Elementary Algebra

SENIOR CONTRIBUTING AUTHORS

LYNN MARECEK, SANTA ANA COLLEGE

MARYANNE ANTHONY-SMITH, FORMERLY OF SANTA ANA COLLEGE

OpenStax
Rice University
6100 Main Street MS-375
Houston, Texas 77005

To learn more about OpenStax, visit https://openstax.org.
Individual print copies and bulk orders can be purchased through our website.

PRINT BOOK ISBN-10	0-9986257-1-X
PRINT BOOK ISBN-13	978-0-9986257-1-3
PDF VERSION ISBN-10	1-947172-25-5
PDF VERSION ISBN-13	978-1-947172-25-8
Revision Number	EA-2017-001(06/17)-LC
Original Publication Year	2017

Printed in Indiana, USA

OPENSTAX

OpenStax provides free, peer-reviewed, openly licensed textbooks for introductory college and Advanced Placement® courses and low-cost, personalized courseware that helps students learn. A nonprofit ed tech initiative based at Rice University, we're committed to helping students access the tools they need to complete their courses and meet their educational goals.

RICE UNIVERSITY

OpenStax, OpenStax CNX, and OpenStax Tutor are initiatives of Rice University. As a leading research university with a distinctive commitment to undergraduate education, Rice University aspires to path-breaking research, unsurpassed teaching, and contributions to the betterment of our world. It seeks to fulfill this mission by cultivating a diverse community of learning and discovery that produces leaders across the spectrum of human endeavor.

FOUNDATION SUPPORT

OpenStax is grateful for the tremendous support of our sponsors. Without their strong engagement, the goal of free access to high-quality textbooks would remain just a dream.

Laura and John Arnold Foundation (LJAF) actively seeks opportunities to invest in organizations and thought leaders that have a sincere interest in implementing fundamental changes that not only yield immediate gains, but also repair broken systems for future generations. LJAF currently focuses its strategic investments on education, criminal justice, research integrity, and public accountability.

The William and Flora Hewlett Foundation has been making grants since 1967 to help solve social and environmental problems at home and around the world. The Foundation concentrates its resources on activities in education, the environment, global development and population, performing arts, and philanthropy, and makes grants to support disadvantaged communities in the San Francisco Bay Area.

Calvin K. Kazanjian was the founder and president of Peter Paul (Almond Joy), Inc. He firmly believed that the more people understood about basic economics the happier and more prosperous they would be. Accordingly, he established the Calvin K. Kazanjian Economics Foundation Inc, in 1949 as a philanthropic, nonpolitical educational organization to support efforts that enhanced economic understanding.

Guided by the belief that every life has equal value, the Bill & Melinda Gates Foundation works to help all people lead healthy, productive lives. In developing countries, it focuses on improving people's health with vaccines and other life-saving tools and giving them the chance to lift themselves out of hunger and extreme poverty. In the United States, it seeks to significantly improve education so that all young people have the opportunity to reach their full potential. Based in Seattle, Washington, the foundation is led by CEO Jeff Raikes and Co-chair William H. Gates Sr., under the direction of Bill and Melinda Gates and Warren Buffett.

The Maxfield Foundation supports projects with potential for high impact in science, education, sustainability, and other areas of social importance.

Our mission at The Michelson 20MM Foundation is to grow access and success by eliminating unnecessary hurdles to affordability. We support the creation, sharing, and proliferation of more effective, more affordable educational content by leveraging disruptive technologies, open educational resources, and new models for collaboration between for-profit, nonprofit, and public entities.

The Bill and Stephanie Sick Fund supports innovative projects in the areas of Education, Art, Science and Engineering.

▤ Table of Contents

Preface   1

1 Foundations   5

1.1 Introduction to Whole Numbers   5
1.2 Use the Language of Algebra   21
1.3 Add and Subtract Integers   40
1.4 Multiply and Divide Integers   61
1.5 Visualize Fractions   76
1.6 Add and Subtract Fractions   92
1.7 Decimals   107
1.8 The Real Numbers   126
1.9 Properties of Real Numbers   142
1.10 Systems of Measurement   160

2 Solving Linear Equations and Inequalities   197

2.1 Solve Equations Using the Subtraction and Addition Properties of Equality   197
2.2 Solve Equations using the Division and Multiplication Properties of Equality   212
2.3 Solve Equations with Variables and Constants on Both Sides   226
2.4 Use a General Strategy to Solve Linear Equations   236
2.5 Solve Equations with Fractions or Decimals   249
2.6 Solve a Formula for a Specific Variable   260
2.7 Solve Linear Inequalities   270

3 Math Models   295

3.1 Use a Problem-Solving Strategy   295
3.2 Solve Percent Applications   312
3.3 Solve Mixture Applications   330
3.4 Solve Geometry Applications: Triangles, Rectangles, and the Pythagorean Theorem   346
3.5 Solve Uniform Motion Applications   369
3.6 Solve Applications with Linear Inequalities   382

4 Graphs   403

4.1 Use the Rectangular Coordinate System   403
4.2 Graph Linear Equations in Two Variables   424
4.3 Graph with Intercepts   444
4.4 Understand Slope of a Line   459
4.5 Use the Slope–Intercept Form of an Equation of a Line   486
4.6 Find the Equation of a Line   512
4.7 Graphs of Linear Inequalities   530

5 Systems of Linear Equations   565

5.1 Solve Systems of Equations by Graphing   565
5.2 Solve Systems of Equations by Substitution   586
5.3 Solve Systems of Equations by Elimination   602
5.4 Solve Applications with Systems of Equations   617
5.5 Solve Mixture Applications with Systems of Equations   635
5.6 Graphing Systems of Linear Inequalities   648

6 Polynomials   673

6.1 Add and Subtract Polynomials   673
6.2 Use Multiplication Properties of Exponents   687
6.3 Multiply Polynomials   701
6.4 Special Products   717
6.5 Divide Monomials   730
6.6 Divide Polynomials   748
6.7 Integer Exponents and Scientific Notation   760

7 Factoring 789

7.1 Greatest Common Factor and Factor by Grouping 789
7.2 Factor Quadratic Trinomials with Leading Coefficient 1 803
7.3 Factor Quadratic Trinomials with Leading Coefficient Other than 1 816
7.4 Factor Special Products 834
7.5 General Strategy for Factoring Polynomials 850
7.6 Quadratic Equations 861

8 Rational Expressions and Equations 883

8.1 Simplify Rational Expressions 883
8.2 Multiply and Divide Rational Expressions 901
8.3 Add and Subtract Rational Expressions with a Common Denominator 914
8.4 Add and Subtract Rational Expressions with Unlike Denominators 923
8.5 Simplify Complex Rational Expressions 937
8.6 Solve Rational Equations 950
8.7 Solve Proportion and Similar Figure Applications 965
8.8 Solve Uniform Motion and Work Applications 981
8.9 Use Direct and Inverse Variation 991

9 Roots and Radicals 1013

9.1 Simplify and Use Square Roots 1013
9.2 Simplify Square Roots 1023
9.3 Add and Subtract Square Roots 1036
9.4 Multiply Square Roots 1046
9.5 Divide Square Roots 1060
9.6 Solve Equations with Square Roots 1074
9.7 Higher Roots 1091
9.8 Rational Exponents 1107

10 Quadratic Equations 1137

10.1 Solve Quadratic Equations Using the Square Root Property 1137
10.2 Solve Quadratic Equations by Completing the Square 1149
10.3 Solve Quadratic Equations Using the Quadratic Formula 1165
10.4 Solve Applications Modeled by Quadratic Equations 1179
10.5 Graphing Quadratic Equations 1190

Index 1309

PREFACE

Welcome to *Elementary Algebra*, an OpenStax resource. This textbook was written to increase student access to high-quality learning materials, maintaining highest standards of academic rigor at little to no cost.

About OpenStax

OpenStax is a nonprofit based at Rice University, and it's our mission to improve student access to education. Our first openly licensed college textbook was published in 2012, and our library has since scaled to over 25 books for college and AP courses used by hundreds of thousands of students. Our adaptive learning technology, designed to improve learning outcomes through personalized educational paths, is being piloted in college courses throughout the country. Through our partnerships with philanthropic foundations and our alliance with other educational resource organizations, OpenStax is breaking down the most common barriers to learning and empowering students and instructors to succeed.

About OpenStax Resources
Customization

Elementary Algebra is licensed under a Creative Commons Attribution 4.0 International (CC BY) license, which means that you can distribute, remix, and build upon the content, as long as you provide attribution to OpenStax and its content contributors.

Because our books are openly licensed, you are free to use the entire book or pick and choose the sections that are most relevant to the needs of your course. Feel free to remix the content by assigning your students certain chapters and sections in your syllabus, in the order that you prefer. You can even provide a direct link in your syllabus to the sections in the web view of your book.

Instructors also have the option of creating a customized version of their OpenStax book. The custom version can be made available to students in low-cost print or digital form through their campus bookstore. Visit your book page on openstax.org for more information.

Errata

All OpenStax textbooks undergo a rigorous review process. However, like any professional-grade textbook, errors sometimes occur. Since our books are web based, we can make updates periodically when deemed pedagogically necessary. If you have a correction to suggest, submit it through the link on your book page on openstax.org. Subject matter experts review all errata suggestions. OpenStax is committed to remaining transparent about all updates, so you will also find a list of past errata changes on your book page on openstax.org.

Format

You can access this textbook for free in web view or PDF through openstax.org, and for a low cost in print.

About *Elementary Algebra*

Elementary Algebra is designed to meet the scope and sequence requirements of a one-semester elementary algebra course. The book's organization makes it easy to adapt to a variety of course syllabi. The text expands on the fundamental concepts of algebra while addressing the needs of students with diverse backgrounds and learning styles. Each topic builds upon previously developed material to demonstrate the cohesiveness and structure of mathematics.

Coverage and Scope

Elementary Algebra follows a nontraditional approach in its presentation of content. Building on the content in *Prealgebra*, the material is presented as a sequence of small steps so that students gain confidence in their ability to succeed in the course. The order of topics was carefully planned to emphasize the logical progression through the course and to facilitate a thorough understanding of each concept. As new ideas are presented, they are explicitly related to previous topics.

Chapter 1: Foundations
Chapter 1 reviews arithmetic operations with whole numbers, integers, fractions, and decimals, to give the student a solid base that will support their study of algebra.

Chapter 2: Solving Linear Equations and Inequalities
In Chapter 2, students learn to verify a solution of an equation, solve equations using the Subtraction and Addition Properties of Equality, solve equations using the Multiplication and Division Properties of Equality, solve equations with variables and constants on both sides, use a general strategy to solve linear equations, solve equations with fractions or decimals, solve a formula for a specific variable, and solve linear inequalities.

Chapter 3: Math Models
Once students have learned the skills needed to solve equations, they apply these skills in Chapter 3 to solve word and number problems.

Chapter 4: Graphs
Chapter 4 covers the rectangular coordinate system, which is the basis for most consumer graphs. Students learn to plot points on a rectangular coordinate system, graph linear equations in two variables, graph with intercepts,

understand slope of a line, use the slope-intercept form of an equation of a line, find the equation of a line, and create graphs of linear inequalities.

Chapter 5: Systems of Linear Equations
Chapter 5 covers solving systems of equations by graphing, substitution, and elimination; solving applications with systems of equations, solving mixture applications with systems of equations, and graphing systems of linear inequalities.

Chapter 6: Polynomials
In Chapter 6, students learn how to add and subtract polynomials, use multiplication properties of exponents, multiply polynomials, use special products, divide monomials and polynomials, and understand integer exponents and scientific notation.

Chapter 7: Factoring
In Chapter 7, students explore the process of factoring expressions and see how factoring is used to solve certain types of equations.

Chapter 8: Rational Expressions and Equations
In Chapter 8, students work with rational expressions, solve rational equations, and use them to solve problems in a variety of applications.

Chapter 9: Roots and Radical
In Chapter 9, students are introduced to and learn to apply the properties of square roots, and extend these concepts to higher order roots and rational exponents.

Chapter 10: Quadratic Equations
In Chapter 10, students study the properties of quadratic equations, solve and graph them. They also learn how to apply them as models of various situations.

All chapters are broken down into multiple sections, the titles of which can be viewed in the **Table of Contents**.

Key Features and Boxes

Examples Each learning objective is supported by one or more worked examples that demonstrate the problem-solving approaches that students must master. Typically, we include multiple Examples for each learning objective to model different approaches to the same type of problem, or to introduce similar problems of increasing complexity.

All Examples follow a simple two- or three-part format. First, we pose a problem or question. Next, we demonstrate the solution, spelling out the steps along the way. Finally (for select Examples), we show students how to check the solution. Most Examples are written in a two-column format, with explanation on the left and math on the right to mimic the way that instructors "talk through" examples as they write on the board in class.

Be Prepared! Each section, beginning with Section 2.1, starts with a few "Be Prepared!" exercises so that students can determine if they have mastered the prerequisite skills for the section. Reference is made to specific Examples from previous sections so students who need further review can easily find explanations. Answers to these exercises can be found in the supplemental resources that accompany this title.

Try It

The Try It feature includes a pair of exercises that immediately follow an Example, providing the student with an immediate opportunity to solve a similar problem. In the Web View version of the text, students can click an Answer link directly below the question to check their understanding. In the PDF, answers to the Try It exercises are located in the Answer Key.

How To

How To feature typically follows the Try It exercises and outlines the series of steps for how to solve the problem in the preceding Example.

Media

The Media icon appears at the conclusion of each section, just prior to the Self Check. This icon marks a list of links to online video tutorials that reinforce the concepts and skills introduced in the section.

Disclaimer: While we have selected tutorials that closely align to our learning objectives, we did not produce these tutorials, nor were they specifically produced or tailored to accompany *Elementary Algebra*.

Self Check The Self Check includes the learning objectives for the section so that students can self-assess their mastery and make concrete plans to improve.

Art Program

Elementary Algebra contains many figures and illustrations. Art throughout the text adheres to a clear, understated style, drawing the eye to the most important information in each figure while minimizing visual distractions.

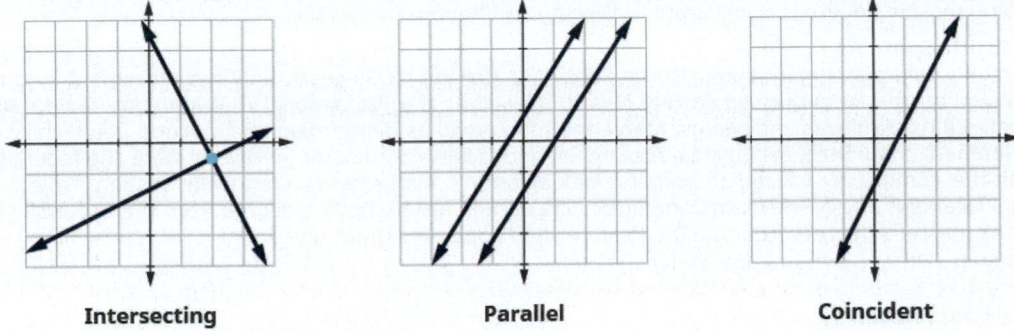

| Intersecting | Parallel | Coincident |

Section Exercises and Chapter Review

Section Exercises Each section of every chapter concludes with a well-rounded set of exercises that can be assigned as homework or used selectively for guided practice. Exercise sets are named *Practice Makes Perfect* to encourage completion of homework assignments.

- Exercises correlate to the learning objectives. This facilitates assignment of personalized study plans based on individual student needs.

- Exercises are carefully sequenced to promote building of skills.

- Values for constants and coefficients were chosen to practice and reinforce arithmetic facts.

- Even and odd-numbered exercises are paired.

- Exercises parallel and extend the text examples and use the same instructions as the examples to help students easily recognize the connection.

- Applications are drawn from many everyday experiences, as well as those traditionally found in college math texts.

- **Everyday Math** highlights practical situations using the concepts from that particular section

- **Writing Exercises** are included in every exercise set to encourage conceptual understanding, critical thinking, and literacy.

Chapter Review Each chapter concludes with a review of the most important takeaways, as well as additional practice problems that students can use to prepare for exams.

- **Key Terms** provide a formal definition for each bold-faced term in the chapter.

- **Key Concepts** summarize the most important ideas introduced in each section, linking back to the relevant Example(s) in case students need to review.

- **Chapter Review Exercises** include practice problems that recall the most important concepts from each section.

- **Practice Test** includes additional problems assessing the most important learning objectives from the chapter.

- **Answer Key** includes the answers to all Try It exercises and every other exercise from the Section Exercises, Chapter Review Exercises, and Practice Test.

Additional Resources
Student and Instructor Resources

We've compiled additional resources for both students and instructors, including Getting Started Guides, manipulative mathematics worksheets, Links to Literacy assignments, and an answer key to Be Prepared Exercises. Instructor resources require a verified instructor account, which can be requested on your openstax.org log-in. Take advantage of these resources to supplement your OpenStax book.

Partner Resources

OpenStax Partners are our allies in the mission to make high-quality learning materials affordable and accessible to students and instructors everywhere. Their tools integrate seamlessly with our OpenStax titles at a low cost. To access the partner resources for your text, visit your book page on openstax.org.

About the Authors
Senior Contributing Authors

Lynn Marecek and MaryAnne Anthony-Smith have been teaching mathematics at Santa Ana College for many years and have worked together on several projects aimed at improving student learning in developmental math courses. They are the authors of *Strategies for Success: Study Skills for the College Math Student*.

Lynn Marecek, Santa Ana College

Lynn Marecek has focused her career on meeting the needs of developmental math students. At Santa Ana College, she has been awarded the Distinguished Faculty Award, Innovation Award, and the Curriculum Development Award four times. She is a Coordinator of Freshman Experience Program, the Department Facilitator for Redesign, and a member of the Student Success and Equity Committee, and the Basic Skills Initiative Task Force. Lynn holds a bachelor's degree from Valparaiso University and master's degrees from Purdue University and National University.

MaryAnne Anthony-Smith, Santa Ana College

MaryAnne Anthony-Smith was a mathematics professor at Santa Ana College for 39 years, until her retirement in June, 2015. She has been awarded the Distinguished Faculty Award, as well as the Professional Development, Curriculum Development, and Professional Achievement awards. MaryAnne has served as department chair, acting dean, chair of the professional development committee, institutional researcher, and faculty coordinator on several state and federally-funded grants. She is the community college coordinator of California's Mathematics Diagnostic Testing Project, a member of AMATYC's Placement and Assessment Committee. She earned her bachelor's degree from the University of California San Diego and master's degrees from San Diego State and Pepperdine Universities.

Reviewers

Jay Abramson, Arizona State University
Bryan Blount, Kentucky Wesleyan College
Gale Burtch, Ivy Tech Community College
Tamara Carter, Texas A&M University
Danny Clarke, Truckee Meadows Community College
Michael Cohen, Hofstra University
Christina Cornejo, Erie Community College
Denise Cutler, Bay de Noc Community College
Lance Hemlow, Raritan Valley Community College
John Kalliongis, Saint Louis Iniversity
Stephanie Krehl, Mid-South Community College
Laurie Lindstrom, Bay de Noc Community College
Beverly Mackie, Lone Star College System
Allen Miller, Northeast Lakeview College
Christian Roldán-Johnson, College of Lake County Community College
Martha Sandoval-Martinez, Santa Ana College
Gowribalan Vamadeva, University of Cincinnati Blue Ash College
Kim Watts, North Lake College
Libby Watts, Tidewater Community College
Allen Wolmer, Atlantic Jewish Academy
John Zarske, Santa Ana College

Figure 1.1 In order to be structurally sound, the foundation of a building must be carefully constructed.

Chapter Outline

1.1 Introduction to Whole Numbers

1.2 Use the Language of Algebra

1.3 Add and Subtract Integers

1.4 Multiply and Divide Integers

1.5 Visualize Fractions

1.6 Add and Subtract Fractions

1.7 Decimals

1.8 The Real Numbers

1.9 Properties of Real Numbers

1.10 Systems of Measurement

Introduction

Just like a building needs a firm foundation to support it, your study of algebra needs to have a firm foundation. To ensure this, we begin this book with a review of arithmetic operations with whole numbers, integers, fractions, and decimals, so that you have a solid base that will support your study of algebra.

1.1 Introduction to Whole Numbers

Learning Objectives

By the end of this section, you will be able to:

› Use place value with whole numbers
› Identify multiples and and apply divisibility tests
› Find prime factorizations and least common multiples

Be Prepared!

A more thorough introduction to the topics covered in this section can be found in *Prealgebra* in the chapters **Whole Numbers** and **The Language of Algebra**.

As we begin our study of elementary algebra, we need to refresh some of our skills and vocabulary. This chapter will focus on whole numbers, integers, fractions, decimals, and real numbers. We will also begin our use of algebraic notation and vocabulary.

Use Place Value with Whole Numbers

The most basic numbers used in algebra are the numbers we use to count objects in our world: 1, 2, 3, 4, and so on. These are called the **counting numbers**. Counting numbers are also called *natural numbers*. If we add zero to the counting numbers, we get the set of **whole numbers**.

Counting Numbers: 1, 2, 3, ...

Whole Numbers: 0, 1, 2, 3, ...

The notation "..." is called ellipsis and means "and so on," or that the pattern continues endlessly.

We can visualize counting numbers and whole numbers on a **number line** (see Figure 1.2).

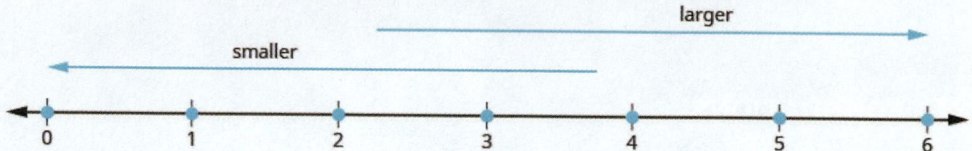

Figure 1.2 The numbers on the number line get larger as they go from left to right, and smaller as they go from right to left. While this number line shows only the whole numbers 0 through 6, the numbers keep going without end.

 MANIPULATIVE MATHEMATICS

Doing the Manipulative Mathematics activity "Number Line-Part 1" will help you develop a better understanding of the counting numbers and the whole numbers.

Our number system is called a place value system, because the value of a digit depends on its position in a number. Figure 1.3 shows the place values. The place values are separated into groups of three, which are called periods. The periods are *ones, thousands, millions, billions, trillions*, and so on. In a written number, commas separate the periods.

Place Value														
Trillions			**Billions**			**Millions**			**Thousands**			**Ones**		
Hundred trillions	Ten trillions	Trillions	Hundred billions	Ten billions	Billions	Hundred millions	Ten millions	Millions	Hundred thousands	Ten thousands	Thousands	Hundreds	Tens	Ones
								5	2	7	8	1	9	4

Figure 1.3 The number 5,278,194 is shown in the chart. The digit 5 is in the millions place. The digit 2 is in the hundred-thousands place. The digit 7 is in the ten-thousands place. The digit 8 is in the thousands place. The digit 1 is in the hundreds place. The digit 9 is in the tens place. The digit 4 is in the ones place.

EXAMPLE 1.1

In the number 63,407,218, find the place value of each digit:

ⓐ 7   ⓑ 0   ⓒ 1   ⓓ 6   ⓔ 3

⊘ **Solution**

Place the number in the place value chart:

Trillions			Billions			Millions			Thousands			Ones		
Hundred trillions	Ten trillions	Trillions	Hundred billions	Ten billions	Billions	Hundred millions	Ten millions	Millions	Hundred thousands	Ten thousands	Thousands	Hundreds	Tens	Ones
						6	3	4	0	7	2	1		8

ⓐ The 7 is in the thousands place.

ⓑ The 0 is in the ten thousands place.

ⓒ The 1 is in the tens place.

ⓓ The 6 is in the ten-millions place.

ⓔ The 3 is in the millions place.

> **TRY IT :: 1.1** For the number 27,493,615, find the place value of each digit:

ⓐ 2 ⓑ 1 ⓒ 4 ⓓ 7 ⓔ 5

> **TRY IT :: 1.2** For the number 519,711,641,328, find the place value of each digit:

ⓐ 9 ⓑ 4 ⓒ 2 ⓓ 6 ⓔ 7

When you write a check, you write out the number in words as well as in digits. To write a number in words, write the number in each period, followed by the name of the period, without the *s* at the end. Start at the left, where the periods have the largest value. The ones period is not named. The commas separate the periods, so wherever there is a comma in the number, put a comma between the words (see **Figure 1.4**). The number 74,218,369 is written as seventy-four million, two hundred eighteen thousand, three hundred sixty-nine.

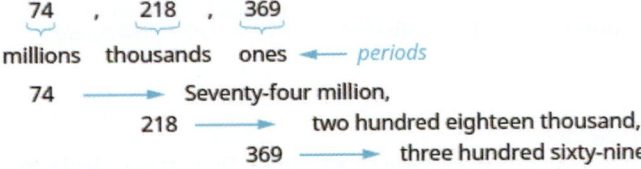

Figure 1.4

  HOW TO :: NAME A WHOLE NUMBER IN WORDS.

Step 1. Start at the left and name the number in each period, followed by the period name.

Step 2. Put commas in the number to separate the periods.

Step 3. Do not name the ones period.

EXAMPLE 1.2

Name the number 8,165,432,098,710 using words.

✓ **Solution**

Name the number in each period, followed by the period name.

$$8 \text{ , } 165 \text{ , } 432 \text{ , } 098 \text{ , } 710$$

trillions   billions   millions   thousands   ones

8 ⟶ Eight trillion,

165 ⟶ One hundred sixty-five billion,

432 ⟶ Four hundred thirty-two million,

098 ⟶ Ninety-eight thousand,

710 ⟶ Seven hundred ten

Put the commas in to separate the periods.

So, 8, 165, 432, 098, 710 is named as eight trillion, one hundred sixty-five billion, four hundred thirty-two million, ninety-eight thousand, seven hundred ten.

> **TRY IT :: 1.3**   Name the number 9, 258, 137, 904, 061 using words.

> **TRY IT :: 1.4**   Name the number 17, 864, 325, 619, 004 using words.

We are now going to reverse the process by writing the digits from the name of the number. To write the number in digits, we first look for the clue words that indicate the periods. It is helpful to draw three blanks for the needed periods and then fill in the blanks with the numbers, separating the periods with commas.

HOW TO :: WRITE A WHOLE NUMBER USING DIGITS.

Step 1.   Identify the words that indicate periods. (Remember, the ones period is never named.)

Step 2.   Draw three blanks to indicate the number of places needed in each period. Separate the periods by commas.

Step 3.   Name the number in each period and place the digits in the correct place value position.

EXAMPLE 1.3

Write *nine billion, two hundred forty-six million, seventy-three thousand, one hundred eighty-nine* as a whole number using digits.

✓ **Solution**

Identify the words that indicate periods.
Except for the first period, all other periods must have three places. Draw three blanks to indicate the number of places needed in each period. Separate the periods by commas.
Then write the digits in each period.

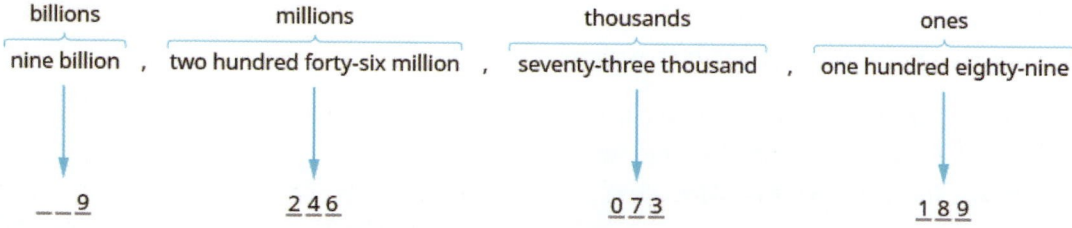

The number is 9,246,073,189.

> **TRY IT :: 1.5**
>
> Write the number two billion, four hundred sixty-six million, seven hundred fourteen thousand, fifty-one as a whole number using digits.

TRY IT : : 1.6

Write the number eleven billion, nine hundred twenty-one million, eight hundred thirty thousand, one hundred six as a whole number using digits.

In 2013, the U.S. Census Bureau estimated the population of the state of New York as 19,651,127. We could say the population of New York was approximately 20 million. In many cases, you don't need the exact value; an approximate number is good enough.

The process of approximating a number is called rounding. Numbers are rounded to a specific place value, depending on how much accuracy is needed. Saying that the population of New York is approximately 20 million means that we rounded to the millions place.

EXAMPLE 1.4   HOW TO ROUND WHOLE NUMBERS

Round 23,658 to the nearest hundred.

⊘ **Solution**

Step 1. Locate the given place value with an arrow. All digits to the left do not change.	Locate the hundreds place in 23,658.	hundredths place ↓ 23,658
Step 2. Underline the digit to the right of the given place value.	Underline the 5, which is to the right of the hundreds place.	hundredths place ↓ 23,6<u>5</u>8
Step 3. Is this digit greater than or equal to 5? Yes–add 1 to the digit in the given place value. No–do <u>not</u> change the digit in the given place value.	Add 1 to the 6 in the hundreds place, since 5 is greater than or equal to 5.	23,658 add 1 ↗
Step 4. Replace all digits to the right of the given place value with zeros.	Replace all digits to the right of the hundreds place with zeros.	23,700 add 1 ↗ ↖ replace with 0s So, 23,700 is rounded to the nearest hundred.

 TRY IT : : 1.7   Round to the nearest hundred: 17,852.

 TRY IT : : 1.8   Round to the nearest hundred: 468,751.

HOW TO : : ROUND WHOLE NUMBERS.

Step 1.   Locate the given place value and mark it with an arrow. All digits to the left of the arrow do not change.

Step 2.   Underline the digit to the right of the given place value.

Step 3.   Is this digit greater than or equal to 5?

  ◦ Yes–add 1 to the digit in the given place value.

  ◦ No–do <u>not</u> change the digit in the given place value.

Step 4.   Replace all digits to the right of the given place value with zeros.

EXAMPLE 1.5

Round 103,978 to the nearest:

    ⓐ hundred   ⓑ thousand   ⓒ ten thousand

✓ **Solution**

ⓐ

Locate the hundreds place in 103,978.

hundreds place
↓
103,978

Underline the digit to the right of the hundreds place.

hundreds place
↓
103,978

Since 7 is greater than or equal to 5, add 1 to the 9. Replace all digits to the right of the hundreds place with zeros.

hundreds place
↓
103,978

add 1 9 + 1 = 10
replace 9 with 0
and carry the 1   replace with 0s

104,000

So, 104,000 is 103,978 rounded to the nearest hundred.

ⓑ

Locate the thousands place and underline the digit to the right of the thousands place.

thousands place
↓
103,978

Since 9 is greater than or equal to 5, add 1 to the 3. Replace all digits to the right of the hundreds place with zeros.

thousands place
↓
103,978

add 1 3 + 1 = 4
replace 3 with 4   replace with 0s

104,000

So, 104,000 is 103,978 rounded to the nearest thousand.

ⓒ

Locate the ten thousands place and underline the digit to the right of the ten thousands place.

ten thousands place
↓
103,978

Since 3 is less than 5, we leave the 0 as is, and then replace the digits to the right with zeros.

100,000

So, 100,000 is 103,978 rounded to the nearest ten thousand.

> **TRY IT : : 1.9**   Round 206,981 to the nearest: ⓐ hundred ⓑ thousand ⓒ ten thousand.

> **TRY IT : : 1.10**   Round 784,951 to the nearest: ⓐ hundred ⓑ thousand ⓒ ten thousand.

Identify Multiples and Apply Divisibility Tests

The numbers 2, 4, 6, 8, 10, and 12 are called **multiples** of 2. A multiple of 2 can be written as the product of a counting number and 2.

$$2, \quad 4, \quad 6, \quad 8, \quad 10, \quad 12, \dots$$
$$2 \cdot 1, \quad 2 \cdot 2, \quad 2 \cdot 3, \quad 2 \cdot 4, \quad 2 \cdot 5, \quad 2 \cdot 6$$

Similarly, a multiple of 3 would be the product of a counting number and 3.

$$3, \quad 6, \quad 9, \quad 12, \quad 15, \quad 18, \dots$$
$$3 \cdot 1, \quad 3 \cdot 2, \quad 3 \cdot 3, \quad 3 \cdot 4, \quad 3 \cdot 5, \quad 3 \cdot 6$$

We could find the multiples of any number by continuing this process.

 MANIPULATIVE MATHEMATICS

Doing the Manipulative Mathematics activity "Multiples" will help you develop a better understanding of multiples.

Table 1.4 shows the multiples of 2 through 9 for the first 12 counting numbers.

Counting Number	1	2	3	4	5	6	7	8	9	10	11	12
Multiples of 2	2	4	6	8	10	12	14	16	18	20	22	24
Multiples of 3	3	6	9	12	15	18	21	24	27	30	33	36
Multiples of 4	4	8	12	16	20	24	28	32	36	40	44	48
Multiples of 5	5	10	15	20	25	30	35	40	45	50	55	60
Multiples of 6	6	12	18	24	30	36	42	48	54	60	66	72
Multiples of 7	7	14	21	28	35	42	49	56	63	70	77	84
Multiples of 8	8	16	24	32	40	48	56	64	72	80	88	96
Multiples of 9	9	18	27	36	45	54	63	72	81	90	99	108
Multiples of 10	10	20	30	40	50	60	0	80	90	100	110	120

Table 1.4

Multiple of a Number

A number is a **multiple** of n if it is the product of a counting number and n.

Another way to say that 15 is a multiple of 3 is to say that 15 is **divisible** by 3. That means that when we divide 3 into 15, we get a counting number. In fact, $15 \div 3$ is 5, so 15 is $5 \cdot 3$.

Divisible by a Number

If a number m is a multiple of n, then m is **divisible** by n.

Look at the multiples of 5 in Table 1.4. They all end in 5 or 0. Numbers with last digit of 5 or 0 are divisible by 5. Looking for other patterns in Table 1.4 that shows multiples of the numbers 2 through 9, we can discover the following divisibility tests:

Divisibility Tests

A number is divisible by:
 	
  • 2 if the last digit is 0, 2, 4, 6, or 8.
  • 3 if the sum of the digits is divisible by 3.
  • 5 if the last digit is 5 or 0.
  • 6 if it is divisible by both 2 and 3.
  • 10 if it ends with 0.

EXAMPLE 1.6

Is 5,625 divisible by 2? By 3? By 5? By 6? By 10?

 **Solution**

Is 5,625 divisible by 2 ?

Does it end in 0, 2, 4, 6, or 8 ?   No.

5,625 is not divisible by 2.

Is 5,625 divisible by 3 ?

What is the sum of the digits? $5 + 6 + 2 + 5 = 18$

Is the sum divisible by 3 ?   Yes. 5,625 is divisible by 3.

Is 5,625 divisible by 5 or 10 ?

What is the last digit? It is 5.   5,625 is divisible by 5 but not by 10.

Is 5,625 divisible by 6 ?

Is it divisible by both 2 and 3 ?   No, 5,625 is not divisible by 2, so 5,625 is

not divisible by 6.

> **TRY IT : : 1.11**   Determine whether 4,962 is divisible by 2, by 3, by 5, by 6, and by 10.

> **TRY IT : : 1.12**   Determine whether 3,765 is divisible by 2, by 3, by 5, by 6, and by 10.

Find Prime Factorizations and Least Common Multiples

In mathematics, there are often several ways to talk about the same ideas. So far, we've seen that if *m* is a multiple of *n*, we can say that *m* is divisible by *n*. For example, since 72 is a multiple of 8, we say 72 is divisible by 8. Since 72 is a multiple of 9, we say 72 is divisible by 9. We can express this still another way.

Since $8 \cdot 9 = 72,$ we say that 8 and 9 are **factors** of 72. When we write $72 = 8 \cdot 9,$ we say we have factored 72.

$$\underbrace{8 \cdot 9}_{factors} = \underbrace{72}_{product}$$

Other ways to factor 72 are $1 \cdot 72, \; 2 \cdot 36, \; 3 \cdot 24, \; 4 \cdot 18,$ and $6 \cdot 12.$ Seventy-two has many factors: 1, 2, 3, 4, 6, 8, 9, 12, 18, 36, and 72.

Factors
If $a \cdot b = m,$ then *a* and *b* are **factors** of *m*.

Some numbers, like 72, have many factors. Other numbers have only two factors.

MANIPULATIVE MATHEMATICS

Doing the Manipulative Mathematics activity "Model Multiplication and Factoring" will help you develop a better understanding of multiplication and factoring.

Prime Number and Composite Number

A **prime number** is a counting number greater than 1, whose only factors are 1 and itself.

A **composite number** is a counting number that is not prime. A composite number has factors other than 1 and itself.

 MANIPULATIVE MATHEMATICS

Doing the Manipulative Mathematics activity "Prime Numbers" will help you develop a better understanding of prime numbers.

The counting numbers from 2 to 19 are listed in **Figure 1.5**, with their factors. Make sure to agree with the "prime" or "composite" label for each!

Number	Factors	Prime or Composite?
2	1,2	Prime
3	1,3	Prime
4	1,2,4	Composite
5	1,5	Prime
6	1,2,3,6	Composite
7	1,7	Prime
8	1,2,4,8	Composite
9	1,3,9	Composite
10	1,2,5,10	Composite

Number	Factors	Prime or Composite?
11	1,11	Prime
12	1,2,3,4,6,12	Composite
13	1,13	Prime
14	1,2,7,14	Composite
15	1,3,5,15	Composite
16	1,2,4,8,16	Composite
17	1,17	Prime
18	1,2,3,6,9,18	Composite
19	1,19	Prime

Figure 1.5

The **prime numbers** less than 20 are 2, 3, 5, 7, 11, 13, 17, and 19. Notice that the only even prime number is 2.

A composite number can be written as a unique product of primes. This is called the **prime factorization** of the number. Finding the prime factorization of a composite number will be useful later in this course.

Prime Factorization

The **prime factorization** of a number is the product of prime numbers that equals the number.

To find the prime factorization of a composite number, find any two factors of the number and use them to create two branches. If a factor is prime, that branch is complete. Circle that prime!

If the factor is not prime, find two factors of the number and continue the process. Once all the branches have circled primes at the end, the factorization is complete. The composite number can now be written as a product of prime numbers.

EXAMPLE 1.7 HOW TO FIND THE PRIME FACTORIZATION OF A COMPOSITE NUMBER

Factor 48.

 **Solution**

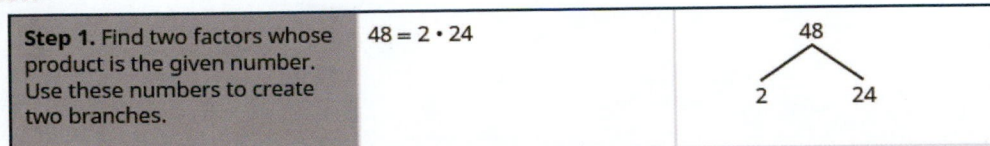

| **Step 1.** Find two factors whose product is the given number. Use these numbers to create two branches. | $48 = 2 \cdot 24$ | |

Step 2. If a factor is prime, that branch is complete. Circle the prime.	2 is prime. Circle the prime.	48 ② 24
Step 3. If a factor is not prime, write it as the product of two factors and continue the process.	24 is not prime. Break it into 2 more factors.	48 ② 24, 24 → 4 6
	4 and 6 are not prime. Break them each into two factors.	48 ② 24, 24 → 4 6
	2 and 3 are prime, so circle them.	48 ② 24, 4 → ②②, 6 → ②③
Step 4. Write the composite number as the product of all the circled primes.		$48 = 2 \cdot 2 \cdot 2 \cdot 2 \cdot 3$

We say $2 \cdot 2 \cdot 2 \cdot 2 \cdot 3$ is the prime factorization of 48. We generally write the primes in ascending order. Be sure to multiply the factors to verify your answer!

If we first factored 48 in a different way, for example as $6 \cdot 8,$ the result would still be the same. Finish the prime factorization and verify this for yourself.

> **TRY IT : : 1.13**   Find the prime factorization of 80.

> **TRY IT : : 1.14**   Find the prime factorization of 60.

HOW TO : : FIND THE PRIME FACTORIZATION OF A COMPOSITE NUMBER.

Step 1.   Find two factors whose product is the given number, and use these numbers to create two branches.

Step 2.   If a factor is prime, that branch is complete. Circle the prime, like a bud on the tree.

Step 3.   If a factor is not prime, write it as the product of two factors and continue the process.

Step 4.   Write the composite number as the product of all the circled primes.

EXAMPLE 1.8

Find the prime factorization of 252.

⊘ Solution

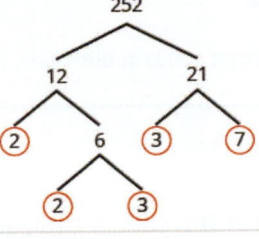

Step 1. Find two factors whose product is 252. 12 and 21 are not prime.

Break 12 and 21 into two more factors. Continue until all primes are factored.

Step 2. Write 252 as the product of all the circled primes.

$$252 = 2 \cdot 2 \cdot 3 \cdot 3 \cdot 7$$

> **TRY IT : : 1.15**   Find the prime factorization of 126.

> **TRY IT : : 1.16**   Find the prime factorization of 294.

One of the reasons we look at multiples and primes is to use these techniques to find the **least common multiple** of two numbers. This will be useful when we add and subtract fractions with different denominators. Two methods are used most often to find the least common multiple and we will look at both of them.

The first method is the Listing Multiples Method. To find the least common multiple of 12 and 18, we list the first few multiples of 12 and 18:

12: 12, 24, **36**, 48, 60, **72**, 84, 96, **108**...

18: 18, **36**, 54, **72**, 90, **108**...

Common Multiples: 36, 72, 108...

Least Common Multiple: 36

Notice that some numbers appear in both lists. They are the **common multiples** of 12 and 18.

We see that the first few common multiples of 12 and 18 are 36, 72, and 108. Since 36 is the smallest of the common multiples, we call it the *least common multiple*. We often use the abbreviation LCM.

Least Common Multiple

The **least common multiple** (LCM) of two numbers is the smallest number that is a multiple of both numbers.

The procedure box lists the steps to take to find the LCM using the prime factors method we used above for 12 and 18.

 HOW TO : : FIND THE LEAST COMMON MULTIPLE BY LISTING MULTIPLES.

Step 1.   List several multiples of each number.

Step 2.   Look for the smallest number that appears on both lists.

Step 3.   This number is the LCM.

EXAMPLE 1.9

Find the least common multiple of 15 and 20 by listing multiples.

⊘ Solution

Make lists of the first few multiples of 15 and of 20, and use them to find the least common multiple.

15: 15, 30, 45, **60**, 75, 90, 105, 120

20: 20, 40, **60**, 80, 100, 120, 140, 160

| Look for the smallest number that appears in both lists. | The first number to appear on both lists is 60, so 60 is the least common multiple of 15 and 20. |

Notice that 120 is in both lists, too. It is a common multiple, but it is not the *least* common multiple.

> **TRY IT : : 1.17**   Find the least common multiple by listing multiples: 9 and 12.

> **TRY IT : : 1.18**   Find the least common multiple by listing multiples: 18 and 24.

Our second method to find the least common multiple of two numbers is to use The Prime Factors Method. Let's find the LCM of 12 and 18 again, this time using their prime factors.

EXAMPLE 1.10   HOW TO FIND THE LEAST COMMON MULTIPLE USING THE PRIME FACTORS METHOD

Find the Least Common Multiple (LCM) of 12 and 18 using the prime factors method.

✓ **Solution**

Step 1. Write each number as a product of primes.		18 12 3 6 3 4 2 3 2 2
Step 2. List the primes of each number. Match primes vertically when possible.	List the primes of 12. List the primes of 18. Line up with the primes of 12 when possible. If not create a new column.	$12 = 2 \cdot 2 \cdot 3$ $18 = 2 \cdot 3 \cdot 3$
Step 3. Bring down the number from each column.		$12 = 2 \cdot 2 \cdot 3$ $18 = 2 \cdot 3 \cdot 3$ $LCM = 2 \cdot 2 \cdot 3 \cdot 3$
Step 4. Multiply the factors.		$LCM = 36$

Notice that the prime factors of 12 $(2 \cdot 2 \cdot 3)$ and the prime factors of 18 $(2 \cdot 3 \cdot 3)$ are included in the LCM $(2 \cdot 2 \cdot 3 \cdot 3)$. So 36 is the least common multiple of 12 and 18.

By matching up the common primes, each common prime factor is used only once. This way you are sure that 36 is the *least* common multiple.

> **TRY IT : : 1.19**   Find the LCM using the prime factors method: 9 and 12.

> **TRY IT : : 1.20**   Find the LCM using the prime factors method: 18 and 24.

 HOW TO : : FIND THE LEAST COMMON MULTIPLE USING THE PRIME FACTORS METHOD.

Step 1.   Write each number as a product of primes.

Step 2.   List the primes of each number. Match primes vertically when possible.

Step 3.   Bring down the columns.

Step 4.   Multiply the factors.

EXAMPLE 1.11

Find the Least Common Multiple (LCM) of 24 and 36 using the prime factors method.

⊘ **Solution**

Find the primes of 24 and 36.
Match primes vertically when possible.

Bring down all columns.

$$24 = 2 \cdot 2 \cdot 2 \cdot 3$$
$$36 = 2 \cdot 2 \cdot \quad 3 \cdot 3$$
$$LCM = 2 \cdot 2 \cdot 2 \cdot 3 \cdot 3$$

Multiply the factors.

$$LCM = 72$$

The LCM of 24 and 36 is 72.

> **TRY IT : : 1.21**   Find the LCM using the prime factors method: 21 and 28.

> **TRY IT : : 1.22**   Find the LCM using the prime factors method: 24 and 32.

▶ **MEDIA : :**

Access this online resource for additional instruction and practice with using whole numbers. You will need to enable Java in your web browser to use the application.

- **Sieve of Eratosthenes (https://openstax.org/l/01sieveoferato)**

1.1 EXERCISES

Practice Makes Perfect

Use Place Value with Whole Numbers

In the following exercises, find the place value of each digit in the given numbers.

1. 51,493
ⓐ 1
ⓑ 4
ⓒ 9
ⓓ 5
ⓔ 3

2. 87,210
ⓐ 2
ⓑ 8
ⓒ 0
ⓓ 7
ⓔ 1

3. 164,285
ⓐ 5
ⓑ 6
ⓒ 1
ⓓ 8
ⓔ 2

4. 395,076
ⓐ 5
ⓑ 3
ⓒ 7
ⓓ 0
ⓔ 9

5. 93,285,170
ⓐ 9
ⓑ 8
ⓒ 7
ⓓ 5
ⓔ 3

6. 36,084,215
ⓐ 8
ⓑ 6
ⓒ 5
ⓓ 4
ⓔ 3

7. 7,284,915,860,132
ⓐ 7
ⓑ 4
ⓒ 5
ⓓ 3
ⓔ 0

8. 2,850,361,159,433
ⓐ 9
ⓑ 8
ⓒ 6
ⓓ 4
ⓔ 2

In the following exercises, name each number using words.

9. 1,078

10. 5,902

11. 364,510

12. 146,023

13. 5,846,103

14. 1,458,398

15. 37,889,005

16. 62,008,465

In the following exercises, write each number as a whole number using digits.

17. four hundred twelve

18. two hundred fifty-three

19. thirty-five thousand, nine hundred seventy-five

20. sixty-one thousand, four hundred fifteen

21. eleven million, forty-four thousand, one hundred sixty-seven

22. eighteen million, one hundred two thousand, seven hundred eighty-three

23. three billion, two hundred twenty-six million, five hundred twelve thousand, seventeen

24. eleven billion, four hundred seventy-one million, thirty-six thousand, one hundred six

In the following, round to the indicated place value.

25. Round to the nearest ten.
ⓐ 386 ⓑ 2,931

26. Round to the nearest ten.
ⓐ 792 ⓑ 5,647

27. Round to the nearest hundred.
ⓐ 13,748 ⓑ 391,794

28. Round to the nearest hundred.

ⓐ 28,166 ⓑ 481,628

29. Round to the nearest ten.

ⓐ 1,492 ⓑ 1,497

30. Round to the nearest ten.

ⓐ 2,791 ⓑ 2,795

31. Round to the nearest hundred.

ⓐ 63,994 ⓑ 63,040

32. Round to the nearest hundred.

ⓐ 49,584 ⓑ 49,548

In the following exercises, round each number to the nearest ⓐ hundred, ⓑ thousand, ⓒ ten thousand.

33. 392,546

34. 619,348

35. 2,586,991

36. 4,287,965

Identify Multiples and Factors

In the following exercises, use the divisibility tests to determine whether each number is divisible by 2, 3, 5, 6, and 10.

37. 84

38. 9,696

39. 75

40. 78

41. 900

42. 800

43. 986

44. 942

45. 350

46. 550

47. 22,335

48. 39,075

Find Prime Factorizations and Least Common Multiples

In the following exercises, find the prime factorization.

49. 86

50. 78

51. 132

52. 455

53. 693

54. 400

55. 432

56. 627

57. 2,160

58. 2,520

In the following exercises, find the least common multiple of the each pair of numbers using the multiples method.

59. 8, 12

60. 4, 3

61. 12, 16

62. 30, 40

63. 20, 30

64. 44, 55

In the following exercises, find the least common multiple of each pair of numbers using the prime factors method.

65. 8, 12

66. 12, 16

67. 28, 40

68. 84, 90

69. 55, 88

70. 60, 72

Everyday Math

71. Writing a Check Jorge bought a car for $24,493. He paid for the car with a check. Write the purchase price in words.

72. Writing a Check Marissa's kitchen remodeling cost $18,549. She wrote a check to the contractor. Write the amount paid in words.

73. Buying a Car Jorge bought a car for $24,493. Round the price to the nearest ⓐ ten ⓑ hundred ⓒ thousand; and ⓓ ten-thousand.

74. Remodeling a Kitchen Marissa's kitchen remodeling cost $18,549, Round the cost to the nearest ⓐ ten ⓑ hundred ⓒ thousand and ⓓ ten-thousand.

75. Population The population of China was 1,339,724,852 on November 1, 2010. Round the population to the nearest ⓐ billion ⓑ hundred-million; and ⓒ million.

76. Astronomy The average distance between Earth and the sun is 149,597,888 kilometers. Round the distance to the nearest ⓐ hundred-million ⓑ ten-million; and ⓒ million.

77. Grocery Shopping Hot dogs are sold in packages of 10, but hot dog buns come in packs of eight. What is the smallest number that makes the hot dogs and buns come out even?

78. Grocery Shopping Paper plates are sold in packages of 12 and party cups come in packs of eight. What is the smallest number that makes the plates and cups come out even?

Writing Exercises

79. Give an everyday example where it helps to round numbers.

80. If a number is divisible by 2 and by 3 why is it also divisible by 6?

81. What is the difference between prime numbers and composite numbers?

82. Explain in your own words how to find the prime factorization of a composite number, using any method you prefer.

Self Check

ⓐ *After completing the exercises, use this checklist to evaluate your mastery of the objectives of this section.*

I can...	Confidently	With some help	No-I don't get it!
use place value with whole numbers.			
identify multiples and apply divisibility tests.			
find prime factorizations and least common multiples.			

ⓑ *If most of your checks were:*

...confidently. Congratulations! You have achieved the objectives in this section. Reflect on the study skills you used so that you can continue to use them. What did you do to become confident of your ability to do these things? Be specific.

...with some help. This must be addressed quickly because topics you do not master become potholes in your road to success. In math, every topic builds upon previous work. It is important to make sure you have a strong foundation before you move on. Who can you ask for help? Your fellow classmates and instructor are good resources. Is there a place on campus where math tutors are available? Can your study skills be improved?

...no—I don't get it! This is a warning sign and you must not ignore it. You should get help right away or you will quickly be overwhelmed. See your instructor as soon as you can to discuss your situation. Together you can come up with a plan to get you the help you need.

1.2 | **Use the Language of Algebra**

Learning Objectives

By the end of this section, you will be able to:

> Use variables and algebraic symbols
> Simplify expressions using the order of operations
> Evaluate an expression
> Identify and combine like terms
> Translate an English phrase to an algebraic expression

Be Prepared!

A more thorough introduction to the topics covered in this section can be found in the *Prealgebra* chapter, **The Language of Algebra**.

Use Variables and Algebraic Symbols

Suppose this year Greg is 20 years old and Alex is 23. You know that Alex is 3 years older than Greg. When Greg was 12, Alex was 15. When Greg is 35, Alex will be 38. No matter what Greg's age is, Alex's age will always be 3 years more, right? In the language of algebra, we say that Greg's age and Alex's age are **variables** and the 3 is a **constant**. The ages change ("vary") but the 3 years between them always stays the same ("constant"). Since Greg's age and Alex's age will always differ by 3 years, 3 is the *constant*.

In algebra, we use letters of the alphabet to represent variables. So if we call Greg's age g, then we could use $g + 3$ to represent Alex's age. See Table 1.8.

Greg's age	Alex's age
12	15
20	23
35	38
g	$g + 3$

Table 1.8

The letters used to represent these changing ages are called *variables*. The letters most commonly used for variables are $x, y, a, b,$ and c.

Variable

A **variable** is a letter that represents a number whose value may change.

Constant

A **constant** is a number whose value always stays the same.

To write algebraically, we need some operation symbols as well as numbers and variables. There are several types of symbols we will be using.

There are four basic arithmetic operations: addition, subtraction, multiplication, and division. We'll list the symbols used to indicate these operations below (Table 1.8). You'll probably recognize some of them.

Operation	Notation	Say:	The result is...
Addition	$a + b$	a plus b	the sum of a and b
Subtraction	$a - b$	a minus b	the difference of a and b
Multiplication	$a \cdot b$, ab, $(a)(b)$, $(a)b$, $a(b)$	a times b	the product of a and b
Division	$a \div b$, a/b, $\frac{a}{b}$, $b\overline{)a}$	a divided by b	the quotient of a and b, a is called the dividend, and b is called the divisor

We perform these operations on two numbers. When translating from symbolic form to English, or from English to symbolic form, pay attention to the words "of" and "and."

- The *difference of* 9 and 2 means subtract 9 and 2, in other words, 9 minus 2, which we write symbolically as $9 - 2$.

- The *product of* 4 and 8 means multiply 4 and 8, in other words 4 times 8, which we write symbolically as $4 \cdot 8$.

In algebra, the cross symbol, $\times$, is not used to show multiplication because that symbol may cause confusion. Does $3xy$ mean $3 \times y$ ('three times y') or $3 \cdot x \cdot y$ (three times x times y)? To make it clear, use $\cdot$ or parentheses for multiplication.

When two quantities have the same value, we say they are equal and connect them with an **equal sign**.

Equality Symbol

$a = b$ is read "a is equal to b"

The symbol "$=$" is called the **equal sign**.

On the number line, the numbers get larger as they go from left to right. The number line can be used to explain the symbols "<" and ">."

Inequality

$a < b$ is read "a is less than b"
a is to the left of b on the number line

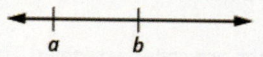

$a > b$ is read "a is greater than b"
a is to the right of b on the number line

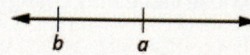

The expressions $a < b$ or $a > b$ can be read from left to right or right to left, though in English we usually read from left to right (Table 1.9). In general, $a < b$ is equivalent to $b > a$. For example 7 < 11 is equivalent to 11 > 7. And $a > b$ is equivalent to $b < a$. For example 17 > 4 is equivalent to 4 < 17.

Inequality Symbols	Words
$a \neq b$	a is *not equal to* b
$a < b$	a is *less than* b
$a \leq b$	a is *less than or equal to* b
$a > b$	a is *greater than* b
$a \geq b$	a is *greater than or equal to* b

Table 1.9

EXAMPLE 1.12

Translate from algebra into English:

ⓐ $17 \leq 26$ ⓑ $8 \neq 17 - 8$ ⓒ $12 > 27 \div 3$ ⓓ $y + 7 < 19$

✓ **Solution**

ⓐ $17 \leq 26$

17 is less than or equal to 26

ⓑ $8 \neq 17 - 8$

8 is not equal to 17 minus 3

ⓒ $12 > 27 \div 3$

12 is greater than 27 divided by 3

ⓓ $y + 7 < 19$

y plus 7 is less than 19

> **TRY IT : : 1.23**   Translate from algebra into English:
>
> ⓐ $14 \leq 27$ ⓑ $19 - 2 \neq 8$ ⓒ $12 > 4 \div 2$ ⓓ $x - 7 < 1$

> **TRY IT : : 1.24**   Translate from algebra into English:
>
> ⓐ $19 \geq 15$ ⓑ $7 = 12 - 5$ ⓒ $15 \div 3 < 8$ ⓓ $y + 3 > 6$

Grouping symbols in algebra are much like the commas, colons, and other punctuation marks in English. They help to make clear which expressions are to be kept together and separate from other expressions. We will introduce three types now.

Grouping Symbols

Parentheses	$()$
Brackets	$[]$
Braces	$\{\}$

Here are some examples of expressions that include grouping symbols. We will simplify expressions like these later in this section.

$$8(14 - 8) \quad 21 - 3[2 + 4(9 - 8)] \quad 24 \div \{13 - 2[1(6 - 5) + 4]\}$$

What is the difference in English between a phrase and a sentence? A phrase expresses a single thought that is incomplete by itself, but a sentence makes a complete statement. "Running very fast" is a phrase, but "The football player was running very fast" is a sentence. A sentence has a subject and a verb. In algebra, we have *expressions* and *equations*.

Expression

An **expression** is a number, a variable, or a combination of numbers and variables using operation symbols.

An **expression** is like an English phrase. Here are some examples of expressions:

Expression	Words	English Phrase
$3 + 5$	3 plus 5	the sum of three and five
$n - 1$	n minus one	the difference of n and one
$6 \cdot 7$	6 times 7	the product of six and seven
$\frac{x}{y}$	x divided by y	the quotient of x and y

Notice that the English phrases do not form a complete sentence because the phrase does not have a verb.

An **equation** is two expressions linked with an equal sign. When you read the words the symbols represent in an equation, you have a complete sentence in English. The equal sign gives the verb.

Equation

An **equation** is two expressions connected by an equal sign.

Here are some examples of equations.

Equation	English Sentence
$3 + 5 = 8$	The sum of three and five is equal to eight.
$n - 1 = 14$	n minus one equals fourteen.
$6 \cdot 7 = 42$	The product of six and seven is equal to forty-two.
$x = 53$	x is equal to fifty-three.
$y + 9 = 2y - 3$	y plus nine is equal to two y minus three.

EXAMPLE 1.13

Determine if each is an expression or an equation:

ⓐ $2(x + 3) = 10$ ⓑ $4(y - 1) + 1$ ⓒ $x \div 25$ ⓓ $y + 8 = 40$

✓ **Solution**

ⓐ $2(x + 3) = 10$ This is an *equation*—two expressions are connected with an equal sign.

ⓑ $4(y - 1) + 1$ This is an *expression*—no equal sign.

ⓒ $x \div 25$ This is an *expression*—no equal sign.

ⓓ $y + 8 = 40$ This is an *equation*—two expressions are connected with an equal sign.

> **TRY IT : : 1.25**   Determine if each is an expression or an equation: ⓐ $3(x - 7) = 27$ ⓑ $5(4y - 2) - 7$.

> **TRY IT : : 1.26**   Determine if each is an expression or an equation: ⓐ $y^3 \div 14$ ⓑ $4x - 6 = 22$.

Suppose we need to multiply 2 nine times. We could write this as $2 \cdot 2 \cdot 2 \cdot 2 \cdot 2 \cdot 2 \cdot 2 \cdot 2 \cdot 2$. This is tedious and it can be hard to keep track of all those 2s, so we use exponents. We write $2 \cdot 2 \cdot 2$ as 2^3 and $2 \cdot 2 \cdot 2 \cdot 2 \cdot 2 \cdot 2 \cdot 2 \cdot 2 \cdot 2$ as 2^9. In expressions such as 2^3, the 2 is called the *base* and the 3 is called the *exponent*. The exponent tells us how many times we need to multiply the base.

$\text{base} \rightarrow 2^3 \leftarrow \text{exponent}$ means multiply 2 by itself, three times, as in $2 \cdot 2 \cdot 2$.

We read 2^3 as "two to the third power" or "two cubed."

We say 2^3 is in *exponential notation* and $2 \cdot 2 \cdot 2$ is in *expanded notation*.

Exponential Notation

a^n means multiply a by itself, n times.

$\text{base} \rightarrow a^n \leftarrow \text{exponent}$

$$a^n = \underbrace{a \cdot a \cdot a \cdot \ldots \cdot a}_{n \text{ factors}}$$

The expression a^n is read a to the nth power.

While we read a^n as "a to the nth power," we usually read:

- a^2 "a squared"
- a^3 "a cubed"

We'll see later why a^2 and a^3 have special names.

Table 1.10 shows how we read some expressions with exponents.

Expression	In Words
7^2	7 to the second power or 7 squared
5^3	5 to the third power or 5 cubed
9^4	9 to the fourth power
12^5	12 to the fifth power

Table 1.10

EXAMPLE 1.14

Simplify: 3^4.

⊘ **Solution**

	3^4
Expand the expression.	$3 \cdot 3 \cdot 3 \cdot 3$
Multiply left to right.	$9 \cdot 3 \cdot 3$
Multiply.	$27 \cdot 3$
Multiply.	81

> **TRY IT :: 1.27**   Simplify: ⓐ 5^3 ⓑ 1^7.

> **TRY IT :: 1.28**   Simplify: ⓐ 7^2 ⓑ 0^5.

Simplify Expressions Using the Order of Operations

To **simplify an expression** means to do all the math possible. For example, to simplify $4 \cdot 2 + 1$ we'd first multiply $4 \cdot 2$ to get 8 and then add the 1 to get 9. A good habit to develop is to work down the page, writing each step of the process below the previous step. The example just described would look like this:

$$4 \cdot 2 + 1$$
$$8 + 1$$
$$9$$

By not using an equal sign when you simplify an expression, you may avoid confusing expressions with equations.

> **Simplify an Expression**
>
> To **simplify an expression**, do all operations in the expression.

We've introduced most of the symbols and notation used in algebra, but now we need to clarify the order of operations. Otherwise, expressions may have different meanings, and they may result in different values. For example, consider the expression:

$$4 + 3 \cdot 7$$

If you simplify this expression, what do you get?

Some students say 49,

	$4 + 3 \cdot 7$
Since $4 + 3$ gives 7.	$7 \cdot 7$
And $7 \cdot 7$ is 49.	49

Others say 25,

	$4 + 3 \cdot 7$
Since $3 \cdot 7$ is 21.	$4 + 21$
And $21 + 4$ makes 25.	25

Imagine the confusion in our banking system if every problem had several different correct answers!

The same expression should give the same result. So mathematicians early on established some guidelines that are called the Order of Operations.

HOW TO :: PERFORM THE ORDER OF OPERATIONS.

Step 1.   Parentheses and Other Grouping Symbols

⚬ Simplify all expressions inside the parentheses or other grouping symbols, working on the innermost parentheses first.

Step 2.   Exponents

⚬ Simplify all expressions with exponents.

Step 3.   Multiplication and Division

⚬ Perform all multiplication and division in order from left to right. These operations have equal priority.

Step 4.   Addition and Subtraction

⚬ Perform all addition and subtraction in order from left to right. These operations have equal priority.

MANIPULATIVE MATHEMATICS

Doing the Manipulative Mathematics activity "Game of 24" give you practice using the order of operations.

Students often ask, "How will I remember the order?" Here is a way to help you remember: Take the first letter of each key word and substitute the silly phrase: "Please Excuse My Dear Aunt Sally."

Parentheses	Please
Exponents	Excuse
Multiplication Division	My Dear
Addition Subtraction	Aunt Sally

It's good that "**My D**ear" goes together, as this reminds us that **m**ultiplication and **d**ivision have equal priority. We do not always do multiplication before division or always do division before multiplication. We do them in order from left to right.

Similarly, "**A**unt **S**ally" goes together and so reminds us that **a**ddition and **s**ubtraction also have equal priority and we do them in order from left to right.

Let's try an example.

___

EXAMPLE 1.15

Simplify: ⓐ $4 + 3 \cdot 7$ ⓑ $(4 + 3) \cdot 7$.

✓ **Solution**

ⓐ

	$4 + 3 \cdot 7$
Are there any **p**arentheses? No.	
Are there any **e**xponents? No.	
Is there any **m**ultiplication or **d**ivision? Yes.	
Multiply first.	$4 + 3 \cdot 7$
Add.	$4 + 21$
	25

ⓑ

	$(4 + 3) \cdot 7$
Are there any **p**arentheses? Yes.	$(4 + 3) \cdot 7$
Simplify inside the parentheses.	$(7)7$
Are there any **e**xponents? No.	
Is there any **m**ultiplication or **d**ivision? Yes.	
Multiply.	49

___

> **TRY IT :: 1.29**   Simplify: ⓐ $12 - 5 \cdot 2$ ⓑ $(12 - 5) \cdot 2$.

> **TRY IT :: 1.30**   Simplify: ⓐ $8 + 3 \cdot 9$ ⓑ $(8 + 3) \cdot 9$.

EXAMPLE 1.16

Simplify: $18 \div 6 + 4(5 - 2)$.

✓ Solution

Parentheses? Yes, subtract first.	$18 \div 6 + 4(5 - 2)$
	$18 \div 6 + 4(3)$
Exponents? No.	
Multiplication or division? Yes.	$18 \div 6 + 4(3)$
Divide first because we multiply and divide left to right.	$3 + 4(3)$
Any other multiplication or division? Yes.	
Multiply.	$3 + 12$
Any other multiplication or division? No.	
Any addition or subtraction? Yes.	15

> **TRY IT :: 1.31**   Simplify: $30 \div 5 + 10(3 - 2)$.

> **TRY IT :: 1.32**   Simplify: $70 \div 10 + 4(6 - 2)$.

When there are multiple grouping symbols, we simplify the innermost parentheses first and work outward.

EXAMPLE 1.17

Simplify: $5 + 2^3 + 3[6 - 3(4 - 2)]$.

✓ Solution

	$5 + 2^3 + 3[6 - 3(4 - 2)]$
Are there any parentheses (or other grouping symbol)? Yes.	
Focus on the parentheses that are inside the brackets.	$5 + 2^3 + 3[6 - 3(4 - 2)]$
Subtract.	$5 + 2^3 + 3[6 - 3(2)]$
Continue inside the brackets and multiply.	$5 + 2^3 + 3[6 - 6]$
Continue inside the brackets and subtract.	$5 + 2^3 + 3[0]$
The expression inside the brackets requires no further simplification.	
Are there any exponents? Yes.	$5 + 2^3 + 3[0]$
Simplify exponents.	$5 + 8 + 3[0]$
Is there any multiplication or division? Yes.	

Multiply.	$5 + 8 + 0$
Is there any addition or subtraction? Yes.	
Add.	$13 + 0$
Add.	13

> **TRY IT : : 1.33**   Simplify: $9 + 5^3 - [4(9 + 3)]$.

> **TRY IT : : 1.34**   Simplify: $7^2 - 2[4(5 + 1)]$.

Evaluate an Expression

In the last few examples, we simplified expressions using the order of operations. Now we'll evaluate some expressions—again following the order of operations. To **evaluate an expression** means to find the value of the expression when the variable is replaced by a given number.

Evaluate an Expression

To **evaluate an expression** means to find the value of the expression when the variable is replaced by a given number.

To evaluate an expression, substitute that number for the variable in the expression and then simplify the expression.

EXAMPLE 1.18

Evaluate $7x - 4$, when ⓐ $x = 5$ and ⓑ $x = 1$.

✓ **Solution**

ⓐ

when $x = 5$	$7x - 4$
	$7(5) - 4$
Multiply.	$35 - 4$
Subtract.	31

ⓑ

when $x = 1$	$7x - 4$
	$7(1) - 4$
Multiply.	$7 - 4$
Subtract.	3

> **TRY IT : : 1.35**   Evaluate $8x - 3$, when ⓐ $x = 2$ and ⓑ $x = 1$.

> **TRY IT : : 1.36**   Evaluate $4y - 4$, when ⓐ $y = 3$ and ⓑ $y = 5$.

EXAMPLE 1.19

Evaluate $x = 4$, when ⓐ x^2 ⓑ 3^x.

✓ **Solution**

ⓐ

	x^2
Replace x with 4.	4^2
Use definition of exponent.	$4 \cdot 4$
Simplify.	16

ⓑ

	3^x
Replace x with 4.	3^4
Use definition of exponent.	$3 \cdot 3 \cdot 3 \cdot 3$
Simplify.	81

> **TRY IT :: 1.37**   Evaluate $x = 3$, when ⓐ x^2 ⓑ 4^x.

> **TRY IT :: 1.38**   Evaluate $x = 6$, when ⓐ x^3 ⓑ 2^x.

EXAMPLE 1.20

Evaluate $2x^2 + 3x + 8$ when $x = 4$.

✓ **Solution**

	$2x^2 + 3x + 8$
Substitute $x = 4$.	$2(4)^2 + 3(4) + 8$
Follow the order of operations.	$2(16) + 3(4) + 8$
	$32 + 12 + 8$
	52

> **TRY IT :: 1.39**   Evaluate $3x^2 + 4x + 1$ when $x = 3$.

> **TRY IT :: 1.40**   Evaluate $6x^2 - 4x - 7$ when $x = 2$.

Indentify and Combine Like Terms

Algebraic expressions are made up of terms. A **term** is a constant, or the product of a constant and one or more variables.

Term

A **term** is a constant, or the product of a constant and one or more variables.

Examples of terms are 7, y, $5x^2$, $9a$, and b^5.

The constant that multiplies the variable is called the **coefficient**.

Coefficient

The **coefficient** of a term is the constant that multiplies the variable in a term.

Think of the coefficient as the number in front of the variable. The coefficient of the term $3x$ is 3. When we write x, the coefficient is 1, since $x = 1 \cdot x$.

EXAMPLE 1.21

Identify the coefficient of each term: ⓐ $14y$ ⓑ $15x^2$ ⓒ a.

⊘ Solution

ⓐ The coefficient of $14y$ is 14.

ⓑ The coefficient of $15x^2$ is 15.

ⓒ The coefficient of a is 1 since $a = 1\,a$.

> **TRY IT : : 1.41**   Identify the coefficient of each term: ⓐ $17x$ ⓑ $41b^2$ ⓒ z.

> **TRY IT : : 1.42**   Identify the coefficient of each term: ⓐ $9p$ ⓑ $13a^3$ ⓒ y^3.

Some terms share common traits. Look at the following 6 terms. Which ones seem to have traits in common?

$$5x \quad 7 \quad n^2 \quad 4 \quad 3x \quad 9n^2$$

The 7 and the 4 are both constant terms.

The 5x and the 3x are both terms with x.

The n^2 and the $9n^2$ are both terms with n^2.

When two terms are constants or have the same variable and exponent, we say they are **like terms**.

- 7 and 4 are like terms.
- 5x and 3x are like terms.
- x^2 and $9x^2$ are like terms.

Like Terms

Terms that are either constants or have the same variables raised to the same powers are called **like terms**.

EXAMPLE 1.22

Identify the like terms: y^3, $7x^2$, 14, 23, $4y^3$, 9x, $5x^2$.

✓ Solution

y^3 and $4y^3$ are like terms because both have y^3; the variable and the exponent match.

$7x^2$ and $5x^2$ are like terms because both have x^2; the variable and the exponent match.

14 and 23 are like terms because both are constants.
There is no other term like $9x$.

> **TRY IT : : 1.43**   Identify the like terms: $9,\ \ 2x^3,\ \ y^2,\ \ 8x^3,\ \ 15,\ \ 9y,\ \ 11y^2$.

> **TRY IT : : 1.44**   Identify the like terms: $4x^3,\ \ 8x^2,\ \ 19,\ 3x^2,\ \ 24,\ 6x^3$.

Adding or subtracting terms forms an expression. In the expression $2x^2 + 3x + 8,$ from **Example 1.20**, the three terms are $2x^2,\ 3x,$ and 8.

EXAMPLE 1.23

Identify the terms in each expression.

ⓐ $9x^2 + 7x + 12$ ⓑ $8x + 3y$

✓ Solution

ⓐ The terms of $9x^2 + 7x + 12$ are $9x^2,\ \ 7x$, and 12.

ⓑ The terms of $8x + 3y$ are $8x$ and $3y$.

> **TRY IT : : 1.45**   Identify the terms in the expression $4x^2 + 5x + 17$.

> **TRY IT : : 1.46**   Identify the terms in the expression $5x + 2y$.

If there are like terms in an expression, you can simplify the expression by combining the like terms. What do you think $4x + 7x + x$ would simplify to? If you thought $12x$, you would be right!

$$4x + 7x + x$$
$$x+x+x+x \quad +x+x+x+x+x+x+x \quad +x$$
$$12x$$

Add the coefficients and keep the same variable. It doesn't matter what x is—if you have 4 of something and add 7 more of the same thing and then add 1 more, the result is 12 of them. For example, 4 oranges plus 7 oranges plus 1 orange is 12 oranges. We will discuss the mathematical properties behind this later.

Simplify: $4x + 7x + x$.

Add the coefficients. $12x$

EXAMPLE 1.24   HOW TO COMBINE LIKE TERMS

Simplify: $2x^2 + 3x + 7 + x^2 + 4x + 5$.

✓ Solution

Step 1. Identify the like terms.	$2x^2 + 3x + 7 + x^2 + 4x + 5$
	$2x^2 + 3x + 7 + x^2 + 4x + 5$

Step 2. Rearrange the expression so the like terms are together.	$2x^2 + x^2 + 3x + 4x + 7 + 5$
Step 3. Combine like terms.	$3x^2 + 7x + 12$

> **TRY IT :: 1.47**   Simplify: $3x^2 + 7x + 9 + 7x^2 + 9x + 8$.

> **TRY IT :: 1.48**   Simplify: $4y^2 + 5y + 2 + 8y^2 + 4y + 5$.

HOW TO :: COMBINE LIKE TERMS.

Step 1.   Identify like terms.

Step 2.   Rearrange the expression so like terms are together.

Step 3.   Add or subtract the coefficients and keep the same variable for each group of like terms.

Translate an English Phrase to an Algebraic Expression

In the last section, we listed many operation symbols that are used in algebra, then we translated expressions and equations into English phrases and sentences. Now we'll reverse the process. We'll translate English phrases into algebraic expressions. The symbols and variables we've talked about will help us do that. Table 1.20 summarizes them.

Operation	Phrase	Expression
Addition	a plus b the sum of a and b a increased by b b more than a the total of a and b b added to a	$a + b$
Subtraction	a minus b the difference of a and b a decreased by b b less than a b subtracted from a	$a - b$
Multiplication	a times b the product of a and b twice a	$a \cdot b$, ab, $a(b)$, $(a)(b)$ $2a$
Division	a divided by b the quotient of a and b the ratio of a and b b divided into a	$a \div b$, a/b, $\frac{a}{b}$, $b\overline{)a}$

Table 1.20

Look closely at these phrases using the four operations:

the **sum** *of a and b*

the **difference** *of a and b*

the **product** *of a and b*

the **quotient** *of a and b*

Each phrase tells us to operate on two numbers. Look for the words *of* and *and* to find the numbers.

EXAMPLE 1.25

Translate each English phrase into an algebraic expression: ⓐ the difference of $17x$ and 5 ⓑ the quotient of $10x^2$ and 7.

✓ **Solution**

ⓐ The key word is *difference*, which tells us the operation is subtraction. Look for the words *of* and *and* to find the numbers to subtract.

the *difference of* 17x *and* 5

17x minus 5

17x – 5

ⓑ The key word is "quotient," which tells us the operation is division.

the *quotient of* 10x² *and* 7

divide 10x² by 7

10x² ÷ 7

This can also be written $10x^2/7$ or $\dfrac{10x^2}{7}$.

> **TRY IT : : 1.49**
>
> Translate the English phrase into an algebraic expression: ⓐ the difference of $14x^2$ and 13 ⓑ the quotient of 12x and 2.

> **TRY IT : : 1.50**
>
> Translate the English phrase into an algebraic expression: ⓐ the sum of $17y^2$ and 19 ⓑ the product of 7 and y.

How old will you be in eight years? What age is eight more years than your age now? Did you add 8 to your present age? Eight "more than" means 8 added to your present age. How old were you seven years ago? This is 7 years less than your age now. You subtract 7 from your present age. Seven "less than" means 7 subtracted from your present age.

EXAMPLE 1.26

Translate the English phrase into an algebraic expression: ⓐ Seventeen more than y ⓑ Nine less than $9x^2$.

✓ **Solution**

ⓐ The key words are *more than*. They tell us the operation is addition. *More than* means "added to."

Seventeen more than y

Seventeen added to y

y + 17

ⓑ The key words are *less than*. They tell us to subtract. *Less than* means "subtracted from."

$$\text{Nine less than } 9x^2$$
$$\text{Nine subtracted from } 9x^2$$
$$9x^2 - 9$$

> **TRY IT : : 1.51**

Translate the English phrase into an algebraic expression: ⓐ Eleven more than x ⓑ Fourteen less than $11a$.

> **TRY IT : : 1.52**

Translate the English phrase into an algebraic expression: ⓐ 13 more than z ⓑ 18 less than $8x$.

EXAMPLE 1.27

Translate the English phrase into an algebraic expression: ⓐ five times the sum of m and n ⓑ the sum of five times m and n.

✓ **Solution**

There are two operation words—*times* tells us to multiply and *sum* tells us to add.

ⓐ Because we are multiplying 5 times the sum we need parentheses around the sum of m and n, $(m + n)$. This forces us to determine the sum first. (Remember the order of operations.)

$$\text{five times the sum of } m \text{ and } n$$
$$5(m + n)$$

ⓑ To take a sum, we look for the words "of" and "and" to see what is being added. Here we are taking the sum *of* five times m and n.

$$\text{the sum of five times } m \text{ and } n$$
$$5m + n$$

> **TRY IT : : 1.53**

Translate the English phrase into an algebraic expression: ⓐ four times the sum of p and q ⓑ the sum of four times p and q.

> **TRY IT : : 1.54**

Translate the English phrase into an algebraic expression: ⓐ the difference of two times x and 8, ⓑ two times the difference of x and 8.

Later in this course, we'll apply our skills in algebra to solving applications. The first step will be to translate an English phrase to an algebraic expression. We'll see how to do this in the next two examples.

EXAMPLE 1.28

The length of a rectangle is 6 less than the width. Let w represent the width of the rectangle. Write an expression for the length of the rectangle.

✓ **Solution**

Write a phrase about the length of the rectangle.	6 less than the width
Substitute w for "the width."	6 less than w
Rewrite "less than" as "subtracted from."	6 subtracted from w
Translate the phrase into algebra.	$w - 6$

> | **TRY IT : : 1.55**

The length of a rectangle is 7 less than the width. Let *w* represent the width of the rectangle. Write an expression for the length of the rectangle.

> | **TRY IT : : 1.56**

The width of a rectangle is 6 less than the length. Let *l* represent the length of the rectangle. Write an expression for the width of the rectangle.

EXAMPLE 1.29

June has dimes and quarters in her purse. The number of dimes is three less than four times the number of quarters. Let *q* represent the number of quarters. Write an expression for the number of dimes.

⊘ **Solution**

Write the phrase about the number of dimes.	three less than four times the number of quarters
Substitute q for the number of quarters.	3 less than 4 times q
Translate "4 times q."	3 less than $4q$
Translate the phrase into algebra.	$4q - 3$

> | **TRY IT : : 1.57**

Geoffrey has dimes and quarters in his pocket. The number of dimes is eight less than four times the number of quarters. Let *q* represent the number of quarters. Write an expression for the number of dimes.

> | **TRY IT : : 1.58**

Lauren has dimes and nickels in her purse. The number of dimes is three more than seven times the number of nickels. Let *n* represent the number of nickels. Write an expression for the number of dimes.

1.2 EXERCISES

Practice Makes Perfect

Use Variables and Algebraic Symbols

In the following exercises, translate from algebra to English.

83. $16 - 9$

84. $3 \cdot 9$

85. $28 \div 4$

86. $x + 11$

87. $(2)(7)$

88. $(4)(8)$

89. $14 < 21$

90. $17 < 35$

91. $36 \geq 19$

92. $6n = 36$

93. $y - 1 > 6$

94. $y - 4 > 8$

95. $2 \leq 18 \div 6$

96. $a \neq 1 \cdot 12$

In the following exercises, determine if each is an expression or an equation.

97. $9 \cdot 6 = 54$

98. $7 \cdot 9 = 63$

99. $5 \cdot 4 + 3$

100. $x + 7$

101. $x + 9$

102. $y - 5 = 25$

Simplify Expressions Using the Order of Operations

In the following exercises, simplify each expression.

103. 5^3

104. 8^3

105. 2^8

106. 10^5

In the following exercises, simplify using the order of operations.

107. ⓐ $3 + 8 \cdot 5$ ⓑ $(3 + 8) \cdot 5$

108. ⓐ $2 + 6 \cdot 3$ ⓑ $(2 + 6) \cdot 3$

109. $2^3 - 12 \div (9 - 5)$

110. $3^2 - 18 \div (11 - 5)$

111. $3 \cdot 8 + 5 \cdot 2$

112. $4 \cdot 7 + 3 \cdot 5$

113. $2 + 8(6 + 1)$

114. $4 + 6(3 + 6)$

115. $4 \cdot 12/8$

116. $2 \cdot 36/6$

117. $(6 + 10) \div (2 + 2)$

118. $(9 + 12) \div (3 + 4)$

119. $20 \div 4 + 6 \cdot 5$

120. $33 \div 3 + 8 \cdot 2$

121. $3^2 + 7^2$

122. $(3 + 7)^2$

123. $3(1 + 9 \cdot 6) - 4^2$

124. $5(2 + 8 \cdot 4) - 7^2$

125. $2[1 + 3(10 - 2)]$

126. $5[2 + 4(3 - 2)]$

Evaluate an Expression

In the following exercises, evaluate the following expressions.

127. $7x + 8$ when $x = 2$

128. $8x - 6$ when $x = 7$

129. x^2 when $x = 12$

130. x^3 when $x = 5$

131. x^5 when $x = 2$

132. 4^x when $x = 2$

133. $x^2 + 3x - 7$ when $x = 4$

134. $6x + 3y - 9$ when
$x = 6, y = 9$

135. $(x - y)^2$ when
$x = 10, y = 7$

136. $(x + y)^2$ when $x = 6, y = 9$

137. $a^2 + b^2$ when $a = 3, b = 8$

138. $r^2 - s^2$ when $r = 12, s = 5$

139. $2l + 2w$ when
$l = 15, w = 12$

140. $2l + 2w$ when
$l = 18, w = 14$

Simplify Expressions by Combining Like Terms

In the following exercises, identify the coefficient of each term.

141. $8a$

142. $13m$

143. $5r^2$

144. $6x^3$

In the following exercises, identify the like terms.

145. $x^3, 8x, 14, 8y, 5, 8x^3$

146. $6z, 3w^2, 1, 6z^2, 4z, w^2$

147. $9a, a^2, 16, 16b^2, 4, 9b^2$

148. $3, 25r^2, 10s, 10r, 4r^2, 3s$

In the following exercises, identify the terms in each expression.

149. $15x^2 + 6x + 2$

150. $11x^2 + 8x + 5$

151. $10y^3 + y + 2$

152. $9y^3 + y + 5$

In the following exercises, simplify the following expressions by combining like terms.

153. $10x + 3x$

154. $15x + 4x$

155. $4c + 2c + c$

156. $6y + 4y + y$

157. $7u + 2 + 3u + 1$

158. $8d + 6 + 2d + 5$

159. $10a + 7 + 5a - 2 + 7a - 4$

160. $7c + 4 + 6c - 3 + 9c - 1$

161. $3x^2 + 12x + 11 + 14x^2 + 8x + 5$

162. $5b^2 + 9b + 10 + 2b^2 + 3b - 4$

Translate an English Phrase to an Algebraic Expression

In the following exercises, translate the phrases into algebraic expressions.

163. the difference of 14 and 9

164. the difference of 19 and 8

165. the product of 9 and 7

166. the product of 8 and 7

167. the quotient of 36 and 9

168. the quotient of 42 and 7

169. the sum of *8x* and *3x*

170. the sum of *13x* and *3x*

171. the quotient of *y* and 3

172. the quotient of *y* and 8

173. eight times the difference of *y* and nine

174. seven times the difference of *y* and one

175. Eric has rock and classical CDs in his car. The number of rock CDs is 3 more than the number of classical CDs. Let c represent the number of classical CDs. Write an expression for the number of rock CDs.

176. The number of girls in a second-grade class is 4 less than the number of boys. Let b represent the number of boys. Write an expression for the number of girls.

177. Greg has nickels and pennies in his pocket. The number of pennies is seven less than twice the number of nickels. Let n represent the number of nickels. Write an expression for the number of pennies.

178. Jeannette has $5 and $10 bills in her wallet. The number of fives is three more than six times the number of tens. Let t represent the number of tens. Write an expression for the number of fives.

Everyday Math

179. Car insurance Justin's car insurance has a $750 deductible per incident. This means that he pays $750 and his insurance company will pay all costs beyond $750. If Justin files a claim for $2,100.

ⓐ how much will he pay?

ⓑ how much will his insurance company pay?

180. Home insurance Armando's home insurance has a $2,500 deductible per incident. This means that he pays $2,500 and the insurance company will pay all costs beyond $2,500. If Armando files a claim for $19,400.

ⓐ how much will he pay?

ⓑ how much will the insurance company pay?

Writing Exercises

181. Explain the difference between an expression and an equation.

182. Why is it important to use the order of operations to simplify an expression?

183. Explain how you identify the like terms in the expression $8a^2 + 4a + 9 - a^2 - 1$.

184. Explain the difference between the phrases "4 times the sum of x and y" and "the sum of 4 times x and y."

Self Check

ⓐ Use this checklist to evaluate your mastery of the objectives of this section.

I can...	Confidently	With some help	No-I don't get it!
use variables and algebraic symbols.			
simplify expressions using the order of operations.			
evaluate an expression.			
identify and combine like terms.			
translate English phrases to algebraic expressions.			

ⓑ After reviewing this checklist, what will you do to become confident for all objectives?

1.3 **Add and Subtract Integers**

Learning Objectives

By the end of this section, you will be able to:

› Use negatives and opposites
› Simplify: expressions with absolute value
› Add integers
› Subtract integers

Be Prepared!

A more thorough introduction to the topics covered in this section can be found in the *Prealgebra* chapter, **Integers**.

Use Negatives and Opposites

Our work so far has only included the counting numbers and the whole numbers. But if you have ever experienced a temperature below zero or accidentally overdrawn your checking account, you are already familiar with negative numbers. **Negative numbers** are numbers less than 0. The negative numbers are to the left of zero on the number line. See **Figure 1.6**.

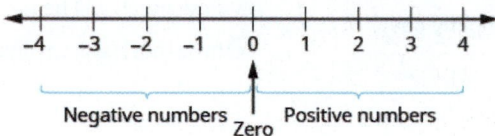

Figure 1.6 The number line shows the location of positive and negative numbers.

The arrows on the ends of the number line indicate that the numbers keep going forever. There is no biggest positive number, and there is no smallest negative number.

Is zero a positive or a negative number? Numbers larger than zero are positive, and numbers smaller than zero are negative. Zero is neither positive nor negative.

Consider how numbers are ordered on the number line. Going from left to right, the numbers increase in value. Going from right to left, the numbers decrease in value. See **Figure 1.7**.

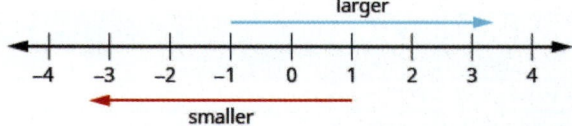

Figure 1.7 The numbers on a number line increase in value going from left to right and decrease in value going from right to left.

 MANIPULATIVE MATHEMATICS

Doing the Manipulative Mathematics activity "Number Line-part 2" will help you develop a better understanding of integers.

Remember that we use the notation:

$a < b$ (read "a is less than b") when a is to the left of b on the number line.

$a > b$ (read "a is greater than b") when a is to the right of b on the number line.

Now we need to extend the number line which showed the whole numbers to include negative numbers, too. The numbers marked by points in **Figure 1.8** are called the integers. The integers are the numbers

$$\ldots -3, -2, -1, 0, 1, 2, 3 \ldots$$

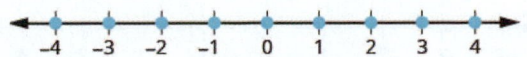

Figure 1.8 All the marked numbers are called *integers*.

EXAMPLE 1.30

Order each of the following pairs of numbers, using < or >: ⓐ 14___6 ⓑ −1___9 ⓒ −1___−4 ⓓ 2___−20.

✓ **Solution**

It may be helpful to refer to the number line shown.

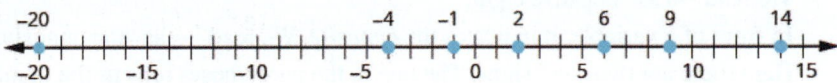

ⓐ

14 is to the right of 6 on the number line.	14___6
	14 > 6

ⓑ

−1 is to the left of 9 on the number line.	−1___9
	−1 < 9

ⓒ

−1 is to the right of −4 on the number line.	−1___−4
	−1 > −4

ⓓ

2 is to the right of −20 on the number line.	2___−20
	2 > −20

> **TRY IT : : 1.59**
>
> Order each of the following pairs of numbers, using < or > : ⓐ 15___7 ⓑ −2___5 ⓒ −3___−7
> ⓓ 5___−17.

> **TRY IT : : 1.60**
>
> Order each of the following pairs of numbers, using < or > : ⓐ 8___13 ⓑ 3___−4 ⓒ −5___−2
> ⓓ 9___−21.

You may have noticed that, on the number line, the negative numbers are a mirror image of the positive numbers, with zero in the middle. Because the numbers 2 and −2 are the same distance from zero, they are called **opposites**. The opposite of 2 is −2, and the opposite of −2 is 2.

Opposite

The **opposite** of a number is the number that is the same distance from zero on the number line but on the opposite side of zero.

Figure 1.9 illustrates the definition.

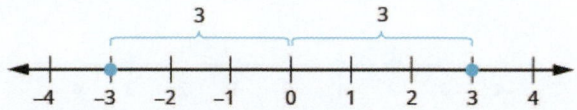

Figure 1.9 The opposite of 3 is -3.

Sometimes in algebra the same symbol has different meanings. Just like some words in English, the specific meaning becomes clear by looking at how it is used. You have seen the symbol "–" used in three different ways.

$10 - 4$	Between two numbers, it indicates the operation of *subtraction*.
	We read $10 - 4$ as "10 minus 4."
-8	In front of a number, it indicates a *negative* number.
	We read -8 as "negative eight."
$-x$	In front of a variable, it indicates the *opposite*. We read $-x$ as "the opposite of x."
$-(-2)$	Here there are two " – " signs. The one in the parentheses tells us the number is negative 2. The one outside the parentheses tells us to take the *opposite* of -2.
	We read $-(-2)$ as "the opposite of negative two."

Opposite Notation

$-a$ means the opposite of the number a.

The notation $-a$ is read as "the opposite of a."

EXAMPLE 1.31

Find: ⓐ the opposite of 7 ⓑ the opposite of -10 ⓒ $-(-6)$.

✓ Solution

ⓐ –7 is the same distance from 0 as 7, but on the opposite side of 0.

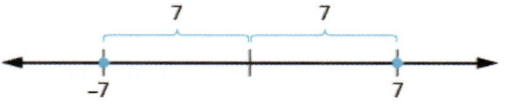

The opposite of 7 is –7.

ⓑ 10 is the same distance from 0 as –10, but on the opposite side of 0.

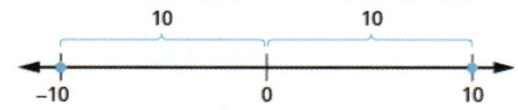

The opposite of –10 is 10.

ⓒ $-(-6)$

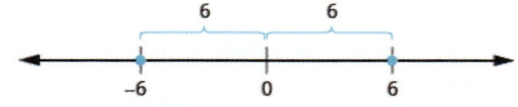

The opposite of –(–6) is –6.

 TRY IT : : 1.61   Find: ⓐ the opposite of 4 ⓑ the opposite of -3 ⓒ $-(-1)$.

 TRY IT : : 1.62   Find: ⓐ the opposite of 8 ⓑ the opposite of -5 ⓒ $-(-5)$.

Our work with opposites gives us a way to define the integers. The whole numbers and their opposites are called the **integers**. The integers are the numbers $\ldots -3, -2, -1, 0, 1, 2, 3\ldots$

> ### Integers
>
> The whole numbers and their opposites are called the **integers**.
>
> The integers are the numbers
>
> $$\ldots -3,\ -2,\ -1,\ 0,\ 1,\ 2,\ 3 \ldots$$

When evaluating the opposite of a variable, we must be very careful. Without knowing whether the variable represents a positive or negative number, we don't know whether $-x$ is positive or negative. We can see this in **Example 1.32**.

EXAMPLE 1.32

Evaluate ⓐ $-x,$ when $x = 8$ ⓑ $-x,$ when $x = -8.$

⊘ **Solution**

ⓐ

To evaluate when $x = 8$ means to substitute 8 for x.

	$-x$
Substitute 8 for x.	$-(8)$
Write the opposite of 8.	-8

ⓑ

To evaluate when $x = -8$ means to substitute –8 for –x.

	$-x$
Substitute –8 for x.	$-(-8)$
Write the opposite of –8.	8

> **TRY IT : : 1.63**   Evaluate $-n,$ when ⓐ $n = 4$ ⓑ $n = -4.$

> **TRY IT : : 1.64**   Evaluate $-m,$ when ⓐ $m = 11$ ⓑ $m = -11.$

Simplify: Expressions with Absolute Value

We saw that numbers such as 2 and -2 are opposites because they are the same distance from 0 on the number line. They are both two units from 0. The distance between 0 and any number on the number line is called the **absolute value** of that number.

> ### Absolute Value
>
> The **absolute value** of a number is its distance from 0 on the number line.
>
> The absolute value of a number n is written as $|n|$.

For example,

- -5 is 5 units away from $0,$ so $|-5| = 5.$

- 5 is 5 units away from $0,$ so $|5| = 5.$

Figure 1.10 illustrates this idea.

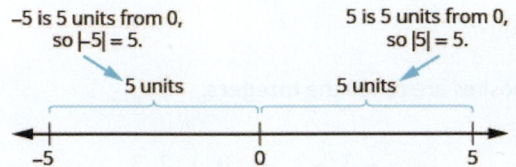

Figure 1.10 The integers 5 and are 5 units away from 0.

The absolute value of a number is never negative (because distance cannot be negative). The only number with absolute value equal to zero is the number zero itself, because the distance from 0 to 0 on the number line is zero units.

Property of Absolute Value

$|n| \geq 0$ for all numbers

Absolute values are always greater than or equal to zero!

Mathematicians say it more precisely, "absolute values are always non-negative." Non-negative means greater than or equal to zero.

EXAMPLE 1.33

Simplify: ⓐ $|3|$ ⓑ $|-44|$ ⓒ $|0|$.

✓ **Solution**

The absolute value of a number is the distance between the number and zero. Distance is never negative, so the absolute value is never negative.

ⓐ $|3|$

3

ⓑ $|-44|$

44

ⓒ $|0|$

0

> **TRY IT : : 1.65**   Simplify: ⓐ $|4|$ ⓑ $|-28|$ ⓒ $|0|$.

> **TRY IT : : 1.66**   Simplify: ⓐ $|-13|$ ⓑ $|47|$ ⓒ $|0|$.

In the next example, we'll order expressions with absolute values. Remember, positive numbers are always greater than negative numbers!

EXAMPLE 1.34

Fill in $<$, $>$, or $=$ for each of the following pairs of numbers:

ⓐ $|-5|$___ $-|-5|$ ⓑ 8___ $-|-8|$ ⓒ -9___ $-|-9|$ ⓓ $-(-16)$___ $-|-16|$

✓ **Solution**

ⓐ

|−5| ___ −|−5|

Simplify.   5 ___ −5

Order.   5 > −5

|−5| > −|−5|

ⓑ

8 ___ −|−8|

Simplify.   8 ___ −8

Order.   8 > −8

8 > −|−8|

ⓒ

9 ___ −|−9|

Simplify.   −9 ___ −9

Order.   −9 = −9

−9 = −|−9|

ⓓ

−(−16) ___ −|−16|

Simplify.   16 ___ −16

Order.   16 > −16

−(−16) > −|−16|

> **TRY IT : : 1.67**

Fill in <, >, or = for each of the following pairs of numbers: ⓐ |−9|___ −|−9| ⓑ 2___ −|−2| ⓒ −8___|−8|

ⓓ −(−9)___ −|−9|.

> **TRY IT : : 1.68**

Fill in <, >, or = for each of the following pairs of numbers: ⓐ 7___ −|−7| ⓑ −(−10)___ −|−10|

ⓒ |−4|___ −|−4| ⓓ −1___|−1|.

We now add absolute value bars to our list of grouping symbols. When we use the order of operations, first we simplify inside the absolute value bars as much as possible, then we take the absolute value of the resulting number.

Grouping Symbols

Parentheses	()	Braces	{ }
Brackets	[]	Absolute value	| |

In the next example, we simplify the expressions inside absolute value bars first, just like we do with parentheses.

EXAMPLE 1.35

Simplify: $24 - |19 - 3(6 - 2)|$.

⊘ Solution

	$24 - |19 - 3(6 - 2)|$
Work inside parentheses first: subtract 2 from 6.	$24 - |19 - 3(4)|$
Multiply 3(4).	$24 - |19 - 12|$
Subtract inside the absolute value bars.	$24 - |7|$
Take the absolute value.	$24 - 7$
Subtract.	17

> **TRY IT ::** 1.69   Simplify: $19 - |11 - 4(3 - 1)|$.

> **TRY IT ::** 1.70   Simplify: $9 - |8 - 4(7 - 5)|$.

EXAMPLE 1.36

Evaluate: ⓐ $|x|$ when $x = -35$ ⓑ $|-y|$ when $y = -20$ ⓒ $-|u|$ when $u = 12$ ⓓ $-|p|$ when $p = -14$.

⊘ Solution

ⓐ $|x|$ when $x = -35$

	$|x|$
Substitute -35 for x.	$|-35|$
Take the absolute value.	35

ⓑ $|-y|$ when $y = -20$

	$|-y|$
Substitute -20 for y.	$|-(-20)|$
Simplify.	$|20|$
Take the absolute value.	20

ⓒ $-|u|$ when $u = 12$

	$-|u|$
Substitute 12 for u.	$-|12|$
Take the absolute value.	-12

ⓓ $-|p|$ when $p = -14$

	$-|p|$
Substitute -14 for p.	$-|-14|$
Take the absolute value.	-14

> **TRY IT : : 1.71**
>
> Evaluate: ⓐ $|x|$ when $x = -17$ ⓑ $|-y|$ when $y = -39$ ⓒ $-|m|$ when $m = 22$ ⓓ $-|p|$ when $p = -11$.

> **TRY IT : : 1.72**
>
> Evaluate: ⓐ $|y|$ when $y = -23$ ⓑ $|-y|$ when $y = -21$ ⓒ $-|n|$ when $n = 37$ ⓓ $-|q|$ when $q = -49$.

Add Integers

Most students are comfortable with the addition and subtraction facts for positive numbers. But doing addition or subtraction with both positive and negative numbers may be more challenging.

 MANIPULATIVE MATHEMATICS

Doing the Manipulative Mathematics activity "Addition of Signed Numbers" will help you develop a better understanding of adding integers."

We will use two color counters to model addition and subtraction of negatives so that you can visualize the procedures instead of memorizing the rules.

We let one color (blue) represent positive. The other color (red) will represent the negatives. If we have one positive counter and one negative counter, the value of the pair is zero. They form a neutral pair. The value of this neutral pair is zero.

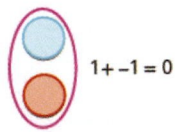

$$1 + -1 = 0$$

We will use the counters to show how to add the four addition facts using the numbers $5, -5$ and $3, -3$.

$$5 + 3 \qquad -5 + (-3) \qquad -5 + 3 \qquad 5 + (-3)$$

To add $5 + 3$, we realize that $5 + 3$ means the sum of 5 and 3.

We start with 5 positives.	5
And then we add 3 positives.	5 3
We now have 8 positives. The sum of 5 and 3 is 8.	8 positives

Now we will add $-5 + (-3)$. Watch for similarities to the last example $5 + 3 = 8$.

To add $-5 + (-3)$, we realize this means the sum of -5 and -3.

We start with 5 negatives.	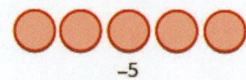
And then we add 3 negatives.	
We now have 8 negatives. The sum of –5 and –3 is –8.	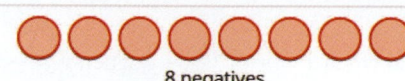

In what ways were these first two examples similar?
- The first example adds 5 positives and 3 positives—both positives.
- The second example adds 5 negatives and 3 negatives—both negatives.

In each case we got 8—either 8 positives or 8 negatives.

When the signs were the same, the counters were all the same color, and so we added them.

EXAMPLE 1.37

Add: ⓐ $1 + 4$ ⓑ $-1 + (-4)$.

✓ **Solution**

ⓐ

1 positive plus 4 positives is 5 positives.

ⓑ

1 negative plus 4 negatives is 5 negatives.

> **TRY IT :: 1.73**   Add: ⓐ $2 + 4$ ⓑ $-2 + (-4)$.

> **TRY IT :: 1.74**   Add: ⓐ $2 + 5$ ⓑ $-2 + (-5)$.

So what happens when the signs are different? Let's add $-5 + 3$. We realize this means the sum of -5 and 3. When the counters were the same color, we put them in a row. When the counters are a different color, we line them up under each other.

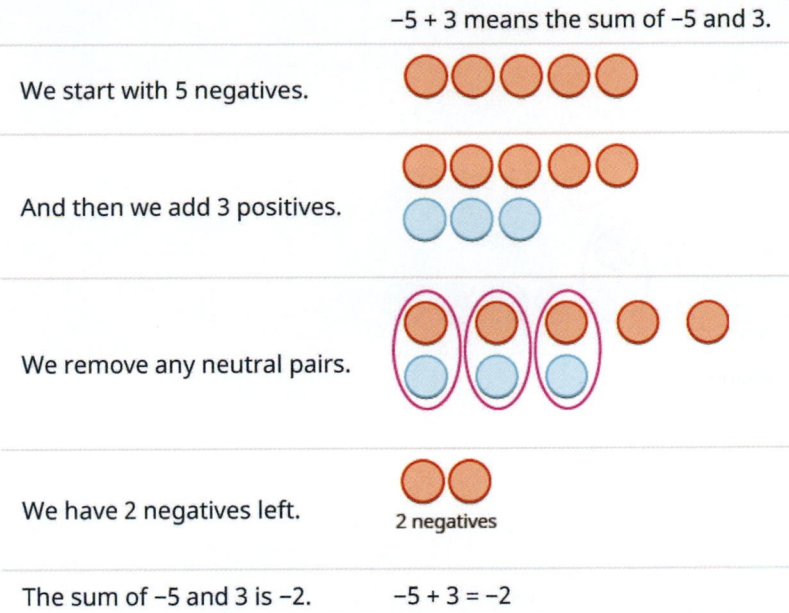

	−5 + 3 means the sum of −5 and 3.
We start with 5 negatives.	
And then we add 3 positives.	
We remove any neutral pairs.	
We have 2 negatives left.	2 negatives
The sum of −5 and 3 is −2.	−5 + 3 = −2

Notice that there were more negatives than positives, so the result was negative. Let's now add the last combination, $5 + (-3)$.

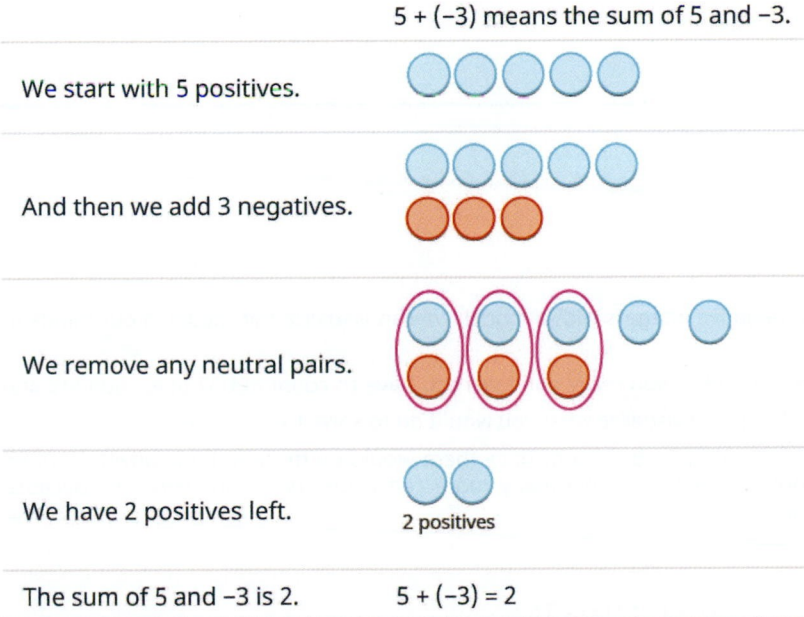

	5 + (−3) means the sum of 5 and −3.
We start with 5 positives.	
And then we add 3 negatives.	
We remove any neutral pairs.	
We have 2 positives left.	2 positives
The sum of 5 and −3 is 2.	5 + (−3) = 2

When we use counters to model addition of positive and negative integers, it is easy to see whether there are more positive or more negative counters. So we know whether the sum will be positive or negative.

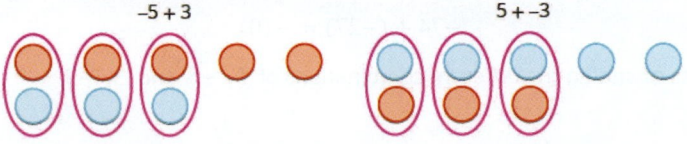

$-5 + 3$ $5 + -3$

More negatives – the sum is negative.   More positives – the sum is positive.

EXAMPLE 1.38

Add: ⓐ $-1 + 5$ ⓑ $1 + (-5)$.

✓ **Solution**

ⓐ

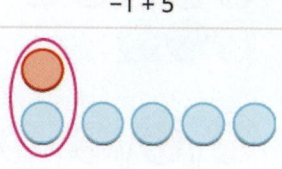

$-1 + 5$

There are more positives, so the sum is positive. 4

ⓑ

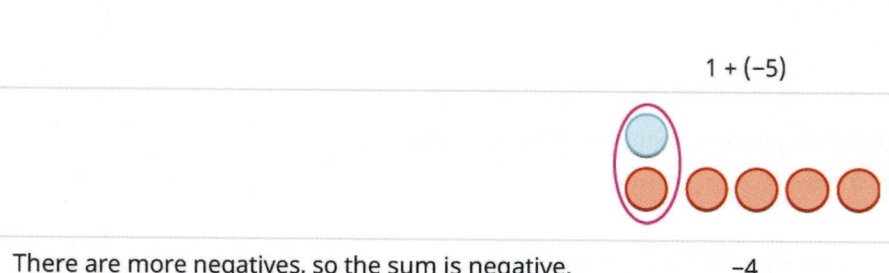

$1 + (-5)$

There are more negatives, so the sum is negative. -4

> **TRY IT : : 1.75** Add: ⓐ $-2 + 4$ ⓑ $2 + (-4)$.

> **TRY IT : : 1.76** Add: ⓐ $-2 + 5$ ⓑ $2 + (-5)$.

Now that we have added small positive and negative integers with a model, we can visualize the model in our minds to simplify problems with any numbers.

When you need to add numbers such as $37 + (-53)$, you really don't want to have to count out 37 blue counters and 53 red counters. With the model in your mind, can you visualize what you would do to solve the problem?

Picture 37 blue counters with 53 red counters lined up underneath. Since there would be more red (negative) counters than blue (positive) counters, the sum would be *negative*. How many more red counters would there be? Because $53 - 37 = 16$, there are 16 more red counters.

Therefore, the sum of $37 + (-53)$ is -16.

$$37 + (-53) = -16$$

Let's try another one. We'll add $-74 + (-27)$. Again, imagine 74 red counters and 27 more red counters, so we'd have 101 red counters. This means the sum is -101.

$$-74 + (-27) = -101$$

Let's look again at the results of adding the different combinations of $5, -5$ and $3, -3$.

Addition of Positive and Negative Integers

$5 + 3$	$-5 + (-3)$
8	-8
both positive, sum positive	both negative, sum negative

When the signs are the same, the counters would be all the same color, so add them.

$-5 + 3$	$5 + (-3)$
-2	2
different signs, more negatives, sum negative	different signs, more positives, sum positive

When the signs are different, some of the counters would make neutral pairs, so subtract to see how many are left.

Visualize the model as you simplify the expressions in the following examples.

EXAMPLE 1.39

Simplify: ⓐ $19 + (-47)$ ⓑ $-14 + (-36)$.

✓ Solution

ⓐ Since the signs are different, we subtract 19 from 47. The answer will be negative because there are more negatives than positives.

$$19 + (-47)$$

Add. -28

ⓑ Since the signs are the same, we add. The answer will be negative because there are only negatives.

$$-14 + (-36)$$

Add. -50

> **TRY IT :: 1.77**   Simplify: ⓐ $-31 + (-19)$ ⓑ $15 + (-32)$.

> **TRY IT :: 1.78**   Simplify: ⓐ $-42 + (-28)$ ⓑ $25 + (-61)$.

The techniques used up to now extend to more complicated problems, like the ones we've seen before. Remember to follow the order of operations!

EXAMPLE 1.40

Simplify: $-5 + 3(-2 + 7)$.

✓ Solution

$$-5 + 3(-2 + 7)$$

Simplify inside the parentheses.	$-5 + 3(5)$
Multiply.	$-5 + 15$
Add left to right.	10

> **TRY IT :: 1.79**   Simplify: $-2 + 5(-4 + 7)$.

> **TRY IT :: 1.80**   Simplify: $-4 + 2(-3 + 5)$.

Subtract Integers

 MANIPULATIVE MATHEMATICS

Doing the Manipulative Mathematics activity "Subtraction of Signed Numbers" will help you develop a better understanding of subtracting integers.

We will continue to use counters to model the subtraction. Remember, the blue counters represent positive numbers and the red counters represent negative numbers.

Perhaps when you were younger, you read "$5 - 3$" as "5 take away 3." When you use counters, you can think of subtraction the same way!

We will model the four subtraction facts using the numbers 5 and 3.

$$5 - 3 \qquad -5 - (-3) \qquad -5 - 3 \qquad 5 - (-3)$$

To subtract $5 - 3$, we restate the problem as "5 take away 3."

We start with 5 positives.	
We 'take away' 3 positives.	
We have 2 positives left.	
The difference of 5 and 3 is 2.	2

Now we will subtract $-5 - (-3)$. Watch for similarities to the last example $5 - 3 = 2$.

To subtract $-5 - (-3)$, we restate this as "−5 take away −3"

We start with 5 negatives.	
We 'take away' 3 negatives.	
We have 2 negatives left.	
The difference of −5 and −3 is −2.	−2

Notice that these two examples are much alike: The first example, we subtract 3 positives from 5 positives and end up with 2 positives.

In the second example, we subtract 3 negatives from 5 negatives and end up with 2 negatives.

Each example used counters of only one color, and the "take away" model of subtraction was easy to apply.

EXAMPLE 1.41

Subtract: ⓐ $7 - 5$ ⓑ $-7 - (-5)$.

⊘ Solution

Take 5 positives from 7 positives and get 2 positives.

$$7 - 5$$
$$2$$

ⓑ

Take 5 negatives from 7 negatives and get 2 negatives.

$$-7 - (-5)$$
$$-2$$

> **TRY IT : :** 1.81   Subtract: ⓐ $6 - 4$ ⓑ $-6 - (-4)$.

> **TRY IT : :** 1.82   Subtract: ⓐ $7 - 4$ ⓑ $-7 - (-4)$.

What happens when we have to subtract one positive and one negative number? We'll need to use both white and red counters as well as some neutral pairs. Adding a neutral pair does not change the value. It is like changing quarters to nickels—the value is the same, but it looks different.

- To subtract $-5 - 3$, we restate it as -5 take away 3.

We start with 5 negatives. We need to take away 3 positives, but we do not have any positives to take away.

Remember, a neutral pair has value zero. If we add 0 to 5 its value is still 5. We add neutral pairs to the 5 negatives until we get 3 positives to take away.

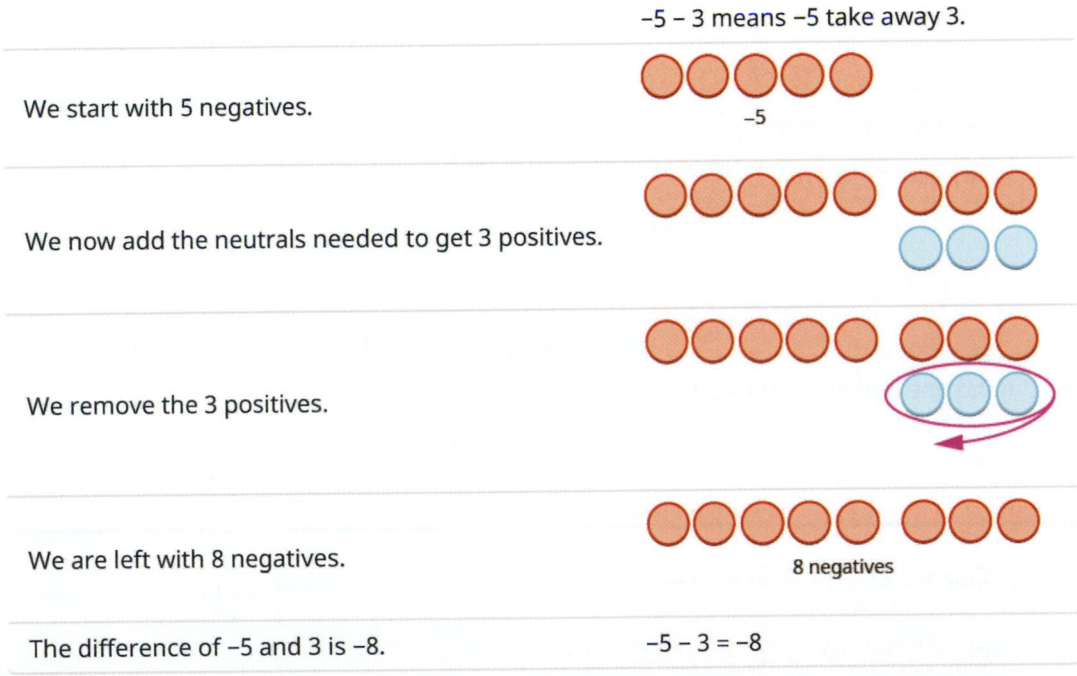

	−5 − 3 means −5 take away 3.
We start with 5 negatives.	−5
We now add the neutrals needed to get 3 positives.	
We remove the 3 positives.	
We are left with 8 negatives.	8 negatives
The difference of −5 and 3 is −8.	$-5 - 3 = -8$

And now, the fourth case, $5 - (-3)$. We start with 5 positives. We need to take away 3 negatives, but there are no negatives to take away. So we add neutral pairs until we have 3 negatives to take away.

5 – (–3) means 5 take away –3.

We start with 5 positives.	
We now add the needed neutrals pairs.	
We remove the 3 negatives.	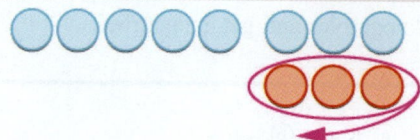
We are left with 8 positives.	 8 positives
The difference of 5 and –3 is 8.	5 – (–3) = 8

EXAMPLE 1.42

Subtract: ⓐ $-3 - 1$ ⓑ $3 - (-1)$.

✓ **Solution**

ⓐ

Take 1 positive from the one added neutral pair.

$-3 - 1$

-4

ⓑ

Take 1 negative from the one added neutral pair.

$3 - (-1)$

4

> **TRY IT : : 1.83**   Subtract: ⓐ $-6 - 4$ ⓑ $6 - (-4)$.

> **TRY IT : : 1.84**   Subtract: ⓐ $-7 - 4$ ⓑ $7 - (-4)$.

Have you noticed that *subtraction of signed numbers can be done by adding the opposite*? In **Example 1.42**, $-3 - 1$ is the same as $-3 + (-1)$ and $3 - (-1)$ is the same as $3 + 1$. You will often see this idea, the **subtraction property**, written as follows:

Subtraction Property

$$a - b = a + (-b)$$

Subtracting a number is the same as adding its opposite.

Look at these two examples.

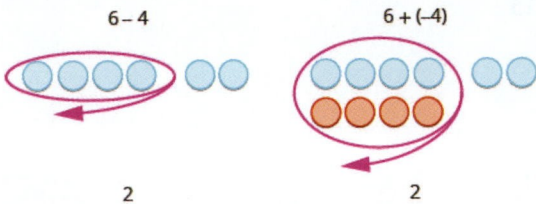

$6 - 4$ gives the same answer as $6 + (-4)$.

Of course, when you have a subtraction problem that has only positive numbers, like $6 - 4$, you just do the subtraction. You already knew how to subtract $6 - 4$ long ago. But *knowing* that $6 - 4$ gives the same answer as $6 + (-4)$ helps when you are subtracting negative numbers. Make sure that you understand how $6 - 4$ and $6 + (-4)$ give the same results!

EXAMPLE 1.43

Simplify: ⓐ $13 - 8$ and $13 + (-8)$ ⓑ $-17 - 9$ and $-17 + (-9)$.

✓ **Solution**

ⓐ

	$13 - 8$	and	$13 + (-8)$
Subtract.	5		5

ⓑ

	$-17 - 9$	and	$-17 + (-9)$
Subtract.	-26		-26

> **TRY IT : : 1.85**   Simplify: ⓐ $21 - 13$ and $21 + (-13)$ ⓑ $-11 - 7$ and $-11 + (-7)$.

> **TRY IT : : 1.86**   Simplify: ⓐ $15 - 7$ and $15 + (-7)$ ⓑ $-14 - 8$ and $-14 + (-8)$.

Look at what happens when we subtract a negative.

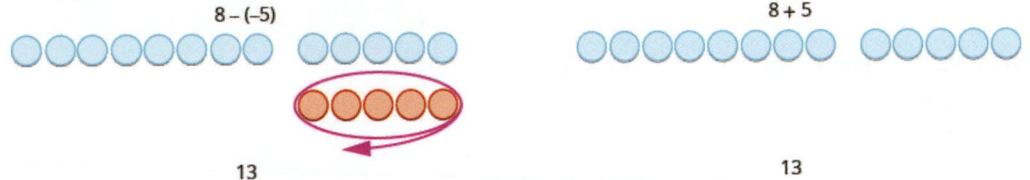

$8 - (-5)$ gives the same answer as $8 + 5$

Subtracting a negative number is like adding a positive!

You will often see this written as $a - (-b) = a + b$.

Does that work for other numbers, too? Let's do the following example and see.

EXAMPLE 1.44

Simplify: ⓐ $9 - (-15)$ and $9 + 15$ ⓑ $-7 - (-4)$ and $-7 + 4$.

Solution

ⓐ

	$9 - (-15)$	$9 + 15$
Subtract.	24	24

ⓑ

	$-7 - (-4)$	$-7 + 4$
Subtract.	-3	-3

> **TRY IT :: 1.87**   Simplify: ⓐ $6 - (-13)$ and $6 + 13$ ⓑ $-5 - (-1)$ and $-5 + 1$.

> **TRY IT :: 1.88**   Simplify: ⓐ $4 - (-19)$ and $4 + 19$ ⓑ $-4 - (-7)$ and $-4 + 7$.

Let's look again at the results of subtracting the different combinations of 5, -5 and 3, -3.

Subtraction of Integers

$5 - 3$	$-5 - (-3)$
2	-2
5 positives take away 3 positives	5 negatives take away 3 negatives
2 positives	2 negatives

When there would be enough counters of the color to take away, subtract.

$-5 - 3$	$5 - (-3)$
-8	8
5 negatives, want to take away 3 positives	5 positives, want to take away 3 negatives
need neutral pairs	need neutral pairs

When there would be not enough counters of the color to take away, add.

What happens when there are more than three integers? We just use the order of operations as usual.

EXAMPLE 1.45

Simplify: $7 - (-4 - 3) - 9$.

Solution

	$7 - (-4 - 3) - 9$
Simplify inside the parentheses first.	$7 - (-7) - 9$
Subtract left to right.	$14 - 9$
Subtract.	5

> **TRY IT :: 1.89**   Simplify: $8 - (-3 - 1) - 9$.

> **TRY IT :: 1.90**   Simplify: $12 - (-9 - 6) - 14$.

▶ **MEDIA ::**

Access these online resources for additional instruction and practice with adding and subtracting integers. You will need to enable Java in your web browser to use the applications.

- **Add Colored Chip (https://openstax.org/l/11AddColorChip)**
- **Subtract Colored Chip (https://openstax.org/l/11SubtrColorChp)**

1.3 EXERCISES

Practice Makes Perfect

Use Negatives and Opposites of Integers

In the following exercises, order each of the following pairs of numbers, using < or >.

185.

ⓐ 9___4

ⓑ −3___6

ⓒ −8___−2

ⓓ 1___−10

186.

ⓐ −7___3

ⓑ −10___−5

ⓒ 2___−6

ⓓ 8___9

In the following exercises, find the opposite of each number.

187.

ⓐ 2

ⓑ −6

188.

ⓐ 9

ⓑ −4

In the following exercises, simplify.

189. $-(-4)$

190. $-(-8)$

191. $-(-15)$

192. $-(-11)$

In the following exercises, evaluate.

193. $-c$ when

ⓐ $c = 12$

ⓑ $c = -12$

194. $-d$ when

ⓐ $d = 21$

ⓑ $d = -21$

Simplify Expressions with Absolute Value

In the following exercises, simplify.

195.

ⓐ $|-32|$

ⓑ $|0|$

ⓒ $|16|$

196.

ⓐ $|0|$

ⓑ $|-40|$

ⓒ $|22|$

In the following exercises, fill in <, >, or = for each of the following pairs of numbers.

197.

ⓐ -6___$|-6|$

ⓑ $-|-3|$___-3

198.

ⓐ $|-5|$___$-|-5|$

ⓑ 9___$-|-9|$

In the following exercises, simplify.

199. $-(-5)$ and $-|-5|$

200. $-|-9|$ and $-(-9)$

201. $8|-7|$

202. $5|-5|$

203. $|15 - 7| - |14 - 6|$

204. $|17 - 8| - |13 - 4|$

205. $18 - |2(8 - 3)|$

206. $18 - |3(8 - 5)|$

In the following exercises, evaluate.

207.

ⓐ $-|p|$ when $p = 19$

ⓑ $-|q|$ when $q = -33$

208.

ⓐ $-|a|$ when $a = 60$

ⓑ $-|b|$ when $b = -12$

Add Integers

In the following exercises, simplify each expression.

209. $-21 + (-59)$

210. $-35 + (-47)$

211. $48 + (-16)$

212. $34 + (-19)$

213. $-14 + (-12) + 4$

214. $-17 + (-18) + 6$

215. $135 + (-110) + 83$

216. $6 - 38 + 27 + (-8) + 126$

217. $19 + 2(-3 + 8)$

218. $24 + 3(-5 + 9)$

Subtract Integers

In the following exercises, simplify.

219. $8 - 2$

220. $-6 - (-4)$

221. $-5 - 4$

222. $-7 - 2$

223. $8 - (-4)$

224. $7 - (-3)$

225.

ⓐ $44 - 28$

ⓑ $44 + (-28)$

226.

ⓐ $35 - 16$

ⓑ $35 + (-16)$

227.

ⓐ $27 - (-18)$

ⓑ $27 + 18$

228.

ⓐ $46 - (-37)$

ⓑ $46 + 37$

In the following exercises, simplify each expression.

229. $15 - (-12)$

230. $14 - (-11)$

231. $48 - 87$

232. $45 - 69$

233. $-17 - 42$

234. $-19 - 46$

235. $-103 - (-52)$

236. $-105 - (-68)$

237. $-45 - (54)$

238. $-58 - (-67)$

239. $8 - 3 - 7$

240. $9 - 6 - 5$

241. $-5 - 4 + 7$

242. $-3 - 8 + 4$

243. $-14 - (-27) + 9$

244. $64 + (-17) - 9$

245. $(2 - 7) - (3 - 8)(2)$

246. $(1 - 8) - (2 - 9)$

247. $-(6 - 8) - (2 - 4)$

248. $-(4 - 5) - (7 - 8)$

249. $25 - [10 - (3 - 12)]$

250. $32 - [5 - (15 - 20)]$

251. $6.3 - 4.3 - 7.2$

252. $5.7 - 8.2 - 4.9$

253. $5^2 - 6^2$

254. $6^2 - 7^2$

Everyday Math

255. Elevation The highest elevation in the United States is Mount McKinley, Alaska, at 20,320 feet above sea level. The lowest elevation is Death Valley, California, at 282 feet below sea level.

Use integers to write the elevation of:

ⓐ Mount McKinley.

ⓑ Death Valley.

256. Extreme temperatures The highest recorded temperature on Earth was $58°$ Celsius, recorded in the Sahara Desert in 1922. The lowest recorded temperature was $90°$ below $0°$ Celsius, recorded in Antarctica in 1983.

Use integers to write the:

ⓐ highest recorded temperature.

ⓑ lowest recorded temperature.

257. State budgets In June, 2011, the state of Pennsylvania estimated it would have a budget surplus of $540 million. That same month, Texas estimated it would have a budget deficit of $27 billion.

Use integers to write the budget of:

ⓐ Pennsylvania.

ⓑ Texas.

258. College enrollments Across the United States, community college enrollment grew by 1,400,000 students from Fall 2007 to Fall 2010. In California, community college enrollment declined by 110,171 students from Fall 2009 to Fall 2010.

Use integers to write the change in enrollment:

ⓐ in the U.S. from Fall 2007 to Fall 2010.

ⓑ in California from Fall 2009 to Fall 2010.

259. Stock Market The week of September 15, 2008 was one of the most volatile weeks ever for the US stock market. The closing numbers of the Dow Jones Industrial Average each day were:

Monday	-504
Tuesday	$+142$
Wednesday	-449
Thursday	$+410$
Friday	$+369$

What was the overall change for the week? Was it positive or negative?

260. Stock Market During the week of June 22, 2009, the closing numbers of the Dow Jones Industrial Average each day were:

Monday	-201
Tuesday	-16
Wednesday	-23
Thursday	$+172$
Friday	-34

What was the overall change for the week? Was it positive or negative?

Writing Exercises

261. Give an example of a negative number from your life experience.

262. What are the three uses of the " $-$ " sign in algebra? Explain how they differ.

263. Explain why the sum of -8 and 2 is negative, but the sum of 8 and -2 is positive.

264. Give an example from your life experience of adding two negative numbers.

Self Check

ⓐ After completing the exercises, use this checklist to evaluate your mastery of the objectives of this section.

I can...	Confidently	With some help	No-I don't get it!
use negatives and opposites of integers.			
simplify expressions with absolute value.			
add integers.			
subtract integers.			

ⓑ *What does this checklist tell you about your mastery of this section? What steps will you take to improve?*

Multiply and Divide Integers

Learning Objectives

By the end of this section, you will be able to:
> Multiply integers
> Divide integers
> Simplify expressions with integers
> Evaluate variable expressions with integers
> Translate English phrases to algebraic expressions
> Use integers in applications

Be Prepared!

A more thorough introduction to the topics covered in this section can be found in the *Prealgebra* chapter, **Integers**.

Multiply Integers

Since multiplication is mathematical shorthand for repeated addition, our model can easily be applied to show multiplication of integers. Let's look at this concrete model to see what patterns we notice. We will use the same examples that we used for addition and subtraction. Here, we will use the model just to help us discover the pattern.

We remember that $a \cdot b$ means add a, b times. Here, we are using the model just to help us discover the pattern.

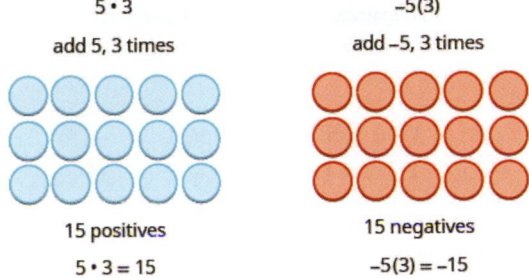

The next two examples are more interesting.

What does it mean to multiply 5 by -3? It means subtract 5, 3 times. Looking at subtraction as "taking away," it means to take away 5, 3 times. But there is nothing to take away, so we start by adding neutral pairs on the workspace. Then we take away 5 three times.

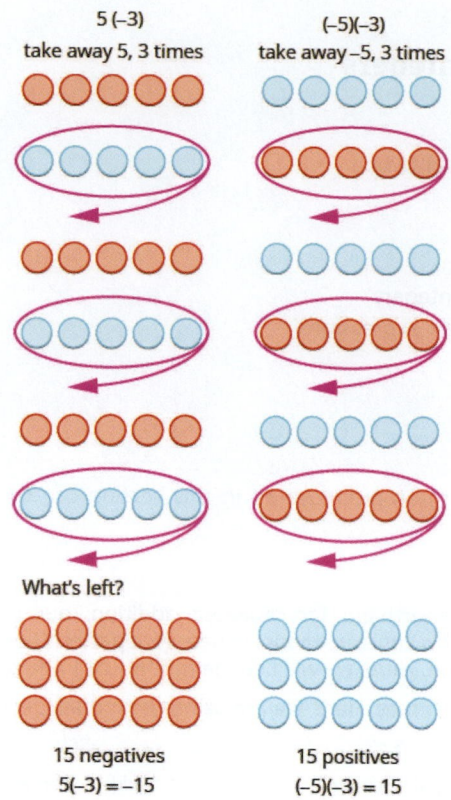

In summary:

$$5 \cdot 3 \;=\; 15 \qquad\quad -5(3) \;=\; -15$$
$$5(-3) \;=\; -15 \qquad (-5)(-3) \;=\; 15$$

Notice that for multiplication of two signed numbers, when the:

- signs are the *same*, the product is *positive*.
- signs are *different*, the product is *negative*.

We'll put this all together in the chart below.

Multiplication of Signed Numbers

For multiplication of two signed numbers:

Same signs	Product	Example
Two positives	Positive	$7 \cdot 4 \;=\; 28$
Two negatives	Positive	$-8(-6) \;=\; 48$

Different signs	Product	Example
Positive · negative	Negative	$7(-9) \;=\; -63$
Negative · positive	Negative	$-5 \cdot 10 \;=\; -50$

EXAMPLE 1.46

Multiply: ⓐ $-9 \cdot 3$ ⓑ $-2(-5)$ ⓒ $4(-8)$ ⓓ $7 \cdot 6$.

✓ **Solution**

ⓐ

$$-9 \cdot 3$$

Multiply, noting that the signs are different so the product is negative.

$$-27$$

ⓑ

$$-2(-5)$$

Multiply, noting that the signs are the same so the product is positive.

$$10$$

ⓒ

$$4(-8)$$

Multiply, with different signs.

$$-32$$

ⓓ

$$7 \cdot 6$$

Multiply, with same signs.

$$42$$

> **TRY IT :: 1.91**   Multiply: ⓐ $-6 \cdot 8$ ⓑ $-4(-7)$ ⓒ $9(-7)$ ⓓ $5 \cdot 12$.

> **TRY IT :: 1.92**   Multiply: ⓐ $-8 \cdot 7$ ⓑ $-6(-9)$ ⓒ $7(-4)$ ⓓ $3 \cdot 13$.

When we multiply a number by 1, the result is the same number. What happens when we multiply a number by -1? Let's multiply a positive number and then a negative number by -1 to see what we get.

$$-1 \cdot 4 \qquad\qquad -1(-3)$$

Multiply.

$$-4 \qquad\qquad\qquad 3$$

-4 is the opposite of 4.   3 is the opposite of -3.

Each time we multiply a number by -1, we get its opposite!

Multiplication by -1

$$-1a = -a$$

Multiplying a number by -1 gives its opposite.

EXAMPLE 1.47

Multiply: ⓐ $-1 \cdot 7$ ⓑ $-1(-11)$.

✓ **Solution**

ⓐ

$$-1 \cdot 7$$

Multiply, noting that the signs are different so the product is negative.

$$-7$$

-7 is the opposite of 7.

ⓑ

Multiply, noting that the signs are the same
so the product is positive.

$$-1(-11)$$
$$11$$

11 is the opposite of -11.

> **TRY IT : : 1.93**   Multiply: ⓐ $-1 \cdot 9$ ⓑ $-1 \cdot (-17)$.

> **TRY IT : : 1.94**   Multiply: ⓐ $-1 \cdot 8$ ⓑ $-1 \cdot (-16)$.

Divide Integers

What about division? Division is the inverse operation of multiplication. So, $15 \div 3 = 5$ because $15 \cdot 3 = 5$. In words, this expression says that 15 can be divided into three groups of five each because adding five three times gives 15. Look at some examples of multiplying integers, to figure out the rules for dividing integers.

$$5 \cdot 3 = 15 \text{ so } 15 \div 3 = 5 \qquad -5(3) = -15 \text{ so } -15 \div 3 = -5$$
$$(-5)(-3) = 15 \text{ so } 15 \div (-3) = -5 \qquad 5(-3) = -15 \text{ so } -15 \div (-3) = 5$$

Division follows the same rules as multiplication!

For division of two signed numbers, when the:

- signs are the *same*, the quotient is *positive*.
- signs are *different*, the quotient is *negative*.

And remember that we can always check the answer of a division problem by multiplying.

Multiplication and Division of Signed Numbers

For multiplication and division of two signed numbers:

- If the signs are the same, the result is positive.
- If the signs are different, the result is negative.

Same signs	Result
Two positives	Positive
Two negatives	Positive
If the signs are the same, the result is positive.	

Different signs	Result
Positive and negative	Negative
Negative and positive	Negative
If the signs are different, the result is negative.	

EXAMPLE 1.48

Divide: ⓐ $-27 \div 3$ ⓑ $-100 \div (-4)$.

✓ Solution

ⓐ

$$-27 \div 3$$

Divide, with different signs, the quotient is negative.

$$-9$$

ⓑ

$$-100 \div (-4)$$

Divide, with signs that are the same the quotient is positive.

$$25$$

> **TRY IT :: 1.95**   Divide: ⓐ $-42 \div 6$ ⓑ $-117 \div (-3)$.

> **TRY IT :: 1.96**   Divide: ⓐ $-63 \div 7$ ⓑ $-115 \div (-5)$.

Simplify Expressions with Integers

What happens when there are more than two numbers in an expression? The order of operations still applies when negatives are included. Remember My Dear Aunt Sally?

Let's try some examples. We'll simplify expressions that use all four operations with integers—addition, subtraction, multiplication, and division. Remember to follow the order of operations.

EXAMPLE 1.49

Simplify: $7(-2) + 4(-7) - 6$.

✓ Solution

$$7(-2) + 4(-7) - 6$$

Multiply first. $-14 + (-28) - 6$

Add. $-42 - 6$

Subtract. -48

> **TRY IT :: 1.97**   Simplify: $8(-3) + 5(-7) - 4$.

> **TRY IT :: 1.98**   Simplify: $9(-3) + 7(-8) - 1$.

EXAMPLE 1.50

Simplify: ⓐ $(-2)^4$ ⓑ -2^4.

✓ Solution

ⓐ

$$(-2)^4$$

Write in expanded form. $(-2)(-2)(-2)(-2)$

Multiply. $4(-2)(-2)$

Multiply. $-8(-2)$

Multiply. 16

ⓑ

	-2^4
Write in expanded form. We are asked to find the opposite of 2^4.	$-(2 \cdot 2 \cdot 2 \cdot 2)$
Multiply.	$-(4 \cdot 2 \cdot 2)$
Multiply.	$-(8 \cdot 2)$
Multiply.	-16

Notice the difference in parts ⓐ and ⓑ. In part ⓐ, the exponent means to raise what is in the parentheses, the (-2) to the 4th power. In part ⓑ, the exponent means to raise just the 2 to the 4th power and then take the opposite.

> **TRY IT : : 1.99**   Simplify: ⓐ $(-3)^4$ ⓑ -3^4.

> **TRY IT : : 1.100**   Simplify: ⓐ $(-7)^2$ ⓑ -7^2.

The next example reminds us to simplify inside parentheses first.

EXAMPLE 1.51

Simplify: $12 - 3(9 - 12)$.

⊘ **Solution**

	$12 - 3(9 - 12)$
Subtract in parentheses first.	$12 - 3(-3)$
Multiply.	$12 - (-9)$
Subtract.	21

> **TRY IT : : 1.101**   Simplify: $17 - 4(8 - 11)$.

> **TRY IT : : 1.102**   Simplify: $16 - 6(7 - 13)$.

EXAMPLE 1.52

Simplify: $8(-9) \div (-2)^3$.

⊘ **Solution**

	$8(-9) \div (-2)^3$
Exponents first.	$8(-9) \div (-8)$
Multiply.	$-72 \div (-8)$
Divide.	9

> **TRY IT : : 1.103**   Simplify: $12(-9) \div (-3)^3$.

> **TRY IT : : 1.104**   Simplify: $18(-4) \div (-2)^3$.

EXAMPLE 1.53

Simplify: $-30 \div 2 + (-3)(-7)$.

✓ Solution

	$-30 \div 2 + (-3)(-7)$
Multiply and divide left to right, so divide first.	$-15 + (-3)(-7)$
Multiply.	$-15 + 21$
Add.	6

> **TRY IT :: 1.105**   Simplify: $-27 \div 3 + (-5)(-6)$.

> **TRY IT :: 1.106**   Simplify: $-32 \div 4 + (-2)(-7)$.

Evaluate Variable Expressions with Integers

Remember that to evaluate an expression means to substitute a number for the variable in the expression. Now we can use negative numbers as well as positive numbers.

EXAMPLE 1.54

When $n = -5$, evaluate: ⓐ $n + 1$ ⓑ $-n + 1$.

✓ Solution

ⓐ

	$n + 1$
Substitute –5 for n.	$-5 + 1$
Simplify.	-4

ⓑ

	$-n + 1$
Substitute –5 for n.	$-(-5) + 1$
Simplify.	$5 + 1$
Add.	6

> **TRY IT :: 1.107**   When $n = -8$, evaluate ⓐ $n + 2$ ⓑ $-n + 2$.

> **TRY IT :: 1.108**   When $y = -9$, evaluate ⓐ $y + 8$ ⓑ $-y + 8$.

EXAMPLE 1.55

Evaluate $(x + y)^2$ when $x = -18$ and $y = 24$.

⊘ Solution

	$(x + y)^2$
Substitute -18 for x and 24 for y.	$(-18 + 24)^2$
Add inside parenthesis.	$(6)^2$
Simplify.	36

> **TRY IT : : 1.109**   Evaluate $(x + y)^2$ when $x = -15$ and $y = 29$.

> **TRY IT : : 1.110**   Evaluate $(x + y)^3$ when $x = -8$ and $y = 10$.

EXAMPLE 1.56

Evaluate $20 - z$ when ⓐ $z = 12$ and ⓑ $z = -12$.

⊘ Solution

ⓐ

	$20 - z$
Substitute 12 for z.	$20 - 12$
Subtract.	8

ⓑ

	$20 - z$
Substitute -12 for z.	$20 - (-12)$
Subtract.	32

> **TRY IT : : 1.111**   Evaluate: $17 - k$ when ⓐ $k = 19$ and ⓑ $k = -19$.

> **TRY IT : : 1.112**   Evaluate: $-5 - b$ when ⓐ $b = 14$ and ⓑ $b = -14$.

EXAMPLE 1.57

Evaluate: $2x^2 + 3x + 8$ when $x = 4$.

⊘ Solution

Substitute 4 for x. Use parentheses to show multiplication.

	$2x^2 + 3x + 8$
Substitute.	$2(4)^2 + 3(4) + 8$
Evaluate exponents.	$2(16) + 3(4) + 8$
Multiply.	$32 + 12 + 8$
Add.	52

> **TRY IT : : 1.113**   Evaluate: $3x^2 - 2x + 6$ when $x = -3$.

> **TRY IT : : 1.114**   Evaluate: $4x^2 - x - 5$ when $x = -2$.

Translate Phrases to Expressions with Integers

Our earlier work translating English to algebra also applies to phrases that include both positive and negative numbers.

EXAMPLE 1.58

Translate and simplify: the sum of 8 and -12, increased by 3.

⊘ Solution

	the **sum** of 8 and -12, increased by 3
Translate.	$[8 + (-12)] + 3$
Simplify. Be careful not to confuse the brackets with an absolute value sign.	$(-4) + 3$
Add.	-1

> **TRY IT : : 1.115**   Translate and simplify the sum of 9 and -16, increased by 4.

> **TRY IT : : 1.116**   Translate and simplify the sum of -8 and -12, increased by 7.

When we first introduced the operation symbols, we saw that the expression may be read in several ways. They are listed in the chart below.

$a - b$
a minus b
the difference of a and b
b subtracted from a
b less than a

Be careful to get a and b in the right order!

EXAMPLE 1.59

Translate and then simplify ⓐ the difference of 13 and -21 ⓑ subtract 24 from -19.

⊘ **Solution**

ⓐ

	the **difference** *of* 13 *and* -21
Translate.	$13 - (-21)$
Simplify.	34

ⓑ

	subtract 24 **from** -19
Translate.	$-19 - 24$
Remember, "subtract b from a means $a - b$.	
Simplify.	-43

> **TRY IT : : 1.117** Translate and simplify ⓐ the difference of 14 and -23 ⓑ subtract 21 from -17.

> **TRY IT : : 1.118** Translate and simplify ⓐ the difference of 11 and -19 ⓑ subtract 18 from -11.

Once again, our prior work translating English to algebra transfers to phrases that include both multiplying and dividing integers. Remember that the key word for multiplication is " product" and for division is " quotient."

EXAMPLE 1.60

Translate to an algebraic expression and simplify if possible: the product of -2 and 14.

⊘ **Solution**

	the product *of* -2 *and* 14
Translate.	$(-2)(14)$
Simplify.	-28

> **TRY IT : : 1.119** Translate to an algebraic expression and simplify if possible: the product of -5 and 12.

> **TRY IT : : 1.120** Translate to an algebraic expression and simplify if possible: the product of 8 and -13.

EXAMPLE 1.61

Translate to an algebraic expression and simplify if possible: the quotient of -56 and -7.

⊘ **Solution**

	the quotient *of* -56 *and* -7
Translate.	$-56 \div (-7)$
Simplify.	8

> **TRY IT : : 1.121** Translate to an algebraic expression and simplify if possible: the quotient of -63 and -9.

> **TRY IT : : 1.122** Translate to an algebraic expression and simplify if possible: the quotient of -72 and -9.

Use Integers in Applications

We'll outline a plan to solve applications. It's hard to find something if we don't know what we're looking for or what to call it! So when we solve an application, we first need to determine what the problem is asking us to find. Then we'll write a phrase that gives the information to find it. We'll translate the phrase into an expression and then simplify the expression to get the answer. Finally, we summarize the answer in a sentence to make sure it makes sense.

EXAMPLE 1.62 HOW TO APPLY A STRATEGY TO SOLVE APPLICATIONS WITH INTEGERS

The temperature in Urbana, Illinois one morning was 11 degrees. By mid-afternoon, the temperature had dropped to -9 degrees. What was the difference of the morning and afternoon temperatures?

✓ Solution

Step 1. Read the problem. Make sure all the words and ideas are understood.	
Step 2. Identify what we are asked to find.	the difference of the morning and afternoon temperatures
Step 3. Write a phrase the gives the information to find it.	the *difference of* 11 *and* -9
Step 4. Translate the phrase to an expression.	$11 - (-9)$
Step 5. Simplify the expression.	20
Step 6. Write a complete sentence that answers the question.	The difference in temperatures was 20 degrees.

> **TRY IT : : 1.123**

The temperature in Anchorage, Alaska one morning was 15 degrees. By mid-afternoon the temperature had dropped to 30 degrees below zero. What was the difference in the morning and afternoon temperatures?

> **TRY IT : : 1.124**

The temperature in Denver was -6 degrees at lunchtime. By sunset the temperature had dropped to -15 degrees. What was the difference in the lunchtime and sunset temperatures?

 HOW TO : : APPLY A STRATEGY TO SOLVE APPLICATIONS WITH INTEGERS.

Step 1.   Read the problem. Make sure all the words and ideas are understood

Step 2.   Identify what we are asked to find.

Step 3.   Write a phrase that gives the information to find it.

Step 4.   Translate the phrase to an expression.

Step 5.   Simplify the expression.

Step 6.   Answer the question with a complete sentence.

EXAMPLE 1.63

The Mustangs football team received three penalties in the third quarter. Each penalty gave them a loss of fifteen yards. What is the number of yards lost?

⊘ Solution

Step 1. Read the problem. Make sure all the words and ideas are understood.

Step 2. Identify what we are asked to find.   the number of yards lost

Step 3. Write a phrase that gives the information to find it.   three times a 15-yard penalty

Step 4. Translate the phrase to an expression. $3(-15)$

Step 5. Simplify the expression. -45

Step 6. Answer the question with a complete sentence.   The team lost 45 yards.

> **TRY IT ::** 1.125

The Bears played poorly and had seven penalties in the game. Each penalty resulted in a loss of 15 yards. What is the number of yards lost due to penalties?

> **TRY IT ::** 1.126

Bill uses the ATM on campus because it is convenient. However, each time he uses it he is charged a $2 fee. Last month he used the ATM eight times. How much was his total fee for using the ATM?

1.4 EXERCISES

Practice Makes Perfect

Multiply Integers

In the following exercises, multiply.

265. $-4 \cdot 8$

266. $-3 \cdot 9$

267. $9(-7)$

268. $13(-5)$

269. -1.6

270. -1.3

271. $-1(-14)$

272. $-1(-19)$

Divide Integers

In the following exercises, divide.

273. $-24 \div 6$

274. $35 \div (-7)$

275. $-52 \div (-4)$

276. $-84 \div (-6)$

277. $-180 \div 15$

278. $-192 \div 12$

Simplify Expressions with Integers

In the following exercises, simplify each expression.

279. $5(-6) + 7(-2) - 3$

280. $8(-4) + 5(-4) - 6$

281. $(-2)^6$

282. $(-3)^5$

283. -4^2

284. -6^2

285. $-3(-5)(6)$

286. $-4(-6)(3)$

287. $(8 - 11)(9 - 12)$

288. $(6 - 11)(8 - 13)$

289. $26 - 3(2 - 7)$

290. $23 - 2(4 - 6)$

291. $65 \div (-5) + (-28) \div (-7)$

292. $52 \div (-4) + (-32) \div (-8)$

293. $9 - 2[3 - 8(-2)]$

294. $11 - 3[7 - 4(-2)]$

295. $(-3)^2 - 24 \div (8 - 2)$

296. $(-4)^2 - 32 \div (12 - 4)$

Evaluate Variable Expressions with Integers

In the following exercises, evaluate each expression.

297. $y + (-14)$ when

ⓐ $y = -33$

ⓑ $y = 30$

298. $x + (-21)$ when

ⓐ $x = -27$

ⓑ $x = 44$

299.

ⓐ $a + 3$ when $a = -7$

ⓑ $-a + 3$ when $a = -7$

300.

ⓐ $d + (-9)$ when $d = -8$

ⓑ $-d + (-9)$ when $d = -8$

301. $m + n$ when
$m = -15,\ n = 7$

302. $p + q$ when
$p = -9,\ q = 17$

303. $r + s$ when $r = -9,\ s = -7$

304. $t + u$ when $t = -6,\ u = -5$

305. $(x + y)^2$ when
$x = -3,\ y = 14$

306. $(y + z)^2$ when
$y = -3$, $z = 15$

307. $-2x + 17$ when
ⓐ $x = 8$
ⓑ $x = -8$

308. $-5y + 14$ when
ⓐ $y = 9$
ⓑ $y = -9$

309. $10 - 3m$ when
ⓐ $m = 5$
ⓑ $m = -5$

310. $18 - 4n$ when
ⓐ $n = 3$
ⓑ $n = -3$

311. $2w^2 - 3w + 7$ when
$w = -2$

312. $3u^2 - 4u + 5$ when $u = -3$

313. $9a - 2b - 8$ when
$a = -6$ and $b = -3$

314. $7m - 4n - 2$ when
$m = -4$ and $n = -9$

Translate English Phrases to Algebraic Expressions

In the following exercises, translate to an algebraic expression and simplify if possible.

315. the sum of 3 and -15, increased by 7

316. the sum of -8 and -9, increased by 23

317. the difference of 10 and -18

318. subtract 11 from -25

319. the difference of -5 and -30

320. subtract -6 from -13

321. the product of -3 and 15

322. the product of -4 and 16

323. the quotient of -60 and -20

324. the quotient of -40 and -20

325. the quotient of -6 and the sum of a and b

326. the quotient of -7 and the sum of m and n

327. the product of -10 and the difference of p and q

328. the product of -13 and the difference of c and d

Use Integers in Applications

In the following exercises, solve.

329. Temperature On January 15, the high temperature in Anaheim, California, was $84°$. That same day, the high temperature in Embarrass, Minnesota was $-12°$. What was the difference between the temperature in Anaheim and the temperature in Embarrass?

330. Temperature On January 21, the high temperature in Palm Springs, California, was $89°$, and the high temperature in Whitefield, New Hampshire was $-31°$. What was the difference between the temperature in Palm Springs and the temperature in Whitefield?

331. Football At the first down, the Chargers had the ball on their 25 yard line. On the next three downs, they lost 6 yards, gained 10 yards, and lost 8 yards. What was the yard line at the end of the fourth down?

332. Football At the first down, the Steelers had the ball on their 30 yard line. On the next three downs, they gained 9 yards, lost 14 yards, and lost 2 yards. What was the yard line at the end of the fourth down?

333. Checking Account Mayra has $124 in her checking account. She writes a check for $152. What is the new balance in her checking account?

334. Checking Account Selina has $165 in her checking account. She writes a check for $207. What is the new balance in her checking account?

335. Checking Account Diontre has a balance of $-$38 in his checking account. He deposits $225 to the account. What is the new balance?

336. Checking Account Reymonte has a balance of $-$49 in his checking account. He deposits $281 to the account. What is the new balance?

Everyday Math

337. Stock market Javier owns 300 shares of stock in one company. On Tuesday, the stock price dropped $12 per share. What was the total effect on Javier's portfolio?

338. Weight loss In the first week of a diet program, eight women lost an average of 3 pounds each. What was the total weight change for the eight women?

Writing Exercises

339. In your own words, state the rules for multiplying integers.

340. In your own words, state the rules for dividing integers.

341. Why is $-2^4 \neq (-2)^4$?

342. Why is $-4^3 = (-4)^3$?

Self Check

ⓐ *After completing the exercises, use this checklist to evaluate your mastery of the objectives of this section.*

I can...	Confidently	With some help	No-I don't get it!
multiply integers.			
divide integers.			
simplify expressions with integers.			
evaluate variable expressions with integers.			
translate English phrases to algebraic expressions.			
use integers in applications.			

ⓑ *On a scale of 1–10, how would you rate your mastery of this section in light of your responses on the checklist? How can you improve this?*

1.5 Visualize Fractions

Learning Objectives

By the end of this section, you will be able to:

› Find equivalent fractions
› Simplify fractions
› Multiply fractions
› Divide fractions
› Simplify expressions written with a fraction bar
› Translate phrases to expressions with fractions

Be Prepared!

A more thorough introduction to the topics covered in this section can be found in the *Prealgebra* chapter, **Fractions**.

Find Equivalent Fractions

Fractions are a way to represent parts of a whole. The fraction $\frac{1}{3}$ means that one whole has been divided into 3 equal parts and each part is one of the three equal parts. See **Figure 1.11**. The fraction $\frac{2}{3}$ represents two of three equal parts.

In the fraction $\frac{2}{3}$, the 2 is called the **numerator** and the 3 is called the **denominator**.

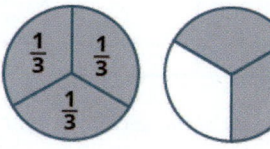

Figure 1.11 The circle on the left has been divided into 3 equal parts. Each part is $\frac{1}{3}$ of the 3 equal parts. In the circle on the right, $\frac{2}{3}$ of the circle is shaded (2 of the 3 equal parts).

 MANIPULATIVE MATHEMATICS

Doing the Manipulative Mathematics activity "Model Fractions" will help you develop a better understanding of fractions, their numerators and denominators.

Fraction

A **fraction** is written $\frac{a}{b}$, where $b \neq 0$ and

- a is the **numerator** and b is the **denominator**.

A fraction represents parts of a whole. The denominator b is the number of equal parts the whole has been divided into, and the numerator a indicates how many parts are included.

If a whole pie has been cut into 6 pieces and we eat all 6 pieces, we ate $\frac{6}{6}$ pieces, or, in other words, one whole pie.

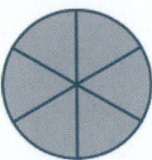

So $\frac{6}{6} = 1$. This leads us to the property of one that tells us that any number, except zero, divided by itself is 1.

Property of One

$$\frac{a}{a} = 1 \quad (a \neq 0)$$

Any number, except zero, divided by itself is one.

 MANIPULATIVE MATHEMATICS

Doing the Manipulative Mathematics activity "Fractions Equivalent to One" will help you develop a better understanding of fractions that are equivalent to one.

If a pie was cut in 6 pieces and we ate all 6, we ate $\frac{6}{6}$ pieces, or, in other words, one whole pie. If the pie was cut into 8 pieces and we ate all 8, we ate $\frac{8}{8}$ pieces, or one whole pie. We ate the same amount—one whole pie.

The fractions $\frac{6}{6}$ and $\frac{8}{8}$ have the same value, 1, and so they are called equivalent fractions. **Equivalent fractions** are fractions that have the same value.

Let's think of pizzas this time. **Figure 1.12** shows two images: a single pizza on the left, cut into two equal pieces, and a second pizza of the same size, cut into eight pieces on the right. This is a way to show that $\frac{1}{2}$ is equivalent to $\frac{4}{8}$. In other words, they are equivalent fractions.

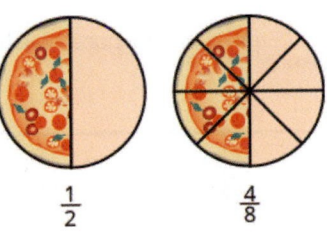

$$\frac{1}{2} \qquad\qquad \frac{4}{8}$$

Figure 1.12 Since the same amount is of each pizza is shaded, we see that $\frac{1}{2}$ is equivalent to $\frac{4}{8}$. They are equivalent fractions.

Equivalent Fractions

Equivalent fractions are fractions that have the same value.

How can we use mathematics to change $\frac{1}{2}$ into $\frac{4}{8}$? How could we take a pizza that is cut into 2 pieces and cut it into 8 pieces? We could cut each of the 2 larger pieces into 4 smaller pieces! The whole pizza would then be cut into 8 pieces instead of just 2. Mathematically, what we've described could be written like this as $\frac{1 \cdot 4}{2 \cdot 4} = \frac{4}{8}$. See **Figure 1.13**.

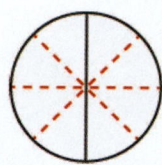

Figure 1.13 Cutting each half of the pizza into 4 pieces, gives us pizza cut into 8 pieces: $\dfrac{1 \cdot 4}{2 \cdot 4} = \dfrac{4}{8}$.

This model leads to the following property:

Equivalent Fractions Property

If a, b, c are numbers where $b \neq 0$, $c \neq 0$, then

$$\frac{a}{b} = \frac{a \cdot c}{b \cdot c}$$

If we had cut the pizza differently, we could get

$$\frac{1 \cdot 2}{2 \cdot 2} = \frac{2}{4} \quad \text{so} \quad \frac{1}{2} = \frac{2}{4}$$

$$\frac{1 \cdot 3}{2 \cdot 3} = \frac{3}{6} \quad \text{so} \quad \frac{1}{2} = \frac{3}{6}$$

$$\frac{1 \cdot 10}{2 \cdot 10} = \frac{10}{20} \quad \text{so} \quad \frac{1}{2} = \frac{10}{20}$$

So, we say $\dfrac{1}{2}$, $\dfrac{2}{4}$, $\dfrac{3}{6}$, and $\dfrac{10}{20}$ are equivalent fractions.

 MANIPULATIVE MATHEMATICS

Doing the Manipulative Mathematics activity "Equivalent Fractions" will help you develop a better understanding of what it means when two fractions are equivalent.

EXAMPLE 1.64

Find three fractions equivalent to $\dfrac{2}{5}$.

⊘ **Solution**

To find a fraction equivalent to $\dfrac{2}{5}$, we multiply the numerator and denominator by the same number. We can choose any number, except for zero. Let's multiply them by 2, 3, and then 5.

$$\frac{2 \cdot 2}{5 \cdot 2} = \frac{4}{10} \qquad \frac{2 \cdot 3}{5 \cdot 3} = \frac{6}{15} \qquad \frac{2 \cdot 5}{5 \cdot 5} = \frac{10}{25}$$

So, $\dfrac{4}{10}$, $\dfrac{6}{15}$, and $\dfrac{10}{25}$ are equivalent to $\dfrac{2}{5}$.

> **TRY IT : : 1.127**   Find three fractions equivalent to $\dfrac{3}{5}$.

> **TRY IT : : 1.128**   Find three fractions equivalent to $\dfrac{4}{5}$.

Simplify Fractions

A fraction is considered **simplified** if there are no common factors, other than 1, in its numerator and denominator.

For example,

- $\frac{2}{3}$ is simplified because there are no common factors of 2 and 3.

- $\frac{10}{15}$ is not simplified because 5 is a common factor of 10 and 15.

Simplified Fraction

A fraction is considered **simplified** if there are no common factors in its numerator and denominator.

The phrase *reduce a fraction* means to simplify the fraction. We simplify, or reduce, a fraction by removing the common factors of the numerator and denominator. A fraction is not simplified until all common factors have been removed. If an expression has fractions, it is not completely simplified until the fractions are simplified.

In **Example 1.64**, we used the equivalent fractions property to find equivalent fractions. Now we'll use the equivalent fractions property in reverse to simplify fractions. We can rewrite the property to show both forms together.

Equivalent Fractions Property

If a, b, c are numbers where $b \neq 0$, $c \neq 0$,

$$\text{then} \quad \frac{a}{b} = \frac{a \cdot c}{b \cdot c} \quad \text{and} \quad \frac{a \cdot c}{b \cdot c} = \frac{a}{b}$$

EXAMPLE 1.65

Simplify: $-\frac{32}{56}$.

⊘ Solution

$$-\frac{32}{56}$$

Rewrite the numerator and denominator showing the common factors.	$-\frac{4 \cdot 8}{7 \cdot 8}$
Simplify using the equivalent fractions property.	$-\frac{4}{7}$

Notice that the fraction $-\frac{4}{7}$ is simplified because there are no more common factors.

> **TRY IT :: 1.129** Simplify: $-\frac{42}{54}$.

> **TRY IT :: 1.130** Simplify: $-\frac{45}{81}$.

Sometimes it may not be easy to find common factors of the numerator and denominator. When this happens, a good idea is to factor the numerator and the denominator into prime numbers. Then divide out the common factors using the equivalent fractions property.

EXAMPLE 1.66   HOW TO SIMPLIFY A FRACTION

Simplify: $-\frac{210}{385}$.

✓ Solution

Step 1. Rewrite the numerator and denominator to show the common factors. If needed, factor the numerator and denominator into prime numbers first.	Rewrite 210 and 385 as the product of the primes.	$-\dfrac{210}{385}$ $-\dfrac{2 \cdot 3 \cdot 5 \cdot 7}{5 \cdot 7 \cdot 11}$
Step 2. Simplify using the equivalent fractions property by dividing out common factors.	Mark the common factors 5 and 7. Divide out the common factors.	$-\dfrac{2 \cdot 3 \cdot \cancel{5} \cdot \cancel{7}}{\cancel{5} \cdot \cancel{7} \cdot 11}$ $-\dfrac{2 \cdot 3}{11}$
Step 3. Multiplify the remaining factors, if necessary.		$-\dfrac{6}{11}$

> **TRY IT : :** 1.131   Simplify: $-\dfrac{69}{120}$.

> **TRY IT : :** 1.132   Simplify: $-\dfrac{120}{192}$.

We now summarize the steps you should follow to simplify fractions.

HOW TO : : SIMPLIFY A FRACTION.

Step 1.   Rewrite the numerator and denominator to show the common factors. If needed, factor the numerator and denominator into prime numbers first.

Step 2.   Simplify using the equivalent fractions property by dividing out common factors.

Step 3.   Multiply any remaining factors, if needed.

EXAMPLE 1.67

Simplify: $\dfrac{5x}{5y}$.

✓ Solution

	$\dfrac{5x}{5y}$
Rewrite showing the common factors, then divide out the common factors.	$\dfrac{\cancel{5} \cdot x}{\cancel{5} \cdot y}$
Simplify.	$\dfrac{x}{y}$

> **TRY IT : :** 1.133   Simplify: $\dfrac{7x}{7y}$.

> **TRY IT : :** 1.134   Simplify: $\dfrac{3a}{3b}$.

Multiply Fractions

Many people find multiplying and dividing fractions easier than adding and subtracting fractions. So we will start with fraction multiplication.

 MANIPULATIVE MATHEMATICS

Doing the Manipulative Mathematics activity "Model Fraction Multiplication" will help you develop a better understanding of multiplying fractions.

We'll use a model to show you how to multiply two fractions and to help you remember the procedure. Let's start with $\frac{3}{4}$.

Now we'll take $\frac{1}{2}$ of $\frac{3}{4}$.

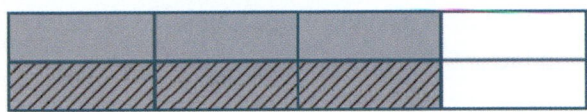

Notice that now, the whole is divided into 8 equal parts. So $\frac{1}{2} \cdot \frac{3}{4} = \frac{3}{8}$.

To multiply fractions, we multiply the numerators and multiply the denominators.

Fraction Multiplication

If a, b, c and d are numbers where $b \neq 0$ and $d \neq 0$, then

$$\frac{a}{b} \cdot \frac{c}{d} = \frac{ac}{bd}$$

To multiply fractions, multiply the numerators and multiply the denominators.

When multiplying fractions, the properties of positive and negative numbers still apply, of course. It is a good idea to determine the sign of the product as the first step. In **Example 1.68**, we will multiply negative and a positive, so the product will be negative.

EXAMPLE 1.68

Multiply: $-\frac{11}{12} \cdot \frac{5}{7}$.

⊘ **Solution**

The first step is to find the sign of the product. Since the signs are the different, the product is negative.

$$-\frac{11}{12} \cdot \frac{5}{7}$$

Determine the sign of the product; multiply.

$$-\frac{11 \cdot 5}{12 \cdot 7}$$

Are there any common factors in the numerator and the denominator? No

$$-\frac{55}{84}$$

> **TRY IT :: 1.135**   Multiply: $-\frac{10}{28} \cdot \frac{8}{15}$.

> **TRY IT ::** 1.136   Multiply: $-\frac{9}{20} \cdot \frac{5}{12}$.

When multiplying a fraction by an integer, it may be helpful to write the integer as a fraction. Any integer, a, can be written as $\frac{a}{1}$. So, for example, $3 = \frac{3}{1}$.

EXAMPLE 1.69

Multiply: $-\frac{12}{5}(-20x)$.

⊘ **Solution**

Determine the sign of the product. The signs are the same, so the product is positive.

$$-\frac{12}{5}(-20x)$$

Write $20x$ as a fraction.	$\frac{12}{5}\left(\frac{20x}{1}\right)$
Multiply.	
Rewrite 20 to show the common factor 5 and divide it out.	$\frac{12 \cdot 4 \cdot \cancel{5}x}{\cancel{5} \cdot 1}$
Simplify.	$48x$

> **TRY IT ::** 1.137   Multiply: $\frac{11}{3}(-9a)$.

> **TRY IT ::** 1.138   Multiply: $\frac{13}{7}(-14b)$.

Divide Fractions

Now that we know how to multiply fractions, we are almost ready to divide. Before we can do that, that we need some vocabulary.

The **reciprocal** of a fraction is found by inverting the fraction, placing the numerator in the denominator and the denominator in the numerator. The reciprocal of $\frac{2}{3}$ is $\frac{3}{2}$.

Notice that $\frac{2}{3} \cdot \frac{3}{2} = 1$. A number and its reciprocal multiply to 1.

To get a product of positive 1 when multiplying two numbers, the numbers must have the same sign. So reciprocals must have the same sign.

The reciprocal of $-\frac{10}{7}$ is $-\frac{7}{10}$, since $-\frac{10}{7}\left(-\frac{7}{10}\right) = 1$.

Reciprocal

The **reciprocal** of $\frac{a}{b}$ is $\frac{b}{a}$.

A number and its reciprocal multiply to one $\frac{a}{b} \cdot \frac{b}{a} = 1$.

MANIPULATIVE MATHEMATICS

Doing the Manipulative Mathematics activity "Model Fraction Division" will help you develop a better understanding of dividing fractions.

To divide fractions, we multiply the first fraction by the reciprocal of the second.

Fraction Division

If a, b, c and d are numbers where $b \neq 0$, $c \neq 0$ and $d \neq 0$, then

$$\frac{a}{b} \div \frac{c}{d} = \frac{a}{b} \cdot \frac{d}{c}$$

To divide fractions, we multiply the first fraction by the reciprocal of the second.

We need to say $b \neq 0$, $c \neq 0$ and $d \neq 0$ to be sure we don't divide by zero!

EXAMPLE 1.70

Divide: $-\frac{2}{3} \div \frac{n}{5}$.

✓ Solution

$$-\frac{2}{3} \div \frac{n}{5}$$

To divide, multiply the first fraction by the reciprocal of the second.

$$-\frac{2}{3} \cdot \frac{5}{n}$$

Multiply.

$$-\frac{10}{3n}$$

> **TRY IT :: 1.139**   Divide: $-\frac{3}{5} \div \frac{p}{7}$.

> **TRY IT :: 1.140**   Divide: $-\frac{5}{8} \div \frac{q}{3}$.

EXAMPLE 1.71

Find the quotient: $-\frac{7}{8} \div \left(-\frac{14}{27}\right)$.

⊘ Solution

$$-\frac{7}{18} \div \left(-\frac{14}{27}\right)$$

To divide, multiply the first fraction by the reciprocal of the second.	$-\frac{7}{18} \cdot -\frac{27}{14}$
Determine the sign of the product, and then multiply..	$\frac{7 \cdot 27}{18 \cdot 14}$
Rewrite showing common factors.	$\frac{7 \cdot 9 \cdot 3}{9 \cdot 2 \cdot 7 \cdot 2}$
Remove common factors.	$\frac{3}{2 \cdot 2}$
Simplify.	$\frac{3}{4}$

> **TRY IT :: 1.141**   Find the quotient: $-\frac{7}{27} \div \left(-\frac{35}{36}\right)$.

> **TRY IT :: 1.142**   Find the quotient: $-\frac{5}{14} \div \left(-\frac{15}{28}\right)$.

There are several ways to remember which steps to take to multiply or divide fractions. One way is to repeat the call outs to yourself. If you do this each time you do an exercise, you will have the steps memorized.

- "To multiply fractions, multiply the numerators and multiply the denominators."
- "To divide fractions, multiply the first fraction by the reciprocal of the second."

Another way is to keep two examples in mind:

One fourth of two pizzas is one half of a pizza.   There are eight quarters in $2.00.

$$2 \cdot \frac{1}{4} \qquad\qquad 2 \div \frac{1}{4}$$

$$\frac{2}{1} \cdot \frac{1}{4} \qquad\qquad \frac{2}{1} \div \frac{1}{4}$$

$$\frac{2}{4} \qquad\qquad \frac{2}{1} \cdot \frac{4}{1}$$

$$\frac{1}{2} \qquad\qquad 8$$

The numerators or denominators of some fractions contain fractions themselves. A fraction in which the numerator or the denominator is a fraction is called a **complex fraction**.

Complex Fraction

A **complex fraction** is a fraction in which the numerator or the denominator contains a fraction.

Some examples of complex fractions are:

$$\frac{\frac{6}{7}}{3} \qquad \frac{\frac{3}{4}}{\frac{5}{8}} \qquad \frac{\frac{x}{2}}{\frac{5}{6}}$$

To simplify a complex fraction, we remember that the fraction bar means division. For example, the complex fraction $\frac{\frac{3}{4}}{\frac{5}{8}}$

means $\frac{3}{4} \div \frac{5}{8}$.

EXAMPLE 1.72

Simplify: $\frac{\frac{3}{4}}{\frac{5}{8}}$.

✓ **Solution**

	$\frac{\frac{3}{4}}{\frac{5}{8}}$
Rewrite as division.	$\frac{3}{4} \div \frac{5}{8}$
Multiply the first fraction by the reciprocal of the second.	$\frac{3}{4} \cdot \frac{8}{5}$
Multiply.	$\frac{3 \cdot 8}{4 \cdot 5}$
Look for common factors.	$\frac{3 \cdot \cancel{4} \cdot 2}{\cancel{4} \cdot 5}$
Divide out common factors and simplify.	$\frac{6}{5}$

> **TRY IT : : 1.143**   Simplify: $\frac{\frac{2}{3}}{\frac{5}{6}}$.

> **TRY IT : : 1.144**   Simplify: $\frac{\frac{3}{7}}{\frac{6}{11}}$.

EXAMPLE 1.73

Simplify: $\frac{\frac{x}{2}}{\frac{xy}{6}}$.

⊘ **Solution**

$$\frac{\frac{x}{2}}{\frac{xy}{6}}$$

Rewrite as division.	$\frac{x}{2} \div \frac{xy}{6}$
Multiply the first fraction by the reciprocal of the second.	$\frac{x}{2} \cdot \frac{6}{xy}$
Multiply.	$\frac{x \cdot 6}{2 \cdot xy}$
Look for common factors.	$\frac{\cancel{x} \cdot 3 \cdot \cancel{2}}{\cancel{2} \cdot \cancel{x} \cdot y}$
Divide out common factors and simplify.	$\frac{3}{y}$

> **TRY IT :: 1.145**

Simplify: $\dfrac{\frac{a}{8}}{\frac{ab}{6}}$.

> **TRY IT :: 1.146**

Simplify: $\dfrac{\frac{p}{2}}{\frac{pq}{8}}$.

Simplify Expressions with a Fraction Bar

The line that separates the numerator from the denominator in a fraction is called a fraction bar. A fraction bar acts as grouping symbol. The order of operations then tells us to simplify the numerator and then the denominator. Then we divide.

To simplify the expression $\frac{5-3}{7+1}$, we first simplify the numerator and the denominator separately. Then we divide.

$$\frac{5-3}{7+1}$$

$$\frac{2}{8}$$

$$\frac{1}{4}$$

 HOW TO :: SIMPLIFY AN EXPRESSION WITH A FRACTION BAR.

Step 1.   Simplify the expression in the numerator. Simplify the expression in the denominator.

Step 2.   Simplify the fraction.

EXAMPLE 1.74

Simplify: $\dfrac{4-2(3)}{2^2+2}$.

✓ Solution

	$\dfrac{4-2(3)}{2^2+2}$
Use the order of operations to simplify the numerator and the denominator.	$\dfrac{4-6}{4+2}$
Simplify the numerator and the denominator.	$\dfrac{-2}{6}$
Simplify. A negative divided by a positive is negative.	$-\dfrac{1}{3}$

> **TRY IT :: 1.147**   Simplify: $\dfrac{6-3(5)}{3^2+3}$.

> **TRY IT :: 1.148**   Simplify: $\dfrac{4-4(6)}{3^2+3}$.

Where does the negative sign go in a fraction? Usually the negative sign is in front of the fraction, but you will sometimes see a fraction with a negative numerator, or sometimes with a negative denominator. Remember that fractions represent division. When the numerator and denominator have different signs, the quotient is negative.

$$\frac{-1}{3}=-\frac{1}{3} \qquad \frac{\text{negative}}{\text{positive}}=\text{negative}$$

$$\frac{1}{-3}=-\frac{1}{3} \qquad \frac{\text{positive}}{\text{negative}}=\text{negative}$$

Placement of Negative Sign in a Fraction

For any positive numbers a and b,

$$\frac{-a}{b}=\frac{a}{-b}=-\frac{a}{b}$$

EXAMPLE 1.75

Simplify: $\dfrac{4(-3)+6(-2)}{-3(2)-2}$.

✓ Solution

The fraction bar acts like a grouping symbol. So completely simplify the numerator and the denominator separately.

	$\dfrac{4(-3)+6(-2)}{-3(2)-2}$
Multiply.	$\dfrac{-12+(-12)}{-6-2}$
Simplify.	$\dfrac{-24}{-8}$
Divide.	3

> **TRY IT :: 1.149**   Simplify: $\dfrac{8(-2)+4(-3)}{-5(2)+3}$.

> **TRY IT :: 1.150**   Simplify: $\dfrac{7(-1)+9(-3)}{-5(3)-2}$.

Translate Phrases to Expressions with Fractions

Now that we have done some work with fractions, we are ready to translate phrases that would result in expressions with fractions.

The English words quotient and ratio are often used to describe fractions. Remember that "quotient" means division. The quotient of a and b is the result we get from dividing a by b, or $\frac{a}{b}$.

EXAMPLE 1.76

Translate the English phrase into an algebraic expression: the quotient of the difference of m and n, and p.

⊘ **Solution**

We are looking for the *quotient of* the difference of m and n, *and p*. This means we want to divide the difference of m and n by p.

$$\frac{m - n}{p}$$

> │ **TRY IT : :** 1.151

Translate the English phrase into an algebraic expression: the quotient of the difference of a and b, and cd.

> │ **TRY IT : :** 1.152

Translate the English phrase into an algebraic expression: the quotient of the sum of p and q, and r

1.5 EXERCISES

Practice Makes Perfect

Find Equivalent Fractions

In the following exercises, find three fractions equivalent to the given fraction. Show your work, using figures or algebra.

343. $\frac{3}{8}$

344. $\frac{5}{8}$

345. $\frac{5}{9}$

346. $\frac{1}{8}$

Simplify Fractions

In the following exercises, simplify.

347. $-\frac{40}{88}$

348. $-\frac{63}{99}$

349. $-\frac{108}{63}$

350. $-\frac{104}{48}$

351. $\frac{120}{252}$

352. $\frac{182}{294}$

353. $-\frac{3x}{12y}$

354. $-\frac{4x}{32y}$

355. $\frac{14x^2}{21y}$

356. $\frac{24a}{32b^2}$

Multiply Fractions

In the following exercises, multiply.

357. $\frac{3}{4} \cdot \frac{9}{10}$

358. $\frac{4}{5} \cdot \frac{2}{7}$

359. $-\frac{2}{3}\left(-\frac{3}{8}\right)$

360. $-\frac{3}{4}\left(-\frac{4}{9}\right)$

361. $-\frac{5}{9} \cdot \frac{3}{10}$

362. $-\frac{3}{8} \cdot \frac{4}{15}$

363. $\left(-\frac{14}{15}\right)\left(\frac{9}{20}\right)$

364. $\left(-\frac{9}{10}\right)\left(\frac{25}{33}\right)$

365. $\left(-\frac{63}{84}\right)\left(-\frac{44}{90}\right)$

366. $\left(-\frac{63}{60}\right)\left(-\frac{40}{88}\right)$

367. $4 \cdot \frac{5}{11}$

368. $5 \cdot \frac{8}{3}$

369. $\frac{3}{7} \cdot 21n$

370. $\frac{5}{6} \cdot 30m$

371. $-8\left(\frac{17}{4}\right)$

372. $(-1)\left(-\frac{6}{7}\right)$

Divide Fractions

In the following exercises, divide.

373. $\frac{3}{4} \div \frac{2}{3}$

374. $\frac{4}{5} \div \frac{3}{4}$

375. $-\frac{7}{9} \div \left(-\frac{7}{4}\right)$

376. $-\frac{5}{6} \div \left(-\frac{5}{6}\right)$

377. $\frac{3}{4} \div \frac{x}{11}$

378. $\frac{2}{5} \div \frac{y}{9}$

379. $\frac{5}{18} \div \left(-\frac{15}{24}\right)$ **380.** $\frac{7}{18} \div \left(-\frac{14}{27}\right)$ **381.** $\frac{8u}{15} \div \frac{12v}{25}$

382. $\frac{12r}{25} \div \frac{18s}{35}$ **383.** $-5 \div \frac{1}{2}$ **384.** $-3 \div \frac{1}{4}$

385. $\frac{3}{4} \div (-12)$ **386.** $-15 \div \left(-\frac{5}{3}\right)$

In the following exercises, simplify.

387. $\dfrac{-\frac{8}{21}}{\frac{12}{35}}$ **388.** $\dfrac{-\frac{9}{16}}{\frac{33}{40}}$ **389.** $\dfrac{-\frac{4}{5}}{2}$

390. $\dfrac{\frac{5}{3}}{10}$ **391.** $\dfrac{\frac{m}{3}}{\frac{n}{2}}$ **392.** $\dfrac{-\frac{3}{8}}{-\frac{y}{12}}$

Simplify Expressions Written with a Fraction Bar

In the following exercises, simplify.

393. $\frac{22+3}{10}$ **394.** $\frac{19-4}{6}$ **395.** $\frac{48}{24-15}$

396. $\frac{46}{4+4}$ **397.** $\frac{-6+6}{8+4}$ **398.** $\frac{-6+3}{17-8}$

399. $\frac{4 \cdot 3}{6 \cdot 6}$ **400.** $\frac{6 \cdot 6}{9 \cdot 2}$ **401.** $\frac{4^2-1}{25}$

402. $\frac{7^2+1}{60}$ **403.** $\frac{8 \cdot 3 + 2 \cdot 9}{14+3}$ **404.** $\frac{9 \cdot 6 - 4 \cdot 7}{22+3}$

405. $\frac{5 \cdot 6 - 3 \cdot 4}{4 \cdot 5 - 2 \cdot 3}$ **406.** $\frac{8 \cdot 9 - 7 \cdot 6}{5 \cdot 6 - 9 \cdot 2}$ **407.** $\frac{5^2 - 3^2}{3-5}$

408. $\frac{6^2 - 4^2}{4-6}$ **409.** $\frac{7 \cdot 4 - 2(8-5)}{9 \cdot 3 - 3 \cdot 5}$ **410.** $\frac{9 \cdot 7 - 3(12-8)}{8 \cdot 7 - 6 \cdot 6}$

411. $\frac{9(8-2) - 3(15-7)}{6(7-1) - 3(17-9)}$ **412.** $\frac{8(9-2) - 4(14-9)}{7(8-3) - 3(16-9)}$

Translate Phrases to Expressions with Fractions

In the following exercises, translate each English phrase into an algebraic expression.

413. the quotient of r and the sum of s and 10

414. the quotient of A and the difference of 3 and B

415. the quotient of the difference of x and y, and -3

416. the quotient of the sum of m and n, and $4q$

Everyday Math

417. Baking. A recipe for chocolate chip cookies calls for $\frac{3}{4}$ cup brown sugar. Imelda wants to double the recipe. ⓐ How much brown sugar will Imelda need? Show your calculation. ⓑ Measuring cups usually come in sets of $\frac{1}{4}, \frac{1}{3}, \frac{1}{2}$, and 1 cup. Draw a diagram to show two different ways that Imelda could measure the brown sugar needed to double the cookie recipe.

418. Baking. Nina is making 4 pans of fudge to serve after a music recital. For each pan, she needs $\frac{2}{3}$ cup of condensed milk. ⓐ How much condensed milk will Nina need? Show your calculation. ⓑ Measuring cups usually come in sets of $\frac{1}{4}, \frac{1}{3}, \frac{1}{2}$, and 1 cup. Draw a diagram to show two different ways that Nina could measure the condensed milk needed for 4 pans of fudge.

419. Portions Don purchased a bulk package of candy that weighs 5 pounds. He wants to sell the candy in little bags that hold $\frac{1}{4}$ pound. How many little bags of candy can he fill from the bulk package?

420. Portions Kristen has $\frac{3}{4}$ yards of ribbon that she wants to cut into 6 equal parts to make hair ribbons for her daughter's 6 dolls. How long will each doll's hair ribbon be?

Writing Exercises

421. Rafael wanted to order half a medium pizza at a restaurant. The waiter told him that a medium pizza could be cut into 6 or 8 slices. Would he prefer 3 out of 6 slices or 4 out of 8 slices? Rafael replied that since he wasn't very hungry, he would prefer 3 out of 6 slices. Explain what is wrong with Rafael's reasoning.

422. Give an example from everyday life that demonstrates how $\frac{1}{2} \cdot \frac{2}{3}$ is $\frac{1}{3}$.

423. Explain how you find the reciprocal of a fraction.

424. Explain how you find the reciprocal of a negative number.

Self Check

ⓐ *After completing the exercises, use this checklist to evaluate your mastery of the objectives of this section.*

I can...	Confidently	With some help	No-I don't get it!
find equivalent fractions.			
simplify fractions.			
multiply fractions.			
divide fractions.			
simplify expressions written with a fraction bar.			
translate phrases to expressions with fractions.			

ⓑ *After looking at the checklist, do you think you are well prepared for the next section? Why or why not?*

1.6   Add and Subtract Fractions

Learning Objectives

By the end of this section, you will be able to:

› Add or subtract fractions with a common denominator
› Add or subtract fractions with different denominators
› Use the order of operations to simplify complex fractions
› Evaluate variable expressions with fractions

Be Prepared!

A more thorough introduction to the topics covered in this section can be found in the *Prealgebra* chapter, **Fractions**.

Add or Subtract Fractions with a Common Denominator

When we multiplied fractions, we just multiplied the numerators and multiplied the denominators right straight across. To add or subtract fractions, they must have a common denominator.

Fraction Addition and Subtraction

If a, b, and c are numbers where $c \neq 0$, then

$$\frac{a}{c} + \frac{b}{c} = \frac{a+b}{c} \quad \text{and} \quad \frac{a}{c} - \frac{b}{c} = \frac{a-b}{c}$$

To add or subtract fractions, add or subtract the numerators and place the result over the common denominator.

 MANIPULATIVE MATHEMATICS

Doing the Manipulative Mathematics activities "Model Fraction Addition" and "Model Fraction Subtraction" will help you develop a better understanding of adding and subtracting fractions.

EXAMPLE 1.77

Find the sum: $\frac{x}{3} + \frac{2}{3}$.

⊘ **Solution**

$$\frac{x}{3} + \frac{2}{3}$$

Add the numerators and place the sum over the common denominator.

$$\frac{x+2}{3}$$

 **TRY IT : : 1.153**   Find the sum: $\frac{x}{4} + \frac{3}{4}$.

 **TRY IT : : 1.154**   Find the sum: $\frac{y}{8} + \frac{5}{8}$.

EXAMPLE 1.78

Find the difference: $-\frac{23}{24} - \frac{13}{24}$.

⊘ Solution

$$-\frac{23}{24} - \frac{13}{24}$$

Subtract the numerators and place the difference over the common denominator.

$$\frac{-23 - 13}{24}$$

Simplify.

$$\frac{-36}{24}$$

Simplify. Remember, $-\frac{a}{b} = \frac{-a}{b}$.

$$-\frac{3}{2}$$

> **TRY IT :: 1.155**　　Find the difference: $-\frac{19}{28} - \frac{7}{28}$.

> **TRY IT :: 1.156**　　Find the difference: $-\frac{27}{32} - \frac{1}{32}$.

EXAMPLE 1.79

Simplify: $-\frac{10}{x} - \frac{4}{x}$.

⊘ Solution

$$-\frac{10}{x} - \frac{4}{x}$$

Subtract the numerators and place the difference over the common denominator.

$$\frac{-14}{x}$$

Rewrite with the sign in front of the fraction.

$$-\frac{14}{x}$$

> **TRY IT :: 1.157**　　Find the difference: $-\frac{9}{x} - \frac{7}{x}$.

> **TRY IT :: 1.158**　　Find the difference: $-\frac{17}{a} - \frac{5}{a}$.

Now we will do an example that has both addition and subtraction.

EXAMPLE 1.80

Simplify: $\frac{3}{8} + \left(-\frac{5}{8}\right) - \frac{1}{8}$.

⊘ Solution

Add and subtract fractions—do they have a common denominator? Yes.

$$\frac{3}{8} + \left(-\frac{5}{8}\right) - \frac{1}{8}$$

Add and subtract the numerators and place the result over the common denominator.

$$\frac{3 + (-5) - 1}{8}$$

Simplify left to right.

$$\frac{-2 - 1}{8}$$

Simplify.

$$-\frac{3}{8}$$

> **TRY IT : : 1.159**   Simplify: $\frac{2}{5} + \left(-\frac{4}{9}\right) - \frac{7}{9}$.

> **TRY IT : : 1.160**   Simplify: $\frac{5}{9} + \left(-\frac{4}{9}\right) - \frac{7}{9}$.

Add or Subtract Fractions with Different Denominators

As we have seen, to add or subtract fractions, their denominators must be the same. The **least common denominator** (LCD) of two fractions is the smallest number that can be used as a common denominator of the fractions. The LCD of the two fractions is the least common multiple (LCM) of their denominators.

Least Common Denominator

The least common denominator (LCD) of two fractions is the least common multiple (LCM) of their denominators.

 MANIPULATIVE MATHEMATICS

Doing the Manipulative Mathematics activity "Finding the Least Common Denominator" will help you develop a better understanding of the LCD.

After we find the least common denominator of two fractions, we convert the fractions to equivalent fractions with the LCD. Putting these steps together allows us to add and subtract fractions because their denominators will be the same!

EXAMPLE 1.81   HOW TO ADD OR SUBTRACT FRACTIONS

Add: $\frac{7}{12} + \frac{5}{18}$.

Solution

Step 1. Do they have a common denominator? No—rewrite each fraction with the LCD (least common denominator).	No. Find the LCD of 12, 18. Change into equivalent fractions with the LCD, 36. Do not simplify the equivalent fractions! If you do, you'll get back to the original fractions and lose the common denominator!	$\begin{array}{l} 12 = 2 \cdot 2 \cdot 3 \\ 18 = 2 \cdot 3 \cdot 3 \\ \hline LCD = 2 \cdot 2 \cdot 3 \cdot 3 \\ LCD = 36 \end{array}$ $\frac{7}{12} + \frac{5}{18}$ $\frac{7 \cdot 3}{12 \cdot 3} + \frac{5 \cdot 2}{18 \cdot 2}$ $\frac{21}{36} + \frac{10}{36}$
Step 2. Add or subtract the fractions.	Add.	$\frac{31}{36}$
Step 3. Simplify, if possible.	Because 31 is a prime number, it has no factors in common with 36. The answer is simplified.	

> **TRY IT : : 1.161**   Add: $\frac{7}{12} + \frac{11}{15}$.

> **TRY IT : : 1.162**   Add: $\frac{13}{15} + \frac{17}{20}$.

HOW TO : : ADD OR SUBTRACT FRACTIONS.

Step 1.   Do they have a common denominator?
- Yes—go to step 2.
- No—rewrite each fraction with the LCD (least common denominator). Find the LCD. Change each fraction into an equivalent fraction with the LCD as its denominator.

Step 2.   Add or subtract the fractions.

Step 3.   Simplify, if possible.

When finding the equivalent fractions needed to create the common denominators, there is a quick way to find the number we need to multiply both the numerator and denominator. This method works if we found the LCD by factoring into primes.

Look at the factors of the LCD and then at each column above those factors. The "missing" factors of each denominator are the numbers we need.

$$\begin{array}{l} \text{missing} \\ \text{factors} \\ 12 = 2 \cdot 2 \cdot 3 \\ 18 = 2 \cdot \quad 3 \cdot 3 \\ \hline \text{LCD} = 2 \cdot 2 \cdot 3 \cdot 3 \\ \text{LCD} = 36 \end{array}$$

In **Example 1.81**, the LCD, 36, has two factors of 2 and two factors of 3.

The numerator 12 has two factors of 2 but only one of 3—so it is "missing" one 3—we multiply the numerator and denominator by 3.

The numerator 18 is missing one factor of 2—so we multiply the numerator and denominator by 2.

We will apply this method as we subtract the fractions in **Example 1.82**.

EXAMPLE 1.82

Subtract: $\dfrac{7}{15} - \dfrac{19}{24}$.

 Solution

Do the fractions have a common denominator? No, so we need to find the LCD.

$$\frac{7}{15} - \frac{19}{24}$$

Find the LCD.	$\begin{array}{l} 15 = \qquad\quad 3 \cdot 5 \\ 24 = 2 \cdot 2 \cdot 2 \cdot 3 \\ \hline \text{LCD} = 2 \cdot 2 \cdot 2 \cdot 3 \cdot 5 \\ \text{LCD} = 120 \end{array}$

Notice, 15 is "missing" three factors of 2 and 24 is "missing" the 5 from the factors of the LCD. So we multiply 8 in the first fraction and 5 in the second fraction to get the LCD.

Rewrite as equivalent fractions with the LCD.	$\dfrac{7 \cdot 8}{15 \cdot 8} - \dfrac{19 \cdot 5}{24 \cdot 5}$
Simplify.	$\dfrac{56}{120} - \dfrac{95}{120}$
Subtract.	$-\dfrac{39}{120}$

Check to see if the answer can be simplified.	$-\dfrac{13 \cdot 3}{40 \cdot 3}$
Both 39 and 120 have a factor of 3.	
Simplify.	$-\dfrac{13}{40}$

Do not simplify the equivalent fractions! If you do, you'll get back to the original fractions and lose the common denominator!

> **TRY IT :: 1.163**   Subtract: $\dfrac{13}{24} - \dfrac{17}{32}$.

> **TRY IT :: 1.164**   Subtract: $\dfrac{21}{32} - \dfrac{9}{28}$.

In the next example, one of the fractions has a variable in its numerator. Notice that we do the same steps as when both numerators are numbers.

EXAMPLE 1.83

Add: $\dfrac{3}{5} + \dfrac{x}{8}$.

⊘ **Solution**

The fractions have different denominators.

$$\dfrac{3}{5} + \dfrac{x}{8}$$

Find the LCD.	$\begin{array}{ll}5 = & \quad\quad 5 \\ 8 = 2 \cdot 2 \cdot 2 & \\ \hline \text{LCD} = 2 \cdot 2 \cdot 2 \cdot 5 \\ \text{LCD} = 40 \end{array}$
Rewrite as equivalent fractions with the LCD.	$\dfrac{3 \cdot 8}{5 \cdot 8} + \dfrac{x \cdot 5}{8 \cdot 5}$
Simplify.	$\dfrac{24}{40} + \dfrac{5x}{40}$
Add.	$\dfrac{24 + 5x}{40}$

Remember, we can only add like terms: 24 and 5x are not like terms.

> **TRY IT :: 1.165**   Add: $\dfrac{y}{6} + \dfrac{7}{9}$.

> **TRY IT :: 1.166**   Add: $\dfrac{x}{6} + \dfrac{7}{15}$.

We now have all four operations for fractions. **Table 1.48** summarizes fraction operations.

Fraction Multiplication	Fraction Division
$\frac{a}{b} \cdot \frac{c}{d} = \frac{ac}{bd}$	$\frac{a}{b} \div \frac{c}{d} = \frac{a}{b} \cdot \frac{d}{c}$
Multiply the numerators and multiply the denominators	Multiply the first fraction by the reciprocal of the second.
Fraction Addition	**Fraction Subtraction**
$\frac{a}{c} + \frac{b}{c} = \frac{a+b}{c}$	$\frac{a}{c} - \frac{b}{c} = \frac{a-b}{c}$
Add the numerators and place the sum over the common denominator.	Subtract the numerators and place the difference over the common denominator.
To multiply or divide fractions, an LCD is NOT needed. To add or subtract fractions, an LCD is needed.	

Table 1.48

EXAMPLE 1.84

Simplify: ⓐ $\frac{5x}{6} - \frac{3}{10}$ ⓑ $\frac{5x}{6} \cdot \frac{3}{10}$.

⊘ Solution

First ask, "What is the operation?" Once we identify the operation that will determine whether we need a common denominator. Remember, we need a common denominator to add or subtract, but not to multiply or divide.

ⓐ What is the operation? The operation is subtraction.

Do the fractions have a common denominator? No.

$$\frac{5x}{6} - \frac{3}{10}$$

Rewrite each fraction as an equivalent fraction with the LCD.

$$\frac{5x \cdot 5}{6 \cdot 5} - \frac{3 \cdot 3}{10 \cdot 3}$$
$$\frac{25x}{30} - \frac{9}{30}$$

Subtract the numerators and place the difference over the common denominators.

$$\frac{25x - 9}{30}$$

Simplify, if possible There are no common factors.
The fraction is simplified.

ⓑ What is the operation? Multiplication.

$$\frac{5x}{6} \cdot \frac{3}{10}$$

To multiply fractions, multiply the numerators and multiply the denominators.

$$\frac{5x \cdot 3}{6 \cdot 10}$$

Rewrite, showing common factors.
Remove common factors.

$$\frac{\cancel{5}x \cdot \cancel{3}}{2 \cdot \cancel{3} \cdot 2 \cdot \cancel{5}}$$

Simplify.

$$\frac{x}{4}$$

Notice we needed an LCD to add $\frac{5x}{6} - \frac{3}{10}$, but not to multiply $\frac{5x}{6} \cdot \frac{3}{10}$.

> **TRY IT : : 1.167**   Simplify: ⓐ $\frac{3a}{4} - \frac{8}{9}$ ⓑ $\frac{3a}{4} \cdot \frac{8}{9}$.

> **TRY IT : : 1.168**   Simplify: ⓐ $\frac{4k}{5} - \frac{1}{6}$ ⓑ $\frac{4k}{5} \cdot \frac{1}{6}$.

Use the Order of Operations to Simplify Complex Fractions

We have seen that a complex fraction is a fraction in which the numerator or denominator contains a fraction. The fraction bar indicates division. We simplified the complex fraction $\frac{\frac{3}{4}}{\frac{5}{8}}$ by dividing $\frac{3}{4}$ by $\frac{5}{8}$.

Now we'll look at complex fractions where the numerator or denominator contains an expression that can be simplified. So we first must completely simplify the numerator and denominator separately using the order of operations. Then we divide the numerator by the denominator.

EXAMPLE 1.85   HOW TO SIMPLIFY COMPLEX FRACTIONS

Simplify: $\dfrac{\left(\frac{1}{2}\right)^2}{4 + 3^2}$.

✓ **Solution**

Step 1. Simplify the numerator. * Remember, $\left(\frac{1}{2}\right)^2$ means $\frac{1}{2} \cdot \frac{1}{2}$.	$\dfrac{\left(\frac{1}{2}\right)^2}{4 + 3^2}$ $\dfrac{\frac{1}{4}}{4 + 3^2}$
Step 2. Simplify the denominator.	$\dfrac{\frac{1}{4}}{4 + 9}$ $\dfrac{\frac{1}{4}}{13}$
Step 3. Divide the numerator by the denominator. Simplify if possible. * Remember, $13 = \frac{13}{1}$	$\frac{1}{4} \div 13$ $\frac{1}{4} \cdot \frac{1}{13}$ $\frac{1}{52}$

> **TRY IT : : 1.169**   Simplify: $\dfrac{\left(\frac{1}{3}\right)^2}{2^3 + 2}$.

> **TRY IT : : 1.170**   Simplify: $\dfrac{1 + 4^2}{\left(\frac{1}{4}\right)^2}$.

  HOW TO : : SIMPLIFY COMPLEX FRACTIONS.

    Step 1.   Simplify the numerator.

    Step 2.   Simplify the denominator.

    Step 3.   Divide the numerator by the denominator. Simplify if possible.

EXAMPLE 1.86

Simplify: $\dfrac{\frac{1}{2}+\frac{2}{3}}{\frac{3}{4}-\frac{1}{6}}$.

⊘ Solution

It may help to put parentheses around the numerator and the denominator.

$$\frac{\left(\frac{1}{2}+\frac{2}{3}\right)}{\left(\frac{3}{4}-\frac{1}{6}\right)}$$

Simplify the numerator (LCD = 6)
and simplify the denominator (LCD = 12).

$$\frac{\left(\frac{3}{6}+\frac{4}{6}\right)}{\left(\frac{9}{12}-\frac{2}{12}\right)}$$

Simplify.

$$\frac{\left(\frac{7}{6}\right)}{\left(\frac{7}{12}\right)}$$

Divide the numerator by the denominator.

$$\frac{7}{6}\div\frac{7}{12}$$

Simplify.

$$\frac{7}{6}\cdot\frac{12}{7}$$

Divide out common factors.

$$\frac{7\cdot6\cdot2}{6\cdot7}$$

Simplify.

$$2$$

> **TRY IT : : 1.171**
>
> Simplify: $\dfrac{\frac{1}{3}+\frac{1}{2}}{\frac{3}{4}-\frac{1}{3}}$.

> **TRY IT : : 1.172**
>
> Simplify: $\dfrac{\frac{2}{3}-\frac{1}{2}}{\frac{1}{4}+\frac{1}{3}}$.

Evaluate Variable Expressions with Fractions

We have evaluated expressions before, but now we can evaluate expressions with fractions. Remember, to evaluate an expression, we substitute the value of the variable into the expression and then simplify.

EXAMPLE 1.87

Evaluate $x+\frac{1}{3}$ when ⓐ $x=-\frac{1}{3}$ ⓑ $x=-\frac{3}{4}$.

✓ Solution

ⓐ To evaluate $x + \frac{1}{3}$ when $x = -\frac{1}{3}$, substitute $-\frac{1}{3}$ for x in the expression.

	$x + \frac{1}{3}$
Substitute $-\frac{1}{3}$ for x.	$-\frac{1}{3} + \frac{1}{3}$
Simplify.	0

ⓑ To evaluate $x + \frac{1}{3}$ when $x = -\frac{3}{4}$, we substitute $-\frac{3}{4}$ for x in the expression.

	$x + \frac{1}{3}$
Substitute $-\frac{3}{4}$ for x.	$-\frac{3}{4} + \frac{1}{3}$
Rewrite as equivalent fractions with the LCD, 12.	$-\frac{3 \cdot 3}{4 \cdot 3} + \frac{1 \cdot 4}{3 \cdot 4}$
Simplify.	$-\frac{9}{12} + \frac{4}{12}$
Add.	$-\frac{5}{12}$

> **TRY IT : : 1.173**   Evaluate $x + \frac{3}{4}$ when ⓐ $x = -\frac{7}{4}$ ⓑ $x = -\frac{5}{4}$.

> **TRY IT : : 1.174**   Evaluate $y + \frac{1}{2}$ when ⓐ $y = \frac{2}{3}$ ⓑ $y = -\frac{3}{4}$.

EXAMPLE 1.88

Evaluate $-\frac{5}{6} - y$ when $y = -\frac{2}{3}$.

✓ Solution

	$-\frac{5}{6} - y$
Substitute $-\frac{2}{3}$ for y.	$-\frac{5}{6} - \left(-\frac{2}{3}\right)$
Rewrite as equivalent fractions with the LCD, 6.	$-\frac{5}{6} - \left(-\frac{4}{6}\right)$
Subtract.	$\frac{-5 - (-4)}{6}$
Simplify.	$-\frac{1}{6}$

> **TRY IT :: 1.175**   Evaluate $-\frac{1}{2} - y$ when $y = -\frac{1}{4}$.

> **TRY IT :: 1.176**   Evaluate $-\frac{3}{8} - y$ when $x = -\frac{5}{2}$.

EXAMPLE 1.89

Evaluate $2x^2 y$ when $x = \frac{1}{4}$ and $y = -\frac{2}{3}$.

⊘ **Solution**

Substitute the values into the expression.

	$2x^2 y$
Substitute $\frac{1}{4}$ for x and $-\frac{2}{3}$ for y.	$2\left(\frac{1}{4}\right)^2\left(-\frac{2}{3}\right)$
Simplify exponents first.	$2\left(\frac{1}{16}\right)\left(-\frac{2}{3}\right)$
Multiply. Divide out the common factors. Notice we write 16 as $2 \cdot 2 \cdot 4$ to make it easy to remove common factors.	$-\dfrac{\cancel{2} \cdot 1 \cdot \cancel{2}}{\cancel{2} \cdot \cancel{2} \cdot 4 \cdot 3}$
Simplify.	$-\frac{1}{12}$

> **TRY IT :: 1.177**   Evaluate $3ab^2$ when $a = -\frac{2}{3}$ and $b = -\frac{1}{2}$.

> **TRY IT :: 1.178**   Evaluate $4c^3 d$ when $c = -\frac{1}{2}$ and $d = -\frac{4}{3}$.

The next example will have only variables, no constants.

EXAMPLE 1.90

Evaluate $\frac{p+q}{r}$ when $p = -4$, $q = -2$, and $r = 8$.

⊘ **Solution**

To evaluate $\frac{p+q}{r}$ when $p = -4$, $q = -2$, and $r = 8$, we substitute the values into the expression.

	$\frac{p+q}{r}$
Substitute –4 for p, –2 for q and 8 for r.	$\frac{-4 + (-2)}{8}$
Add in the numerator first.	$\frac{-6}{8}$
Simplify.	$-\frac{3}{4}$

> **TRY IT : : 1.179**   Evaluate $\dfrac{a+b}{c}$ when $a = -8$, $b = -7$, and $c = 6$.

> **TRY IT : : 1.180**   Evaluate $\dfrac{x+y}{z}$ when $x = 9$, $y = -18$, and $z = -6$.

1.6 EXERCISES

Practice Makes Perfect

Add and Subtract Fractions with a Common Denominator

In the following exercises, add.

425. $\frac{6}{13} + \frac{5}{13}$

426. $\frac{4}{15} + \frac{7}{15}$

427. $\frac{x}{4} + \frac{3}{4}$

428. $\frac{8}{9} + \frac{6}{9}$

429. $-\frac{3}{16} + \left(-\frac{7}{16}\right)$

430. $-\frac{5}{16} + \left(-\frac{9}{16}\right)$

431. $-\frac{8}{17} + \frac{15}{17}$

432. $-\frac{9}{19} + \frac{17}{19}$

433. $\frac{6}{13} + \left(-\frac{10}{13}\right) + \left(-\frac{12}{13}\right)$

434. $\frac{5}{12} + \left(-\frac{7}{12}\right) + \left(-\frac{11}{12}\right)$

In the following exercises, subtract.

435. $\frac{11}{15} - \frac{7}{15}$

436. $\frac{9}{13} - \frac{4}{13}$

437. $\frac{11}{12} - \frac{5}{12}$

438. $\frac{7}{12} - \frac{5}{12}$

439. $\frac{19}{21} - \frac{4}{21}$

440. $\frac{17}{21} - \frac{8}{21}$

441. $\frac{5y}{8} - \frac{7}{8}$

442. $\frac{11z}{13} - \frac{8}{13}$

443. $-\frac{23}{u} - \frac{15}{u}$

444. $-\frac{29}{v} - \frac{26}{v}$

445. $-\frac{3}{5} - \left(-\frac{4}{5}\right)$

446. $-\frac{3}{7} - \left(-\frac{5}{7}\right)$

447. $-\frac{7}{9} - \left(-\frac{5}{9}\right)$

448. $-\frac{8}{11} - \left(-\frac{5}{11}\right)$

Mixed Practice

In the following exercises, simplify.

449. $-\frac{5}{18} \cdot \frac{9}{10}$

450. $-\frac{3}{14} \cdot \frac{7}{12}$

451. $\frac{n}{5} - \frac{4}{5}$

452. $\frac{6}{11} - \frac{s}{11}$

453. $-\frac{7}{24} + \frac{2}{24}$

454. $-\frac{5}{18} + \frac{1}{18}$

455. $\frac{8}{15} \div \frac{12}{5}$

456. $\frac{7}{12} \div \frac{9}{28}$

Add or Subtract Fractions with Different Denominators

In the following exercises, add or subtract.

457. $\frac{1}{2} + \frac{1}{7}$

458. $\frac{1}{3} + \frac{1}{8}$

459. $\frac{1}{3} - \left(-\frac{1}{9}\right)$

460. $\frac{1}{4} - \left(-\frac{1}{8}\right)$

461. $\frac{7}{12} + \frac{5}{8}$

462. $\frac{5}{12} + \frac{3}{8}$

463. $\frac{7}{12} - \frac{9}{16}$

464. $\frac{7}{16} - \frac{5}{12}$

465. $\frac{2}{3} - \frac{3}{8}$

466. $\frac{5}{6} - \frac{3}{4}$

467. $-\frac{11}{30} + \frac{27}{40}$

468. $-\frac{9}{20} + \frac{17}{30}$

469. $-\frac{13}{30} + \frac{25}{42}$

470. $-\frac{23}{30} + \frac{5}{48}$

471. $-\frac{39}{56} - \frac{22}{35}$

472. $-\frac{33}{49} - \frac{18}{35}$

473. $-\frac{2}{3} - \left(-\frac{3}{4}\right)$

474. $-\frac{3}{4} - \left(-\frac{4}{5}\right)$

475. $1 + \frac{7}{8}$

476. $1 - \frac{3}{10}$

477. $\frac{x}{3} + \frac{1}{4}$

478. $\frac{y}{2} + \frac{2}{3}$

479. $\frac{y}{4} - \frac{3}{5}$

480. $\frac{x}{5} - \frac{1}{4}$

Mixed Practice

In the following exercises, simplify.

481. ⓐ $\frac{2}{3} + \frac{1}{6}$ ⓑ $\frac{2}{3} \div \frac{1}{6}$

482. ⓐ $-\frac{2}{5} - \frac{1}{8}$ ⓑ $-\frac{2}{5} \cdot \frac{1}{8}$

483. ⓐ $\frac{5n}{6} \div \frac{8}{15}$ ⓑ $\frac{5n}{6} - \frac{8}{15}$

484. ⓐ $\frac{3a}{8} \div \frac{7}{12}$ ⓑ $\frac{3a}{8} - \frac{7}{12}$

485. $-\frac{3}{8} \div \left(-\frac{3}{10}\right)$

486. $-\frac{5}{12} \div \left(-\frac{5}{9}\right)$

487. $-\frac{3}{8} + \frac{5}{12}$

488. $-\frac{1}{8} + \frac{7}{12}$

489. $\frac{5}{6} - \frac{1}{9}$

490. $\frac{5}{9} - \frac{1}{6}$

491. $-\frac{7}{15} - \frac{y}{4}$

492. $-\frac{3}{8} - \frac{x}{11}$

493. $\frac{11}{12a} \cdot \frac{9a}{16}$

494. $\frac{10y}{13} \cdot \frac{8}{15y}$

Use the Order of Operations to Simplify Complex Fractions

In the following exercises, simplify.

495. $\dfrac{2^3 + 4^2}{\left(\frac{2}{3}\right)^2}$

496. $\dfrac{3^3 - 3^2}{\left(\frac{3}{4}\right)^2}$

497. $\dfrac{\left(\frac{3}{5}\right)^2}{\left(\frac{3}{7}\right)^2}$

498. $\dfrac{\left(\frac{3}{4}\right)^2}{\left(\frac{5}{8}\right)^2}$

499. $\dfrac{2}{\frac{1}{3} + \frac{1}{5}}$

500. $\dfrac{5}{\frac{1}{4} + \frac{1}{3}}$

501. $\dfrac{\frac{7}{8} - \frac{2}{3}}{\frac{1}{2} + \frac{3}{8}}$

502. $\dfrac{\frac{3}{4} - \frac{3}{5}}{\frac{1}{4} + \frac{2}{5}}$

503. $\frac{1}{2} + \frac{2}{3} \cdot \frac{5}{12}$

504. $\frac{1}{3} + \frac{2}{5} \cdot \frac{3}{4}$

505. $1 - \frac{3}{5} \div \frac{1}{10}$

506. $1 - \frac{5}{6} \div \frac{1}{12}$

507. $\frac{2}{3} + \frac{1}{6} + \frac{3}{4}$

508. $\frac{2}{3} + \frac{1}{4} + \frac{3}{5}$

509. $\frac{3}{8} - \frac{1}{6} + \frac{3}{4}$

510. $\frac{2}{5} + \frac{5}{8} - \frac{3}{4}$

511. $12\left(\frac{9}{20} - \frac{4}{15}\right)$

512. $8\left(\frac{15}{16} - \frac{5}{6}\right)$

513. $\dfrac{\frac{5}{8} + \frac{1}{6}}{\frac{19}{24}}$

514. $\dfrac{\frac{1}{6} + \frac{3}{10}}{\frac{14}{30}}$

515. $\left(\frac{5}{9} + \frac{1}{6}\right) \div \left(\frac{2}{3} - \frac{1}{2}\right)$

516. $\left(\frac{3}{4} + \frac{1}{6}\right) \div \left(\frac{5}{8} - \frac{1}{3}\right)$

Evaluate Variable Expressions with Fractions

In the following exercises, evaluate.

517. $x + \left(-\frac{5}{6}\right)$ when

ⓐ $x = \frac{1}{3}$

ⓑ $x = -\frac{1}{6}$

518. $x + \left(-\frac{11}{12}\right)$ when

ⓐ $x = \frac{11}{12}$

ⓑ $x = \frac{3}{4}$

519. $x - \frac{2}{5}$ when

ⓐ $x = \frac{3}{5}$

ⓑ $x = -\frac{3}{5}$

520. $x - \frac{1}{3}$ when

ⓐ $x = \frac{2}{3}$

ⓑ $x = -\frac{2}{3}$

521. $\frac{7}{10} - w$ when

ⓐ $w = \frac{1}{2}$

ⓑ $w = -\frac{1}{2}$

522. $\frac{5}{12} - w$ when

ⓐ $w = \frac{1}{4}$

ⓑ $w = -\frac{1}{4}$

523. $2x^2 y^3$ when $x = -\frac{2}{3}$ and $y = -\frac{1}{2}$

524. $8u^2 v^3$ when $u = -\frac{3}{4}$ and $v = -\frac{1}{2}$

525. $\frac{a+b}{a-b}$ when $a = -3$, $b = 8$

526. $\frac{r-s}{r+s}$ when $r = 10$, $s = -5$

Everyday Math

527. Decorating Laronda is making covers for the throw pillows on her sofa. For each pillow cover, she needs $\frac{1}{2}$ yard of print fabric and $\frac{3}{8}$ yard of solid fabric. What is the total amount of fabric Laronda needs for each pillow cover?

528. Baking Vanessa is baking chocolate chip cookies and oatmeal cookies. She needs $\frac{1}{2}$ cup of sugar for the chocolate chip cookies and $\frac{1}{4}$ of sugar for the oatmeal cookies. How much sugar does she need altogether?

Writing Exercises

529. Why do you need a common denominator to add or subtract fractions? Explain.

530. How do you find the LCD of 2 fractions?

Self Check

ⓐ *After completing the exercises, use this checklist to evaluate your mastery of the objectives of this section.*

I can...	Confidently	With some help	No-I don't get it!
add and subtract fractions with different denominators.			
identify and use fraction operations.			
use the order of operations to simplify complex fractions.			
evaluate variable expressions with fractions.			

ⓑ *After looking at the checklist, do you think you are well-prepared for the next chapter? Why or why not?*

1.7 Decimals

Learning Objectives

By the end of this section, you will be able to:
- › Name and write decimals
- › Round decimals
- › Add and subtract decimals
- › Multiply and divide decimals
- › Convert decimals, fractions, and percents

Be Prepared!

A more thorough introduction to the topics covered in this section can be found in the *Prealgebra* chapter, **Decimals**.

Name and Write Decimals

Decimals are another way of writing fractions whose denominators are powers of 10.

$$0.1 = \frac{1}{10} \qquad \text{0.1 is "one tenth"}$$

$$0.01 = \frac{1}{100} \qquad \text{0.01 is "one hundredth"}$$

$$0.001 = \frac{1}{1,000} \qquad \text{0.001 is "one thousandth"}$$

$$0.0001 = \frac{1}{10,000} \qquad \text{0.0001 is "one ten-thousandth"}$$

Notice that "ten thousand" is a number larger than one, but "one ten-thousand**th**" is a number smaller than one. The "th" at the end of the name tells you that the number is smaller than one.

When we name a whole number, the name corresponds to the place value based on the powers of ten. We read 10,000 as "ten thousand" and 10,000,000 as "ten million." Likewise, the names of the decimal places correspond to their fraction values. **Figure 1.14** shows the names of the place values to the left and right of the decimal point.

Place Value										
Hundred thousands	Ten thousands	Thousands	Hundreds	Tens	Ones	Tenths	Hundredths	Thousandths	Ten-thousandths	Hundred-thousandths

Figure 1.14 Place value of decimal numbers are shown to the left and right of the decimal point.

EXAMPLE 1.91	HOW TO NAME DECIMALS

Name the decimal 4.3.

 Solution

Step 1. Name the number to the left of the decimal point.	4 is to the left of the decimal point.	4.3 four _____
Step 2. Write 'and' for the decimal point.		four and _____

Step 3. Name the 'number' part to the right of the decimal point as if it were a whole number.	3 is to the right of the decimal point.	four and three _____
Step 4. Name the decimal place.		four and three tenths

> **TRY IT : : 1.181**   Name the decimal: 6.7.

> **TRY IT : : 1.182**   Name the decimal: 5.8.

We summarize the steps needed to name a decimal below.

HOW TO : : NAME A DECIMAL.

Step 1.   Name the number to the left of the decimal point.

Step 2.   Write "and" for the decimal point.

Step 3.   Name the "number" part to the right of the decimal point as if it were a whole number.

Step 4.   Name the decimal place of the last digit.

EXAMPLE 1.92

Name the decimal: −15.571.

⊘ **Solution**

−15.571

Name the number to the left of the decimal point.	negative fifteen _____
Write "and" for the decimal point.	negative fifteen and _____
Name the number to the right of the decimal point.	negative fifteen and five hundred seventy-one _____
The 1 is in the thousandths place.	negative fifteen and five hundred seventy-one thousandths

> **TRY IT : : 1.183**   Name the decimal: −13.461.

> **TRY IT : : 1.184**   Name the decimal: −2.053.

When we write a check we write both the numerals and the name of the number. Let's see how to write the decimal from the name.

EXAMPLE 1.93   HOW TO WRITE DECIMALS

Write "fourteen and twenty-four thousandths" as a decimal.

⊘ **Solution**

| Step 1. Look for the word 'and'; it locates the decimal point. Place a decimal point under the word 'and'.

Translate the words before 'and' into the whole number and place to the left of the decimal point. | | fourteen and twenty-four thousandths
fourteen <u>and</u> twenty-four thousandths

_____ . _____

14 . _____ |

Step 2. Mark the number of decimal places needed to the right of the decimal point by noting the place value indicated by the last word.	The last word is 'thousandths'.	14.____ ____ ____ tenths hundredths thousandths
Step 3. Translate the words after 'and' into the number to the right of the decimal point. Write the number in the spaces – putting the final digit in the last place.		14.____ <u>2</u> <u>4</u>
Step 4. Fill in zeros for empty place holders as needed.	Zeros are needed in the tenths place.	14.<u>0</u> <u>2</u> <u>4</u> Fourteen and twenty-four thousandths is written 14.024.

> **TRY IT : : 1.185**   Write as a decimal: thirteen and sixty-eight thousandths.

> **TRY IT : : 1.186**   Write as a decimal: five and ninety-four thousandths.

We summarize the steps to writing a decimal.

HOW TO : : WRITE A DECIMAL.

Step 1.   Look for the word "and"—it locates the decimal point.

   ◦   Place a decimal point under the word "and." Translate the words before "and" into the whole number and place it to the left of the decimal point.

   ◦   If there is no "and," write a "0" with a decimal point to its right.

Step 2.   Mark the number of decimal places needed to the right of the decimal point by noting the place value indicated by the last word.

Step 3.   Translate the words after "and" into the number to the right of the decimal point. Write the number in the spaces—putting the final digit in the last place.

Step 4.   Fill in zeros for place holders as needed.

Round Decimals

Rounding decimals is very much like rounding whole numbers. We will round decimals with a method based on the one we used to round whole numbers.

EXAMPLE 1.94   HOW TO ROUND DECIMALS

Round 18.379 to the nearest hundredth.

✓ **Solution**

Step 1. Locate the given place value and mark it with an arrow.		hundredths place ↓ 18.379
Step 2. Underline the digit to the right of the given place value.		hundredths place ↓ 18.379

Step 3. Is this digit greater than or equal to 5? **Yes:** Add 1 to the digit in the given place value. **No:** Do <u>not</u> change the digit in the given place value.	Because 9 is greater than or equal to 5, add 1 to the 7.	18.37 9 add 1 delete
Step 4. Rewrite the number, removing all digits to the right of the rounding digit.		18.38 18.38 is 18.379 rounded to the nearest hundredth.

> **TRY IT ∷ 1.187**   Round to the nearest hundredth: 1.047.

> **TRY IT ∷ 1.188**   Round to the nearest hundredth: 9.173.

We summarize the steps for rounding a decimal here.

HOW TO ∷ ROUND DECIMALS.

Step 1.   Locate the given place value and mark it with an arrow.

Step 2.   Underline the digit to the right of the place value.

Step 3.   Is this digit greater than or equal to 5?

  ◦ Yes—add 1 to the digit in the given place value.

  ◦ No—do <u>not</u> change the digit in the given place value.

Step 4.   Rewrite the number, deleting all digits to the right of the rounding digit.

EXAMPLE 1.95

Round 18.379 to the nearest ⓐ tenth ⓑ whole number.

⊘ **Solution**

Round 18.379

ⓐ to the nearest tenth

Locate the tenths place with an arrow.	tenths place ↓ 18.379
Underline the digit to the right of the given place value.	tenths place ↓ 18.3<u>7</u>9
Because 7 is greater than or equal to 5, add 1 to the 3.	18.379 add 1 ↗ ⌣ delete
Rewrite the number, deleting all digits to the right of the rounding digit.	18.4
Notice that the deleted digits were NOT replaced with zeros.	So, 18.379 rounded to the nearest tenth is 18.4.

ⓑ to the nearest whole number

Locate the ones place with an arrow.	ones place ↓ 18.379
Underline the digit to the right of the given place value.	ones place ↓ 18.<u>3</u>79
Since 3 is not greater than or equal to 5, do not add 1 to the 8.	18.379 do not add 1 ↗ ⌣ delete
Rewrite the number, deleting all digits to the right of the rounding digit.	18
	So, 18.379 rounded to the nearest whole number is 18.

> **TRY IT ∷ 1.189**   Round 6.582 to the nearest ⓐ hundredth ⓑ tenth ⓒ whole number.

> **TRY IT ∷ 1.190**   Round 15.2175 to the nearest ⓐ thousandth ⓑ hundredth ⓒ tenth.

Add and Subtract Decimals

To add or subtract decimals, we line up the decimal points. By lining up the decimal points this way, we can add or subtract the corresponding place values. We then add or subtract the numbers as if they were whole numbers and then place the decimal point in the sum.

HOW TO :: ADD OR SUBTRACT DECIMALS.

Step 1.   Write the numbers so the decimal points line up vertically.

Step 2.   Use zeros as place holders, as needed.

Step 3.   Add or subtract the numbers as if they were whole numbers. Then place the decimal point in the answer under the decimal points in the given numbers.

EXAMPLE 1.96

Add: $23.5 + 41.38$.

✓ Solution

Write the numbers so the decimal points line up vertically.

$$\begin{array}{r} 23.5 \\ +41.38 \\ \hline \end{array}$$

Put 0 as a placeholder after the 5 in 23.5. Remember, $\frac{5}{10} = \frac{50}{100}$ so $0.5 = 0.50$.

$$\begin{array}{r} 23.50 \\ +41.38 \\ \hline \end{array}$$

Add the numbers as if they were whole numbers. Then place the decimal point in the sum.

$$\begin{array}{r} 23.50 \\ +41.38 \\ \hline 64.88 \end{array}$$

> **TRY IT :: 1.191**   Add: $4.8 + 11.69$.

> **TRY IT :: 1.192**   Add: $5.123 + 18.47$.

EXAMPLE 1.97

Subtract: $20 - 14.65$.

✓ Solution

Write the numbers so the decimal points line up vertically.
Remember, 20 is a whole number, so place the decimal point after the 0.

$$\begin{array}{c} 20 - 14.65 \\ \begin{array}{r} 20. \\ -14.65 \\ \hline \end{array} \end{array}$$

Put in zeros to the right as placeholders.

$$\begin{array}{r} 20.00 \\ -14.65 \\ \hline \end{array}$$

Subtract and place the decimal point in the answer.

$$\begin{array}{r} \overset{\scriptscriptstyle 1}{\cancel{2}}\overset{\scriptscriptstyle 9}{\cancel{0}}.\overset{\scriptscriptstyle 9}{\cancel{0}}\,\overset{\scriptscriptstyle 10}{\cancel{0}} \\ -14.65 \\ \hline 5.35 \end{array}$$

> **TRY IT :: 1.193**   Subtract: $10 - 9.58$.

> **TRY IT :: 1.194**   Subtract: $50 - 37.42$.

Multiply and Divide Decimals

Multiplying decimals is very much like multiplying whole numbers—we just have to determine where to place the decimal point. The procedure for multiplying decimals will make sense if we first convert them to fractions and then multiply.

So let's see what we would get as the product of decimals by converting them to fractions first. We will do two examples side-by-side. Look for a pattern!

	(0.3) (0.7)	(0.2) (0.46)
	1 place 1 place	1 place 2 places
Convert to fractions.	$\dfrac{3}{10} \cdot \dfrac{7}{10}$	$\dfrac{2}{10} \cdot \dfrac{46}{100}$
Multiply.	$\dfrac{21}{100}$	$\dfrac{92}{1000}$
Convert to decimals.	0.21	0.092
	2 places	3 places

Notice, in the first example, we multiplied two numbers that each had one digit after the decimal point and the product had two decimal places. In the second example, we multiplied a number with one decimal place by a number with two decimal places and the product had three decimal places.

We multiply the numbers just as we do whole numbers, temporarily ignoring the decimal point. We then count the number of decimal points in the factors and that sum tells us the number of decimal places in the product.

The rules for multiplying positive and negative numbers apply to decimals, too, of course!

When *multiplying* two numbers,

- if their signs are the *same* the product is *positive*.
- if their signs are *different* the product is *negative*.

When we multiply signed decimals, first we determine the sign of the product and then multiply as if the numbers were both positive. Finally, we write the product with the appropriate sign.

HOW TO :: MULTIPLY DECIMALS.

Step 1. Determine the sign of the product.

Step 2. Write in vertical format, lining up the numbers on the right. Multiply the numbers as if they were whole numbers, temporarily ignoring the decimal points.

Step 3. Place the decimal point. The number of decimal places in the product is the sum of the number of decimal places in the factors.

Step 4. Write the product with the appropriate sign.

EXAMPLE 1.98

Multiply: $(-3.9)(4.075)$.

Solution

	$(-3.9)(4.075)$
The signs are different. The product will be negative.	
Write in vertical format, lining up the numbers on the right.	4.075 × 3.9
Multiply.	4.075 × 3.9 36675 12225 158925
Add the number of decimal places in the factors (1 + 3). (−3.9) (4.075) 1 place 3 places	4.075 × 3.9 36675 12225 15.8925
Place the decimal point 4 places from the right.	4 places
The signs are different, so the product is negative.	$(-3.9)(4.075) = -15.8925$

> **TRY IT :: 1.195**   Multiply: $-4.5(6.107)$.

> **TRY IT :: 1.196**   Multiply: $-10.79(8.12)$.

In many of your other classes, especially in the sciences, you will multiply decimals by powers of 10 (10, 100, 1000, etc.). If you multiply a few products on paper, you may notice a pattern relating the number of zeros in the power of 10 to number of decimal places we move the decimal point to the right to get the product.

HOW TO :: MULTIPLY A DECIMAL BY A POWER OF TEN.

Step 1.   Move the decimal point to the right the same number of places as the number of zeros in the power of 10.

Step 2.   Add zeros at the end of the number as needed.

EXAMPLE 1.99

Multiply 5.63 ⓐ by 10 ⓑ by 100 ⓒ by 1,000.

Solution

By looking at the number of zeros in the multiple of ten, we see the number of places we need to move the decimal to the right.

ⓐ

	$5.63(10)$
There is 1 zero in 10, so move the decimal point 1 place to the right.	5.63 56.3

ⓑ

	5.63(100)
There are 2 zeros in 100, so move the decimal point 2 places to the right.	5.63
	563

ⓒ

	5.63(1,000)
There are 3 zeros in 1,000, so move the decimal point 3 places to the right.	5.63
A zero must be added at the end.	5,630

> | **TRY IT : : 1.197**   Multiply 2.58 ⓐ by 10 ⓑ by 100 ⓒ by 1,000.

> | **TRY IT : : 1.198**   Multiply 14.2 ⓐ by 10 ⓑ by 100 ⓒ by 1,000.

Just as with multiplication, division of decimals is very much like dividing whole numbers. We just have to figure out where the decimal point must be placed.

To divide decimals, determine what power of 10 to multiply the denominator by to make it a whole number. Then multiply the numerator by that same power of $10.$ Because of the equivalent fractions property, we haven't changed the value of the fraction! The effect is to move the decimal points in the numerator and denominator the same number of places to the right. For example:

$$\frac{0.8}{0.4}$$

$$\frac{0.8(10)}{0.4(10)}$$

$$\frac{8}{4}$$

We use the rules for dividing positive and negative numbers with decimals, too. When dividing signed decimals, first determine the sign of the quotient and then divide as if the numbers were both positive. Finally, write the quotient with the appropriate sign.

We review the notation and vocabulary for division:

$$\underset{\text{dividend}}{a} \div \underset{\text{divisor}}{b} = \underset{\text{quotient}}{c} \qquad \underset{\text{divisor}}{b}\overline{\smash{)}\underset{\text{dividend}}{a}}\;\overset{\text{quotient}}{c}$$

We'll write the steps to take when dividing decimals, for easy reference.

HOW TO : : DIVIDE DECIMALS.

Step 1.   Determine the sign of the quotient.

Step 2.   Make the divisor a whole number by "moving" the decimal point all the way to the right. "Move" the decimal point in the dividend the same number of places—adding zeros as needed.

Step 3.   Divide. Place the decimal point in the quotient above the decimal point in the dividend.

Step 4.   Write the quotient with the appropriate sign.

EXAMPLE 1.100

Divide: $-25.56 \div (-0.06)$.

✓ Solution

Remember, you can "move" the decimals in the divisor and dividend because of the Equivalent Fractions Property.

$$-25.65 \div (-0.06)$$

The signs are the same.	The quotient is positive.
Make the divisor a whole number by "moving" the decimal point all the way to the right.	
"Move" the decimal point in the dividend the same number of places.	$0.06\overline{)25.65}$
Divide. Place the decimal point in the quotient above the decimal point in the dividend.	$\begin{array}{r} 427.5 \\ 006.\overline{)2565.0} \\ \underline{-24} \\ 16 \\ \underline{-12} \\ 45 \\ \underline{-42} \\ 30 \\ \underline{30} \end{array}$
Write the quotient with the appropriate sign.	$-25.65 \div (-0.06) = 427.5$

> **TRY IT : : 1.199**   Divide: $-23.492 \div (-0.04)$.

> **TRY IT : : 1.200**   Divide: $-4.11 \div (-0.12)$.

A common application of dividing whole numbers into decimals is when we want to find the price of one item that is sold as part of a multi-pack. For example, suppose a case of 24 water bottles costs $3.99. To find the price of one water bottle, we would divide $3.99 by 24. We show this division in **Example 1.101**. In calculations with money, we will round the answer to the nearest cent (hundredth).

EXAMPLE 1.101

Divide: $\$3.99 \div 24$.

✓ Solution

$$\$3.99 \div 24$$

Place the decimal point in the quotient above the decimal point in the dividend.	
Divide as usual. When do we stop? Since this division involves money, we round it to the nearest cent (hundredth.) To do this, we must carry the division to the thousandths place.	$\begin{array}{r} 0.166 \\ 24\overline{)3.990} \\ \underline{24} \\ 159 \\ \underline{144} \\ 150 \\ \underline{144} \\ 6 \end{array}$
Round to the nearest cent.	$\$0.166 \approx \$0.17$
	$\$3.99 \div 24 \approx \$0.17$

> **TRY IT : : 1.201**   Divide: $\$6.99 \div 36$.

> **TRY IT : : 1.202**   Divide: $\$4.99 \div 12$.

Convert Decimals, Fractions, and Percents

We convert decimals into fractions by identifying the place value of the last (farthest right) digit. In the decimal 0.03 the 3 is in the hundredths place, so 100 is the denominator of the fraction equivalent to 0.03.

$$0\,0.03 = \frac{3}{100}$$

Notice, when the number to the left of the decimal is zero, we get a fraction whose numerator is less than its denominator. Fractions like this are called proper fractions.

The steps to take to convert a decimal to a fraction are summarized in the procedure box.

HOW TO : : CONVERT A DECIMAL TO A PROPER FRACTION.

Step 1.   Determine the place value of the final digit.

Step 2.   Write the fraction.

  ◦ numerator—the "numbers" to the right of the decimal point

  ◦ denominator—the place value corresponding to the final digit

---

EXAMPLE 1.102

Write 0.374 as a fraction.

✓ **Solution**

0.374

	0.374
Determine the place value of the final digit.	0.3 7 4 tenths hundredths thousandths
Write the fraction for 0.374: • The numerator is 374. • The denominator is 1,000.	$\dfrac{374}{1000}$
Simplify the fraction.	$\dfrac{2 \cdot 187}{2 \cdot 500}$
Divide out the common factors.	$\dfrac{187}{500}$ so, $0.374 = \dfrac{187}{500}$

Did you notice that the number of zeros in the denominator of $\dfrac{374}{1,000}$ is the same as the number of decimal places in 0.374?

> **TRY IT : : 1.203**   Write 0.234 as a fraction.

> **TRY IT : : 1.204**   Write 0.024 as a fraction.

We've learned to convert decimals to fractions. Now we will do the reverse—convert fractions to decimals. Remember that the fraction bar means division. So $\frac{4}{5}$ can be written $4 \div 5$ or $5\overline{)4}$. This leads to the following method for converting a fraction to a decimal.

 HOW TO : : CONVERT A FRACTION TO A DECIMAL.

To convert a fraction to a decimal, divide the numerator of the fraction by the denominator of the fraction.

EXAMPLE 1.103

Write $-\dfrac{5}{8}$ as a decimal.

⊘ **Solution**

Since a fraction bar means division, we begin by writing $\dfrac{5}{8}$ as $8\overline{)5}$. Now divide.

$$
\begin{array}{r}
0.625 \\
8\overline{)5.000} \\
\underline{48} \\
20 \\
\underline{16} \\
40 \\
\underline{40}
\end{array}
$$

so, $-\dfrac{5}{8} = -0.625$

> **TRY IT : :** 1.205   Write $-\dfrac{7}{8}$ as a decimal.

> **TRY IT : :** 1.206   Write $-\dfrac{3}{8}$ as a decimal.

When we divide, we will not always get a zero remainder. Sometimes the quotient ends up with a decimal that repeats. A **repeating decimal** is a decimal in which the last digit or group of digits repeats endlessly. A bar is placed over the repeating block of digits to indicate it repeats.

Repeating Decimal

A **repeating decimal** is a decimal in which the last digit or group of digits repeats endlessly.

A bar is placed over the repeating block of digits to indicate it repeats.

EXAMPLE 1.104

Write $\dfrac{43}{22}$ as a decimal.

✓ Solution

$$\frac{43}{22}$$

Divide 43 by 22.

```
        1.95454
  22)43.00000
     22
     ─
     210
     198
     ───
     120  ←──── 120 repeats
     110
     ───
     100
      88
      ──
     120
     110
     ───
     100
      88
      ──
      ...
```

The pattern repeats, so the numbers in the quotient will repeat as well.

100 repeats

so, $\frac{43}{22} = 1.9\overline{54}$

> **TRY IT :: 1.207**   Write $\frac{27}{11}$ as a decimal.

> **TRY IT :: 1.208**   Write $\frac{51}{22}$ as a decimal.

Sometimes we may have to simplify expressions with fractions and decimals together.

EXAMPLE 1.105

Simplify: $\frac{7}{8} + 6.4$.

✓ Solution

First we must change one number so both numbers are in the same form. We can change the fraction to a decimal, or change the decimal to a fraction. Usually it is easier to change the fraction to a decimal.

$$\frac{7}{8} + 6.4$$

Change $\frac{7}{8}$ to a decimal.	$\begin{array}{r} 0.875 \\ 8)\overline{7.000} \\ \underline{64} \\ 60 \\ \underline{56} \\ 40 \\ \underline{40} \\ 0 \end{array}$
Add.	$0.875 + 6.4$
	7.275
	So, $\frac{7}{8} + 6.4 = 7.275$

> **TRY IT :: 1.209**   Simplify: $\frac{3}{8} + 4.9$.

> **TRY IT :: 1.210**   Simplify: $5.7 + \frac{13}{20}$.

A **percent** is a ratio whose denominator is 100. Percent means per hundred. We use the percent symbol, %, to show percent.

Percent
A **percent** is a ratio whose denominator is 100.

Since a percent is a ratio, it can easily be expressed as a fraction. Percent means per 100, so the denominator of the fraction is 100. We then change the fraction to a decimal by dividing the numerator by the denominator.

	6%	78%	135%
Write as a ratio with denominator 100.	$\frac{6}{100}$	$\frac{78}{100}$	$\frac{135}{100}$
Change the fraction to a decimal by dividing the numerator by the denominator.	0.06	0.78	1.35

Do you see the pattern? *To convert a percent number to a decimal number, we move the decimal point two places to the left.*

6%   78%   2.7%   135%

0.06   0.78   0.027   1.35

EXAMPLE 1.106

Convert each percent to a decimal: ⓐ 62% ⓑ 135% ⓒ 35.7%.

✓ **Solution**

ⓐ

62%

| Move the decimal point two places to the left. | 0.62 |

ⓑ

135%

| Move the decimal point two places to the left. | 1.35 |

ⓒ

5.7%

| Move the decimal point two places to the left. | 0.057 |

> **TRY IT : : 1.211**   Convert each percent to a decimal: ⓐ 9% ⓑ 87% ⓒ 3.9%.

> **TRY IT : : 1.212**   Convert each percent to a decimal: ⓐ 3% ⓑ 91% ⓒ 8.3%.

Converting a decimal to a percent makes sense if we remember the definition of percent and keep place value in mind.

To convert a decimal to a percent, remember that percent means per hundred. If we change the decimal to a fraction whose denominator is 100, it is easy to change that fraction to a percent.

	0.83	1.05	0.075
Write as a fraction.	$\frac{83}{100}$	$1\frac{5}{100}$	$\frac{75}{1000}$
The denominator is 100.		$\frac{105}{100}$	$\frac{7.5}{100}$
Write the ratio as a percent.	83%	105%	7.5%

Recognize the pattern? To convert a decimal to a percent, we move the decimal point two places to the right and then add the percent sign.

0.05	0.83	1.05	0.075	0.3
5%	83%	105%	7.5%	30%

EXAMPLE 1.107

Convert each decimal to a percent: ⓐ 0.51 ⓑ 1.25 ⓒ 0.093.

✓ **Solution**

ⓐ

	0.51
Move the decimal point two places to the right.	51%

ⓑ

	1.25
Move the decimal point two places to the right.	125%

ⓒ

	0.093
Move the decimal point two places to the right.	9.3%

> **TRY IT : : 1.213**   Convert each decimal to a percent: ⓐ 0.17 ⓑ 1.75 ⓒ 0.0825.

> **TRY IT : : 1.214**   Convert each decimal to a percent: ⓐ 0.41 ⓑ 2.25 ⓒ 0.0925.

1.7 EXERCISES

Practice Makes Perfect

Name and Write Decimals

In the following exercises, write as a decimal.

531. Twenty-nine and eighty-one hundredths

532. Sixty-one and seventy-four hundredths

533. Seven tenths

534. Six tenths

535. Twenty-nine thousandth

536. Thirty-five thousandths

537. Negative eleven and nine ten-thousandths

538. Negative fifty-nine and two ten-thousandths

In the following exercises, name each decimal.

539. 5.5

540. 14.02

541. 8.71

542. 2.64

543. 0.002

544. 0.479

545. -17.9

546. -31.4

Round Decimals

In the following exercises, round each number to the nearest tenth.

547. 0.67

548. 0.49

549. 2.84

550. 4.63

In the following exercises, round each number to the nearest hundredth.

551. 0.845

552. 0.761

553. 0.299

554. 0.697

555. 4.098

556. 7.096

In the following exercises, round each number to the nearest ⓐ hundredth ⓑ tenth ⓒ whole number.

557. 5.781

558. 1.6381

559. 63.479

560. 84.281

Add and Subtract Decimals

In the following exercises, add or subtract.

561. $16.92 + 7.56$

562. $248.25 - 91.29$

563. $21.76 - 30.99$

564. $38.6 + 13.67$

565. $-16.53 - 24.38$

566. $-19.47 - 32.58$

567. $-38.69 + 31.47$

568. $29.83 + 19.76$

569. $72.5 - 100$

570. $86.2 - 100$

571. $15 + 0.73$

572. $27 + 0.87$

573. $91.95 - (-10.462)$

574. $94.69 - (-12.678)$

575. $55.01 - 3.7$

576. $59.08 - 4.6$

577. $2.51 - 7.4$

578. $3.84 - 6.1$

Multiply and Divide Decimals

In the following exercises, multiply.

579. $(0.24)(0.6)$

580. $(0.81)(0.3)$

581. $(5.9)(7.12)$

582. $(2.3)(9.41)$

583. $(-4.3)(2.71)$

584. $(-8.5)(1.69)$

585. $(-5.18)(-65.23)$

586. $(-9.16)(-68.34)$

587. $(0.06)(21.75)$

588. $(0.08)(52.45)$

589. $(9.24)(10)$

590. $(6.531)(10)$

591. $(55.2)(1000)$

592. $(99.4)(1000)$

In the following exercises, divide.

593. $4.75 \div 25$

594. $12.04 \div 43$

595. $\$117.25 \div 48$

596. $\$109.24 \div 36$

597. $0.6 \div 0.2$

598. $0.8 \div 0.4$

599. $1.44 \div (-0.3)$

600. $1.25 \div (-0.5)$

601. $-1.75 \div (-0.05)$

602. $-1.15 \div (-0.05)$

603. $5.2 \div 2.5$

604. $6.5 \div 3.25$

605. $11 \div 0.55$

606. $14 \div 0.35$

Convert Decimals, Fractions and Percents

In the following exercises, write each decimal as a fraction.

607. 0.04

608. 0.19

609. 0.52

610. 0.78

611. 1.25

612. 1.35

613. 0.375

614. 0.464

615. 0.095

616. 0.085

In the following exercises, convert each fraction to a decimal.

617. $\frac{17}{20}$

618. $\frac{13}{20}$

619. $\frac{11}{4}$

620. $\frac{17}{4}$

621. $-\frac{310}{25}$

622. $-\frac{284}{25}$

623. $\frac{15}{11}$

624. $\frac{18}{11}$

625. $\frac{15}{111}$

626. $\frac{25}{111}$

627. $2.4 + \frac{5}{8}$

628. $3.9 + \frac{9}{20}$

In the following exercises, convert each percent to a decimal.

629. 1%

630. 2%

631. 63%

632. 71%

633. 150%

634. 250%

635. 21.4%

636. 39.3%

637. 7.8%

638. 6.4%

In the following exercises, convert each decimal to a percent.

639. 0.01

640. 0.03

641. 1.35

642. 1.56

643. 3

644. 4

645. 0.0875

646. 0.0625

647. 2.254

648. 2.317

Everyday Math

649. Salary Increase Danny got a raise and now makes $58,965.95 a year. Round this number to the nearest

ⓐ dollar

ⓑ thousand dollars

ⓒ ten thousand dollars.

650. New Car Purchase Selena's new car cost $23,795.95. Round this number to the nearest

ⓐ dollar

ⓑ thousand dollars

ⓒ ten thousand dollars.

651. Sales Tax Hyo Jin lives in San Diego. She bought a refrigerator for $1,624.99 and when the clerk calculated the sales tax it came out to exactly $142.186625. Round the sales tax to the nearest

ⓐ penny and

ⓑ dollar.

652. Sales Tax Jennifer bought a $1,038.99 dining room set for her home in Cincinnati. She calculated the sales tax to be exactly $67.53435. Round the sales tax to the nearest

ⓐ penny and

ⓑ dollar.

653. Paycheck Annie has two jobs. She gets paid $14.04 per hour for tutoring at City College and $8.75 per hour at a coffee shop. Last week she tutored for 8 hours and worked at the coffee shop for 15 hours.

ⓐ How much did she earn?

ⓑ If she had worked all 23 hours as a tutor instead of working both jobs, how much more would she have earned?

654. Paycheck Jake has two jobs. He gets paid $7.95 per hour at the college cafeteria and $20.25 at the art gallery. Last week he worked 12 hours at the cafeteria and 5 hours at the art gallery.

ⓐ How much did he earn?

ⓑ If he had worked all 17 hours at the art gallery instead of working both jobs, how much more would he have earned?

Writing Exercises

655. How does knowing about US money help you learn about decimals?

656. Explain how you write "three and nine hundredths" as a decimal.

657. Without solving the problem "44 is 80% of what number" think about what the solution might be. Should it be a number that is greater than 44 or less than 44? Explain your reasoning.

658. When the Szetos sold their home, the selling price was 500% of what they had paid for the house 30 years ago. Explain what 500% means in this context.

Self Check

ⓐ *After completing the exercises, use this checklist to evaluate your mastery of the objectives of this section.*

I can...	Confidently	With some help	No-I don't get it!
name and write decimals.			
round decimals.			
add and subtract decimals.			
multiply and divide decimals.			
convert decimals, fractions, and percents.			

ⓑ What does this checklist tell you about your mastery of this section? What steps will you take to improve?

1.8 The Real Numbers

Learning Objectives

By the end of this section, you will be able to:
› Simplify expressions with square roots
› Identify integers, rational numbers, irrational numbers, and real numbers
› Locate fractions on the number line
› Locate decimals on the number line

Be Prepared!

A more thorough introduction to the topics covered in this section can be found in the *Prealgebra* chapters, **Decimals** and **Properties of Real Numbers**.

Simplify Expressions with Square Roots

Remember that when a number n is multiplied by itself, we write n^2 and read it "n squared." The result is called the **square** of n. For example,

$$8^2 \quad \text{read '8 squared'}$$
$$64 \quad \text{64 is called the } square \text{ of 8.}$$

Similarly, 121 is the square of 11, because 11^2 is 121.

Square of a Number

If $n^2 = m,$ then m is the **square** of n.

 MANIPULATIVE MATHEMATICS

Doing the Manipulative Mathematics activity "Square Numbers" will help you develop a better understanding of perfect square numbers.

Complete the following table to show the squares of the counting numbers 1 through 15.

Number	n	1	2	3	4	5	6	7	8	9	10	11	12	13	14	15
Square	n^2								64			121				

The numbers in the second row are called perfect square numbers. It will be helpful to learn to recognize the perfect square numbers.

The squares of the counting numbers are positive numbers. What about the squares of negative numbers? We know that when the signs of two numbers are the same, their product is positive. So the square of any negative number is also positive.

$$(-3)^2 = 9 \quad (-8)^2 = 64 \quad (-11)^2 = 121 \quad (-15)^2 = 225$$

Did you notice that these squares are the same as the squares of the positive numbers?

Sometimes we will need to look at the relationship between numbers and their squares in reverse. Because $10^2 = 100,$ we say 100 is the square of 10. We also say that 10 is a *square root* of 100. A number whose square is m is called a **square root** of m.

Square Root of a Number

If $n^2 = m,$ then n is a **square root** of m.

Notice $(-10)^2 = 100$ also, so -10 is also a square root of 100. Therefore, both 10 and -10 are square roots of 100.

So, every positive number has two square roots—one positive and one negative. What if we only wanted the positive square root of a positive number? The radical sign, $\sqrt{m}$, denotes the positive square root. The positive square root is called the principal square root. When we use the radical sign that always means we want the principal square root.

We also use the radical sign for the square root of zero. Because $0^2 = 0$, $\sqrt{0} = 0$. Notice that zero has only one square root.

Square Root Notation

$\sqrt{m}$ is read "the square root of m"

$$\text{radical sign} \longrightarrow \sqrt{m} \longleftarrow \text{radicand}$$

If $m = n^2$, then $\sqrt{m} = n$, for $n \geq 0$.

The square root of m, $\sqrt{m}$, is the positive number whose square is m.

Since 10 is the principal square root of 100, we write $\sqrt{100} = 10$. You may want to complete the following table to help you recognize square roots.

$\sqrt{1}$	$\sqrt{4}$	$\sqrt{9}$	$\sqrt{16}$	$\sqrt{25}$	$\sqrt{36}$	$\sqrt{49}$	$\sqrt{64}$	$\sqrt{81}$	$\sqrt{100}$	$\sqrt{121}$	$\sqrt{144}$	$\sqrt{169}$	$\sqrt{196}$	$\sqrt{225}$
									10					

EXAMPLE 1.108

Simplify: ⓐ $\sqrt{25}$ ⓑ $\sqrt{121}$.

⊘ Solution

ⓐ

$$\sqrt{25}$$

Since $5^2 = 25$ 5

ⓑ

$$\sqrt{121}$$

Since $11^2 = 121$ 11

> **TRY IT :: 1.215**   Simplify: ⓐ $\sqrt{36}$ ⓑ $\sqrt{169}$.

> **TRY IT :: 1.216**   Simplify: ⓐ $\sqrt{16}$ ⓑ $\sqrt{196}$.

We know that every positive number has two square roots and the radical sign indicates the positive one. We write $\sqrt{100} = 10$. If we want to find the negative square root of a number, we place a negative in front of the radical sign. For example, $-\sqrt{100} = -10$. We read $-\sqrt{100}$ as "the opposite of the square root of 10."

EXAMPLE 1.109

Simplify: ⓐ $-\sqrt{9}$ ⓑ $-\sqrt{144}$.

⊘ Solution

ⓐ

$$-\sqrt{9}$$

The negative is in front of the radical sign. -3

ⓑ

$$-\sqrt{144}$$
The negative is in front of the radical sign. -12

> **TRY IT :: 1.217**   Simplify: ⓐ $-\sqrt{4}$ ⓑ $-\sqrt{225}$.

> **TRY IT :: 1.218**   Simplify: ⓐ $-\sqrt{81}$ ⓑ $-\sqrt{100}$.

Identify Integers, Rational Numbers, Irrational Numbers, and Real Numbers

We have already described numbers as *counting numbers*, *whole numbers*, and *integers*. What is the difference between these types of numbers?

Counting numbers	1, 2, 3, 4, …
Whole numbers	0, 1, 2, 3, 4, …
Integers	…−3, −2, −1, 0, 1, 2, 3, …

What type of numbers would we get if we started with all the integers and then included all the fractions? The numbers we would have form the set of rational numbers. A **rational number** is a number that can be written as a ratio of two integers.

Rational Number

A **rational number** is a number of the form $\frac{p}{q}$, where p and q are integers and $q \neq 0$.

A rational number can be written as the ratio of two integers.

All signed fractions, such as $\frac{4}{5}$, $-\frac{7}{8}$, $\frac{13}{4}$, $-\frac{20}{3}$ are rational numbers. Each numerator and each denominator is an integer.

Are integers rational numbers? To decide if an integer is a rational number, we try to write it as a ratio of two integers. Each integer can be written as a ratio of integers in many ways. For example, 3 is equivalent to $\frac{3}{1}$, $\frac{6}{2}$, $\frac{9}{3}$, $\frac{12}{4}$, $\frac{15}{5}$ …

An easy way to write an integer as a ratio of integers is to write it as a fraction with denominator one.

$$3 = \frac{3}{1} \qquad -8 = -\frac{8}{1} \qquad 0 = \frac{0}{1}$$

Since any integer can be written as the ratio of two integers, *all integers are rational numbers*! Remember that the counting numbers and the whole numbers are also integers, and so they, too, are rational.

What about decimals? Are they rational? Let's look at a few to see if we can write each of them as the ratio of two integers. We've already seen that integers are rational numbers. The integer -8 could be written as the decimal -8.0. So, clearly, some decimals are rational.

Think about the decimal 7.3. Can we write it as a ratio of two integers? Because 7.3 means $7\frac{3}{10}$, we can write it as an improper fraction, $\frac{73}{10}$. So 7.3 is the ratio of the integers 73 and 10. It is a rational number.

In general, any decimal that ends after a number of digits (such as 7.3 or -1.2684) is a rational number. We can use the place value of the last digit as the denominator when writing the decimal as a fraction.

EXAMPLE 1.110

Write as the ratio of two integers: ⓐ -27 ⓑ 7.31.

✓ **Solution**

ⓐ

$$-27$$

Write it as a fraction with denominator 1.　　$\dfrac{-27}{1}$

ⓑ

$$7.31$$

Write is as a mixed number. Remember.
7 is the whole number and the decimal
part, 0.31, indicates hundredths.　　$7\dfrac{31}{100}$

Convert to an improper fraction.　　$\dfrac{731}{100}$

So we see that -27 and 7.31 are both rational numbers, since they can be written as the ratio of two integers.

> **TRY IT :: 1.219**　　Write as the ratio of two integers: ⓐ -24 ⓑ 3.57.

> **TRY IT :: 1.220**　　Write as the ratio of two integers: ⓐ -19 ⓑ 8.41.

Let's look at the decimal form of the numbers we know are rational.

We have seen that *every integer is a rational number*, since $a = \dfrac{a}{1}$ for any integer, a. We can also change any integer to a decimal by adding a decimal point and a zero.

Integer	-2	-1	0	1	2	3
Decimal form	-2.0	-1.0	0.0	1.0	2.0	3.0

These decimal numbers stop.

We have also seen that *every fraction is a rational number*. Look at the decimal form of the fractions we considered above.

Ratio of integers	$\dfrac{4}{5}$	$-\dfrac{7}{8}$	$\dfrac{13}{4}$	$-\dfrac{20}{3}$
The decimal form	0.8	-0.875	3.25	$-6.666...$ $-6.\overline{6}$

These decimals either stop or repeat.

What do these examples tell us?

Every rational number can be written both as a ratio of integers, ($\dfrac{p}{q}$, where p and q are integers and $q \neq 0$), and as a decimal that either stops or repeats.

Here are the numbers we looked at above expressed as a ratio of integers and as a decimal:

	Fractions				Integers					
Number	$\dfrac{4}{5}$	$-\dfrac{7}{8}$	$\dfrac{13}{4}$	$-\dfrac{20}{3}$	-2	-1	0	1	2	3
Ratio of Integers	$\dfrac{4}{5}$	$-\dfrac{7}{8}$	$\dfrac{13}{4}$	$-\dfrac{20}{3}$	$-\dfrac{2}{1}$	$-\dfrac{1}{1}$	$\dfrac{0}{1}$	$\dfrac{1}{1}$	$\dfrac{2}{1}$	$\dfrac{3}{1}$
Decimal Form	0.8	-0.875	3.25	$-6.\overline{6}$	-2.0	-1.0	0.0	1.0	2.0	3.0

Rational Number

A **rational number** is a number of the form $\frac{p}{q}$, where p and q are integers and $q \neq 0$.

Its decimal form stops or repeats.

Are there any decimals that do not stop or repeat? Yes!

The number π (the Greek letter *pi*, pronounced "pie"), which is very important in describing circles, has a decimal form that does not stop or repeat.

$$\pi = 3.141592654...$$

We can even create a decimal pattern that does not stop or repeat, such as

$$2.01001000100001...$$

Numbers whose decimal form does not stop or repeat cannot be written as a fraction of integers. We call these numbers irrational.

Irrational Number

An **irrational number** is a number that cannot be written as the ratio of two integers.

Its decimal form does not stop and does not repeat.

Let's summarize a method we can use to determine whether a number is rational or irrational.

Rational or Irrational?

If the decimal form of a number

- *repeats or stops*, the number is **rational**.
- *does not repeat and does not stop*, the number is **irrational**.

EXAMPLE 1.111

Given the numbers $0.58\overline{3}, 0.47, 3.605551275...$ list the ⓐ rational numbers ⓑ irrational numbers.

✓ Solution

ⓐ

Look for decimals that repeat or stop.	The 3 repeats in $0.58\overline{3}$.
	The decimal 0.47 stops after the 7.
	So $0.58\overline{3}$ and 0.47 are rational.

ⓑ

| Look for decimals that neither stop nor repeat. | $3.605551275...$ has no repeating block of digits and it does not stop. |
| | So $3.605551275...$ is irrational. |

> **TRY IT : : 1.221**

For the given numbers list the ⓐ rational numbers ⓑ irrational numbers: $0.29, 0.81\overline{6}, 2.515115111....$

> **TRY IT : : 1.222**

For the given numbers list the ⓐ rational numbers ⓑ irrational numbers: $2.6\overline{3}, 0.125, 0.418302...$

EXAMPLE 1.112

For each number given, identify whether it is rational or irrational: ⓐ $\sqrt{36}$ ⓑ $\sqrt{44}$.

⊘ **Solution**

ⓐ Recognize that 36 is a perfect square, since $6^2 = 36$. So $\sqrt{36} = 6$, therefore $\sqrt{36}$ is rational.

ⓑ Remember that $6^2 = 36$ and $7^2 = 49$, so 44 is not a perfect square. Therefore, the decimal form of $\sqrt{44}$ will never repeat and never stop, so $\sqrt{44}$ is irrational.

> **TRY IT : : 1.223**   For each number given, identify whether it is rational or irrational: ⓐ $\sqrt{81}$ ⓑ $\sqrt{17}$.

> **TRY IT : : 1.224**   For each number given, identify whether it is rational or irrational: ⓐ $\sqrt{116}$ ⓑ $\sqrt{121}$.

We have seen that all counting numbers are whole numbers, all whole numbers are integers, and all integers are rational numbers. The irrational numbers are numbers whose decimal form does not stop and does not repeat. When we put together the rational numbers and the irrational numbers, we get the set of **real numbers**.

Real Number

A **real number** is a number that is either rational or irrational.

All the numbers we use in elementary algebra are real numbers. **Figure 1.15** illustrates how the number sets we've discussed in this section fit together.

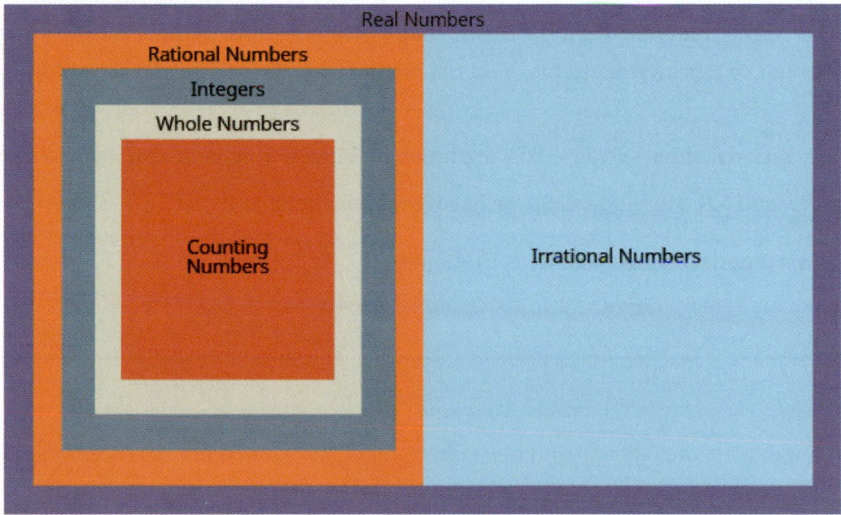

Figure 1.15 This chart shows the number sets that make up the set of real numbers. Does the term "real numbers" seem strange to you? Are there any numbers that are not "real," and, if so, what could they be?

Can we simplify $\sqrt{-25}$? Is there a number whose square is -25?

$$(\quad)^2 = -25?$$

None of the numbers that we have dealt with so far has a square that is -25. Why? Any positive number squared is positive. Any negative number squared is positive. So we say there is no real number equal to $\sqrt{-25}$.

The square root of a negative number is not a real number.

EXAMPLE 1.113

For each number given, identify whether it is a real number or not a real number: ⓐ $\sqrt{-169}$ ⓑ $-\sqrt{64}$.

⊘ Solution

ⓐ There is no real number whose square is -169. Therefore, $\sqrt{-169}$ is not a real number.

ⓑ Since the negative is in front of the radical, $-\sqrt{64}$ is -8, Since -8 is a real number, $-\sqrt{64}$ is a real number.

> **TRY IT : : 1.225**

For each number given, identify whether it is a real number or not a real number: ⓐ $\sqrt{-196}$ ⓑ $-\sqrt{81}$.

> **TRY IT : : 1.226**

For each number given, identify whether it is a real number or not a real number: ⓐ $-\sqrt{49}$ ⓑ $\sqrt{-121}$.

EXAMPLE 1.114

Given the numbers $-7, \dfrac{14}{5}, 8, \sqrt{5}, 5.9, -\sqrt{64},$ list the ⓐ whole numbers ⓑ integers ⓒ rational numbers ⓓ irrational numbers ⓔ real numbers.

⊘ Solution

ⓐ Remember, the whole numbers are $0, 1, 2, 3, \ldots$ and 8 is the only whole number given.

ⓑ The integers are the whole numbers, their opposites, and 0. So the whole number 8 is an integer, and -7 is the opposite of a whole number so it is an integer, too. Also, notice that 64 is the square of 8 so $-\sqrt{64} = -8$. So the integers are $-7, 8, -\sqrt{64}$.

ⓒ Since all integers are rational, then $-7, 8, -\sqrt{64}$ are rational. Rational numbers also include fractions and decimals that repeat or stop, so $\dfrac{14}{5}$ and 5.9 are rational. So the list of rational numbers is $-7, \dfrac{14}{5}, 8, 5.9, -\sqrt{64}$.

ⓓ Remember that 5 is not a perfect square, so $\sqrt{5}$ is irrational.

ⓔ All the numbers listed are real numbers.

> **TRY IT : : 1.227**

For the given numbers, list the ⓐ whole numbers ⓑ integers ⓒ rational numbers ⓓ irrational numbers ⓔ real numbers: $-3, -\sqrt{2}, 0.\overline{3}, \dfrac{9}{5}, 4, \sqrt{49}$.

> **TRY IT : : 1.228**

For the given numbers, list the ⓐ whole numbers ⓑ integers ⓒ rational numbers ⓓ irrational numbers ⓔ real numbers: $-\sqrt{25}, -\dfrac{3}{8}, -1, 6, \sqrt{121}, 2.041975\ldots$

Locate Fractions on the Number Line

The last time we looked at the number line, it only had positive and negative integers on it. We now want to include fractions and decimals on it.

 MANIPULATIVE MATHEMATICS

Doing the Manipulative Mathematics activity "Number Line Part 3" will help you develop a better understanding of the location of fractions on the number line.

Let's start with fractions and locate $\frac{1}{5}$, $-\frac{4}{5}$, 3, $\frac{7}{4}$, $-\frac{9}{2}$, -5, and $\frac{8}{3}$ on the number line.

We'll start with the whole numbers 3 and -5. because they are the easiest to plot. See **Figure 1.16**.

The proper fractions listed are $\frac{1}{5}$ and $-\frac{4}{5}$. We know the proper fraction $\frac{1}{5}$ has value less than one and so would be located between 0 and 1. The denominator is 5, so we divide the unit from 0 to 1 into 5 equal parts $\frac{1}{5}, \frac{2}{5}, \frac{3}{5}, \frac{4}{5}$. We plot $\frac{1}{5}$. See **Figure 1.16**.

Similarly, $-\frac{4}{5}$ is between 0 and -1. After dividing the unit into 5 equal parts we plot $-\frac{4}{5}$. See **Figure 1.16**.

Finally, look at the improper fractions $\frac{7}{4}$, $-\frac{9}{2}$, $\frac{8}{3}$. These are fractions in which the numerator is greater than the denominator. Locating these points may be easier if you change each of them to a mixed number. See **Figure 1.16**.

$$\frac{7}{4} = 1\frac{3}{4} \qquad -\frac{9}{2} = -4\frac{1}{2} \qquad \frac{8}{3} = 2\frac{2}{3}$$

Figure 1.16 shows the number line with all the points plotted.

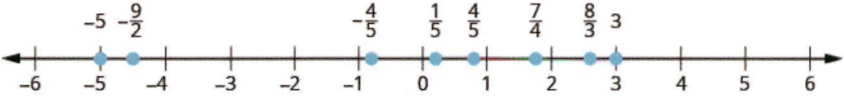

Figure 1.16

EXAMPLE 1.115

Locate and label the following on a number line: 4, $\frac{3}{4}$, $-\frac{1}{4}$, -3, $\frac{6}{5}$, $-\frac{5}{2}$, and $\frac{7}{3}$.

⊘ Solution

Locate and plot the integers, 4, -3.

Locate the proper fraction $\frac{3}{4}$ first. The fraction $\frac{3}{4}$ is between 0 and 1. Divide the distance between 0 and 1 into four equal parts then, we plot $\frac{3}{4}$. Similarly plot $-\frac{1}{4}$.

Now locate the improper fractions $\frac{6}{5}$, $-\frac{5}{2}$, $\frac{7}{3}$. It is easier to plot them if we convert them to mixed numbers and then plot them as described above: $\frac{6}{5} = 1\frac{1}{5}$, $-\frac{5}{2} = -2\frac{1}{2}$, $\frac{7}{3} = 2\frac{1}{3}$.

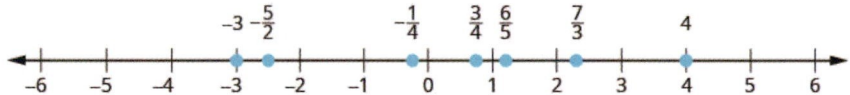

> **TRY IT :: 1.229**   Locate and label the following on a number line: -1, $\frac{1}{3}$, $\frac{6}{5}$, $-\frac{7}{4}$, $\frac{9}{2}$, 5, $-\frac{8}{3}$.

> **TRY IT :: 1.230**   Locate and label the following on a number line: -2, $\frac{2}{3}$, $\frac{7}{5}$, $-\frac{7}{4}$, $\frac{7}{2}$, 3, $-\frac{7}{3}$.

In **Example 1.116**, we'll use the inequality symbols to order fractions. In previous chapters we used the number line to order numbers.

- $a < b$ "a is less than b" when a is to the left of b on the number line
- $a > b$ "a is greater than b" when a is to the right of b on the number line

As we move from left to right on a number line, the values increase.

EXAMPLE 1.116

Order each of the following pairs of numbers, using < or >. It may be helpful to refer **Figure 1.17**.

ⓐ $-\frac{2}{3}$___-1 ⓑ $-3\frac{1}{2}$___-3 ⓒ $-\frac{3}{4}$___$-\frac{1}{4}$ ⓓ -2___$-\frac{8}{3}$

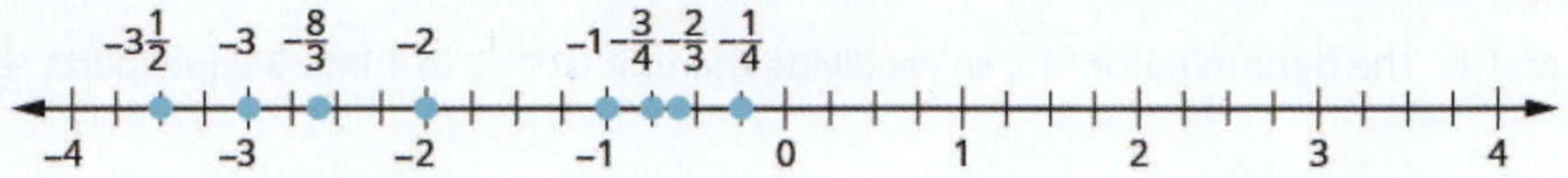

Figure 1.17

✓ **Solution**

Be careful when ordering negative numbers.

ⓐ

$-\frac{2}{3}$ is to the right of -1 on the number line.

$$-\frac{2}{3}\_\_\_-1$$
$$-\frac{2}{3} > -1$$

ⓑ

$-3\frac{1}{2}$ is to the left of -3 on the number line.

$$-3\frac{1}{2}\_\_\_-3$$
$$-3\frac{1}{2} < -3$$

ⓒ

$-\frac{3}{4}$ is to the left of $-\frac{1}{4}$ on the number line.

$$-\frac{3}{4}\_\_\_-\frac{1}{4}$$
$$-\frac{3}{4} < -\frac{1}{4}$$

ⓓ

-2 is to the right of $-\frac{8}{3}$ on the number line.

$$-2\_\_\_-\frac{8}{3}$$
$$-2 > -\frac{8}{3}$$

> **TRY IT :: 1.231**   Order each of the following pairs of numbers, using < or >:
>
> ⓐ $-\frac{1}{3}$___-1 ⓑ $-1\frac{1}{2}$___-2 ⓒ $-\frac{2}{3}$___$-\frac{1}{3}$ ⓓ -3___$-\frac{7}{3}$.

> **TRY IT :: 1.232**   Order each of the following pairs of numbers, using < or >:
>
> ⓐ -1___$-\frac{2}{3}$ ⓑ $-2\frac{1}{4}$___-2 ⓒ $-\frac{3}{5}$___$-\frac{4}{5}$ ⓓ -4___$-\frac{10}{3}$.

Locate Decimals on the Number Line

Since decimals are forms of fractions, locating decimals on the number line is similar to locating fractions on the number line.

EXAMPLE 1.117

Locate 0.4 on the number line.

⊘ Solution

A proper fraction has value less than one. The decimal number 0.4 is equivalent to $\frac{4}{10}$, a proper fraction, so 0.4 is located

between 0 and 1. On a number line, divide the interval between 0 and 1 into 10 equal parts. Now label the parts 0.1, 0.2, 0.3, 0.4, 0.5, 0.6, 0.7, 0.8, 0.9, 1.0. We write 0 as 0.0 and 1 and 1.0, so that the numbers are consistently in tenths. Finally, mark 0.4 on the number line. See **Figure 1.18**.

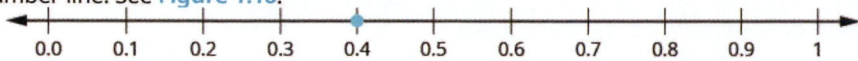

Figure 1.18

> **TRY IT : : 1.233**   Locate on the number line: 0.6.

> **TRY IT : : 1.234**   Locate on the number line: 0.9.

EXAMPLE 1.118

Locate -0.74 on the number line.

⊘ Solution

The decimal -0.74 is equivalent to $-\frac{74}{100}$, so it is located between 0 and -1. On a number line, mark off and label

the hundredths in the interval between 0 and -1. See **Figure 1.19**.

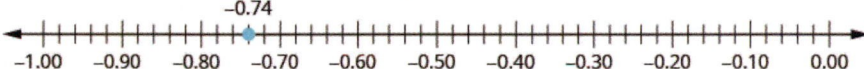

Figure 1.19

> **TRY IT : : 1.235**   Locate on the number line: -0.6.

> **TRY IT : : 1.236**   Locate on the number line: -0.7.

Which is larger, 0.04 or 0.40? If you think of this as money, you know that $0.40 (forty cents) is greater than $0.04 (four cents). So,

$0.40 > 0.04$

Again, we can use the number line to order numbers.

- $a < b$ "a is less than b" when a is to the left of b on the number line
- $a > b$ "a is greater than b" when a is to the right of b on the number line

Where are 0.04 and 0.40 located on the number line? See **Figure 1.20**.

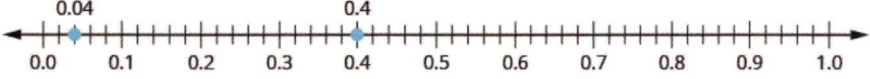

Figure 1.20

We see that 0.40 is to the right of 0.04 on the number line. This is another way to demonstrate that 0.40 > 0.04.

How does 0.31 compare to 0.308? This doesn't translate into money to make it easy to compare. But if we convert 0.31 and 0.308 into fractions, we can tell which is larger.

	0.31	0.308
Convert to fractions.	$\dfrac{31}{100}$	$\dfrac{308}{1000}$
We need a common denominator to compare them.	$\dfrac{31 \cdot 10}{100 \cdot 10}$	$\dfrac{308}{1000}$
	$\dfrac{310}{1000}$	$\dfrac{308}{1000}$

Because 310 > 308, we know that $\dfrac{310}{1000} > \dfrac{308}{1000}$. Therefore, 0.31 > 0.308.

Notice what we did in converting 0.31 to a fraction—we started with the fraction $\dfrac{31}{100}$ and ended with the equivalent fraction $\dfrac{310}{1000}$. Converting $\dfrac{310}{1000}$ back to a decimal gives 0.310. So 0.31 is equivalent to 0.310. Writing zeros at the end of a decimal does not change its value!

$$\frac{31}{100} = \frac{310}{1000} \quad \text{and} \quad 0.31 = 0.310$$

We say 0.31 and 0.310 are **equivalent decimals**.

Equivalent Decimals

Two decimals are equivalent if they convert to equivalent fractions.

We use equivalent decimals when we order decimals.

The steps we take to order decimals are summarized here.

HOW TO :: ORDER DECIMALS.

Step 1. Write the numbers one under the other, lining up the decimal points.

Step 2. Check to see if both numbers have the same number of digits. If not, write zeros at the end of the one with fewer digits to make them match.

Step 3. Compare the numbers as if they were whole numbers.

Step 4. Order the numbers using the appropriate inequality sign.

EXAMPLE 1.119

Order 0.64___0.6 using < or > .

⊘ **Solution**

Write the numbers one under the other, lining up the decimal points.	0.64 0.6
Add a zero to 0.6 to make it a decimal with 2 decimal places. Now they are both hundredths.	0.64 0.60
64 is greater than 60.	64 > 60
64 hundredths is greater than 60 hundredths.	0.64 > 0.60
	0.64 > 0.6

> | **TRY IT :: 1.237**   Order each of the following pairs of numbers, using $<$ or $>$: 0.42___0.4.

> | **TRY IT :: 1.238**   Order each of the following pairs of numbers, using $<$ or $>$: 0.18___0.1.

EXAMPLE 1.120

Order 0.83___0.803 using $<$ or $>$.

⊘ **Solution**

	0.83___0.803
Write the numbers one under the other, lining up the decimals.	0.83 0.803
They do not have the same number of digits. Write one zero at the end of 0.83.	0.830 0.803
Since 830 > 803, 830 thousandths is greater than 803 thousandths.	0.830 > 0.803
	0.83 > 0.803

> | **TRY IT :: 1.239**   Order the following pair of numbers, using $<$ or $>$: 0.76___0.706.

> | **TRY IT :: 1.240**   Order the following pair of numbers, using $<$ or $>$: 0.305___0.35.

When we order negative decimals, it is important to remember how to order negative integers. Recall that larger numbers are to the right on the number line. For example, because -2 lies to the right of -3 on the number line, we know that $-2 > -3$. Similarly, smaller numbers lie to the left on the number line. For example, because -9 lies to the left of -6 on the number line, we know that $-9 < -6$. See **Figure 1.21**.

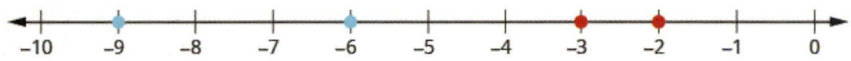

Figure 1.21

If we zoomed in on the interval between 0 and -1, as shown in **Example 1.121**, we would see in the same way that $-0.2 > -0.3$ and $-0.9 < -0.6$.

EXAMPLE 1.121

Use $<$ or $>$ to order -0.1___-0.8.

⊘ **Solution**

	-0.1___-0.8
Write the numbers one under the other, lining up the decimal points. They have the same number of digits.	-0.1 -0.8
Since $-1 > -8$, -1 tenth is greater than -8 tenths.	$-0.1 > -0.8$

> **TRY IT : :** 1.241   Order the following pair of numbers, using < or >: -0.3___-0.5.

> **TRY IT : :** 1.242   Order the following pair of numbers, using < or >: -0.6___-0.7.

1.8 EXERCISES

Practice Makes Perfect

Simplify Expressions with Square Roots

In the following exercises, simplify.

659. $\sqrt{36}$

660. $\sqrt{4}$

661. $\sqrt{64}$

662. $\sqrt{169}$

663. $\sqrt{9}$

664. $\sqrt{16}$

665. $\sqrt{100}$

666. $\sqrt{144}$

667. $-\sqrt{4}$

668. $-\sqrt{100}$

669. $-\sqrt{1}$

670. $-\sqrt{121}$

Identify Integers, Rational Numbers, Irrational Numbers, and Real Numbers

In the following exercises, write as the ratio of two integers.

671. ⓐ 5 ⓑ 3.19

672. ⓐ 8 ⓑ 1.61

673. ⓐ -12 ⓑ 9.279

674. ⓐ -16 ⓑ 4.399

In the following exercises, list the ⓐ rational numbers, ⓑ irrational numbers

675. $0.75, 0.22\bar{3}, 1.39174$

676. $0.36, 0.94729..., 2.52\bar{8}$

677. $0.4\bar{5}, 1.919293..., 3.59$

678. $0.1\bar{3}, 0.42982..., 1.875$

In the following exercises, identify whether each number is rational or irrational.

679. ⓐ $\sqrt{25}$ ⓑ $\sqrt{30}$

680. ⓐ $\sqrt{44}$ ⓑ $\sqrt{49}$

681. ⓐ $\sqrt{164}$ ⓑ $\sqrt{169}$

682. ⓐ $\sqrt{225}$ ⓑ $\sqrt{216}$

In the following exercises, identify whether each number is a real number or not a real number.

683. ⓐ $-\sqrt{81}$ ⓑ $\sqrt{-121}$

684. ⓐ $-\sqrt{64}$ ⓑ $\sqrt{-9}$

685. ⓐ $\sqrt{-36}$ ⓑ $-\sqrt{144}$

686. ⓐ $\sqrt{-49}$ ⓑ $-\sqrt{144}$

In the following exercises, list the ⓐ whole numbers, ⓑ integers, ⓒ rational numbers, ⓓ irrational numbers, ⓔ real numbers for each set of numbers.

687.
$-8, 0, 1.95286..., \frac{12}{5}, \sqrt{36}, 9$

688.
$-9, -3\frac{4}{9}, -\sqrt{9}, 0.40\bar{9}, \frac{11}{6}, 7$

689.
$-\sqrt{100}, -7, -\frac{8}{3}, -1, 0.77, 3\frac{1}{4}$

690.
$-6, -\frac{5}{2}, 0, 0.\overline{714285}, 2\frac{1}{5}, \sqrt{14}$

Locate Fractions on the Number Line

In the following exercises, locate the numbers on a number line.

691. $\frac{3}{4}, \frac{8}{5}, \frac{10}{3}$

692. $\frac{1}{4}, \frac{9}{5}, \frac{11}{3}$

693. $\frac{3}{10}, \frac{7}{2}, \frac{11}{6}, 4$

694. $\frac{7}{10}, \frac{5}{2}, \frac{13}{8}, 3$

695. $\frac{2}{5}, -\frac{2}{5}$

696. $\frac{3}{4}, -\frac{3}{4}$

697. $\frac{3}{4}, -\frac{3}{4}, 1\frac{2}{3}, -1\frac{2}{3}, \frac{5}{2}, -\frac{5}{2}$

698. $\frac{1}{5}, -\frac{2}{5}, 1\frac{3}{4}, -1\frac{3}{4}, \frac{8}{3}, -\frac{8}{3}$

In the following exercises, order each of the pairs of numbers, using < or >.

699. $-1 \underline{\quad} -\frac{1}{4}$

700. $-1 \underline{\quad} -\frac{1}{3}$

701. $-2\frac{1}{2} \underline{\quad} -3$

702. $-1\frac{3}{4} \underline{\quad} -2$

703. $-\frac{5}{12} \underline{\quad} -\frac{7}{12}$

704. $-\frac{9}{10} \underline{\quad} -\frac{3}{10}$

705. $-3 \underline{\quad} -\frac{13}{5}$

706. $-4 \underline{\quad} -\frac{23}{6}$

Locate Decimals on the Number Line *In the following exercises, locate the number on the number line.*

707. 0.8

708. -0.9

709. -1.6

710. 3.1

In the following exercises, order each pair of numbers, using < or >.

711. 0.37 ___ 0.63

712. 0.86 ___ 0.69

713. 0.91 ___ 0.901

714. 0.415 ___ 0.41

715. -0.5 ___ -0.3

716. -0.1 ___ -0.4

717. -0.62 ___ -0.619

718. -7.31 ___ -7.3

Everyday Math

719. Field trip All the 5th graders at Lincoln Elementary School will go on a field trip to the science museum. Counting all the children, teachers, and chaperones, there will be 147 people. Each bus holds 44 people.

ⓐ How many busses will be needed?

ⓑ Why must the answer be a whole number?

ⓒ Why shouldn't you round the answer the usual way, by choosing the whole number closest to the exact answer?

720. Child care Serena wants to open a licensed child care center. Her state requires there be no more than 12 children for each teacher. She would like her child care center to serve 40 children.

ⓐ How many teachers will be needed?

ⓑ Why must the answer be a whole number?

ⓒ Why shouldn't you round the answer the usual way, by choosing the whole number closest to the exact answer?

Writing Exercises

721. In your own words, explain the difference between a rational number and an irrational number.

722. Explain how the sets of numbers (counting, whole, integer, rational, irrationals, reals) are related to each other.

Self Check

ⓐ *After completing the exercises, use this checklist to evaluate your mastery of the objective of this section.*

I can...	Confidently	With some help	No-I don't get it!
simplify expressions with square roots.			
identify integers, rational numbers, irrational numbers, and real numbers.			
locate fractions on the number line.			
locate decimals on the number line.			

ⓑ *On a scale of* $1 - 10,$ *how would you rate your mastery of this section in light of your responses on the checklist? How can you improve this?*

1.9 Properties of Real Numbers

Learning Objectives

By the end of this section, you will be able to:

› Use the commutative and associative properties
› Use the identity and inverse properties of addition and multiplication
› Use the properties of zero
› Simplify expressions using the distributive property

Be Prepared!

A more thorough introduction to the topics covered in this section can be found in the *Prealgebra* chapter, **The Properties of Real Numbers**.

Use the Commutative and Associative Properties

Think about adding two numbers, say 5 and 3. The order we add them doesn't affect the result, does it?

$$5 + 3 \qquad 3 + 5$$
$$8 \qquad 8$$
$$5 + 3 = 3 + 5$$

The results are the same.

As we can see, the order in which we add does not matter!

What about multiplying 5 and 3?

$$5 \cdot 3 \qquad 3 \cdot 5$$
$$15 \qquad 15$$
$$5 \cdot 3 = 3 \cdot 5$$

Again, the results are the same!

The order in which we multiply does not matter!

These examples illustrate the **commutative property**. When adding or multiplying, changing the *order* gives the same result.

Commutative Property

of Addition	If a, b are real numbers, then	$a + b = b + a$
of Multiplication	If a, b are real numbers, then	$a \cdot b = b \cdot a$

When adding or multiplying, changing the *order* gives the same result.

The commutative property has to do with order. If you change the order of the numbers when adding or multiplying, the result is the same.

What about subtraction? Does order matter when we subtract numbers? Does $7 - 3$ give the same result as $3 - 7$?

$$7 - 3 \qquad 3 - 7$$
$$4 \qquad -4$$
$$4 \neq -4$$
$$7 - 3 \neq 3 - 7$$

The results are not the same.

Since changing the order of the subtraction did not give the same result, we know that *subtraction is not commutative*.

Let's see what happens when we divide two numbers. Is division commutative?

$$12 \div 4 \qquad 4 \div 12$$

$$\frac{12}{4} \qquad \frac{4}{12}$$

$$3 \qquad \frac{1}{3}$$

$$3 \neq \frac{1}{3}$$

$$12 \div 4 \neq 4 \div 12$$

The results are not the same.

Since changing the order of the division did not give the same result, *division is not commutative*. The commutative properties only apply to addition and multiplication!

- Addition and multiplication *are* commutative.
- Subtraction and Division *are not* commutative.

If you were asked to simplify this expression, how would you do it and what would your answer be?

$$7 + 8 + 2$$

Some people would think $7 + 8$ is 15 and then $15 + 2$ is 17. Others might start with $8 + 2$ makes 10 and then $7 + 10$ makes 17.

Either way gives the same result. Remember, we use parentheses as grouping symbols to indicate which operation should be done first.

$$(7 + 8) + 2$$

Add $7 + 8$.	$15 + 2$
Add.	17

$$7 + (8 + 2)$$

Add $8 + 2$.	$7 + 10$
Add.	17

$$(7 + 8) + 2 = 7 + (8 + 2)$$

When adding three numbers, changing the grouping of the numbers gives the same result.

This is true for multiplication, too.

$$\left(5 \cdot \frac{1}{3}\right) \cdot 3$$

Multiply.	$5 \cdot \frac{1}{3}$	$\frac{5}{3} \cdot 3$
Multiply.		5

$$5 \cdot \left(\frac{1}{3} \cdot 3\right)$$

Multiply.	$\frac{1}{3} \cdot 3$	$5 \cdot 1$
Multiply.		5

$$\left(5 \cdot \frac{1}{3}\right) \cdot 3 = 5 \cdot \left(\frac{1}{3} \cdot 3\right)$$

When multiplying three numbers, changing the grouping of the numbers gives the same result.

You probably know this, but the terminology may be new to you. These examples illustrate the **associative property**.

Associative Property

of Addition	If a, b, c are real numbers, then $(a + b) + c = a + (b + c)$
of Multiplication	If a, b, c are real numbers, then $(a \cdot b) \cdot c = a \cdot (b \cdot c)$

When adding or multiplying, changing the *grouping* gives the same result.

Let's think again about multiplying $5 \cdot \frac{1}{3} \cdot 3$. We got the same result both ways, but which way was easier? Multiplying $\frac{1}{3}$ and 3 first, as shown above on the right side, eliminates the fraction in the first step. Using the associative property can make the math easier!

The associative property has to do with grouping. If we change how the numbers are grouped, the result will be the same. Notice it is the same three numbers in the same order—the only difference is the grouping.

We saw that subtraction and division were not commutative. They are not associative either.

When simplifying an expression, it is always a good idea to plan what the steps will be. In order to combine like terms in the next example, we will use the commutative property of addition to write the like terms together.

EXAMPLE 1.122

Simplify: $18p + 6q + 15p + 5q$.

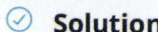 **Solution**

$$18p + 6q + 15p + 5q$$

Use the commutative property of addition to re-order so that like terms are together. $18p + 15p + 6q + 5q$

Add like terms. $33p + 11q$

> **TRY IT :: 1.243**   Simplify: $23r + 14s + 9r + 15s$.

> **TRY IT :: 1.244**   Simplify: $37m + 21n + 4m - 15n$.

When we have to simplify algebraic expressions, we can often make the work easier by applying the commutative or associative property first, instead of automatically following the order of operations. When adding or subtracting fractions, combine those with a common denominator first.

EXAMPLE 1.123

Simplify: $\left(\frac{5}{13} + \frac{3}{4}\right) + \frac{1}{4}$.

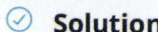 **Solution**

$$\left(\frac{5}{13} + \frac{3}{4}\right) + \frac{1}{4}$$

Notice that the last 2 terms have a common denominator, so change the grouping. $\frac{5}{13} + \left(\frac{3}{4} + \frac{1}{4}\right)$

Add in parentheses first. $\frac{5}{13} + \left(\frac{4}{4}\right)$

Simplify the fraction. $\frac{5}{13} + 1$

Add. $1\frac{5}{13}$

Convert to an improper fraction. $\frac{18}{13}$

> **TRY IT : : 1.245**   Simplify: $\left(\frac{7}{15}+\frac{5}{8}\right)+\frac{3}{8}$.

> **TRY IT : : 1.246**   Simplify: $\left(\frac{2}{9}+\frac{7}{12}\right)+\frac{5}{12}$.

EXAMPLE 1.124

Use the associative property to simplify $6(3x)$.

⊘ **Solution**

Use the associative property of multiplication, $(a \cdot b) \cdot c = a \cdot (b \cdot c)$, to change the grouping.

$$6(3x)$$

Change the grouping. $(6 \cdot 3)x$

Multiply in the parentheses. $18x$

Notice that we can multiply $6 \cdot 3$ but we could not multiply $3x$ without having a value for x.

> **TRY IT : : 1.247**   Use the associative property to simplify $8(4x)$.

> **TRY IT : : 1.248**   Use the associative property to simplify $-9(7y)$.

Use the Identity and Inverse Properties of Addition and Multiplication

What happens when we add 0 to any number? Adding 0 doesn't change the value. For this reason, we call 0 the **additive identity**.

For example,

$$13+0 \qquad -14+0 \qquad 0+(-8)$$
$$13 \qquad\quad -14 \qquad\quad -8$$

These examples illustrate the **Identity Property of Addition** that states that for any real number a, $a+0=a$ and $0+a=a$.

What happens when we multiply any number by one? Multiplying by 1 doesn't change the value. So we call 1 the **multiplicative identity**.

For example,

$$43 \cdot 1 \qquad -27 \cdot 1 \qquad 1 \cdot \frac{3}{5}$$
$$43 \qquad\quad -27 \qquad\quad \frac{3}{5}$$

These examples illustrate the **Identity Property of Multiplication** that states that for any real number a, $a \cdot 1 = a$ and $1 \cdot a = a$.

We summarize the Identity Properties below.

Identity Property

of addition For any real number a:	$a+0=a \quad 0+a=a$
0 is the **additive identity**	
of multiplication For any real number a:	$a \cdot 1 = a \quad 1 \cdot a = a$
1 is the **multiplicative identity**	

What number added to 5 gives the additive identity, 0?

$$5 + \underline{} = 0 \qquad \text{We know } 5 + (-5) = 0$$

What number added to –6 gives the additive identity, 0?

$$-6 + \underline{} = 0 \qquad \text{We know } -6 + 6 = 0$$

Notice that in each case, the missing number was the opposite of the number!

We call $-a.$ the **additive inverse** of a. *The opposite of a number is its additive inverse.* A number and its opposite add to zero, which is the additive identity. This leads to the **Inverse Property of Addition** that states for any real number a, $a + (-a) = 0$. Remember, a number and its opposite add to zero.

What number multiplied by $\frac{2}{3}$ gives the multiplicative identity, 1? In other words, $\frac{2}{3}$ times what results in 1?

$$\frac{2}{3} \cdot \underline{} = 1 \qquad \text{We know } \frac{2}{3} \cdot \frac{3}{2} = 1$$

What number multiplied by 2 gives the multiplicative identity, 1? In other words 2 times what results in 1?

$$2 \cdot \underline{} = 1 \qquad \text{We know } 2 \cdot \frac{1}{2} = 1$$

Notice that in each case, the missing number was the reciprocal of the number!

We call $\frac{1}{a}$ the **multiplicative inverse** of a. *The reciprocal of a number is its multiplicative inverse.* A number and its reciprocal multiply to one, which is the multiplicative identity. This leads to the **Inverse Property of Multiplication** that states that for any real number a, $a \neq 0$, $a \cdot \frac{1}{a} = 1$.

We'll formally state the inverse properties here:

Inverse Property

of addition	For any real number a, $-a.$ is the **additive inverse** of a. A number and its opposite add to zero.	$a + (-a) = 0$
of multiplication	For any real number a, $a \neq 0$ $\frac{1}{a}$ is the **multiplicative inverse** of a. A number and its reciprocal multiply to one.	$a \cdot \frac{1}{a} = 1$

EXAMPLE 1.125

Find the additive inverse of ⓐ $\frac{5}{8}$ ⓑ 0.6 ⓒ -8 ⓓ $-\frac{4}{3}$.

✓ **Solution**

To find the additive inverse, we find the opposite.

ⓐ The additive inverse of $\frac{5}{8}$ is the opposite of $\frac{5}{8}$. The additive inverse of $\frac{5}{8}$ is $-\frac{5}{8}$.

ⓑ The additive inverse of 0.6 is the opposite of 0.6. The additive inverse of 0.6 is -0.6.

ⓒ The additive inverse of -8 is the opposite of -8. We write the opposite of -8 as $-(-8)$, and then simplify it to 8. Therefore, the additive inverse of -8 is 8.

ⓓ The additive inverse of $-\frac{4}{3}$ is the opposite of $-\frac{4}{3}$. We write this as $-\left(-\frac{4}{3}\right)$, and then simplify to $\frac{4}{3}$. Thus, the additive inverse of $-\frac{4}{3}$ is $\frac{4}{3}$.

> **TRY IT : : 1.249** Find the additive inverse of: ⓐ $\frac{7}{9}$ ⓑ 1.2 ⓒ -14 ⓓ $-\frac{9}{4}$.

> **TRY IT : : 1.250** Find the additive inverse of: ⓐ $\frac{7}{13}$ ⓑ 8.4 ⓒ -46 ⓓ $-\frac{5}{2}$.

EXAMPLE 1.126

Find the multiplicative inverse of ⓐ 9 ⓑ $-\frac{1}{9}$ ⓒ 0.9.

⊘ **Solution**

To find the multiplicative inverse, we find the reciprocal.

ⓐ The multiplicative inverse of 9 is the reciprocal of 9, which is $\frac{1}{9}$. Therefore, the multiplicative inverse of 9 is

$\frac{1}{9}$.

ⓑ The multiplicative inverse of $-\frac{1}{9}$ is the reciprocal of $-\frac{1}{9}$, which is -9. Thus, the multiplicative inverse of

$-\frac{1}{9}$ is -9.

ⓒ To find the multiplicative inverse of 0.9, we first convert 0.9 to a fraction, $\frac{9}{10}$. Then we find the reciprocal of

the fraction. The reciprocal of $\frac{9}{10}$ is $\frac{10}{9}$. So the multiplicative inverse of 0.9 is $\frac{10}{9}$.

> **TRY IT : : 1.251** Find the multiplicative inverse of ⓐ 4 ⓑ $-\frac{1}{7}$ ⓒ 0.3

> **TRY IT : : 1.252** Find the multiplicative inverse of ⓐ 18 ⓑ $-\frac{4}{5}$ ⓒ 0.6.

Use the Properties of Zero

The identity property of addition says that when we add 0 to any number, the result is that same number. What happens when we multiply a number by 0? Multiplying by 0 makes the product equal zero.

Multiplication by Zero

For any real number a.

$$a \cdot 0 = 0 \qquad\qquad 0 \cdot a = 0$$

The product of any real number and 0 is 0.

What about division involving zero? What is $0 \div 3$? Think about a real example: If there are no cookies in the cookie jar and 3 people are to share them, how many cookies does each person get? There are no cookies to share, so each person gets 0 cookies. So,

$$0 \div 3 = 0$$

We can check division with the related multiplication fact.

$$12 \div 6 = 2 \text{ because } 2 \cdot 6 = 12.$$

So we know $0 \div 3 = 0$ because $0 \cdot 3 = 0$.

Division of Zero

For any real number a, except 0, $\frac{0}{a} = 0$ and $0 \div a = 0$.

Zero divided by any real number except zero is zero.

Now think about dividing *by* zero. What is the result of dividing 4 by 0? Think about the related multiplication fact:

$4 \div 0 = ?$ means $? \cdot 0 = 4$. Is there a number that multiplied by 0 gives 4? Since any real number multiplied by 0 gives 0, there is no real number that can be multiplied by 0 to obtain 4.

We conclude that there is no answer to $4 \div 0$ and so we say that division by 0 is undefined.

Division by Zero

For any real number a, except 0, $\frac{a}{0}$ and $a \div 0$ are undefined.

Division by zero is undefined.

We summarize the properties of zero below.

Properties of Zero

Multiplication by Zero: For any real number a,

$\quad a \cdot 0 = 0 \quad 0 \cdot a = 0 \quad$ The product of any number and 0 is 0.

Division of Zero, Division by Zero: For any real number a, $a \neq 0$

$\quad \frac{0}{a} = 0 \quad\quad$ Zero divided by any real number, except itself is zero.

$\quad \frac{a}{0}$ is undefined $\quad$ Division by zero is undefined.

EXAMPLE 1.127

Simplify: ⓐ $-8 \cdot 0$ ⓑ $\frac{0}{-2}$ ⓒ $\frac{-32}{0}$.

✓ **Solution**

ⓐ

$$-8 \cdot 0$$

The product of any real number and 0 is 0. $\quad\quad 0$

ⓑ

$$\frac{0}{-2}$$

Zero divided by any real number, except itself, is 0. $\quad\quad 0$

ⓒ

$$\frac{-32}{0}$$

Division by 0 is undefined. $\quad\quad$ Undefined

> **TRY IT :: 1.253** $\quad$ Simplify: ⓐ $-14 \cdot 0$ ⓑ $\frac{0}{-6}$ ⓒ $\frac{-2}{0}$.

> **TRY IT :: 1.254** $\quad$ Simplify: ⓐ $0(-17)$ ⓑ $\frac{0}{-10}$ ⓒ $\frac{-5}{0}$.

We will now practice using the properties of identities, inverses, and zero to simplify expressions.

EXAMPLE 1.128

Simplify: ⓐ $\dfrac{0}{n+5}$, where $n \neq -5$ ⓑ $\dfrac{10-3p}{0}$, where $10-3p \neq 0$.

✓ **Solution**

ⓐ

$$\dfrac{0}{n+5}$$

Zero divided by any real number except itself is 0.

$$0$$

ⓑ

$$\dfrac{10-3p}{0}$$

Division by 0 is undefined.

Undefined

EXAMPLE 1.129

Simplify: $-84n + (-73n) + 84n$.

✓ **Solution**

$$-84n + (-73n) + 84n$$

Notice that the first and third terms are opposites; use the commutative property of addition to re-order the terms.

$$-84n + 84n + (-73n)$$

Add left to right.

$$0 + (-73)$$

Add.

$$-73n$$

> **TRY IT :: 1.255**   Simplify: $-27a + (-48a) + 27a$.

> **TRY IT :: 1.256**   Simplify: $39x + (-92x) + (-39x)$.

Now we will see how recognizing reciprocals is helpful. Before multiplying left to right, look for reciprocals—their product is 1.

EXAMPLE 1.130

Simplify: $\dfrac{7}{15} \cdot \dfrac{8}{23} \cdot \dfrac{15}{7}$.

✓ **Solution**

$$\frac{7}{15} \cdot \frac{8}{23} \cdot \frac{15}{7}$$

Notice the first and third terms are reciprocals, so use the commutative property of multiplication to re-order the factors.

$$\frac{7}{15} \cdot \frac{15}{7} \cdot \frac{8}{23}$$

Multiply left to right.

$$1 \cdot \frac{8}{23}$$

Multiply.

$$\frac{8}{23}$$

> **TRY IT :: 1.257**   Simplify: $\frac{9}{16} \cdot \frac{5}{49} \cdot \frac{16}{9}$.

> **TRY IT :: 1.258**   Simplify: $\frac{6}{17} \cdot \frac{11}{25} \cdot \frac{17}{6}$.

> **TRY IT :: 1.259**   Simplify: ⓐ $\frac{0}{m+7}$, where $m \neq -7$ ⓑ $\frac{18 - 6c}{0}$, where $18 - 6c \neq 0$.

> **TRY IT :: 1.260**   Simplify: ⓐ $\frac{0}{d-4}$, where $d \neq 4$ ⓑ $\frac{15 - 4q}{0}$, where $15 - 4q \neq 0$.

EXAMPLE 1.131

Simplify: $\frac{3}{4} \cdot \frac{4}{3}(6x + 12)$.

✓ **Solution**

$$\frac{3}{4} \cdot \frac{4}{3}(6x + 12)$$

There is nothing to do in the parentheses, so multiply the two fractions first—notice, they are reciprocals.

$$1(6x + 12)$$

Simplify by recognizing the multiplicative identity.

$$6x + 12$$

> **TRY IT :: 1.261**   Simplify: $\frac{2}{5} \cdot \frac{5}{2}(20y + 50)$.

> **TRY IT :: 1.262**   Simplify: $\frac{3}{8} \cdot \frac{8}{3}(12z + 16)$.

Simplify Expressions Using the Distributive Property

Suppose that three friends are going to the movies. They each need $9.25—that's 9 dollars and 1 quarter—to pay for their tickets. How much money do they need all together?

You can think about the dollars separately from the quarters. They need 3 times $9 so $27, and 3 times 1 quarter, so 75 cents. In total, they need \$27.75. If you think about doing the math in this way, you are using the **distributive property**.

Distributive Property

If a, b, c are real numbers, then $\quad a(b+c) = ab + ac$

Also, $\quad (b+c)a = ba + ca$
$$a(b-c) = ab - ac$$
$$(b-c)a = ba - ca$$

Back to our friends at the movies, we could find the total amount of money they need like this:

$$3(9.25)$$

$$3(9 \ + \ 0.25)$$
$$3(9) \ + \ 3(0.25)$$
$$27 \ + \ 0.75$$
$$27.75$$

In algebra, we use the **distributive property** to remove parentheses as we simplify expressions.

For example, if we are asked to simplify the expression $3(x+4)$, the order of operations says to work in the parentheses first. But we cannot add x and 4, since they are not like terms. So we use the distributive property, as shown in **Example 1.132**.

EXAMPLE 1.132

Simplify: $3(x+4)$.

⊘ Solution

$$3(x+4)$$

| Distribute. | $3 \cdot x + 3 \cdot 4$ |
| Multiply. | $3x + 12$ |

> **TRY IT : : 1.263**   Simplify: $4(x+2)$.

> **TRY IT : : 1.264**   Simplify: $6(x+7)$.

Some students find it helpful to draw in arrows to remind them how to use the distributive property. Then the first step in **Example 1.132** would look like this:

$$3(x+4)$$

EXAMPLE 1.133

Simplify: $8\left(\frac{3}{8}x + \frac{1}{4}\right)$.

⊘ Solution

$$8\left(\frac{3}{8}x + \frac{1}{4}\right)$$

| Distribute. | $8 \cdot \frac{3}{8}x + 8 \cdot \frac{1}{4}$ |
| Multiply. | $3x + 2$ |

> **TRY IT : : 1.265**   Simplify: $6\left(\frac{5}{6}y + \frac{1}{2}\right)$.

> **TRY IT : : 1.266**   Simplify: $12\left(\frac{1}{3}n + \frac{3}{4}\right)$.

Using the distributive property as shown in **Example 1.134** will be very useful when we solve money applications in later chapters.

EXAMPLE 1.134

Simplify: $100(0.3 + 0.25q)$.

⊘ **Solution**

	$100(0.3 + 0.25q)$
Distribute.	$100(0.3) + 100(0.25q)$
Multiply.	$30 + 25q$

> **TRY IT : : 1.267**   Simplify: $100(0.7 + 0.15p)$.

> **TRY IT : : 1.268**   Simplify: $100(0.04 + 0.35d)$.

When we distribute a negative number, we need to be extra careful to get the signs correct!

EXAMPLE 1.135

Simplify: $-2(4y + 1)$.

⊘ **Solution**

	$-2(4y + 1)$
Distribute.	$-2 \cdot 4y + (-2) \cdot 1$
Multiply.	$-8y - 2$

> **TRY IT : : 1.269**   Simplify: $-3(6m + 5)$.

> **TRY IT : : 1.270**   Simplify: $-6(8n + 11)$.

EXAMPLE 1.136

Simplify: $-11(4 - 3a)$.

⊘ Solution

Distribute.	$-11(4 - 3a)$
Multiply.	$-11 \cdot 4 - (-11) \cdot 3a$ $-44 - (-33a)$
Simplify.	$-44 + 33a$

Notice that you could also write the result as $33a - 44$. Do you know why?

> **TRY IT :: 1.271**   Simplify: $-5(2 - 3a)$.

> **TRY IT :: 1.272**   Simplify: $-7(8 - 15y)$.

Example 1.137 will show how to use the distributive property to find the opposite of an expression.

EXAMPLE 1.137

Simplify: $-(y + 5)$.

⊘ Solution

$$-(y + 5)$$

Multiplying by -1 results in the opposite.	$-1(y + 5)$
Distribute.	$-1 \cdot y + (-1) \cdot 5$
Simplify.	$-y + (-5)$
	$-y - 5$

> **TRY IT :: 1.273**   Simplify: $-(z - 11)$.

> **TRY IT :: 1.274**   Simplify: $-(x - 4)$.

There will be times when we'll need to use the distributive property as part of the order of operations. Start by looking at the parentheses. If the expression inside the parentheses cannot be simplified, the next step would be multiply using the distributive property, which removes the parentheses. The next two examples will illustrate this.

EXAMPLE 1.138

Simplify: $8 - 2(x + 3)$.

Be sure to follow the order of operations. Multiplication comes before subtraction, so we will distribute the 2 first and then subtract.

⊘ **Solution**

$$8 - 2(x + 3)$$

Distribute. $8 - 2 \cdot x - 2 \cdot 3$

Multiply. $8 - 2x - 6$

Combine like terms. $-2x + 2$

> | **TRY IT ::** 1.275   Simplify: $9 - 3(x + 2)$.

> | **TRY IT ::** 1.276   Simplify: $7x - 5(x + 4)$.

EXAMPLE 1.139

Simplify: $4(x - 8) - (x + 3)$.

⊘ **Solution**

$$4(x - 8) - (x + 3)$$

Distribute. $4x - 32 - x - 3$

Combine like terms. $3x - 35$

> | **TRY IT ::** 1.277   Simplify: $6(x - 9) - (x + 12)$.

> | **TRY IT ::** 1.278   Simplify: $8(x - 1) - (x + 5)$.

All the properties of real numbers we have used in this chapter are summarized in Table 1.74.

Commutative Property	
of addition If a, b are real numbers, then	$a + b = b + a$
of multiplication If a, b are real numbers, then	$a \cdot b = b \cdot a$
Associative Property	
of addition If a, b, c are real numbers, then	$(a + b) + c = a + (b + c)$
of multiplication If a, b, c are real numbers, then	$(a \cdot b) \cdot c = a \cdot (b \cdot c)$
Distributive Property	
If a, b, c are real numbers, then	$a(b + c) = ab + ac$
Identity Property	
of addition For any real number a: 0 is the **additive identity**	$a + 0 = a$ $0 + a = a$
of multiplication For any real number a: **1** is the **multiplicative identity**	$a \cdot 1 = a$ $1 \cdot a = a$
Inverse Property	
of addition For any real number a, $-a$ is the **additive inverse** of a	$a + (-a) = 0$
of multiplication For any real number a, $a \neq 0$ $\frac{1}{a}$ is the **multiplicative inverse** of a.	$a \cdot \frac{1}{a} = 1$
Properties of Zero	
For any real number a,	$a \cdot 0 = 0$ $0 \cdot a = 0$
For any real number a, $a \neq 0$	$\frac{0}{a} = 0$
For any real number a, $a \neq 0$	$\frac{a}{0}$ is undefined

Table 1.74

1.9 EXERCISES

Practice Makes Perfect

Use the Commutative and Associative Properties

In the following exercises, use the associative property to simplify.

723. $3(4x)$

724. $4(7m)$

725. $(y + 12) + 28$

726. $(n + 17) + 33$

In the following exercises, simplify.

727. $\frac{1}{2} + \frac{7}{8} + \left(-\frac{1}{2}\right)$

728. $\frac{2}{5} + \frac{5}{12} + \left(-\frac{2}{5}\right)$

729. $\frac{3}{20} \cdot \frac{49}{11} \cdot \frac{20}{3}$

730. $\frac{13}{18} \cdot \frac{25}{7} \cdot \frac{18}{13}$

731. $-24.7 \cdot \frac{3}{8}$

732. $-36 \cdot 11 \cdot \frac{4}{9}$

733. $\left(\frac{5}{6} + \frac{8}{15}\right) + \frac{7}{15}$

734. $\left(\frac{11}{12} + \frac{4}{9}\right) + \frac{5}{9}$

735. $17(0.25)(4)$

736. $36(0.2)(5)$

737. $[2.48(12)](0.5)$

738. $[9.731(4)](0.75)$

739. $7(4a)$

740. $9(8w)$

741. $-15(5m)$

742. $-23(2n)$

743. $12\left(\frac{5}{6}p\right)$

744. $20\left(\frac{3}{5}q\right)$

745.
$43m + (-12n) + (-16m) + (-9n)$

746.
$-22p + 17q + (-35p) + (-27q)$

747. $\frac{3}{8}g + \frac{1}{12}h + \frac{7}{8}g + \frac{5}{12}h$

748. $\frac{5}{6}a + \frac{3}{10}b + \frac{1}{6}a + \frac{9}{10}b$

749. $6.8p + 9.14q + (-4.37p) + (-0.88q)$

750. $9.6m + 7.22n + (-2.19m) + (-0.65n)$

Use the Identity and Inverse Properties of Addition and Multiplication

In the following exercises, find the additive inverse of each number.

751.
ⓐ $\frac{2}{5}$

ⓑ 4.3

ⓒ -8

ⓓ $-\frac{10}{3}$

752.
ⓐ $\frac{5}{9}$

ⓑ 2.1

ⓒ -3

ⓓ $-\frac{9}{5}$

753.
ⓐ $-\frac{7}{6}$

ⓑ -0.075

ⓒ 23

ⓓ $\frac{1}{4}$

754.

ⓐ $-\dfrac{8}{3}$

ⓑ -0.019

ⓒ 52

ⓓ $\dfrac{5}{6}$

In the following exercises, find the multiplicative inverse of each number.

755. ⓐ 6 ⓑ $-\dfrac{3}{4}$ ⓒ 0.7

756. ⓐ 12 ⓑ $-\dfrac{9}{2}$ ⓒ 0.13

757. ⓐ $\dfrac{11}{12}$ ⓑ -1.1 ⓒ -4

758. ⓐ $\dfrac{17}{20}$ ⓑ -1.5 ⓒ -3

Use the Properties of Zero

In the following exercises, simplify.

759. $\dfrac{0}{6}$

760. $\dfrac{3}{0}$

761. $0 \div \dfrac{11}{12}$

762. $\dfrac{6}{0}$

763. $\dfrac{0}{3}$

764. $0 \cdot \dfrac{8}{15}$

765. $(-3.14)(0)$

766. $\dfrac{\frac{1}{10}}{0}$

Mixed Practice

In the following exercises, simplify.

767. $19a + 44 - 19a$

768. $27c + 16 - 27c$

769. $10(0.1d)$

770. $100(0.01p)$

771. $\dfrac{0}{u - 4.99}$, where $u \neq 4.99$

772. $\dfrac{0}{v - 65.1}$, where $v \neq 65.1$

773. $0 \div \left(x - \dfrac{1}{2}\right)$, where $x \neq \dfrac{1}{2}$

774. $0 \div \left(y - \dfrac{1}{6}\right)$, where $x \neq \dfrac{1}{6}$

775. $\dfrac{32 - 5a}{0}$, where $32 - 5a \neq 0$

776. $\dfrac{28 - 9b}{0}$, where $28 - 9b \neq 0$

777. $\left(\dfrac{3}{4} + \dfrac{9}{10}m\right) \div 0$ where $\dfrac{3}{4} + \dfrac{9}{10}m \neq 0$

778. $\left(\dfrac{5}{16}n - \dfrac{3}{7}\right) \div 0$ where $\dfrac{5}{16}n - \dfrac{3}{7} \neq 0$

779. $15 \cdot \dfrac{3}{5}(4d + 10)$

780. $18 \cdot \dfrac{5}{6}(15h + 24)$

Simplify Expressions Using the Distributive Property

In the following exercises, simplify using the distributive property.

781. $8(4y + 9)$

782. $9(3w + 7)$

783. $6(c - 13)$

784. $7(y - 13)$

785. $\dfrac{1}{4}(3q + 12)$

786. $\dfrac{1}{5}(4m + 20)$

787. $9\left(\frac{5}{9}y - \frac{1}{3}\right)$

788. $10\left(\frac{3}{10}x - \frac{2}{5}\right)$

789. $12\left(\frac{1}{4} + \frac{2}{3}r\right)$

790. $12\left(\frac{1}{6} + \frac{3}{4}s\right)$

791. $r(s - 18)$

792. $u(v - 10)$

793. $(y + 4)p$

794. $(a + 7)x$

795. $-7(4p + 1)$

796. $-9(9a + 4)$

797. $-3(x - 6)$

798. $-4(q - 7)$

799. $-(3x - 7)$

800. $-(5p - 4)$

801. $16 - 3(y + 8)$

802. $18 - 4(x + 2)$

803. $4 - 11(3c - 2)$

804. $9 - 6(7n - 5)$

805. $22 - (a + 3)$

806. $8 - (r - 7)$

807. $(5m - 3) - (m + 7)$

808. $(4y - 1) - (y - 2)$

809. $5(2n + 9) + 12(n - 3)$

810. $9(5u + 8) + 2(u - 6)$

811. $9(8x - 3) - (-2)$

812. $4(6x - 1) - (-8)$

813. $14(c - 1) - 8(c - 6)$

814. $11(n - 7) - 5(n - 1)$

815. $6(7y + 8) - (30y - 15)$

816. $7(3n + 9) - (4n - 13)$

Everyday Math

817. Insurance copayment Carrie had to have 5 fillings done. Each filling cost $80. Her dental insurance required her to pay 20% of the cost as a copay. Calculate Carrie's copay:

ⓐ First, by multiplying 0.20 by 80 to find her copay for each filling and then multiplying your answer by 5 to find her total copay for 5 fillings.

ⓑ Next, by multiplying [5(0.20)](80)

ⓒ Which of the properties of real numbers says that your answers to parts (a), where you multiplied 5[(0.20)(80)] and (b), where you multiplied [5(0.20)](80), should be equal?

818. Cooking time Helen bought a 24-pound turkey for her family's Thanksgiving dinner and wants to know what time to put the turkey in to the oven. She wants to allow 20 minutes per pound cooking time. Calculate the length of time needed to roast the turkey:

ⓐ First, by multiplying $24 \cdot 20$ to find the total number of minutes and then multiplying the answer by $\frac{1}{60}$ to convert minutes into hours.

ⓑ Next, by multiplying $24\left(20 \cdot \frac{1}{60}\right)$.

ⓒ Which of the properties of real numbers says that your answers to parts (a), where you multiplied $(24 \cdot 20)\frac{1}{60}$, and (b), where you multiplied $24\left(20 \cdot \frac{1}{60}\right)$, should be equal?

819. Buying by the case Trader Joe's grocery stores sold a bottle of wine they called "Two Buck Chuck" for $1.99. They sold a case of 12 bottles for $23.88. To find the cost of 12 bottles at $1.99, notice that 1.99 is $2 - 0.01$.

ⓐ Multiply 12(1.99) by using the distributive property to multiply $12(2 - 0.01)$.

ⓑ Was it a bargain to buy "Two Buck Chuck" by the case?

820. Multi-pack purchase Adele's shampoo sells for $3.99 per bottle at the grocery store. At the warehouse store, the same shampoo is sold as a 3 pack for $10.49. To find the cost of 3 bottles at $3.99, notice that 3.99 is $4 - 0.01$.

ⓐ Multiply 3(3.99) by using the distributive property to multiply $3(4 - 0.01)$.

ⓑ How much would Adele save by buying 3 bottles at the warehouse store instead of at the grocery store?

Writing Exercises

821. In your own words, state the commutative property of addition.

822. What is the difference between the additive inverse and the multiplicative inverse of a number?

823. Simplify $8\left(x - \frac{1}{4}\right)$ using the distributive property and explain each step.

824. Explain how you can multiply 4($5.97) without paper or calculator by thinking of $5.97 as $6 - 0.03$ and then using the distributive property.

Self Check

ⓐ *After completing the exercises, use this checklist to evaluate your mastery of the objectives of this section.*

I can...	Confidently	With some help	No-I don't get it!
use the commutative and associative properties.			
use the identity and inverse properties of addition and multiplication.			
use the properties of zero.			
simplify expressions using the distributive property.			

ⓑ *After reviewing this checklist, what will you do to become confident for all objectives?*

1.10 ## Systems of Measurement

Learning Objectives

By the end of this section, you will be able to:

- Make unit conversions in the US system
- Use mixed units of measurement in the US system
- Make unit conversions in the metric system
- Use mixed units of measurement in the metric system
- Convert between the US and the metric systems of measurement
- Convert between Fahrenheit and Celsius temperatures

Be Prepared!

A more thorough introduction to the topics covered in this section can be found in the *Prealgebra* chapter, **The Properties of Real Numbers**.

Make Unit Conversions in the U.S. System

There are two systems of measurement commonly used around the world. Most countries use the metric system. The U.S. uses a different system of measurement, usually called the **U.S. system**. We will look at the U.S. system first.

The U.S. system of measurement uses units of inch, foot, yard, and mile to measure length and pound and ton to measure weight. For capacity, the units used are cup, pint, quart, and gallons. Both the U.S. system and the metric system measure time in seconds, minutes, and hours.

The equivalencies of measurements are shown in Table 1.75. The table also shows, in parentheses, the common abbreviations for each measurement.

U.S. System of Measurement			
Length	1 foot (ft.) = 12 inches (in.) 1 yard (yd.) = 3 feet (ft.) 1 mile (mi.) = 5,280 feet (ft.)	**Volume**	3 teaspoons (t) = 1 tablespoon (T) 16 tablespoons (T) = 1 cup (C) 1 cup (C) = 8 fluid ounces (fl. oz.) 1 pint (pt.) = 2 cups (C) 1 quart (qt.) = 2 pints (pt.) 1 gallon (gal) = 4 quarts (qt.)
Weight	1 pound (lb.) = 16 ounces (oz.) 1 ton = 2000 pounds (lb.)	**Time**	1 minute (min) = 60 seconds (sec) 1 hour (hr) = 60 minutes (min) 1 day = 24 hours (hr) 1 week (wk) = 7 days 1 year (yr) = 365 days

Table 1.75

In many real-life applications, we need to convert between units of measurement, such as feet and yards, minutes and seconds, quarts and gallons, etc. We will use the identity property of multiplication to do these conversions. We'll restate the identity property of multiplication here for easy reference.

Identity Property of Multiplication

For any real number a : $\qquad a \cdot 1 = a \qquad 1 \cdot a = a$

1 is the **multiplicative identity**

To use the identity property of multiplication, we write 1 in a form that will help us convert the units. For example, suppose we want to change inches to feet. We know that 1 foot is equal to 12 inches, so we will write 1 as the fraction $\dfrac{1 \text{ foot}}{12 \text{ inches}}$.

When we multiply by this fraction we do not change the value, but just change the units.

But $\frac{12 \text{ inches}}{1 \text{ foot}}$ also equals 1. How do we decide whether to multiply by $\frac{1 \text{ foot}}{12 \text{ inches}}$ or $\frac{12 \text{ inches}}{1 \text{ foot}}$? We choose the fraction that will make the units we want to convert *from* divide out. Treat the unit words like factors and "divide out" common units like we do common factors. If we want to convert 66 inches to feet, which multiplication will eliminate the inches?

$$66 \text{ inches} \cdot \frac{1 \text{ foot}}{12 \text{ inches}} \quad \text{or} \quad \cancel{66 \text{ inches} \cdot \frac{12 \text{ inches}}{1 \text{ foot}}}$$

The first form works since $66 \text{ inches} \cdot \frac{1 \text{ foot}}{12 \text{ inches}}$.

The inches divide out and leave only feet. The second form does not have any units that will divide out and so will not help us.

EXAMPLE 1.140 HOW TO MAKE UNIT CONVERSIONS

MaryAnne is 66 inches tall. Convert her height into feet.

✓ **Solution**

Step 1. Multiply the measurement to be converted by 1; write 1 as a fraction relating the units given and the units needed.	Multiply 66 inches by 1, writing 1 as a fraction relating inches and feet. We need inches in the denominator so that the inches will divide out!	$66 \text{ inches} \cdot 1$ $66 \text{ inches} \cdot \frac{1 \text{ foot}}{12 \text{ inches}}$
Step 2. Multiply.	Think of 66 inches as $\frac{66 \text{ inches}}{1}$.	$\frac{66 \text{ inches} \cdot 1 \text{ foot}}{12 \text{ inches}}$
Step 3. Simplify the fraction.	Notice: inches divide out.	$\frac{66 \cancel{\text{ inches}} \cdot 1 \text{ foot}}{12 \cancel{\text{ inches}}}$ $\frac{66 \text{ feet}}{12}$
Step 4. Simplify.	Divide 66 by 12.	5.5 feet

> **TRY IT :: 1.279**   Lexie is 30 inches tall. Convert her height to feet.

> **TRY IT :: 1.280**   Rene bought a hose that is 18 yards long. Convert the length to feet.

HOW TO :: MAKE UNIT CONVERSIONS.

Step 1. Multiply the measurement to be converted by 1; write 1 as a fraction relating the units given and the units needed.

Step 2. Multiply.

Step 3. Simplify the fraction.

Step 4. Simplify.

When we use the identity property of multiplication to convert units, we need to make sure the units we want to change from will divide out. Usually this means we want the conversion fraction to have those units in the denominator.

EXAMPLE 1.141

Ndula, an elephant at the San Diego Safari Park, weighs almost 3.2 tons. Convert her weight to pounds.

✓ **Solution**

We will convert 3.2 tons into pounds. We will use the identity property of multiplication, writing 1 as the fraction

$$\frac{2000 \text{ pounds}}{1 \text{ ton}}.$$

	3.2 tons
Multiply the measurement to be converted, by 1.	$3.2 \text{ tons} \cdot 1$
Write 1 as a fraction relating tons and pounds.	$3.2 \text{ tons} \cdot \dfrac{2,000 \text{ pounds}}{1 \text{ ton}}$
Simplify.	$\dfrac{3.2 \cancel{\text{ tons}} \cdot 2,000 \text{ pounds}}{1 \cancel{\text{ ton}}}$
Multiply.	6,400 pounds
	Ndula weighs almost 6,400 pounds.

> **TRY IT : : 1.281**   Arnold's SUV weighs about 4.3 tons. Convert the weight to pounds.

> **TRY IT : : 1.282**   The Carnival *Destiny* cruise ship weighs 51,000 tons. Convert the weight to pounds.

Sometimes, to convert from one unit to another, we may need to use several other units in between, so we will need to multiply several fractions.

EXAMPLE 1.142

Juliet is going with her family to their summer home. She will be away from her boyfriend for 9 weeks. Convert the time to minutes.

⊘ **Solution**

To convert weeks into minutes we will convert weeks into days, days into hours, and then hours into minutes. To do this we will multiply by conversion factors of 1.

	9 weeks
Write 1 as $\dfrac{7 \text{ days}}{1 \text{ week}}$, $\dfrac{24 \text{ hours}}{1 \text{ day}}$, and $\dfrac{60 \text{ minutes}}{1 \text{ hour}}$.	$\dfrac{9 \text{ wk}}{1} \cdot \dfrac{7 \text{ days}}{1 \text{ wk}} \cdot \dfrac{24 \text{ hr}}{1 \text{ day}} \cdot \dfrac{60 \text{ min}}{1 \text{ hr}}$
Divide out the common units.	$\dfrac{9 \cancel{\text{ wk}}}{1} \cdot \dfrac{7 \text{ days}}{1 \cancel{\text{ wk}}} \cdot \dfrac{24 \cancel{\text{ hr}}}{1 \cancel{\text{ day}}} \cdot \dfrac{60 \text{ min}}{1 \cancel{\text{ hr}}}$
Multiply.	$\dfrac{9 \cdot 7 \cdot 24 \cdot 60 \text{ min}}{1 \cdot 1 \cdot 1 \cdot 1}$
Multiply.	90,720 min

Juliet and her boyfriend will be apart for 90,720 minutes (although it may seem like an eternity!).

> **TRY IT : : 1.283**
>
> The distance between the earth and the moon is about 250,000 miles. Convert this length to yards.

> **TRY IT : : 1.284**
>
> The astronauts of Expedition 28 on the International Space Station spend 15 weeks in space. Convert the time to minutes.

EXAMPLE 1.143

How many ounces are in 1 gallon?

✓ Solution

We will convert gallons to ounces by multiplying by several conversion factors. Refer to Table 1.75.

	1 gallon
Multiply the measurement to be converted by 1.	$\dfrac{1 \text{ gallon}}{1} \cdot \dfrac{4 \text{ quarts}}{1 \text{ gallon}} \cdot \dfrac{2 \text{ pints}}{1 \text{ quart}} \cdot \dfrac{2 \text{ cups}}{1 \text{ pint}} \cdot \dfrac{8 \text{ ounces}}{1 \text{ cup}}$
Use conversion factors to get to the right unit. Simplify.	$\dfrac{1 \ \cancel{\text{gallon}}}{1} \cdot \dfrac{4 \ \cancel{\text{quarts}}}{1 \ \cancel{\text{gallon}}} \cdot \dfrac{2 \ \cancel{\text{pints}}}{1 \ \cancel{\text{quart}}} \cdot \dfrac{2 \ \cancel{\text{cups}}}{1 \ \cancel{\text{pint}}} \cdot \dfrac{8 \text{ ounces}}{1 \ \cancel{\text{cup}}}$
Multiply.	$\dfrac{1 \cdot 4 \cdot 2 \cdot 2 \cdot 8 \text{ ounces}}{1 \cdot 1 \cdot 1 \cdot 1 \cdot 1}$
Simplify.	128 ounces There are 128 ounces in a gallon.

> **TRY IT : : 1.285**   How many cups are in 1 gallon?

> **TRY IT : : 1.286**   How many teaspoons are in 1 cup?

Use Mixed Units of Measurement in the U.S. System

We often use mixed units of measurement in everyday situations. Suppose Joe is 5 feet 10 inches tall, stays at work for 7 hours and 45 minutes, and then eats a 1 pound 2 ounce steak for dinner—all these measurements have mixed units.

Performing arithmetic operations on measurements with mixed units of measures requires care. Be sure to add or subtract like units!

EXAMPLE 1.144

Seymour bought three steaks for a barbecue. Their weights were 14 ounces, 1 pound 2 ounces and 1 pound 6 ounces. How many total pounds of steak did he buy?

✓ Solution

We will add the weights of the steaks to find the total weight of the steaks.

Add the ounces. Then add the pounds.	$\begin{array}{rl} & 14 \text{ ounces} \\ 1 \text{ pound} & 2 \text{ ounces} \\ +\,1 \text{ pound} & 6 \text{ ounces} \\ \hline 2 \text{ pounds} & 22 \text{ ounces} \end{array}$
Convert 22 ounces to pounds and ounces.	2 pounds + 1 pound, 6 ounces
Add the pounds.	3 pounds, 6 ounces
	Seymour bought 3 pounds 6 ounces of steak.

> **TRY IT : : 1.287**
>
> Laura gave birth to triplets weighing 3 pounds 3 ounces, 3 pounds 3 ounces, and 2 pounds 9 ounces. What was the total birth weight of the three babies?

> | **TRY IT ::** 1.288

Stan cut two pieces of crown molding for his family room that were 8 feet 7 inches and 12 feet 11 inches. What was the total length of the molding?

EXAMPLE 1.145

Anthony bought four planks of wood that were each 6 feet 4 inches long. What is the total length of the wood he purchased?

✓ **Solution**

We will multiply the length of one plank to find the total length.

Multiply the inches and then the feet.	6 feet 4 inches × 4 24 feet 16 inches
Convert the 16 inches to feet. Add the feet.	24 feet + 1 foot 4 inches 25 feet 4 inches
	Anthony bought 25 feet and 4 inches of wood.

> | **TRY IT ::** 1.289

Henri wants to triple his spaghetti sauce recipe that uses 1 pound 8 ounces of ground turkey. How many pounds of ground turkey will he need?

> | **TRY IT ::** 1.290

Joellen wants to double a solution of 5 gallons 3 quarts. How many gallons of solution will she have in all?

Make Unit Conversions in the Metric System

In the **metric system**, units are related by powers of 10. The roots words of their names reflect this relation. For example, the basic unit for measuring length is a meter. One kilometer is 1,000 meters; the prefix *kilo* means *thousand*. One centimeter is $\frac{1}{100}$ of a meter, just like one cent is $\frac{1}{100}$ of one dollar.

The equivalencies of measurements in the metric system are shown in Table 1.81. The common abbreviations for each measurement are given in parentheses.

Metric System of Measurement		
Length	**Mass**	**Capacity**
1 kilometer (km) = 1,000 m	1 kilogram (kg) = 1,000 g	1 kiloliter (kL) = 1,000 L
1 hectometer (hm) = 100 m	1 hectogram (hg) = 100 g	1 hectoliter (hL) = 100 L
1 dekameter (dam) = 10 m	1 dekagram (dag) = 10 g	1 dekaliter (daL) = 10 L
1 meter (m) = 1 m	1 gram (g) = 1 g	1 liter (L) = 1 L
1 decimeter (dm) = 0.1 m	1 decigram (dg) = 0.1 g	1 deciliter (dL) = 0.1 L
1 centimeter (cm) = 0.01 m	1 centigram (cg) = 0.01 g	1 centiliter (cL) = 0.01 L
1 millimeter (mm) = 0.001 m	1 milligram (mg) = 0.001 g	1 milliliter (mL) = 0.001 L
1 meter = 100 centimeters	1 gram = 100 centigrams	1 liter = 100 centiliters
1 meter = 1,000 millimeters	1 gram = 1,000 milligrams	1 liter = 1,000 milliliters

Table 1.81

To make conversions in the metric system, we will use the same technique we did in the US system. Using the identity property of multiplication, we will multiply by a conversion factor of one to get to the correct units.

Have you ever run a 5K or 10K race? The length of those races are measured in kilometers. The metric system is commonly used in the United States when talking about the length of a race.

EXAMPLE 1.146

Nick ran a 10K race. How many meters did he run?

✓ **Solution**

We will convert kilometers to meters using the identity property of multiplication.

	10 kilometers
Multiply the measurement to be converted by 1.	10 kilometers · 1
Write 1 as a fraction relating kilometers and meters.	10 kilometers · $\frac{1,000 \text{ meters}}{1 \text{ kilometers}}$
Simplify.	$\frac{10 \text{ kilometers} \cdot 1,000 \text{ m}}{1 \text{ kilometers}}$
Multiply.	10,000 meters
	Nick ran 10,000 meters.

> **TRY IT :: 1.291**   Sandy completed her first 5K race! How many meters did she run?

> **TRY IT :: 1.292**   Herman bought a rug 2.5 meters in length. How many centimeters is the length?

EXAMPLE 1.147

Eleanor's newborn baby weighed 3,200 grams. How many kilograms did the baby weigh?

⊘ **Solution**

We will convert grams into kilograms.

	3,200 grams
Multiply the measurement to be converted by 1.	3,200 grams · 1
Write 1 as a function relating kilograms and grams.	$3,200 \text{ grams} \cdot \dfrac{1 \text{ kg}}{1,000 \text{ grams}}$
Simplify.	$3,200 \text{ grams} \cdot \dfrac{1 \text{ kg}}{1,000 \text{ grams}}$
Multiply.	$\dfrac{3,200 \text{ kilograms}}{1,000}$
Divide.	3.2 kilograms The baby weighed 3.2 kilograms.

> **TRY IT : : 1.293** Kari's newborn baby weighed 2,800 grams. How many kilograms did the baby weigh?

> **TRY IT : : 1.294**
>
> Anderson received a package that was marked 4,500 grams. How many kilograms did this package weigh?

As you become familiar with the metric system you may see a pattern. Since the system is based on multiples of ten, the calculations involve multiplying by multiples of ten. We have learned how to simplify these calculations by just moving the decimal.

To multiply by 10, 100, or 1,000, we move the decimal to the right one, two, or three places, respectively. To multiply by 0.1, 0.01, or 0.001, we move the decimal to the left one, two, or three places, respectively.

We can apply this pattern when we make measurement conversions in the metric system. In **Example 1.147**, we changed 3,200 grams to kilograms by multiplying by $\dfrac{1}{1000}$ (or 0.001). This is the same as moving the decimal three places to the left.

$$3,200 \cdot \dfrac{1}{1,000} \qquad 3,200.$$
$$3.2 \qquad\qquad 3.2$$

EXAMPLE 1.148

Convert ⓐ 350 L to kiloliters ⓑ 4.1 L to milliliters.

⊘ **Solution**

ⓐ We will convert liters to kiloliters. In **Table 1.81**, we see that 1 kiloliter = 1,000 liters.

	350 L
Multiply by 1, writing 1 as a fraction relating liters to kiloliters.	$350 \text{ L} \cdot \dfrac{1 \text{ kL}}{1,000 \text{ L}}$
Simplify.	$350 \text{ L} \cdot \dfrac{1 \text{ kL}}{1,000 \text{ L}}$
Move the decimal 3 units to the left. (350.)	0.35 kL

ⓑ We will convert liters to milliliters. From **Table 1.81** we see that $1 \text{ liter} = 1,000 \text{ milliliters}$.

	4.1 L
Multiply by 1, writing 1 as a fraction relating liters to milliliters.	$4.1 \text{ L} \cdot \dfrac{1,000 \text{ mL}}{1 \text{ L}}$
Simplify.	$4.1 \, \cancel{\text{L}} \cdot \dfrac{1,000 \text{ mL}}{1 \, \cancel{\text{L}}}$
Move the decimal 3 units to the right.	4.100 mL 
	4,100 mL

> **TRY IT : : 1.295** Convert: ⓐ 725 L to kiloliters ⓑ 6.3 L to milliliters

> **TRY IT : : 1.296** Convert: ⓐ 350 hL to liters ⓑ 4.1 L to centiliters

Use Mixed Units of Measurement in the Metric System

Performing arithmetic operations on measurements with mixed units of measures in the metric system requires the same care we used in the US system. But it may be easier because of the relation of the units to the powers of 10. Make sure to add or subtract like units.

EXAMPLE 1.149

Ryland is 1.6 meters tall. His younger brother is 85 centimeters tall. How much taller is Ryland than his younger brother?

✓ Solution

We can convert both measurements to either centimeters or meters. Since meters is the larger unit, we will subtract the lengths in meters. We convert 85 centimeters to meters by moving the decimal 2 places to the left.

Write the 85 centimeters as meters.
$$\begin{array}{r} 1.60 \text{ m} \\ -0.85 \text{ m} \\ \hline 0.75 \text{ m} \end{array}$$

Ryland is 0.75 m taller than his brother.

> **TRY IT : : 1.297**
>
> Mariella is 1.58 meters tall. Her daughter is 75 centimeters tall. How much taller is Mariella than her daughter? Write the answer in centimeters.

> **TRY IT : : 1.298**
>
> The fence around Hank's yard is 2 meters high. Hank is 96 centimeters tall. How much shorter than the fence is Hank? Write the answer in meters.

EXAMPLE 1.150

Dena's recipe for lentil soup calls for 150 milliliters of olive oil. Dena wants to triple the recipe. How many liters of olive oil will she need?

✓ Solution

We will find the amount of olive oil in millileters then convert to liters.

	Triple 150 mL
Translate to algebra.	$3 \cdot 150$ mL
Multiply.	450 mL
Convert to liters.	$450 \cdot \dfrac{0.001\ L}{1\ mL}$
Simplify.	0.45 L

Dena needs 0.45 liters of olive oil.

> **TRY IT : : 1.299**

A recipe for Alfredo sauce calls for 250 milliliters of milk. Renata is making pasta with Alfredo sauce for a big party and needs to multiply the recipe amounts by 8. How many liters of milk will she need?

> **TRY IT : : 1.300**

To make one pan of baklava, Dorothea needs 400 grams of filo pastry. If Dorothea plans to make 6 pans of baklava, how many kilograms of filo pastry will she need?

Convert Between the U.S. and the Metric Systems of Measurement

Many measurements in the United States are made in metric units. Our soda may come in 2-liter bottles, our calcium may come in 500-mg capsules, and we may run a 5K race. To work easily in both systems, we need to be able to convert between the two systems.

Table 1.86 shows some of the most common conversions.

Conversion Factors Between U.S. and Metric Systems		
Length	**Mass**	**Capacity**
1 in. = 2.54 cm 1 ft. = 0.305 m 1 yd. = 0.914 m 1 mi. = 1.61 km 1 m = 3.28 ft.	1 lb. = 0.45 kg 1 oz. = 28 g 1 kg = 2.2 lb.	1 qt. = 0.95 L 1 fl. oz. = 30 mL 1 L = 1.06 qt.

Table 1.86

Figure 1.22 shows how inches and centimeters are related on a ruler.

Figure 1.22 This ruler shows inches and centimeters.

Figure 1.23 shows the ounce and milliliter markings on a measuring cup.

Figure 1.23 This measuring cup shows ounces and milliliters.

Figure 1.24 shows how pounds and kilograms marked on a bathroom scale.

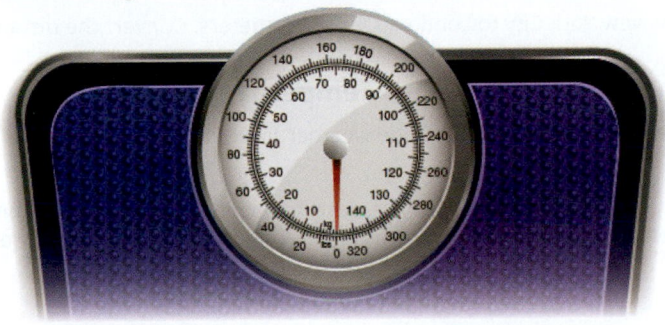

Figure 1.24 This scale shows pounds and kilograms.

We make conversions between the systems just as we do within the systems—by multiplying by unit conversion factors.

EXAMPLE 1.151

Lee's water bottle holds 500 mL of water. How many ounces are in the bottle? Round to the nearest tenth of an ounce.

✓ Solution

$$500 \text{ mL}$$

Multiply by a unit conversion factor relating mL and ounces.

$$500 \text{ milliliters} \cdot \frac{1 \text{ ounce}}{30 \text{ milliliters}}$$

Simplify.

$$\frac{50 \text{ ounce}}{30}$$

Divide.

$$16.7 \text{ ounces.}$$
The water bottle has 16.7 ounces.

> | **TRY IT :: 1.301**   How many quarts of soda are in a 2-L bottle?

> | **TRY IT :: 1.302**   How many liters are in 4 quarts of milk?

EXAMPLE 1.152

Soleil was on a road trip and saw a sign that said the next rest stop was in 100 kilometers. How many miles until the next rest stop?

✓ **Solution**

	100 kilometers
Multiply by a unit conversion factor relating km and mi.	$100 \text{ kilometers} \cdot \dfrac{1 \text{ mile}}{1.61 \text{ kilometer}}$
Simplify.	$\dfrac{100 \text{ miles}}{1.61}$
Divide.	62 miles
	Soleil will travel 62 miles.

> **TRY IT :: 1.303**   The height of Mount Kilimanjaro is 5,895 meters. Convert the height to feet.

> **TRY IT :: 1.304**

The flight distance from New York City to London is 5,586 kilometers. Convert the distance to miles.

Convert between Fahrenheit and Celsius Temperatures

Have you ever been in a foreign country and heard the weather forecast? If the forecast is for $22°C$, what does that mean?

The U.S. and metric systems use different scales to measure temperature. The U.S. system uses degrees Fahrenheit, written $°F$. The metric system uses degrees Celsius, written $°C$. Figure 1.25 shows the relationship between the two systems.

Celsius (°C)   **Fahrenheit (°F)**

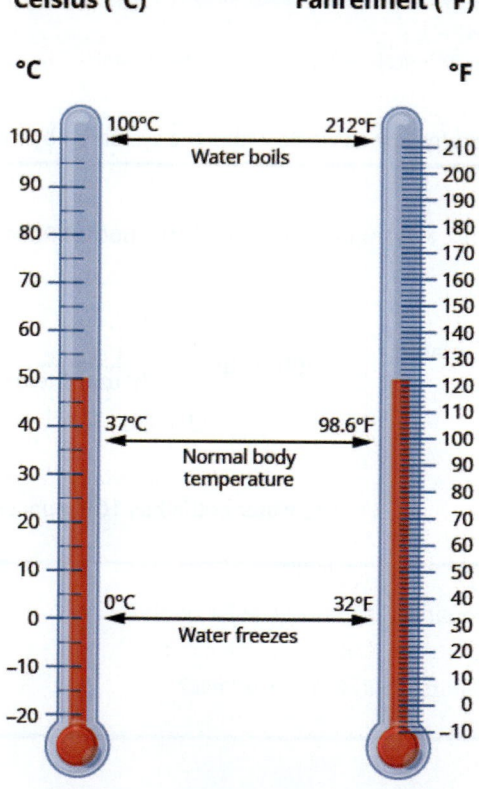

Figure 1.25 The diagram shows normal body temperature, along with the freezing and boiling temperatures of water in degrees Fahrenheit and degrees Celsius.

Temperature Conversion

To convert from Fahrenheit temperature, F, to Celsius temperature, C, use the formula

$$C = \frac{5}{9}(F - 32).$$

To convert from Celsius temperature, C, to Fahrenheit temperature, F, use the formula

$$F = \frac{9}{5}C + 32.$$

EXAMPLE 1.153

Convert $50°$ Fahrenheit into degrees Celsius.

⊘ Solution

We will substitute $50°F$ into the formula to find C.

$$C = \frac{5}{9}(F - 32)$$

Substitute 50 for F.	$C = \frac{5}{9}(50 - 32)$
Simplify in parentheses.	$C = \frac{5}{9}(18)$
Multiply.	$C = 10$
	So we found that 50°F is equivalent to 10°C.

> **TRY IT : : 1.305**   Convert the Fahrenheit temperature to degrees Celsius: $59°$ Fahrenheit.

> **TRY IT : : 1.306**   Convert the Fahrenheit temperature to degrees Celsius: $41°$ Fahrenheit.

EXAMPLE 1.154

While visiting Paris, Woody saw the temperature was $20°$ Celsius. Convert the temperature into degrees Fahrenheit.

⊘ Solution

We will substitute $20°C$ into the formula to find F.

$$F = \frac{9}{5}C + 32$$

Substitute 20 for C.	$F = \frac{9}{5}(20) + 32$
Multiply.	$F = 36 + 32$
Add.	$F = 68$
	So we found that 20°C is equivalent to 68°F.

> **TRY IT : : 1.307**
>
> Convert the Celsius temperature to degrees Fahrenheit: the temperature in Helsinki, Finland, was $15°$ Celsius.

> **TRY IT : : 1.308**
>
> Convert the Celsius temperature to degrees Fahrenheit: the temperature in Sydney, Australia, was $10°$ Celsius.

1.10 EXERCISES

Practice Makes Perfect

Make Unit Conversions in the U.S. System

In the following exercises, convert the units.

825. A park bench is 6 feet long. Convert the length to inches.

826. A floor tile is 2 feet wide. Convert the width to inches.

827. A ribbon is 18 inches long. Convert the length to feet.

828. Carson is 45 inches tall. Convert his height to feet.

829. A football field is 160 feet wide. Convert the width to yards.

830. On a baseball diamond, the distance from home plate to first base is 30 yards. Convert the distance to feet.

831. Ulises lives 1.5 miles from school. Convert the distance to feet.

832. Denver, Colorado, is 5,183 feet above sea level. Convert the height to miles.

833. A killer whale weighs 4.6 tons. Convert the weight to pounds.

834. Blue whales can weigh as much as 150 tons. Convert the weight to pounds.

835. An empty bus weighs 35,000 pounds. Convert the weight to tons.

836. At take-off, an airplane weighs 220,000 pounds. Convert the weight to tons.

837. Rocco waited $1\frac{1}{2}$ hours for his appointment. Convert the time to seconds.

838. Misty's surgery lasted $2\frac{1}{4}$ hours. Convert the time to seconds.

839. How many teaspoons are in a pint?

840. How many tablespoons are in a gallon?

841. JJ's cat, Posy, weighs 14 pounds. Convert her weight to ounces.

842. April's dog, Beans, weighs 8 pounds. Convert his weight to ounces.

843. Crista will serve 20 cups of juice at her son's party. Convert the volume to gallons.

844. Lance needs 50 cups of water for the runners in a race. Convert the volume to gallons.

845. Jon is 6 feet 4 inches tall. Convert his height to inches.

846. Faye is 4 feet 10 inches tall. Convert her height to inches.

847. The voyage of the *Mayflower* took 2 months and 5 days. Convert the time to days.

848. Lynn's cruise lasted 6 days and 18 hours. Convert the time to hours.

849. Baby Preston weighed 7 pounds 3 ounces at birth. Convert his weight to ounces.

850. Baby Audrey weighted 6 pounds 15 ounces at birth. Convert her weight to ounces.

Use Mixed Units of Measurement in the U.S. System

In the following exercises, solve.

851. Eli caught three fish. The weights of the fish were 2 pounds 4 ounces, 1 pound 11 ounces, and 4 pounds 14 ounces. What was the total weight of the three fish?

852. Judy bought 1 pound 6 ounces of almonds, 2 pounds 3 ounces of walnuts, and 8 ounces of cashews. How many pounds of nuts did Judy buy?

853. One day Anya kept track of the number of minutes she spent driving. She recorded 45, 10, 8, 65, 20, and 35. How many hours did Anya spend driving?

854. Last year Eric went on 6 business trips. The number of days of each was 5, 2, 8, 12, 6, and 3. How many weeks did Eric spend on business trips last year?

855. Renee attached a 6 feet 6 inch extension cord to her computer's 3 feet 8 inch power cord. What was the total length of the cords?

856. Fawzi's SUV is 6 feet 4 inches tall. If he puts a 2 feet 10 inch box on top of his SUV, what is the total height of the SUV and the box?

857. Leilani wants to make 8 placemats. For each placemat she needs 18 inches of fabric. How many yards of fabric will she need for the 8 placemats?

858. Mireille needs to cut 24 inches of ribbon for each of the 12 girls in her dance class. How many yards of ribbon will she need altogether?

Make Unit Conversions in the Metric System

In the following exercises, convert the units.

859. Ghalib ran 5 kilometers. Convert the length to meters.

860. Kitaka hiked 8 kilometers. Convert the length to meters.

861. Estrella is 1.55 meters tall. Convert her height to centimeters.

862. The width of the wading pool is 2.45 meters. Convert the width to centimeters.

863. Mount Whitney is 3,072 meters tall. Convert the height to kilometers.

864. The depth of the Mariana Trench is 10,911 meters. Convert the depth to kilometers.

865. June's multivitamin contains 1,500 milligrams of calcium. Convert this to grams.

866. A typical ruby-throated hummingbird weights 3 grams. Convert this to milligrams.

867. One stick of butter contains 91.6 grams of fat. Convert this to milligrams.

868. One serving of gourmet ice cream has 25 grams of fat. Convert this to milligrams.

869. The maximum mass of an airmail letter is 2 kilograms. Convert this to grams.

870. Dimitri's daughter weighed 3.8 kilograms at birth. Convert this to grams.

871. A bottle of wine contained 750 milliliters. Convert this to liters.

872. A bottle of medicine contained 300 milliliters. Convert this to liters.

Use Mixed Units of Measurement in the Metric System

In the following exercises, solve.

873. Matthias is 1.8 meters tall. His son is 89 centimeters tall. How much taller is Matthias than his son?

874. Stavros is 1.6 meters tall. His sister is 95 centimeters tall. How much taller is Stavros than his sister?

875. A typical dove weighs 345 grams. A typical duck weighs 1.2 kilograms. What is the difference, in grams, of the weights of a duck and a dove?

876. Concetta had a 2-kilogram bag of flour. She used 180 grams of flour to make biscotti. How many kilograms of flour are left in the bag?

877. Harry mailed 5 packages that weighed 420 grams each. What was the total weight of the packages in kilograms?

878. One glass of orange juice provides 560 milligrams of potassium. Linda drinks one glass of orange juice every morning. How many grams of potassium does Linda get from her orange juice in 30 days?

879. Jonas drinks 200 milliliters of water 8 times a day. How many liters of water does Jonas drink in a day?

880. One serving of whole grain sandwich bread provides 6 grams of protein. How many milligrams of protein are provided by 7 servings of whole grain sandwich bread?

Convert Between the U.S. and the Metric Systems of Measurement

In the following exercises, make the unit conversions. Round to the nearest tenth.

881. Bill is 75 inches tall. Convert his height to centimeters.

882. Frankie is 42 inches tall. Convert his height to centimeters.

883. Marcus passed a football 24 yards. Convert the pass length to meters

884. Connie bought 9 yards of fabric to make drapes. Convert the fabric length to meters.

885. Each American throws out an average of 1,650 pounds of garbage per year. Convert this weight to kilograms.

886. An average American will throw away 90,000 pounds of trash over his or her lifetime. Convert this weight to kilograms.

887. A 5K run is 5 kilometers long. Convert this length to miles.

888. Kathryn is 1.6 meters tall. Convert her height to feet.

889. Dawn's suitcase weighed 20 kilograms. Convert the weight to pounds.

890. Jackson's backpack weighed 15 kilograms. Convert the weight to pounds.

891. Ozzie put 14 gallons of gas in his truck. Convert the volume to liters.

892. Bernard bought 8 gallons of paint. Convert the volume to liters.

Convert between Fahrenheit and Celsius Temperatures

In the following exercises, convert the Fahrenheit temperatures to degrees Celsius. Round to the nearest tenth.

893. $86°$ Fahrenheit

894. $77°$ Fahrenheit

895. $104°$ Fahrenheit

896. $14°$ Fahrenheit

897. $72°$ Fahrenheit

898. $4°$ Fahrenheit

899. $0°$ Fahrenheit

900. $120°$ Fahrenheit

In the following exercises, convert the Celsius temperatures to degrees Fahrenheit. Round to the nearest tenth.

901. $5°$ Celsius

902. $25°$ Celsius

903. $-10°$ Celsius

904. $-15°$ Celsius

905. $22°$ Celsius

906. $8°$ Celsius

907. $43°$ Celsius

908. $16°$ Celsius

Everyday Math

909. Nutrition Julian drinks one can of soda every day. Each can of soda contains 40 grams of sugar. How many kilograms of sugar does Julian get from soda in 1 year?

910. Reflectors The reflectors in each lane-marking stripe on a highway are spaced 16 yards apart. How many reflectors are needed for a one mile long lane-marking stripe?

Writing Exercises

911. Some people think that $65°$ to $75°$ Fahrenheit is the ideal temperature range.

ⓐ What is your ideal temperature range? Why do you think so?

ⓑ Convert your ideal temperatures from Fahrenheit to Celsius.

912.

ⓐ Did you grow up using the U.S. or the metric system of measurement?

ⓑ Describe two examples in your life when you had to convert between the two systems of measurement.

Self Check

ⓐ After completing the exercises, use this checklist to evaluate your mastery of the objectives of this section.

I can...	Confidently	With some help	No-I don't get it!
define U.S. units of measurement and convert from one unit to another.			
use U.S units of measurement.			
define metric units of measurement and convert from one unit to another.			
use metric units of measurement.			
convert between the U.S. and the metric system of measurement.			
convert between Fahrenheit and Celsius temperatures.			

ⓑ Overall, after looking at the checklist, do you think you are well-prepared for the next Chapter? Why or why not?

CHAPTER 1 REVIEW

KEY TERMS

absolute value The absolute value of a number is its distance from 0 on the number line. The absolute value of a number n is written as $|n|$.

additive identity The additive identity is the number 0; adding 0 to any number does not change its value.

additive inverse The opposite of a number is its additive inverse. A number and it additive inverse add to 0.

coefficient The coefficient of a term is the constant that multiplies the variable in a term.

complex fraction A complex fraction is a fraction in which the numerator or the denominator contains a fraction.

composite number A composite number is a counting number that is not prime. A composite number has factors other than 1 and itself.

constant A constant is a number whose value always stays the same.

counting numbers The counting numbers are the numbers 1, 2, 3, ...

decimal A decimal is another way of writing a fraction whose denominator is a power of ten.

denominator The denominator is the value on the bottom part of the fraction that indicates the number of equal parts into which the whole has been divided.

divisible by a number If a number m is a multiple of n, then m is divisible by n. (If 6 is a multiple of 3, then 6 is divisible by 3.)

equality symbol The symbol " $=$ " is called the equal sign. We read $a = b$ as " a is equal to b."

equation An equation is two expressions connected by an equal sign.

equivalent decimals Two decimals are equivalent if they convert to equivalent fractions.

equivalent fractions Equivalent fractions are fractions that have the same value.

evaluate an expression To evaluate an expression means to find the value of the expression when the variable is replaced by a given number.

expression An expression is a number, a variable, or a combination of numbers and variables using operation symbols.

factors If $a \cdot b = m$, then a and b are factors of m. Since $3 \cdot 4 = 12$, then 3 and 4 are factors of 12.

fraction A fraction is written $\frac{a}{b}$, where $b \neq 0$ a is the numerator and b is the denominator. A fraction represents parts of a whole. The denominator b is the number of equal parts the whole has been divided into, and the numerator a indicates how many parts are included.

integers The whole numbers and their opposites are called the integers: ...–3, –2, –1, 0, 1, 2, 3...

irrational number An irrational number is a number that cannot be written as the ratio of two integers. Its decimal form does not stop and does not repeat.

least common denominator The least common denominator (LCD) of two fractions is the Least common multiple (LCM) of their denominators.

least common multiple The least common multiple of two numbers is the smallest number that is a multiple of both numbers.

like terms Terms that are either constants or have the same variables raised to the same powers are called like terms.

multiple of a number A number is a multiple of n if it is the product of a counting number and n.

multiplicative identity The multiplicative identity is the number 1; multiplying 1 by any number does not change the value of the number.

multiplicative inverse The reciprocal of a number is its multiplicative inverse. A number and its multiplicative inverse multiply to one.

number line A number line is used to visualize numbers. The numbers on the number line get larger as they go from left to right, and smaller as they go from right to left.

numerator The numerator is the value on the top part of the fraction that indicates how many parts of the whole are included.

opposite The opposite of a number is the number that is the same distance from zero on the number line but on the opposite side of zero: $-a$ means the opposite of the number. The notation $-a$ is read "the opposite of a."

origin The origin is the point labeled 0 on a number line.

percent A percent is a ratio whose denominator is 100.

prime factorization The prime factorization of a number is the product of prime numbers that equals the number.

prime number A prime number is a counting number greater than 1, whose only factors are 1 and itself.

radical sign A radical sign is the symbol $\sqrt{m}$ that denotes the positive square root.

rational number A rational number is a number of the form $\frac{p}{q}$, where p and q are integers and $q \neq 0$. A rational number can be written as the ratio of two integers. Its decimal form stops or repeats.

real number A real number is a number that is either rational or irrational.

reciprocal The reciprocal of $\frac{a}{b}$ is $\frac{b}{a}$. A number and its reciprocal multiply to one: $\frac{a}{b} \cdot \frac{b}{a} = 1$.

repeating decimal A repeating decimal is a decimal in which the last digit or group of digits repeats endlessly.

simplified fraction A fraction is considered simplified if there are no common factors in its numerator and denominator.

simplify an expression To simplify an expression, do all operations in the expression.

square and square root If $n^2 = m$, then m is the square of n and n is a square root of m.

term A term is a constant or the product of a constant and one or more variables.

variable A variable is a letter that represents a number whose value may change.

whole numbers The whole numbers are the numbers 0, 1, 2, 3, ...

KEY CONCEPTS

1.1 Introduction to Whole Numbers

- **Place Value** as in **Figure 1.3**.
- **Name a Whole Number in Words**
  Step 1. Start at the left and name the number in each period, followed by the period name.
  Step 2. Put commas in the number to separate the periods.
  Step 3. Do not name the ones period.
- **Write a Whole Number Using Digits**
  Step 1. Identify the words that indicate periods. (Remember the ones period is never named.)
  Step 2. Draw 3 blanks to indicate the number of places needed in each period. Separate the periods by commas.
  Step 3. Name the number in each period and place the digits in the correct place value position.
- **Round Whole Numbers**
  Step 1. Locate the given place value and mark it with an arrow. All digits to the left of the arrow do not change.
  Step 2. Underline the digit to the right of the given place value.
  Step 3. Is this digit greater than or equal to 5?
    - Yes—add 1 to the digit in the given place value.
    - No—do not change the digit in the given place value.
  Step 4. Replace all digits to the right of the given place value with zeros.
- **Divisibility Tests:** A number is divisible by:
  - 2 if the last digit is 0, 2, 4, 6, or 8.
  - 3 if the sum of the digits is divisible by 3.
  - 5 if the last digit is 5 or 0.
  - 6 if it is divisible by both 2 and 3.
  - 10 if it ends with 0.
- **Find the Prime Factorization of a Composite Number**
  Step 1. Find two factors whose product is the given number, and use these numbers to create two branches.
  Step 2. If a factor is prime, that branch is complete. Circle the prime, like a bud on the tree.
  Step 3. If a factor is not prime, write it as the product of two factors and continue the process.
  Step 4. Write the composite number as the product of all the circled primes.

- **Find the Least Common Multiple by Listing Multiples**

  Step 1. List several multiples of each number.

  Step 2. Look for the smallest number that appears on both lists.

  Step 3. This number is the LCM.

- **Find the Least Common Multiple Using the Prime Factors Method**

  Step 1. Write each number as a product of primes.

  Step 2. List the primes of each number. Match primes vertically when possible.

  Step 3. Bring down the columns.

  Step 4. Multiply the factors.

1.2 Use the Language of Algebra

- **Notation**   **The result is...**

  - $a + b$ the sum of a and b

  - $a - b$ the difference of a and b

  - $a \cdot b$, ab, $(a)(b)$ $(a)b$, $a(b)$ the product of a and b

  - $a \div b$, a/b, $\frac{a}{b}$, $b\overline{)a}$ the quotient of a and b

- **Inequality**

  - $a < b$ is read "a is less than b" a is to the left of b on the number line

  - $a > b$ is read "a is greater than b" a is to the right of b on the number line

- **Inequality Symbols**   **Words**

  - $a \neq b$ a is **not equal to** b

  - $a < b$ a is **less than** b

  - $a \leq b$ a is **less than or equal to** b

  - $a > b$ a is **greater than** b

  - $a \geq b$ a is **greater than or equal to** b

- **Grouping Symbols**

  - Parentheses ( )

  - Brackets [ ]

  - Braces { }

- **Exponential Notation**

  - a^n means multiply a by itself, n times. The expression a^n is read a to the nth power.

- **Order of Operations:** When simplifying mathematical expressions perform the operations in the following order:

  Step 1. Parentheses and other Grouping Symbols: Simplify all expressions inside the parentheses or other grouping symbols, working on the innermost parentheses first.

  Step 2. Exponents: Simplify all expressions with exponents.

  Step 3. Multiplication and Division: Perform all multiplication and division in order from left to right. These operations have equal priority.

  Step 4. Addition and Subtraction: Perform all addition and subtraction in order from left to right. These operations have equal priority.

- **Combine Like Terms**

Step 1. Identify like terms.

Step 2. Rearrange the expression so like terms are together.

Step 3. Add or subtract the coefficients and keep the same variable for each group of like terms.

1.3 Add and Subtract Integers

- **Addition of Positive and Negative Integers**

$5 + 3$	$-5 + (-3)$
8	-8

both positive,	both negative,
sum positive	sum negative

$-5 + 3$	$5 + (-3)$
-2	2

different signs,	different signs,
more negatives	more positives
sum negative	sum positive

- **Property of Absolute Value**: $|n| \geq 0$ for all numbers. Absolute values are always greater than or equal to zero!

- **Subtraction of Integers**

$5 - 3$	$-5 - (-3)$
2	-2

5 positives	5 negatives
take away 3 positives	take away 3 negatives
2 positives	2 negatives

$-5 - 3$	$5 - (-3)$
-8	8

5 negatives, want to	5 positives, want to
subtract 3 positives	subtract 3 negatives
need neutral pairs	need neutral pairs

- **Subtraction Property:** Subtracting a number is the same as adding its opposite.

1.4 Multiply and Divide Integers

- **Multiplication and Division of Two Signed Numbers**
  - Same signs—Product is positive
  - Different signs—Product is negative
- **Strategy for Applications**

Step 1. Identify what you are asked to find.

Step 2. Write a phrase that gives the information to find it.

Step 3. Translate the phrase to an expression.

Step 4. Simplify the expression.

Step 5. Answer the question with a complete sentence.

1.5 Visualize Fractions

- **Equivalent Fractions Property:** If a, b, c are numbers where $b \neq 0$, $c \neq 0$, then
$$\frac{a}{b} = \frac{a \cdot c}{b \cdot c} \text{ and } \frac{a \cdot c}{b \cdot c} = \frac{a}{b}.$$

- **Fraction Division:** If a, b, c and d are numbers where $b \neq 0$, $c \neq 0$, and $d \neq 0$, then $\frac{a}{b} \div \frac{c}{d} = \frac{a}{b} \cdot \frac{d}{c}$. To divide fractions, multiply the first fraction by the reciprocal of the second.

- **Fraction Multiplication:** If a, b, c and d are numbers where $b \neq 0$, and $d \neq 0$, then $\frac{a}{b} \cdot \frac{c}{d} = \frac{ac}{bd}$. To multiply fractions, multiply the numerators and multiply the denominators.

- **Placement of Negative Sign in a Fraction:** For any positive numbers a and b, $\frac{-a}{b} = \frac{a}{-b} = -\frac{a}{b}$.

- **Property of One:** $\frac{a}{a} = 1$; Any number, except zero, divided by itself is one.

- **Simplify a Fraction**

  Step 1. Rewrite the numerator and denominator to show the common factors. If needed, factor the numerator and denominator into prime numbers first.

  Step 2. Simplify using the equivalent fractions property by dividing out common factors.

  Step 3. Multiply any remaining factors.

- **Simplify an Expression with a Fraction Bar**

  Step 1. Simplify the expression in the numerator. Simplify the expression in the denominator.

  Step 2. Simplify the fraction.

1.6 Add and Subtract Fractions

- **Fraction Addition and Subtraction:** If a, b, and c are numbers where $c \neq 0$, then

  $\frac{a}{c} + \frac{b}{c} = \frac{a+b}{c}$ and $\frac{a}{c} - \frac{b}{c} = \frac{a-b}{c}$.

  To add or subtract fractions, add or subtract the numerators and place the result over the common denominator.

- **Strategy for Adding or Subtracting Fractions**

  Step 1. Do they have a common denominator?
  Yes—go to step 2.
  No—Rewrite each fraction with the LCD (Least Common Denominator). Find the LCD. Change each fraction into an equivalent fraction with the LCD as its denominator.

  Step 2. Add or subtract the fractions.

  Step 3. Simplify, if possible. To multiply or divide fractions, an LCD IS NOT needed. To add or subtract fractions, an LCD IS needed.

- **Simplify Complex Fractions**

  Step 1. Simplify the numerator.

  Step 2. Simplify the denominator.

  Step 3. Divide the numerator by the denominator. Simplify if possible.

1.7 Decimals

- **Name a Decimal**

  Step 1. Name the number to the left of the decimal point.

  Step 2. Write "and" for the decimal point.

  Step 3. Name the "number" part to the right of the decimal point as if it were a whole number.

  Step 4. Name the decimal place of the last digit.

- **Write a Decimal**

  Step 1. Look for the word 'and'—it locates the decimal point. Place a decimal point under the word 'and.' Translate the words before 'and' into the whole number and place it to the left of the decimal point. If there is no "and," write a "0" with a decimal point to its right.

  Step 2. Mark the number of decimal places needed to the right of the decimal point by noting the place value indicated by the last word.

  Step 3. Translate the words after 'and' into the number to the right of the decimal point. Write the number in the spaces—putting the final digit in the last place.

  Step 4. Fill in zeros for place holders as needed.

- **Round a Decimal**

  Step 1. Locate the given place value and mark it with an arrow.

  Step 2. Underline the digit to the right of the place value.

  Step 3. Is this digit greater than or equal to 5? Yes—add 1 to the digit in the given place value. No—do <u>not</u> change the digit in the given place value.

  Step 4. Rewrite the number, deleting all digits to the right of the rounding digit.

- **Add or Subtract Decimals**

  Step 1. Write the numbers so the decimal points line up vertically.

  Step 2. Use zeros as place holders, as needed.

  Step 3. Add or subtract the numbers as if they were whole numbers. Then place the decimal in the answer under the decimal points in the given numbers.

- **Multiply Decimals**

  Step 1. Determine the sign of the product.

  Step 2. Write in vertical format, lining up the numbers on the right. Multiply the numbers as if they were whole numbers, temporarily ignoring the decimal points.

  Step 3. Place the decimal point. The number of decimal places in the product is the sum of the decimal places in the factors.

  Step 4. Write the product with the appropriate sign.

- **Multiply a Decimal by a Power of Ten**

  Step 1. Move the decimal point to the right the same number of places as the number of zeros in the power of 10.

  Step 2. Add zeros at the end of the number as needed.

- **Divide Decimals**

  Step 1. Determine the sign of the quotient.

  Step 2. Make the divisor a whole number by "moving" the decimal point all the way to the right. "Move" the decimal point in the dividend the same number of places - adding zeros as needed.

  Step 3. Divide. Place the decimal point in the quotient above the decimal point in the dividend.

  Step 4. Write the quotient with the appropriate sign.

- **Convert a Decimal to a Proper Fraction**

  Step 1. Determine the place value of the final digit.

  Step 2. Write the fraction: numerator—the 'numbers' to the right of the decimal point; denominator—the place value corresponding to the final digit.

- **Convert a Fraction to a Decimal** Divide the numerator of the fraction by the denominator.

1.8 The Real Numbers

- **Square Root Notation**

  $\sqrt{m}$ is read 'the square root of m.' If $m = n^2$, then $\sqrt{m} = n$, for $n \geq 0$.

- **Order Decimals**

  Step 1. Write the numbers one under the other, lining up the decimal points.

  Step 2. Check to see if both numbers have the same number of digits. If not, write zeros at the end of the one with fewer digits to make them match.

  Step 3. Compare the numbers as if they were whole numbers.

  Step 4. Order the numbers using the appropriate inequality sign.

1.9 Properties of Real Numbers

- **Commutative Property of**

  ◦ **Addition:** If a, b are real numbers, then $a + b = b + a$.

  ◦ **Multiplication:** If a, b are real numbers, then $a \cdot b = b \cdot a$. When adding or multiplying, changing the *order* gives the same result.

- **Associative Property of**

- **Addition:** If a, b, c are real numbers, then $(a + b) + c = a + (b + c)$.
  - **Multiplication:** If a, b, c are real numbers, then $(a \cdot b) \cdot c = a \cdot (b \cdot c)$.
    When adding or multiplying, changing the *grouping* gives the same result.
- **Distributive Property:** If a, b, c are real numbers, then
  - $a(b + c) = ab + ac$
  - $(b + c)a = ba + ca$
  - $a(b - c) = ab - ac$
  - $(b - c)a = ba - ca$
- **Identity Property**
  - **of Addition:** For any real number a: $a + 0 = a$ $0 + a = a$
    0 is the **additive identity**
  - **of Multiplication:** For any real number a: $a \cdot 1 = a$ $1 \cdot a = a$
    1 is the **multiplicative identity**
- **Inverse Property**
  - **of Addition:** For any real number a, $a + (-a) = 0$. A number and its *opposite* add to zero. $-a$ is the **additive inverse** of a.
  - **of Multiplication:** For any real number a, $(a \neq 0)$ $a \cdot \frac{1}{a} = 1$. A number and its *reciprocal* multiply to one. $\frac{1}{a}$ is the **multiplicative inverse** of a.
- **Properties of Zero**
  - For any real number a,
    $a \cdot 0 = 0$ $0 \cdot a = 0$ – The product of any real number and 0 is 0.
  - $\frac{0}{a} = 0$ for $a \neq 0$ – Zero divided by any real number except zero is zero.
  - $\frac{a}{0}$ is undefined – Division by zero is undefined.

1.10 Systems of Measurement

- **Metric System of Measurement**
  - **Length**

1 kilometer (km)	=	1,000 m
1 hectometer (hm)	=	100 m
1 dekameter (dam)	=	10 m
1 meter (m)	=	1 m
1 decimeter (dm)	=	0.1 m
1 centimeter (cm)	=	0.01 m
1 millimeter (mm)	=	0.001 m
1 meter	=	100 centimeters
1 meter	=	1,000 millimeters

  - **Mass**

$$
\begin{array}{rcl}
1 \text{ kilogram (kg)} & = & 1{,}000 \text{ g} \\
1 \text{ hectogram (hg)} & = & 100 \text{ g} \\
1 \text{ dekagram (dag)} & = & 10 \text{ g} \\
1 \text{ gram (g)} & = & 1 \text{ g} \\
1 \text{ decigram (dg)} & = & 0.1 \text{ g} \\
1 \text{ centigram (cg)} & = & 0.01 \text{ g} \\
1 \text{ milligram (mg)} & = & 0.001 \text{ g} \\
1 \text{ gram} & = & 100 \text{ centigrams} \\
1 \text{ gram} & = & 1{,}000 \text{ milligrams}
\end{array}
$$

◦ **Capacity**

$$
\begin{array}{rcl}
1 \text{ kiloliter (kL)} & = & 1{,}000 \text{ L} \\
1 \text{ hectoliter (hL)} & = & 100 \text{ L} \\
1 \text{ dekaliter (daL)} & = & 10 \text{ L} \\
1 \text{ liter (L)} & = & 1 \text{ L} \\
1 \text{ deciliter (dL)} & = & 0.1 \text{ L} \\
1 \text{ centiliter (cL)} & = & 0.01 \text{ L} \\
1 \text{ milliliter (mL)} & = & 0.001 \text{ L} \\
1 \text{ liter} & = & 100 \text{ centiliters} \\
1 \text{ liter} & = & 1{,}000 \text{ milliliters}
\end{array}
$$

- **Temperature Conversion**

  ◦ To convert from Fahrenheit temperature, F, to Celsius temperature, C, use the formula $C = \frac{5}{9}(F - 32)$

  ◦ To convert from Celsius temperature, C, to Fahrenheit temperature, F, use the formula $F = \frac{9}{5}C + 32$

REVIEW EXERCISES

1.1 Introduction to Whole Numbers

Use Place Value with Whole Number

In the following exercises find the place value of each digit.

913. 26,915

ⓐ 1
ⓑ 2
ⓒ 9
ⓓ 5
ⓔ 6

914. 359,417

ⓐ 9
ⓑ 3
ⓒ 4
ⓓ 7
ⓔ 1

915. 58,129,304

ⓐ 5
ⓑ 0
ⓒ 1
ⓓ 8
ⓔ 2

916. 9,430,286,157

ⓐ 6
ⓑ 4
ⓒ 9
ⓓ 0
ⓔ 5

In the following exercises, name each number.

917. 6,104

918. 493,068

919. 3,975,284

920. 85,620,435

In the following exercises, write each number as a whole number using digits.

921. three hundred fifteen

922. sixty-five thousand, nine hundred twelve

923. ninety million, four hundred twenty-five thousand, sixteen

924. one billion, forty-three million, nine hundred twenty-two thousand, three hundred eleven

In the following exercises, round to the indicated place value.

925. Round to the nearest ten.

ⓐ 407 ⓑ 8,564

926. Round to the nearest hundred.

ⓐ 25,846 ⓑ 25,864

In the following exercises, round each number to the nearest ⓐ hundred ⓑ thousand ⓒ ten thousand.

927. 864,951

928. 3,972,849

Identify Multiples and Factors

In the following exercises, use the divisibility tests to determine whether each number is divisible by 2, by 3, by 5, by 6, and by 10.

929. 168

930. 264

931. 375

932. 750

933. 1430

934. 1080

Find Prime Factorizations and Least Common Multiples

In the following exercises, find the prime factorization.

935. 420

936. 115

937. 225

938. 2475

939. 1560

940. 56

941. 72

942. 168

943. 252

944. 391

In the following exercises, find the least common multiple of the following numbers using the multiples method.

945. 6,15

946. 60, 75

In the following exercises, find the least common multiple of the following numbers using the prime factors method.

947. 24, 30

948. 70, 84

1.2 Use the Language of Algebra

Use Variables and Algebraic Symbols

In the following exercises, translate the following from algebra to English.

949. $25 - 7$

950. $5 \cdot 6$

951. $45 \div 5$

952. $x + 8$

953. $42 \geq 27$

954. $3n = 24$

955. $3 \leq 20 \div 4$

956. $a \neq 7 \cdot 4$

In the following exercises, determine if each is an expression or an equation.

957. $6 \cdot 3 + 5$

958. $y - 8 = 32$

Simplify Expressions Using the Order of Operations

In the following exercises, simplify each expression.

959. 3^5

960. 10^8

In the following exercises, simplify

961. $6 + 10/2 + 2$

962. $9 + 12/3 + 4$

963. $20 \div (4 + 6) \cdot 5$

964. $33 \div (3 + 8) \cdot 2$

965. $4^2 + 5^2$

966. $(4 + 5)^2$

Evaluate an Expression

In the following exercises, evaluate the following expressions.

967. $9x + 7$ when $x = 3$

968. $5x - 4$ when $x = 6$

969. x^4 when $x = 3$

970. 3^x when $x = 3$

971. $x^2 + 5x - 8$ when $x = 6$

972. $2x + 4y - 5$ when $x = 7, y = 8$

Simplify Expressions by Combining Like Terms

In the following exercises, identify the coefficient of each term.

973. $12n$

974. $9x^2$

In the following exercises, identify the like terms.

975. $3n, n^2, 12, 12p^2, 3, 3n^2$

976. $5, 18r^2, 9s, 9r, 5r^2, 5s$

In the following exercises, identify the terms in each expression.

977. $11x^2 + 3x + 6$

978. $22y^3 + y + 15$

In the following exercises, simplify the following expressions by combining like terms.

979. $17a + 9a$

980. $18z + 9z$

981. $9x + 3x + 8$

982. $8a + 5a + 9$

983. $7p + 6 + 5p - 4$

984. $8x + 7 + 4x - 5$

Translate an English Phrase to an Algebraic Expression

In the following exercises, translate the following phrases into algebraic expressions.

985. the sum of 8 and 12

986. the sum of 9 and 1

987. the difference of x and 4

988. the difference of x and 3

989. the product of 6 and y

990. the product of 9 and y

991. Adele bought a skirt and a blouse. The skirt cost $15 more than the blouse. Let b represent the cost of the blouse. Write an expression for the cost of the skirt.

992. Marcella has 6 fewer boy cousins than girl cousins. Let g represent the number of girl cousins. Write an expression for the number of boy cousins.

1.3 Add and Subtract Integers

Use Negatives and Opposites of Integers

In the following exercises, order each of the following pairs of numbers, using < or >.

993.
ⓐ 6___2

ⓑ −7___4

ⓒ −9___−1

ⓓ 9___−3

994.
ⓐ −5___1

ⓑ −4___−9

ⓒ 6___10

ⓓ 3___−8

In the following exercises,, find the opposite of each number.

995. ⓐ −8 ⓑ 1

996. ⓐ −2 ⓑ 6

In the following exercises, simplify.

997. $-(-19)$

998. $-(-53)$

In the following exercises, simplify.

999. $-m$ when

ⓐ $m = 3$

ⓑ $m = -3$

1000. $-p$ when

ⓐ $p = 6$

ⓑ $p = -6$

Simplify Expressions with Absolute Value

In the following exercises,, simplify.

1001. ⓐ $|7|$ ⓑ $|-25|$ ⓒ $|0|$

1002. ⓐ $|5|$ ⓑ $|0|$ ⓒ $|-19|$

In the following exercises, fill in <, >, or = for each of the following pairs of numbers.

1003.
ⓐ -8___$|-8|$

ⓑ $-|-2|$___-2

1004.
ⓐ $|-3|$___$-|-3|$

ⓑ 4___$-|-4|$

In the following exercises, simplify.

1005. $|8 - 4|$

1006. $|9 - 6|$

1007. $8(14 - 2|-2|)$

1008. $6(13 - 4|-2|)$

In the following exercises, evaluate.

1009. ⓐ $|x|$ when $x = -28$ ⓑ

1010.
ⓐ $|y|$ when $y = -37$

ⓑ $|-z|$ when $z = -24$

Add Integers

In the following exercises, simplify each expression.

1011. $-200 + 65$

1012. $-150 + 45$

1013. $2 + (-8) + 6$

1014. $4 + (-9) + 7$

1015. $140 + (-75) + 67$

1016. $-32 + 24 + (-6) + 10$

Subtract Integers
In the following exercises, simplify.

1017. $9 - 3$

1018. $-5 - (-1)$

1019. ⓐ $15 - 6$ ⓑ $15 + (-6)$

1020. ⓐ $12 - 9$ ⓑ $12 + (-9)$

1021. ⓐ $8 - (-9)$ ⓑ $8 + 9$

1022. ⓐ $4 - (-4)$ ⓑ $4 + 4$

In the following exercises, simplify each expression.

1023. $10 - (-19)$

1024. $11 - (-18)$

1025. $31 - 79$

1026. $39 - 81$

1027. $-31 - 11$

1028. $-32 - 18$

1029. $-15 - (-28) + 5$

1030. $71 + (-10) - 8$

1031. $-16 - (-4 + 1) - 7$

1032. $-15 - (-6 + 4) - 3$

Multiply Integers
In the following exercises, multiply.

1033. $-5(7)$

1034. $-8(6)$

1035. $-18(-2)$

1036. $-10(-6)$

Divide Integers
In the following exercises, divide.

1037. $-28 \div 7$

1038. $56 \div (-7)$

1039. $-120 \div (-20)$

1040. $-200 \div 25$

Simplify Expressions with Integers
In the following exercises, simplify each expression.

1041. $-8(-2) - 3(-9)$

1042. $-7(-4) - 5(-3)$

1043. $(-5)^3$

1044. $(-4)^3$

1045. $-4 \cdot 2 \cdot 11$

1046. $-5 \cdot 3 \cdot 10$

1047. $-10(-4) \div (-8)$

1048. $-8(-6) \div (-4)$

1049. $31 - 4(3 - 9)$

1050. $24 - 3(2 - 10)$

Evaluate Variable Expressions with Integers
In the following exercises, evaluate each expression.

1051. $x + 8$ when

ⓐ $x = -26$

ⓑ $x = -95$

1052. $y + 9$ when

ⓐ $y = -29$

ⓑ $y = -84$

1053. When $b = -11$, evaluate:

ⓐ $b + 6$

ⓑ $-b + 6$

1054. When $c = -9$, evaluate:

ⓐ $c + (-4)$

ⓑ $-c + (-4)$

1055. $p^2 - 5p + 2$ when $p = -1$

1056. $q^2 - 2q + 9$ when $q = -2$

1057. $6x - 5y + 15$ when $x = 3$ and $y = -1$

1058. $3p - 2q + 9$ when $p = 8$ and $q = -2$

Translate English Phrases to Algebraic Expressions

In the following exercises, translate to an algebraic expression and simplify if possible.

1059. the sum of -4 and -17, increased by 32

1060. ⓐ the difference of 15 and -7 ⓑ subtract 15 from -7

1061. the quotient of -45 and -9

1062. the product of -12 and the difference of c and d

Use Integers in Applications

In the following exercises, solve.

1063. Temperature The high temperature one day in Miami Beach, Florida, was $76°$. That same day, the high temperature in Buffalo, New York was $-8°$. What was the difference between the temperature in Miami Beach and the temperature in Buffalo?

1064. Checking Account Adrianne has a balance of $-\$22$ in her checking account. She deposits $\$301$ to the account. What is the new balance?

1.5 Visualize Fractions

Find Equivalent Fractions

In the following exercises, find three fractions equivalent to the given fraction. Show your work, using figures or algebra.

1065. $\dfrac{1}{4}$

1066. $\dfrac{1}{3}$

1067. $\dfrac{5}{6}$

1068. $\dfrac{2}{7}$

Simplify Fractions

In the following exercises, simplify.

1069. $\dfrac{7}{21}$

1070. $\dfrac{8}{24}$

1071. $\dfrac{15}{20}$

1072. $\dfrac{12}{18}$

1073. $-\dfrac{168}{192}$

1074. $-\dfrac{140}{224}$

1075. $\dfrac{11x}{11y}$

1076. $\dfrac{15a}{15b}$

Multiply Fractions

In the following exercises, multiply.

1077. $\dfrac{2}{5} \cdot \dfrac{1}{3}$

1078. $\dfrac{1}{2} \cdot \dfrac{3}{8}$

1079. $\dfrac{7}{12}\left(-\dfrac{8}{21}\right)$

1080. $\frac{5}{12}\left(-\frac{8}{15}\right)$

1081. $-28p\left(-\frac{1}{4}\right)$

1082. $-51q\left(-\frac{1}{3}\right)$

1083. $\frac{14}{5}(-15)$

1084. $-1\left(-\frac{3}{8}\right)$

Divide Fractions

In the following exercises, divide.

1085. $\frac{1}{2} \div \frac{1}{4}$

1086. $\frac{1}{2} \div \frac{1}{8}$

1087. $-\frac{4}{5} \div \frac{4}{7}$

1088. $-\frac{3}{4} \div \frac{3}{5}$

1089. $\frac{5}{8} \div \frac{a}{10}$

1090. $\frac{5}{6} \div \frac{c}{15}$

1091. $\frac{7p}{12} \div \frac{21p}{8}$

1092. $\frac{5q}{12} \div \frac{15q}{8}$

1093. $\frac{2}{5} \div (-10)$

1094. $-18 \div -\left(\frac{9}{2}\right)$

In the following exercises, simplify.

1095. $\dfrac{\frac{2}{3}}{\frac{8}{9}}$

1096. $\dfrac{\frac{4}{5}}{\frac{8}{15}}$

1097. $\dfrac{-\frac{9}{10}}{3}$

1098. $\dfrac{\frac{2}{5}}{8}$

1099. $\dfrac{\frac{r}{5}}{\frac{s}{3}}$

1100. $\dfrac{-\frac{x}{6}}{-\frac{8}{9}}$

Simplify Expressions Written with a Fraction Bar

In the following exercises, simplify.

1101. $\frac{4+11}{8}$

1102. $\frac{9+3}{7}$

1103. $\frac{30}{7-12}$

1104. $\frac{15}{4-9}$

1105. $\frac{22-14}{19-13}$

1106. $\frac{15+9}{18+12}$

1107. $\frac{5 \cdot 8}{-10}$

1108. $\frac{3 \cdot 4}{-24}$

1109. $\frac{15 \cdot 5 - 5^2}{2 \cdot 10}$

1110. $\frac{12 \cdot 9 - 3^2}{3 \cdot 18}$

1111. $\frac{2+4(3)}{-3-2^2}$

1112. $\frac{7+3(5)}{-2-3^2}$

Translate Phrases to Expressions with Fractions

In the following exercises, translate each English phrase into an algebraic expression.

1113. the quotient of c and the sum of d and 9.

1114. the quotient of the difference of h and k, and -5.

1.6 Add and Subtract Fractions

Add and Subtract Fractions with a Common Denominator

In the following exercises, add.

1115. $\frac{4}{9} + \frac{1}{9}$

1116. $\frac{2}{9} + \frac{5}{9}$

1117. $\frac{y}{3} + \frac{2}{3}$

1118. $\frac{7}{p} + \frac{9}{p}$

1119. $-\frac{1}{8} + \left(-\frac{3}{8}\right)$

1120. $-\frac{1}{8} + \left(-\frac{5}{8}\right)$

In the following exercises, subtract.

1121. $\frac{4}{5} - \frac{1}{5}$

1122. $\frac{4}{5} - \frac{3}{5}$

1123. $\frac{y}{17} - \frac{9}{17}$

1124. $\frac{x}{19} - \frac{8}{19}$

1125. $-\frac{8}{d} - \frac{3}{d}$

1126. $-\frac{7}{c} - \frac{7}{c}$

Add or Subtract Fractions with Different Denominators

In the following exercises, add or subtract.

1127. $\frac{1}{3} + \frac{1}{5}$

1128. $\frac{1}{4} + \frac{1}{5}$

1129. $\frac{1}{5} - \left(-\frac{1}{10}\right)$

1130. $\frac{1}{2} - \left(-\frac{1}{6}\right)$

1131. $\frac{2}{3} + \frac{3}{4}$

1132. $\frac{3}{4} + \frac{2}{5}$

1133. $\frac{11}{12} - \frac{3}{8}$

1134. $\frac{5}{8} - \frac{7}{12}$

1135. $-\frac{9}{16} - \left(-\frac{4}{5}\right)$

1136. $-\frac{7}{20} - \left(-\frac{5}{8}\right)$

1137. $1 + \frac{5}{6}$

1138. $1 - \frac{5}{9}$

Use the Order of Operations to Simplify Complex Fractions

In the following exercises, simplify.

1139. $\dfrac{\left(\frac{1}{5}\right)^2}{2 + 3^2}$

1140. $\dfrac{\left(\frac{1}{3}\right)^2}{5 + 2^2}$

1141. $\dfrac{\frac{2}{3} + \frac{1}{2}}{\frac{3}{4} - \frac{2}{3}}$

1142. $\dfrac{\frac{3}{4} + \frac{1}{2}}{\frac{5}{6} - \frac{2}{3}}$

Evaluate Variable Expressions with Fractions

In the following exercises, evaluate.

1143. $x + \frac{1}{2}$ when

ⓐ $x = -\frac{1}{8}$

ⓑ $x = -\frac{1}{2}$

1144. $x + \frac{2}{3}$ when

ⓐ $x = -\frac{1}{6}$

ⓑ $x = -\frac{5}{3}$

1145. $4p^2q$ when $p = -\frac{1}{2}$ and $q = \frac{5}{9}$

1146. $5m^2n$ when $m = -\frac{2}{5}$ and $n = \frac{1}{3}$

1147. $\frac{u + v}{w}$ when $u = -4, v = -8, w = 2$

1148. $\frac{m + n}{p}$ when $m = -6, n = -2, p = 4$

1.7 Decimals

Name and Write Decimals

In the following exercises, write as a decimal.

1149. Eight and three hundredths

1150. Nine and seven hundredths

1151. One thousandth

1152. Nine thousandths

In the following exercises, name each decimal.

1153. 7.8

1154. 5.01

1155. 0.005

1156. 0.381

Round Decimals

In the following exercises, round each number to the nearest ⓐ hundredth ⓑ tenth ⓒ whole number.

1157. 5.7932

1158. 3.6284

1159. 12.4768

1160. 25.8449

Add and Subtract Decimals

In the following exercises, add or subtract.

1161. $18.37 + 9.36$

1162. $256.37 - 85.49$

1163. $15.35 - 20.88$

1164. $37.5 + 12.23$

1165. $-4.2 + (-9.3)$

1166. $-8.6 + (-8.6)$

1167. $100 - 64.2$

1168. $100 - 65.83$

1169. $2.51 + 40$

1170. $9.38 + 60$

Multiply and Divide Decimals

In the following exercises, multiply.

1171. $(0.3)(0.4)$

1172. $(0.6)(0.7)$

1173. $(8.52)(3.14)$

1174. $(5.32)(4.86)$

1175. $(0.09)(24.78)$

1176. $(0.04)(36.89)$

In the following exercises, divide.

1177. $0.15 \div 5$

1178. $0.27 \div 3$

1179. $\$8.49 \div 12$

1180. $\$16.99 \div 9$

1181. $12 \div 0.08$

1182. $5 \div 0.04$

Convert Decimals, Fractions, and Percents

In the following exercises, write each decimal as a fraction.

1183. 0.08

1184. 0.17

1185. 0.425

1186. 0.184

1187. 1.75

1188. 0.035

In the following exercises, convert each fraction to a decimal.

1189. $\frac{2}{5}$

1190. $\frac{4}{5}$

1191. $-\frac{3}{8}$

1192. $-\dfrac{5}{8}$ **1193.** $\dfrac{5}{9}$ **1194.** $\dfrac{2}{9}$

1195. $\dfrac{1}{2} + 6.5$ **1196.** $\dfrac{1}{4} + 10.75$

In the following exercises, convert each percent to a decimal.

1197. 5%   **1198.** 9%   **1199.** 40%

1200. 50%   **1201.** 115%   **1202.** 125%

In the following exercises, convert each decimal to a percent.

1203. 0.18   **1204.** 0.15   **1205.** 0.009

1206. 0.008   **1207.** 1.5   **1208.** 2.2

1.8 The Real Numbers

Simplify Expressions with Square Roots

In the following exercises, simplify.

1209. $\sqrt{64}$ **1210.** $\sqrt{144}$ **1211.** $-\sqrt{25}$

1212. $-\sqrt{81}$

Identify Integers, Rational Numbers, Irrational Numbers, and Real Numbers

In the following exercises, write as the ratio of two integers.

1213. ⓐ 9 ⓑ 8.47   **1214.** ⓐ -15 ⓑ 3.591

In the following exercises, list the ⓐ rational numbers, ⓑ irrational numbers.

1215. 0.84, 0.79132..., $1.\overline{3}$ **1216.** $2.3\overline{8}$, 0.572, 4.93814...

In the following exercises, identify whether each number is rational or irrational.

1217. ⓐ $\sqrt{121}$ ⓑ $\sqrt{48}$ **1218.** ⓐ $\sqrt{56}$ ⓑ $\sqrt{16}$

In the following exercises, identify whether each number is a real number or not a real number.

1219. ⓐ $\sqrt{-9}$ ⓑ $-\sqrt{169}$ **1220.** ⓐ $\sqrt{-64}$ ⓑ $-\sqrt{81}$

In the following exercises, list the ⓐ whole numbers, ⓑ integers, ⓒ rational numbers, ⓓ irrational numbers, ⓔ real numbers for each set of numbers.

1221.
$-4,\ 0,\ \dfrac{5}{6},\ \sqrt{16},\ \sqrt{18},\ 5.2537...$

1222.
$-\sqrt{4},\ 0.\overline{36},\ \dfrac{13}{3},\ 6.9152...,\ \sqrt{48},\ 10\dfrac{1}{2}$

Locate Fractions on the Number Line

In the following exercises, locate the numbers on a number line.

1223. $\dfrac{2}{3},\ \dfrac{5}{4},\ \dfrac{12}{5}$ **1224.** $\dfrac{1}{3},\ \dfrac{7}{4},\ \dfrac{13}{5}$ **1225.** $2\dfrac{1}{3},\ -2\dfrac{1}{3}$

1226. $1\frac{3}{5}, -1\frac{3}{5}$

In the following exercises, order each of the following pairs of numbers, using < or >.

1227. $-1\underline{\quad} -\frac{1}{8}$

1228. $-3\frac{1}{4}\underline{\quad}-4$

1229. $-\frac{7}{9}\underline{\quad}-\frac{4}{9}$

1230. $-2\underline{\quad}-\frac{19}{8}$

Locate Decimals on the Number Line

In the following exercises, locate on the number line.

1231. 0.3

1232. −0.2

1233. −2.5

1234. 2.7

In the following exercises, order each of the following pairs of numbers, using < or >.

1235. 0.9___0.6

1236. 0.7___0.8

1237. −0.6___−0.59

1238. −0.27___−0.3

1.9 Properties of Real Numbers

Use the Commutative and Associative Properties

In the following exercises, use the Associative Property to simplify.

1239. $-12(4m)$

1240. $30\left(\frac{5}{6}q\right)$

1241. $(a+16)+31$

1242. $(c+0.2)+0.7$

In the following exercises, simplify.

1243. $6y+37+(-6y)$

1244. $\frac{1}{4}+\frac{11}{15}+\left(-\frac{1}{4}\right)$

1245. $\frac{14}{11}\cdot\frac{35}{9}\cdot\frac{14}{11}$

1246. $-18\cdot15\cdot\frac{2}{9}$

1247. $\left(\frac{7}{12}+\frac{4}{5}\right)+\frac{1}{5}$

1248. $(3.98d+0.75d)+1.25d$

1249. $11x+8y+16x+15y$

1250.
$52m+(-20n)+(-18m)+(-5n)$

Use the Identity and Inverse Properties of Addition and Multiplication

In the following exercises, find the additive inverse of each number.

1251.

ⓐ $\frac{1}{3}$

ⓑ 5.1

ⓒ −14

ⓓ $-\frac{8}{5}$

1252.

ⓐ $-\frac{7}{8}$

ⓑ −0.03

ⓒ 17

ⓓ $\frac{12}{5}$

In the following exercises, find the multiplicative inverse of each number.

1253. ⓐ 10 ⓑ $-\frac{4}{9}$ ⓒ 0.6

1254. ⓐ $-\frac{9}{2}$ ⓑ -7 ⓒ 2.1

Use the Properties of Zero

In the following exercises, simplify.

1255. $83 \cdot 0$

1256. $\frac{0}{9}$

1257. $\frac{5}{0}$

1258. $0 \div \frac{2}{3}$

In the following exercises, simplify.

1259. $43 + 39 + (-43)$

1260. $(n + 6.75) + 0.25$

1261. $\frac{5}{13} \cdot 57 \cdot \frac{13}{5}$

1262. $\frac{1}{6} \cdot 17 \cdot 12$

1263. $\frac{2}{3} \cdot 28 \cdot \frac{3}{7}$

1264. $9(6x - 11) + 15$

Simplify Expressions Using the Distributive Property

In the following exercises, simplify using the Distributive Property.

1265. $7(x + 9)$

1266. $9(u - 4)$

1267. $-3(6m - 1)$

1268. $-8(-7a - 12)$

1269. $\frac{1}{3}(15n - 6)$

1270. $(y + 10) \cdot p$

1271. $(a - 4) - (6a + 9)$

1272. $4(x + 3) - 8(x - 7)$

1.10 Systems of Measurement

1.1 Define U.S. Units of Measurement and Convert from One Unit to Another

In the following exercises, convert the units. Round to the nearest tenth.

1273. A floral arbor is 7 feet tall. Convert the height to inches.

1274. A picture frame is 42 inches wide. Convert the width to feet.

1275. Kelly is 5 feet 4 inches tall. Convert her height to inches.

1276. A playground is 45 feet wide. Convert the width to yards.

1277. The height of Mount Shasta is 14,179 feet. Convert the height to miles.

1278. Shamu weights 4.5 tons. Convert the weight to pounds.

1279. The play lasted $1\frac{3}{4}$ hours. Convert the time to minutes.

1280. How many tablespoons are in a quart?

1281. Naomi's baby weighed 5 pounds 14 ounces at birth. Convert the weight to ounces.

1282. Trinh needs 30 cups of paint for her class art project. Convert the volume to gallons.

Use Mixed Units of Measurement in the U.S. System.

In the following exercises, solve.

1283. John caught 4 lobsters. The weights of the lobsters were 1 pound 9 ounces, 1 pound 12 ounces, 4 pounds 2 ounces, and 2 pounds 15 ounces. What was the total weight of the lobsters?

1284. Every day last week Pedro recorded the number of minutes he spent reading. The number of minutes were 50, 25, 83, 45, 32, 60, 135. How many hours did Pedro spend reading?

1285. Fouad is 6 feet 2 inches tall. If he stands on a rung of a ladder 8 feet 10 inches high, how high off the ground is the top of Fouad's head?

1286. Dalila wants to make throw pillow covers. Each cover takes 30 inches of fabric. How many yards of fabric does she need for 4 covers?

Make Unit Conversions in the Metric System

In the following exercises, convert the units.

1287. Donna is 1.7 meters tall. Convert her height to centimeters.

1288. Mount Everest is 8,850 meters tall. Convert the height to kilometers.

1289. One cup of yogurt contains 488 milligrams of calcium. Convert this to grams.

1290. One cup of yogurt contains 13 grams of protein. Convert this to milligrams.

1291. Sergio weighed 2.9 kilograms at birth. Convert this to grams.

1292. A bottle of water contained 650 milliliters. Convert this to liters.

Use Mixed Units of Measurement in the Metric System

In the following exerices, solve.

1293. Minh is 2 meters tall. His daughter is 88 centimeters tall. How much taller is Minh than his daughter?

1294. Selma had a 1 liter bottle of water. If she drank 145 milliliters, how much water was left in the bottle?

1295. One serving of cranberry juice contains 30 grams of sugar. How many kilograms of sugar are in 30 servings of cranberry juice?

1296. One ounce of tofu provided 2 grams of protein. How many milligrams of protein are provided by 5 ounces of tofu?

Convert between the U.S. and the Metric Systems of Measurement

In the following exercises, make the unit conversions. Round to the nearest tenth.

1297. Majid is 69 inches tall. Convert his height to centimeters.

1298. A college basketball court is 84 feet long. Convert this length to meters.

1299. Caroline walked 2.5 kilometers. Convert this length to miles.

1300. Lucas weighs 78 kilograms. Convert his weight to pounds.

1301. Steve's car holds 55 liters of gas. Convert this to gallons.

1302. A box of books weighs 25 pounds. Convert the weight to kilograms.

Convert between Fahrenheit and Celsius Temperatures

In the following exercises, convert the Fahrenheit temperatures to degrees Celsius. Round to the nearest tenth.

1303. 95° Fahrenheit

1304. 23° Fahrenheit

1305. 20° Fahrenheit

1306. 64° Fahrenheit

In the following exercises, convert the Celsius temperatures to degrees Fahrenheit. Round to the nearest tenth.

1307. 30° Celsius

1308. −5° Celsius

1309. −12° Celsius

1310. 24° Celsius

PRACTICE TEST

1311. Write as a whole number using digits: two hundred five thousand, six hundred seventeen.

1312. Find the prime factorization of 504.

1313. Find the Least Common Multiple of 18 and 24.

1314. Combine like terms: $5n + 8 + 2n - 1$.

In the following exercises, evaluate.

1315. $-|x|$ when $x = -2$

1316. $11 - a$ when $a = -3$

1317. Translate to an algebraic expression and simplify: twenty less than negative 7.

1318. Monique has a balance of $-\$18$ in her checking account. She deposits $152 to the account. What is the new balance?

1319. Round 677.1348 to the nearest hundredth.

1320. Convert $\frac{4}{5}$ to a decimal.

1321. Convert 1.85 to a percent.

1322. Locate $\frac{2}{3}$, -1.5, and $\frac{9}{4}$ on a number line.

In the following exercises, simplify each expression.

1323. $4 + 10(3 + 9) - 5^2$

1324. $-85 + 42$

1325. $-19 - 25$

1326. $(-2)^4$

1327. $-5(-9) \div 15$

1328. $\frac{3}{8} \cdot \frac{11}{12}$

1329. $\frac{4}{5} \div \frac{9}{20}$

1330. $\frac{12 + 3 \cdot 5}{15 - 6}$

1331. $\frac{m}{7} + \frac{10}{7}$

1332. $\frac{7}{12} - \frac{3}{8}$

1333. $-5.8 + (-4.7)$

1334. $100 - 64.25$

1335. $(0.07)(31.95)$

1336. $9 \div 0.05$

1337. $-14\left(\frac{5}{7}p\right)$

1338. $(u + 8) - 9$

1339. $6x + (-4y) + 9x + 8y$

1340. $\frac{0}{23}$

1341. $\frac{75}{0}$

1342. $-2(13q - 5)$

1343. A movie lasted $1\frac{2}{3}$ hours. How many minutes did it last? *(1 hour = 60 minutes)*

1344. Mike's SUV is 5 feet 11 inches tall. He wants to put a rooftop cargo bag on the the SUV. The cargo bag is 1 foot 6 inches tall. What will the total height be of the SUV with the cargo bag on the roof? *(1 foot = 12 inches)*

1345. Jennifer ran 2.8 miles. Convert this length to kilometers. *(1 mile = 1.61 kilometers)*

Figure 2.1 The rocks in this formation must remain perfectly balanced around the center for the formation to hold its shape.

Chapter Outline

2.1 Solve Equations Using the Subtraction and Addition Properties of Equality

2.2 Solve Equations using the Division and Multiplication Properties of Equality

2.3 Solve Equations with Variables and Constants on Both Sides

2.4 Use a General Strategy to Solve Linear Equations

2.5 Solve Equations with Fractions or Decimals

2.6 Solve a Formula for a Specific Variable

2.7 Solve Linear Inequalities

Introduction

If we carefully placed more rocks of equal weight on both sides of this formation, it would still balance. Similarly, the expressions in an equation remain balanced when we add the same quantity to both sides of the equation. In this chapter, we will solve equations, remembering that what we do to one side of the equation, we must also do to the other side.

2.1   Solve Equations Using the Subtraction and Addition Properties of Equality

Learning Objectives

By the end of this section, you will be able to:

› Verify a solution of an equation
› Solve equations using the Subtraction and Addition Properties of Equality
› Solve equations that require simplification
› Translate to an equation and solve
› Translate and solve applications

Be Prepared!

Before you get started, take this readiness quiz.

1. Evaluate $x + 4$ when $x = -3$.
   If you missed this problem, review **Example 1.54**.

2. Evaluate $15 - y$ when $y = -5$.
   If you missed this problem, review **Example 1.56**.

3. Simplify $4(4n + 1) - 15n$.
   If you missed this problem, review **Example 1.138**.

4.   Translate into algebra "5 is less than x."
     If you missed this problem, review **Example 1.26**.

Verify a Solution of an Equation

Solving an equation is like discovering the answer to a puzzle. The purpose in solving an equation is to find the value or values of the variable that make each side of the equation the same – so that we end up with a true statement. Any value of the variable that makes the equation true is called a solution to the equation. It is the answer to the puzzle!

Solution of an equation

A **solution of an equation** is a value of a variable that makes a true statement when substituted into the equation.

HOW TO :: TO DETERMINE WHETHER A NUMBER IS A SOLUTION TO AN EQUATION.

Step 1.   Substitute the number in for the variable in the equation.

Step 2.   Simplify the expressions on both sides of the equation.

Step 3.   Determine whether the resulting equation is true (the left side is equal to the right side)

   ◦   If it is true, the number is a solution.

   ◦   If it is not true, the number is not a solution.

EXAMPLE 2.1

Determine whether $x = \frac{3}{2}$ is a solution of $4x - 2 = 2x + 1$.

✓ Solution

Since a solution to an equation is a value of the variable that makes the equation true, begin by substituting the value of the solution for the variable.

	$4x - 2 = 2x + 1$
Substitute $\frac{3}{2}$ for x.	$4\left(\frac{3}{2}\right) - 3 \overset{?}{=} 2\left(\frac{3}{2}\right) + 1$
Multiply.	$6 - 2 \overset{?}{=} 3 + 1$
Subtract.	$4 = 4 \checkmark$

Since $x = \frac{3}{2}$ results in a true equation (4 is in fact equal to 4), $\frac{3}{2}$ is a solution to the equation $4x - 2 = 2x + 1$.

> **TRY IT :: 2.1**   Is $y = \frac{4}{3}$ a solution of $9y + 2 = 6y + 3$?

> **TRY IT :: 2.2**   Is $y = \frac{7}{5}$ a solution of $5y + 3 = 10y - 4$?

Solve Equations Using the Subtraction and Addition Properties of Equality

We are going to use a model to clarify the process of solving an equation. An envelope represents the variable – since its contents are unknown – and each counter represents one. We will set out one envelope and some counters on our workspace, as shown in **Figure 2.2**. Both sides of the workspace have the same number of counters, but some counters are "hidden" in the envelope. Can you tell how many counters are in the envelope?

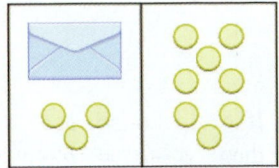

Figure 2.2 The illustration shows a model of an equation with one variable. On the left side of the workspace is an unknown (envelope) and three counters, while on the right side of the workspace are eight counters.

What are you thinking? What steps are you taking in your mind to figure out how many counters are in the envelope?

Perhaps you are thinking: "I need to remove the 3 counters at the bottom left to get the envelope by itself. The 3 counters on the left can be matched with 3 on the right and so I can take them away from both sides. That leaves five on the right—so there must be 5 counters in the envelope." See **Figure 2.3** for an illustration of this process.

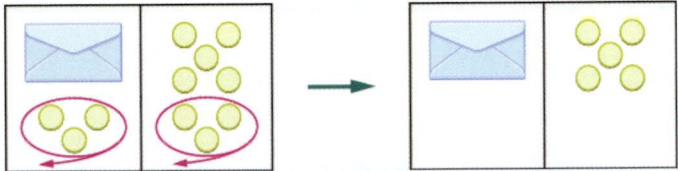

Figure 2.3 The illustration shows a model for solving an equation with one variable. On both sides of the workspace remove three counters, leaving only the unknown (envelope) and five counters on the right side. The unknown is equal to five counters.

What algebraic equation would match this situation? In **Figure 2.4** each side of the workspace represents an expression and the center line takes the place of the equal sign. We will call the contents of the envelope x.

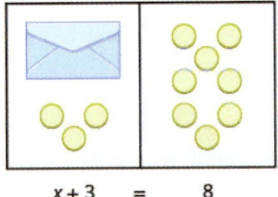

$$x + 3 \quad = \quad 8$$

Figure 2.4 The illustration shows a model for the equation $x + 3 = 8$.

Let's write algebraically the steps we took to discover how many counters were in the envelope:

$$x + 3 = 8$$

First, we took away three from each side. $\qquad x + 3 - 3 = 8 - 3$

Then we were left with five. $\qquad x = 5$

Check:

Five in the envelope plus three more does equal eight!

$$5 + 3 = 8$$

Our model has given us an idea of what we need to do to solve one kind of equation. The goal is to isolate the variable by itself on one side of the equation. To solve equations such as these mathematically, we use the **Subtraction Property of Equality**.

Subtraction Property of Equality

For any numbers a, b, and c,

$$\text{If} \quad a = b,$$
$$\text{then} \quad a - c = b - c$$

When you subtract the same quantity from both sides of an equation, you still have equality.

 MANIPULATIVE MATHEMATICS

Doing the Manipulative Mathematics activity "Subtraction Property of Equality" will help you develop a better understanding of how to solve equations by using the Subtraction Property of Equality.

Let's see how to use this property to solve an equation. Remember, the goal is to isolate the variable on one side of the equation. And we check our solutions by substituting the value into the equation to make sure we have a true statement.

EXAMPLE 2.2

Solve: $y + 37 = -13$.

⊘ Solution

To get y by itself, we will undo the addition of 37 by using the Subtraction Property of Equality.

$$y + 37 = -13$$

Subtract 37 from each side to 'undo' the addition.	$y + 37 - 37 = -13 - 37$
Simplify.	$y = -50$
Check:	$y + 37 = -13$
Substitute $y = -50$	$-50 + 37 = -13$
	$-13 \overset{?}{=} -13 \checkmark$

Since $y = -50$ makes $y + 37 = -13$ a true statement, we have the solution to this equation.

> **TRY IT :: 2.3**   Solve: $x + 19 = -27$.

> **TRY IT :: 2.4**   Solve: $x + 16 = -34$.

What happens when an equation has a number subtracted from the variable, as in the equation $x - 5 = 8$? We use another property of equations to solve equations where a number is subtracted from the variable. We want to isolate the variable, so to 'undo' the subtraction we will add the number to both sides. We use the **Addition Property of Equality**.

Addition Property of Equality

For any numbers a, b, and c,

$$\text{If} \quad a = b,$$
$$\text{then} \quad a + c = b + c$$

When you add the same quantity to both sides of an equation, you still have equality.

In Example 2.2, 37 was added to the y and so we subtracted 37 to 'undo' the addition. In Example 2.3, we will need to 'undo' subtraction by using the Addition Property of Equality.

EXAMPLE 2.3

Solve: $a - 28 = -37$.

✓ Solution

$$a - 28 = -37$$

Add 28 to each side to 'undo' the subtraction.	$a - 28 + 28 = -37 + 28$
Simplify.	$a = -9$
Check:	$a - 28 = -37$
Substitute $a = -9$	$-9 - 28 = -37$
	$-37 \overset{?}{=} -37 \checkmark$
	The solution to $a - 28 = -37$ is $a = -9$.

> **TRY IT :: 2.5**   Solve: $n - 61 = -75$.

> **TRY IT :: 2.6**   Solve: $p - 41 = -73$.

EXAMPLE 2.4

Solve: $x - \frac{5}{8} = \frac{3}{4}$.

✓ Solution

$$x - \frac{5}{8} = \frac{3}{4}$$

Use the Addition Property of Equality.	$x - \frac{5}{8} + \frac{5}{8} = \frac{3}{4} + \frac{5}{8}$
Find the LCD to add the fractions on the right.	$x - \frac{5}{8} + \frac{5}{8} = \frac{6}{8} + \frac{5}{8}$
Simplify.	$x = \frac{11}{8}$
Check:	$x - \frac{5}{8} = \frac{3}{4}$
Substitute $x = \frac{11}{8}$.	$\frac{11}{8} - \frac{5}{8} \overset{?}{=} \frac{3}{4}$
Subtract.	$\frac{6}{8} \overset{?}{=} \frac{3}{4}$
Simplify.	$\frac{3}{4} = \frac{3}{4} \checkmark$
	The solution to $x - \frac{5}{8} = \frac{3}{4}$ is $x = \frac{11}{8}$.

> **TRY IT :: 2.7**   Solve: $p - \frac{2}{3} = \frac{5}{6}$.

> **TRY IT :: 2.8**   Solve: $q - \frac{1}{2} = \frac{5}{6}$.

The next example will be an equation with decimals.

EXAMPLE 2.5

Solve: $n - 0.63 = -4.2$.

✓ **Solution**

	$n - 0.63 = -4.2$
Use the Addition Property of Equality.	$n - 0.63 + 0.63 = -4.2 + 0.63$
Add.	$n = -3.57$
Check:	$n = -3.57$
Let $n = -3.57$.	$-3.57 - 0.63 \overset{?}{=} -4.2$
	$-4.2 = -4.2 ✓$

> **TRY IT :: 2.9**   Solve: $b - 0.47 = -2.1$.

> **TRY IT :: 2.10**   Solve: $c - 0.93 = -4.6$.

Solve Equations That Require Simplification

In the previous examples, we were able to isolate the variable with just one operation. Most of the equations we encounter in algebra will take more steps to solve. Usually, we will need to simplify one or both sides of an equation before using the Subtraction or Addition Properties of Equality.

You should always simplify as much as possible before you try to isolate the variable. Remember that to simplify an expression means to do all the operations in the expression. Simplify one side of the equation at a time. Note that simplification is different from the process used to solve an equation in which we apply an operation to both sides.

EXAMPLE 2.6   HOW TO SOLVE EQUATIONS THAT REQUIRE SIMPLIFICATION

Solve: $9x - 5 - 8x - 6 = 7$.

✓ **Solution**

Step 1. Simplify the expressions on each side as much as possible.	Rearrange the terms, using the Commutative Property of Addition. Combine like terms. Notice that each side is now simplified as much as possible.	$9x - 5 - 8x - 6 = 7$ $9x - 8x - 5 - 6 = 7$ $x - 11 = 7$
Step 2. Isolate the variable.	Now isolate x. Undo subtraction by adding 11 to both sides.	$x - 11 + 11 = 7 + 11$
Step 3. Simplify the expressions on both sides of the equation.		$x = 18$

Step 4. Check the solution.		**Check:** Substitute $x = 18$.
		$9x - 5 - 8x - 6 = 7$
		$9(18) - 5 - 8(18) - 6 \overset{?}{=} 7$
		$162 - 5 - 144 - 6 \overset{?}{=} 7$
		$157 - 144 - 6 \overset{?}{=} 7$
		$13 - 6 \overset{?}{=} 7$
		$7 = 7 \checkmark$
		The solution to $9x - 5 - 8x - 6 = 7$ is $x = 18$.

> **TRY IT :: 2.11**   Solve: $8y - 4 - 7y - 7 = 4$.

> **TRY IT :: 2.12**   Solve: $6z + 5 - 5z - 4 = 3$.

EXAMPLE 2.7

Solve: $5(n - 4) - 4n = -8$.

⊘ **Solution**

We simplify both sides of the equation as much as possible before we try to isolate the variable.

	$5(n - 4) - 4n = -8$
Distribute on the left.	$5n - 20 - 4n = -8$
Use the Commutative Property to rearrange terms.	$5n - 4n - 20 = -8$
Combine like terms.	$n - 20 = -8$
Each side is as simplified as possible. Next, isolate n.	
Undo subtraction by using the Addition Property of Equality.	$n - 20 + 20 = -8 + 20$
Add.	$n = 12$

Check. Substitute $n = 12$.

$$5(n - 4) - 4n = -8$$
$$5(12 - 4) - 4(12) \overset{?}{=} -8z$$
$$5(8) - 48 \overset{?}{=} -8$$
$$40 - 48 \overset{?}{=} -8$$
$$-8 = -8 \checkmark$$

The solution to $5(n - 4) - 4n = -8$ is $n = 12$.

> **TRY IT :: 2.13**   Solve: $5(p - 3) - 4p = -10$.

> **TRY IT :: 2.14**   Solve: $4(q + 2) - 3q = -8$.

EXAMPLE 2.8

Solve: $3(2y - 1) - 5y = 2(y + 1) - 2(y + 3)$.

⊘ Solution

We simplify both sides of the equation before we isolate the variable.

$$3(2y-1) - 5y = 2(y+1) - 2(y+3)$$

Distribute on both sides.	$6y - 3 - 5y = 2y + 2 - 2y - 6$
Use the Commutative Property of Addition.	$6y - 5y - 3 = 2y - 2y + 2 - 6$
Combine like terms.	$y - 3 = -4$
Each side is as simplified as possible. Next, isolate y.	
Undo subtraction by using the Addition Property of Equality.	$y - 3 + 3 = -4 + 3$
Add.	$y = -1$

Check. Let $y = -1$.

$$3(2y - 1) - 5y = 2(y + 1) - 2(y + 3)$$

$$3(2(-1) - 1) - 5(-1) \stackrel{?}{=} 2(-1 + 1) - 2(-1 + 3)$$

$$3(-2 - 1) + 5 \stackrel{?}{=} 2(0) - 2(2)$$

$$3(-3) + 5 \stackrel{?}{=} -4$$

$$-9 + 5 \stackrel{?}{=} -4$$

$$-4 = -4 \checkmark$$

The solution to $3(2y - 1) - 5y = 2(y + 1) - 2(y + 3)$ is $y = -1$.

> **TRY IT :: 2.15**   Solve: $4(2h - 3) - 7h = 6(h - 2) - 6(h - 1)$.

> **TRY IT :: 2.16**   Solve: $2(5x + 2) - 9x = 3(x - 2) - 3(x - 4)$.

Translate to an Equation and Solve

To solve applications algebraically, we will begin by translating from English sentences into equations. Our first step is to look for the word (or words) that would translate to the equals sign. Table 2.8 shows us some of the words that are commonly used.

Equals =
is
is equal to
is the same as
the result is
gives
was
will be

Table 2.8

The steps we use to translate a sentence into an equation are listed below.

HOW TO :: TRANSLATE AN ENGLISH SENTENCE TO AN ALGEBRAIC EQUATION.

Step 1.   Locate the "equals" word(s). Translate to an equals sign (=).

Step 2.   Translate the words to the left of the "equals" word(s) into an algebraic expression.

Step 3.   Translate the words to the right of the "equals" word(s) into an algebraic expression.

EXAMPLE 2.9

Translate and solve: Eleven more than x is equal to 54.

⊘ **Solution**

	Eleven more than x	is equal to	54
Translate.	$x + 11$	$=$	54
Subtract 11 from both sides.	$x + 11 - 11$	$=$	$54 - 11$
Simplify.	x	$=$	43

Check: Is 54 eleven more than 43?

$$43 + 11 \overset{?}{=} 54$$
$$54 = 54 \checkmark$$

> **TRY IT :: 2.17**   Translate and solve: Ten more than x is equal to 41.

> **TRY IT :: 2.18**   Translate and solve: Twelve less than x is equal to 51.

EXAMPLE 2.10

Translate and solve: The difference of $12t$ and $11t$ is -14.

⊘ **Solution**

	The difference of $12t$ and $11t$ is -14
Translate.	$12t - 11t = -14$
Simplify.	$t = -14$

Check:

$$12(-14) - 11(-14) \overset{?}{=} -14$$

$$-168 + 154 \overset{?}{=} -14$$

$$-14 = -14 \checkmark$$

> **TRY IT : : 2.19**   Translate and solve: The difference of $4x$ and $3x$ is 14.

> **TRY IT : : 2.20**   Translate and solve: The difference of $7a$ and $6a$ is -8.

Translate and Solve Applications

Most of the time a question that requires an algebraic solution comes out of a real life question. To begin with that question is asked in English (or the language of the person asking) and not in math symbols. Because of this, it is an important skill to be able to translate an everyday situation into algebraic language.

We will start by restating the problem in just one sentence, assign a variable, and then translate the sentence into an equation to solve. When assigning a variable, choose a letter that reminds you of what you are looking for. For example, you might use q for the number of quarters if you were solving a problem about coins.

| EXAMPLE 2.11 | HOW TO SOLVE TRANSLATE AND SOLVE APPLICATIONS |

The MacIntyre family recycled newspapers for two months. The two months of newspapers weighed a total of 57 pounds. The second month, the newspapers weighed 28 pounds. How much did the newspapers weigh the first month?

⊘ **Solution**

Step 1. Read the problem. Make sure all the words and ideas are understood.	The problem is about the weight of newspapers.	
Step 2. Identify what we are asked to find.	What are we asked to find?	"How much did the newspapers weigh the 2ⁿᵈ month?"
Step 3. Name what we are looking for. Choose a variable to represent that quantity.	Choose a variable.	Let w = weight of the newspapers the 1ˢᵗ month
Step 4. Translate into an equation. It may be helpful to restate the problem in one sentence with the important information.	Restate the problem. We know the weight of the newspapers the second month is 28 pounds. Translate into an equation, using the variable w.	Weight of newspapers the 1ˢᵗ month plus the weight of the newspapers the 2ⁿᵈ month equals 57 pounds. Weight from 1ˢᵗ month plus 28 equals 57. $w + 28 = 57$
Step 5. Solve the equation using good algebra techniques.	Solve.	$w + 28 - 28 = 57 - 28$ $w = 29$
Step 6. Check the answer in the problem and make sure it makes sense.	Does 1ˢᵗ month's weight plus 2ⁿᵈ month's weight equal 57 pounds?	**Check:** Does 1ˢᵗ month's weight plus 2ⁿᵈ month's weight equal 57 pounds? $29 + 28 \overset{?}{=} 57$ $57 = 57 \checkmark$

Step 7. Answer the question with a complete sentence.	Write a sentence to answer "How much did the newspapers weigh the 2nd month?"	The 2nd month the newspapers weighed 29 pounds.

> **TRY IT : : 2.21**

Translate into an algebraic equation and solve:

The Pappas family has two cats, Zeus and Athena. Together, they weigh 23 pounds. Zeus weighs 16 pounds. How much does Athena weigh?

> **TRY IT : : 2.22**

Translate into an algebraic equation and solve:

Sam and Henry are roommates. Together, they have 68 books. Sam has 26 books. How many books does Henry have?

HOW TO : : SOLVE AN APPLICATION.

Step 1.   **Read** the problem. Make sure all the words and ideas are understood.

Step 2.   **Identify** what we are looking for.

Step 3.   **Name** what we are looking for. Choose a variable to represent that quantity.

Step 4.   **Translate** into an equation. It may be helpful to restate the problem in one sentence with the important information.

Step 5.   **Solve** the equation using good algebra techniques.

Step 6.   **Check** the answer in the problem and make sure it makes sense.

Step 7.   **Answer** the question with a complete sentence.

EXAMPLE 2.12

Randell paid $28,675 for his new car. This was $875 less than the sticker price. What was the sticker price of the car?

⊘ **Solution**

Step 1. Read the problem.

Step 2. Identify what we are looking for.   "What was the sticker price of the car?"

Step 3. Name what we are looking for.
Choose a variable to represent that quantity.   Let s = the sticker price of the car.

Step 4. Translate into an equation. Restate
the problem in one sentence. $28,675 is $875 less than the sticker price

                                                       $28,675 is $875 less than s
                                                       $$28,675 = s - 875$$
Step 5. Solve the equation. $$28,675 + 875 = s - 875 + 875$$
                                                       $$29,550 = s$$

Step 6. Check the answer.
Is $875 less than $29,550 equal to $28,675?
$$29,550 - 875 \overset{?}{=} 28,675$$
$$28,675 = 28,675 \checkmark$$

Step 7. Answer the question with
a complete sentence.   The sticker price of the car was $29,550.

> **TRY IT :: 2.23**

Translate into an algebraic equation and solve:

Eddie paid $19,875 for his new car. This was $1,025 less than the sticker price. What was the sticker price of the car?

> **TRY IT :: 2.24**

Translate into an algebraic equation and solve:

The admission price for the movies during the day is $7.75. This is $3.25 less the price at night. How much does the movie cost at night?

2.1 EXERCISES
Practice Makes Perfect

Verify a Solution of an Equation
In the following exercises, determine whether the given value is a solution to the equation.

1. Is $y = \frac{5}{3}$ a solution of

$6y + 10 = 12y$?

2. Is $x = \frac{9}{4}$ a solution of

$4x + 9 = 8x$?

3. Is $u = -\frac{1}{2}$ a solution of

$8u - 1 = 6u$?

4. Is $v = -\frac{1}{3}$ a solution of

$9v - 2 = 3v$?

Solve Equations using the Subtraction and Addition Properties of Equality
In the following exercises, solve each equation using the Subtraction and Addition Properties of Equality.

5. $x + 24 = 35$

6. $x + 17 = 22$

7. $y + 45 = -66$

8. $y + 39 = -83$

9. $b + \frac{1}{4} = \frac{3}{4}$

10. $a + \frac{2}{5} = \frac{4}{5}$

11. $p + 2.4 = -9.3$

12. $m + 7.9 = 11.6$

13. $a - 45 = 76$

14. $a - 30 = 57$

15. $m - 18 = -200$

16. $m - 12 = -12$

17. $x - \frac{1}{3} = 2$

18. $x - \frac{1}{5} = 4$

19. $y - 3.8 = 10$

20. $y - 7.2 = 5$

21. $x - 165 = -420$

22. $z - 101 = -314$

23. $z + 0.52 = -8.5$

24. $x + 0.93 = -4.1$

25. $q + \frac{3}{4} = \frac{1}{2}$

26. $p + \frac{1}{3} = \frac{5}{6}$

27. $p - \frac{2}{5} = \frac{2}{3}$

28. $y - \frac{3}{4} = \frac{3}{5}$

Solve Equations that Require Simplification
In the following exercises, solve each equation.

29. $c + 31 - 10 = 46$

30. $m + 16 - 28 = 5$

31. $9x + 5 - 8x + 14 = 20$

32. $6x + 8 - 5x + 16 = 32$

33. $-6x - 11 + 7x - 5 = -16$

34. $-8n - 17 + 9n - 4 = -41$

35. $5(y - 6) - 4y = -6$

36. $9(y - 2) - 8y = -16$

37. $8(u + 1.5) - 7u = 4.9$

38. $5(w + 2.2) - 4w = 9.3$

39. $6a - 5(a - 2) + 9 = -11$

40. $8c - 7(c - 3) + 4 = -16$

41. $6(y - 2) - 5y = 4(y + 3)$
$-4(y - 1)$

42. $9(x - 1) - 8x = -3(x + 5)$
$+3(x - 5)$

43. $3(5n - 1) - 14n + 9$
$= 10(n - 4) - 6n - 4(n + 1)$

44. $2(8m + 3) - 15m - 4$
$= 9(m + 6) - 2(m - 1) - 7m$

45. $-(j + 2) + 2j - 1 = 5$

46. $-(k + 7) + 2k + 8 = 7$

47. $-\left(\frac{1}{4}a - \frac{3}{4}\right) + \frac{5}{4}a = -2$ **48.** $-\left(\frac{2}{3}d - \frac{1}{3}\right) + \frac{5}{3}d = -4$ **49.** $8(4x + 5) - 5(6x) - x$
$= 53 - 6(x + 1) + 3(2x + 2)$

50. $6(9y - 1) - 10(5y) - 3y$
$= 22 - 4(2y - 12) + 8(y - 6)$

Translate to an Equation and Solve

In the following exercises, translate to an equation and then solve it.

51. Nine more than x is equal to 52.

52. The sum of x and -15 is 23.

53. Ten less than m is -14.

54. Three less than y is -19.

55. The sum of y and -30 is 40.

56. Twelve more than p is equal to 67.

57. The difference of $9x$ and $8x$ is 107.

58. The difference of $5c$ and $4c$ is 602.

59. The difference of n and $\frac{1}{6}$ is $\frac{1}{2}$.

60. The difference of f and $\frac{1}{3}$ is $\frac{1}{12}$.

61. The sum of $-4n$ and $5n$ is -82.

62. The sum of $-9m$ and $10m$ is -95.

Translate and Solve Applications

In the following exercises, translate into an equation and solve.

63. Distance Avril rode her bike a total of 18 miles, from home to the library and then to the beach. The distance from Avril's house to the library is 7 miles. What is the distance from the library to the beach?

64. Reading Jeff read a total of 54 pages in his History and Sociology textbooks. He read 41 pages in his History textbook. How many pages did he read in his Sociology textbook?

65. Age Eva's daughter is 15 years younger than her son. Eva's son is 22 years old. How old is her daughter?

66. Age Pablo's father is 3 years older than his mother. Pablo's mother is 42 years old. How old is his father?

67. Groceries For a family birthday dinner, Celeste bought a turkey that weighed 5 pounds less than the one she bought for Thanksgiving. The birthday turkey weighed 16 pounds. How much did the Thanksgiving turkey weigh?

68. Weight Allie weighs 8 pounds less than her twin sister Lorrie. Allie weighs 124 pounds. How much does Lorrie weigh?

69. Health Connor's temperature was 0.7 degrees higher this morning than it had been last night. His temperature this morning was 101.2 degrees. What was his temperature last night?

70. Health The nurse reported that Tricia's daughter had gained 4.2 pounds since her last checkup and now weighs 31.6 pounds. How much did Tricia's daughter weigh at her last checkup?

71. Salary Ron's paycheck this week was $17.43 less than his paycheck last week. His paycheck this week was $103.76. How much was Ron's paycheck last week?

72. Textbooks Melissa's math book cost $22.85 less than her art book cost. Her math book cost $93.75. How much did her art book cost?

Everyday Math

73. Construction Miguel wants to drill a hole for a $\frac{5}{8}$ inch screw. The hole should be $\frac{1}{12}$ inch smaller than the screw. Let d equal the size of the hole he should drill. Solve the equation $d - \frac{1}{12} = \frac{5}{8}$ to see what size the hole should be.

74. Baking Kelsey needs $\frac{2}{3}$ cup of sugar for the cookie recipe she wants to make. She only has $\frac{3}{8}$ cup of sugar and will borrow the rest from her neighbor. Let s equal the amount of sugar she will borrow. Solve the equation $\frac{3}{8} + s = \frac{2}{3}$ to find the amount of sugar she should ask to borrow.

Writing Exercises

75. Is -8 a solution to the equation $3x = 16 - 5x$? How do you know?

76. What is the first step in your solution to the equation $10x + 2 = 4x + 26$?

Self Check

ⓐ *After completing the exercises, use this checklist to evaluate your mastery of the objectives of this section.*

I can...	Confidently	With some help	No-I don't get it!
verify a solution of an equation.			
solve equations using the subtraction and addition properties of equality.			
solve equations that require simplification.			
translate to an equation and solve.			
translate and solve applications.			

ⓑ *If most of your checks were:*

...confidently. *Congratulations! You have achieved your goals in this section! Reflect on the study skills you used so that you can continue to use them. What did you do to become confident of your ability to do these things? Be specific!*

...with some help. *This must be addressed quickly as topics you do not master become potholes in your road to success. Math is sequential - every topic builds upon previous work. It is important to make sure you have a strong foundation before you move on. Who can you ask for help? Your fellow classmates and instructor are good resources. Is there a place on campus where math tutors are available? Can your study skills be improved?*

...no - I don't get it! *This is critical and you must not ignore it. You need to get help immediately or you will quickly be overwhelmed. See your instructor as soon as possible to discuss your situation. Together you can come up with a plan to get you the help you need.*

2.2 | Solve Equations using the Division and Multiplication Properties of Equality

Learning Objectives

By the end of this section, you will be able to:

> Solve equations using the Division and Multiplication Properties of Equality
> Solve equations that require simplification
> Translate to an equation and solve
> Translate and solve applications

Be Prepared!

Before you get started, take this readiness quiz.

1. Simplify: $-7\left(\frac{1}{-7}\right)$.
   If you missed this problem, review **Example 1.68**.
2. Evaluate $9x + 2$ when $x = -3$.
   If you missed this problem, review **Example 1.57**.

Solve Equations Using the Division and Multiplication Properties of Equality

You may have noticed that all of the equations we have solved so far have been of the form $x + a = b$ or $x - a = b$. We were able to isolate the variable by adding or subtracting the constant term on the side of the equation with the variable. Now we will see how to solve equations that have a variable multiplied by a constant and so will require division to isolate the variable.

Let's look at our puzzle again with the envelopes and counters in **Figure 2.5**.

Figure 2.5 The illustration shows a model of an equation with one variable multiplied by a constant. On the left side of the workspace are two instances of the unknown (envelope), while on the right side of the workspace are six counters.

In the illustration there are two identical envelopes that contain the same number of counters. Remember, the left side of the workspace must equal the right side, but the counters on the left side are "hidden" in the envelopes. So how many counters are in each envelope?

How do we determine the number? We have to separate the counters on the right side into two groups of the same size to correspond with the two envelopes on the left side. The 6 counters divided into 2 equal groups gives 3 counters in each group (since $6 \div 2 = 3$).

What equation models the situation shown in **Figure 2.6**? There are two envelopes, and each contains x counters. Together, the two envelopes must contain a total of 6 counters.

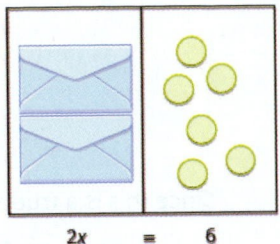

Figure 2.6 The illustration shows a model of the equation $2x = 6$.

	$2x = 6$
If we divide both sides of the equation by 2, as we did with the envelopes and counters,	$\dfrac{2x}{2} = \dfrac{6}{2}$
we get:	$x = 3$

We found that each envelope contains 3 counters. Does this check? We know $2 \cdot 3 = 6$, so it works! Three counters in each of two envelopes does equal six!

This example leads to the **Division Property of Equality**.

The Division Property of Equality

For any numbers a, b, and c, and $c \neq 0$,

$$\text{If} \quad a = b,$$
$$\text{then} \quad \frac{a}{c} = \frac{b}{c}$$

When you divide both sides of an equation by any non-zero number, you still have equality.

 MANIPULATIVE MATHEMATICS

Doing the Manipulative Mathematics activity " Division Property of Equality" will help you develop a better understanding of how to solve equations by using the Division Property of Equality.

The goal in solving an equation is to 'undo' the operation on the variable. In the next example, the variable is multiplied by 5, so we will divide both sides by 5 to 'undo' the multiplication.

EXAMPLE 2.13

Solve: $5x = -27$.

✓ **Solution**

To isolate x, "undo" the multiplication by 5.	$5x = -27$
Divide to 'undo' the multiplication.	$\dfrac{5x}{5} = \dfrac{-27}{5}$
Simplify.	$x = -\dfrac{27}{5}$
Check:	$5x = -27$

Substitute $-\frac{27}{5}$ for x. $5\left(-\frac{27}{5}\right) \overset{?}{=} -27$

$$-27 = -27 \checkmark$$

Since this is a true statement, $x = -\frac{27}{5}$
is the solution to $5x = -27$.

> **TRY IT :: 2.25**   Solve: $3y = -41$.

> **TRY IT :: 2.26**   Solve: $4z = -55$.

Consider the equation $\frac{x}{4} = 3$. We want to know what number divided by 4 gives 3. So to "undo" the division, we will need
to multiply by 4. The **Multiplication Property of Equality** will allow us to do this. This property says that if we start with
two equal quantities and multiply both by the same number, the results are equal.

The Multiplication Property of Equality

For any numbers a, b, and c,

$$\text{If} \quad a = b,$$
$$\text{then} \quad ac = bc$$

If you multiply both sides of an equation by the same number, you still have equality.

EXAMPLE 2.14

Solve: $\frac{y}{-7} = -14$.

⊘ **Solution**

Here y is divided by -7. We must multiply by -7 to isolate y.

	$\frac{y}{-7} = -14$
Multiply both sides by -7.	$-7\left(\frac{y}{-7}\right) = -7(-14)$
Multiply.	$\frac{-7y}{7} = 98$
Simplify.	$y = 98$
Check: $\frac{y}{-7} = -14$	
Substitute $y = 98$.	$\frac{98}{-7} \overset{?}{=} -14$
Divide.	$-14 = -14 \checkmark$

> **TRY IT :: 2.27**   Solve: $\frac{a}{-7} = -42$.

> **TRY IT :: 2.28**   Solve: $\frac{b}{-6} = -24$.

EXAMPLE 2.15

Solve: $-n = 9$.

✓ Solution

		$-n = 9$
Remember $-n$ is equivalent to $-1n$.		$-1n = 9$
Divide both sides by -1.		$\dfrac{-1n}{-1} = \dfrac{9}{-1}$
Divide.		$n = -9$

Notice that there are two other ways to solve $-n = 9$. We can also solve this equation by multiplying both sides by -1 and also by taking the opposite of both sides.

Check:		$-n = 9$
Substitute $n = -9$.		$-(-9) \overset{?}{=} 9$
Simplify.		$9 = 9 \checkmark$

> **TRY IT :: 2.29**   Solve: $-k = 8$.

> **TRY IT :: 2.30**   Solve: $-g = 3$.

EXAMPLE 2.16

Solve: $\dfrac{3}{4}x = 12$.

✓ Solution

Since the product of a number and its reciprocal is 1, our strategy will be to isolate x by multiplying by the reciprocal of $\dfrac{3}{4}$.

		$\dfrac{3}{4}x = 12$
Multiply by the reciprocal of $\dfrac{3}{4}$.		$\dfrac{4}{3} \cdot \dfrac{3}{4}x = \dfrac{4}{3} \cdot 12$
Reciprocals multiply to 1.		$1x = \dfrac{4}{3} \cdot \dfrac{12}{1}$
Multiply.		$x = 16$

Notice that we could have divided both sides of the equation $\dfrac{3}{4}x = 12$ by $\dfrac{3}{4}$ to isolate x. While this would work, most people would find multiplying by the reciprocal easier.

Check:		$\dfrac{3}{4}x = 12$
Substitute $x = 16$.		$\dfrac{3}{4} \cdot 16 \overset{?}{=} 12$
		$12 = 12 \checkmark$

> **TRY IT :: 2.31**   Solve: $\frac{2}{5}n = 14$.

> **TRY IT :: 2.32**   Solve: $\frac{5}{6}y = 15$.

In the next example, all the variable terms are on the right side of the equation. As always, our goal in solving the equation is to isolate the variable.

EXAMPLE 2.17

Solve: $\frac{8}{15} = -\frac{4}{5}x$.

⊘ **Solution**

$$\frac{8}{15} = -\frac{4}{5}x$$

Multiply by the reciprocal of $-\frac{4}{5}$.	$\left(-\frac{5}{4}\right)\left(\frac{8}{15}\right) = \left(-\frac{5}{4}\right)\left(-\frac{4}{5}x\right)$
Reciprocals multiply to 1.	$-\frac{5 \cdot 4 \cdot 2}{4 \cdot 3 \cdot 5} = 1x$
Multiply.	$-\frac{2}{3} = x$
Check:	$\frac{8}{15} = -\frac{4}{5}x$
Let $x = -\frac{2}{3}$.	$\frac{8}{15} = -\frac{4}{5}\left(-\frac{2}{3}\right)$
	$\frac{8}{15} = \frac{8}{15}$ ✓

> **TRY IT :: 2.33**   Solve: $\frac{9}{25} = -\frac{4}{5}z$.

> **TRY IT :: 2.34**   Solve: $\frac{5}{6} = -\frac{8}{3}r$.

Solve Equations That Require Simplification

Many equations start out more complicated than the ones we have been working with.

With these more complicated equations the first step is to simplify both sides of the equation as much as possible. This usually involves combining like terms or using the distributive property.

EXAMPLE 2.18

Solve: $14 - 23 = 12y - 4y - 5y$.

⊘ **Solution**

Begin by simplifying each side of the equation.

	$14 - 23 \stackrel{?}{=} -36 + 12 + 15$
Simplify each side.	$-9 = 3y$
Divide both sides by 3 to isolate y.	$\dfrac{-9}{3} = \dfrac{3y}{3}$
Divide.	$-3 = y$
Check:	$14 - 23 = 12y - 4y - 5y$
Substitute $y = -3$.	$14 - 23 \stackrel{?}{=} 12(-3) - 4(-3) - 5(-3)$
	$14 - 23 \stackrel{?}{=} -36 + 12 + 15$
	$-9 = -9 ✓$

> **TRY IT : : 2.35**   Solve: $18 - 27 = 15c - 9c - 3c$.

> **TRY IT : : 2.36**   Solve: $18 - 22 = 12x - x - 4x$.

EXAMPLE 2.19

Solve: $-4(a - 3) - 7 = 25$.

⊘ Solution

Here we will simplify each side of the equation by using the distributive property first.

	$-4(a - 3) - 7 = 25$
Distribute.	$-4a + 12 - 7 = 25$
Simplify.	$-4a + 5 = 25$
Simplify.	$-4a = 20$
Divide both sides by -4 to isolate a.	$\dfrac{-4a}{-4} = \dfrac{20}{-4}$
Divide.	$a = -5$
Check:	$-4(a - 3) - 7 = 25$
Substitute $a = -5$.	$-4(-5 - 3) - 7 \stackrel{?}{=} 25$
	$-4(-8) - 7 \stackrel{?}{=} 25$
	$32 - 7 \stackrel{?}{=} 25$
	$25 = 25 ✓$

> **TRY IT : : 2.37**   Solve: $-4(q - 2) - 8 = 24$.

> **TRY IT : : 2.38**   Solve: $-6(r - 2) - 12 = 30$.

Now we have covered all four properties of equality—subtraction, addition, division, and multiplication. We'll list them all together here for easy reference.

Properties of Equality

Subtraction Property of Equality	Addition Property of Equality
For any real numbers a, b, and c,	For any real numbers a, b, and c,
if $\quad a = b,$	if $\quad a = b,$
then $a - c = b - c.$	then $a + c = b + c.$

Division Property of Equality	Multiplication Property of Equality
For any numbers a, b, and c, and $c \neq 0$,	For any numbers a, b, and c,
if $\quad a = b,$	if $\quad a = b,$
then $\frac{a}{c} = \frac{b}{c}.$	then $ac = bc.$

When you add, subtract, multiply, or divide the same quantity from both sides of an equation, you still have equality.

Translate to an Equation and Solve

In the next few examples, we will translate sentences into equations and then solve the equations. You might want to review the translation table in the previous chapter.

EXAMPLE 2.20

Translate and solve: The number 143 is the product of -11 and y.

✓ Solution

Begin by translating the sentence into an equation.

	The number 143 is the product of -11 and y.
Translate.	$143 = -11y$
Divide by -11.	$\dfrac{143}{-11} = \dfrac{-11y}{-11}$
Simplify.	$-13 = y$

Check:

$$143 = -11y$$
$$143 \overset{?}{=} -11(-13)$$
$$143 = 143 \checkmark$$

> **TRY IT :: 2.39**   Translate and solve: The number 132 is the product of *−12* and *y*.

> **TRY IT :: 2.40**   Translate and solve: The number 117 is the product of *−13* and *z*.

EXAMPLE 2.21

Translate and solve: n divided by 8 is -32.

Solution

Begin by translating the sentence into an equation. Translate.	n divided by 8 is -32. $\dfrac{n}{8} = -32$
Multiple both sides by 8.	$8 \cdot \dfrac{n}{8} = 8(-32)$
Simplify.	$n = -256$
Check:	Is n divided by 8 equal to -32?
Let $n = -256$.	Is -256 divided by 8 equal to -32?
Translate.	$\dfrac{-256}{8} \overset{?}{=} -32$
Simplify.	$-32 = -32$ ✓

> **TRY IT :: 2.41**   Translate and solve: n divided by 7 is equal to -21.

> **TRY IT :: 2.42**   Translate and solve: n divided by 8 is equal to -56.

EXAMPLE 2.22

Translate and solve: The quotient of y and -4 is 68.

Solution

Begin by translating the sentence into an equation.

Translate.	The quotient of y and -4 is 68. $\dfrac{y}{-4} = 68$
Multiply both sides by -4.	$-4\left(\dfrac{y}{-4}\right) = -4(68)$
Simplify.	$y = -272$
Check:	Is the quotient of y and -4 equal to 68?
Let $y = -272$.	Is the quotient of -272 and -4 equal to 68?
Translate.	$\dfrac{-272}{-4} \overset{?}{=} 68$
Simplify.	$68 = 68$ ✓

> **TRY IT :: 2.43**   Translate and solve: The quotient of q and -8 is 72.

> **TRY IT :: 2.44**   Translate and solve: The quotient of p and -9 is 81.

EXAMPLE 2.23

Translate and solve: Three-fourths of p is 18.

✓ Solution

Begin by translating the sentence into an equation. Remember, "of" translates into multiplication.

	Three-fourths of p is 18.
Translate.	$\frac{3}{4}p = 18$
Multiply both sides by $\frac{4}{3}$.	$\frac{4}{3} \cdot \frac{3}{4}p = \frac{4}{3} \cdot 18$
Simplify.	$p = 24$
Check:	Is three-fourths of p equal to 18?
Let $p = 24$.	Is three-fourths of 24 equal to 18?
Translate.	$\frac{3}{4} \cdot 24 \overset{?}{=} 18$
Simplify.	$18 = 18 \checkmark$

> **TRY IT : : 2.45**   Translate and solve: Two-fifths of f is 16.

> **TRY IT : : 2.46**   Translate and solve: Three-fourths of f is 21.

EXAMPLE 2.24

Translate and solve: The sum of three-eighths and x is one-half.

✓ Solution

Begin by translating the sentence into an equation.

	The sum of three – eighths and x is $\frac{1}{2}$
Translate.	$\frac{3}{8} + x = \frac{1}{2}$
Subtract $\frac{3}{8}$ from each side.	$\frac{3}{8} - \frac{3}{8} + x = \frac{1}{2} - \frac{3}{8}$
Simplify and rewrite fractions with common denominators.	$x = \frac{4}{8} - \frac{3}{8}$
Simplify.	$x = \frac{1}{8}$
Check:	Is the sum of three-eighths and x equal to one-half?
Let $x = \frac{1}{8}$.	Is the sum of three-eighths and one-eighth equal to one-half?
Translate.	$\frac{3}{8} + \frac{1}{8} \overset{?}{=} \frac{1}{2}$

Simplify.	$\frac{4}{8} \overset{?}{=} \frac{1}{2}$
Simplify.	$\frac{1}{2} = \frac{1}{2} \checkmark$

> **TRY IT : : 2.47**   Translate and solve: The sum of five-eighths and x is one-fourth.

> **TRY IT : : 2.48**   Translate and solve: The sum of three-fourths and x is five-sixths.

Translate and Solve Applications

To solve applications using the Division and Multiplication Properties of Equality, we will follow the same steps we used in the last section. We will restate the problem in just one sentence, assign a variable, and then translate the sentence into an equation to solve.

EXAMPLE 2.25

Denae bought 6 pounds of grapes for $10.74. What was the cost of one pound of grapes?

⊘ **Solution**

What are you asked to find?	The cost of 1 pound of grapes
Assign a variable.	Let $c =$ the cost of one pound.
Write a sentence that gives the information to find it.	The cost of 6 pounds is $10.74.
Translate into an equation.	$6c = 10.74$
Solve.	$\frac{6c}{6} = \frac{10.74}{6}$
	$c = 1.79$

The grapes cost $1.79 per pound.

Check: If one pound costs $1.79, do 6 pounds cost $10.74?

$$6(1.79) \overset{?}{=} 10.74$$
$$10.74 = 10.74 \checkmark$$

> **TRY IT : : 2.49**   Translate and solve:
>
> Arianna bought a 24-pack of water bottles for $9.36. What was the cost of one water bottle?

> **TRY IT : : 2.50**   Translate and solve:
>
> At JB's Bowling Alley, 6 people can play on one lane for $34.98. What is the cost for each person?

EXAMPLE 2.26

Andreas bought a used car for \$12,000. Because the car was 4-years old, its price was $\frac{3}{4}$ of the original price, when the car was new. What was the original price of the car?

✓ **Solution**

What are you asked to find?	The original price of the car
Assign a variable.	Let $p =$ the original price.
Write a sentence that gives the information to find it.	\$12,000 is $\frac{3}{4}$ of the original price.
Translate into an equation.	$12{,}000 = \frac{3}{4}p$
Solve.	$\frac{4}{3}(12{,}000) = \frac{4}{3} \cdot \frac{3}{4}p$ $16{,}000 = p$ The original cost of the car was \$16,000.

Check: Is $\frac{3}{4}$ of \$16,000 equal to \$12,000?

$$\frac{3}{4} \cdot 16{,}000 \overset{?}{=} 12{,}000$$

$$12{,}000 = 12{,}000 \checkmark$$

> **TRY IT :: 2.51**
>
> Translate and solve:
>
> The annual property tax on the Mehta's house is \$1,800, calculated as $\frac{15}{1{,}000}$ of the assessed value of the house. What is the assessed value of the Mehta's house?

> **TRY IT :: 2.52**
>
> Translate and solve:
>
> Stella planted 14 flats of flowers in $\frac{2}{3}$ of her garden. How many flats of flowers would she need to fill the whole garden?

2.2 EXERCISES

Practice Makes Perfect

Solve Equations Using the Division and Multiplication Properties of Equality

In the following exercises, solve each equation using the Division and Multiplication Properties of Equality and check the solution.

77. $8x = 56$

78. $7p = 63$

79. $-5c = 55$

80. $-9x = -27$

81. $-809 = 15y$

82. $-731 = 19y$

83. $-37p = -541$

84. $-19m = -586$

85. $0.25z = 3.25$

86. $0.75a = 11.25$

87. $-13x = 0$

88. $24x = 0$

89. $\frac{x}{4} = 35$

90. $\frac{z}{2} = 54$

91. $-20 = \frac{q}{-5}$

92. $\frac{c}{-3} = -12$

93. $\frac{y}{9} = -16$

94. $\frac{q}{6} = -38$

95. $\frac{m}{-12} = 45$

96. $-24 = \frac{p}{-20}$

97. $-y = 6$

98. $-u = 15$

99. $-v = -72$

100. $-x = -39$

101. $\frac{2}{3}y = 48$

102. $\frac{3}{5}r = 75$

103. $-\frac{5}{8}w = 40$

104. $24 = -\frac{3}{4}x$

105. $-\frac{2}{5} = \frac{1}{10}a$

106. $-\frac{1}{3}q = -\frac{5}{6}$

107. $-\frac{7}{10}x = -\frac{14}{3}$

108. $\frac{3}{8}y = -\frac{1}{4}$

109. $\frac{7}{12} = -\frac{3}{4}p$

110. $\frac{11}{18} = -\frac{5}{6}q$

111. $-\frac{5}{18} = -\frac{10}{9}u$

112. $-\frac{7}{20} = -\frac{7}{4}v$

Solve Equations That Require Simplification

In the following exercises, solve each equation requiring simplification.

113. $100 - 16 = 4p - 10p - p$

114. $-18 - 7 = 5t - 9t - 6t$

115. $\frac{7}{8}n - \frac{3}{4}n = 9 + 2$

116. $\frac{5}{12}q + \frac{1}{2}q = 25 - 3$

117. $0.25d + 0.10d = 6 - 0.75$

118. $0.05p - 0.01p = 2 + 0.24$

119. $-10(q - 4) - 57 = 93$

120. $-12(d - 5) - 29 = 43$

121. $-10(x + 4) - 19 = 85$

122. $-15(z + 9) - 11 = 75$

Mixed Practice

In the following exercises, solve each equation.

123. $\frac{9}{10}x = 90$

124. $\frac{5}{12}y = 60$

125. $y + 46 = 55$

126. $x + 33 = 41$

127. $\frac{w}{-2} = 99$

128. $\frac{s}{-3} = -60$

129. $27 = 6a$

130. $-a = 7$

131. $-x = 2$

132. $z - 16 = -59$

133. $m - 41 = -14$

134. $0.04r = 52.60$

135. $63.90 = 0.03p$

136. $-15x = -120$

137. $84 = -12z$

138. $19.36 = x - 0.2x$

139. $c - 0.3c = 35.70$

140. $-y = -9$

141. $-x = -8$

Translate to an Equation and Solve

In the following exercises, translate to an equation and then solve.

142. 187 is the product of -17 and m.

143. 133 is the product of -19 and n.

144. -184 is the product of 23 and p.

145. -152 is the product of 8 and q.

146. u divided by 7 is equal to -49.

147. r divided by 12 is equal to -48.

148. h divided by -13 is equal to -65.

149. j divided by -20 is equal to -80.

150. The quotient c and -19 is 38.

151. The quotient of b and -6 is 18.

152. The quotient of h and 26 is -52.

153. The quotient k and 22 is -66.

154. Five-sixths of y is 15.

155. Three-tenths of x is 15.

156. Four-thirds of w is 36.

157. Five-halves of v is 50.

158. The sum of nine-tenths and g is two-thirds.

159. The sum of two-fifths and f is one-half.

160. The difference of p and one-sixth is two-thirds.

161. The difference of q and one-eighth is three-fourths.

Translate and Solve Applications

In the following exercises, translate into an equation and solve.

162. Kindergarten Connie's kindergarten class has 24 children. She wants them to get into 4 equal groups. How many children will she put in each group?

163. Balloons Ramona bought 18 balloons for a party. She wants to make 3 equal bunches. How many balloons did she use in each bunch?

164. Tickets Mollie paid $36.25 for 5 movie tickets. What was the price of each ticket?

165. Shopping Serena paid $12.96 for a pack of 12 pairs of sport socks. What was the price of pair of sport socks?

166. Sewing Nancy used 14 yards of fabric to make flags for one-third of the drill team. How much fabric, would Nancy need to make flags for the whole team?

167. MPG John's SUV gets 18 miles per gallon (mpg). This is half as many mpg as his wife's hybrid car. How many miles per gallon does the hybrid car get?

168. Height Aiden is 27 inches tall. He is $\frac{3}{8}$ as tall as his father. How tall is his father?

169. Real estate Bea earned $11,700 commission for selling a house, calculated as $\frac{6}{100}$ of the selling price. What was the selling price of the house?

Everyday Math

170. Commission Every week Perry gets paid $150 plus 12% of his total sales amount. Solve the equation $840 = 150 + 0.12(a - 1250)$ for a, to find the total amount Perry must sell in order to be paid $840 one week.

171. Stamps Travis bought $9.45 worth of 49-cent stamps and 21-cent stamps. The number of 21-cent stamps was 5 less than the number of 49-cent stamps. Solve the equation $0.49s + 0.21(s - 5) = 9.45$ for s, to find the number of 49-cent stamps Travis bought.

Writing Exercises

172. Frida started to solve the equation $-3x = 36$ by adding 3 to both sides. Explain why Frida's method will not solve the equation.

173. Emiliano thinks $x = 40$ is the solution to the equation $\frac{1}{2}x = 80$. Explain why he is wrong.

Self Check

ⓐ *After completing the exercises, use this checklist to evaluate your mastery of the objectives of this section.*

I can...	Confidently	With some help	No-I don't get it!
solve equations using the Division and Multiplication Properties of equality.			
solve equations that require simplification.			
translate to an equation and solve.			
translate and solve applications.			

ⓑ *What does this checklist tell you about your mastery of this section? What steps will you take to improve?*

2.3 Solve Equations with Variables and Constants on Both Sides

Learning Objectives

By the end of this section, you will be able to:

> Solve an equation with constants on both sides
> Solve an equation with variables on both sides
> Solve an equation with variables and constants on both sides

Be Prepared!

Before you get started, take this readiness quiz.

1. Simplify: $4y - 9 + 9$.

    If you missed this problem, review **Example 1.129**.

Solve Equations with Constants on Both Sides

In all the equations we have solved so far, all the variable terms were on only one side of the equation with the constants on the other side. This does not happen all the time—so now we will learn to solve equations in which the variable terms, or constant terms, or both are on both sides of the equation.

Our strategy will involve choosing one side of the equation to be the "variable side", and the other side of the equation to be the "constant side." Then, we will use the Subtraction and Addition Properties of Equality to get all the variable terms together on one side of the equation and the constant terms together on the other side.

By doing this, we will transform the equation that began with variables and constants on both sides into the form $ax = b$. We already know how to solve equations of this form by using the Division or Multiplication Properties of Equality.

EXAMPLE 2.27

Solve: $7x + 8 = -13$.

 Solution

In this equation, the variable is found only on the left side. It makes sense to call the left side the "variable" side. Therefore, the right side will be the "constant" side. We will write the labels above the equation to help us remember what goes where.

$$\overset{\text{variable}}{7x} + 8 = \overset{\text{constant}}{-13}$$

Since the left side is the " x ", or variable side, the 8 is out of place. We must "undo" adding 8 by subtracting 8, and to keep the equality we must subtract 8 from both sides.

	$\overset{\text{variable}}{7x} + 8 = \overset{\text{constant}}{-14}\ 3$
Use the Subtraction Property of Equality.	$7x + 8 - 8 = -13 - 8$
Simplify.	$7x = -21$
Now all the variables are on the left and the constant on the right. The equation looks like those you learned to solve earlier.	
Use the Division Property of Equality.	$\dfrac{7x}{7} = \dfrac{-21}{7}$
Simplify.	$x = -3$
Check:	$7x + 8 = -13$
Let $x = -3$.	$7(-3) + 8 \overset{?}{=} -13$

$$-21 + 8 \cdot 2 \stackrel{?}{=} -13$$

$$-13 = -13 \checkmark$$

> **TRY IT :: 2.53**   Solve: $3x + 4 = -8$.

> **TRY IT :: 2.54**   Solve: $5a + 3 = -37$.

EXAMPLE 2.28

Solve: $8y - 9 = 31$.

⊘ Solution

Notice, the variable is only on the left side of the equation, so we will call this side the "variable" side, and the right side will be the "constant" side. Since the left side is the "variable" side, the 9 is out of place. It is subtracted from the $8y$, so to "undo" subtraction, add 9 to both sides. Remember, whatever you do to the left, you must do to the right.

	variable constant
	$8y - 9 = 31$
Add 9 to both sides.	$8y - 9 + 9 = 31 + 9$
Simplify.	$8y = 40$
	The variables are now on one side and the constants on the other. We continue from here as we did earlier.
Divide both sides by 8.	$\dfrac{8y}{8} = \dfrac{40}{8}$
Simplify.	$y = 5$
Check:	$8y - 9 = 31$
Let $y = 5$.	$8 \cdot 5 - 9 \stackrel{?}{=} 31$
	$40 - 9 \stackrel{?}{=} 31$
	$31 = 31 \checkmark$

> **TRY IT :: 2.55**   Solve: $5y - 9 = 16$.

> **TRY IT :: 2.56**   Solve: $3m - 8 = 19$.

Solve Equations with Variables on Both Sides

What if there are variables on both sides of the equation? For equations like this, begin as we did above—choose a "variable" side and a "constant" side, and then use the subtraction and addition properties of equality to collect all variables on one side and all constants on the other side.

EXAMPLE 2.29

Solve: $9x = 8x - 6$.

⊘ Solution

Here the variable is on both sides, but the constants only appear on the right side, so let's make the right side the "constant" side. Then the left side will be the "variable" side.

	variable constant
	$9x = 8x - 6$
We don't want any x's on the right, so subtract the $8x$ from both sides.	$9x - 8x = 8x - 8x - 6$
Simplify.	$x = -6$

We succeeded in getting the variables on one side and the constants on the other, and have obtained the solution.

Check:	$9x = 8x - 6$
Let $x = -6$.	$9(-6) \overset{?}{=} 8(-6) - 6$
	$-54 \overset{?}{=} -48 - 6$
	$-54 = -54\ \checkmark$

> **TRY IT : : 2.57**   Solve: $6n = 5n - 10$.

> **TRY IT : : 2.58**   Solve: $-6c = -7c - 1$.

EXAMPLE 2.30

Solve: $5y - 9 = 8y$.

⊘ **Solution**

The only constant is on the left and the y's are on both sides. Let's leave the constant on the left and get the variables to the right.

	constant variable
	$5y - 9 = 8y$
Subtract $5y$ from both sides.	$5y - 5y - 9 = 8y - 5y$
Simplify.	$-9 = 3y$
We have the y's on the right and the constants on the left. Divide both sides by 3.	$\dfrac{-9}{3} = \dfrac{3y}{3}$
Simplify.	$-3 = y$
Check:	$5y - 9 = 8y$
Let $y = -3$.	$5(-3) - 9 \overset{?}{=} 8(-3)$
	$-15 - 9 \overset{?}{=} -24$
	$-24 = -24\ \checkmark$

> **TRY IT : : 2.59**   Solve: $3p - 14 = 5p$.

> **TRY IT : : 2.60**   Solve: $8m + 9 = 5m$.

EXAMPLE 2.31

Solve: $12x = -x + 26$.

⊘ Solution

The only constant is on the right, so let the left side be the "variable" side.

	variable constant
	$12x = -x + 26$
Remove the $-x$ from the right side by adding x to both sides.	$12x + x = -x + x + 26$
Simplify.	$13x = 26$
All the x's are on the left and the constants are on the right. Divide both sides by 13.	$\dfrac{13x}{13} = \dfrac{26}{13}$
Simplify.	$x = 2$

> **TRY IT :: 2.61**   Solve: $12j = -4j + 32$.

> **TRY IT :: 2.62**   Solve: $8h = -4h + 12$.

Solve Equations with Variables and Constants on Both Sides

The next example will be the first to have variables and constants on both sides of the equation. It may take several steps to solve this equation, so we need a clear and organized strategy.

EXAMPLE 2.32   HOW TO SOLVE EQUATIONS WITH VARIABLES AND CONSTANTS ON BOTH SIDES

Solve: $7x + 5 = 6x + 2$.

⊘ Solution

Step 1. Choose which side will be the "variable" side—the other side will be the "constant" side.	The variable terms are 7x and 6x. Since 7 is greater than 6, we will make the left side the "x" side. The right side will be the "constant" side.	variable constant $7x + 5 = 6x + 2$
Step 2. Collect the variable terms to the "variable" side of the equation, using the addition or subtraction property of equality.	With the right side as the "constant" side, the 6x is out of place, so subtract 6x from both sides. Combine like terms. Now, the variable is only on the left side!	$7x - 6x + 5 = 6x - 6x + 2$ $x + 5 = 2$
Step 3. Collect all the constants to the other side of the equation, using the addition or subtraction property of equality.	The right side is the "constant" side, so the 5 is out of place. Subtract 5 from both sides. Simplify.	$x + 5 - 5 = 2 - 5$ $x = -3$
Step 4. Make the coefficient of the variable equal 1, using the multiplication or division property of equality.	The coefficient of x is one. The equation is solved.	

Step 5. Check.	Let $x = -3$	**Check:**
		$7x + 6 = 6x + 2$
	Simplify.	$(-3) + 5 = 6(-3) + 2$
	Add.	$-21 + 5 = -18 + 2$
		$-16 = -16$ ✓

> **TRY IT :: 2.63**   Solve: $12x + 8 = 6x + 2$.

> **TRY IT :: 2.64**   Solve: $9y + 4 = 7y + 12$.

We'll list the steps below so you can easily refer to them. But we'll call this the 'Beginning Strategy' because we'll be adding some steps later in this chapter.

HOW TO :: BEGINNING STRATEGY FOR SOLVING EQUATIONS WITH VARIABLES AND CONSTANTS ON BOTH SIDES OF THE EQUATION.

Step 1.   Choose which side will be the "variable" side—the other side will be the "constant" side.

Step 2.   Collect the variable terms to the "variable" side of the equation, using the Addition or Subtraction Property of Equality.

Step 3.   Collect all the constants to the other side of the equation, using the Addition or Subtraction Property of Equality.

Step 4.   Make the coefficient of the variable equal 1, using the Multiplication or Division Property of Equality.

Step 5.   Check the solution by substituting it into the original equation.

In Step 1, a helpful approach is to make the "variable" side the side that has the variable with the larger coefficient. This usually makes the arithmetic easier.

EXAMPLE 2.33

Solve: $8n - 4 = -2n + 6$.

⊘ **Solution**

In the first step, choose the variable side by comparing the coefficients of the variables on each side.

Since $8 > -2$, make the left side the "variable" side.	variable constant $8n - 4 = -2n + 6$
We don't want variable terms on the right side—add $2n$ to both sides to leave only constants on the right.	$8n + 2n - 4 = -2n + 2n + 6$
Combine like terms.	$10n - 4 = 6$
We don't want any constants on the left side, so add 4 to both sides.	$10n - 4 + 4 = 6 + 4$
Simplify.	$10n = 10$
The variable term is on the left and the constant term is on the right. To get the coefficient of n to be one, divide both sides by 10.	$\dfrac{10n}{10} = \dfrac{10}{10}$
Simplify.	$n = 1$
Check:	$8n - 4 = -2n + 6$

Let $n = 1$.

$$8 \cdot 1 - 4 \overset{?}{=} -2 \cdot 1 + 6$$

$$8 - 4 \overset{?}{=} -2 + 6$$

$$4 = 4 \checkmark$$

> **TRY IT :: 2.65**   Solve: $8q - 5 = -4q + 7$.

> **TRY IT :: 2.66**   Solve: $7n - 3 = n + 3$.

EXAMPLE 2.34

Solve: $7a - 3 = 13a + 7$.

⊘ **Solution**

In the first step, choose the variable side by comparing the coefficients of the variables on each side.

Since $13 > 7$, make the right side the "variable" side and the left side the "constant" side.

	constant variable
	$7a - 3 = 13a + 7$
Subtract $7a$ from both sides to remove the variable term from the left.	$7a - 7a - 3 = 13a - 7a + 7$
Combine like terms.	$-3 = 6a + 7$
Subtract 7 from both sides to remove the constant from the right.	$-3 - 7 = 6a + 7 - 7$
Simplify.	$-10 = 6a$
Divide both sides by 6 to make 1 the coefficient of a.	$\dfrac{-10}{6} = \dfrac{6a}{6}$
Simplify.	$-\dfrac{5}{3} = a$
Check:	$7a - 3 = 13a + 7$
Let $a = -\dfrac{5}{3}$.	$7\left(-\dfrac{5}{3}\right) - 3 \overset{?}{=} 13\left(-\dfrac{5}{3}\right) + 7$
	$-\dfrac{35}{3} - \dfrac{9}{3} \overset{?}{=} -\dfrac{65}{3} + \dfrac{21}{3}$
	$-\dfrac{54}{3} = -\dfrac{54}{3} \checkmark$

> **TRY IT :: 2.67**   Solve: $2a - 2 = 6a + 18$.

> **TRY IT :: 2.68**   Solve: $4k - 1 = 7k + 17$.

In the last example, we could have made the left side the "variable" side, but it would have led to a negative coefficient on the variable term. (Try it!) While we could work with the negative, there is less chance of errors when working with positives. The strategy outlined above helps avoid the negatives!

To solve an equation with fractions, we just follow the steps of our strategy to get the solution!

EXAMPLE 2.35

Solve: $\frac{5}{4}x + 6 = \frac{1}{4}x - 2$.

⊘ Solution

Since $\frac{5}{4} > \frac{1}{4}$, make the left side the "variable" side and the right side the "constant" side.

	variable constant
	$\frac{5}{4}x + 6 = \frac{1}{4}x - 2$
Subtract $\frac{1}{4}x$ from both sides.	$\frac{5}{4}x - \frac{1}{4}x + 6 = \frac{1}{4}x - \frac{1}{4}x - 2$
Combine like terms.	$x + 6 = -2$
Subtract 6 from both sides.	$x + 6 - 6 = -2 - 6$
Simplify.	$x = -8$

Check:
$$\frac{5}{4}x + 6 = \frac{1}{4}x - 2$$
Let $x = -8$,
$$\frac{5}{4}(-8) + 6 \overset{?}{=} \frac{1}{4}(-8) - 2$$
$$-10 + 6 \overset{?}{=} -2 - 2$$
$$-4 = -4 ✓$$

> **TRY IT :: 2.69**   Solve: $\frac{7}{8}x - 12 = -\frac{1}{8}x - 2$.

> **TRY IT :: 2.70**   Solve: $\frac{7}{6}y + 11 = \frac{1}{6}y + 8$.

We will use the same strategy to find the solution for an equation with decimals.

EXAMPLE 2.36

Solve: $7.8x + 4 = 5.4x - 8$.

⊘ Solution

Since $7.8 > 5.4$, make the left side the "variable" side and the right side the "constant" side.

	variable side constant side
	$7.8x + 4 = 5.4x - 8$
Subtract $5.4x$ from both sides.	$7.8x - 5.4x + 4 = 5.4x - 5.4x - 8$
Combine like terms.	$2.4x + 4 = -8$
Subtract 4 from both sides.	$2.4x + 4 - 4 = -8 - 4$
Simplify.	$2.4x = -12$
Use the Division Propery of Equality.	$\frac{2.4x}{2.4} = \frac{-12}{2.4}$
Simplify.	$x = -5$
Check:	$7.8x + 4 = 5.4x - 5$

Let $x = -5$.
$$7.8(-5) + 4 = 5.4(-5) - 8$$
$$-39 + 4 \overset{?}{=} -27 - 8$$
$$-35 = -35 \checkmark$$

> **TRY IT :: 2.71**   Solve: $2.8x + 12 = -1.4x - 9$.

> **TRY IT :: 2.72**   Solve: $3.6y + 8 = 1.2y - 4$.

2.3 EXERCISES

Practice Makes Perfect

Solve Equations with Constants on Both Sides

In the following exercises, solve the following equations with constants on both sides.

174. $9x - 3 = 60$

175. $12x - 8 = 64$

176. $14w + 5 = 117$

177. $15y + 7 = 97$

178. $2a + 8 = -28$

179. $3m + 9 = -15$

180. $-62 = 8n - 6$

181. $-77 = 9b - 5$

182. $35 = -13y + 9$

183. $60 = -21x - 24$

184. $-12p - 9 = 9$

185. $-14q - 2 = 16$

Solve Equations with Variables on Both Sides

In the following exercises, solve the following equations with variables on both sides.

186. $19z = 18z - 7$

187. $21k = 20k - 11$

188. $9x + 36 = 15x$

189. $8x + 27 = 11x$

190. $c = -3c - 20$

191. $b = -4b - 15$

192. $9q = 44 - 2q$

193. $5z = 39 - 8z$

194. $6y + \frac{1}{2} = 5y$

195. $4x + \frac{3}{4} = 3x$

196. $-18a - 8 = -22a$

197. $-11r - 8 = -7r$

Solve Equations with Variables and Constants on Both Sides

In the following exercises, solve the following equations with variables and constants on both sides.

198. $8x - 15 = 7x + 3$

199. $6x - 17 = 5x + 2$

200. $26 + 13d = 14d + 11$

201. $21 + 18f = 19f + 14$

202. $2p - 1 = 4p - 33$

203. $12q - 5 = 9q - 20$

204. $4a + 5 = -a - 40$

205. $8c + 7 = -3c - 37$

206. $5y - 30 = -5y + 30$

207. $7x - 17 = -8x + 13$

208. $7s + 12 = 5 + 4s$

209. $9p + 14 = 6 + 4p$

210. $2z - 6 = 23 - z$

211. $3y - 4 = 12 - y$

212. $\frac{5}{3}c - 3 = \frac{2}{3}c - 16$

213. $\frac{7}{4}m - 7 = \frac{3}{4}m - 13$

214. $8 - \frac{2}{5}q = \frac{3}{5}q + 6$

215. $11 - \frac{1}{5}a = \frac{4}{5}a + 4$

216. $\frac{4}{3}n + 9 = \frac{1}{3}n - 9$

217. $\frac{5}{4}a + 15 = \frac{3}{4}a - 5$

218. $\frac{1}{4}y + 7 = \frac{3}{4}y - 3$

219. $\frac{3}{5}p + 2 = \frac{4}{5}p - 1$

220. $14n + 8.25 = 9n + 19.60$

221. $13z + 6.45 = 8z + 23.75$

222. $2.4w - 100 = 0.8w + 28$

223. $2.7w - 80 = 1.2w + 10$

224. $5.6r + 13.1 = 3.5r + 57.2$

225. $6.6x - 18.9 = 3.4x + 54.7$

Everyday Math

226. Concert tickets At a school concert the total value of tickets sold was $1506. Student tickets sold for $6 and adult tickets sold for $9. The number of adult tickets sold was 5 less than 3 times the number of student tickets. Find the number of student tickets sold, s, by solving the equation $6s + 27s - 45 = 1506$
.

227. Making a fence Jovani has 150 feet of fencing to make a rectangular garden in his backyard. He wants the length to be 15 feet more than the width. Find the width, w, by solving the equation $150 = 2w + 30 + 2w$.

Writing Exercises

228. Solve the equation $\frac{6}{5}y - 8 = \frac{1}{5}y + 7$ explaining all the steps of your solution as in the examples in this section.

229. Solve the equation $10x + 14 = -2x + 38$ explaining all the steps of your solution as in the examples in this section.

230. When solving an equation with variables on both sides, why is it usually better to choose the side with the larger coefficient of x to be the "variable" side?

231. Is $x = -2$ a solution to the equation $5 - 2x = -4x + 1$? How do you know?

Self Check

ⓐ *After completing the exercises, use this checklist to evaluate your mastery of the objectives of this section.*

I can...	Confidently	With some help	No-I don't get it!
solve an equation with constants on both sides.			
solve an equation with variables on both sides.			
solve an equation with variables and constants on both sides.			

ⓑ *What does this checklist tell you about your mastery of this section? What steps will you take to improve?*

2.4 Use a General Strategy to Solve Linear Equations

Learning Objectives

By the end of this section, you will be able to:

› Solve equations using a general strategy
› Classify equations

Be Prepared!

Before you get started, take this readiness quiz.

1. Simplify: $-(a - 4)$.

   If you missed this problem, review **Example 1.137**.

2. Multiply: $\frac{3}{2}(12x + 20)$.

   If you missed this problem, review **Example 1.133**.

3. Simplify: $5 - 2(n + 1)$.

   If you missed this problem, review **Example 1.138**.

4. Multiply: $3(7y + 9)$.

   If you missed this problem, review **Example 1.132**.

5. Multiply: $(2.5)(6.4)$.

   If you missed this problem, review **Example 1.97**.

Solve Equations Using the General Strategy

Until now we have dealt with solving one specific form of a linear equation. It is time now to lay out one overall strategy that can be used to solve any linear equation. Some equations we solve will not require all these steps to solve, but many will.

Beginning by simplifying each side of the equation makes the remaining steps easier.

EXAMPLE 2.37   HOW TO SOLVE LINEAR EQUATIONS USING THE GENERAL STRATEGY

Solve: $-6(x + 3) = 24$.

 **Solution**

Step 1. Simplify each side of the equation as much as possible.	Use the Distributive Property.	$-6(x + 3) = 24$
	Notice that each side of the equation is simplified as much as possible.	$-6x - 18 = 24$
Step 2. Collect all variable terms on one side of the equation.	Nothing to do – all x's are on the left side.	
Step 3. Collect constant terms on the other side of the equation.	To get constants only on the right, add 18 to each side. Simplify.	$-6x - 18 + 18 = 24 + 18$ $-6x = 42$
Step 4. Make the coefficient of the variable term to equal to 1.	Divide each side by -6.	$\dfrac{-6x}{-6} = \dfrac{42}{-6}$
	Simplify.	$x = -7$

Step 5. Check the solution.	Let $x = -7$	**Check:**
		$-6(x + 3) = 24$
	Simplify.	$-6(-7 + 3) \stackrel{?}{=} 24$
	Multiply.	$-6(-4) \stackrel{?}{=} 24$
		$24 = 24 \checkmark$

> **TRY IT ::** 2.73   Solve: $5(x + 3) = 35$.

> **TRY IT ::** 2.74   Solve: $6(y - 4) = -18$.

HOW TO :: GENERAL STRATEGY FOR SOLVING LINEAR EQUATIONS.

Step 1.   **Simplify each side of the equation as much as possible.**
Use the Distributive Property to remove any parentheses.
Combine like terms.

Step 2.   **Collect all the variable terms on one side of the equation.**
Use the Addition or Subtraction Property of Equality.

Step 3.   **Collect all the constant terms on the other side of the equation.**
Use the Addition or Subtraction Property of Equality.

Step 4.   **Make the coefficient of the variable term to equal to 1.**
Use the Multiplication or Division Property of Equality.
State the solution to the equation.

Step 5.   **Check the solution.** Substitute the solution into the original equation to make sure the result
is a true statement.

EXAMPLE 2.38

Solve: $-(y + 9) = 8$.

 **Solution**

	$-(y + 9) = 8$
Simplify each side of the equation as much as possible by distributing.	$-y - 9 = 8$
The only y term is on the left side, so all variable terms are on the left side of the equation.	
Add 9 to both sides to get all constant terms on the right side of the equation.	$-y - 9 + 9 = 8 + 9$
Simplify.	$-y = 17$
Rewrite $-y$ as $-1y$.	$-1y = 17$
Make the coefficient of the variable term to equal to 1 by dividing both sides by -1.	$\dfrac{-1y}{-1} = \dfrac{17}{-1}$
Simplify.	$y = -17$
Check:	$-(y + 9) = 8$

Let $y = -17$. $-(-17 + 9) \overset{?}{=} 8$

$-(-8) \overset{?}{=} 8$

$8 = 8 \checkmark$

> **TRY IT : : 2.75**   Solve: $-(y + 8) = -2$.

> **TRY IT : : 2.76**   Solve: $-(z + 4) = -12$.

EXAMPLE 2.39

Solve: $5(a - 3) + 5 = -10$.

⊘ **Solution**

	$5(a - 3) + 5 = -10$
Simplify each side of the equation as much as possible.	
Distribute.	$5a - 15 + 5 = -10$
Combine like terms.	$5a - 10 = -10$
The only a term is on the left side, so all variable terms are on one side of the equation.	
Add 10 to both sides to get all constant terms on the other side of the equation.	$5a - 10 + 10 = -10 + 10$
Simplify.	$5a = 0$
Make the coefficient of the variable term to equal to 1 by dividing both sides by 5.	$\dfrac{5a}{5} = \dfrac{0}{5}$
Simplify.	$a = 0$
Check:	$5(a - 3) + 5 = -10$
Let $a = 0$.	$5(0 - 3) + 5 \overset{?}{=} -10$
	$5(-3) + 5 \overset{?}{=} -10$
	$-15 + 5 \overset{?}{=} -10$
	$-10 = -10 \checkmark$

> **TRY IT : : 2.77**   Solve: $2(m - 4) + 3 = -1$.

> **TRY IT : : 2.78**   Solve: $7(n - 3) - 8 = -15$.

EXAMPLE 2.40

Solve: $\frac{2}{3}(6m - 3) = 8 - m$.

✓ Solution

$$\frac{2}{3}(6m - 3) = 8 - m$$

Distribute.	$4m - 2 = 8 - m$
Add m to get the variables only to the left.	$4m + m - 2 = 8 - m + m$
Simplify.	$5m - 2 = 8$
Add 2 to get constants only on the right.	$5m - 2 + 2 = 8 + 2$
Simplify.	$5m = 10$
Divide by 5.	$\dfrac{5m}{5} = \dfrac{10}{5}$
Simplify.	$m = 2$
Check:	$\frac{2}{3}(6m - 3) = 8 - m$
Let $m = 2$.	$\frac{2}{3}(6 \cdot 2 - 3) \overset{?}{=} 8 - 2$
	$\frac{2}{3}(12 - 3) \overset{?}{=} 6$
	$\frac{2}{3}(9) \overset{?}{=} 6$
	$6 = 6 \checkmark$

> **TRY IT :: 2.79**   Solve: $\frac{1}{3}(6u + 3) = 7 - u$.

> **TRY IT :: 2.80**   Solve: $\frac{2}{3}(9x - 12) = 8 + 2x$.

EXAMPLE 2.41

Solve: $8 - 2(3y + 5) = 0$.

✓ Solution

$$8 - 2(3y + 5) = 0$$

Simplify—use the Distributive Property.	$8 - 6y - 10 = 0$
Combine like terms.	$-6y - 2 = 0$
Add 2 to both sides to collect constants on the right.	$-6y - 2 + 2 = 0 + 2$
Simplify.	$-6y = 2$
Divide both sides by -6.	$\dfrac{-6y}{-6} = \dfrac{2}{-6}$
Simplify.	$y = -\dfrac{1}{3}$

Check: Let $y = -\frac{1}{3}$.

$$8 - 2(3y + 5) = 0$$

$$8 - 2\left[3\left(-\frac{1}{3}\right) + 5\right] = 0$$

$$8 - 2(-1 + 5) \stackrel{?}{=} 0$$

$$8 - 2(4) \stackrel{?}{=} 0$$

$$8 - 8 \stackrel{?}{=} 0$$

$$0 = 0 \checkmark$$

> **TRY IT ::** 2.81   Solve: $12 - 3(4j + 3) = -17$.

> **TRY IT ::** 2.82   Solve: $-6 - 8(k - 2) = -10$.

EXAMPLE 2.42

Solve: $4(x - 1) - 2 = 5(2x + 3) + 6$.

⊘ **Solution**

	$4(x - 1) - 2 = 5(2x + 3) + 6$
Distribute.	$4x - 4 - 2 = 10x + 15 + 6$
Combine like terms.	$4x - 6 = 10x + 21$
Subtract $4x$ to get the variables only on the right side since $10 > 4$.	$4x - 4x - 6 = 10x - 4x + 21$
Simplify.	$-6 = 6x + 21$
Subtract 21 to get the constants on left.	$-6 - 21 = 6x + 21 - 21$
Simplify.	$-27 = 6x$
Divide by 6.	$\dfrac{-27}{6} = \dfrac{6x}{6}$
Simplify.	$-\dfrac{9}{2} = x$
Check:	$4(x - 1) - 2 = 5(2x + 3) + 6$
Let $x = -\dfrac{9}{2}$.	$4\left(-\dfrac{9}{2} - 1\right) - 2 \stackrel{?}{=} 5\left[2\left(-\dfrac{9}{2}\right) + 3\right] + 6$
	$4\left(-\dfrac{11}{2}\right) - 2 \stackrel{?}{=} 5(-9 + 3) + 6$
	$-22 - 2 \stackrel{?}{=} 5(-6) + 6$
	$-24 \stackrel{?}{=} -30 + 6$
	$-24 = -24 \checkmark$

> **TRY IT :: 2.83**   Solve: $6(p-3)-7 = 5(4p+3)-12$.

> **TRY IT :: 2.84**   Solve: $8(q+1)-5 = 3(2q-4)-1$.

EXAMPLE 2.43

Solve: $10[3 - 8(2s-5)] = 15(40-5s)$.

⊘ **Solution**

	$10[3 - 8(2s - 5)] = 15(40 - 5s)$
Simplify from the innermost parentheses first.	$10[3 - 16s + 40] = 15(40 - 5s)$
Combine like terms in the brackets.	$10[43 - 16s] = 15(40 - 5s)$
Distribute.	$430 - 160s = 600 - 75s$
Add $160s$ to get the s's to the right.	$430 - 160s + 160s = 600 - 75s + 160s$
Simplify.	$430 = 600 + 85s$
Subtract 600 to get the constants to the left.	$430 - 600 = 600 + 85s - 600$
Simplify.	$-170 = 85s$
Divide.	$\dfrac{-170}{85} = \dfrac{85s}{85}$
Simplify.	$-2 = s$
Check:	$10[3 - 8(2s - 5)] = 15(40 - 5s)$
Substitute $s = -2$.	$10[3 - 8(2(-2) - 5)] \overset{?}{=} 15(40 - 5(-2))$
	$10[3 - 8(-4 - 5)] \overset{?}{=} 15(40 + 10)$
	$10[3 - 8(-9)] \overset{?}{=} 15(50)$
	$10[3 + 72] \overset{?}{=} 750$
	$10[75] \overset{?}{=} 750$
	$750 = 750$ ✓

> **TRY IT :: 2.85**   Solve: $6[4 - 2(7y - 1)] = 8(13 - 8y)$.

> **TRY IT :: 2.86**   Solve: $12[1 - 5(4z - 1)] = 3(24 + 11z)$.

EXAMPLE 2.44

Solve: $0.36(100n + 5) = 0.6(30n + 15)$.

⊘ **Solution**

$$0.36(100n + 5) = 0.6(30n + 15)$$

Distribute.	$36n + 1.8 = 18n + 9$
Subtract $18n$ to get the variables to the left.	$36n - 18n + 1.8 = 18n - 18n + 9$
Simplify.	$18n + 1.8 = 9$
Subtract 1.8 to get the constants to the right.	$18n + 1.8 - 1.8 = 9 - 1.8$
Simplify.	$18n = 7.2$
Divide.	$\dfrac{18n}{18} = \dfrac{7.2}{18}$
Simplify.	$n = 0.4$
Check:	$0.36(100n + 5) = 0.6(30n + 15)$
Let $n = 0.4$.	$0.36(100(0.4) + 5) \overset{?}{=} 0.6(30(0.4) + 15)$
	$0.36(40 + 5) \overset{?}{=} 0.6(12 + 15)$
	$0.36(45) \overset{?}{=} 0.6(27)$
	$16.2 = 16.2 \checkmark$

> **TRY IT :: 2.87**   Solve: $0.55(100n + 8) = 0.6(85n + 14)$.

> **TRY IT :: 2.88**   Solve: $0.15(40m - 120) = 0.5(60m + 12)$.

Classify Equations

Consider the equation we solved at the start of the last section, $7x + 8 = -13$. The solution we found was $x = -3$. This means the equation $7x + 8 = -13$ is true when we replace the variable, x, with the value -3. We showed this when we checked the solution $x = -3$ and evaluated $7x + 8 = -13$ for $x = -3$.

$$7(-3) + 8 \overset{?}{=} -13$$
$$-21 + 8 \overset{?}{=} -13$$
$$-13 = -13 \checkmark$$

If we evaluate $7x + 8$ for a different value of x, the left side will not be -13.

The equation $7x + 8 = -13$ is true when we replace the variable, x, with the value -3, but not true when we replace x with any other value. Whether or not the equation $7x + 8 = -13$ is true depends on the value of the variable. Equations like this are called conditional equations.

All the equations we have solved so far are conditional equations.

Conditional equation

An equation that is true for one or more values of the variable and false for all other values of the variable is a **conditional equation**.

Now let's consider the equation $2y + 6 = 2(y + 3)$. Do you recognize that the left side and the right side are equivalent? Let's see what happens when we solve for y.

$$2y + 6 = 2(y + 3)$$

Distribute.	$2y + 6 = 2y + 6$
Subtract $2y$ to get the y's to one side.	$2y - 2y + 6 = 2y - 2y + 6$
Simplify—the y's are gone!	$6 = 6$

But $6 = 6$ is true.

This means that the equation $2y + 6 = 2(y + 3)$ is true for any value of y. We say the solution to the equation is all of the real numbers. An equation that is true for any value of the variable like this is called an identity.

Identity

An equation that is true for any value of the variable is called an **identity**.

The solution of an identity is all real numbers.

What happens when we solve the equation $5z = 5z - 1$?

	$5z = 5z - 1$
Subtract $5z$ to get the constant alone on the right.	$5z - 5z = 5z - 5z - 1$
Simplify—the z's are gone!	$0 \neq -1$

But $0 \neq -1$.

Solving the equation $5z = 5z - 1$ led to the false statement $0 = -1$. The equation $5z = 5z - 1$ will not be true for any value of z. It has no solution. An equation that has no solution, or that is false for all values of the variable, is called a contradiction.

Contradiction

An equation that is false for all values of the variable is called a **contradiction**.

A contradiction has no solution.

EXAMPLE 2.45

Classify the equation as a conditional equation, an identity, or a contradiction. Then state the solution.

$6(2n - 1) + 3 = 2n - 8 + 5(2n + 1)$

✓ **Solution**

	$6(2n - 1) + 3 = 2n - 8 + 5(2n + 1)$
Distribute.	$12n - 6 + 3 = 2n - 8 + 10n + 5$
Combine like terms.	$12n - 3 = 12n - 3$
Subtract $12n$ to get the n's to one side.	$12n - 12n - 3 = 12n - 12n - 3$
Simplify.	$-3 = -3$
This is a true statement.	The equation is an identity. The solution is all real numbers.

> **TRY IT ::** 2.89

Classify the equation as a conditional equation, an identity, or a contradiction and then state the solution:

$4 + 9(3x - 7) = -42x - 13 + 23(3x - 2)$

> **TRY IT ::** 2.90

Classify the equation as a conditional equation, an identity, or a contradiction and then state the solution:

$8(1 - 3x) + 15(2x + 7) = 2(x + 50) + 4(x + 3) + 1$

EXAMPLE 2.46

Classify as a conditional equation, an identity, or a contradiction. Then state the solution.

$10 + 4(p - 5) = 0$

✓ **Solution**

	$10 + 4(p - 5) = 0$
Distribute.	$10 + 4p - 20 = 0$
Combine like terms.	$4p - 10 = 0$
Add 10 to both sides.	$4p - 10 + 10 = 0 + 10$
Simplify.	$4p = 10$
Divide.	$\dfrac{4p}{4} = \dfrac{10}{4}$
Simplify.	$p = \dfrac{5}{2}$
The equation is true when $p = \dfrac{5}{2}$.	This is a conditional equation. The solution is $p = \dfrac{5}{2}$.

> **TRY IT ::** 2.91

Classify the equation as a conditional equation, an identity, or a contradiction and then state the solution:

$11(q + 3) - 5 = 19$

> **TRY IT ::** 2.92

Classify the equation as a conditional equation, an identity, or a contradiction and then state the solution:

$6 + 14(k - 8) = 95$

EXAMPLE 2.47

Classify the equation as a conditional equation, an identity, or a contradiction. Then state the solution.

$5m + 3(9 + 3m) = 2(7m - 11)$

✓ Solution

$$5m + 3(9 + 3m) = 2(7m - 11)$$

Distribute.	$5m + 27 + 9m = 14m - 22$
Combine like terms.	$14m + 27 = 14m - 22$
Subtract $14m$ from both sides.	$14m + 27 - 14m = 14m - 22 - 14m$
Simplify.	$27 \neq -22$
But $27 \neq -22$.	The equation is a contradiction. It has no solution.

> **TRY IT :: 2.93**
>
> Classify the equation as a conditional equation, an identity, or a contradiction and then state the solution:
>
> $12c + 5(5 + 3c) = 3(9c - 4)$

> **TRY IT :: 2.94**
>
> Classify the equation as a conditional equation, an identity, or a contradiction and then state the solution:
>
> $4(7d + 18) = 13(3d - 2) - 11d$

Type of equation	What happens when you solve it?	Solution
Conditional Equation	True for one or more values of the variables and false for all other values	One or more values
Identity	**True** for any value of the variable	All real numbers
Contradiction	**False** for all values of the variable	No solution

Table 2.42

2.4 EXERCISES

Practice Makes Perfect

Solve Equations Using the General Strategy for Solving Linear Equations

In the following exercises, solve each linear equation.

232. $15(y - 9) = -60$

233. $21(y - 5) = -42$

234. $-9(2n + 1) = 36$

235. $-16(3n + 4) = 32$

236. $8(22 + 11r) = 0$

237. $5(8 + 6p) = 0$

238. $-(w - 12) = 30$

239. $-(t - 19) = 28$

240. $9(6a + 8) + 9 = 81$

241. $8(9b - 4) - 12 = 100$

242. $32 + 3(z + 4) = 41$

243. $21 + 2(m - 4) = 25$

244. $51 + 5(4 - q) = 56$

245. $-6 + 6(5 - k) = 15$

246. $2(9s - 6) - 62 = 16$

247. $8(6t - 5) - 35 = -27$

248. $3(10 - 2x) + 54 = 0$

249. $-2(11 - 7x) + 54 = 4$

250. $\frac{2}{3}(9c - 3) = 22$

251. $\frac{3}{5}(10x - 5) = 27$

252. $\frac{1}{5}(15c + 10) = c + 7$

253. $\frac{1}{4}(20d + 12) = d + 7$

254. $18 - (9r + 7) = -16$

255. $15 - (3r + 8) = 28$

256. $5 - (n - 1) = 19$

257. $-3 - (m - 1) = 13$

258. $11 - 4(y - 8) = 43$

259. $18 - 2(y - 3) = 32$

260. $24 - 8(3v + 6) = 0$

261. $35 - 5(2w + 8) = -10$

262. $4(a - 12) = 3(a + 5)$

263. $-2(a - 6) = 4(a - 3)$

264. $2(5 - u) = -3(2u + 6)$

265. $5(8 - r) = -2(2r - 16)$

266. $3(4n - 1) - 2 = 8n + 3$

267. $9(2m - 3) - 8 = 4m + 7$

268.
$12 + 2(5 - 3y) = -9(y - 1) - 2$

269.
$-15 + 4(2 - 5y) = -7(y - 4) + 4$

270. $8(x - 4) - 7x = 14$

271. $5(x - 4) - 4x = 14$

272.
$5 + 6(3s - 5) = -3 + 2(8s - 1)$

273.
$-12 + 8(x - 5) = -4 + 3(5x - 2)$

274.
$4(u - 1) - 8 = 6(3u - 2) - 7$

275. $7(2n - 5) = 8(4n - 1) - 9$

276. $4(p - 4) - (p + 7) = 5(p - 3)$

277.
$3(a - 2) - (a + 6) = 4(a - 1)$

278. $-(9y + 5) - (3y - 7)$
$= 16 - (4y - 2)$

279. $-(7m + 4) - (2m - 5)$
$= 14 - (5m - 3)$

280. $4[5 - 8(4c - 3)]$
$= 12(1 - 13c) - 8$

281. $5[9 - 2(6d - 1)]$
$= 11(4 - 10d) - 139$

282. $3[-9 + 8(4h - 3)]$
$= 2(5 - 12h) - 19$

283. $3[-14 + 2(15k - 6)]$
$= 8(3 - 5k) - 24$

284. $5[2(m + 4) + 8(m - 7)]$
$= 2[3(5 + m) - (21 - 3m)]$

285. $10[5(n + 1) + 4(n - 1)]$
$= 11[7(5 + n) - (25 - 3n)]$

286. $5(1.2u - 4.8) = -12$

287. $4(2.5v - 0.6) = 7.6$

288. $0.25(q - 6) = 0.1(q + 18)$

289. $0.2(p - 6) = 0.4(p + 14)$

290. $0.2(30n + 50) = 28$

291. $0.5(16m + 34) = -15$

Classify Equations

In the following exercises, classify each equation as a conditional equation, an identity, or a contradiction and then state the solution.

292. $23z + 19 = 3(5z - 9) + 8z + 46$

293. $15y + 32 = 2(10y - 7) - 5y + 46$

294. $5(b - 9) + 4(3b + 9) = 6(4b - 5) - 7b + 21$

295. $9(a - 4) + 3(2a + 5) = 7(3a - 4) - 6a + 7$

296. $18(5j - 1) + 29 = 47$

297. $24(3d - 4) + 100 = 52$

298. $22(3m - 4) = 8(2m + 9)$

299. $30(2n - 1) = 5(10n + 8)$

300. $7v + 42 = 11(3v + 8) - 2(13v - 1)$

301. $18u - 51 = 9(4u + 5) - 6(3u - 10)$

302. $3(6q - 9) + 7(q + 4) = 5(6q + 8) - 5(q + 1)$

303. $5(p + 4) + 8(2p - 1) = 9(3p - 5) - 6(p - 2)$

304. $12(6h - 1) = 8(8h + 5) - 4$

305. $9(4k - 7) = 11(3k + 1) + 4$

306. $45(3y - 2) = 9(15y - 6)$

307. $60(2x - 1) = 15(8x + 5)$

308. $16(6n + 15) = 48(2n + 5)$

309. $36(4m + 5) = 12(12m + 15)$

310. $9(14d + 9) + 4d = 13(10d + 6) + 3$

311. $11(8c + 5) - 8c = 2(40c + 25) + 5$

Everyday Math

312. Fencing Micah has 44 feet of fencing to make a dog run in his yard. He wants the length to be 2.5 feet more than the width. Find the length, L, by solving the equation $2L + 2(L - 2.5) = 44$.

313. Coins Rhonda has $1.90 in nickels and dimes. The number of dimes is one less than twice the number of nickels. Find the number of nickels, n, by solving the equation $0.05n + 0.10(2n - 1) = 1.90$.

Writing Exercises

314. Using your own words, list the steps in the general strategy for solving linear equations.

315. Explain why you should simplify both sides of an equation as much as possible before collecting the variable terms to one side and the constant terms to the other side.

316. What is the first step you take when solving the equation $3 - 7(y - 4) = 38$? Why is this your first step?

317. Solve the equation $\frac{1}{4}(8x + 20) = 3x - 4$ explaining all the steps of your solution as in the examples in this section.

Self Check

ⓐ *After completing the exercises, use this checklist to evaluate your mastery of the objective of this section.*

I can...	Confidently	With some help	No-I don't get it!
solve equations using the general strategy for solving linear equations.			
classify equations.			

ⓑ *On a scale of 1-10, how would you rate your mastery of this section in light of your responses on the checklist? How can you improve this?*

2.5 Solve Equations with Fractions or Decimals

Learning Objectives

By the end of this section, you will be able to:

> Solve equations with fraction coefficients
> Solve equations with decimal coefficients

Be Prepared!

Before you get started, take this readiness quiz.

1. Multiply: $8 \cdot \frac{3}{8}$.

   If you missed this problem, review **Example 1.69**.

2. Find the LCD of $\frac{5}{6}$ and $\frac{1}{4}$.

   If you missed this problem, review **Example 1.82**.

3. Multiply 4.78 by 100.
   If you missed this problem, review **Example 1.98**.

Solve Equations with Fraction Coefficients

Let's use the general strategy for solving linear equations introduced earlier to solve the equation, $\frac{1}{8}x + \frac{1}{2} = \frac{1}{4}$.

$$\frac{1}{8}x + \frac{1}{2} = \frac{1}{4}$$

To isolate the x term, subtract $\frac{1}{2}$ from both sides.	$\frac{1}{8}x + \frac{1}{2} - \frac{1}{2} = \frac{1}{4} - \frac{1}{2}$
Simplify the left side.	$\frac{1}{8}x = \frac{1}{4} - \frac{1}{2}$
Change the constants to equivalent fractions with the LCD.	$\frac{1}{8}x = \frac{1}{4} - \frac{2}{4}$
Subtract.	$\frac{1}{8}x = -\frac{1}{4}$
Multiply both sides by the reciprocal of $\frac{1}{8}$.	$\frac{8}{1} \cdot \frac{1}{8}x = \frac{8}{1}\left(-\frac{1}{4}\right)$
Simplify.	$x = -2$

This method worked fine, but many students do not feel very confident when they see all those fractions. So, we are going to show an alternate method to solve equations with fractions. This alternate method eliminates the fractions.

We will apply the Multiplication Property of Equality and multiply both sides of an equation by the least common denominator of all the fractions in the equation. The result of this operation will be a new equation, equivalent to the first, but without fractions. This process is called "clearing" the equation of fractions.

Let's solve a similar equation, but this time use the method that eliminates the fractions.

EXAMPLE 2.48   HOW TO SOLVE EQUATIONS WITH FRACTION COEFFICIENTS

Solve: $\frac{1}{6}y - \frac{1}{3} = \frac{5}{6}$.

✓ **Solution**

Step 1. Find the least common denominator of *all* the fractions in the equation.	What is the LCD of $\frac{1}{6}$, $\frac{1}{3}$, and $\frac{5}{6}$?	$\frac{1}{6}y - \frac{1}{3} = \frac{5}{6}$ LCD = 6
Step 2. Multiply both sides of the equation by that LCD. This clears the fractions.	Multiply both sides of the equation by the LCD 6.	$6\left(\frac{1}{6}y - \frac{1}{3}\right) = 6\left(\frac{5}{6}\right)$
	Use the Distributive Property.	$6 \cdot \frac{1}{6}y - 6 \cdot \frac{1}{3} = 6 \cdot \frac{5}{6}$
	Simplify – and notice, no more fractions!	$y - 2 = 5$
Step 3. Solve using the General Strategy for Solving Linear Equations.	To isolate the "y" term, add 2.	$y - 2 + 2 = 5 + 2$
	Simplify.	$y = 7$

> **TRY IT :: 2.95**   Solve: $\frac{1}{4}x + \frac{1}{2} = \frac{5}{8}$.

> **TRY IT :: 2.96**   Solve: $\frac{1}{8}x + \frac{1}{2} = \frac{1}{4}$.

Notice in **Example 2.48**, once we cleared the equation of fractions, the equation was like those we solved earlier in this chapter. We changed the problem to one we already knew how to solve! We then used the General Strategy for Solving Linear Equations.

HOW TO :: STRATEGY TO SOLVE EQUATIONS WITH FRACTION COEFFICIENTS.

Step 1.   Find the least common denominator of *all* the fractions in the equation.

Step 2.   Multiply both sides of the equation by that LCD. This clears the fractions.

Step 3.   Solve using the General Strategy for Solving Linear Equations.

EXAMPLE 2.49

Solve: $6 = \frac{1}{2}v + \frac{2}{5}v - \frac{3}{4}v$.

✓ **Solution**

We want to clear the fractions by multiplying both sides of the equation by the LCD of all the fractions in the equation.

Find the LCD of all fractions in the equation.	$6 = \frac{1}{2}v + \frac{2}{5}v - \frac{3}{4}v$
The LCD is 20.	
Multiply both sides of the equation by 20.	$20(6) = 20 \cdot \left(\frac{1}{2}v + \frac{2}{5}v - \frac{3}{4}v\right)$
Distribute.	$20(6) = 20 \cdot \frac{1}{2}v + 20 \cdot \frac{2}{5}v - 20 \cdot \frac{3}{4}v$
Simplify—notice, no more fractions!	$120 = 10v + 8v - 15v$
Combine like terms.	$120 = 3v$

Divide by 3.	$\dfrac{120}{3} = \dfrac{3v}{3}$
Simplify.	$40 = v$
Check:	$6 = \dfrac{1}{2}v + \dfrac{2}{5}v - \dfrac{3}{4}v$
Let $v = 40$.	$6 \overset{?}{=} \dfrac{1}{2}(40) + \dfrac{2}{5}(40) - \dfrac{3}{4}(40)$
	$6 \overset{?}{=} 20 + 16 - 30$
	$6 = 6 ✓$

> **TRY IT :: 2.97**   Solve: $7 = \dfrac{1}{2}x + \dfrac{3}{4}x - \dfrac{2}{3}x$.

> **TRY IT :: 2.98**   Solve: $-1 = \dfrac{1}{2}u + \dfrac{1}{4}u - \dfrac{2}{3}u$.

In the next example, we again have variables on both sides of the equation.

EXAMPLE 2.50

Solve: $a + \dfrac{3}{4} = \dfrac{3}{8}a - \dfrac{1}{2}$.

⊘ **Solution**

	$a + \dfrac{3}{4} = \dfrac{3}{8}a - \dfrac{1}{2}$
Find the LCD of all fractions in the equation. The LCD is 8.	
Multiply both sides by the LCD.	$8\left(a + \dfrac{3}{4}\right) = 8\left(\dfrac{3}{8}a - \dfrac{1}{2}\right)$
Distribute.	$8 \cdot a + 8 \cdot \dfrac{3}{4} = 8 \cdot \dfrac{3}{8}a - 8 \cdot \dfrac{1}{2}$
Simplify—no more fractions.	$8a + 6 = 3a - 4$
Subtract $3a$ from both sides.	$8a - 3a + 6 = 3a - 3a - 4$
Simplify.	$5a + 6 = -4$
Subtract 6 from both sides.	$5a + 6 - 6 = -4 - 6$
Simplify.	$5a = -10$
Divide by 5.	$\dfrac{5a}{5} = \dfrac{-10}{5}$
Simplify.	$a = -2$
Check:	$a + \dfrac{3}{4} = \dfrac{3}{8}a - \dfrac{1}{2}$
Let $a = -2$.	$-2 + \dfrac{3}{4} \overset{?}{=} \dfrac{3}{8}(-2) - \dfrac{1}{2}$
	$-\dfrac{8}{4} + \dfrac{3}{4} \overset{?}{=} -\dfrac{16}{8} - \dfrac{4}{8}$

$$-\frac{5}{4} = -\frac{10}{8}$$

$$-\frac{5}{4} = -\frac{5}{4} \checkmark$$

> **TRY IT :: 2.99**   Solve: $x + \frac{1}{3} = \frac{1}{6}x - \frac{1}{2}$.

> **TRY IT :: 2.100**   Solve: $c + \frac{3}{4} = \frac{1}{2}c - \frac{1}{4}$.

In the next example, we start by using the Distributive Property. This step clears the fractions right away.

EXAMPLE 2.51

Solve: $-5 = \frac{1}{4}(8x + 4)$.

⊘ **Solution**

	$-5 = \frac{1}{4}(8x + 4)$
Distribute.	$-5 = \frac{1}{4} \cdot 8x + \frac{1}{4} \cdot 4$
Simplify. Now there are no fractions.	$-5 = 2x + 1$
Subtract 1 from both sides.	$-5 - 1 = 2x + 1 - 1$
Simplify.	$-6 = 2x$
Divide by 2.	$\frac{-6}{2} = \frac{2x}{2}$
Simplify.	$-3 = x$
Check:	$-5 = \frac{1}{4}(8x + 4)$
Let $x = -3$.	$-5 \stackrel{?}{=} \frac{1}{2}(4(-3) + 2)$
	$-5 \stackrel{?}{=} \frac{1}{2}(-12 + 2)$
	$-5 \stackrel{?}{=} \frac{1}{2}(-10)$
	$-5 = -5 \checkmark$

> **TRY IT :: 2.101**   Solve: $-11 = \frac{1}{2}(6p + 2)$.

> **TRY IT :: 2.102**   Solve: $8 = \frac{1}{3}(9q + 6)$.

In the next example, even after distributing, we still have fractions to clear.

EXAMPLE 2.52

Solve: $\frac{1}{2}(y-5) = \frac{1}{4}(y-1)$.

✓ **Solution**

$$\frac{1}{2}(y-5) = \frac{1}{4}(y-1)$$

Distribute.	$\frac{1}{2}\cdot y - \frac{1}{2}\cdot 5 = \frac{1}{4}\cdot y - \frac{1}{4}\cdot 1$
Simplify.	$\frac{1}{2}y - \frac{5}{2} = \frac{1}{4}y - \frac{1}{4}$
Multiply by the LCD, 4.	$4\left(\frac{1}{2}y - \frac{5}{2}\right) = 4\left(\frac{1}{4}y - \frac{1}{4}\right)$
Distribute.	$4\cdot\frac{1}{2}y - 4\cdot\frac{5}{2} = 4\cdot\frac{1}{4}y - 4\cdot\frac{1}{4}$
Simplify.	$2y - 10 = y - 1$
Collect the variables to the left.	$2y - y - 10 = y - y - 1$
Simplify.	$y - 10 = -1$
Collect the constants to the right.	$y - 10 + 10 = -1 + 10$
Simplify.	$y = 9$
Check:	$\frac{1}{2}(y-5) = \frac{1}{4}(y-1)$
Let $y = 9$.	$\frac{1}{2}(9-5) \overset{?}{=} \frac{1}{4}(9-1)$
Finish the check on your own.	

> **TRY IT :: 2.103** Solve: $\frac{1}{5}(n+3) = \frac{1}{4}(n+2)$.

> **TRY IT :: 2.104** Solve: $\frac{1}{2}(m-3) = \frac{1}{4}(m-7)$.

EXAMPLE 2.53

Solve: $\frac{5x-3}{4} = \frac{x}{2}$.

⊘ Solution

$$\frac{5x-3}{4}=\frac{x}{2}$$

Multiply by the LCD, 4.	$4\left(\frac{5x-3}{4}\right)=4\left(\frac{x}{2}\right)$
Simplify.	$5x-3=2x$
Collect the variables to the right.	$5x-5x-3=2x-5x$
Simplify.	$-3=-3x$
Divide.	$\frac{-3}{-3}=\frac{-3x}{-3}$
Simplify.	$1=x$
Check:	$\frac{5x-3}{4}=\frac{x}{2}$
Let $x=1$.	$\frac{5(1)-3}{4}\stackrel{?}{=}\frac{1}{2}$
	$\frac{2}{4}\stackrel{?}{=}\frac{1}{2}$
	$\frac{1}{2}=\frac{1}{2}\checkmark$

> **TRY IT :: 2.105**   Solve: $\frac{4y-7}{3}=\frac{y}{6}$.

> **TRY IT :: 2.106**   Solve: $\frac{-2z-5}{4}=\frac{z}{8}$.

EXAMPLE 2.54

Solve: $\frac{a}{6}+2=\frac{a}{4}+3$.

⊘ Solution

$$\frac{a}{6}+2=\frac{a}{4}+3$$

Multiply by the LCD, 12.	$12\left(\frac{a}{6}+2\right)=12\left(\frac{a}{4}+3\right)$
Distribute.	$12\cdot\frac{a}{6}+12\cdot2=12\cdot\frac{a}{4}+12\cdot3$
Simplify.	$2a+24=3a+36$
Collect the variables to the right.	$2a-2a+24=3a-2a+36$
Simplify.	$24=a+36$
Collect the constants to the left.	$24-36=a+36-36$
Simplify.	$a=-12$
Check:	$\frac{a}{6}+2=\frac{a}{4}+3$

Let $a = -12$. $\dfrac{-12}{6} + 2 \overset{?}{=} \dfrac{-12}{4} + 3$

$-2 + 2 \overset{?}{=} -3 + 3$

$0 = 0 \checkmark$

> **TRY IT : : 2.107**   Solve: $\dfrac{b}{10} + 2 = \dfrac{b}{4} + 5$.

> **TRY IT : : 2.108**   Solve: $\dfrac{c}{6} + 3 = \dfrac{c}{3} + 4$.

EXAMPLE 2.55

Solve: $\dfrac{4q + 3}{2} + 6 = \dfrac{3q + 5}{4}$.

⊘ **Solution**

	$\dfrac{4q + 3}{2} + 6 = \dfrac{3q + 5}{4}$
Multiply by the LCD, 4.	$4\left(\dfrac{4q + 3}{2} + 6\right) = 4\left(\dfrac{3q + 5}{4}\right)$
Distribute.	$4\left(\dfrac{4q + 3}{2}\right) + 4 \cdot 6 = 4 \cdot \left(\dfrac{3q + 5}{4}\right)$
	$2(4q + 3) + 24 = 3q + 5$
Simplify.	$8q + 6 + 24 = 3q + 5$
	$8q + 30 = 3q + 5$
Collect the variables to the left.	$8q - 3q + 30 = 3q - 3q + 5$
Simplify.	$5q + 30 = 5$
Collect the constants to the right.	$5q + 30 - 30 = 5 - 30$
Simplify.	$5q = -25$
Divide by 5.	$\dfrac{5q}{5} = \dfrac{-25}{5}$
Simplify.	$q = -5$
Check:	$\dfrac{4q + 3}{2} + 6 = \dfrac{3q + 5}{4}$
Let $q = -5$.	$\dfrac{4(-5) + 3}{2} + 6 \overset{?}{=} \dfrac{3(-5) + 5}{4}$

Finish the check on your own.

> **TRY IT : : 2.109**   Solve: $\dfrac{3r + 5}{6} + 1 = \dfrac{4r + 3}{3}$.

> **TRY IT : : 2.110**   Solve: $\dfrac{2s + 3}{2} + 1 = \dfrac{3s + 2}{4}$.

Solve Equations with Decimal Coefficients

Some equations have decimals in them. This kind of equation will occur when we solve problems dealing with money or percentages. But decimals can also be expressed as fractions. For example, $0.3 = \frac{3}{10}$ and $0.17 = \frac{17}{100}$. So, with an equation with decimals, we can use the same method we used to clear fractions—multiply both sides of the equation by the least common denominator.

EXAMPLE 2.56

Solve: $0.06x + 0.02 = 0.25x - 1.5$.

⊘ **Solution**

Look at the decimals and think of the equivalent fractions.

$$0.06 = \frac{6}{100} \qquad 0.02 = \frac{2}{100} \qquad 0.25 = \frac{25}{100} \qquad 1.5 = 1\frac{5}{10}$$

Notice, the LCD is 100.

By multiplying by the LCD, we will clear the decimals from the equation.

	$0.06x + 0.02 = 0.25x - 1.5$
Multiply both sides by 100.	$100(0.06x + 0.02) = 100(0.25x - 1.5)$
Distribute.	$100(0.06x) + 100(0.02) = 100(0.25x) - 100(1.5)$
Multiply, and now we have no more decimals.	$6x + 2 = 25x - 150$
Collect the variables to the right.	$6x - 6x + 2 = 25x - 6x - 150$
Simplify.	$2 = 19x - 150$
Collect the constants to the left.	$2 + 150 = 19x - 150 + 150$
Simplify.	$152 = 19x$
Divide by 19.	$\frac{152}{19} = \frac{19x}{19}$
Simplify.	$8 = x$

Check: Let $x = 8$.

$$0.06(8) + 0.02 \overset{?}{=} 0.25(8) - 1.5$$
$$0.48 + 0.02 \overset{?}{=} 2.00 - 1.5$$
$$0.50 = 0.50 \checkmark$$

> **TRY IT :: 2.111**   Solve: $0.14h + 0.12 = 0.35h - 2.4$.

> **TRY IT :: 2.112**   Solve: $0.65k - 0.1 = 0.4k - 0.35$.

The next example uses an equation that is typical of the money applications in the next chapter. Notice that we distribute the decimal before we clear all the decimals.

EXAMPLE 2.57

Solve: $0.25x + 0.05(x + 3) = 2.85$.

⊘ Solution

$$0.25x + 0.05(x + 3) = 2.85$$

Distribute first.	$0.25x + 0.05x + 0.15 = 2.85$
Combine like terms.	$0.30x + 0.15 = 2.85$
To clear decimals, multiply by 100.	$100(0.30x + 0.15) = 100(2.85)$
Distribute.	$30x + 15 = 285$
Subtract 15 from both sides.	$30x + 15 - 15 = 285 - 15$
Simplify.	$30x = 270$
Divide by 30.	$\dfrac{30x}{30} = \dfrac{270}{30}$
Simplify.	$x = 9$
Check it yourself by substituting $x = 9$ into the original equation.	

> **TRY IT :: 2.113**   Solve: $0.25n + 0.05(n + 5) = 2.95$.

> **TRY IT :: 2.114**   Solve: $0.10d + 0.05(d - 5) = 2.15$.

2.5 EXERCISES

Practice Makes Perfect

Solve Equations with Fraction Coefficients

In the following exercises, solve each equation with fraction coefficients.

318. $\frac{1}{4}x - \frac{1}{2} = -\frac{3}{4}$

319. $\frac{3}{4}x - \frac{1}{2} = \frac{1}{4}$

320. $\frac{5}{6}y - \frac{2}{3} = -\frac{3}{2}$

321. $\frac{5}{6}y - \frac{1}{3} = -\frac{7}{6}$

322. $\frac{1}{2}a + \frac{3}{8} = \frac{3}{4}$

323. $\frac{5}{8}b + \frac{1}{2} = -\frac{3}{4}$

324. $2 = \frac{1}{3}x - \frac{1}{2}x + \frac{2}{3}x$

325. $2 = \frac{3}{5}x - \frac{1}{3}x + \frac{2}{5}x$

326. $\frac{1}{4}m - \frac{4}{5}m + \frac{1}{2}m = -1$

327. $\frac{5}{6}n - \frac{1}{4}n - \frac{1}{2}n = -2$

328. $x + \frac{1}{2} = \frac{2}{3}x - \frac{1}{2}$

329. $x + \frac{3}{4} = \frac{1}{2}x - \frac{5}{4}$

330. $\frac{1}{3}w + \frac{5}{4} = w - \frac{1}{4}$

331. $\frac{3}{2}z + \frac{1}{3} = z - \frac{2}{3}$

332. $\frac{1}{2}x - \frac{1}{4} = \frac{1}{12}x + \frac{1}{6}$

333. $\frac{1}{2}a - \frac{1}{4} = \frac{1}{6}a + \frac{1}{12}$

334. $\frac{1}{3}b + \frac{1}{5} = \frac{2}{5}b - \frac{3}{5}$

335. $\frac{1}{3}x + \frac{2}{5} = \frac{1}{5}x - \frac{2}{5}$

336. $1 = \frac{1}{6}(12x - 6)$

337. $1 = \frac{1}{5}(15x - 10)$

338. $\frac{1}{4}(p - 7) = \frac{1}{3}(p + 5)$

339. $\frac{1}{5}(q + 3) = \frac{1}{2}(q - 3)$

340. $\frac{1}{2}(x + 4) = \frac{3}{4}$

341. $\frac{1}{3}(x + 5) = \frac{5}{6}$

342. $\frac{5q - 8}{5} = \frac{2q}{10}$

343. $\frac{4m + 2}{6} = \frac{m}{3}$

344. $\frac{4n + 8}{4} = \frac{n}{3}$

345. $\frac{3p + 6}{3} = \frac{p}{2}$

346. $\frac{u}{3} - 4 = \frac{u}{2} - 3$

347. $\frac{v}{10} + 1 = \frac{v}{4} - 2$

348. $\frac{c}{15} + 1 = \frac{c}{10} - 1$

349. $\frac{d}{6} + 3 = \frac{d}{8} + 2$

350. $\frac{3x + 4}{2} + 1 = \frac{5x + 10}{8}$

351. $\frac{10y - 2}{3} + 3 = \frac{10y + 1}{9}$

352. $\frac{7u - 1}{4} - 1 = \frac{4u + 8}{5}$

353. $\frac{3v - 6}{2} + 5 = \frac{11v - 4}{5}$

Solve Equations with Decimal Coefficients

In the following exercises, solve each equation with decimal coefficients.

354. $0.6y + 3 = 9$

355. $0.4y - 4 = 2$

356. $3.6j - 2 = 5.2$

357. $2.1k + 3 = 7.2$

358. $0.4x + 0.6 = 0.5x - 1.2$

359. $0.7x + 0.4 = 0.6x + 2.4$

360.
$0.23x + 1.47 = 0.37x - 1.05$

361.
$0.48x + 1.56 = 0.58x - 0.64$

362. $0.9x - 1.25 = 0.75x + 1.75$

363. $1.2x - 0.91 = 0.8x + 2.29$

364. $0.05n + 0.10(n + 8) = 2.15$

365. $0.05n + 0.10(n + 7) = 3.55$

366. $0.10d + 0.25(d + 5) = 4.05$ **367.** $0.10d + 0.25(d + 7) = 5.25$ **368.** $0.05(q - 5) + 0.25q = 3.05$

369. $0.05(q - 8) + 0.25q = 4.10$

Everyday Math

370. Coins Taylor has $2.00 in dimes and pennies. The number of pennies is 2 more than the number of dimes. Solve the equation $0.10d + 0.01(d + 2) = 2$ for d, the number of dimes.

371. Stamps Paula bought $22.82 worth of 49-cent stamps and 21-cent stamps. The number of 21-cent stamps was 8 less than the number of 49-cent stamps. Solve the equation $0.49s + 0.21(s - 8) = 22.82$ for s, to find the number of 49-cent stamps Paula bought.

Writing Exercises

372. Explain how you find the least common denominator of $\frac{3}{8}$, $\frac{1}{6}$, and $\frac{2}{3}$.

373. If an equation has several fractions, how does multiplying both sides by the LCD make it easier to solve?

374. If an equation has fractions only on one side, why do you have to multiply both sides of the equation by the LCD?

375. In the equation $0.35x + 2.1 = 3.85$ what is the LCD? How do you know?

Self Check

ⓐ *After completing the exercises, use this checklist to evaluate your mastery of the objectives of this section.*

I can...	Confidently	With some help	No-I don't get it!
solve equations with fraction coefficients.			
solve equations with decimal coefficients.			

ⓑ *Overall, after looking at the checklist, do you think you are well-prepared for the next section? Why or why not?*

2.6   # Solve a Formula for a Specific Variable

Learning Objectives

By the end of this section, you will be able to:
> Use the Distance, Rate, and Time formula
> Solve a formula for a specific variable

Be Prepared!

Before you get started, take this readiness quiz.

1. Solve: $15t = 120$.
   If you missed this problem, review **Example 2.13**.

2. Solve: $6x + 24 = 96$.
   If you missed this problem, review **Example 2.27**.

Use the Distance, Rate, and Time Formula

One formula you will use often in algebra and in everyday life is the formula for distance traveled by an object moving at a constant rate. Rate is an equivalent word for "speed." The basic idea of rate may already familiar to you. Do you know what distance you travel if you drive at a steady rate of 60 miles per hour for 2 hours? (This might happen if you use your car's cruise control while driving on the highway.) If you said 120 miles, you already know how to use this formula!

Distance, Rate, and Time

For an object moving at a uniform (constant) rate, the distance traveled, the elapsed time, and the rate are related by the formula:

$$d = rt \qquad \text{where} \quad \begin{aligned} d &= \text{distance} \\ r &= \text{rate} \\ t &= \text{time} \end{aligned}$$

We will use the Strategy for Solving Applications that we used earlier in this chapter. When our problem requires a formula, we change Step 4. In place of writing a sentence, we write the appropriate formula. We write the revised steps here for reference.

 HOW TO :: SOLVE AN APPLICATION (WITH A FORMULA).

Step 1.   **Read** the problem. Make sure all the words and ideas are understood.

Step 2.   **Identify** what we are looking for.

Step 3.   **Name** what we are looking for. Choose a variable to represent that quantity.

Step 4.   **Translate** into an equation. Write the appropriate formula for the situation. Substitute in the given information.

Step 5.   **Solve** the equation using good algebra techniques.

Step 6.   **Check** the answer in the problem and make sure it makes sense.

Step 7.   **Answer** the question with a complete sentence.

You may want to create a mini-chart to summarize the information in the problem. See the chart in this first example.

EXAMPLE 2.58

Jamal rides his bike at a uniform rate of 12 miles per hour for $3\frac{1}{2}$ hours. What distance has he traveled?

✓ Solution

Step 1. Read the problem.

Step 2. Identify what you are looking for.	distance traveled
Step 3. Name. Choose a variable to represent it.	Let d = distance.
Step 4. Translate: Write the appropriate formula.	$d = rt$

$$\boxed{\begin{array}{l} d = ? \\ r = 12 \text{ mph} \\ t = 3\frac{1}{2} \text{ hours} \end{array}}$$

Substitute in the given information.	$d = 12 \cdot 3\frac{1}{2}$
Step 5. Solve the equation.	$d = 42$ miles

Step 6. Check

Does 42 miles make sense?

Jamal rides:

12 miles in 1 hour,

24 miles in 2 hours,

36 miles in 3 hours, — 42 miles in $3\frac{1}{2}$ hours is reasonable

48 miles in 4 hours.

Step 7. Answer the question with a complete sentence.	Jamal rode 42 miles.

> **TRY IT : : 2.115**   Lindsay drove for $5\frac{1}{2}$ hours at 60 miles per hour. How much distance did she travel?

> **TRY IT : : 2.116**   Trinh walked for $2\frac{1}{3}$ hours at 3 miles per hour. How far did she walk?

EXAMPLE 2.59

Rey is planning to drive from his house in San Diego to visit his grandmother in Sacramento, a distance of 520 miles. If he can drive at a steady rate of 65 miles per hour, how many hours will the trip take?

✓ Solution

Step 1. Read the problem.

Step 2. Identify what you are looking for.	How many hours (time)
Step 3. Name. Choose a variable to represent it.	Let t = time.

$$\boxed{\begin{array}{l} d = 520 \text{ miles} \\ r = 65 \text{ mph} \\ t = ? \text{ hours} \end{array}}$$

Step 4. Translate.
Write the appropriate formula. $d = rt$

Substitute in the given information. $520 = 65t$

Step 5. Solve the equation. $t = 8$

Step 6. Check. Substitute the numbers into
the formula and make sure the result is a
true statement.

$$
\begin{aligned}
d &= rt \\
520 &\stackrel{?}{=} 65 \cdot 8 \\
520 &= 520 \checkmark
\end{aligned}
$$

Step 7. Answer the question with a complete sentence. Rey's trip will take 8 hours.

> **TRY IT ::** 2.117

> Lee wants to drive from Phoenix to his brother's apartment in San Francisco, a distance of 770 miles. If he drives
> at a steady rate of 70 miles per hour, how many hours will the trip take?

> **TRY IT ::** 2.118

> Yesenia is 168 miles from Chicago. If she needs to be in Chicago in 3 hours, at what rate does she need to drive?

Solve a Formula for a Specific Variable

You are probably familiar with some geometry formulas. A formula is a mathematical description of the relationship
between variables. Formulas are also used in the sciences, such as chemistry, physics, and biology. In medicine they are
used for calculations for dispensing medicine or determining body mass index. Spreadsheet programs rely on formulas
to make calculations. It is important to be familiar with formulas and be able to manipulate them easily.

In **Example 2.58** and **Example 2.59**, we used the formula $d = rt$. This formula gives the value of d, distance, when you
substitute in the values of r and t, the rate and time. But in **Example 2.59**, we had to find the value of t. We substituted
in values of d and r and then used algebra to solve for t. If you had to do this often, you might wonder why there is not a
formula that gives the value of t when you substitute in the values of d and r. We can make a formula like this by solving
the formula $d = rt$ for t.

To solve a formula for a specific variable means to isolate that variable on one side of the equals sign with a coefficient of
1. All other variables and constants are on the other side of the equals sign. To see how to solve a formula for a specific
variable, we will start with the distance, rate and time formula.

EXAMPLE 2.60

Solve the formula $d = rt$ for t:

      ⓐ when $d = 520$ and $r = 65$ ⓑ in general

⊘ **Solution**

We will write the solutions side-by-side to demonstrate that solving a formula in general uses the same steps as when we
have numbers to substitute.

ⓐ when $d = 520$ and $r = 65$

		ⓑ in general	
Write the formula.	$d = rt$	Write the formula.	$d = rt$
Substitute.	$520 = 65t$		
Divide, to isolate t.	$\dfrac{520}{65} = \dfrac{65t}{65}$	Divide, to isolate t.	$\dfrac{d}{r} = \dfrac{rt}{r}$
Simplify.	$8 = t$	Simplify.	$\dfrac{d}{r} = t$

We say the formula $t = \dfrac{d}{r}$ is solved for t.

> **TRY IT :: 2.119**   Solve the formula $d = rt$ for r:
>
> ⓐ when $d = 180$ and $t = 4$ ⓑ in general

> **TRY IT :: 2.120**   Solve the formula $d = rt$ for r:
>
> ⓐ when $d = 780$ and $t = 12$ ⓑ in general

EXAMPLE 2.61

Solve the formula $A = \dfrac{1}{2}bh$ for h:

ⓐ when $A = 90$ and $b = 15$ ⓑ in general

✓ **Solution**

ⓐ when $A = 90$ and $b = 15$

		ⓑ in general	
Write the formula.	$A = \dfrac{1}{2}bh$	Write the formula.	$A = \dfrac{1}{2}bh$
Substitute.	$90 = \dfrac{1}{2} \cdot 15 \cdot h$		
Clear the fractions.	$2 \cdot 90 = 2 \cdot \dfrac{1}{2}15h$	Clear the fractions.	$2 \cdot A = 2 \cdot \dfrac{1}{2}bh$
Simplify.	$180 = 15h$	Simplify.	$2A = bh$
Solve for h.	$12 = h$	Solve for h.	$\dfrac{2A}{b} = h$

We can now find the height of a triangle, if we know the area and the base, by using the formula $h = \dfrac{2A}{b}$.

> **TRY IT :: 2.121**   Use the formula $A = \dfrac{1}{2}bh$ to solve for h:
>
> ⓐ when $A = 170$ and $b = 17$ ⓑ in general

> **TRY IT :: 2.122**   Use the formula $A = \frac{1}{2}bh$ to solve for b:

ⓐ when $A = 62$ and $h = 31$ ⓑ in general

The formula $I = Prt$ is used to calculate simple interest, I, for a principal, P, invested at rate, r, for t years.

EXAMPLE 2.62

Solve the formula $I = Prt$ to find the principal, P:

ⓐ when $I = \$5,600$, $r = 4\%$, $t = 7$ years ⓑ in general

⊘ **Solution**

ⓐ $I = \$5,600$, $r = 4\%$, $t = 7$ years

ⓑ in general

Write the formula.	$I = Prt$	Write the formula.	$I = Prt$
Substitute.	$5600 = P(0.04)(7)$		
Simplify.	$5600 = P(0.28)$	Simplify.	$I = P(rt)$
Divide, to isolate P.	$\dfrac{5600}{0.28} = \dfrac{P(0.28)}{0.28}$	Divide, to isolate P.	$\dfrac{I}{rt} = \dfrac{P(rt)}{rt}$
Simplify.	$20,000 = P$	Simplify.	$\dfrac{I}{rt} = P$
The principal is	$\$20,000$		$P = \dfrac{I}{rt}$

> **TRY IT :: 2.123**   Use the formula $I = Prt$ to find the principal, P:

ⓐ when $I = \$2,160$, $r = 6\%$, $t = 3$ years ⓑ in general

> **TRY IT :: 2.124**   Use the formula $I = Prt$ to find the principal, P:

ⓐ when $I = \$5,400$, $r = 12\%$, $t = 5$ years ⓑ in general

Later in this class, and in future algebra classes, you'll encounter equations that relate two variables, usually x and y. You might be given an equation that is solved for y and need to solve it for x, or vice versa. In the following example, we're given an equation with both x and y on the same side and we'll solve it for y.

EXAMPLE 2.63

Solve the formula $3x + 2y = 18$ for y:

ⓐ when $x = 4$ ⓑ in general

✓ Solution

ⓐ when $x = 4$

	$3x + 2y = 18$
Substitute.	$3(4) + 2y = 18$
Subtract to isolate the y-term.	$12 - 12 + 2y = 18 - 12$
Divide.	$\frac{2y}{2} = \frac{6}{2}$
Simplify.	$y = 3$

ⓑ in general

	$3x + 2y = 18$
Subtract to isolate the y-term.	$3x - 3x + 2y = 18 - 3x$
Divide.	$\frac{2y}{2} = \frac{18}{2} - \frac{3x}{2}$
Simplify.	$y = -\frac{3x}{2} + 9$

> **TRY IT :: 2.125** Solve the formula $3x + 4y = 10$ for y:
>
> ⓐ when $x = \frac{14}{3}$ ⓑ in general

> **TRY IT :: 2.126** Solve the formula $5x + 2y = 18$ for y:
>
> ⓐ when $x = 4$ ⓑ in general

In Examples 1.60 through 1.64 we used the numbers in part ⓐ as a guide to solving in general in part ⓑ. Now we will solve a formula in general without using numbers as a guide.

EXAMPLE 2.64

Solve the formula $P = a + b + c$ for a.

✓ Solution

We will isolate a on one side of the equation.	$P = a + b + c$
Both b and c are added to a, so we subtract them from both sides of the equation.	$P - b - c = a + b + c - b - c$
Simplify.	$P - b - c = a$
	$a = P - b - c$

> **TRY IT :: 2.127** Solve the formula $P = a + b + c$ for b.

> **TRY IT :: 2.128** Solve the formula $P = a + b + c$ for c.

EXAMPLE 2.65

Solve the formula $6x + 5y = 13$ for y.

⊘ Solution

	$6x + 5y = 13$
Subtract $6x$ from both sides to isolate the term with y.	$6x - 6x + 5y = 13 - 6x$
Simplify.	$5y = 13 - 6x$
Divide by 5 to make the coefficient 1.	$\dfrac{5y}{5} = \dfrac{13 - 6x}{5}$
Simplify.	$y = \dfrac{13 - 6x}{5}$

The fraction is simplified. We cannot divide $13 - 6x$ by 5.

>	**TRY IT :: 2.129**	Solve the formula $4x + 7y = 9$ for y.

>	**TRY IT :: 2.130**	Solve the formula $5x + 8y = 1$ for y.

2.6 EXERCISES

Practice Makes Perfect

Use the Distance, Rate, and Time Formula

In the following exercises, solve.

376. Steve drove for $8\frac{1}{2}$ hours at 72 miles per hour. How much distance did he travel?

377. Socorro drove for $4\frac{5}{6}$ hours at 60 miles per hour. How much distance did she travel?

378. Yuki walked for $1\frac{3}{4}$ hours at 4 miles per hour. How far did she walk?

379. Francie rode her bike for $2\frac{1}{2}$ hours at 12 miles per hour. How far did she ride?

380. Connor wants to drive from Tucson to the Grand Canyon, a distance of 338 miles. If he drives at a steady rate of 52 miles per hour, how many hours will the trip take?

381. Megan is taking the bus from New York City to Montreal. The distance is 380 miles and the bus travels at a steady rate of 76 miles per hour. How long will the bus ride be?

382. Aurelia is driving from Miami to Orlando at a rate of 65 miles per hour. The distance is 235 miles. To the nearest tenth of an hour, how long will the trip take?

383. Kareem wants to ride his bike from St. Louis to Champaign, Illinois. The distance is 180 miles. If he rides at a steady rate of 16 miles per hour, how many hours will the trip take?

384. Javier is driving to Bangor, 240 miles away. If he needs to be in Bangor in 4 hours, at what rate does he need to drive?

385. Alejandra is driving to Cincinnati, 450 miles away. If she wants to be there in 6 hours, at what rate does she need to drive?

386. Aisha took the train from Spokane to Seattle. The distance is 280 miles and the trip took 3.5 hours. What was the speed of the train?

387. Philip got a ride with a friend from Denver to Las Vegas, a distance of 750 miles. If the trip took 10 hours, how fast was the friend driving?

Solve a Formula for a Specific Variable

In the following exercises, use the formula $d = rt$.

388. Solve for t

ⓐ when $d = 350$ and $r = 70$

ⓑ in general

389. Solve for t

ⓐ when $d = 240$ and $r = 60$

ⓑ in general

390. Solve for t

ⓐ when $d = 510$ and $r = 60$

ⓑ in general

391. Solve for t

ⓐ when $d = 175$ and $r = 50$

ⓑ in general

392. Solve for r

ⓐ when $d = 204$ and $t = 3$

ⓑ in general

393. Solve for r

ⓐ when $d = 420$ and $t = 6$

ⓑ in general

394. Solve for r

ⓐ when $d = 160$ and $t = 2.5$

ⓑ in general

395. Solve for r

ⓐ when $d = 180$ and $t = 4.5$

ⓑ in general

In the following exercises, use the formula $A = \frac{1}{2}bh$.

396. Solve for b

ⓐ when $A = 126$ and $h = 18$

ⓑ in general

397. Solve for h

ⓐ when $A = 176$ and $b = 22$

ⓑ in general

398. Solve for h

ⓐ when $A = 375$ and $b = 25$

ⓑ in general

399. Solve for b

ⓐ when $A = 65$ and $h = 13$

ⓑ in general

In the following exercises, use the formula I = Prt.

400. Solve for the principal, *P* for

ⓐ $I = \$5,480$, $r = 4\%$, $t = 7$ years

ⓑ in general

401. Solve for the principal, *P* for

ⓐ $I = \$3,950$, $r = 6\%$, $t = 5$ years

ⓑ in general

402. Solve for the time, *t* for

ⓐ $I = \$2,376$, $P = \$9,000$, $r = 4.4\%$

ⓑ in general

403. Solve for the time, *t* for

ⓐ $I = \$624$, $P = \$6,000$, $r = 5.2\%$

ⓑ in general

In the following exercises, solve.

404. Solve the formula $2x + 3y = 12$ for y

ⓐ when $x = 3$

ⓑ in general

405. Solve the formula $5x + 2y = 10$ for y

ⓐ when $x = 4$

ⓑ in general

406. Solve the formula $3x - y = 7$ for y

ⓐ when $x = -2$

ⓑ in general

407. Solve the formula $4x + y = 5$ for y

ⓐ when $x = -3$

ⓑ in general

408. Solve $a + b = 90$ for b.

409. Solve $a + b = 90$ for a.

410. Solve $180 = a + b + c$ for a.

411. Solve $180 = a + b + c$ for c.

412. Solve the formula $8x + y = 15$ for y.

413. Solve the formula $9x + y = 13$ for y.

414. Solve the formula $-4x + y = -6$ for y.

415. Solve the formula $-5x + y = -1$ for y.

416. Solve the formula $4x + 3y = 7$ for y.

417. Solve the formula $3x + 2y = 11$ for y.

418. Solve the formula $x - y = -4$ for y.

419. Solve the formula $x - y = -3$ for y.

420. Solve the formula $P = 2L + 2W$ for L.

421. Solve the formula $P = 2L + 2W$ for W.

422. Solve the formula $C = \pi d$ for d.

423. Solve the formula $C = \pi d$ for π.

424. Solve the formula $V = LWH$ for L.

425. Solve the formula $V = LWH$ for H.

Everyday Math

426. Converting temperature While on a tour in Greece, Tatyana saw that the temperature was 40° Celsius. Solve for F in the formula $C = \frac{5}{9}(F - 32)$ to find the Fahrenheit temperature.

427. Converting temperature Yon was visiting the United States and he saw that the temperature in Seattle one day was 50° Fahrenheit. Solve for C in the formula $F = \frac{9}{5}C + 32$ to find the Celsius temperature.

Writing Exercises

428. Solve the equation $2x + 3y = 6$ for y

ⓐ when $x = -3$

ⓑ in general

ⓒ Which solution is easier for you, ⓐ or ⓑ? Why?

429. Solve the equation $5x - 2y = 10$ for x

ⓐ when $y = 10$

ⓑ in general

ⓒ Which solution is easier for you, ⓐ or ⓑ? Why?

Self Check

ⓐ After completing the exercises, use this checklist to evaluate your mastery of the objectives of this section.

I can...	Confidently	With some help	No-I don't get it!
use the distance, rate, and time formula.			
solve a formula for a specific variable.			

ⓑ What does this checklist tell you about your mastery of this section? What steps will you take to improve?

2.7 Solve Linear Inequalities

Learning Objectives

By the end of this section, you will be able to:

> Graph inequalities on the number line
> Solve inequalities using the Subtraction and Addition Properties of inequality
> Solve inequalities using the Division and Multiplication Properties of inequality
> Solve inequalities that require simplification
> Translate to an inequality and solve

Be Prepared!

Before you get started, take this readiness quiz.

1. Translate from algebra to English: $15 > x$.
   If you missed this problem, review **Example 1.12**.

2. Solve: $n - 9 = -42$.
   If you missed this problem, review **Example 2.3**.

3. Solve: $-5p = -23$.
   If you missed this problem, review **Example 2.13**.

4. Solve: $3a - 12 = 7a - 20$.
   If you missed this problem, review **Example 2.34**.

Graph Inequalities on the Number Line

Do you remember what it means for a number to be a solution to an equation? A solution of an equation is a value of a variable that makes a true statement when substituted into the equation.

What about the solution of an inequality? What number would make the inequality $x > 3$ true? Are you thinking, 'x could be 4'? That's correct, but x could be 5 too, or 20, or even 3.001. Any number greater than 3 is a solution to the inequality $x > 3$.

We show the solutions to the inequality $x > 3$ on the number line by shading in all the numbers to the right of 3, to show that all numbers greater than 3 are solutions. Because the number 3 itself is not a solution, we put an open parenthesis at 3. The graph of $x > 3$ is shown in **Figure 2.7**. Please note that the following convention is used: light blue arrows point in the positive direction and dark blue arrows point in the negative direction.

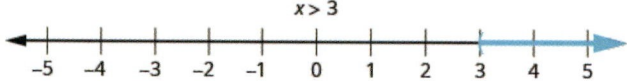

Figure 2.7 The inequality $x > 3$ is graphed on this number line.

The graph of the inequality $x \geq 3$ is very much like the graph of $x > 3$, but now we need to show that 3 is a solution, too. We do that by putting a bracket at $x = 3$, as shown in **Figure 2.8**.

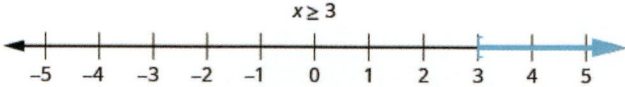

Figure 2.8 The inequality $x \geq 3$ is graphed on this number line.

Notice that the open parentheses symbol, (, shows that the endpoint of the inequality is not included. The open bracket symbol, [, shows that the endpoint is included.

EXAMPLE 2.66

Graph on the number line:

ⓐ $x \leq 1$ ⓑ $x < 5$ ⓒ $x > -1$

✓ Solution

ⓐ $x \leq 1$

This means all numbers less than or equal to 1. We shade in all the numbers on the number line to the left of 1 and put a bracket at $x = 1$ to show that it is included.

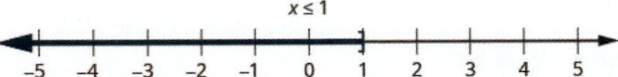

ⓑ $x < 5$

This means all numbers less than 5, but not including 5. We shade in all the numbers on the number line to the left of 5 and put a parenthesis at $x = 5$ to show it is not included.

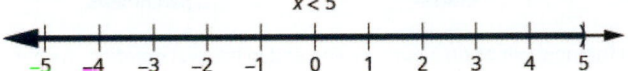

ⓒ $x > -1$

This means all numbers greater than -1, but not including -1. We shade in all the numbers on the number line to the right of -1, then put a parenthesis at $x = -1$ to show it is not included.

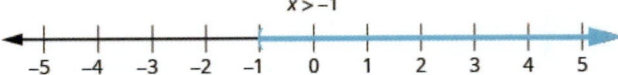

> **TRY IT :: 2.131**   Graph on the number line: ⓐ $x \leq -1$ ⓑ $x > 2$ ⓒ $x < 3$

> **TRY IT :: 2.132**   Graph on the number line: ⓐ $x > -2$ ⓑ $x < -3$ ⓒ $x \geq -1$

We can also represent inequalities using *interval notation*. As we saw above, the inequality $x > 3$ means all numbers greater than 3. There is no upper end to the solution to this inequality. In interval notation, we express $x > 3$ as $(3, \infty)$. The symbol ∞ is read as 'infinity'. It is not an actual number. **Figure 2.9** shows both the number line and the interval notation.

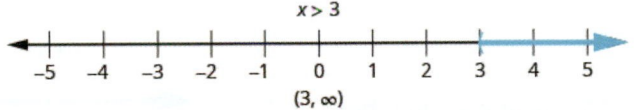

Figure 2.9 The inequality $x > 3$ is graphed on this number line and written in interval notation.

The inequality $x \leq 1$ means all numbers less than or equal to 1. There is no lower end to those numbers. We write $x \leq 1$ in interval notation as $(-\infty, 1]$. The symbol $-\infty$ is read as 'negative infinity'. **Figure 2.10** shows both the number line and interval notation.

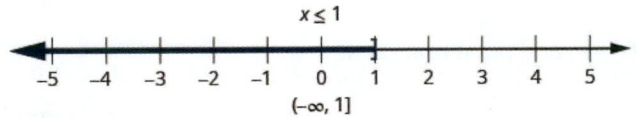

Figure 2.10 The inequality $x \leq 1$ is graphed on this number line and written in interval notation.

Inequalities, Number Lines, and Interval Notation

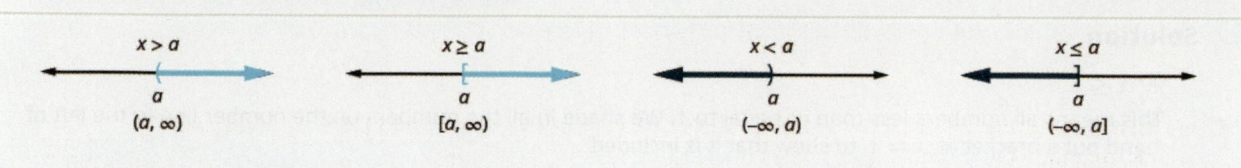

Did you notice how the parenthesis or bracket in the interval notation matches the symbol at the endpoint of the arrow? These relationships are shown in **Figure 2.11**.

Figure 2.11 The notation for inequalities on a number line and in interval notation use similar symbols to express the endpoints of intervals.

EXAMPLE 2.67

Graph on the number line and write in interval notation.

ⓐ $x \geq -3$ ⓑ $x < 2.5$ ⓒ $x \leq -\frac{3}{5}$

✓ **Solution**

ⓐ

	$x \geq -3$
Shade to the right of -3, and put a bracket at -3.	
Write in interval notation.	$[-3, \infty)$

ⓑ

	$x < 2.5$
Shade to the left of 2.5, and put a parenthesis at 2.5.	
Write in interval notation.	$(-\infty, 2.5)$

ⓒ

	$x \leq -\frac{3}{5}$
Shade to the left of $-\frac{3}{5}$, and put a bracket at $-\frac{3}{5}$.	
Write in interval notation.	$\left(-\infty, -\frac{3}{5}\right]$

> **TRY IT ::** 2.133   Graph on the number line and write in interval notation:

ⓐ $x > 2$ ⓑ $x \leq -1.5$ ⓒ $x \geq \frac{3}{4}$

> **TRY IT : : 2.134**   Graph on the number line and write in interval notation:

ⓐ $x \le -4$ ⓑ $x \ge 0.5$ ⓒ $x < -\frac{2}{3}$

Solve Inequalities using the Subtraction and Addition Properties of Inequality

The Subtraction and Addition Properties of Equality state that if two quantities are equal, when we add or subtract the same amount from both quantities, the results will be equal.

Properties of Equality

Subtraction Property of Equality	**Addition Property of Equality**
For any numbers a, b, and c,	For any numbers a, b, and c,
if $a = b$,	if $a = b$,
then $a - c = b - c$.	then $a + c = b + c$.

Similar properties hold true for inequalities.

For example, we know that −4 is less than 2.	$-4 < 2$
If we subtract 5 from both quantities, is the left side still less than the right side?	$-4 - 5 \,?\, 2 - 5$
We get −9 on the left and −3 on the right.	$-9 \,?\, -3$
And we know −9 is less than −3.	$-9 < -3$

The inequality sign stayed the same.

Similarly we could show that the inequality also stays the same for addition.

This leads us to the Subtraction and Addition Properties of Inequality.

Properties of Inequality

Subtraction Property of Inequality	**Addition Property of Inequality**
For any numbers a, b, and c,	For any numbers a, b, and c,
if $a < b$	if $a < b$
then $a - c < b - c$.	then $a + c < b + c$.
if $a > b$	if $a > b$
then $a - c > b - c$.	then $a + c > b + c$.

We use these properties to solve inequalities, taking the same steps we used to solve equations. Solving the inequality $x + 5 > 9$, the steps would look like this:

$$x + 5 > 9$$

Subtract 5 from both sides to isolate x. $\quad x + 5 - 5 > 9 - 5$

Simplify. $\qquad\qquad\qquad\qquad\qquad\qquad\quad x > 4$

Any number greater than 4 is a solution to this inequality.

EXAMPLE 2.68

Solve the inequality $n - \frac{1}{2} \le \frac{5}{8}$, graph the solution on the number line, and write the solution in interval notation.

⊘ Solution

$$n - \frac{1}{2} \leq \frac{5}{8}$$

Add $\frac{1}{2}$ to both sides of the inequality.	$n - \frac{1}{2} + \frac{1}{2} \leq \frac{5}{8} + \frac{1}{2}$
Simplify.	$n \leq \frac{9}{8}$
Graph the solution on the number line.	
Write the solution in interval notation.	$\left(-\infty, \frac{9}{8}\right]$

> **TRY IT : : 2.135**

Solve the inequality, graph the solution on the number line, and write the solution in interval notation.

$$p - \frac{3}{4} \geq \frac{1}{6}$$

> **TRY IT : : 2.136**

Solve the inequality, graph the solution on the number line, and write the solution in interval notation.

$$r - \frac{1}{3} \leq \frac{7}{12}$$

Solve Inequalities using the Division and Multiplication Properties of Inequality

The Division and Multiplication Properties of Equality state that if two quantities are equal, when we divide or multiply both quantities by the same amount, the results will also be equal (provided we don't divide by 0).

Properties of Equality

Division Property of Equality	Multiplication Property of Equality
For any numbers a, b, c, and $c \neq 0$, if $\quad a = b$, then $\frac{a}{c} = \frac{b}{c}$.	For any real numbers a, b, c, if $\quad a = b$, then $ac = bc$.

Are there similar properties for inequalities? What happens to an inequality when we divide or multiply both sides by a constant?

Consider some numerical examples.

	$10 < 15$		$10 < 15$
Divide both sides by 5.	$\frac{10}{5}$? $\frac{15}{5}$	Multiply both sides by 5.	$10(5)$? $15(5)$
Simplify.	$2 ? 3$		$50 ? 75$
Fill in the inequality signs.	$2 < 3$		$50 < 75$

The inequality signs stayed the same.

Does the inequality stay the same when we divide or multiply by a negative number?

	10 < 15		10 < 15
Divide both sides by –5.	$\frac{10}{-5}$? $\frac{15}{-5}$	Multiply both sides by –5.	10(–5) ? 15(–5)
Simplify.	–2 ? –3		–50 ? –75
Fill in the inequality signs.	–2 > –3		–50 > –75

The inequality signs reversed their direction.

When we divide or multiply an inequality by a positive number, the inequality sign stays the same. When we divide or multiply an inequality by a negative number, the inequality sign reverses.

Here are the Division and Multiplication Properties of Inequality for easy reference.

Division and Multiplication Properties of Inequality

For any real numbers a, b, c

if $a < b$ and $c > 0$, then $\frac{a}{c} < \frac{b}{c}$ and $ac < bc$.

if $a > b$ and $c > 0$, then $\frac{a}{c} > \frac{b}{c}$ and $ac > bc$.

if $a < b$ and $c < 0$, then $\frac{a}{c} > \frac{b}{c}$ and $ac > bc$.

if $a > b$ and $c < 0$, then $\frac{a}{c} < \frac{b}{c}$ and $ac < bc$.

When we **divide or multiply** an inequality by a:

- **positive** number, the inequality stays the **same**.
- **negative** number, the inequality **reverses**.

EXAMPLE 2.69

Solve the inequality $7y < 42$, graph the solution on the number line, and write the solution in interval notation.

⊘ Solution

	$7y < 42$
Divide both sides of the inequality by 7. Since $7 > 0$, the inequality stays the same.	$\frac{7y}{7} < \frac{42}{7}$
Simplify.	$y < 6$
Graph the solution on the number line.	
Write the solution in interval notation.	$(-\infty, 6)$

> **TRY IT : : 2.137**

Solve the inequality, graph the solution on the number line, and write the solution in interval notation.

$9c > 72$

> **TRY IT :: 2.138**

   Solve the inequality, graph the solution on the number line, and write the solution in interval notation.

   $12d \leq 60$

EXAMPLE 2.70

Solve the inequality $-10a \geq 50$, graph the solution on the number line, and write the solution in interval notation.

⊘ **Solution**

	$-10a \geq 50$
Divide both sides of the inequality by -10. Since $-10 < 0$, the inequality reverses.	$\dfrac{-10a}{-10} \leq \dfrac{50}{-10}$
Simplify.	$a \leq -5$
Graph the solution on the number line.	
Write the solution in interval notation.	$(-\infty, -5]$

> **TRY IT :: 2.139**

   Solve each inequality, graph the solution on the number line, and write the solution in interval notation.

   $-8q < 32$

> **TRY IT :: 2.140**

   Solve each inequality, graph the solution on the number line, and write the solution in interval notation.

   $-7r \leq -70$

Solving Inequalities

Sometimes when solving an inequality, the variable ends up on the right. We can rewrite the inequality in reverse to get the variable to the left.

$$x > a \text{ has the same meaning as } a < x$$

Think about it as "If Xavier is taller than Alex, then Alex is shorter than Xavier."

EXAMPLE 2.71

Solve the inequality $-20 < \frac{4}{5}u$, graph the solution on the number line, and write the solution in interval notation.

⊘ **Solution**

	$-20 < \dfrac{4}{5}u$
Multiply both sides of the inequality by $\frac{5}{4}$. Since $\frac{5}{4} > 0$, the inequality stays the same.	$\dfrac{5}{4}(-20) < \dfrac{5}{4}\left(\dfrac{4}{5}u\right)$
Simplify.	$-25 < u$

Rewrite the variable on the left.	$u > -25$
Graph the solution on the number line.	
Write the solution in interval notation.	$(-25, \infty)$

> **TRY IT : : 2.141**
>
> Solve the inequality, graph the solution on the number line, and write the solution in interval notation.
>
> $24 \le \frac{3}{8}m$

> **TRY IT : : 2.142**
>
> Solve the inequality, graph the solution on the number line, and write the solution in interval notation.
>
> $-24 < \frac{4}{3}n$

EXAMPLE 2.72

Solve the inequality $\frac{t}{-2} \ge 8$, graph the solution on the number line, and write the solution in interval notation.

✓ **Solution**

$$\frac{t}{-2} \ge 8$$

Multiply both sides of the inequality by -2. Since $-2 < 0$, the inequality reverses.	$-2\left(\frac{t}{-2}\right) \le -2(8)$
Simplify.	$t \le -16$
Graph the solution on the number line.	
Write the solution in interval notation.	$(-\infty, -16]$

> **TRY IT : : 2.143**
>
> Solve the inequality, graph the solution on the number line, and write the solution in interval notation.
>
> $\frac{k}{-12} \le 15$

> **TRY IT : : 2.144**
>
> Solve the inequality, graph the solution on the number line, and write the solution in interval notation.
>
> $\frac{u}{-4} \ge -16$

Solve Inequalities That Require Simplification

Most inequalities will take more than one step to solve. We follow the same steps we used in the general strategy for solving linear equations, but be sure to pay close attention during multiplication or division.

EXAMPLE 2.73

Solve the inequality $4m \leq 9m + 17$, graph the solution on the number line, and write the solution in interval notation.

⊘ **Solution**

	$4m \leq 9m + 17$
Subtract $9m$ from both sides to collect the variables on the left.	$4m - 9m \leq 9m - 9m + 17$
Simplify.	$-5m \leq 17$
Divide both sides of the inequality by −5, and reverse the inequality.	$\dfrac{-5m}{-5} \geq \dfrac{17}{-5}$
Simplify.	$m \geq -\dfrac{17}{5}$
Graph the solution on the number line.	
Write the solution in interval notation.	$\left[-\dfrac{17}{5}, \infty \right)$

> **TRY IT : : 2.145**
>
> Solve the inequality $3q \geq 7q - 23$, graph the solution on the number line, and write the solution in interval notation.

> **TRY IT : : 2.146**
>
> Solve the inequality $6x < 10x + 19$, graph the solution on the number line, and write the solution in interval notation.

EXAMPLE 2.74

Solve the inequality $8p + 3(p - 12) > 7p - 28$, graph the solution on the number line, and write the solution in interval notation.

⊘ **Solution**

Simplify each side as much as possible.	$8p + 3(p - 12) > 7p - 28$
Distribute.	$8p + 3p - 36 > 7p - 28$
Combine like terms.	$11p - 36 > 7p - 28$
Subtract $7p$ from both sides to collect the variables on the left.	$11p - 36 - 7p > 7p - 28 - 7p$
Simplify.	$4p - 36 > -28$
Add 36 to both sides to collect the constants on the right.	$4p - 36 + 36 > -28 + 36$
Simplify.	$4p > 8$
Divide both sides of the inequality by 4; the inequality stays the same.	$\dfrac{4p}{4} > \dfrac{8}{4}$

Simplify.	$p > 2$
Graph the solution on the number line.	
Write the solution in interal notation.	$(2, \infty)$

> **TRY IT : : 2.147**
>
> Solve the inequality $9y + 2(y + 6) > 5y - 24$, graph the solution on the number line, and write the solution in interval notation.

> **TRY IT : : 2.148**
>
> Solve the inequality $6u + 8(u - 1) > 10u + 32$, graph the solution on the number line, and write the solution in interval notation.

Just like some equations are identities and some are contradictions, inequalities may be identities or contradictions, too. We recognize these forms when we are left with only constants as we solve the inequality. If the result is a true statement, we have an identity. If the result is a false statement, we have a contradiction.

EXAMPLE 2.75

Solve the inequality $8x - 2(5 - x) < 4(x + 9) + 6x$, graph the solution on the number line, and write the solution in interval notation.

✓ **Solution**

Simplify each side as much as possible.	$8x - 2(5 - x) < 4(x + 9) + 6x$
Distribute.	$8x - 10 + 2x < 4x + 36 + 6x$
Combine like terms.	$10x - 10 < 10x + 36$
Subtract $10x$ from both sides to collect the variables on the left.	$10x - 10 - 10x < 10x + 36 - 10x$
Simplify.	$-10 < 36$
The x's are gone, and we have a true statement.	The inequality is an identity. The solution is all real numbers.
Graph the solution on the number line.	
Write the solution in interval notation.	$(-\infty, \infty)$

> **TRY IT : : 2.149**
>
> Solve the inequality $4b - 3(3 - b) > 5(b - 6) + 2b$, graph the solution on the number line, and write the solution in interval notation.

> **TRY IT : : 2.150**
>
> Solve the inequality $9h - 7(2 - h) < 8(h + 11) + 8h$, graph the solution on the number line, and write the solution in interval notation.

EXAMPLE 2.76

Solve the inequality $\frac{1}{3}a - \frac{1}{8}a > \frac{5}{24}a + \frac{3}{4}$, graph the solution on the number line, and write the solution in interval notation.

✓ **Solution**

	$\frac{1}{3}a - \frac{1}{8}a > \frac{5}{24}a + \frac{3}{4}$
Multiply both sides by the LCD, 24, to clear the fractions.	$24\left(\frac{1}{3}a - \frac{1}{8}a\right) > 24\left(\frac{5}{24}a + \frac{3}{4}\right)$
Simplify.	$8a - 3a > 5a + 18$
Combine like terms.	$5a > 5a + 18$
Subtract $5a$ from both sides to collect the variables on the left.	$5a - 5a > 5a - 5a + 18$
Simplify.	$0 > 18$
The statement is false!	The inequality is a contradiction.
	There is no solution.
Graph the solution on the number line.	
Write the solution in interval notation.	There is no solution.

> **TRY IT :: 2.151**
>
> Solve the inequality $\frac{1}{4}x - \frac{1}{12}x > \frac{1}{6}x + \frac{7}{8}$, graph the solution on the number line, and write the solution in interval notation.

> **TRY IT :: 2.152**
>
> Solve the inequality $\frac{2}{5}z - \frac{1}{3}z < \frac{1}{15}z - \frac{3}{5}$, graph the solution on the number line, and write the solution in interval notation.

Translate to an Inequality and Solve

To translate English sentences into inequalities, we need to recognize the phrases that indicate the inequality. Some words are easy, like 'more than' and 'less than'. But others are not as obvious.

Think about the phrase 'at least' – what does it mean to be 'at least 21 years old'? It means 21 or more. The phrase 'at least' is the same as 'greater than or equal to'.

Table 2.72 shows some common phrases that indicate inequalities.

$>$	$\geq$	$<$	$\leq$
is greater than	is greater than or equal to	is less than	is less than or equal to
is more than	is at least	is smaller than	is at most
is larger than	is no less than	has fewer than	is no more than
exceeds	is the minimum	is lower than	is the maximum

Table 2.72

EXAMPLE 2.77

Translate and solve. Then write the solution in interval notation and graph on the number line.

Twelve times c is no more than 96.

✓ Solution

	Twelve times c is no more than 96
Translate.	$12c \le 96$
Solve—divide both sides by 12.	$\dfrac{12c}{12} \le \dfrac{96}{12}$
Simplify.	$c \le 8$
Write in interval notation.	$(-\infty, -8]$
Graph on the number line.	

> **TRY IT :: 2.153**
>
> Translate and solve. Then write the solution in interval notation and graph on the number line.
>
> Twenty times y is at most 100

> **TRY IT :: 2.154**
>
> Translate and solve. Then write the solution in interval notation and graph on the number line.
>
> Nine times z is no less than 135

EXAMPLE 2.78

Translate and solve. Then write the solution in interval notation and graph on the number line.

Thirty less than x is at least 45.

✓ Solution

	Thirty less than x is at least 45.
Translate.	$x - 30 \ge 45$
Solve—add 30 to both sides.	$x - 30 + 30 \ge 45 + 30$
Simplify.	$x \ge 75$
Write in interval notation.	$[75, \infty)$
Graph on the number line.	

> **TRY IT :: 2.155**
>
> Translate and solve. Then write the solution in interval notation and graph on the number line.
>
> Nineteen less than p is no less than 47

> **TRY IT :: 2.156**
>
> Translate and solve. Then write the solution in interval notation and graph on the number line.
>
> Four more than a is at most 15.

2.7 EXERCISES

Practice Makes Perfect

Graph Inequalities on the Number Line

In the following exercises, graph each inequality on the number line.

430.
ⓐ $x \leq 2$

ⓑ $x > -1$

ⓒ $x < 0$

431.
ⓐ $x > 1$

ⓑ $x < -2$

ⓒ $x \geq -3$

432.
ⓐ $x \geq -3$

ⓑ $x < 4$

ⓒ $x \leq -2$

433.
ⓐ $x \leq 0$

ⓑ $x > -4$

ⓒ $x \geq -1$

In the following exercises, graph each inequality on the number line and write in interval notation.

434.
ⓐ $x < -2$

ⓑ $x \geq -3.5$

ⓒ $x \leq \frac{2}{3}$

435.
ⓐ $x > 3$

ⓑ $x \leq -0.5$

ⓒ $x \geq \frac{1}{3}$

436.
ⓐ $x \geq -4$

ⓑ $x < 2.5$

ⓒ $x > -\frac{3}{2}$

437.
ⓐ $x \leq 5$

ⓑ $x \geq -1.5$

ⓒ $x < -\frac{7}{3}$

Solve Inequalities using the Subtraction and Addition Properties of Inequality

In the following exercises, solve each inequality, graph the solution on the number line, and write the solution in interval notation.

438. $n - 11 < 33$

439. $m - 45 \leq 62$

440. $u + 25 > 21$

441. $v + 12 > 3$

442. $a + \frac{3}{4} \geq \frac{7}{10}$

443. $b + \frac{7}{8} \geq \frac{1}{6}$

444. $f - \frac{13}{20} < -\frac{5}{12}$

445. $g - \frac{11}{12} < -\frac{5}{18}$

Solve Inequalities using the Division and Multiplication Properties of Inequality

In the following exercises, solve each inequality, graph the solution on the number line, and write the solution in interval notation.

446. $8x > 72$

447. $6y < 48$

448. $7r \leq 56$

449. $9s \geq 81$

450. $-5u \geq 65$

451. $-8v \leq 96$

452. $-9c < 126$

453. $-7d > 105$

454. $20 > \frac{2}{5}h$

455. $40 < \frac{5}{8}k$

456. $\frac{7}{6}j \geq 42$

457. $\frac{9}{4}g \leq 36$

458. $\frac{a}{-3} \leq 9$

459. $\frac{b}{-10} \geq 30$

460. $-25 < \frac{p}{-5}$

461. $-18 > \frac{q}{-6}$

462. $9t \geq -27$

463. $7s < -28$

464. $\frac{2}{3}y > -36$

465. $\frac{3}{5}x \leq -45$

Solve Inequalities That Require Simplification

In the following exercises, solve each inequality, graph the solution on the number line, and write the solution in interval notation.

466. $4v \geq 9v - 40$

467. $5u \leq 8u - 21$

468. $13q < 7q - 29$

469. $9p > 14p - 18$

470. $12x + 3(x + 7) > 10x - 24$

471. $9y + 5(y + 3) < 4y - 35$

472. $6h - 4(h - 1) \leq 7h - 11$

473. $4k - (k - 2) \geq 7k - 26$

474.
$8m - 2(14 - m) \geq 7(m - 4) + 3m$

475.
$6n - 12(3 - n) \leq 9(n - 4) + 9n$

476. $\frac{3}{4}b - \frac{1}{3}b < \frac{5}{12}b - \frac{1}{2}$

477.
$9u + 5(2u - 5) \geq 12(u - 1) + 7u$

478.
$\frac{2}{3}g - \frac{1}{2}(g - 14) \leq \frac{1}{6}(g + 42)$

479. $\frac{5}{6}a - \frac{1}{4}a > \frac{7}{12}a + \frac{2}{3}$

480.
$\frac{4}{5}h - \frac{2}{3}(h - 9) \geq \frac{1}{15}(2h + 90)$

481.
$12v + 3(4v - 1) \leq 19(v - 2) + 5v$

Mixed practice

In the following exercises, solve each inequality, graph the solution on the number line, and write the solution in interval notation.

482. $15k \leq -40$

483. $35k \geq -77$

484.
$23p - 2(6 - 5p) > 3(11p - 4)$

485.
$18q - 4(10 - 3q) < 5(6q - 8)$

486. $-\frac{9}{4}x \geq -\frac{5}{12}$

487. $-\frac{21}{8}y \leq -\frac{15}{28}$

488. $c + 34 < -99$

489. $d + 29 > -61$

490. $\frac{m}{18} \geq -4$

491. $\frac{n}{13} \leq -6$

Translate to an Inequality and Solve

In the following exercises, translate and solve .Then write the solution in interval notation and graph on the number line.

492. Fourteen times d is greater than 56.

493. Ninety times c is less than 450.

494. Eight times z is smaller than -40 .

495. Ten times y is at most -110.

496. Three more than h is no less than 25.

497. Six more than k exceeds 25.

498. Ten less than w is at least 39.

499. Twelve less than x is no less than 21.

500. Negative five times r is no more than 95.

501. Negative two times s is lower than 56.

502. Nineteen less than b is at most -22.

503. Fifteen less than a is at least -7.

Everyday Math

504. Safety A child's height, h, must be at least 57 inches for the child to safely ride in the front seat of a car. Write this as an inequality.

505. Fighter pilots The maximum height, h, of a fighter pilot is 77 inches. Write this as an inequality.

506. Elevators The total weight, w, of an elevator's passengers can be no more than 1,200 pounds. Write this as an inequality.

507. Shopping The number of items, n, a shopper can have in the express check-out lane is at most 8. Write this as an inequality.

Writing Exercises

508. Give an example from your life using the phrase 'at least'.

509. Give an example from your life using the phrase 'at most'.

510. Explain why it is necessary to reverse the inequality when solving $-5x > 10$.

511. Explain why it is necessary to reverse the inequality when solving $\frac{n}{-3} < 12$.

Self Check

ⓐ *After completing the exercises, use this checklist to evaluate your mastery of the objectives of this section.*

I can...	Confidently	With some help	No-I don't get it!
graph inequalities on the number line.			
solve inequalities using the Subtraction and Addition Properties of Inequality.			
solve inequalities using the Division and Multiplication Properties of Inequality.			
solve inequalities that require simplification.			
translate to an inequality and solve.			

ⓑ *What does this checklist tell you about your mastery of this section? What steps will you take to improve?*

CHAPTER 2 REVIEW

KEY TERMS

conditional equation An equation that is true for one or more values of the variable and false for all other values of the variable is a conditional equation.

contradiction An equation that is false for all values of the variable is called a contradiction. A contradiction has no solution.

identity An equation that is true for any value of the variable is called an identity. The solution of an identity is all real numbers.

solution of an equation A solution of an equation is a value of a variable that makes a true statement when substituted into the equation.

KEY CONCEPTS

2.1 Solve Equations Using the Subtraction and Addition Properties of Equality

- **To Determine Whether a Number is a Solution to an Equation**

  Step 1. **Substitute the number in for the variable in the equation.**

  Step 2. **Simplify the expressions on both sides of the equation.**

  Step 3. **Determine whether the resulting statement is true.**
    - If it is true, the number is a solution.
    - If it is not true, the number is not a solution.

- **Addition Property of Equality**
  - For any numbers a, b, and c, if $a = b$, then $a + c = b + c$.

- **Subtraction Property of Equality**
  - For any numbers a, b, and c, if $a = b$, then $a - c = b - c$.

- **To Translate a Sentence to an Equation**

  Step 1. Locate the "equals" word(s). Translate to an equal sign (=).

  Step 2. Translate the words to the left of the "equals" word(s) into an algebraic expression.

  Step 3. Translate the words to the right of the "equals" word(s) into an algebraic expression.

- **To Solve an Application**

  Step 1. Read the problem. Make sure all the words and ideas are understood.

  Step 2. Identify what we are looking for.

  Step 3. Name what we are looking for. Choose a variable to represent that quantity.

  Step 4. Translate into an equation. It may be helpful to restate the problem in one sentence with the important information.

  Step 5. Solve the equation using good algebra techniques.

  Step 6. Check the answer in the problem and make sure it makes sense.

  Step 7. Answer the question with a complete sentence.

2.2 Solve Equations using the Division and Multiplication Properties of Equality

- **The Division Property of Equality**—For any numbers a, b, and c, and $c \neq 0$, if $a = b$, then $\frac{a}{c} = \frac{b}{c}$.

  When you divide both sides of an equation by any non-zero number, you still have equality.

- **The Multiplication Property of Equality**—For any numbers a, b, and c, if $a = b$, then $ac = bc$.

  If you multiply both sides of an equation by the same number, you still have equality.

2.3 Solve Equations with Variables and Constants on Both Sides

- **Beginning Strategy for Solving an Equation with Variables and Constants on Both Sides of the Equation**

  Step 1. Choose which side will be the "variable" side—the other side will be the "constant" side.

Step 2.   Collect the variable terms to the "variable" side of the equation, using the Addition or Subtraction Property of Equality.

Step 3.   Collect all the constants to the other side of the equation, using the Addition or Subtraction Property of Equality.

Step 4.   Make the coefficient of the variable equal 1, using the Multiplication or Division Property of Equality.

Step 5.   Check the solution by substituting it into the original equation.

2.4 Use a General Strategy to Solve Linear Equations

- **General Strategy for Solving Linear Equations**

  Step 1.   Simplify each side of the equation as much as possible.
  Use the Distributive Property to remove any parentheses.
  Combine like terms.

  Step 2.   Collect all the variable terms on one side of the equation.
  Use the Addition or Subtraction Property of Equality.

  Step 3.   Collect all the constant terms on the other side of the equation.
  Use the Addition or Subtraction Property of Equality.

  Step 4.   Make the coefficient of the variable term to equal to 1.
  Use the Multiplication or Division Property of Equality.
  State the solution to the equation.

  Step 5.   Check the solution.
  Substitute the solution into the original equation.

2.5 Solve Equations with Fractions or Decimals

- **Strategy to Solve an Equation with Fraction Coefficients**

  Step 1.   Find the least common denominator of all the fractions in the equation.

  Step 2.   Multiply both sides of the equation by that LCD. This clears the fractions.

  Step 3.   Solve using the General Strategy for Solving Linear Equations.

2.6 Solve a Formula for a Specific Variable

- **To Solve an Application (with a formula)**

  Step 1.   **Read** the problem. Make sure all the words and ideas are understood.

  Step 2.   **Identify** what we are looking for.

  Step 3.   **Name** what we are looking for. Choose a variable to represent that quantity.

  Step 4.   **Translate** into an equation. Write the appropriate formula for the situation. Substitute in the given information.

  Step 5.   **Solve** the equation using good algebra techniques.

  Step 6.   **Check** the answer in the problem and make sure it makes sense.

  Step 7.   **Answer** the question with a complete sentence.

- **Distance, Rate and Time**
  For an object moving at a uniform (constant) rate, the distance traveled, the elapsed time, and the rate are related by the formula: $d = rt$ where d = distance, r = rate, t = time.

- **To solve a formula for a specific variable** means to get that variable by itself with a coefficient of 1 on one side of the equation and all other variables and constants on the other side.

2.7 Solve Linear Inequalities

- **Subtraction Property of Inequality**
  For any numbers a, b, and c,
  if $a < b$ then $a - c < b - c$ and
  if $a > b$ then $a - c > b - c$.

- **Addition Property of Inequality**
  For any numbers a, b, and c,

if $a < b$ then $a + c < b + c$ and

if $a > b$ then $a + c > b + c$.

- **Division and Multiplication Properties of Inequality**
  For any numbers a, b, and c,

  if $a < b$ and $c > 0$, then $\frac{a}{c} < \frac{b}{c}$ and $ac > bc$.

  if $a > b$ and $c > 0$, then $\frac{a}{c} > \frac{b}{c}$ and $ac > bc$.

  if $a < b$ and $c < 0$, then $\frac{a}{c} > \frac{b}{c}$ and $ac > bc$.

  if $a > b$ and $c < 0$, then $\frac{a}{c} < \frac{b}{c}$ and $ac < bc$.

- When we **divide or multiply** an inequality by a:

  ◦ **positive** number, the inequality stays the **same**.

  ◦ **negative** number, the inequality **reverses**.

REVIEW EXERCISES

2.1 Section 2.1 Solve Equations using the Subtraction and Addition Properties of Equality

Verify a Solution of an Equation

In the following exercises, determine whether each number is a solution to the equation.

512. $10x - 1 = 5x;\ x = \frac{1}{5}$

513. $w + 2 = \frac{5}{8};\ w = \frac{3}{8}$

514. $-12n + 5 = 8n;\ n = -\frac{5}{4}$

515. $6a - 3 = -7a,\ a = \frac{3}{13}$

Solve Equations using the Subtraction and Addition Properties of Equality

In the following exercises, solve each equation using the Subtraction Property of Equality.

516. $x + 7 = 19$

517. $y + 2 = -6$

518. $a + \frac{1}{3} = \frac{5}{3}$

519. $n + 3.6 = 5.1$

In the following exercises, solve each equation using the Addition Property of Equality.

520. $u - 7 = 10$

521. $x - 9 = -4$

522. $c - \frac{3}{11} = \frac{9}{11}$

523. $p - 4.8 = 14$

In the following exercises, solve each equation.

524. $n - 12 = 32$

525. $y + 16 = -9$

526. $f + \frac{2}{3} = 4$

527. $d - 3.9 = 8.2$

Solve Equations That Require Simplification

In the following exercises, solve each equation.

528. $y + 8 - 15 = -3$

529. $7x + 10 - 6x + 3 = 5$

530. $6(n - 1) - 5n = -14$

531. $8(3p + 5) - 23(p - 1) = 35$

Translate to an Equation and Solve

In the following exercises, translate each English sentence into an algebraic equation and then solve it.

532. The sum of -6 and m is 25.

533. Four less than n is 13.

Translate and Solve Applications

In the following exercises, translate into an algebraic equation and solve.

534. Rochelle's daughter is 11 years old. Her son is 3 years younger. How old is her son?

535. Tan weighs 146 pounds. Minh weighs 15 pounds more than Tan. How much does Minh weigh?

536. Peter paid $9.75 to go to the movies, which was $46.25 less than he paid to go to a concert. How much did he pay for the concert?

537. Elissa earned $152.84 this week, which was $21.65 more than she earned last week. How much did she earn last week?

2.2 Section 2.2 Solve Equations using the Division and Multiplication Properties of Equality

Solve Equations Using the Division and Multiplication Properties of Equality

In the following exercises, solve each equation using the division and multiplication properties of equality and check the solution.

538. $8x = 72$

539. $13a = -65$

540. $0.25p = 5.25$

541. $-y = 4$

542. $\frac{n}{6} = 18$

543. $\frac{y}{-10} = 30$

544. $36 = \frac{3}{4}x$

545. $\frac{5}{8}u = \frac{15}{16}$

546. $-18m = -72$

547. $\frac{c}{9} = 36$

548. $0.45x = 6.75$

549. $\frac{11}{12} = \frac{2}{3}y$

Solve Equations That Require Simplification

In the following exercises, solve each equation requiring simplification.

550. $5r - 3r + 9r = 35 - 2$

551. $24x + 8x - 11x = -7 - 14$

552. $\frac{11}{12}n - \frac{5}{6}n = 9 - 5$

553. $-9(d - 2) - 15 = -24$

Translate to an Equation and Solve

In the following exercises, translate to an equation and then solve.

554. 143 is the product of -11 and y.

555. The quotient of b and and 9 is -27.

556. The sum of q and one-fourth is one.

557. The difference of s and one-twelfth is one fourth.

Translate and Solve Applications

In the following exercises, translate into an equation and solve.

558. Ray paid $21 for 12 tickets at the county fair. What was the price of each ticket?

559. Janet gets paid $24 per hour. She heard that this is $\frac{3}{4}$ of what Adam is paid. How much is Adam paid per hour?

2.3 Section 2.3 Solve Equations with Variables and Constants on Both Sides

Solve an Equation with Constants on Both Sides

In the following exercises, solve the following equations with constants on both sides.

560. $8p + 7 = 47$

561. $10w - 5 = 65$

562. $3x + 19 = -47$

563. $32 = -4 - 9n$

Solve an Equation with Variables on Both Sides

In the following exercises, solve the following equations with variables on both sides.

564. $7y = 6y - 13$

565. $5a + 21 = 2a$

566. $k = -6k - 35$

567. $4x - \frac{3}{8} = 3x$

Solve an Equation with Variables and Constants on Both Sides

In the following exercises, solve the following equations with variables and constants on both sides.

568. $12x - 9 = 3x + 45$

569. $5n - 20 = -7n - 80$

570. $4u + 16 = -19 - u$

571. $\frac{5}{8}c - 4 = \frac{3}{8}c + 4$

2.4 Section 2.4 Use a General Strategy for Solving Linear Equations

Solve Equations Using the General Strategy for Solving Linear Equations

In the following exercises, solve each linear equation.

572. $6(x + 6) = 24$

573. $9(2p - 5) = 72$

574. $-(s + 4) = 18$

575. $8 + 3(n - 9) = 17$

576. $23 - 3(y - 7) = 8$

577. $\frac{1}{3}(6m + 21) = m - 7$

578. $4(3.5y + 0.25) = 365$

579. $0.25(q - 8) = 0.1(q + 7)$

580. $8(r - 2) = 6(r + 10)$

581. $5 + 7(2 - 5x) = 2(9x + 1)$
$-(13x - 57)$

582. $(9n + 5) - (3n - 7)$
$= 20 - (4n - 2)$

583. $2[-16 + 5(8k - 6)]$
$= 8(3 - 4k) - 32$

Classify Equations

In the following exercises, classify each equation as a conditional equation, an identity, or a contradiction and then state the solution.

584. $17y - 3(4 - 2y) = 11(y - 1)$
$+ 12y - 1$

585. $9u + 32 = 15(u - 4)$
$- 3(2u + 21)$

586. $-8(7m + 4) = -6(8m + 9)$

587. $21(c - 1) - 19(c + 1)$
$= 2(c - 20)$

2.5 Section 2.5 Solve Equations with Fractions and Decimals

Solve Equations with Fraction Coefficients

In the following exercises, solve each equation with fraction coefficients.

588. $\frac{2}{5}n - \frac{1}{10} = \frac{7}{10}$

589. $\frac{1}{3}x + \frac{1}{5}x = 8$

590. $\frac{3}{4}a - \frac{1}{3} = \frac{1}{2}a - \frac{5}{6}$

591. $\frac{1}{2}(k - 3) = \frac{1}{3}(k + 16)$

592. $\frac{3x - 2}{5} = \frac{3x + 4}{8}$

593. $\frac{5y - 1}{3} + 4 = \frac{-8y + 4}{6}$

Solve Equations with Decimal Coefficients

In the following exercises, solve each equation with decimal coefficients.

594. $0.8x - 0.3 = 0.7x + 0.2$

595. $0.36u + 2.55 = 0.41u + 6.8$

596. $0.6p - 1.9 = 0.78p + 1.7$

597. $0.6p - 1.9 = 0.78p + 1.7$

2.6 Section 2.6 Solve a Formula for a Specific Variable

Use the Distance, Rate, and Time Formula

In the following exercises, solve.

598. Natalie drove for $7\frac{1}{2}$ hours at 60 miles per hour. How much distance did she travel?

599. Mallory is taking the bus from St. Louis to Chicago. The distance is 300 miles and the bus travels at a steady rate of 60 miles per hour. How long will the bus ride be?

600. Aaron's friend drove him from Buffalo to Cleveland. The distance is 187 miles and the trip took 2.75 hours. How fast was Aaron's friend driving?

601. Link rode his bike at a steady rate of 15 miles per hour for $2\frac{1}{2}$ hours. How much distance did he travel?

Solve a Formula for a Specific Variable

In the following exercises, solve.

602. Use the formula. $d = rt$ to solve for t
ⓐ when $d = 510$ and $r = 60$
ⓑ in general

603. Use the formula. $d = rt$ to solve for r
ⓐ when when $d = 451$ and $t = 5.5$
ⓑ in general

604. Use the formula $A = \frac{1}{2}bh$ to solve for b
ⓐ when $A = 390$ and $h = 26$
ⓑ in general

605. Use the formula $A = \frac{1}{2}bh$ to solve for h

ⓐ when $A = 153$ and $b = 18$

ⓑ in general

606. Use the formula $I = Prt$ to solve for the principal, P for

ⓐ $I = \$2,501,\ r = 4.1\%,$ $t = 5$ years

ⓑ in general

607. Solve the formula $4x + 3y = 6$ for y

ⓐ when $x = -2$

ⓑ in general

608. Solve $180 = a + b + c$ for c
.

609. Solve the formula $V = LWH$ for H.

2.7 Section 2.7 Solve Linear Inequalities

Graph Inequalities on the Number Line

In the following exercises, graph each inequality on the number line.

610.

ⓐ $x \le 4$

ⓑ $x > -2$

ⓒ $x < 1$

611.

ⓐ $x > 0$

ⓑ $x < -3$

ⓒ $x \ge -1$

In the following exercises, graph each inequality on the number line and write in interval notation.

612.

ⓐ $x < -1$

ⓑ $x \ge -2.5$

ⓒ $x \le \frac{5}{4}$

613.

ⓐ $x > 2$

ⓑ $x \le -1.5$

ⓒ $x \ge \frac{5}{3}$

Solve Inequalities using the Subtraction and Addition Properties of Inequality

In the following exercises, solve each inequality, graph the solution on the number line, and write the solution in interval notation.

614. $n - 12 \le 23$

615. $m + 14 \le 56$

616. $a + \frac{2}{3} \ge \frac{7}{12}$

617. $b - \frac{7}{8} \ge -\frac{1}{2}$

Solve Inequalities using the Division and Multiplication Properties of Inequality

In the following exercises, solve each inequality, graph the solution on the number line, and write the solution in interval notation.

618. $9x > 54$

619. $-12d \le 108$

620. $\frac{5}{2}j < -60$

621. $\frac{q}{-2} \ge -24$

Solve Inequalities That Require Simplification

In the following exercises, solve each inequality, graph the solution on the number line, and write the solution in interval notation.

622. $6p > 15p - 30$

623. $9h - 7(h - 1) \le 4h - 23$

624. $5n - 15(4 - n) < 10(n - 6) + 10n$

625. $\frac{3}{8}a - \frac{1}{12}a > \frac{5}{12}a + \frac{3}{4}$

Translate to an Inequality and Solve

In the following exercises, translate and solve. Then write the solution in interval notation and graph on the number line.

626. Five more than z is at most 19.

627. Three less than c is at least 360.

628. Nine times n exceeds 42.

629. Negative two times a is no more than 8.

Everyday Math

630. Describe how you have used two topics from this chapter in your life outside of your math class during the past month.

PRACTICE TEST

631. Determine whether each number is a solution to the equation $6x - 3 = x + 20$.

ⓐ 5

ⓑ $\frac{23}{5}$

In the following exercises, solve each equation.

632. $n - \frac{2}{3} = \frac{1}{4}$

633. $\frac{9}{2}c = 144$

634. $4y - 8 = 16$

635. $-8x - 15 + 9x - 1 = -21$

636. $-15a = 120$

637. $\frac{2}{3}x = 6$

638. $x - 3.8 = 8.2$

639. $10y = -5y - 60$

640. $8n - 2 = 6n - 12$

641. $9m - 2 - 4m - m = 42 - 8$

642. $-5(2x - 1) = 45$

643. $-(d - 9) = 23$

644.
$\frac{1}{4}(12m - 28) = 6 - 2(3m - 1)$

645. $2(6x - 5) - 8 = -22$

646.
$8(3a - 5) - 7(4a - 3) = 20 - 3a$

647. $\frac{1}{4}p - \frac{1}{3} = \frac{1}{2}$

648. $0.1d + 0.25(d + 8) = 4.1$

649.
$14n - 3(4n + 5) = -9 + 2(n - 8)$

650. $9(3u - 2) - 4[6 - 8(u - 1)]$
$= 3(u - 2)$

651. Solve the formula
$x - 2y = 5$ for y

ⓐ when $x = -3$

ⓑ in general

In the following exercises, graph on the number line and write in interval notation.

652. $x \geq -3.5$

653. $x < \frac{11}{4}$

In the following exercises,, solve each inequality, graph the solution on the number line, and write the solution in interval notation.

654. $8k \geq 5k - 120$

655. $3c - 10(c - 2) < 5c + 16$

In the following exercises, translate to an equation or inequality and solve.

656. 4 less than twice x is 16.

657. Fifteen more than n is at least 48.

658. Samuel paid $25.82 for gas this week, which was $3.47 less than he paid last week. How much had he paid last week?

659. Jenna bought a coat on sale for $120, which was $\frac{2}{3}$ of the original price. What was the original price of the coat?

660. Sean took the bus from Seattle to Boise, a distance of 506 miles. If the trip took $7\frac{2}{3}$ hours, what was the speed of the bus?

3 MATH MODELS

Figure 3.1 Sophisticated mathematical models are used to predict traffic patterns on our nation's highways.

Chapter Outline

3.1 Use a Problem-Solving Strategy

3.2 Solve Percent Applications

3.3 Solve Mixture Applications

3.4 Solve Geometry Applications: Triangles, Rectangles, and the Pythagorean Theorem

3.5 Solve Uniform Motion Applications

3.6 Solve Applications with Linear Inequalities

✎ Introduction

Mathematical formulas model phenomena in every facet of our lives. They are used to explain events and predict outcomes in fields such as transportation, business, economics, medicine, chemistry, engineering, and many more. In this chapter, we will apply our skills in solving equations to solve problems in a variety of situations.

3.1 Use a Problem-Solving Strategy

Learning Objectives

By the end of this section, you will be able to:
- Approach word problems with a positive attitude
- Use a problem-solving strategy for word problems
- Solve number problems

Be Prepared!

Before you get started, take this readiness quiz.

1. Translate "6 less than twice *x*" into an algebraic expression.
   If you missed this problem, review **Example 1.26**.

2. Solve: $\frac{2}{3}x = 24$.
   If you missed this problem, review **Example 2.16**.

3. Solve: $3x + 8 = 14$.
   If you missed this problem, review **Example 2.27**.

Approach Word Problems with a Positive Attitude

"If you think you can... or think you can't... you're right."—Henry Ford

The world is full of word problems! Will my income qualify me to rent that apartment? How much punch do I need to make for the party? What size diamond can I afford to buy my girlfriend? Should I fly or drive to my family reunion?

How much money do I need to fill the car with gas? How much tip should I leave at a restaurant? How many socks should I pack for vacation? What size turkey do I need to buy for Thanksgiving dinner, and then what time do I need to put it in the oven? If my sister and I buy our mother a present, how much does each of us pay?

Now that we can solve equations, we are ready to apply our new skills to word problems. Do you know anyone who has had negative experiences in the past with word problems? Have you ever had thoughts like the student below?

Figure 3.2 Negative thoughts can be barriers to success.

When we feel we have no control, and continue repeating negative thoughts, we set up barriers to success. We need to calm our fears and change our negative feelings.

Start with a fresh slate and begin to think positive thoughts. If we take control and believe we can be successful, we will be able to master word problems! Read the positive thoughts in Figure 3.3 and say them out loud.

Figure 3.3 Thinking positive thoughts is a first step towards success.

Think of something, outside of school, that you can do now but couldn't do 3 years ago. Is it driving a car? Snowboarding? Cooking a gourmet meal? Speaking a new language? Your past experiences with word problems happened when you were younger—now you're older and ready to succeed!

Use a Problem-Solving Strategy for Word Problems

We have reviewed translating English phrases into algebraic expressions, using some basic mathematical vocabulary and symbols. We have also translated English sentences into algebraic equations and solved some word problems. The word problems applied math to everyday situations. We restated the situation in one sentence, assigned a variable, and then wrote an equation to solve the problem. This method works as long as the situation is familiar and the math is not too complicated.

Now, we'll expand our strategy so we can use it to successfully solve any word problem. We'll list the strategy here, and then we'll use it to solve some problems. We summarize below an effective strategy for problem solving.

HOW TO :: USE A PROBLEM-SOLVING STRATEGY TO SOLVE WORD PROBLEMS.

Step 1.   **Read** the problem. Make sure all the words and ideas are understood.

Step 2.   **Identify** what we are looking for.

Step 3.   **Name** what we are looking for. Choose a variable to represent that quantity.

Step 4.   **Translate** into an equation. It may be helpful to restate the problem in one sentence with all the important information. Then, translate the English sentence into an algebraic equation.

Step 5.   **Solve** the equation using good algebra techniques.

Step 6.   **Check** the answer in the problem and make sure it makes sense.

Step 7.   **Answer** the question with a complete sentence.

EXAMPLE 3.1

Pilar bought a purse on sale for $18, which is one-half of the original price. What was the original price of the purse?

✓ **Solution**

Step 1. Read the problem. Read the problem two or more times if necessary. Look up any unfamiliar words in a dictionary or on the internet.
  • *In this problem, is it clear what is being discussed? Is every word familiar?*

Step 2. Identify what you are looking for. Did you ever go into your bedroom to get something and then forget what you were looking for? It's hard to find something if you are not sure what it is! Read the problem again and look for words that tell you what you are looking for!
  • *In this problem, the words "what was the original price of the purse" tell us what we need to find.*

Step 3. Name what we are looking for. Choose a variable to represent that quantity. We can use any letter for the variable, but choose one that makes it easy to remember what it represents.

  • Let $p = $ the original price of the purse.

Step 4. Translate into an equation. It may be helpful to restate the problem in one sentence with all the important information. Translate the English sentence into an algebraic equation.

Reread the problem carefully to see how the given information is related. Often, there is one sentence that gives this information, or it may help to write one sentence with all the important information. Look for clue words to help translate the sentence into algebra. Translate the sentence into an equation.

Restate the problem in one sentence with all the important information.	18	is	one-half the original price.
Translate into an equation.	18	=	$\frac{1}{2} \cdot p$

Step 5. Solve the equation using good algebraic techniques. Even if you know the solution right away, using good algebraic techniques here will better prepare you to solve problems that do not have obvious answers.

Solve the equation.	$18 = \frac{1}{2}p$
Multiply both sides by 2.	$2 \cdot 18 = 2 \cdot \frac{1}{2}p$
Simplify.	$36 = p$

Step 6. Check the answer in the problem to make sure it makes sense. We solved the equation and found that $p = 36$,

which means "the original price" was $36.

- *Does $36 make sense in the problem? Yes, because 18 is one-half of 36, and the purse was on sale at half the original price.*

Step 7. Answer the question with a complete sentence. The problem asked "What was the original price of the purse?"

- *The answer to the question is: "The original price of the purse was $36."*

If this were a homework exercise, our work might look like this:

Pilar bought a purse on sale for $18, which is one-half the original price. What was the original price of the purse?

	Let $p =$ the original price.
	18 is one-half the original price.
	$18 = \frac{1}{2}p$
Multiply both sides by 2.	$2 \cdot 18 = 2 \cdot \frac{1}{2}p$
Simplify.	$36 = p$
Check. Is $36 a reasonable price for a purse?	
Yes.	
Is 18 one half of 36?	
$18 \overset{?}{=} \frac{1}{2} \cdot 36$	
$18 = 18 \checkmark$	
	The original price of the purse was $36.

> **TRY IT ::** 3.1

Joaquin bought a bookcase on sale for $120, which was two-thirds of the original price. What was the original price of the bookcase?

> **TRY IT ::** 3.2

Two-fifths of the songs in Mariel's playlist are country. If there are 16 country songs, what is the total number of songs in the playlist?

Let's try this approach with another example.

EXAMPLE 3.2

Ginny and her classmates formed a study group. The number of girls in the study group was three more than twice the number of boys. There were 11 girls in the study group. How many boys were in the study group?

⊘ **Solution**

Step 1. **Read** the problem.	
Step 2. **Identify** what we are looking for.	How many boys were in the study group?
Step 3. **Name.** Choose a variable to represent the number of boys.	Let $n =$ the number of boys.

Step 4. Translate. Restate the problem in one sentence with all the important information.	The number of girls (11)	was	three more than twice the number of boys
Translate into an equation.	11	=	$2b + 3$
Step 5. Solve the equation.	$11 = 2b + 3$		
Subtract 3 from each side.	$11 - 3 = 2b + 3 - 3$		
Simplify.	$8 = 2b$		
Divide each side by 2.	$\dfrac{8}{2} = \dfrac{2b}{2}$		
Simplify.	$4 = b$		
Step 6. Check. First, is our answer reasonable? Yes, having 4 boys in a study group seems OK. The problem says the number of girls was 3 more than twice the number of boys. If there are four boys, does that make eleven girls? Twice 4 boys is 8. Three more than 8 is 11.			
Step 7. Answer the question.	There were 4 boys in the study group.		

> **TRY IT : : 3.3**

> Guillermo bought textbooks and notebooks at the bookstore. The number of textbooks was 3 more than twice the number of notebooks. He bought 7 textbooks. How many notebooks did he buy?

> **TRY IT : : 3.4**

> Gerry worked Sudoku puzzles and crossword puzzles this week. The number of Sudoku puzzles he completed is eight more than twice the number of crossword puzzles. He completed 22 Sudoku puzzles. How many crossword puzzles did he do?

Solve Number Problems

Now that we have a problem solving strategy, we will use it on several different types of word problems. The first type we will work on is "number problems." Number problems give some clues about one or more numbers. We use these clues to write an equation. Number problems don't usually arise on an everyday basis, but they provide a good introduction to practicing the problem solving strategy outlined above.

EXAMPLE 3.3

The difference of a number and six is 13. Find the number.

⊘ **Solution**

Step 1. Read the problem. Are all the words familiar?	
Step 2. Identify what we are looking for.	the number
Step 3. Name. Choose a variable to represent the number.	Let $n =$ the number.
Step 4. Translate. Remember to look for clue words like "difference... of... and..."	
Restate the problem as one sentence.	The difference of the number and 6 is 13
Translate into an equation.	$n - 6$ = 13

Step 5. **Solve** the equation.	$n - 6 = 13$
Simplify.	$n = 19$

Step 6. Check.

The difference of 19 and 6 is 13. It checks!

Step 7. **Answer** the question.	The number is 19.

> **TRY IT :: 3.5**   The difference of a number and eight is 17. Find the number.

> **TRY IT :: 3.6**   The difference of a number and eleven is -7. Find the number.

EXAMPLE 3.4

The sum of twice a number and seven is 15. Find the number.

⊘ **Solution**

Step 1. Read the problem.

Step 2. **Identify** what we are looking for.	the number

Step 3. **Name.** Choose a variable to represent the number.	Let $n =$ the number.

Step 4. Translate.

Restate the problem as one sentence.	The sum of twice a number and 7 is 15
Translate into an equation.	$2n + 7$ = 15

Step 5. **Solve** the equation.	$2n + 7 = 15$
Subtract 7 from each side and simplify.	$2n = 8$
Divide each side by 2 and simplify.	$n = 4$

Step 6. Check.

Is the sum of twice 4 and 7 equal to 15?

$$2 \cdot 4 + 7 \overset{?}{=} 15$$
$$15 = 15 \checkmark$$

Step 7. **Answer** the question.	The number is 4.

Did you notice that we left out some of the steps as we solved this equation? If you're not yet ready to leave out these steps, write down as many as you need.

> **TRY IT :: 3.7**   The sum of four times a number and two is 14. Find the number.

> **TRY IT :: 3.8**   The sum of three times a number and seven is 25. Find the number.

Some number word problems ask us to find two or more numbers. It may be tempting to name them all with different variables, but so far we have only solved equations with one variable. In order to avoid using more than one variable, we will define the numbers in terms of the same variable. Be sure to read the problem carefully to discover how all the numbers relate to each other.

EXAMPLE 3.5

One number is five more than another. The sum of the numbers is 21. Find the numbers.

⊘ **Solution**

Step 1. Read the problem.	
Step 2. Identify what we are looking for.	We are looking for two numbers.
Step 3. Name. We have two numbers to name and need a name for each.	
Choose a variable to represent the first number.	Let $n = 1^{st}$ number.
What do we know about the second number?	One number is five more than another.
	$n + 5 = 2^{nd}$ number
Step 4. Translate. Restate the problem as one sentence with all the important information.	The sum of the 1st number and the 2nd number is 21.
Translate into an equation.	1st number $+ \ 2^{nd}$ number $= \ 21$
Substitute the variable expressions.	$n \quad + \quad n + 5 \quad = \quad 21$
Step 5. Solve the equation.	$n + n + 5 = 21$
Combine like terms.	$2n + 5 = 21$
Subtract 5 from both sides and simplify.	$2n = 16$
Divide by 2 and simplify.	$n = 8 \quad 1^{st}$ number
Find the second number, too.	$n + 5 \quad 2^{nd}$ number
	$8 + 5$
	13
Step 6. Check.	
Do these numbers check in the problem?	
Is one number 5 more than the other?	$13 \overset{?}{=} 8 + 5$
Is thirteen 5 more than 8? Yes.	$13 = 13 \ \checkmark$
Is the sum of the two numbers 21?	$8 + 13 \overset{?}{=} 21$
	$21 = 21 \ \checkmark$
Step 7. Answer the question.	The numbers are 8 and 13.

> **TRY IT : : 3.9**

One number is six more than another. The sum of the numbers is twenty-four. Find the numbers.

> **TRY IT : : 3.10**

The sum of two numbers is fifty-eight. One number is four more than the other. Find the numbers.

EXAMPLE 3.6

The sum of two numbers is negative fourteen. One number is four less than the other. Find the numbers.

⊘ **Solution**

Step 1. Read the problem.

Step 2. Identify what we are looking for.	We are looking for two numbers.

Step 3. Name.

Choose a variable.	Let $n = 1^{st}$ number.
One number is 4 less than the other.	$n - 4 = 2^{nd}$ number

Step 4. Translate.

Write as one sentence.	The sum of the 2 numbers is negative 14.
Translate into an equation.	$\underbrace{1^{st} \text{ number}} + \underbrace{2^{nd} \text{ number}}$ $\underbrace{\text{is}}$ $\underbrace{\text{negative fourteen}}$

Step 5. Solve the equation.	$n \quad + \quad n-4 \quad = \quad -14$
Combine like terms.	$n + n - 4 = -14$
Add 4 to each side and simplify.	$2n - 4 = -14$
Simplify.	$2n = -10$
	$n = -5 \quad 1^{st} \text{ number}$
	$n - 4 \quad 2^{nd} \text{ number}$
	${\color{red}-5} - 4$
	-9

Step 6. Check.

Is –9 four less than –5?	$-5 - 4 \overset{?}{=} -9$
	$-9 = -9 \checkmark$
Is their sum –14?	$-5 + (-9) \overset{?}{=} -14$
	$-14 = -14 \checkmark$

Step 7. Answer the question.	The numbers are –5 and –9.

> **TRY IT : : 3.11**

The sum of two numbers is negative twenty-three. One number is seven less than the other. Find the numbers.

> **TRY IT : : 3.12** The sum of two numbers is -18. One number is 40 more than the other. Find the numbers.

EXAMPLE 3.7

One number is ten more than twice another. Their sum is one. Find the numbers.

✓ Solution

Step 1. Read the problem.	
Step 2. Identify what you are looking for.	We are looking for two numbers.
Step 3. Name.	
Choose a variable.	Let $x = 1^{st}$ number.
One number is 10 more than twice another.	$2x + 10 = 2^{nd}$ number
Step 4. Translate.	
Restate as one sentence.	Their sum is one.
	The sum of the two numbers is 1.
Translate into an equation.	$x + 2x + 10 = 1$
Step 5. Solve the equation.	
Combine like terms.	$x + 2x + 10 = 1$
Subtract 10 from each side.	$3x + 10 = 1$
Divide each side by 3.	$3x = -9$
	$x = -3$ 1st number
	$2x + 10$ 2nd number
	$2(-3) + 10$
	4
Step 6. Check.	
Is ten more than twice –3 equal to 4?	$2(-3) + 10 \overset{?}{=} 4$
	$-6 + 10 \overset{?}{=} 4$
	$4 = 4 \checkmark$
Is their sum 1?	$-3 + 4 \overset{?}{=} 1$
	$1 = 1 \checkmark$
Step 7. Answer the question.	The numbers are –3 and –4.

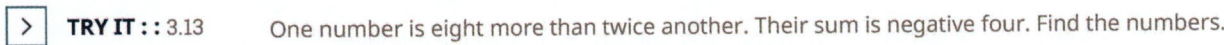

> **TRY IT :: 3.13**   One number is eight more than twice another. Their sum is negative four. Find the numbers.

> **TRY IT :: 3.14**   One number is three more than three times another. Their sum is -5. Find the numbers.

Some number problems involve consecutive integers. *Consecutive integers* are integers that immediately follow each other. Examples of consecutive integers are:

$$1, 2, 3, 4$$
$$-10, -9, -8, -7$$
$$150, 151, 152, 153$$

Notice that each number is one more than the number preceding it. So if we define the first integer as n, the next consecutive integer is $n + 1$. The one after that is one more than $n + 1$, so it is $n + 1 + 1$, which is $n + 2$.

$$n \qquad 1^{st} \text{ integer}$$
$$n + 1 \qquad 2^{nd} \text{ consecutive integer}$$
$$n + 2 \qquad 3^{rd} \text{ consecutive integer} \ldots \text{ etc.}$$

EXAMPLE 3.8

The sum of two consecutive integers is 47. Find the numbers.

✓ **Solution**

Step 1. Read the problem.	
Step 2. Identify what you are looking for.	two consecutive integers
Step 3. Name each number.	Let $n = 1^{st}$ integer.
	$n + 1 =$ next consecutive integer
Step 4. Translate.	
Restate as one sentence.	The sum of the integers is 47.
Translate into an equation.	$n + n + 1 = 47$
Step 5. Solve the equation.	$n + n + 1 = 47$
Combine like terms.	$2n + 1 = 47$
Subtract 1 from each side.	$2n = 46$
Divide each side by 2.	$n = 23 \quad 1^{st}$ integer
	$n + 1 \qquad$ next consecutive integer
	$23 + 1$
	24
Step 6. Check.	

$$23 + 24 \overset{?}{=} 47$$
$$47 = 47 \checkmark$$

Step 7. Answer the question.	The two consecutive integers are 23 and 24.

> **TRY IT : : 3.15**   The sum of two consecutive integers is 95. Find the numbers.

> **TRY IT : : 3.16**   The sum of two consecutive integers is −31. Find the numbers.

EXAMPLE 3.9

Find three consecutive integers whose sum is −42.

✓ Solution

Step 1. Read the problem.

Step 2. Identify what we are looking for.	three consecutive integers
Step 3. Name each of the three numbers.	Let $n = 1^{st}$ integer.
	$n + 1 = $ 2nd consecutive integer
	$n + 2 = $ 3rd consecutive integer

Step 4. Translate.

Restate as one sentence.	The sum of the three integers is −42.
Translate into an equation.	$n + n + 1 + n + 2 = -42$
Step 5. Solve the equation.	$n + n + 1 + n + 2 = -42$
Combine like terms.	$3n + 3 = -42$
Subtract 3 from each side.	$3n = -45$
Divide each side by 3.	$n = -15$ 1st integer
	$n + 1$ 2nd integer
	$-15 + 1$
	-14
	$n + 2$ 3rd integer
	$-15 + 2$
	-13

Step 6. Check.

$$-13 + (-14) + (-15) \overset{?}{=} -42$$
$$-42 = -42 \checkmark$$

Step 7. Answer the question.	The three consecutive integers are −13, −14, and −15.

> **TRY IT :: 3.17**   Find three consecutive integers whose sum is -96.

> **TRY IT :: 3.18**   Find three consecutive integers whose sum is -36.

Now that we have worked with consecutive integers, we will expand our work to include consecutive even integers and consecutive odd integers. *Consecutive even integers* are even integers that immediately follow one another. Examples of consecutive even integers are:

$$18, 20, 22$$

$$64, 66, 68$$
$$-12, -10, -8$$

Notice each integer is 2 more than the number preceding it. If we call the first one n, then the next one is $n + 2$. The next one would be $n + 2 + 2$ or $n + 4$.

$$n \qquad 1^{st} \text{ even integer}$$

$$n + 2 \qquad 2^{nd} \text{ consecutive even integer}$$

$$n + 4 \qquad 3^{rd} \text{ consecutive even integer} \dots \text{etc.}$$

Consecutive odd integers are odd integers that immediately follow one another. Consider the consecutive odd integers 77, 79, and 81.

$$77, 79, 81$$

$$n, n + 2, n + 4$$

$$n \qquad 1^{st} \text{ odd integer}$$

$$n + 2 \qquad 2^{nd} \text{ consecutive odd integer}$$

$$n + 4 \qquad 3^{rd} \text{ consecutive odd integer} \dots \text{etc.}$$

Does it seem strange to add 2 (an even number) to get from one odd integer to the next? Do you get an odd number or an even number when we add 2 to 3? to 11? to 47?

Whether the problem asks for consecutive even numbers or odd numbers, you don't have to do anything different. The pattern is still the same—to get from one odd or one even integer to the next, add 2.

EXAMPLE 3.10

Find three consecutive even integers whose sum is 84.

⊘ Solution

Step 1. Read the problem.

Step 2. Identify what we are looking for.   three consecutive even integers

Step 3. Name the integers.   Let $n = 1^{st}$ even integer.

$$n + 2 = 2^{nd} \text{ consecutive even integer}$$

$$n + 4 = 3^{rd} \text{ consecutive even integer}$$

Step 4. Translate.

Restate as one sentence.   The sum of the three even integers is 84.

Translate into an equation. $n + n + 2 + n + 4 = 84$

Step 5. Solve the equation.

Combine like terms. $n + n + 2 + n + 4 = 84$

Subtract 6 from each side. $3n + 6 = 84$

Divide each side by 3. $3n = 78$

$$n = 26 \quad 1^{st} \text{ integer}$$

$$\begin{array}{ccc} n & + & 2 \quad 2^{nd} \text{ integer} \\ 26 & + & 2 \\ & 28 \end{array}$$

$$\begin{array}{ccc} n & + & 4 \quad 3^{rd} \text{ integer} \\ 26 & + & 4 \\ & 30 \end{array}$$

Step 6. Check.

$$26 + 28 + 30 \stackrel{?}{=} 84$$

$$84 = 84 \checkmark$$

Step 7. Answer the question.   The three consecutive integers are 26, 28, and 30.

 **TRY IT : : 3.19**   Find three consecutive even integers whose sum is 102.

> **TRY IT : : 3.20**   Find three consecutive even integers whose sum is -24.

EXAMPLE 3.11

A married couple together earns $110,000 a year. The wife earns $16,000 less than twice what her husband earns. What does the husband earn?

⊘ **Solution**

Step 1. Read the problem.	
Step 2. Identify what we are looking for.	How much does the husband earn?
Step 3. Name.	
Choose a variable to represent the amount the husband earns.	Let $h =$ the amount the husband earns.
The wife earns $16,000 less than twice that.	$2h - 16,000$ the amount the wife earns.
Step 4. Translate.	Together the husband and wife earn $110,000.
Restate the problem in one sentence with all the important information.	The amount the husband earns plus the amount the wife earns is $110,000
Translate into an equation.	$h \quad + \quad 2h - 16,000 \quad = \quad 110,000$
Step 5. Solve the equation.	$h + 2h - 16,000 = 110,000$
Combine like terms.	$3h - 16,000 = 110,000$
Add 16,000 to both sides and simplify.	$3h = 126,000$
Divide each side by 3.	$h = 42,000$
	$42,000$ amount husband earns
	$2h - 16,000$ amount wife earns
	$2(42,000) - 16,000$
	$84,000 - 16,000$
	$68,000$
Step 6. Check.	
If the wife earns $68,000 and the husband earns $42,000 is the total $110,000? Yes!	
Step 7. Answer the question.	The husband earns $42,000 a year.

> **TRY IT : : 3.21**

> According to the National Automobile Dealers Association, the average cost of a car in 2014 was $28,500. This was $1,500 less than 6 times the cost in 1975. What was the average cost of a car in 1975?

> **TRY IT : :** 3.22

U.S. Census data shows that the median price of new home in the United States in November 2014 was $280,900. This was $10,700 more than 14 times the price in November 1964. What was the median price of a new home in November 1964?

3.1 EXERCISES

Practice Makes Perfect

Use the Approach Word Problems with a Positive Attitude

In the following exercises, prepare the lists described.

1. List five positive thoughts you can say to yourself that will help you approach word problems with a positive attitude. You may want to copy them on a sheet of paper and put it in the front of your notebook, where you can read them often.

2. List five negative thoughts that you have said to yourself in the past that will hinder your progress on word problems. You may want to write each one on a small piece of paper and rip it up to symbolically destroy the negative thoughts.

Use a Problem-Solving Strategy for Word Problems

In the following exercises, solve using the problem solving strategy for word problems. Remember to write a complete sentence to answer each question.

3. Two-thirds of the children in the fourth-grade class are girls. If there are 20 girls, what is the total number of children in the class?

4. Three-fifths of the members of the school choir are women. If there are 24 women, what is the total number of choir members?

5. Zachary has 25 country music CDs, which is one-fifth of his CD collection. How many CDs does Zachary have?

6. One-fourth of the candies in a bag of M&M's are red. If there are 23 red candies, how many candies are in the bag?

7. There are 16 girls in a school club. The number of girls is four more than twice the number of boys. Find the number of boys.

8. There are 18 Cub Scouts in Pack 645. The number of scouts is three more than five times the number of adult leaders. Find the number of adult leaders.

9. Huong is organizing paperback and hardback books for her club's used book sale. The number of paperbacks is 12 less than three times the number of hardbacks. Huong had 162 paperbacks. How many hardback books were there?

10. Jeff is lining up children's and adult bicycles at the bike shop where he works. The number of children's bicycles is nine less than three times the number of adult bicycles. There are 42 adult bicycles. How many children's bicycles are there?

11. Philip pays $1,620 in rent every month. This amount is $120 more than twice what his brother Paul pays for rent. How much does Paul pay for rent?

12. Marc just bought an SUV for $54,000. This is $7,400 less than twice what his wife paid for her car last year. How much did his wife pay for her car?

13. Laurie has $46,000 invested in stocks and bonds. The amount invested in stocks is $8,000 less than three times the amount invested in bonds. How much does Laurie have invested in bonds?

14. Erica earned a total of $50,450 last year from her two jobs. The amount she earned from her job at the store was $1,250 more than three times the amount she earned from her job at the college. How much did she earn from her job at the college?

Solve Number Problems

In the following exercises, solve each number word problem.

15. The sum of a number and eight is 12. Find the number.

16. The sum of a number and nine is 17. Find the number.

17. The difference of a number and 12 is three. Find the number.

18. The difference of a number and eight is four. Find the number.

19. The sum of three times a number and eight is 23. Find the number.

20. The sum of twice a number and six is 14. Find the number.

21. The difference of twice a number and seven is 17. Find the number.

22. The difference of four times a number and seven is 21. Find the number.

23. Three times the sum of a number and nine is 12. Find the number.

24. Six times the sum of a number and eight is 30. Find the number.

25. One number is six more than the other. Their sum is 42. Find the numbers.

26. One number is five more than the other. Their sum is 33. Find the numbers.

27. The sum of two numbers is 20. One number is four less than the other. Find the numbers.

28. The sum of two numbers is 27. One number is seven less than the other. Find the numbers.

29. The sum of two numbers is -45. One number is nine more than the other. Find the numbers.

30. The sum of two numbers is -61. One number is 35 more than the other. Find the numbers.

31. The sum of two numbers is -316. One number is 94 less than the other. Find the numbers.

32. The sum of two numbers is -284. One number is 62 less than the other. Find the numbers.

33. One number is 14 less than another. If their sum is increased by seven, the result is 85. Find the numbers.

34. One number is 11 less than another. If their sum is increased by eight, the result is 71. Find the numbers.

35. One number is five more than another. If their sum is increased by nine, the result is 60. Find the numbers.

36. One number is eight more than another. If their sum is increased by 17, the result is 95. Find the numbers.

37. One number is one more than twice another. Their sum is -5. Find the numbers.

38. One number is six more than five times another. Their sum is six. Find the numbers.

39. The sum of two numbers is 14. One number is two less than three times the other. Find the numbers.

40. The sum of two numbers is zero. One number is nine less than twice the other. Find the numbers.

41. The sum of two consecutive integers is 77. Find the integers.

42. The sum of two consecutive integers is 89. Find the integers.

43. The sum of two consecutive integers is -23. Find the integers.

44. The sum of two consecutive integers is -37. Find the integers.

45. The sum of three consecutive integers is 78. Find the integers.

46. The sum of three consecutive integers is 60. Find the integers.

47. Find three consecutive integers whose sum is -36.

48. Find three consecutive integers whose sum is -3.

49. Find three consecutive even integers whose sum is 258.

50. Find three consecutive even integers whose sum is 222.

51. Find three consecutive odd integers whose sum is 171.

52. Find three consecutive odd integers whose sum is 291.

53. Find three consecutive even integers whose sum is -36.

54. Find three consecutive even integers whose sum is -84.

55. Find three consecutive odd integers whose sum is -213.

56. Find three consecutive odd integers whose sum is -267.

Everyday Math

57. Sale Price Patty paid $35 for a purse on sale for $10 off the original price. What was the original price of the purse?

58. Sale Price Travis bought a pair of boots on sale for $25 off the original price. He paid $60 for the boots. What was the original price of the boots?

59. Buying in Bulk Minh spent $6.25 on five sticker books to give his nephews. Find the cost of each sticker book.

60. Buying in Bulk Alicia bought a package of eight peaches for $3.20. Find the cost of each peach.

61. Price before Sales Tax Tom paid $1,166.40 for a new refrigerator, including $86.40 tax. What was the price of the refrigerator?

62. Price before Sales Tax Kenji paid $2,279 for a new living room set, including $129 tax. What was the price of the living room set?

Writing Exercises

63. What has been your past experience solving word problems?

64. When you start to solve a word problem, how do you decide what to let the variable represent?

65. What are consecutive odd integers? Name three consecutive odd integers between 50 and 60.

66. What are consecutive even integers? Name three consecutive even integers between -50 and -40.

Self Check

ⓐ *After completing the exercises, use this checklist to evaluate your mastery of the objectives of this section.*

I can...	Confidently	With some help	No-I don't get it!
approach word problems with a positive attitude.			
use a problem solving strategy for word problems.			
solve number problems.			

ⓑ *If most of your checks were:*

...confidently. Congratulations! You have achieved your goals in this section! Reflect on the study skills you used so that you can continue to use them. What did you do to become confident of your ability to do these things? Be specific!

...with some help. This must be addressed quickly as topics you do not master become potholes in your road to success. Math is sequential—every topic builds upon previous work. It is important to make sure you have a strong foundation before you move on. Who can you ask for help? Your fellow classmates and instructor are good resources. Is there a place on campus where math tutors are available? Can your study skills be improved?

...no—I don't get it! This is critical and you must not ignore it. You need to get help immediately or you will quickly be overwhelmed. See your instructor as soon as possible to discuss your situation. Together you can come up with a plan to get you the help you need.

3.2 Solve Percent Applications

Learning Objectives

By the end of this section, you will be able to:

› Translate and solve basic percent equations
› Solve percent applications
› Find percent increase and percent decrease
› Solve simple interest applications
› Solve applications with discount or mark-up

Be Prepared!

Before you get started, take this readiness quiz.

1. Convert 4.5% to a decimal.
   If you missed this problem, review **Example 1.26**.

2. Convert 0.6 to a percent.
   If you missed this problem, review **Example 1.26**.

3. Round 0.875 to the nearest hundredth.
   If you missed this problem, review **Example 1.26**.

4. Multiply (4.5)(2.38).
   If you missed this problem, review **Example 1.26**.

5. Solve $3.5 = 0.7n$.
   If you missed this problem, review **Example 1.26**.

6. Subtract $50 - 37.45$.
   If you missed this problem, review **Example 1.26**.

Translate and Solve Basic Percent Equations

We will solve percent equations using the methods we used to solve equations with fractions or decimals. Without the tools of algebra, the best method available to solve percent problems was by setting them up as proportions. Now as an algebra student, you can just translate English sentences into algebraic equations and then solve the equations.

We can use any letter you like as a variable, but it is a good idea to choose a letter that will remind us of what you are looking for. We must be sure to change the given percent to a decimal when we put it in the equation.

EXAMPLE 3.12

Translate and solve: What number is 35% of 90?

 **Solution**

	What number is 35% of 90?
Translate into algebra. Let n = the number.	$n = 0.35 \cdot 90$
Remember "of" means multiply, "is" means equals.	
Multiply.	$n = 31.5$
	31.5 is 35% of 90

 TRY IT : : 3.23   Translate and solve:

What number is 45% of 80?

> **TRY IT : : 3.24**   Translate and solve:

What number is 55% of 60?

We must be very careful when we translate the words in the next example. The unknown quantity will not be isolated at first, like it was in **Example 3.12**. We will again use direct translation to write the equation.

EXAMPLE 3.13

Translate and solve: 6.5% of what number is $1.17?

⊘ **Solution**

	6.5%	of	what number	is	$1.17?
Translate. Let $n =$ the number.	0.065	•	n	=	1.17
Multiply.			$0.065n$	=	1.17
Divide both sides by 0.065 and simplify.				$n = 18$	

6.5% of $18 is $1.17

> **TRY IT : : 3.25**   Translate and solve:

7.5% of what number is $1.95?

> **TRY IT : : 3.26**   Translate and solve:

8.5% of what number is $3.06?

In the next example, we are looking for the percent.

EXAMPLE 3.14

Translate and solve: 144 is what percent of 96?

⊘ **Solution**

	144	is	what percent	of	96?
Translate into algebra. Let $p =$ the percent.	144	=	p	•	96
Multiply.		$144 = 96\,p$			
Divide by 96 and simplify.		$1.5 = p$			
Convert to percent.		$150\% = p$			

144 is 150% of 96

Note that we are asked to find percent, so we must have our final result in percent form.

> **TRY IT : : 3.27**   Translate and solve:

110 is what percent of 88?

> **TRY IT : : 3.28**   Translate and solve:

126 is what percent of 72?

Solve Applications of Percent

Many applications of percent—such as tips, sales tax, discounts, and interest—occur in our daily lives. To solve these applications we'll translate to a basic percent equation, just like those we solved in previous examples. Once we translate the sentence into a percent equation, we know how to solve it.

We will restate the problem solving strategy we used earlier for easy reference.

HOW TO :: USE A PROBLEM-SOLVING STRATEGY TO SOLVE AN APPLICATION.

Step 1. **Read** the problem. Make sure all the words and ideas are understood.

Step 2. **Identify** what we are looking for.

Step 3. **Name** what we are looking for. Choose a variable to represent that quantity.

Step 4. **Translate** into an equation. It may be helpful to restate the problem in one sentence with all the important information. Then, translate the English sentence into an algebraic equation.

Step 5. **Solve** the equation using good algebra techniques.

Step 6. **Check** the answer in the problem and make sure it makes sense.

Step 7. **Answer** the question with a complete sentence.

Now that we have the strategy to refer to, and have practiced solving basic percent equations, we are ready to solve percent applications. Be sure to ask yourself if your final answer makes sense—since many of the applications will involve everyday situations, you can rely on your own experience.

EXAMPLE 3.15

Dezohn and his girlfriend enjoyed a nice dinner at a restaurant and his bill was $68.50. He wants to leave an 18% tip. If the tip will be 18% of the total bill, how much tip should he leave?

✓ **Solution**

Step 1. Read the problem.	
Step 2. Identify what we are looking for.	the amount of tip should Dezohn leave
Step 3. Name what we are looking for.	
Choose a variable to represent it.	Let t = amount of tip.
Step 4. Translate into an equation.	The tip is 18% of the total bill.
Write a sentence that gives the information to find it.	The tip is 18% of $68.50
Translate the sentence into an equation.	$t = 0.18 \cdot 68.50$
Step 5. Solve the equation. Multiply.	$t = 12.33$
Step 6. Check. Does this make sense?	
Yes, 20% of $70 is $14.	
Step 7. Answer the question with a complete sentence.	Dezohn should leave a tip of $12.33.

Notice that we used t to represent the unknown tip.

> **TRY IT :: 3.29**

Cierra and her sister enjoyed a dinner in a restaurant and the bill was $81.50. If she wants to leave 18% of the total bill as her tip, how much should she leave?

TRY IT : : 3.30

Kimngoc had lunch at her favorite restaurant. She wants to leave 15% of the total bill as her tip. If her bill was $14.40, how much will she leave for the tip?

EXAMPLE 3.16

The label on Masao's breakfast cereal said that one serving of cereal provides 85 milligrams (mg) of potassium, which is 2% of the recommended daily amount. What is the total recommended daily amount of potassium?

✓ **Solution**

Step 1. Read the problem.	
Step 2. Identify what we are looking for.	the total amount of potassium that is recommended
Step 3. Name what we are looking for.	
Choose a variable to represent it.	Let $a =$ total amount of potassium.
Step 4. Translate. Write a sentence that gives the information to find it.	85 mg is 2% of the total amount
Translate into an equation.	$85 = 0.02 \cdot a$
Step 5. Solve the equation.	$4{,}250 = a$
Step 6. Check. Does this make sense?	
Yes, 2% is a small percent and 85 is a small part of 4,250.	
Step 7. Answer the question with a complete sentence.	The amount of potassium that is recommended is 4,250 mg.

 TRY IT : : 3.31

One serving of wheat square cereal has seven grams of fiber, which is 28% of the recommended daily amount. What is the total recommended daily amount of fiber?

 TRY IT : : 3.32

One serving of rice cereal has 190 mg of sodium, which is 8% of the recommended daily amount. What is the total recommended daily amount of sodium?

EXAMPLE 3.17

Mitzi received some gourmet brownies as a gift. The wrapper said each brownie was 480 calories, and had 240 calories of fat. What percent of the total calories in each brownie comes from fat?

✓ **Solution**

Step 1. Read the problem.	
Step 2. Identify what we are looking for.	the percent of the total calories from fat
Step 3. Name what we are looking for.	

Choose a variable to represent it.	Let $p =$ percent of fat.
Step 4. Translate. Write a sentence that gives the information to find it.	What percent of 480 is 240?
Translate into an equation.	$p \cdot 480 = 240$
Step 5. Solve the equation.	$480p = 240$
Divide by 480.	$p = 0.5$
Put in a percent form.	$p = 50\%$
Step 6. Check. Does this make sense?	
Yes, 240 is half of 480, so 50% makes sense.	
Step 7. Answer the question with a complete sentence.	Of the total calories in each brownie, 50% is fat.

> **TRY IT :: 3.33**
>
> Solve. Round to the nearest whole percent.
>
> Veronica is planning to make muffins from a mix. The package says each muffin will be 230 calories and 60 calories will be from fat. What percent of the total calories is from fat?

> **TRY IT :: 3.34**
>
> Solve. Round to the nearest whole percent.
>
> The mix Ricardo plans to use to make brownies says that each brownie will be 190 calories, and 76 calories are from fat. What percent of the total calories are from fat?

Find Percent Increase and Percent Decrease

People in the media often talk about how much an amount has increased or decreased over a certain period of time. They usually express this increase or decrease as a percent.

To find the percent increase, first we find the amount of increase, the difference of the new amount and the original amount. Then we find what percent the amount of increase is of the original amount.

HOW TO :: FIND THE PERCENT INCREASE.

Step 1.   Find the amount of increase.
        new amount − original amount = increase

Step 2.   Find the percent increase.
        The increase is what percent of the original amount?

EXAMPLE 3.18

In 2011, the California governor proposed raising community college fees from $26 a unit to $36 a unit. Find the percent increase. (Round to the nearest tenth of a percent.)

⊘ **Solution**

Step 1. Read the problem.

Step 2. Identify what we are looking for.	the percent increase
Step 3. Name what we are looking for.	
Choose a variable to represent it.	Let $p =$ the percent.
Step 4. Translate. Write a sentence that gives the information to find it.	
First find the amount of increase.	new amount – original amount = increase
	$36 - 26 = 10$
Find the percent.	Increase is what percent of the original amount?
	10 is what percent of 26?
Translate into an equation.	$10 = p \cdot 26$
Step 5. Solve the equation.	$10 = 26\,p$
Divide by 26.	$0.384 = p$
Change to percent form; round to the nearest tenth.	$38.4\% = p$
Step 6. Check. Does this make sense?	
Yes, 38.4% is close to $\frac{1}{3}$, and 10 is close to $\frac{1}{3}$ of 26.	
Step 7. Answer the question with a complete sentence.	The new fees represent a 38.4% increase over the old fees.

Notice that we rounded the division to the nearest thousandth in order to round the percent to the nearest tenth.

> **TRY IT : : 3.35** Find the percent increase. (Round to the nearest tenth of a percent.)

In 2011, the IRS increased the deductible mileage cost to 55.5 cents from 51 cents.

> **TRY IT : : 3.36**

Find the percent increase.

In 1995, the standard bus fare in Chicago was $1.50. In 2008, the standard bus fare was $2.25.

Finding the percent decrease is very similar to finding the percent increase, but now the amount of decrease is the difference of the original amount and the new amount. Then we find what percent the amount of decrease is of the original amount.

HOW TO : : FIND THE PERCENT DECREASE.

Step 1.   Find the amount of decrease.
original amount − new amount = decrease

Step 2.   Find the percent decrease.
Decrease is what percent of the original amount?

EXAMPLE 3.19

The average price of a gallon of gas in one city in June 2014 was $3.71. The average price in that city in July was $3.64. Find the percent decrease.

✓ **Solution**

Step 1. Read the problem.	
Step 2. Identify what we are looking for.	the percent decrease
Step 3. Name what we are looking for.	
Choose a variable to represent that quantity.	Let $p =$ the percent decrease.
Step 4. Translate. Write a sentence that gives the information to find it.	
First find the amount of decrease.	$3.71 - 3.64 = 0.07$
Find the percent.	Decrease is what percent of the original amaount?
	0.07 is what percent of 3.71?
Translate into an equation.	$0.07 \quad = \quad p \quad \cdot \quad 3.71$
Step 5. Solve the equation.	$0.07 = 3.71\,p$
Divide by 3.71.	$0.019 = p$
Change to percent form; round to the nearest tenth.	$1.9\% = p$
Step 6. Check. Does this make sense?	
Yes, if the original price was $4, a 2% decrease would be 8 cents.	
Step 7. Answer the question with a complete sentence.	The price of gas decreased 1.9%.

> **TRY IT : : 3.37**

Find the percent decrease. (Round to the nearest tenth of a percent.)

The population of North Dakota was about 672,000 in 2010. The population is projected to be about 630,000 in 2020.

> **TRY IT : : 3.38**

Find the percent decrease.

Last year, Sheila's salary was $42,000. Because of furlough days, this year, her salary was $37,800.

Solve Simple Interest Applications

Do you know that banks pay you to keep your money? The money a customer puts in the bank is called the **principal**, P, and the money the bank pays the customer is called the **interest**. The interest is computed as a certain percent of the principal; called the **rate of interest**, r. We usually express rate of interest as a percent per year, and we calculate it by using the decimal equivalent of the percent. The variable t, (for *time*) represents the number of years the money is in the account.

To find the interest we use the simple interest formula, $I = Prt$.

Simple Interest

If an amount of money, P, called the principal, is invested for a period of t years at an annual interest rate r, the amount of interest, I, earned is

$$I = Prt \quad \text{where} \quad \begin{aligned} I &= \text{interest} \\ P &= \text{principal} \\ r &= \text{rate} \\ t &= \text{time} \end{aligned}$$

Interest earned according to this formula is called **simple interest**.

Interest may also be calculated another way, called compound interest. This type of interest will be covered in later math classes.

The formula we use to calculate simple interest is $I = Prt$. To use the formula, we substitute in the values the problem gives us for the variables, and then solve for the unknown variable. It may be helpful to organize the information in a chart.

EXAMPLE 3.20

Nathaly deposited $12,500 in her bank account where it will earn 4% interest. How much interest will Nathaly earn in 5 years?

$$\begin{aligned} I &= ? \\ P &= \$12{,}500 \\ r &= 4\% \\ t &= 5 \text{ years} \end{aligned}$$

 **Solution**

Step 1. Read the problem.

Step 2. Identify what we are looking for.

the amount of interest earned

Step 3. Name what we are looking for. Choose a variable to represent that quantity

Let I = the amount of interest.

Step 4. Translate into an equation.
  Write the formula.

$I = Prt$

  Substitute in the given information.

$I = (12{,}500)(.04)(5)$

Step 5. Solve the equation.

$I = 2{,}500$

Step 6. Check: Does this make sense?
  Is $2,500 is a reasonable interest on
  $12,500? Yes.

Step 7. Answer the question with a complete sentence.

The interest is $2,500.

> **TRY IT : : 3.39**

Areli invested a principal of $950 in her bank account with interest rate 3%. How much interest did she earn in 5 years?

> **TRY IT : : 3.40**

Susana invested a principal of $36,000 in her bank account with interest rate 6.5%. How much interest did she earn in 3 years?

There may be times when we know the amount of interest earned on a given principal over a certain length of time, but we don't know the rate. To find the rate, we use the simple interest formula, substitute in the given values for the principal and time, and then solve for the rate.

EXAMPLE 3.21

Loren loaned his brother $3,000 to help him buy a car. In 4 years his brother paid him back the $3,000 plus $660 in interest. What was the rate of interest?

$$
\begin{aligned}
I &= \$660 \\
P &= \$3,000 \\
r &= \ ? \\
t &= 4 \text{ years}
\end{aligned}
$$

✓ **Solution**

Step 1. Read the problem.

Step 2. Identify what we are looking for.   the rate of interest

Step 3. Name what we are looking for. Choose a variable to represent that quantity.   Let r = rate of interest.

Step 4. Translate into an equation.
Write the formula.
Substitute in the given information.

$$
\begin{aligned}
I &= Prt \\
660 &= (3,000)r(4)
\end{aligned}
$$

Step 5. Solve the equation.
Divide.
Change to percent form.

$$
\begin{aligned}
660 &= (12,000)r \\
0.055 &= r \\
5.5\% &= r
\end{aligned}
$$

Step 6. Check: Does this make sense?

$$I = Prt$$

$$660 \overset{?}{=} (3,000)(0.055)(4)$$

$$660 = 660 ✓$$

Step 7. Answer the question with a complete sentence.   The rate of interest was 5.5%.

Notice that in this example, Loren's brother paid Loren interest, just like a bank would have paid interest if Loren invested his money there.

> **TRY IT : : 3.41**
>
> Jim loaned his sister $5,000 to help her buy a house. In 3 years, she paid him the $5,000, plus $900 interest. What was the rate of interest?

> **TRY IT : : 3.42**
>
> Hang borrowed $7,500 from her parents to pay her tuition. In 5 years, she paid them $1,500 interest in addition to the $7,500 she borrowed. What was the rate of interest?

EXAMPLE 3.22

Eduardo noticed that his new car loan papers stated that with a 7.5% interest rate, he would pay $6,596.25 in interest over 5 years. How much did he borrow to pay for his car?

⊘ Solution

Step 1. Read the problem.

Step 2. Identify what we are looking for.	the amount borrowed (the principal)
Step 3. Name what we are looking for. Choose a variable to represent that quantity.	Let P = principal borrowed.

Step 4. Translate into an equation.
Write the formula.
Substitute in the given information.

$$I = Prt$$
$$6{,}596.25 = P(0.075)(5)$$

Step 5. Solve the equation.

$$6{,}596.25 = 0.375P$$
$$17{,}590 = P$$

Divide.

Step 6. Check: Does this make sense?

$$I = Prt$$
$$6{,}596.25 \overset{?}{=} (17{,}590)(0.075)(5)$$
$$6{,}596.25 = 6{,}596.25 \checkmark$$

Step 7. Answer the question with a complete sentence.	The principal was \$17,590.

> **TRY IT : : 3.43**
>
> Sean's new car loan statement said he would pay \$4,866.25 in interest from an interest rate of 8.5% over 5 years. How much did he borrow to buy his new car?

> **TRY IT : : 3.44**
>
> In 5 years, Gloria's bank account earned \$2,400 interest at 5%. How much had she deposited in the account?

Solve Applications with Discount or Mark-up

Applications of discount are very common in retail settings. When you buy an item on sale, the original price has been discounted by some dollar amount. The **discount rate**, usually given as a percent, is used to determine the amount of the discount. To determine the **amount of discount**, we multiply the discount rate by the original price.

We summarize the discount model in the box below.

Discount

amount of discount = discount rate × original price

sale price = original price − amount of discount

Keep in mind that the sale price should always be less than the original price.

EXAMPLE 3.23

Elise bought a dress that was discounted 35% off of the original price of \$140. What was ⓐ the amount of discount and ⓑ the sale price of the dress?

⊘ Solution

ⓐ

Original price = $140
Discount rate = 35%
Discount = ?

Step 1. Read the problem.

Step 2. Identify what we are looking for.   the amount of discount

Step 3. Name what we are looking for.
Choose a variable to represent that quantity.   Let d = the amount of discount.

Step 4. Translate into an equation. Write a
sentence that gives the information to find it.   The discount is 35% of $140.
Translate into an equation. $d = 0.35(140)$

Step 5. Solve the equation. $d = 49$

Step 6. Check: Does this make sense?
    Is a $49 discount reasonable for a
    $140 dress? Yes.

Step 7. Write a complete sentence to answer
the question.   The amount of discount was $49.

ⓑ
Read the problem again.

Step 1. Identify what we are looking for.	the sale price of the dress
Step 2. Name what we are looking for.	
Choose a variable to represent that quantity.	Let $s =$ the sale price.
Step 3. Translate into an equation.	
Write a sentence that gives the information to find it.	The sale price is the $140 minus the $49 discount
Translate into an equation.	s $=$ 140 $-$ 49
Step 4. Solve the equation.	$s = 91$
Step 5. Check. Does this make sense?	
Is the sale price less than the original price?	
Yes, $91 is less than $140.	
Step 6. Answer the question with a complete sentence.	The sale price of the dress was $91.

> **TRY IT : : 3.45**   Find ⓐ the amount of discount and ⓑ the sale price:

Sergio bought a belt that was discounted 40% from an original price of $29.

> **TRY IT : : 3.46**   Find ⓐ the amount of discount and ⓑ the sale price:

Oscar bought a barbecue that was discounted 65% from an original price of $395.

There may be times when we know the original price and the sale price, and we want to know the discount rate. To find the discount rate, first we will find the amount of discount and then use it to compute the rate as a percent of the original price. **Example 3.24** will show this case.

EXAMPLE 3.24

Jeannette bought a swimsuit at a sale price of $13.95. The original price of the swimsuit was $31. Find the ⓐ amount of discount and ⓑ discount rate.

✓ **Solution**

ⓐ

$$\text{Original price} = \$31$$
$$\text{Discount} = ?$$
$$\text{Sale Price} = \$13.95$$

Step 1. Read the problem.

Step 2. Identify what we are looking for.   the amount of discount

Step 3. Name what we are looking for.
Choose a variable to represent that quantity.   Let d = the amount of discount.

Step 4. Translate into an equation.
    Write a sentence that gives the   The discount is the difference between the original
    information to find it.   price and the sale price.

    Translate into an equation. $d = 31 - 13.95$

Step 5. Solve the equation. $d = 17.05$

Step 6. Check: Does this make sense?
    Is 17.05 less than 31? Yes.

Step 7. Answer the question with a complete sentence.   The amount of discount was $17.05.

ⓑ

Read the problem again.

Step 1. **Identify** what we are looking for.	the discount rate
Step 2. **Name** what we are looking for.	
Choose a variable to represent it.	Let r = the discount rate.
Step 3. **Translate** into an equation.	
Write a sentence that gives the information to find it.	The discount of $17.05 is what percent of $31?
Translate into an equation.	$17.05 = r \cdot 31$
Step 4. **Solve** the equation.	$17.05 = 31r$

| Divide both sides by 31. | $0.55 = r$ |
| Change to percent form. | $r = 55\%$ |

Step 5. Check. Does this make sense?

Is $17.05 equal to 55% of $31?

$$17.05 \overset{?}{=} 0.55(31)$$

$$17.05 = 17.05 \checkmark$$

| **Step 6. Answer** the question with a complete sentence. | The rate of discount was 55%. |

> **TRY IT :: 3.47**
>
> Find ⓐ the amount of discount and ⓑ the discount rate.
>
> Lena bought a kitchen table at the sale price of $375.20. The original price of the table was $560.

> **TRY IT :: 3.48**
>
> Find ⓐ the amount of discount and ⓑ the discount rate.
>
> Nick bought a multi-room air conditioner at a sale price of $340. The original price of the air conditioner was $400.

Applications of mark-up are very common in retail settings. The price a retailer pays for an item is called the **original cost**. The retailer then adds a **mark-up** to the original cost to get the **list price**, the price he sells the item for. The mark-up is usually calculated as a percent of the original cost. To determine the amount of mark-up, multiply the mark-up rate by the original cost.

We summarize the mark-up model in the box below.

Mark-Up

$$\text{amount of mark-up} = \text{mark-up rate} \times \text{original cost}$$
$$\text{list price} = \text{original cost} + \text{amount of mark up}$$

Keep in mind that the list price should always be more than the original cost.

EXAMPLE 3.25

Adam's art gallery bought a photograph at original cost $250. Adam marked the price up 40%. Find the ⓐ amount of mark-up and ⓑ the list price of the photograph.

⊘ **Solution**

ⓐ

Step 1. Read the problem.	
Step 2. Identify what we are looking for.	the amount of mark-up
Step 3. Name what we are looking for.	
Choose a variable to represent it.	Let $m =$ the amount of markup.
Step 4. Translate into an equation.	

Write a sentence that gives the information to find it.	The mark-up is 40% of the $250 original cost
Translate into an equation.	$m = 0.40 \cdot 250$
Step 5. Solve the equation.	$m = 100$
Step 6. Check. Does this make sense?	
Yes, 40% is less than one-half and 100 is less than half of 250.	
Step 7. Answer the question with a complete sentence.	The mark-up on the phtograph was $100.

 ⓑ

Step 1. Read the problem again.	
Step 2. Identify what we are looking for.	the list price
Step 3. Name what we are looking for.	
Choose a variable to represent it.	Let $p =$ the list price.
Step 4. Translate into an equation.	
Write a sentence that gives the information to find it.	The list price is original cost plus the mark-up
Translate into an equation.	$p = 250 + 100$
Step 5. Solve the equation.	$p = 350$
Step 6. Check. Does this make sense?	
Is the list price more than the net price? Is $350 more than $250? Yes.	
Step 7. Answer the question with a complete sentence.	The list price of the photograph was $350.

> **TRY IT :: 3.49**   Find ⓐ the amount of mark-up and ⓑ the list price.

Jim's music store bought a guitar at original cost $1,200. Jim marked the price up 50%.

> **TRY IT :: 3.50**   Find ⓐ the amount of mark-up and ⓑ the list price.

The Auto Resale Store bought Pablo's Toyota for $8,500. They marked the price up 35%.

3.2 EXERCISES
Practice Makes Perfect

Translate and Solve Basic Percent Equations

In the following exercises, translate and solve.

67. What number is 45% of 120?

68. What number is 65% of 100?

69. What number is 24% of 112?

70. What number is 36% of 124?

71. 250% of 65 is what number?

72. 150% of 90 is what number?

73. 800% of 2250 is what number?

74. 600% of 1740 is what number?

75. 28 is 25% of what number?

76. 36 is 25% of what number?

77. 81 is 75% of what number?

78. 93 is 75% of what number?

79. 8.2% of what number is $2.87?

80. 6.4% of what number is $2.88?

81. 11.5% of what number is $108.10?

82. 12.3% of what number is $92.25?

83. What percent of 260 is 78?

84. What percent of 215 is 86?

85. What percent of 1500 is 540?

86. What percent of 1800 is 846?

87. 30 is what percent of 20?

88. 50 is what percent of 40?

89. 840 is what percent of 480?

90. 790 is what percent of 395?

Solve Percent Applications

In the following exercises, solve.

91. Geneva treated her parents to dinner at their favorite restaurant. The bill was $74.25. Geneva wants to leave 16% of the total bill as a tip. How much should the tip be?

92. When Hiro and his co-workers had lunch at a restaurant near their work, the bill was $90.50. They want to leave 18% of the total bill as a tip. How much should the tip be?

93. Trong has 12% of each paycheck automatically deposited to his savings account. His last paycheck was $2165. How much money was deposited to Trong's savings account?

94. Cherise deposits 8% of each paycheck into her retirement account. Her last paycheck was $1,485. How much did Cherise deposit into her retirement account?

95. One serving of oatmeal has eight grams of fiber, which is 33% of the recommended daily amount. What is the total recommended daily amount of fiber?

96. One serving of trail mix has 67 grams of carbohydrates, which is 22% of the recommended daily amount. What is the total recommended daily amount of carbohydrates?

97. A bacon cheeseburger at a popular fast food restaurant contains 2070 milligrams (mg) of sodium, which is 86% of the recommended daily amount. What is the total recommended daily amount of sodium?

98. A grilled chicken salad at a popular fast food restaurant contains 650 milligrams (mg) of sodium, which is 27% of the recommended daily amount. What is the total recommended daily amount of sodium?

99. After 3 months on a diet, Lisa had lost 12% of her original weight. She lost 21 pounds. What was Lisa's original weight?

100. Tricia got a 6% raise on her weekly salary. The raise was $30 per week. What was her original salary?

101. Yuki bought a dress on sale for $72. The sale price was 60% of the original price. What was the original price of the dress?

102. Kim bought a pair of shoes on sale for $40.50. The sale price was 45% of the original price. What was the original price of the shoes?

103. Tim left a $9 tip for a $50 restaurant bill. What percent tip did he leave?

104. Rashid left a $15 tip for a $75 restaurant bill. What percent tip did he leave?

105. The nutrition fact sheet at a fast food restaurant says the fish sandwich has 380 calories, and 171 calories are from fat. What percent of the total calories is from fat?

106. The nutrition fact sheet at a fast food restaurant says a small portion of chicken nuggets has 190 calories, and 114 calories are from fat. What percent of the total calories is from fat?

107. Emma gets paid $3,000 per month. She pays $750 a month for rent. What percent of her monthly pay goes to rent?

108. Dimple gets paid $3,200 per month. She pays $960 a month for rent. What percent of her monthly pay goes to rent?

Find Percent Increase and Percent Decrease

In the following exercises, solve.

109. Tamanika got a raise in her hourly pay, from $15.50 to $17.36. Find the percent increase.

110. Ayodele got a raise in her hourly pay, from $24.50 to $25.48. Find the percent increase.

111. Annual student fees at the University of California rose from about $4,000 in 2000 to about $12,000 in 2010. Find the percent increase.

112. The price of a share of one stock rose from $12.50 to $50. Find the percent increase.

113. According to *Time* magazine annual global seafood consumption rose from 22 pounds per person in the 1960s to 38 pounds per person in 2011. Find the percent increase. (Round to the nearest tenth of a percent.)

114. In one month, the median home price in the Northeast rose from $225,400 to $241,500. Find the percent increase. (Round to the nearest tenth of a percent.)

115. A grocery store reduced the price of a loaf of bread from $2.80 to $2.73. Find the percent decrease.

116. The price of a share of one stock fell from $8.75 to $8.54. Find the percent decrease.

117. Hernando's salary was $49,500 last year. This year his salary was cut to $44,055. Find the percent decrease.

118. In 10 years, the population of Detroit fell from 950,000 to about 712,500. Find the percent decrease.

119. In 1 month, the median home price in the West fell from $203,400 to $192,300. Find the percent decrease. (Round to the nearest tenth of a percent.)

120. Sales of video games and consoles fell from $1,150 million to $1,030 million in 1 year. Find the percent decrease. (Round to the nearest tenth of a percent.)

Solve Simple Interest Applications

In the following exercises, solve.

121. Casey deposited $1,450 in a bank account with interest rate 4%. How much interest was earned in two years?

122. Terrence deposited $5,720 in a bank account with interest rate 6%. How much interest was earned in 4 years?

123. Robin deposited $31,000 in a bank account with interest rate 5.2%. How much interest was earned in 3 years?

124. Carleen deposited $16,400 in a bank account with interest rate 3.9%. How much interest was earned in 8 years?

125. Hilaria borrowed $8,000 from her grandfather to pay for college. Five years later, she paid him back the $8,000, plus $1,200 interest. What was the rate of interest?

126. Kenneth loaned his niece $1,200 to buy a computer. Two years later, she paid him back the $1,200, plus $96 interest. What was the rate of interest?

127. Lebron loaned his daughter $20,000 to help her buy a condominium. When she sold the condominium four years later, she paid him the $20,000, plus $3,000 interest. What was the rate of interest?

128. Pablo borrowed $50,000 to start a business. Three years later, he repaid the $50,000, plus $9,375 interest. What was the rate of interest?

129. In 10 years, a bank account that paid 5.25% earned $18,375 interest. What was the principal of the account?

130. In 25 years, a bond that paid 4.75% earned $2,375 interest. What was the principal of the bond?

131. Joshua's computer loan statement said he would pay $1,244.34 in interest for a 3-year loan at 12.4%. How much did Joshua borrow to buy the computer?

132. Margaret's car loan statement said she would pay $7,683.20 in interest for a 5-year loan at 9.8%. How much did Margaret borrow to buy the car?

Solve Applications with Discount or Mark-up

In the following exercises, find the sale price.

133. Perla bought a cell phone that was on sale for $50 off. The original price of the cell phone was $189.

134. Sophie saw a dress she liked on sale for $15 off. The original price of the dress was $96.

135. Rick wants to buy a tool set with original price $165. Next week the tool set will be on sale for $40 off.

136. Angelo's store is having a sale on televisions. One television, with original price $859, is selling for $125 off.

In the following exercises, find ⓐ the amount of discount and ⓑ the sale price.

137. Janelle bought a beach chair on sale at 60% off. The original price was $44.95.

138. Errol bought a skateboard helmet on sale at 40% off. The original price was $49.95.

139. Kathy wants to buy a camera that lists for $389. The camera is on sale with a 33% discount.

140. Colleen bought a suit that was discounted 25% from an original price of $245.

141. Erys bought a treadmill on sale at 35% off. The original price was $949.95 (round to the nearest cent.)

142. Jay bought a guitar on sale at 45% off. The original price was $514.75 (round to the nearest cent.)

In the following exercises, find ⓐ the amount of discount and ⓑ the discount rate. (Round to the nearest tenth of a percent if needed.)

143. Larry and Donna bought a sofa at the sale price of $1,344. The original price of the sofa was $1,920.

144. Hiroshi bought a lawnmower at the sale price of $240. The original price of the lawnmower is $300.

145. Patty bought a baby stroller on sale for $301.75. The original price of the stroller was $355.

146. Bill found a book he wanted on sale for $20.80. The original price of the book was $32.

147. Nikki bought a patio set on sale for $480. The original price was $850. To the nearest tenth of a percent, what was the rate of discount?

148. Stella bought a dinette set on sale for $725. The original price was $1,299. To the nearest tenth of a percent, what was the rate of discount?

In the following exercises, find ⓐ the amount of the mark-up and ⓑ the list price.

149. Daria bought a bracelet at original cost $16 to sell in her handicraft store. She marked the price up 45%.

150. Regina bought a handmade quilt at original cost $120 to sell in her quilt store. She marked the price up 55%.

151. Tom paid $0.60 a pound for tomatoes to sell at his produce store. He added a 33% mark-up.

152. Flora paid her supplier $0.74 a stem for roses to sell at her flower shop. She added an 85% mark-up.

153. Alan bought a used bicycle for $115. After re-conditioning it, he added 225% mark-up and then advertised it for sale.

154. Michael bought a classic car for $8,500. He restored it, then added 150% mark-up before advertising it for sale.

Everyday Math

155. Leaving a Tip At the campus coffee cart, a medium coffee costs $1.65. MaryAnne brings $2.00 with her when she buys a cup of coffee and leaves the change as a tip. What percent tip does she leave?

156. Splitting a Bill Four friends went out to lunch and the bill came to $53.75. They decided to add enough tip to make a total of $64, so that they could easily split the bill evenly among themselves. What percent tip did they leave?

Writing Exercises

157. Without solving the problem "44 is 80% of what number" think about what the solution might be. Should it be a number that is greater than 44 or less than 44? Explain your reasoning.

158. Without solving the problem "What is 20% of 300?" think about what the solution might be. Should it be a number that is greater than 300 or less than 300? Explain your reasoning.

159. After returning from vacation, Alex said he should have packed 50% fewer shorts and 200% more shirts. Explain what Alex meant.

160. Because of road construction in one city, commuters were advised to plan that their Monday morning commute would take 150% of their usual commuting time. Explain what this means.

Self Check

ⓐ After completing the exercises, use this checklist to evaluate your mastery of the objectives of this section.

I can...	Confidently	With some help	No-I don't get it!
translate and solve basic percent equations.			
solve percent applications.			
find percent increase and percent decrease.			
solve simple interest applications.			
solve applications with discount or mark-up.			

ⓑ After reviewing this checklist, what will you do to become confident for all goals?

Solve Mixture Applications

Learning Objectives

By the end of this section, you will be able to:

› Solve coin word problems
› Solve ticket and stamp word problems
› Solve mixture word problems
› Use the mixture model to solve investment problems using simple interest

---

Be Prepared!

Before you get started, take this readiness quiz.

1. Multiply: 14(0.25).
   If you missed this problem, review **Example 1.97**.

2. Solve: $0.25x + 0.10(x + 4) = 2.5$.

   If you missed this problem, review **Example 2.44**.

3. The number of dimes is three more than the number of quarters. Let q represent the number of quarters. Write an expression for the number of dimes.
   If you missed this problem, review **Example 1.26**.

Solve Coin Word Problems

In **mixture problems**, we will have two or more items with different values to combine together. The mixture model is used by grocers and bartenders to make sure they set fair prices for the products they sell. Many other professionals, like chemists, investment bankers, and landscapers also use the mixture model.

 MANIPULATIVE MATHEMATICS

Doing the Manipulative Mathematics activity *Coin Lab* will help you develop a better understanding of mixture word problems.

We will start by looking at an application everyone is familiar with—money!

Imagine that we take a handful of coins from a pocket or purse and place them on a desk. How would we determine the value of that pile of coins? If we can form a step-by-step plan for finding the total value of the coins, it will help us as we begin solving coin word problems.

So what would we do? To get some order to the mess of coins, we could separate the coins into piles according to their value. Quarters would go with quarters, dimes with dimes, nickels with nickels, and so on. To get the total value of all the coins, we would add the total value of each pile.

How would we determine the value of each pile? Think about the dime pile—how much is it worth? If we count the number of dimes, we'll know how many we have—the *number* of dimes.

But this does not tell us the *value* of all the dimes. Say we counted 17 dimes, how much are they worth? Each dime is worth $0.10—that is the *value* of one dime. To find the total value of the pile of 17 dimes, multiply 17 by $0.10 to get $1.70. This is the total value of all 17 dimes. This method leads to the following model.

Total Value of Coins

For the same type of coin, the total value of a number of coins is found by using the model

$$number \cdot value = total\ value$$

where

number is the number of coins

value is the value of each coin

total value is the total value of all the coins

The number of dimes times the value of each dime equals the total value of the dimes.

$$number \cdot value\ =\ total\ value$$
$$17 \cdot \$0.10\ =\ \$1.70$$

We could continue this process for each type of coin, and then we would know the total value of each type of coin. To get the total value of *all* the coins, add the total value of each type of coin.

Let's look at a specific case. Suppose there are 14 quarters, 17 dimes, 21 nickels, and 39 pennies.

Type	Number	•	Value($)	=	Total Value($)
Quarters	14		0.25		3.50
Dimes	17		0.10		1.70
Nickels	21		0.05		1.05
Pennies	39		0.01		0.39
					6.64

The total value of all the coins is $6.64.

Notice how the chart helps organize all the information! Let's see how we use this method to solve a coin word problem.

EXAMPLE 3.26

Adalberto has $2.25 in dimes and nickels in his pocket. He has nine more nickels than dimes. How many of each type of coin does he have?

✓ **Solution**

Step 1. Read the problem. Make sure all the words and ideas are understood.
- Determine the types of coins involved.
  Think about the strategy we used to find the value of the handful of coins. The first thing we need is to notice what types of coins are involved. Adalberto has dimes and nickels.
- **Create a table** to organize the information. See chart below.
  - Label the columns "type," "number," "value," "total value."
  - List the types of coins.
  - Write in the value of each type of coin.
  - Write in the total value of all the coins.

We can work this problem all in cents or in dollars. Here we will do it in dollars and put in the dollar sign ($) in the table as a reminder.

The value of a dime is $0.10 and the value of a nickel is $0.05. The total value of all the coins is $2.25. The table below shows this information.

Type	Number	•	Value($)	=	Total Value($)
Dimes			0.10		
Nickels			0.05		
					2.25

Step 2. Identify what we are looking for.
- We are asked to find the number of dimes and nickels Adalberto has.

Step 3. Name what we are looking for. Choose a variable to represent that quantity.
- Use variable expressions to represent the number of each type of coin and write them in the table.
- Multiply the number times the value to get the total value of each type of coin.

Next we counted the number of each type of coin. In this problem we cannot count each type of coin—that is what you are looking for—but we have a clue. There are nine more nickels than dimes. The number of nickels is nine more than the number of dimes.

$$\text{Let } d \; = \; \text{number of dimes.}$$
$$d + 9 \; = \; \text{number of nickels}$$

Fill in the "number" column in the table to help get everything organized.

Type	Number •	Value($) =	Total Value($)
Dimes	d	0.10	
Nickels	$d + 9$	0.05	
			2.25

Now we have all the information we need from the problem!

We multiply the number times the value to get the total value of each type of coin. While we do not know the actual number, we do have an expression to represent it.

And so now multiply $number \cdot value = total\ value$. See how this is done in the table below.

Type	Number •	Value($) =	Total Value($)
Dimes	d	0.10	$0.10d$
Nickels	$d + 9$	0.05	$0.05(d + 9)$
			2.25

Notice that we made the heading of the table show the model.

Step 4. Translate into an equation. It may be helpful to restate the problem in one sentence. Translate the English sentence into an algebraic equation.

Write the equation by adding the total values of all the types of coins.

$$\underbrace{\text{Value of dimes}} \; + \; \underbrace{\text{value of nickels}} \; = \; \underbrace{\text{total value of coins}}$$

Translate to an equation. $\qquad 0.10d \qquad + \qquad 0.05(d + 9) \qquad = \qquad 2.25$

Step 5. Solve the equation using good algebra techniques.

Now solve this equation.	$0.10d + 0.05(d + 9) = 2.25$
Distribute.	$0.10d + 0.05d + 0.45 = 2.25$
Combine like terms.	$0.15d + 0.45 = 2.25$
Subtract 0.45 from each side.	$0.15d = 1.80$
Divide.	$d = 12$
So there are 12 dimes.	
The number of nickels is $d + 9$.	$d + 9$
	$12 + 9$
	21

Step 6. Check the answer in the problem and make sure it makes sense.

Does this check?

$$12 \text{ dimes} \qquad 12(0.10) = 1.20$$
$$21 \text{ nickels} \qquad 21(0.05) = \underline{1.05}$$
$$\$2.25 \checkmark$$

Step 7. Answer the question with a complete sentence.

   • Adalberto has twelve dimes and twenty-one nickels.

If this were a homework exercise, our work might look like the following.

Adalberto has \$2.25 in dimes and nickels in his pocket. He has nine more nickels than dimes.

How many of each type does he have?

Type	Number • Value(\$)		= Total Value (\$)
Dimes	d	0.10	0.10d
Nickels	d + 9	0.05	0.05(d + 9)
			2.25 ✓

12 dimes   12(0.10) = 1.20

21 nickels   21(0.05) = 1.05

$2.25 ✓

Adalberto has twelve dimes and twenty-one nickels.

$$0.10d + 0.05d + 0.45 = 2.25 \qquad d + 9$$

$$0.15d + 0.45 = 2.25 \qquad 12 + 9$$

$$0.15d = 1.80 \qquad 21 \text{ nickels}$$

$$d = 12 \text{ dimes}$$

> **TRY IT : : 3.51**

Michaela has \$2.05 in dimes and nickels in her change purse. She has seven more dimes than nickels. How many coins of each type does she have?

> **TRY IT : : 3.52**

Liliana has \$2.10 in nickels and quarters in her backpack. She has 12 more nickels than quarters. How many coins of each type does she have?

HOW TO :: SOLVE COIN WORD PROBLEMS.

Step 1.   **Read** the problem. Make sure all the words and ideas are understood.

    ◦ Determine the types of coins involved.

    ◦ Create a table to organize the information.

    ◦ Label the columns "type," "number," "value," "total value."

    ◦ List the types of coins.

    ◦ Write in the value of each type of coin.

    ◦ Write in the total value of all the coins.

Type	Number	•	Value($)	=	Total Value($)

Step 2.   **Identify** what we are looking for.

Step 3.   **Name** what we are looking for. Choose a variable to represent that quantity.

    ◦ Use variable expressions to represent the number of each type of coin and write them in the table.

    ◦ Multiply the number times the value to get the total value of each type of coin.

Step 4.   **Translate** into an equation.
It may be helpful to restate the problem in one sentence with all the important information. Then, translate the sentence into an equation.
Write the equation by adding the total values of all the types of coins.

Step 5.   **Solve** the equation using good algebra techniques.

Step 6.   **Check** the answer in the problem and make sure it makes sense.

Step 7.   **Answer** the question with a complete sentence.

EXAMPLE 3.27

Maria has $2.43 in quarters and pennies in her wallet. She has twice as many pennies as quarters. How many coins of each type does she have?

 **Solution**

Step 1. Read the problem.

Determine the types of coins involved.

We know that Maria has quarters and pennies.

Create a table to organize the information.

- Label the columns "type," "number," "value," "total value."
- List the types of coins.
- Write in the value of each type of coin.
- Write in the total value of all the coins.

Type	Number	•	Value($)	=	Total Value($)
Quarters			0.25		
Pennies			0.01		
					2.43

Step 2. Identify what you are looking for.

- We are looking for the number of quarters and pennies.

Step 3. Name. Represent the number of quarters and pennies using variables.

- We know Maria has twice as many pennies as quarters. The number of pennies is defined in terms of quarters.
- Let q represent the number of quarters.
- Then the number of pennies is $2q$.

Type	Number	•	Value($)	=	Total Value($)
Quarters	q		0.25		
Pennies	$2q$		0.01		
					2.43

Multiply the 'number' and the 'value' to get the 'total value' of each type of coin.

Type	Number	•	Value($)	=	Total Value($)
Quarters	q		0.25		$0.25q$
Pennies	$2q$		0.01		$0.01(2q)$
					2.43

Step 4. Translate. Write the equation by adding the 'total value' of all the types of coins.

Step 5. Solve the equation.

Multiply.

Combine like terms.

Divide by 0.27

$$0.25q + 0.01(2q) = 2.43$$
$$0.25q + 0.02q = 2.43$$
$$0.27q = 2.43$$
$$q = 9 \text{ quarters}$$

The number of pennies is $2q$.

$2q$

$2 \cdot 9$

18 pennies

Step 6. Check the answer in the problem.

Maria has 9 quarters and 18 pennies. Does this make $2.43?

9 quarters	$9(0.25)$	$=$	2.25
18 pennies	$18(0.01)$	$=$	0.18
Total			$2.43 ✓

Step 7. Answer the question.   Maria has nine quarters and eighteen pennies.

> **TRY IT : : 3.53**

Sumanta has $4.20 in nickels and dimes in her piggy bank. She has twice as many nickels as dimes. How many coins of each type does she have?

> **TRY IT : : 3.54**

Alison has three times as many dimes as quarters in her purse. She has $9.35 altogether. How many coins of each type does she have?

In the next example, we'll show only the completed table—remember the steps we take to fill in the table.

EXAMPLE 3.28

Danny has $2.14 worth of pennies and nickels in his piggy bank. The number of nickels is two more than ten times the number of pennies. How many nickels and how many pennies does Danny have?

⊘ Solution

Step 1. Read the problem.

Determine the types of coins involved.	pennies and nickels
Create a table.	
Write in the value of each type of coin.	Pennies are worth $0.01. Nickels are worth $0.05.
Step 2. Identify what we are looking for.	the number of pennies and nickels

Step 3. Name. Represent the number of each type of coin using variables.

The number of nickels is defined in terms of the number of pennies, so start with pennies.	Let $p =$ number of pennies.
The number of nickels is two more than ten times the number of pennies.	$10p + 2 =$ number of nickels.

Multiply the number and the value to get the total value of each type of coin.

Type	Number •	Value ($) =	Total Value ($)
pennies	p	0.01	$0.01p$
nickels	$10p + 2$	0.05	$0.05(10p + 2)$
			$2.14

Step 4. Translate. Write the equation by adding the total value of all the types of coins.	$0.01p + 0.05(10p + 2) = 2.14$
Step 5. Solve the equation.	$0.01p + 0.50p + 0.10 = 2.14$ $0.51p + 0.10 = 2.14$ $0.51p = 2.04$ $p = 4$ pennies
How many nickels?	$10p + 2$ $10(4) + 2$ 42 nickels

Step 6. Check the answer in the problem and make sure it makes sense
Danny has four pennies and 42 nickels.
Is the total value $2.14?
$$4(0.01) + 42(0.05) \overset{?}{=} 2.14$$
$$2.14 = 2.14 \checkmark$$

Step 7. Answer the question.	Danny has four pennies and 42 nickels.

> **TRY IT : : 3.55**
>
> Jesse has $6.55 worth of quarters and nickels in his pocket. The number of nickels is five more than two times the number of quarters. How many nickels and how many quarters does Jesse have?

TRY IT : : 3.56

Elane has $7.00 total in dimes and nickels in her coin jar. The number of dimes that Elane has is seven less than three times the number of nickels. How many of each coin does Elane have?

Solve Ticket and Stamp Word Problems

Problems involving tickets or stamps are very much like coin problems. Each type of ticket and stamp has a value, just like each type of coin does. So to solve these problems, we will follow the same steps we used to solve coin problems.

EXAMPLE 3.29

At a school concert, the total value of tickets sold was $1,506. Student tickets sold for $6 each and adult tickets sold for $9 each. The number of adult tickets sold was five less than three times the number of student tickets sold. How many student tickets and how many adult tickets were sold?

 **Solution**

Step 1. Read the problem.
- Determine the types of tickets involved. There are student tickets and adult tickets.
- Create a table to organize the information.

Type	Number •	Value($) =	Total Value($)
Student		6	
Adult		9	
			1506

Step 2. Identify what we are looking for.
- We are looking for the number of student and adult tickets.

Step 3. Name. Represent the number of each type of ticket using variables.

We know the number of adult tickets sold was five less than three times the number of student tickets sold.
- Let s be the number of student tickets.
- Then $3s - 5$ is the number of adult tickets

Multiply the number times the value to get the total value of each type of ticket.

Type	Number •	Value($) =	Total Value($)
Student	s	6	$6s$
Adult	$3s - 5$	9	$9(3s - 5)$
			1506

Step 4. Translate. Write the equation by adding the total values of each type of ticket.
$$6s + 9(3s - 5) = 1506$$

Step 5. Solve the equation.
$$\begin{aligned} 6s + 27s - 45 &= 1506 \\ 33s - 45 &= 1506 \\ 33s &= 1551 \\ s &= 47 \text{ student tickets} \\ 3s - 5 \\ 3(47) - 5 \\ 136 \text{ adult tickets} \end{aligned}$$

Step 6. Check the answer.

There were 47 student tickets at $6 each and 136 adult tickets at $9 each. Is the total value $1,506? We find the total value of each type of ticket by multiplying the number of tickets times its value then add to get the total value of all the tickets sold.

$$47 \cdot 6 = 282$$
$$136 \cdot 9 = 1{,}224$$
$$1{,}506 \checkmark$$

Step 7. Answer the question. They sold 47 student tickets and 136 adult tickets.

> **TRY IT :: 3.57**

The first day of a water polo tournament the total value of tickets sold was $17,610. One-day passes sold for $20 and tournament passes sold for $30. The number of tournament passes sold was 37 more than the number of day passes sold. How many day passes and how many tournament passes were sold?

> **TRY IT :: 3.58**

At the movie theater, the total value of tickets sold was $2,612.50. Adult tickets sold for $10 each and senior/child tickets sold for $7.50 each. The number of senior/child tickets sold was 25 less than twice the number of adult tickets sold. How many senior/child tickets and how many adult tickets were sold?

We have learned how to find the total number of tickets when the number of one type of ticket is based on the number of the other type. Next, we'll look at an example where we know the total number of tickets and have to figure out how the two types of tickets relate.

Suppose Bianca sold a total of 100 tickets. Each ticket was either an adult ticket or a child ticket. If she sold 20 child tickets, how many adult tickets did she sell?

- *Did you say '80'? How did you figure that out? Did you subtract 20 from 100?*

If she sold 45 child tickets, how many adult tickets did she sell?

- *Did you say '55'? How did you find it? By subtracting 45 from 100?*

What if she sold 75 child tickets? How many adult tickets did she sell?

- *The number of adult tickets must be $100 - 75$. She sold 25 adult tickets.*

Now, suppose Bianca sold x child tickets. Then how many adult tickets did she sell? To find out, we would follow the same logic we used above. In each case, we subtracted the number of child tickets from 100 to get the number of adult tickets. We now do the same with x.

We have summarized this below.

Child tickets	Adult tickets
20	80
45	55
75	25
x	$100 - x$

We can apply these techniques to other examples

EXAMPLE 3.30

Galen sold 810 tickets for his church's carnival for a total of $2,820. Children's tickets cost $3 each and adult tickets cost $5 each. How many children's tickets and how many adult tickets did he sell?

✓ **Solution**

Step 1. Read the problem.
- Determine the types of tickets involved. There are children tickets and adult tickets.
- Create a table to organize the information.

Type	Number	•	Value($)	=	Total Value($)
Children			3		
Adult			5		
					2820

Step 2. Identify what we are looking for.
- We are looking for the number of children and adult tickets.

Step 3. Name. Represent the number of each type of ticket using variables.
- We know the total number of tickets sold was 810. This means the number of children's tickets plus the number of adult tickets must add up to 810.
- Let c be the number of children tickets.
- Then $810 - c$ is the number of adult tickets.
- Multiply the number times the value to get the total value of each type of ticket.

Type	Number •	Value($) =	Total Value($)
Children	c	3	$3c$
Adult	$810 - c$	5	$5(810 - c)$
			2820

Step 4. Translate.
- Write the equation by adding the total values of each type of ticket.

Step 5. Solve the equation.

$$3c + 5(810 - c) = 2,820$$
$$3c + 4,050 - 5c = 2,820$$
$$-2c = -1,230$$
$$c = 615 \text{ children tickets}$$

How many adults?

$$810 - c$$
$$810 - 615$$
$$195 \text{ adult tickets}$$

Step 6. Check the answer. There were 615 children's tickets at $3 each and 195 adult tickets at $5 each. Is the total value $2,820?

$$615 \cdot 3 = 1845$$
$$195 \cdot 5 = \underline{975}$$
$$2,820 \checkmark$$

Step 7. Answer the question. Galen sold 615 children's tickets and 195 adult tickets.

> **TRY IT : : 3.59**
>
> During her shift at the museum ticket booth, Leah sold 115 tickets for a total of $1,163. Adult tickets cost $12 and student tickets cost $5. How many adult tickets and how many student tickets did Leah sell?

> **TRY IT : : 3.60**
>
> A whale-watching ship had 40 paying passengers on board. The total collected from tickets was $1,196. Full-fare passengers paid $32 each and reduced-fare passengers paid $26 each. How many full-fare passengers and how many reduced-fare passengers were on the ship?

Now, we'll do one where we fill in the table all at once.

EXAMPLE 3.31

Monica paid $8.36 for stamps. The number of 41-cent stamps was four more than twice the number of two-cent stamps. How many 41-cent stamps and how many two-cent stamps did Monica buy?

⊘ **Solution**

The types of stamps are 41-cent stamps and two-cent stamps. Their names also give the value!

"The number of 41-cent stamps was four more than twice the number of two-cent stamps."

Let x = number of 2-cent stamps.

$2x + 4$ = number of 41-cent stamps

Type	Number	•	Value($)	=	Total Value($)
41 cent stamps	$2x + 4$		0.41		$0.41(2x + 4)$
2 cent stamps	x		0.02		$0.02x$
					8.36

Write the equation from the total values.

$$0.41(2x + 4) + 0.02x = 8.36$$
$$0.82x + 1.64 + 0.02x = 8.36$$

Solve the equation.

$$0.84x + 1.64 = 8.36$$
$$0.84x = 6.72$$
$$x = 8$$

Monica bought eight two-cent stamps.

Find the number of 41-cent stamps she bought by evaluating

$2x + 4$ for $x = 8$.
$2x + 4$
$2(8) + 4$
20

Check.

$$8(0.02) + 20(0.41) \stackrel{?}{=} 8.36$$
$$0.16 + 8.20 \stackrel{?}{=} 8.36$$
$$8.36 = 8.36 ✓$$

Monica bought eight two-cent stamps and 20 41-cent stamps.

> **TRY IT : : 3.61**

Eric paid $13.36 for stamps. The number of 41-cent stamps was eight more than twice the number of two-cent stamps. How many 41-cent stamps and how many two-cent stamps did Eric buy?

> **TRY IT : : 3.62**

Kailee paid $12.66 for stamps. The number of 41-cent stamps was four less than three times the number of 20-cent stamps. How many 41-cent stamps and how many 20-cent stamps did Kailee buy?

Solve Mixture Word Problems

Now we'll solve some more general applications of the mixture model. Grocers and bartenders use the mixture model to set a fair price for a product made from mixing two or more ingredients. Financial planners use the mixture model when they invest money in a variety of accounts and want to find the overall interest rate. Landscape designers use the mixture model when they have an assortment of plants and a fixed budget, and event coordinators do the same when choosing appetizers and entrees for a banquet.

Our first mixture word problem will be making trail mix from raisins and nuts.

EXAMPLE 3.32

Henning is mixing raisins and nuts to make 10 pounds of trail mix. Raisins cost $2 a pound and nuts cost $6 a pound. If Henning wants his cost for the trail mix to be $5.20 a pound, how many pounds of raisins and how many pounds of nuts should he use?

⊘ **Solution**

As before, we fill in a chart to organize our information.

The 10 pounds of trail mix will come from mixing raisins and nuts.

Let x = number of pounds of raisins.

$10 - x$ = number of pounds of nuts

We enter the price per pound for each item.

We multiply the number times the value to get the total value.

Type	Number of pounds	• Price per pound ($)	= Total Value ($)
Raisins	x	2	$2x$
Nuts	$10 - x$	6	$6(10 - x)$
Trail mix	10	5.20	$10(5.20)$

Notice that the last line in the table gives the information for the total amount of the mixture.

We know the value of the raisins plus the value of the nuts will be the value of the trail mix.

Write the equation from the total values.	$2x + 6(10 + x) = 10(5.20)$
Solve the equation.	$2x + 60 - 6x = 52$
	$-4x = -8$
	$x = 2$ pounds of raisins
Find the number of pounds of nuts.	$10 - x$
	$10 - 2$
	8 pounds of nuts

Check.

$$2(\$2) + 8(\$6) \stackrel{?}{=} 10(\$5.20)$$
$$\$4 + \$48 \stackrel{?}{=} \$52$$
$$\$52 = \$52 \checkmark$$

Henning mixed two pounds of raisins with eight pounds of nuts.

> **TRY IT : : 3.63**

Orlando is mixing nuts and cereal squares to make a party mix. Nuts sell for $7 a pound and cereal squares sell for $4 a pound. Orlando wants to make 30 pounds of party mix at a cost of $6.50 a pound, how many pounds of nuts and how many pounds of cereal squares should he use?

> **TRY IT : : 3.64**

Becca wants to mix fruit juice and soda to make a punch. She can buy fruit juice for $3 a gallon and soda for $4 a gallon. If she wants to make 28 gallons of punch at a cost of $3.25 a gallon, how many gallons of fruit juice and how many gallons of soda should she buy?

We can also use the mixture model to solve investment problems using simple interest. We have used the simple interest formula, $I = Prt$, where t represented the number of years. When we just need to find the interest for one year, $t = 1$, so then $I = Pr$.

EXAMPLE 3.33

Stacey has $20,000 to invest in two different bank accounts. One account pays interest at 3% per year and the other account pays interest at 5% per year. How much should she invest in each account if she wants to earn 4.5% interest per year on the total amount?

⊘ Solution

We will fill in a chart to organize our information. We will use the simple interest formula to find the interest earned in the different accounts.

The interest on the mixed investment will come from adding the interest from the account earning 3% and the interest from the account earning 5% to get the total interest on the $20,000.

$$\text{Let } x = \text{amount invested at 3\%.}$$
$$20,000 - x = \text{amount invested at 5\%}$$

The amount invested is the *principal* for each account.

We enter the interest rate for each account.

We multiply the amount invested times the rate to get the interest.

Type	Amount Invested	• Rate =	Interest
3%	x	0.03	0.03x
5%	20,000 − x	0.05	0.05(20,000 − x)
4.5%	20,000	0.045	0.045(20,000)

Notice that the total amount invested, 20,000, is the sum of the amount invested at 3% and the amount invested at 5%. And the total interest, 0.045(20,000), is the sum of the interest earned in the 3% account and the interest earned in the 5% account.

As with the other mixture applications, the last column in the table gives us the equation to solve.

Write the equation from the interest earned. $0.03x + 0.05(20,000 - x) = 0.045(20,000)$

Solve the equation.

$$0.03x + 1,000 - 0.05x = 900$$
$$-0.02x + 1,000 = 900$$
$$-0.02x = -100$$
$$x = 5,000$$

amount invested at 3%

Find the amount invested at 5%.

20,000 − x
20,000 − 5,000
15,000 = amount invested at 5%

Check.

$$0.03x + 0.05(15,000 + x) \overset{?}{=} 0.045(20,000)$$
$$150 + 750 \overset{?}{=} 900$$
$$900 = 900 ✓$$

Stacey should invest $5,000 in the account that earns 3% and $15,000 in the account that earns 5%.

> **TRY IT :: 3.65**

Remy has $14,000 to invest in two mutual funds. One fund pays interest at 4% per year and the other fund pays interest at 7% per year. How much should she invest in each fund if she wants to earn 6.1% interest on the total amount?

> **TRY IT :: 3.66**

Marco has $8,000 to save for his daughter's college education. He wants to divide it between one account that pays 3.2% interest per year and another account that pays 8% interest per year. How much should he invest in each account if he wants the interest on the total investment to be 6.5%?

3.3 EXERCISES

Practice Makes Perfect

Solve Coin Word Problems

In the following exercises, solve each coin word problem.

161. Jaime has $2.60 in dimes and nickels. The number of dimes is 14 more than the number of nickels. How many of each coin does he have?

162. Lee has $1.75 in dimes and nickels. The number of nickels is 11 more than the number of dimes. How many of each coin does he have?

163. Ngo has a collection of dimes and quarters with a total value of $3.50. The number of dimes is seven more than the number of quarters. How many of each coin does he have?

164. Connor has a collection of dimes and quarters with a total value of $6.30. The number of dimes is 14 more than the number of quarters. How many of each coin does he have?

165. A cash box of $1 and $5 bills is worth $45. The number of $1 bills is three more than the number of $5 bills. How many of each bill does it contain?

166. Joe's wallet contains $1 and $5 bills worth $47. The number of $1 bills is five more than the number of $5 bills. How many of each bill does he have?

167. Rachelle has $6.30 in nickels and quarters in her coin purse. The number of nickels is twice the number of quarters. How many coins of each type does she have?

168. Deloise has $1.20 in pennies and nickels in a jar on her desk. The number of pennies is three times the number of nickels. How many coins of each type does she have?

169. Harrison has $9.30 in his coin collection, all in pennies and dimes. The number of dimes is three times the number of pennies. How many coins of each type does he have?

170. Ivan has $8.75 in nickels and quarters in his desk drawer. The number of nickels is twice the number of quarters. How many coins of each type does he have?

171. In a cash drawer there is $125 in $5 and $10 bills. The number of $10 bills is twice the number of $5 bills. How many of each are in the drawer?

172. John has $175 in $5 and $10 bills in his drawer. The number of $5 bills is three times the number of $10 bills. How many of each are in the drawer?

173. Carolyn has $2.55 in her purse in nickels and dimes. The number of nickels is nine less than three times the number of dimes. Find the number of each type of coin.

174. Julio has $2.75 in his pocket in nickels and dimes. The number of dimes is 10 less than twice the number of nickels. Find the number of each type of coin.

175. Chi has $11.30 in dimes and quarters. The number of dimes is three more than three times the number of quarters. How many of each are there?

176. Tyler has $9.70 in dimes and quarters. The number of quarters is eight more than four times the number of dimes. How many of each coin does he have?

177. Mukul has $3.75 in quarters, dimes and nickels in his pocket. He has five more dimes than quarters and nine more nickels than quarters. How many of each coin are in his pocket?

178. Vina has $4.70 in quarters, dimes and nickels in her purse. She has eight more dimes than quarters and six more nickels than quarters. How many of each coin are in her purse?

Solve Ticket and Stamp Word Problems

In the following exercises, solve each ticket or stamp word problem.

179. The school play sold $550 in tickets one night. The number of $8 adult tickets was 10 less than twice the number of $5 child tickets. How many of each ticket were sold?

180. If the number of $8 child tickets is seventeen less than three times the number of $12 adult tickets and the theater took in $584, how many of each ticket were sold?

181. The movie theater took in $1,220 one Monday night. The number of $7 child tickets was ten more than twice the number of $9 adult tickets. How many of each were sold?

182. The ball game sold $1,340 in tickets one Saturday. The number of $12 adult tickets was 15 more than twice the number of $5 child tickets. How many of each were sold?

183. The ice rink sold 95 tickets for the afternoon skating session, for a total of $828. General admission tickets cost $10 each and youth tickets cost $8 each. How many general admission tickets and how many youth tickets were sold?

184. For the 7:30 show time, 140 movie tickets were sold. Receipts from the $13 adult tickets and the $10 senior tickets totaled $1,664. How many adult tickets and how many senior tickets were sold?

185. The box office sold 360 tickets to a concert at the college. The total receipts were $4170. General admission tickets cost $15 and student tickets cost $10. How many of each kind of ticket was sold?

186. Last Saturday, the museum box office sold 281 tickets for a total of $3954. Adult tickets cost $15 and student tickets cost $12. How many of each kind of ticket was sold?

187. Julie went to the post office and bought both $0.41 stamps and $0.26 postcards. She spent $51.40. The number of stamps was 20 more than twice the number of postcards. How many of each did she buy?

188. Jason went to the post office and bought both $0.41 stamps and $0.26 postcards and spent $10.28. The number of stamps was four more than twice the number of postcards. How many of each did he buy?

189. Maria spent $12.50 at the post office. She bought three times as many $0.41 stamps as $0.02 stamps. How many of each did she buy?

190. Hector spent $33.20 at the post office. He bought four times as many $0.41 stamps as $0.02 stamps. How many of each did he buy?

191. Hilda has $210 worth of $10 and $12 stock shares. The numbers of $10 shares is five more than twice the number of $12 shares. How many of each does she have?

192. Mario invested $475 in $45 and $25 stock shares. The number of $25 shares was five less than three times the number of $45 shares. How many of each type of share did he buy?

Solve Mixture Word Problems

In the following exercises, solve each mixture word problem.

193. Lauren in making 15 liters of mimosas for a brunch banquet. Orange juice costs her $1.50 per liter and champagne costs her $12 per liter. How many liters of orange juice and how many liters of champagne should she use for the mimosas to cost Lauren $5 per liter?

194. Macario is making 12 pounds of nut mixture with macadamia nuts and almonds. Macadamia nuts cost $9 per pound and almonds cost $5.25 per pound. How many pounds of macadamia nuts and how many pounds of almonds should Macario use for the mixture to cost $6.50 per pound to make?

195. Kaapo is mixing Kona beans and Maui beans to make 25 pounds of coffee blend. Kona beans cost Kaapo $15 per pound and Maui beans cost $24 per pound. How many pounds of each coffee bean should Kaapo use for his blend to cost him $17.70 per pound?

196. Estelle is making 30 pounds of fruit salad from strawberries and blueberries. Strawberries cost $1.80 per pound and blueberries cost $4.50 per pound. If Estelle wants the fruit salad to cost her $2.52 per pound, how many pounds of each berry should she use?

197. Carmen wants to tile the floor of his house. He will need 1000 square feet of tile. He will do most of the floor with a tile that costs $1.50 per square foot, but also wants to use an accent tile that costs $9.00 per square foot. How many square feet of each tile should he plan to use if he wants the overall cost to be $3 per square foot?

198. Riley is planning to plant a lawn in his yard. He will need nine pounds of grass seed. He wants to mix Bermuda seed that costs $4.80 per pound with Fescue seed that costs $3.50 per pound. How much of each seed should he buy so that the overall cost will be $4.02 per pound?

199. Vartan was paid $25,000 for a cell phone app that he wrote and wants to invest it to save for his son's education. He wants to put some of the money into a bond that pays 4% annual interest and the rest into stocks that pay 9% annual interest. If he wants to earn 7.4% annual interest on the total amount, how much money should he invest in each account?

200. Vern sold his 1964 Ford Mustang for $55,000 and wants to invest the money to earn him 5.8% interest per year. He will put some of the money into Fund A that earns 3% per year and the rest in Fund B that earns 10% per year. How much should he invest into each fund if he wants to earn 5.8% interest per year on the total amount?

201. Stephanie inherited $40,000. She wants to put some of the money in a certificate of deposit that pays 2.1% interest per year and the rest in a mutual fund account that pays 6.5% per year. How much should she invest in each account if she wants to earn 5.4% interest per year on the total amount?

202. Avery and Caden have saved $27,000 towards a down payment on a house. They want to keep some of the money in a bank account that pays 2.4% annual interest and the rest in a stock fund that pays 7.2% annual interest. How much should they put into each account so that they earn 6% interest per year?

203. Dominic pays 7% interest on his $15,000 college loan and 12% interest on his $11,000 car loan. What average interest rate does he pay on the total $26,000 he owes? (Round your answer to the nearest tenth of a percent.)

204. Liam borrowed a total of $35,000 to pay for college. He pays his parents 3% interest on the $8,000 he borrowed from them and pays the bank 6.8% on the rest. What average interest rate does he pay on the total $35,000? (Round your answer to the nearest tenth of a percent.)

Everyday Math

205. As the treasurer of her daughter's Girl Scout troop, Laney collected money for some girls and adults to go to a 3-day camp. Each girl paid $75 and each adult paid $30. The total amount of money collected for camp was $765. If the number of girls is three times the number of adults, how many girls and how many adults paid for camp?

206. Laurie was completing the treasurer's report for her son's Boy Scout troop at the end of the school year. She didn't remember how many boys had paid the $15 full-year registration fee and how many had paid the $10 partial-year fee. She knew that the number of boys who paid for a full-year was ten more than the number who paid for a partial-year. If $250 was collected for all the registrations, how many boys had paid the full-year fee and how many had paid the partial-year fee?

Writing Exercises

207. Suppose you have six quarters, nine dimes, and four pennies. Explain how you find the total value of all the coins.

208. Do you find it helpful to use a table when solving coin problems? Why or why not?

209. In the table used to solve coin problems, one column is labeled "number" and another column is labeled "value." What is the difference between the "number" and the "value?"

210. What similarities and differences did you see between solving the coin problems and the ticket and stamp problems?

Self Check

ⓐ After completing the exercises, use this checklist to evaluate your mastery of the objectives of this section.

I can...	Confidently	With some help	No-I don't get it!
solve coin word problems.			
solve ticket and stamp word problems.			
solve mixture word problems.			

ⓑ After reviewing this checklist, what will you do to become confident for all objectives?

3.4 Solve Geometry Applications: Triangles, Rectangles, and the Pythagorean Theorem

Learning Objectives

By the end of this section, you will be able to:

› Solve applications using properties of triangles
› Use the Pythagorean Theorem
› Solve applications using rectangle properties

Be Prepared!

Before you get started, take this readiness quiz.

1. Simplify: $\frac{1}{2}(6h)$.

   If you missed this problem, review **Example 1.122**.

2. The length of a rectangle is three less than the width. Let w represent the width. Write an expression for the length of the rectangle.
   If you missed this problem, review **Example 1.26**.

3. Solve: $A = \frac{1}{2}bh$ for b when $A = 260$ and $h = 52$.

   If you missed this problem, review **Example 2.61**.

4. Simplify: $\sqrt{144}$.
   If you missed this problem, review **Example 1.111**.

Solve Applications Using Properties of Triangles

In this section we will use some common geometry formulas. We will adapt our problem-solving strategy so that we can solve geometry applications. The geometry formula will name the variables and give us the equation to solve. In addition, since these applications will all involve shapes of some sort, most people find it helpful to draw a figure and label it with the given information. We will include this in the first step of the problem solving strategy for geometry applications.

HOW TO :: SOLVE GEOMETRY APPLICATIONS.

Step 1. **Read** the problem and make sure all the words and ideas are understood. Draw the figure and label it with the given information.

Step 2. **Identify** what we are looking for.

Step 3. **Label** what we are looking for by choosing a variable to represent it.

Step 4. **Translate** into an equation by writing the appropriate formula or model for the situation. Substitute in the given information.

Step 5. **Solve** the equation using good algebra techniques.

Step 6. **Check** the answer by substituting it back into the equation solved in step 5 and by making sure it makes sense in the context of the problem.

Step 7. **Answer** the question with a complete sentence.

We will start geometry applications by looking at the properties of triangles. Let's review some basic facts about triangles. Triangles have three sides and three interior angles. Usually each side is labeled with a lowercase letter to match the uppercase letter of the opposite vertex.

The plural of the word *vertex* is *vertices*. All triangles have three vertices. Triangles are named by their vertices: The triangle in **Figure 3.4** is called $\triangle ABC$.

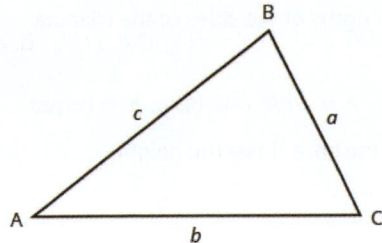

Figure 3.4 Triangle ABC has vertices A, B, and C. The lengths of the sides are a, b, and c.

The three angles of a triangle are related in a special way. The sum of their measures is $180°$. Note that we read $m\angle A$ as "the measure of angle A." So in $\triangle ABC$ in **Figure 3.4**,

$$m\angle A + m\angle B + m\angle C = 180°$$

Because the perimeter of a figure is the length of its boundary, the perimeter of $\triangle ABC$ is the sum of the lengths of its three sides.

$$P = a + b + c$$

To find the area of a triangle, we need to know its base and height. The height is a line that connects the base to the opposite vertex and makes a $90°$ angle with the base. We will draw $\triangle ABC$ again, and now show the height, h. See **Figure 3.5**.

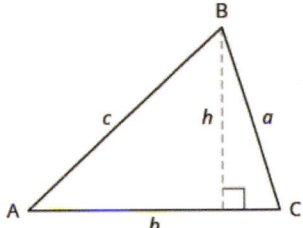

Figure 3.5 The formula for the area of $\triangle ABC$ is $A = \frac{1}{2}bh$, where b is the base and h is the height.

Triangle Properties

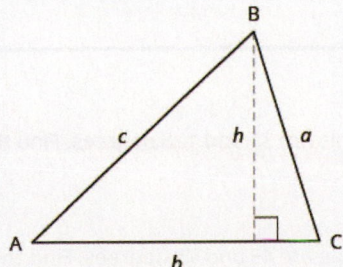

For $\triangle ABC$

Angle measures:

$$m\angle A + m\angle B + m\angle C = 180$$

- The sum of the measures of the angles of a triangle is $180°$.

Perimeter:

$$P = a + b + c$$

- The perimeter is the sum of the lengths of the sides of the triangle.

Area:

$$A = \frac{1}{2}bh, \; b = \text{base}, \; h = \text{height}$$

- The area of a triangle is one-half the base times the height.

EXAMPLE 3.34

The measures of two angles of a triangle are 55 and 82 degrees. Find the measure of the third angle.

✓ **Solution**

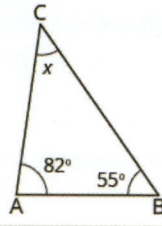

Step 1. Read the problem. Draw the figure and label it with the given information.

Step 2. Identify what you are looking for.	the measure of the third angle in a triangle
Step 3. Name. Choose a variable to represent it.	Let $x =$ the measure of the angle.
Step 4. Translate.	
Write the appropriate formula and substitute.	$m \angle A + m \angle B + m \angle C = 180$
Step 5. Solve the equation.	$\begin{aligned} 55 + 82 + x &= 180 \\ 137 + x &= 180 \\ x &= 43 \end{aligned}$

Step 6. Check.

$$55 + 82 + 43 \overset{?}{=} 180$$
$$180 = 180 \checkmark$$

Step 7. Answer the question.	The measure of the third angle is 43 degrees.

> **TRY IT : : 3.67**
>
> The measures of two angles of a triangle are 31 and 128 degrees. Find the measure of the third angle.

> **TRY IT : : 3.68**
>
> The measures of two angles of a triangle are 49 and 75 degrees. Find the measure of the third angle.

EXAMPLE 3.35

The perimeter of a triangular garden is 24 feet. The lengths of two sides are four feet and nine feet. How long is the third side?

✓ **Solution**

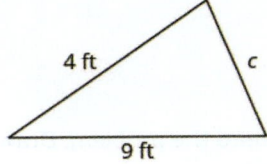

Step 1. Read the problem. Draw the figure and label it with the given information.

$$P = 24 \text{ ft}$$

Step 2. Identify what you are looking for.	length of the third side of a triangle
Step 3. Name. Choose a variable to represent it.	Let $c =$ the third side.

Step 4. Translate.

Write the appropriate formula and substitute.

$$P = a + b + c$$

Substitute in the given information.

$$24 \text{ ft} = 4 \text{ ft} + 9 \text{ ft} + c$$

$$24 = 13 + c$$

Step 5. Solve the equation.

$$11 = c$$

Step 6. Check.

$$P = a + b + c$$
$$24 \overset{?}{=} 4 + 9 + 11$$
$$24 = 24 \checkmark$$

Step 7. Answer the question.	The third side is 11 feet long.

> **TRY IT :: 3.69**

The perimeter of a triangular garden is 48 feet. The lengths of two sides are 18 feet and 22 feet. How long is the third side?

> **TRY IT :: 3.70**

The lengths of two sides of a triangular window are seven feet and five feet. The perimeter is 18 feet. How long is the third side?

EXAMPLE 3.36

The area of a triangular church window is 90 square meters. The base of the window is 15 meters. What is the window's height?

⊘ **Solution**

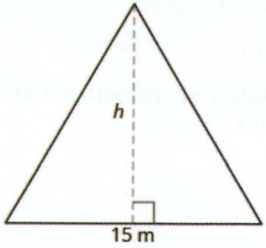

Step 1. Read the problem. Draw the figure and label it with the given information.

Area $= 90m^2$

| **Step 2. Identify** what you are looking for. | height of a triangle |
| **Step 3. Name.** Choose a variable to represent it. | Let $h =$ the height. |

Step 4. Translate.

Write the appropriate formula.

$$\underbrace{A}\ =\ \underbrace{\frac{1}{2}}\ \cdot\ \underbrace{b}\ \cdot\ \underbrace{h}$$

Substitute in the given information.

$$90 \text{ m}^2\ =\ \frac{1}{2}\ \cdot\ 15 \text{ m}\ \cdot\ h$$

Step 5. Solve the equation.

$$90 = \frac{15}{2}h$$

$$12 = h$$

Step 6. Check.

$$A\ =\ \frac{1}{2}bh$$

$$90\ \overset{?}{=}\ \frac{1}{2}\cdot 15 \cdot 12$$

$$90\ =\ 90\ \checkmark$$

Step 7. Answer the question.

The height of the triangle is 12 meters.

 **TRY IT ::** 3.71

The area of a triangular painting is 126 square inches. The base is 18 inches. What is the height?

> **TRY IT ::** 3.72 A triangular tent door has area 15 square feet. The height is five feet. What is the base?

The triangle properties we used so far apply to all triangles. Now we will look at one specific type of triangle—a right triangle. A right triangle has one $90°$ angle, which we usually mark with a small square in the corner.

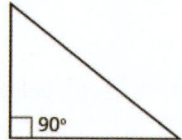

Right Triangle

A **right triangle** has one $90°$ angle, which is often marked with a square at the vertex.

EXAMPLE 3.37

One angle of a right triangle measures $28°$. What is the measure of the third angle?

✓ **Solution**

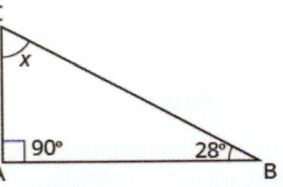

Step 1. Read the problem. Draw the figure and label it with the given information.	
Step 2. Identify what you are looking for.	the measure of an angle
Step 3. Name. Choose a variable to represent it.	Let $x =$ the measure of an angle.
Step 4. Translate.	$m \angle A + m \angle B + m \angle C = 180$
Write the appropriate formula and substitute.	$x + 90 + 28 = 180$
Step 5. Solve the equation.	$\begin{aligned} x + 118 &= 180 \\ x &= 62 \end{aligned}$
Step 6. Check.	

$$180 \overset{?}{=} 90 + 28 + 62$$
$$180 = 180 ✓$$

Step 7. Answer the question.	The measure of the third angle is $62°$.

> **TRY IT :: 3.73** One angle of a right triangle measures $56°$. What is the measure of the other small angle?

> **TRY IT :: 3.74** One angle of a right triangle measures $45°$. What is the measure of the other small angle?

In the examples we have seen so far, we could draw a figure and label it directly after reading the problem. In the next example, we will have to define one angle in terms of another. We will wait to draw the figure until we write expressions for all the angles we are looking for.

EXAMPLE 3.38

The measure of one angle of a right triangle is 20 degrees more than the measure of the smallest angle. Find the measures of all three angles.

✓ **Solution**

Step 1. Read the problem.	
Step 2. Identify what you are looking for.	the measures of all three angles
Step 3. Name. Choose a variable to represent it.	Let $a = 1^{st}$ angle. $a + 20 = 2^{nd}$ angle $90 = 3^{rd}$ angle (the right angle)

Draw the figure and label it with the given information

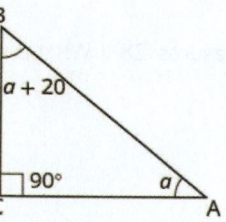

Step 4. Translate	$m\angle A + m\angle B + m\angle C = 180$
Write the appropriate formula. Substitute into the formula.	$a + (a + 20) + 90 = 180$

Step 5. Solve the equation.

$$2a + 110 = 180$$
$$2a = 70$$
$$a = 35 \text{ first angle}$$
$$a + 20 \text{ second angle}$$
$$35 + 20$$
$$55$$
$$90 \text{ third angle}$$

Step 6. Check.

$$35 + 55 + 90 \overset{?}{=} 180$$
$$180 = 180 \checkmark$$

Step 7. Answer the question.   The three angles measure 35°, 55°, and 90°.

> **TRY IT ∷ 3.75**

The measure of one angle of a right triangle is 50° more than the measure of the smallest angle. Find the measures of all three angles.

> **TRY IT ∷ 3.76**

The measure of one angle of a right triangle is 30° more than the measure of the smallest angle. Find the measures of all three angles.

Use the Pythagorean Theorem

We have learned how the measures of the angles of a triangle relate to each other. Now, we will learn how the lengths of the sides relate to each other. An important property that describes the relationship among the lengths of the three sides of a right triangle is called the Pythagorean Theorem. This theorem has been used around the world since ancient times. It is named after the Greek philosopher and mathematician, Pythagoras, who lived around 500 BC.

Before we state the Pythagorean Theorem, we need to introduce some terms for the sides of a triangle. Remember that a right triangle has a 90° angle, marked with a small square in the corner. The side of the triangle opposite the 90° angle is called the *hypotenuse* and each of the other sides are called *legs*.

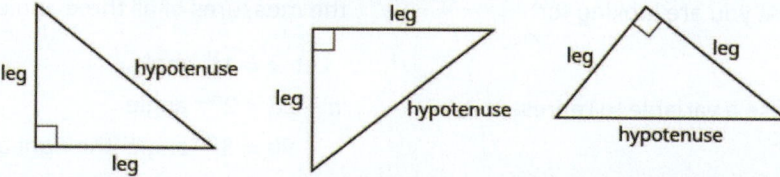

The Pythagorean Theorem tells how the lengths of the three sides of a right triangle relate to each other. It states that in any right triangle, the sum of the squares of the lengths of the two legs equals the square of the length of the hypotenuse.

In symbols we say: in any right triangle, $a^2 + b^2 = c^2$, where a and b are the lengths of the legs and c is the length of the hypotenuse.

Writing the formula in every exercise and saying it aloud as you write it, may help you remember the Pythagorean Theorem.

The Pythagorean Theorem

In any right triangle, $a^2 + b^2 = c^2$.

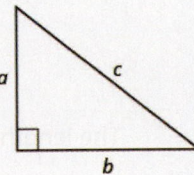

where a and b are the lengths of the legs, c is the length of the hypotenuse.

To solve exercises that use the Pythagorean Theorem, we will need to find square roots. We have used the notation $\sqrt{m}$ and the definition:

If $m = n^2$, then $\sqrt{m} = n$, for $n \geq 0$.

For example, we found that $\sqrt{25}$ is 5 because $25 = 5^2$.

Because the Pythagorean Theorem contains variables that are squared, to solve for the length of a side in a right triangle, we will have to use square roots.

EXAMPLE 3.39

Use the Pythagorean Theorem to find the length of the hypotenuse shown below.

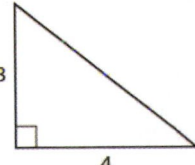

⊘ **Solution**

Step 1. Read the problem.	
Step 2. Identify what you are looking for.	the length of the hypotenuse of the triangle
Step 3. Name. Choose a variable to represent it. Label side c on the figure.	Let c = the length of the hypotenuse.

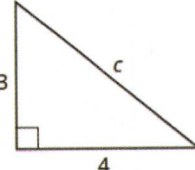

Step 4. Translate.	
Write the appropriate formula.	$a^2 + b^2 = c^2$
Substitute.	$3^2 + 4^2 = c^2$
Step 5. Solve the equation.	$9 + 16 = c^2$

Simplify.	$25 = c^2$
Use the definition of square root.	$\sqrt{25} = c$
Simplify.	$5 = c$

Step 6. Check.

$3^2 + 4^2 \overset{?}{=} 5^2$

$9 + 16 \overset{?}{=} 25$

$25 = 25 \checkmark$

Step 7. Answer the question.	The length of the hypotenuse is 5.

 TRY IT : : 3.77

Use the Pythagorean Theorem to find the length of the hypotenuse in the triangle shown below.

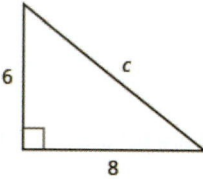

 TRY IT : : 3.78

Use the Pythagorean Theorem to find the length of the hypotenuse in the triangle shown below.

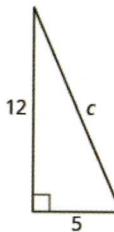

EXAMPLE 3.40

Use the Pythagorean Theorem to find the length of the leg shown below.

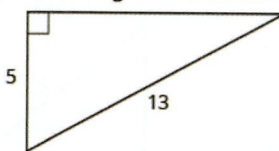

⊘ **Solution**

Step 1. Read the problem.	
Step 2. Identify what you are looking for.	the length of the leg of the triangle
Step 3. Name. Choose a variable to represent it.	Let b = the leg of the triangle.

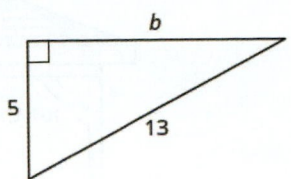

Lable side b.

Step 4. Translate

Write the appropriate formula.	$a^2 + b^2 = c^2$
Substitute.	$5^2 + b^2 = 13^2$

Step 5. Solve the equation.

	$25 + b^2 = 169$
Isolate the variable term.	$b^2 = 144$
Use the definition of square root.	$b^2 = \sqrt{144}$
Simplify.	$b = 12$

Step 6. Check.

$5^2 + 12^2 \overset{?}{=} 13^2$
$25 + 144 \overset{?}{=} 169$
$169 = 169 \checkmark$

Step 7. Answer the question.

The length of the leg is 12.

> **TRY IT : : 3.79**   Use the Pythagorean Theorem to find the length of the leg in the triangle shown below.

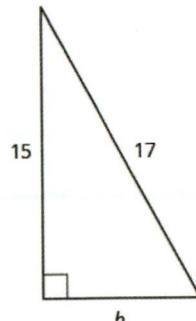

> **TRY IT : : 3.80**   Use the Pythagorean Theorem to find the length of the leg in the triangle shown below.

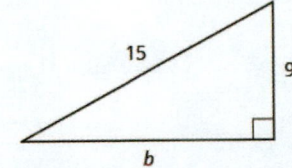

EXAMPLE 3.41

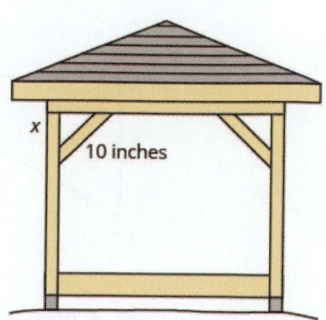

Kelvin is building a gazebo and wants to brace each corner by placing a 10″ piece of wood diagonally as shown above.

If he fastens the wood so that the ends of the brace are the same distance from the corner, what is the length of the legs of the right triangle formed? Approximate to the nearest tenth of an inch.

✓ **Solution**

Step 1. Read the problem.

Step 2. Identify what we are looking for.

the distance from the corner that the bracket should be attached

Step 3. Name. Choose a variable to represent it.

Let x = the distance from the corner.

Step 4. Translate.
Write the appropriate formula and substitute.

$$a^2 + b^2 = c^2$$
$$x^2 + x^2 = 10^2$$

Step 5. Solve the equation.
Isolate the variable.
Use the definition of square root.
Simplify. Approximate to the nearest tenth.

$$2x^2 = 100$$
$$x^2 = 50$$
$$x = \sqrt{50}$$
$$x \approx 7.1$$

Step 6. Check.

$$a^2 + b^2 = c^2$$

$$(7.1)^2 + (7.1)^2 \approx 10^2 \text{ Yes.}$$

Step 7. Answer the question.

Kelvin should fasten each piece of wood approximately 7.1″ from the corner.

> **TRY IT : : 3.81**

John puts the base of a 13-foot ladder five feet from the wall of his house as shown below. How far up the wall does the ladder reach?

> **TRY IT :: 3.82**

Randy wants to attach a 17 foot string of lights to the top of the 15 foot mast of his sailboat, as shown below. How far from the base of the mast should he attach the end of the light string?

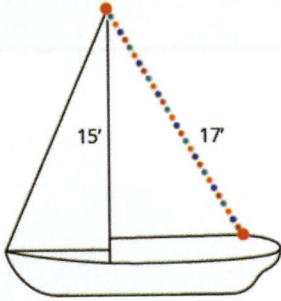

Solve Applications Using Rectangle Properties

You may already be familiar with the properties of rectangles. Rectangles have four sides and four right $(90°)$ angles. The opposite sides of a rectangle are the same length. We refer to one side of the rectangle as the length, *L*, and its adjacent side as the width, *W*.

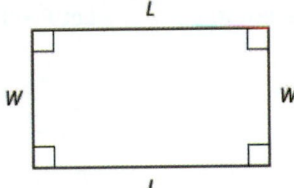

The distance around this rectangle is $L + W + L + W$, or $2L + 2W$. This is the perimeter, *P*, of the rectangle.

$$P = 2L + 2W$$

What about the area of a rectangle? Imagine a rectangular rug that is 2-feet long by 3-feet wide. Its area is 6 square feet. There are six squares in the figure.

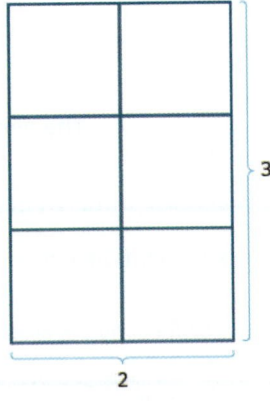

$$A = 6$$
$$A = 2 \cdot 3$$
$$A = L \cdot W$$

The area is the length times the width.

The formula for the area of a rectangle is $A = LW$.

Properties of Rectangles

Rectangles have four sides and four right $(90°)$ angles.

The lengths of opposite sides are equal.

The perimeter of a rectangle is the sum of twice the length and twice the width.

$$P = 2L + 2W$$

The area of a rectangle is the product of the length and the width.

$$A = L \cdot W$$

EXAMPLE 3.42

The length of a rectangle is 32 meters and the width is 20 meters. What is the perimeter?

✓ **Solution**

Step 1. Read the problem.
Draw the figure and label it with the given information.

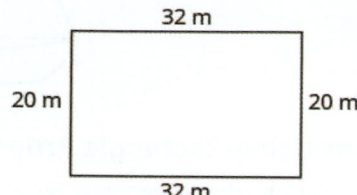

Step 2. Identify what you are looking for.	the perimeter of a rectangle
Step 3. Name. Choose a variable to represent it.	Let P = the perimeter.
Step 4. Translate.	
Write the appropriate formula.	$P = 2L + 2W$
Substitute.	$P = 2(32\text{ m}) + 2(20\text{ m})$
Step 5. Solve the equation.	$P = 64 + 40$ $P = 104$

Step 6. Check.

$$
\begin{aligned}
P &\overset{?}{=} 104 \\
20 + 32 + 20 + 32 &\overset{?}{=} 104 \\
104 &= 104 \checkmark
\end{aligned}
$$

Step 7. Answer the question.	The perimeter of the rectangle is 104 meters.

> **TRY IT : : 3.83**   The length of a rectangle is 120 yards and the width is 50 yards. What is the perimeter?

> **TRY IT : : 3.84**   The length of a rectangle is 62 feet and the width is 48 feet. What is the perimeter?

EXAMPLE 3.43

The area of a rectangular room is 168 square feet. The length is 14 feet. What is the width?

⊘ Solution

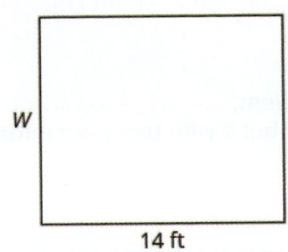

Step 1. Read the problem.
Draw the figure and label it with the given information.

Step 2. Identify what you are looking for.	the width of a rectangular room
Step 3. Name. Choose a variable to represent it.	Let W = the width.
Step 4. Translate.	
Write the appropriate formula.	$A = LW$
Substitute.	$168 = 14W$
Step 5. Solve the equation.	$\dfrac{168}{14} = \dfrac{14W}{14}$ $12 = W$

Step 6. Check.

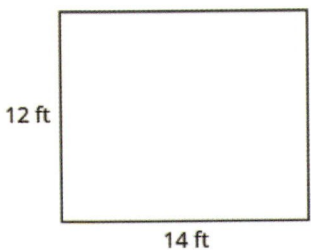

$$A = LW$$
$$168 \overset{?}{=} 14 \cdot 12$$
$$168 = 168 \checkmark$$

Step 7. Answer the question.	The width of the room is 12 feet.

> **TRY IT : : 3.85**   The area of a rectangle is 598 square feet. The length is 23 feet. What is the width?

> **TRY IT : : 3.86**   The width of a rectangle is 21 meters. The area is 609 square meters. What is the length?

EXAMPLE 3.44

Find the length of a rectangle with perimeter 50 inches and width 10 inches.

✓ **Solution**

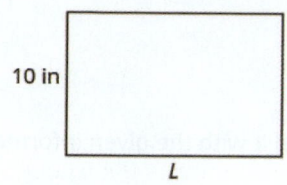

Step 1. **Read** the problem. Draw the figure and label it with the given information.	
Step 2. **Identify** what you are looking for.	the length of the rectangle
Step 3. **Name.** Choose a variable to represent it.	Let L = the length.
Step 4. **Translate.**	
Write the appropriate formula.	$P = 2L + 2W$
Substitute.	$50 = 2L + 2(10)$
Step 5. **Solve** the equation.	$50 - 20 = 2L + 20 - 20$
	$30 = 2L$
	$\dfrac{30}{2} = \dfrac{2L}{2}$
	$15 = L$

Step 6. Check.

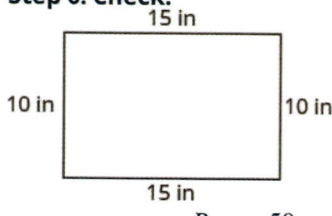

$$
\begin{array}{rcl}
P & = & 50 \\
15 + 10 + 15 + 10 & \overset{?}{=} & 50 \\
50 & = & 50 \ \checkmark
\end{array}
$$

Step 7. **Answer** the question.	The length is 15 inches.

> **TRY IT :: 3.87**   Find the length of a rectangle with: perimeter 80 and width 25.

> **TRY IT :: 3.88**   Find the length of a rectangle with: perimeter 30 and width 6.

We have solved problems where either the length or width was given, along with the perimeter or area; now we will learn how to solve problems in which the width is defined in terms of the length. We will wait to draw the figure until we write an expression for the width so that we can label one side with that expression.

EXAMPLE 3.45

The width of a rectangle is two feet less than the length. The perimeter is 52 feet. Find the length and width.

✓ **Solution**

Step 1. Read the problem.

Step 2. Identify what you are looking for.

the length and width of a rectangle

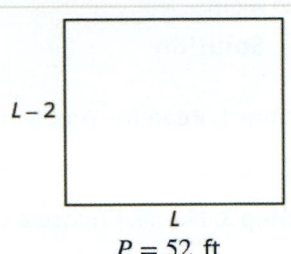

$P = 52$ ft

Step 3. Name. Choose a variable to represent it.
Since the width is defined in terms of the length, we let L = length. The width is two feet less than the length, so we let $L − 2$ = width.

Step 4. Translate.

Write the appropriate formula. The formula for the perimeter of a rectangle relates all the information.	$P = 2L + 2W$
Substitute in the given information.	$52 = 2L + 2(L − 2)$
Step 5. Solve the equation.	$52 = 2L + 2L − 4$
Combine like terms.	$52 = 4L − 4$
Add 4 to each side.	$56 = 4L$
Divide by 4.	$\dfrac{56}{4} = \dfrac{4L}{4}$ $14 = L$ The length is 14 feet.
Now we need to find the width.	The width is $L − 2$. $L − 2$ $14 − 2$ 12 The width is 12 feet.

Step 6. Check.
Since $14 + 12 + 14 + 12 = 52$, this works!

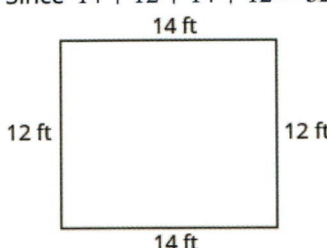

Step 7. Answer the question.

The length is 14 feet and the width is 12 feet.

> **TRY IT :: 3.89**

The width of a rectangle is seven meters less than the length. The perimeter is 58 meters. Find the length and width.

> **TRY IT :: 3.90**

The length of a rectangle is eight feet more than the width. The perimeter is 60 feet. Find the length and width.

EXAMPLE 3.46

The length of a rectangle is four centimeters more than twice the width. The perimeter is 32 centimeters. Find the length and width.

✅ **Solution**

Step 1. Read the problem.	
Step 2. Identify what you are looking for.	the length and the width
Step 3. Name. Choose a variable to represent the width.	Let W = width
The length is four more than twice the width.	$2W + 4$ = length

$$2W + 4$$
$$P = 32 \text{ cm}$$

Step 4. Translate	
Write the appropriate formula.	$P = 2L + 2W$
Substitute in the given information.	$32 = 2(2W + 4) + 2W$
Step 5. Solve the equation.	$32 = 4W + 8 + 2W$
	$32 = 6W + 8$
	$24 = 6W$
	$4 = W$ (width)
	$2W + 4$ (length)
	$2(\textcolor{red}{4}) + 4$
	12
	The length is 12 cm.

Step 6. Check.

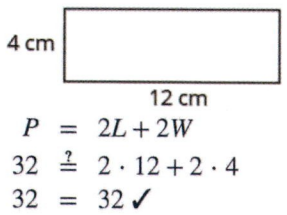

$$\begin{aligned} P &= 2L + 2W \\ 32 &\overset{?}{=} 2 \cdot 12 + 2 \cdot 4 \\ 32 &= 32 ✓ \end{aligned}$$

Step 7. Answer the question.	The length is 12 cm and the width is 4 cm.

> **TRY IT : : 3.91**

> The length of a rectangle is eight more than twice the width. The perimeter is 64. Find the length and width.

> **TRY IT : : 3.92**

> The width of a rectangle is six less than twice the length. The perimeter is 18. Find the length and width.

EXAMPLE 3.47

The perimeter of a rectangular swimming pool is 150 feet. The length is 15 feet more than the width. Find the length and width.

⊘ **Solution**

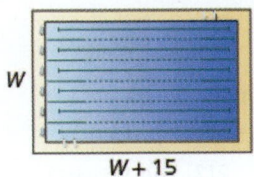

Step 1. Read the problem.
Draw the figure and label it with the given
information.

W

$W + 15$

$P = 150$ ft

Step 2. Identify what you are looking for.	the length and the width of the pool

Step 3. Name. Choose a variable to represent the width. The length is 15 feet more than the width.	Let W = width $W + 15$ = length

Step 4. Translate

Write the appropriate formula.	$P = 2L + 2W$
Substitute.	$150 = 2(W + 15) + 2W$

Step 5. Solve the equation.	$150 = 2W + 30 + 2W$
	$150 = 4W + 30$
	$120 = 4W$
	$30 = W$ (the width of the pool)
	$W + 15$ (the length of the pool)
	$30 + 15$
	45

Step 6. Check.

$$
\begin{aligned}
P &= 2L + 2W \\
150 &\overset{?}{=} 2(45) + 2(30) \\
150 &= 150 \checkmark
\end{aligned}
$$

Step 7. Answer the question.	The length of the pool is 45 feet and the width is 30 feet.

> **TRY IT : : 3.93**

The perimeter of a rectangular swimming pool is 200 feet. The length is 40 feet more than the width. Find the
length and width.

> **TRY IT : : 3.94**

The length of a rectangular garden is 30 yards more than the width. The perimeter is 300 yards. Find the length
and width.

3.4 EXERCISES

Practice Makes Perfect

Solving Applications Using Triangle Properties

In the following exercises, solve using triangle properties.

211. The measures of two angles of a triangle are 26 and 98 degrees. Find the measure of the third angle.

212. The measures of two angles of a triangle are 61 and 84 degrees. Find the measure of the third angle.

213. The measures of two angles of a triangle are 105 and 31 degrees. Find the measure of the third angle.

214. The measures of two angles of a triangle are 47 and 72 degrees. Find the measure of the third angle.

215. The perimeter of a triangular pool is 36 yards. The lengths of two sides are 10 yards and 15 yards. How long is the third side?

216. A triangular courtyard has perimeter 120 meters. The lengths of two sides are 30 meters and 50 meters. How long is the third side?

217. If a triangle has sides 6 feet and 9 feet and the perimeter is 23 feet, how long is the third side?

218. If a triangle has sides 14 centimeters and 18 centimeters and the perimeter is 49 centimeters, how long is the third side?

219. A triangular flag has base one foot and height 1.5 foot. What is its area?

220. A triangular window has base eight feet and height six feet. What is its area?

221. What is the base of a triangle with area 207 square inches and height 18 inches?

222. What is the height of a triangle with area 893 square inches and base 38 inches?

223. One angle of a right triangle measures 33 degrees. What is the measure of the other small angle?

224. One angle of a right triangle measures 51 degrees. What is the measure of the other small angle?

225. One angle of a right triangle measures 22.5 degrees. What is the measure of the other small angle?

226. One angle of a right triangle measures 36.5 degrees. What is the measure of the other small angle?

227. The perimeter of a triangle is 39 feet. One side of the triangle is one foot longer than the second side. The third side is two feet longer than the second side. Find the length of each side.

228. The perimeter of a triangle is 35 feet. One side of the triangle is five feet longer than the second side. The third side is three feet longer than the second side. Find the length of each side.

229. One side of a triangle is twice the shortest side. The third side is five feet more than the shortest side. The perimeter is 17 feet. Find the lengths of all three sides.

230. One side of a triangle is three times the shortest side. The third side is three feet more than the shortest side. The perimeter is 13 feet. Find the lengths of all three sides.

231. The two smaller angles of a right triangle have equal measures. Find the measures of all three angles.

232. The measure of the smallest angle of a right triangle is 20° less than the measure of the next larger angle. Find the measures of all three angles.

233. The angles in a triangle are such that one angle is twice the smallest angle, while the third angle is three times as large as the smallest angle. Find the measures of all three angles.

234. The angles in a triangle are such that one angle is 20° more than the smallest angle, while the third angle is three times as large as the smallest angle. Find the measures of all three angles.

Use the Pythagorean Theorem

In the following exercises, use the Pythagorean Theorem to find the length of the hypotenuse.

235.

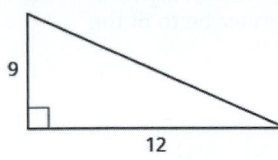

236.

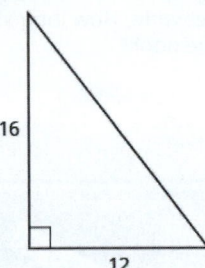

237.

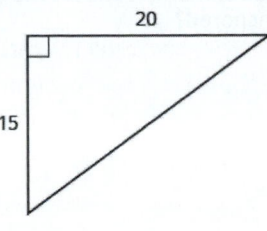

238.

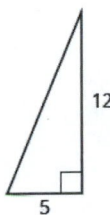

In the following exercises, use the Pythagorean Theorem to find the length of the leg. Round to the nearest tenth, if necessary.

239.

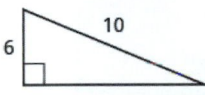

240.

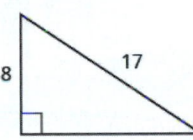

241.

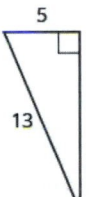

242.

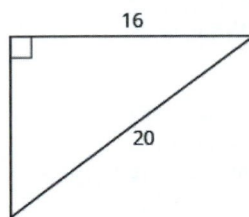

243.

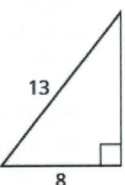

244.

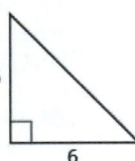

245.

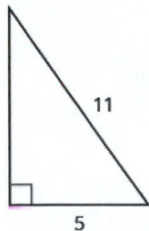

246.

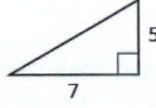

In the following exercises, solve using the Pythagorean Theorem. Approximate to the nearest tenth, if necessary.

247. A 13-foot string of lights will be attached to the top of a 12-foot pole for a holiday display, as shown below. How far from the base of the pole should the end of the string of lights be anchored?

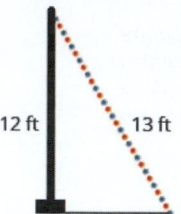

12 ft   13 ft

248. Pam wants to put a banner across her garage door, as shown below, to congratulate her son for his college graduation. The garage door is 12 feet high and 16 feet wide. How long should the banner be to fit the garage door?

249. Chi is planning to put a path of paving stones through her flower garden, as shown below. The flower garden is a square with side 10 feet. What will the length of the path be?

10'

250. Brian borrowed a 20 foot extension ladder to use when he paints his house. If he sets the base of the ladder 6 feet from the house, as shown below, how far up will the top of the ladder reach?

20'

6'

Solve Applications Using Rectangle Properties

In the following exercises, solve using rectangle properties.

251. The length of a rectangle is 85 feet and the width is 45 feet. What is the perimeter?

252. The length of a rectangle is 26 inches and the width is 58 inches. What is the perimeter?

253. A rectangular room is 15 feet wide by 14 feet long. What is its perimeter?

254. A driveway is in the shape of a rectangle 20 feet wide by 35 feet long. What is its perimeter?

255. The area of a rectangle is 414 square meters. The length is 18 meters. What is the width?

256. The area of a rectangle is 782 square centimeters. The width is 17 centimeters. What is the length?

257. The width of a rectangular window is 24 inches. The area is 624 square inches. What is the length?

258. The length of a rectangular poster is 28 inches. The area is 1316 square inches. What is the width?

259. Find the length of a rectangle with perimeter 124 and width 38.

260. Find the width of a rectangle with perimeter 92 and length 19.

261. Find the width of a rectangle with perimeter 16.2 and length 3.2.

262. Find the length of a rectangle with perimeter 20.2 and width 7.8.

263. The length of a rectangle is nine inches more than the width. The perimeter is 46 inches. Find the length and the width.

264. The width of a rectangle is eight inches more than the length. The perimeter is 52 inches. Find the length and the width.

265. The perimeter of a rectangle is 58 meters. The width of the rectangle is five meters less than the length. Find the length and the width of the rectangle.

266. The perimeter of a rectangle is 62 feet. The width is seven feet less than the length. Find the length and the width.

267. The width of the rectangle is 0.7 meters less than the length. The perimeter of a rectangle is 52.6 meters. Find the dimensions of the rectangle.

268. The length of the rectangle is 1.1 meters less than the width. The perimeter of a rectangle is 49.4 meters. Find the dimensions of the rectangle.

269. The perimeter of a rectangle is 150 feet. The length of the rectangle is twice the width. Find the length and width of the rectangle.

270. The length of a rectangle is three times the width. The perimeter of the rectangle is 72 feet. Find the length and width of the rectangle.

271. The length of a rectangle is three meters less than twice the width. The perimeter of the rectangle is 36 meters. Find the dimensions of the rectangle.

272. The length of a rectangle is five inches more than twice the width. The perimeter is 34 inches. Find the length and width.

273. The perimeter of a rectangular field is 560 yards. The length is 40 yards more than the width. Find the length and width of the field.

274. The perimeter of a rectangular atrium is 160 feet. The length is 16 feet more than the width. Find the length and width of the atrium.

275. A rectangular parking lot has perimeter 250 feet. The length is five feet more than twice the width. Find the length and width of the parking lot.

276. A rectangular rug has perimeter 240 inches. The length is 12 inches more than twice the width. Find the length and width of the rug.

Everyday Math

277. Christa wants to put a fence around her triangular flowerbed. The sides of the flowerbed are six feet, eight feet and 10 feet. How many feet of fencing will she need to enclose her flowerbed?

278. Jose just removed the children's playset from his back yard to make room for a rectangular garden. He wants to put a fence around the garden to keep out the dog. He has a 50 foot roll of fence in his garage that he plans to use. To fit in the backyard, the width of the garden must be 10 feet. How long can he make the other length?

Writing Exercises

279. If you need to put tile on your kitchen floor, do you need to know the perimeter or the area of the kitchen? Explain your reasoning.

280. If you need to put a fence around your backyard, do you need to know the perimeter or the area of the backyard? Explain your reasoning.

281. Look at the two figures below.

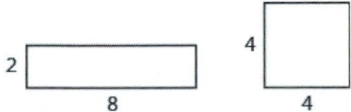

ⓐ Which figure looks like it has the larger area?

ⓑ Which looks like it has the larger perimeter?

ⓒ Now calculate the area and perimeter of each figure.

ⓓ Which has the larger area?

ⓔ Which has the larger perimeter?

282. Write a geometry word problem that relates to your life experience, then solve it and explain all your steps.

Self Check

ⓐ *After completing the exercises, use this checklist to evaluate your mastery of the objectives of this section.*

I can...	Confidently	With some help	No-I don't get it!
solve applications using triangle properties.			
use the Pythagorean Theorem.			
solve applications using rectangle properties.			

ⓑ *What does this checklist tell you about your mastery of this section? What steps will you take to improve?*

3.5 | Solve Uniform Motion Applications

Learning Objectives

By the end of this section, you will be able to:

> Solve uniform motion applications

Be Prepared!

Before you get started, take this readiness quiz.

1. Find the distance travelled by a car going 70 miles per hour for 3 hours.
   If you missed this problem, review **Example 2.58**.

2. Solve $x + 1.2(x - 10) = 98$.

   If you missed this problem, review **Example 2.39**.

3. Convert 90 minutes to hours.
   If you missed this problem, review **Example 1.140**.

Solve Uniform Motion Applications

When planning a road trip, it often helps to know how long it will take to reach the destination or how far to travel each day. We would use the distance, rate, and time formula, $D = rt$, which we have already seen.

In this section, we will use this formula in situations that require a little more algebra to solve than the ones we saw earlier. Generally, we will be looking at comparing two scenarios, such as two vehicles travelling at different rates or in opposite directions. When the speed of each vehicle is constant, we call applications like this *uniform motion problems*.

Our problem-solving strategies will still apply here, but we will add to the first step. The first step will include drawing a diagram that shows what is happening in the example. Drawing the diagram helps us understand what is happening so that we will write an appropriate equation. Then we will make a table to organize the information, like we did for the money applications.

The steps are listed here for easy reference:

HOW TO :: USE A PROBLEM-SOLVING STRATEGY IN DISTANCE, RATE, AND TIME APPLICATIONS.

Step 1.   **Read** the problem. Make sure all the words and ideas are understood.
- Draw a diagram to illustrate what it happening.
- Create a table to organize the information.
- Label the columns rate, time, distance.
- List the two scenarios.
- Write in the information you know.

	Rate	•	Time	=	Distance

Step 2.   **Identify** what we are looking for.

Step 3.   **Name** what we are looking for. Choose a variable to represent that quantity.
- Complete the chart.
- Use variable expressions to represent that quantity in each row.
- Multiply the rate times the time to get the distance.

Step 4.   **Translate** into an equation.
- Restate the problem in one sentence with all the important information.
- Then, translate the sentence into an equation.

Step 5.   **Solve** the equation using good algebra techniques.

Step 6.   **Check** the answer in the problem and make sure it makes sense.

Step 7.   **Answer** the question with a complete sentence.

EXAMPLE 3.48

An express train and a local train leave Pittsburgh to travel to Washington, D.C. The express train can make the trip in 4 hours and the local train takes 5 hours for the trip. The speed of the express train is 12 miles per hour faster than the speed of the local train. Find the speed of both trains.

✓ **Solution**

Step 1. Read the problem. Make sure all the words and ideas are understood.
- Draw a diagram to illustrate what it happening. Shown below is a sketch of what is happening in the example.

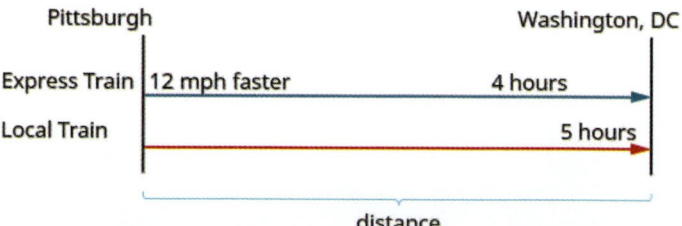

- Create a table to organize the information.
- Label the columns "Rate," "Time," and "Distance."

- List the two scenarios.
- Write in the information you know.

Step 2. Identify what we are looking for.

- We are asked to find the speed of both trains.
- Notice that the distance formula uses the word "rate," but it is more common to use "speed" when we talk about vehicles in everyday English.

Step 3. Name what we are looking for. Choose a variable to represent that quantity.

- Complete the chart
- Use variable expressions to represent that quantity in each row.
- We are looking for the speed of the trains. Let's let r represent the speed of the local train. Since the speed of the express train is 12 mph faster, we represent that as $r + 12$.

$$r = \text{speed of the local train}$$
$$r + 12 = \text{speed of the express train}$$

Fill in the speeds into the chart.

	Rate (mph)	•	Time (hrs)	=	Distance (miles)
Express	$r + 12$		4		
Local	r		5		

Multiply the rate times the time to get the distance.

	Rate (mph)	•	Time (hrs)	=	Distance (miles)
Express	$r + 12$		4		$4(r + 12)$
Local	r		5		$5r$

Step 4. Translate into an equation.

- Restate the problem in one sentence with all the important information.
- Then, translate the sentence into an equation.
- The equation to model this situation will come from the relation between the distances. Look at the diagram we drew above. How is the distance travelled by the express train related to the distance travelled by the local train?
- Since both trains leave from Pittsburgh and travel to Washington, D.C. they travel the same distance. So we write:

distance traveled by express train $=$ distance traveled by local train

Translate to an equation. $4(r + 12)$ $=$ $5r$

Step 5. Solve the equation using good algebra techniques.

Now solve this equation.

$$4(r + 12) = 5r$$
$$4r + 48 = 5r$$
$$48 = r$$

So the speed of the local train is 48 mph.

Find the speed of the express train.

$$r + 12$$
$$48 + 12$$
$$60$$

The speed of the express train is 60 mph.

Step 6. Check the answer in the problem and make sure it makes sense.

express train   60 mph (4 hours) = 240 miles

local train   48 mph (5 hours) = 240 miles ✓

Step 7. Answer the question with a complete sentence.

- The speed of the local train is 48 mph and the speed of the express train is 60 mph.

> **TRY IT : : 3.95**

Wayne and Dennis like to ride the bike path from Riverside Park to the beach. Dennis's speed is seven miles per hour faster than Wayne's speed, so it takes Wayne 2 hours to ride to the beach while it takes Dennis 1.5 hours for the ride. Find the speed of both bikers.

> **TRY IT : : 3.96**

Jeromy can drive from his house in Cleveland to his college in Chicago in 4.5 hours. It takes his mother 6 hours to make the same drive. Jeromy drives 20 miles per hour faster than his mother. Find Jeromy's speed and his mother's speed.

In **Example 3.48**, the last example, we had two trains traveling the same distance. The diagram and the chart helped us write the equation we solved. Let's see how this works in another case.

EXAMPLE 3.49

Christopher and his parents live 115 miles apart. They met at a restaurant between their homes to celebrate his mother's birthday. Christopher drove 1.5 hours while his parents drove 1 hour to get to the restaurant. Christopher's average speed was 10 miles per hour faster than his parents' average speed. What were the average speeds of Christopher and of his parents as they drove to the restaurant?

⊘ **Solution**

Step 1. Read the problem. Make sure all the words and ideas are understood.

- Draw a diagram to illustrate what it happening. Below shows a sketch of what is happening in the example.

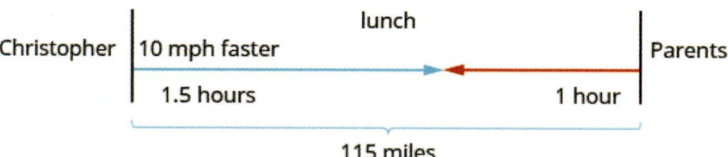

- Create a table to organize the information.
- Label the columns rate, time, distance.
- List the two scenarios.
- Write in the information you know.

	Rate (mph) •	Time (hrs) =	Distance (miles)
Christopher		1.5	
Parents		1	
			115

Step 2. Identify what we are looking for.

- We are asked to find the average speeds of Christopher and his parents.

Step 3. Name what we are looking for. Choose a variable to represent that quantity.

- Complete the chart.
- Use variable expressions to represent that quantity in each row.
- We are looking for their average speeds. Let's let r represent the average speed of the parents. Since the Christopher's speed is 10 mph faster, we represent that as $r + 10$.

Fill in the speeds into the chart.

	Rate (mph) $\cdot$	Time (hrs) $=$	Distance (miles)
Christopher	$r + 10$	1.5	$1.5(r + 10)$
Parents	r	1	r
			115

Multiply the rate times the time to get the distance.

Step 4. Translate into an equation.

- Restate the problem in one sentence with all the important information.
- Then, translate the sentence into an equation.
- Again, we need to identify a relationship between the distances in order to write an equation. Look at the diagram we created above and notice the relationship between the distance Christopher traveled and the distance his parents traveled.

The distance Christopher travelled plus the distance his parents travel must add up to 115 miles. So we write:

$$\underbrace{\text{distance traveled by Christopher}} + \underbrace{\text{distance traveled by his parents}} = 115$$

Translate to an equation. $\qquad 1.5(r + 10) \qquad + \qquad r \qquad = 115$

Step 5. Solve the equation using good algebra techniques.

Now solve this equation.
$$
\begin{aligned}
1.5(r + 10) + r &= 115 \\
1.5r + 15 + r &= 115 \\
2.5r + 15 &= 115 \\
2.5r &= 100 \\
r &= 40
\end{aligned}
$$
So the parents' speed was 40 mph.

Christopher's speed is $r + 10$.
$$
\begin{aligned}
& r + 10 \\
& 40 + 10 \\
& 50
\end{aligned}
$$
Christopher's speed was 50 mph.

Step 6. Check the answer in the problem and make sure it makes sense.

Christopher drove $\quad$ 50 mph (1.5 hours) $\quad = \quad$ 75 miles

His parents drove $\quad$ 40 mph (1 hours) $\quad = \quad$ <u>40 miles</u>

$\qquad\qquad\qquad\qquad\qquad\qquad$ 115 miles

Step 7. Answer the question with a complete sentence. $\qquad$ Christopher's speed was 50 mph.
$\qquad\qquad\qquad\qquad\qquad\qquad\qquad\qquad\qquad\qquad\qquad$ His parents' speed was 40 mph.

> **TRY IT : : 3.97**

Carina is driving from her home in Anaheim to Berkeley on the same day her brother is driving from Berkeley to Anaheim, so they decide to meet for lunch along the way in Buttonwillow. The distance from Anaheim to Berkeley is 410 miles. It takes Carina 3 hours to get to Buttonwillow, while her brother drives 4 hours to get there. The average speed Carina's brother drove was 15 miles per hour faster than Carina's average speed. Find Carina's and her brother's average speeds.

> **TRY IT : : 3.98**

Ashley goes to college in Minneapolis, 234 miles from her home in Sioux Falls. She wants her parents to bring her more winter clothes, so they decide to meet at a restaurant on the road between Minneapolis and Sioux Falls. Ashley and her parents both drove 2 hours to the restaurant. Ashley's average speed was seven miles per hour faster than her parents' average speed. Find Ashley's and her parents' average speed.

As you read the next example, think about the relationship of the distances traveled. Which of the previous two examples is more similar to this situation?

EXAMPLE 3.50

Two truck drivers leave a rest area on the interstate at the same time. One truck travels east and the other one travels west. The truck traveling west travels at 70 mph and the truck traveling east has an average speed of 60 mph. How long will they travel before they are 325 miles apart?

⊘ **Solution**

Step 1. Read the problem. Make sure all the words and ideas are understood.

- Draw a diagram to illustrate what it happening.

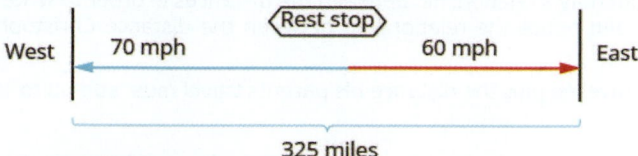

- Create a table to organize the information.

	Rate (mph)	•	Time (hrs)	=	Distance (miles)
West	70				
East	60				
					325

Step 2. Identify what we are looking for.

- We are asked to find the amount of time the trucks will travel until they are 325 miles apart.

Step 3. Name what we are looking for. Choose a variable to represent that quantity.

- We are looking for the time travelled. Both trucks will travel the same amount of time. Let's call the time t. Since their speeds are different, they will travel different distances.
- Complete the chart.

	Rate (mph)	•	Time (hrs)	=	Distance (miles)
West	70		t		$70t$
East	60		t		$60t$
					325

Step 4. Translate into an equation.

- We need to find a relation between the distances in order to write an equation. Looking at the diagram, what is the relationship between the distance each of the trucks will travel?
- The distance traveled by the truck going west plus the distance travelled by the truck going east must add up to 325 miles. So we write:

$$\text{distance traveled by westbound truck} + \text{distance traveled by eastbound truck} = 325$$

Translate to an equation. $70t$ $+$ $60t$ $= 325$

Step 5. Solve the equation using good algebra techniques.

Now solve this equation.
$$70t + 60t = 325$$
$$130t = 325$$
$$t = 2.5$$

So it will take the trucks 2.5 hours to be 325 miles apart.

Step 6. Check the answer in the problem and make sure it makes sense.

Truck going West   70 mph (2.5 hours) = 175 miles
Truck going East   60 mph (2.5 hours) = 150 miles
                                                     325 miles

Step 7. Answer the question with a complete sentence.   It will take the trucks 2.5 hours to be 325 miles apart.

 TRY IT : : 3.99

Pierre and Monique leave their home in Portland at the same time. Pierre drives north on the turnpike at a speed of 75 miles per hour while Monique drives south at a speed of 68 miles per hour. How long will it take them to be 429 miles apart?

 TRY IT : : 3.100

Thanh and Nhat leave their office in Sacramento at the same time. Thanh drives north on I-5 at a speed of 72 miles per hour. Nhat drives south on I-5 at a speed of 76 miles per hour. How long will it take them to be 330 miles apart?

Matching Units in Problems

It is important to make sure the units match when we use the distance rate and time formula. For instance, if the rate is in miles per hour, then the time must be in hours.

EXAMPLE 3.51

When Katie Mae walks to school, it takes her 30 minutes. If she rides her bike, it takes her 15 minutes. Her speed is three miles per hour faster when she rides her bike than when she walks. What are her walking speed and her speed riding her bike?

⊘ **Solution**

First, we draw a diagram that represents the situation to help us see what is happening.

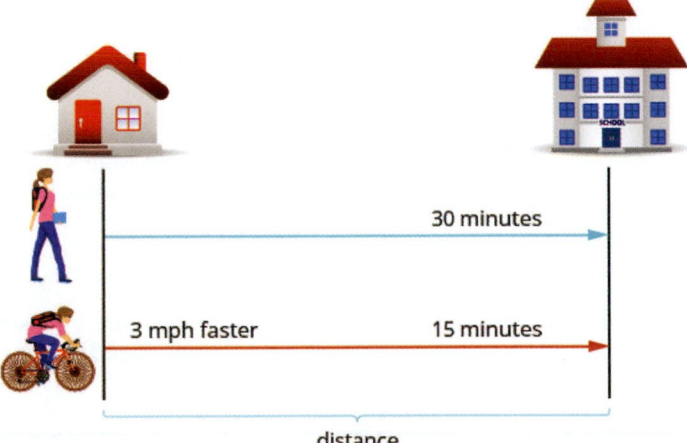

distance

We are asked to find her speed walking and riding her bike. Let's call her walking speed r. Since her biking speed is three miles per hour faster, we will call that speed $r + 3$. We write the speeds in the chart.

The speed is in miles per hour, so we need to express the times in hours, too, in order for the units to be the same. Remember, one hour is 60 minutes. So:

$$30 \text{ minutes is } \frac{30}{60} \text{ or } \frac{1}{2} \text{ hour}$$

$$15 \text{ minutes is } \frac{15}{60} \text{ or } \frac{1}{4} \text{ hour}$$

Next, we multiply rate times time to fill in the distance column.

	Rate (mph) $\cdot$	Time (hrs) =	Distance (miles)
Walk	r	$\frac{1}{2}$	$\frac{1}{2}r$
Bike	$r+3$	$\frac{1}{4}$	$\frac{1}{4}(r+3)$

The equation will come from the fact that the distance from Katie Mae's home to her school is the same whether she is walking or riding her bike.

So we say:

	distance walked $=$ distance covered by bike
Translate into an equation.	$\frac{1}{2}r \quad = \quad \frac{1}{4}(r+3)$
Solve this equation.	$\frac{1}{2}r = \frac{1}{4}(r+3)$
Clear the fractions by multiplying by the LCD of all the fractions in the equation.	$8 \cdot \frac{1}{2}r = 8 \cdot \frac{1}{4}(r+3)$
Simplify.	$4r = 2(r+3)$
	$4r = 2r + 6$
	$2r = 6$
	$r = 3$ mph (Katie Mae's walking speed)
	$r + 3$ biking speed
	$3 + 3$
	6 mph (Katie Mae's biking speed)
Let's check if this works. Walk 3 mph (0.5 hour) = 1.5 miles Bike 6 mph (0.25 hour) = 1.5 miles	
Yes, either way Katie Mae travels 1.5 miles to school.	Katie Mae's walking speed is 3 mph. Her speed riding her bike is 6 mph.

> **TRY IT : :** 3.101

Suzy takes 50 minutes to hike uphill from the parking lot to the lookout tower. It takes her 30 minutes to hike back down to the parking lot. Her speed going downhill is 1.2 miles per hour faster than her speed going uphill. Find Suzy's uphill and downhill speeds.

> **TRY IT : :** 3.102

Llewyn takes 45 minutes to drive his boat upstream from the dock to his favorite fishing spot. It takes him 30 minutes to drive the boat back downstream to the dock. The boat's speed going downstream is four miles per hour faster than its speed going upstream. Find the boat's upstream and downstream speeds.

In the distance, rate, and time formula, time represents the actual amount of elapsed time (in hours, minutes, etc.). If a problem gives us starting and ending times as clock times, we must find the elapsed time in order to use the formula.

EXAMPLE 3.52

Hamilton loves to travel to Las Vegas, 255 miles from his home in Orange County. On his last trip, he left his house at 2:00 pm. The first part of his trip was on congested city freeways. At 4:00 pm, the traffic cleared and he was able to drive through the desert at a speed 1.75 times faster than when he drove in the congested area. He arrived in Las Vegas at 6:30 pm. How fast was he driving during each part of his trip?

⊘ Solution

A diagram will help us model this trip.

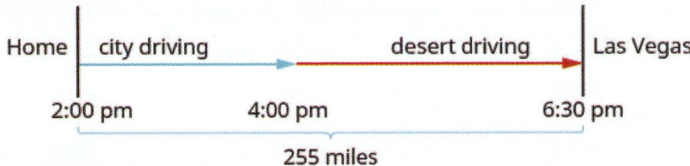

Next, we create a table to organize the information.

We know the total distance is 255 miles. We are looking for the rate of speed for each part of the trip. The rate in the desert is 1.75 times the rate in the city. If we let $r =$ the rate in the city, then the rate in the desert is $1.75r$.

The times here are given as clock times. Hamilton started from home at 2:00 pm and entered the desert at 4:30 pm. So he spent two hours driving the congested freeways in the city. Then he drove faster from 4:00 pm until 6:30 pm in the desert. So he drove 2.5 hours in the desert.

Now, we multiply the rates by the times.

	Rate (mph)	·	Time (hrs)	=	Distance (miles)
City	r		2		$2r$
Desert	$1.75r$		2.5		$2.5(1.75r)$
					255

By looking at the diagram below, we can see that the sum of the distance driven in the city and the distance driven in the desert is 255 miles.

distance driven in the city $+$ distance driven in desert $= 255$

Translate into an equation.	$2r$	$+$	$2.5(1.75r)$	$= 255$

Solve this equation.

$$2r + 2.5(1.75r) = 255$$

$$2r + 4.375r = 255$$

$$6.375r = 255$$

$$r = 40 \text{ mph city}$$

$$1.75r \text{ desert speed}$$

$$1.75(40)$$

$$70 \text{ mph}$$

Check.

$$\text{City } 40 \text{ mph } (2 \text{ hours}) = 80 \text{ miles}$$
$$\text{Desert } 70 \text{ mph } (2.5 \text{ hours}) = \underline{175 \text{ miles}}$$
$$255 \text{ miles}$$

Hamilton drove 40 mph in the city and 70 mph in the desert.

> **TRY IT : :** 3.103

Cruz is training to compete in a triathlon. He left his house at 6:00 and ran until 7:30. Then he rode his bike until 9:45. He covered a total distance of 51 miles. His speed when biking was 1.6 times his speed when running. Find Cruz's biking and running speeds.

> **TRY IT : :** 3.104

Phuong left home on his bicycle at 10:00. He rode on the flat street until 11:15, then rode uphill until 11:45. He rode a total of 31 miles. His speed riding uphill was 0.6 times his speed on the flat street. Find his speed biking uphill and on the flat street.

3.5 EXERCISES

Practice Makes Perfect

Solve Uniform Motion Applications

In the following exercises, solve.

283. Lilah is moving from Portland to Seattle. It takes her three hours to go by train. Mason leaves the train station in Portland and drives to the train station in Seattle with all Lilah's boxes in his car. It takes him 2.4 hours to get to Seattle, driving at 15 miles per hour faster than the speed of the train. Find Mason's speed and the speed of the train.

284. Kathy and Cheryl are walking in a fundraiser. Kathy completes the course in 4.8 hours and Cheryl completes the course in 8 hours. Kathy walks two miles per hour faster than Cheryl. Find Kathy's speed and Cheryl's speed.

285. Two busses go from Sacramento for San Diego. The express bus makes the trip in 6.8 hours and the local bus takes 10.2 hours for the trip. The speed of the express bus is 25 mph faster than the speed of the local bus. Find the speed of both busses.

286. A commercial jet and a private airplane fly from Denver to Phoenix. It takes the commercial jet 1.1 hours for the flight, and it takes the private airplane 1.8 hours. The speed of the commercial jet is 210 miles per hour faster than the speed of the private airplane. Find the speed of both airplanes.

287. Saul drove his truck 3 hours from Dallas towards Kansas City and stopped at a truck stop to get dinner. At the truck stop he met Erwin, who had driven 4 hours from Kansas City towards Dallas. The distance between Dallas and Kansas City is 542 miles, and Erwin's speed was eight miles per hour slower than Saul's speed. Find the speed of the two truckers.

288. Charlie and Violet met for lunch at a restaurant between Memphis and New Orleans. Charlie had left Memphis and drove 4.8 hours towards New Orleans. Violet had left New Orleans and drove 2 hours towards Memphis, at a speed 10 miles per hour faster than Charlie's speed. The distance between Memphis and New Orleans is 394 miles. Find the speed of the two drivers.

289. Sisters Helen and Anne live 332 miles apart. For Thanksgiving, they met at their other sister's house partway between their homes. Helen drove 3.2 hours and Anne drove 2.8 hours. Helen's average speed was four miles per hour faster than Anne's. Find Helen's average speed and Anne's average speed.

290. Ethan and Leo start riding their bikes at the opposite ends of a 65-mile bike path. After Ethan has ridden 1.5 hours and Leo has ridden 2 hours, they meet on the path. Ethan's speed is six miles per hour faster than Leo's speed. Find the speed of the two bikers.

291. Elvira and Aletheia live 3.1 miles apart on the same street. They are in a study group that meets at a coffee shop between their houses. It took Elvira half an hour and Aletheia two-thirds of an hour to walk to the coffee shop. Aletheia's speed is 0.6 miles per hour slower than Elvira's speed. Find both women's walking speeds.

292. DaMarcus and Fabian live 23 miles apart and play soccer at a park between their homes. DaMarcus rode his bike for three-quarters of an hour and Fabian rode his bike for half an hour to get to the park. Fabian's speed was six miles per hour faster than DaMarcus' speed. Find the speed of both soccer players.

293. Cindy and Richard leave their dorm in Charleston at the same time. Cindy rides her bicycle north at a speed of 18 miles per hour. Richard rides his bicycle south at a speed of 14 miles per hour. How long will it take them to be 96 miles apart?

294. Matt and Chris leave their uncle's house in Phoenix at the same time. Matt drives west on I-60 at a speed of 76 miles per hour. Chris drives east on I-60 at a speed of 82 miles per hour. How many hours will it take them to be 632 miles apart?

295. Two busses leave Billings at the same time. The Seattle bus heads west on I-90 at a speed of 73 miles per hour while the Chicago bus heads east at a speed of 79 miles an hour. How many hours will it take them to be 532 miles apart?

296. Two boats leave the same dock in Cairo at the same time. One heads north on the Mississippi River while the other heads south. The northbound boat travels four miles per hour. The southbound boat goes eight miles per hour. How long will it take them to be 54 miles apart?

297. Lorena walks the path around the park in 30 minutes. If she jogs, it takes her 20 minutes. Her jogging speed is 1.5 miles per hour faster than her walking speed. Find Lorena's walking speed and jogging speed.

298. Julian rides his bike uphill for 45 minutes, then turns around and rides back downhill. It takes him 15 minutes to get back to where he started. His uphill speed is 3.2 miles per hour slower than his downhill speed. Find Julian's uphill and downhill speed.

299. Cassius drives his boat upstream for 45 minutes. It takes him 30 minutes to return downstream. His speed going upstream is three miles per hour slower than his speed going downstream. Find his upstream and downstream speeds.

300. It takes Darline 20 minutes to drive to work in light traffic. To come home, when there is heavy traffic, it takes her 36 minutes. Her speed in light traffic is 24 miles per hour faster than her speed in heavy traffic. Find her speed in light traffic and in heavy traffic.

301. At 1:30 Marlon left his house to go to the beach, a distance of 7.6 miles. He rode his skateboard until 2:15, then walked the rest of the way. He arrived at the beach at 3:00. Marlon's speed on his skateboard is 2.5 times his walking speed. Find his speed when skateboarding and when walking.

302. Aaron left at 9:15 to drive to his mountain cabin 108 miles away. He drove on the freeway until 10:45, and then he drove on the mountain road. He arrived at 11:05. His speed on the freeway was three times his speed on the mountain road. Find Aaron's speed on the freeway and on the mountain road.

303. Marisol left Los Angeles at 2:30 to drive to Santa Barbara, a distance of 95 miles. The traffic was heavy until 3:20. She drove the rest of the way in very light traffic and arrived at 4:20. Her speed in heavy traffic was 40 miles per hour slower than her speed in light traffic. Find her speed in heavy traffic and in light traffic.

304. Lizette is training for a marathon. At 7:00 she left her house and ran until 8:15, then she walked until 11:15. She covered a total distance of 19 miles. Her running speed was five miles per hour faster than her walking speed. Find her running and walking speeds.

Everyday Math

305. John left his house in Irvine at 8:35 am to drive to a meeting in Los Angeles, 45 miles away. He arrived at the meeting at 9:50. At 3:30 pm, he left the meeting and drove home. He arrived home at 5:18.

ⓐ What was his average speed on the drive from Irvine to Los Angeles?

ⓑ What was his average speed on the drive from Los Angeles to Irvine?

ⓒ What was the total time he spent driving to and from this meeting?

ⓓ John drove a total of 90 miles roundtrip. Find his average speed. (Round to the nearest tenth.)

306. Sarah wants to arrive at her friend's wedding at 3:00. The distance from Sarah's house to the wedding is 95 miles. Based on usual traffic patterns, Sarah predicts she can drive the first 15 miles at 60 miles per hour, the next 10 miles at 30 miles per hour, and the remainder of the drive at 70 miles per hour.

ⓐ How long will it take Sarah to drive the first 15 miles?

ⓑ How long will it take Sarah to drive the next 10 miles?

ⓒ How long will it take Sarah to drive the rest of the trip?

ⓓ What time should Sarah leave her house?

Writing Exercises

307. When solving a uniform motion problem, how does drawing a diagram of the situation help you?

308. When solving a uniform motion problem, how does creating a table help you?

Self Check

ⓐ *After completing the exercises, use this checklist to evaluate your mastery of the objectives of this section.*

I can...	Confidently	With some help	No-I don't get it!
solve uniform motion applications.			

ⓑ *What does this checklist tell you about your mastery of this section? What steps will you take to improve?*

3.6 **Solve Applications with Linear Inequalities**

Learning Objectives

By the end of this section, you will be able to:

› Solve applications with linear inequalities

Be Prepared!

Before you get started, take this readiness quiz.

1. Write as an inequality: *x* is at least 30.
   If you missed this problem, review **Example 2.77**.

2. Solve $8 - 3y < 41$.
   If you missed this problem, review **Example 2.73**.

Solve Applications with Linear Inequalities

Many real-life situations require us to solve inequalities. In fact, inequality applications are so common that we often do not even realize we are doing algebra. For example, how many gallons of gas can be put in the car for $20? Is the rent on an apartment affordable? Is there enough time before class to go get lunch, eat it, and return? How much money should each family member's holiday gift cost without going over budget?

The method we will use to solve applications with linear inequalities is very much like the one we used when we solved applications with equations. We will read the problem and make sure all the words are understood. Next, we will identify what we are looking for and assign a variable to represent it. We will restate the problem in one sentence to make it easy to translate into an inequality. Then, we will solve the inequality.

EXAMPLE 3.53

Emma got a new job and will have to move. Her monthly income will be $5,265. To qualify to rent an apartment, Emma's monthly income must be at least three times as much as the rent. What is the highest rent Emma will qualify for?

 **Solution**

Step 1. Read the problem.

Step 2. Identify what we are looking for.   the highest rent Emma will qualify for

Step 3. Name what we are looking for.

   Choose a variable to represent that quantity.   Let r = the rent.

Step 4. Translate into an inequality.

   First write a sentence that gives the information   Emma's monthly income must be at least
   to find it.   three times the rent.

Step 5. Solve the inequality.

   Remember, $a > x$ has the same meaning $5,625 \ge 3r$
   as $x < a$. $1,755 \ge r$
                                                          $r \le 1,755$

Step 6. Check the answer in the problem
and make sure it makes sense.

   A maximum rent of $1,755 seems
   reasonable for an income of $5,625.

Step 7. Answer the question with a
complete sentence.   The maximum rent is $1,755.

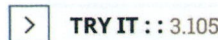 **TRY IT :: 3.105**

Alan is loading a pallet with boxes that each weighs 45 pounds. The pallet can safely support no more than 900 pounds. How many boxes can he safely load onto the pallet?

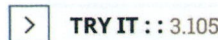 **TRY IT :: 3.106**

The elevator in Yehire's apartment building has a sign that says the maximum weight is 2,100 pounds. If the average weight of one person is 150 pounds, how many people can safely ride the elevator?

Sometimes an application requires the solution to be a whole number, but the algebraic solution to the inequality is not a whole number. In that case, we must round the algebraic solution to a whole number. The context of the application will determine whether we round up or down. To check applications like this, we will round our answer to a number that is easy to compute with and make sure that number makes the inequality true.

EXAMPLE 3.54

Dawn won a mini-grant of $4,000 to buy tablet computers for her classroom. The tablets she would like to buy cost $254.12 each, including tax and delivery. What is the maximum number of tablets Dawn can buy?

⊘ Solution

Step 1. Read the problem.

Step 2. Identify what we are looking for. — the maximum number of tablets Dawn can buy

Step 3. Name what we are looking for.

Choose a variable to represent that quantity. — Let n = the number of tablets.

Step 4. Translate. Write a sentence that gives the information to find it. — $254.12 times the number of tablets is no more than $4,000.

Translate into an inequality.
$$254.12n \leq 4,000$$

Step 5. Solve the inequality.
$$n \leq 15.74$$

But n must be a whole number of tablets,

so round to 15.
$$n \leq 15$$

Step 6. Check the answer in the problem and make sure it makes sense.

Rounding down the price to $250, 15 tablets would cost $3,750, while 16 tablets would be $4,000. So a maximum of 15 tablets at $254.12 seems reasonable.

Step 7. Answer the question with a complete sentence. — Dawn can buy a maximum of 15 tablets.

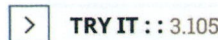 **TRY IT :: 3.107**

Angie has $20 to spend on juice boxes for her son's preschool picnic. Each pack of juice boxes costs $2.63. What is the maximum number of packs she can buy?

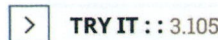 **TRY IT :: 3.108**

Daniel wants to surprise his girlfriend with a birthday party at her favorite restaurant. It will cost $42.75 per person for dinner, including tip and tax. His budget for the party is $500. What is the maximum number of people Daniel can have at the party?

EXAMPLE 3.55

Pete works at a computer store. His weekly pay will be either a fixed amount, $925, or $500 plus 12% of his total sales. How much should his total sales be for his variable pay option to exceed the fixed amount of $925?

✓ **Solution**

Step 1. Read the problem.

Step 2. Identify what we are looking for.

the total sales needed for his variable pay option to exceed the fixed amount of $925

Step 3. Name what we are looking for.

Choose a variable to represent that quantity.

Let s = the total sales.

Step 4. Translate Write a sentence that gives the information to find it.

$500 plus 12% of total sales is more than $925.

Translate into an inequality. Remember to convert the percent to a decimal.

$$500 + 0.12s > 925$$

Step 5. Solve the inequality.

$$0.12s > 425$$
$$s > 3{,}541.\overline{66}$$

Step 6. Check the answer in the problem and make sure it makes sense.

If we round the total sales up to $4,000, we see that $500 + 0.12(4{,}000) = 980$, which is more than $925.

Step 7. Answer the question with a complete sentence.

The total sales must be more than $3,541.67.

> **TRY IT : : 3.109**

Tiffany just graduated from college and her new job will pay her $20,000 per year plus 2% of all sales. She wants to earn at least $100,000 per year. For what total sales will she be able to achieve her goal?

> **TRY IT : : 3.110**

Christian has been offered a new job that pays $24,000 a year plus 3% of sales. For what total sales would this new job pay more than his current job which pays $60,000?

EXAMPLE 3.56

Sergio and Lizeth have a very tight vacation budget. They plan to rent a car from a company that charges $75 a week plus $0.25 a mile. How many miles can they travel and still keep within their $200 budget?

⊘ **Solution**

Step 1. Read the problem.

Step 2. Identify what we are looking for.   the number of miles Sergio and Lizeth can travel

Step 3. Name what we are looking for.

    Choose a variable to represent that quantity.   Let $m =$ the number of miles.

Step 4. Translate Write a sentence that $75 plus 0.25 times the number of miles is
gives the information to find it.   less than or equal to $200.

    Translate into an inequality. $$75 + 0.25m \le 200$$

Step 5. Solve the inequality. $$0.25m \;\le\; 125$$
$$m \;\le\; 500 \text{ miles}$$

Step 6. Check the answer in the problem
and make sure it makes sense.
    Yes, $75 + 0.25(500) = 200$.

Step 7. Write a sentence that answers the question.   Sergio and Lizeth can travel 500 miles
and still stay on budget.

> **TRY IT : : 3.111**
>
> Taleisha's phone plan costs her $28.80 a month plus $0.20 per text message. How many text messages can she use and keep her monthly phone bill no more than $50?

> **TRY IT : : 3.112**
>
> Rameen's heating bill is $5.42 per month plus $1.08 per therm. How many therms can Rameen use if he wants his heating bill to be a maximum of $87.50?

A common goal of most businesses is to make a profit. *Profit* is the money that remains when the expenses have been subtracted from the money earned. In the next example, we will find the number of jobs a small businessman needs to do every month in order to make a certain amount of profit.

EXAMPLE 3.57

Elliot has a landscape maintenance business. His monthly expenses are $1,100. If he charges $60 per job, how many jobs must he do to earn a profit of at least $4,000 a month?

⊘ **Solution**

Step 1. Read the problem.

Step 2. Identify what we are looking for.

the number of jobs Elliot needs

Step 3. Name what we are looking for. Choose a variable to represent it.

Let j = the number of jobs.

Step 4. Translate Write a sentence that gives the information to find it.

Translate into an inequality.

$60 times the number of jobs minus $1,100 is at least $4,000.

$$60j - 1100 \geq 4,000$$
$$60j \geq 5,100$$
$$j \geq 85 \text{ jobs}$$

Step 5. Solve the inequality.

Step 6. Check the answer in the problem and make sure it makes sense.

If Elliot did 90 jobs, his profit would be $60(90) - 1,100$, or $4,300. This is more than $4,000.

Step 7. Write a sentence that answers the question.

Elliot must work at least 85 jobs.

> **TRY IT :: 3.113**

Caleb has a pet sitting business. He charges $32 per hour. His monthly expenses are $2,272. How many hours must he work in order to earn a profit of at least $800 per month?

> **TRY IT :: 3.114**

Felicity has a calligraphy business. She charges $2.50 per wedding invitation. Her monthly expenses are $650. How many invitations must she write to earn a profit of at least $2,800 per month?

Sometimes life gets complicated! There are many situations in which several quantities contribute to the total expense. We must make sure to account for all the individual expenses when we solve problems like this.

EXAMPLE 3.58

Brenda's best friend is having a destination wedding and the event will last 3 days. Brenda has $500 in savings and can earn $15 an hour babysitting. She expects to pay $350 airfare, $375 for food and entertainment and $60 a night for her share of a hotel room. How many hours must she babysit to have enough money to pay for the trip?

⊘ Solution

Step 1. Read the problem.

Step 2. Identify what we are looking for.	the number of hours Brenda must babysit
Step 3. Name what we are looking for. Choose a variable to represent that quantity.	Let $h =$ the number of hours.
Step 4. Translate Write a sentence that gives the information to find it.	The expenses must be less than or equal to the income. The cost of airfare plus the cost of food and entertainment and the hotel bill must be less than or equal to the savings plus the amount earned babysitting.
Translate into an inequality.	$\$350 + \$375 + \$60(3) \leq \$500 + \$15h$

Step 5. Solve the inequality.

$$
\begin{aligned}
905 &\leq 500 + 15h \\
405 &\leq 15h \\
27 &\leq h \\
h &\geq 27
\end{aligned}
$$

Step 6. Check the answer in the problem and make sure it makes sense.

We substitute 27 into the inequality.

$905 \leq 500 + 15h$

$905 \leq 500 + 15(27)$

$905 \leq 905$

Step 7. Write a sentence that answers the question.	Brenda must babysit at least 27 hours.

> **TRY IT : : 3.115**

Malik is planning a 6-day summer vacation trip. He has $840 in savings, and he earns $45 per hour for tutoring. The trip will cost him $525 for airfare, $780 for food and sightseeing, and $95 per night for the hotel. How many hours must he tutor to have enough money to pay for the trip?

> **TRY IT : : 3.116**

Josue wants to go on a 10-day road trip next spring. It will cost him $180 for gas, $450 for food, and $49 per night for a motel. He has $520 in savings and can earn $30 per driveway shoveling snow. How many driveways must he shovel to have enough money to pay for the trip?

3.6 EXERCISES

Practice Makes Perfect

Solve Applications with Linear Inequalities

In the following exercises, solve.

309. Mona is planning her son's birthday party and has a budget of $285. The Fun Zone charges $19 per child. How many children can she have at the party and stay within her budget?

310. Carlos is looking at apartments with three of his friends. They want the monthly rent to be no more than $2360. If the roommates split the rent evenly among the four of them, what is the maximum rent each will pay?

311. A water taxi has a maximum load of 1,800 pounds. If the average weight of one person is 150 pounds, how many people can safely ride in the water taxi?

312. Marcela is registering for her college classes, which cost $105 per unit. How many units can she take to have a maximum cost of $1,365?

313. Arleen got a $20 gift card for the coffee shop. Her favorite iced drink costs $3.79. What is the maximum number of drinks she can buy with the gift card?

314. Teegan likes to play golf. He has budgeted $60 next month for the driving range. It costs him $10.55 for a bucket of balls each time he goes. What is the maximum number of times he can go to the driving range next month?

315. Joni sells kitchen aprons online for $32.50 each. How many aprons must she sell next month if she wants to earn at least $1,000?

316. Ryan charges his neighbors $17.50 to wash their car. How many cars must he wash next summer if his goal is to earn at least $1,500?

317. Keshad gets paid $2,400 per month plus 6% of his sales. His brother earns $3,300 per month. For what amount of total sales will Keshad's monthly pay be higher than his brother's monthly pay?

318. Kimuyen needs to earn $4,150 per month in order to pay all her expenses. Her job pays her $3,475 per month plus 4% of her total sales. What is the minimum Kimuyen's total sales must be in order for her to pay all her expenses?

319. Andre has been offered an entry-level job. The company offered him $48,000 per year plus 3.5% of his total sales. Andre knows that the average pay for this job is $62,000. What would Andre's total sales need to be for his pay to be at least as high as the average pay for this job?

320. Nataly is considering two job offers. The first job would pay her $83,000 per year. The second would pay her $66,500 plus 15% of her total sales. What would her total sales need to be for her salary on the second offer be higher than the first?

321. Jake's water bill is $24.80 per month plus $2.20 per ccf (hundred cubic feet) of water. What is the maximum number of ccf Jake can use if he wants his bill to be no more than $60?

322. Kiyoshi's phone plan costs $17.50 per month plus $0.15 per text message. What is the maximum number of text messages Kiyoshi can use so the phone bill is no more than $56.50?

323. Marlon's TV plan costs $49.99 per month plus $5.49 per first-run movie. How many first-run movies can he watch if he wants to keep his monthly bill to be a maximum of $100?

324. Kellen wants to rent a banquet room in a restaurant for her cousin's baby shower. The restaurant charges $350 for the banquet room plus $32.50 per person for lunch. How many people can Kellen have at the shower if she wants the maximum cost to be $1,500?

325. Moshde runs a hairstyling business from her house. She charges $45 for a haircut and style. Her monthly expenses are $960. She wants to be able to put at least $1,200 per month into her savings account order to open her own salon. How many "cut & styles" must she do to save at least $1,200 per month?

326. Noe installs and configures software on home computers. He charges $125 per job. His monthly expenses are $1,600. How many jobs must he work in order to make a profit of at least $2,400?

327. Katherine is a personal chef. She charges $115 per four-person meal. Her monthly expenses are $3,150. How many four-person meals must she sell in order to make a profit of at least $1,900?

328. Melissa makes necklaces and sells them online. She charges $88 per necklace. Her monthly expenses are $3745. How many necklaces must she sell if she wants to make a profit of at least $1,650?

329. Five student government officers want to go to the state convention. It will cost them $110 for registration, $375 for transportation and food, and $42 per person for the hotel. There is $450 budgeted for the convention in the student government savings account. They can earn the rest of the money they need by having a car wash. If they charge $5 per car, how many cars must they wash in order to have enough money to pay for the trip?

330. Cesar is planning a 4-day trip to visit his friend at a college in another state. It will cost him $198 for airfare, $56 for local transportation, and $45 per day for food. He has $189 in savings and can earn $35 for each lawn he mows. How many lawns must he mow to have enough money to pay for the trip?

331. Alonzo works as a car detailer. He charges $175 per car. He is planning to move out of his parents' house and rent his first apartment. He will need to pay $120 for application fees, $950 for security deposit, and first and last months' rent at $1,140 per month. He has $1,810 in savings. How many cars must he detail to have enough money to rent the apartment?

332. Eun-Kyung works as a tutor and earns $60 per hour. She has $792 in savings. She is planning an anniversary party for her parents. She would like to invite 40 guests. The party will cost her $1,520 for food and drinks and $150 for the photographer. She will also have a favor for each of the guests, and each favor will cost $7.50. How many hours must she tutor to have enough money for the party?

Everyday Math

333. Maximum Load on a Stage In 2014, a high school stage collapsed in Fullerton, California, when 250 students got on stage for the finale of a musical production. Two dozen students were injured. The stage could support a maximum of 12,750 pounds. If the average weight of a student is assumed to be 140 pounds, what is the maximum number of students who could safely be on the stage?

334. Maximum Weight on a Boat In 2004, a water taxi sank in Baltimore harbor and five people drowned. The water taxi had a maximum capacity of 3,500 pounds (25 people with average weight 140 pounds). The average weight of the 25 people on the water taxi when it sank was 168 pounds per person. What should the maximum number of people of this weight have been?

335. Wedding Budget Adele and Walter found the perfect venue for their wedding reception. The cost is $9,850 for up to 100 guests, plus $38 for each additional guest. How many guests can attend if Adele and Walter want the total cost to be no more than $12,500?

336. Shower Budget Penny is planning a baby shower for her daughter-in-law. The restaurant charges $950 for up to 25 guests, plus $31.95 for each additional guest. How many guests can attend if Penny wants the total cost to be no more than $1,500?

Writing Exercises

337. Find your last month's phone bill and the hourly salary you are paid at your job. (If you do not have a job, use the hourly salary you would realistically be paid if you had a job.) Calculate the number of hours of work it would take you to earn at least enough money to pay your phone bill by writing an appropriate inequality and then solving it.

338. Find out how many units you have left, after this term, to achieve your college goal and estimate the number of units you can take each term in college. Calculate the number of terms it will take you to achieve your college goal by writing an appropriate inequality and then solving it.

Self Check

ⓐ *After completing the exercises, use this checklist to evaluate your mastery of the objectives of this section.*

I can...	Confidently	With some help	No–I don't get it!
solve applications with linear inequalities.			

ⓑ *What does this checklist tell you about your mastery of this section? What steps will you take to improve?*

CHAPTER 3 REVIEW

KEY TERMS

amount of discount The amount of discount is the amount resulting when a discount rate is multiplied by the original price of an item.

discount rate The discount rate is the percent used to determine the amount of a discount, common in retail settings.

interest Interest is the money that a bank pays its customers for keeping their money in the bank.

list price The list price is the price a retailer sells an item for.

mark-up A mark-up is a percentage of the original cost used to increase the price of an item.

mixture problems Mixture problems combine two or more items with different values together.

original cost The original cost in a retail setting, is the price that a retailer pays for an item.

principal The principal is the original amount of money invested or borrowed for a period of time at a specific interest rate.

rate of interest The rate of interest is a percent of the principal, usually expressed as a percent per year.

simple interest Simple interest is the interest earned according to the formula $I = Prt$.

KEY CONCEPTS

3.1 Use a Problem-Solving Strategy

- **Problem-Solving Strategy**

  Step 1. **Read** the problem. Make sure all the words and ideas are understood.

  Step 2. **Identify** what we are looking for.

  Step 3. **Name** what we are looking for. Choose a variable to represent that quantity.

  Step 4. **Translate** into an equation. It may be helpful to restate the problem in one sentence with all the important information. Then, translate the English sentence into an algebra equation.

  Step 5. **Solve** the equation using good algebra techniques.

  Step 6. **Check** the answer in the problem and make sure it makes sense.

  Step 7. **Answer** the question with a complete sentence.

- **Consecutive Integers**
  Consecutive integers are integers that immediately follow each other.

$$n \qquad 1^{st} \text{ integer}$$
$$n + 1 \qquad 2^{nd} \text{ integer consecutive integer}$$
$$n + 2 \qquad 3^{rd} \text{ consecutive integer . . . etc.}$$

  Consecutive even integers are even integers that immediately follow one another.

$$n \qquad 1^{st} \text{ integer}$$
$$n + 2 \qquad 2^{nd} \text{ integer consecutive integer}$$
$$n + 4 \qquad 3^{rd} \text{ consecutive integer . . . etc.}$$

  Consecutive odd integers are odd integers that immediately follow one another.

$$n \qquad 1^{st} \text{ integer}$$
$$n + 2 \qquad 2^{nd} \text{ integer consecutive integer}$$
$$n + 4 \qquad 3^{rd} \text{ consecutive integer . . . etc.}$$

3.2 Solve Percent Applications

- **Percent Increase** To find the percent increase:

Step 1. Find the amount of increase. increase = new amount − original amount

Step 2. Find the percent increase. Increase is what percent of the original amount?

- **Percent Decrease** To find the percent decrease:

Step 1. Find the amount of decrease. decrease = original amount − new amount

Step 2. Find the percent decrease. Decrease is what percent of the original amount?

- **Simple Interest** If an amount of money, *P*, called the principal, is invested for a period of *t* years at an annual interest rate *r*, the amount of interest, *I*, earned is

$$I = Prt$$

$$\text{where} \quad \begin{aligned} I &= \text{interest} \\ P &= \text{principal} \\ r &= \text{rate} \\ t &= \text{time} \end{aligned}$$

- **Discount**
  - ○ amount of discount is discount rate · original price
  - ○ sale price is original price − discount

- **Mark-up**
  - ○ amount of mark-up is mark-up rate · original cost
  - ○ list price is original cost + mark up

3.3 Solve Mixture Applications

- **Total Value of Coins** For the same type of coin, the total value of a number of coins is found by using the model. *number · value = total value* where *number* is the number of coins and *value* is the value of each coin; *total value* is the total value of all the coins

- **Problem-Solving Strategy—Coin Word Problems**

Step 1. **Read** the problem. Make all the words and ideas are understood. Determine the types of coins involved.
   - ▪ Create a table to organize the information.
   - ▪ Label the columns type, number, value, total value.
   - ▪ List the types of coins.
   - ▪ Write in the value of each type of coin.
   - ▪ Write in the total value of all the coins.

Step 2. **Identify** what we are looking for.

Step 3. **Name** what we are looking for. Choose a variable to represent that quantity.
   Use variable expressions to represent the number of each type of coin and write them in the table.
   Multiply the number times the value to get the total value of each type of coin.

Step 4. **Translate** into an equation. It may be helpful to restate the problem in one sentence with all the important information. Then, translate the sentence into an equation.
   Write the equation by adding the total values of all the types of coins.

Step 5. **Solve** the equation using good algebra techniques.

Step 6. **Check** the answer in the problem and make sure it makes sense.

Step 7. **Answer** the question with a complete sentence.

3.4 Solve Geometry Applications: Triangles, Rectangles, and the Pythagorean Theorem

- **Problem-Solving Strategy for Geometry Applications**

Step 1. **Read** the problem and make all the words and ideas are understood. Draw the figure and label it with the given information.

Step 2. **Identify** what we are looking for.

Step 3. **Name** what we are looking for by choosing a variable to represent it.

Step 4. **Translate** into an equation by writing the appropriate formula or model for the situation. Substitute in the given information.

Step 5. **Solve** the equation using good algebra techniques.

Step 6. **Check** the answer in the problem and make sure it makes sense.

Step 7. **Answer** the question with a complete sentence.

- **Triangle Properties For** $\triangle ABC$

Angle measures:

  ◦ $m\angle A + m\angle B + m\angle C = 180$

Perimeter:

  ◦ $P = a + b + c$

Area:

  ◦ $A = \frac{1}{2}bh,\ \ b = \text{base},\ h = \text{height}$

A right triangle has one $90°$ angle.

- **The Pythagorean Theorem** In any right triangle, $a^2 + b^2 = c^2$ where c is the length of the hypotenuse and a and b are the lengths of the legs.

- **Properties of Rectangles**

  ◦ Rectangles have four sides and four right (90°) angles.

  ◦ The lengths of opposite sides are equal.

  ◦ The perimeter of a rectangle is the sum of twice the length and twice the width: $P = 2L + 2W$. The area of a rectangle is the length times the width: $A = LW$.

3.5 Solve Uniform Motion Applications

- **Distance, Rate, and Time**

  ◦ $D = rt$ where D = distance, r = rate, t = time

- **Problem-Solving Strategy—Distance, Rate, and Time Applications**

Step 1. **Read** the problem. Make sure all the words and ideas are understood.
Draw a diagram to illustrate what it happening.
Create a table to organize the information: Label the columns rate, time, distance. List the two scenarios. Write in the information you know.

Step 2. **Identify** what we are looking for.

Step 3. **Name** what we are looking for. Choose a variable to represent that quantity.
Complete the chart.
Use variable expressions to represent that quantity in each row.
Multiply the rate times the time to get the distance.

Step 4. **Translate** into an equation.
Restate the problem in one sentence with all the important information.
Then, translate the sentence into an equation.

Step 5. **Solve** the equation using good algebra techniques.

Step 6. **Check** the answer in the problem and make sure it makes sense.

Step 7. **Answer** the question with a complete sentence.

3.6 Solve Applications with Linear Inequalities

- **Solving inequalities**

Step 1. **Read** the problem.

Step 2. **Identify** what we are looking for.

Step 3. **Name** what we are looking for. Choose a variable to represent that quantity.

Step 4. **Translate.** Write a sentence that gives the information to find it. Translate into an inequality.

Step 5. **Solve** the inequality.

Step 6. **Check** the answer in the problem and make sure it makes sense.

Step 7. **Answer** the question with a complete sentence.

REVIEW EXERCISES

3.1 3.1 Using a Problem Solving Strategy

Approach Word Problems with a Positive Attitude

In the following exercises, reflect on your approach to word problems.

339. How has your attitude towards solving word problems changed as a result of working through this chapter? Explain.

340. Did the problem-solving strategy help you solve word problems in this chapter? Explain.

Use a Problem-Solving Strategy for Word Problems

In the following exercises, solve using the problem-solving strategy for word problems. Remember to write a complete sentence to answer each question.

341. Three-fourths of the people at a concert are children. If there are 87 children, what is the total number of people at the concert?

342. There are nine saxophone players in the band. The number of saxophone players is one less than twice the number of tuba players. Find the number of tuba players.

Solve Number Problems

In the following exercises, solve each number word problem.

343. The sum of a number and three is forty-one. Find the number.

344. Twice the difference of a number and ten is fifty-four. Find the number.

345. One number is nine less than another. Their sum is negative twenty-seven. Find the numbers.

346. One number is eleven more than another. If their sum is increased by seventeen, the result is 90. Find the numbers.

347. One number is two more than four times another. Their sum is -13. Find the numbers.

348. The sum of two consecutive integers is -135. Find the numbers.

349. Find three consecutive integers whose sum is -141.

350. Find three consecutive even integers whose sum is 234.

351. Find three consecutive odd integers whose sum is 51.

352. Koji has $5,502 in his savings account. This is $30 less than six times the amount in his checking account. How much money does Koji have in his checking account?

3.2 3.2 Solve Percent Applications

Translate and Solve Basic Percent Equations

In the following exercises, translate and solve.

353. What number is 67% of 250?

354. 300% of 82 is what number?

355. 12.5% of what number is 20?

356. 72 is 30% of what number?

357. What percent of 125 is 150?

358. 127.5 is what percent of 850?

Solve Percent Applications

In the following exercises, solve.

359. The bill for Dino's lunch was $19.45. He wanted to leave 20% of the total bill as a tip. How much should the tip be?

360. Reza was very sick and lost 15% of his original weight. He lost 27 pounds. What was his original weight?

361. Dolores bought a crib on sale for $350. The sale price was 40% of the original price. What was the original price of the crib?

362. Jaden earns $2,680 per month. He pays $938 a month for rent. What percent of his monthly pay goes to rent?

Find Percent Increase and Percent Decrease

In the following exercises, solve.

363. Angel's got a raise in his annual salary from $55,400 to $56,785. Find the percent increase.

364. Rowena's monthly gasoline bill dropped from $83.75 last month to $56.95 this month. Find the percent decrease.

Solve Simple Interest Applications

In the following exercises, solve.

365. Winston deposited $3,294 in a bank account with interest rate 2.6%. How much interest was earned in 5 years?

366. Moira borrowed $4,500 from her grandfather to pay for her first year of college. Three years later, she repaid the $4,500 plus $243 interest. What was the rate of interest?

367. Jaime's refrigerator loan statement said he would pay $1,026 in interest for a 4-year loan at 13.5%. How much did Jaime borrow to buy the refrigerator?

368. In 12 years, a bond that paid 6.35% interest earned $7,620 interest. What was the principal of the bond?

Solve Applications with Discount or Mark-up

In the following exercises, find the sale price.

369. The original price of a handbag was $84. Carole bought it on sale for $21 off.

370. Marian wants to buy a coffee table that costs $495. Next week the coffee table will be on sale for $149 off.

In the following exercises, find ⓐ the amount of discount and ⓑ the sale price.

371. Emmett bought a pair of shoes on sale at 40% off from an original price of $138.

372. Anastasia bought a dress on sale at 75% off from an original price of $280.

In the following exercises, find ⓐ the amount of discount and ⓑ the discount rate. (Round to the nearest tenth of a percent, if needed.)

373. Zack bought a printer for his office that was on sale for $380. The original price of the printer was $450.

374. Lacey bought a pair of boots on sale for $95. The original price of the boots was $200.

In the following exercises, find ⓐ the amount of the mark-up and ⓑ the list price.

375. Nga and Lauren bought a chest at a flea market for $50. They re-finished it and then added a 350% mark-up.

376. Carly bought bottled water for $0.24 per bottle at the discount store. She added a 75% mark-up before selling them at the football game.

3.3 3.3 Solve Mixture Applications

Solve Coin Word Problems

In the following exercises, solve each coin word problem.

377. Francie has $4.35 in dimes and quarters. The number of dimes is five more than the number of quarters. How many of each coin does she have?

378. Scott has $0.39 in pennies and nickels. The number of pennies is eight times the number of nickels. How many of each coin does he have?

379. Paulette has $140 in $5 and $10 bills. The number of $10 bills is one less than twice the number of $5 bills. How many of each does she have?

380. Lenny has $3.69 in pennies, dimes, and quarters. The number of pennies is three more than the number of dimes. The number of quarters is twice the number of dimes. How many of each coin does he have?

Solve Ticket and Stamp Word Problems

In the following exercises, solve each ticket or stamp word problem.

381. A church luncheon made $842. Adult tickets cost $10 each and children's tickets cost $6 each. The number of children was 12 more than twice the number of adults. How many of each ticket were sold?

382. Tickets for a basketball game cost $2 for students and $5 for adults. The number of students was three less than 10 times the number of adults. The total amount of money from ticket sales was $619. How many of each ticket were sold?

383. 125 tickets were sold for the jazz band concert for a total of $1,022. Student tickets cost $6 each and general admission tickets cost $10 each. How many of each kind of ticket were sold?

384. One afternoon the water park sold 525 tickets for a total of $13,545. Child tickets cost $19 each and adult tickets cost $40 each. How many of each kind of ticket were sold?

385. Ana spent $4.06 buying stamps. The number of $0.41 stamps she bought was five more than the number of $0.26 stamps. How many of each did she buy?

386. Yumi spent $34.15 buying stamps. The number of $0.56 stamps she bought was 10 less than four times the number of $0.41 stamps. How many of each did she buy?

Solve Mixture Word Problems

In the following exercises, solve each mixture word problem.

387. Marquese is making 10 pounds of trail mix from raisins and nuts. Raisins cost $3.45 per pound and nuts cost $7.95 per pound. How many pounds of raisins and how many pounds of nuts should Marquese use for the trail mix to cost him $6.96 per pound?

388. Amber wants to put tiles on the backsplash of her kitchen counters. She will need 36 square feet of tile. She will use basic tiles that cost $8 per square foot and decorator tiles that cost $20 per square foot. How many square feet of each tile should she use so that the overall cost of the backsplash will be $10 per square foot?

389. Shawn has $15,000 to invest. She will put some of it into a fund that pays 4.5% annual interest and the rest in a certificate of deposit that pays 1.8% annual interest. How much should she invest in each account if she wants to earn 4.05% annual interest on the total amount?

390. Enrique borrowed $23,500 to buy a car. He pays his uncle 2% interest on the $4,500 he borrowed from him, and he pays the bank 11.5% interest on the rest. What average interest rate does he pay on the total $23,500? (Round your answer to the nearest tenth of a percent.)

3.4 3.4 Solve Geometry Applications: Triangles, Rectangles and the Pythagorean Theorem

Solve Applications Using Triangle Properties

In the following exercises, solve using triangle properties.

391. The measures of two angles of a triangle are 22 and 85 degrees. Find the measure of the third angle.

392. The playground at a shopping mall is a triangle with perimeter 48 feet. The lengths of two sides are 19 feet and 14 feet. How long is the third side?

393. A triangular road sign has base 30 inches and height 40 inches. What is its area?

394. What is the height of a triangle with area 67.5 square meters and base 9 meters?

395. One angle of a triangle is 30° more than the smallest angle. The largest angle is the sum of the other angles. Find the measures of all three angles.

396. One angle of a right triangle measures 58°. What is the measure of the other angles of the triangle?

397. The measure of the smallest angle in a right triangle is 45° less than the measure of the next larger angle. Find the measures of all three angles.

398. The perimeter of a triangle is 97 feet. One side of the triangle is eleven feet more than the smallest side. The third side is six feet more than twice the smallest side. Find the lengths of all sides.

Use the Pythagorean Theorem

In the following exercises, use the Pythagorean Theorem to find the length of the hypotenuse.

399.

24, 10

400.
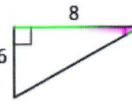
8, 6

In the following exercises, use the Pythagorean Theorem to find the length of the missing side. Round to the nearest tenth, if necessary.

401.
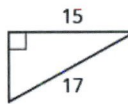
15, 17

402.
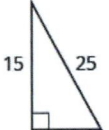
15, 25

403.

7, 4

404.
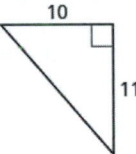
10, 11

In the following exercises, solve. Approximate to the nearest tenth, if necessary.

405. Sergio needs to attach a wire to hold the antenna to the roof of his house, as shown in the figure. The antenna is 8 feet tall and Sergio has 10 feet of wire. How far from the base of the antenna can he attach the wire?

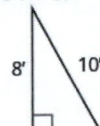
8′, 10′

406. Seong is building shelving in his garage. The shelves are 36 inches wide and 15 inches tall. He wants to put a diagonal brace across the back to stabilize the shelves, as shown. How long should the brace be?

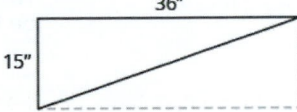
36″, 15″

Solve Applications Using Rectangle Properties

In the following exercises, solve using rectangle properties.

407. The length of a rectangle is 36 feet and the width is 19 feet. Find the ⓐ perimeter ⓑ area.

408. A sidewalk in front of Kathy's house is in the shape of a rectangle four feet wide by 45 feet long. Find the ⓐ perimeter ⓑ area.

409. The area of a rectangle is 2356 square meters. The length is 38 meters. What is the width?

410. The width of a rectangle is 45 centimeters. The area is 2,700 square centimeters. What is the length?

411. The length of a rectangle is 12 cm more than the width. The perimeter is 74 cm. Find the length and the width.

412. The width of a rectangle is three more than twice the length. The perimeter is 96 inches. Find the length and the width.

3.5 3.5 Solve Uniform Motion Applications

Solve Uniform Motion Applications

In the following exercises, solve.

413. When Gabe drives from Sacramento to Redding it takes him 2.2 hours. It takes Elsa 2 hours to drive the same distance. Elsa's speed is seven miles per hour faster than Gabe's speed. Find Gabe's speed and Elsa's speed.

414. Louellen and Tracy met at a restaurant on the road between Chicago and Nashville. Louellen had left Chicago and drove 3.2 hours towards Nashville. Tracy had left Nashville and drove 4 hours towards Chicago, at a speed one mile per hour faster than Louellen's speed. The distance between Chicago and Nashville is 472 miles. Find Louellen's speed and Tracy's speed.

415. Two busses leave Amarillo at the same time. The Albuquerque bus heads west on the I-40 at a speed of 72 miles per hour, and the Oklahoma City bus heads east on the I-40 at a speed of 78 miles per hour. How many hours will it take them to be 375 miles apart?

416. Kyle rowed his boat upstream for 50 minutes. It took him 30 minutes to row back downstream. His speed going upstream is two miles per hour slower than his speed going downstream. Find Kyle's upstream and downstream speeds.

417. At 6:30, Devon left her house and rode her bike on the flat road until 7:30. Then she started riding uphill and rode until 8:00. She rode a total of 15 miles. Her speed on the flat road was three miles per hour faster than her speed going uphill. Find Devon's speed on the flat road and riding uphill.

418. Anthony drove from New York City to Baltimore, a distance of 192 miles. He left at 3:45 and had heavy traffic until 5:30. Traffic was light for the rest of the drive, and he arrived at 7:30. His speed in light traffic was four miles per hour more than twice his speed in heavy traffic. Find Anthony's driving speed in heavy traffic and light traffic.

3.6 3.6 Solve Applications with Linear Inequalities

Solve Applications with Linear Inequalities

In the following exercises, solve.

419. Julianne has a weekly food budget of $231 for her family. If she plans to budget the same amount for each of the seven days of the week, what is the maximum amount she can spend on food each day?

420. Rogelio paints watercolors. He got a $100 gift card to the art supply store and wants to use it to buy $12'' \times 16''$ canvases. Each canvas costs $10.99. What is the maximum number of canvases he can buy with his gift card?

421. Briana has been offered a sales job in another city. The offer was for $42,500 plus 8% of her total sales. In order to make it worth the move, Briana needs to have an annual salary of at least $66,500. What would her total sales need to be for her to move?

422. Renee's car costs her $195 per month plus $0.09 per mile. How many miles can Renee drive so that her monthly car expenses are no more than $250?

423. Costa is an accountant. During tax season, he charges $125 to do a simple tax return. His expenses for buying software, renting an office, and advertising are $6,000. How many tax returns must he do if he wants to make a profit of at least $8,000?

424. Jenna is planning a 5-day resort vacation with three of her friends. It will cost her $279 for airfare, $300 for food and entertainment, and $65 per day for her share of the hotel. She has $550 saved towards her vacation and can earn $25 per hour as an assistant in her uncle's photography studio. How many hours must she work in order to have enough money for her vacation?

PRACTICE TEST

425. Four-fifths of the people on a hike are children. If there are 12 children, what is the total number of people on the hike?

426. One number is three more than twice another. Their sum is −63. Find the numbers.

427. The sum of two consecutive odd integers is −96. Find the numbers.

428. Marla's breakfast was 525 calories. This was 35% of her total calories for the day. How many calories did she have that day?

429. Humberto's hourly pay increased from $16.25 to $17.55. Find the percent increase.

430. Melinda deposited $5,985 in a bank account with an interest rate of 1.9%. How much interest was earned in 2 years?

431. Dotty bought a freezer on sale for $486.50. The original price of the freezer was $695. Find ⓐ the amount of discount and ⓑ the discount rate.

432. Bonita has $2.95 in dimes and quarters in her pocket. If she has five more dimes than quarters, how many of each coin does she have?

433. At a concert, $1,600 in tickets were sold. Adult tickets were $9 each and children's tickets were $4 each. If the number of adult tickets was 30 less than twice the number of children's tickets, how many of each kind were sold?

434. Kim is making eight gallons of punch from fruit juice and soda. The fruit juice costs $6.04 per gallon and the soda costs $4.28 per gallon. How much fruit juice and how much soda should she use so that the punch costs $5.71 per gallon?

435. The measure of one angle of a triangle is twice the measure of the smallest angle. The measure of the third angle is 14 more than the measure of the smallest angle. Find the measures of all three angles.

436. What is the height of a triangle with area 277.2 square inches and base 44 inches?

In the following exercises, use the Pythagorean Theorem to find the length of the missing side. Round to the nearest tenth, if necessary.

437.

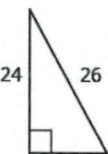

438.

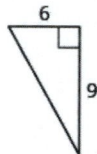

439. A baseball diamond is really a square with sides of 90 feet. How far is it from home plate to second base, as shown?

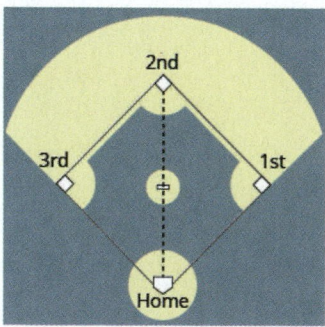

440. The length of a rectangle is two feet more than five times the width. The perimeter is 40 feet. Find the dimensions of the rectangle.

441. Two planes leave Dallas at the same time. One heads east at a speed of 428 miles per hour. The other plane heads west at a speed of 382 miles per hour. How many hours will it take them to be 2,025 miles apart?

442. Leon drove from his house in Cincinnati to his sister's house in Cleveland, a distance of 252 miles. It took him $4\frac{1}{2}$ hours. For the first half hour he had heavy traffic, and the rest of the time his speed was five miles per hour less than twice his speed in heavy traffic. What was his speed in heavy traffic?

443. Chloe has a budget of $800 for costumes for the 18 members of her musical theater group. What is the maximum she can spend for each costume?

444. Frank found a rental car deal online for $49 per week plus $0.24 per mile. How many miles could he drive if he wants the total cost for one week to be no more than $150?

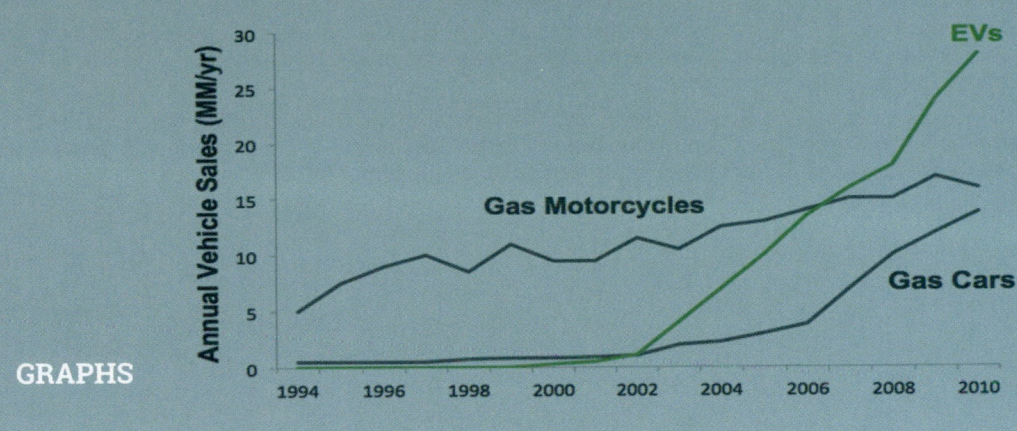

4   GRAPHS

Figure 4.1 This graph illustrates the annual vehicle sales of gas motorcycles, gas cars, and electric vehicles from 1994 to 2010. It is a line graph with x- and y-axes, one of the most common types of graphs. (credit: Steve Jurvetson, Flickr)

Chapter Outline

4.1 Use the Rectangular Coordinate System

4.2 Graph Linear Equations in Two Variables

4.3 Graph with Intercepts

4.4 Understand Slope of a Line

4.5 Use the Slope–Intercept Form of an Equation of a Line

4.6 Find the Equation of a Line

4.7 Graphs of Linear Inequalities

✎ Introduction

Graphs are found in all areas of our lives—from commercials showing you which cell phone carrier provides the best coverage, to bank statements and news articles, to the boardroom of major corporations. In this chapter, we will study the rectangular coordinate system, which is the basis for most consumer graphs. We will look at linear graphs, slopes of lines, equations of lines, and linear inequalities.

4.1   Use the Rectangular Coordinate System

Learning Objectives

By the end of this section, you will be able to:

> Plot points in a rectangular coordinate system
> Verify solutions to an equation in two variables
> Complete a table of solutions to a linear equation
> Find solutions to a linear equation in two variables

Be Prepared!

Before you get started, take this readiness quiz.

1. Evaluate $x + 3$ when $x = -1$.
   If you missed this problem, review **Example 1.54**.

2. Evaluate $2x - 5y$ when $x = 3$ and $y = -2$.
   If you missed this problem, review **Example 1.55**.

3. Solve for y: $40 - 4y = 20$.
   If you missed this problem, review **Example 2.27**.

Plot Points on a Rectangular Coordinate System

Just like maps use a grid system to identify locations, a grid system is used in algebra to show a relationship between two variables in a **rectangular coordinate system**. The rectangular coordinate system is also called the *xy*-plane or the 'coordinate plane'.

The horizontal number line is called the *x-axis*. The vertical number line is called the *y-axis*. The *x*-axis and the *y*-axis together form the rectangular coordinate system. These axes divide a plane into four regions, called **quadrants**. The quadrants are identified by Roman numerals, beginning on the upper right and proceeding counterclockwise. See **Figure 4.2**.

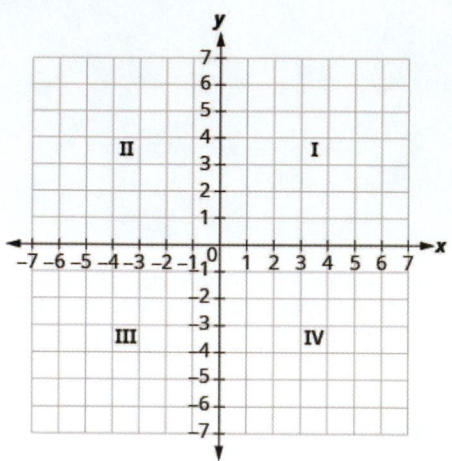

Figure 4.2 'Quadrant' has the root 'quad,' which means 'four.'

In the rectangular coordinate system, every point is represented by an *ordered pair*. The first number in the ordered pair is the **x-coordinate** of the point, and the second number is the **y-coordinate** of the point.

Ordered Pair

An **ordered pair**, (x, y), gives the coordinates of a point in a rectangular coordinate system.

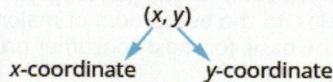

The first number is the *x*-coordinate.

The second number is the *y*-coordinate.

The phrase 'ordered pair' means the order is important. What is the ordered pair of the point where the axes cross? At that point both coordinates are zero, so its ordered pair is $(0, 0)$. The point $(0, 0)$ has a special name. It is called the origin.

The Origin

The point $(0, 0)$ is called the **origin**. It is the point where the *x*-axis and *y*-axis intersect.

We use the coordinates to locate a point on the *xy*-plane. Let's plot the point $(1, 3)$ as an example. First, locate 1 on the *x*-axis and lightly sketch a vertical line through $x = 1$. Then, locate 3 on the *y*-axis and sketch a horizontal line through $y = 3$. Now, find the point where these two lines meet—that is the point with coordinates $(1, 3)$.

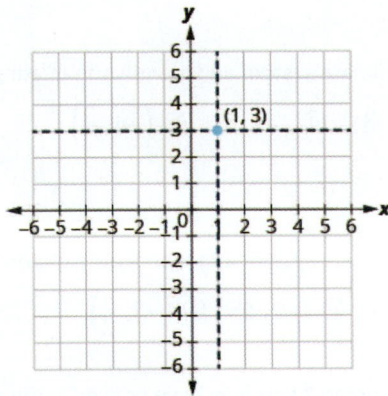

Notice that the vertical line through $x = 1$ and the horizontal line through $y = 3$ are not part of the graph. We just used them to help us locate the point $(1, 3)$.

EXAMPLE 4.1

Plot each point in the rectangular coordinate system and identify the quadrant in which the point is located:

ⓐ $(-5, 4)$ ⓑ $(-3, -4)$ ⓒ $(2, -3)$ ⓓ $(-2, 3)$ ⓔ $\left(3, \dfrac{5}{2}\right)$.

✓ **Solution**

The first number of the coordinate pair is the x-coordinate, and the second number is the y-coordinate.

ⓐ Since $x = -5$, the point is to the left of the y-axis. Also, since $y = 4$, the point is above the x-axis. The point $(-5, 4)$ is in Quadrant II.

ⓑ Since $x = -3$, the point is to the left of the y-axis. Also, since $y = -4$, the point is below the x-axis. The point $(-3, -4)$ is in Quadrant III.

ⓒ Since $x = 2$, the point is to the right of the y-axis. Since $y = -3$, the point is below the x-axis. The point $(2, -3)$ is in Quadrant IV.

ⓓ Since $x = -2$, the point is to the left of the y-axis. Since $y = 3$, the point is above the x-axis. The point $(-2, 3)$ is in Quadrant II.

ⓔ Since $x = 3$, the point is to the right of the y-axis. Since $y = \dfrac{5}{2}$, the point is above the x-axis. (It may be helpful to write $\dfrac{5}{2}$ as a mixed number or decimal.) The point $\left(3, \dfrac{5}{2}\right)$ is in Quadrant I.

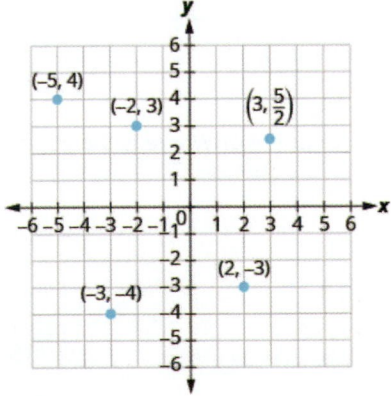

Figure 4.3

> **TRY IT : : 4.1**

Plot each point in a rectangular coordinate system and identify the quadrant in which the point is located:

ⓐ $(-2, 1)$ ⓑ $(-3, -1)$ ⓒ $(4, -4)$ ⓓ $(-4, 4)$ ⓔ $\left(-4, \frac{3}{2}\right)$.

> **TRY IT : : 4.2**

Plot each point in a rectangular coordinate system and identify the quadrant in which the point is located:

ⓐ $(-4, 1)$ ⓑ $(-2, 3)$ ⓒ $(2, -5)$ ⓓ $(-2, 5)$ ⓔ $\left(-3, \frac{5}{2}\right)$.

How do the signs affect the location of the points? You may have noticed some patterns as you graphed the points in the previous example.

For the point in **Figure 4.3** in Quadrant IV, what do you notice about the signs of the coordinates? What about the signs of the coordinates of points in the third quadrant? The second quadrant? The first quadrant?

Can you tell just by looking at the coordinates in which quadrant the point $(-2, 5)$ is located? In which quadrant is $(2, -5)$ located?

Quadrants

We can summarize sign patterns of the quadrants in this way.

Quadrant I	Quadrant II	Quadrant III	Quadrant IV
(x, y)	(x, y)	(x, y)	(x, y)
$(+, +)$	$(-, +)$	$(-, -)$	$(+, -)$

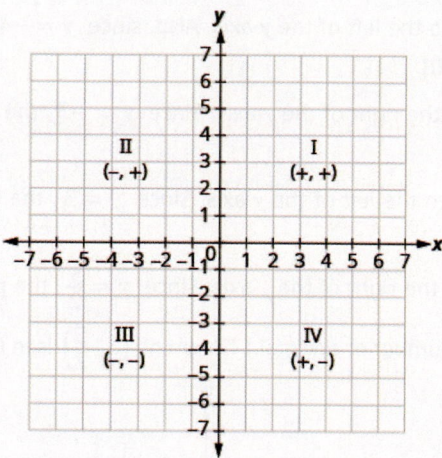

What if one coordinate is zero as shown in **Figure 4.4**? Where is the point $(0, 4)$ located? Where is the point $(-2, 0)$ located?

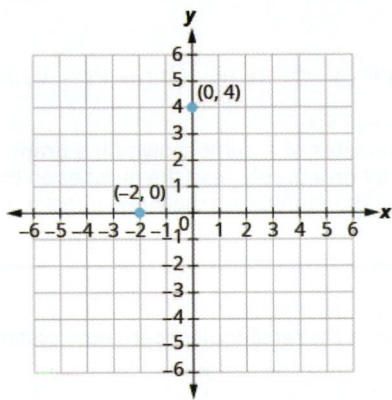

Figure 4.4

The point $(0, 4)$ is on the *y*-axis and the point $(-2, 0)$ is on the *x*-axis.

Points on the Axes

Points with a *y*-coordinate equal to 0 are on the *x*-axis, and have coordinates $(a, 0)$.

Points with an *x*-coordinate equal to 0 are on the *y*-axis, and have coordinates $(0, b)$.

EXAMPLE 4.2

Plot each point:

ⓐ $(0, 5)$ ⓑ $(4, 0)$ ⓒ $(-3, 0)$ ⓓ $(0, 0)$ ⓔ $(0, -1)$.

✓ **Solution**

ⓐ Since $x = 0$, the point whose coordinates are $(0, 5)$ is on the *y*-axis.

ⓑ Since $y = 0$, the point whose coordinates are $(4, 0)$ is on the *x*-axis.

ⓒ Since $y = 0$, the point whose coordinates are $(-3, 0)$ is on the *x*-axis.

ⓓ Since $x = 0$ and $y = 0$, the point whose coordinates are $(0, 0)$ is the origin.

ⓔ Since $x = 0$, the point whose coordinates are $(0, -1)$ is on the *y*-axis.

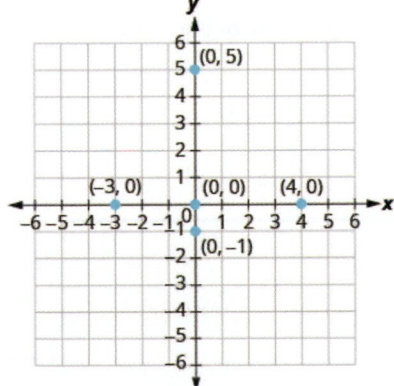

> **TRY IT : : 4.3**   Plot each point:

ⓐ $(4, 0)$ ⓑ $(-2, 0)$ ⓒ $(0, 0)$ ⓓ $(0, 2)$ ⓔ $(0, -3)$.

> **TRY IT : : 4.4**   Plot each point:

ⓐ (−5, 0)   ⓑ (3, 0)   ⓒ (0, 0)   ⓓ (0, −1)   ⓔ (0, 4).

In algebra, being able to identify the coordinates of a point shown on a graph is just as important as being able to plot points. To identify the *x*-coordinate of a point on a graph, read the number on the *x*-axis directly above or below the point. To identify the *y*-coordinate of a point, read the number on the *y*-axis directly to the left or right of the point. Remember, when you write the ordered pair use the correct order, (*x*, *y*).

EXAMPLE 4.3

Name the ordered pair of each point shown in the rectangular coordinate system.

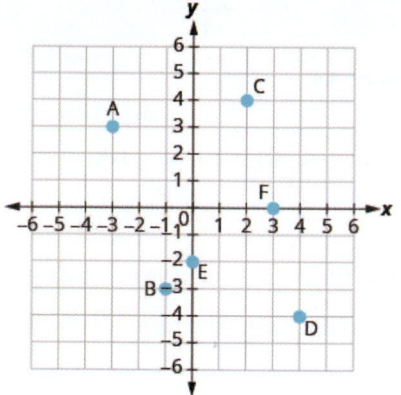

✓ **Solution**

Point A is above −3 on the *x*-axis, so the *x*-coordinate of the point is −3.

- The point is to the left of 3 on the *y*-axis, so the *y*-coordinate of the point is 3.
- The coordinates of the point are (−3, 3).

Point B is below −1 on the *x*-axis, so the *x*-coordinate of the point is −1.

- The point is to the left of −3 on the *y*-axis, so the *y*-coordinate of the point is −3.
- The coordinates of the point are (−1, −3).

Point C is above 2 on the *x*-axis, so the *x*-coordinate of the point is 2.

- The point is to the right of 4 on the *y*-axis, so the *y*-coordinate of the point is 4.
- The coordinates of the point are (2, 4).

Point D is below 4 on the *x*-axis, so the *x*-coordinate of the point is 4.

- The point is to the right of −4 on the *y*-axis, so the *y*-coordinate of the point is −4.
- The coordinates of the point are (4, −4).

Point E is on the *y*-axis at $y = -2$. The coordinates of point E are (0, −2).

Point F is on the *x*-axis at $x = 3$. The coordinates of point F are (3, 0).

> **TRY IT ::** 4.5   Name the ordered pair of each point shown in the rectangular coordinate system.

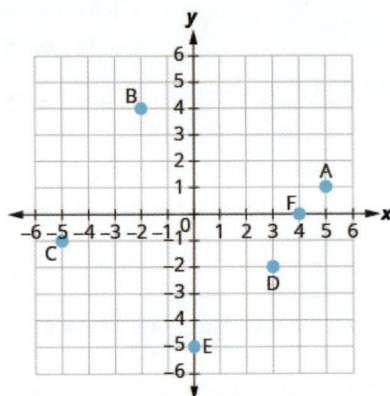

> **TRY IT ::** 4.6   Name the ordered pair of each point shown in the rectangular coordinate system.

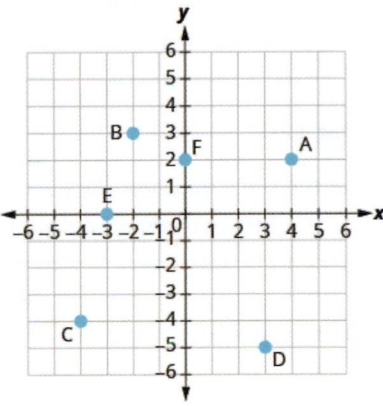

Verify Solutions to an Equation in Two Variables

Up to now, all the equations you have solved were equations with just one variable. In almost every case, when you solved the equation you got exactly one solution. The process of solving an equation ended with a statement like $x = 4$. (Then, you checked the solution by substituting back into the equation.)

Here's an example of an equation in one variable, and its one solution.

$$
\begin{aligned}
3x + 5 &= 17 \\
3x &= 12 \\
x &= 4
\end{aligned}
$$

But equations can have more than one variable. Equations with two variables may be of the form $Ax + By = C$. Equations of this form are called **linear equations in two variables**.

Linear Equation

An equation of the form $Ax + By = C$, where A and B are not both zero, is called a **linear equation in two variables**.

Notice the word *line* in **linear**. Here is an example of a linear equation in two variables, x and y.

$$Ax + By = C$$

$$x + 4y = 8$$

$$A = 1, \; B = 4, \; C = 8$$

The equation $y = -3x + 5$ is also a linear equation. But it does not appear to be in the form $Ax + By = C$. We can use the Addition Property of Equality and rewrite it in $Ax + By = C$ form.

Add to both sides.

Simplify.

Use the Commutative Property to put it in
$Ax + By = C$ form.

$$y = -3x + 5$$
$$y + 3x = -3x + 5 + 3x$$
$$y + 3x = 5$$
$$3x + y = 5$$

By rewriting $y = -3x + 5$ as $3x + y = 5$, we can easily see that it is a linear equation in two variables because it is of the form $Ax + By = C$. When an equation is in the form $Ax + By = C$, we say it is in *standard form*.

Standard Form of Linear Equation

A linear equation is in standard form when it is written $Ax + By = C$.

Most people prefer to have A, B, and C be integers and $A \geq 0$ when writing a linear equation in standard form, although it is not strictly necessary.

Linear equations have infinitely many solutions. For every number that is substituted for x there is a corresponding y value. This pair of values is a *solution* to the linear equation and is represented by the ordered pair (x, y). When we substitute these values of x and y into the equation, the result is a true statement, because the value on the left side is equal to the value on the right side.

Solution of a Linear Equation in Two Variables

An **ordered pair** (x, y) is a **solution** of the linear equation $Ax + By = C$, if the equation is a true statement when the x- and y-values of the ordered pair are substituted into the equation.

EXAMPLE 4.4

Determine which ordered pairs are solutions to the equation $x + 4y = 8$.

ⓐ $(0, 2)$ ⓑ $(2, -4)$ ⓒ $(-4, 3)$

⊘ Solution

Substitute the x- and y-values from each ordered pair into the equation and determine if the result is a true statement.

(a)	(b)	(c)
$(0, 2)$	$(2, -4)$	$(-4, 3)$
$x = 0, y = 2$	$x = 2, y = -4$	$x = -4, y = 3$
$x + 4y = 8$	$x + 4y = 8$	$x + 4y = 8$
$0 + 4 \cdot 2 \overset{?}{=} 8$	$2 + 4(-4) \overset{?}{=} 8$	$-4 + 4 \cdot 3 \overset{?}{=} 8$
$0 + 8 \overset{?}{=} 8$	$2 + (-16) \overset{?}{=} 8$	$-4 + 12 \overset{?}{=} 8$
$8 = 8 ✓$	$-14 \neq 8$	$8 = 8 ✓$
$(0, 2)$ is a solution.	$(2, -4)$ is not a solution.	$(-4, 3)$ is a solution.

> **TRY IT :: 4.7**   Which of the following ordered pairs are solutions to $2x + 3y = 6$?
>
> ⓐ $(3, 0)$ ⓑ $(2, 0)$ ⓒ $(6, -2)$

> **TRY IT :: 4.8**   Which of the following ordered pairs are solutions to the equation $4x - y = 8$?
>
> ⓐ $(0, 8)$ ⓑ $(2, 0)$ ⓒ $(1, -4)$

EXAMPLE 4.5

Which of the following ordered pairs are solutions to the equation $y = 5x - 1$?

ⓐ $(0, -1)$ ⓑ $(1, 4)$ ⓒ $(-2, -7)$

✓ Solution

Substitute the x- and y-values from each ordered pair into the equation and determine if it results in a true statement.

(a)	(b)	(c)
$(0, -1)$	$(1, 4)$	$(-2, -7)$
$x = 0, y = -1$	$x = 1, y = 4$	$x = -2, y = -7$
$y = 5x - 1$	$y = 5x - 1$	$y = 5x - 1$
$-1 \overset{?}{=} 5(0) - 1$	$4 \overset{?}{=} 5(1) - 1$	$-7 \overset{?}{=} 5(-2) - 1$
$-1 \overset{?}{=} 0 - 1$	$4 \overset{?}{=} 5 - 1$	$-7 \overset{?}{=} -10 - 1$
$-1 = -1 \checkmark$	$4 = 4 \checkmark$	$-7 \neq -11$
$(0, -1)$ is a solution.	$(1, 4)$ is a solution.	$(-2, -7)$ is not a solution.

> **TRY IT :: 4.9**   Which of the following ordered pairs are solutions to the equation $y = 4x - 3$?
>
> ⓐ $(0, 3)$ ⓑ $(1, 1)$ ⓒ $(-1, -1)$

> **TRY IT :: 4.10**   Which of the following ordered pairs are solutions to the equation $y = -2x + 6$?
>
> ⓐ $(0, 6)$ ⓑ $(1, 4)$ ⓒ $(-2, -2)$

Complete a Table of Solutions to a Linear Equation in Two Variables

In the examples above, we substituted the x- and y-values of a given ordered pair to determine whether or not it was a solution to a linear equation. But how do you find the ordered pairs if they are not given? It's easier than you might think—you can just pick a value for x and then solve the equation for y. Or, pick a value for y and then solve for x.

We'll start by looking at the solutions to the equation $y = 5x - 1$ that we found in **Example 4.5**. We can summarize this information in a table of solutions, as shown in **Table 4.1**.

$y = 5x - 1$		
x	y	(x, y)
0	-1	$(0, -1)$
1	4	$(1, 4)$

Table 4.1

To find a third solution, we'll let $x = 2$ and solve for y.

	$y = 5x - 1$
Substitute $x = 2$.	$y = 5(2) - 1$
Multiply.	$y = 10 - 1$
Simplify.	$y = 9$

The ordered pair $(2, 9)$ is a solution to $y = 5x - 1$. We will add it to **Table 4.2**.

$y = 5x - 1$		
x	y	(x, y)
0	-1	$(0, -1)$
1	4	$(1, 4)$
2	9	$(2, 9)$

Table 4.2

We can find more solutions to the equation by substituting in any value of x or any value of y and solving the resulting equation to get another ordered pair that is a solution. There are infinitely many solutions of this equation.

EXAMPLE 4.6

Complete Table 4.3 to find three solutions to the equation $y = 4x - 2$.

$y = 4x - 2$		
x	y	(x, y)
0		
−1		
2		

Table 4.3

⊘ **Solution**

Substitute $x = 0$, $x = -1$, and $x = 2$ into $y = 4x - 2$.

$x = 0$	$x = -1$	$x = 2$
$y = 4x - 2$	$y = 4x - 2$	$y = 4x - 2$
$y = 4 \cdot 0 - 2$	$y = 4(-1) - 2$	$y = 4 \cdot 2 - 2$
$y = 0 - 2$	$y = -4 - 2$	$y = 8 - 2$
$y = -2$	$y = -6$	$y = 6$
$(0, -2)$	$(-1, -6)$	$(2, 6)$

The results are summarized in Table 4.4.

$y = 4x - 2$		
x	y	(x, y)
0	−2	$(0, -2)$
−1	−6	$(-1, -6)$
2	6	$(2, 6)$

Table 4.4

> **TRY IT :: 4.11**   Complete the table to find three solutions to this equation: $y = 3x - 1$.

$y = 3x - 1$		
x	y	(x, y)
0		
−1		
2		

> **TRY IT : : 4.12**   Complete the table to find three solutions to this equation: $y = 6x + 1$.

$y = 6x + 1$		
x	y	(x, y)
0		
1		
-2		

EXAMPLE 4.7

Complete Table 4.5 to find three solutions to the equation $5x - 4y = 20$.

$5x - 4y = 20$		
x	y	(x, y)
0		
	0	
	5	

Table 4.5

⊘ **Solution**

Substitute the given value into the equation $5x - 4y = 20$ and solve for the other variable. Then, fill in the values in the table.

$x = 0$	$y = 0$	$y = 5$
$5x - 4y = 20$	$5x - 4y = 20$	$5x - 4y = 20$
$5 \cdot 0 - 4y = 20$	$5x - 4 \cdot 0 = 20$	$5x - 4 \cdot 5 = 20$
$0 - 4y = 20$	$5x - 0 = 20$	$5x - 20 = 20$
$-4y = 20$	$5x = 20$	$5x = 40$
$y = -5$	$x = 4$	$x = 8$
$(0, -5)$	$(4, 0)$	$(8, 5)$

The results are summarized in Table 4.6.

$5x - 4y = 20$		
x	y	(x, y)
0	-5	$(0, -5)$
4	0	$(4, 0)$
8	5	$(8, 5)$

Table 4.6

> **TRY IT :: 4.13**   Complete the table to find three solutions to this equation: $2x - 5y = 20$.

$2x - 5y = 20$		
x	y	(x, y)
0		
	0	
−5		

> **TRY IT :: 4.14**   Complete the table to find three solutions to this equation: $3x - 4y = 12$.

$3x - 4y = 12$		
x	y	(x, y)
0		
	0	
−4		

Find Solutions to a Linear Equation

To find a solution to a linear equation, you really can pick *any* number you want to substitute into the equation for x or y. But since you'll need to use that number to solve for the other variable it's a good idea to choose a number that's easy to work with.

When the equation is in *y*-form, with the *y* by itself on one side of the equation, it is usually easier to choose values of x and then solve for y.

EXAMPLE 4.8

Find three solutions to the equation $y = -3x + 2$.

⊘ **Solution**

We can substitute any value we want for x or any value for y. Since the equation is in *y*-form, it will be easier to substitute in values of x. Let's pick $x = 0$, $x = 1$, and $x = -1$.

	$x = 0$	$x = 1$	$x = -1$
	$y = -3x + 2$	$y = -3x + 2$	$y = -3x + 2$
Substitute the value into the equation.	$y = -3 \cdot 0 + 2$	$y = -3 \cdot 1 + 2$	$y = -3(-1) + 2$
Simplify.	$y = 0 + 2$	$y = -3 + 2$	$y = 3 + 2$
Simplify.	$y = 2$	$y = -1$	$y = 5$
Write the ordered pair.	$(0, 2)$	$(1, -1)$	$(-1, 5)$
Check.			

$y = -3x + 2$	$y = -3x + 2$	$y = -3x + 2$
$2 \stackrel{?}{=} -3 \cdot 0 + 2$	$-1 \stackrel{?}{=} -3 \cdot 1 + 2$	$5 \stackrel{?}{=} -3(-1) + 2$
$2 \stackrel{?}{=} 0 + 2$	$-1 \stackrel{?}{=} -3 + 2$	$5 \stackrel{?}{=} 3 + 2$
$2 = 2 ✓$	$-1 = -1 ✓$	$5 = 5 ✓$

So, $(0, 2)$, $(1, -1)$ and $(-1, 5)$ are all solutions to $y = -3x + 2$. We show them in Table 4.8.

$y = -3x + 2$		
x	y	(x, y)
0	2	$(0, 2)$
1	-1	$(1, -1)$
-1	5	$(-1, 5)$

Table 4.8

> **TRY IT :: 4.15**   Find three solutions to this equation: $y = -2x + 3$.

> **TRY IT :: 4.16**   Find three solutions to this equation: $y = -4x + 1$.

We have seen how using zero as one value of x makes finding the value of y easy. When an equation is in standard form, with both the x and y on the same side of the equation, it is usually easier to first find one solution when $x = 0$ find a second solution when $y = 0$, and then find a third solution.

EXAMPLE 4.9

Find three solutions to the equation $3x + 2y = 6$.

⊘ **Solution**

We can substitute any value we want for x or any value for y. Since the equation is in standard form, let's pick first $x = 0$, then $y = 0$, and then find a third point.

	$x = 0$	$y = 0$	$x = 1$
	$3x + 2y = 6$	$3x + 2y = 6$	$3x + 2y = 6$
Substitute the value into the equation.	$3(0) + 2y = 6$	$3x + 2(0) = 6$	$3(1) + 2y = 6$
Simplify.	$0 + 2y = 6$	$3x + 0 = 6$	$3 + 2y = 6$
Solve.	$2y = 6$	$3x = 6$	$2y = 3$
	$y = 3$	$x = 2$	$y = \frac{3}{2}$
Write the ordered pair.	$(0, 3)$	$(2, 0)$	$\left(1, \frac{3}{2}\right)$

Check.

$3x + 2y = 6$	$3x + 2y = 6$	$3x + 2y = 6$
$3 \cdot 0 + 2 \cdot 3 \stackrel{?}{=} 6$	$3 \cdot 2 + 2 \cdot 0 \stackrel{?}{=} 6$	$3 \cdot 1 + 2 \cdot \frac{3}{2} \stackrel{?}{=} 6$
$0 + 6 \stackrel{?}{=} 6$	$6 + 0 \stackrel{?}{=} 6$	$3 + 3 \stackrel{?}{=} 6$
$6 = 6 \checkmark$	$6 = 6 \checkmark$	$6 = 6 \checkmark$

So $(0, 3)$, $(2, 0)$, and $\left(1, \frac{3}{2}\right)$ are all solutions to the equation $3x + 2y = 6$. We can list these three solutions in Table 4.10.

$3x + 2y = 6$		
x	y	(x, y)
0	3	$(0, 3)$
2	0	$(2, 0)$
1	$\frac{3}{2}$	$\left(1, \frac{3}{2}\right)$

Table 4.10

> **TRY IT : : 4.17** Find three solutions to the equation $2x + 3y = 6$.

> **TRY IT : : 4.18** Find three solutions to the equation $4x + 2y = 8$.

4.1 EXERCISES

Practice Makes Perfect

Plot Points in a Rectangular Coordinate System

In the following exercises, plot each point in a rectangular coordinate system and identify the quadrant in which the point is located.

1.
ⓐ $(-4, 2)$
ⓑ $(-1, -2)$
ⓒ $(3, -5)$
ⓓ $(-3, 5)$
ⓔ $\left(\frac{5}{3}, 2\right)$

2.
ⓐ $(-2, -3)$
ⓑ $(3, -3)$
ⓒ $(-4, 1)$
ⓓ $(4, -1)$
ⓔ $\left(\frac{3}{2}, 1\right)$

3.
ⓐ $(3, -1)$
ⓑ $(-3, 1)$
ⓒ $(-2, 2)$
ⓓ $(-4, -3)$
ⓔ $\left(1, \frac{14}{5}\right)$

4.
ⓐ $(-1, 1)$
ⓑ $(-2, -1)$
ⓒ $(2, 1)$
ⓓ $(1, -4)$
ⓔ $\left(3, \frac{7}{2}\right)$

In the following exercises, plot each point in a rectangular coordinate system.

5.
ⓐ $(-2, 0)$
ⓑ $(-3, 0)$
ⓒ $(0, 0)$
ⓓ $(0, 4)$
ⓔ $(0, 2)$

6.
ⓐ $(0, 1)$
ⓑ $(0, -4)$
ⓒ $(-1, 0)$
ⓓ $(0, 0)$
ⓔ $(5, 0)$

7.
ⓐ $(0, 0)$
ⓑ $(0, -3)$
ⓒ $(-4, 0)$
ⓓ $(1, 0)$
ⓔ $(0, -2)$

8.
ⓐ $(-3, 0)$
ⓑ $(0, 5)$
ⓒ $(0, -2)$
ⓓ $(2, 0)$
ⓔ $(0, 0)$

In the following exercises, name the ordered pair of each point shown in the rectangular coordinate system.

9.

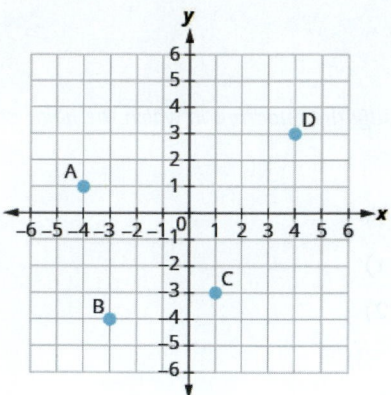

10.

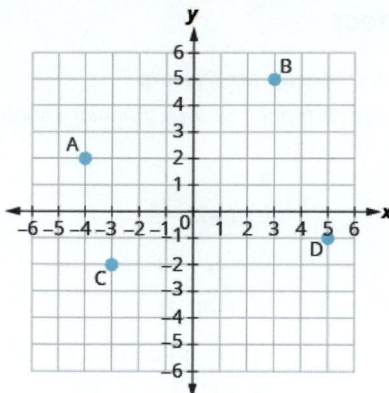

11.

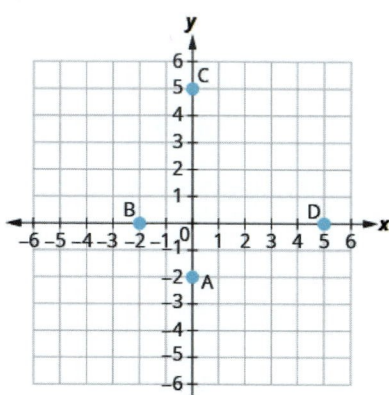

12.

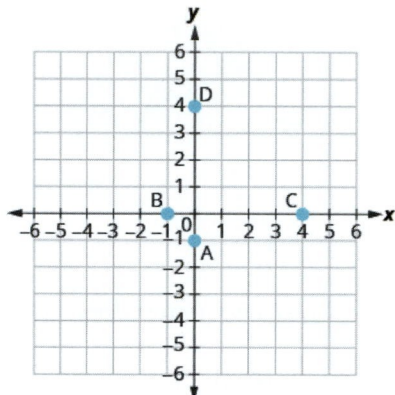

Verify Solutions to an Equation in Two Variables

In the following exercises, which ordered pairs are solutions to the given equations?

13. $2x + y = 6$

ⓐ $(1, 4)$

ⓑ $(3, 0)$

ⓒ $(2, 3)$

14. $x + 3y = 9$

ⓐ $(0, 3)$

ⓑ $(6, 1)$

ⓒ $(-3, -3)$

15. $4x - 2y = 8$

ⓐ $(3, 2)$

ⓑ $(1, 4)$

ⓒ $(0, -4)$

16. $3x - 2y = 12$

ⓐ $(4, 0)$

ⓑ $(2, -3)$

ⓒ $(1, 6)$

17. $y = 4x + 3$

ⓐ $(4, 3)$

ⓑ $(-1, -1)$

ⓒ $\left(\frac{1}{2}, 5\right)$

18. $y = 2x - 5$

ⓐ $(0, -5)$

ⓑ $(2, 1)$

ⓒ $\left(\frac{1}{2}, -4\right)$

19. $y = \frac{1}{2}x - 1$

ⓐ $(2, 0)$

ⓑ $(-6, -4)$

ⓒ $(-4, -1)$

20. $y = \frac{1}{3}x + 1$

ⓐ $(-3, 0)$

ⓑ $(9, 4)$

ⓒ $(-6, -1)$

Complete a Table of Solutions to a Linear Equation

In the following exercises, complete the table to find solutions to each linear equation.

21. $y = 2x - 4$

x	y	(x, y)
0		
2		
-1		

22. $y = 3x - 1$

x	y	(x, y)
0		
2		
-1		

23. $y = -x + 5$

x	y	(x, y)
0		
3		
-2		

24. $y = -x + 2$

x	y	(x, y)
0		
3		
-2		

25. $y = \frac{1}{3}x + 1$

x	y	(x, y)
0		
3		
6		

26. $y = \frac{1}{2}x + 4$

x	y	(x, y)
0		
2		
4		

27. $y = -\frac{3}{2}x - 2$

x	y	(x, y)
0		
2		
-2		

28. $y = -\frac{2}{3}x - 1$

x	y	(x, y)
0		
3		
-3		

29. $x + 3y = 6$

x	y	(x, y)
0		
3		
	0	

30. $x + 2y = 8$

x	y	(x, y)
0		
4		
	0	

31. $2x - 5y = 10$

x	y	(x, y)
0		
10		
	0	

32. $3x - 4y = 12$

x	y	(x, y)
0		
8		
	0	

Find Solutions to a Linear Equation

In the following exercises, find three solutions to each linear equation.

33. $y = 5x - 8$

34. $y = 3x - 9$

35. $y = -4x + 5$

36. $y = -2x + 7$

37. $x + y = 8$

38. $x + y = 6$

39. $x + y = -2$

40. $x + y = -1$

41. $3x + y = 5$

42. $2x + y = 3$

43. $4x - y = 8$

44. $5x - y = 10$

45. $2x + 4y = 8$

46. $3x + 2y = 6$

47. $5x - 2y = 10$

48. $4x - 3y = 12$

Everyday Math

49. Weight of a baby. Mackenzie recorded her baby's weight every two months. The baby's age, in months, and weight, in pounds, are listed in the table below, and shown as an ordered pair in the third column.

ⓐ Plot the points on a coordinate plane.

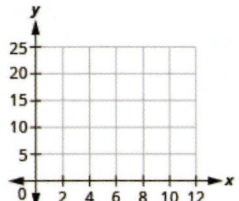

ⓑ Why is only Quadrant I needed?

Age x	Weight y	(x, y)
0	7	(0, 7)
2	11	(2, 11)
4	15	(4, 15)
6	16	(6, 16)
8	19	(8, 19)
10	20	(10, 20)
12	21	(12, 21)

50. Weight of a child. Latresha recorded her son's height and weight every year. His height, in inches, and weight, in pounds, are listed in the table below, and shown as an ordered pair in the third column.

ⓐ Plot the points on a coordinate plane.

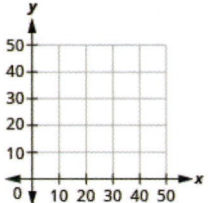

ⓑ Why is only Quadrant I needed?

Height x	Weight y	(x, y)
28	22	(28, 22)
31	27	(31, 27)
33	33	(33, 33)
37	35	(37, 35)
40	41	(40, 41)
42	45	(42, 45)

Writing Exercises

51. Explain in words how you plot the point $(4, -2)$ in a rectangular coordinate system.

52. How do you determine if an ordered pair is a solution to a given equation?

53. Is the point $(-3, 0)$ on the *x*-axis or *y*-axis? How do you know?

54. Is the point $(0, 8)$ on the *x*-axis or *y*-axis? How do you know?

Self Check

ⓐ *After completing the exercises, use this checklist to evaluate your mastery of the objectives of this section.*

I can...	Confidently	With some help	No-I don't get it!
plot points in a rectangular coordinate system.			
identify points on a graph.			
verify solutions to an equation in two variables.			
complete a table of solutions to a linear equation.			
find solutions to a linear equation.			

ⓑ *If most of your checks were:*

...confidently. *Congratulations! You have achieved the objectives in this section. Reflect on the study skills you used so that you can continue to use them. What did you do to become confident of your ability to do these things? Be specific.*

...with some help. *This must be addressed quickly because topics you do not master become potholes in your road to success. In math every topic builds upon previous work. It is important to make sure you have a strong foundation before you move on. Who can you ask for help? Your fellow classmates and instructor are good resources. Is there a place on campus where math tutors are available? Can your study skills be improved?*

...no, I don't get it. *This is a warning sign and you must not ignore it. You should get help right away or you will quickly be overwhelmed. See your instructor as soon as you can to discuss your situation. Together you can come up with a plan to get you the help you need.*

4.2 Graph Linear Equations in Two Variables

Learning Objectives

By the end of this section, you will be able to:

> Recognize the relationship between the solutions of an equation and its graph.
> Graph a linear equation by plotting points.
> Graph vertical and horizontal lines.

Be Prepared!

Before you get started, take this readiness quiz.

1. Evaluate $3x + 2$ when $x = -1$.
   If you missed this problem, review **Example 1.57**.
2. Solve $3x + 2y = 12$ for y in general.
   If you missed this problem, review **Example 2.63**.

Recognize the Relationship Between the Solutions of an Equation and its Graph

In the previous section, we found several solutions to the equation $3x + 2y = 6$. They are listed in **Table 4.11**. So, the ordered pairs $(0, 3)$, $(2, 0)$, and $\left(1, \frac{3}{2}\right)$ are some solutions to the equation $3x + 2y = 6$. We can plot these solutions in the rectangular coordinate system as shown in **Figure 4.5**.

$3x + 2y = 6$		
x	y	(x, y)
0	3	$(0, 3)$
2	0	$(2, 0)$
1	$\frac{3}{2}$	$\left(1, \frac{3}{2}\right)$

Table 4.11

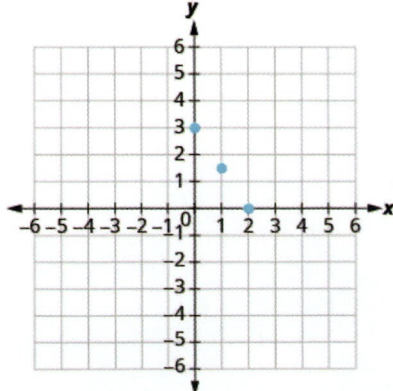

Figure 4.5

Notice how the points line up perfectly? We connect the points with a line to get the graph of the equation $3x + 2y = 6$. See **Figure 4.6**. Notice the arrows on the ends of each side of the line. These arrows indicate the line continues.

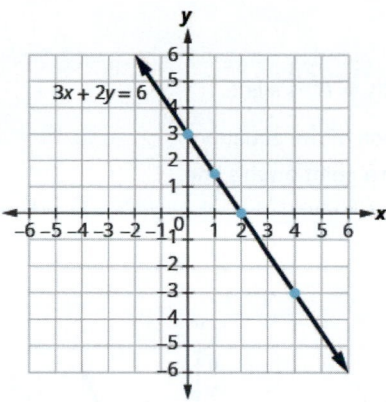

Figure 4.6

Every point on the line is a solution of the equation. Also, every solution of this equation is a point on this line. Points *not* on the line are not solutions.

Notice that the point whose coordinates are $(-2, 6)$ is on the line shown in **Figure 4.7**. If you substitute $x = -2$ and $y = 6$ into the equation, you find that it is a solution to the equation.

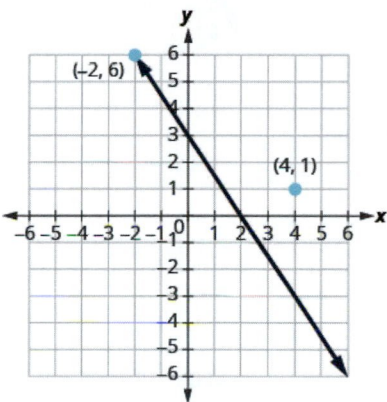

Figure 4.7

Test $(-2, 6)$

$$3x + 2y = 6$$

$$3(-2) + 2(6) = 6$$

$$-6 + 12 = 6$$

$$6 = 6 \checkmark$$

So the point $(-2, 6)$ is a solution to the equation $3x + 2y = 6$. (The phrase "the point whose coordinates are $(-2, 6)$" is often shortened to "the point $(-2, 6)$.")

What about $(4, 1)$?

$$3x + 2y = 6$$

$$3 \cdot 4 + 2 \cdot 1 = 6$$

$$12 + 2 \stackrel{?}{=} 6$$

$$14 \neq 6$$

So $(4, 1)$ is not a solution to the equation $3x + 2y = 6$. Therefore, the point $(4, 1)$ is not on the line. See **Figure 4.6**. This is an example of the saying, "A picture is worth a thousand words." The line shows you *all* the solutions to the equation. Every point on the line is a solution of the equation. And, every solution of this equation is on this line. This line is called the *graph* of the equation $3x + 2y = 6$.

Graph of a Linear Equation

The **graph of a linear equation** $Ax + By = C$ is a line.

- Every point on the line is a solution of the equation.
- Every solution of this equation is a point on this line.

EXAMPLE 4.10

The graph of $y = 2x - 3$ is shown.

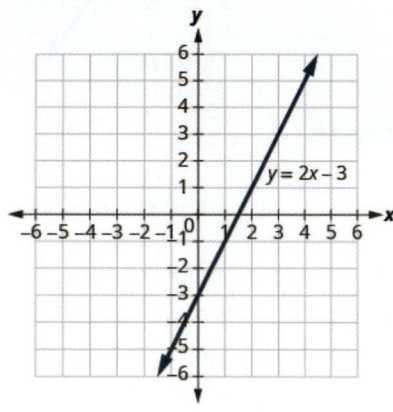

For each ordered pair, decide:

ⓐ Is the ordered pair a solution to the equation?

ⓑ Is the point on the line?

A $(0, -3)$ B $(3, 3)$ C $(2, -3)$ D $(-1, -5)$

⊘ **Solution**

Substitute the *x*- and *y*- values into the equation to check if the ordered pair is a solution to the equation.

ⓐ

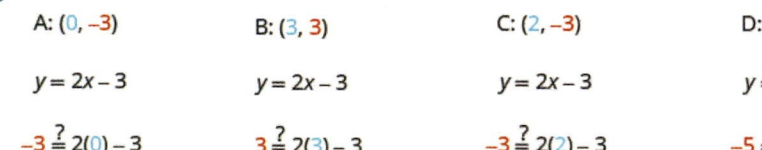

A: $(0, -3)$	B: $(3, 3)$	C: $(2, -3)$	D: $(-1, -5)$
$y = 2x - 3$	$y = 2x - 3$	$y = 2x - 3$	$y = 2x - 3$
$-3 \stackrel{?}{=} 2(0) - 3$	$3 \stackrel{?}{=} 2(3) - 3$	$-3 \stackrel{?}{=} 2(2) - 3$	$-5 \stackrel{?}{=} 2(-1) - 3$
$-3 = -3 \checkmark$	$3 = 3 \checkmark$	$-3 \neq 1$	$-5 = -5 \checkmark$
$(0, -3)$ is a solution.	$(3, 3)$ is a solution.	$(2, -3)$ is not a solution.	$(-1, -5)$ is a solution.

ⓑ Plot the points A $(0, 3)$, B $(3, 3)$, C $(2, -3)$, and D $(-1, -5)$.

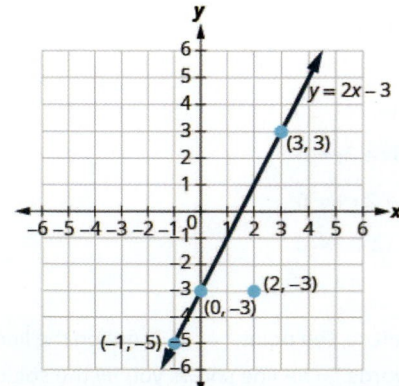

The points $(0, 3)$, $(3, 3)$, and $(-1, -5)$ are on the line $y = 2x - 3$, and the point $(2, -3)$ is not on the line.

The points that are solutions to $y = 2x - 3$ are on the line, but the point that is not a solution is not on the line.

> **TRY IT :: 4.19**   Use the graph of $y = 3x - 1$ to decide whether each ordered pair is:

- a solution to the equation.
- on the line.

ⓐ $(0, -1)$ ⓑ $(2, 5)$

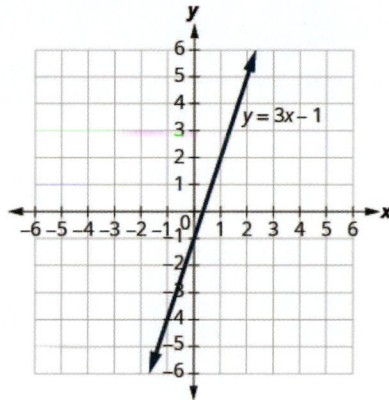

> **TRY IT :: 4.20**   Use graph of $y = 3x - 1$ to decide whether each ordered pair is:

- a solution to the equation
- on the line

ⓐ $(3, -1)$ ⓑ $(-1, -4)$

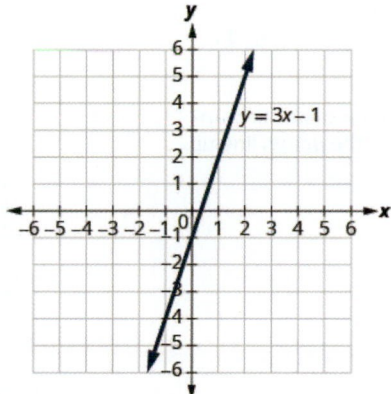

Graph a Linear Equation by Plotting Points

There are several methods that can be used to graph a linear equation. The method we used to graph $3x + 2y = 6$ is called plotting points, or the Point–Plotting Method.

EXAMPLE 4.11   HOW TO GRAPH AN EQUATION BY PLOTTING POINTS

Graph the equation $y = 2x + 1$ by plotting points.

⊘ **Solution**

Step 1. Find three points whose coordinates are solutions to the equation.	You can choose any values for x or y. In this case, since y is isolated on the left side of the equation, it is easier to choose values for x.	$y = 2x + 1$ $x = 0$ $y = 2x + 1$ $y = 2 \cdot 0 + 1$ $y = 0 + 1$ $y = 1$ $x = 1$ $y = 2x + 1$ $y = 2 \cdot 1 + 1$ $y = 2 + 1$ $y = 3$ $x = -2$ $y = 2x + 1$ $y = 2(-2) + 1$ $y = -4 + 1$ $y = -3$
Organize the solutions in a table.	Put the three solutions in a table.	<table><tr><td colspan="3">$y = 2x + 1$</td></tr><tr><td>x</td><td>y</td><td>(x, y)</td></tr><tr><td>0</td><td>1</td><td>(0, 1)</td></tr><tr><td>1</td><td>3</td><td>(1, 3)</td></tr><tr><td>−2</td><td>−3</td><td>(−2, −3)</td></tr></table>
Step 2. Plot the points in a rectangular coordinate system. Check that the points line up. If they do not, carefully check your work!	Plot: (0, 1), (1, 3), (−2, −3). Do the points line up? Yes, the points line up.	

| Step 3. Draw the line through the three points. Extend the line to fill the grid and put arrows on both ends of the line. | This line is the graph of $y = 2x + 1$. | 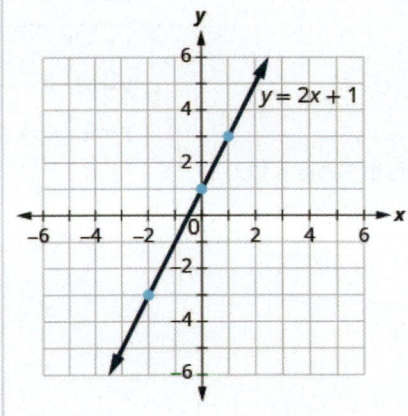 |

 TRY IT :: 4.21   Graph the equation by plotting points: $y = 2x - 3$.

 TRY IT :: 4.22   Graph the equation by plotting points: $y = -2x + 4$.

The steps to take when graphing a linear equation by plotting points are summarized below.

HOW TO :: GRAPH A LINEAR EQUATION BY PLOTTING POINTS.

Step 1.   Find three points whose coordinates are solutions to the equation. Organize them in a table.

Step 2.   Plot the points in a rectangular coordinate system. Check that the points line up. If they do not, carefully check your work.

Step 3.   Draw the line through the three points. Extend the line to fill the grid and put arrows on both ends of the line.

It is true that it only takes two points to determine a line, but it is a good habit to use three points. If you only plot two points and one of them is incorrect, you can still draw a line but it will not represent the solutions to the equation. It will be the wrong line.

If you use three points, and one is incorrect, the points will not line up. This tells you something is wrong and you need to check your work. Look at the difference between part (a) and part (b) in **Figure 4.8**.

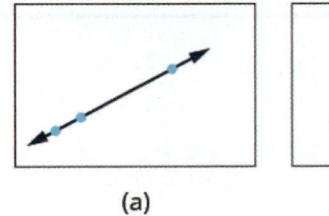

(a)   (b)

Figure 4.8

Let's do another example. This time, we'll show the last two steps all on one grid.

EXAMPLE 4.12

Graph the equation $y = -3x$.

 Solution

Find three points that are solutions to the equation. Here, again, it's easier to choose values for x. Do you see why?

$$x = 0 \qquad\qquad x = 1 \qquad\qquad x = -2$$

$$y = -3x \qquad\qquad y = -3x \qquad\qquad y = -3x$$

$$y = -3 \cdot 0 \qquad\qquad y = -3 \cdot 1 \qquad\qquad y = -3(-2)$$

$$y = 0 \qquad\qquad y = -3 \qquad\qquad y = 6$$

We list the points in **Table 4.12**.

$y = -3x$		
x	y	(x, y)
0	0	$(0, 0)$
1	-3	$(1, -3)$
-2	6	$(-2, 6)$

Table 4.12

Plot the points, check that they line up, and draw the line.

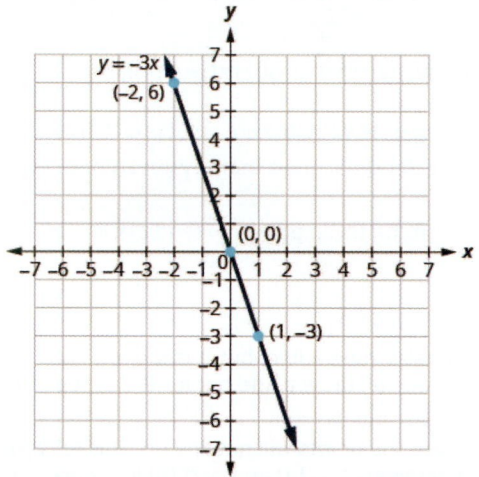

> **TRY IT :: 4.23**   Graph the equation by plotting points: $y = -4x$.

> **TRY IT :: 4.24**   Graph the equation by plotting points: $y = x$.

When an equation includes a fraction as the coefficient of x, we can still substitute any numbers for x. But the math is easier if we make 'good' choices for the values of x. This way we will avoid fraction answers, which are hard to graph precisely.

EXAMPLE 4.13

Graph the equation $y = \frac{1}{2}x + 3$.

⊘ **Solution**

Find three points that are solutions to the equation. Since this equation has the fraction $\frac{1}{2}$ as a coefficient of x, we will choose values of x carefully. We will use zero as one choice and multiples of 2 for the other choices. Why are multiples of

2 a good choice for values of x?

$$x = 0 \qquad\qquad x = 2 \qquad\qquad x = 4$$

$$y = \tfrac{1}{2}x + 3 \qquad y = \tfrac{1}{2}x + 3 \qquad y = \tfrac{1}{2}x + 3$$

$$y = \tfrac{1}{2}(0) + 3 \qquad y = \tfrac{1}{2}(2) + 3 \qquad y = \tfrac{1}{2}(4) + 3$$

$$y = 0 + 3 \qquad\quad y = 1 + 3 \qquad\quad y = 2 + 3$$

$$y = 3 \qquad\qquad y = 4 \qquad\qquad y = 5$$

The points are shown in **Table 4.13**.

$y = \tfrac{1}{2}x + 3$		
x	y	(x, y)
0	3	(0, 3)
2	4	(2, 4)
4	5	(4, 5)

Table 4.13

Plot the points, check that they line up, and draw the line.

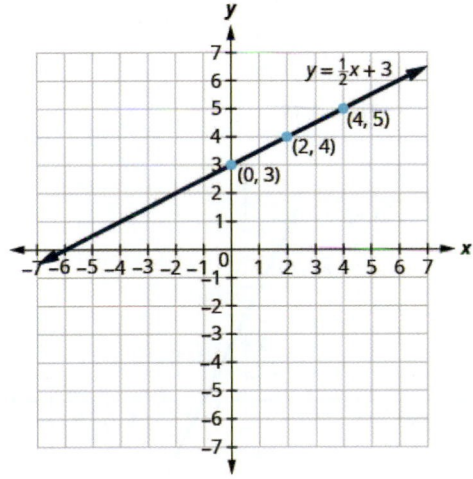

> **TRY IT : : 4.25**   Graph the equation $y = \tfrac{1}{3}x - 1$.

> **TRY IT : : 4.26**   Graph the equation $y = \tfrac{1}{4}x + 2$.

So far, all the equations we graphed had y given in terms of x. Now we'll graph an equation with x and y on the same side. Let's see what happens in the equation $2x + y = 3$. If $y = 0$ what is the value of x?

$$y = 0$$
$$2x + y = 3$$
$$2x + 0 = 3$$
$$2x = 3$$
$$x = \frac{3}{2}$$
$$\left(\frac{3}{2}, 0\right)$$

This point has a fraction for the x- coordinate and, while we could graph this point, it is hard to be precise graphing fractions. Remember in the example $y = \frac{1}{2}x + 3$, we carefully chose values for x so as not to graph fractions at all. If we solve the equation $2x + y = 3$ for y, it will be easier to find three solutions to the equation.

$$2x + y = 3$$
$$y = -2x + 3$$

The solutions for $x = 0$, $x = 1$, and $x = -1$ are shown in the Table 4.14. The graph is shown in Figure 4.9.

$2x + y = 3$		
x	y	(x, y)
0	3	$(0, 3)$
1	1	$(1, 1)$
-1	5	$(-1, 5)$

Table 4.14

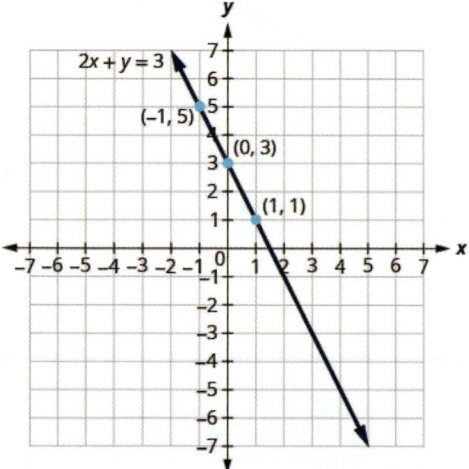

Figure 4.9

Can you locate the point $\left(\frac{3}{2}, 0\right)$, which we found by letting $y = 0$, on the line?

EXAMPLE 4.14

Graph the equation $3x + y = -1$.

✓ Solution

Find three points that are solutions to the equation. $\qquad 3x + y = -1$

First solve the equation for y. $\qquad\qquad\qquad y = -3x - 1$

We'll let x be 0, 1, and -1 to find 3 points. The ordered pairs are shown in Table 4.16. Plot the points, check that they line up, and draw the line. See Figure 4.10.

$3x + y = -1$		
x	y	(x, y)
0	-1	$(0, -1)$
1	-4	$(1, -4)$
-1	2	$(-1, 2)$

Table 4.15

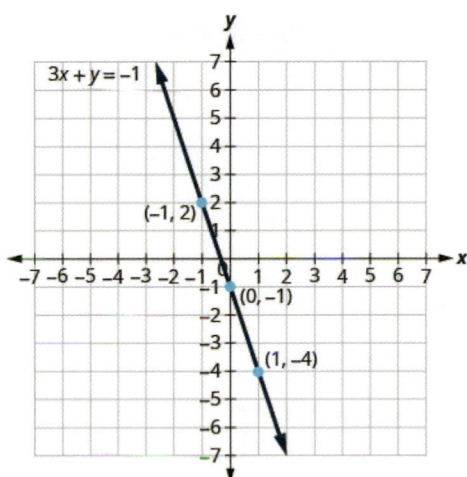

Figure 4.10

> **TRY IT : : 4.27**   Graph the equation $2x + y = 2$.

> **TRY IT : : 4.28**   Graph the equation $4x + y = -3$.

If you can choose any three points to graph a line, how will you know if your graph matches the one shown in the answers in the book? If the points where the graphs cross the x- and y-axis are the same, the graphs match!

The equation in Example 4.14 was written in standard form, with both x and y on the same side. We solved that equation for y in just one step. But for other equations in standard form it is not that easy to solve for y, so we will leave them in standard form. We can still find a first point to plot by letting $x = 0$ and solving for y. We can plot a second point by letting $y = 0$ and then solving for x. Then we will plot a third point by using some other value for x or y.

EXAMPLE 4.15

Graph the equation $2x - 3y = 6$.

⊘ Solution

Find three points that are solutions to the equation.

$$2x - 3y = 6$$

First let $x = 0$.

$$2(0) - 3y = 6$$

Solve for y.

$$-3y = 6$$
$$y = -2$$

Now let $y = 0$.

$$2x - 3(0) = 6$$

Solve for x.

$$2x = 6$$
$$x = 3$$

We need a third point. Remember, we can choose any value for x or y. We'll let $x = 6$.

$$2(6) - 3y = 6$$

Solve for y.

$$12 - 3y = 6$$
$$-3y = -6$$
$$y = 2$$

We list the ordered pairs in Table 4.17. Plot the points, check that they line up, and draw the line. See Figure 4.11.

$2x - 3y = 6$		
x	y	(x, y)
0	−2	(0, −2)
3	0	(3, 0)
6	2	(6, 2)

Table 4.16

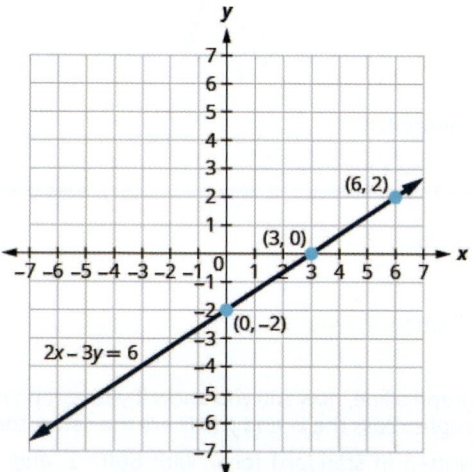

Figure 4.11

> **TRY IT : :** 4.29   Graph the equation $4x + 2y = 8$.

> **TRY IT : :** 4.30   Graph the equation $2x - 4y = 8$.

Graph Vertical and Horizontal Lines

Can we graph an equation with only one variable? Just x and no y, or just y without an x? How will we make a table of values to get the points to plot?

Let's consider the equation $x = -3$. This equation has only one variable, x. The equation says that x is *always* equal to -3, so its value does not depend on y. No matter what y is, the value of x is always -3.

So to make a table of values, write -3 in for all the x values. Then choose any values for y. Since x does not depend on y, you can choose any numbers you like. But to fit the points on our coordinate graph, we'll use 1, 2, and 3 for the y-coordinates. See Table 4.17.

$x = -3$		
x	y	(x, y)
-3	1	$(-3, 1)$
-3	2	$(-3, 2)$
-3	3	$(-3, 3)$

Table 4.17

Plot the points from Table 4.17 and connect them with a straight line. Notice in Figure 4.12 that we have graphed a *vertical line*.

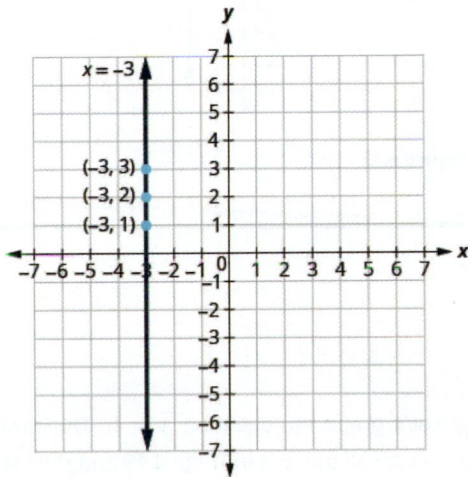

Figure 4.12

Vertical Line

A **vertical line** is the graph of an equation of the form $x = a$.

The line passes through the x-axis at $(a, 0)$.

EXAMPLE 4.16

Graph the equation $x = 2$.

⊘ **Solution**

The equation has only one variable, x, and x is always equal to 2. We create Table 4.18 where x is always 2 and then put in any values for y. The graph is a vertical line passing through the x-axis at 2. See **Figure 4.13**.

x = 2		
x	**y**	**(x, y)**
2	1	(2, 1)
2	2	(2, 2)
2	3	(2, 3)

Table 4.18

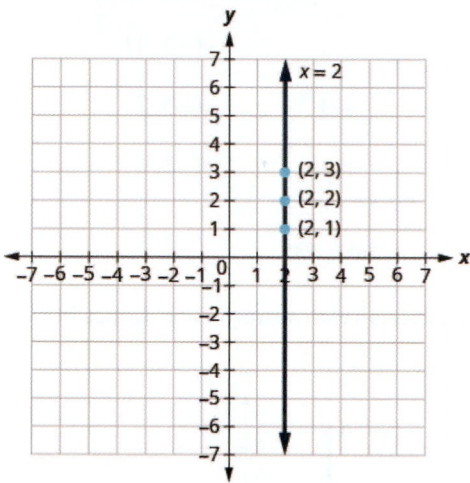

Figure 4.13

> **TRY IT :: 4.31**   Graph the equation $x = 5$.

> **TRY IT :: 4.32**   Graph the equation $x = -2$.

What if the equation has y but no x? Let's graph the equation $y = 4$. This time the y- value is a constant, so in this equation, y does not depend on x. Fill in 4 for all the y's in **Table 4.19** and then choose any values for x. We'll use 0, 2, and 4 for the x-coordinates.

y = 4		
x	**y**	**(x, y)**
0	4	(0, 4)
2	4	(2, 4)
4	4	(4, 4)

Table 4.19

The graph is a horizontal line passing through the y-axis at 4. See **Figure 4.14**.

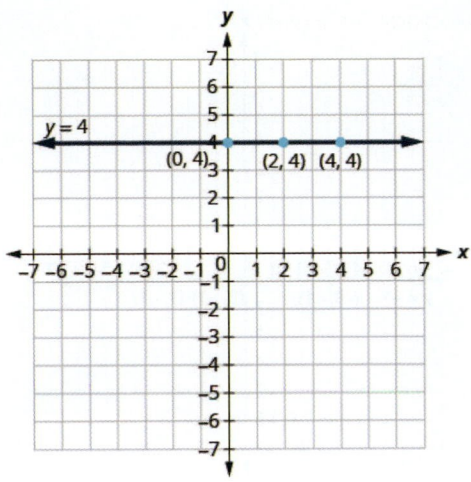

Figure 4.14

Horizontal Line

A **horizontal line** is the graph of an equation of the form $y = b$.

The line passes through the y-axis at $(0, b)$.

EXAMPLE 4.17

Graph the equation $y = -1$.

⊘ **Solution**

The equation $y = -1$ has only one variable, y. The value of y is constant. All the ordered pairs in **Table 4.20** have the same y-coordinate. The graph is a horizontal line passing through the y-axis at -1, as shown in **Figure 4.15**.

	$y = -1$	
x	y	(x, y)
0	-1	$(0, -1)$
3	-1	$(3, -1)$
-3	-1	$(-3, -1)$

Table 4.20

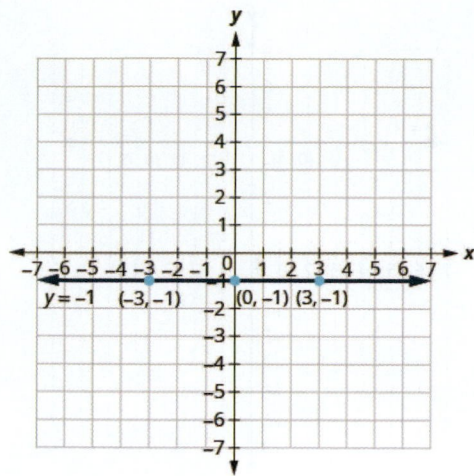

Figure 4.15

> **TRY IT : :** 4.33   Graph the equation $y = -4$.

> **TRY IT : :** 4.34   Graph the equation $y = 3$.

The equations for vertical and horizontal lines look very similar to equations like $y = 4x$. What is the difference between the equations $y = 4x$ and $y = 4$?

The equation $y = 4x$ has both x and y. The value of y depends on the value of x. The y-coordinate changes according to the value of x. The equation $y = 4$ has only one variable. The value of y is constant. The y-coordinate is always 4. It does not depend on the value of x. See Table 4.21.

$y = 4x$				$y = 4$		
x	y	(x, y)		x	y	(x, y)
0	0	$(0, 0)$		0	4	$(0, 4)$
1	4	$(1, 4)$		1	4	$(1, 4)$
2	8	$(2, 8)$		2	4	$(2, 4)$

Table 4.21

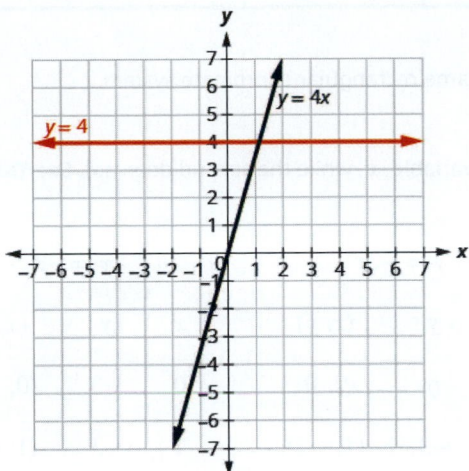

Figure 4.16

Notice, in **Figure 4.16**, the equation $y = 4x$ gives a slanted line, while $y = 4$ gives a horizontal line.

EXAMPLE 4.18

Graph $y = -3x$ and $y = -3$ in the same rectangular coordinate system.

✓ **Solution**

Notice that the first equation has the variable x, while the second does not. See Table 4.22. The two graphs are shown in Figure 4.17.

$y = -3x$				$y = -3$		
x	y	(x, y)		x	y	(x, y)
0	0	$(0, 0)$		0	-3	$(0, -3)$
1	-3	$(1, -3)$		1	-3	$(1, -3)$
2	-6	$(2, -6)$		2	-3	$(2, -3)$

Table 4.22

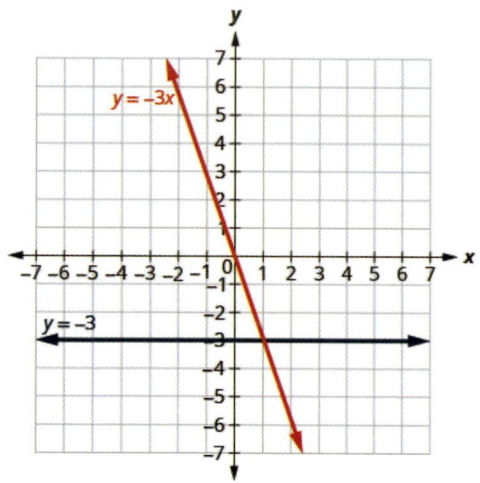

Figure 4.17

> **TRY IT :: 4.35**   Graph $y = -4x$ and $y = -4$ in the same rectangular coordinate system.

> **TRY IT :: 4.36**   Graph $y = 3$ and $y = 3x$ in the same rectangular coordinate system.

4.2 EXERCISES

Practice Makes Perfect

Recognize the Relationship Between the Solutions of an Equation and its Graph

In the following exercises, for each ordered pair, decide:

ⓐ *Is the ordered pair a solution to the equation?*   ⓑ *Is the point on the line?*

55. $y = x + 2$

ⓐ (0, 2)

ⓑ (1, 2)

ⓒ (−1, 1)

ⓓ (−3, −1)

56. $y = x − 4$

ⓐ (0, −4)

ⓑ (3, −1)

ⓒ (2, 2)

ⓓ (1, −5)

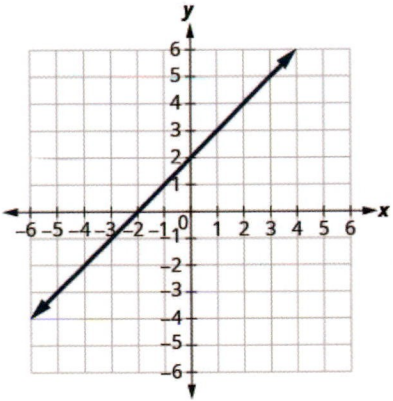

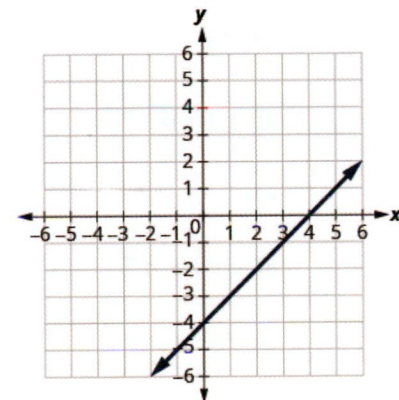

57. $y = \frac{1}{2}x − 3$

ⓐ (0, −3)

ⓑ (2, −2)

ⓒ (−2, −4)

ⓓ (4, 1)

58. $y = \frac{1}{3}x + 2$

ⓐ (0, 2)

ⓑ (3, 3)

ⓒ (−3, 2)

ⓓ (−6, 0)

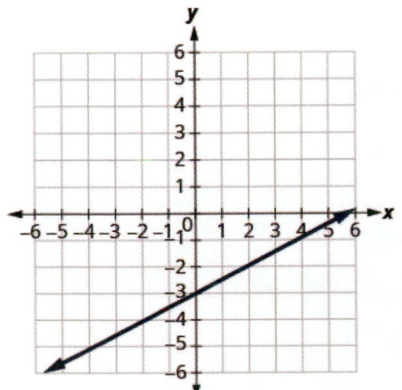

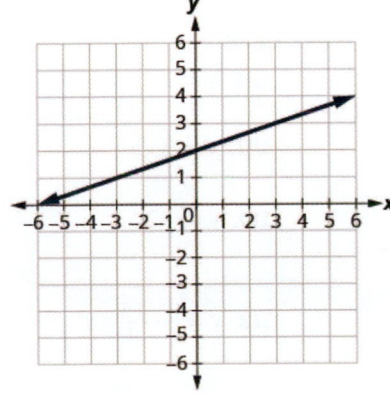

Graph a Linear Equation by Plotting Points

In the following exercises, graph by plotting points.

59. $y = 3x - 1$

60. $y = 2x + 3$

61. $y = -2x + 2$

62. $y = -3x + 1$

63. $y = x + 2$

64. $y = x - 3$

65. $y = -x - 3$

66. $y = -x - 2$

67. $y = 2x$

68. $y = 3x$

69. $y = -4x$

70. $y = -2x$

71. $y = \frac{1}{2}x + 2$

72. $y = \frac{1}{3}x - 1$

73. $y = \frac{4}{3}x - 5$

74. $y = \frac{3}{2}x - 3$

75. $y = -\frac{2}{5}x + 1$

76. $y = -\frac{4}{5}x - 1$

77. $y = -\frac{3}{2}x + 2$

78. $y = -\frac{5}{3}x + 4$

79. $x + y = 6$

80. $x + y = 4$

81. $x + y = -3$

82. $x + y = -2$

83. $x - y = 2$

84. $x - y = 1$

85. $x - y = -1$

86. $x - y = -3$

87. $3x + y = 7$

88. $5x + y = 6$

89. $2x + y = -3$

90. $4x + y = -5$

91. $\frac{1}{3}x + y = 2$

92. $\frac{1}{2}x + y = 3$

93. $\frac{2}{5}x - y = 4$

94. $\frac{3}{4}x - y = 6$

95. $2x + 3y = 12$

96. $4x + 2y = 12$

97. $3x - 4y = 12$

98. $2x - 5y = 10$

99. $x - 6y = 3$

100. $x - 4y = 2$

101. $5x + 2y = 4$

102. $3x + 5y = 5$

Graph Vertical and Horizontal Lines

In the following exercises, graph each equation.

103. $x = 4$

104. $x = 3$

105. $x = -2$

106. $x = -5$

107. $y = 3$

108. $y = 1$

109. $y = -5$

110. $y = -2$

111. $x = \frac{7}{3}$

112. $x = \frac{5}{4}$

113. $y = -\frac{15}{4}$

114. $y = -\frac{5}{3}$

In the following exercises, graph each pair of equations in the same rectangular coordinate system.

115. $y = 2x$ and $y = 2$

116. $y = 5x$ and $y = 5$

117. $y = -\frac{1}{2}x$ and $y = -\frac{1}{2}$

118. $y = -\frac{1}{3}x$ and $y = -\frac{1}{3}$

Mixed Practice

In the following exercises, graph each equation.

119. $y = 4x$

120. $y = 2x$

121. $y = -\frac{1}{2}x + 3$

122. $y = \frac{1}{4}x - 2$

123. $y = -x$

124. $y = x$

125. $x - y = 3$

126. $x + y = -5$

127. $4x + y = 2$

128. $2x + y = 6$

129. $y = -1$

130. $y = 5$

131. $2x + 6y = 12$

132. $5x + 2y = 10$

133. $x = 3$

134. $x = -4$

Everyday Math

135. Motor home cost. The Robinsons rented a motor home for one week to go on vacation. It cost them $594 plus $0.32 per mile to rent the motor home, so the linear equation $y = 594 + 0.32x$ gives the cost, y, for driving x miles. Calculate the rental cost for driving 400, 800, and 1200 miles, and then graph the line.

136. Weekly earnings. At the art gallery where he works, Salvador gets paid $200 per week plus 15% of the sales he makes, so the equation $y = 200 + 0.15x$ gives the amount, y, he earns for selling x dollars of artwork. Calculate the amount Salvador earns for selling $900, $1600, and $2000, and then graph the line.

Writing Exercises

137. Explain how you would choose three *x*- values to make a table to graph the line $y = \frac{1}{5}x - 2$.

138. What is the difference between the equations of a vertical and a horizontal line?

Self Check

ⓐ *After completing the exercises, use this checklist to evaluate your mastery of the objectives of this section.*

I can...	Confidently	With some help	No-I don't get it!
recognize the relation between the solutions of an equation and its graph.			
graph a linear equation by plotting points.			
graph vertical and horizontal lines.			

ⓑ *After reviewing this checklist, what will you do to become confident for all goals?*

4.3 Graph with Intercepts

Learning Objectives

By the end of this section, you will be able to:

> Identify the x- and y- intercepts on a graph

> Find the x- and y- intercepts from an equation of a line

> Graph a line using the intercepts

Be Prepared!

Before you get started, take this readiness quiz.

1. Solve: $3 \cdot 0 + 4y = -2$.

    If you missed this problem, review **Example 2.17**.

Identify the *x*- and *y*- Intercepts on a Graph

Every linear equation can be represented by a unique line that shows all the solutions of the equation. We have seen that when graphing a line by plotting points, you can use any three solutions to graph. This means that two people graphing the line might use different sets of three points.

At first glance, their two lines might not appear to be the same, since they would have different points labeled. But if all the work was done correctly, the lines should be exactly the same. One way to recognize that they are indeed the same line is to look at where the line crosses the *x*- axis and the *y*- axis. These points are called the *intercepts* of the line.

Intercepts of a Line

The points where a line crosses the *x*- axis and the *y*- axis are called the **intercepts of a line**.

Let's look at the graphs of the lines in **Figure 4.18**.

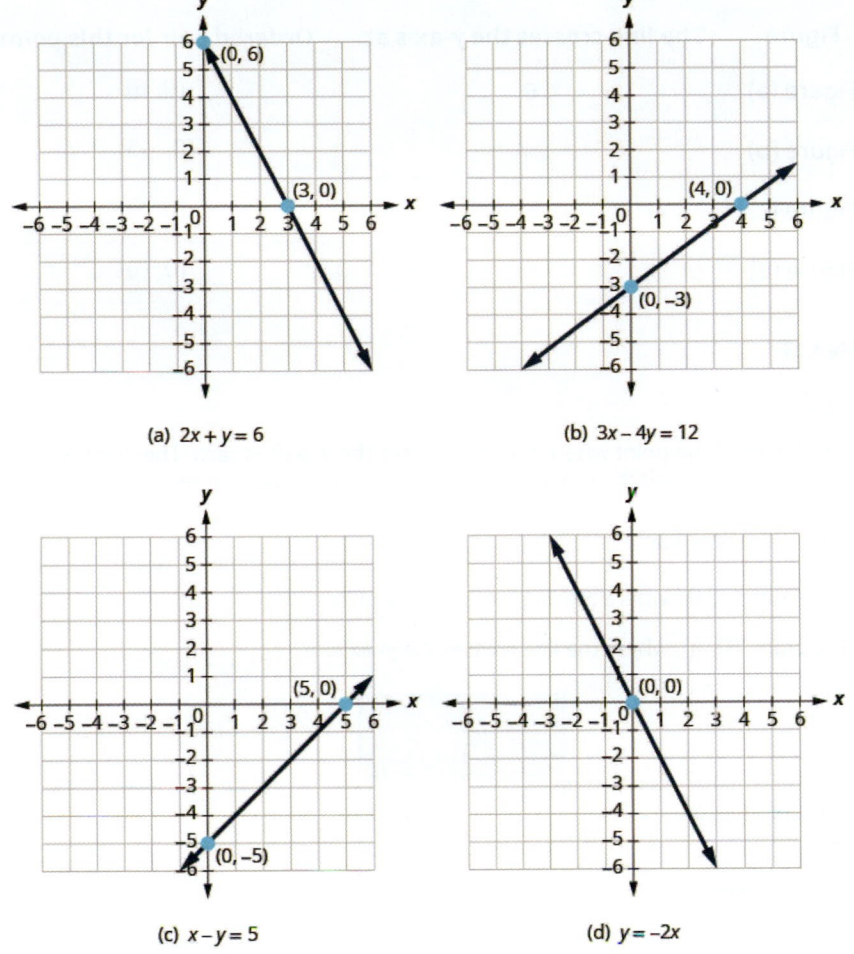

(a) $2x + y = 6$

(b) $3x - 4y = 12$

(c) $x - y = 5$

(d) $y = -2x$

Figure 4.18 Examples of graphs crossing the x-negative axis.

First, notice where each of these lines crosses the x negative axis. See **Figure 4.18**.

Figure	The line crosses the x- axis at:	Ordered pair of this point
Figure (a)	3	$(3, 0)$
Figure (b)	4	$(4, 0)$
Figure (c)	5	$(5, 0)$
Figure (d)	0	$(0, 0)$

Table 4.23

Do you see a pattern?

For each row, the y- coordinate of the point where the line crosses the x- axis is zero. The point where the line crosses the x- axis has the form $(a, 0)$ and is called the **x- intercept of a line**. The x- intercept occurs when y is zero.

Now, let's look at the points where these lines cross the y- axis. See **Table 4.24**.

Figure	The line crosses the y-axis at:	Ordered pair for this point
Figure (a)	6	(0, 6)
Figure (b)	−3	(0, −3)
Figure (c)	−5	(0, 5)
Figure (d)	0	(0, 0)

Table 4.24

What is the pattern here?

In each row, the x- coordinate of the point where the line crosses the y- axis is zero. The point where the line crosses the y- axis has the form $(0, b)$ and is called the y- *intercept* of the line. The y- intercept occurs when x is zero.

x- intercept and y- intercept of a line

The x- intercept is the point $(a, 0)$ where the line crosses the x- axis.

The y- intercept is the point $(0, b)$ where the line crosses the y- axis.

x	y
a	0
0	b

- The x-intercept occurs when y is zero.

- The y-intercept occurs when x is zero.

EXAMPLE 4.19

Find the x- and y- intercepts on each graph.

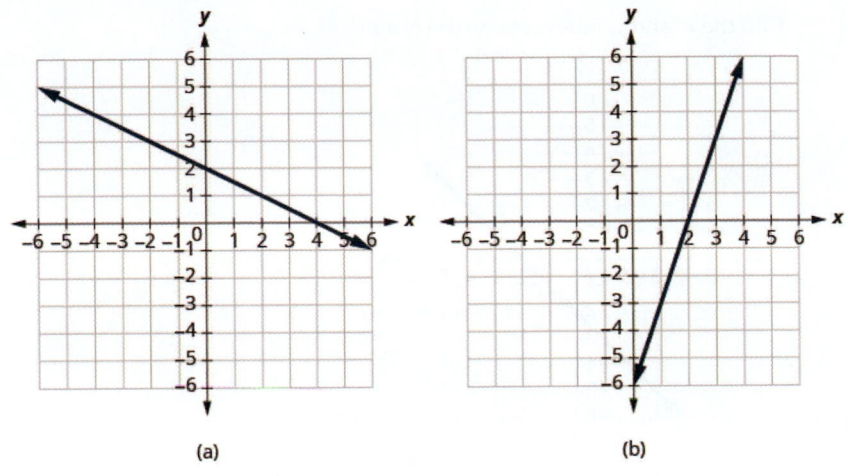

(a)

(b)

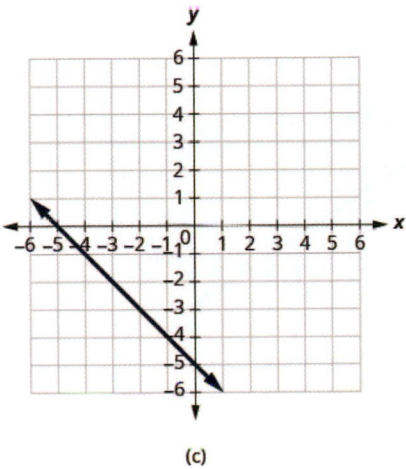

(c)

✓ **Solution**

ⓐ The graph crosses the x- axis at the point $(4, 0)$. The x- intercept is $(4, 0)$.
The graph crosses the y- axis at the point $(0, 2)$. The y- intercept is $(0, 2)$.

ⓑ The graph crosses the x- axis at the point $(2, 0)$. The x- intercept is $(2, 0)$
The graph crosses the y- axis at the point $(0, -6)$. The y- intercept is $(0, -6)$.

ⓒ The graph crosses the x- axis at the point $(-5, 0)$. The x- intercept is $(-5, 0)$.
The graph crosses the y- axis at the point $(0, -5)$. The y- intercept is $(0, -5)$.

> **TRY IT :: 4.37**   Find the *x*- and *y*- intercepts on the graph.

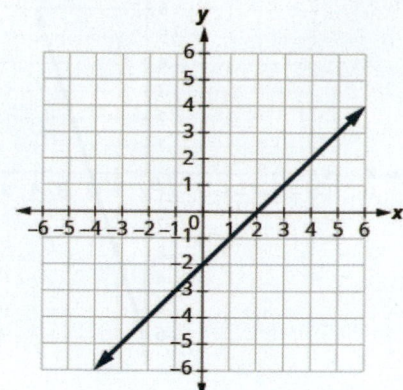

> **TRY IT :: 4.38**   Find the *x*- and *y*- intercepts on the graph.

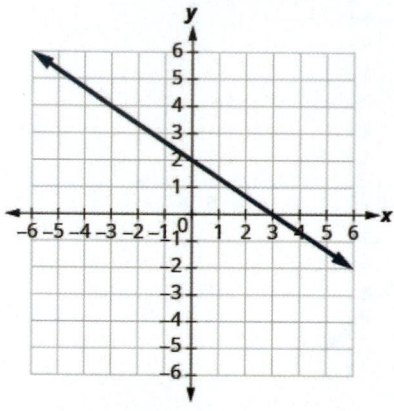

Find the *x*- and *y*- Intercepts from an Equation of a Line

Recognizing that the *x*- intercept occurs when *y* is zero and that the *y*- intercept occurs when *x* is zero, gives us a method to find the intercepts of a line from its equation. To find the *x*- intercept, let $y = 0$ and solve for *x*. To find the *y*- intercept, let $x = 0$ and solve for *y*.

Find the *x*- and *y*- Intercepts from the Equation of a Line

Use the equation of the line. To find:

- the *x*- intercept of the line, let $y = 0$ and solve for x.

- the *y*- intercept of the line, let $x = 0$ and solve for y.

EXAMPLE 4.20

Find the intercepts of $2x + y = 6$.

⊘ **Solution**

We will let $y = 0$ to find the *x*- intercept, and let $x = 0$ to find the *y*- intercept. We will fill in the table, which reminds us of what we need to find.

2x + y = 6	
x	y
	0
0	

x-intercept (row with y = 0)
y-intercept (row with x = 0)

To find the *x*- intercept, let $y = 0$.

	$2x + y = 6$
Let y = 0.	$2x + 0 = 6$
Simplify.	$2x = 6$
	$x = 3$
The *x*-intercept is	$(3, 0)$
To find the *y*-intercept, let x = 0.	
	$2x + y = 6$
Let x = 0.	$2 \cdot 0 + y = 6$
Simplify.	$0 + y = 6$
	$y = 6$
The *y*-intercept is	$(0, 6)$

The intercepts are the points $(3, 0)$ and $(0, 6)$ as shown in Table 4.26.

$2x + y = 6$	
x	y
3	0
0	6

Table 4.26

> **TRY IT :: 4.39** Find the intercepts of $3x + y = 12$.

> **TRY IT :: 4.40** Find the intercepts of $x + 4y = 8$.

EXAMPLE 4.21

Find the intercepts of $4x - 3y = 12$.

✓ **Solution**

To find the *x*-intercept, let *y* = 0.

	$4x - 3y = 12$
Let $y = 0$.	$4x - 3 \cdot 0 = 12$
Simplify.	$4x - 0 = 12$
	$4x = 12$
	$x = 3$
The *x*-intercept is	$(3, 0)$
To find the *y*-intercept, let *x* = 0.	
	$4x - 3y = 12$
Let $x = 0$.	$4 \cdot 0 - 3y = 12$
Simplify.	$0 - 3y = 12$
	$-3y = 12$
	$y = -4$
The *y*-intercept is	$(0, -4)$

The intercepts are the points (3, 0) and (0, –4) as shown in **Table 4.28**.

$4x - 3y = 12$	
x	y
3	0
0	−4

> **TRY IT : : 4.41** Find the intercepts of $3x - 4y = 12$.

> **TRY IT : : 4.42** Find the intercepts of $2x - 4y = 8$.

Graph a Line Using the Intercepts

To graph a linear equation by plotting points, you need to find three points whose coordinates are solutions to the equation. You can use the *x*- and *y*- intercepts as two of your three points. Find the intercepts, and then find a third point to ensure accuracy. Make sure the points line up—then draw the line. This method is often the quickest way to graph a line.

EXAMPLE 4.22   HOW TO GRAPH A LINE USING INTERCEPTS

Graph $-x + 2y = 6$ using the intercepts.

⊘ Solution

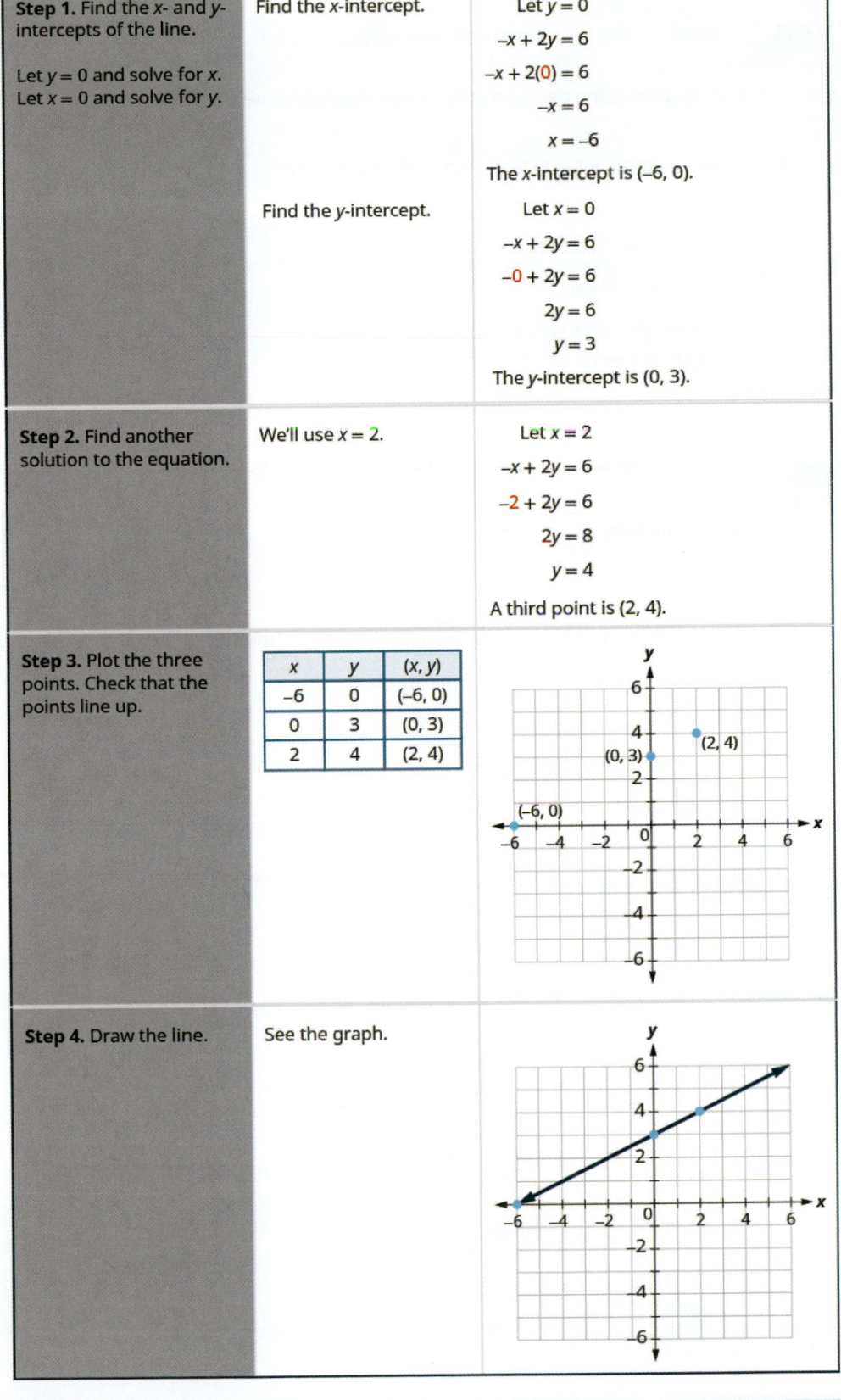

Step 1. Find the *x*- and *y*-intercepts of the line. Let *y* = 0 and solve for *x*. Let *x* = 0 and solve for *y*.	Find the *x*-intercept.	Let $y = 0$ $-x + 2y = 6$ $-x + 2(0) = 6$ $-x = 6$ $x = -6$ The *x*-intercept is (–6, 0).
	Find the *y*-intercept.	Let $x = 0$ $-x + 2y = 6$ $-0 + 2y = 6$ $2y = 6$ $y = 3$ The *y*-intercept is (0, 3).
Step 2. Find another solution to the equation.	We'll use *x* = 2.	Let $x = 2$ $-x + 2y = 6$ $-2 + 2y = 6$ $2y = 8$ $y = 4$ A third point is (2, 4).

Step 3. Plot the three points. Check that the points line up.

x	y	(x, y)
–6	0	(–6, 0)
0	3	(0, 3)
2	4	(2, 4)

Step 4. Draw the line.   See the graph.

> **TRY IT :: 4.43**   Graph $x - 2y = 4$ using the intercepts.

> **TRY IT :: 4.44**   Graph $-x + 3y = 6$ using the intercepts.

The steps to graph a linear equation using the intercepts are summarized below.

HOW TO :: GRAPH A LINEAR EQUATION USING THE INTERCEPTS.

Step 1.   Find the x- and y- intercepts of the line.
- Let $y = 0$ and solve for x
- Let $x = 0$ and solve for y.

Step 2.   Find a third solution to the equation.

Step 3.   Plot the three points and check that they line up.

Step 4.   Draw the line.

EXAMPLE 4.23

Graph $4x - 3y = 12$ using the intercepts.

⊘ **Solution**

Find the intercepts and a third point.

x-intercept, let $y = 0$	y-intercept, let $x = 0$	third point, let $y = 4$
$4x - 3y = 12$	$4x - 3y = 12$	$4x - 3y = 12$
$4x - 3(0) = 12$	$4(0) - 3y = 12$	$4x - 3(4) = 12$
$4x = 12$	$-3y = 12$	$4x - 12 = 12$
$x = 3$	$y = -4$	$4x = 24$
		$x = 6$

We list the points in **Table 4.29** and show the graph below.

$4x - 3y = 12$		
x	y	(x, y)
3	0	$(3, 0)$
0	−4	$(0, -4)$
6	4	$(6, 4)$

Table 4.29

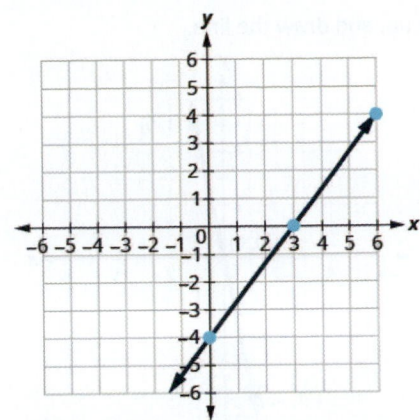

> **TRY IT ::** 4.45   Graph $5x - 2y = 10$ using the intercepts.

> **TRY IT ::** 4.46   Graph $3x - 4y = 12$ using the intercepts.

EXAMPLE 4.24

Graph $y = 5x$ using the intercepts.

✓ **Solution**

x-intercept	y-intercept
Let $y = 0$.	Let $x = 0$.
$y = 5x$	$y = 5x$
$0 = 5x$	$y = 5 \cdot 0$
$0 = x$	$y = 0$
$(0, 0)$	$(0, 0)$

This line has only one intercept. It is the point $(0, 0)$.

To ensure accuracy we need to plot three points. Since the x- and y- intercepts are the same point, we need *two* more points to graph the line.

Let $x = 1$.	Let $x = -1$.
$y = 5x$	$y = 5x$
$y = 5 \cdot 1$	$y = 5(-1)$
$y = 5$	$y = -5$

See Table 4.30.

$y = 5x$		
x	y	(x, y)
0	0	$(0, 0)$
1	5	$(1, 5)$
-1	-5	$(-1, -5)$

Table 4.30

Plot the three points, check that they line up, and draw the line.

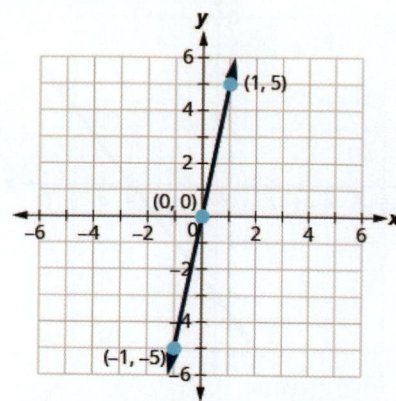

> **TRY IT : : 4.47**   Graph $y = 4x$ using the intercepts.

> **TRY IT : : 4.48**   Graph $y = -x$ the intercepts.

4.3 EXERCISES

Practice Makes Perfect

Identify the *x*- and *y*- Intercepts on a Graph

In the following exercises, find the x- and y- intercepts on each graph.

139.

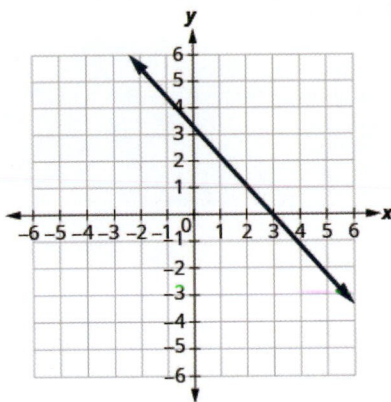

140.

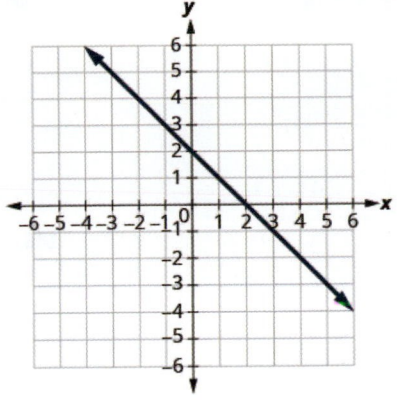

141.

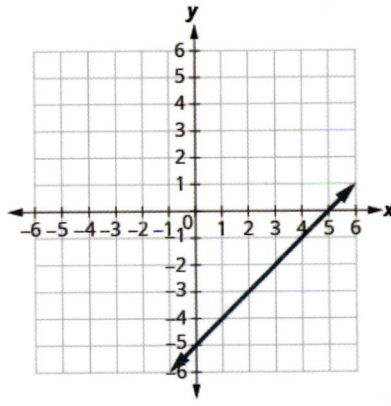

142.

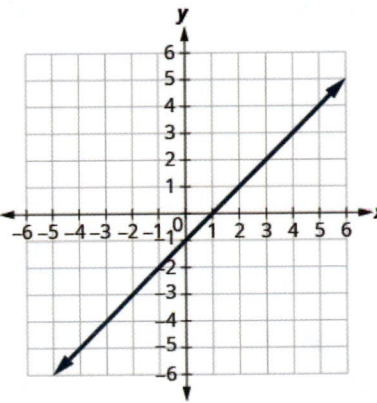

143.

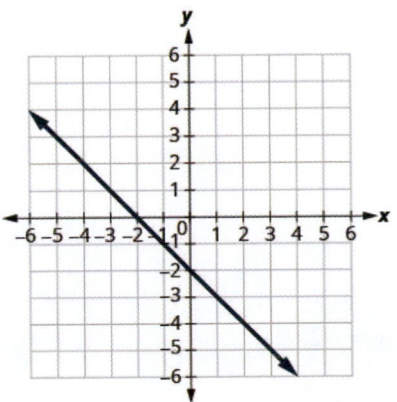

144.

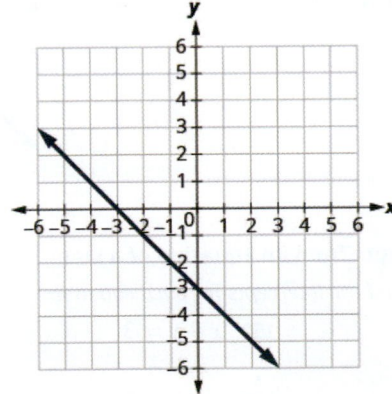

145.

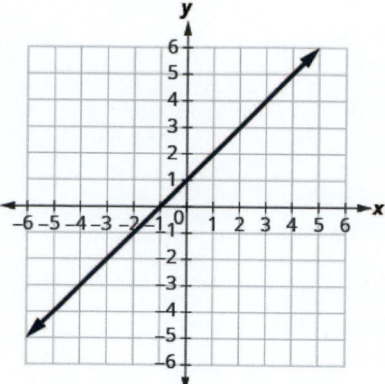

146.

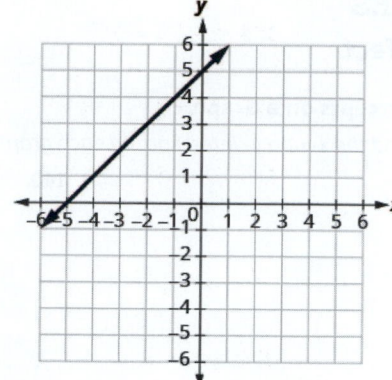

147.

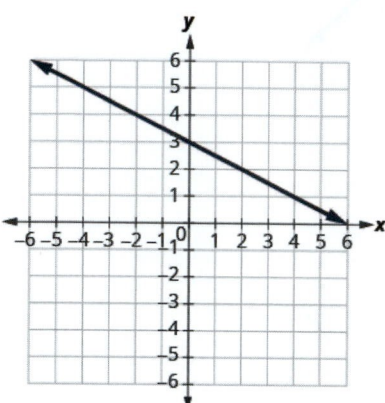

148.

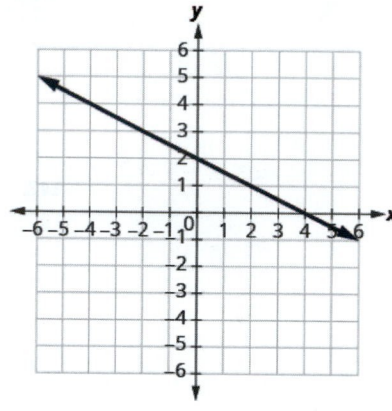

149.

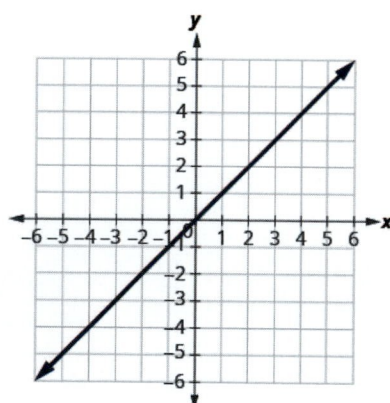

150.

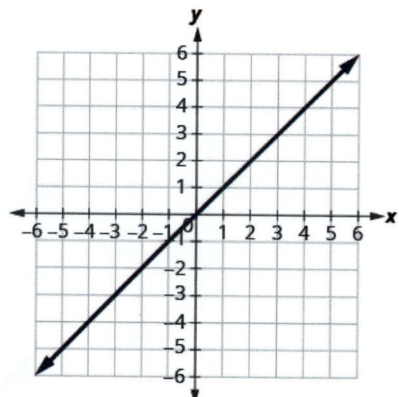

Find the *x*- and *y*- Intercepts from an Equation of a Line

In the following exercises, find the intercepts for each equation.

151. $x + y = 4$

152. $x + y = 3$

153. $x + y = -2$

154. $x + y = -5$

155. $x - y = 5$

156. $x - y = 1$

157. $x - y = -3$

158. $x - y = -4$

159. $x + 2y = 8$

160. $x + 2y = 10$

161. $3x + y = 6$

162. $3x + y = 9$

163. $x - 3y = 12$

164. $x - 2y = 8$

165. $4x - y = 8$

166. $5x - y = 5$

167. $2x + 5y = 10$

168. $2x + 3y = 6$

169. $3x - 2y = 12$

170. $3x - 5y = 30$

171. $y = \frac{1}{3}x + 1$

172. $y = \frac{1}{4}x - 1$

173. $y = \frac{1}{5}x + 2$

174. $y = \frac{1}{3}x + 4$

175. $y = 3x$

176. $y = -2x$

177. $y = -4x$

178. $y = 5x$

Graph a Line Using the Intercepts

In the following exercises, graph using the intercepts.

179. $-x + 5y = 10$

180. $-x + 4y = 8$

181. $x + 2y = 4$

182. $x + 2y = 6$

183. $x + y = 2$

184. $x + y = 5$

185. $x + y = -3$

186. $x + y = -1$

187. $x - y = 1$

188. $x - y = 2$

189. $x - y = -4$

190. $x - y = -3$

191. $4x + y = 4$

192. $3x + y = 3$

193. $2x + 4y = 12$

194. $3x + 2y = 12$

195. $3x - 2y = 6$

196. $5x - 2y = 10$

197. $2x - 5y = -20$

198. $3x - 4y = -12$

199. $3x - y = -6$

200. $2x - y = -8$

201. $y = -2x$

202. $y = -4x$

203. $y = x$

204. $y = 3x$

Everyday Math

205. Road trip. Damien is driving from Chicago to Denver, a distance of 1000 miles. The x- axis on the graph below shows the time in hours since Damien left Chicago. The y- axis represents the distance he has left to drive.

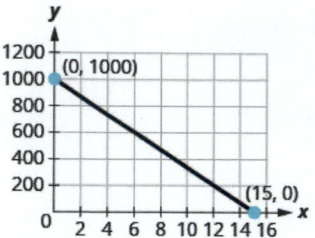

ⓐ Find the x- and y- intercepts.

ⓑ Explain what the x- and y- intercepts mean for Damien.

206. Road trip. Ozzie filled up the gas tank of his truck and headed out on a road trip. The x- axis on the graph below shows the number of miles Ozzie drove since filling up. The y- axis represents the number of gallons of gas in the truck's gas tank.

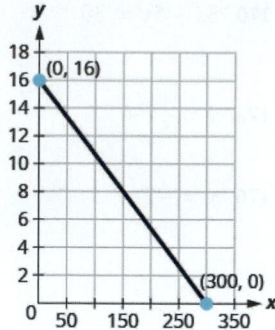

ⓐ Find the x- and y- intercepts.

ⓑ Explain what the x- and y- intercepts mean for Ozzie.

Writing Exercises

207. How do you find the x- intercept of the graph of $3x - 2y = 6$?

208. Do you prefer to use the method of plotting points or the method using the intercepts to graph the equation $4x + y = -4$? Why?

209. Do you prefer to use the method of plotting points or the method using the intercepts to graph the equation $y = \frac{2}{3}x - 2$? Why?

210. Do you prefer to use the method of plotting points or the method using the intercepts to graph the equation $y = 6$? Why?

Self Check

ⓐ After completing the exercises, use this checklist to evaluate your mastery of the objectives of this section.

I can...	Confidently	With some help	No-I don't get it!
identify the x and y intercepts on a graph.			
find the x and y intercepts from an equation of a line.			
graph a line using the intercepts.			

ⓑ What does this checklist tell you about your mastery of this section? What steps will you take to improve?

4.4 Understand Slope of a Line

Learning Objectives

By the end of this section, you will be able to:

› Use geoboards to model slope
› Use $m = \frac{rise}{run}$ to find the slope of a line from its graph
› Find the slope of horizontal and vertical lines
› Use the slope formula to find the slope of a line between two points
› Graph a line given a point and the slope
› Solve slope applications

Be Prepared!

Before you get started, take this readiness quiz.

1. Simplify: $\frac{1-4}{8-2}$.
   If you missed this problem, review **Example 1.74**.

2. Divide: $\frac{0}{4}, \frac{4}{0}$.
   If you missed this problem, review **Example 1.127**.

3. Simplify: $\frac{15}{-3}, \frac{-15}{3}, \frac{-15}{-3}$.
   If you missed this problem, review **Example 1.65**.

When you graph linear equations, you may notice that some lines tilt up as they go from left to right and some lines tilt down. Some lines are very steep and some lines are flatter. What determines whether a line tilts up or down or if it is steep or flat?

In mathematics, the 'tilt' of a line is called the *slope* of the line. The concept of slope has many applications in the real world. The pitch of a roof, grade of a highway, and a ramp for a wheelchair are some examples where you literally see slopes. And when you ride a bicycle, you feel the slope as you pump uphill or coast downhill.

In this section, we will explore the concept of slope.

Use Geoboards to Model Slope

A **geoboard** is a board with a grid of pegs on it. Using rubber bands on a geoboard gives us a concrete way to model lines on a coordinate grid. By stretching a rubber band between two pegs on a geoboard, we can discover how to find the slope of a line.

 MANIPULATIVE MATHEMATICS

Doing the Manipulative Mathematics activity "Exploring Slope" will help you develop a better understanding of the slope of a line. (Graph paper can be used instead of a geoboard, if needed.)

We'll start by stretching a rubber band between two pegs as shown in **Figure 4.19**.

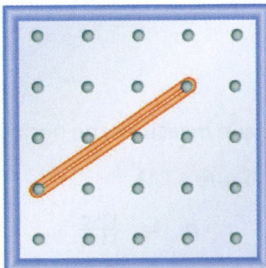

Figure 4.19

Doesn't it look like a line?

Now we stretch one part of the rubber band straight up from the left peg and around a third peg to make the sides of a right triangle, as shown in Figure 4.20

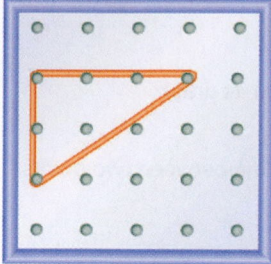

Figure 4.20

We carefully make a 90° angle around the third peg, so one of the newly formed lines is vertical and the other is horizontal.

To find the slope of the line, we measure the distance along the vertical and horizontal sides of the triangle. The vertical distance is called the rise and the horizontal distance is called the run, as shown in Figure 4.21.

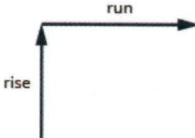

Figure 4.21

If our geoboard and rubber band look just like the one shown in Figure 4.22, the rise is 2. The rubber band goes up 2 units. (Each space is one unit.)

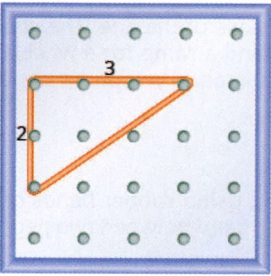

Figure 4.22 The rise on this geoboard is 2, as the rubber band goes up two units.

What is the run?

The rubber band goes across 3 units. The run is 3 (see Figure 4.22).

The slope of a line is the ratio of the rise to the run. In mathematics, it is always referred to with the letter m.

Slope of a Line

The **slope of a line** of a line is $m = \frac{\text{rise}}{\text{run}}$.

The **rise** measures the vertical change and the **run** measures the horizontal change between two points on the line.

What is the slope of the line on the geoboard in Figure 4.22?

$$m = \frac{\text{rise}}{\text{run}}$$

$$m = \frac{2}{3}$$

The line has slope $\frac{2}{3}$. This means that the line rises 2 units for every 3 units of run.

When we work with geoboards, it is a good idea to get in the habit of starting at a peg on the left and connecting to a peg to the right. If the rise goes up it is positive and if it goes down it is negative. The run will go from left to right and be positive.

EXAMPLE 4.25

What is the slope of the line on the geoboard shown?

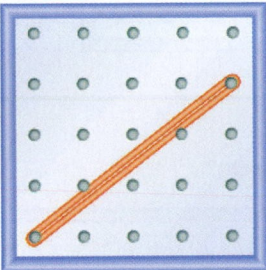

⊘ **Solution**

Use the definition of slope: $m = \frac{\text{rise}}{\text{run}}$.

Start at the left peg and count the spaces up and to the right to reach the second peg.

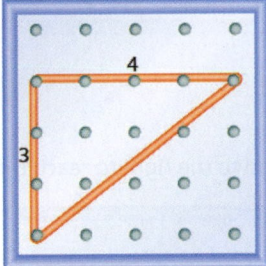

The rise is 3. $m = \frac{3}{\text{run}}$

The run is 4. $m = \frac{3}{4}$

The slope is $\frac{3}{4}$.

This means that the line rises 3 units for every 4 units of run.

> **TRY IT : : 4.49**   What is the slope of the line on the geoboard shown?

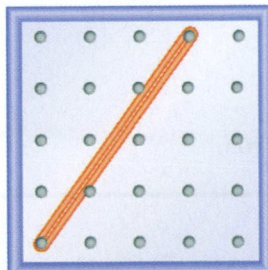

> **TRY IT :: 4.50**   What is the slope of the line on the geoboard shown?

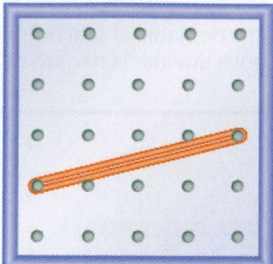

EXAMPLE 4.26

What is the slope of the line on the geoboard shown?

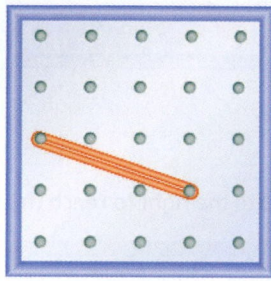

✓ **Solution**

Use the definition of slope: $m = \frac{\text{rise}}{\text{run}}$.

Start at the left peg and count the units down and to the right to reach the second peg.

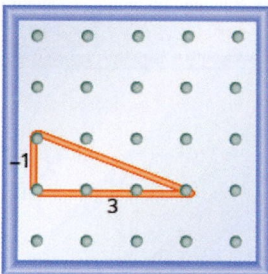

The rise is -1. $m = \frac{-1}{\text{run}}$

The run is 3. $m = \frac{-1}{3}$

$$m = -\frac{1}{3}$$

The slope is $-\frac{1}{3}$.

This means that the line drops 1 unit for every 3 units of run.

> **TRY IT : : 4.51**   What is the slope of the line on the geoboard?

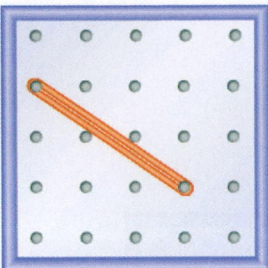

> **TRY IT : : 4.52**   What is the slope of the line on the geoboard?

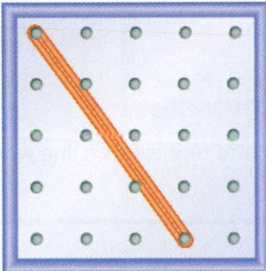

Notice that in **Example 4.25** the slope is positive and in **Example 4.26** the slope is negative. Do you notice any difference in the two lines shown in **Figure 4.23**(a) and **Figure 4.23**(b)?

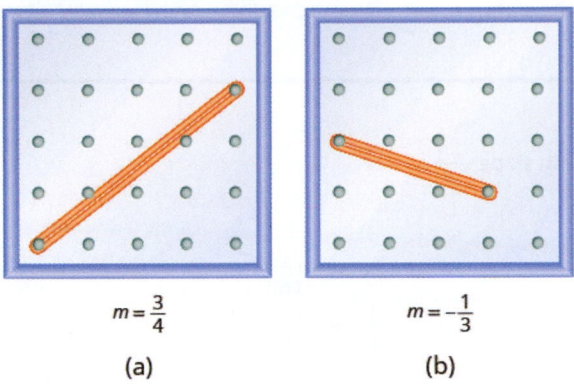

$$m = \frac{3}{4}$$ $$m = -\frac{1}{3}$$

(a)   (b)

Figure 4.23

We 'read' a line from left to right just like we read words in English. As you read from left to right, the line in **Figure 4.23**(a) is going up; it has **positive slope**. The line in **Figure 4.23**(b) is going down; it has **negative slope**.

Positive and Negative Slopes

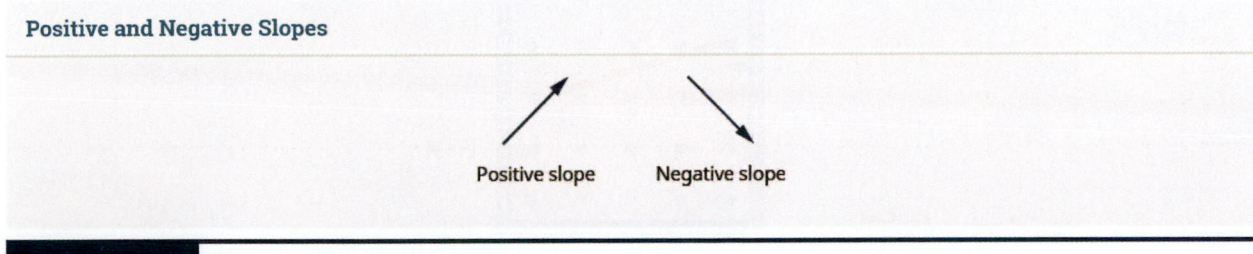

Positive slope   Negative slope

EXAMPLE 4.27

Use a geoboard to model a line with slope $\frac{1}{2}$.

⊘ **Solution**

To model a line on a geoboard, we need the rise and the run.

Use the slope formula. $m = \dfrac{\text{rise}}{\text{run}}$

Replace m with $\dfrac{1}{2}$. $\dfrac{1}{2} = \dfrac{\text{rise}}{\text{run}}$

So, the rise is 1 and the run is 2.

Start at a peg in the lower left of the geoboard.

Stretch the rubber band up 1 unit, and then right 2 units.

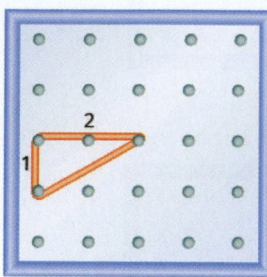

The hypotenuse of the right triangle formed by the rubber band represents a line whose slope is $\dfrac{1}{2}$.

> **TRY IT : : 4.53**   Model the slope $m = \dfrac{1}{3}$. Draw a picture to show your results.

> **TRY IT : : 4.54**   Model the slope $m = \dfrac{3}{2}$. Draw a picture to show your results.

EXAMPLE 4.28

Use a geoboard to model a line with slope $\dfrac{-1}{4}$.

✓ **Solution**

Use the slope formula. $m = \dfrac{\text{rise}}{\text{run}}$

Replace m with $\dfrac{-1}{4}$. $\dfrac{-1}{4} = \dfrac{\text{rise}}{\text{run}}$

So, the rise is -1 and the run is 4.

Since the rise is negative, we choose a starting peg on the upper left that will give us room to count down.

We stretch the rubber band down 1 unit, then go to the right 4 units, as shown.

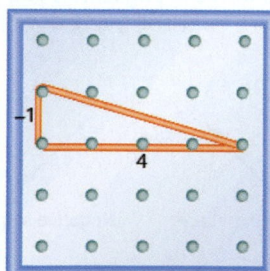

The hypotenuse of the right triangle formed by the rubber band represents a line whose slope is $\dfrac{-1}{4}$.

> **TRY IT : : 4.55**   Model the slope $m = \dfrac{-2}{3}$. Draw a picture to show your results.

> | **TRY IT :: 4.56** Model the slope $m = \frac{-1}{3}$. Draw a picture to show your results.

Use $m = \frac{\text{rise}}{\text{run}}$ to Find the Slope of a Line from its Graph

Now, we'll look at some graphs on the xy-coordinate plane and see how to find their slopes. The method will be very similar to what we just modeled on our geoboards.

To find the slope, we must count out the rise and the run. But where do we start?

We locate two points on the line whose coordinates are integers. We then start with the point on the left and sketch a right triangle, so we can count the rise and run.

EXAMPLE 4.29   HOW TO USE $m = \frac{\text{rise}}{\text{run}}$ TO FIND THE SLOPE OF A LINE FROM ITS GRAPH

Find the slope of the line shown.

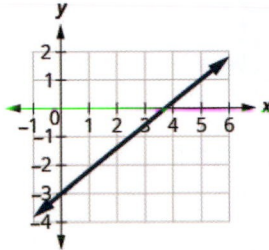

⊘   **Solution**

Step 1. Locate two points on the graph whose coordinates are integers.	Mark (0, –3) and (5, 1).	
Step 2. Starting with the point on the left, sketch a right triangle, going from the first point to the second point.	Starting at (0, –3), sketch a right triangle to (5, 1).	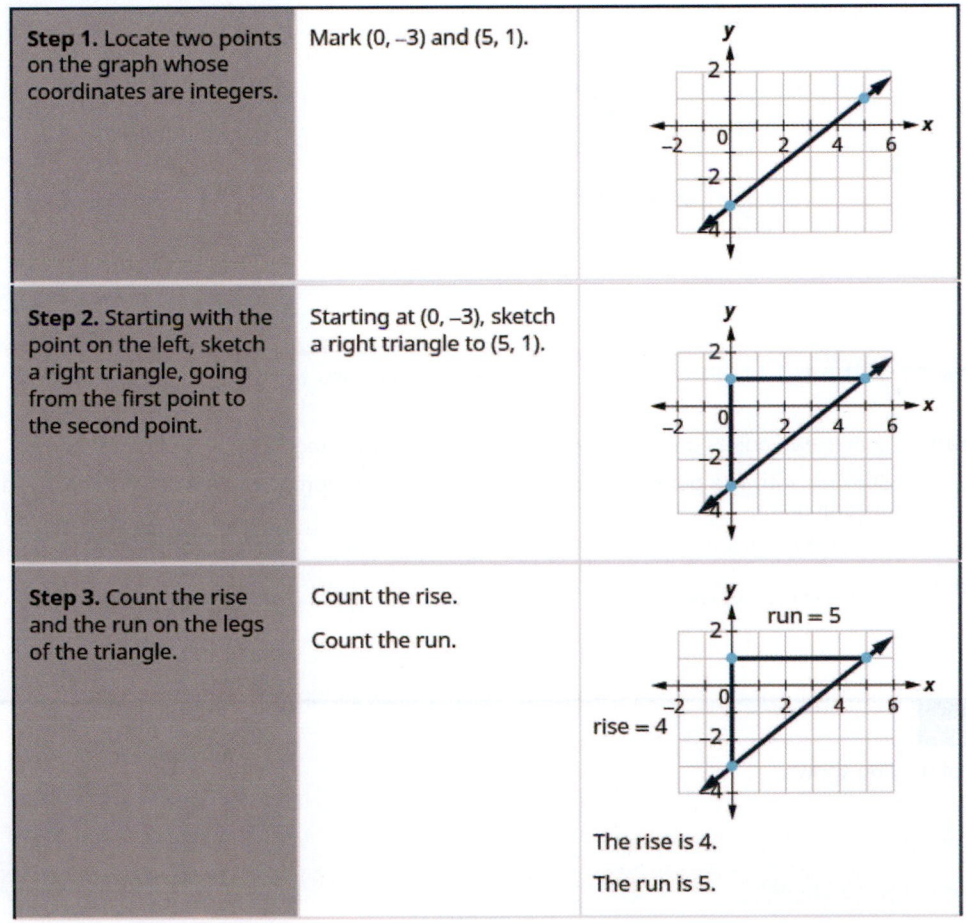
Step 3. Count the rise and the run on the legs of the triangle.	Count the rise. Count the run. The rise is 4. The run is 5.	

Step 4. Take the ratio of rise to run to find the slope. $m = \dfrac{\text{rise}}{\text{run}}$	Use the slope formula.	$m = \dfrac{\text{rise}}{\text{run}}$
	Substitute the values of the rise and run.	$m = \dfrac{4}{5}$ The slope of the line is $\dfrac{4}{5}$. This means that y increases 4 units as x increases 5 units.

> **TRY IT : : 4.57**   Find the slope of the line shown.

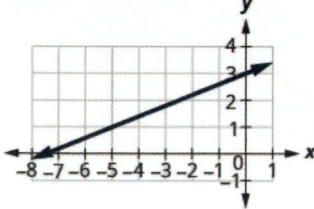

> **TRY IT : : 4.58**   Find the slope of the line shown.

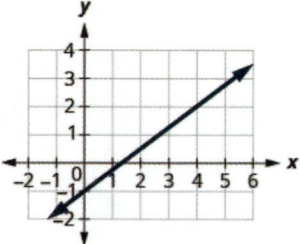

HOW TO : : FIND THE SLOPE OF A LINE FROM ITS GRAPH USING $m = \dfrac{\text{rise}}{\text{run}}$.

Step 1. Locate two points on the line whose coordinates are integers.

Step 2. Starting with the point on the left, sketch a right triangle, going from the first point to the second point.

Step 3. Count the rise and the run on the legs of the triangle.

Step 4. Take the ratio of rise to run to find the slope, $m = \dfrac{\text{rise}}{\text{run}}$.

EXAMPLE 4.30

Find the slope of the line shown.

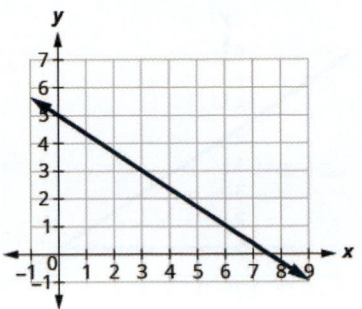

⊘ **Solution**

Locate two points on the graph whose coordinates are integers.	$(0, 5)$ and $(3, 3)$
Which point is on the left?	$(0, 5)$

Starting at $(0, 5)$, sketch a right triangle to $(3, 3)$.

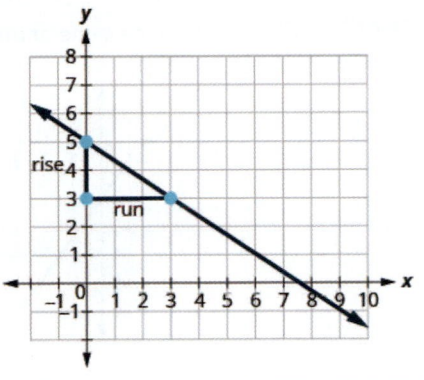

Count the rise—it is negative.	The rise is -2.
Count the run.	The run is 3.
Use the slope formula.	$m = \dfrac{\text{rise}}{\text{run}}$
Substitute the values of the rise and run.	$m = \dfrac{-2}{3}$
Simplify.	$m = -\dfrac{2}{3}$
	The slope of the line is $-\dfrac{2}{3}$.

So y increases by 3 units as x decreases by 2 units.

What if we used the points $(-3, 7)$ and $(6, 1)$ to find the slope of the line?

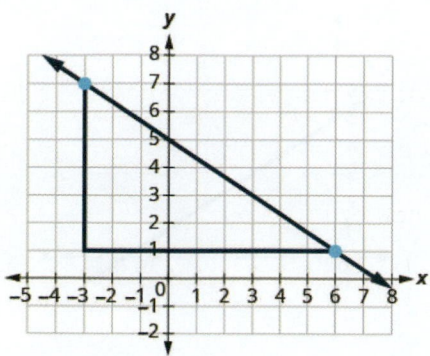

The rise would be -6 and the run would be 9. Then $m = \frac{-6}{9}$, and that simplifies to $m = -\frac{2}{3}$. Remember, it does not matter which points you use—the slope of the line is always the same.

> **TRY IT : : 4.59**   Find the slope of the line shown.

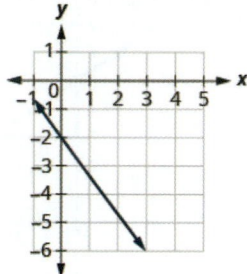

> **TRY IT : : 4.60**   Find the slope of the line shown.

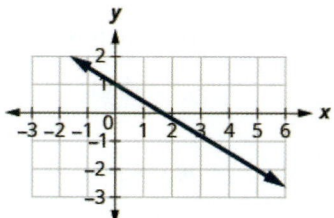

In the last two examples, the lines had y-intercepts with integer values, so it was convenient to use the y-intercept as one of the points to find the slope. In the next example, the y-intercept is a fraction. Instead of using that point, we'll look for two other points whose coordinates are integers. This will make the slope calculations easier.

EXAMPLE 4.31

Find the slope of the line shown.

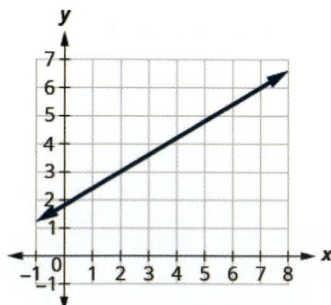

⊘ **Solution**

Locate two points on the graph whose coordinates are integers.	$(2, 3)$ and $(7, 6)$
Which point is on the left?	$(2, 3)$

Starting at $(2, 3)$, sketch a right triangle to $(7, 6)$.

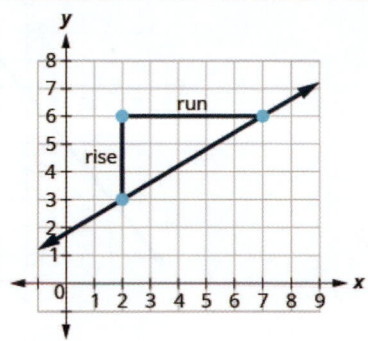

Count the rise.	The rise is 3.
Count the run.	The run is 5.
Use the slope formula.	$m = \dfrac{\text{rise}}{\text{run}}$
Substitute the values of the rise and run.	$m = \dfrac{3}{5}$
	The slope of the line is $\dfrac{3}{5}$.

This means that y increases 5 units as x increases 3 units.

When we used geoboards to introduce the concept of slope, we said that we would always start with the point on the left and count the rise and the run to get to the point on the right. That way the run was always positive and the rise determined whether the slope was positive or negative.

What would happen if we started with the point on the right?

Let's use the points $(2, 3)$ and $(7, 6)$ again, but now we'll start at $(7, 6)$.

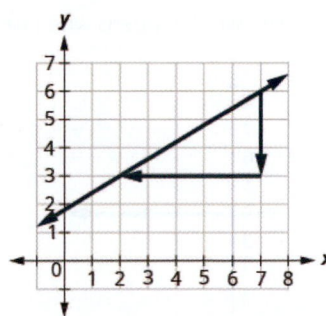

Count the rise.	The rise is -3.
Count the run. It goes from right to left, so it is negative.	The run is -5.
Use the slope formula.	$m = \dfrac{\text{rise}}{\text{run}}$
Substitute the values of the rise and run.	$m = \dfrac{-3}{-5}$
	The slope of the line is $\dfrac{3}{5}$.

It does not matter where you start—the slope of the line is always the same.

> **TRY IT : :** 4.61   Find the slope of the line shown.

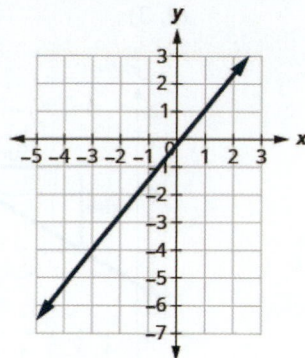

> **TRY IT : :** 4.62   Find the slope of the line shown.

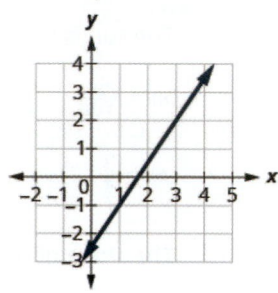

Find the Slope of Horizontal and Vertical Lines

Do you remember what was special about horizontal and vertical lines? Their equations had just one variable.

Horizontal line $y = b$ **Vertical line** $x = a$

y-coordinates are the same. x-coordinates are the same.

So how do we find the slope of the horizontal line $y = 4$? One approach would be to graph the horizontal line, find two points on it, and count the rise and the run. Let's see what happens when we do this.

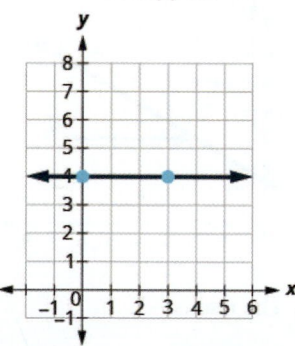

What is the rise?   The rise is 0.
What is the run?   The run is 3.

$$m = \frac{\text{rise}}{\text{run}}$$

What is the slope? $m = \frac{0}{3}$

$$m = 0$$

The slope of the horizontal line $y = 4$ is 0.

All horizontal lines have slope 0. When the *y*-coordinates are the same, the rise is 0.

Slope of a Horizontal Line

The slope of a horizontal line, $y = b$, is 0.

The floor of your room is horizontal. Its slope is 0. If you carefully placed a ball on the floor, it would not roll away. Now, we'll consider a vertical line, the line.

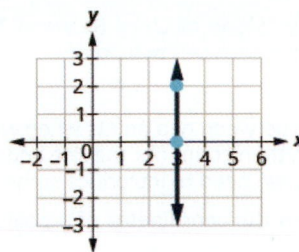

What is the rise?   The rise is 2.
What is the run?   The run is 0.

$$m = \frac{\text{rise}}{\text{run}}$$

What is the slope?

$$m = \frac{2}{0}$$

But we can't divide by 0. Division by 0 is not defined. So we say that the slope of the vertical line $x = 3$ is undefined.

The slope of any vertical line is undefined. When the *x*-coordinates of a line are all the same, the run is 0.

Slope of a Vertical Line

The slope of a vertical line, $x = a$, is undefined.

EXAMPLE 4.32

Find the slope of each line:

ⓐ $x = 8$ ⓑ $y = -5$.

✓ Solution

ⓐ $x = 8$
This is a vertical line.
Its slope is undefined.

ⓑ $y = -5$
This is a horizontal line.
It has slope 0.

> **TRY IT : : 4.63**   Find the slope of the line: $x = -4$.

> **TRY IT : : 4.64**   Find the slope of the line: $y = 7$.

Quick Guide to the Slopes of Lines

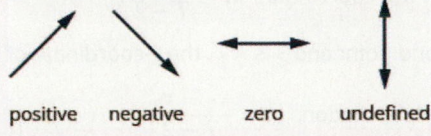

positive   negative   zero   undefined

Remember, we 'read' a line from left to right, just like we read written words in English.

Use the Slope Formula to find the Slope of a Line Between Two Points

 MANIPULATIVE MATHEMATICS

Doing the Manipulative Mathematics activity "Slope of Lines Between Two Points" will help you develop a better understanding of how to find the slope of a line between two points.

Sometimes we'll need to find the slope of a line between two points when we don't have a graph to count out the rise and the run. We could plot the points on grid paper, then count out the rise and the run, but as we'll see, there is a way to find the slope without graphing. Before we get to it, we need to introduce some algebraic notation.

We have seen that an ordered pair (x, y) gives the coordinates of a point. But when we work with slopes, we use two points. How can the same symbol (x, y) be used to represent two different points? Mathematicians use subscripts to distinguish the points.

$$(x_1, y_1) \quad \text{read ' } x \text{ sub 1, } y \text{ sub 1'}$$
$$(x_2, y_2) \quad \text{read ' } x \text{ sub 2, } y \text{ sub 2'}$$

The use of subscripts in math is very much like the use of last name initials in elementary school. Maybe you remember Laura C. and Laura M. in your third grade class?

We will use (x_1, y_1) to identify the first point and (x_2, y_2) to identify the second point.

If we had more than two points, we could use (x_3, y_3), (x_4, y_4), and so on.

Let's see how the rise and run relate to the coordinates of the two points by taking another look at the slope of the line between the points $(2, 3)$ and $(7, 6)$.

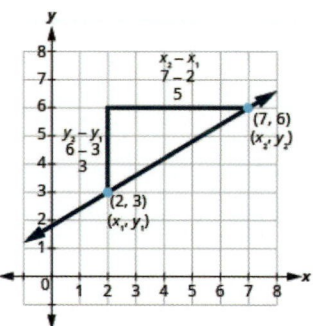

Since we have two points, we will use subscript notation, $\begin{pmatrix} x_1, y_1 \\ 2, 3 \end{pmatrix} \begin{pmatrix} x_2, y_2 \\ 7, 6 \end{pmatrix}$.

On the graph, we counted the rise of 3 and the run of 5.

Notice that the rise of 3 can be found by subtracting the y-coordinates 6 and 3.

$$3 = 6 - 3$$

And the run of 5 can be found by subtracting the x-coordinates 7 and 2.

$$5 = 7 - 2$$

We know $m = \frac{\text{rise}}{\text{run}}$. So $m = \frac{3}{5}$.

We rewrite the rise and run by putting in the coordinates $m = \frac{6 - 3}{7 - 2}$.

But 6 is y_2, the y-coordinate of the second point and 3 is y_1, the y-coordinate of the first point.

So we can rewrite the slope using subscript notation. $m = \frac{y_2 - y_1}{7 - 2}$

Also, 7 is x_2, the x-coordinate of the second point and 2 is x_1, the x-coordinate of the first point.

So, again, we rewrite the slope using subscript notation. $m = \frac{y_2 - y_1}{x_2 - x_1}$

We've shown that $m = \frac{y_2 - y_1}{x_2 - x_1}$ is really another version of $m = \frac{\text{rise}}{\text{run}}$. We can use this formula to find the slope of a line when we have two points on the line.

Slope Formula

The slope of the line between two points (x_1, y_1) and (x_2, y_2) is

$$m = \frac{y_2 - y_1}{x_2 - x_1}$$

This is the **slope formula**.

The slope is:

y of the second point minus y of the first point

over

x of the second point minus x of the first point.

EXAMPLE 4.33

Use the slope formula to find the slope of the line between the points $(1, 2)$ and $(4, 5)$.

✓ Solution

We'll call $(1, 2)$ point #1 and $(4, 5)$ point #2.

$$\begin{pmatrix} x_1, y_1 \\ 1, 2 \end{pmatrix} \begin{pmatrix} x_2, y_2 \\ 4, 5 \end{pmatrix}$$

Use the slope formula.

$$m = \frac{y_2 - y_1}{x_2 - x_1}$$

Substitute the values.

y of the second point minus y of the first point

$$m = \frac{5 - 2}{x_2 - x_1}$$

x of the second point minus x of the first point

$$m = \frac{5 - 2}{4 - 1}$$

Simplify the numerator and the denominator.

$$m = \frac{3}{3}$$

Simplify.

$$m = 1$$

Let's confirm this by counting out the slope on a graph using $m = \frac{\text{rise}}{\text{run}}$.

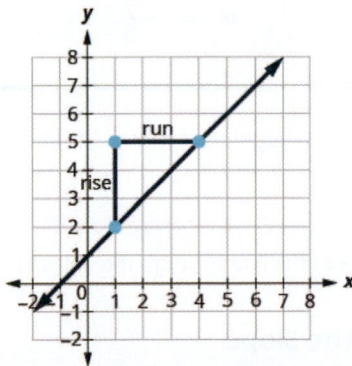

It doesn't matter which point you call point #1 and which one you call point #2. The slope will be the same. Try the calculation yourself.

> **TRY IT :: 4.65**   Use the slope formula to find the slope of the line through the points: $(8, 5)$ and $(6, 3)$.

> **TRY IT : : 4.66**   Use the slope formula to find the slope of the line through the points: $(1, 5)$ and $(5, 9)$.

EXAMPLE 4.34

Use the slope formula to find the slope of the line through the points $(-2, -3)$ and $(-7, 4)$.

⊘ **Solution**

We'll call $(-2, -3)$ point #1 and $(-7, 4)$ point #2.

$$\left(\overset{x_1,\ y_1}{-2,\ -3}\right)\left(\overset{x_2,\ y_2}{-7,\ 4}\right)$$

Use the slope formula.

$$m = \frac{y_2 - y_1}{x_2 - x_1}$$

Substitute the values.

y of the second point minus y of the first point

$$m = \frac{4 - (-3)}{x_2 - x_1}$$

x of the second point minus x of the first point

$$m = \frac{4 - (-3)}{-7 - (-2)}$$

Simplify.

$$m = \frac{7}{-5}$$

$$m = -\frac{7}{5}$$

Let's verify this slope on the graph shown.

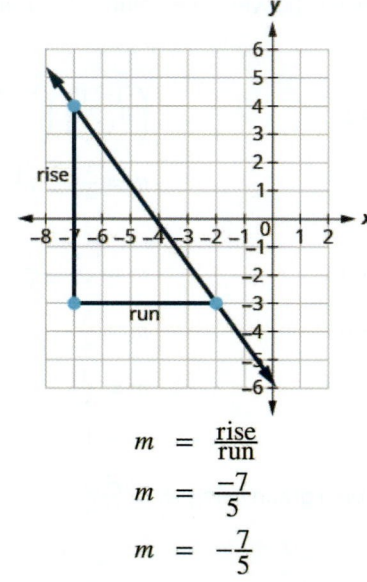

$$m = \frac{\text{rise}}{\text{run}}$$

$$m = \frac{-7}{5}$$

$$m = -\frac{7}{5}$$

> **TRY IT : : 4.67**   Use the slope formula to find the slope of the line through the points: $(-3, 4)$ and $(2, -1)$.

> **TRY IT : : 4.68**
>
> Use the slope formula to find the slope of the line through the pair of points: $(-2, 6)$ and $(-3, -4)$.

Graph a Line Given a Point and the Slope

Up to now, in this chapter, we have graphed lines by plotting points, by using intercepts, and by recognizing horizontal and vertical lines.

One other method we can use to graph lines is called the **point–slope method**. We will use this method when we know one point and the slope of the line. We will start by plotting the point and then use the definition of slope to draw the graph of the line.

EXAMPLE 4.35   HOW TO GRAPH A LINE GIVEN A POINT AND THE SLOPE

Graph the line passing through the point $(1, -1)$ whose slope is $m = \frac{3}{4}$.

✓ **Solution**

Step 1. Plot the given point.	Plot $(1, -1)$.	
Step 2. Use the slope formula $m = \dfrac{\text{rise}}{\text{run}}$ to identify the rise and the run.	Identify the rise and the run.	$m = \dfrac{3}{4}$ $\dfrac{\text{rise}}{\text{run}} = \dfrac{3}{4}$ rise $= 3$ run $= 4$
Step 3. Starting at the given point, count out the rise and run to mark the second point.	Start at $(1, -1)$ and count the rise and the run. Up 3 units, right 4 units.	
Step 4. Connect the points with a line.	Connect the two points with a line.	

> **TRY IT : : 4.69**   Graph the line passing through the point $(2, -2)$ with the slope $m = \frac{4}{3}$.

> **TRY IT : : 4.70**   Graph the line passing through the point $(-2, 3)$ with the slope $m = \frac{1}{4}$.

HOW TO :: GRAPH A LINE GIVEN A POINT AND THE SLOPE.

Step 1.   Plot the given point.

Step 2.   Use the slope formula $m = \frac{\text{rise}}{\text{run}}$ to identify the rise and the run.

Step 3.   Starting at the given point, count out the rise and run to mark the second point.

Step 4.   Connect the points with a line.

EXAMPLE 4.36

Graph the line with y-intercept 2 whose slope is $m = -\frac{2}{3}$.

✓ **Solution**

Plot the given point, the y-intercept, $(0, 2)$.

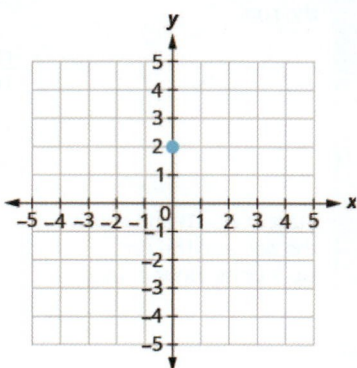

Identify the rise and the run.

$$m = -\frac{2}{3}$$
$$\frac{\text{rise}}{\text{run}} = \frac{-2}{3}$$
$$\text{rise} = -2$$
$$\text{run} = 3$$

Count the rise and the run. Mark the second point.

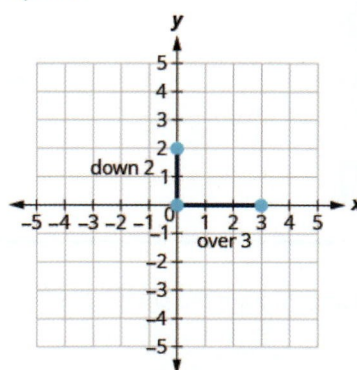

Connect the two points with a line.

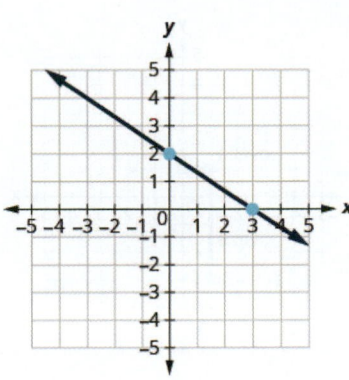

You can check your work by finding a third point. Since the slope is $m = -\frac{2}{3}$, it can be written as $m = \frac{2}{-3}$. Go back to $(0, 2)$ and count out the rise, 2, and the run, -3.

> **TRY IT :: 4.71**   Graph the line with the y-intercept 4 and slope $m = -\frac{5}{2}$.

> **TRY IT :: 4.72**   Graph the line with the x-intercept -3 and slope $m = -\frac{3}{4}$.

EXAMPLE 4.37

Graph the line passing through the point $(-1, -3)$ whose slope is $m = 4$.

⊘ **Solution**

Plot the given point.

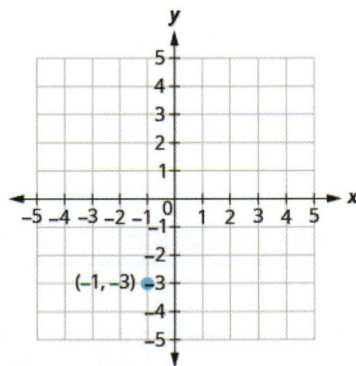

Identify the rise and the run.	m	$=$	4
Write 4 as a fraction.	$\frac{\text{rise}}{\text{run}}$	$=$	$\frac{4}{1}$
	rise	$=$	4 run $= 1$

Count the rise and run and mark the second point.

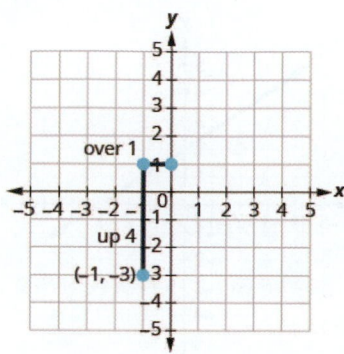

Connect the two points with a line.

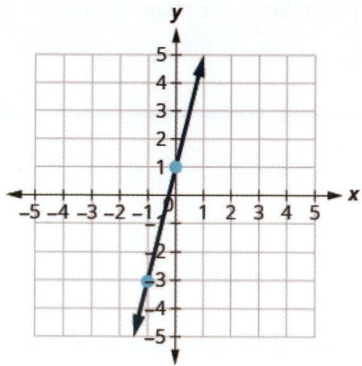

You can check your work by finding a third point. Since the slope is $m = 4$, it can be written as $m = \frac{-4}{-1}$. Go back to $(-1, -3)$ and count out the rise, -4, and the run, -1.

> | **TRY IT : : 4.73**   Graph the line with the point $(-2, 1)$ and slope $m = 3$.

> | **TRY IT : : 4.74**   Graph the line with the point $(4, -2)$ and slope $m = -2$.

Solve Slope Applications

At the beginning of this section, we said there are many applications of slope in the real world. Let's look at a few now.

EXAMPLE 4.38

The 'pitch' of a building's roof is the slope of the roof. Knowing the pitch is important in climates where there is heavy snowfall. If the roof is too flat, the weight of the snow may cause it to collapse. What is the slope of the roof shown?

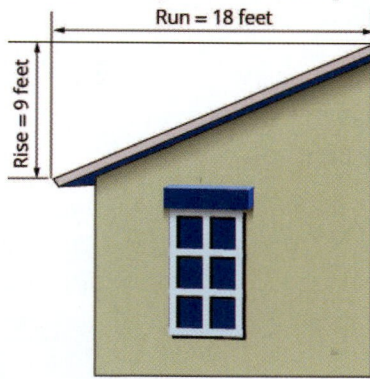

✓ Solution

Use the slope formula. $$m = \frac{\text{rise}}{\text{run}}$$

Substitute the values for rise and run. $$m = \frac{9}{18}$$

Simplify. $$m = \frac{1}{2}$$

The slope of the roof is $\frac{1}{2}$.

The roof rises 1 foot for every 2 feet of horizontal run.

> **TRY IT : : 4.75**   Use **Example 4.38**, substituting the rise = 14 and run = 24.

> **TRY IT : : 4.76**   Use **Example 4.38**, substituting rise = 15 and run = 36.

EXAMPLE 4.39

Have you ever thought about the sewage pipes going from your house to the street? They must slope down $\frac{1}{4}$ inch per foot in order to drain properly. What is the required slope?

✓ Solution

Use the slope formula. $$m = \frac{\text{rise}}{\text{run}}$$

$$m = \frac{-\frac{1}{4}\text{inch}}{1 \text{ foot}}$$

$$m = \frac{-\frac{1}{4}\text{inch}}{12 \text{ inches}}$$

Simplify. $$m = -\frac{1}{48}$$

The slope of the pipe is $-\frac{1}{48}$.

The pipe drops 1 inch for every 48 inches of horizontal run.

> **TRY IT : : 4.77**   Find the slope of a pipe that slopes down $\frac{1}{3}$ inch per foot.

> **TRY IT : : 4.78**   Find the slope of a pipe that slopes down $\frac{3}{4}$ inch per yard.

▶ **MEDIA : :**

Access these online resources for additional instruction and practice with understanding slope of a line.

- **Practice Slope with a Virtual Geoboard (https://openstax.org/l/25Geoboard)**
- **Small, Medium, and Large Virtual Geoboards (https://openstax.org/l/25VirtualGeo)**
- **Explore Area and Perimeter with a Geoboard (https://openstax.org/l/25APGeoboard)**

4.4 EXERCISES

Practice Makes Perfect

Use Geoboards to Model Slope

In the following exercises, find the slope modeled on each geoboard.

211.

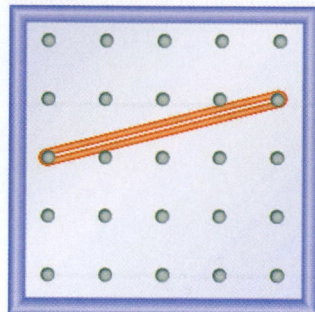

212.

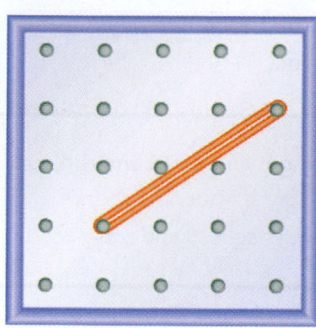

213.

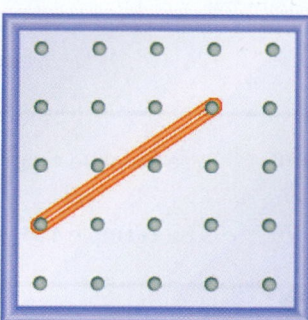

214.

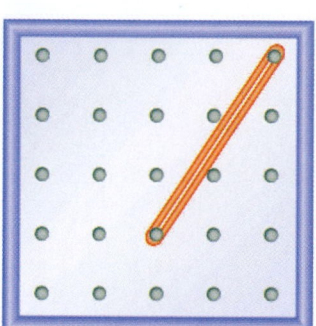

215.

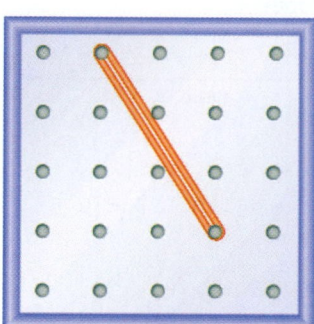

216.

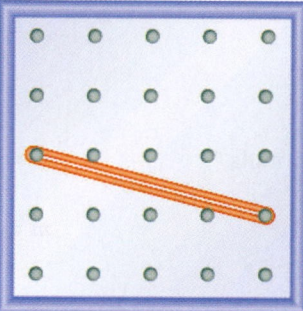

217.

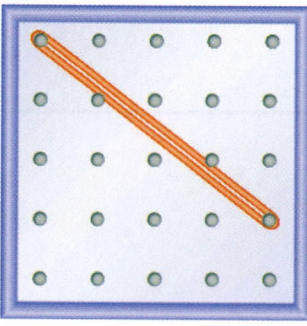

218.

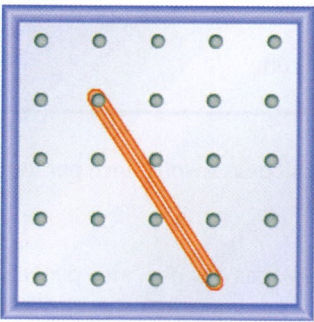

In the following exercises, model each slope. Draw a picture to show your results.

219. $\frac{2}{3}$

220. $\frac{3}{4}$

221. $\frac{1}{4}$

222. $\frac{4}{3}$

223. $-\frac{1}{2}$

224. $-\frac{3}{4}$

225. $-\frac{2}{3}$

226. $-\frac{3}{2}$

Use $m = \frac{rise}{run}$ **to find the Slope of a Line from its Graph**

In the following exercises, find the slope of each line shown.

227.

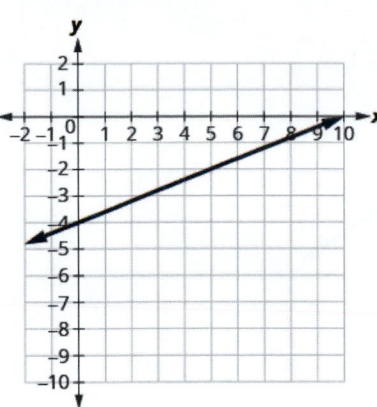

228.

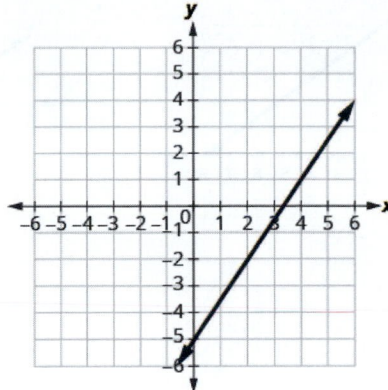

229.

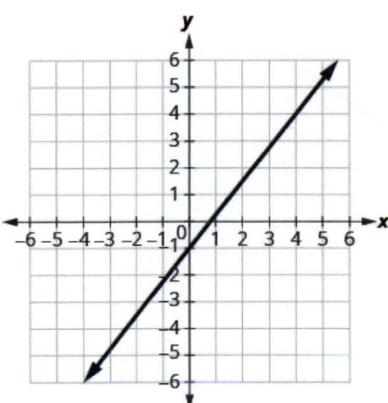

230.

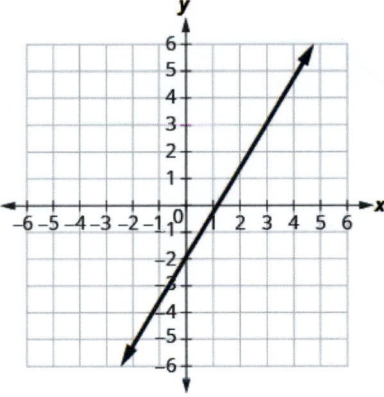

231.

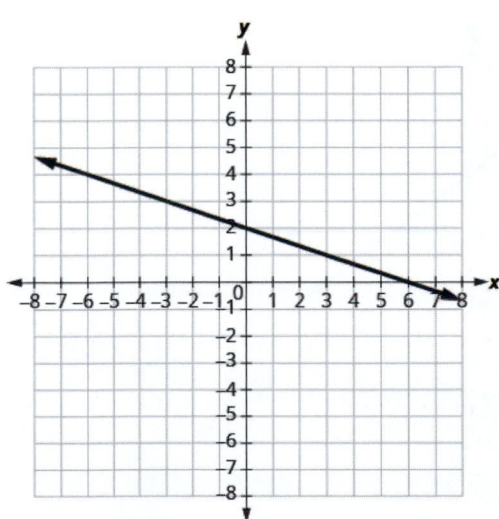

232.

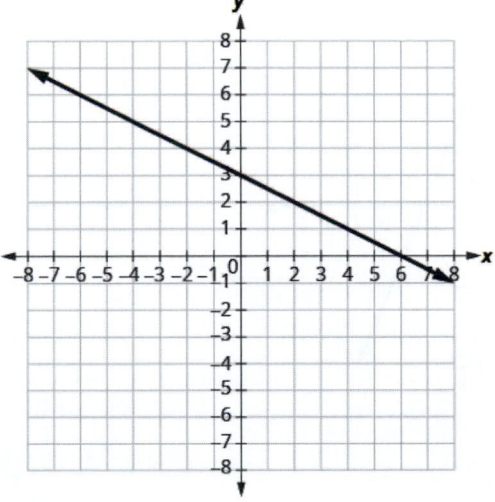

233.

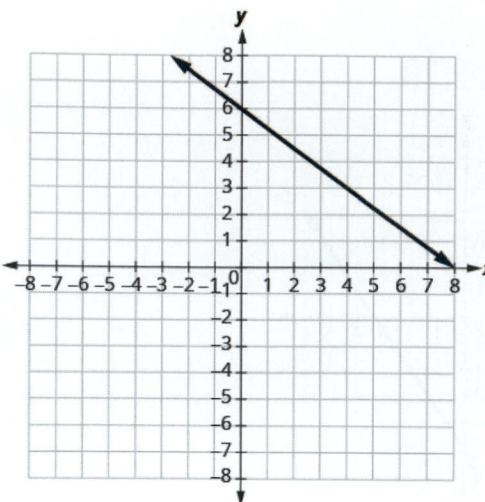

234.

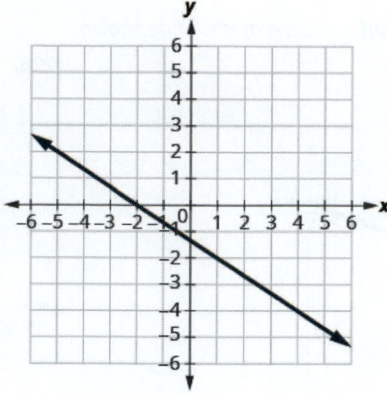

235.

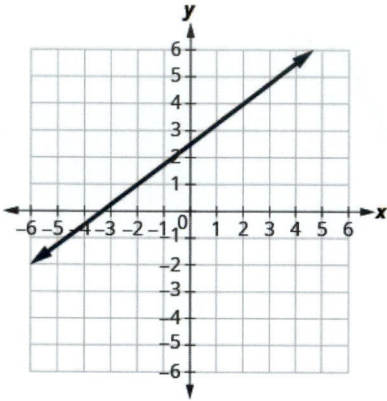

236.

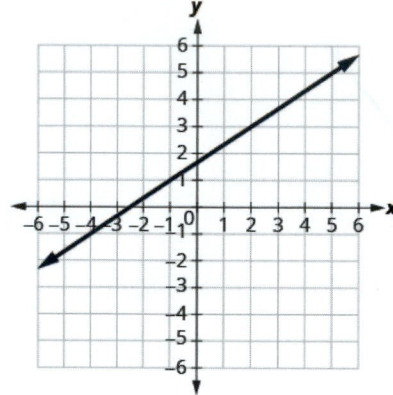

237.

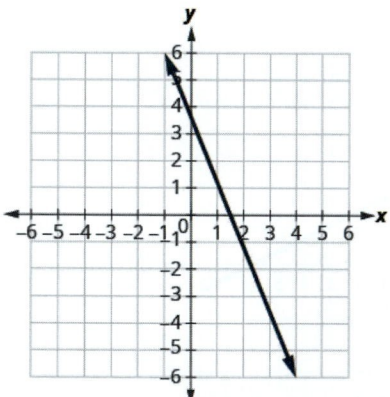

238.

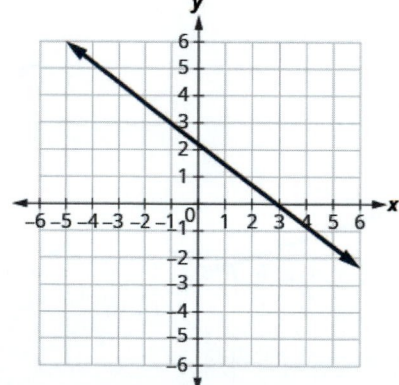

239.

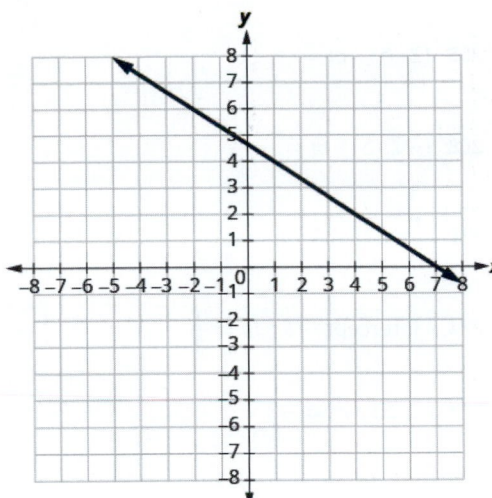

240.

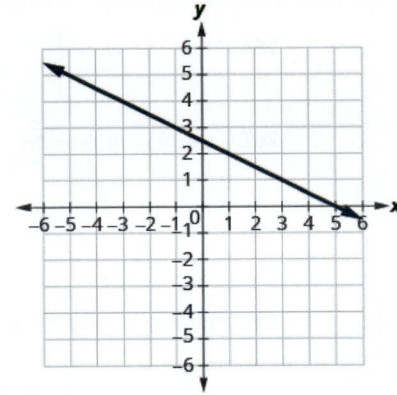

241.

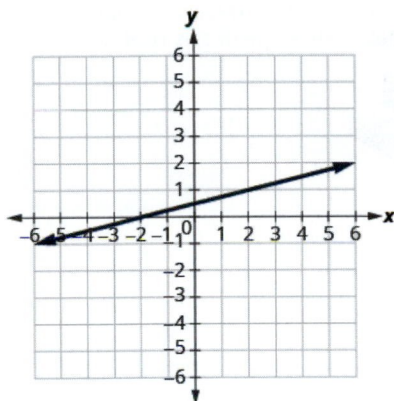

242.

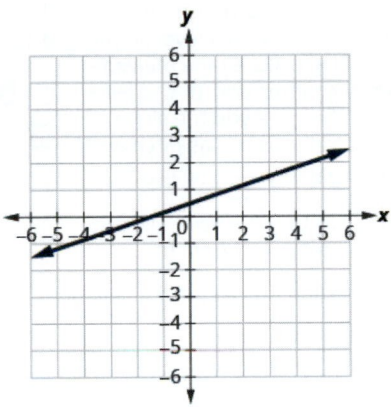

Find the Slope of Horizontal and Vertical Lines

In the following exercises, find the slope of each line.

243. $y = 3$

244. $y = 1$

245. $x = 4$

246. $x = 2$

247. $y = -2$

248. $y = -3$

249. $x = -5$

250. $x = -4$

Use the Slope Formula to find the Slope of a Line between Two Points

In the following exercises, use the slope formula to find the slope of the line between each pair of points.

251. $(1, 4), (3, 9)$

252. $(2, 3), (5, 7)$

253. $(0, 3), (4, 6)$

254. $(0, 1), (5, 4)$

255. $(2, 5), (4, 0)$

256. $(3, 6), (8, 0)$

257. $(-3, 3), (4, -5)$

258. $(-2, 4), (3, -1)$

259. $(-1, -2), (2, 5)$

260. $(-2, -1), (6, 5)$

261. $(4, -5), (1, -2)$

262. $(3, -6), (2, -2)$

Graph a Line Given a Point and the Slope

In the following exercises, graph each line with the given point and slope.

263. $(1, -2)$; $m = \frac{3}{4}$

264. $(1, -1)$; $m = \frac{2}{3}$

265. $(2, 5)$; $m = -\frac{1}{3}$

266. $(1, 4)$; $m = -\frac{1}{2}$

267. $(-3, 4)$; $m = -\frac{3}{2}$

268. $(-2, 5)$; $m = -\frac{5}{4}$

269. $(-1, -4)$; $m = \frac{4}{3}$

270. $(-3, -5)$; $m = \frac{3}{2}$

271. y-intercept 3; $m = -\frac{2}{5}$

272. y-intercept 5; $m = -\frac{4}{3}$

273. x-intercept -2; $m = \frac{3}{4}$

274. x-intercept -1; $m = \frac{1}{5}$

275. $(-3, 3)$; $m = 2$

276. $(-4, 2)$; $m = 4$

277. $(1, 5)$; $m = -3$

278. $(2, 3)$; $m = -1$

Everyday Math

279. Slope of a roof. An easy way to determine the slope of a roof is to set one end of a 12 inch level on the roof surface and hold it level. Then take a tape measure or ruler and measure from the other end of the level down to the roof surface. This will give you the slope of the roof. Builders, sometimes, refer to this as pitch and state it as an " x 12 pitch" meaning $\frac{x}{12}$, where x is the measurement from the roof to the level—the rise. It is also sometimes stated as an " x-in-12 pitch".

ⓐ What is the slope of the roof in this picture?

ⓑ What is the pitch in construction terms?

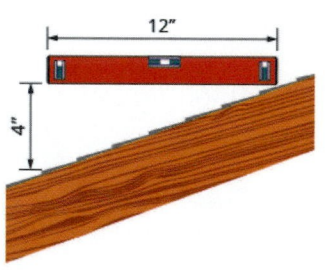

280. The slope of the roof shown here is measured with a 12″ level and a ruler. What is the slope of this roof?

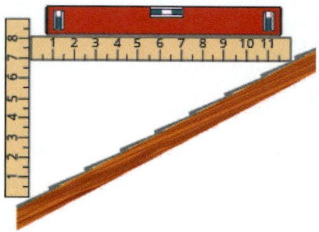

281. Road grade. A local road has a grade of 6%. The grade of a road is its slope expressed as a percent. Find the slope of the road as a fraction and then simplify. What rise and run would reflect this slope or grade?

282. Highway grade. A local road rises 2 feet for every 50 feet of highway.

ⓐ What is the slope of the highway?

ⓑ The grade of a highway is its slope expressed as a percent. What is the grade of this highway?

283. Wheelchair ramp. The rules for wheelchair ramps require a maximum 1-inch rise for a 12-inch run.

ⓐ How long must the ramp be to accommodate a 24-inch rise to the door?

ⓑ Create a model of this ramp.

284. Wheelchair ramp. A 1-inch rise for a 16-inch run makes it easier for the wheelchair rider to ascend a ramp.

ⓐ How long must a ramp be to easily accommodate a 24-inch rise to the door?

ⓑ Create a model of this ramp.

Writing Exercises

285. What does the sign of the slope tell you about a line?

286. How does the graph of a line with slope $m = \frac{1}{2}$ differ from the graph of a line with slope $m = 2$?

287. Why is the slope of a vertical line "undefined"?

Self Check

ⓐ *After completing the exercises, use this checklist to evaluate your mastery of the objectives of this section.*

I can...	Confidently	With some help	No-I don't get it
use geoboards to model slope.			
use $m = \frac{rise}{run}$ to find the slope of a line from its graph.			
find the slope of horizontal and vertical lines.			
use the slope formula to find the slope of a line between two points.			
graph a line given a point and the slope.			
solve slope applications.			

ⓑ *On a scale of 1–10, how would you rate your mastery of this section in light of your responses on the checklist? How can you improve this?*

4.5 | Use the Slope–Intercept Form of an Equation of a Line

Learning Objectives

By the end of this section, you will be able to:

> Recognize the relation between the graph and the slope–intercept form of an equation of a line
> Identify the slope and y-intercept form of an equation of a line
> Graph a line using its slope and intercept
> Choose the most convenient method to graph a line
> Graph and interpret applications of slope–intercept
> Use slopes to identify parallel lines
> Use slopes to identify perpendicular lines

Be Prepared!

Before you get started, take this readiness quiz.

1. Add: $\frac{x}{4} + \frac{1}{4}$.
   If you missed this problem, review **Example 1.77**.

2. Find the reciprocal of $\frac{3}{7}$.
   If you missed this problem, review **Example 1.70**.

3. Solve $2x - 3y = 12$ for y.
   If you missed this problem, review **Example 2.63**.

Recognize the Relation Between the Graph and the Slope–Intercept Form of an Equation of a Line

We have graphed linear equations by plotting points, using intercepts, recognizing horizontal and vertical lines, and using the point-slope method. Once we see how an equation in slope–intercept form and its graph are related, we'll have one more method we can use to graph lines.

In **Graph Linear Equations in Two Variables**, we graphed the line of the equation $y = \frac{1}{2}x + 3$ by plotting points. See **Figure 4.24**. Let's find the slope of this line.

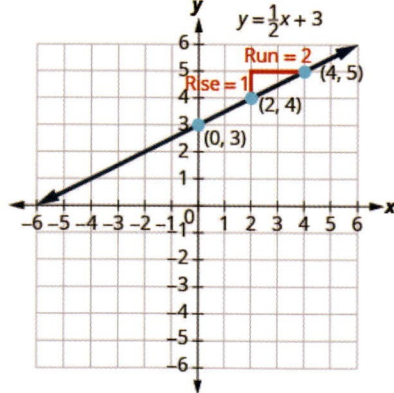

Figure 4.24

The red lines show us the rise is 1 and the run is 2. Substituting into the slope formula:

$$m = \frac{\text{rise}}{\text{run}}$$
$$m = \frac{1}{2}$$

What is the y-intercept of the line? The y-intercept is where the line crosses the y-axis, so y-intercept is $(0, 3)$. The equation of this line is:

$$y = \tfrac{1}{2}x + 3$$

Notice, the line has:

$$\text{slope } m = \tfrac{1}{2} \text{ and } y\text{-intercept } (0, 3)$$

When a linear equation is solved for y, the coefficient of the x term is the slope and the constant term is the y-coordinate of the y-intercept. We say that the equation $y = \tfrac{1}{2}x + 3$ is in slope–intercept form.

$$m = \tfrac{1}{2}; y\text{-intercept is } (0, 3)$$

$$y = \tfrac{1}{2}x + 3$$

$$y = mx + b$$

Slope-Intercept Form of an Equation of a Line

The **slope–intercept form** of an equation of a line with slope m and y-intercept, $(0, b)$ is,

$$y = mx + b$$

Sometimes the slope–intercept form is called the "y-form."

EXAMPLE 4.40

Use the graph to find the slope and y-intercept of the line, $y = 2x + 1$.

Compare these values to the equation $y = mx + b$.

⊘ Solution

To find the slope of the line, we need to choose two points on the line. We'll use the points $(0, 1)$ and $(1, 3)$.

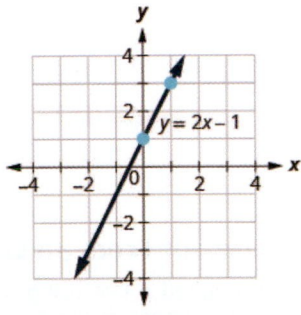

Find the rise and run.	$m = \dfrac{\text{rise}}{\text{run}}$
	$m = \dfrac{2}{1}$
	$m = 2$
Find the y-intercept of the line.	The y-intercept is the point $(0, 1)$.
We found slope $m = 2$ and y-intercept $(0, 1)$.	$y = 2x + 1$ $y = mx + b$

The slope is the same as the coefficient of x and the y-coordinate of the y-intercept is the same as the constant term.

> **TRY IT : :** 4.79

Use the graph to find the slope and y-intercept of the line $y = \frac{2}{3}x - 1$. Compare these values to the equation $y = mx + b$.

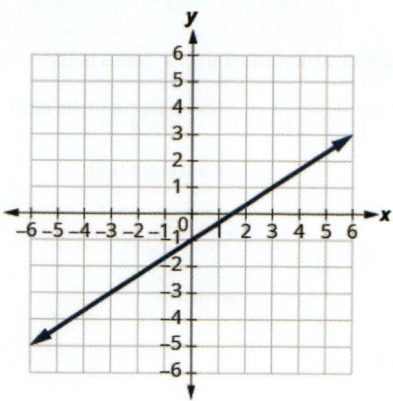

> **TRY IT : :** 4.80

Use the graph to find the slope and y-intercept of the line $y = \frac{1}{2}x + 3$. Compare these values to the equation $y = mx + b$.

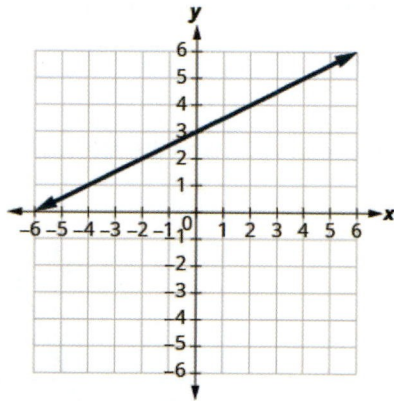

Identify the Slope and y-Intercept From an Equation of a Line

In **Understand Slope of a Line**, we graphed a line using the slope and a point. When we are given an equation in slope–intercept form, we can use the y-intercept as the point, and then count out the slope from there. Let's practice finding the values of the slope and y-intercept from the equation of a line.

EXAMPLE 4.41

Identify the slope and y-intercept of the line with equation $y = -3x + 5$.

⊘ **Solution**

We compare our equation to the slope–intercept form of the equation.

	$y = mx + b$
Write the equation of the line.	$y = -3x + 5$
Identify the slope.	$m = -3$

Identify the *y*-intercept. *y*-intercept is (0, 5)

> **TRY IT :: 4.81** Identify the slope and *y*-intercept of the line $y = \frac{2}{5}x - 1$.

> **TRY IT :: 4.82** Identify the slope and *y*-intercept of the line $y = -\frac{4}{3}x + 1$.

When an equation of a line is not given in slope–intercept form, our first step will be to solve the equation for y.

EXAMPLE 4.42

Identify the slope and *y*-intercept of the line with equation $x + 2y = 6$.

✓ **Solution**

This equation is not in slope–intercept form. In order to compare it to the slope–intercept form we must first solve the equation for y.

Solve for *y*.	$x + 2y = 6$
Subtract *x* from each side.	$2y = -x + 6$
Divide both sides by 2.	$\frac{2y}{2} = \frac{-x + 6}{2}$
Simplify.	$\frac{2y}{2} = \frac{-x}{2} + \frac{6}{2}$
$\left(\text{Remember: } \frac{a+b}{c} = \frac{a}{c} + \frac{b}{c}\right)$	
Simplify.	$y = -\frac{1}{2}x + 3$
Write the slope–intercept form of the equation of the line.	$y = mx + b$
Write the equation of the line.	$y = -\frac{1}{2}x + 3$
Identify the slope.	$m = -\frac{1}{2}$
Identify the *y*-intercept.	*y*-intercept is (0, 3)

> **TRY IT :: 4.83** Identify the slope and *y*-intercept of the line $x + 4y = 8$.

> **TRY IT :: 4.84** Identify the slope and *y*-intercept of the line $3x + 2y = 12$.

Graph a Line Using its Slope and Intercept

Now that we know how to find the slope and *y*-intercept of a line from its equation, we can graph the line by plotting the *y*-intercept and then using the slope to find another point.

EXAMPLE 4.43 HOW TO GRAPH A LINE USING ITS SLOPE AND INTERCEPT

Graph the line of the equation $y = 4x - 2$ using its slope and *y*-intercept.

⊘ **Solution**

Step 1. Find the slope–intercept form of the equation.	This equation is in slope–intercept form.	$y = 4x - 2$
Step 2. Identify the slope and y-intercept.	Use $y = mx + b$ Find the slope. Find the y-intercept.	$y = mx + b$ $y = 4x + (-2)$ $m = 4$ $b = -2, (0, -2)$
Step 3. Plot the y-intercept.	Plot $(0, -2)$.	
Step 4. Use the slope formula $m = \dfrac{\text{rise}}{\text{run}}$ to identify the rise and the run.	Identify the rise and the run.	$m = 4$ $\dfrac{\text{rise}}{\text{run}} = \dfrac{4}{1}$ $\text{rise} = 4$ $\text{run} = 1$
Step 5. Starting at the y-intercept, count out the rise and run to mark the second point.	Start at $(0, -2)$ and count the rise and the run. Up 4, right 1.	

Step 6. Connect the points with a line.	Connect the two points with a line.	

> **TRY IT : : 4.85**   Graph the line of the equation $y = 4x + 1$ using its slope and y-intercept.

> **TRY IT : : 4.86**   Graph the line of the equation $y = 2x - 3$ using its slope and y-intercept.

HOW TO : : GRAPH A LINE USING ITS SLOPE AND Y-INTERCEPT.

Step 1.   Find the slope-intercept form of the equation of the line.

Step 2.   Identify the slope and y-intercept.

Step 3.   Plot the y-intercept.

Step 4.   Use the slope formula $m = \frac{\text{rise}}{\text{run}}$ to identify the rise and the run.

Step 5.   Starting at the y-intercept, count out the rise and run to mark the second point.

Step 6.   Connect the points with a line.

EXAMPLE 4.44

Graph the line of the equation $y = -x + 4$ using its slope and y-intercept.

✓ **Solution**

	$y = mx + b$
The equation is in slope–intercept form.	$y = -x + 4$
Identify the slope and y-intercept.	$m = -1$
	y-intercept is $(0, 4)$
Plot the y-intercept.	See graph below.
Identify the rise and the run.	$m = \frac{-1}{1}$
Count out the rise and run to mark the second point.	rise -1, run 1

Draw the line.

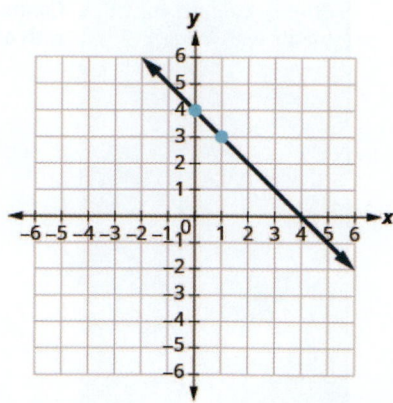

To check your work, you can find another point on the line and make sure it is a solution of the equation. In the graph we see the line goes through (4, 0).

Check.

$$y = -x + 4$$

$$0 \overset{?}{=} -4 + 4$$

$$0 = 0 ✓$$

> **TRY IT : : 4.87**   Graph the line of the equation $y = -x - 3$ using its slope and y-intercept.

> **TRY IT : : 4.88**   Graph the line of the equation $y = -x - 1$ using its slope and y-intercept.

EXAMPLE 4.45

Graph the line of the equation $y = -\frac{2}{3}x - 3$ using its slope and y-intercept.

⊘ **Solution**

$$y = mx + b$$

The equation is in slope–intercept form.	$y = -\frac{2}{3}x - 3$
Identify the slope and y-intercept.	$m = -\frac{2}{3}$; y-intercept is (0, –3)
Plot the y-intercept.	See graph below.
Identify the rise and the run.	
Count out the rise and run to mark the second point.	

Draw the line.

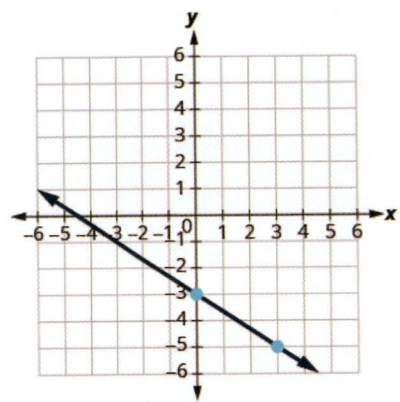

> **TRY IT : : 4.89**   Graph the line of the equation $y = -\frac{5}{2}x + 1$ using its slope and y-intercept.

> **TRY IT : : 4.90**   Graph the line of the equation $y = -\frac{3}{4}x - 2$ using its slope and y-intercept.

EXAMPLE 4.46

Graph the line of the equation $4x - 3y = 12$ using its slope and y-intercept.

✓ **Solution**

$$4x - 3y = 12$$

Find the slope–intercept form of the equation.	$-3y = -4x + 12$
	$-\dfrac{3y}{3} = \dfrac{-4x + 12}{-3}$
The equation is now in slope–intercept form.	$y = \dfrac{4}{3}x - 4$
Identify the slope and y-intercept.	$m = \dfrac{4}{3}$
	y-intercept is $(0, -4)$
Plot the y-intercept.	See graph below.
Identify the rise and the run; count out the rise and run to mark the second point.	

Draw the line.

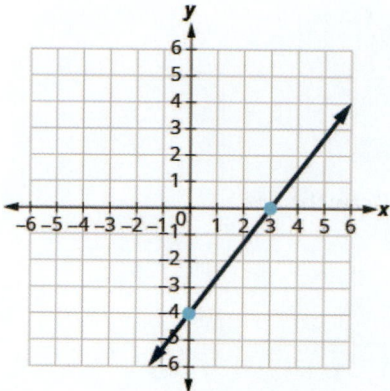

> TRY IT : : 4.91   Graph the line of the equation $2x - y = 6$ using its slope and y-intercept.

> TRY IT : : 4.92   Graph the line of the equation $3x - 2y = 8$ using its slope and y-intercept.

We have used a grid with x and y both going from about -10 to 10 for all the equations we've graphed so far. Not all linear equations can be graphed on this small grid. Often, especially in applications with real-world data, we'll need to extend the axes to bigger positive or smaller negative numbers.

EXAMPLE 4.47

Graph the line of the equation $y = 0.2x + 45$ using its slope and y-intercept.

⊘ **Solution**

We'll use a grid with the axes going from about -80 to 80.

	$y = mx + b$
The equation is in slope–intercept form.	$y = 0.2x + 45$
Identify the slope and y-intercept.	$m = 0.2$
	The y-intercept is (0, 45)
Plot the y-intercept.	See graph below.

Count out the rise and run to mark the second point. The slope is $m = 0.2$; in fraction form this means $m = \frac{2}{10}$. Given the scale of our graph, it would be easier to use the equivalent fraction $m = \frac{10}{50}$.

Draw the line.

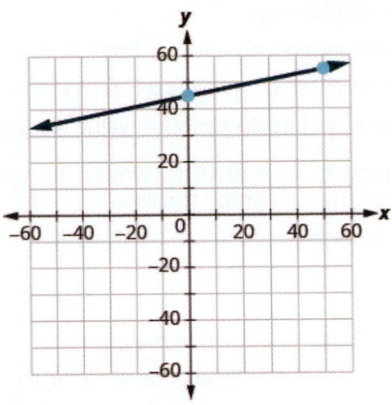

> **TRY IT : : 4.93**   Graph the line of the equation $y = 0.5x + 25$ using its slope and y-intercept.

> **TRY IT : : 4.94**   Graph the line of the equation $y = 0.1x - 30$ using its slope and y-intercept.

Now that we have graphed lines by using the slope and y-intercept, let's summarize all the methods we have used to graph lines. See **Figure 4.25**.

Methods to Graph Lines			
Point Plotting	**Slope–Intercept**	**Intercepts**	**Recognize Vertical and Horizontal Lines**
$x \mid y$	$y = mx + b$	$x \mid y$ 0 0	
Find three points. Plot the points, make sure they line up, then draw the line.	Find the slope and y-intercept. Start at the y-intercept, then count the slope to get a second point.	Find the intercepts and a third point. Plot the points, make sure they line up, then draw the line.	The equation has only one variable. $x = a$ vertical $y = b$ horizontal

Figure 4.25

Choose the Most Convenient Method to Graph a Line

Now that we have seen several methods we can use to graph lines, how do we know which method to use for a given equation?

While we could plot points, use the slope–intercept form, or find the intercepts for *any* equation, if we recognize the most convenient way to graph a certain type of equation, our work will be easier. Generally, plotting points is not the most efficient way to graph a line. We saw better methods in sections 4.3, 4.4, and earlier in this section. Let's look for some patterns to help determine the most convenient method to graph a line.

Here are six equations we graphed in this chapter, and the method we used to graph each of them.

	Equation	**Method**
#1	$x = 2$	Vertical line
#2	$y = 4$	Horizontal line
#3	$-x + 2y = 6$	Intercepts
#4	$4x - 3y = 12$	Intercepts
#5	$y = 4x - 2$	Slope–intercept
#6	$y = -x + 4$	Slope–intercept

Equations #1 and #2 each have just one variable. Remember, in equations of this form the value of that one variable is constant; it does not depend on the value of the other variable. Equations of this form have graphs that are vertical or horizontal lines.

In equations #3 and #4, both x and y are on the same side of the equation. These two equations are of the form $Ax + By = C$. We substituted $y = 0$ to find the x-intercept and $x = 0$ to find the y-intercept, and then found a third point by choosing another value for x or y.

Equations #5 and #6 are written in slope–intercept form. After identifying the slope and y-intercept from the equation we used them to graph the line.

This leads to the following strategy.

Strategy for Choosing the Most Convenient Method to Graph a Line

Consider the form of the equation.
- If it only has one variable, it is a vertical or horizontal line.
  - $x = a$ is a vertical line passing through the x-axis at a.
  - $y = b$ is a horizontal line passing through the y-axis at b.
- If y is isolated on one side of the equation, in the form $y = mx + b$, graph by using the slope and y-intercept.
  - Identify the slope and y-intercept and then graph.
- If the equation is of the form $Ax + By = C$, find the intercepts.
  - Find the x- and y-intercepts, a third point, and then graph.

EXAMPLE 4.48

Determine the most convenient method to graph each line.

ⓐ $y = -6$ ⓑ $5x - 3y = 15$ ⓒ $x = 7$ ⓓ $y = \frac{2}{5}x - 1$.

✓ Solution

ⓐ $y = -6$

This equation has only one variable, y. Its graph is a horizontal line crossing the y-axis at -6.

ⓑ $5x - 3y = 15$

This equation is of the form $Ax + By = C$. The easiest way to graph it will be to find the intercepts and one more point.

ⓒ $x = 7$

There is only one variable, x. The graph is a vertical line crossing the x-axis at 7.

ⓓ $y = \frac{2}{5}x - 1$

Since this equation is in $y = mx + b$ form, it will be easiest to graph this line by using the slope and y-intercept.

> **TRY IT : : 4.95**

Determine the most convenient method to graph each line: ⓐ $3x + 2y = 12$ ⓑ $y = 4$ ⓒ $y = \frac{1}{5}x - 4$ ⓓ $x = -7$.

> **TRY IT : : 4.96**

Determine the most convenient method to graph each line: ⓐ $x = 6$ ⓑ $y = -\frac{3}{4}x + 1$ ⓒ $y = -8$ ⓓ $4x - 3y = -1$.

Graph and Interpret Applications of Slope–Intercept

Many real-world applications are modeled by linear equations. We will take a look at a few applications here so you can see how equations written in slope–intercept form relate to real-world situations.

Usually when a linear equation models a real-world situation, different letters are used for the variables, instead of *x* and *y*. The variable names remind us of what quantities are being measured.

EXAMPLE 4.49

The equation $F = \frac{9}{5}C + 32$ is used to convert temperatures, C, on the Celsius scale to temperatures, F, on the Fahrenheit scale.

ⓐ Find the Fahrenheit temperature for a Celsius temperature of 0.

ⓑ Find the Fahrenheit temperature for a Celsius temperature of 20.

ⓒ Interpret the slope and *F*-intercept of the equation.

ⓓ Graph the equation.

⊘ Solution

ⓐ

Find the Fahrenheit temperature for a Celsius temperature of 0.	$F = \frac{9}{5}C + 32$
Find F when $C = 0$.	$F = \frac{9}{5}(0) + 32$
Simplify.	$F = 32$

ⓑ

Find the Fahrenheit temperature for a Celsius temperature of 20.	$F = \frac{9}{5}C + 32$
Find F when $C = 20$.	$F = \frac{9}{5}(20) + 32$
Simplify.	$F = 36 + 32$
Simplify.	$F = 68$

ⓒ Interpret the slope and *F*-intercept of the equation.

Even though this equation uses F and C, it is still in slope–intercept form.

$$y = mx + b$$

$$F = mC + b$$

$$F = \frac{9}{5}C + 32$$

The slope, $\frac{9}{5}$, means that the temperature Fahrenheit (*F*) increases 9 degrees when the temperature Celsius (*C*) increases 5 degrees.

The *F*-intercept means that when the temperature is $0°$ on the Celsius scale, it is $32°$ on the Fahrenheit scale.

ⓓ Graph the equation.

We'll need to use a larger scale than our usual. Start at the *F*-intercept $(0, 32)$ then count out the rise of 9 and the run of 5 to get a second point. See Figure 4.26.

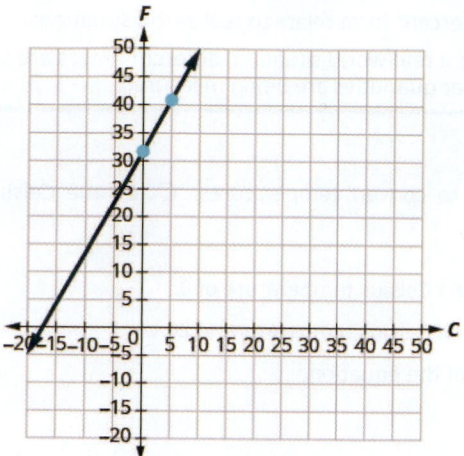

Figure 4.26

> **TRY IT : : 4.97**

The equation $h = 2s + 50$ is used to estimate a woman's height in inches, h, based on her shoe size, s.

ⓐ Estimate the height of a child who wears women's shoe size 0.

ⓑ Estimate the height of a woman with shoe size 8.

ⓒ Interpret the slope and h-intercept of the equation.

ⓓ Graph the equation.

> **TRY IT : : 4.98**

The equation $T = \frac{1}{4}n + 40$ is used to estimate the temperature in degrees Fahrenheit, T, based on the number

of cricket chirps, n, in one minute.

ⓐ Estimate the temperature when there are no chirps.

ⓑ Estimate the temperature when the number of chirps in one minute is 100.

ⓒ Interpret the slope and T-intercept of the equation.

ⓓ Graph the equation.

The cost of running some types business has two components—a *fixed cost* and a *variable cost*. The fixed cost is always the same regardless of how many units are produced. This is the cost of rent, insurance, equipment, advertising, and other items that must be paid regularly. The variable cost depends on the number of units produced. It is for the material and labor needed to produce each item.

EXAMPLE 4.50

Stella has a home business selling gourmet pizzas. The equation $C = 4p + 25$ models the relation between her weekly cost, C, in dollars and the number of pizzas, p, that she sells.

ⓐ Find Stella's cost for a week when she sells no pizzas.

ⓑ Find the cost for a week when she sells 15 pizzas.

ⓒ Interpret the slope and C-intercept of the equation.

ⓓ Graph the equation.

✓ Solution

ⓐ Find Stella's cost for a week when she sells no pizzas.	$C = 4p + 25$
Find C when $p = 0$.	$C = 4(0) + 25$
Simplify.	$C = 25$
	Stella's fixed cost is $25 when she sells no pizzas.
ⓑ Find the cost for a week when she sells 15 pizzas.	$C = 4p + 25$
Find C when $p = 15$.	$C = 4(15) + 25$
Simplify.	$C = 60 + 25$
	$C = 85$
	Stella's costs are $85 when she sells 15 pizzas.
ⓒ Interpret the slope and C-intercept of the equation.	$y = mx + b$ $C = 4p + 25$
	The slope, 4, means that the cost increases by $4 for each pizza Stella sells. The C-intercept means that even when Stella sells no pizzas, her costs for the week are $25.
ⓓ Graph the equation. We'll need to use a larger scale than our usual. Start at the C-intercept (0, 25) then count out the rise of 4 and the run of 1 to get a second point.	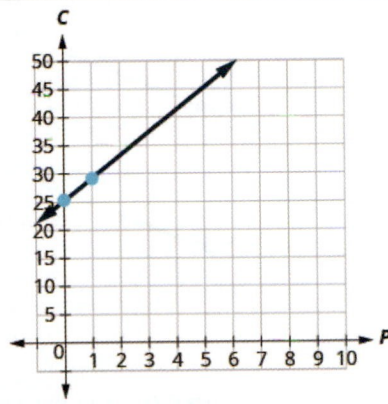

> **TRY IT : : 4.99**
>
> Sam drives a delivery van. The equation $C = 0.5m + 60$ models the relation between his weekly cost, C, in dollars and the number of miles, m, that he drives.
>
> ⓐ Find Sam's cost for a week when he drives 0 miles.
> ⓑ Find the cost for a week when he drives 250 miles.
> ⓒ Interpret the slope and C-intercept of the equation.
> ⓓ Graph the equation.

> **TRY IT : : 4.100**

Loreen has a calligraphy business. The equation $C = 1.8n + 35$ models the relation between her weekly cost, C, in dollars and the number of wedding invitations, n, that she writes.

ⓐ Find Loreen's cost for a week when she writes no invitations.
ⓑ Find the cost for a week when she writes 75 invitations.
ⓒ Interpret the slope and C-intercept of the equation.
ⓓ Graph the equation.

Use Slopes to Identify Parallel Lines

The slope of a line indicates how steep the line is and whether it rises or falls as we read it from left to right. Two lines that have the same slope are called parallel lines. Parallel lines never intersect.

We say this more formally in terms of the rectangular coordinate system. Two lines that have the same slope and different y-intercepts are called **parallel lines**. See **Figure 4.27**.

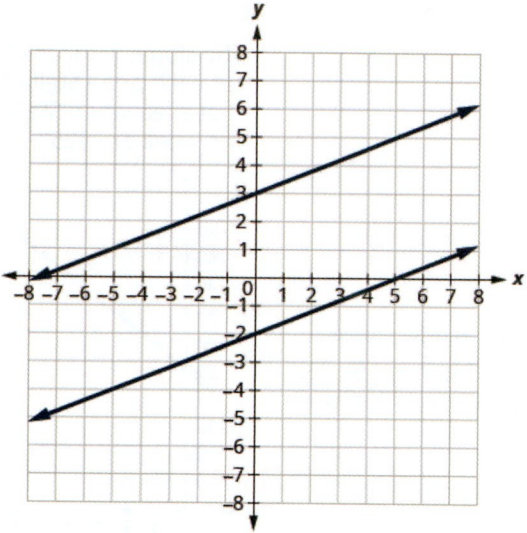

Figure 4.27 Verify that both lines have the same slope, $m = \dfrac{2}{5}$, and different y-intercepts.

What about vertical lines? The slope of a vertical line is undefined, so vertical lines don't fit in the definition above. We say that vertical lines that have different x-intercepts are parallel. See **Figure 4.28**.

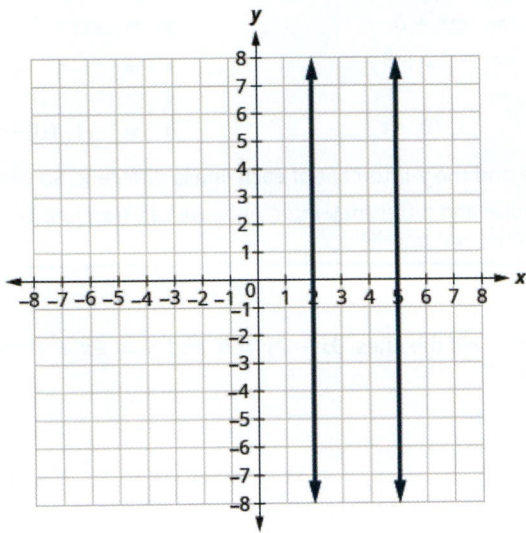

Figure 4.28 Vertical lines with diferent x-intercepts are parallel.

Parallel Lines

Parallel lines are lines in the same plane that do not intersect.

- Parallel lines have the same slope and different y-intercepts.
- If m_1 and m_2 are the slopes of two parallel lines then $m_1 = m_2$.
- Parallel vertical lines have different x-intercepts.

Let's graph the equations $y = -2x + 3$ and $2x + y = -1$ on the same grid. The first equation is already in slope–intercept form: $y = -2x + 3$. We solve the second equation for y:

$$\begin{aligned} 2x + y &= -1 \\ y &= -2x - 1 \end{aligned}$$

Graph the lines.

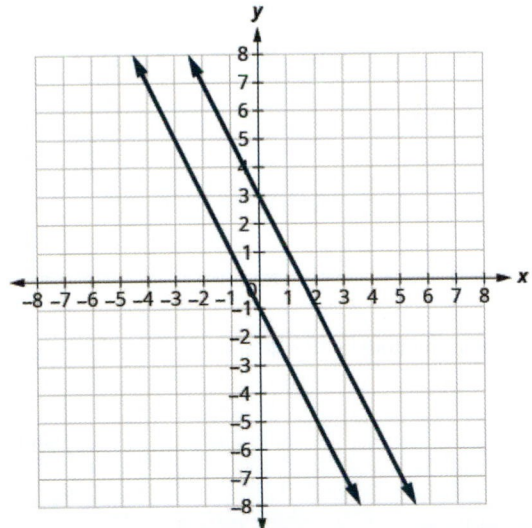

Notice the lines look parallel. What is the slope of each line? What is the y-intercept of each line?

$$
\begin{aligned}
y &= mx + b \\
y &= -2x + 3 \\
m &= -2 \\
b &= 3, (0, 3)
\end{aligned}
\qquad
\begin{aligned}
y &= mx + b \\
y &= -2x - 1 \\
m &= -2 \\
b &= -1, (0, -1)
\end{aligned}
$$

The slopes of the lines are the same and the y-intercept of each line is different. So we know these lines are parallel.

Since parallel lines have the same slope and different y-intercepts, we can now just look at the slope–intercept form of the equations of lines and decide if the lines are parallel.

EXAMPLE 4.51

Use slopes and y-intercepts to determine if the lines $3x - 2y = 6$ and $y = \frac{3}{2}x + 1$ are parallel.

⊘ **Solution**

$$3x - 2y = 6 \qquad \text{and} \qquad y = \frac{3}{2}x + 1$$

Solve the first equation for y.

$$-2y = -3x + 6$$

$$\frac{-2y}{-2} = \frac{-3x + 6}{-2}$$

The equation is now in slope–intercept form.

$$y = \frac{3}{2}x - 3$$

The equation of the second line is already in slope–intercept form.

$$y = \frac{3}{2}x + 1$$

Identify the slope and y-intercept of both lines.

$$
\begin{aligned}
y &= \tfrac{3}{2}x - 3 \\
y &= mx + b \\
m &= \tfrac{3}{2}
\end{aligned}
\qquad
\begin{aligned}
y &= \tfrac{3}{2}x + 1 \\
y &= mx + b \\
m &= \tfrac{3}{2}
\end{aligned}
$$

y-intercept is $(0, -3)$ \qquad y-intercept is $(0, 1)$

The lines have the same slope and different y-intercepts and so they are parallel. You may want to graph the lines to confirm whether they are parallel.

> **TRY IT :: 4.101**
>
> Use slopes and y-intercepts to determine if the lines $2x + 5y = 5$ and $y = -\frac{2}{5}x - 4$ are parallel.

> **TRY IT :: 4.102**
>
> Use slopes and y-intercepts to determine if the lines $4x - 3y = 6$ and $y = \frac{4}{3}x - 1$ are parallel.

EXAMPLE 4.52

Use slopes and y-intercepts to determine if the lines $y = -4$ and $y = 3$ are parallel.

⊘ **Solution**

$$y = -4 \qquad \text{and} \qquad y = 3$$
$$y = 0x - 4 \qquad\qquad\quad y = 0x + 3$$

Write each equation in slope–intercept form.

Since there is no x term we write $0x$.

Identify the slope and y-intercept of both lines.

$$
\begin{aligned}
y &= 0x - 4 \\
y &= mx + b \\
m &= 0
\end{aligned}
\qquad
\begin{aligned}
y &= 0x + 3 \\
y &= mx + b \\
m &= 0
\end{aligned}
$$

y-intercept is $(0, 4)$ \qquad y-intercept is $(0, 3)$

The lines have the same slope and different y-intercepts and so they are parallel.

There is another way you can look at this example. If you recognize right away from the equations that these are horizontal lines, you know their slopes are both 0. Since the horizontal lines cross the y-axis at $y = -4$ and at $y = 3$, we know the y-intercepts are $(0, -4)$ and $(0, 3)$. The lines have the same slope and different y-intercepts and so they are parallel.

> **TRY IT : : 4.103**   Use slopes and y-intercepts to determine if the lines $y = 8$ and $y = -6$ are parallel.

> **TRY IT : : 4.104**   Use slopes and y-intercepts to determine if the lines $y = 1$ and $y = -5$ are parallel.

EXAMPLE 4.53

Use slopes and y-intercepts to determine if the lines $x = -2$ and $x = -5$ are parallel.

⊘ **Solution**

$$x = -2 \text{ and } x = -5$$

Since there is no y, the equations cannot be put in slope–intercept form. But we recognize them as equations of vertical lines. Their x-intercepts are -2 and -5. Since their x-intercepts are different, the vertical lines are parallel.

> **TRY IT : : 4.105**   Use slopes and y-intercepts to determine if the lines $x = 1$ and $x = -5$ are parallel.

> **TRY IT : : 4.106**   Use slopes and y-intercepts to determine if the lines $x = 8$ and $x = -6$ are parallel.

EXAMPLE 4.54

Use slopes and y-intercepts to determine if the lines $y = 2x - 3$ and $-6x + 3y = -9$ are parallel. You may want to graph these lines, too, to see what they look like.

⊘ **Solution**

$$y = 2x - 3 \quad \text{and} \quad -6x + 3y = -9$$

The first equation is already in slope–intercept form.
Solve the second equation for y.

$$y = 2x - 3$$
$$-6x + 3y = -9$$
$$3y = 6x - 9$$
$$\frac{3y}{3} = \frac{6x - 9}{3}$$

The second equation is now in slope–intercept form.
Identify the slope and y-intercept of both lines.

$$y = 2x - 3$$
$$y = 2x - 3$$
$$y = mx + b$$
$$m = 2$$

y-intercept is $(0, -3)$

$$y = 2x - 3$$
$$y = mx + b$$
$$m = 2$$

y-intercept is $(0, -3)$

The lines have the same slope, but they also have the same y-intercepts. Their equations represent the same line. They are not parallel; they are the same line.

> **TRY IT : : 4.107**
>
> Use slopes and y-intercepts to determine if the lines $y = -\frac{1}{2}x - 1$ and $x + 2y = 2$ are parallel.

> **TRY IT : :** 4.108

Use slopes and y-intercepts to determine if the lines $y = \frac{3}{4}x - 3$ and $3x - 4y = 12$ are parallel.

Use Slopes to Identify Perpendicular Lines

Let's look at the lines whose equations are $y = \frac{1}{4}x - 1$ and $y = -4x + 2$, shown in **Figure 4.29**.

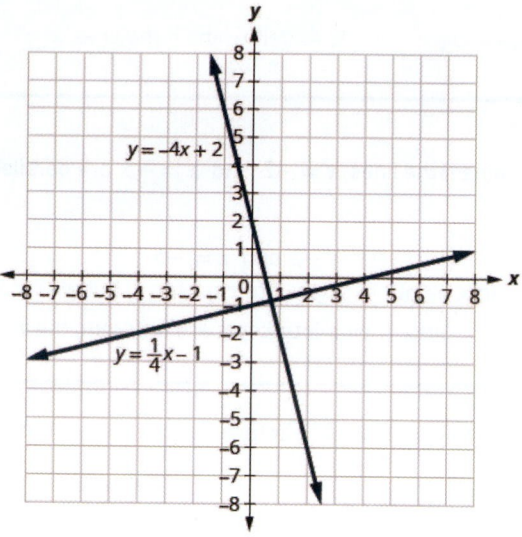

Figure 4.29

These lines lie in the same plane and intersect in right angles. We call these lines **perpendicular**.

What do you notice about the slopes of these two lines? As we read from left to right, the line $y = \frac{1}{4}x - 1$ rises, so its slope is positive. The line $y = -4x + 2$ drops from left to right, so it has a negative slope. Does it make sense to you that the slopes of two perpendicular lines will have opposite signs?

If we look at the slope of the first line, $m_1 = \frac{1}{4}$, and the slope of the second line, $m_2 = -4$, we can see that they are *negative reciprocals* of each other. If we multiply them, their product is -1.

$$m_1 \cdot m_2$$
$$\frac{1}{4}(-4)$$
$$-1$$

This is always true for perpendicular lines and leads us to this definition.

Perpendicular Lines

Perpendicular lines are lines in the same plane that form a right angle.

If m_1 and m_2 are the slopes of two perpendicular lines, then:

$$m_1 \cdot m_2 = -1 \quad \text{and} \quad m_1 = \frac{-1}{m_2}$$

Vertical lines and horizontal lines are always perpendicular to each other.

We were able to look at the slope–intercept form of linear equations and determine whether or not the lines were parallel. We can do the same thing for perpendicular lines.

We find the slope–intercept form of the equation, and then see if the slopes are negative reciprocals. If the product of the slopes is -1, the lines are perpendicular. Perpendicular lines may have the same y-intercepts.

EXAMPLE 4.55

Use slopes to determine if the lines, $y = -5x - 4$ and $x - 5y = 5$ are perpendicular.

✓ **Solution**

The first equation is in slope–intercept form.	$y = -5x - 4$	
Solve the second equation for y.	$x - 5y = 5$	
	$-5y = -x + 5$	
	$\dfrac{-5y}{-5} = \dfrac{-x + 5}{-5}$	
	$y = \frac{1}{5}x - 1$	
Identify the slope of each line.	$y = -5x - 4$	$y = \frac{1}{5}x - 1$
	$y = mx + b$	$y = mx + b$
	$m_1 = -5$	$m_2 = \frac{1}{5}$

The slopes are negative reciprocals of each other, so the lines are perpendicular. We check by multiplying the slopes,

$$m_1 \cdot m_2$$
$$-5\left(\frac{1}{5}\right)$$
$$-1 \checkmark$$

> **TRY IT :: 4.109**   Use slopes to determine if the lines $y = -3x + 2$ and $x - 3y = 4$ are perpendicular.

> **TRY IT :: 4.110**   Use slopes to determine if the lines $y = 2x - 5$ and $x + 2y = -6$ are perpendicular.

EXAMPLE 4.56

Use slopes to determine if the lines, $7x + 2y = 3$ and $2x + 7y = 5$ are perpendicular.

✓ **Solution**

Solve the equations for y.	$7x + 2y = 3$	$2x + 7y = 5$
	$2y = -7x + 3$	$7y = -2x + 5$
	$\dfrac{2y}{2} = \dfrac{-7x + 3}{2}$	$\dfrac{7y}{7} = \dfrac{-2x + 5}{7}$
	$y = -\frac{7}{2}x + \frac{3}{2}$	$y = -\frac{2}{7}x + \frac{5}{7}$
Identify the slope of each line.	$y = mx + b$	$y = mx + b$
	$m_1 = -\frac{7}{2}$	$m_2 = -\frac{2}{7}$

The slopes are reciprocals of each other, but they have the same sign. Since they are not negative reciprocals, the lines are not perpendicular.

> **TRY IT :: 4.111**   Use slopes to determine if the lines $5x + 4y = 1$ and $4x + 5y = 3$ are perpendicular.

> **TRY IT :: 4.112**   Use slopes to determine if the lines $2x - 9y = 3$ and $9x - 2y = 1$ are perpendicular.

▶ | **MEDIA : :**

Access this online resource for additional instruction and practice with graphs.

- **Explore the Relation Between a Graph and the Slope–Intercept Form of an Equation of a Line (https://openstax.org/l/25GraphPractice)**

4.5 EXERCISES

Practice Makes Perfect

Recognize the Relation Between the Graph and the Slope–Intercept Form of an Equation of a Line

In the following exercises, use the graph to find the slope and y-intercept of each line. Compare the values to the equation $y = mx + b$.

288.

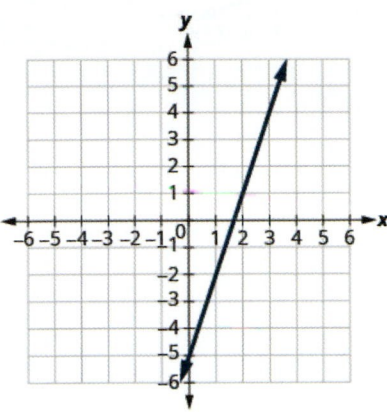

$y = 3x - 5$

289.

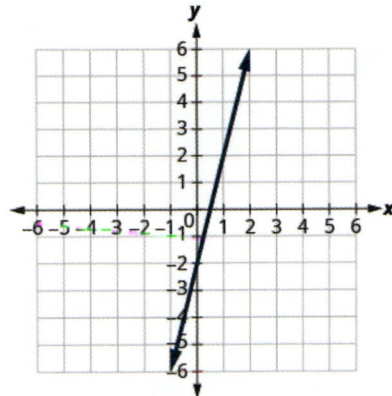

$y = 4x - 2$

290.

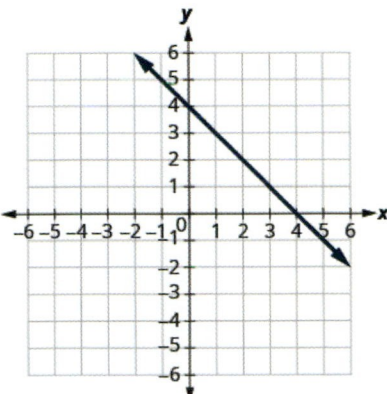

$y = -x + 4$

291.

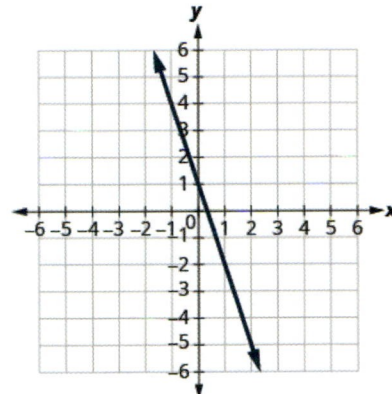

$y = -3x + 1$

292.

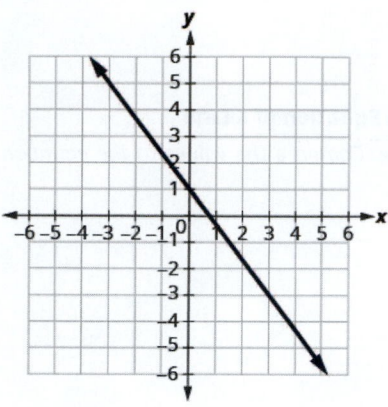

$$y = -\frac{4}{3}x + 1$$

293.

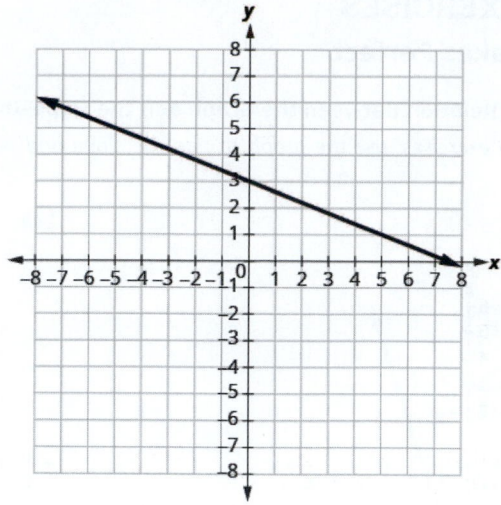

$$y = -\frac{2}{5}x + 3$$

Identify the Slope and y-Intercept From an Equation of a Line

In the following exercises, identify the slope and y-intercept of each line.

294. $y = -7x + 3$

295. $y = -9x + 7$

296. $y = 6x - 8$

297. $y = 4x - 10$

298. $3x + y = 5$

299. $4x + y = 8$

300. $6x + 4y = 12$

301. $8x + 3y = 12$

302. $5x - 2y = 6$

303. $7x - 3y = 9$

Graph a Line Using Its Slope and Intercept

In the following exercises, graph the line of each equation using its slope and y-intercept.

304. $y = x + 3$

305. $y = x + 4$

306. $y = 3x - 1$

307. $y = 2x - 3$

308. $y = -x + 2$

309. $y = -x + 3$

310. $y = -x - 4$

311. $y = -x - 2$

312. $y = -\frac{3}{4} - 1$

313. $y = -\frac{2}{5} - 3$

314. $y = -\frac{3}{5} + 2$

315. $y = -\frac{2}{3} + 1$

316. $3x - 4y = 8$

317. $4x - 3y = 6$

318. $y = 0.1x + 15$

319. $y = 0.3x + 25$

Choose the Most Convenient Method to Graph a Line

In the following exercises, determine the most convenient method to graph each line.

320. $x = 2$

321. $y = 4$

322. $y = 5$

323. $x = -3$

324. $y = -3x + 4$

325. $y = -5x + 2$

326. $x - y = 5$

327. $x - y = 1$

328. $y = \frac{2}{3}x - 1$

329. $y = \frac{4}{5}x - 3$

330. $y = -3$

331. $y = -1$

332. $3x - 2y = -12$

333. $2x - 5y = -10$

334. $y = -\frac{1}{4} + 3$

335. $y = -\frac{1}{3}x + 5$

Graph and Interpret Applications of Slope–Intercept

336. The equation $P = 31 + 1.75w$ models the relation between the amount of Tuyet's monthly water bill payment, P, in dollars, and the number of units of water, w, used.

ⓐ Find Tuyet's payment for a month when 0 units of water are used.

ⓑ Find Tuyet's payment for a month when 12 units of water are used.

ⓒ Interpret the slope and P-intercept of the equation.

ⓓ Graph the equation.

337. The equation $P = 28 + 2.54w$ models the relation between the amount of Randy's monthly water bill payment, P, in dollars, and the number of units of water, w, used.

ⓐ Find the payment for a month when Randy used 0 units of water.

ⓑ Find the payment for a month when Randy used 15 units of water.

ⓒ Interpret the slope and P-intercept of the equation.

ⓓ Graph the equation.

338. Bruce drives his car for his job. The equation $R = 0.575m + 42$ models the relation between the amount in dollars, R, that he is reimbursed and the number of miles, m, he drives in one day.

ⓐ Find the amount Bruce is reimbursed on a day when he drives 0 miles.

ⓑ Find the amount Bruce is reimbursed on a day when he drives 220 miles.

ⓒ Interpret the slope and R-intercept of the equation.

ⓓ Graph the equation.

339. Janelle is planning to rent a car while on vacation. The equation $C = 0.32m + 15$ models the relation between the cost in dollars, C, per day and the number of miles, m, she drives in one day.

ⓐ Find the cost if Janelle drives the car 0 miles one day.

ⓑ Find the cost on a day when Janelle drives the car 400 miles.

ⓒ Interpret the slope and C-intercept of the equation.

ⓓ Graph the equation.

340. Cherie works in retail and her weekly salary includes commission for the amount she sells. The equation $S = 400 + 0.15c$ models the relation between her weekly salary, S, in dollars and the amount of her sales, c, in dollars.

ⓐ Find Cherie's salary for a week when her sales were 0.

ⓑ Find Cherie's salary for a week when her sales were 3600.

ⓒ Interpret the slope and S-intercept of the equation.

ⓓ Graph the equation.

341. Patel's weekly salary includes a base pay plus commission on his sales. The equation $S = 750 + 0.09c$ models the relation between his weekly salary, S, in dollars and the amount of his sales, c, in dollars.

ⓐ Find Patel's salary for a week when his sales were 0.

ⓑ Find Patel's salary for a week when his sales were 18,540.

ⓒ Interpret the slope and S-intercept of the equation.

ⓓ Graph the equation.

342. Costa is planning a lunch banquet. The equation $C = 450 + 28g$ models the relation between the cost in dollars, C, of the banquet and the number of guests, g.

ⓐ Find the cost if the number of guests is 40.

ⓑ Find the cost if the number of guests is 80.

ⓒ Interpret the slope and C-intercept of the equation.

ⓓ Graph the equation.

343. Margie is planning a dinner banquet. The equation $C = 750 + 42g$ models the relation between the cost in dollars, C of the banquet and the number of guests, g.

ⓐ Find the cost if the number of guests is 50.

ⓑ Find the cost if the number of guests is 100.

ⓒ Interpret the slope and C-intercept of the equation.

ⓓ Graph the equation.

Use Slopes to Identify Parallel Lines

In the following exercises, use slopes and y-intercepts to determine if the lines are parallel.

344. $y = \frac{3}{4}x - 3; \quad 3x - 4y = -2$

345. $y = \frac{2}{3}x - 1; \quad 2x - 3y = -2$

346. $2x - 5y = -3; \quad y = \frac{2}{5}x + 1$

347. $3x - 4y = -2; \quad y = \frac{3}{4}x - 3$

348. $2x - 4y = 6; \quad x - 2y = 3$

349. $6x - 3y = 9; \quad 2x - y = 3$

350. $4x + 2y = 6; \quad 6x + 3y = 3$

351. $8x + 6y = 6; \quad 12x + 9y = 12$

352. $x = 5; \quad x = -6$

353. $x = 7; \quad x = -8$

354. $x = -4; \quad x = -1$

355. $x = -3; \quad x = -2$

356. $y = 2; \quad y = 6$

357. $y = 5; \quad y = 1$

358. $y = -4; \quad y = 3$

359. $y = -1; \quad y = 2$

360. $x - y = 2; \quad 2x - 2y = 4$

361. $4x + 4y = 8; \quad x + y = 2$

362. $x - 3y = 6; \quad 2x - 6y = 12$

363. $5x - 2y = 11; \quad 5x - y = 7$

364. $3x - 6y = 12; \quad 6x - 3y = 3$

365. $4x - 8y = 16; \quad x - 2y = 4$

366. $9x - 3y = 6; \quad 3x - y = 2$

367. $x - 5y = 10; \quad 5x - y = -10$

368. $7x - 4y = 8; \quad 4x + 7y = 14$

369. $9x - 5y = 4; \quad 5x + 9y = -1$

Use Slopes to Identify Perpendicular Lines

In the following exercises, use slopes and y-intercepts to determine if the lines are perpendicular.

370. $3x - 2y = 8; 2x + 3y = 6$

371. $x - 4y = 8; 4x + y = 2$

372. $2x + 5y = 3; 5x - 2y = 6$

373. $2x + 3y = 5; 3x - 2y = 7$

374. $3x - 2y = 1; 2x - 3y = 2$

375. $3x - 4y = 8; 4x - 3y = 6$

376. $5x + 2y = 6; 2x + 5y = 8$

377. $2x + 4y = 3; 6x + 3y = 2$

378. $4x - 2y = 5; 3x + 6y = 8$

379. $2x - 6y = 4; 12x + 4y = 9$

380. $6x - 4y = 5; 8x + 12y = 3$

381. $8x - 2y = 7; 3x + 12y = 9$

Everyday Math

382. The equation $C = \frac{5}{9}F - 17.8$ can be used to convert temperatures F, on the Fahrenheit scale to temperatures, C, on the Celsius scale.

ⓐ Explain what the slope of the equation means.

ⓑ Explain what the C–intercept of the equation means.

383. The equation $n = 4T - 160$ is used to estimate the number of cricket chirps, n, in one minute based on the temperature in degrees Fahrenheit, T.

ⓐ Explain what the slope of the equation means.

ⓑ Explain what the n–intercept of the equation means. Is this a realistic situation?

Writing Exercises

384. Explain in your own words how to decide which method to use to graph a line.

385. Why are all horizontal lines parallel?

Self Check

ⓐ *After completing the exercises, use this checklist to evaluate your mastery of the objectives of this section.*

I can...	Confidently	With some help	No-I don't get it!
recognize the relation between the graph and the slope-intercept form of an equation of a line.			
identify the slope and y-intercept from an equation of a line.			
graph a line using its slope and intercept.			
choose the most convenient method to graph a line.			
graph and interpret applications of slope-intercept.			
use slopes to identify parallel lines.			

ⓑ *After looking at the checklist, do you think you are well-prepared for the next section? Why or why not?*

4.6 | Find the Equation of a Line

Learning Objectives

By the end of this section, you will be able to:

> Find an equation of the line given the slope and y-intercept

> Find an equation of the line given the slope and a point
> Find an equation of the line given two points
> Find an equation of a line parallel to a given line
> Find an equation of a line perpendicular to a given line

Be Prepared!

Before you get started, take this readiness quiz.

1. Solve: $\frac{2}{3} = \frac{x}{5}$.

   If you missed this problem, review **Example 2.14**.

2. Simplify: $-\frac{2}{5}(x - 15)$.

   If you missed this problem, review **Example 1.133**.

How do online retailers know that 'you may also like' a particular item based on something you just ordered? How can economists know how a rise in the minimum wage will affect the unemployment rate? How do medical researchers create drugs to target cancer cells? How can traffic engineers predict the effect on your commuting time of an increase or decrease in gas prices? It's all mathematics.

You are at an exciting point in your mathematical journey as the mathematics you are studying has interesting applications in the real world.

The physical sciences, social sciences, and the business world are full of situations that can be modeled with linear equations relating two variables. Data is collected and graphed. If the data points appear to form a straight line, an equation of that line can be used to predict the value of one variable based on the value of the other variable.

To create a mathematical model of a linear relation between two variables, we must be able to find the equation of the line. In this section we will look at several ways to write the equation of a line. The specific method we use will be determined by what information we are given.

Find an Equation of the Line Given the Slope and *y*-Intercept

We can easily determine the slope and intercept of a line if the equation was written in slope–intercept form, $y = mx + b$.

Now, we will do the reverse—we will start with the slope and *y*-intercept and use them to find the equation of the line.

EXAMPLE 4.57

Find an equation of a line with slope -7 and *y*-intercept $(0, -1)$.

 **Solution**

Since we are given the slope and *y*-intercept of the line, we can substitute the needed values into the slope–intercept form, $y = mx + b$.

Name the slope.	$m = -7$
Name the *y*-intercept.	*y*-intercept $(0, -1)$
Substitute the values into $y = mx + b$.	$y = mx + b$
	$y = -7x + (-1)$
	$y = -7x + -1$

> **TRY IT : : 4.113**   Find an equation of a line with slope $\frac{2}{5}$ and y-intercept $(0, 4)$.

> **TRY IT : : 4.114**   Find an equation of a line with slope -1 and y-intercept $(0, -3)$.

Sometimes, the slope and intercept need to be determined from the graph.

EXAMPLE 4.58

Find the equation of the line shown.

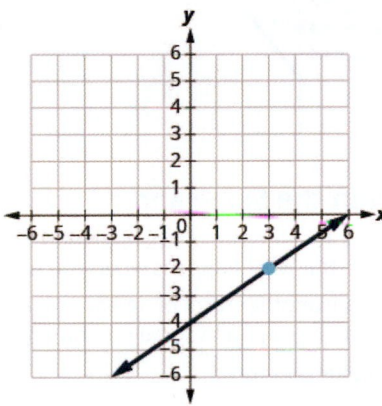

⊘ **Solution**

We need to find the slope and y-intercept of the line from the graph so we can substitute the needed values into the slope–intercept form, $y = mx + b$.

To find the slope, we choose two points on the graph.

The y-intercept is $(0, -4)$ and the graph passes through $(3, -2)$.

Find the slope by counting the rise and run.	$m = \dfrac{\text{rise}}{\text{run}}$
	$m = \dfrac{2}{3}$
Find the y-intercept.	y-intercept $(0, -4)$
Substitute the values into $y = mx + b$.	$y = mx + b$
	$y = \dfrac{2}{3}x - 4$

> **TRY IT : : 4.115**   Find the equation of the line shown in the graph.

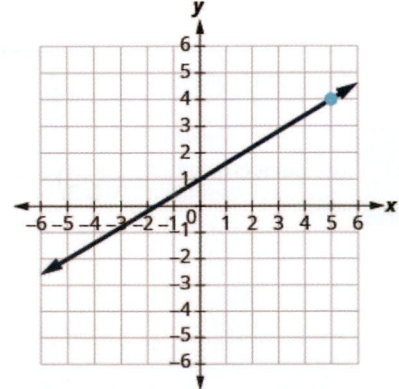

> **TRY IT : : 4.116**   Find the equation of the line shown in the graph.

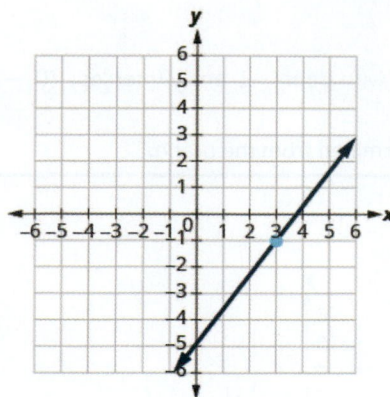

Find an Equation of the Line Given the Slope and a Point

Finding an equation of a line using the slope–intercept form of the equation works well when you are given the slope and y-intercept or when you read them off a graph. But what happens when you have another point instead of the y-intercept?

We are going to use the slope formula to derive another form of an equation of the line. Suppose we have a line that has slope m and that contains some specific point (x_1, y_1) and some other point, which we will just call (x, y). We can write the slope of this line and then change it to a different form.

$$m = \frac{y - y_1}{x - x_1}$$

Multiply both sides of the equation by $x - x_1$.
$$m(x - x_1) = \left(\frac{y - y_1}{x - x_1}\right)(x - x_1)$$

Simplify.
$$m(x - x_1) = y - y_1$$

Rewrite the equation with the y terms on the left.
$$y - y_1 = m(x - x_1)$$

This format is called the point–slope form of an equation of a line.

Point–slope Form of an Equation of a Line

The **point–slope form** of an equation of a line with slope m and containing the point (x_1, y_1) is

$$y - y_1 = m(x - x_1)$$

We can use the point–slope form of an equation to find an equation of a line when we are given the slope and one point. Then we will rewrite the equation in slope–intercept form. Most applications of linear equations use the the slope–intercept form.

EXAMPLE 4.59   FIND AN EQUATION OF A LINE GIVEN THE SLOPE AND A POINT

Find an equation of a line with slope $m = \frac{2}{5}$ that contains the point $(10, 3)$. Write the equation in slope–intercept form.

✓ **Solution**

Step 1. Identify the slope.	The slope is given.	$m = \frac{2}{5}$
Step 2. Identify the point.	The point is given.	$\begin{pmatrix} x, & y, \\ 10, & 3 \end{pmatrix}$

Step 3. Substitute the values into the point–slope form, $y - y_1 = m(x - x_1)$.		$y - y_1 = m(x - x_1)$ $y - 3 = \frac{2}{5}(x - 10)$
	Simplify.	$y - 3 = \frac{2}{5}x - 4$
Step 4. Write the equation in slope–intercept form.		$y = \frac{2}{5}x - 1$

> **TRY IT : : 4.117**   Find an equation of a line with slope $m = \frac{5}{6}$ and containing the point $(6, 3)$.

> **TRY IT : : 4.118**   Find an equation of a line with slope $m = \frac{2}{3}$ and containing thepoint $(9, 2)$.

HOW TO : : FIND AN EQUATION OF A LINE GIVEN THE SLOPE AND A POINT.

Step 1.   Identify the slope.

Step 2.   Identify the point.

Step 3.   Substitute the values into the point-slope form, $y - y_1 = m(x - x_1)$.

Step 4.   Write the equation in slope–intercept form.

EXAMPLE 4.60

Find an equation of a line with slope $m = -\frac{1}{3}$ that contains the point $(6, -4)$. Write the equation in slope–intercept form.

✓ **Solution**

Since we are given a point and the slope of the line, we can substitute the needed values into the point–slope form, $y - y_1 = m(x - x_1)$.

Identify the slope.	$m = -\frac{1}{3}$
Identify the point.	$\left(\overset{x_1}{6}, \overset{y_1}{-4}\right)$
Substitute the values into $y - y_1 = m(x - x_1)$.	$y - y_1 = m(x - x_1)$
	$y - (-4) = -\frac{1}{3}(x - 6)$
Simplify.	$y + 4 = -\frac{1}{3}x + 2$
Write in slope–intercept form.	$y = -\frac{1}{3}x - 2$

> **TRY IT : : 4.119**   Find an equation of a line with slope $m = -\frac{2}{5}$ and containing the point $(10, -5)$.

> **TRY IT :: 4.120**   Find an equation of a line with slope $m = -\frac{3}{4}$, and containing the point $(4, -7)$.

EXAMPLE 4.61

Find an equation of a horizontal line that contains the point $(-1, 2)$. Write the equation in slope–intercept form.

⊘ **Solution**

Every horizontal line has slope 0. We can substitute the slope and points into the point–slope form, $y - y_1 = m(x - x_1)$.

Identify the slope.	$m = 0$
Identify the point.	$\begin{pmatrix} x_1 & y_1 \\ -1, & 2 \end{pmatrix}$
Substitute the values into $y - y_1 = m(x - x_1)$.	$y - y_1 = m(x - x_1)$
	$y - 2 = 0(x - (-1))$
Simplify.	$y - 2 = 0(x + 1)$
	$y - 2 = 0$
	$y = 2$
Write in slope–intercept form.	It is in y-form, but could be written $y = 0x + 2$.

Did we end up with the form of a horizontal line, $y = a$?

> **TRY IT :: 4.121**   Find an equation of a horizontal line containing the point $(-3, 8)$.

> **TRY IT :: 4.122**   Find an equation of a horizontal line containing the point $(-1, 4)$.

Find an Equation of the Line Given Two Points

When real-world data is collected, a linear model can be created from two data points. In the next example we'll see how to find an equation of a line when just two points are given.

We have two options so far for finding an equation of a line: slope–intercept or point–slope. Since we will know two points, it will make more sense to use the point–slope form.

But then we need the slope. Can we find the slope with just two points? Yes. Then, once we have the slope, we can use it and one of the given points to find the equation.

EXAMPLE 4.62   FIND AN EQUATION OF A LINE GIVEN TWO POINTS

Find an equation of a line that contains the points $(5, 4)$ and $(3, 6)$. Write the equation in slope–intercept form.

⊘ **Solution**

Step 1. Find the slope using the given points.	To use the point-slope form, we first find the slope.	$m = \dfrac{y_2 - y_1}{x_2 - x_1}$ $m = \dfrac{6 - 4}{3 - 5}$ $m = \dfrac{2}{-2}$ $m = -1$

Step 2. Choose one point.	Choose either point.	$\left(\overset{x_1}{5},\ \overset{y_1}{4}\right)$
Step 3. Substitute the values into the point-slope form, $y - y_1 = m(x - x_1)$.	Simplify.	$y - y_1 = m(x - x_1)$ $y - 4 = -1(x - 5)$ $y - 4 = -1x + 5$
Step 4. Write the equation in slope–intercept form.		$y = -1x + 9$

Use the point $(3,\ 6)$ and see that you get the same equation.

> **TRY IT : :** 4.123   Find an equation of a line containing the points $(3,\ 1)$ and $(5,\ 6)$.

> **TRY IT : :** 4.124   Find an equation of a line containing the points $(1,\ 4)$ and $(6,\ 2)$.

HOW TO : : FIND AN EQUATION OF A LINE GIVEN TWO POINTS.

Step 1.   Find the slope using the given points.

Step 2.   Choose one point.

Step 3.   Substitute the values into the point-slope form, $y - y_1 = m(x - x_1)$.

Step 4.   Write the equation in slope–intercept form.

EXAMPLE 4.63

Find an equation of a line that contains the points $(-3,\ -1)$ and $(2,\ -2)$. Write the equation in slope–intercept form.

⊘ **Solution**

Since we have two points, we will find an equation of the line using the point–slope form. The first step will be to find the slope.

Find the slope of the line through $(-3, -1)$ and $(2, -2)$.	$m = \dfrac{y_2 - y_1}{x_2 - x_1}$
	$m = \dfrac{-2 - (-1)}{2 - (-3)}$
	$m = \dfrac{-1}{5}$
	$m = -\dfrac{1}{5}$
Choose either point.	$\begin{pmatrix} x_1 & y_1 \\ 2 & -2 \end{pmatrix}$
Substitute the values into $y - y_1 = m(x - x_1)$.	$y - y_1 = m(x - x_1)$
	$y - (-2) = -\dfrac{1}{5}(x - 2)$
	$y + 2 = -\dfrac{1}{5}x + \dfrac{2}{5}$
Write in slope–intercept form.	$y = -\dfrac{1}{5}x - \dfrac{8}{5}$

> **TRY IT :: 4.125**   Find an equation of a line containing the points $(-2, -4)$ and $(1, -3)$.

> **TRY IT :: 4.126**   Find an equation of a line containing the points $(-4, -3)$ and $(1, -5)$.

EXAMPLE 4.64

Find an equation of a line that contains the points $(-2, 4)$ and $(-2, -3)$. Write the equation in slope–intercept form.

⊘ **Solution**

Again, the first step will be to find the slope.

Find the slope of the line through $(-2, 4)$ and $(-2, -3)$.

$$m = \dfrac{y_2 - y_1}{x_2 - x_1}$$

$$m = \dfrac{-3 - 4}{-2 - (-2)}$$

$$m = \dfrac{-7}{0}$$

The slope is undefined.

This tells us it is a vertical line. Both of our points have an x-coordinate of -2. So our equation of the line is $x = -2$. Since there is no y, we cannot write it in slope–intercept form.

You may want to sketch a graph using the two given points. Does the graph agree with our conclusion that this is a vertical line?

> **TRY IT :: 4.127**   Find an equation of a line containing the points $(5, 1)$ and $(5, -4)$.

> **TRY IT :: 4.128**   Find an equaion of a line containing the points $(-4, 4)$ and $(-4, 3)$.

We have seen that we can use either the slope–intercept form or the point–slope form to find an equation of a line. Which form we use will depend on the information we are given. This is summarized in Table 4.46.

To Write an Equation of a Line		
If given:	Use:	Form:
Slope and y-intercept	slope–intercept	$y = mx + b$
Slope and a point	point–slope	$y - y_1 = m(x - x_1)$
Two points	point–slope	$y - y_1 = m(x - x_1)$

Table 4.46

Find an Equation of a Line Parallel to a Given Line

Suppose we need to find an equation of a line that passes through a specific point and is parallel to a given line. We can use the fact that parallel lines have the same slope. So we will have a point and the slope—just what we need to use the point–slope equation.

First let's look at this graphically.

The graph shows the graph of $y = 2x - 3$. We want to graph a line parallel to this line and passing through the point $(-2, 1)$.

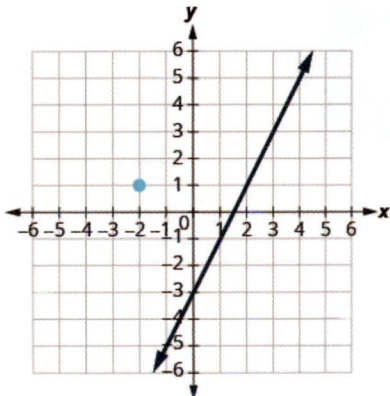

We know that parallel lines have the same slope. So the second line will have the same slope as $y = 2x - 3$. That slope is $m_{\parallel} = 2$. We'll use the notation $m_{\parallel}$ to represent the slope of a line parallel to a line with slope m. (Notice that the subscript $\parallel$ looks like two parallel lines.)

The second line will pass through $(-2, 1)$ and have $m = 2$. To graph the line, we start at $(-2, 1)$ and count out the rise and run. With $m = 2$ (or $m = \frac{2}{1}$), we count out the rise 2 and the run 1. We draw the line.

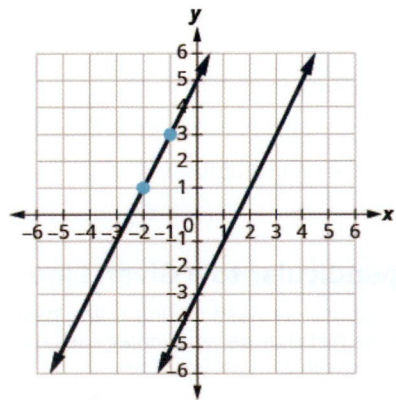

Do the lines appear parallel? Does the second line pass through $(-2, 1)$?

Now, let's see how to do this algebraically.

We can use either the slope–intercept form or the point-slope form to find an equation of a line. Here we know one point and can find the slope. So we will use the point–slope form.

EXAMPLE 4.65 HOW TO FIND AN EQUATION OF A LINE PARALLEL TO A GIVEN LINE

Find an equation of a line parallel to $y = 2x - 3$ that contains the point $(-2, 1)$. Write the equation in slope–intercept form.

✓ **Solution**

Step 1. Find the slope of the given line.	The line is in slope–intercept form, $y = 2x - 3$.	$m = 2$
Step 2. Find the slope of the parallel line.	Parallel lines have the same slope.	$m_{\shortparallel} = 2$
Step 3. Identify the point.	The given point is, $(-2, 1)$.	$\begin{pmatrix} x_1, & y_1 \\ -2, & 1 \end{pmatrix}$
Step 4. Substitute the values into the point-slope form, $y - y_1 = m(x - x_1)$.	Simplify.	$y - y_1 = m(x - x_1)$ $y - 1 = 2(x - (-2))$ $y - 1 = 2(x + 2)$ $y - 1 = 2x + 4$
Step 5. Write the equation in slope–intercept form.		$y = 2x + 5$

Does this equation make sense? What is the y-intercept of the line? What is the slope?

> **TRY IT :: 4.129**
>
> Find an equation of a line parallel to the line $y = 3x + 1$ that contains the point $(4, 2)$. Write the equation in slope–intercept form.

> **TRY IT :: 4.130**   Find an equation of a line parallel to the line $y = \frac{1}{2}x - 3$ that contains the point $(6, 4)$.

HOW TO :: FIND AN EQUATION OF A LINE PARALLEL TO A GIVEN LINE.

Step 1.   Find the slope of the given line.

Step 2.   Find the slope of the parallel line.

Step 3.   Identify the point.

Step 4.   Substitute the values into the point–slope form, $y - y_1 = m(x - x_1)$.

Step 5.   Write the equation in slope–intercept form.

Find an Equation of a Line Perpendicular to a Given Line

Now, let's consider perpendicular lines. Suppose we need to find a line passing through a specific point and which is perpendicular to a given line. We can use the fact that perpendicular lines have slopes that are negative reciprocals. We will again use the point–slope equation, like we did with parallel lines.

The graph shows the graph of $y = 2x - 3$. Now, we want to graph a line perpendicular to this line and passing through $(-2, 1)$.

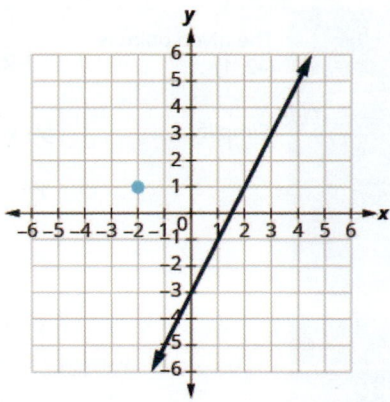

We know that perpendicular lines have slopes that are negative reciprocals. We'll use the notation $m_\perp$ to represent the slope of a line perpendicular to a line with slope m. (Notice that the subscript $\perp$ looks like the right angles made by two perpendicular lines.)

$$y = 2x - 3 \quad \text{perpendicular line}$$

$$m = 2 \qquad m_\perp = -\frac{1}{2}$$

We now know the perpendicular line will pass through $(-2, 1)$ with $m_\perp = -\frac{1}{2}$.

To graph the line, we will start at $(-2, 1)$ and count out the rise -1 and the run 2. Then we draw the line.

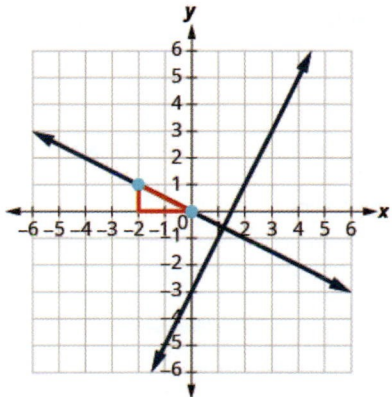

Do the lines appear perpendicular? Does the second line pass through $(-2, 1)$?

Now, let's see how to do this algebraically. We can use either the slope–intercept form or the point–slope form to find an equation of a line. In this example we know one point, and can find the slope, so we will use the point–slope form.

EXAMPLE 4.66   HOW TO FIND AN EQUATION OF A LINE PERPENDICULAR TO A GIVEN LINE

Find an equation of a line perpendicular to $y = 2x - 3$ that contains the point $(-2, 1)$. Write the equation in slope–intercept form.

✓ **Solution**

Step 1. Find the slope of the given line.	The line is in slope–intercept form, $y = 2x - 3$.	$m = 2$
Step 2. Find the slope of the perpendicular line.	The slopes of perpendicular lines are negative reciprocals.	$m_\perp = -\dfrac{1}{2}$

Step 3. Identify the point.	The given point is, $(-2, 1)$	$\begin{pmatrix} x, & y_1 \\ -2, & 1 \end{pmatrix}$
Step 4. Substitute the values into the point–slope form, $y - y_1 = m(x - x_1)$.	Simplify.	$y - y_1 = m(x - x_1)$ $y - 1 = -\frac{1}{2}(x - (-2))$ $y - 1 = -\frac{1}{2}(x + 2)$ $y - 1 = -\frac{1}{2}x - 1$
Step 5. Write the equation in slope–intercept form.		$y = -\frac{1}{2}x$

> **TRY IT : : 4.131**
>
> Find an equation of a line perpendicular to the line $y = 3x + 1$ that contains the point $(4, 2)$. Write the equation in slope–intercept form.

> **TRY IT : : 4.132**
>
> Find an equation of a line perpendicular to the line $y = \frac{1}{2}x - 3$ that contains the point $(6, 4)$.

HOW TO : : FIND AN EQUATION OF A LINE PERPENDICULAR TO A GIVEN LINE.

Step 1.   Find the slope of the given line.

Step 2.   Find the slope of the perpendicular line.

Step 3.   Identify the point.

Step 4.   Substitute the values into the point–slope form, $y - y_1 = m(x - x_1)$.

Step 5.   Write the equation in slope–intercept form.

EXAMPLE 4.67

Find an equation of a line perpendicular to $x = 5$ that contains the point $(3, -2)$. Write the equation in slope–intercept form.

⊘ **Solution**

Again, since we know one point, the point–slope option seems more promising than the slope–intercept option. We need the slope to use this form, and we know the new line will be perpendicular to $x = 5$. This line is vertical, so its perpendicular will be horizontal. This tells us the $m_\perp = 0$.

Identify the point.	$(3, -2)$
Identify the slope of the perpendicular line.	$m_\perp = 0$
Substitute the values into $y - y_1 = m(x - x_1)$.	$y - y_1 = m(x - x_1)$ $y - (-2) = 0(x - 3)$
Simplify.	$y + 2 = 0$ $y = -2$

Sketch the graph of both lines. Do they appear to be perpendicular?

> **TRY IT : : 4.133**

Find an equation of a line that is perpendicular to the line $x = 4$ that contains the point $(4, -5)$. Write the equation in slope–intercept form.

> **TRY IT : : 4.134**

Find an equation of a line that is perpendicular to the line $x = 2$ that contains the point $(2, -1)$. Write the equation in slope–intercept form.

In **Example 4.67**, we used the point–slope form to find the equation. We could have looked at this in a different way. We want to find a line that is perpendicular to $x = 5$ that contains the point $(3, -2)$. The graph shows us the line $x = 5$ and the point $(3, -2)$.

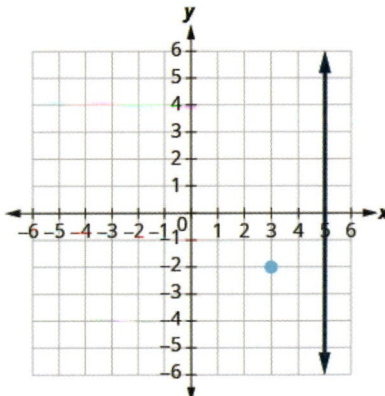

We know every line perpendicular to a vetical line is horizontal, so we will sketch the horizontal line through $(3, -2)$.

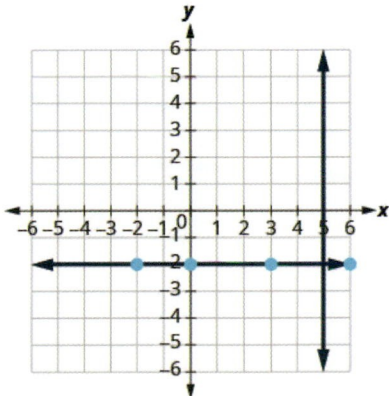

Do the lines appear perpendicular?

If we look at a few points on this horizontal line, we notice they all have y-coordinates of -2. So, the equation of the line perpendicular to the vertical line $x = 5$ is $y = -2$.

EXAMPLE 4.68

Find an equation of a line that is perpendicular to $y = -4$ that contains the point $(-4, 2)$. Write the equation in slope–intercept form.

⊘ **Solution**

The line $y = -4$ is a horizontal line. Any line perpendicular to it must be vertical, in the form $x = a$. Since the perpendicular line is vertical and passes through $(-4, 2)$, every point on it has an x-coordinate of -4. The equation of the perpendicular line is $x = -4$. You may want to sketch the lines. Do they appear perpendicular?

> | **TRY IT : :** 4.135

Find an equation of a line that is perpendicular to the line $y = 1$ that contains the point $(-5, 1)$. Write the equation in slope–intercept form.

> | **TRY IT : :** 4.136

Find an equation of a line that is perpendicular to the line $y = -5$ that contains the point $(-4, -5)$.

▶ | **MEDIA : :**

Access this online resource for additional instruction and practice with finding the equation of a line.

- **Use the Point-Slope Form of an Equation of a Line (https://openstax.org/l/25PointSlopeForm)**

4.6 EXERCISES

Practice Makes Perfect

Find an Equation of the Line Given the Slope and *y*-Intercept

In the following exercises, find the equation of a line with given slope and y-intercept. Write the equation in slope–intercept form.

386. slope 3 and *y*-intercept $(0, 5)$

387. slope 4 and *y*-intercept $(0, 1)$

388. slope 6 and *y*-intercept $(0, -4)$

389. slope 8 and *y*-intercept $(0, -6)$

390. slope -1 and *y*-intercept $(0, 3)$

391. slope -1 and *y*-intercept $(0, 7)$

392. slope -2 and *y*-intercept $(0, -3)$

393. slope -3 and *y*-intercept $(0, -1)$

394. slope $\frac{3}{5}$ and *y*-intercept $(0, -1)$

395. slope $\frac{1}{5}$ and *y*-intercept $(0, -5)$

396. slope $-\frac{3}{4}$ and *y*-intercept $(0, -2)$

397. slope $-\frac{2}{3}$ and *y*-intercept $(0, -3)$

398. slope 0 and *y*-intercept $(0, -1)$

399. slope 0 and *y*-intercept $(0, 2)$

400. slope -3 and *y*-intercept $(0, 0)$

401. slope -4 and *y*-intercept $(0, 0)$

In the following exercises, find the equation of the line shown in each graph. Write the equation in slope–intercept form.

402.

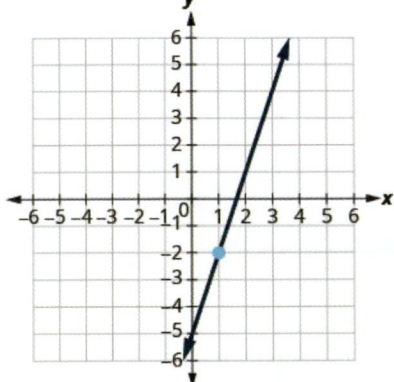

403.

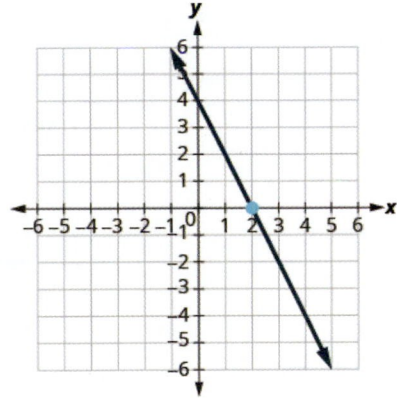

404.

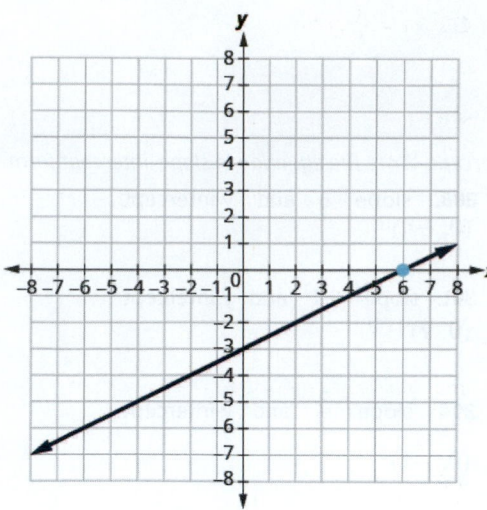

405.

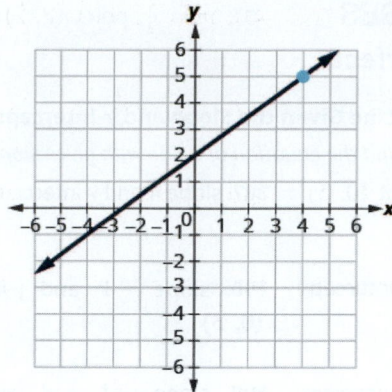

406.

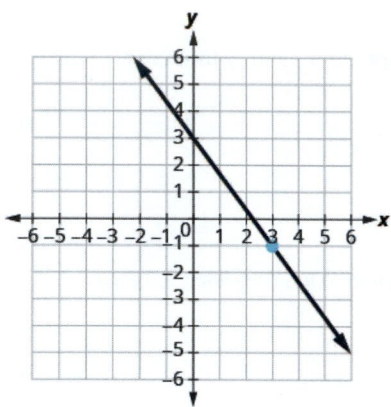

407.

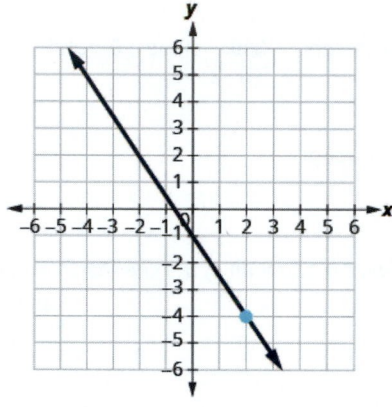

408.

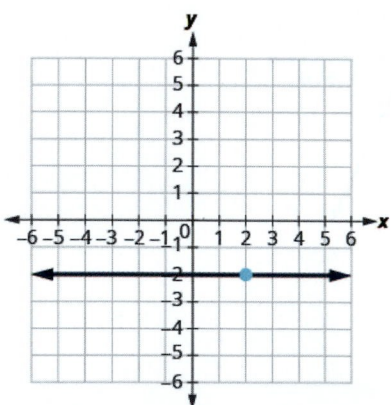

409.

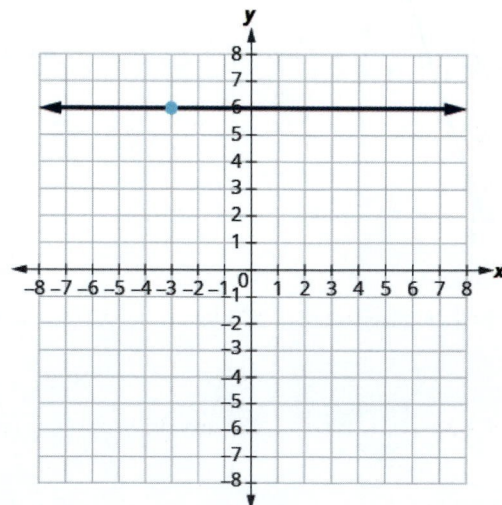

Find an Equation of the Line Given the Slope and a Point

In the following exercises, find the equation of a line with given slope and containing the given point. Write the equation in slope-intercept form.

410. $m = \frac{5}{8}$, point $(8, 3)$

411. $m = \frac{3}{8}$, point $(8, 2)$

412. $m = \frac{1}{6}$, point $(6, 1)$

413. $m = \frac{5}{6}$, point $(6, 7)$

414. $m = -\frac{3}{4}$, point $(8, -5)$

415. $m = -\frac{3}{5}$, point $(10, -5)$

416. $m = -\frac{1}{4}$, point $(-12, -6)$

417. $m = -\frac{1}{3}$, point $(-9, -8)$

418. Horizontal line containing $(-2, 5)$

419. Horizontal line containing $(-1, 4)$

420. Horizontal line containing $(-2, -3)$

421. Horizontal line containing $(-1, -7)$

422. $m = -\frac{3}{2}$, point $(-4, -3)$

423. $m = -\frac{5}{2}$, point $(-8, -2)$

424. $m = -7$, point $(-1, -3)$

425. $m = -4$, point $(-2, -3)$

426. Horizontal line containing $(2, -3)$

427. Horizontal line containing $(4, -8)$

Find an Equation of the Line Given Two Points

In the following exercises, find the equation of a line containing the given points. Write the equation in slope–intercept form.

428. $(2, 6)$ and $(5, 3)$

429. $(3, 1)$ and $(2, 5)$

430. $(4, 3)$ and $(8, 1)$

431. $(2, 7)$ and $(3, 8)$

432. $(-3, -4)$ and $(5 - 2)$

433. $(-5, -3)$ and $(4, -6)$

434. $(-1, 3)$ and $(-6, -7)$

435. $(-2, 8)$ and $(-4, -6)$

436. $(6, -4)$ and $(-2, 5)$

437. $(3, -2)$ and $(-4, 4)$

438. $(0, 4)$ and $(2, -3)$

439. $(0, -2)$ and $(-5, -3)$

440. $(7, 2)$ and $(7, -2)$

441. $(4, 2)$ and $(4, -3)$

442. $(-7, -1)$ and $(-7, -4)$

443. $(-2, 1)$ and $(-2, -4)$

444. $(6, 1)$ and $(0, 1)$

445. $(6, 2)$ and $(-3, 2)$

446. $(3, -4)$ and $(5, -4)$

447. $(-6, -3)$ and $(-1, -3)$

448. $(4, 3)$ and $(8, 0)$

449. $(0, 0)$ and $(1, 4)$

450. $(-2, -3)$ and $(-5, -6)$

451. $(-3, 0)$ and $(-7, -2)$

452. $(8, -1)$ and $(8, -5)$

453. $(3, 5)$ and $(-7, 5)$

Find an Equation of a Line Parallel to a Given Line

In the following exercises, find an equation of a line parallel to the given line and contains the given point. Write the equation in slope–intercept form.

454. line $y = 4x + 2$, point $(1, 2)$

455. line $y = 3x + 4$, point $(2, 5)$

456. line $y = -2x - 3$, point $(-1, 3)$

457. line $y = -3x - 1$, point $(2, -3)$

458. line $3x - y = 4$, point $(3, 1)$

459. line $2x - y = 6$, point $(3, 0)$

460. line $4x + 3y = 6$, point $(0, -3)$

461. line $2x + 3y = 6$, point $(0, 5)$

462. line $x = -3$, point $(-2, -1)$

463. line $x = -4$, point $(-3, -5)$ **464.** line $x - 2 = 0$, point $(1, -2)$ **465.** line $x - 6 = 0$, point $(4, -3)$

466. line $y = 5$, point $(2, -2)$ **467.** line $y = 1$, point $(3, -4)$ **468.** line $y + 2 = 0$, point $(3, -3)$

469. line $y + 7 = 0$, point $(1, -1)$

Find an Equation of a Line Perpendicular to a Given Line

In the following exercises, find an equation of a line perpendicular to the given line and contains the given point. Write the equation in slope–intercept form.

470. line $y = -2x + 3$, point $(2, 2)$ **471.** line $y = -x + 5$, point $(3, 3)$ **472.** line $y = \frac{3}{4}x - 2$, point $(-3, 4)$

473. line $y = \frac{2}{3}x - 4$, point $(2, -4)$ **474.** line $2x - 3y = 8$, point $(4, -1)$ **475.** line $4x - 3y = 5$, point $(-3, 2)$

476. line $2x + 5y = 6$, point $(0, 0)$ **477.** line $4x + 5y = -3$, point $(0, 0)$ **478.** line $y - 3 = 0$, point $(-2, -4)$

479. line $y - 6 = 0$, point $(-5, -3)$ **480.** line y-axis, point $(3, 4)$ **481.** line y-axis, point $(2, 1)$

Mixed Practice

In the following exercises, find the equation of each line. Write the equation in slope–intercept form.

482. Containing the points $(4, 3)$ and $(8, 1)$ **483.** Containing the points $(2, 7)$ and $(3, 8)$

484. $m = \frac{1}{6}$, containing point $(6, 1)$ **485.** $m = \frac{5}{6}$, containing point $(6, 7)$

486. Parallel to the line $4x + 3y = 6$, containing point $(0, -3)$ **487.** Parallel to the line $2x + 3y = 6$, containing point $(0, 5)$

488. $m = -\frac{3}{4}$, containing point $(8, -5)$ **489.** $m = -\frac{3}{5}$, containing point $(10, -5)$

490. Perpendicular to the line $y - 1 = 0$, point $(-2, 6)$ **491.** Perpendicular to the line y-axis, point $(-6, 2)$

492. Containing the points $(4, 3)$ and $(8, 1)$ **493.** Containing the points $(-2, 0)$ and $(-3, -2)$

494. Parallel to the line $x = -3$, containing point $(-2, -1)$ **495.** Parallel to the line $x = -4$, containing point $(-3, -5)$

496. Containing the points $(-3, -4)$ and $(2, -5)$ **497.** Containing the points $(-5, -3)$ and $(4, -6)$

498. Perpendicular to the line $x - 2y = 5$, containing point $(-2, 2)$

499. Perpendicular to the line $4x + 3y = 1$, containing point $(0, 0)$

Everyday Math

500. Cholesterol. The age, x, and LDL cholesterol level, y, of two men are given by the points $(18, 68)$ and $(27, 122)$. Find a linear equation that models the relationship between age and LDL cholesterol level.

501. Fuel consumption. The city mpg, x, and highway mpg, y, of two cars are given by the points $(29, 40)$ and $(19, 28)$. Find a linear equation that models the relationship between city mpg and highway mpg.

Writing Exercises

502. Why are all horizontal lines parallel?

503. Explain in your own words why the slopes of two perpendicular lines must have opposite signs.

Self Check

ⓐ After completing the exercises, use this checklist to evaluate your mastery of the objectives of this section.

I can...	Confidently	With some help	No-I don't get it!
find the equation of the line given the slope and y-intercept.			
find an equation of the line given the slope and a point.			
find an equation of the line given two points.			
find an equation of a line parallel to a given line.			
find an equation of a line perpendicular to a given line.			

ⓑ On a scale of 1-10, how would you rate your mastery of this section in light of your responses on the checklist? How can you improve this?

4.7 Graphs of Linear Inequalities

Learning Objectives

By the end of this section, you will be able to:
> Verify solutions to an inequality in two variables
> Recognize the relation between the solutions of an inequality and its graph
> Graph linear inequalities

Be Prepared!

Before you get started, take this readiness quiz.

1. Solve: $4x + 3 > 23$.
   If you missed this problem, review **Example 2.73**.

2. Translate from algebra to English: $x < 5$.
   If you missed this problem, review **Example 1.12**.

3. Evaluate $3x - 2y$ when $x = 1$, $y = -2$.
   If you missed this problem, review **Example 1.55**.

Verify Solutions to an Inequality in Two Variables

We have learned how to solve inequalities in one variable. Now, we will look at inequalities in two variables. Inequalities in two variables have many applications. If you ran a business, for example, you would want your revenue to be greater than your costs—so that your business would make a profit.

Linear Inequality

A **linear inequality** is an inequality that can be written in one of the following forms:
$$Ax + By > C \quad Ax + By \geq C \quad Ax + By < C \quad Ax + By \leq C$$
where A and B are not both zero.

Do you remember that an inequality with one variable had many solutions? The solution to the inequality $x > 3$ is any number greater than 3. We showed this on the number line by shading in the number line to the right of 3, and putting an open parenthesis at 3. See **Figure 4.30**.

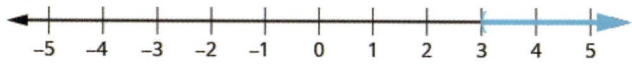

Figure 4.30

Similarly, inequalities in two variables have many solutions. Any ordered pair (x, y) that makes the inequality true when we substitute in the values is a solution of the inequality.

Solution of a Linear Inequality

An ordered pair (x, y) is a **solution of a linear inequality** if the inequality is true when we substitute the values of x and y.

EXAMPLE 4.69

Determine whether each ordered pair is a solution to the inequality $y > x + 4$:

ⓐ $(0, 0)$ ⓑ $(1, 6)$ ⓒ $(2, 6)$ ⓓ $(-5, -15)$ ⓔ $(-8, 12)$

✓ **Solution**

ⓐ

(0, 0)	$y > x + 4$
Substitute 0 for x and 0 for y.	$0 \overset{?}{>} 0 + 4$
	$0 \not> 4$
Simplify.	So, $(0, 0)$ is not a solution to $y > x + 4$.

ⓑ

(1, 6)	$y > x + 4$
Substitute 1 for x and 6 for y.	$6 \overset{?}{>} 1 + 4$
	$6 > 5$
Simplify.	So, $(1, 6)$ is a solution to $y > x + 4$.

ⓒ

(2, 6)	$y > x + 4$
Substitute 2 for x and 6 for y.	$6 \overset{?}{>} 2 + 4$
	$6 \not> 6$
Simplify.	So, $(2, 6)$ is not a solution to $y > x + 4$.

ⓓ

(−5, −15)	$y > x + 4$
Substitute −5 for x and −15 for y.	$-15 \overset{?}{>} -5 + 4$
	$-15 \not> -1$
Simplify.	So, $(-5, -15)$ is not a solution to $y > x + 4$.

ⓔ

(−8, 12)	$y > x + 4$
Substitute −8 for x and 12 for y.	$12 \overset{?}{>} -8 + 4$
	$12 > -4$
Simplify.	So, $(-8, 12)$ is a solution to $y > x + 4$.

> **TRY IT :: 4.137**   Determine whether each ordered pair is a solution to the inequality $y > x - 3$:

ⓐ $(0, 0)$ ⓑ $(4, 9)$ ⓒ $(-2, 1)$ ⓓ $(-5, -3)$ ⓔ $(5, 1)$

> **TRY IT :: 4.138**   Determine whether each ordered pair is a solution to the inequality $y < x + 1$:

ⓐ $(0, 0)$ ⓑ $(8, 6)$ ⓒ $(-2, -1)$ ⓓ $(3, 4)$ ⓔ $(-1, -4)$

Recognize the Relation Between the Solutions of an Inequality and its Graph

Now, we will look at how the solutions of an inequality relate to its graph.

Let's think about the number line in **Figure 4.30** again. The point $x = 3$ separated that number line into two parts. On one side of 3 are all the numbers less than 3. On the other side of 3 all the numbers are greater than 3. See **Figure 4.31**.

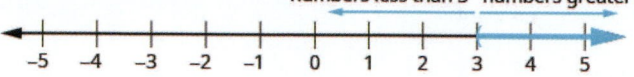

Figure 4.31

The solution to $x > 3$ is the shaded part of the number line to the right of $x = 3$.

Similarly, the line $y = x + 4$ separates the plane into two regions. On one side of the line are points with $y < x + 4$. On the other side of the line are the points with $y > x + 4$. We call the line $y = x + 4$ a boundary line.

Boundary Line

The line with equation $Ax + By = C$ is the **boundary line** that separates the region where $Ax + By > C$ from the region where $Ax + By < C$.

For an inequality in one variable, the endpoint is shown with a parenthesis or a bracket depending on whether or not a is included in the solution:

Similarly, for an inequality in two variables, the boundary line is shown with a solid or dashed line to indicate whether or not it the line is included in the solution. This is summarized in **Table 4.52**

$Ax + By < C$	$Ax + By \leq C$
$Ax + By > C$	$Ax + By \geq C$
Boundary line is not included in solution.	Boundary line is included in solution.
Boundary line is dashed.	**Boundary line is solid.**

Table 4.52

Now, let's take a look at what we found in **Example 4.69**. We'll start by graphing the line $y = x + 4$, and then we'll plot the five points we tested. See **Figure 4.32**.

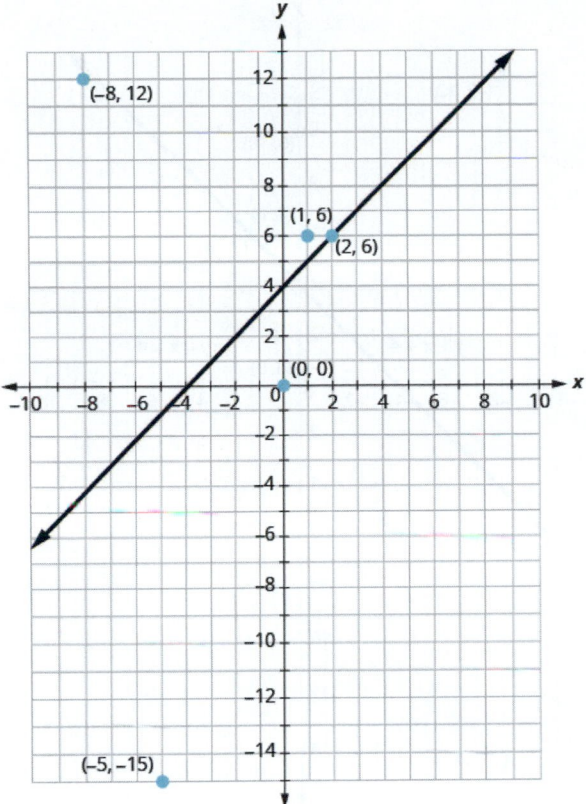

Figure 4.32

In **Example 4.69** we found that some of the points were solutions to the inequality $y > x + 4$ and some were not.

Which of the points we plotted are solutions to the inequality $y > x + 4$? The points $(1, 6)$ and $(-8, 12)$ are solutions to the inequality $y > x + 4$. Notice that they are both on the same side of the boundary line $y = x + 4$.

The two points $(0, 0)$ and $(-5, -15)$ are on the other side of the boundary line $y = x + 4$, and they are not solutions to the inequality $y > x + 4$. For those two points, $y < x + 4$.

What about the point $(2, 6)$? Because $6 = 2 + 4$, the point is a solution to the equation $y = x + 4$. So the point $(2, 6)$ is on the boundary line.

Let's take another point on the left side of the boundary line and test whether or not it is a solution to the inequality $y > x + 4$. The point $(0, 10)$ clearly looks to be to the left of the boundary line, doesn't it? Is it a solution to the inequality?

$$y > x + 4$$
$$10 \overset{?}{>} 0 + 4$$
$$10 > 4 \qquad \text{So, } (0, 10) \text{ is a solution to } y > x + 4.$$

Any point you choose on the left side of the boundary line is a solution to the inequality $y > x + 4$. All points on the left are solutions.

Similarly, all points on the right side of the boundary line, the side with $(0, 0)$ and $(-5, -15)$, are not solutions to $y > x + 4$. See **Figure 4.33**.

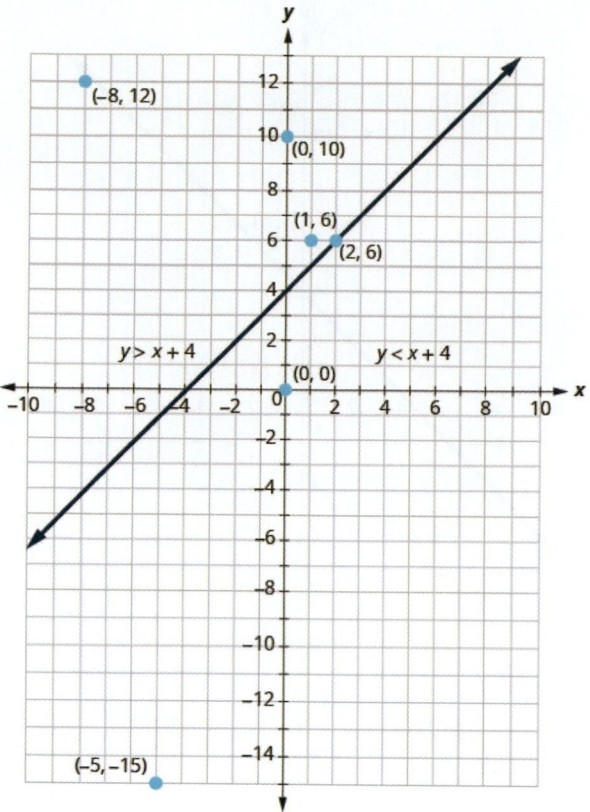

Figure 4.33

The graph of the inequality $y > x + 4$ is shown in **Figure 4.34** below. The line $y = x + 4$ divides the plane into two regions. The shaded side shows the solutions to the inequality $y > x + 4$.

The points on the boundary line, those where $y = x + 4$, are not solutions to the inequality $y > x + 4$, so the line itself is not part of the solution. We show that by making the line dashed, not solid.

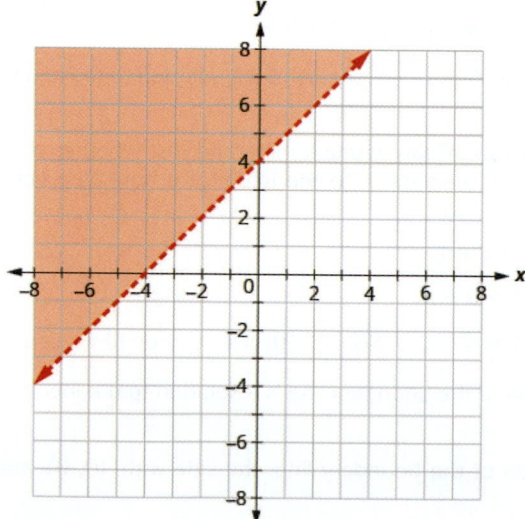

Figure 4.34 The graph of the inequality $y > x + 4$.

EXAMPLE 4.70

The boundary line shown is $y = 2x - 1$. Write the inequality shown by the graph.

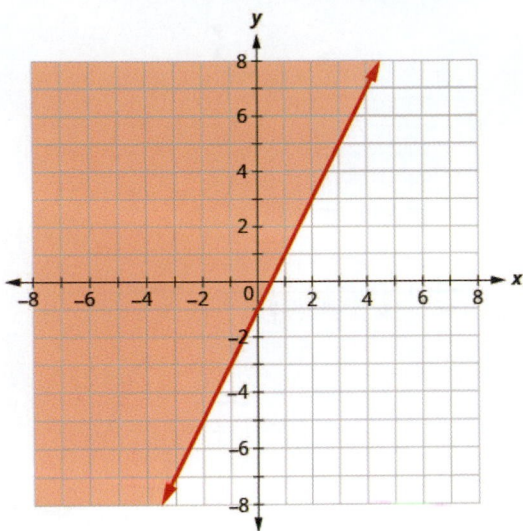

⊘ Solution

The line $y = 2x - 1$ is the boundary line. On one side of the line are the points with $y > 2x - 1$ and on the other side of the line are the points with $y < 2x - 1$.

Let's test the point $(0, 0)$ and see which inequality describes its side of the boundary line.

At $(0, 0)$, which inequality is true:

$$y > 2x - 1 \quad \text{or} \quad y < 2x - 1\,?$$
$$y > 2x - 1 \qquad\qquad y < 2x - 1$$
$$0 \overset{?}{>} 2 \cdot 0 - 1 \qquad\quad 0 \overset{?}{<} 2 \cdot 0 - 1$$
$$0 > -1 \text{ True} \qquad 0 < -1 \text{ False}$$

Since, $y > 2x - 1$ is true, the side of the line with $(0, 0)$, is the solution. The shaded region shows the solution of the inequality $y > 2x - 1$.

Since the boundary line is graphed with a solid line, the inequality includes the equal sign.

The graph shows the inequality $y \geq 2x - 1$.

We could use any point as a test point, provided it is not on the line. Why did we choose $(0, 0)$? Because it's the easiest to evaluate. You may want to pick a point on the other side of the boundary line and check that $y < 2x - 1$.

> **TRY IT :: 4.139**   Write the inequality shown by the graph with the boundary line $y = -2x + 3$.

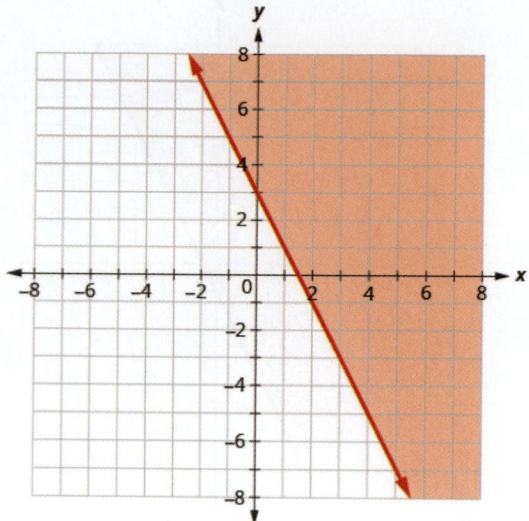

> **TRY IT :: 4.140**   Write the inequality shown by the graph with the boundary line $y = \frac{1}{2}x - 4$.

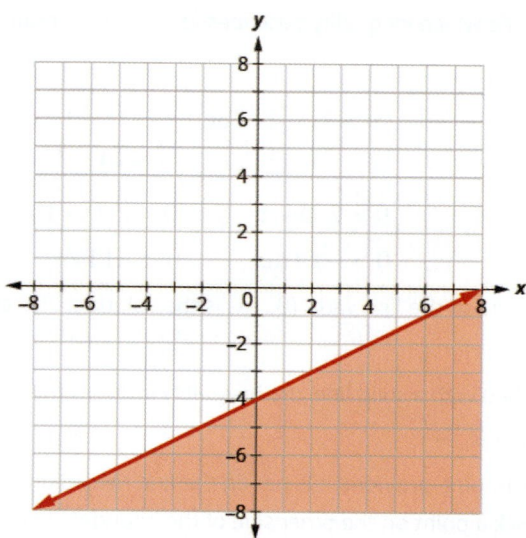

EXAMPLE 4.71

The boundary line shown is $2x + 3y = 6$. Write the inequality shown by the graph.

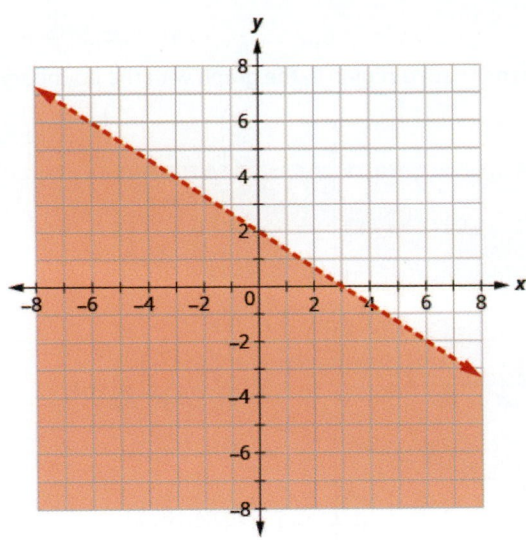

✓ **Solution**

The line $2x + 3y = 6$ is the boundary line. On one side of the line are the points with $2x + 3y > 6$ and on the other side of the line are the points with $2x + 3y < 6$.

Let's test the point $(0, 0)$ and see which inequality describes its side of the boundary line.

At $(0, 0)$, which inequality is true:

$$
\begin{array}{ccl}
2x + 3y &>& 6 \\
2x + 3y &>& 6 \\
2(0) + 3(0) &\overset{?}{>}& 6 \\
0 &>& 6 \text{ False}
\end{array}
\qquad \text{or} \qquad
\begin{array}{ccl}
2x + 3y &<& 6? \\
2x + 3y &<& 6 \\
2(0) + 3(0) &\overset{?}{<}& 6 \\
0 &<& 6 \text{ True}
\end{array}
$$

So the side with $(0, 0)$ is the side where $2x + 3y < 6$.

(You may want to pick a point on the other side of the boundary line and check that $2x + 3y > 6$.)

Since the boundary line is graphed as a dashed line, the inequality does not include an equal sign.

The graph shows the solution to the inequality $2x + 3y < 6$.

> **TRY IT : :** 4.141

Write the inequality shown by the shaded region in the graph with the boundary line $x - 4y = 8$.

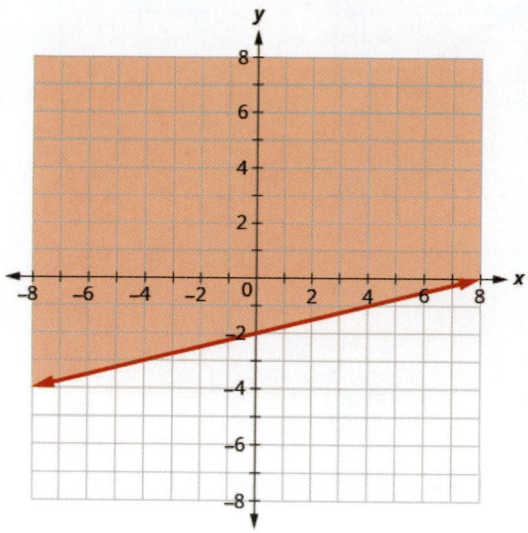

> **TRY IT : :** 4.142

Write the inequality shown by the shaded region in the graph with the boundary line $3x - y = 6$.

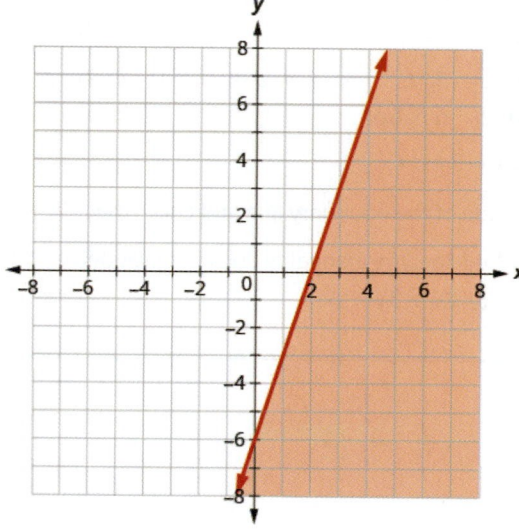

Graph Linear Inequalities

Now, we're ready to put all this together to graph linear inequalities.

EXAMPLE 4.72   HOW TO GRAPH LINEAR INEQUALITIES

Graph the linear inequality $y \geq \frac{3}{4}x - 2$.

⊘ **Solution**

Step 1. Identify and graph the boundary line.	Replace the inequality sign with an equal sign to find the boundary line.	
• If the inequality is ≤ or ≥, the boundary line is solid.	Graph the boundary line $y = \frac{3}{4}x - 2$.	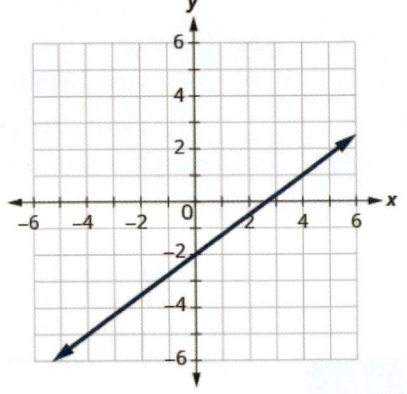
• If the inequality is < or >, the boundary line is dashed.	The inequality sign is ≥, so we draw a solid line.	
Step 2. Test a point that is not on the boundary line. Is it a solution of the inequality?	We'll test (0, 0). Is it a solution of the inequality?	At (0, 0), is $y \geq \frac{3}{4}x - 2$? $0 \overset{?}{\geq} \frac{3}{4}(0) - 2$ $0 \geq -2$ So, (0, 0) is a solution.
Step 3. Shade in one side of the boundary line. • If the test point is a solution, shade in the side that includes the point. • If the test point is not a solution, shade in the opposite side.	The test point (0, 0), is a solution to $y \geq \frac{3}{4}x - 2$. So we shade in that side.	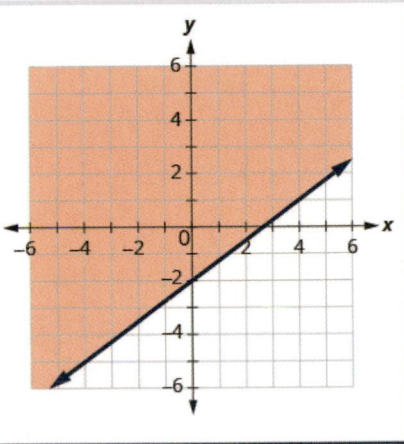

> **TRY IT : : 4.143**   Graph the linear inequality $y \geq \frac{5}{2}x - 4$.

> **TRY IT : : 4.144**   Graph the linear inequality $y < \frac{2}{3}x - 5$.

The steps we take to graph a linear inequality are summarized here.

HOW TO :: GRAPH A LINEAR INEQUALITY.

Step 1.   Identify and graph the boundary line.

  ∘   If the inequality is ≤ or ≥ , the boundary line is solid.

  ∘   If the inequality is < or >, the boundary line is dashed.

Step 2.   Test a point that is not on the boundary line. Is it a solution of the inequality?

Step 3.   Shade in one side of the boundary line.

  ∘   If the test point is a solution, shade in the side that includes the point.

  ∘   If the test point is not a solution, shade in the opposite side.

EXAMPLE 4.73

Graph the linear inequality $x - 2y < 5$.

✅ **Solution**

First we graph the boundary line $x - 2y = 5$. The inequality is < so we draw a dashed line.

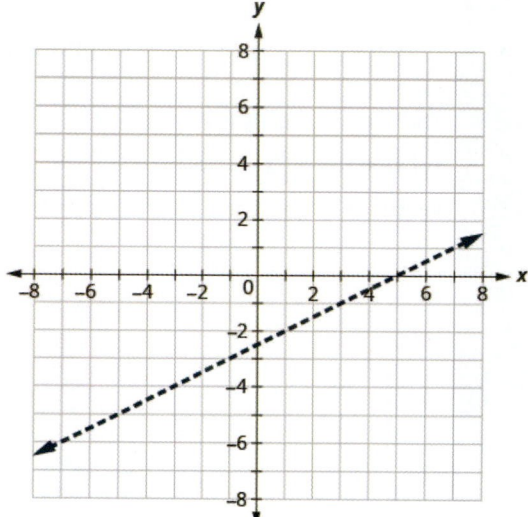

Then we test a point. We'll use $(0, 0)$ again because it is easy to evaluate and it is not on the boundary line.

Is $(0, 0)$ a solution of $x - 2y < 5$?

$$0 - 2(0) \overset{?}{<} 5$$

$$0 - 0 \overset{?}{<} 5$$

$$0 < 5$$

The point $(0, 0)$ is a solution of $x - 2y < 5$, so we shade in that side of the boundary line.

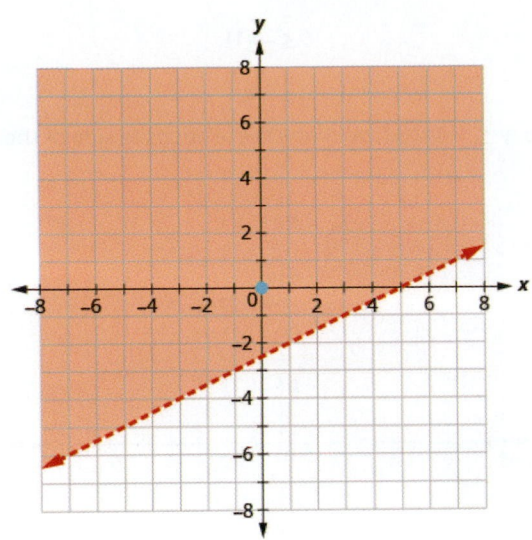

> **TRY IT :: 4.145**   Graph the linear inequality $2x - 3y \le 6$.

> **TRY IT :: 4.146**   Graph the linear inequality $2x - y > 3$.

What if the boundary line goes through the origin? Then we won't be able to use $(0, 0)$ as a test point. No problem—we'll just choose some other point that is not on the boundary line.

EXAMPLE 4.74

Graph the linear inequality $y \le -4x$.

⊘ Solution

First we graph the boundary line $y = -4x$. It is in slope-intercept form, with $m = -4$ and $b = 0$. The inequality is $\le$ so we draw a solid line.

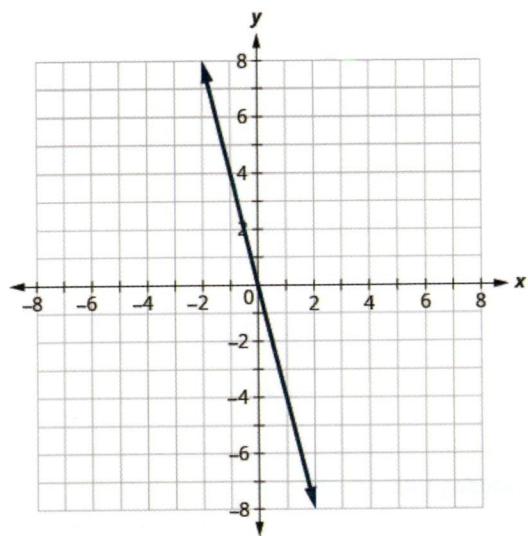

Now, we need a test point. We can see that the point $(1, 0)$ is not on the boundary line.

Is $(1, 0)$ a solution of $y \le -4x$?

$$0 \overset{?}{\leq} -4(1)$$

$$0 \nleq -4$$

The point $(1, 0)$ is not a solution to $y \leq -4x$, so we shade in the opposite side of the boundary line. See Figure 4.35.

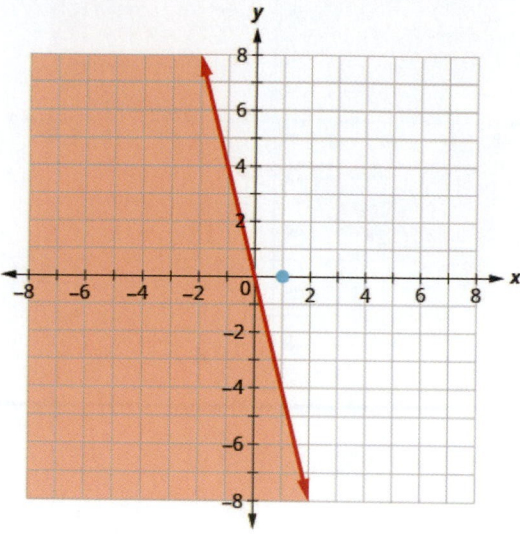

Figure 4.35

> **TRY IT : : 4.147**   Graph the linear inequality $y > -3x$.

> **TRY IT : : 4.148**   Graph the linear inequality $y \geq -2x$.

Some linear inequalities have only one variable. They may have an x but no y, or a y but no x. In these cases, the boundary line will be either a vertical or a horizontal line. Do you remember?

$$x = a \quad \text{vertical line}$$
$$y = b \quad \text{horizontal line}$$

EXAMPLE 4.75

Graph the linear inequality $y > 3$.

✓ Solution

First we graph the boundary line $y = 3$. It is a horizontal line. The inequality is > so we draw a dashed line.

We test the point $(0, 0)$.

$$y > 3$$

$$0 \not> 3$$

$(0, 0)$ is not a solution to $y > 3$.

So we shade the side that does not include (0, 0).

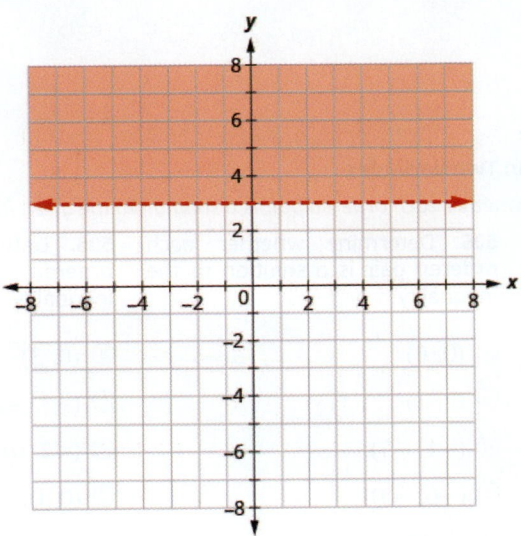

> **TRY IT : : 4.149**   Graph the linear inequality $y < 5$.

> **TRY IT : : 4.150**   Graph the linear inequality $y \leq -1$.

4.7 EXERCISES

Practice Makes Perfect

Verify Solutions to an Inequality in Two Variables

In the following exercises, determine whether each ordered pair is a solution to the given inequality.

504. Determine whether each ordered pair is a solution to the inequality $y > x - 1$:

ⓐ $(0, 1)$

ⓑ $(-4, -1)$

ⓒ $(4, 2)$

ⓓ $(3, 0)$

ⓔ $(-2, -3)$

505. Determine whether each ordered pair is a solution to the inequality $y > x - 3$:

ⓐ $(0, 0)$

ⓑ $(2, 1)$

ⓒ $(-1, -5)$

ⓓ $(-6, -3)$

ⓔ $(1, 0)$

506. Determine whether each ordered pair is a solution to the inequality $y < x + 2$:

ⓐ $(0, 3)$

ⓑ $(-3, -2)$

ⓒ $(-2, 0)$

ⓓ $(0, 0)$

ⓔ $(-1, 4)$

507. Determine whether each ordered pair is a solution to the inequality $y < x + 5$:

ⓐ $(-3, 0)$

ⓑ $(1, 6)$

ⓒ $(-6, -2)$

ⓓ $(0, 1)$

ⓔ $(5, -4)$

508. Determine whether each ordered pair is a solution to the inequality $x + y > 4$:

ⓐ $(5, 1)$

ⓑ $(-2, 6)$

ⓒ $(3, 2)$

ⓓ $(10, -5)$

ⓔ $(0, 0)$

509. Determine whether each ordered pair is a solution to the inequality $x + y > 2$:

ⓐ $(1, 1)$

ⓑ $(4, -3)$

ⓒ $(0, 0)$

ⓓ $(-8, 12)$

ⓔ $(3, 0)$

Recognize the Relation Between the Solutions of an Inequality and its Graph

In the following exercises, write the inequality shown by the shaded region.

510. Write the inequality shown by the graph with the boundary line $y = 3x - 4$.

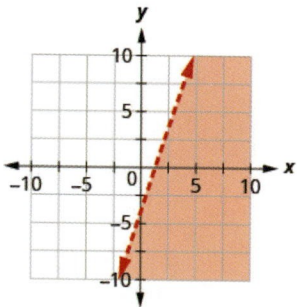

511. Write the inequality shown by the graph with the boundary line $y = 2x - 4$.

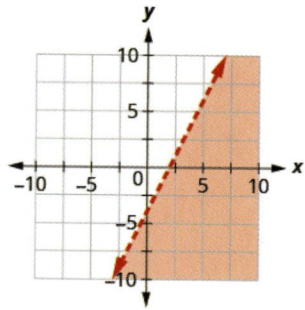

512. Write the inequality shown by the graph with the boundary line $y = -\frac{1}{2}x + 1$.

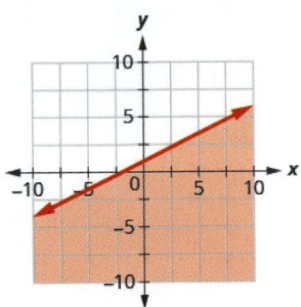

513. Write the inequality shown by the graph with the boundary line $y = -\frac{1}{3}x - 2$.

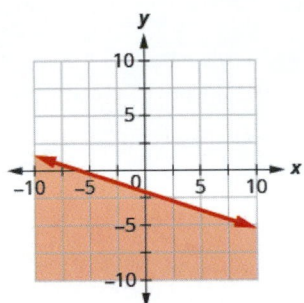

514. Write the inequality shown by the shaded region in the graph with the boundary line $x + y = 5$.

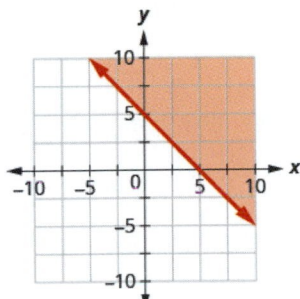

515. Write the inequality shown by the shaded region in the graph with the boundary line $x + y = 3$.

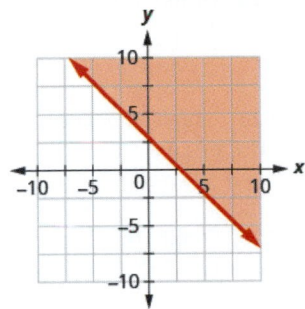

516. Write the inequality shown by the shaded region in the graph with the boundary line $2x + y = -4$.

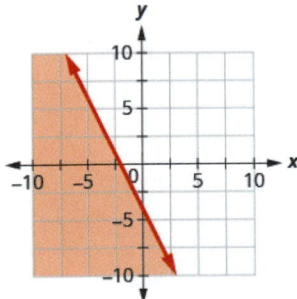

517. Write the inequality shown by the shaded region in the graph with the boundary line $x + 2y = -2$.

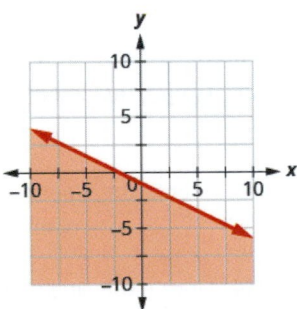

518. Write the inequality shown by the shaded region in the graph with the boundary line $3x - y = 6$.

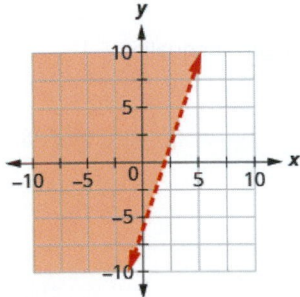

519. Write the inequality shown by the shaded region in the graph with the boundary line $2x - y = 4$.

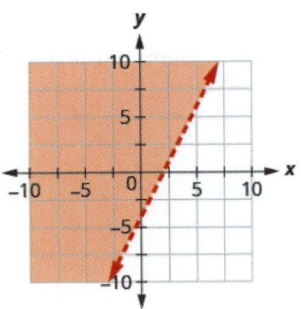

520. Write the inequality shown by the shaded region in the graph with the boundary line $2x - 5y = 10$.

521. Write the inequality shown by the shaded region in the graph with the boundary line $4x - 3y = 12$.

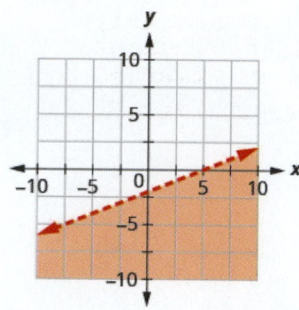

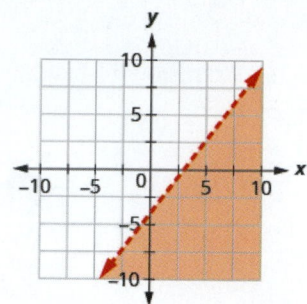

Graph Linear Inequalities

In the following exercises, graph each linear inequality.

522. Graph the linear inequality $y > \frac{2}{3}x - 1$.

523. Graph the linear inequality $y < \frac{3}{5}x + 2$.

524. Graph the linear inequality $y \leq -\frac{1}{2}x + 4$.

525. Graph the linear inequality $y \geq -\frac{1}{3}x - 2$.

526. Graph the linear inequality $x - y \leq 3$.

527. Graph the linear inequality $x - y \geq -2$.

528. Graph the linear inequality $4x + y > -4$.

529. Graph the linear inequality $x + 5y < -5$.

530. Graph the linear inequality $3x + 2y \geq -6$.

531. Graph the linear inequality $4x + 2y \geq -8$.

532. Graph the linear inequality $y > 4x$.

533. Graph the linear inequality $y > x$.

534. Graph the linear inequality $y \leq -x$.

535. Graph the linear inequality $y \leq -3x$.

536. Graph the linear inequality $y \geq -2$.

537. Graph the linear inequality $y < -1$.

538. Graph the linear inequality $y < 4$.

539. Graph the linear inequality $y \geq 2$.

540. Graph the linear inequality $x \leq 5$.

541. Graph the linear inequality $x > -2$.

542. Graph the linear inequality $x > -3$.

543. Graph the linear inequality $x \leq 4$.

544. Graph the linear inequality $x - y < 4$.

545. Graph the linear inequality $x - y < -3$.

546. Graph the linear inequality $y \geq \frac{3}{2}x$.

547. Graph the linear inequality $y \leq \frac{5}{4}x$.

548. Graph the linear inequality $y > -2x + 1$.

549. Graph the linear inequality $y < -3x - 4$.

550. Graph the linear inequality $x \leq -1$.

551. Graph the linear inequality $x \geq 0$.

Everyday Math

552. Money. Gerry wants to have a maximum of $100 cash at the ticket booth when his church carnival opens. He will have $1 bills and $5 bills. If x is the number of $1 bills and y is the number of $5 bills, the inequality $x + 5y \le 100$ models the situation.

  ⓐ Graph the inequality.

  ⓑ List three solutions to the inequality $x + 5y \le 100$ where both x and y are integers.

553. Shopping. Tula has $20 to spend at the used book sale. Hardcover books cost $2 each and paperback books cost $0.50 each. If x is the number of hardcover books Tula can buy and y is the number of paperback books she can buy, the inequality $2x + \frac{1}{2}y \le 20$ models the situation.

  ⓐ Graph the inequality.

  ⓑ List three solutions to the inequality $2x + \frac{1}{2}y \le 20$ where both x and y are whole numbers.

Writing Exercises

554. Lester thinks that the solution of any inequality with a > sign is the region above the line and the solution of any inequality with a < sign is the region below the line. Is Lester correct? Explain why or why not.

555. Explain why in some graphs of linear inequalities the boundary line is solid but in other graphs it is dashed.

Self Check

ⓐ After completing the exercises, use this checklist to evaluate your mastery of the objectives of this section.

I can...	Confidently	With some help	No-I don't get it!
verify solutions to an inequality in two variables.			
recognize the relation between the solutions of an inequality and its graph.			
graph linear inequalities.			

ⓑ What does this checklist tell you about your mastery of this section? What steps will you take to improve?

CHAPTER 4 REVIEW

KEY TERMS

boundary line The line with equation $Ax + By = C$ that separates the region where $Ax + By > C$ from the region where $Ax + By < C$.

geoboard A geoboard is a board with a grid of pegs on it.

graph of a linear equation The graph of a linear equation $Ax + By = C$ is a straight line. Every point on the line is a solution of the equation. Every solution of this equation is a point on this line.

horizontal line A horizontal line is the graph of an equation of the form $y = b$. The line passes through the y-axis at $(0, b)$.

intercepts of a line The points where a line crosses the x- axis and the y- axis are called the intercepts of the line.

linear equation A linear equation is of the form $Ax + By = C$, where A and B are not both zero, is called a linear equation in two variables.

linear inequality An inequality that can be written in one of the following forms:
$$Ax + By > C \quad Ax + By \geq C \quad Ax + By < C \quad Ax + By \leq C$$

where A and B are not both zero.

negative slope A negative slope of a line goes down as you read from left to right.

ordered pair An ordered pair (x, y) gives the coordinates of a point in a rectangular coordinate system.

origin The point $(0, 0)$ is called the origin. It is the point where the x-axis and y-axis intersect.

parallel lines Lines in the same plane that do not intersect.

perpendicular lines Lines in the same plane that form a right angle.

point–slope form The point–slope form of an equation of a line with slope m and containing the point (x_1, y_1) is $y - y_1 = m(x - x_1)$.

positive slope A positive slope of a line goes up as you read from left to right.

quadrant The x-axis and the y-axis divide a plane into four regions, called quadrants.

rectangular coordinate system A grid system is used in algebra to show a relationship between two variables; also called the xy-plane or the 'coordinate plane'.

rise The rise of a line is its vertical change.

run The run of a line is its horizontal change.

slope formula The slope of the line between two points (x_1, y_1) and (x_2, y_2) is $m = \frac{y_2 - y_1}{x_2 - x_1}$.

slope of a line The slope of a line is $m = \frac{\text{rise}}{\text{run}}$. The rise measures the vertical change and the run measures the horizontal change.

slope-intercept form of an equation of a line The slope–intercept form of an equation of a line with slope m and y-intercept, $(0, b)$ is, $y = mx + b$.

solution of a linear inequality An ordered pair (x, y) is a solution to a linear inequality the inequality is true when we substitute the values of x and y.

vertical line A vertical line is the graph of an equation of the form $x = a$. The line passes through the x-axis at $(a, 0)$.

x- intercept The point $(a, 0)$ where the line crosses the x- axis; the x- intercept occurs when y is zero.

x-coordinate The first number in an ordered pair (x, y).

y-coordinate The second number in an ordered pair (x, y).

y-intercept The point $(0, b)$ where the line crosses the y- axis; the y- intercept occurs when x is zero.

KEY CONCEPTS

4.1 Use the Rectangular Coordinate System

- **Sign Patterns of the Quadrants**

Quadrant I	Quadrant II	Quadrant III	Quadrant IV
(x, y)	(x, y)	(x, y)	(x, y)
$(+, +)$	$(-, +)$	$(-, -)$	$(+, -)$

- **Points on the Axes**
    - On the x-axis, $y = 0$. Points with a y-coordinate equal to 0 are on the x-axis, and have coordinates $(a, 0)$.

    - On the y-axis, $x = 0$. Points with an x-coordinate equal to 0 are on the y-axis, and have coordinates $(0, b)$.

- **Solution of a Linear Equation**
    - An ordered pair (x, y) is a solution of the linear equation $Ax + By = C$, if the equation is a true statement when the x- and y- values of the ordered pair are substituted into the equation.

4.2 Graph Linear Equations in Two Variables

- **Graph a Linear Equation by Plotting Points**

  Step 1. Find three points whose coordinates are solutions to the equation. Organize them in a table.

  Step 2. Plot the points in a rectangular coordinate system. Check that the points line up. If they do not, carefully check your work!

  Step 3. Draw the line through the three points. Extend the line to fill the grid and put arrows on both ends of the line.

4.3 Graph with Intercepts

- **Find the x- and y- Intercepts from the Equation of a Line**
    - Use the equation of the line to find the x- intercept of the line, let $y = 0$ and solve for x.

    - Use the equation of the line to find the y- intercept of the line, let $x = 0$ and solve for y.

- **Graph a Linear Equation using the Intercepts**

  Step 1. Find the x- and y- intercepts of the line.

  Let $y = 0$ and solve for x.

  Let $x = 0$ and solve for y.

  Step 2. Find a third solution to the equation.

  Step 3. Plot the three points and then check that they line up.

  Step 4. Draw the line.

- **Strategy for Choosing the Most Convenient Method to Graph a Line:**
    - Consider the form of the equation.
    - If it only has one variable, it is a vertical or horizontal line.
      $x = a$ is a vertical line passing through the x- axis at a
      $y = b$ is a horizontal line passing through the y- axis at b.

    - If y is isolated on one side of the equation, graph by plotting points.
    - Choose any three values for x and then solve for the corresponding y- values.
    - If the equation is of the form $ax + by = c$, find the intercepts. Find the x- and y- intercepts and then a third point.

4.4 Understand Slope of a Line

- **Find the Slope of a Line from its Graph using** $m = \frac{\text{rise}}{\text{run}}$

  Step 1. Locate two points on the line whose coordinates are integers.

  Step 2. Starting with the point on the left, sketch a right triangle, going from the first point to the second point.

  Step 3. Count the rise and the run on the legs of the triangle.

  Step 4. Take the ratio of rise to run to find the slope.

- **Graph a Line Given a Point and the Slope**

  Step 1. Plot the given point.

  Step 2. Use the slope formula $m = \frac{\text{rise}}{\text{run}}$ to identify the rise and the run.

  Step 3. Starting at the given point, count out the rise and run to mark the second point.

  Step 4. Connect the points with a line.

- **Slope of a Horizontal Line**

  - The slope of a horizontal line, $y = b$, is 0.

- **Slope of a vertical line**

  - The slope of a vertical line, $x = a$, is undefined

4.5 Use the Slope–Intercept Form of an Equation of a Line

- The slope–intercept form of an equation of a line with slope m and y-intercept, $(0, b)$ is, $y = mx + b$.

- **Graph a Line Using its Slope and y-Intercept**

  Step 1. Find the slope-intercept form of the equation of the line.

  Step 2. Identify the slope and y-intercept.

  Step 3. Plot the y-intercept.

  Step 4. Use the slope formula $m = \frac{\text{rise}}{\text{run}}$ to identify the rise and the run.

  Step 5. Starting at the y-intercept, count out the rise and run to mark the second point.

  Step 6. Connect the points with a line.

- **Strategy for Choosing the Most Convenient Method to Graph a Line:** Consider the form of the equation.

  - If it only has one variable, it is a vertical or horizontal line.
    $x = a$ is a vertical line passing through the x-axis at a.
    $y = b$ is a horizontal line passing through the y-axis at b.

  - If y is isolated on one side of the equation, in the form $y = mx + b$, graph by using the slope and y-intercept.
    Identify the slope and y-intercept and then graph.

  - If the equation is of the form $Ax + By = C$, find the intercepts.
    Find the x- and y-intercepts, a third point, and then graph.

- Parallel lines are lines in the same plane that do not intersect.

  - Parallel lines have the same slope and different y-intercepts.

  - If m_1 and m_2 are the slopes of two parallel lines then $m_1 = m_2$.

  - Parallel vertical lines have different x-intercepts.

- Perpendicular lines are lines in the same plane that form a right angle.

- If m_1 and m_2 are the slopes of two perpendicular lines, then $m_1 \cdot m_2 = -1$ and $m_1 = \frac{-1}{m_2}$.

- Vertical lines and horizontal lines are always perpendicular to each other.

4.6 Find the Equation of a Line

- **To Find an Equation of a Line Given the Slope and a Point**
  Step 1. Identify the slope.

  Step 2. Identify the point.

  Step 3. Substitute the values into the point-slope form, $y - y_1 = m(x - x_1)$.

  Step 4. Write the equation in slope-intercept form.

- **To Find an Equation of a Line Given Two Points**
  Step 1. Find the slope using the given points.

  Step 2. Choose one point.

  Step 3. Substitute the values into the point-slope form, $y - y_1 = m(x - x_1)$.

  Step 4. Write the equation in slope-intercept form.

- **To Write and Equation of a Line**
  - If given slope and y-intercept, use slope–intercept form $y = mx + b$.

  - If given slope and a point, use point–slope form $y - y_1 = m(x - x_1)$.

  - If given two points, use point–slope form $y - y_1 = m(x - x_1)$.

- **To Find an Equation of a Line Parallel to a Given Line**
  Step 1. Find the slope of the given line.

  Step 2. Find the slope of the parallel line.

  Step 3. Identify the point.

  Step 4. Substitute the values into the point-slope form, $y - y_1 = m(x - x_1)$.

  Step 5. Write the equation in slope-intercept form.

- **To Find an Equation of a Line Perpendicular to a Given Line**

  Step 1. Find the slope of the given line.

  Step 2. Find the slope of the perpendicular line.

  Step 3. Identify the point.

  Step 4. Substitute the values into the point-slope form, $y - y_1 = m(x - x_1)$.

  Step 5. Write the equation in slope-intercept form.

4.7 Graphs of Linear Inequalities

- **To Graph a Linear Inequality**

  Step 1. Identify and graph the boundary line.
  If the inequality is $\leq$ or $\geq$, the boundary line is solid.
  If the inequality is < or >, the boundary line is dashed.

  Step 2. Test a point that is not on the boundary line. Is it a solution of the inequality?

Step 3. Shade in one side of the boundary line.
         If the test point is a solution, shade in the side that includes the point.
         If the test point is not a solution, shade in the opposite side.

REVIEW EXERCISES

4.1 Rectangular Coordinate System

Plot Points in a Rectangular Coordinate System

In the following exercises, plot each point in a rectangular coordinate system.

556.
ⓐ $(-1, -5)$

ⓑ $(-3, 4)$

ⓒ $(2, -3)$

ⓓ $\left(1, \frac{5}{2}\right)$

557.
ⓐ $(4, 3)$

ⓑ $(-4, 3)$

ⓒ $(-4, -3)$

ⓓ $(4, -3)$

558.
ⓐ $(-2, 0)$

ⓑ $(0, -4)$

ⓒ $(0, 5)$

ⓓ $(3, 0)$

559.
ⓐ $\left(2, \frac{3}{2}\right)$

ⓑ $\left(3, \frac{4}{3}\right)$

ⓒ $\left(\frac{1}{3}, -4\right)$

ⓓ $\left(\frac{1}{2}, -5\right)$

Identify Points on a Graph

In the following exercises, name the ordered pair of each point shown in the rectangular coordinate system.

560.

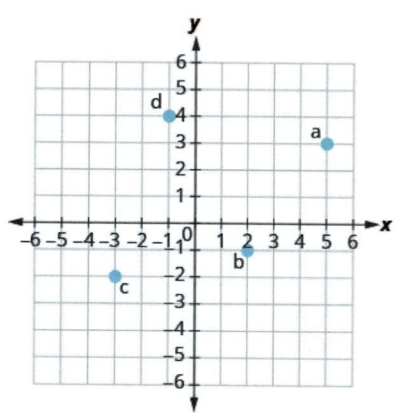

561.

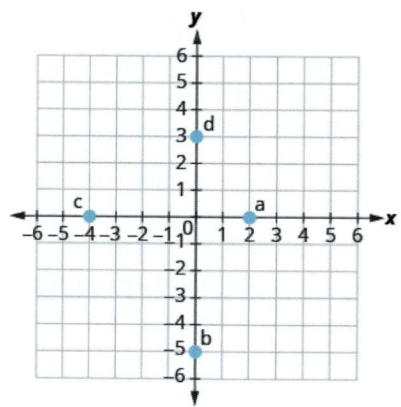

Verify Solutions to an Equation in Two Variables

In the following exercises, which ordered pairs are solutions to the given equations?

562. $5x + y = 10$

ⓐ $(5, 1)$

ⓑ $(2, 0)$

ⓒ $(4, -10)$

563. $y = 6x - 2$

ⓐ $(1, 4)$

ⓑ $\left(\frac{1}{3}, 0\right)$

ⓒ $(6, -2)$

Complete a Table of Solutions to a Linear Equation in Two Variables

In the following exercises, complete the table to find solutions to each linear equation.

564. $y = 4x - 1$

x	y	(x, y)
0		
1		
−2		

565. $y = -\frac{1}{2}x + 3$

x	y	(x, y)
0		
4		
−2		

566. $x + 2y = 5$

x	y	(x, y)
	0	
1		
−1		

567. $3x + 2y = 6$

x	y	(x, y)
0		
	0	
−2		

Find Solutions to a Linear Equation in Two Variables

In the following exercises, find three solutions to each linear equation.

568. $x + y = 3$

569. $x + y = -4$

570. $y = 3x + 1$

571. $y = -x - 1$

4.2 Graphing Linear Equations

Recognize the Relation Between the Solutions of an Equation and its Graph

In the following exercises, for each ordered pair, decide:

ⓐ Is the ordered pair a solution to the equation?   ⓑ Is the point on the line?

572. $y = -x + 4$

$(0, 4)$ $(-1, 3)$

$(2, 2)$ $(-2, 6)$

573. $y = \frac{2}{3}x - 1$

$(0, -1)$ $(3, 1)$

$(-3, -3)$ $(6, 4)$

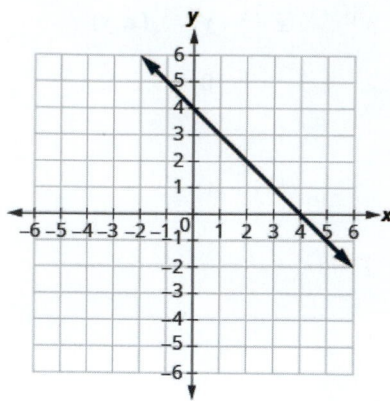

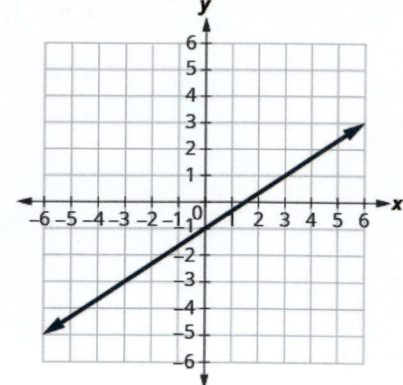

Graph a Linear Equation by Plotting Points

In the following exercises, graph by plotting points.

574. $y = 4x - 3$

575. $y = -3x$

576. $y = \frac{1}{2}x + 3$

577. $x - y = 6$

578. $2x + y = 7$

579. $3x - 2y = 6$

Graph Vertical and Horizontal lines

In the following exercises, graph each equation.

580. $y = -2$

581. $x = 3$

In the following exercises, graph each pair of equations in the same rectangular coordinate system.

582. $y = -2x$ and $y = -2$

583. $y = \frac{4}{3}x$ and $y = \frac{4}{3}$

4.3 Graphing with Intercepts

Identify the *x*- and *y*-Intercepts on a Graph

In the following exercises, find the x- and y-intercepts.

584.

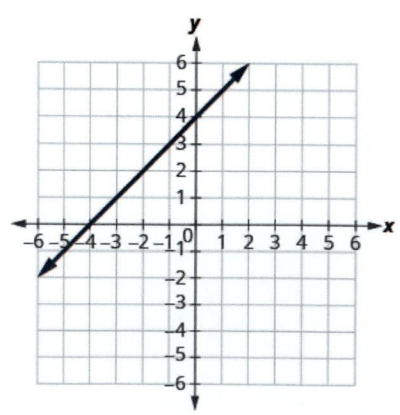

585.

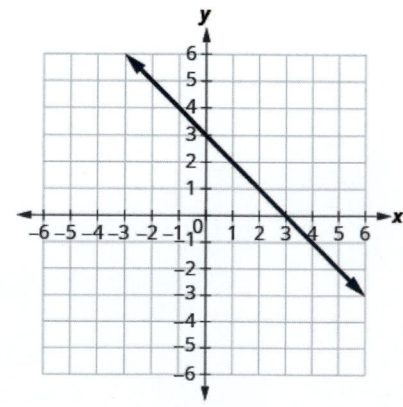

Find the *x*- and *y*-Intercepts from an Equation of a Line

In the following exercises, find the intercepts of each equation.

586. $x + y = 5$

587. $x - y = -1$

588. $x + 2y = 6$

589. $2x + 3y = 12$

590. $y = \frac{3}{4}x - 12$

591. $y = 3x$

Graph a Line Using the Intercepts

In the following exercises, graph using the intercepts.

592. $-x + 3y = 3$

593. $x + y = -2$

594. $x - y = 4$

595. $2x - y = 5$

596. $2x - 4y = 8$

597. $y = 2x$

4.4 Slope of a Line

Use Geoboards to Model Slope

In the following exercises, find the slope modeled on each geoboard.

598.

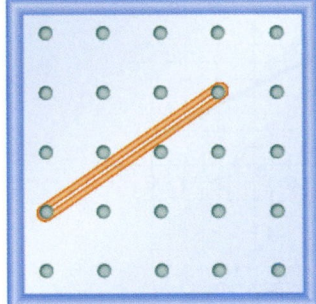

599.

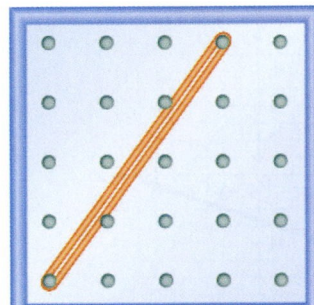

600.

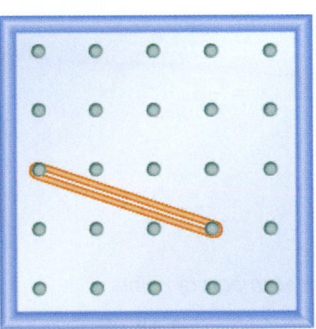

601.

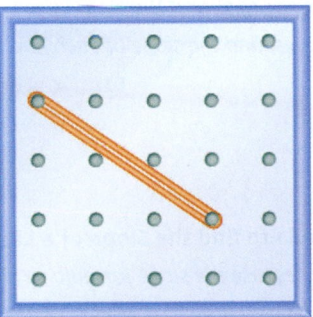

In the following exercises, model each slope. Draw a picture to show your results.

602. $\frac{1}{3}$

603. $\frac{3}{2}$

604. $-\frac{2}{3}$

605. $-\frac{1}{2}$

Use $m = \frac{\text{rise}}{\text{run}}$ to find the Slope of a Line from its Graph

In the following exercises, find the slope of each line shown.

606.

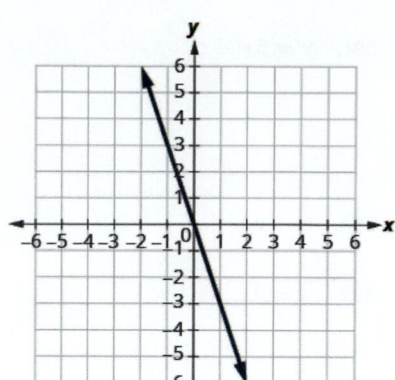

607.

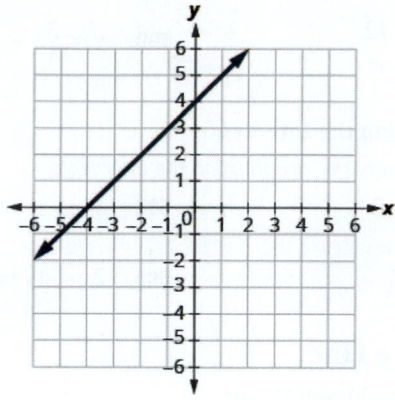

608.

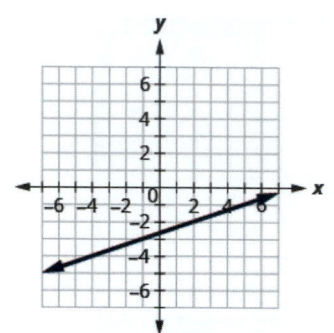

609.

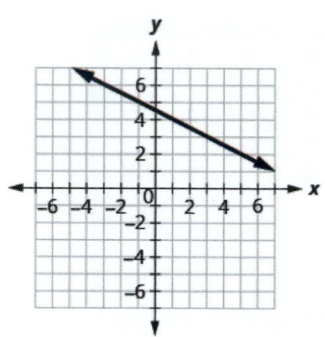

Find the Slope of Horizontal and Vertical Lines

In the following exercises, find the slope of each line.

610. $y = 2$

611. $x = 5$

612. $x = -3$

613. $y = -1$

Use the Slope Formula to find the Slope of a Line between Two Points

In the following exercises, use the slope formula to find the slope of the line between each pair of points.

614. $(-1, -1), (0, 5)$

615. $(3, 5), (4, -1)$

616. $(-5, -2), (3, 2)$

617. $(2, 1), (4, 6)$

Graph a Line Given a Point and the Slope

In the following exercises, graph each line with the given point and slope.

618. $(2, -2)$; $m = \frac{5}{2}$

619. $(-3, 4)$; $m = -\frac{1}{3}$

620. x-intercept -4; $m = 3$

621. y-intercept 1; $m = -\frac{3}{4}$

Solve Slope Applications

In the following exercises, solve these slope applications.

622. The roof pictured below has a rise of 10 feet and a run of 15 feet. What is its slope?

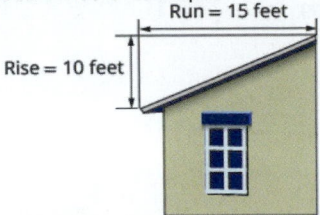

623. A mountain road rises 50 feet for a 500-foot run. What is its slope?

4.5 Intercept Form of an Equation of a Line

Recognize the Relation Between the Graph and the Slope–Intercept Form of an Equation of a Line

In the following exercises, use the graph to find the slope and y-intercept of each line. Compare the values to the equation $y = mx + b$.

624.

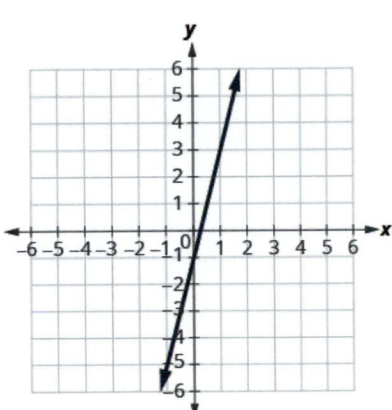

$y = 4x - 1$

625.

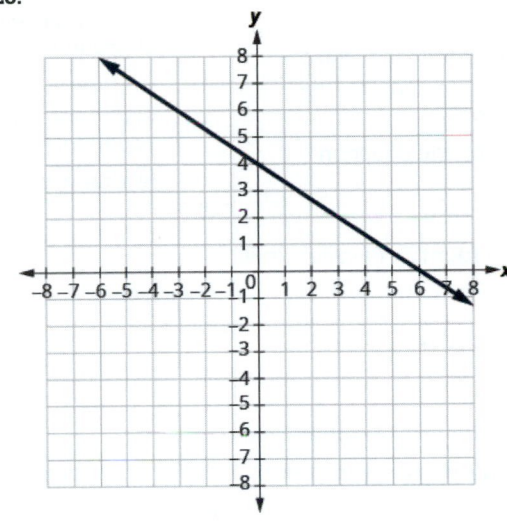

$y = -\frac{2}{3}x + 4$

Identify the Slope and y-Intercept from an Equation of a Line

In the following exercises, identify the slope and y-intercept of each line.

626. $y = -4x + 9$

627. $y = \frac{5}{3}x - 6$

628. $5x + y = 10$

629. $4x - 5y = 8$

Graph a Line Using Its Slope and Intercept

In the following exercises, graph the line of each equation using its slope and y-intercept.

630. $y = 2x + 3$

631. $y = -x - 1$

632. $y = -\frac{2}{5}x + 3$

633. $4x - 3y = 12$

In the following exercises, determine the most convenient method to graph each line.

634. $x = 5$

635. $y = -3$

636. $2x + y = 5$

637. $x - y = 2$

638. $y = x + 2$

639. $y = \frac{3}{4}x - 1$

Graph and Interpret Applications of Slope–Intercept

640. Katherine is a private chef. The equation $C = 6.5m + 42$ models the relation between her weekly cost, C, in dollars and the number of meals, m, that she serves.

ⓐ Find Katherine's cost for a week when she serves no meals.

ⓑ Find the cost for a week when she serves 14 meals.

ⓒ Interpret the slope and C-intercept of the equation.

ⓓ Graph the equation.

641. Marjorie teaches piano. The equation $P = 35h - 250$ models the relation between her weekly profit, P, in dollars and the number of student lessons, s, that she teaches.

ⓐ Find Marjorie's profit for a week when she teaches no student lessons.

ⓑ Find the profit for a week when she teaches 20 student lessons.

ⓒ Interpret the slope and P-intercept of the equation.

ⓓ Graph the equation.

Use Slopes to Identify Parallel Lines

In the following exercises, use slopes and y-intercepts to determine if the lines are parallel.

642. $4x - 3y = -1$; $y = \frac{4}{3}x - 3$

643. $2x - y = 8$; $x - 2y = 4$

Use Slopes to Identify Perpendicular Lines

In the following exercises, use slopes and y-intercepts to determine if the lines are perpendicular.

644. $y = 5x - 1$; $10x + 2y = 0$

645. $3x - 2y = 5$; $2x + 3y = 6$

4.6 Find the Equation of a Line

Find an Equation of the Line Given the Slope and *y*-Intercept

In the following exercises, find the equation of a line with given slope and y-intercept. Write the equation in slope–intercept form.

646. slope $\frac{1}{3}$ and *y*-intercept $(0, -6)$

647. slope -5 and *y*-intercept $(0, -3)$

648. slope 0 and *y*-intercept $(0, 4)$

649. slope -2 and *y*-intercept $(0, 0)$

In the following exercises, find the equation of the line shown in each graph. Write the equation in slope–intercept form.

650.

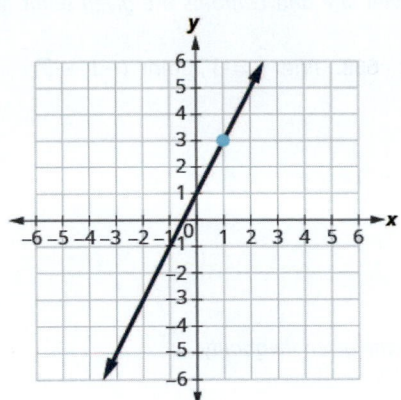

651.

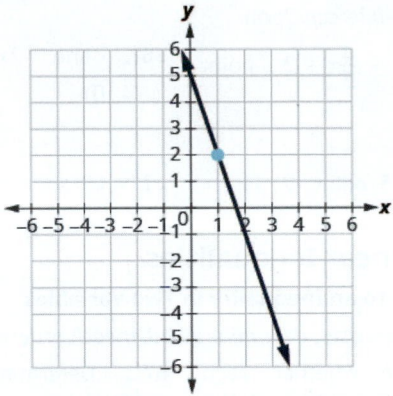

652.

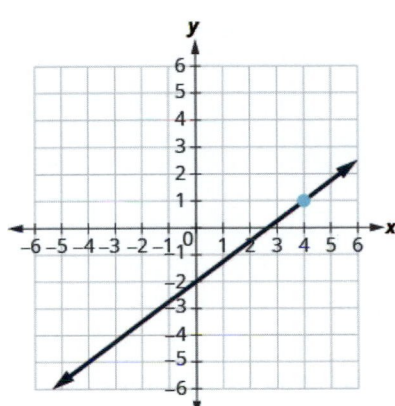

653.

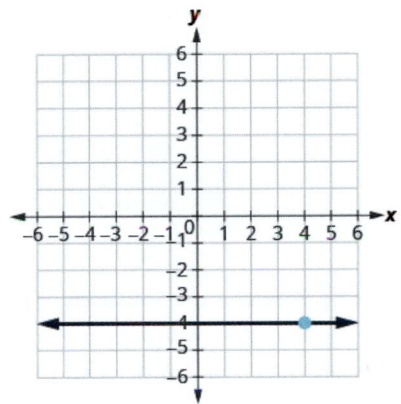

Find an Equation of the Line Given the Slope and a Point

In the following exercises, find the equation of a line with given slope and containing the given point. Write the equation in slope–intercept form.

654. $m = -\frac{1}{4}$, point $(-8, 3)$

655. $m = \frac{3}{5}$, point $(10, 6)$

656. Horizontal line containing $(-2, 7)$

657. $m = -2$, point $(-1, -3)$

Find an Equation of the Line Given Two Points

In the following exercises, find the equation of a line containing the given points. Write the equation in slope–intercept form.

658. $(2, 10)$ and $(-2, -2)$

659. $(7, 1)$ and $(5, 0)$

660. $(3, 8)$ and $(3, -4)$.

661. $(5, 2)$ and $(-1, 2)$

Find an Equation of a Line Parallel to a Given Line

In the following exercises, find an equation of a line parallel to the given line and contains the given point. Write the equation in slope–intercept form.

662. line $y = -3x + 6$, point $(1, -5)$

663. line $2x + 5y = -10$, point $(10, 4)$

664. line $x = 4$, point $(-2, -1)$

665. line $y = -5$, point $(-4, 3)$

Find an Equation of a Line Perpendicular to a Given Line

In the following exercises, find an equation of a line perpendicular to the given line and contains the given point. Write the equation in slope–intercept form.

666. line $y = -\frac{4}{5}x + 2$, point $(8, 9)$

667. line $2x - 3y = 9$, point $(-4, 0)$

668. line $y = 3$, point $(-1, -3)$

669. line $x = -5$ point $(2, 1)$

4.7 Graph Linear Inequalities

Verify Solutions to an Inequality in Two Variables

In the following exercises, determine whether each ordered pair is a solution to the given inequality.

670. Determine whether each ordered pair is a solution to the inequality $y < x - 3$:

ⓐ $(0, 1)$

ⓑ $(-2, -4)$

ⓒ $(5, 2)$

ⓓ $(3, -1)$

ⓔ $(-1, -5)$

671. Determine whether each ordered pair is a solution to the inequality $x + y > 4$:

ⓐ $(6, 1)$

ⓑ $(-3, 6)$

ⓒ $(3, 2)$

ⓓ $(-5, 10)$

ⓔ $(0, 0)$

Recognize the Relation Between the Solutions of an Inequality and its Graph

In the following exercises, write the inequality shown by the shaded region.

672. Write the inequality shown by the graph with the boundary line $y = -x + 2$.

673. Write the inequality shown by the graph with the boundary line $y = \frac{2}{3}x - 3$.

674. Write the inequality shown by the shaded region in the graph with the boundary line $x + y = -4$.

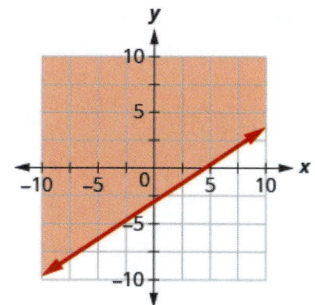

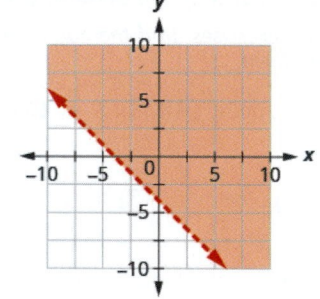

675. Write the inequality shown by the shaded region in the graph with the boundary line $x - 2y = 6$.

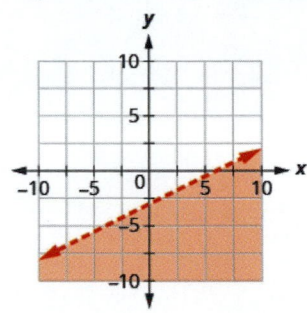

Graph Linear Inequalities

In the following exercises, graph each linear inequality.

676. Graph the linear inequality $y > \frac{2}{5}x - 4$.

677. Graph the linear inequality $y \leq -\frac{1}{4}x + 3$.

678. Graph the linear inequality $x - y \leq 5$.

679. Graph the linear inequality $3x + 2y > 10$.

680. Graph the linear inequality $y \leq -3x$.

681. Graph the linear inequality $y < 6$.

PRACTICE TEST

682. Plot each point in a rectangular coordinate system.

ⓐ (2, 5)

ⓑ (−1, −3)

ⓒ (0, 2)

ⓓ $\left(-4, \frac{3}{2}\right)$

ⓔ (5, 0)

683. Which of the given ordered pairs are solutions to the equation $3x − y = 6$?

ⓐ (3, 3)

ⓑ (2, 0)

ⓒ (4, −6)

684. Find three solutions to the linear equation $y = −2x − 4$.

685. Find the x- and y-intercepts of the equation $4x − 3y = 12$.

Find the slope of each line shown.

686.

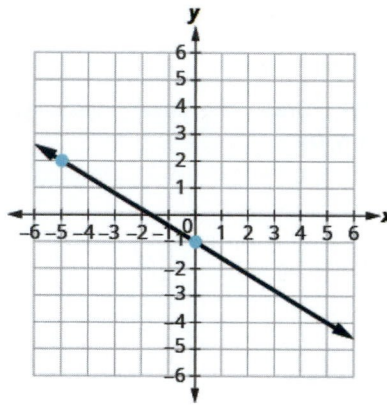

687.

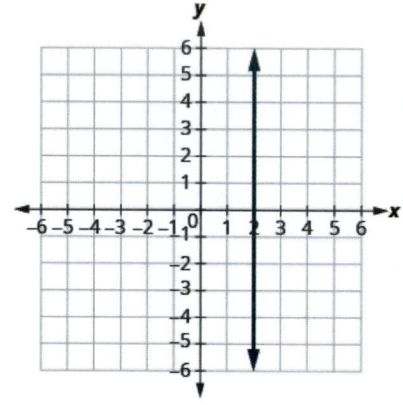

688.

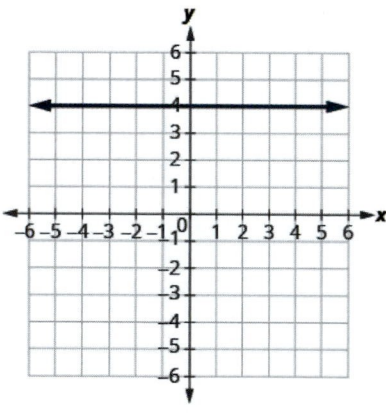

689. Find the slope of the line between the points (5, 2) and (−1, −4).

690. Graph the line with slope $\frac{1}{2}$ containing the point (−3, −4).

Graph the line for each of the following equations.

691. $y = \frac{5}{3}x − 1$

692. $y = −x$

693. $x − y = 2$

694. $4x + 2y = -8$

695. $y = 2$

696. $x = -3$

Find the equation of each line. Write the equation in slope–intercept form.

697. slope $-\frac{3}{4}$ and y-intercept $(0, -2)$

698. $m = 2$, point $(-3, -1)$

699. containing $(10, 1)$ and $(6, -1)$

700. parallel to the line $y = -\frac{2}{3}x - 1$, containing the point $(-3, 8)$

701. perpendicular to the line $y = \frac{5}{4}x + 2$, containing the point $(-10, 3)$

702. Write the inequality shown by the graph with the boundary line $y = -x - 3$.

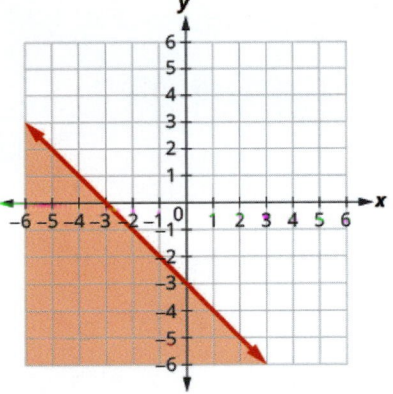

Graph each linear inequality.

703. $y > \frac{3}{2}x + 5$

704. $x - y \geq -4$

705. $y \leq -5x$

706. $y < 3$

5   SYSTEMS OF LINEAR EQUATIONS

Figure 5.1 Designing the number and sizes of windows in a home can pose challenges for an architect.

Chapter Outline

5.1 Solve Systems of Equations by Graphing

5.2 Solve Systems of Equations by Substitution

5.3 Solve Systems of Equations by Elimination

5.4 Solve Applications with Systems of Equations

5.5 Solve Mixture Applications with Systems of Equations

5.6 Graphing Systems of Linear Inequalities

Introduction

An architect designing a home may have restrictions on both the area and perimeter of the windows because of energy and structural concerns. The length and width chosen for each window would have to satisfy two equations: one for the area and the other for the perimeter. Similarly, a banker may have a fixed amount of money to put into two investment funds. A restaurant owner may want to increase profits, but in order to do that he will need to hire more staff. A job applicant may compare salary and costs of commuting for two job offers.

In this chapter, we will look at methods to solve situations like these using equations with two variables.

5.1   Solve Systems of Equations by Graphing

Learning Objectives

By the end of this section, you will be able to:

› Determine whether an ordered pair is a solution of a system of equations
› Solve a system of linear equations by graphing
› Determine the number of solutions of linear system
› Solve applications of systems of equations by graphing

Be Prepared!

Before you get started, take this readiness quiz.

1. For the equation $y = \frac{2}{3}x - 4$

    ⓐ is $(6, 0)$ a solution? ⓑ is $(-3, -2)$ a solution?

    If you missed this problem, review **Example 2.1**.

2. Find the slope and y-intercept of the line $3x - y = 12$.

    If you missed this problem, review **Example 4.42**.

3. Find the x- and y-intercepts of the line $2x - 3y = 12$.
   If you missed this problem, review **Example 4.21**.

Determine Whether an Ordered Pair is a Solution of a System of Equations

In **Solving Linear Equations and Inequalities** we learned how to solve linear equations with one variable. Remember that the solution of an equation is a value of the variable that makes a true statement when substituted into the equation.

Now we will work with **systems of linear equations**, two or more linear equations grouped together.

System of Linear Equations

When two or more linear equations are grouped together, they form a system of linear equations.

We will focus our work here on systems of two linear equations in two unknowns. Later, you may solve larger systems of equations.

An example of a system of two linear equations is shown below. We use a brace to show the two equations are grouped together to form a system of equations.

$$\begin{cases} 2x + y = 7 \\ x - 2y = 6 \end{cases}$$

A linear equation in two variables, like $2x + y = 7$, has an infinite number of solutions. Its graph is a line. Remember, every point on the line is a solution to the equation and every solution to the equation is a point on the line.

To solve a system of two linear equations, we want to find the values of the variables that are solutions to both equations. In other words, we are looking for the ordered pairs (x, y) that make both equations true. These are called the solutions to a system of equations.

Solutions of a System of Equations

Solutions of a system of equations are the values of the variables that make all the equations true. A solution of a system of two linear equations is represented by an ordered pair (x, y).

To determine if an ordered pair is a solution to a system of two equations, we substitute the values of the variables into each equation. If the ordered pair makes both equations true, it is a solution to the system.

Let's consider the system below:

$$\begin{cases} 3x - y = 7 \\ x - 2y = 4 \end{cases}$$

Is the ordered pair $(2, -1)$ a solution?

We substitute $x = 2$ and $y = -1$ into both equations.

$$3x - y = 7 \qquad\qquad x - 2y = 4$$
$$3(2) - (-1) \overset{?}{=} 7 \qquad\qquad 2 - 2(-1) \overset{?}{=} 4$$
$$7 = 7 \text{ true} \qquad\qquad 4 = 4 \text{ true}$$

The ordered pair (2, –1) made both equations true. Therefore (2, –1) is a solution to this system.

Let's try another ordered pair. Is the ordered pair (3, 2) a solution?

We substitute $x = 3$ and $y = 2$ into both equations.

$$3x - y = 7 \qquad\qquad x - 2y = 4$$
$$3(3) - 2 \overset{?}{=} 7 \qquad\qquad 2 - 2(2) \overset{?}{=} 4$$
$$7 = 7 \text{ true} \qquad\qquad -2 = 4 \text{ false}$$

The ordered pair (3, 2) made one equation true, but it made the other equation false. Since it is not a solution to **both** equations, it is not a solution to this system.

EXAMPLE 5.1

Determine whether the ordered pair is a solution to the system: $\begin{cases} x - y = -1 \\ 2x - y = -5 \end{cases}$

ⓐ $(-2, -1)$ ⓑ $(-4, -3)$

✓ **Solution**

ⓐ

$$\begin{cases} x - y = -1 \\ 2x - y = -5 \end{cases}$$

We substitute $x = -2$ and $y = -1$ into both equations.

$$x - y = -1 \qquad\qquad 2x - y = -5$$
$$-2 - (-1) \overset{?}{=} -1 \qquad 2(-2) - (-1) \overset{?}{=} -5$$
$$-1 = -1 \checkmark \qquad\qquad 5 \neq -5$$

$(-2, -1)$ does not make both equations true. $(-2, -1)$ is not a solution.

ⓑ

We substitute $x = -4$ and $y = -3$ into both equations.

$$x - y = -1 \qquad\qquad 2x - y = -5$$
$$-4 - (-3) \overset{?}{=} -1 \qquad 2(-4) - (-3) \overset{?}{=} -5$$
$$-1 = -1 \checkmark \qquad\qquad -5 = -5 \checkmark$$

$(-4, -3)$ does not make both equations true. $(-4, -3)$ is a solution.

> **TRY IT :: 5.1**
>
> Determine whether the ordered pair is a solution to the system: $\begin{cases} 3x + y = 0 \\ x + 2y = -5 \end{cases}$.
>
> ⓐ $(1, -3)$ ⓑ $(0, 0)$

> **TRY IT :: 5.2**
>
> Determine whether the ordered pair is a solution to the system: $\begin{cases} x - 3y = -8 \\ -3x - y = 4 \end{cases}$.
>
> ⓐ $(2, -2)$ ⓑ $(-2, 2)$

Solve a System of Linear Equations by Graphing

In this chapter we will use three methods to solve a system of linear equations. The first method we'll use is graphing.

The graph of a linear equation is a line. Each point on the line is a solution to the equation. For a system of two equations, we will graph two lines. Then we can see all the points that are solutions to each equation. And, by finding what the lines have in common, we'll find the solution to the system.

Most linear equations in one variable have one solution, but we saw that some equations, called contradictions, have no solutions and for other equations, called identities, all numbers are solutions.

Similarly, when we solve a system of two linear equations represented by a graph of two lines in the same plane, there are three possible cases, as shown in **Figure 5.2**:

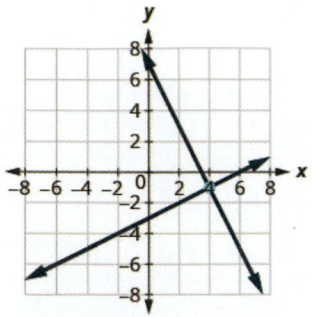 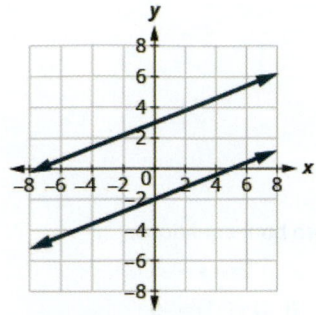 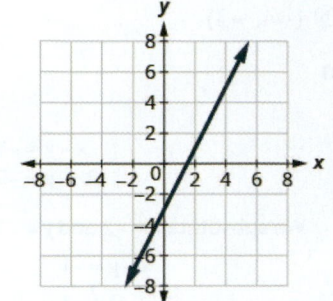

The lines intersect.
Intersecting lines have one point in common.
There is one solution to this system.

The lines are parallel.
Parallel lines have no points in common.
There is no solution to this system.

Both equations give the same line.
Because we have just one line, there are infinitely many solutions.

Figure 5.2

For the first example of solving a system of linear equations in this section and in the next two sections, we will solve the same system of two linear equations. But we'll use a different method in each section. After seeing the third method, you'll decide which method was the most convenient way to solve this system.

EXAMPLE 5.2 HOW TO SOLVE A SYSTEM OF LINEAR EQUATIONS BY GRAPHING

Solve the system by graphing: $\begin{cases} 2x + y = 7 \\ x - 2y = 6 \end{cases}$.

✓ **Solution**

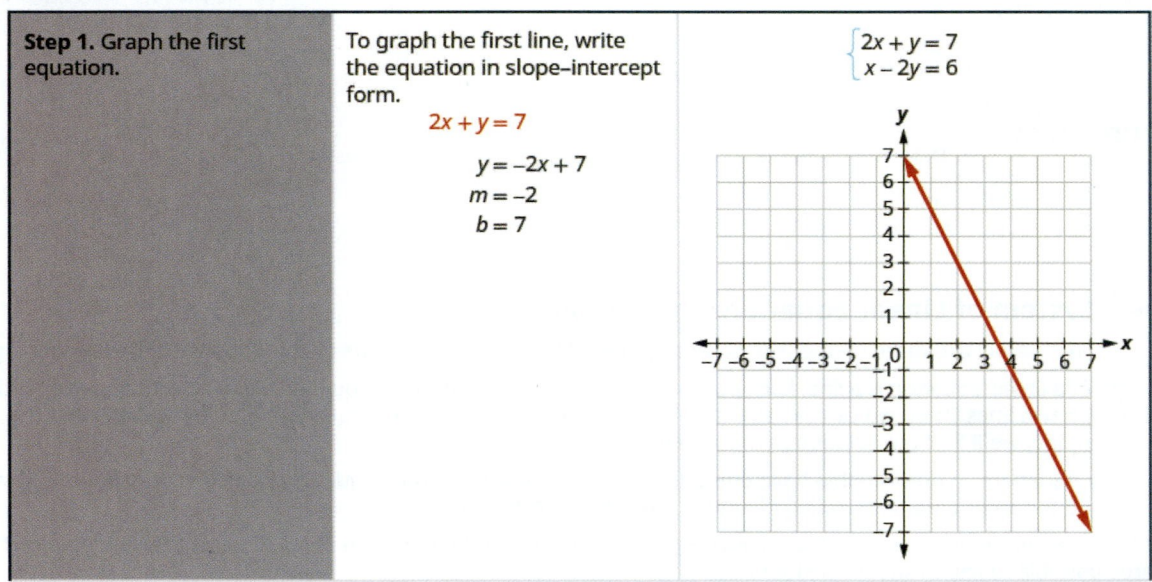

Step 1. Graph the first equation.	To graph the first line, write the equation in slope–intercept form. $2x + y = 7$ $y = -2x + 7$ $m = -2$ $b = 7$	$\begin{cases} 2x + y = 7 \\ x - 2y = 6 \end{cases}$

Step 2. Graph the second equation on the same rectangular coordinate system.	To graph the second line, use intercepts. $x - 2y = 6$ $(0, -3)$ $(6, 0)$	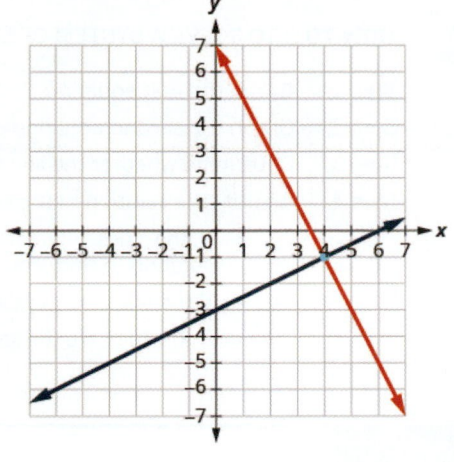
Step 3. Determine whether the lines intersect, are parallel, or are the same line.	Look at the graph of the lines.	The lines intersect.
Step 4. Identify the solution to the system. If the lines intersect, identify the point of intersection. Check to make sure it is a solution to both equations. This is the solution to the system. If the lines are parallel, the system has no solution If the lines are the same, the system has an infinite number of solutions.	Since the lines intersect, find the point of intersection. Check the point in both equations.	The lines intersect at $(4, -1)$. $2x + y = 7$ $2(4) + (-1) \overset{?}{=} 7$ $8 - 1 \overset{?}{=} 7$ $7 = 7 \checkmark$ $x - 2y = 6$ $4 - 2(-1) \overset{?}{=} 6$ $6 = 6 \checkmark$ The solution is $(4, -1)$.

> **TRY IT :: 5.3**   Solve each system by graphing: $\begin{cases} x - 3y = -3 \\ x + y = 5 \end{cases}$.

> **TRY IT :: 5.4**   Solve each system by graphing: $\begin{cases} -x + y = 1 \\ 3x + 2y = 12 \end{cases}$.

The steps to use to solve a system of linear equations by graphing are shown below.

HOW TO : : TO SOLVE A SYSTEM OF LINEAR EQUATIONS BY GRAPHING.

Step 1.   Graph the first equation.

Step 2.   Graph the second equation on the same rectangular coordinate system.

Step 3.   Determine whether the lines intersect, are parallel, or are the same line.

Step 4.   Identify the solution to the system.

- If the lines intersect, identify the point of intersection. Check to make sure it is a solution to both equations. This is the solution to the system.
- If the lines are parallel, the system has no solution.
- If the lines are the same, the system has an infinite number of solutions.

EXAMPLE 5.3

Solve the system by graphing: $\begin{cases} y = 2x + 1 \\ y = 4x - 1 \end{cases}$.

⊘ Solution

Both of the equations in this system are in slope-intercept form, so we will use their slopes and y-intercepts to graph them. $\begin{cases} y = 2x + 1 \\ y = 4x - 1 \end{cases}$

Find the slope and y-intercept of the first equation.	$y = 2x + 1$ $m = 2$ $b = 1$
Find the slope and y-intercept of the first equation.	$y = 4x - 1$ $m = 4$ $b = -1$
Graph the two lines.	
Determine the point of intersection.	The lines intersect at $(1, 3)$.

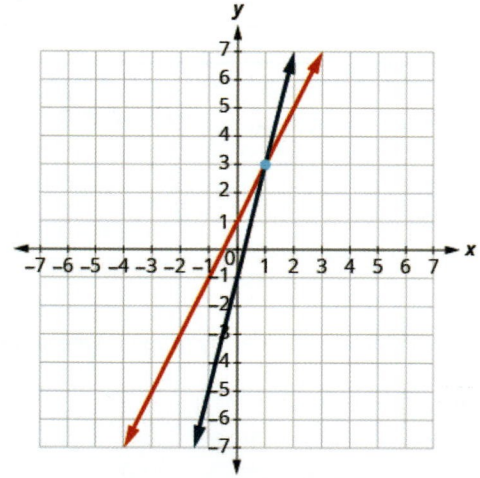

Check the solution in both equations.	$\begin{array}{ll} y = 2x + 1 & y = 4x - 1 \\ 3 \overset{?}{=} 2 \cdot 1 + 1 & 3 \overset{?}{=} 4 \cdot 1 - 1 \\ 3 = 3 \checkmark & 3 = 3 \checkmark \end{array}$

The solution is $(1, 3)$.

> **TRY IT :: 5.5**

Solve each system by graphing: $\begin{cases} y = 2x + 2 \\ y = -x - 4 \end{cases}$.

> **TRY IT :: 5.6**

Solve each system by graphing: $\begin{cases} y = 3x + 3 \\ y = -x + 7 \end{cases}$.

Both equations in **Example 5.3** were given in slope–intercept form. This made it easy for us to quickly graph the lines. In the next example, we'll first re-write the equations into slope–intercept form.

EXAMPLE 5.4

Solve the system by graphing: $\begin{cases} 3x + y = -1 \\ 2x + y = 0 \end{cases}$.

⊘ **Solution**

We'll solve both of these equations for y so that we can easily graph them using their slopes and y-intercepts.

$\begin{cases} 3x + y = -1 \\ 2x + y = 0 \end{cases}$

Solve the first equation for y.	$\begin{aligned} 3x + y &= -1 \\ y &= -3x - 1 \end{aligned}$
Find the slope and y-intercept.	$\begin{aligned} m &= -3 \\ b &= -1 \end{aligned}$
Solve the second equation for y.	$\begin{aligned} 2x + y &= 0 \\ y &= -2x \end{aligned}$
Find the slope and y-intercept.	$\begin{aligned} m &= -2 \\ b &= 0 \end{aligned}$

Graph the lines.

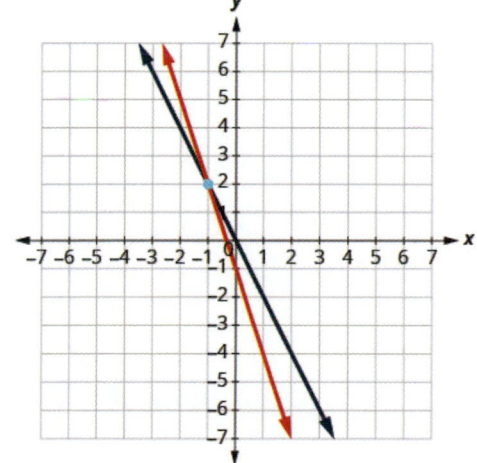

Determine the point of intersection.	The lines intersect at $(-1, 2)$.
Check the solution in both equations.	$\begin{aligned} 3x + y &= -1 \\ 3(-1) + 2 &\overset{?}{=} -1 \\ -1 &= -1 ✓ \end{aligned}$ $\begin{aligned} 2x + y &= 0 \\ 2(-1) + 2 &\overset{?}{=} 0 \\ 0 &= 0 ✓ \end{aligned}$
	The solution is $(-1, 2)$.

> **TRY IT ::** 5.7

Solve each system by graphing: $\begin{cases} -x + y = 1 \\ 2x + y = 10 \end{cases}$.

> **TRY IT ::** 5.8

Solve each system by graphing: $\begin{cases} 2x + y = 6 \\ x + y = 1 \end{cases}$.

Usually when equations are given in standard form, the most convenient way to graph them is by using the intercepts. We'll do this in **Example 5.5**.

EXAMPLE 5.5

Solve the system by graphing: $\begin{cases} x + y = 2 \\ x - y = 4 \end{cases}$.

⊘ **Solution**

We will find the x- and y-intercepts of both equations and use them to graph the lines.

$$x + y = 2$$

To find the intercepts, let $x = 0$ and solve for y, then let $y = 0$ and solve for x.	$\begin{aligned} x + y &= 2 \\ 0 + y &= 2 \\ y &= 2 \end{aligned}$	$\begin{aligned} x + y &= 2 \\ x + 0 &= 2 \\ x &= 2 \end{aligned}$	x	y
			0	2
			2	0

$$x - y = 4$$

To find the intercepts, let $x = 0$ then let $y = 0$.	$\begin{aligned} x - y &= 4 \\ 0 - y &= 4 \\ -y &= 4 \\ y &= -4 \end{aligned}$	$\begin{aligned} x - y &= 4 \\ x - 0 &= 4 \\ x &= 4 \end{aligned}$	x	y
			0	-4
			4	0

Graph the line.

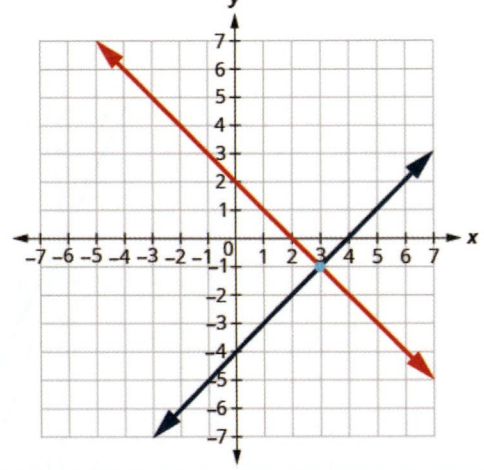

Determine the point of intersection.	The lines intersect at (3, –1).

Check the solution in both equations.	$\begin{aligned} x + y &= 2 \\ 3 + (-1) &\overset{?}{=} 2 \\ 2 &= 2 ✓ \end{aligned}$ $\qquad$ $\begin{aligned} x - y &= 4 \\ 3 - (-1) &\overset{?}{=} 4 \\ 4 &= 4 ✓ \end{aligned}$ The solution is (3, –1).

> **TRY IT : : 5.9** Solve each system by graphing: $\begin{cases} x + y = 6 \\ x - y = 2 \end{cases}$.

> **TRY IT : : 5.10** Solve each system by graphing: $\begin{cases} x + y = 2 \\ x - y = -8 \end{cases}$.

Do you remember how to graph a linear equation with just one variable? It will be either a vertical or a horizontal line.

EXAMPLE 5.6

Solve the system by graphing: $\begin{cases} y = 6 \\ 2x + 3y = 12 \end{cases}$.

⊘ **Solution**

	$\begin{cases} y = 6 \\ 2x + 3y = 12 \end{cases}$
We know the first equation represents a horizontal line whose y-intercept is 6.	$y = 6$
The second equation is most conveniently graphed using intercepts.	$2x + 3y = 12$
To find the intercepts, let $x = 0$ and then $y = 0$.	<table><tr><td>x</td><td>y</td></tr><tr><td>0</td><td>4</td></tr><tr><td>6</td><td>0</td></tr></table>
Graph the lines.	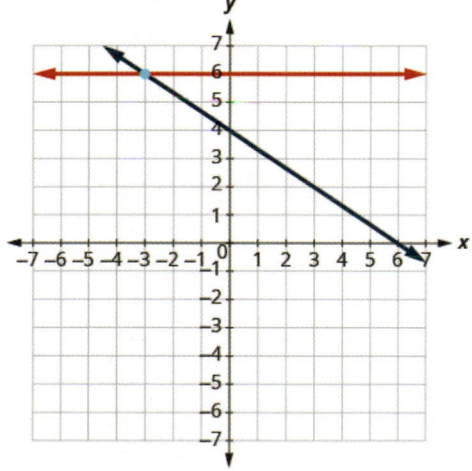
Determine the point of intersection.	The lines intersect at $(-3, 6)$.
Check the solution to both equations.	$\begin{array}{ccc} y &=& 6 \\ 6 &\overset{?}{=}& 6\ \checkmark \\ 2 &=& 2 \end{array}$ $\qquad$ $\begin{array}{ccc} 2x + 3y &=& 12 \\ 2(-3) + 3(6) &\overset{?}{=}& 12 \\ -6 + 18 &\overset{?}{=}& 12 \\ 12 &=& 12\ \checkmark \end{array}$
	The solution is $(-3, 6)$.

> **TRY IT :: 5.11**

Solve each system by graphing: $\begin{cases} y = -1 \\ x + 3y = 6 \end{cases}$.

> **TRY IT :: 5.12**

Solve each system by graphing: $\begin{cases} x = 4 \\ 3x - 2y = 24 \end{cases}$.

In all the systems of linear equations so far, the lines intersected and the solution was one point. In the next two examples, we'll look at a system of equations that has no solution and at a system of equations that has an infinite number of solutions.

EXAMPLE 5.7

Solve the system by graphing: $\begin{cases} y = \frac{1}{2}x - 3 \\ x - 2y = 4 \end{cases}$.

⊘ **Solution**

$$\begin{cases} y = \frac{1}{2}x - 3 \\ x - 2y = 4 \end{cases}$$

To graph the first equation, we will use its slope and y-intercept.	$y = \frac{1}{2}x - 3$
	$m = \frac{1}{2}$
	$b = -3$
To graph the second equation, we will use the intercepts.	$x - 2y = 4$

x	y
0	–2
4	0

Graph the lines.

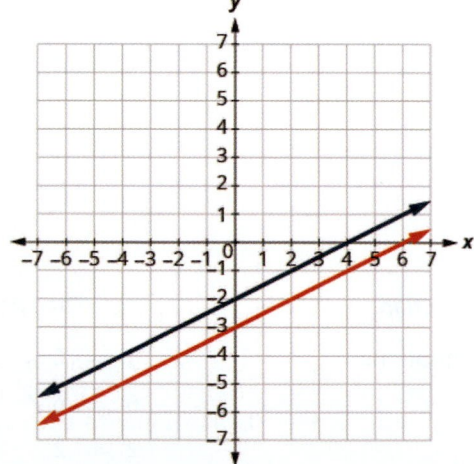

Determine the point of intersection.	The lines are parallel.
	Since no point is on both lines, there is no ordered pair that makes both equations true. There is no solution to this system.

> **TRY IT :: 5.13**

Solve each system by graphing: $\begin{cases} y = -\frac{1}{4}x + 2 \\ x + 4y = -8 \end{cases}$.

> **TRY IT :: 5.14**

Solve each system by graphing: $\begin{cases} y = 3x - 1 \\ 6x - 2y = 6 \end{cases}$.

EXAMPLE 5.8

Solve the system by graphing: $\begin{cases} y = 2x - 3 \\ -6x + 3y = -9 \end{cases}$.

⊘ **Solution**

$$\begin{cases} y = 2x - 3 \\ -6x + 3y = -9 \end{cases}$$

Find the slope and y-intercept of the first equation.	$y = 2x - 3$ $m = 2$ $b = -3$
Find the intercepts of the second equation.	$-6x + 3y = -9$

x	y
0	-3
$\frac{3}{2}$	0

Graph the lines.

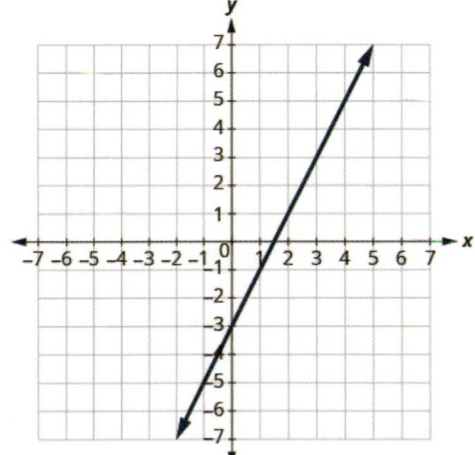

Determine the point of intersection.	The lines are the same!
	Since every point on the line makes both equations true, there are infinitely many ordered pairs that make both equations true.
	There are infinitely many solutions to this system.

> **TRY IT :: 5.15**

Solve each system by graphing: $\begin{cases} y = -3x - 6 \\ 6x + 2y = -12 \end{cases}$.

> **TRY IT :: 5.16**

Solve each system by graphing: $\begin{cases} y = \frac{1}{2}x - 4 \\ 2x - 4y = 16 \end{cases}$.

If you write the second equation in **Example 5.8** in slope-intercept form, you may recognize that the equations have the same slope and same y-intercept.

When we graphed the second line in the last example, we drew it right over the first line. We say the two lines are coincident. Coincident lines have the same slope and same y-intercept.

Coincident Lines

Coincident lines have the same slope and same y-intercept.

Determine the Number of Solutions of a Linear System

There will be times when we will want to know how many solutions there will be to a system of linear equations, but we might not actually have to find the solution. It will be helpful to determine this without graphing.

We have seen that two lines in the same plane must either intersect or are parallel. The systems of equations in **Example 5.2** through **Example 5.6** all had two intersecting lines. Each system had one solution.

A system with parallel lines, like **Example 5.7**, has no solution. What happened in **Example 5.8**? The equations have coincident lines, and so the system had infinitely many solutions.

We'll organize these results in **Figure 5.3** below:

Graph	Number of solutions
2 intersecting lines	1
Parallel lines	None
Same line	Infinitely many

Figure 5.3

Parallel lines have the same slope but different y-intercepts. So, if we write both equations in a system of linear equations in slope–intercept form, we can see how many solutions there will be without graphing! Look at the system we solved in **Example 5.7**.

$$\begin{cases} y = \frac{1}{2}x - 3 \\ x - 2y = 4 \end{cases}$$

The first line is in slope–intercept form.   If we solve the second equation for y, we get

$$y = \frac{1}{2}x - 3$$

$$\begin{aligned} x - 2y &= 4 \\ -2y &= -x + 4 \\ y &= \frac{1}{2}x - 2 \end{aligned}$$

$$m = \frac{1}{2},\ b = -3$$ $$m = \frac{1}{2},\ b = -2$$

The two lines have the same slope but different y-intercepts. They are parallel lines.

Figure 5.4 shows how to determine the number of solutions of a linear system by looking at the slopes and intercepts.

Number of Solutions of a Linear System of Equations			
Slopes	Intercepts	Type of Lines	Number of Solutions
Different		Intersecting	1 point
Same	Different	Parallel	No solution
Same	Same	Coincident	Infinitely many solutions

Figure 5.4

Let's take one more look at our equations in **Example 5.7** that gave us parallel lines.

$$\begin{cases} y = \frac{1}{2}x - 3 \\ x - 2y = 4 \end{cases}$$

When both lines were in slope-intercept form we had:

$$y = \frac{1}{2}x - 3 \qquad y = \frac{1}{2}x - 2$$

Do you recognize that it is impossible to have a single ordered pair (x, y) that is a solution to both of those equations?

We call a system of equations like this an inconsistent system. It has no solution.

A system of equations that has at least one solution is called a consistent system.

Consistent and Inconsistent Systems

A **consistent system** of equations is a system of equations with at least one solution.

An **inconsistent system** of equations is a system of equations with no solution.

We also categorize the equations in a system of equations by calling the equations *independent* or *dependent*. If two equations are **independent equations**, they each have their own set of solutions. Intersecting lines and parallel lines are independent.

If two equations are dependent, all the solutions of one equation are also solutions of the other equation. When we graph two dependent equations, we get coincident lines.

Independent and Dependent Equations

Two equations are **independent** if they have different solutions.

Two equations are **dependent** if all the solutions of one equation are also solutions of the other equation.

Let's sum this up by looking at the graphs of the three types of systems. See **Figure 5.5** and **Figure 5.6**.

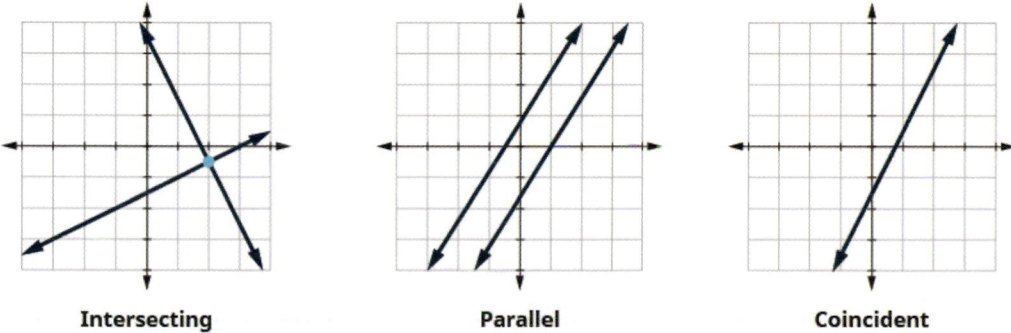

| Intersecting | Parallel | Coincident |

Figure 5.5

Lines	Intersecting	Parallel	Coincident
Number of solutions	1 point	No solution	Infinitely many
Consistent/inconsistent	Consistent	Inconsistent	Consistent
Dependent/independent	Independent	Independent	Dependent

Figure 5.6

EXAMPLE 5.9

Without graphing, determine the number of solutions and then classify the system of equations: $\begin{cases} y = 3x - 1 \\ 6x - 2y = 12 \end{cases}$.

⊘ Solution

We will compare the slopes and intercepts of the two lines.

$$\begin{cases} y & = & 3x - 1 \\ 6x - 2y & = & 12 \end{cases}$$

The first equation is already in slope-intercept form.

$$y = 3x - 1$$

Write the second equation in slope–intercept form.

$$6x - 2y = 12$$
$$-2y = -6x + 12$$
$$\frac{-2y}{-2} = \frac{-6x + 12}{-2}$$
$$y = 3x - 6$$

Find the slope and intercept of each line.

$$\begin{array}{ll} y = 3x - 1 & y = 3x - 6 \\ m = 3 & m = 3 \\ b = -1 & b = -6 \end{array}$$

Since the slopes are the same and y-intercepts are different, the lines are parallel.

A system of equations whose graphs are parallel lines has no solution and is inconsistent and independent.

> **TRY IT :: 5.17**
>
> Without graphing, determine the number of solutions and then classify the system of equations.
>
> $$\begin{cases} y = -2x - 4 \\ 4x + 2y = 9 \end{cases}$$

> **TRY IT :: 5.18**
>
> Without graphing, determine the number of solutions and then classify the system of equations.
>
> $$\begin{cases} y = \frac{1}{3}x - 5 \\ x - 3y = 6 \end{cases}$$

EXAMPLE 5.10

Without graphing, determine the number of solutions and then classify the system of equations: $\begin{cases} 2x + y = -3 \\ x - 5y = 5 \end{cases}$.

✓ Solution

We will compare the slope and intercepts of the two lines.

$$\begin{cases} 2x + y = -3 \\ x - 5y = 5 \end{cases}$$

Write both equations in slope–intercept form.

$$\begin{array}{ll} 2x + y = -3 & x - 5y = 5 \\ y = -2x - 3 & -5y = -x + 5 \\ & \dfrac{-5y}{-5} = \dfrac{-x+5}{-5} \\ & y = \tfrac{1}{5}x - 1 \end{array}$$

Find the slope and intercept of each line.

$$\begin{array}{ll} y = -2x - 3 & y = \tfrac{1}{5}x - 1 \\ m = -2 & m = \tfrac{1}{5} \\ b = -3 & b = -1 \end{array}$$

Since the slopes are different, the lines intersect.

A system of equations whose graphs are intersect has 1 solution and is consistent and independent.

> **TRY IT : : 5.19**
>
> Without graphing, determine the number of solutions and then classify the system of equations.
>
> $$\begin{cases} 3x + 2y = 2 \\ 2x + y = 1 \end{cases}$$

> **TRY IT : : 5.20**
>
> Without graphing, determine the number of solutions and then classify the system of equations.
>
> $$\begin{cases} x + 4y = 12 \\ -x + y = 3 \end{cases}$$

EXAMPLE 5.11

Without graphing, determine the number of solutions and then classify the system of equations. $\begin{cases} 3x - 2y = 4 \\ y = \tfrac{3}{2}x - 2 \end{cases}$

✓ Solution

We will compare the slope and intercepts of the two lines.

$$\begin{cases} 3x - 2y = 4 \\ \quad\; y = \frac{3}{2}x - 2 \end{cases}$$

Write the first equation in slope–intercept form.

$$
\begin{aligned}
3x - 2y &= 4 \\
-2y &= -3x + 4 \\
\frac{-2y}{-2} &= \frac{-3x + 4}{-2} \\
y &= \tfrac{3}{2}x - 2
\end{aligned}
$$

The second equation is already in slope–intercept form.

$$y = \tfrac{3}{2}x - 2$$

Since the equations are the same, they have the same slope and same y-intercept and so the lines are coincident.

A system of equations whose graphs are coincident lines has infinitely many solutions and is consistent and dependent.

> **TRY IT ::** 5.21
>
> Without graphing, determine the number of solutions and then classify the system of equations.
>
> $$\begin{cases} 4x - 5y = 20 \\ y = \frac{4}{5}x - 4 \end{cases}$$

> **TRY IT ::** 5.22
>
> Without graphing, determine the number of solutions and then classify the system of equations.
>
> $$\begin{cases} -2x - 4y = 8 \\ y = -\frac{1}{2}x - 2 \end{cases}$$

Solve Applications of Systems of Equations by Graphing

We will use the same problem solving strategy we used in Math Models to set up and solve applications of systems of linear equations. We'll modify the strategy slightly here to make it appropriate for systems of equations.

 HOW TO :: USE A PROBLEM SOLVING STRATEGY FOR SYSTEMS OF LINEAR EQUATIONS.

Step 1.   **Read** the problem. Make sure all the words and ideas are understood.

Step 2.   **Identify** what we are looking for.

Step 3.   **Name** what we are looking for. Choose variables to represent those quantities.

Step 4.   **Translate** into a system of equations.

Step 5.   **Solve** the system of equations using good algebra techniques.

Step 6.   **Check** the answer in the problem and make sure it makes sense.

Step 7.   **Answer** the question with a complete sentence.

Step 5 is where we will use the method introduced in this section. We will graph the equations and find the solution.

EXAMPLE 5.12

Sondra is making 10 quarts of punch from fruit juice and club soda. The number of quarts of fruit juice is 4 times the number of quarts of club soda. How many quarts of fruit juice and how many quarts of club soda does Sondra need?

⊘ Solution

Step 1. Read the problem.

Step 2. Identify what we are looking for.

We are looking for the number of quarts of fruit juice and the number of quarts of club soda that Sondra will need.

Step 3. Name what we are looking for. Choose variables to represent those quantities.

Let $f =$ number of quarts of fruit juice.

$c =$ number of quarts of club soda

Step 4. Translate into a system of equations.

The number of quarts of fruit juice and the number of quarts of club soda is 10

$$f \quad + \quad c \quad = 10$$

The number of quarts of fruit juice is four times the number of quarts of club soda

$$f \quad = \quad 4c$$

We now have the system. $\begin{cases} f + c = 10 \\ f = 4c \end{cases}$

Step 5. Solve the system of equations using good algebra techniques.

$$
\begin{array}{ll}
f = 4c & f + c = 10 \\
m = 4 & f = -c + 10 \\
b = 0 & m = -1 \\
& b = 10
\end{array}
$$

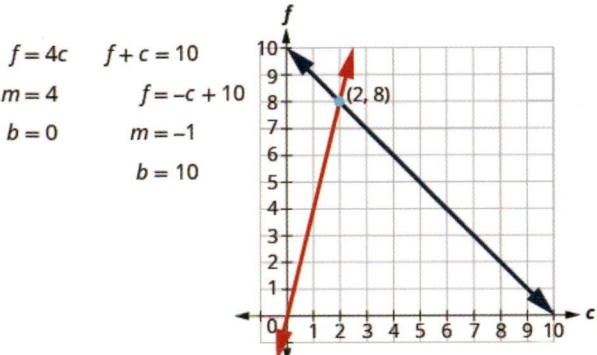

The point of intersection (2, 8) is the solution. This means Sondra needs 2 quarts of club soda and 8 quarts of fruit juice.

Step 6. Check the answer in the problem and make sure it makes sense.

Does this make sense in the problem?

Yes, the number of quarts of fruit juice, 8 is 4 times the number of quarts of club soda, 2.

Yes, 10 quarts of punch is 8 quarts of fruit juice plus 2 quarts of club soda.

Step 7. Answer the question with a complete sentence.

Sondra needs 8 quarts of fruit juice and 2 quarts of soda.

> **TRY IT ::** 5.23

Manny is making 12 quarts of orange juice from concentrate and water. The number of quarts of water is 3 times the number of quarts of concentrate. How many quarts of concentrate and how many quarts of water does Manny need?

> **TRY IT ::** 5.24

Alisha is making an 18 ounce coffee beverage that is made from brewed coffee and milk. The number of ounces of brewed coffee is 5 times greater than the number of ounces of milk. How many ounces of coffee and how many ounces of milk does Alisha need?

▶ **MEDIA : :**

Access these online resources for additional instruction and practice with solving systems of equations by graphing.

- **Instructional Video Solving Linear Systems by Graphing (http://www.openstax.org/l/25linsysGraph)**
- **Instructional Video Solve by Graphing (http://www.openstax.org/l/25solvesbyGraph)**

5.1 EXERCISES
Practice Makes Perfect

Determine Whether an Ordered Pair is a Solution of a System of Equations. *In the following exercises, determine if the following points are solutions to the given system of equations.*

1. $\begin{cases} 2x - 6y = 0 \\ 3x - 4y = 5 \end{cases}$

ⓐ (3, 1) ⓑ (−3, 4)

2. $\begin{cases} 7x - 4y = -1 \\ -3x - 2y = 1 \end{cases}$

ⓐ   ⓑ (1, −2)

3. $\begin{cases} 2x + y = 5 \\ x + y = 1 \end{cases}$

ⓐ (4, −3) ⓑ (2, 0)

4. $\begin{cases} -3x + y = 8 \\ -x + 2y = -9 \end{cases}$

ⓐ (−5, −7) ⓑ (−5, 7)

5. $\begin{cases} x + y = 2 \\ y = \frac{3}{4}x \end{cases}$

ⓐ $\left(\frac{8}{7}, \frac{6}{7}\right)$ ⓑ $\left(1, \frac{3}{4}\right)$

6. $\begin{cases} x + y = 1 \\ y = \frac{2}{5}x \end{cases}$

ⓐ $\left(\frac{5}{7}, \frac{2}{7}\right)$ ⓑ (5, 2)

7. $\begin{cases} x + 5y = 10 \\ y = \frac{3}{5}x + 1 \end{cases}$

ⓐ (−10, 4) ⓑ $\left(\frac{5}{4}, \frac{7}{4}\right)$

8. $\begin{cases} x + 3y = 9 \\ y = \frac{2}{3}x - 2 \end{cases}$

ⓐ (−6, 5) ⓑ $\left(5, \frac{4}{3}\right)$

Solve a System of Linear Equations by Graphing *In the following exercises, solve the following systems of equations by graphing.*

9. $\begin{cases} 3x + y = -3 \\ 2x + 3y = 5 \end{cases}$

10. $\begin{cases} -x + y = 2 \\ 2x + y = -4 \end{cases}$

11. $\begin{cases} -3x + y = -1 \\ 2x + y = 4 \end{cases}$

12. $\begin{cases} -2x + 3y = -3 \\ x + y = 4 \end{cases}$

13. $\begin{cases} y = x + 2 \\ y = -2x + 2 \end{cases}$

14. $\begin{cases} y = x - 2 \\ y = -3x + 2 \end{cases}$

15. $\begin{cases} y = \frac{3}{2}x + 1 \\ y = -\frac{1}{2}x + 5 \end{cases}$

16. $\begin{cases} y = \frac{2}{3}x - 2 \\ y = -\frac{1}{3}x - 5 \end{cases}$

17. $\begin{cases} -x + y = -3 \\ 4x + 4y = 4 \end{cases}$

18. $\begin{cases} x - y = 3 \\ 2x - y = 4 \end{cases}$

19. $\begin{cases} -3x + y = -1 \\ 2x + y = 4 \end{cases}$

20. $\begin{cases} -3x + y = -2 \\ 4x - 2y = 6 \end{cases}$

21. $\begin{cases} x + y = 5 \\ 2x - y = 4 \end{cases}$

22. $\begin{cases} x - y = 2 \\ 2x - y = 6 \end{cases}$

23. $\begin{cases} x + y = 2 \\ x - y = 0 \end{cases}$

24. $\begin{cases} x + y = 6 \\ x - y = -8 \end{cases}$

25. $\begin{cases} x + y = -5 \\ x - y = 3 \end{cases}$

26. $\begin{cases} x + y = 4 \\ x - y = 0 \end{cases}$

27. $\begin{cases} x + y = -4 \\ -x + 2y = -2 \end{cases}$

28. $\begin{cases} -x + 3y = 3 \\ x + 3y = 3 \end{cases}$

29. $\begin{cases} -2x + 3y = 3 \\ x + 3y = 12 \end{cases}$

30. $\begin{cases} 2x - y = 4 \\ 2x + 3y = 12 \end{cases}$

31. $\begin{cases} 2x + 3y = 6 \\ y = -2 \end{cases}$

32. $\begin{cases} -2x + y = 2 \\ y = 4 \end{cases}$

33. $\begin{cases} x - 3y = -3 \\ y = 2 \end{cases}$

34. $\begin{cases} 2x - 2y = 8 \\ y = -3 \end{cases}$

35. $\begin{cases} 2x - y = -1 \\ x = 1 \end{cases}$

36. $\begin{cases} x + 2y = 2 \\ x = -2 \end{cases}$

37. $\begin{cases} x - 3y = -6 \\ x = -3 \end{cases}$

38. $\begin{cases} x + y = 4 \\ x = 1 \end{cases}$

39. $\begin{cases} 4x - 3y = 8 \\ 8x - 6y = 14 \end{cases}$

40. $\begin{cases} x + 3y = 4 \\ -2x - 6y = 3 \end{cases}$

41. $\begin{cases} -2x + 4y = 4 \\ y = \frac{1}{2}x \end{cases}$

42. $\begin{cases} 3x + 5y = 10 \\ y = -\frac{3}{5}x + 1 \end{cases}$

43. $\begin{cases} x = -3y + 4 \\ 2x + 6y = 8 \end{cases}$

44. $\begin{cases} 4x = 3y + 7 \\ 8x - 6y = 14 \end{cases}$

45. $\begin{cases} 2x + y = 6 \\ -8x - 4y = -24 \end{cases}$

46. $\begin{cases} 5x + 2y = 7 \\ -10x - 4y = -14 \end{cases}$

47. $\begin{cases} x + 3y = -6 \\ 4y = -\frac{4}{3}x - 8 \end{cases}$

48. $\begin{cases} -x + 2y = -6 \\ y = -\frac{1}{2}x - 1 \end{cases}$

49. $\begin{cases} -3x + 2y = -2 \\ y = -x + 4 \end{cases}$

50. $\begin{cases} -x + 2y = -2 \\ y = -x - 1 \end{cases}$

Determine the Number of Solutions of a Linear System *Without graphing the following systems of equations, determine the number of solutions and then classify the system of equations.*

51. $\begin{cases} y = \frac{2}{3}x + 1 \\ -2x + 3y = 5 \end{cases}$

52. $\begin{cases} y = \frac{1}{3}x + 2 \\ x - 3y = 9 \end{cases}$

53. $\begin{cases} y = -2x + 1 \\ 4x + 2y = 8 \end{cases}$

54. $\begin{cases} y = 3x + 4 \\ 9x - 3y = 18 \end{cases}$

55. $\begin{cases} y = \frac{2}{3}x + 1 \\ 2x - 3y = 7 \end{cases}$

56. $\begin{cases} 3x + 4y = 12 \\ y = -3x - 1 \end{cases}$

57. $\begin{cases} 4x + 2y = 10 \\ 4x - 2y = -6 \end{cases}$

58. $\begin{cases} 5x + 3y = 4 \\ 2x - 3y = 5 \end{cases}$

59. $\begin{cases} y = -\frac{1}{2}x + 5 \\ x + 2y = 10 \end{cases}$

60. $\begin{cases} y = x + 1 \\ -x + y = 1 \end{cases}$

61. $\begin{cases} y = 2x + 3 \\ 2x - y = -3 \end{cases}$

62. $\begin{cases} 5x - 2y = 10 \\ y = \frac{5}{2}x - 5 \end{cases}$

Solve Applications of Systems of Equations by Graphing *In the following exercises, solve.*

63. Molly is making strawberry infused water. For each ounce of strawberry juice, she uses three times as many ounces of water. How many ounces of strawberry juice and how many ounces of water does she need to make 64 ounces of strawberry infused water?

64. Jamal is making a snack mix that contains only pretzels and nuts. For every ounce of nuts, he will use 2 ounces of pretzels. How many ounces of pretzels and how many ounces of nuts does he need to make 45 ounces of snack mix?

65. Enrique is making a party mix that contains raisins and nuts. For each ounce of nuts, he uses twice the amount of raisins. How many ounces of nuts and how many ounces of raisins does he need to make 24 ounces of party mix?

66. Owen is making lemonade from concentrate. The number of quarts of water he needs is 4 times the number of quarts of concentrate. How many quarts of water and how many quarts of concentrate does Owen need to make 100 quarts of lemonade?

Everyday Math

67. Leo is planning his spring flower garden. He wants to plant tulip and daffodil bulbs. He will plant 6 times as many daffodil bulbs as tulip bulbs. If he wants to plant 350 bulbs, how many tulip bulbs and how many daffodil bulbs should he plant?

68. A marketing company surveys 1,200 people. They surveyed twice as many females as males. How many males and females did they survey?

Writing Exercises

69. In a system of linear equations, the two equations have the same slope. Describe the possible solutions to the system.

70. In a system of linear equations, the two equations have the same intercepts. Describe the possible solutions to the system.

Self Check

After completing the exercises, use this checklist to evaluate your mastery of the objectives of this section.

I can...	Confidently	With some help	No-I don't get it!
determine whether an ordered pair is a solution of a system of equations.			
solve a system of linear equations by graphing.			
determine the number of solutions of a linear system.			
solve applications of systems of equations by graphing.			

If most of your checks were:

*...**confidently.** Congratulations! You have achieved the objectives in this section. Reflect on the study skills you used so that you can continue to use them. What did you do to become confident of your ability to do these things? Be specific.*

*...**with some help.** This must be addressed quickly because topics you do not master become potholes in your road to success. In math every topic builds upon previous work. It is important to make sure you have a strong foundation before you move on. Who can you ask for help? Your fellow classmates and instructor are good resources. Is there a place on campus where math tutors are available? Can your study skills be improved?*

*...**no - I don't get it!** This is a warning sign and you must not ignore it. You should get help right away or you will quickly be overwhelmed. See your instructor as soon as you can to discuss your situation. Together you can come up with a plan to get you the help you need.*

5.2 **Solve Systems of Equations by Substitution**

Learning Objectives

By the end of this section, you will be able to:
> Solve a system of equations by substitution
> Solve applications of systems of equations by substitution

Be Prepared!

Before you get started, take this readiness quiz.

1. Simplify $-5(3 - x)$.
   If you missed this problem, review **Example 1.136**.

2. Simplify $4 - 2(n + 5)$.
   If you missed this problem, review **Example 1.123**.

3. Solve for y. $8y - 8 = 32 - 2y$
   If you missed this problem, review **Example 2.34**.

4. Solve for x. $3x - 9y = -3$
   If you missed this problem, review **Example 2.65**.

Solving systems of linear equations by graphing is a good way to visualize the types of solutions that may result. However, there are many cases where solving a system by graphing is inconvenient or imprecise. If the graphs extend beyond the small grid with x and y both between −10 and 10, graphing the lines may be cumbersome. And if the solutions to the system are not integers, it can be hard to read their values precisely from a graph.

In this section, we will solve systems of linear equations by the substitution method.

Solve a System of Equations by Substitution

We will use the same system we used first for graphing.

$$\begin{cases} 2x + y = 7 \\ x - 2y = 6 \end{cases}$$

We will first solve one of the equations for either x or y. We can choose either equation and solve for either variable—but we'll try to make a choice that will keep the work easy.

Then we substitute that expression into the other equation. The result is an equation with just one variable—and we know how to solve those!

After we find the value of one variable, we will substitute that value into one of the original equations and solve for the other variable. Finally, we check our solution and make sure it makes both equations true.

We'll fill in all these steps now in **Example 5.13**.

EXAMPLE 5.13   HOW TO SOLVE A SYSTEM OF EQUATIONS BY SUBSTITUTION

Solve the system by substitution. $\begin{cases} 2x + y = 7 \\ x - 2y = 6 \end{cases}$

⊘ **Solution**

Step 1. Solve one of the equations for either variable.	We'll solve the first equation for y.	$2x + y = 7$ $y = 7 - 2x$

Step 2. Substitute the expression from Step 1 into the other equation.	We replace y in the second equation with the expression $7 - 2x$.	$x - 2y = 6$ $x - 2(7 - 2x) = 6$
Step 3. Solve the resulting equation.	Now we have an equation with just 1 variable. We know how to solve this!	$x - 2(7 - 2x) = 6$ $x - 14 + 4x = 6$ $5x = 20$ $x = 4$
Step 4. Substitute the solution in Step 3 into one of the original equations to find the other variable.	We'll use the first equation and replace x with 4.	$2x + y = 7$ $2(4) + y = 7$ $8 + y = 7$ $y = -1$
Step 5. Write the solution as an ordered pair.	The ordered pair is (x, y).	$(4, -1)$
Step 6. Check that the ordered pair is a solution to **both** original equations.	Substitute $(4, -1)$ into both equations and make sure they are both true.	$2x + y = 7 \qquad x - 2y = 6$ $2(4) + (-1) \overset{?}{=} 7 \qquad 4 - 2(-1) \overset{?}{=} 6$ $7 = 7 \checkmark \qquad 6 = 6 \checkmark$ Both equations are true. $(4, -1)$ is the solution to the system.

> **TRY IT :: 5.25**   Solve the system by substitution. $\begin{cases} -2x + y = -11 \\ x + 3y = 9 \end{cases}$

> **TRY IT :: 5.26**   Solve the system by substitution. $\begin{cases} x + 3y = 10 \\ 4x + y = 18 \end{cases}$

HOW TO :: SOLVE A SYSTEM OF EQUATIONS BY SUBSTITUTION.

Step 1.   Solve one of the equations for either variable.

Step 2.   Substitute the expression from Step 1 into the other equation.

Step 3.   Solve the resulting equation.

Step 4.   Substitute the solution in Step 3 into one of the original equations to find the other variable.

Step 5.   Write the solution as an ordered pair.

Step 6.   Check that the ordered pair is a solution to **both** original equations.

If one of the equations in the system is given in slope–intercept form, Step 1 is already done! We'll see this in **Example 5.14**.

EXAMPLE 5.14

Solve the system by substitution.

$$\begin{cases} x + y = -1 \\ y = x + 5 \end{cases}$$

⊘ Solution

The second equation is already solved for *y*. We will substitute the expression in place of *y* in the first equation.

$$\begin{cases} x + y = -1 \\ \quad y = x + 5 \end{cases}$$

The second equation is already solved for *y*. We will substitute into the first equation.	
Replace the *y* with *x* + 5.	$y = x + 5$ $x + y = -1$
Solve the resulting equation for *x*.	$x + x + 5 = -1$
	$2x + 5 = -1$
	$2x = -6$
	$x = -3$
Substitute *x* = −3 into *y* = *x* + 5 to find *y*.	$y = x + 5$
	$y = -3 + 5$
The ordered pair is (−3, 2).	$y = 2$

Check the ordered pair in both equations:

$$\begin{array}{rcl} x + y &=& -1 \\ -3 + 2 &\overset{?}{=}& -1 \\ -1 &=& -1 \checkmark \end{array} \qquad \begin{array}{rcl} y &=& x + 5 \\ 2 &\overset{?}{=}& -3 + 5 \\ 2 &=& 2 \checkmark \end{array}$$

The solution is (−3, 2).

> **TRY IT : : 5.27** Solve the system by substitution. $\begin{cases} x + y = 6 \\ y = 3x - 2 \end{cases}$

> **TRY IT : : 5.28** Solve the system by substitution. $\begin{cases} 2x - y = 1 \\ y = -3x - 6 \end{cases}$

If the equations are given in standard form, we'll need to start by solving for one of the variables. In this next example, we'll solve the first equation for *y*.

EXAMPLE 5.15

Solve the system by substitution. $\begin{cases} 3x + y = 5 \\ 2x + 4y = -10 \end{cases}$

⊘ Solution

We need to solve one equation for one variable. Then we will substitute that expression into the other equation.

	$3x + y = 5$
Solve for y.	$y = \boxed{-3x + 5}$
Substitute into the other equation.	$2x + 4y = -10$
Replace the y with $-3x + 5$.	$2x + 4(-3x + 5) = -10$
Solve the resulting equation for x.	$2x - 12x + 20 = -10$
	$-10x + 20 = -10$
	$-10x = -30$
	$\boxed{x = -3}$
Substitute $x = 3$ into $3x + y = 5$ to find y.	$3x + y = 5$
	$3(3) + y = 5$
	$9 + y = 5$
The ordered pair is (3, –4).	$y = -4$

Check the ordered pair in both equations:

$$3x + y \;=\; 5 \qquad\qquad 2x + 4y \;=\; -10$$
$$3 \cdot 3 + (-4) \overset{?}{=} 5 \qquad 2 \cdot 3 + 4(-4) \;=\; -10$$
$$9 - 4 \overset{?}{=} 5 \qquad\qquad 6 - 16 \overset{?}{=} -10$$
$$5 = 5 \checkmark \qquad\qquad -10 = -10 \checkmark$$

The solution is (3, –4).

> **TRY IT :: 5.29**　Solve the system by substitution. $\begin{cases} 4x + y = 2 \\ 3x + 2y = -1 \end{cases}$

> **TRY IT :: 5.30**　Solve the system by substitution. $\begin{cases} -x + y = 4 \\ 4x - y = 2 \end{cases}$

In **Example 5.15** it was easiest to solve for y in the first equation because it had a coefficient of 1. In **Example 5.16** it will be easier to solve for x.

EXAMPLE 5.16

Solve the system by substitution. $\begin{cases} x - 2y = -2 \\ 3x + 2y = 34 \end{cases}$

⊘ **Solution**

We will solve the first equation for x and then substitute the expression into the second equation.

	$x - 2y = -2$
Solve for x.	$x = \boxed{2y - 2}$
Substitute into the other equation.	$3x + 2y = 34$
Replace the x with $2y - 2$.	$3(2y - 2) + 2y = 34$

Solve the resulting equation for y.	$6y - 6 + 2y = 34$
	$8y - 6 = 34$
	$8y = 40$
	$y = 5$
Substitute $y = 5$ into $x - 2y = -2$ to find x.	$x - 2y = -2$
	$x - 2 \cdot 5 = -2$
	$x - 10 = -2$
	$x = 8$

The ordered pair is (8, 5).

Check the ordered pair in both equations:

$$x - 2y = -2 \qquad\qquad 3x + 2y = 34$$
$$8 - 2 \cdot 5 \overset{?}{=} -2 \qquad\qquad 3 \cdot 8 + 2 \cdot 5 \overset{?}{=} 34$$
$$8 - 10 \overset{?}{=} -2 \qquad\qquad 24 + 10 \overset{?}{=} 34$$
$$-2 = -2 \checkmark \qquad\qquad 34 = 34 \checkmark$$

The solution is (8, 5).

> **TRY IT : : 5.31**
>
> Solve the system by substitution. $\begin{cases} x - 5y = 13 \\ 4x - 3y = 1 \end{cases}$

> **TRY IT : : 5.32**
>
> Solve the system by substitution. $\begin{cases} x - 6y = -6 \\ 2x - 4y = 4 \end{cases}$

When both equations are already solved for the same variable, it is easy to substitute!

EXAMPLE 5.17

Solve the system by substitution. $\begin{cases} y = -2x + 5 \\ y = \frac{1}{2}x \end{cases}$

⊘ Solution

Since both equations are solved for y, we can substitute one into the other.

Substitute $\frac{1}{2}x$ for y in the first equation.	$y = \frac{1}{2}x$ $y = -2x + 5$
Replace the y with $\frac{1}{2}x$.	$\frac{1}{2}x = -2x + 5$
Solve the resulting equation. Start by clearing the fraction.	$2\left(\frac{1}{2}x\right) = 2(-2x + 5)$
Solve for x.	$x = -4x + 10$
	$5x = 10$

Substitute $x = 2$ into $y = \frac{1}{2}x$ to find y.

$$\boxed{x = 2}$$
$$y = \frac{1}{2}x$$
$$y = \frac{1}{2} \cdot 2$$
$$y = 1$$

The ordered pair is (2,1).

Check the ordered pair in both equations:

$$\begin{aligned} y &= \tfrac{1}{2}x & y &= -2x + 5 \\ 1 &\overset{?}{=} \tfrac{1}{2} \cdot 2 & 1 &\overset{?}{=} -2 \cdot 2 + 5 \\ & & 1 &= -4 + 5 \\ 1 &= 1 \checkmark & 1 &= 1 \checkmark \end{aligned}$$

The solution is (2,1).

> **TRY IT :: 5.33**
> Solve the system by substitution. $\begin{cases} y = 3x - 16 \\ y = \frac{1}{3}x \end{cases}$

> **TRY IT :: 5.34**
> Solve the system by substitution. $\begin{cases} y = -x + 10 \\ y = \frac{1}{4}x \end{cases}$

Be very careful with the signs in the next example.

EXAMPLE 5.18

Solve the system by substitution. $\begin{cases} 4x + 2y = 4 \\ 6x - y = 8 \end{cases}$

⊘ **Solution**

We need to solve one equation for one variable. We will solve the first equation for y.

$$4x + 2y = 4$$

Solve the first equation for y.	$2y = -4x + 4$
	$y = \boxed{-2x + 2}$
Substitute $-2x + 2$ for y in the second equation.	$6x - y = 8$
Replace the y with $-2x + 2$.	$6x - (-2x + 2) = 8$
Solve the equation for x.	$6x + 2x - 2 = 8$
	$8x - 2 = 8$
	$8x = 10$

Substitute $x = \frac{5}{4}$ into $4x + 2y = 4$ to find y.

$$\boxed{x = \frac{5}{4}}$$
$$4x + 2y = 4$$
$$4\left(\frac{5}{4}\right) + 2y = 4$$
$$5 + 2y = 4$$
$$2y = -1$$
$$y = -\frac{1}{2}$$

The ordered pair is $\left(\frac{5}{4}, -\frac{1}{2}\right)$.

Check the ordered pair in both equations.

$$
\begin{array}{rcl}
4x + 2y &=& 4 \\
4\left(\frac{5}{4}\right) + 2\left(-\frac{1}{2}\right) &\overset{?}{=}& 4 \\
5 - 1 &\overset{?}{=}& 4 \\
4 &=& 4 \checkmark
\end{array}
\qquad
\begin{array}{rcl}
6x - y &=& 8 \\
6\left(\frac{5}{4}\right) - \left(-\frac{1}{2}\right) &\overset{?}{=}& 8 \\
\frac{15}{4} - \left(-\frac{1}{2}\right) &\overset{?}{=}& 8 \\
\frac{16}{2} &\overset{?}{=}& 8 \\
8 &=& 8 \checkmark
\end{array}
$$

The solution is $\left(\frac{5}{4}, -\frac{1}{2}\right)$.

> **TRY IT :: 5.35**
>
> Solve the system by substitution. $\begin{cases} x - 4y = -4 \\ -3x + 4y = 0 \end{cases}$

> **TRY IT :: 5.36**
>
> Solve the system by substitution. $\begin{cases} 4x - y = 0 \\ 2x - 3y = 5 \end{cases}$

In **Example 5.19**, it will take a little more work to solve one equation for x or y.

EXAMPLE 5.19

Solve the system by substitution. $\begin{cases} 4x - 3y = 6 \\ 15y - 20x = -30 \end{cases}$

⊘ **Solution**

We need to solve one equation for one variable. We will solve the first equation for x.

$$4x - 3y = 6$$

Solve the first equation for x.
$$4x = 3y + 6$$

Substitute $\frac{3}{4}y + \frac{3}{2}$ for x in the second equation.

$$\boxed{x = \frac{3}{4}y + \frac{3}{2}}$$
$$15y - 20x = -30$$

Replace the x with $\frac{3}{4}y + \frac{3}{2}$.
$$15y - 20\left(\frac{3}{4}y + \frac{3}{2}\right) = -30$$

Solve for y.	$15y - 15y - 30 = -30$
	$0 - 30 = -30$
	$0 = 0$

Since $0 = 0$ is a true statement, the system is consistent. The equations are dependent. The graphs of these two equations would give the same line. The system has infinitely many solutions.

> **TRY IT :: 5.37** Solve the system by substitution. $\begin{cases} 2x - 3y = 12 \\ -12y + 8x = 48 \end{cases}$

> **TRY IT :: 5.38** Solve the system by substitution. $\begin{cases} 5x + 2y = 12 \\ -4y - 10x = -24 \end{cases}$

Look back at the equations in **Example 5.19**. Is there any way to recognize that they are the same line?

Let's see what happens in the next example.

EXAMPLE 5.20

Solve the system by substitution. $\begin{cases} 5x - 2y = -10 \\ y = \frac{5}{2}x \end{cases}$

✓ **Solution**

The second equation is already solved for y, so we can substitute for y in the first equation.

Substitute x for y in the first equation.	$5x - 2y = -10$
Replace the y with $\frac{5}{2}x$.	$5x - 2\left(\frac{5}{2}x\right) = -10$
Solve for x.	$5x - 5x = -10$
	$0 \neq -10$

Since $0 = -10$ is a false statement the equations are inconsistent. The graphs of the two equation would be parallel lines. The system has no solutions.

> **TRY IT :: 5.39** Solve the system by substitution. $\begin{cases} 3x + 2y = 9 \\ y = -\frac{3}{2}x + 1 \end{cases}$

> **TRY IT :: 5.40** Solve the system by substitution. $\begin{cases} 5x - 3y = 2 \\ y = \frac{5}{3}x - 4 \end{cases}$

Solve Applications of Systems of Equations by Substitution

We'll copy here the problem solving strategy we used in the **Solving Systems of Equations by Graphing** section for solving systems of equations. Now that we know how to solve systems by substitution, that's what we'll do in Step 5.

HOW TO :: HOW TO USE A PROBLEM SOLVING STRATEGY FOR SYSTEMS OF LINEAR EQUATIONS.

Step 1.   **Read** the problem. Make sure all the words and ideas are understood.

Step 2.   **Identify** what we are looking for.

Step 3.   **Name** what we are looking for. Choose variables to represent those quantities.

Step 4.   **Translate** into a system of equations.

Step 5.   **Solve** the system of equations using good algebra techniques.

Step 6.   **Check** the answer in the problem and make sure it makes sense.

Step 7.   **Answer** the question with a complete sentence.

Some people find setting up word problems with two variables easier than setting them up with just one variable. Choosing the variable names is easier when all you need to do is write down two letters. Think about this in the next example—how would you have done it with just one variable?

EXAMPLE 5.21

The sum of two numbers is zero. One number is nine less than the other. Find the numbers.

 Solution

Step 1. Read the problem.	
Step 2. Identify what we are looking for.	We are looking for two numbers.
Step 3. Name what we are looking for.	Let $n =$ the first number Let $m =$ the second number
Step 4. Translate into a system of equations.	The sum of two numbers is zero.
	$$n + m = 0$$
	One number is nine less than the other.
	$$n = m - 9$$
The system is:	$$\begin{cases} n + m = 0 \\ \quad n = m - 9 \end{cases}$$
Step 5. Solve the system of equations. We will use substitution since the second equation is solved for n.	
	$$n = \boxed{m - 9}$$
Substitute $m - 9$ for n in the first equation.	$$n + m = 0$$
Solve for m.	$$m - 9 + m = 0$$
	$$2m - 9 = 0$$
	$$2m = 9$$
Substitute $m = \frac{9}{2}$ into the second equation and then solve for n.	$$m = \boxed{\frac{9}{2}}$$
	$$n = m - 9$$
	$$m = \frac{9}{2} - 9$$

$$m = \frac{9}{2} - \frac{18}{2}$$

$$n = -\frac{9}{2}$$

Step 6. Check the answer in the problem.	Do these numbers make sense in the problem? We will leave this to you!
Step 7. Answer the question.	The numbers are $\frac{9}{2}$ and $-\frac{9}{2}$.

> **TRY IT : : 5.41**   The sum of two numbers is 10. One number is 4 less than the other. Find the numbers.

> **TRY IT : : 5.42**   The sum of two number is –6. One number is 10 less than the other. Find the numbers.

In the **Example 5.22**, we'll use the formula for the perimeter of a rectangle, $P = 2L + 2W$.

EXAMPLE 5.22

The perimeter of a rectangle is 88. The length is five more than twice the width. Find the length and the width.

⊘ **Solution**

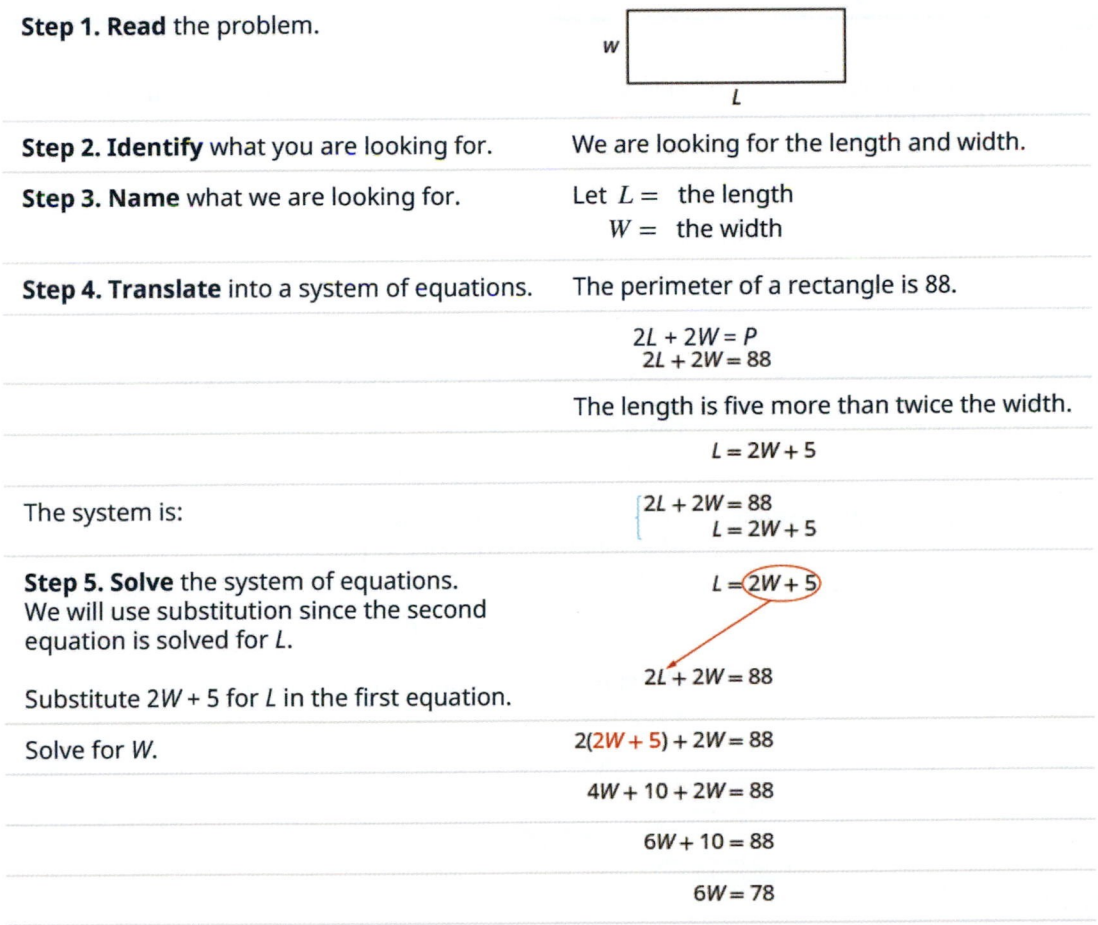

Step 1. Read the problem.	
Step 2. Identify what you are looking for.	We are looking for the length and width.
Step 3. Name what we are looking for.	Let $L =$ the length $W =$ the width
Step 4. Translate into a system of equations.	The perimeter of a rectangle is 88.
	$2L + 2W = P$ $2L + 2W = 88$
	The length is five more than twice the width.
	$L = 2W + 5$
The system is:	$\begin{cases} 2L + 2W = 88 \\ L = 2W + 5 \end{cases}$
Step 5. Solve the system of equations. We will use substitution since the second equation is solved for L.	$L = \boxed{2W + 5}$
Substitute $2W + 5$ for L in the first equation.	$2L + 2W = 88$
Solve for W.	$2(2W + 5) + 2W = 88$
	$4W + 10 + 2W = 88$
	$6W + 10 = 88$
	$6W = 78$

Substitute $W = 13$ into the second equation and then solve for L.	$W = \boxed{13}$ $L = 2W + 5$ $L = 2 \cdot 13 + 5$ $L = 31$
Step 6. Check the answer in the problem.	Does a rectangle with length 31 and width 13 have perimeter 88? Yes.
Step 7. Answer the equation.	The length is 31 and the width is 13.

> **TRY IT :: 5.43**
>
> The perimeter of a rectangle is 40. The length is 4 more than the width. Find the length and width of the rectangle.

> **TRY IT :: 5.44**
>
> The perimeter of a rectangle is 58. The length is 5 more than three times the width. Find the length and width of the rectangle.

For **Example 5.23** we need to remember that the sum of the measures of the angles of a triangle is 180 degrees and that a right triangle has one 90 degree angle.

EXAMPLE 5.23

The measure of one of the small angles of a right triangle is ten more than three times the measure of the other small angle. Find the measures of both angles.

⊘ **Solution**

We will draw and label a figure.

Step 1. Read the problem.	
Step 2. Identify what you are looking for.	We are looking for the measures of the angles.
Step 3. Name what we are looking for.	Let $a =$ the measure of the 1st angle $b =$ the measure of the 2nd angle
Step 4. Translate into a system of equations.	The measure of one of the small angles of a right triangle is ten more than three times the measure of the other small angle.
	$a = 3b + 10$
	The sum of the measures of the angles of a triangle is 180.
	$a + b + 90 = 180$
The system is:	$\begin{cases} a = 3b + 10 \\ a + b + 90 = 180 \end{cases}$

Step 5. Solve the system of equations. We will use substitution since the first equation is solved for a.

$a = \boxed{3b + 10}$

$a + b + 90 = 180$

Substitute $3b + 10$ for a in the second equation.	$(3b + 10) + b + 90 = 180$
Solve for b.	$4b + 100 = 180$
	$4b = 80$
	$b = \boxed{20}$
	$a = 3b + 10$
Substitute $b = 20$ into the first equation and then solve for a.	$a = 3 \cdot 20 + 10$
	$a = 70$
Step 6. Check the answer in the problem.	We will leave this to you!
Step 7. Answer the question.	The measures of the small angles are 20 and 70.

> **TRY IT : : 5.45**
>
> The measure of one of the small angles of a right triangle is 2 more than 3 times the measure of the other small angle. Find the measure of both angles.

> **TRY IT : : 5.46**
>
> The measure of one of the small angles of a right triangle is 18 less than twice the measure of the other small angle. Find the measure of both angles.

EXAMPLE 5.24

Heather has been offered two options for her salary as a trainer at the gym. Option A would pay her $25,000 plus $15 for each training session. Option B would pay her $10,000 + $40 for each training session. How many training sessions would make the salary options equal?

⊘ **Solution**

Step 1. Read the problem.

Step 2. Identify what you are looking for.	We are looking for the number of training sessions that would make the pay equal.
Step 3. Name what we are looking for.	Let $s =$ Heather's salary. $n =$ the number of training sessions
Step 4. Translate into a system of equations.	Option A would pay her $25,000 plus $15 for each training session.
	$s = 25,000 + 15n$
	Option B would pay her $10,000 + $40 for each training session

$$s = 10,000 + 40n$$

The system is:	$\begin{cases} s = 25,000 + 15n \\ s = 10,000 + 40n \end{cases}$

Step 5. Solve the system of equations. We will use substitution.

$$s = \boxed{25,000 + 15n}$$
$$s = 10,000 + 40n$$

Substitute $25,000 + 15n$ for s in the second equation.

$$25,000 + 15n = 10,000 + 40n$$

Solve for n.

$$25,000 = 10,000 + 25n$$

$$15,000 = 25n$$

$$600 = n$$

Step 6. Check the answer.

Are 600 training sessions a year reasonable?
Are the two options equal when $n = 600$?

Step 7. Answer the question.

The salary options would be equal for 600 training sessions.

> **TRY IT : : 5.47**

Geraldine has been offered positions by two insurance companies. The first company pays a salary of $12,000 plus a commission of $100 for each policy sold. The second pays a salary of $20,000 plus a commission of $50 for each policy sold. How many policies would need to be sold to make the total pay the same?

> **TRY IT : : 5.48**

Kenneth currently sells suits for company A at a salary of $22,000 plus a $10 commission for each suit sold. Company B offers him a position with a salary of $28,000 plus a $4 commission for each suit sold. How many suits would Kenneth need to sell for the options to be equal?

> **MEDIA : :**

Access these online resources for additional instruction and practice with solving systems of equations by substitution.

- **Instructional Video-Solve Linear Systems by Substitution (http://www.openstax.org/l/25SolvingLinear)**
- **Instructional Video-Solve by Substitution (http://www.openstax.org/l/25Substitution)**

5.2 EXERCISES
Practice Makes Perfect

Solve a System of Equations by Substitution

In the following exercises, solve the systems of equations by substitution.

71. $\begin{cases} 2x + y = -4 \\ 3x - 2y = -6 \end{cases}$

72. $\begin{cases} 2x + y = -2 \\ 3x - y = 7 \end{cases}$

73. $\begin{cases} x - 2y = -5 \\ 2x - 3y = -4 \end{cases}$

74. $\begin{cases} x - 3y = -9 \\ 2x + 5y = 4 \end{cases}$

75. $\begin{cases} 5x - 2y = -6 \\ y = 3x + 3 \end{cases}$

76. $\begin{cases} -2x + 2y = 6 \\ y = -3x + 1 \end{cases}$

77. $\begin{cases} 2x + 3y = 3 \\ y = -x + 3 \end{cases}$

78. $\begin{cases} 2x + 5y = -14 \\ y = -2x + 2 \end{cases}$

79. $\begin{cases} 2x + 5y = 1 \\ y = \frac{1}{3}x - 2 \end{cases}$

80. $\begin{cases} 3x + 4y = 1 \\ y = -\frac{2}{5}x + 2 \end{cases}$

81. $\begin{cases} 3x - 2y = 6 \\ y = \frac{2}{3}x + 2 \end{cases}$

82. $\begin{cases} -3x - 5y = 3 \\ y = \frac{1}{2}x - 5 \end{cases}$

83. $\begin{cases} 2x + y = 10 \\ -x + y = -5 \end{cases}$

84. $\begin{cases} -2x + y = 10 \\ -x + 2y = 16 \end{cases}$

85. $\begin{cases} 3x + y = 1 \\ -4x + y = 15 \end{cases}$

86. $\begin{cases} x + y = 0 \\ 2x + 3y = -4 \end{cases}$

87. $\begin{cases} x + 3y = 1 \\ 3x + 5y = -5 \end{cases}$

88. $\begin{cases} x + 2y = -1 \\ 2x + 3y = 1 \end{cases}$

89. $\begin{cases} 2x + y = 5 \\ x - 2y = -15 \end{cases}$

90. $\begin{cases} 4x + y = 10 \\ x - 2y = -20 \end{cases}$

91. $\begin{cases} y = -2x - 1 \\ y = -\frac{1}{3}x + 4 \end{cases}$

92. $\begin{cases} y = x - 6 \\ y = -\frac{3}{2}x + 4 \end{cases}$

93. $\begin{cases} y = 2x - 8 \\ y = \frac{3}{5}x + 6 \end{cases}$

94. $\begin{cases} y = -x - 1 \\ y = x + 7 \end{cases}$

95. $\begin{cases} 4x + 2y = 8 \\ 8x - y = 1 \end{cases}$

96. $\begin{cases} -x - 12y = -1 \\ 2x - 8y = -6 \end{cases}$

97. $\begin{cases} 15x + 2y = 6 \\ -5x + 2y = -4 \end{cases}$

98. $\begin{cases} 2x - 15y = 7 \\ 12x + 2y = -4 \end{cases}$

99. $\begin{cases} y = 3x \\ 6x - 2y = 0 \end{cases}$

100. $\begin{cases} x = 2y \\ 4x - 8y = 0 \end{cases}$

101. $\begin{cases} 2x + 16y = 8 \\ -x - 8y = -4 \end{cases}$

102. $\begin{cases} 15x + 4y = 6 \\ -30x - 8y = -12 \end{cases}$

103. $\begin{cases} y = -4x \\ 4x + y = 1 \end{cases}$

104. $\begin{cases} y = -\frac{1}{4}x \\ x + 4y = 8 \end{cases}$

105. $\begin{cases} y = \frac{7}{8}x + 4 \\ -7x + 8y = 6 \end{cases}$

106. $\begin{cases} y = -\frac{2}{3}x + 5 \\ 2x + 3y = 11 \end{cases}$

Solve Applications of Systems of Equations by Substitution

In the following exercises, translate to a system of equations and solve.

107. The sum of two numbers is 15. One number is 3 less than the other. Find the numbers.

108. The sum of two numbers is 30. One number is 4 less than the other. Find the numbers.

109. The sum of two numbers is −26. One number is 12 less than the other. Find the numbers.

110. The perimeter of a rectangle is 50. The length is 5 more than the width. Find the length and width.

111. The perimeter of a rectangle is 60. The length is 10 more than the width. Find the length and width.

112. The perimeter of a rectangle is 58. The length is 5 more than three times the width. Find the length and width.

113. The perimeter of a rectangle is 84. The length is 10 more than three times the width. Find the length and width.

114. The measure of one of the small angles of a right triangle is 14 more than 3 times the measure of the other small angle. Find the measure of both angles.

115. The measure of one of the small angles of a right triangle is 26 more than 3 times the measure of the other small angle. Find the measure of both angles.

116. The measure of one of the small angles of a right triangle is 15 less than twice the measure of the other small angle. Find the measure of both angles.

117. The measure of one of the small angles of a right triangle is 45 less than twice the measure of the other small angle. Find the measure of both angles.

118. Maxim has been offered positions by two car dealers. The first company pays a salary of $10,000 plus a commission of $1,000 for each car sold. The second pays a salary of $20,000 plus a commission of $500 for each car sold. How many cars would need to be sold to make the total pay the same?

119. Jackie has been offered positions by two cable companies. The first company pays a salary of $ 14,000 plus a commission of $100 for each cable package sold. The second pays a salary of $20,000 plus a commission of $25 for each cable package sold. How many cable packages would need to be sold to make the total pay the same?

120. Amara currently sells televisions for company A at a salary of $17,000 plus a $100 commission for each television she sells. Company B offers her a position with a salary of $29,000 plus a $20 commission for each television she sells. How many televisions would Amara need to sell for the options to be equal?

121. Mitchell currently sells stoves for company A at a salary of $12,000 plus a $150 commission for each stove he sells. Company B offers him a position with a salary of $24,000 plus a $50 commission for each stove he sells. How many stoves would Mitchell need to sell for the options to be equal?

Everyday Math

122. When Gloria spent 15 minutes on the elliptical trainer and then did circuit training for 30 minutes, her fitness app says she burned 435 calories. When she spent 30 minutes on the elliptical trainer and 40 minutes circuit training she burned 690 calories. Solve the system $\begin{cases} 15e + 30c = 435 \\ 30e + 40c = 690 \end{cases}$ for e, the number of calories she burns for each minute on the elliptical trainer, and c, the number of calories she burns for each minute of circuit training.

123. Stephanie left Riverside, California, driving her motorhome north on Interstate 15 towards Salt Lake City at a speed of 56 miles per hour. Half an hour later, Tina left Riverside in her car on the same route as Stephanie, driving 70 miles per hour. Solve the system $\begin{cases} 56s = 70t \\ s = t + \frac{1}{2} \end{cases}$.

ⓐ for t to find out how long it will take Tina to catch up to Stephanie.

ⓑ what is the value of s, the number of hours Stephanie will have driven before Tina catches up to her?

Writing Exercises

124. Solve the system of equations
$$\begin{cases} x + y = 10 \\ x - y = 6 \end{cases}$$

ⓐ by graphing.

ⓑ by substitution.

ⓒ Which method do you prefer? Why?

125. Solve the system of equations
$$\begin{cases} 3x + y = 12 \\ x = y - 8 \end{cases}$$
by substitution and explain all your steps in words.

Self Check

ⓐ *After completing the exercises, use this checklist to evaluate your mastery of the objectives of this section.*

I can...	Confidently	With some help	No-I don't get it!
solve a system of equations by substitution.			
solve applications of systems of equations by substitution.			

ⓑ *After reviewing this checklist, what will you do to become confident for all objectives?*

5.3   Solve Systems of Equations by Elimination

Learning Objectives

By the end of this section, you will be able to:

> Solve a system of equations by elimination
> Solve applications of systems of equations by elimination
> Choose the most convenient method to solve a system of linear equations

---

Be Prepared!

Before you get started, take this readiness quiz.

1. Simplify $-5(6 - 3a)$.
   If you missed this problem, review **Example 1.136**.

2. Solve the equation $\frac{1}{3}x + \frac{5}{8} = \frac{31}{24}$.
   If you missed this problem, review **Example 2.48**.

We have solved systems of linear equations by graphing and by substitution. Graphing works well when the variable coefficients are small and the solution has integer values. Substitution works well when we can easily solve one equation for one of the variables and not have too many fractions in the resulting expression.

The third method of solving systems of linear equations is called the Elimination Method. When we solved a system by substitution, we started with two equations and two variables and reduced it to one equation with one variable. This is what we'll do with the elimination method, too, but we'll have a different way to get there.

Solve a System of Equations by Elimination

The Elimination Method is based on the Addition Property of Equality. The Addition Property of Equality says that when you add the same quantity to both sides of an equation, you still have equality. We will extend the Addition Property of Equality to say that when you add equal quantities to both sides of an equation, the results are equal.

For any expressions a, b, c, and d,

$$\begin{aligned} \text{if} \quad a &= b \\ \text{and} \quad c &= d \\ \text{then} \quad a + c &= b + d \end{aligned}$$

To solve a system of equations by elimination, we start with both equations in standard form. Then we decide which variable will be easiest to eliminate. How do we decide? We want to have the coefficients of one variable be opposites, so that we can add the equations together and eliminate that variable.

Notice how that works when we add these two equations together:

$$\begin{aligned} 3x + y &= 5 \\ 2x - y &= 0 \\ \hline 5x \quad\;\; &= 5 \end{aligned}$$

The y's add to zero and we have one equation with one variable.

Let's try another one:

$$\begin{cases} x + 4y = 2 \\ 2x + 5y = -2 \end{cases}$$

This time we don't see a variable that can be immediately eliminated if we add the equations.

But if we multiply the first equation by –2, we will make the coefficients of x opposites. We must multiply every term on both sides of the equation by –2.

$$\begin{cases} -2(x + 4y) = -2(2) \\ \;\; 2x + 5y = -2 \end{cases}$$

$$\begin{cases} -2x - 8y = -4 \\ \;\; 2x + 5y = -2 \end{cases}$$

Now we see that the coefficients of the *x* terms are opposites, so *x* will be eliminated when we add these two equations. Add the equations yourself—the result should be −3*y* = −6. And that looks easy to solve, doesn't it? Here is what it would look like.

$$\begin{cases} -2x - 8y = -4 \\ \underline{2x + 5y = -2} \\ -3y = -6 \end{cases}$$

We'll do one more:

$$\begin{cases} 4x - 3y = 10 \\ 3x + 5y = -7 \end{cases}$$

It doesn't appear that we can get the coefficients of one variable to be opposites by multiplying one of the equations by a constant, unless we use fractions. So instead, we'll have to multiply both equations by a constant.

We can make the coefficients of *x* be opposites if we multiply the first equation by 3 and the second by −4, so we get 12*x* and −12*x*.

$$3(4x - 3y) = 3(10)$$

$$-4(3x + 5y) = -4(-7)$$

This gives us these two new equations:

$$\begin{cases} 12x - 9y = 30 \\ -12x - 20y = 28 \end{cases}$$

When we add these equations,

$$\begin{cases} 12x - 9y = 30 \\ \underline{-12x - 20y = 28} \\ -29y = 58 \end{cases}$$

the *x*'s are eliminated and we just have −29*y* = 58.

Once we get an equation with just one variable, we solve it. Then we substitute that value into one of the original equations to solve for the remaining variable. And, as always, we check our answer to make sure it is a solution to both of the original equations.

Now we'll see how to use elimination to solve the same system of equations we solved by graphing and by substitution.

EXAMPLE 5.25   HOW TO SOLVE A SYSTEM OF EQUATIONS BY ELIMINATION

Solve the system by elimination. $\begin{cases} 2x + y = 7 \\ x - 2y = 6 \end{cases}$

✓ **Solution**

Step 1. Write both equations in standard form. If any coefficients are fractions, clear them.	Both equations are in standard form, $Ax + By = C$. There are no fractions.	$\begin{cases} 2x + y = 7 \\ x - 2y = 6 \end{cases}$
Step 2. Make the coefficients of one variable opposites. Decide which variable you will eliminate. Multiply one or both equations so that the coefficients of that variable are opposites.	We can eliminate the *y*'s by multiplying the first equation by 2. Multiply both sides of $2x + y = 7$ by 2.	$\begin{cases} 2x + y = 7 \\ x - 2y = 6 \end{cases}$ $\begin{cases} 2(2x + y) = 2(7) \\ x - 2y = 6 \end{cases}$
Step 3. Add the equations resulting from Step 2 to eliminate one variable.	We add the *x*'s, *y*'s, and constants.	$\begin{cases} 4x + 2y = 14 \\ \underline{x - 2y = 6} \\ 5x = 20 \end{cases}$

Step 4. Solve for the remaining variable.	Solve for x.	$x = 4$
Step 5. Substitute the solution from Step 4 into one of the original equations. Then solve for the other variable.	Substitute $x = 4$ into the second equation, $x - 2y = 6$. Then solve for y.	$x - 2y = 6$ $4 - 2y = 6$ $-2y = 2$ $y = -1$
Step 6. Write the solution as an ordered pair.	Write it as (x, y).	$(4, -1)$
Step 7. Check that the ordered pair is a solution to **both** original equations.	Substitute $(4, -1)$ into $2x + y = 7$ and $x - 2y = 6$ Do they make both equations true? Yes!	$2x + y = 7$ $\qquad$ $x - 2y = 6$ $2(4) + (-1) \overset{?}{=} 7$ $\quad$ $4 - 2(-1) \overset{?}{=} 6$ $\qquad 7 = 7 \checkmark$ $\qquad\qquad 6 = 6 \checkmark$ The solution is $(4, -1)$.

> **TRY IT :: 5.49**　Solve the system by elimination. $\begin{cases} 3x + y = 5 \\ 2x - 3y = 7 \end{cases}$

> **TRY IT :: 5.50**　Solve the system by elimination. $\begin{cases} 4x + y = -5 \\ -2x - 2y = -2 \end{cases}$

The steps are listed below for easy reference.

HOW TO :: HOW TO SOLVE A SYSTEM OF EQUATIONS BY ELIMINATION.

Step 1.　Write both equations in standard form. If any coefficients are fractions, clear them.

Step 2.　Make the coefficients of one variable opposites.
- Decide which variable you will eliminate.
- Multiply one or both equations so that the coefficients of that variable are opposites.

Step 3.　Add the equations resulting from Step 2 to eliminate one variable.

Step 4.　Solve for the remaining variable.

Step 5.　Substitute the solution from Step 4 into one of the original equations. Then solve for the other variable.

Step 6.　Write the solution as an ordered pair.

Step 7.　Check that the ordered pair is a solution to **both** original equations.

First we'll do an example where we can eliminate one variable right away.

EXAMPLE 5.26

Solve the system by elimination. $\begin{cases} x + y = 10 \\ x - y = 12 \end{cases}$

✓ Solution

$$\begin{cases} x + y = 10 \\ x - y = 12 \end{cases}$$

Both equations are in standard form.	
The coefficients of y are already opposites.	
Add the two equations to eliminate y. The resulting equation has only 1 variable, x.	$\begin{cases} x + y = 10 \\ \underline{x - y = 12} \\ 2x \quad\; = 22 \end{cases}$
Solve for x, the remaining variable.	$x = \boxed{11}$
Substitute $x = 11$ into one of the original equations.	$x + y = 10$
	$11 + y = 10$
Solve for the other variable, y.	$y = -1$
Write the solution as an ordered pair.	The ordered pair is $(11, -1)$.

Check that the ordered pair is a solution to **both** original equations.

$$\begin{aligned} x + y &= 10 & x - y &= 12 \\ 11 + (-1) &\overset{?}{=} 10 & 11 - (-1) &\overset{?}{=} 12 \\ 10 &= 10 \;\checkmark & 12 &= 12 \;\checkmark \end{aligned}$$

The solution is $(11, -1)$.

> **TRY IT : : 5.51**
>
> Solve the system by elimination. $\begin{cases} 2x + y = 5 \\ x - y = 4 \end{cases}$

> **TRY IT : : 5.52**
>
> Solve the system by elimination. $\begin{cases} x + y = 3 \\ -2x - y = -1 \end{cases}$

In **Example 5.27**, we will be able to make the coefficients of one variable opposites by multiplying one equation by a constant.

EXAMPLE 5.27

Solve the system by elimination. $\begin{cases} 3x - 2y = -2 \\ 5x - 6y = 10 \end{cases}$

✓ Solution

$$\begin{cases} 3x - 2y = -2 \\ 5x - 6y = 10 \end{cases}$$

Both equations are in standard form.	
None of the coefficients are opposites.	
We can make the coefficients of y opposites by multiplying the first equation by -3.	$\begin{cases} -3(3x - 2y) = -3(-2) \\ 5x - 6y = 10 \end{cases}$

Simplify.	$\begin{cases} -9x + 6y = 6 \\ 5x - 6y = 10 \end{cases}$
Add the two equations to eliminate y.	$\begin{array}{r} -9x + 6y = 6 \\ 5x - 6y = 10 \\ \hline -4x \quad\quad = 16 \end{array}$
Solve for the remaining variable, x. Substitute $x = -4$ into one of the original equations.	$x = -4$ $3x - 2y = -2$
	$3(-4) - 2y = -2$
Solve for y.	$-12 - 2y = -2$
	$-2y = 10$
	$y = -5$
Write the solution as an ordered pair.	The ordered pair is $(-4, -5)$.

Check that the ordered pair is a solution to
both original equations.

$$\begin{array}{ll}
3x - 2y = -2 & 5x - 6y = 10 \\
3(-4) - 2(-5) \overset{?}{=} -2 & 3(-4) - 6(-5) \overset{?}{=} 10 \\
-12 + 10 \overset{?}{=} -2 & -20 + 30 \overset{?}{=} 10 \\
-2y = -2 \checkmark & 10 = 10 \checkmark
\end{array}$$

The solution is $(-4, -5)$.

> **TRY IT :: 5.53**
>
> Solve the system by elimination. $\begin{cases} 4x - 3y = 1 \\ 5x - 9y = -4 \end{cases}$

> **TRY IT :: 5.54**
>
> Solve the system by elimination. $\begin{cases} 3x + 2y = 2 \\ 6x + 5y = 8 \end{cases}$

Now we'll do an example where we need to multiply both equations by constants in order to make the coefficients of one variable opposites.

EXAMPLE 5.28

Solve the system by elimination. $\begin{cases} 4x - 3y = 9 \\ 7x + 2y = -6 \end{cases}$

⊘ **Solution**

In this example, we cannot multiply just one equation by any constant to get opposite coefficients. So we will strategically multiply both equations by a constant to get the opposites.

$$\begin{cases} 4x - 3y = 9 \\ 7x + 2y = -6 \end{cases}$$

Both equations are in standard form. To get opposite coefficients of y, we will multiply the first equation by 2 and the second equation by 3.	$\begin{cases} 2(4x - 3y) = 2(9) \\ 3(7x + 2y) = 3(-6) \end{cases}$

Simplify.	$\begin{cases} 8x - 6y = 18 \\ 21x + 6y = -18 \end{cases}$
Add the two equations to eliminate y.	$\begin{array}{r} 8x - 6y = 18 \\ 21x + 6y = -18 \\ \hline 39x = 0 \end{array}$
Solve for x.	$x = \boxed{0}$
Substitute $x = 0$ into one of the original equations.	$7x + 2y = -6$
	$7 \cdot 0 + 2y = -6$
Solve for y.	$2y = -6$
	$y = -3$
Write the solution as an ordered pair.	The ordered pair is $(0, -3)$.

Check that the ordered pair is a solution to **both** original equations.

$$\begin{array}{rcl} 4x - 3y & = & 9 \\ 4(0) - 3(-3) & \overset{?}{=} & 9 \\ 9 & = & 9 \checkmark \end{array} \qquad \begin{array}{rcl} 7x + 2y & = & -6 \\ 7(0) + 2(-3) & \overset{?}{=} & -6 \\ -6 & = & -6 \checkmark \end{array}$$

The solution is $(0, -3)$.

What other constants could we have chosen to eliminate one of the variables? Would the solution be the same?

> **TRY IT :: 5.55** Solve the system by elimination. $\begin{cases} 3x - 4y = -9 \\ 5x + 3y = 14 \end{cases}$

> **TRY IT :: 5.56** Solve the system by elimination. $\begin{cases} 7x + 8y = 4 \\ 3x - 5y = 27 \end{cases}$

When the system of equations contains fractions, we will first clear the fractions by multiplying each equation by its LCD.

EXAMPLE 5.29

Solve the system by elimination. $\begin{cases} x + \frac{1}{2}y = 6 \\ \frac{3}{2}x + \frac{2}{3}y = \frac{17}{2} \end{cases}$

⊘ **Solution**

In this example, both equations have fractions. Our first step will be to multiply each equation by its LCD to clear the fractions.

$$\begin{cases} x + \frac{1}{2}y = 6 \\ \frac{3}{2}x + \frac{2}{3}y = \frac{17}{2} \end{cases}$$

To clear the fractions, multiply each equation by its LCD.	$\begin{cases} 2\left(x + \frac{1}{2}y\right) = 2(6) \\ 6\left(\frac{3}{2}x + \frac{2}{3}y\right) = 6\left(\frac{17}{2}\right) \end{cases}$

Simplify.	$\begin{cases} 2x + y = 12 \\ 9x + 4y = 51 \end{cases}$

Now we are ready to eliminate one of the variables. Notice that both equations are in standard form.

We can eliminate y multiplying the top equation by -4.	$\begin{cases} -4(2x + y) = -4(12) \\ 9x + 4y = 51 \end{cases}$

Simplify and add.	$\begin{array}{r} -8x - 4y = -48 \\ 9x + 4y = 51 \\ \hline x = \boxed{3} \end{array}$
	$x + \frac{1}{2}y = 6$

Substitute $x = 3$ into one of the original equations.	
Solve for y.	$3 + \frac{1}{2}y = 6$
	$\frac{1}{2}y = 3$
	$y = 6$

Write the solution as an ordered pair.	The ordered pair is (3, 6).

Check that the ordered pair is a solution to **both** original equations.

$$x + \frac{1}{2}y = 6 \qquad\qquad \frac{3}{2}x + \frac{2}{3}y = \frac{17}{2}$$
$$3 + \frac{1}{2}(6) \stackrel{?}{=} 6 \qquad\qquad \frac{3}{2}(3) + \frac{2}{3}(6) \stackrel{?}{=} \frac{17}{2}$$
$$3 + 6 \stackrel{?}{=} 6 \qquad\qquad \frac{9}{2} + 4 \stackrel{?}{=} \frac{17}{2}$$
$$6 = 6 \checkmark \qquad\qquad \frac{9}{2} + \frac{8}{2} \stackrel{?}{=} \frac{17}{2}$$
$$\frac{17}{2} = \frac{17}{2} \checkmark$$

	The solution is (3, 6).

> **TRY IT :: 5.57**

Solve the system by elimination. $\begin{cases} \frac{1}{3}x - \frac{1}{2}y = 1 \\ \frac{3}{4}x - y = \frac{5}{2} \end{cases}$

> **TRY IT :: 5.58**

Solve the system by elimination. $\begin{cases} x + \frac{3}{5}y = -\frac{1}{5} \\ -\frac{1}{2}x - \frac{2}{3}y = \frac{5}{6} \end{cases}$

In the **Solving Systems of Equations by Graphing** we saw that not all systems of linear equations have a single ordered pair as a solution. When the two equations were really the same line, there were infinitely many solutions. We called that a consistent system. When the two equations described parallel lines, there was no solution. We called that an inconsistent system.

EXAMPLE 5.30

Solve the system by elimination. $\begin{cases} 3x + 4y = 12 \\ y = 3 - \frac{3}{4}x \end{cases}$

✓ **Solution**

$$\begin{cases} 3x + 4y &= 12 \\ y &= 3 - \frac{3}{4}x \end{cases}$$

Write the second equation in standard form.

$$\begin{cases} 3x + 4y &= 12 \\ \frac{3}{4}x + y &= 3 \end{cases}$$

Clear the fractions by multiplying the second equation by 4.

$$\begin{cases} 3x + 4y &= 12 \\ 4\left(\frac{3}{4}x + y\right) &= 4(3) \end{cases}$$

Simplify.

$$\begin{cases} 3x + 4y &= 12 \\ 3x + 4y &= 12 \end{cases}$$

To eliminate a variable, we multiply the second equation by -1.

$$\begin{cases} 3x + 4y &= 12 \\ -3x - 4y &= -12 \end{cases}$$
$$0 = 0$$

Simplify and add.

This is a true statement. The equations are consistent but dependent. Their graphs would be the same line. The system has infinitely many solutions.

After we cleared the fractions in the second equation, did you notice that the two equations were the same? That means we have coincident lines.

> **TRY IT : : 5.59**

Solve the system by elimination. $\begin{cases} 5x - 3y = 15 \\ y = -5 + \frac{5}{3}x \end{cases}$

> **TRY IT : : 5.60**

Solve the system by elimination. $\begin{cases} x + 2y = 6 \\ y = -\frac{1}{2}x + 3 \end{cases}$

EXAMPLE 5.31

Solve the system by elimination. $\begin{cases} -6x + 15y = 10 \\ 2x - 5y = -5 \end{cases}$

⊘ **Solution**

The equations are in standard form.

$$\begin{cases} -6x + 15y = 10 \\ 2x - 5y = -5 \end{cases}$$

Multiply the second equation by 3 to eliminate a variable.

$$\begin{cases} -6x + 15y = 10 \\ 3(2x - 5y) = 3(-5) \end{cases}$$

Simplify and add.

$$\begin{cases} -6x + 15y = 10 \\ \underline{6x - 15y = -15} \end{cases}$$
$$0 \neq -5$$

This statement is false. The equations are inconsistent and so their graphs would be parallel lines. The system does not have a solution.

> **TRY IT : : 5.61**
> Solve the system by elimination. $\begin{cases} -3x + 2y = 8 \\ 9x - 6y = 13 \end{cases}$

> **TRY IT : : 5.62**
> Solve the system by elimination. $\begin{cases} 7x - 3y = -2 \\ -14x + 6y = 8 \end{cases}$

Solve Applications of Systems of Equations by Elimination

Some applications problems translate directly into equations in standard form, so we will use the elimination method to solve them. As before, we use our Problem Solving Strategy to help us stay focused and organized.

EXAMPLE 5.32

The sum of two numbers is 39. Their difference is 9. Find the numbers.

⊘ Solution

Step 1. Read the problem

Step 2. Identify what we are looking for. We are looking for two numbers.

Step 3. Name what we are looking for. Let $n = $ the first number.
$m = $ the second number

Step 4. Translate into a system of equations. The sum of two numbers is 39.
$$n + m = 39$$
Their difference is 9.
$$n - m = 9$$
$$\begin{cases} n + m = 39 \\ n - m = 9 \end{cases}$$

The system is:

Step 5. Solve the system of equations.
To solve the system of equations, use elimination. The equations are in standard form and the coefficients of m are opposites. Add.

$$\begin{cases} n + m = 39 \\ \underline{n - m = 9} \end{cases}$$
$$2n \quad\quad = 48$$

Solve for n.
$$n = 24$$

Substitute $n = 24$ into one of the original equations and solve for m.
$$n + m = 39$$
$$24 + m = 39$$
$$m = 15$$

Step 6. Check the answer. Since $24 + 15 = 39$ and
$24 - 15 = 9,$ the answers check.

Step 7. Answer the question. The numbers are 24 and 15.

> **TRY IT : : 5.63**   The sum of two numbers is 42. Their difference is 8. Find the numbers.

> **TRY IT : : 5.64**   The sum of two numbers is −15. Their difference is −35. Find the numbers.

EXAMPLE 5.33

Joe stops at a burger restaurant every day on his way to work. Monday he had one order of medium fries and two small sodas, which had a total of 620 calories. Tuesday he had two orders of medium fries and one small soda, for a total of 820 calories. How many calories are there in one order of medium fries? How many calories in one small soda?

⊘ Solution

Step 1. Read the problem.

Step 2. Identify what we are looking for.	We are looking for the number of calories in one order of medium fries and in one small soda.
Step 3. Name what we are looking for.	Let f = the number of calories in 1 order of medium fries. s = the number of calories in 1 small soda.
Step 4. Translate into a system of equations:	one medium fries and two small sodas had a total of 620 calories
	$f + 2s = 620$

	two medium fries and one small soda had a total of 820 calories.
	$2f + s = 820$
Our system is:	$\begin{cases} f + 2s = 620 \\ 2f + s = 820 \end{cases}$
Step 5. Solve the system of equations. To solve the system of equations, use elimination. The equations are in standard form. To get opposite coefficients of f, multiply the top equation by –2.	$\begin{cases} -2(f + 2s) = -2(620) \\ 2f + s = 820 \end{cases}$
Simplify and add.	$\begin{array}{r} -2f - 4s = -1240 \\ 2f + s = 820 \\ \hline -3s = -420 \end{array}$
Solve for s.	$s = 140$
Substitute $s = 140$ into one of the original equations and then solve for f.	$f + 2s = 620$
	$f + 2 \cdot 140 = 620$
	$f + 280 = 620$
	$f = 340$
Step 6. Check the answer.	Verify that these numbers make sense in the problem and that they are solutions to both equations. We leave this to you!
Step 7. Answer the question.	The small soda has 140 calories and the fries have 340 calories.

> **TRY IT : :** 5.65

Malik stops at the grocery store to buy a bag of diapers and 2 cans of formula. He spends a total of $37. The next week he stops and buys 2 bags of diapers and 5 cans of formula for a total of $87. How much does a bag of diapers cost? How much is one can of formula?

> **TRY IT : :** 5.66

To get her daily intake of fruit for the day, Sasha eats a banana and 8 strawberries on Wednesday for a calorie count of 145. On the following Wednesday, she eats two bananas and 5 strawberries for a total of 235 calories for the fruit. How many calories are there in a banana? How many calories are in a strawberry?

Choose the Most Convenient Method to Solve a System of Linear Equations

When you will have to solve a system of linear equations in a later math class, you will usually not be told which method to use. You will need to make that decision yourself. So you'll want to choose the method that is easiest to do and minimizes your chance of making mistakes.

Graphing	Substitution	Elimination
Use when you need a picture of the situation.	Use when one equation is already solved for one variable.	Use when the equations are in standard form.

EXAMPLE 5.34

For each system of linear equations decide whether it would be more convenient to solve it by substitution or elimination. Explain your answer.

(a) $\begin{cases} 3x + 8y = 40 \\ 7x - 4y = -32 \end{cases}$ (b) $\begin{cases} 5x + 6y = 12 \\ y = \frac{2}{3}x - 1 \end{cases}$

⊘ Solution

(a) $\begin{cases} 3x + 8y = 40 \\ 7x - 4y = -32 \end{cases}$

Since both equations are in standard form, using elimination will be most convenient.

(b) $\begin{cases} 5x + 6y = 12 \\ y = \frac{2}{3}x - 1 \end{cases}$

Since one equation is already solved for y, using substitution will be most convenient.

> **TRY IT : : 5.67**

For each system of linear equations, decide whether it would be more convenient to solve it by substitution or elimination. Explain your answer.

(a) $\begin{cases} 4x - 5y = -32 \\ 3x + 2y = -1 \end{cases}$ (b) $\begin{cases} x = 2y - 1 \\ 3x - 5y = -7 \end{cases}$

> **TRY IT : : 5.68**

For each system of linear equations, decide whether it would be more convenient to solve it by substitution or elimination. Explain your answer.

(a) $\begin{cases} y = 2x - 1 \\ 3x - 4y = -6 \end{cases}$ (b) $\begin{cases} 6x - 2y = 12 \\ 3x + 7y = -13 \end{cases}$

▶ **MEDIA : :**

Access these online resources for additional instruction and practice with solving systems of linear equations by elimination.

- **Instructional Video-Solving Systems of Equations by Elimination (http://www.openstax.org/l/25Elimination1)**
- **Instructional Video-Solving by Elimination (http://www.openstax.org/l/25Elimination2)**
- **Instructional Video-Solving Systems by Elimination (http://www.openstax.org/l/25Elimination3)**

5.3 EXERCISES

Practice Makes Perfect

Solve a System of Equations by Elimination

In the following exercises, solve the systems of equations by elimination.

126. $\begin{cases} 5x + 2y = 2 \\ -3x - y = 0 \end{cases}$

127. $\begin{cases} -3x + y = -9 \\ x - 2y = -12 \end{cases}$

128. $\begin{cases} 6x - 5y = -1 \\ 2x + y = 13 \end{cases}$

129. $\begin{cases} 3x - y = -7 \\ 4x + 2y = -6 \end{cases}$

130. $\begin{cases} x + y = -1 \\ x - y = -5 \end{cases}$

131. $\begin{cases} x + y = -8 \\ x - y = -6 \end{cases}$

132. $\begin{cases} 3x - 2y = 1 \\ -x + 2y = 9 \end{cases}$

133. $\begin{cases} -7x + 6y = -10 \\ x - 6y = 22 \end{cases}$

134. $\begin{cases} 3x + 2y = -3 \\ -x - 2y = -19 \end{cases}$

135. $\begin{cases} 5x + 2y = 1 \\ -5x - 4y = -7 \end{cases}$

136. $\begin{cases} 6x + 4y = -4 \\ -6x - 5y = 8 \end{cases}$

137. $\begin{cases} 3x - 4y = -11 \\ x - 2y = -5 \end{cases}$

138. $\begin{cases} 5x - 7y = 29 \\ x + 3y = -3 \end{cases}$

139. $\begin{cases} 6x - 5y = -75 \\ -x - 2y = -13 \end{cases}$

140. $\begin{cases} -x + 4y = 8 \\ 3x + 5y = 10 \end{cases}$

141. $\begin{cases} 2x - 5y = 7 \\ 3x - y = 17 \end{cases}$

142. $\begin{cases} 5x - 3y = -1 \\ 2x - y = 2 \end{cases}$

143. $\begin{cases} 7x + y = -4 \\ 13x + 3y = 4 \end{cases}$

144. $\begin{cases} -3x + 5y = -13 \\ 2x + y = -26 \end{cases}$

145. $\begin{cases} 3x - 5y = -9 \\ 5x + 2y = 16 \end{cases}$

146. $\begin{cases} 4x - 3y = 3 \\ 2x + 5y = -31 \end{cases}$

147. $\begin{cases} 4x + 7y = 14 \\ -2x + 3y = 32 \end{cases}$

148. $\begin{cases} 5x + 2y = 21 \\ 7x - 4y = 9 \end{cases}$

149. $\begin{cases} 3x + 8y = -3 \\ 2x + 5y = -3 \end{cases}$

150. $\begin{cases} 11x + 9y = -5 \\ 7x + 5y = -1 \end{cases}$

151. $\begin{cases} 3x + 8y = 67 \\ 5x + 3y = 60 \end{cases}$

152. $\begin{cases} 2x + 9y = -4 \\ 3x + 13y = -7 \end{cases}$

153. $\begin{cases} \frac{1}{3}x - y = -3 \\ x + \frac{5}{2}y = 2 \end{cases}$

154. $\begin{cases} x + \frac{1}{2}y = \frac{3}{2} \\ \frac{1}{5}x - \frac{1}{5}y = 3 \end{cases}$

155. $\begin{cases} x + \frac{1}{3}y = -1 \\ \frac{1}{2}x - \frac{1}{3}y = -2 \end{cases}$

156. $\begin{cases} \frac{1}{3}x - y = -3 \\ \frac{2}{3}x + \frac{5}{2}y = 3 \end{cases}$

157. $\begin{cases} 2x + y = 3 \\ 6x + 3y = 9 \end{cases}$

158. $\begin{cases} x - 4y = -1 \\ -3x + 12y = 3 \end{cases}$

159. $\begin{cases} -3x - y = 8 \\ 6x + 2y = -16 \end{cases}$

160. $\begin{cases} 4x + 3y = 2 \\ 20x + 15y = 10 \end{cases}$

161. $\begin{cases} 3x + 2y = 6 \\ -6x - 4y = -12 \end{cases}$

162. $\begin{cases} 5x - 8y = 12 \\ 10x - 16y = 20 \end{cases}$

163. $\begin{cases} -11x + 12y = 60 \\ -22x + 24y = 90 \end{cases}$

164. $\begin{cases} 7x - 9y = 16 \\ -21x + 27y = -24 \end{cases}$

165. $\begin{cases} 5x - 3y = 15 \\ y = \frac{5}{3}x - 2 \end{cases}$

166. $\begin{cases} 2x + 4y = 7 \\ y = -\frac{1}{2}x - 4 \end{cases}$

Solve Applications of Systems of Equations by Elimination

In the following exercises, translate to a system of equations and solve.

167. The sum of two numbers is 65. Their difference is 25. Find the numbers.

168. The sum of two numbers is 37. Their difference is 9. Find the numbers.

169. The sum of two numbers is −27. Their difference is −59. Find the numbers.

170. The sum of two numbers is −45. Their difference is −89. Find the numbers.

171. Andrea is buying some new shirts and sweaters. She is able to buy 3 shirts and 2 sweaters for $114 or she is able to buy 2 shirts and 4 sweaters for $164. How much does a shirt cost? How much does a sweater cost?

172. Peter is buying office supplies. He is able to buy 3 packages of paper and 4 staplers for $40 or he is able to buy 5 packages of paper and 6 staplers for $62. How much does a package of paper cost? How much does a stapler cost?

173. The total amount of sodium in 2 hot dogs and 3 cups of cottage cheese is 4720 mg. The total amount of sodium in 5 hot dogs and 2 cups of cottage cheese is 6300 mg. How much sodium is in a hot dog? How much sodium is in a cup of cottage cheese?

174. The total number of calories in 2 hot dogs and 3 cups of cottage cheese is 960 calories. The total number of calories in 5 hot dogs and 2 cups of cottage cheese is 1190 calories. How many calories are in a hot dog? How many calories are in a cup of cottage cheese?

Choose the Most Convenient Method to Solve a System of Linear Equations

In the following exercises, decide whether it would be more convenient to solve the system of equations by substitution or elimination.

175.

ⓐ $\begin{cases} 8x - 15y = -32 \\ 6x + 3y = -5 \end{cases}$

ⓑ $\begin{cases} x = 4y - 3 \\ 4x - 2y = -6 \end{cases}$

176.

ⓐ $\begin{cases} y = 7x - 5 \\ 3x - 2y = 16 \end{cases}$

ⓑ $\begin{cases} 12x - 5y = -42 \\ 3x + 7y = -15 \end{cases}$

177.

ⓐ $\begin{cases} y = 4x + 9 \\ 5x - 2y = -21 \end{cases}$

ⓑ $\begin{cases} 9x - 4y = 24 \\ 3x + 5y = -14 \end{cases}$

178.

ⓐ $\begin{cases} 14x - 15y = -30 \\ 7x + 2y = 10 \end{cases}$

ⓑ $\begin{cases} x = 9y - 11 \\ 2x - 7y = -27 \end{cases}$

Everyday Math

179. Norris can row 3 miles upstream against the current in the same amount of time it takes him to row 5 miles downstream, with the current. Solve the system. $\begin{cases} r - c = 3 \\ r + c = 5 \end{cases}$

ⓐ for r, his rowing speed in still water.

ⓑ Then solve for c, the speed of the river current.

180. Josie wants to make 10 pounds of trail mix using nuts and raisins, and she wants the total cost of the trail mix to be $54. Nuts cost $6 per pound and raisins cost $3 per pound. Solve the system $\begin{cases} n + r = 10 \\ 6n + 3r = 54 \end{cases}$

to find n, the number of pounds of nuts, and r, the number of pounds of raisins she should use.

Writing Exercises

181. Solve the system
$$\begin{cases} x + y = 10 \\ 5x + 8y = 56 \end{cases}$$

ⓐ by substitution

ⓑ by graphing

ⓒ Which method do you prefer? Why?

182. Solve the system
$$\begin{cases} x + y = -12 \\ y = 4 - \frac{1}{2}x \end{cases}$$

ⓐ by substitution

ⓑ by graphing

ⓒ Which method do you prefer? Why?

Self Check

ⓐ *After completing the exercises, use this checklist to evaluate your mastery of the objectives of this section.*

I can...	Confidently	With some help	No-I don't get it!
solve a system of equations by elimination.			
solve applications of systems of equations by elimination.			
choose the most convenient method to solve a system of linear equations.			

ⓑ *What does this checklist tell you about your mastery of this section? What steps will you take to improve?*

5.4 Solve Applications with Systems of Equations

Learning Objectives

By the end of this section, you will be able to:

› Translate to a system of equations
› Solve direct translation applications
› Solve geometry applications
› Solve uniform motion applications

Be Prepared!

Before you get started, take this readiness quiz.

1. The sum of twice a number and nine is 31. Find the number.
   If you missed this problem, review **Example 3.4**.

2. Twins Jon and Ron together earned $96,000 last year. Ron earned $8,000 more than three times what Jon earned. How much did each of the twins earn?
   If you missed this problem, review **Example 3.11**.

3. Alessio rides his bike $3\frac{1}{2}$ hours at a rate of 10 miles per hour. How far did he ride?

   If you missed this problem, review **Example 2.58**.

Previously in this chapter we solved several applications with systems of linear equations. In this section, we'll look at some specific types of applications that relate two quantities. We'll translate the words into linear equations, decide which is the most convenient method to use, and then solve them.

We will use our Problem Solving Strategy for Systems of Linear Equations.

HOW TO : : USE A PROBLEM SOLVING STRATEGY FOR SYSTEMS OF LINEAR EQUATIONS.

Step 1. **Read** the problem. Make sure all the words and ideas are understood.

Step 2. **Identify** what we are looking for.

Step 3. **Name** what we are looking for. Choose variables to represent those quantities.

Step 4. **Translate** into a system of equations.

Step 5. **Solve** the system of equations using good algebra techniques.

Step 6. **Check** the answer in the problem and make sure it makes sense.

Step 7. **Answer** the question with a complete sentence.

Translate to a System of Equations

Many of the problems we solved in earlier applications related two quantities. Here are two of the examples from the chapter on **Math Models**.

• The sum of two numbers is negative fourteen. One number is four less than the other. Find the numbers.

• A married couple together earns $110,000 a year. The wife earns $16,000 less than twice what her husband earns. What does the husband earn?

In that chapter we translated each situation into one equation using only one variable. Sometimes it was a bit of a challenge figuring out how to name the two quantities, wasn't it?

Let's see how we can translate these two problems into a system of equations with two variables. We'll focus on Steps 1 through 4 of our Problem Solving Strategy.

EXAMPLE 5.35 HOW TO TRANSLATE TO A SYSTEM OF EQUATIONS

Translate to a system of equations:

The sum of two numbers is negative fourteen. One number is four less than the other. Find the numbers.

✓ **Solution**

Step 1. **Read** the problem. Make sure you understand all the words and ideas.	This is a number problem.	The sum of two numbers is negative fourteen. One number is four less than the other. Find the numbers.
Step 2. **Identify** what you are looking for.	"Find the numbers."	We are looking for 2 numbers.
Step 3. **Name** what you are looking for. Choose variables to represent those quantities.	We will use two variables, m and n.	Let $m =$ one number $n =$ second number
Step 4. **Translate** into a system of equations.	We will write one equation for each sentence.	The sum of the numbers is -14 $m + n = -14$ One number is four less than the other $m = n - 4$ The system is: $\begin{cases} m + n = -14 \\ m = n - 4 \end{cases}$

> **TRY IT : : 5.69**

Translate to a system of equations:

The sum of two numbers is negative twenty-three. One number is 7 less than the other. Find the numbers.

> **TRY IT : : 5.70**

Translate to a system of equations:

The sum of two numbers is negative eighteen. One number is 40 more than the other. Find the numbers.

We'll do another example where we stop after we write the system of equations.

EXAMPLE 5.36

Translate to a system of equations:

A married couple together earns $110,000 a year. The wife earns $16,000 less than twice what her husband earns. What does the husband earn?

✓ **Solution**

We are looking for the amount that the husband and wife each earn.

Let $h =$ the amount the husband earns. $w =$ the amount the wife earns

Translate.

A married couple together earns $110,000.
$$w + h = 110{,}000$$
The wife earns $16,000 less than twice what husband earns.
$$w = 2h - 16{,}000$$

The system of equations is:

$$\begin{cases} w + h = 110{,}000 \\ w = 2h - 16{,}000 \end{cases}$$

> **TRY IT : :** 5.71

Translate to a system of equations:

A couple has a total household income of $84,000. The husband earns $18,000 less than twice what the wife earns. How much does the wife earn?

> **TRY IT : :** 5.72

Translate to a system of equations:

A senior employee makes $5 less than twice what a new employee makes per hour. Together they make $43 per hour. How much does each employee make per hour?

Solve Direct Translation Applications

We set up, but did not solve, the systems of equations in **Example 5.35** and **Example 5.36** Now we'll translate a situation to a system of equations and then solve it.

EXAMPLE 5.37

Translate to a system of equations and then solve:

Devon is 26 years older than his son Cooper. The sum of their ages is 50. Find their ages.

⊘ **Solution**

Step 1. Read the problem.	
Step 2. Identify what we are looking for.	We are looking for the ages of Devon and Cooper.
Step 3. Name what we are looking for.	Let $d =$ Devon's age. $c =$ Cooper's age
Step 4. Translate into a system of equations.	Devon is 26 years older than Cooper.
	$d = c + 26$
	The sum of their ages is 50.
	$d + c = 50$
The system is:	$\begin{cases} d = c + 26 \\ d + c = 50 \end{cases}$
Step 5. Solve the system of equations. Solve by substitution.	$\begin{cases} d = \boxed{c + 26} \\ d + c = 50 \end{cases}$ $d + c = 50$
Substitute $c + 26$ into the second equation.	$c + 26 + c = 50$
Solve for c.	$2c + 26 = 50$
	$2c = 24$
	$c = \boxed{12}$ $d = c + 26$
Substitute $c = 12$ into the first equation and then solve for d.	$d = 12 + 26$
	$d = 38$

Step 6. Check the answer in the problem.	Is Devon's age 26 more than Cooper's? Yes, 38 is 26 more than 12. Is the sum of their ages 50? Yes, 38 plus 12 is 50.
Step 7. Answer the question.	Devon is 38 and Cooper is 12 years old.

> **TRY IT : : 5.73**
>
> Translate to a system of equations and then solve:
>
> Ali is 12 years older than his youngest sister, Jameela. The sum of their ages is 40. Find their ages.

> **TRY IT : : 5.74**   Translate to a system of equations and then solve:
>
> Jake's dad is 6 more than 3 times Jake's age. The sum of their ages is 42. Find their ages.

EXAMPLE 5.38

Translate to a system of equations and then solve:

When Jenna spent 10 minutes on the elliptical trainer and then did circuit training for 20 minutes, her fitness app says she burned 278 calories. When she spent 20 minutes on the elliptical trainer and 30 minutes circuit training she burned 473 calories. How many calories does she burn for each minute on the elliptical trainer? How many calories does she burn for each minute of circuit training?

⊘ **Solution**

Step 1. Read the problem.

Step 2. Identify what we are looking for.	We are looking for the number of calories burned each minute on the elliptical trainer and each minute of circuit training.
Step 3. Name what we are looking for.	Let $e =$ number of calories burned per minute on the elliptical trainer. $c =$ number of calories burned per minute while circuit training
Step 4. Translate into a system of equations.	10 minutes on the elliptical and circuit training for 20 minutes, burned 278 calories
	$10e + 20c = 278$
	20 minutes on the elliptical and 30 minutes of circuit training burned 473 calories
	$20e + 30c = 473$
The system is:	$\begin{cases} 10e + 20c = 278 \\ 20e + 30c = 473 \end{cases}$
Step 5. Solve the system of equations.	
Multiply the first equation by –2 to get opposite coefficients of e.	$\begin{cases} -2(10e + 20c) = -2(278) \\ \quad 20e + 30c\ = 473 \end{cases}$

Simplify and add the equations.	$\begin{cases} -20e - 40c = -556 \\ 20e + 30c = 473 \end{cases}$
Solve for c.	$-10c = -83$ $c = 8.3$
Substitute $c = 8.3$ into one of the original equations to solve for e.	$10e + 20c = 278$
	$10e + 20(8.3) = 278$
	$10e + 166 = 278$
	$10e = 112$
	$e = 11.2$
Step 6. Check the answer in the problem.	Check the math on your own.

$\begin{cases} 10(11.2) + 20(8.3) \overset{?}{=} 278 \\ 20(11.2) + 30(8.3) \overset{?}{=} 473 \end{cases}$

Step 7. Answer the question.	Jenna burns 8.3 calories per minute circuit training and 11.2 calories per minute while on the elliptical trainer.

> **TRY IT : : 5.75**

Translate to a system of equations and then solve:

Mark went to the gym and did 40 minutes of Bikram hot yoga and 10 minutes of jumping jacks. He burned 510 calories. The next time he went to the gym, he did 30 minutes of Bikram hot yoga and 20 minutes of jumping jacks burning 470 calories. How many calories were burned for each minute of yoga? How many calories were burned for each minute of jumping jacks?

> **TRY IT : : 5.76**

Translate to a system of equations and then solve:

Erin spent 30 minutes on the rowing machine and 20 minutes lifting weights at the gym and burned 430 calories. During her next visit to the gym she spent 50 minutes on the rowing machine and 10 minutes lifting weights and burned 600 calories. How many calories did she burn for each minutes on the rowing machine? How many calories did she burn for each minute of weight lifting?

Solve Geometry Applications

When we learned about Math Models, we solved geometry applications using properties of triangles and rectangles. Now we'll add to our list some properties of angles.

The measures of two complementary angles add to 90 degrees. The measures of two supplementary angles add to 180 degrees.

Complementary and Supplementary Angles

Two angles are **complementary** if the sum of the measures of their angles is 90 degrees.

Two angles are **supplementary** if the sum of the measures of their angles is 180 degrees.

If two angles are complementary, we say that *one angle is the complement of the other.*

If two angles are supplementary, we say that *one angle is the supplement of the other.*

EXAMPLE 5.39

Translate to a system of equations and then solve:

The difference of two complementary angles is 26 degrees. Find the measures of the angles.

✓ Solution

Step 1. Read the problem.

Step 2. Identify what we are looking for.   We are looking for the measure of each angle.

Step 3. Name what we are looking for.   Let $x =$ the measure of the first angle.
                                                                 $y =$ the measure of the second angle

Step 4. Translate into a system of equations.   The angles are complementary.
$$x + y = 90$$
The difference of the two angles is 26 degrees.
$$x - y = 26$$

The system is
$$\begin{cases} x + y = 90 \\ x - y = 26 \end{cases}$$

Step 5. Solve the system of equations by elimination.
$$\begin{cases} x + y = 90 \\ x - y = 26 \end{cases}$$
$$2x \quad\;\; = 116$$
$$x = 58$$
$$x + y = 90$$
Substitute $x = 58$ into the first equation.
$$58 + y = 90$$
$$y = 32$$

Step 6. Check the answer in the problem.
$$58 + 42 = 90 \checkmark$$
$$58 - 32 = 26 \checkmark$$

Step 7. Answer the question.   The angle measures are 58 degrees and 42 degrees.

> **TRY IT :: 5.77**   Translate to a system of equations and then solve:

The difference of two complementary angles is 20 degrees. Find the measures of the angles.

> **TRY IT :: 5.78**   Translate to a system of equations and then solve:

The difference of two complementary angles is 80 degrees. Find the measures of the angles.

EXAMPLE 5.40

Translate to a system of equations and then solve:

Two angles are supplementary. The measure of the larger angle is twelve degrees less than five times the measure of the smaller angle. Find the measures of both angles.

✓ Solution

Step 1. Read the problem.	
Step 2. Identify what we are looking for.	We are looking for the measure of each angle.
Step 3. Name what we are looking for.	Let $x =$ the measure of the first angle. $y =$ the measure of the second angle

Step 4. Translate into a system of equations.

The angles are supplementary.

$$x + y = 180$$

The larger angle is twelve less than five times the smaller angle

$$y = 5x - 12$$

The system is:

$$\begin{cases} x + y = 180 \\ y = \boxed{5x - 12} \end{cases}$$

Step 5. Solve the system of equations substitution.

$$x + y = 180$$

Substitute $5x - 12$ for y in the first equation.

$$x + 5x - 12 = 180$$

Solve for x.

$$6x - 12 = 180$$

$$6x = 192$$

$$x = \boxed{32}$$
$$y = 5x - 12$$

Substitute 32 for in the second equation, then solve for y.

$$y = 5 \cdot 32 - 12$$

$$y = 160 - 12$$

$$y = 148$$

Step 6. Check the answer in the problem.

$$32 + 158 = 180 \checkmark$$
$$5 \cdot 32 - 12 = 147 \checkmark$$

Step 7. Answer the question.

The angle measures are 148 and 32.

> **TRY IT : : 5.79**
>
> Translate to a system of equations and then solve:
>
> Two angles are supplementary. The measure of the larger angle is 12 degrees more than three times the smaller angle. Find the measures of the angles.

> **TRY IT : : 5.80**
>
> Translate to a system of equations and then solve:
>
> Two angles are supplementary. The measure of the larger angle is 18 less than twice the measure of the smaller angle. Find the measures of the angles.

EXAMPLE 5.41

Translate to a system of equations and then solve:

Randall has 125 feet of fencing to enclose the rectangular part of his backyard adjacent to his house. He will only need to fence around three sides, because the fourth side will be the wall of the house. He wants the length of the fenced yard (parallel to the house wall) to be 5 feet more than four times as long as the width. Find the length and the width.

⊘ Solution

Step 1. Read the problem.

Step 2. Identify what you are looking for.	We are looking for the length and width.

Step 3. Name what we are looking for.	Let $L =$ the length of the fenced yard. $\quad W =$ the width of the fenced yard
Step 4. Translate into a system of equations.	One length and two widths equal 125.
	$L + 2W = 125$
	The length will be 5 feet more than four times the width.
	$L = 4W + 5$
The system is:	$\begin{cases} L + 2W = 125 \\ L = 4W + 5 \end{cases}$
Step 5. Solve the system of equations by substitution.	$L + 2W = 125$
Substitute $L = 4W + 5$ into the first equation, then solve for W.	$4W + 5 + 2W = 125$
	$6W + 5 = 125$
	$6W = 120$
Substitute 20 for W in the second equation, then solve for L.	$W = 20$
	$L = 4W + 5$
	$L = 4 \cdot 20 + 5$
	$L = 80 + 5$
	$L = 85$

Step 6. Check the answer in the problem.

$$20 + 28 + 20 = 125 \checkmark$$
$$85 = 4 \cdot 20 + 5 \checkmark$$

Step 7. Answer the equation.	The length is 85 feet and the width is 20 feet.

> **TRY IT :: 5.81**

Translate to a system of equations and then solve:

Mario wants to put a rectangular fence around the pool in his backyard. Since one side is adjacent to the house, he will only need to fence three sides. There are two long sides and the one shorter side is parallel to the house. He needs 155 feet of fencing to enclose the pool. The length of the long side is 10 feet less than twice the width. Find the length and width of the pool area to be enclosed.

TRY IT : : 5.82

Translate to a system of equations and then solve:

Alexis wants to build a rectangular dog run in her yard adjacent to her neighbor's fence. She will use 136 feet of fencing to completely enclose the rectangular dog run. The length of the dog run along the neighbor's fence will be 16 feet less than twice the width. Find the length and width of the dog run.

Solve Uniform Motion Applications

We used a table to organize the information in uniform motion problems when we introduced them earlier. We'll continue using the table here. The basic equation was $D = rt$ where D is the distance travelled, r is the rate, and t is the time.

Our first example of a uniform motion application will be for a situation similar to some we have already seen, but now we can use two variables and two equations.

EXAMPLE 5.42

Translate to a system of equations and then solve:

Joni left St. Louis on the interstate, driving west towards Denver at a speed of 65 miles per hour. Half an hour later, Kelly left St. Louis on the same route as Joni, driving 78 miles per hour. How long will it take Kelly to catch up to Joni?

⊘ **Solution**

A diagram is useful in helping us visualize the situation.

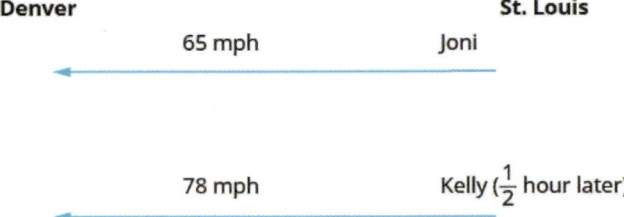

Identify and name what we are looking for.
A chart will help us organize the data.
We know the rates of both Joni and Kelly, and so we enter them in the chart.

We are looking for the length of time Kelly, k, and Joni, j, will each drive.
Since $D = r \cdot t$ we can fill in the Distance column.

Type	Rate	•	Time	=	Distance
Joni	65		j		$65j$
Kelly	78		k		$78k$

Translate into a system of equations.
To make the system of equations, we must recognize that Kelly and Joni will drive the same distance. So,
$65j = 78k$.

Also, since Kelly left later, her time will be $\frac{1}{2}$ hour less than Joni's time.

So, $k = j - \frac{1}{2}$.

Now we have the system.

$$\begin{cases} k = j - \frac{1}{2} \\ 65j = 78k \end{cases}$$

Solve the system of equations by substitution.

$$65j = 78k$$

Substitute $k = j - \frac{1}{2}$ into the second equation, then solve for j.

$$65j = 78\left(j - \frac{1}{2}\right)$$

$$65j = 78j - 39$$

$$-13j = -39$$

$$j = 3$$

| To find Kelly's time, substitute $j = 3$ into the first equation, then solve for k. | $k = j - \dfrac{1}{2}$ |

$$k = 3 - \dfrac{1}{2}$$

$$k = \dfrac{5}{2} \text{ or } k = 2\dfrac{1}{2}$$

Check the answer in the problem.

   Joni 3 hours (65 mph) = 195 miles.

   Kelly $2\dfrac{1}{2}$ hours (78 mph) = 195 miles.

   Yes, they will have traveled the same distance when they meet.

Answer the question.

Kelly will catch up to Joni in $2\dfrac{1}{2}$ hours.

By then, Joni will have traveled 3 hours.

> **TRY IT : :** 5.83

Translate to a system of equations and then solve: Mitchell left Detroit on the interstate driving south towards Orlando at a speed of 60 miles per hour. Clark left Detroit 1 hour later traveling at a speed of 75 miles per hour, following the same route as Mitchell. How long will it take Clark to catch Mitchell?

> **TRY IT : :** 5.84

Translate to a system of equations and then solve: Charlie left his mother's house traveling at an average speed of 36 miles per hour. His sister Sally left 15 minutes (1/4 hour) later traveling the same route at an average speed of 42 miles per hour. How long before Sally catches up to Charlie?

Many real-world applications of uniform motion arise because of the effects of currents—of water or air—on the actual speed of a vehicle. Cross-country airplane flights in the United States generally take longer going west than going east because of the prevailing wind currents.

Let's take a look at a boat travelling on a river. Depending on which way the boat is going, the current of the water is either slowing it down or speeding it up.

Figure 5.7 and **Figure 5.8** show how a river current affects the speed at which a boat is actually travelling. We'll call the speed of the boat in still water b and the speed of the river current c.

In **Figure 5.7** the boat is going downstream, in the same direction as the river current. The current helps push the boat, so the boat's actual speed is faster than its speed in still water. The actual speed at which the boat is moving is $b + c$.

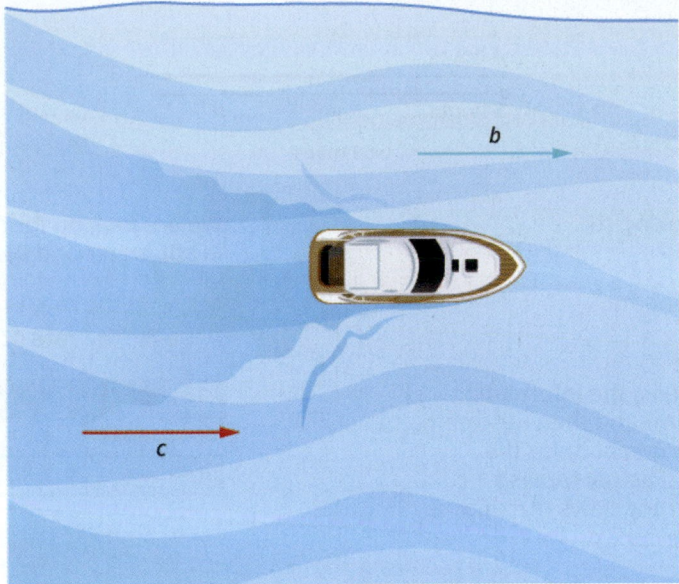

Figure 5.7

In **Figure 5.8** the boat is going upstream, opposite to the river current. The current is going against the boat, so the boat's actual speed is slower than its speed in still water. The actual speed of the boat is $b - c$.

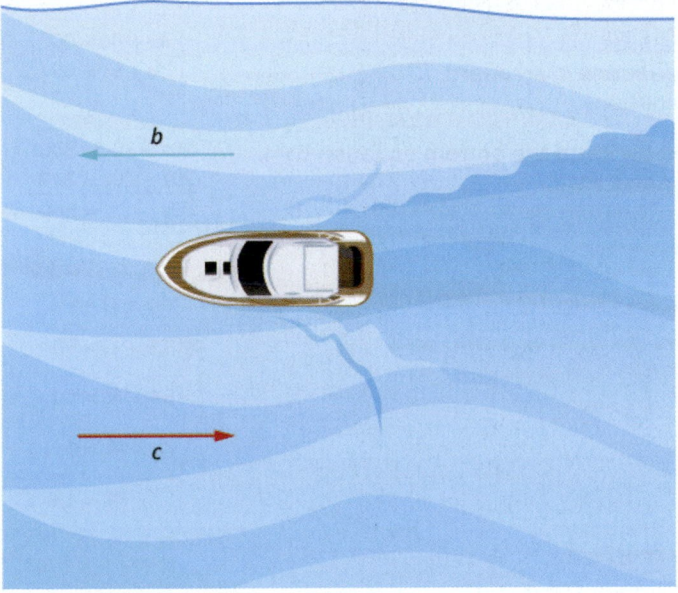

Figure 5.8

We'll put some numbers to this situation in **Example 5.43**.

EXAMPLE 5.43

Translate to a system of equations and then solve:

A river cruise ship sailed 60 miles downstream for 4 hours and then took 5 hours sailing upstream to return to the dock. Find the speed of the ship in still water and the speed of the river current.

 **Solution**

Read the problem.

This is a uniform motion problem and a picture will help us visualize the situation.

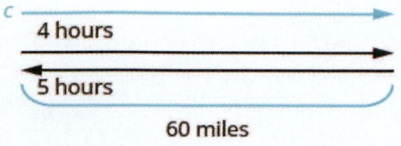

		Rate	**• Time =**	**Distance**
Identify what we are looking for.				We are looking for the speed of the ship in still water and the speed of the current.

Identify what we are looking for.

We are looking for the speed of the ship in still water and the speed of the current.

Name what we are looking for.

Let $s =$ the rate of the ship in still water.

$c =$ the rate of the current

A chart will help us organize the information. The ship goes downstream and then upstream. Going downstream, the current helps the ship; therefore, the ship's actual rate is $s + c$. Going upstream, the current slows the ship; therefore, the actual rate is $s - c$.

	Rate	**• Time =**	**Distance**
downstream	$s + c$	4	60
upstream	$s - c$	5	60

Downstream it takes 4 hours.
Upstream it takes 5 hours.
Each way the distance is 60 miles.

Translate into a system of equations. Since rate times time is distance, we can write the system of equations.

$$\begin{cases} 4(s + c) = 60 \\ 5(s - c) = 60 \end{cases}$$

Solve the system of equations. Distribute to put both equations in standard form, then solve by elimination.

$$\begin{cases} 4s + 4c = 60 \\ 5s - 5c = 60 \end{cases}$$

Multiply the top equation by 5 and the bottom equation by 4. Add the equations, then solve for s.

$$\begin{cases} 20s + 20c = 300 \\ 20s - 20c = 240 \end{cases}$$
$$40s \quad\quad = 540$$

$$s \quad = 13.5$$

Substitute $s = 13.5$ into one of the original equations.

$$4(s + c) = 60$$

$$4(13.5 + c) = 60$$

$$54 + 4c = 60$$

$$4c = 6$$

$$4c = 1.5$$

Check the answer in the problem.

The downstream rate would be
13.5 + 1.5 = 15 mph.
In 4 hours the ship would travel
15 · 4 = 60 miles.
The upstream rate would be
13.5 – 1.5 = 12 mph.
In 5 hours the ship would travel
12 · 5 = 60 miles.

Answer the question.

The rate of the ship is 13.5 mph and the rate of the current is 1.5 mph.

> **TRY IT : : 5.85**

Translate to a system of equations and then solve: A Mississippi river boat cruise sailed 120 miles upstream for 12 hours and then took 10 hours to return to the dock. Find the speed of the river boat in still water and the speed of the river current.

> **TRY IT : : 5.86**

Translate to a system of equations and then solve: Jason paddled his canoe 24 miles upstream for 4 hours. It took him 3 hours to paddle back. Find the speed of the canoe in still water and the speed of the river current.

Wind currents affect airplane speeds in the same way as water currents affect boat speeds. We'll see this in **Example 5.44**. A wind current in the same direction as the plane is flying is called a *tailwind*. A wind current blowing against the direction of the plane is called a *headwind*.

EXAMPLE 5.44

Translate to a system of equations and then solve:

A private jet can fly 1095 miles in three hours with a tailwind but only 987 miles in three hours into a headwind. Find the speed of the jet in still air and the speed of the wind.

⊘ **Solution**

Read the problem.

This is a uniform motion problem and a picture will help us visualize.

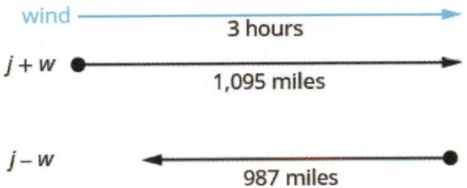

Identify what we are looking for.		We are looking for the speed of the jet in still air and the speed of the wind.	

Name what we are looking for.

Let $j =$ the speed of the jet in still air.
$w =$ the speed of the wind

A chart will help us organize the information. The jet makes two trips-one in a tailwind and one in a headwind.
In a tailwind, the wind helps the jet and so the rate is $j + w$.
In a headwind, the wind slows the jet and so the rate is $j - w$.

	Rate	• Time	= Distance
tailwind	$j + w$	3	1095
headwind	$j - w$	3	987

Each trip takes 3 hours.
In a tailwind the jet flies 1095 miles.
In a headwind the jet flies 987 miles.

Translate into a system of equations.
Since rate times time is distance, we get the system of equations.

$$\begin{cases} 3(j + w) = 1095 \\ 3(j - w) = 987 \end{cases}$$

Solve the system of equations.
Distribute, then solve by elimination.

$$\begin{array}{l} 3j + 3w = 1095 \\ \underline{3j - 3w = 987} \\ 6j \qquad\;\; = 2082 \end{array}$$

Add, and solve for *j*.	$j = 347$
	$3(j + w) = 1095$
Substitute *j* = 347 into one of the original equations, then solve for *w*.	

$$3(347 + w) = 1095$$

$$1041 + 3w = 1095$$

$$3w = 54$$

$$w = 18$$

Check the answer in the problem.

With the tailwind, the actual rate of the jet would be
    347 + 18 = 365 mph.
In 3 hours the jet would travel
    365 · 3 = 1095 miles.
Going into the headwind, the jet's actual rate would be
    347 − 18 = 329 mph.
In 3 hours the jet would travel
    329 · 3 = 987 miles.

Answer the question.   The rate of the jet is 347 mph and the rate of the wind is 18 mph.

> **TRY IT : : 5.87**

Translate to a system of equations and then solve: A small jet can fly 1,325 miles in 5 hours with a tailwind but only 1035 miles in 5 hours into a headwind. Find the speed of the jet in still air and the speed of the wind.

> **TRY IT : : 5.88**

Translate to a system of equations and then solve: A commercial jet can fly 1728 miles in 4 hours with a tailwind but only 1536 miles in 4 hours into a headwind. Find the speed of the jet in still air and the speed of the wind.

5.4 EXERCISES
Practice Makes Perfect

Translate to a System of Equations

In the following exercises, translate to a system of equations. Do not solve the system.

183. The sum of two numbers is fifteen. One number is three less than the other. Find the numbers.

184. The sum of two numbers is twenty-five. One number is five less than the other. Find the numbers.

185. The sum of two numbers is negative thirty. One number is five times the other. Find the numbers.

186. The sum of two numbers is negative sixteen. One number is seven times the other. Find the numbers.

187. Twice a number plus three times a second number is twenty-two. Three times the first number plus four times the second is thirty-one. Find the numbers.

188. Six times a number plus twice a second number is four. Twice the first number plus four times the second number is eighteen. Find the numbers.

189. Three times a number plus three times a second number is fifteen. Four times the first plus twice the second number is fourteen. Find the numbers.

190. Twice a number plus three times a second number is negative one. The first number plus four times the second number is two. Find the numbers.

191. A married couple together earn $75,000. The husband earns $15,000 more than five times what his wife earns. What does the wife earn?

192. During two years in college, a student earned $9,500. The second year she earned $500 more than twice the amount she earned the first year. How much did she earn the first year?

193. Daniela invested a total of $50,000, some in a certificate of deposit (CD) and the remainder in bonds. The amount invested in bonds was $5000 more than twice the amount she put into the CD. How much did she invest in each account?

194. Jorge invested $28,000 into two accounts. The amount he put in his money market account was $2,000 less than twice what he put into a CD. How much did he invest in each account?

195. In her last two years in college, Marlene received $42,000 in loans. The first year she received a loan that was $6,000 less than three times the amount of the second year's loan. What was the amount of her loan for each year?

196. Jen and David owe $22,000 in loans for their two cars. The amount of the loan for Jen's car is $2000 less than twice the amount of the loan for David's car. How much is each car loan?

Solve Direct Translation Applications

In the following exercises, translate to a system of equations and solve.

197. Alyssa is twelve years older than her sister, Bethany. The sum of their ages is forty-four. Find their ages.

198. Robert is 15 years older than his sister, Helen. The sum of their ages is sixty-three. Find their ages.

199. The age of Noelle's dad is six less than three times Noelle's age. The sum of their ages is seventy-four. Find their ages.

200. The age of Mark's dad is 4 less than twice Marks's age. The sum of their ages is ninety-five. Find their ages.

201. Two containers of gasoline hold a total of fifty gallons. The big container can hold ten gallons less than twice the small container. How many gallons does each container hold?

202. June needs 48 gallons of punch for a party and has two different coolers to carry it in. The bigger cooler is five times as large as the smaller cooler. How many gallons can each cooler hold?

203. Shelly spent 10 minutes jogging and 20 minutes cycling and burned 300 calories. The next day, Shelly swapped times, doing 20 minutes of jogging and 10 minutes of cycling and burned the same number of calories. How many calories were burned for each minute of jogging and how many for each minute of cycling?

204. Drew burned 1800 calories Friday playing one hour of basketball and canoeing for two hours. Saturday he spent two hours playing basketball and three hours canoeing and burned 3200 calories. How many calories did he burn per hour when playing basketball?

205. Troy and Lisa were shopping for school supplies. Each purchased different quantities of the same notebook and thumb drive. Troy bought four notebooks and five thumb drives for $116. Lisa bought two notebooks and three thumb dives for $68. Find the cost of each notebook and each thumb drive.

206. Nancy bought seven pounds of oranges and three pounds of bananas for $17. Her husband later bought three pounds of oranges and six pounds of bananas for $12. What was the cost per pound of the oranges and the bananas?

Solve Geometry Applications *In the following exercises, translate to a system of equations and solve.*

207. The difference of two complementary angles is 30 degrees. Find the measures of the angles.

208. The difference of two complementary angles is 68 degrees. Find the measures of the angles.

209. The difference of two supplementary angles is 70 degrees. Find the measures of the angles.

210. The difference of two supplementary angles is 24 degrees. Find the measure of the angles.

211. The difference of two supplementary angles is 8 degrees. Find the measures of the angles.

212. The difference of two supplementary angles is 88 degrees. Find the measures of the angles.

213. The difference of two complementary angles is 55 degrees. Find the measures of the angles.

214. The difference of two complementary angles is 17 degrees. Find the measures of the angles.

215. Two angles are supplementary. The measure of the larger angle is four more than three times the measure of the smaller angle. Find the measures of both angles.

216. Two angles are supplementary. The measure of the larger angle is five less than four times the measure of the smaller angle. Find the measures of both angles.

217. Two angles are complementary. The measure of the larger angle is twelve less than twice the measure of the smaller angle. Find the measures of both angles.

218. Two angles are complementary. The measure of the larger angle is ten more than four times the measure of the smaller angle. Find the measures of both angles.

219. Wayne is hanging a string of lights 45 feet long around the three sides of his rectangular patio, which is adjacent to his house. The length of his patio, the side along the house, is five feet longer than twice its width. Find the length and width of the patio.

220. Darrin is hanging 200 feet of Christmas garland on the three sides of fencing that enclose his rectangular front yard. The length is five feet less than five times the width. Find the length and width of the fencing.

221. A frame around a rectangular family portrait has a perimeter of 60 inches. The length is fifteen less than twice the width. Find the length and width of the frame.

222. The perimeter of a rectangular toddler play area is 100 feet. The length is ten more than three times the width. Find the length and width of the play area.

Solve Uniform Motion Applications *In the following exercises, translate to a system of equations and solve.*

223. Sarah left Minneapolis heading east on the interstate at a speed of 60 mph. Her sister followed her on the same route, leaving two hours later and driving at a rate of 70 mph. How long will it take for Sarah's sister to catch up to Sarah?

224. College roommates John and David were driving home to the same town for the holidays. John drove 55 mph, and David, who left an hour later, drove 60 mph. How long will it take for David to catch up to John?

225. At the end of spring break, Lucy left the beach and drove back towards home, driving at a rate of 40 mph. Lucy's friend left the beach for home 30 minutes (half an hour) later, and drove 50 mph. How long did it take Lucy's friend to catch up to Lucy?

226. Felecia left her home to visit her daughter driving 45 mph. Her husband waited for the dog sitter to arrive and left home twenty minutes (1/3 hour) later. He drove 55 mph to catch up to Felecia. How long before he reaches her?

227. The Jones family took a 12 mile canoe ride down the Indian River in two hours. After lunch, the return trip back up the river took three hours. Find the rate of the canoe in still water and the rate of the current.

228. A motor boat travels 60 miles down a river in three hours but takes five hours to return upstream. Find the rate of the boat in still water and the rate of the current.

229. A motor boat traveled 18 miles down a river in two hours but going back upstream, it took 4.5 hours due to the current. Find the rate of the motor boat in still water and the rate of the current. (Round to the nearest hundredth.).

230. A river cruise boat sailed 80 miles down the Mississippi River for four hours. It took five hours to return. Find the rate of the cruise boat in still water and the rate of the current. (Round to the nearest hundredth.).

231. A small jet can fly 1,072 miles in 4 hours with a tailwind but only 848 miles in 4 hours into a headwind. Find the speed of the jet in still air and the speed of the wind.

232. A small jet can fly 1,435 miles in 5 hours with a tailwind but only 1215 miles in 5 hours into a headwind. Find the speed of the jet in still air and the speed of the wind.

233. A commercial jet can fly 868 miles in 2 hours with a tailwind but only 792 miles in 2 hours into a headwind. Find the speed of the jet in still air and the speed of the wind.

234. A commercial jet can fly 1,320 miles in 3 hours with a tailwind but only 1,170 miles in 3 hours into a headwind. Find the speed of the jet in still air and the speed of the wind.

Everyday Math

235. At a school concert, 425 tickets were sold. Student tickets cost $5 each and adult tickets cost $8 each. The total receipts for the concert were $2,851. Solve the system

$$\begin{cases} s + a = 425 \\ 5s + 8a = 2,851 \end{cases}$$

to find s, the number of student tickets and a, the number of adult tickets.

236. The first graders at one school went on a field trip to the zoo. The total number of children and adults who went on the field trip was 115. The number of adults was $\frac{1}{4}$ the number of children. Solve the system

$$\begin{cases} c + a = 115 \\ a = \frac{1}{4}c \end{cases}$$

to find c, the number of children and a, the number of adults.

Writing Exercises

237. Write an application problem similar to **Example 5.37** using the ages of two of your friends or family members. Then translate to a system of equations and solve it.

238. Write a uniform motion problem similar to **Example 5.42** that relates to where you live with your friends or family members. Then translate to a system of equations and solve it.

Self Check

ⓐ *After completing the exercises, use this checklist to evaluate your mastery of the objectives of this section.*

I can...	Confidently	With some help	No-I don't get it!
translate to a system of equations.			
solve direct translation applications.			
solve geometry applications.			
solve uniform motion applications.			

ⓑ *On a scale of 1-10, how would you rate your mastery of this section in light of your responses on the checklist? How can you improve this?*

5.5 Solve Mixture Applications with Systems of Equations

Learning Objectives

By the end of this section, you will be able to:

› Solve mixture applications
› Solve interest applications

Be Prepared!

Before you get started, take this readiness quiz.

1. Multiply 4.025(1,562).
   If you missed this problem, review **Example 1.98**.

2. Write 8.2% as a decimal.
   If you missed this problem, review **Example 1.106**.

3. Earl's dinner bill came to $32.50 and he wanted to leave an 18% tip. How much should the tip be?
   If you missed this problem, review **Example 3.15**.

Solve Mixture Applications

When we solved mixture applications with coins and tickets earlier, we started by creating a table so we could organize the information. For a coin example with nickels and dimes, the table looked like this:

Type	Number •	Value($) =	Total Value($)
nickels		0.05	
dimes		0.10	

Using one variable meant that we had to relate the number of nickels and the number of dimes. We had to decide if we were going to let n be the number of nickels and then write the number of dimes in terms of n, or if we would let d be the number of dimes and write the number of nickels in terms of d.

Now that we know how to solve systems of equations with two variables, we'll just let n be the number of nickels and d be the number of dimes. We'll write one equation based on the total value column, like we did before, and the other equation will come from the number column.

For the first example, we'll do a ticket problem where the ticket prices are in whole dollars, so we won't need to use decimals just yet.

EXAMPLE 5.45

Translate to a system of equations and solve:

The box office at a movie theater sold 147 tickets for the evening show, and receipts totaled $1,302. How many $11 adult and how many $8 child tickets were sold?

⊘ **Solution**

Step 1. Read the problem.	We will create a table to organize the information.
Step 2. Identify what we are looking for.	We are looking for the number of adult tickets and the number of child tickets sold.
Step 3. Name what we are looking for.	Let $a =$ the number of adult tickets. $c =$ the number of child tickets
A table will help us organize the data. We have two types of tickets: adult and child.	Write a and c for the number of tickets.
Write the total number of tickets sold at the bottom of the Number column.	Altogether 147 were sold.

Write the value of each type of ticket in the Value column.	The value of each adult ticket is $11. The value of each child tickets is $8.	

The number times the value gives the total value, so the total value of adult tickets is $a \cdot 11 = 11a$, and the total value of child tickets is $c \cdot 8 = 8c$.

Type	Number	• Value ($)	= Total Value ($)
adult	a	11	$11a$
child	c	8	$8c$
	147		1302

Altogether the total value of the tickets was $1,302.

Fill in the Total Value column.

Step 4. Translate into a system of equations.

The Number column and the Total Value column give us the system of equations. We will use the elimination method to solve this system.

$$\begin{cases} a + c = 147 \\ 11a + 8c = 1302 \end{cases}$$

Multiply the first equation by –8.

$$\begin{cases} -8(a + c) = -8(147) \\ 11a + 8c = 1302 \end{cases}$$

Simplify and add, then solve for a.

$$\begin{aligned} -8a + 8c &= -1176 \\ \underline{11a + 8c = 1302} \\ 3a \quad\quad = 126 \end{aligned}$$

$$a \quad\quad = 42$$

$$a + c = 147$$

Substitute a = 42 into the first equation, then solve for c.

$$42 + c = 147$$

$$c = 105$$

Step 5. Check the answer in the problem.

42 adult tickets at $11 per ticket makes $462
105 child tickets at $8 per ticket makes $840.
The total receipts are $1,302. ✓

Step 6. Answer the question.

The movie theater sold 42 adult tickets and 105 child tickets.

> **TRY IT :: 5.89**

Translate to a system of equations and solve:

The ticket office at the zoo sold 553 tickets one day. The receipts totaled $3,936. How many $9 adult tickets and how many $6 child tickets were sold?

> **TRY IT :: 5.90**

Translate to a system of equations and solve:

A science center sold 1,363 tickets on a busy weekend. The receipts totaled $12,146. How many $12 adult tickets and how many $7 child tickets were sold?

In **Example 5.46** we'll solve a coin problem. Now that we know how to work with systems of two variables, naming the variables in the 'number' column will be easy.

EXAMPLE 5.46

Translate to a system of equations and solve:

Priam has a collection of nickels and quarters, with a total value of $7.30. The number of nickels is six less than three times the number of quarters. How many nickels and how many quarters does he have?

☑ **Solution**

Step 1. Read the problem.	We will create a table to organize the information.
Step 2. Identify what we are looking for.	We are looking for the number of nickels and the number of quarters.
Step 3. Name what we are looking for.	Let $n =$ the number of nickels. $q =$ the number of quarters
A table will help us organize the data. We have two types of coins, nickels and quarters.	Write n and q for the number of each type of coin.
Fill in the Value column with the value of each type of coin.	The value of each nickel is $0.05. The value of each quarter is $0.25.

The number times the value gives the total value, so, the total value of the nickels is $n(0.05) = 0.05n$ and the total value of quarters is $q(0.25) = 0.25q$. Altogether the total value of the coins is $7.30.

Type	Number	• Value ($)	= Total Value ($)
nickels	n	0.05	$0.05n$
quarters	q	0.25	$0.25q$
			7.30

Step 4. Translate into a system of equations.

The Total value column gives one equation.	$0.05n + 0.25q = 7.30$
We also know the number of nickels is six less than three times the number of quarters. Translate to get the second equation.	$n = 3q - 6$
Now we have the system to solve.	$\begin{cases} 0.05n + 0.25q = 7.30 \\ n = 3q - 6 \end{cases}$

Step 5. Solve the system of equations
We will use the substitution method.
Substitute $n = 3q - 6$ into the first equation.
Simplify and solve for q.

$$0.05n + 0.25q = 7.30$$

$$0.05(3q - 6) + 0.25q = 7.3$$

$$0.15q - 0.3 + 0.25q = 7.3$$

$$0.4q - 0.3 = 7.3$$

$$0.4q = 7.6$$

$$q = 19$$

To find the number of nickels, substitute
$q = 19$ into the second equation.

$$n = 3q - 6$$

$$n = 3 \cdot 19 - 6$$

$$n = 51$$

Step 6. Check the answer in the problem.

$$
\begin{aligned}
19 \text{ quarters at } \$ 0.25 &= \$ 4.75 \\
51 \text{ nickels at } \$ 0.05 &= \$ 2.55 \\
\text{Total} &= \$ 7.30 \checkmark \\
3 \cdot 19 - 16 &= 51 \checkmark
\end{aligned}
$$

Step 7. Answer the question. Priam has 19 quarters and 51 nickels.

> **TRY IT : : 5.91**

Translate to a system of equations and solve:

Matilda has a handful of quarters and dimes, with a total value of $8.55. The number of quarters is 3 more than twice the number of dimes. How many dimes and how many quarters does she have?

> **TRY IT : : 5.92**

Translate to a system of equations and solve:

Juan has a pocketful of nickels and dimes. The total value of the coins is $8.10. The number of dimes is 9 less than twice the number of nickels. How many nickels and how many dimes does Juan have?

Some mixture applications involve combining foods or drinks. Example situations might include combining raisins and nuts to make a trail mix or using two types of coffee beans to make a blend.

EXAMPLE 5.47

Translate to a system of equations and solve:

Carson wants to make 20 pounds of trail mix using nuts and chocolate chips. His budget requires that the trail mix costs him $7.60 per pound. Nuts cost $9.00 per pound and chocolate chips cost $2.00 per pound. How many pounds of nuts and how many pounds of chocolate chips should he use?

⊘ **Solution**

Step 1. Read the problem.	We will create a table to organize the information.
Step 2. Identify what we are looking for.	We are looking for the number of pounds of nuts and the number of pounds of chocolate chips.
Step 3. Name what we are looking for.	Let $n =$ the number of pound of nuts. $c =$ the number of pounds of chips

Carson will mix nuts and chocolate chips to get trail mix.
Write in n and c for the number of pounds of nuts and chocolate chips.

There will be 20 pounds of trail mix.
Put the price per pound of each item in the Value column.
Fill in the last column using

Number · Value = Total Value

Type	Number of pounds	· Value ($) =	Total Value ($)
nuts	n	9.00	$9n$
chocolate chips	c	2.00	$2c$
trail mix	20	7.60	$7.60(20) = 152$

Step 4. Translate into a system of equations.
We get the equations from the Number and Total Value columns.

$$
\begin{cases}
n + c = 20 \\
9n + 2c = 152
\end{cases}
$$

Step 5. Solve the system of equations
We will use elimination to solve the system.

Multiply the first equation by –2 to eliminate *c*.	$\begin{cases} -2(n+c) = -2(20) \\ 9n + 2c = 152 \end{cases}$
Simplify and add. Solve for *n*.	$\begin{aligned} -2n - 2c &= -40 \\ \underline{9n + 2c} &= \underline{152} \\ 7n &= 112 \end{aligned}$
	$n = 16$
To find the number of pounds of chocolate chips, substitute *n* = 16 into the first equation, then solve for *c*.	$n + c = 20$ $16 + c = 20$
	$c = 4$

Step 6. Check the answer in the problem.

$$16 + 4 = 20 \checkmark$$
$$9 \cdot 16 + 2 \cdot 4 = 152 \checkmark$$

Step 7. Answer the question.	Carson should mix 16 pounds of nuts with 4 pounds of chocolate chips to create the trail mix.

 TRY IT : : 5.93

Translate to a system of equations and solve:

Greta wants to make 5 pounds of a nut mix using peanuts and cashews. Her budget requires the mixture to cost her $6 per pound. Peanuts are $4 per pound and cashews are $9 per pound. How many pounds of peanuts and how many pounds of cashews should she use?

 TRY IT : : 5.94

Translate to a system of equations and solve:

Sammy has most of the ingredients he needs to make a large batch of chili. The only items he lacks are beans and ground beef. He needs a total of 20 pounds combined of beans and ground beef and has a budget of $3 per pound. The price of beans is $1 per pound and the price of ground beef is $5 per pound. How many pounds of beans and how many pounds of ground beef should he purchase?

Another application of mixture problems relates to concentrated cleaning supplies, other chemicals, and mixed drinks. The concentration is given as a percent. For example, a 20% concentrated household cleanser means that 20% of the total amount is cleanser, and the rest is water. To make 35 ounces of a 20% concentration, you mix 7 ounces (20% of 35) of the cleanser with 28 ounces of water.

For these kinds of mixture problems, we'll use percent instead of value for one of the columns in our table.

EXAMPLE 5.48

Translate to a system of equations and solve:

Sasheena is a lab assistant at her community college. She needs to make 200 milliliters of a 40% solution of sulfuric acid for a lab experiment. The lab has only 25% and 50% solutions in the storeroom. How much should she mix of the 25% and the 50% solutions to make the 40% solution?

⊘ **Solution**

Step 1. Read the problem.	A figure may help us visualize the situation, then we will create a table to organize the information.

Sasheena must mix some of the 25% solution and some of the 50% solution together to get 200 ml of the 40% solution.	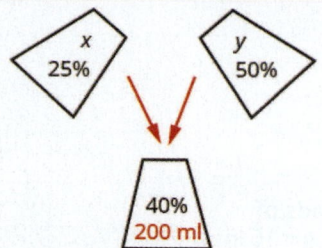

Step 2. Identify what we are looking for.	We are looking for how much of each solution she needs.

Step 3. Name what we are looking for.	Let $x =$ number of ml of 25% solution. $y =$ number of ml of 50% solution

A table will help us organize the data.

She will mix x ml of 25% with y ml of 50% to get 200 ml of 40% solution.

We write the percents as decimals in the chart.

We multiply the number of units times the concentration to get the total amount of sulfuric acid in each solution.

Type	Number of units	· Concentration %	= Amount
25%	x	0.25	$0.25x$
50%	y	0.50	$0.50y$
40%	200	0.40	0.40(200)

Step 4. Translate into a system of equations. We get the equations from the Number column and the Amount column.

Now we have the system.	$\begin{cases} x + y = 200 \\ 0.25x + 0.50y = 0.40(200) \end{cases}$

Step 5. Solve the system of equations. We will solve the system by elimination. Multiply the first equation by −0.5 to eliminate y.	$\begin{cases} -0.5(x + y) = -0.5(200) \\ 0.25x + 0.50y = 80 \end{cases}$

Simplify and add to solve for x.	$\begin{cases} -0.5x - 0.5y = -100 \\ 0.25x + 0.5y = 80 \end{cases}$ $\begin{aligned} -0.25x &= -20 \\ x &= 80 \end{aligned}$

To solve for y, substitute $x = 80$ into the first equation.	$x + y = 200$
	$80 + y = 200$
	$y = 120$

Step 6. Check the answer in the problem.

$$80 + 120 = 120 \checkmark$$
$$0.25(80) + 0.50(120) = 80 \checkmark$$
$$\text{Yes!}$$

| **Step 7. Answer** the question. | Sasheena should mix 80 ml of the 25% solution with 120 ml of the 50% solution to get the 200 ml of the 40% solution. |

> **TRY IT : :** 5.95

Translate to a system of equations and solve:

LeBron needs 150 milliliters of a 30% solution of sulfuric acid for a lab experiment but only has access to a 25% and a 50% solution. How much of the 25% and how much of the 50% solution should he mix to make the 30% solution?

> **TRY IT : :** 5.96

Translate to a system of equations and solve:

Anatole needs to make 250 milliliters of a 25% solution of hydrochloric acid for a lab experiment. The lab only has a 10% solution and a 40% solution in the storeroom. How much of the 10% and how much of the 40% solutions should he mix to make the 25% solution?

Solve Interest Applications

The formula to model interest applications is $I = Prt$. Interest, I, is the product of the principal, P, the rate, r, and the time, t. In our work here, we will calculate the interest earned in one year, so t will be 1.

We modify the column titles in the mixture table to show the formula for interest, as you'll see in **Example 5.49**.

EXAMPLE 5.49

Translate to a system of equations and solve:

Adnan has $40,000 to invest and hopes to earn 7.1% interest per year. He will put some of the money into a stock fund that earns 8% per year and the rest into bonds that earns 3% per year. How much money should he put into each fund?

⊘ **Solution**

Step 1. Read the problem.	A chart will help us organize the information.
Step 2. Identify what we are looking for.	We are looking for the amount to invest in each fund.
Step 3. Name what we are looking for.	Let $s =$ the amount invested in stocks. $b =$ the amount invested in bonds.

Write the interest rate as a decimal for each fund.
Multiply:
Principal · Rate · Time
to get the Interest.

Account	Principal ·	Rate ·	Time =	Interest
stock fund	s	0.08	1	0.08s
bonds	b	0.03	1	0.03b
Total	40,000	0.071		0.071(40,000)

Step 4. Translate into a system of equations.
We get our system of equations from the Principal column and the Interest column.

$$\begin{cases} s + b = 40{,}000 \\ 0.08s + 0.03b = 0.071(40{,}000) \end{cases}$$

Step 5. Solve the system of equations Solve by elimination. Multiply the top equation by −0.03.	$\begin{cases} -0.03(s+b) = -0.03(40{,}000) \\ 0.08s + 0.03b = 2{,}840 \end{cases}$
Simplify and add to solve for s.	$\begin{cases} -0.03s - 0.03b = -1{,}200 \\ \underline{0.08s + 0.03b = \ \ 2{,}840} \\ \qquad\quad 0.05s = \ \ 1{,}640 \end{cases}$
	$s = 32{,}800$
To find b, substitute $s = 32{,}800$ into the first equation.	$s + b = 40{,}000$
	$32{,}800 + b = 40{,}000$
	$b = 7{,}200$
Step 6. Check the answer in the problem.	We leave the check to you.
Step 7. Answer the question.	Adnan should invest $32,000 in stock and $7,200 in bonds.

Did you notice that the Principal column represents the total amount of money invested while the Interest column represents only the interest earned? Likewise, the first equation in our system, $s + b = 40{,}000$, represents the total amount of money invested and the second equation, $0.08s + 0.03b = 0.071(40{,}000)$, represents the interest earned.

 TRY IT : : 5.97

Translate to a system of equations and solve:

Leon had $50,000 to invest and hopes to earn 6.2 % interest per year. He will put some of the money into a stock fund that earns 7% per year and the rest in to a savings account that earns 2% per year. How much money should he put into each fund?

 TRY IT : : 5.98

Translate to a system of equations and solve:

Julius invested $7,000 into two stock investments. One stock paid 11% interest and the other stock paid 13% interest. He earned 12.5% interest on the total investment. How much money did he put in each stock?

EXAMPLE 5.50

Translate to a system of equations and solve:

Rosie owes $21,540 on her two student loans. The interest rate on her bank loan is 10.5% and the interest rate on the federal loan is 5.9%. The total amount of interest she paid last year was $1,669.68. What was the principal for each loan?

 **Solution**

Step 1. Read the problem.	A chart will help us organize the information.
Step 2. Identify what we are looking for.	We are looking for the principal of each loan.
Step 3. Name what we are looking for.	Let $b = $ the principal for the bank loan. $f = $ the principal on the federal loan
The total loans are $21,540.	

Record the interest rates as decimals in the chart.

Account	Principal	•	Rate	•	Time	=	Interest
bank	b		0.105		1		$0.105b$
federal	f		0.059		1		$0.059f$
Total	21,540						1669.68

Multiply using the formula $I = Pr\,t$ to get the Interest.

Step 4. Translate into a system of equations.
The system of equations comes from the Principal column and the Interest column.

$$\begin{cases} b + f = 21,540 \\ 0.105b + 0.059f = 1669.68 \end{cases}$$

Step 5. Solve the system of equations
We will use substitution to solve.
Solve the first equation for b.

$$b + f = 21,540$$
$$b = -f + 21,540$$

Substitute $b = -f + 21,540$ into the second equation.

$$0.105b + 0.059f = 1669.68$$
$$0.105(-f + 21,540) + 0.059f = 1669.68$$

Simplify and solve for f.

$$-0.105f + 2261.70 + 0.059f = 1669.68$$

$$-0.046f + 2261.70 = 1669.68$$

$$-0.046f = -592.02$$

$$f = 12,870$$

To find b, substitute $f = 12,870$ into the first equation.

$$b + f = 21,540$$

$$12,870 + f = 21,540$$

$$f = 8,670$$

Step 6. Check the answer in the problem.

We leave the check to you.

Step 7. Answer the question.

The principal of the bank loan is $12,870 and the principal for the federal loan is $8,670.

> **TRY IT : : 5.99**

Translate to a system of equations and solve:

Laura owes $18,000 on her student loans. The interest rate on the bank loan is 2.5% and the interest rate on the federal loan is 6.9 %. The total amount of interest she paid last year was $1,066. What was the principal for each loan?

> **TRY IT : : 5.100**

Translate to a system of equations and solve:

Jill's Sandwich Shoppe owes $65,200 on two business loans, one at 4.5% interest and the other at 7.2% interest. The total amount of interest owed last year was $3,582. What was the principal for each loan?

> ▶ **MEDIA : :**
>
> Access these online resources for additional instruction and practice with solving application problems with systems of linear equations.
>
> - **Cost and Mixture Word Problems (http://www.openstax.org/l/25LinEqu1)**
> - **Mixture Problems (http://www.openstax.org/l/25EqMixture)**

5.5 EXERCISES

Practice Makes Perfect

Solve Mixture Applications

In the following exercises, translate to a system of equations and solve.

239. Tickets to a Broadway show cost $35 for adults and $15 for children. The total receipts for 1650 tickets at one performance were $47,150. How many adult and how many child tickets were sold?

240. Tickets for a show are $70 for adults and $50 for children. One evening performance had a total of 300 tickets sold and the receipts totaled $17,200. How many adult and how many child tickets were sold?

241. Tickets for a train cost $10 for children and $22 for adults. Josie paid $1,200 for a total of 72 tickets. How many children's tickets and how many adult tickets did Josie buy?

242. Tickets for a baseball game are $69 for Main Level seats and $39 for Terrace Level seats. A group of sixteen friends went to the game and spent a total of $804 for the tickets. How many of Main Level and how many Terrace Level tickets did they buy?

243. Tickets for a dance recital cost $15 for adults and $7 for children. The dance company sold 253 tickets and the total receipts were $2,771. How many adult tickets and how many child tickets were sold?

244. Tickets for the community fair cost $12 for adults and $5 dollars for children. On the first day of the fair, 312 tickets were sold for a total of $2,204. How many adult tickets and how many child tickets were sold?

245. Brandon has a cup of quarters and dimes with a total value of $3.80. The number of quarters is four less than twice the number of quarters. How many quarters and how many dimes does Brandon have?

246. Sherri saves nickels and dimes in a coin purse for her daughter. The total value of the coins in the purse is $0.95. The number of nickels is two less than five times the number of dimes. How many nickels and how many dimes are in the coin purse?

247. Peter has been saving his loose change for several days. When he counted his quarters and dimes, he found they had a total value $13.10. The number of quarters was fifteen more than three times the number of dimes. How many quarters and how many dimes did Peter have?

248. Lucinda had a pocketful of dimes and quarters with a value of $ 6.20. The number of dimes is eighteen more than three times the number of quarters. How many dimes and how many quarters does Lucinda have?

249. A cashier has 30 bills, all of which are $10 or $20 bills. The total value of the money is $460. How many of each type of bill does the cashier have?

250. A cashier has 54 bills, all of which are $10 or $20 bills. The total value of the money is $910. How many of each type of bill does the cashier have?

251. Marissa wants to blend candy selling for $1.80 per pound with candy costing $1.20 per pound to get a mixture that costs her $1.40 per pound to make. She wants to make 90 pounds of the candy blend. How many pounds of each type of candy should she use?

252. How many pounds of nuts selling for $6 per pound and raisins selling for $3 per pound should Kurt combine to obtain 120 pounds of trail mix that cost him $5 per pound?

253. Hannah has to make twenty-five gallons of punch for a potluck. The punch is made of soda and fruit drink. The cost of the soda is $1.79 per gallon and the cost of the fruit drink is $2.49 per gallon. Hannah's budget requires that the punch cost $2.21 per gallon. How many gallons of soda and how many gallons of fruit drink does she need?

254. Joseph would like to make 12 pounds of a coffee blend at a cost of $6.25 per pound. He blends Ground Chicory at $4.40 a pound with Jamaican Blue Mountain at $8.84 per pound. How much of each type of coffee should he use?

255. Julia and her husband own a coffee shop. They experimented with mixing a City Roast Columbian coffee that cost $7.80 per pound with French Roast Columbian coffee that cost $8.10 per pound to make a 20 pound blend. Their blend should cost them $7.92 per pound. How much of each type of coffee should they buy?

256. Melody wants to sell bags of mixed candy at her lemonade stand. She will mix chocolate pieces that cost $4.89 per bag with peanut butter pieces that cost $3.79 per bag to get a total of twenty-five bags of mixed candy. Melody wants the bags of mixed candy to cost her $4.23 a bag to make. How many bags of chocolate pieces and how many bags of peanut butter pieces should she use?

257. Jotham needs 70 liters of a 50% alcohol solution. He has a 30% and an 80% solution available. How many liters of the 30% and how many liters of the 80% solutions should he mix to make the 50% solution?

258. Joy is preparing 15 liters of a 25% saline solution. She only has 40% and 10% solution in her lab. How many liters of the 40% and how many liters of the 10% should she mix to make the 25% solution?

259. A scientist needs 65 liters of a 15% alcohol solution. She has available a 25% and a 12% solution. How many liters of the 25% and how many liters of the 12% solutions should she mix to make the 15% solution?

260. A scientist needs 120 milliliters of a 20% acid solution for an experiment. The lab has available a 25% and a 10% solution. How many liters of the 25% and how many liters of the 10% solutions should the scientist mix to make the 20% solution?

261. A 40% antifreeze solution is to be mixed with a 70% antifreeze solution to get 240 liters of a 50% solution. How many liters of the 40% and how many liters of the 70% solutions will be used?

262. A 90% antifreeze solution is to be mixed with a 75% antifreeze solution to get 360 liters of a 85% solution. How many liters of the 90% and how many liters of the 75% solutions will be used?

Solve Interest Applications

In the following exercises, translate to a system of equations and solve.

263. Hattie had $3,000 to invest and wants to earn 10.6% interest per year. She will put some of the money into an account that earns 12% per year and the rest into an account that earns 10% per year. How much money should she put into each account?

264. Carol invested $2,560 into two accounts. One account paid 8% interest and the other paid 6% interest. She earned 7.25% interest on the total investment. How much money did she put in each account?

265. Sam invested $48,000, some at 6% interest and the rest at 10%. How much did he invest at each rate if he received $4,000 in interest in one year?

266. Arnold invested $64,000, some at 5.5% interest and the rest at 9%. How much did he invest at each rate if he received $4,500 in interest in one year?

267. After four years in college, Josie owes $65,800 in student loans. The interest rate on the federal loans is 4.5% and the rate on the private bank loans is 2%. The total interest she owed for one year was $2,878.50. What is the amount of each loan?

268. Mark wants to invest $10,000 to pay for his daughter's wedding next year. He will invest some of the money in a short term CD that pays 12% interest and the rest in a money market savings account that pays 5% interest. How much should he invest at each rate if he wants to earn $1,095 in interest in one year?

269. A trust fund worth $25,000 is invested in two different portfolios. This year, one portfolio is expected to earn 5.25% interest and the other is expected to earn 4%. Plans are for the total interest on the fund to be $1150 in one year. How much money should be invested at each rate?

270. A business has two loans totaling $85,000. One loan has a rate of 6% and the other has a rate of 4.5%. This year, the business expects to pay $4650 in interest on the two loans. How much is each loan?

Everyday Math

In the following exercises, translate to a system of equations and solve.

271. Laurie was completing the treasurer's report for her son's Boy Scout troop at the end of the school year. She didn't remember how many boys had paid the $15 full-year registration fee and how many had paid the $10 partial-year fee. She knew that the number of boys who paid for a full-year was ten more than the number who paid for a partial-year. If $250 was collected for all the registrations, how many boys had paid the full-year fee and how many had paid the partial-year fee?

272. As the treasurer of her daughter's Girl Scout troop, Laney collected money for some girls and adults to go to a three-day camp. Each girl paid $75 and each adult paid $30. The total amount of money collected for camp was $765. If the number of girls is three times the number of adults, how many girls and how many adults paid for camp?

Writing Exercises

273. Take a handful of two types of coins, and write a problem similar to **Example 5.46** relating the total number of coins and their total value. Set up a system of equations to describe your situation and then solve it.

274. In **Example 5.50** we solved the system of equations
$$\begin{cases} b + f = 21{,}540 \\ 0.105b + 0.059f = 1669.68 \end{cases}$$
by substitution. Would you have used substitution or elimination to solve this system? Why?

Self Check

After completing the exercises, use this checklist to evaluate your mastery of the objectives of this section.

I can...	Confidently	With some help	No-I don't get it!
solve mixture applications.			
solve interest applications.			

After looking at the checklist, do you think you are well-prepared for the next section? Why or why not?

5.6 Graphing Systems of Linear Inequalities

Learning Objectives

By the end of this section, you will be able to:

> Determine whether an ordered pair is a solution of a system of linear inequalities
> Solve a system of linear inequalities by graphing
> Solve applications of systems of inequalities

Be Prepared!

Before you get started, take this readiness quiz.

1. Graph $x > 2$ on a number line.
   If you missed this problem, review **Example 2.66**.

2. Solve the inequality $2a < 5a + 12$.
   If you missed this problem, review **Example 2.73**.

3. Determine whether the ordered pair $\left(3, \frac{1}{2}\right)$ is a solution to the system $\begin{cases} x + 2y = 4 \\ y = 6x \end{cases}$.

   If you missed this problem, review **Example 5.1**.

Determine Whether an Ordered Pair is a Solution of a System of Linear Inequalities

The definition of a system of linear inequalities is very similar to the definition of a system of linear equations.

System of Linear Inequalities

Two or more linear inequalities grouped together form a **system of linear inequalities**.

A system of linear inequalities looks like a system of linear equations, but it has inequalities instead of equations. A system of two linear inequalities is shown below.

$$\begin{cases} x + 4y \geq 10 \\ 3x - 2y < 12 \end{cases}$$

To solve a system of linear inequalities, we will find values of the variables that are solutions to both inequalities. We solve the system by using the graphs of each inequality and show the solution as a graph. We will find the region on the plane that contains all ordered pairs (x, y) that make both inequalities true.

Solutions of a System of Linear Inequalities

Solutions of a system of linear inequalities are the values of the variables that make all the inequalities true.

The solution of a system of linear inequalities is shown as a shaded region in the x-y coordinate system that includes all the points whose ordered pairs make the inequalities true.

To determine if an ordered pair is a solution to a system of two inequalities, we substitute the values of the variables into each inequality. If the ordered pair makes both inequalities true, it is a solution to the system.

EXAMPLE 5.51

Determine whether the ordered pair is a solution to the system. $\begin{cases} x + 4y \geq 10 \\ 3x - 2y < 12 \end{cases}$

ⓐ (−2, 4) ⓑ (3,1)

⊘ Solution

ⓐ Is the ordered pair (−2, 4) a solution?
We substitute $x = -2$ and $y = 4$ into both inequalities.

$$x + 4y \geq 10 \qquad\qquad\qquad 3x - 2y < 12$$

$$-2 + 4(4) \overset{?}{\geq} 10 \qquad\qquad 3(-2) - 2(4) \overset{?}{<} 12$$

$$14 \geq 10 \text{ true} \qquad\qquad -14 < 12 \text{ true}$$

The ordered pair (−2, 4) made both inequalities true. Therefore (−2, 4) is a solution to this system.

ⓑ Is the ordered pair (3,1) a solution?
We substitute $x = 3$ and $y = 1$ into both inequalities.

$$x + 4y \geq 10 \qquad\qquad\qquad 3x - 2y < 12$$

$$3 + 4(1) \overset{?}{\geq} 10 \qquad\qquad 3(3) - 2(1) \overset{?}{<} 12$$

$$7 \geq 10 \text{ false} \qquad\qquad 7 < 12 \text{ true}$$

The ordered pair (3,1) made one inequality true, but the other one false. Therefore (3,1) is not a solution to this system.

> **TRY IT :: 5.101**   Determine whether the ordered pair is a solution to the system.
$$\begin{cases} x - 5y > 10 \\ 2x + 3y > -2 \end{cases}$$

ⓐ $(3, -1)$ ⓑ $(6, -3)$

> **TRY IT :: 5.102**   Determine whether the ordered pair is a solution to the system.
$$\begin{cases} y > 4x - 2 \\ 4x - y < 20 \end{cases}$$

ⓐ $(2, 1)$ ⓑ $(4, -1)$

Solve a System of Linear Inequalities by Graphing

The solution to a single linear inequality is the region on one side of the boundary line that contains all the points that make the inequality true. The solution to a system of two linear inequalities is a region that contains the solutions to both inequalities. To find this region, we will graph each inequality separately and then locate the region where they are both true. The solution is always shown as a graph.

EXAMPLE 5.52   HOW TO SOLVE A SYSTEM OF LINEAR INEQUALITIES

Solve the system by graphing.
$$\begin{cases} y \geq 2x - 1 \\ y < x + 1 \end{cases}$$

⊘ **Solution**

Step 1. Graph the first inequality.	We will graph $y \geq 2x - 1$.	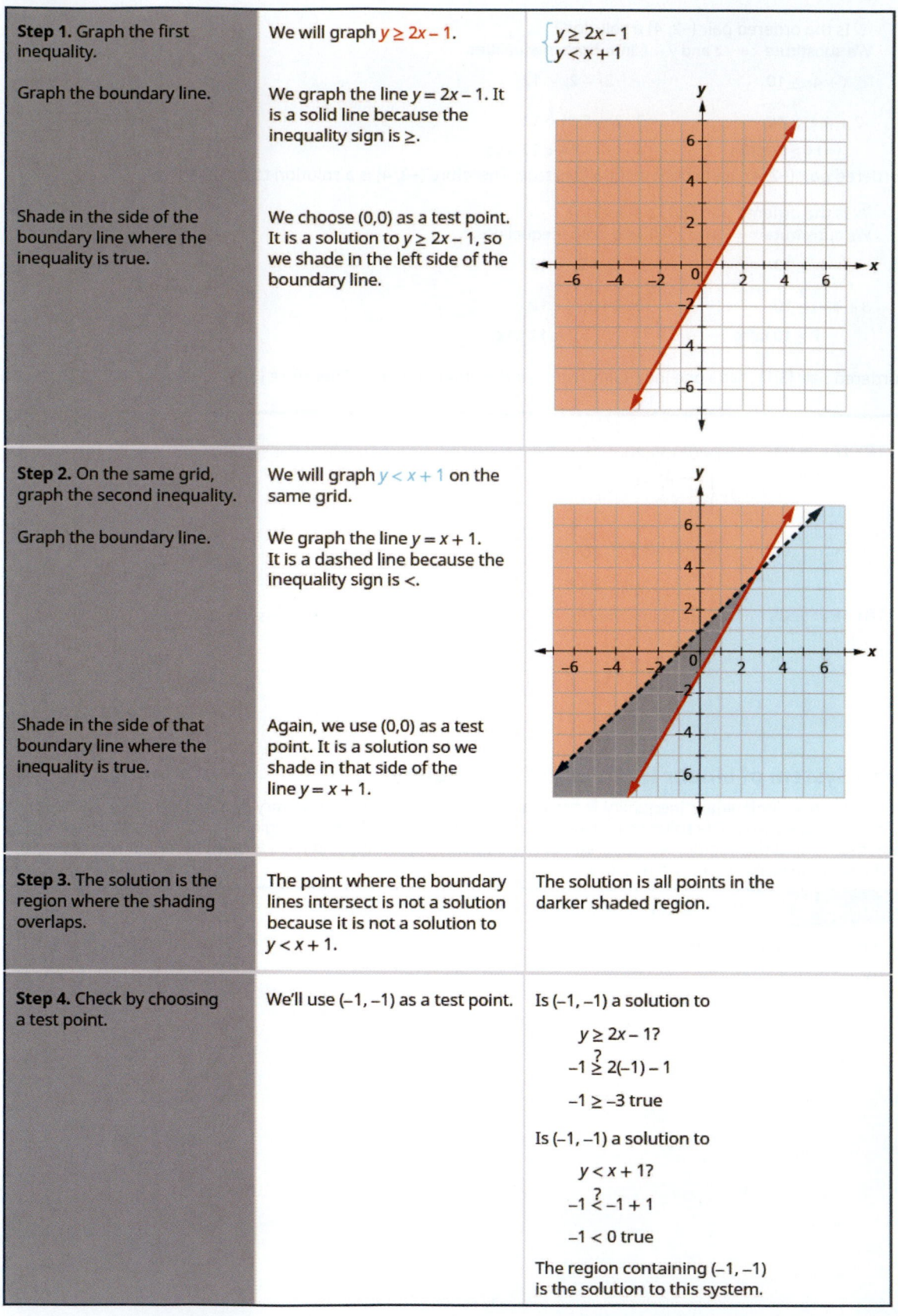
Graph the boundary line.	We graph the line $y = 2x - 1$. It is a solid line because the inequality sign is $\geq$.	
Shade in the side of the boundary line where the inequality is true.	We choose (0,0) as a test point. It is a solution to $y \geq 2x - 1$, so we shade in the left side of the boundary line.	
Step 2. On the same grid, graph the second inequality.	We will graph $y < x + 1$ on the same grid.	
Graph the boundary line.	We graph the line $y = x + 1$. It is a dashed line because the inequality sign is $<$.	
Shade in the side of that boundary line where the inequality is true.	Again, we use (0,0) as a test point. It is a solution so we shade in that side of the line $y = x + 1$.	
Step 3. The solution is the region where the shading overlaps.	The point where the boundary lines intersect is not a solution because it is not a solution to $y < x + 1$.	The solution is all points in the darker shaded region.
Step 4. Check by choosing a test point.	We'll use (–1, –1) as a test point.	Is (–1, –1) a solution to $$y \geq 2x - 1?$$ $$-1 \overset{?}{\geq} 2(-1) - 1$$ $$-1 \geq -3 \text{ true}$$ Is (–1, –1) a solution to $$y < x + 1?$$ $$-1 \overset{?}{<} -1 + 1$$ $$-1 < 0 \text{ true}$$ The region containing (–1, –1) is the solution to this system.

TRY IT :: 5.103

Solve the system by graphing. $\begin{cases} y < 3x + 2 \\ y > -x - 1 \end{cases}$

 TRY IT :: 5.104

Solve the system by graphing. $\begin{cases} y < -\frac{1}{2}x + 3 \\ y < 3x - 4 \end{cases}$

HOW TO :: SOLVE A SYSTEM OF LINEAR INEQUALITIES BY GRAPHING.

Step 1. Graph the first inequality.
- Graph the boundary line.
- Shade in the side of the boundary line where the inequality is true.

Step 2. On the same grid, graph the second inequality.
- Graph the boundary line.
- Shade in the side of that boundary line where the inequality is true.

Step 3. The solution is the region where the shading overlaps.

Step 4. Check by choosing a test point.

EXAMPLE 5.53

Solve the system by graphing. $\begin{cases} x - y > 3 \\ y < -\frac{1}{5}x + 4 \end{cases}$

Solution

Graph $x - y > 3$, by graphing $x - y = 3$ and testing a point.

The intercepts are $x = 3$ and $y = -3$ and the boundary line will be dashed.

Test $(0, 0)$. It makes the inequality false. So, shade the side that does not contain $(0, 0)$ red.

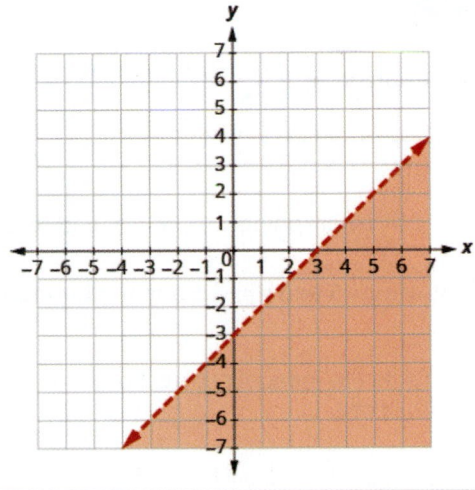

Graph $y < -\frac{1}{5}x + 4$ by graphing $y = -\frac{1}{5}x + 4$

using the slope $m = -\frac{1}{5}$ and y–intercept

b = 4. The boundary line will be dashed.

Test (0, 0). It makes the inequality true, so shade the side that contains (0, 0) blue.

Choose a test point in the solution and verify that it is a solution to both inequalities.

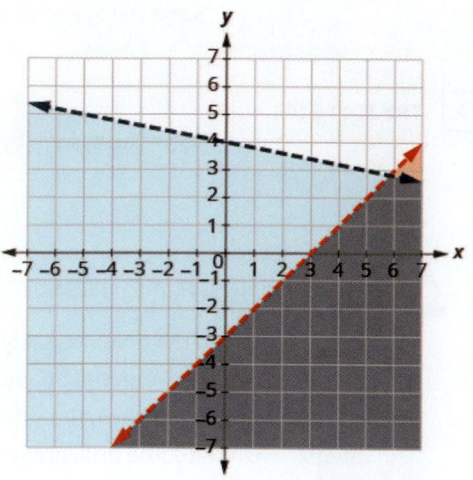

The point of intersection of the two lines is not included as both boundary lines were dashed. The solution is the area shaded twice which is the darker-shaded region.

> **TRY IT : : 5.105**

Solve the system by graphing. $\begin{cases} x + y \leq 2 \\ y \geq \frac{2}{3}x - 1 \end{cases}$

> **TRY IT : : 5.106**

Solve the system by graphing. $\begin{cases} 3x - 2y \leq 6 \\ y > -\frac{1}{4}x + 5 \end{cases}$

EXAMPLE 5.54

Solve the system by graphing. $\begin{cases} x - 2y < 5 \\ y > -4 \end{cases}$

⊘ **Solution**

Graph $x - 2y < 5$, by graphing $x - 2y = 5$ and testing a point.
The intercepts are x = 5 and y = –2.5 and the boundary line will be dashed.

Test (0, 0). It makes the inequality true. So, shade the side that contains (0, 0) red.

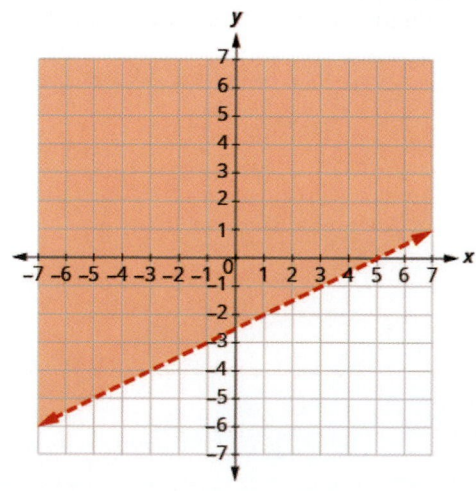

Graph $y > -4$, by graphing $y = -4$ and recognizing that it is a horizontal line through $y = -4$. The boundary line will be dashed.

Test $(0, 0)$. It makes the inequality true. So, shade (blue) the side that contains $(0, 0)$ blue.

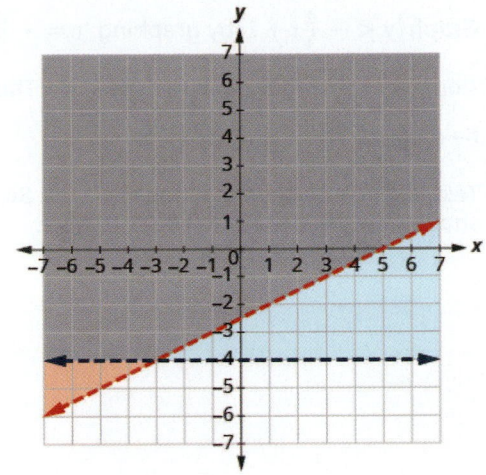

The point $(0, 0)$ is in the solution and we have already found it to be a solution of each inequality. The point of intersection of the two lines is not included as both boundary lines were dashed.

The solution is the area shaded twice which is the darker-shaded region.

> **TRY IT : : 5.107**   Solve the system by graphing. $\begin{cases} y \ge 3x - 2 \\ y < -1 \end{cases}$

> **TRY IT : : 5.108**   Solve the system by graphing. $\begin{cases} x > -4 \\ x - 2y \le -4 \end{cases}$

Systems of linear inequalities where the boundary lines are parallel might have no solution. We'll see this in **Example 5.55**.

EXAMPLE 5.55

Solve the system by graphing. $\begin{cases} 4x + 3y \ge 12 \\ y < -\frac{4}{3}x + 1 \end{cases}$

⊘ **Solution**

Graph $4x + 3y \ge 12$, by graphing $4x + 3y = 12$ and testing a point.
The intercepts are $x = 3$ and $y = 4$ and the boundary line will be solid.

Test $(0, 0)$. It makes the inequality false. So, shade the side that does not contain $(0, 0)$ red.

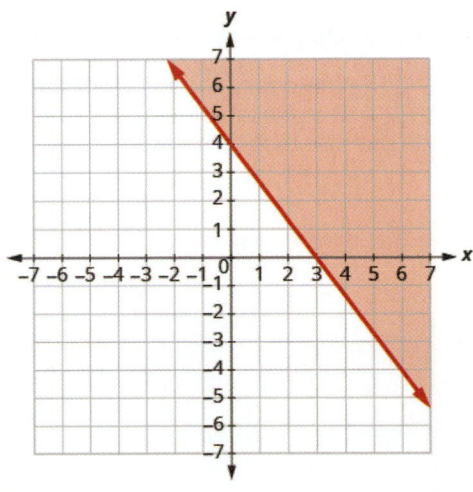

Graph $y < -\frac{4}{3}x + 1$ by graphing $y = -\frac{4}{3}x + 1$ using the slope $m = \frac{4}{3}$ and the y-intercept $b = 1$. The boundary line will be dashed.

Test $(0, 0)$. It makes the inequality true. So, shade the side that contains $(0, 0)$ blue.

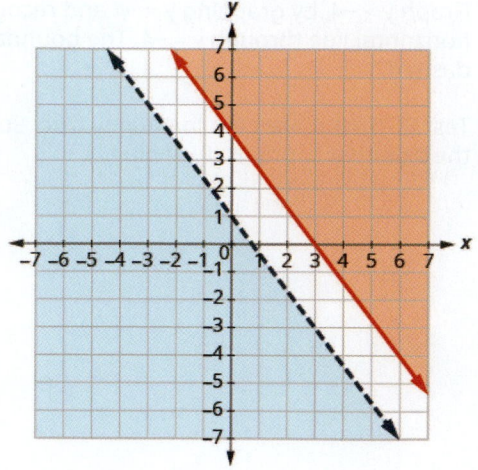

There is no point in both shaded regions, so the system has no solution. This system has no solution.

> **TRY IT : : 5.109**

Solve the system by graphing. $\begin{cases} 3x - 2y \le 12 \\ y \ge \frac{3}{2}x + 1 \end{cases}$

> **TRY IT : : 5.110**

Solve the system by graphing. $\begin{cases} x + 3y > 8 \\ y < -\frac{1}{3}x - 2 \end{cases}$

EXAMPLE 5.56

Solve the system by graphing. $\begin{cases} y > \frac{1}{2}x - 4 \\ x - 2y < -4 \end{cases}$

☑ **Solution**

Graph $y > \frac{1}{2}x - 4$ by graphing $y = \frac{1}{2}x - 4$

using the slope $m = \frac{1}{2}$ and the intercept

$b = -4$. The boundary line will be dashed. Test $(0, 0)$. It makes the inequality true. So, shade the side that contains $(0, 0)$ red.

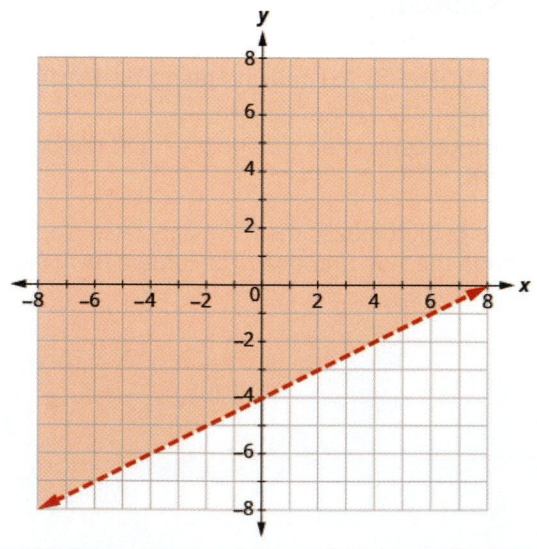

Graph $x - 2y < -4$ by graphing $x - 2y = -4$ and

testing a point.
The intercepts are $x = -4$ and $y = 2$ and the boundary line will be dashed.

Choose a test point in the solution and verify that it is a solution to both inequalities.

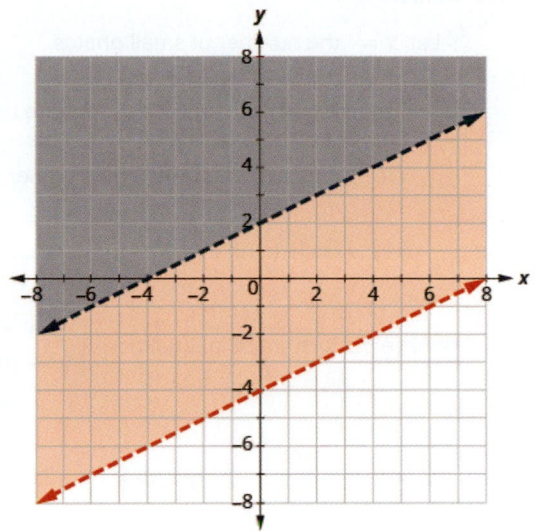

No point on the boundary lines is included in the solution as both lines are dashed.

The solution is the region that is shaded twice, which is also the solution to $x - 2y < -4$.

> **TRY IT :: 5.111**
> Solve the system by graphing. $\begin{cases} y \geq 3x + 1 \\ -3x + y \geq -4 \end{cases}$

> **TRY IT :: 5.112**
> Solve the system by graphing. $\begin{cases} y \leq -\frac{1}{4}x + 2 \\ x + 4y \leq 4 \end{cases}$

Solve Applications of Systems of Inequalities

The first thing we'll need to do to solve applications of systems of inequalities is to translate each condition into an inequality. Then we graph the system as we did above to see the region that contains the solutions. Many situations will be realistic only if both variables are positive, so their graphs will only show Quadrant I.

EXAMPLE 5.57

Christy sells her photographs at a booth at a street fair. At the start of the day, she wants to have at least 20 photos to display at her booth. Each small photo she displays costs her $4 and each large photo costs her $10. She doesn't want to spend more than $200 on photos to display.

ⓐ Write a system of inequalities to model this situation.

ⓑ Graph the system.

ⓒ Could she display 15 small and 5 large photos?

ⓓ Could she display 3 large and 22 small photos?

✓ **Solution**

ⓐ Let $x =$ the number of small photos.

$y =$ the number of large photos

To find the system of inequalities, translate the information.

She wants to have at least 25 photos.

The number of small plus the number of large should be at least 25.

$$x + y \geq 25$$

$4 for each small and $10 for each large must be no more than $200

$$4x + 10y \leq 200$$

We have our system of inequalities. $\begin{cases} x + y \geq 25 \\ 4x + 10y \leq 200 \end{cases}$

ⓑ

To graph $x + y \geq 25$, graph $x + y = 25$ as a solid line.

Choose (0, 0) as a test point. Since it does not make the inequality
true, shade the side that does not include the point (0, 0) red.

To graph $4x + 10y \leq 200$, graph $4x + 10y = 200$ as a
solid line.
Choose (0, 0) as a test point. Since it does not make the inequality
true, shade the side that includes the point (0, 0) blue.

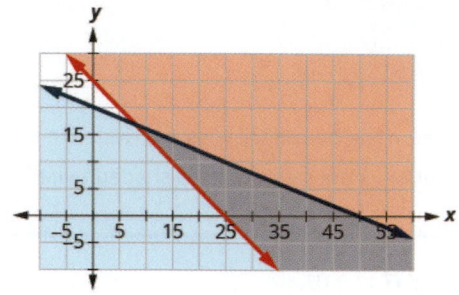

The solution of the system is the region of the graph that is double shaded and so is shaded darker.

ⓒ To determine if 10 small and 20 large photos would work, we see if the point (10, 20) is in the solution region.
It is not. Christy would not display 10 small and 20 large photos.

ⓓ To determine if 20 small and 10 large photos would work, we see if the point (20, 10) is in the solution region.
It is. Christy could choose to display 20 small and 10 large photos.
Notice that we could also test the possible solutions by substituting the values into each inequality.

> **TRY IT :: 5.113**

A trailer can carry a maximum weight of 160 pounds and a maximum volume of 15 cubic feet. A microwave oven
weighs 30 pounds and has 2 cubic feet of volume, while a printer weighs 20 pounds and has 3 cubic feet of space.

ⓐ Write a system of inequalities to model this situation.

ⓑ Graph the system.

ⓒ Could 4 microwaves and 2 printers be carried on this trailer?

ⓓ Could 7 microwaves and 3 printers be carried on this trailer?

> **TRY IT :: 5.114**

Mary needs to purchase supplies of answer sheets and pencils for a standardized test to be given to the juniors
at her high school. The number of the answer sheets needed is at least 5 more than twice the number of pencils.
The pencils cost $2 and the answer sheets cost $1. Mary's budget for these supplies allows for a maximum cost of
$400.

ⓐ Write a system of inequalities to model this situation.

ⓑ Graph the system.

ⓒ Could Mary purchase 100 pencils and 100 answer sheets?

ⓓ Could Mary purchase 150 pencils and 150 answer sheets?

EXAMPLE 5.58

Omar needs to eat at least 800 calories before going to his team practice. All he wants is hamburgers and cookies, and he doesn't want to spend more than $5. At the hamburger restaurant near his college, each hamburger has 240 calories and costs $1.40. Each cookie has 160 calories and costs $0.50.

ⓐ Write a system of inequalities to model this situation.

ⓑ Graph the system.

ⓒ Could he eat 3 hamburgers and 1 cookie?

ⓓ Could he eat 2 hamburgers and 4 cookies?

✓ Solution

ⓐ Let $h =$ the number of hamburgers.
       $c =$ the number of cookies

To find the system of inequalities, translate the information.
The calories from hamburgers at 240 calories each, plus the calories from cookies at 160 calories each must be more that 800.

$$240h + 160c \geq 800$$

The amount spent on hamburgers at $1.40 each, plus the amount spent on cookies at $0.50 each must be no more than $5.00.

$$1.40h + 0.50c \leq 5$$

We have our system of inequalities. $\begin{cases} 240h + 160c \geq 800 \\ 1.40h + 0.50c \leq 5 \end{cases}$

ⓑ

To graph $240h + 160c \geq 800$ graph $240h + 160c = 800$ as a solid line.
Choose (0, 0) as a test point. it does not make the inequality true.
So, shade (red) the side that does not include the point (0, 0).

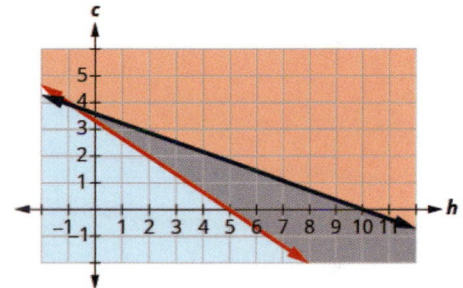

To graph $1.40h + 0.50c \leq 5$, graph $1.40h + 0.50c = 5$ as a solid line.
Choose (0,0) as a test point. It makes the inequality true. So, shade
(blue) the side that includes the point.

The solution of the system is the region of the graph that is double shaded and so is shaded darker.

ⓒ To determine if 3 hamburgers and 2 cookies would meet Omar's criteria, we see if the point (3, 2) is in the solution region. It is. He might choose to eat 3 hamburgers and 2 cookies.

ⓓ To determine if 2 hamburgers and 4 cookies would meet Omar's criteria, we see if the point (2, 4) is in the solution region. It is not. Omar would not choose to eat 2 hamburgers and 4 cookies.

We could also test the possible solutions by substituting the values into each inequality.

> **TRY IT : : 5.115**

Tension needs to eat at least an extra 1,000 calories a day to prepare for running a marathon. He has only $25 to spend on the extra food he needs and will spend it on $0.75 donuts which have 360 calories each and $2 energy drinks which have 110 calories.

ⓐ Write a system of inequalities that models this situation.
ⓑ Graph the system.
ⓒ Can he buy 8 donuts and 4 energy drinks?
ⓓ Can he buy 1 donut and 3 energy drinks?

> **TRY IT : : 5.116**

Philip's doctor tells him he should add at least 1000 more calories per day to his usual diet. Philip wants to buy protein bars that cost $1.80 each and have 140 calories and juice that costs $1.25 per bottle and have 125 calories. He doesn't want to spend more than $12.

ⓐ Write a system of inequalities that models this situation.
ⓑ Graph the system.
ⓒ Can he buy 3 protein bars and 5 bottles of juice?
ⓓ Can he buy 5 protein bars and 3 bottles of juice?

▶ **MEDIA : :**

Access these online resources for additional instruction and practice with graphing systems of linear inequalities.

- **Graphical System of Inequalities (http://www.openstax.org/l/25GSInequal1)**
- **Systems of Inequalities (http://www.openstax.org/l/25GSInequal2)**
- **Solving Systems of Linear Inequalities by Graphing (http://www.openstax.org/l/25GSInequal3)**

5.6 EXERCISES

Practice Makes Perfect

Determine Whether an Ordered Pair is a Solution of a System of Linear Inequalities

In the following exercises, determine whether each ordered pair is a solution to the system.

275. $\begin{cases} 3x + y > 5 \\ 2x - y \le 10 \end{cases}$

ⓐ $(3, -3)$ ⓑ $(7, 1)$

276. $\begin{cases} 4x - y < 10 \\ -2x + 2y > -8 \end{cases}$

ⓐ $(5, -2)$ ⓑ $(-1, 3)$

277. $\begin{cases} y > \frac{2}{3}x - 5 \\ x + \frac{1}{2}y \le 4 \end{cases}$

ⓐ $(6, -4)$ ⓑ $(3, 0)$

278. $\begin{cases} y < \frac{3}{2}x + 3 \\ \frac{3}{4}x - 2y < 5 \end{cases}$

ⓐ $(-4, -1)$ ⓑ $(8, 3)$

279. $\begin{cases} 7x + 2y > 14 \\ 5x - y \le 8 \end{cases}$

ⓐ $(2, 3)$ ⓑ $(7, -1)$

280. $\begin{cases} 6x - 5y < 20 \\ -2x + 7y > -8 \end{cases}$

ⓐ $(1, -3)$ ⓑ $(-4, 4)$

281. $\begin{cases} 2x + 3y \ge 2 \\ 4x - 6y < -1 \end{cases}$

ⓐ $\left(\frac{3}{2}, \frac{4}{3}\right)$ ⓑ $\left(\frac{1}{4}, \frac{7}{6}\right)$

282. $\begin{cases} 5x - 3y < -2 \\ 10x + 6y > 4 \end{cases}$

ⓐ $\left(\frac{1}{5}, \frac{2}{3}\right)$ ⓑ $\left(-\frac{3}{10}, \frac{7}{6}\right)$

Solve a System of Linear Inequalities by Graphing

In the following exercises, solve each system by graphing.

283. $\begin{cases} y \le 3x + 2 \\ y > x - 1 \end{cases}$

284. $\begin{cases} y < -2x + 2 \\ y \ge -x - 1 \end{cases}$

285. $\begin{cases} y < 2x - 1 \\ y \le -\frac{1}{2}x + 4 \end{cases}$

286. $\begin{cases} y \ge -\frac{2}{3}x + 2 \\ y > 2x - 3 \end{cases}$

287. $\begin{cases} x - y > 1 \\ y < -\frac{1}{4}x + 3 \end{cases}$

288. $\begin{cases} x + 2y < 4 \\ y < x - 2 \end{cases}$

289. $\begin{cases} 3x - y \le 6 \\ y \ge -\frac{1}{2}x \end{cases}$

290. $\begin{cases} 2x + 4y \ge 8 \\ y \le \frac{3}{4}x \end{cases}$

291. $\begin{cases} 2x - 5y < 10 \\ 3x + 4y \ge 12 \end{cases}$

292. $\begin{cases} 3x - 2y \le 6 \\ -4x - 2y > 8 \end{cases}$

293. $\begin{cases} 2x + 2y > -4 \\ -x + 3y \ge 9 \end{cases}$

294. $\begin{cases} 2x + y > -6 \\ -x + 2y \ge -4 \end{cases}$

295. $\begin{cases} x - 2y < 3 \\ y \le 1 \end{cases}$

296. $\begin{cases} x - 3y > 4 \\ y \le -1 \end{cases}$

297. $\begin{cases} y \ge -\frac{1}{2}x - 3 \\ x \le 2 \end{cases}$

298. $\begin{cases} y \le -\frac{2}{3}x + 5 \\ x \ge 3 \end{cases}$

299. $\begin{cases} y \ge \frac{3}{4}x - 2 \\ y < 2 \end{cases}$

300. $\begin{cases} y \le -\frac{1}{2}x + 3 \\ y < 1 \end{cases}$

301. $\begin{cases} 3x - 4y < 8 \\ x < 1 \end{cases}$

302. $\begin{cases} -3x + 5y > 10 \\ x > -1 \end{cases}$

303. $\begin{cases} x \ge 3 \\ y \le 2 \end{cases}$

304. $\begin{cases} x \le -1 \\ y \ge 3 \end{cases}$

305. $\begin{cases} 2x + 4y > 4 \\ y \le -\frac{1}{2}x - 2 \end{cases}$

306. $\begin{cases} x - 3y \ge 6 \\ y > \frac{1}{3}x + 1 \end{cases}$

307. $\begin{cases} -2x + 6y < 0 \\ 6y > 2x + 4 \end{cases}$

308. $\begin{cases} -3x + 6y > 12 \\ 4y \le 2x - 4 \end{cases}$

309. $\begin{cases} y \ge -3x + 2 \\ 3x + y > 5 \end{cases}$

310. $\begin{cases} y \ge \frac{1}{2}x - 1 \\ -2x + 4y \ge 4 \end{cases}$

311. $\begin{cases} y \le -\frac{1}{4}x - 2 \\ x + 4y < 6 \end{cases}$

312. $\begin{cases} y \ge 3x - 1 \\ -3x + y > -4 \end{cases}$

313. $\begin{cases} 3y > x + 2 \\ -2x + 6y > 8 \end{cases}$

314. $\begin{cases} y < \frac{3}{4}x - 2 \\ -3x + 4y < 7 \end{cases}$

Solve Applications of Systems of Inequalities

In the following exercises, translate to a system of inequalities and solve.

315. Caitlyn sells her drawings at the county fair. She wants to sell at least 60 drawings and has portraits and landscapes. She sells the portraits for $15 and the landscapes for $10. She needs to sell at least $800 worth of drawings in order to earn a profit.

ⓐ Write a system of inequalities to model this situation.

ⓑ Graph the system.

ⓒ Will she make a profit if she sells 20 portraits and 35 landscapes?

ⓓ Will she make a profit if she sells 50 portraits and 20 landscapes?

316. Jake does not want to spend more than $50 on bags of fertilizer and peat moss for his garden. Fertilizer costs $2 a bag and peat moss costs $5 a bag. Jake's van can hold at most 20 bags.

ⓐ Write a system of inequalities to model this situation.

ⓑ Graph the system.

ⓒ Can he buy 15 bags of fertilizer and 4 bags of peat moss?

ⓓ Can he buy 10 bags of fertilizer and 10 bags of peat moss?

317. Reiko needs to mail her Christmas cards and packages and wants to keep her mailing costs to no more than $500. The number of cards is at least 4 more than twice the number of packages. The cost of mailing a card (with pictures enclosed) is $3 and for a package the cost is $7.

ⓐ Write a system of inequalities to model this situation.

ⓑ Graph the system.

ⓒ Can she mail 60 cards and 26 packages?

ⓓ Can she mail 90 cards and 40 packages?

318. Juan is studying for his final exams in Chemistry and Algebra. He knows he only has 24 hours to study, and it will take him at least three times as long to study for Algebra than Chemistry.

ⓐ Write a system of inequalities to model this situation.

ⓑ Graph the system.

ⓒ Can he spend 4 hours on Chemistry and 20 hours on Algebra?

ⓓ Can he spend 6 hours on Chemistry and 18 hours on Algebra?

319. Jocelyn is pregnant and needs to eat at least 500 more calories a day than usual. When buying groceries one day with a budget of $15 for the extra food, she buys bananas that have 90 calories each and chocolate granola bars that have 150 calories each. The bananas cost $0.35 each and the granola bars cost $2.50 each.

ⓐ Write a system of inequalities to model this situation.

ⓑ Graph the system.

ⓒ Could she buy 5 bananas and 6 granola bars?

ⓓ Could she buy 3 bananas and 4 granola bars?

320. Mark is attempting to build muscle mass and so he needs to eat at least an additional 80 grams of protein a day. A bottle of protein water costs $3.20 and a protein bar costs $1.75. The protein water supplies 27 grams of protein and the bar supplies 16 gram. If he has $ 10 dollars to spend

ⓐ Write a system of inequalities to model this situation.

ⓑ Graph the system.

ⓒ Could he buy 3 bottles of protein water and 1 protein bar?

ⓓ Could he buy no bottles of protein water and 5 protein bars?

321. Jocelyn desires to increase both her protein consumption and caloric intake. She desires to have at least 35 more grams of protein each day and no more than an additional 200 calories daily. An ounce of cheddar cheese has 7 grams of protein and 110 calories. An ounce of parmesan cheese has 11 grams of protein and 22 calories.

ⓐ Write a system of inequalities to model this situation.

ⓑ Graph the system.

ⓒ Could she eat 1 ounce of cheddar cheese and 3 ounces of parmesan cheese?

ⓓ Could she eat 2 ounces of cheddar cheese and 1 ounce of parmesan cheese?

322. Mark is increasing his exercise routine by running and walking at least 4 miles each day. His goal is to burn a minimum of 1,500 calories from this exercise. Walking burns 270 calories/mile and running burns 650 calories.

ⓐ Write a system of inequalities to model this situation.

ⓑ Graph the system.

ⓒ Could he meet his goal by walking 3 miles and running 1 mile?

ⓓ Could he meet his goal by walking 2 miles and running 2 mile?

Everyday Math

323. Tickets for an American Baseball League game for 3 adults and 3 children cost less than $75, while tickets for 2 adults and 4 children cost less than $62.

ⓐ Write a system of inequalities to model this problem.

ⓑ Graph the system.

ⓒ Could the tickets cost $20 for adults and $8 for children?

ⓓ Could the tickets cost $15 for adults and $5 for children?

324. Grandpa and Grandma are treating their family to the movies. Matinee tickets cost $4 per child and $4 per adult. Evening tickets cost $6 per child and $8 per adult. They plan on spending no more than $80 on the matinee tickets and no more than $100 on the evening tickets.

ⓐ Write a system of inequalities to model this situation.

ⓑ Graph the system.

ⓒ Could they take 9 children and 4 adults to both shows?

ⓓ Could they take 8 children and 5 adults to both shows?

Writing Exercises

325. Graph the inequality $x - y \geq 3$. How do you know which side of the line $x - y = 3$ should be shaded?

326. Graph the system $\begin{cases} x + 2y \leq 6 \\ y \geq -\frac{1}{2}x - 4 \end{cases}$. What does the solution mean?

Self Check

ⓐ *After completing the exercises, use this checklist to evaluate your mastery of the objectives of this section.*

I can...	Confidently	With some help	No-I don't get it!
determine whether an ordered pair is a solution of a system of linear inequalities.			
solve a system of linear inequalities by graphing.			
solve applications of systems of inequalities.			

ⓑ *After reviewing this checklist, what will you do to become confident for all objectives?*

CHAPTER 5 REVIEW

KEY TERMS

coincident lines Coincident lines are lines that have the same slope and same *y*-intercept.

complementary angles Two angles are complementary if the sum of the measures of their angles is 90 degrees.

consistent system A consistent system of equations is a system of equations with at least one solution.

dependent equations Two equations are dependent if all the solutions of one equation are also solutions of the other equation.

inconsistent system An inconsistent system of equations is a system of equations with no solution.

independent equations Two equations are independent if they have different solutions.

solutions of a system of equations Solutions of a system of equations are the values of the variables that make all the equations true. A solution of a system of two linear equations is represented by an ordered pair (*x, y*).

supplementary angles Two angles are supplementary if the sum of the measures of their angles is 180 degrees.

system of linear equations When two or more linear equations are grouped together, they form a system of linear equations.

system of linear inequalities Two or more linear inequalities grouped together form a system of linear inequalities.

KEY CONCEPTS

5.1 Solve Systems of Equations by Graphing

- **To solve a system of linear equations by graphing**

  Step 1. Graph the first equation.

  Step 2. Graph the second equation on the same rectangular coordinate system.

  Step 3. Determine whether the lines intersect, are parallel, or are the same line.

  Step 4. Identify the solution to the system.
  If the lines intersect, identify the point of intersection. Check to make sure it is a solution to both equations. This is the solution to the system.
  If the lines are parallel, the system has no solution.
  If the lines are the same, the system has an infinite number of solutions.

  Step 5. Check the solution in both equations.

- Determine the number of solutions from the graph of a linear system

Graph	Number of solutions
2 intersecting lines	1
Parallel lines	None
Same line	Infinitely many

- Determine the number of solutions of a linear system by looking at the slopes and intercepts

Number of Solutions of a Linear System of Equations			
Slopes	Intercepts	Type of Lines	Number of Solutions
Different		Intersecting	1 point
Same	Different	Parallel	No solution
Same	Same	Coincident	Infinitely many solutions

- Determine the number of solutions and how to classify a system of equations

Lines	Intersecting	Parallel	Coincident
Number of solutions	1 point	No solution	Infinitely many
Consistent/inconsistent	Consistent	Inconsistent	Consistent
Dependent/independent	Independent	Independent	Dependent

- **Problem Solving Strategy for Systems of Linear Equations**

  Step 1. **Read** the problem. Make sure all the words and ideas are understood.

  Step 2. **Identify** what we are looking for.

  Step 3. **Name** what we are looking for. Choose variables to represent those quantities.

  Step 4. **Translate** into a system of equations.

  Step 5. **Solve** the system of equations using good algebra techniques.

  Step 6. **Check** the answer in the problem and make sure it makes sense.

  Step 7. **Answer** the question with a complete sentence.

5.2 Solve Systems of Equations by Substitution

- **Solve a system of equations by substitution**

  Step 1. Solve one of the equations for either variable.

  Step 2. Substitute the expression from Step 1 into the other equation.

  Step 3. Solve the resulting equation.

  Step 4. Substitute the solution in Step 3 into one of the original equations to find the other variable.

  Step 5. Write the solution as an ordered pair.

  Step 6. Check that the ordered pair is a solution to both original equations.

5.3 Solve Systems of Equations by Elimination

- **To Solve a System of Equations by Elimination**

  Step 1. Write both equations in standard form. If any coefficients are fractions, clear them.

  Step 2. Make the coefficients of one variable opposites.

  - Decide which variable you will eliminate.

  - Multiply one or both equations so that the coefficients of that variable are opposites.

  Step 3. Add the equations resulting from Step 2 to eliminate one variable.

  Step 4. Solve for the remaining variable.

  Step 5. Substitute the solution from Step 4 into one of the original equations. Then solve for the other variable.

  Step 6. Write the solution as an ordered pair.

  Step 7. Check that the ordered pair is a solution to **both** original equations.

5.5 Solve Mixture Applications with Systems of Equations

- **Table for coin and mixture applications**

Type	Number	• Value($) =	Total Value($)
Total			

- **Table for concentration applications**

Type	Number of units	• Concentration %	= Amount
Total			

- **Table for interest applications**

Account	Principal	•	Rate	•	Time	=	Interest
					1		
					1		
Total							

5.6 Graphing Systems of Linear Inequalities

- **To Solve a System of Linear Inequalities by Graphing**

  Step 1. Graph the first inequality.
  - Graph the boundary line.
  - Shade in the side of the boundary line where the inequality is true.

  Step 2. On the same grid, graph the second inequality.
  - Graph the boundary line.
  - Shade in the side of that boundary line where the inequality is true.

  Step 3. The solution is the region where the shading overlaps.

  Step 4. Check by choosing a test point.

REVIEW EXERCISES

5.1 Section 5.1 Solve Systems of Equations by Graphing

Determine Whether an Ordered Pair is a Solution of a System of Equations.

In the following exercises, determine if the following points are solutions to the given system of equations.

327. $\begin{cases} x + 3y = -9 \\ 2x - 4y = 12 \end{cases}$

ⓐ $(-3, -2)$ ⓑ $(0, -3)$

328. $\begin{cases} x + y = 8 \\ y = x - 4 \end{cases}$

ⓐ $(6, 2)$ ⓑ $(9, -1)$

Solve a System of Linear Equations by Graphing

In the following exercises, solve the following systems of equations by graphing.

329. $\begin{cases} 3x + y = 6 \\ x + 3y = -6 \end{cases}$

330. $\begin{cases} y = x - 2 \\ y = -2x - 2 \end{cases}$

331. $\begin{cases} 2x - y = 6 \\ y = 4 \end{cases}$

332. $\begin{cases} x + 4y = -1 \\ x = 3 \end{cases}$

333. $\begin{cases} 2x - y = 5 \\ 4x - 2y = 10 \end{cases}$

334. $\begin{cases} -x + 2y = 4 \\ y = \frac{1}{2}x - 3 \end{cases}$

Determine the Number of Solutions of a Linear System

In the following exercises, without graphing determine the number of solutions and then classify the system of equations.

335. $\begin{cases} y = \frac{2}{5}x + 2 \\ -2x + 5y = 10 \end{cases}$

336. $\begin{cases} 3x + 2y = 6 \\ y = -3x + 4 \end{cases}$

337. $\begin{cases} 5x - 4y = 0 \\ y = \frac{5}{4}x - 5 \end{cases}$

338. $\begin{cases} y = -\frac{3}{4}x + 1 \\ 6x + 8y = 8 \end{cases}$

Solve Applications of Systems of Equations by Graphing

339. LaVelle is making a pitcher of caffe mocha. For each ounce of chocolate syrup, she uses five ounces of coffee. How many ounces of chocolate syrup and how many ounces of coffee does she need to make 48 ounces of caffe mocha?

340. Eli is making a party mix that contains pretzels and chex. For each cup of pretzels, he uses three cups of chex. How many cups of pretzels and how many cups of chex does he need to make 12 cups of party mix?

5.2 Section 5.2 Solve Systems of Equations by Substitution

Solve a System of Equations by Substitution

In the following exercises, solve the systems of equations by substitution.

341. $\begin{cases} 3x - y = -5 \\ y = 2x + 4 \end{cases}$

342. $\begin{cases} 3x - 2y = 2 \\ y = \frac{1}{2}x + 3 \end{cases}$

343. $\begin{cases} x - y = 0 \\ 2x + 5y = -14 \end{cases}$

344. $\begin{cases} y = -2x + 7 \\ y = \frac{2}{3}x - 1 \end{cases}$

345. $\begin{cases} y = -5x \\ 5x + y = 6 \end{cases}$

346. $\begin{cases} y = -\frac{1}{3}x + 2 \\ x + 3y = 6 \end{cases}$

Solve Applications of Systems of Equations by Substitution

In the following exercises, translate to a system of equations and solve.

347. The sum of two number is 55. One number is 11 less than the other. Find the numbers.

348. The perimeter of a rectangle is 128. The length is 16 more than the width. Find the length and width.

349. The measure of one of the small angles of a right triangle is 2 less than 3 times the measure of the other small angle. Find the measure of both angles.

350. Gabriela works for an insurance company that pays her a salary of $32,000 plus a commission of $100 for each policy she sells. She is considering changing jobs to a company that would pay a salary of $40,000 plus a commission of $80 for each policy sold. How many policies would Gabriela need to sell to make the total pay the same?

5.3 Section 5.3 Solve Systems of Equations by Elimination

Solve a System of Equations by Elimination *In the following exercises, solve the systems of equations by elimination.*

351. $\begin{cases} x + y = 12 \\ x - y = -10 \end{cases}$

352. $\begin{cases} 4x + 2y = 2 \\ -4x - 3y = -9 \end{cases}$

353. $\begin{cases} 3x - 8y = 20 \\ x + 3y = 1 \end{cases}$

354. $\begin{cases} 3x - 2y = 6 \\ 4x + 3y = 8 \end{cases}$

355. $\begin{cases} 9x + 4y = 2 \\ 5x + 3y = 5 \end{cases}$

356. $\begin{cases} -x + 3y = 8 \\ 2x - 6y = -20 \end{cases}$

Solve Applications of Systems of Equations by Elimination

In the following exercises, translate to a system of equations and solve.

357. The sum of two numbers is -90. Their difference is 16. Find the numbers.

358. Omar stops at a donut shop every day on his way to work. Last week he had 8 donuts and 5 cappuccinos, which gave him a total of 3,000 calories. This week he had 6 donuts and 3 cappuccinos, which was a total of 2,160 calories. How many calories are in one donut? How many calories are in one cappuccino?

Choose the Most Convenient Method to Solve a System of Linear Equations

In the following exercises, decide whether it would be more convenient to solve the system of equations by substitution or elimination.

359. $\begin{cases} 6x - 5y = 27 \\ 3x + 10y = -24 \end{cases}$

360. $\begin{cases} y = 3x - 9 \\ 4x - 5y = 23 \end{cases}$

5.4 Section 5.4 Solve Applications with Systems of Equations

Translate to a System of Equations

In the following exercises, translate to a system of equations. Do not solve the system.

361. The sum of two numbers is -32. One number is two less than twice the other. Find the numbers.

362. Four times a number plus three times a second number is -9. Twice the first number plus the second number is three. Find the numbers.

363. Last month Jim and Debbie earned $7,200. Debbie earned $1,600 more than Jim earned. How much did they each earn?

364. Henri has $24,000 invested in stocks and bonds. The amount in stocks is $6,000 more than three times the amount in bonds. How much is each investment?

Solve Direct Translation Applications

In the following exercises, translate to a system of equations and solve.

365. Pam is 3 years older than her sister, Jan. The sum of their ages is 99. Find their ages.

366. Mollie wants to plant 200 bulbs in her garden. She wantsall irises and tulips. She wants to plant three times as many tulips as irises. How many irises and how many tulips should she plant?

Solve Geometry Applications

In the following exercises, translate to a system of equations and solve.

367. The difference of two supplementary angles is 58 degrees. Find the measures of the angles.

368. Two angles are complementary. The measure of the larger angle is five more than four times the measure of the smaller angle. Find the measures of both angles.

369. Becca is hanging a 28 foot floral garland on the two sides and top of a pergola to prepare for a wedding. The height is four feet less than the width. Find the height and width of the pergola.

370. The perimeter of a city rectangular park is 1428 feet. The length is 78 feet more than twice the width. Find the length and width of the park.

Solve Uniform Motion Applications

In the following exercises, translate to a system of equations and solve.

371. Sheila and Lenore were driving to their grandmother's house. Lenore left one hour after Sheila. Sheila drove at a rate of 45 mph, and Lenore drove at a rate of 60 mph. How long will it take for Lenore to catch up to Sheila?

372. Bob left home, riding his bike at a rate of 10 miles per hour to go to the lake. Cheryl, his wife, left 45 minutes ($\frac{3}{4}$ hour) later, driving her car at a rate of 25 miles per hour. How long will it take Cheryl to catch up to Bob?

373. Marcus can drive his boat 36 miles down the river in three hours but takes four hours to return upstream. Find the rate of the boat in still water and the rate of the current.

374. A passenger jet can fly 804 miles in 2 hours with a tailwind but only 776 miles in 2 hours into a headwind. Find the speed of the jet in still air and the speed of the wind.

5.5 Section 5.5 Solve Mixture Applications with Systems of Equations

Solve Mixture Applications

In the following exercises, translate to a system of equations and solve.

375. Lynn paid a total of $2,780 for 261 tickets to the theater. Student tickets cost $10 and adult tickets cost $15. How many student tickets and how many adult tickets did Lynn buy?

376. Priam has dimes and pennies in a cup holder in his car. The total value of the coins is $4.21. The number of dimes is three less than four times the number of pennies. How many dimes and how many pennies are in the cup?

377. Yumi wants to make 12 cups of party mix using candies and nuts. Her budget requires the party mix to cost her $1.29 per cup. The candies are $2.49 per cup and the nuts are $0.69 per cup. How many cups of candies and how many cups of nuts should she use?

378. A scientist needs 70 liters of a 40% solution of alcohol. He has a 30% and a 60% solution available. How many liters of the 30% and how many liters of the 60% solutions should he mix to make the 40% solution?

Solve Interest Applications

In the following exercises, translate to a system of equations and solve.

379. Jack has $12,000 to invest and wants to earn 7.5% interest per year. He will put some of the money into a savings account that earns 4% per year and the rest into CD account that earns 9% per year. How much money should he put into each account?

380. When she graduates college, Linda will owe $43,000 in student loans. The interest rate on the federal loans is 4.5% and the rate on the private bank loans is 2%. The total interest she owes for one year was $1585. What is the amount of each loan?

5.6 Section 5.6 Graphing Systems of Linear Inequalities

Determine Whether an Ordered Pair is a Solution of a System of Linear Inequalities

In the following exercises, determine whether each ordered pair is a solution to the system.

381. $\begin{cases} 4x + y > 6 \\ 3x - y \leq 12 \end{cases}$

ⓐ $(2, -1)$ ⓑ $(3, -2)$

382. $\begin{cases} y > \frac{1}{3}x + 2 \\ x - \frac{1}{4}y \leq 10 \end{cases}$

ⓐ $(6, 5)$ ⓑ $(15, 8)$

Solve a System of Linear Inequalities by Graphing

In the following exercises, solve each system by graphing.

383. $\begin{cases} y < 3x + 1 \\ y \geq -x - 2 \end{cases}$

384. $\begin{cases} x - y > -1 \\ y < \frac{1}{3}x - 2 \end{cases}$

385. $\begin{cases} 2x - 3y < 6 \\ 3x + 4y \geq 12 \end{cases}$

386. $\begin{cases} y \leq -\frac{3}{4}x + 1 \\ x \geq -5 \end{cases}$

387. $\begin{cases} x + 3y < 5 \\ y \geq -\frac{1}{3}x + 6 \end{cases}$

388. $\begin{cases} y \geq 2x - 5 \\ -6x + 3y > -4 \end{cases}$

Solve Applications of Systems of Inequalities

In the following exercises, translate to a system of inequalities and solve.

389. Roxana makes bracelets and necklaces and sells them at the farmers' market. She sells the bracelets for $12 each and the necklaces for $18 each. At the market next weekend she will have room to display no more than 40 pieces, and she needs to sell at least $500 worth in order to earn a profit.

ⓐ Write a system of inequalities to model this situation.

ⓑ Graph the system.

ⓒ Should she display 26 bracelets and 14 necklaces?

ⓓ Should she display 39 bracelets and 1 necklace?

390. Annie has a budget of $600 to purchase paperback books and hardcover books for her classroom. She wants the number of hardcover to be at least 5 more than three times the number of paperback books. Paperback books cost $4 each and hardcover books cost $15 each.

ⓐ Write a system of inequalities to model this situation.

ⓑ Graph the system.

ⓒ Can she buy 8 paperback books and 40 hardcover books?

ⓓ Can she buy 10 paperback books and 37 hardcover books?

PRACTICE TEST

391. $\begin{cases} x - 4y = -8 \\ 2x + 5y = 10 \end{cases}$

ⓐ $(0, 2)$ ⓑ $(4, 3)$

In the following exercises, solve the following systems by graphing.

392. $\begin{cases} x - y = 5 \\ x + 2y = -4 \end{cases}$

393. $\begin{cases} x - y > -2 \\ y \le 3x + 1 \end{cases}$

In the following exercises, solve each system of equations. Use either substitution or elimination.

394. $\begin{cases} 3x - 2y = 3 \\ y = 2x - 1 \end{cases}$

395. $\begin{cases} x + y = -3 \\ x - y = 11 \end{cases}$

396. $\begin{cases} 4x - 3y = 7 \\ 5x - 2y = 0 \end{cases}$

397. $\begin{cases} y = -\dfrac{4}{5}x + 1 \\ 8x + 10y = 10 \end{cases}$

398. $\begin{cases} 2x + 3y = 12 \\ -4x + 6y = -16 \end{cases}$

In the following exercises, translate to a system of equations and solve.

399. The sum of two numbers is −24. One number is 104 less than the other. Find the numbers.

400. Ramon wants to plant cucumbers and tomatoes in his garden. He has room for 16 plants, and he wants to plant three times as many cucumbers as tomatoes. How many cucumbers and how many tomatoes should he plant?

401. Two angles are complementary. The measure of the larger angle is six more than twice the measure of the smaller angle. Find the measures of both angles.

402. On Monday, Lance ran for 30 minutes and swam for 20 minutes. His fitness app told him he had burned 610 calories. On Wednesday, the fitness app told him he burned 695 calories when he ran for 25 minutes and swam for 40 minutes. How many calories did he burn for one minute of running? How many calories did he burn for one minute of swimming?

403. Kathy left home to walk to the mall, walking quickly at a rate of 4 miles per hour. Her sister Abby left home 15 minutes later and rode her bike to the mall at a rate of 10 miles per hour. How long will it take Abby to catch up to Kathy?

404. It takes $5\dfrac{1}{2}$ hours for a jet to fly 2,475 miles with a headwind from San Jose, California to Lihue, Hawaii. The return flight from Lihue to San Jose with a tailwind, takes 5 hours. Find the speed of the jet in still air and the speed of the wind.

405. Liz paid $160 for 28 tickets to take the Brownie troop to the science museum. Children's tickets cost $5 and adult tickets cost $9. How many children's tickets and how many adult tickets did Liz buy?

406. A pharmacist needs 20 liters of a 2% saline solution. He has a 1% and a 5% solution available. How many liters of the 1% and how many liters of the 5% solutions should she mix to make the 2% solution?

407. Translate to a system of inequalities and solve.

Andi wants to spend no more than $50 on Halloween treats. She wants to buy candy bars that cost $1 each and lollipops that cost $0.50 each, and she wants the number of lollipops to be at least three times the number of candy bars.

ⓐ Write a system of inequalities to model this situation.

ⓑ Graph the system.

ⓒ Can she buy 20 candy bars and 70 lollipops?

ⓓ Can she buy 15 candy bars and 65 lollipops?

Figure 6.1 Architects use polynomials to design curved shapes such as this suspension bridge, the Silver Jubilee bridge in Halton, England.

Chapter Outline

6.1 Add and Subtract Polynomials

6.2 Use Multiplication Properties of Exponents

6.3 Multiply Polynomials

6.4 Special Products

6.5 Divide Monomials

6.6 Divide Polynomials

6.7 Integer Exponents and Scientific Notation

Introduction

We have seen that the graphs of linear equations are straight lines. Graphs of other types of equations, called polynomial equations, are curves, like the outline of this suspension bridge. Architects use polynomials to design the shape of a bridge like this and to draw the blueprints for it. Engineers use polynomials to calculate the stress on the bridge's supports to ensure they are strong enough for the intended load. In this chapter, you will explore operations with and properties of polynomials.

6.1 Add and Subtract Polynomials

Learning Objectives

By the end of this section, you will be able to:

› Identify polynomials, monomials, binomials, and trinomials
› Determine the degree of polynomials
› Add and subtract monomials
› Add and subtract polynomials
› Evaluate a polynomial for a given value

Be Prepared!

Before you get started, take this readiness quiz.

1. Simplify: $8x + 3x$.
    If you missed this problem, review **Example 1.24**.

2. Subtract: $(5n + 8) - (2n - 1)$.
    If you missed this problem, review **Example 1.139**.

3. Write in expanded form: a^5.
   If you missed this problem, review **Example 1.14**.

Identify Polynomials, Monomials, Binomials and Trinomials

You have learned that a *term* is a constant or the product of a constant and one or more variables. When it is of the form ax^m, where a is a constant and m is a whole number, it is called a monomial. Some examples of monomial are 8, $-2x^2$, $4y^3$, and $11z^7$.

Monomials

A **monomial** is a term of the form ax^m, where a is a constant and m is a positive whole number.

A monomial, or two or more monomials combined by addition or subtraction, is a polynomial. Some polynomials have special names, based on the number of terms. A monomial is a polynomial with exactly one term. A binomial has exactly two terms, and a trinomial has exactly three terms. There are no special names for polynomials with more than three terms.

Polynomials

polynomial—A monomial, or two or more monomials combined by addition or subtraction, is a polynomial.

- **monomial**—A polynomial with exactly one term is called a monomial.
- **binomial**—A polynomial with exactly two terms is called a binomial.
- **trinomial**—A polynomial with exactly three terms is called a trinomial.

Here are some examples of polynomials.

Polynomial	$b + 1$	$4y^2 - 7y + 2$	$4x^4 + x^3 + 8x^2 - 9x + 1$	
Monomial	14	$8y^2$	$-9x^3 y^5$	-13
Binomial	$a + 7$	$4b - 5$	$y^2 - 16$	$3x^3 - 9x^2$
Trinomial	$x^2 - 7x + 12$	$9y^2 + 2y - 8$	$6m^4 - m^3 + 8m$	$z^4 + 3z^2 - 1$

Notice that every monomial, binomial, and trinomial is also a polynomial. They are just special members of the "family" of polynomials and so they have special names. We use the words *monomial*, *binomial*, and *trinomial* when referring to these special polynomials and just call all the rest *polynomials*.

EXAMPLE 6.1

Determine whether each polynomial is a monomial, binomial, trinomial, or other polynomial.

ⓐ $4y^2 - 8y - 6$ ⓑ $-5a^4 b^2$ ⓒ $2x^5 - 5x^3 - 9x^2 + 3x + 4$ ⓓ $13 - 5m^3$ ⓔ q

✓ **Solution**

	Polynomial	Number of terms	Type
(a)	$4y^2 - 8y - 6$	3	Trinomial
(b)	$-5a^4 b^2$	1	Monomial
(c)	$2x^5 - 5x^3 - 9x^2 + 3x + 4$	5	Polynomial
(d)	$13 - 5m^3$	2	Binomial
(e)	q	1	Monomial

> **TRY IT :: 6.1**   Determine whether each polynomial is a monomial, binomial, trinomial, or other polynomial:

ⓐ $5b$ ⓑ $8y^3 - 7y^2 - y - 3$ ⓒ $-3x^2 - 5x + 9$ ⓓ $81 - 4a^2$ ⓔ $-5x^6$

> **TRY IT : : 6.2** Determine whether each polynomial is a monomial, binomial, trinomial, or other polynomial:

ⓐ $27z^3 - 8$ ⓑ $12m^3 - 5m^2 - 2m$ ⓒ $\frac{5}{6}$ ⓓ $8x^4 - 7x^2 - 6x - 5$ ⓔ $-n^4$

Determine the Degree of Polynomials

The degree of a polynomial and the degree of its terms are determined by the exponents of the variable.

A monomial that has no variable, just a constant, is a special case. The degree of a constant is 0—it has no variable.

Degree of a Polynomial

The **degree of a term** is the sum of the exponents of its variables.

The **degree of a constant** is 0.

The **degree of a polynomial** is the highest degree of all its terms.

Let's see how this works by looking at several polynomials. We'll take it step by step, starting with monomials, and then progressing to polynomials with more terms.

Monomial	14	$8y^2$	$-9x^3y^5$	$-13a$
Degree	0	2	8	1

Binomial	$a + 7$	$4b^2 - 5b$	$x^2y^2 - 16$	$3n^3 - 9n^2$
Degree of each term	0 1	2 1	4 0	3 2
Degree of polynomial	1	2	4	3

Trinomial	$x^2 - 7x + 12$	$9a^2 + 6ab + b^2$	$6m^4 - m^3n^2 + 8mn^5$	$z^4 + 3z^2 - 1$
Degree of each term	2 1 0	2 2 2	4 5 6	4 2 0
Degree of polynomial	2	2	6	4

Polynomial	$b + 1$	$4y^2 - 7y + 2$	$4x^4 + x^3 + 8x^2 - 9x + 1$	
Degree of each term	1 0	2 1 0	4 3 2 1 0	
Degree of polynomial	1	2	4	

A polynomial is in **standard form** when the terms of a polynomial are written in descending order of degrees. Get in the habit of writing the term with the highest degree first.

EXAMPLE 6.2

Find the degree of the following polynomials.

ⓐ $10y$ ⓑ $4x^3 - 7x + 5$ ⓒ -15 ⓓ $-8b^2 + 9b - 2$ ⓔ $8xy^2 + 2y$

⊘ **Solution**

ⓐ

	$10y$
The exponent of y is one. $y = y^1$	The degree is 1.

ⓑ

	$4x^3 - 7x + 5$
The highest degree of all the terms is 3.	The degree is 3.

ⓒ

	-15
The degree of a constant is 0.	The degree is 0.

ⓓ

	$-8b^2 + 9b - 2$
The highest degree of all the terms is 2.	The degree is 2.

ⓔ

	$8xy^2 + 2y$
The highest degree of all the terms is 3.	The degree is 3.

> **TRY IT :: 6.3**   Find the degree of the following polynomials:
>
> ⓐ $-15b$ ⓑ $10z^4 + 4z^2 - 5$ ⓒ $12c^5 d^4 + 9c^3 d^9 - 7$ ⓓ $3x^2 y - 4x$ ⓔ -9

> **TRY IT :: 6.4**   Find the degree of the following polynomials:
>
> ⓐ 52 ⓑ $a^4 b - 17a^4$ ⓒ $5x + 6y + 2z$ ⓓ $3x^2 - 5x + 7$ ⓔ $-a^3$

Add and Subtract Monomials

You have learned how to simplify expressions by combining like terms. Remember, like terms must have the same variables with the same exponent. Since monomials are terms, adding and subtracting monomials is the same as combining like terms. If the monomials are like terms, we just combine them by adding or subtracting the coefficient.

EXAMPLE 6.3

Add: $25y^2 + 15y^2$.

⊘ **Solution**

	$25y^2 + 15y^2$
Combine like terms.	$40y^2$

> **TRY IT :: 6.5**   Add: $12q^2 + 9q^2$.

> **TRY IT :: 6.6**   Add: $-15c^2 + 8c^2$.

EXAMPLE 6.4

Subtract: $16p - (-7p)$.

⊘ **Solution**

	$16p - (-7p)$
Combine like terms.	$23p$

> **TRY IT :: 6.7**   Subtract: $8m - (-5m)$.

> **TRY IT :: 6.8**   Subtract: $-15z^3 - (-5z^3)$.

Remember that like terms must have the same variables with the same exponents.

EXAMPLE 6.5

Simplify: $c^2 + 7d^2 - 6c^2$.

✓ **Solution**

$$c^2 + 7d^2 - 6c^2$$

Combine like terms. $\qquad -5c^2 + 7d^2$

> | **TRY IT :: 6.9** $\qquad$ Add: $8y^2 + 3z^2 - 3y^2$.

> | **TRY IT :: 6.10** $\qquad$ Add: $3m^2 + n^2 - 7m^2$.

EXAMPLE 6.6

Simplify: $u^2v + 5u^2 - 3v^2$.

✓ **Solution**

$$u^2v + 5u^2 - 3v^2$$

There are no like terms to combine. $\qquad u^2v + 5u^2 - 3v^2$

> | **TRY IT :: 6.11** $\qquad$ Simplify: $m^2n^2 - 8m^2 + 4n^2$.

> | **TRY IT :: 6.12** $\qquad$ Simplify: $pq^2 - 6p - 5q^2$.

Add and Subtract Polynomials

We can think of adding and subtracting polynomials as just adding and subtracting a series of monomials. Look for the like terms—those with the same variables and the same exponent. The Commutative Property allows us to rearrange the terms to put like terms together.

EXAMPLE 6.7

Find the sum: $\left(5y^2 - 3y + 15\right) + \left(3y^2 - 4y - 11\right)$.

✓ **Solution**

Identify like terms.	$(\underline{5y^2} - \underline{3y} + 15) + (\underline{3y^2} - \underline{4y} - 11)$
Rearrange to get the like terms together.	$\underline{5y^2} + \underline{3y^2} - \underline{3y} - \underline{4y} + 15 - 11$
Combine like terms.	$8y^2 - 7y + 4$

> | **TRY IT :: 6.13** $\qquad$ Find the sum: $\left(7x^2 - 4x + 5\right) + \left(x^2 - 7x + 3\right)$.

> | **TRY IT :: 6.14** $\qquad$ Find the sum: $\left(14y^2 + 6y - 4\right) + \left(3y^2 + 8y + 5\right)$.

EXAMPLE 6.8

Find the difference: $\left(9w^2 - 7w + 5\right) - \left(2w^2 - 4\right)$.

✓ Solution

	$(9w^2 - 7w + 5) - (2w^2 - 4)$
Distribute and identify like terms.	$9w^2 - 7w + 5 - 2w^2 + 4$
Rearrange the terms.	$9w^2 - 2w^2 - 7w + 5 + 4$
Combine like terms.	$7w^2 - 7w + 9$

> **TRY IT : :** 6.15   Find the difference: $\left(8x^2 + 3x - 19\right) - \left(7x^2 - 14\right)$.

> **TRY IT : :** 6.16   Find the difference: $\left(9b^2 - 5b - 4\right) - \left(3b^2 - 5b - 7\right)$.

EXAMPLE 6.9

Subtract: $\left(c^2 - 4c + 7\right)$ from $\left(7c^2 - 5c + 3\right)$.

✓ Solution

	Subtract $(c^2 - 4c + 7)$ from $(7c^2 - 5c + 3)$.
	$(7c^2 - 5c + 3) - (c^2 - 4c + 7)$
Distribute and identify like terms.	$7c^2 - 5c + 3 - c^2 + 4c - 7$
Rearrange the terms.	$7c^2 - c^2 - 5c + 4c + 3 - 7$
Combine like terms.	$6c^2 - c - 4$

> **TRY IT : :** 6.17   Subtract: $\left(5z^2 - 6z - 2\right)$ from $\left(7z^2 + 6z - 4\right)$.

> **TRY IT : :** 6.18   Subtract: $\left(x^2 - 5x - 8\right)$ from $\left(6x^2 + 9x - 1\right)$.

EXAMPLE 6.10

Find the sum: $\left(u^2 - 6uv + 5v^2\right) + \left(3u^2 + 2uv\right)$.

✓ Solution

	$\left(u^2 - 6uv + 5v^2\right) + \left(3u^2 + 2uv\right)$
Distribute.	$u^2 - 6uv + 5v^2 + 3u^2 + 2uv$
Rearrange the terms, to put like terms together.	$u^2 + 3u^2 - 6uv + 2uv + 5v^2$
Combine like terms.	$4u^2 - 4uv + 5v^2$

> **TRY IT ::** 6.19   Find the sum: $\left(3x^2 - 4xy + 5y^2\right) + \left(2x^2 - xy\right)$.

> **TRY IT ::** 6.20   Find the sum: $\left(2x^2 - 3xy - 2y^2\right) + \left(5x^2 - 3xy\right)$.

EXAMPLE 6.11

Find the difference: $\left(p^2 + q^2\right) - \left(p^2 + 10pq - 2q^2\right)$.

⊘ **Solution**

$$\left(p^2 + q^2\right) - \left(p^2 + 10pq - 2q^2\right)$$

Distribute. $\qquad\qquad p^2 + q^2 - p^2 - 10pq + 2q^2$

Rearrange the terms, to put like terms together. $\qquad p^2 - p^2 - 10pq + q^2 + 2q^2$

Combine like terms. $\qquad\qquad -10pq^2 + 3q^2$

> **TRY IT ::** 6.21   Find the difference: $\left(a^2 + b^2\right) - \left(a^2 + 5ab - 6b^2\right)$.

> **TRY IT ::** 6.22   Find the difference: $\left(m^2 + n^2\right) - \left(m^2 - 7mn - 3n^2\right)$.

EXAMPLE 6.12

Simplify: $\left(a^3 - a^2 b\right) - \left(ab^2 + b^3\right) + \left(a^2 b + ab^2\right)$.

⊘ **Solution**

$$\left(a^3 - a^2 b\right) - \left(ab^2 + b^3\right) + \left(a^2 b + ab^2\right)$$

Distribute. $\qquad\qquad a^3 - a^2 b - ab^2 - b^3 + a^2 b + ab^2$

Rearrange the terms, to put like terms together. $\qquad a^3 - a^2 b + a^2 b - ab^2 + ab^2 - b^3$

Combine like terms. $\qquad\qquad a^3 - b^3$

> **TRY IT ::** 6.23   Simplify: $\left(x^3 - x^2 y\right) - \left(xy^2 + y^3\right) + \left(x^2 y + xy^2\right)$.

> **TRY IT ::** 6.24   Simplify: $\left(p^3 - p^2 q\right) + \left(pq^2 + q^3\right) - \left(p^2 q + pq^2\right)$.

Evaluate a Polynomial for a Given Value

We have already learned how to evaluate expressions. Since polynomials are expressions, we'll follow the same procedures to evaluate a polynomial. We will substitute the given value for the variable and then simplify using the order of operations.

EXAMPLE 6.13

Evaluate $5x^2 - 8x + 4$ when

ⓐ $x = 4$ ⓑ $x = -2$ ⓒ $x = 0$

✓ Solution

ⓐ $x = 4$

	$5x^2 - 8x + 4$
Substitute 4 for x.	$5(4)^2 - 8(4) + 4$
Simplify the exponents.	$5 \cdot 16 - 8(4) + 4$
Multiply.	$80 - 32 + 4$
Simplify.	52

ⓑ $x = -2$

	$5x^2 - 8x + 4$
Substitute –2 for x.	$5(-2)^2 - 8(-2) + 4$
Simplify the exponents.	$5 \cdot 4 - 8(-2) + 4$
Multiply.	$20 + 16 + 4$
Simplify.	40

ⓒ $x = 0$

	$5x^2 - 8x + 4$
Substitute 0 for x.	$5(0)^2 - 8(0) + 4$
Simplify the exponents.	$5 \cdot 0 - 8(0) + 4$
Multiply.	$0 + 0 + 4$
Simplify.	4

> **TRY IT ∷ 6.25**   Evaluate: $3x^2 + 2x - 15$ when
>
> ⓐ $x = 3$ ⓑ $x = -5$ ⓒ $x = 0$

> **TRY IT ∷ 6.26**   Evaluate: $5z^2 - z - 4$ when
>
> ⓐ $z = -2$ ⓑ $z = 0$ ⓒ $z = 2$

EXAMPLE 6.14

The polynomial $-16t^2 + 250$ gives the height of a ball t seconds after it is dropped from a 250 foot tall building. Find the height after $t = 2$ seconds.

✅ **Solution**

$$-16t^2 + 250$$

Substitute $t = 2$.	$-16(2)^2 + 250$
Simplify.	$-16 \cdot 4 + 250$
Simplify.	$-64 + 250$
Simplify.	186

After 2 seconds the height of the ball is 186 feet.

> **TRY IT : : 6.27**
>
> The polynomial $-16t^2 + 250$ gives the height of a ball t seconds after it is dropped from a 250-foot tall building. Find the height after $t = 0$ seconds.

> **TRY IT : : 6.28**
>
> The polynomial $-16t^2 + 250$ gives the height of a ball t seconds after it is dropped from a 250-foot tall building. Find the height after $t = 3$ seconds.

EXAMPLE 6.15

The polynomial $6x^2 + 15xy$ gives the cost, in dollars, of producing a rectangular container whose top and bottom are squares with side x feet and sides of height y feet. Find the cost of producing a box with $x = 4$ feet and $y = 6$ feet.

✅ **Solution**

$$6x^2 + 15xy$$

Substitute $x = 4$, $y = 6$.	$6(4)^2 + 15(4)(6)$
Simplify.	$6 \cdot 16 + 15(4)(6)$
Simplify.	$96 + 360$
Simplify.	456

The cost of producing the box is $456.

> **TRY IT : : 6.29**
>
> The polynomial $6x^2 + 15xy$ gives the cost, in dollars, of producing a rectangular container whose top and bottom are squares with side x feet and sides of height y feet. Find the cost of producing a box with $x = 6$ feet and $y = 4$ feet.

> **TRY IT : : 6.30**
>
> The polynomial $6x^2 + 15xy$ gives the cost, in dollars, of producing a rectangular container whose top and bottom are squares with side x feet and sides of height y feet. Find the cost of producing a box with $x = 5$ feet and $y = 8$ feet.

▶ **MEDIA : :**

Access these online resources for additional instruction and practice with adding and subtracting polynomials.

- **Add and Subtract Polynomials 1 (https://openstax.org/l/25Addsubtrpoly1)**
- **Add and Subtract Polynomials 2 (https://openstax.org/l/25Addsubtrpoly2)**
- **Add and Subtract Polynomial 3 (https://openstax.org/l/25Addsubtrpoly3)**
- **Add and Subtract Polynomial 4 (https://openstax.org/l/25Addsubtrpoly4)**

6.1 EXERCISES

Practice Makes Perfect

Identify Polynomials, Monomials, Binomials, and Trinomials

In the following exercises, determine if each of the following polynomials is a monomial, binomial, trinomial, or other polynomial.

1.
ⓐ $81b^5 - 24b^3 + 1$
ⓑ $5c^3 + 11c^2 - c - 8$
ⓒ $\frac{14}{15}y + \frac{1}{7}$
ⓓ 5
ⓔ $4y + 17$

2.
ⓐ $x^2 - y^2$
ⓑ $-13c^4$
ⓒ $x^2 + 5x - 7$
ⓓ $x^2y^2 - 2xy + 8$
ⓔ 19

3.
ⓐ $8 - 3x$
ⓑ $z^2 - 5z - 6$
ⓒ $y^3 - 8y^2 + 2y - 16$
ⓓ $81b^5 - 24b^3 + 1$
ⓔ -18

4.
ⓐ $11y^2$
ⓑ -73
ⓒ $6x^2 - 3xy + 4x - 2y + y^2$
ⓓ $4y + 17$
ⓔ $5c^3 + 11c^2 - c - 8$

Determine the Degree of Polynomials

In the following exercises, determine the degree of each polynomial.

5.
ⓐ $6a^2 + 12a + 14$
ⓑ $18xy^2z$
ⓒ $5x + 2$
ⓓ $y^3 - 8y^2 + 2y - 16$
ⓔ -24

6.
ⓐ $9y^3 - 10y^2 + 2y - 6$
ⓑ $-12p^4$
ⓒ $a^2 + 9a + 18$
ⓓ $20x^2y^2 - 10a^2b^2 + 30$
ⓔ 17

7.
ⓐ $14 - 29x$
ⓑ $z^2 - 5z - 6$
ⓒ $y^3 - 8y^2 + 2y - 16$
ⓓ $23ab^2 - 14$
ⓔ -3

8.
ⓐ $62y^2$
ⓑ 15
ⓒ $6x^2 - 3xy + 4x - 2y + y^2$
ⓓ $10 - 9x$
ⓔ $m^4 + 4m^3 + 6m^2 + 4m + 1$

Add and Subtract Monomials

In the following exercises, add or subtract the monomials.

9. $7x^2 + 5x^2$

10. $4y^3 + 6y^3$

11. $-12w + 18w$

12. $-3m + 9m$

13. $4a - 9a$

14. $-y - 5y$

15. $28x - (-12x)$

16. $13z - (-4z)$

17. $-5b - 17b$

18. $-10x - 35x$

19. $12a + 5b - 22a$

20. $14x - 3y - 13x$

21. $2a^2 + b^2 - 6a^2$

22. $5u^2 + 4v^2 - 6u^2$

23. $xy^2 - 5x - 5y^2$

24. $pq^2 - 4p - 3q^2$

25. $a^2 b - 4a - 5ab^2$

26. $x^2 y - 3x + 7xy^2$

27. $12a + 8b$

28. $19y + 5z$

29. Add: $4a, -3b, -8a$

30. Add: $4x, 3y, -3x$

31. Subtract $5x^6$ from $-12x^6$.

32. Subtract $2p^4$ from $-7p^4$.

Add and Subtract Polynomials

In the following exercises, add or subtract the polynomials.

33. $(5y^2 + 12y + 4) + (6y^2 - 8y + 7)$

34. $(4y^2 + 10y + 3) + (8y^2 - 6y + 5)$

35. $(x^2 + 6x + 8) + (-4x^2 + 11x - 9)$

36. $(y^2 + 9y + 4) + (-2y^2 - 5y - 1)$

37. $(8x^2 - 5x + 2) + (3x^2 + 3)$

38. $(7x^2 - 9x + 2) + (6x^2 - 4)$

39. $(5a^2 + 8) + (a^2 - 4a - 9)$

40. $(p^2 - 6p - 18) + (2p^2 + 11)$

41. $(4m^2 - 6m - 3) - (2m^2 + m - 7)$

42. $(3b^2 - 4b + 1) - (5b^2 - b - 2)$

43. $(a^2 + 8a + 5) - (a^2 - 3a + 2)$

44. $(b^2 - 7b + 5) - (b^2 - 2b + 9)$

45. $(12s^2 - 15s) - (s - 9)$

46. $(10r^2 - 20r) - (r - 8)$

47. Subtract $(9x^2 + 2)$ from $(12x^2 - x + 6)$.

48. Subtract $(5y^2 - y + 12)$ from $(10y^2 - 8y - 20)$.

49. Subtract $(7w^2 - 4w + 2)$ from $(8w^2 - w + 6)$.

50. Subtract $(5x^2 - x + 12)$ from $(9x^2 - 6x - 20)$.

51. Find the sum of $(2p^3 - 8)$ and $(p^2 + 9p + 18)$.

52. Find the sum of $(q^2 + 4q + 13)$ and $(7q^3 - 3)$.

53. Find the sum of $(8a^3 - 8a)$ and $(a^2 + 6a + 12)$.

54. Find the sum of $(b^2 + 5b + 13)$ and $(4b^3 - 6)$.

55. Find the difference of $(w^2 + w - 42)$ and $(w^2 - 10w + 24)$.

56. Find the difference of $(z^2 - 3z - 18)$ and $(z^2 + 5z - 20)$.

57. Find the difference of $(c^2 + 4c - 33)$ and $(c^2 - 8c + 12)$.

58. Find the difference of $(t^2 - 5t - 15)$ and $(t^2 + 4t - 17)$.

59. $\left(7x^2 - 2xy + 6y^2\right) + \left(3x^2 - 5xy\right)$

60. $\left(-5x^2 - 4xy - 3y^2\right) + \left(2x^2 - 7xy\right)$

61. $\left(7m^2 + mn - 8n^2\right) + \left(3m^2 + 2mn\right)$

62. $\left(2r^2 - 3rs - 2s^2\right) + \left(5r^2 - 3rs\right)$

63. $\left(a^2 - b^2\right) - \left(a^2 + 3ab - 4b^2\right)$

64. $\left(m^2 + 2n^2\right) - \left(m^2 - 8mn - n^2\right)$

65. $\left(u^2 - v^2\right) - \left(u^2 - 4uv - 3v^2\right)$

66. $\left(j^2 - k^2\right) - \left(j^2 - 8jk - 5k^2\right)$

67. $\left(p^3 - 3p^2q\right) + \left(2pq^2 + 4q^3\right) - \left(3p^2q + pq^2\right)$

68. $\left(a^3 - 2a^2b\right) + \left(ab^2 + b^3\right) - \left(3a^2b + 4ab^2\right)$

69. $\left(x^3 - x^2y\right) - \left(4xy^2 - y^3\right) + \left(3x^2y - xy^2\right)$

70. $\left(x^3 - 2x^2y\right) - \left(xy^2 - 3y^3\right) - \left(x^2y - 4xy^2\right)$

Evaluate a Polynomial for a Given Value

In the following exercises, evaluate each polynomial for the given value.

71. Evaluate $8y^2 - 3y + 2$ when:

ⓐ $y = 5$

ⓑ $y = -2$

ⓒ $y = 0$

72. Evaluate $5y^2 - y - 7$ when:

ⓐ $y = -4$

ⓑ $y = 1$

ⓒ $y = 0$

73. Evaluate $4 - 36x$ when:

ⓐ $x = 3$

ⓑ $x = 0$

ⓒ $x = -1$

74. Evaluate $16 - 36x^2$ when:

ⓐ $x = -1$

ⓑ $x = 0$

ⓒ $x = 2$

75. A painter drops a brush from a platform 75 feet high. The polynomial $-16t^2 + 75$ gives the height of the brush t seconds after it was dropped. Find the height after $t = 2$ seconds.

76. A girl drops a ball off a cliff into the ocean. The polynomial $-16t^2 + 250$ gives the height of a ball t seconds after it is dropped from a 250-foot tall cliff. Find the height after $t = 2$ seconds.

77. A manufacturer of stereo sound speakers has found that the revenue received from selling the speakers at a cost of p dollars each is given by the polynomial $-4p^2 + 420p$. Find the revenue received when $p = 60$ dollars.

78. A manufacturer of the latest basketball shoes has found that the revenue received from selling the shoes at a cost of p dollars each is given by the polynomial $-4p^2 + 420p$. Find the revenue received when $p = 90$ dollars.

Everyday Math

79. Fuel Efficiency The fuel efficiency (in miles per gallon) of a car going at a speed of x miles per hour is given by the polynomial $-\frac{1}{150}x^2 + \frac{1}{3}x$. Find the fuel efficiency when $x = 30$ mph.

80. Stopping Distance The number of feet it takes for a car traveling at x miles per hour to stop on dry, level concrete is given by the polynomial $0.06x^2 + 1.1x$. Find the stopping distance when $x = 40$ mph.

81. Rental Cost The cost to rent a rug cleaner for d days is given by the polynomial $5.50d + 25$. Find the cost to rent the cleaner for 6 days.

82. Height of Projectile The height (in feet) of an object projected upward is given by the polynomial $-16t^2 + 60t + 90$ where t represents time in seconds. Find the height after $t = 2.5$ seconds.

83. Temperature Conversion The temperature in degrees Fahrenheit is given by the polynomial $\frac{9}{5}c + 32$ where c represents the temperature in degrees Celsius. Find the temperature in degrees Fahrenheit when $c = 65°$.

Writing Exercises

84. Using your own words, explain the difference between a monomial, a binomial, and a trinomial.

85. Using your own words, explain the difference between a polynomial with five terms and a polynomial with a degree of 5.

86. Ariana thinks the sum $6y^2 + 5y^4$ is $11y^6$. What is wrong with her reasoning?

87. Jonathan thinks that $\frac{1}{3}$ and $\frac{1}{x}$ are both monomials. What is wrong with his reasoning?

Self Check

ⓐ *After completing the exercises, use this checklist to evaluate your mastery of the objectives of this section.*

I can...	Confidently	With some help	No-I don't get it!
identify polynomials, monomials, binomials, and trinomials.			
determine the degree of polynomials.			
add and subtract monomials.			
add and subtract polynomials.			
evaluate a polynomial for a given value.			

ⓑ *If most of your checks were:*

...confidently. *Congratulations! You have achieved the objectives in this section. Reflect on the study skills you used so that you can continue to use them. What did you do to become confident of your ability to do these things? Be specific.*

...with some help. *This must be addressed quickly because topics you do not master become potholes in your road to success. In math every topic builds upon previous work. It is important to make sure you have a strong foundation before you move on. Who can you ask for help? Your fellow classmates and instructor are good resources. Is there a place on campus where math tutors are available? Can your study skills be improved?*

...no - I don't get it! *This is a warning sign and you must not ignore it. You should get help right away or you will quickly be overwhelmed. See your instructor as soon as you can to discuss your situation. Together you can come up with a plan to get you the help you need.*

6.2 Use Multiplication Properties of Exponents

Learning Objectives

By the end of this section, you will be able to:

› Simplify expressions with exponents
› Simplify expressions using the Product Property for Exponents
› Simplify expressions using the Power Property for Exponents
› Simplify expressions using the Product to a Power Property
› Simplify expressions by applying several properties
› Multiply monomials

Be Prepared!

Before you get started, take this readiness quiz.

1. Simplify: $\frac{3}{4} \cdot \frac{3}{4}$.

   If you missed this problem, review **Example 1.68**.

2. Simplify: $(-2)(-2)(-2)$.

   If you missed this problem, review **Example 1.50**.

Simplify Expressions with Exponents

Remember that an exponent indicates repeated multiplication of the same quantity. For example, 2^4 means to multiply 2 by itself 4 times, so 2^4 means $2 \cdot 2 \cdot 2 \cdot 2$.

Let's review the vocabulary for expressions with exponents.

Exponential Notation

$a^m \leftarrow$ exponent a^m means multiply m factors of a

base $a^m = a \cdot a \cdot a \cdot ..... \cdot a$

          m factors

This is read a to the mth power.

In the expression a^m, the *exponent* m tells us how many times we use the *base* a as a factor.

4^3 $(-9)^5$

$4 \cdot 4 \cdot 4$ $(-9)(-9)(-9)(-9)(-9)$

3 factors   5 factors

Before we begin working with variable expressions containing exponents, let's simplify a few expressions involving only numbers.

EXAMPLE 6.16

Simplify: ⓐ 4^3 ⓑ 7^1 ⓒ $\left(\frac{5}{6}\right)^2$ ⓓ $(0.63)^2$.

✓ **Solution**

	4^3
Multiply three factors of 4.	$4 \cdot 4 \cdot 4$
Simplify.	64

ⓑ

	7^1
Multiply one factor of 7.	7

ⓒ

	$\left(\dfrac{5}{6}\right)^2$
Multiply two factors.	$\left(\dfrac{5}{6}\right)\left(\dfrac{5}{6}\right)$
Simplify.	$\dfrac{25}{36}$

ⓓ

	$(0.63)^2$
Multiply two factors.	$(0.63)(0.63)$
Simplify.	0.3969

> **TRY IT ::** 6.31

Simplify: ⓐ 6^3 ⓑ 15^1 ⓒ $\left(\dfrac{3}{7}\right)^2$ ⓓ $(0.43)^2$.

> **TRY IT ::** 6.32

Simplify: ⓐ 2^5 ⓑ 21^1 ⓒ $\left(\dfrac{2}{5}\right)^3$ ⓓ $(0.218)^2$.

EXAMPLE 6.17

Simplify: ⓐ $(-5)^4$ ⓑ -5^4.

✓ **Solution**

ⓐ

	$(-5)^4$
Multiply four factors of -5.	$(-5)(-5)(-5)(-5)$
Simplify.	625

ⓑ

	-5^4
Multiply four factors of 5.	$-(5 \cdot 5 \cdot 5 \cdot 5)$
Simplify.	-625

> **TRY IT ::** 6.33

Simplify: ⓐ $(-3)^4$ ⓑ -3^4.

> **TRY IT ::** 6.34

Simplify: ⓐ $(-13)^2$ ⓑ -13^2.

Notice the similarities and differences in **Example 6.17**ⓐ and **Example 6.17**ⓑ! Why are the answers different? As we follow the order of operations in part ⓐ the parentheses tell us to raise the (-5) to the 4th power. In part ⓑ we raise just the 5 to the 4th power and then take the opposite.

Simplify Expressions Using the Product Property for Exponents

You have seen that when you combine like terms by adding and subtracting, you need to have the same base with the same exponent. But when you multiply and divide, the exponents may be different, and sometimes the bases may be different, too.

We'll derive the properties of exponents by looking for patterns in several examples.

First, we will look at an example that leads to the Product Property.

$$x^2 \cdot x^3$$

What does this mean? How many factors altogether?	$\underbrace{x \cdot x}_{\text{2 factors}} \cdot \underbrace{x \cdot x \cdot x}_{\text{3 factors}}$ $\underbrace{\qquad\qquad\qquad}_{\text{5 factors}}$
So, we have	x^5
Notice that 5 is the sum of the exponents, 2 and 3.	$x^2 \cdot x^3$ is x^{2+3}, or x^5

We write:

$$x^2 \cdot x^3$$
$$x^{2+3}$$
$$x^5$$

The base stayed the same and we added the exponents. This leads to the **Product Property for Exponents**.

Product Property for Exponents

If a is a real number, and m and n are counting numbers, then

$$a^m \cdot a^n = a^{m+n}$$

To multiply with like bases, add the exponents.

An example with numbers helps to verify this property.

$$2^2 \cdot 2^3 \stackrel{?}{=} 2^{2+3}$$
$$4 \cdot 8 \stackrel{?}{=} 2^5$$
$$32 = 32 \checkmark$$

EXAMPLE 6.18

Simplify: $y^5 \cdot y^6$.

✓ **Solution**

	$y^5 \cdot y^6$
Use the product property, $a^m \cdot a^n = a^{m+n}$.	y^{5+6}
Simplify.	y^{11}

> **TRY IT : : 6.35**   Simplify: $b^9 \cdot b^8$.

> **TRY IT : : 6.36**   Simplify: $x^{12} \cdot x^4$.

EXAMPLE 6.19

Simplify: ⓐ $2^5 \cdot 2^9$ ⓑ $3 \cdot 3^4$.

✓ **Solution**

ⓐ

	$2^5 \cdot 2^9$
Use the product property, $a^m \cdot a^n = a^{m+n}$.	2^{5+9}
Simplify.	2^{14}

ⓑ

	$3 \cdot 3^4$
Use the product property, $a^m \cdot a^n = a^{m+n}$.	3^{1+4}
Simplify.	3^5

> **TRY IT : : 6.37**   Simplify: ⓐ $5 \cdot 5^5$ ⓑ $4^9 \cdot 4^9$.

> **TRY IT : : 6.38**   Simplify: ⓐ $7^6 \cdot 7^8$ ⓑ $10 \cdot 10^{10}$.

EXAMPLE 6.20

Simplify: ⓐ $a^7 \cdot a$ ⓑ $x^{27} \cdot x^{13}$.

✓ **Solution**

ⓐ

	$a^7 \cdot a$
Rewrite, $a = a^1$.	$a^7 \cdot a^1$
Use the product property, $a^m \cdot a^n = a^{m+n}$.	a^{7+1}
Simplify.	a^8

ⓑ

	$x^{27} \cdot x^{13}$
Notice, the bases are the same, so add the exponents.	x^{27+13}
Simplify.	x^{40}

> **TRY IT :: 6.39** Simplify: ⓐ $p^5 \cdot p$ ⓑ $y^{14} \cdot y^{29}$.

> **TRY IT :: 6.40** Simplify: ⓐ $z \cdot z^7$ ⓑ $b^{15} \cdot b^{34}$.

We can extend the Product Property for Exponents to more than two factors.

EXAMPLE 6.21

Simplify: $d^4 \cdot d^5 \cdot d^2$.

✓ **Solution**

	$d^4 \cdot d^5 \cdot d^2$
Add the exponents, since bases are the same.	d^{4+5+2}
Simplify.	d^{11}

> **TRY IT :: 6.41** Simplify: $x^6 \cdot x^4 \cdot x^8$.

> **TRY IT :: 6.42** Simplify: $b^5 \cdot b^9 \cdot b^5$.

Simplify Expressions Using the Power Property for Exponents

Now let's look at an exponential expression that contains a power raised to a power. See if you can discover a general property.

$$(x^2)^3$$

What does this mean? How many factors altogether?	$x^2 \quad \cdot \quad x^2 \quad \cdot \quad x^2$ $\underbrace{x \cdot x} \quad \cdot \quad \underbrace{x \cdot x} \quad \cdot \quad \underbrace{x \cdot x}$ 2 *factors* 2 *factors* 2 *factors* $\underbrace{\qquad\qquad\qquad\qquad}$ 6 *factors*
So we have	x^6
Notice that 6 is the product of the exponents, 2 and 3.	$(x^2)^3$ is $x^{2\cdot3}$ or x^6

We write:

$$\left(x^2\right)^3$$
$$x^{2\cdot3}$$
$$x^6$$

We multiplied the exponents. This leads to the **Power Property for Exponents.**

Power Property for Exponents

If a is a real number, and m and n are whole numbers, then

$$(a^m)^n = a^{m\cdot n}$$

To raise a power to a power, multiply the exponents.

An example with numbers helps to verify this property.

$$\left(3^2\right)^3 \overset{?}{=} 3^{2\cdot3}$$
$$(9)^3 \overset{?}{=} 3^6$$
$$729 = 729 \checkmark$$

EXAMPLE 6.22

Simplify: ⓐ $(y^5)^9$ ⓑ $(4^4)^7$.

✓ **Solution**

ⓐ

	$(y^5)^9$
Use the power property, $(a^m)^n = a^{m\cdot n}$.	$y^{5\cdot9}$
Simplify.	y^{45}

ⓑ

	$(4^4)^7$
Use the power property.	$4^{4 \cdot 7}$
Simplify.	4^{28}

> **TRY IT :: 6.43**　　Simplify: ⓐ $(b^7)^5$ ⓑ $(5^4)^3$.

> **TRY IT :: 6.44**　　Simplify: ⓐ $(z^6)^9$ ⓑ $(3^7)^7$.

Simplify Expressions Using the Product to a Power Property

We will now look at an expression containing a product that is raised to a power. Can you find this pattern?

	$(2x)^3$
What does this mean?	$2x \cdot 2x \cdot 2x$
We group the like factors together.	$2 \cdot 2 \cdot 2 \cdot x \cdot x \cdot x$
How many factors of 2 and of x?	$2^3 \cdot x^3$

Notice that each factor was raised to the power and $(2x)^3$ is $2^3 \cdot x^3$.

We write:　　　　　　$(2x)^3$

$2^3 \cdot x^3$

The exponent applies to each of the factors! This leads to the **Product to a Power Property for Exponents.**

Product to a Power Property for Exponents

If a and b are real numbers and m is a whole number, then

$$(ab)^m = a^m b^m$$

To raise a product to a power, raise each factor to that power.

An example with numbers helps to verify this property:

$$(2 \cdot 3)^2 \overset{?}{=} 2^2 \cdot 3^2$$
$$6^2 \overset{?}{=} 4 \cdot 9$$
$$36 = 36 \checkmark$$

EXAMPLE 6.23

Simplify: ⓐ $(-9d)^2$ ⓑ $(3mn)^3$.

⊘ **Solution**

ⓐ

	$(-9d)^2$
Use Power of a Product Property, $(ab)^m = a^m b^m$.	$(-9)^2 d^2$
Simplify.	$81d^2$

ⓑ

	$(3mn)^3$
Use Power of a Product Property, $(ab)^m = a^m b^m$.	$(3)^3 m^3 n^3$
Simplify.	$27m^3 n^3$

> **TRY IT : : 6.45** Simplify: ⓐ $(-12y)^2$ ⓑ $(2wx)^5$.

> **TRY IT : : 6.46** Simplify: ⓐ $(5wx)^3$ ⓑ $(-3y)^3$.

Simplify Expressions by Applying Several Properties

We now have three properties for multiplying expressions with exponents. Let's summarize them and then we'll do some examples that use more than one of the properties.

Properties of Exponents

If a and b are real numbers, and m and n are whole numbers, then

Product Property	$a^m \cdot a^n = a^{m+n}$
Power Property	$(a^m)^n = a^{m \cdot n}$
Product to a Power	$(ab)^m = a^m b^m$

All exponent properties hold true for any real numbers m and n. Right now, we only use whole number exponents.

EXAMPLE 6.24

Simplify: ⓐ $(y^3)^6 (y^5)^4$ ⓑ $\left(-6x^4 y^5\right)^2$.

⊘ **Solution**

ⓐ

	$(y^3)^6 (y^5)^4$
Use the Power Property.	$y^{15} \cdot y^{20}$
Add the exponents.	y^{35}

ⓑ

	$\left(-6x^4 y^5\right)^2$
Use the Product to a Power Property.	$(-6)^2 (x^4)^2 (y^5)^2$
Use the Power Property.	$(-6)^2 (x^8)(y^{10})$
Simplify.	$36x^8 y^{10}$

> **TRY IT : : 6.47** Simplify: ⓐ $(a^4)^5 (a^7)^4$ ⓑ $\left(-2c^4 d^2\right)^3$.

> **TRY IT : : 6.48** Simplify: ⓐ $\left(-3x^6 y^7\right)^4$ ⓑ $(q^4)^5 (q^3)^3$.

EXAMPLE 6.25

Simplify: ⓐ $(5m)^2(3m^3)$ ⓑ $(3x^2y)^4(2xy^2)^3$.

✓ **Solution**

ⓐ

$$(5m)^2(3m^3)$$

Raise $5m$ to the second power.	$5^2m^2 \cdot 3m^3$
Simplify.	$25m^2 \cdot 3m^3$
Use the Commutative Property.	$25 \cdot 3 \cdot m^2 \cdot m^3$
Multiply the constants and add the exponents.	$75m^5$

ⓑ

$$(3x^2y)^4(2xy^2)^3$$

Use the Product to a Power Property.	$(3^4x^8y^4)(2^3x^3y^6)$
Simplify.	$(81x^8y^4)(8x^3y^6)$
Use the Commutative Property.	$81 \cdot 8 \cdot x^8 \cdot x^3 \cdot y^4 \cdot y^6$
Multiply the constants and add the exponents.	$648x^{11}y^{10}$

> **TRY IT :: 6.49**   Simplify: ⓐ $(5n)^2(3n^{10})$ ⓑ $(c^4d^2)^5(3cd^5)^4$.

> **TRY IT :: 6.50**   Simplify: ⓐ $(a^3b^2)^6(4ab^3)^4$ ⓑ $(2x)^3(5x^7)$.

Multiply Monomials

Since a monomial is an algebraic expression, we can use the properties of exponents to multiply monomials.

EXAMPLE 6.26

Multiply: $(3x^2)(-4x^3)$.

✓ **Solution**

$$(3x^2)(-4x^3)$$

Use the Commutative Property to rearrange the terms.	$3 \cdot (-4) \cdot x^2 \cdot x^3$
Multiply.	$-12x^5$

> **TRY IT :: 6.51**   Multiply: $(5y^7)(-7y^4)$.

> **TRY IT :: 6.52**   Multiply: $(-6b^4)(-9b^5)$.

EXAMPLE 6.27

Multiply: $\left(\frac{5}{6}x^3y\right)(12xy^2)$.

✓ Solution

$$\left(\tfrac{5}{6}x^3 y\right)\left(12xy^2\right)$$

Use the Commutative Property to rearrange the terms. $\tfrac{5}{6} \cdot 12 \cdot x^3 \cdot x \cdot y \cdot y^2$

Multiply. $10x^4 y^3$

> **TRY IT : : 6.53**   Multiply: $\left(\tfrac{2}{5}a^4 b^3\right)\left(15ab^3\right)$.

> **TRY IT : : 6.54**   Multiply: $\left(\tfrac{2}{3}r^5 s\right)\left(12r^6 s^7\right)$.

▶ **MEDIA : :**

Access these online resources for additional instruction and practice with using multiplication properties of exponents:

- **Multiplication Properties of Exponents (https://openstax.org/l/25MultiPropExp)**

6.2 EXERCISES

Practice Makes Perfect

Simplify Expressions with Exponents

In the following exercises, simplify each expression with exponents.

88.
ⓐ 3^5
ⓑ 9^1
ⓒ $\left(\frac{1}{3}\right)^2$
ⓓ $(0.2)^4$

89.
ⓐ 10^4
ⓑ 17^1
ⓒ $\left(\frac{2}{9}\right)^2$
ⓓ $(0.5)^3$

90.
ⓐ 2^6
ⓑ 14^1
ⓒ $\left(\frac{2}{5}\right)^3$
ⓓ $(0.7)^2$

91.
ⓐ 8^3
ⓑ 8^1
ⓒ $\left(\frac{3}{4}\right)^3$
ⓓ $(0.4)^3$

92.
ⓐ $(-6)^4$
ⓑ -6^4

93.
ⓐ $(-2)^6$
ⓑ -2^6

94.
ⓐ $-\left(\frac{1}{4}\right)^4$
ⓑ $\left(-\frac{1}{4}\right)^4$

95.
ⓐ $-\left(\frac{2}{3}\right)^2$
ⓑ $\left(-\frac{2}{3}\right)^2$

96.
ⓐ -0.5^2
ⓑ $(-0.5)^2$

97.
ⓐ -0.1^4
ⓑ $(-0.1)^4$

Simplify Expressions Using the Product Property for Exponents

In the following exercises, simplify each expression using the Product Property for Exponents.

98. $d^3 \cdot d^6$

99. $x^4 \cdot x^2$

100. $n^{19} \cdot n^{12}$

101. $q^{27} \cdot q^{15}$

102. ⓐ $4^5 \cdot 4^9$ ⓑ $8^9 \cdot 8$

103. ⓐ $3^{10} \cdot 3^6$ ⓑ $5 \cdot 5^4$

104. ⓐ $y \cdot y^3$ ⓑ $z^{25} \cdot z^8$

105. ⓐ $w^5 \cdot w$ ⓑ $u^{41} \cdot u^{53}$

106. $w \cdot w^2 \cdot w^3$

107. $y \cdot y^3 \cdot y^5$

108. $a^4 \cdot a^3 \cdot a^9$

109. $c^5 \cdot c^{11} \cdot c^2$

110. $m^x \cdot m^3$

111. $n^y \cdot n^2$

112. $y^a \cdot y^b$

113. $x^p \cdot x^q$

Simplify Expressions Using the Power Property for Exponents

In the following exercises, simplify each expression using the Power Property for Exponents.

114. ⓐ $(m^4)^2$ ⓑ $(10^3)^6$

115. ⓐ $(b^2)^7$ ⓑ $(3^8)^2$

116. ⓐ $(y^3)^x$ ⓑ $(5^x)^y$

117. ⓐ $(x^2)^y$ ⓑ $(7^a)^b$

Simplify Expressions Using the Product to a Power Property

In the following exercises, simplify each expression using the Product to a Power Property.

118. ⓐ $(6a)^2$ ⓑ $(3xy)^2$

119. ⓐ $(5x)^2$ ⓑ $(4ab)^2$

120. ⓐ $(-4m)^3$ ⓑ $(5ab)^3$

121. ⓐ $(-7n)^3$ ⓑ $(3xyz)^4$

Simplify Expressions by Applying Several Properties

In the following exercises, simplify each expression.

122.
ⓐ $(y^2)^4 \cdot (y^3)^2$
ⓑ $(10a^2 b)^3$

123.
ⓐ $(w^4)^3 \cdot (w^5)^2$
ⓑ $(2xy^4)^5$

124.
ⓐ $(-2r^3 s^2)^4$
ⓑ $(m^5)^3 \cdot (m^9)^4$

125.
ⓐ $(-10q^2 p^4)^3$
ⓑ $(n^3)^{10} \cdot (n^5)^2$

126.
ⓐ $(3x)^2(5x)$
ⓑ $(5t^2)^3 (3t)^2$

127.
ⓐ $(2y)^3(6y)$
ⓑ $(10k^4)^3 (5k^6)^2$

128.
ⓐ $(5a)^2 (2a)^3$
ⓑ $(\frac{1}{2}y^2)^3 (\frac{2}{3}y)^2$

129.
ⓐ $(4b)^2 (3b)^3$
ⓑ $(\frac{1}{2}j^2)^5 (\frac{2}{5}j^3)^2$

130.
ⓐ $(\frac{2}{5}x^2 y)^3$
ⓑ $(\frac{8}{9}xy^4)^2$

131.
ⓐ $(2r^2)^3 (4r)^2$
ⓑ $(3x^3)^3 (x^5)^4$

132.
ⓐ $(m^2 n)^2 (2mn^5)^4$
ⓑ $(3pq^4)^2 (6p^6 q)^2$

Multiply Monomials

In the following exercises, multiply the monomials.

133. $(6y^7)(-3y^4)$

134. $(-10x^5)(-3x^3)$

135. $(-8u^6)(-9u)$

136. $(-6c^4)(-12c)$

137. $(\frac{1}{5}f^8)(20f^3)$

138. $(\frac{1}{4}d^5)(36d^2)$

139. $(4a^3 b)(9a^2 b^6)$

140. $(6m^4 n^3)(7mn^5)$

141. $(\frac{4}{7}rs^2)(14rs^3)$

142. $(\frac{5}{8}x^3 y)(24x^5 y)$

143. $(\frac{2}{3}x^2 y)(\frac{3}{4}xy^2)$

144. $(\frac{3}{5}m^3 n^2)(\frac{5}{9}m^2 n^3)$

Mixed Practice

In the following exercises, simplify each expression.

145. $\left(x^2\right)^4 \cdot \left(x^3\right)^2$

146. $\left(y^4\right)^3 \cdot \left(y^5\right)^2$

147. $\left(a^2\right)^6 \cdot \left(a^3\right)^8$

148. $\left(b^7\right)^5 \cdot \left(b^2\right)^6$

149. $(2m^6)^3$

150. $(3y^2)^4$

151. $(10x^2y)^3$

152. $(2mn^4)^5$

153. $(-2a^3b^2)^4$

154. $(-10u^2v^4)^3$

155. $\left(\frac{2}{3}x^2y\right)^3$

156. $\left(\frac{7}{9}pq^4\right)^2$

157. $(8a^3)^2(2a)^4$

158. $(5r^2)^3(3r)^2$

159. $(10p^4)^3(5p^6)^2$

160. $(4x^3)^3(2x^5)^4$

161. $\left(\frac{1}{2}x^2y^3\right)^4\left(4x^5y^3\right)^2$

162. $\left(\frac{1}{3}m^3n^2\right)^4\left(9m^8n^3\right)^2$

163. $(3m^2n)^2(2mn^5)^4$

164. $(2pq^4)^3(5p^6q)^2$

Everyday Math

165. Email Kate emails a flyer to ten of her friends and tells them to forward it to ten of their friends, who forward it to ten of their friends, and so on. The number of people who receive the email on the second round is 10^2, on the third round is 10^3, as shown in the table below. How many people will receive the email on the sixth round? Simplify the expression to show the number of people who receive the email.

Round	Number of people
1	10
2	10^2
3	10^3
...	...
6	?

166. Salary Jamal's boss gives him a 3% raise every year on his birthday. This means that each year, Jamal's salary is 1.03 times his last year's salary. If his original salary was $35,000, his salary after 1 year was $35,000(1.03)$, after 2 years was $35,000(1.03)^2$, after 3 years was $35,000(1.03)^3$, as shown in the table below. What will Jamal's salary be after 10 years? Simplify the expression, to show Jamal's salary in dollars.

Year	Salary
1	$35,000(1.03)$
2	$35,000(1.03)^2$
3	$35,000(1.03)^3$
...	...
10	?

167. Clearance A department store is clearing out merchandise in order to make room for new inventory. The plan is to mark down items by 30% each week. This means that each week the cost of an item is 70% of the previous week's cost. If the original cost of a sofa was $1,000, the cost for the first week would be $1,000(0.70) and the cost of the item during the second week would be $1,000(0.70)^2$. Complete the table shown below. What will be the cost of the sofa during the fifth week? Simplify the expression, to show the cost in dollars.

Week	Cost
1	$1,000(0.70)
2	$1,000(0.70)^2$
3	
...	...
8	?

168. Depreciation Once a new car is driven away from the dealer, it begins to lose value. Each year, a car loses 10% of its value. This means that each year the value of a car is 90% of the previous year's value. If a new car was purchased for $20,000, the value at the end of the first year would be $20,000(0.90) and the value of the car after the end of the second year would be $20,000(0.90)^2$. Complete the table shown below. What will be the value of the car at the end of the eighth year? Simplify the expression, to show the value in dollars.

Week	Cost
1	$20,000(0.90)
2	$20,000(0.90)^2$
3	
4	...
5	?

Writing Exercises

169. Use the Product Property for Exponents to explain why $x \cdot x = x^2$.

170. Explain why $-5^3 = (-5)^3$ but $-5^4 \neq (-5)^4$.

171. Jorge thinks $\left(\frac{1}{2}\right)^2$ is 1. What is wrong with his reasoning?

172. Explain why $x^3 \cdot x^5$ is x^8, and not x^{15}.

Self Check

ⓐ After completing the exercises, use this checklist to evaluate your mastery of the objectives of this section.

I can...	Confidently	With some help	No-I don't get it!
simplify expressions with exponents.			
simplify expressions using the Product Property for Exponents.			
simplify expressions using the Power Property for Exponents.			
simplify expressions using the Product to a Power Property.			
simplify expressions by applying several properties.			
multiply monomials.			

ⓑ After reviewing this checklist, what will you do to become confident for all goals?

6.3 Multiply Polynomials

Learning Objectives

By the end of this section, you will be able to:

> Multiply a polynomial by a monomial
> Multiply a binomial by a binomial
> Multiply a trinomial by a binomial

Be Prepared!

Before you get started, take this readiness quiz.

1. Distribute: $2(x + 3)$.
   If you missed this problem, review **Example 1.132**.

2. Combine like terms: $x^2 + 9x + 7x + 63$.
   If you missed this problem, review **Example 1.24**.

Multiply a Polynomial by a Monomial

We have used the Distributive Property to simplify expressions like $2(x - 3)$. You multiplied both terms in the parentheses, x and 3, by 2, to get $2x - 6$. With this chapter's new vocabulary, you can say you were multiplying a binomial, $x - 3$, by a monomial, 2.

Multiplying a binomial by a monomial is nothing new for you! Here's an example:

EXAMPLE 6.28

Multiply: $4(x + 3)$.

⊘ Solution

$$4(x + 3)$$

Distribute.	$4 \cdot x + 4 \cdot 3$
Simplify.	$4x + 12$

>	**TRY IT : :** 6.55	Multiply: $5(x + 7)$.

>	**TRY IT : :** 6.56	Multiply: $3(y + 13)$.

EXAMPLE 6.29

Multiply: $y(y - 2)$.

⊘ Solution

$$y(y - 2)$$

Distribute.	$y \cdot y - y \cdot 2$
Simplify.	$y^2 - 2y$

> **TRY IT ::** 6.57   Multiply: $x(x - 7)$.

> **TRY IT ::** 6.58   Multiply: $d(d - 11)$.

EXAMPLE 6.30

Multiply: $7x(2x + y)$.

⊘ **Solution**

$$7x(2x + y)$$

Distribute.	$7x \cdot 2x + 7x \cdot y$
Simplify.	$14x^2 + 7xy$

> **TRY IT ::** 6.59   Multiply: $5x(x + 4y)$.

> **TRY IT ::** 6.60   Multiply: $2p(6p + r)$.

EXAMPLE 6.31

Multiply: $-2y\left(4y^2 + 3y - 5\right)$.

⊘ **Solution**

$$-2y(4y^2 + 3y - 5)$$

Distribute.	$-2y \cdot 4y^2 + (-2y) \cdot 3y - (-2y) \cdot 5$
Simplify.	$-8y^3 - 6y^2 + 10y$

> **TRY IT ::** 6.61   Multiply: $-3y\left(5y^2 + 8y - 7\right)$.

> **TRY IT ::** 6.62   Multiply: $4x^2(2x^2 - 3x + 5)$.

EXAMPLE 6.32

Multiply: $2x^3(x^2 - 8x + 1)$.

Solution

$$2x^3(x^2 - 8x + 1)$$

Distribute.	$2x^3 \cdot x^2 + (2x^3) \cdot (-8x) + (2x^3) \cdot 1$
Simplify.	$2x^5 - 16x^4 + 2x^3$

> **TRY IT :: 6.63**   Multiply: $4x(3x^2 - 5x + 3)$.

> **TRY IT :: 6.64**   Multiply: $-6a^3(3a^2 - 2a + 6)$.

EXAMPLE 6.33

Multiply: $(x + 3)p$.

Solution

The monomial is the second factor.	$(x + 3)p$
Distribute.	$x \cdot p + 3 \cdot p$
Simplify.	$xp + 3p$

> **TRY IT :: 6.65**   Multiply: $(x + 8)p$.

> **TRY IT :: 6.66**   Multiply: $(a + 4)p$.

Multiply a Binomial by a Binomial

Just like there are different ways to represent multiplication of numbers, there are several methods that can be used to multiply a binomial times a binomial. We will start by using the Distributive Property.

Multiply a Binomial by a Binomial Using the Distributive Property

Look at Example 6.33, where we multiplied a binomial by a monomial.

$$(x + 3)p$$

We distributed the p to get:	$xp + 3p$
What if we have $(x + 7)$ instead of p?	$(x + 3)(x + 7)$
Distribute $(x + 7)$.	$x(x + 7) + 3(x + 7)$
Distribute again.	$x^2 + 7x + 3x + 21$
Combine like terms.	$x^2 + 10x + 21$

Notice that before combining like terms, you had four terms. You multiplied the two terms of the first binomial by the two terms of the second binomial—four multiplications.

EXAMPLE 6.34

Multiply: $(y + 5)(y + 8)$.

⊘ Solution

$$(y + 5)(y + 8)$$

Distribute $(y + 8)$.	$y(y + 8) + 5(y + 8)$
Distribute again	$y^2 + 8y + 5y + 40$
Combine like terms.	$y^2 + 13y + 40$

> **TRY IT : :** 6.67   Multiply: $(x + 8)(x + 9)$.

> **TRY IT : :** 6.68   Multiply: $(5x + 9)(4x + 3)$.

EXAMPLE 6.35

Multiply: $(2y + 5)(3y + 4)$.

✓ Solution

$$(2y + 5)(3y + 4)$$

Distribute $(3y + 4)$.	$2y(3y + 4) + 5(3y + 4)$
Distribute again	$6y^2 + 8y + 15y + 20$
Combine like terms.	$6y^2 + 23y + 20$

> **TRY IT :: 6.69**   Multiply: $(3b + 5)(4b + 6)$.

> **TRY IT :: 6.70**   Multiply: $(a + 10)(a + 7)$.

EXAMPLE 6.36

Multiply: $(4y + 3)(2y - 5)$.

✓ Solution

$$(4y + 3)(2y - 5)$$

Distribute.	$4y(2y - 5) + 3(2y - 5)$
Distribute again.	$8y^2 - 20y + 6y - 15$
Combine like terms.	$8y^2 - 14y - 15$

> **TRY IT :: 6.71**   Multiply: $(5y + 2)(6y - 3)$.

> **TRY IT :: 6.72**   Multiply: $(3c + 4)(5c - 2)$.

EXAMPLE 6.37

Multiply: $(x + 2)(x - y)$.

✓ Solution

$$(x - 2)(x - y)$$

Distribute.	$x(x - y) - 2(x - y)$
Distribute again.	$x^2 - xy - 2x + 2y$
There are no like terms to combine.	

> **TRY IT :: 6.73**   Multiply: $(a + 7)(a - b)$.

> **TRY IT :: 6.74**   Multiply: $(x + 5)(x - y)$.

Multiply a Binomial by a Binomial Using the FOIL Method

Remember that when you multiply a binomial by a binomial you get four terms. Sometimes you can combine like terms to get a trinomial, but sometimes, like in Example 6.37, there are no like terms to combine.

Let's look at the last example again and pay particular attention to how we got the four terms.

$$(x - 2)(x - y)$$
$$x^2 - xy - 2x + 2y$$

Where did the first term, x^2, come from?

It is the product of x and x, the *first* terms in $(x - 2)$ and $(x - y)$.

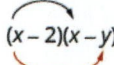

$(x - 2)(x - y)$

First

The next term, $-xy$, is the product of x and $-y$, the two *outer* terms.

$(x - 2)(x - y)$

Outer

The third term, $-2x$, is the product of -2 and x, the two *inner* terms.

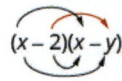

$(x - 2)(x - y)$

Inner

And the last term, $+2y$, came from multiplying the two *last* terms, -2 and $-y$.

$(x - 2)(x - y)$

Last

We abbreviate "First, Outer, Inner, Last" as FOIL. The letters stand for 'First, **O**uter, **I**nner, **L**ast'. The word FOIL is easy to remember and ensures we find all four products.

$$(x - 2)(x - y)$$
$$\underset{F}{x^2} - \underset{O}{xy} - \underset{I}{2x} + \underset{L}{2y}$$

Let's look at $(x + 3)(x + 7)$.

Distibutive Property	FOIL
$(x + 3)(x + 7)$	$(x + 3)(x + 7)$
$x(x + 7) + 3(x + 7)$	
$\underset{F}{x^2} + \underset{O}{7x} + \underset{I}{3x} + \underset{L}{21}$	$\underset{F}{x^2} + \underset{O}{7x} + \underset{I}{3x} + \underset{L}{21}$
$x^2 + 10x + 21$	$x^2 + 10x + 21$

Notice how the terms in third line fit the FOIL pattern.

Now we will do an example where we use the FOIL pattern to multiply two binomials.

EXAMPLE 6.38   HOW TO MULTIPLY A BINOMIAL BY A BINOMIAL USING THE FOIL METHOD

Multiply using the FOIL method: $(x + 5)(x + 9)$.

 Solution

Step 1. Multiply the *First* terms.	$(x + 5)(x + 9)$	
	$(x + 5)(x + 9)$	$x^2 + \_\_ + \_\_ + \_\_$ F O I L
Step 2. Multiply the *Outer* terms.	$(x + 5)(x + 9)$	$x^2 + 9x + \_\_ + \_\_$ F O I L
Step 3. Multiply the *Inner* terms.	$(x + 5)(x + 9)$	$x^2 + 9x + 5x + \_\_$ F O I L
Step 4. Multiply the *Last* terms.	$(x + 5)(x + 9)$	$x^2 + 9x + 5x + 45$ F O I L
Step 5. Combine like terms, when possible.		$x^2 + 14x + 45$

> **TRY IT :: 6.75**   Multiply using the FOIL method: $(x + 6)(x + 8)$.

> **TRY IT :: 6.76**   Multiply using the FOIL method: $(y + 17)(y + 3)$.

We summarize the steps of the FOIL method below. The FOIL method only applies to multiplying binomials, not other polynomials!

HOW TO :: MULTIPLY TWO BINOMIALS USING THE FOIL METHOD

Step 1. Multiply the *First* terms.

Step 2. Multiply the *Outer* terms.

Step 3. Multiply the *Inner* terms.

Step 4. Multiply the *Last* terms.

Step 5. Combine like terms, when possible.

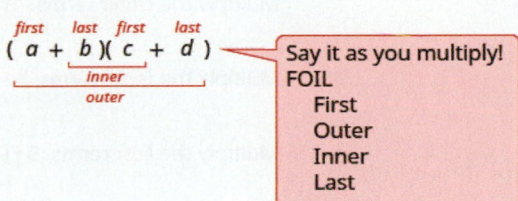

When you multiply by the FOIL method, drawing the lines will help your brain focus on the pattern and make it easier to apply.

EXAMPLE 6.39

Multiply: $(y - 7)(y + 4)$.

⊘ **Solution**

$$(y - 7)(y + 4)$$

Multiply the *First* terms.	$(y - 7)(y + 4)$	$y^2 + \_ + \_ + \_$ F O I L
Multiply the *Outer* terms.	$(y - 7)(y + 4)$	$y^2 + 4y + \_ + \_$ F O I L
Multiply the *Inner* terms.	$(y - 7)(y + 4)$	$y^2 + 4y - 7y + \_$ F O I L
Multiply the *Last* terms.	$(y - 7)(y + 4)$	$y^2 + 4y - 7y -28$ F O I L
Combine like terms.		$y^2 - 3y - 28$

> **TRY IT : : 6.77**   Multiply: $(x - 7)(x + 5)$.

> **TRY IT : : 6.78**   Multiply: $(b - 3)(b + 6)$.

EXAMPLE 6.40

Multiply: $(4x + 3)(2x - 5)$.

⊘ **Solution**

$$(4x - 3)(2x - 5)$$

$$(4x + 3)(2x - 5)$$

Multiply the *First* terms, $4x \cdot 2x$.		$8x^2 + \_ + \_ + \_$ F O I L
Multiply the *Outer* terms, $4x \cdot (-5)$.		$8x^2 - 20x + \_ + \_$ F O I L
Multiply the *Inner* terms, $3 \cdot 2x$.		$8x^2 - 2x + 6x + \_$ F O I L
Multiply the *Last* terms, $3 \cdot (-5)$.		$8x^2 - 20x + 6x -15$ F O I L
Combine like terms.		$8x^2 - 14x - 15$

> **TRY IT : : 6.79**   Multiply: $(3x + 7)(5x - 2)$.

> **TRY IT : : 6.80**   Multiply: $(4y + 5)(4y - 10)$.

The final products in the last four examples were trinomials because we could combine the two middle terms. This is not always the case.

EXAMPLE 6.41

Multiply: $(3x - y)(2x - 5)$.

⊘ Solution

	$(3x - y)(2x - 5)$
	$(3x - y)(2x - 5)$
Multiply the *First*.	$6x^2 + \_ + \_ + \_$ F O I L
Multiply the *Outer*.	$6x^2 - 15x + \_ + \_$ F O I L
Multiply the *Inner*.	$6x^2 - 15x - 2xy + \_$ F O I L
Multiply the *Last*.	$6x^2 - 15x - 2xy + 5y$ F O I L
Combine like terms—there are none.	$6x^2 - 15x - 2xy + 5y$

> **TRY IT :: 6.81**   Multiply: $(10c - d)(c - 6)$.

> **TRY IT :: 6.82**   Multiply: $(7x - y)(2x - 5)$.

Be careful of the exponents in the next example.

EXAMPLE 6.42

Multiply: $\left(n^2 + 4\right)(n - 1)$.

⊘ Solution

	$(n^2 + 4)(n - 1)$
	$(n^2 + 4)(n - 1)$
Multiply the *First*.	$n^3 + \_ + \_ + \_$ F O I L
Multiply the *Outer*.	$n^3 - n^2 + \_ + \_$ F O I L
Multiply the *Inner*.	$n^3 - n^2 + 4n + \_$ F O I L
Multiply the *Last*.	$n^3 - n^2 + 4n - 4$ F O I L
Combine like terms—there are none.	$n^3 - n^2 + 4n - 4$

> **TRY IT :: 6.83**   Multiply: $\left(x^2 + 6\right)(x - 8)$.

> **TRY IT :: 6.84**   Multiply: $\left(y^2 + 7\right)(y - 9)$.

EXAMPLE 6.43

Multiply: $(3pq + 5)(6pq - 11)$.

⊘ **Solution**

	$(3pq + 5)(6pq - 11)$	
Multiply the *First*.	$\underset{F}{18p^2q^2} + \underset{O}{\_} + \underset{I}{\_} + \underset{L}{\_}$	
Multiply the *Outer*.	$\underset{F}{18p^2q^2} - \underset{O}{33pq} + \underset{I}{\_} + \underset{L}{\_}$	$(3pq + 5)(6pq - 11)$
Multiply the *Inner*.	$\underset{F}{18p^2q^2} - \underset{O}{33pq} + \underset{I}{30pq} + \underset{L}{\_}$	
Multiply the *Last*.	$\underset{F}{18p^2q^2} - \underset{O}{33pq} + \underset{I}{30pq} - \underset{L}{55}$	
Combine like terms—there are none.	$18p^2q^2 - 3pq - 55$	

> **TRY IT :: 6.85**   Multiply: $(2ab + 5)(4ab - 4)$.

> **TRY IT :: 6.86**   Multiply: $(2xy + 3)(4xy - 5)$.

Multiply a Binomial by a Binomial Using the Vertical Method

The FOIL method is usually the quickest method for multiplying two binomials, but it *only* works for binomials. You can use the Distributive Property to find the product of any two polynomials. Another method that works for all polynomials is the Vertical Method. It is very much like the method you use to multiply whole numbers. Look carefully at this example of multiplying two-digit numbers.

$$\begin{array}{r} 23 \\ \times 46 \\ \hline 138 \\ 92 \\ \hline 1058 \end{array}$$

138 partial product   Start by multiplying 23 by 6 to get 138.
92   partial product   Next, multiply 23 by 4, lining up the partial product in the correct columns.
1058 product   Last you add the partial products.

Now we'll apply this same method to multiply two binomials.

EXAMPLE 6.44

Multiply using the Vertical Method: $(3y - 1)(2y - 6)$.

⊘ **Solution**

It does not matter which binomial goes on the top.

Multiply $3y - 1$ by -6. $3y - 1$
Multiply $3y - 1$ by $2y$. $\times\ 2y - 6$
                                  ─────────────
                                  $-18y + 6$ partial product
                                  $6y^2 - 2y$ partial product
                                  ─────────────
Add like terms. $6y^2 - 20y + 6$ product

Notice the partial products are the same as the terms in the FOIL method.

$$
\begin{array}{r}
3y - 1 \\
\times\ 2y - 6 \\
\hline
-18y + 6 \\
6y^2 - 2y \\
\hline
6y^2 - 20x + 6
\end{array}
$$

$$(3y - 1)(2y - 6)$$
$$6y^2 - 2y - 18y + 6$$
$$6y^2 - 20y + 6$$

> **TRY IT : : 6.87**   Multiply using the Vertical Method: $(5m - 7)(3m - 6)$.

> **TRY IT : : 6.88**   Multiply using the Vertical Method: $(6b - 5)(7b - 3)$.

We have now used three methods for multiplying binomials. Be sure to practice each method, and try to decide which one you prefer. The methods are listed here all together, to help you remember them.

Multiplying Two Binomials

To multiply binomials, use the:

- Distributive Property
- FOIL Method
- Vertical Method

Remember, FOIL only works when multiplying two binomials.

Multiply a Trinomial by a Binomial

We have multiplied monomials by monomials, monomials by polynomials, and binomials by binomials. Now we're ready to multiply a trinomial by a binomial. Remember, FOIL will not work in this case, but we can use either the Distributive Property or the Vertical Method. We first look at an example using the Distributive Property.

EXAMPLE 6.45

Multiply using the Distributive Property: $(b + 3)(2b^2 - 5b + 8)$.

⊘ **Solution**

Distribute.	$b(2b^2 - 5b + 8) + 3(2b^2 - 5b + 8)$
Multiply.	$2b^3 - 5b^2 + 8b + 6b^2 - 15b + 24$
Combine like terms.	$2b^3 + b^2 - 7b + 24$

> **TRY IT : : 6.89**   Multiply using the Distributive Property: $(y - 3)(y^2 - 5y + 2)$.

> **TRY IT : : 6.90**   Multiply using the Distributive Property: $(x + 4)(2x^2 - 3x + 5)$.

Now let's do this same multiplication using the Vertical Method.

EXAMPLE 6.46

Multiply using the Vertical Method: $(b + 3)(2b^2 - 5b + 8)$.

⊘ **Solution**

It is easier to put the polynomial with fewer terms on the bottom because we get fewer partial products this way.

$$2b^2 - 5b + 8$$
$$\underline{\times \qquad b + 3}$$

Multiply $(2b^2 - 5b + 8)$ by 3. $6b^2 - 15b + 24$

$$\underline{2b^3 - 5b^2 + 8b}$$

Multiply $(2b^2 - 5b + 8)$ by b. $2b^3 + b^2 - 7b + 24$

Add like terms.

> **TRY IT : : 6.91**   Multiply using the Vertical Method: $(y - 3)(y^2 - 5y + 2)$.

> **TRY IT : : 6.92**   Multiply using the Vertical Method: $(x + 4)(2x^2 - 3x + 5)$.

We have now seen two methods you can use to multiply a trinomial by a binomial. After you practice each method, you'll probably find you prefer one way over the other. We list both methods are listed here, for easy reference.

Multiplying a Trinomial by a Binomial

To multiply a trinomial by a binomial, use the:
- Distributive Property
- Vertical Method

▶ **MEDIA : :**

Access these online resources for additional instruction and practice with multiplying polynomials:
- **Multiplying Exponents 1 (https://openstax.org/l/25MultiplyExp1)**
- **Multiplying Exponents 2 (https://openstax.org/l/25MultiplyExp2)**
- **Multiplying Exponents 3 (https://openstax.org/l/25MultiplyExp3)**

6.3 EXERCISES
Practice Makes Perfect

Multiply a Polynomial by a Monomial
In the following exercises, multiply.

173. $4(w + 10)$

174. $6(b + 8)$

175. $-3(a + 7)$

176. $-5(p + 9)$

177. $2(x - 7)$

178. $7(y - 4)$

179. $-3(k - 4)$

180. $-8(j - 5)$

181. $q(q + 5)$

182. $k(k + 7)$

183. $-b(b + 9)$

184. $-y(y + 3)$

185. $-x(x - 10)$

186. $-p(p - 15)$

187. $6r(4r + s)$

188. $5c(9c + d)$

189. $12x(x - 10)$

190. $9m(m - 11)$

191. $-9a(3a + 5)$

192. $-4p(2p + 7)$

193. $3(p^2 + 10p + 25)$

194. $6(y^2 + 8y + 16)$

195. $-8x(x^2 + 2x - 15)$

196. $-5t(t^2 + 3t - 18)$

197. $5q^3\left(q^3 - 2q + 6\right)$

198. $4x^3\left(x^4 - 3x + 7\right)$

199. $-8y(y^2 + 2y - 15)$

200. $-5m(m^2 + 3m - 18)$

201. $5q^3(q^2 - 2q + 6)$

202. $9r^3(r^2 - 3r + 5)$

203. $-4z^2(3z^2 + 12z - 1)$

204. $-3x^2(7x^2 + 10x - 1)$

205. $(2m - 9)m$

206. $(8j - 1)j$

207. $(w - 6) \cdot 8$

208. $(k - 4) \cdot 5$

209. $4(x + 10)$

210. $6(y + 8)$

211. $15(r - 24)$

212. $12(v - 30)$

213. $-3(m + 11)$

214. $-4(p + 15)$

215. $-8(z - 5)$

216. $-3(x - 9)$

217. $u(u + 5)$

218. $q(q + 7)$

219. $n(n^2 - 3n)$

220. $s(s^2 - 6s)$

221. $6x(4x + y)$

222. $5a(9a + b)$

223. $5p(11p - 5q)$

224. $12u(3u - 4v)$

225. $3(v^2 + 10v + 25)$

226. $6(x^2 + 8x + 16)$

227. $2n(4n^2 - 4n + 1)$

228. $3r(2r^2 - 6r + 2)$

229. $-8y(y^2 + 2y - 15)$

230. $-5m(m^2 + 3m - 18)$

231. $5q^3(q^2 - 2q + 6)$

232. $9r^3(r^2 - 3r + 5)$

233. $-4z^2(3z^2 + 12z - 1)$ **234.** $-3x^2(7x^2 + 10x - 1)$ **235.** $(2y - 9)y$

236. $(8b - 1)b$

Multiply a Binomial by a Binomial

In the following exercises, multiply the following binomials using: ⓐ the Distributive Property ⓑ the FOIL method ⓒ the Vertical Method.

237. $(w + 5)(w + 7)$ **238.** $(y + 9)(y + 3)$ **239.** $(p + 11)(p - 4)$

240. $(q + 4)(q - 8)$

In the following exercises, multiply the binomials. Use any method.

241. $(x + 8)(x + 3)$ **242.** $(y + 7)(y + 4)$ **243.** $(y - 6)(y - 2)$

244. $(x - 7)(x - 2)$ **245.** $(w - 4)(w + 7)$ **246.** $(q - 5)(q + 8)$

247. $(p + 12)(p - 5)$ **248.** $(m + 11)(m - 4)$ **249.** $(6p + 5)(p + 1)$

250. $(7m + 1)(m + 3)$ **251.** $(2t - 9)(10t + 1)$ **252.** $(3r - 8)(11r + 1)$

253. $(5x - y)(3x - 6)$ **254.** $(10a - b)(3a - 4)$ **255.** $(a + b)(2a + 3b)$

256. $(r + s)(3r + 2s)$ **257.** $(4z - y)(z - 6)$ **258.** $(5x - y)(x - 4)$

259. $(x^2 + 3)(x + 2)$ **260.** $(y^2 - 4)(y + 3)$ **261.** $(x^2 + 8)(x^2 - 5)$

262. $(y^2 - 7)(y^2 - 4)$ **263.** $(5ab - 1)(2ab + 3)$ **264.** $(2xy + 3)(3xy + 2)$

265. $(6pq - 3)(4pq - 5)$ **266.** $(3rs - 7)(3rs - 4)$

Multiply a Trinomial by a Binomial

In the following exercises, multiply using ⓐ the Distributive Property ⓑ the Vertical Method.

267. $(x + 5)(x^2 + 4x + 3)$ **268.** $(u + 4)(u^2 + 3u + 2)$ **269.** $(y + 8)(4y^2 + y - 7)$

270. $(a + 10)(3a^2 + a - 5)$

In the following exercises, multiply. Use either method.

271. $(w - 7)(w^2 - 9w + 10)$ **272.** $(p - 4)(p^2 - 6p + 9)$ **273.** $(3q + 1)(q^2 - 4q - 5)$

274. $(6r + 1)(r^2 - 7r - 9)$

Mixed Practice

275. $(10y - 6) + (4y - 7)$ **276.** $(15p - 4) + (3p - 5)$ **277.**
$$\left(x^2 - 4x - 34\right) - \left(x^2 + 7x - 6\right)$$

278.
$$\left(j^2 - 8j - 27\right) - \left(j^2 + 2j - 12\right)$$

279. $5q(3q^2 - 6q + 11)$

280. $8t(2t^2 - 5t + 6)$

281. $(s - 7)(s + 9)$

282. $(x - 5)(x + 13)$

283. $\left(y^2 - 2y\right)(y + 1)$

284. $\left(a^2 - 3a\right)(4a + 5)$

285. $(3n - 4)\left(n^2 + n - 7\right)$

286. $(6k - 1)\left(k^2 + 2k - 4\right)$

287. $(7p + 10)(7p - 10)$

288. $(3y + 8)(3y - 8)$

289. $(4m^2 - 3m - 7)m^2$

290. $(15c^2 - 4c + 5)c^4$

291. $(5a + 7b)(5a + 7b)$

292. $(3x - 11y)(3x - 11y)$

293. $(4y + 12z)(4y - 12z)$

Everyday Math

294. Mental math You can use binomial multiplication to multiply numbers without a calculator. Say you need to multiply 13 times 15. Think of 13 as $10 + 3$ and 15 as $10 + 5$.

  ⓐ Multiply $(10 + 3)(10 + 5)$ by the FOIL method.

  ⓑ Multiply $13 \cdot 15$ without using a calculator.

  ⓒ Which way is easier for you? Why?

295. Mental math You can use binomial multiplication to multiply numbers without a calculator. Say you need to multiply 18 times 17. Think of 18 as $20 - 2$ and 17 as $20 - 3$.

  ⓐ Multiply $(20 - 2)(20 - 3)$ by the FOIL method.

  ⓑ Multiply $18 \cdot 17$ without using a calculator.

  ⓒ Which way is easier for you? Why?

Writing Exercises

296. Which method do you prefer to use when multiplying two binomials: the Distributive Property, the FOIL method, or the Vertical Method? Why?

297. Which method do you prefer to use when multiplying a trinomial by a binomial: the Distributive Property or the Vertical Method? Why?

298. Multiply the following:

$(x + 2)(x - 2)$

$(y + 7)(y - 7)$

$(w + 5)(w - 5)$

Explain the pattern that you see in your answers.

299. Multiply the following:

$(m - 3)(m + 3)$

$(n - 10)(n + 10)$

$(p - 8)(p + 8)$

Explain the pattern that you see in your answers.

300. Multiply the following:

$(p + 3)(p + 3)$

$(q + 6)(q + 6)$

$(r + 1)(r + 1)$

Explain the pattern that you see in your answers.

301. Multiply the following:

$(x - 4)(x - 4)$

$(y - 1)(y - 1)$

$(z - 7)(z - 7)$

Explain the pattern that you see in your answers.

Self Check

(a) *After completing the exercises, use this checklist to evaluate your mastery of the objectives of this section.*

I can...	Confidently	With some help	No-I don't get it!
multiply a polynomial by a monomial.			
multiply a binomial by a binomial.			
multiply a trinomial by a binomial.			

(b) *What does this checklist tell you about your mastery of this section? What steps will you take to improve?*

6.4 Special Products

Learning Objectives

By the end of this section, you will be able to:

> Square a binomial using the Binomial Squares Pattern
> Multiply conjugates using the Product of Conjugates Pattern
> Recognize and use the appropriate special product pattern

Be Prepared!

Before you get started, take this readiness quiz.

1. Simplify: ⓐ 9^2 ⓑ $(-9)^2$ ⓒ -9^2.
   If you missed this problem, review **Example 1.50**.

Square a Binomial Using the Binomial Squares Pattern

Mathematicians like to look for patterns that will make their work easier. A good example of this is squaring binomials. While you can always get the product by writing the binomial twice and using the methods of the last section, there is less work to do if you learn to use a pattern.

Let's start by looking at $(x + 9)^2$.

What does this mean?	$(x + 9)^2$
It means to multiply $(x + 9)$ by itself.	$(x + 9)(x + 9)$
Then, using FOIL, we get:	$x^2 + 9x + 9x + 81$
Combining like terms gives:	$x^2 + 18x + 81$

Here's another one:	$(y - 7)^2$
Multiply $(y - 7)$ by itself.	$(y - 7)(y - 7)$
Using FOIL, we get:	$y^2 - 7y - 7y + 49$
And combining like terms:	$y^2 - 14y + 49$

And one more:	$(2x + 3)^2$
Multiply.	$(2x + 3)(2x + 3)$
Use FOIL:	$4x^2 + 6x + 6x + 9$
Combine like terms.	$4x^2 + 12x + 9$

Look at these results. Do you see any patterns?

What about the number of terms? In each example we squared a binomial and the result was a trinomial.

$$(a + b)^2 = \underline{} + \underline{} + \underline{}$$

Now look at the **first term** in each result. Where did it come from?

$(x + 9)^2$ $\qquad$ $(y - 7)^2$ $\qquad$ $(2x + 3)^2$

$(x + 9)(x + 9)$ $\qquad$ $(y - 7)(y - 7)$ $\qquad$ $(2x + 3)(2x + 3)$

$x^2 + 9x + 9x + 81$ $\qquad$ $y^2 - 7y - 7y + 49$ $\qquad$ $4x^2 + 6x + 6x + 9$

$x^2 + 18x + 81$ $\qquad$ $y^2 - 14y + 49$ $\qquad$ $4x^2 + 12x + 9$

The first term is the product of the first terms of each binomial. Since the binomials are identical, it is just the square of the first term!

$$(a + b)^2 = a^2 + \underline{\quad} + \underline{\quad}$$

To get the **first term** of the product, **square the first term.**

Where did the **last term** come from? Look at the examples and find the pattern.

The last term is the product of the last terms, which is the square of the last term.

$$(a + b)^2 = \underline{\quad} + \underline{\quad} + b^2$$

To get the **last term** of the product, **square the last term.**

Finally, look at the **middle term**. Notice it came from adding the "outer" and the "inner" terms—which are both the same! So the middle term is double the product of the two terms of the binomial.

$$(a + b)^2 = \underline{\quad} + 2ab + \underline{\quad}$$
$$(a - b)^2 = \underline{\quad} - 2ab + \underline{\quad}$$

To get the **middle term** of the product, **multiply the terms and double their product.**

Putting it all together:

Binomial Squares Pattern

If a and b are real numbers,

$$(a + b)^2 = a^2 + 2ab + b^2$$
$$(a - b)^2 = a^2 - 2ab + b^2$$

$(a + b)^2$	=	a^2	+	$2ab$	+	b^2
(binomial)²		(first term)²		2(product of terms)		(last term)²

To square a binomial:

- square the first term
- square the last term
- double their product

A number example helps verify the pattern.

$$(10 + 4)^2$$

Square the first term.	$10^2 + \underline{\quad} + \underline{\quad}$
Square the last term.	$10^2 + \underline{\quad} + 4^2$
Double their product.	$10^2 + 2 \cdot 10 \cdot 4 + 4^2$
Simplify.	$100 + 80 + 16$
Simplify.	196

To multiply $(10 + 4)^2$ usually you'd follow the Order of Operations.

$$(10 + 4)^2$$
$$(14)^2$$
$$196$$

The pattern works!

EXAMPLE 6.47

Multiply: $(x + 5)^2$.

⊘ **Solution**

$$\left(\overset{a+b}{x+5}\right)^2$$

Square the first term.	$\overset{a^2 + 2ab + b^2}{x^2 + \underline{\quad} + \underline{\quad}}$
Square the last term.	$\overset{a^2 + 2ab + b^2}{x^2 + \underline{\quad} + 5^2}$
Double the product.	$\overset{a^2 + 2 \cdot a \cdot b + b^2}{x^2 + 2 \cdot x \cdot 5 + 5^2}$
Simplify.	$x^2 + 10x + 25$

> **TRY IT :: 6.93** Multiply: $(x + 9)^2$.

> **TRY IT :: 6.94** Multiply: $(y + 11)^2$.

EXAMPLE 6.48

Multiply: $(y - 3)^2$.

⊘ **Solution**

$$\left(\overset{a-b}{y-3}\right)^2$$

Square the first term.	$\overset{a^2 - 2ab + b^2}{y^2 - \underline{\quad} + \underline{\quad}}$
Square the last term.	$\overset{a^2 - 2ab + b^2}{y^2 - \underline{\quad} + 3^2}$
Double the product.	$\overset{a^2 - 2 \cdot a \cdot b + b^2}{y^2 - 2 \cdot y \cdot 3 + 3^2}$
Simplify.	$y^2 - 6y + 9$

> **TRY IT :: 6.95** Multiply: $(x - 9)^2$.

> **TRY IT :: 6.96** Multiply: $(p - 13)^2$.

EXAMPLE 6.49

Multiply: $(4x + 6)^2$.

⊘ Solution

$$\left(\overset{a\ +\ b}{4x + 6}\right)^2$$

Use the pattern.	$\overset{a^2\ +\ 2\ \cdot\ a\ \cdot\ b\ +\ b^2}{(4x)^2 + 2 \cdot 4x \cdot 6 + 6^2}$
Simplify.	$16x^2 + 48x + 36$

> **TRY IT :: 6.97**   Multiply: $(6x + 3)^2$.

> **TRY IT :: 6.98**   Multiply: $(4x + 9)^2$.

EXAMPLE 6.50

Multiply: $(2x - 3y)^2$.

⊘ Solution

$$\left(\overset{a\ -\ b}{2x - 3y}\right)^2$$

Use the pattern.	$\overset{a^2\ -\ 2\ \cdot\ a\ \cdot\ b\ +\ b^2}{(2x)^2 - 2 \cdot 2x \cdot 3y + (3y)^2}$
Simplify.	$4x^2 - 12xy + 9y^2$

> **TRY IT :: 6.99**   Multiply: $(2c - d)^2$.

> **TRY IT :: 6.100**   Multiply: $(4x - 5y)^2$.

EXAMPLE 6.51

Multiply: $\left(4u^3 + 1\right)^2$.

⊘ Solution

$$\left(\overset{a\ +\ b}{4u^3 + 1}\right)^2$$

Use the pattern.	$\overset{a^2\ +\ 2\ \cdot\ a\ \cdot\ b\ +\ b^2}{(4u^3)^2 + 2 \cdot 4u^3 \cdot 1 + (1)^2}$
Simplify.	$16u^6 + 8u^3 + 1$

> **TRY IT :: 6.101**   Multiply: $\left(2x^2 + 1\right)^2$.

 **TRY IT :: 6.102**   Multiply: $\left(3y^3 + 2\right)^2$.

Multiply Conjugates Using the Product of Conjugates Pattern

We just saw a pattern for squaring binomials that we can use to make multiplying some binomials easier. Similarly, there is a pattern for another product of binomials. But before we get to it, we need to introduce some vocabulary.

What do you notice about these pairs of binomials?

$$(x - 9)(x + 9) \qquad (y - 8)(y + 8) \qquad (2x - 5)(2x + 5)$$

Look at the first term of each binomial in each pair.

$$(x - 9)(x + 9) \qquad (y - 8)(y + 8) \qquad (2x - 5)(2x + 5)$$

Notice the first terms are the same in each pair.

Look at the last terms of each binomial in each pair.

$$(x - 9)(x + 9) \qquad (y - 8)(y + 8) \qquad (2x - 5)(2x + 5)$$

Notice the last terms are the same in each pair.

Notice how each pair has one sum and one difference.

$$\left(\underset{\text{Difference}}{x - 9}\right)\left(\underset{\text{Sum}}{x + 9}\right) \qquad \left(\underset{\text{Difference}}{y - 8}\right)\left(\underset{\text{Sum}}{y + 8}\right) \qquad \left(\underset{\text{Difference}}{2x - 5}\right)\left(\underset{\text{Sum}}{2x + 5}\right)$$

A pair of binomials that each have the same first term and the same last term, but one is a sum and one is a difference has a special name. It is called a *conjugate pair* and is of the form $(a - b), (a + b)$.

Conjugate Pair

A **conjugate pair** is two binomials of the form

$$(a - b), (a + b).$$

The pair of binomials each have the same first term and the same last term, but one binomial is a sum and the other is a difference.

There is a nice pattern for finding the product of conjugates. You could, of course, simply FOIL to get the product, but using the pattern makes your work easier.

Let's look for the pattern by using FOIL to multiply some conjugate pairs.

$$\begin{array}{ccc} (x - 9)(x + 9) & (y - 8)(y + 8) & (2x - 5)(2x + 5) \\ x^2 + 9x - 9x - 81 & y^2 + 8y - 8y - 64 & 4x^2 + 10x - 10x - 25 \\ x^2 - 81 & y^2 - 64 & 4x^2 - 25 \end{array}$$

$$\begin{array}{ccc} (x + 9)(x - 9) & (y - 8)(y + 8) & (2x - 5)(2x + 5) \end{array}$$

$$\begin{array}{ccc} x^2 - 9x + 9x - 81 & y^2 + 8y - 8y - 64 & 4x^2 + 10x - 10x - 25 \end{array}$$

$$\begin{array}{ccc} x^2 - 81 & y^2 - 64 & 4x^2 - 25 \end{array}$$

Each **first term** is the product of the first terms of the binomials, and since they are identical it is the square of the first term.

$$(a + b)(a - b) = a^2 - \underline{\quad\quad}$$

To get the **first term, square the first term**.

The **last term** came from multiplying the last terms, the square of the last term.

$$(a + b)(a - b) = a^2 - b^2$$

To get the **last term, square the last term**.

What do you observe about the products?

The product of the two binomials is also a binomial! Most of the products resulting from FOIL have been trinomials.

Why is there no middle term? Notice the two middle terms you get from FOIL combine to 0 in every case, the result of one addition and one subtraction.

The product of conjugates is always of the form $a^2 - b^2$. This is called a difference of squares.

This leads to the pattern:

Product of Conjugates Pattern

If a and b are real numbers,

$$(a - b)(a + b) = a^2 - b^2 \qquad \underbrace{(a - b)(a + b)}_{conjugates} = \overbrace{a^2 \quad - \quad b^2}^{\substack{difference \\ squares}}$$

The product is called a difference of squares.

To multiply conjugates, square the first term, square the last term, and write the product as a difference of squares.

Let's test this pattern with a numerical example.

$$(10 - 2)(10 + 2)$$

It is the product of conjugates, so the result will be the difference of two squares.

$$\underline{\quad\quad} - \underline{\quad\quad}$$

Square the first term.

$$10^2 - \underline{\quad\quad}$$

Square the last term.

$$10^2 - 2^2$$

Simplify.

$$100 - 4$$

Simplify.

$$96$$

What do you get using the Order of Operations?

$$(10 - 2)(10 + 2)$$
$$(8)(12)$$
$$96$$

Notice, the result is the same!

EXAMPLE 6.52

Multiply: $(x - 8)(x + 8)$.

⊘ **Solution**

First, recognize this as a product of conjugates. The binomials have the same first terms, and the same last terms, and one binomial is a sum and the other is a difference.

It fits the pattern.	$\left(\overset{a\ -\ b}{x - 8}\right)\left(\overset{a\ +\ b}{x + 8}\right)$
Square the first term, x.	$\overset{a^2\ -\ b^2}{x^2 - \underline{\quad}}$
Square the last term, 8.	$\overset{a^2\ -\ b^2}{x^2 - 8^2}$
The product is a difference of squares.	$\overset{a^2\ -\ b^2}{x^2 - 64}$

> **TRY IT ::** 6.103   Multiply: $(x - 5)(x + 5)$.

> **TRY IT ::** 6.104   Multiply: $(w - 3)(w + 3)$.

EXAMPLE 6.53

Multiply: $(2x + 5)(2x - 5)$.

✓ **Solution**

Are the binomials conjugates?

It is the product of conjugates.	$\left(\overset{a\ +\ b}{2x + 5}\right)\left(\overset{a\ -\ b}{2x - 5}\right)$
Square the first term, $2x$.	$\overset{a^2\ -\ b^2}{(2x)^2 - \underline{}}$
Square the last term, 5.	$\overset{a^2\ -\ b^2}{(2x)^2 - 5^2}$
Simplify. The product is a difference of squares.	$\overset{a^2\ -\ b^2}{4x^2 - 25}$

> **TRY IT : : 6.105**   Multiply: $(6x + 5)(6x - 5)$.

> **TRY IT : : 6.106**   Multiply: $(2x + 7)(2x - 7)$.

The binomials in the next example may look backwards – the variable is in the second term. But the two binomials are still conjugates, so we use the same pattern to multiply them.

EXAMPLE 6.54

Find the product: $(3 + 5x)(3 - 5x)$.

✓ **Solution**

It is the product of conjugates.	$\left(\overset{a\ -\ b}{3 + 5x}\right)\left(\overset{a\ +\ b}{3 - 5x}\right)$
Use the pattern.	$\overset{a^2\ -\ b^2}{3^2 - (5x)^2}$
Simplify.	$9 - 25x^2$

> **TRY IT : : 6.107**   Multiply: $(7 + 4x)(7 - 4x)$.

> **TRY IT : : 6.108**   Multiply: $(9 - 2y)(9 + 2y)$.

Now we'll multiply conjugates that have two variables.

EXAMPLE 6.55

Find the product: $(5m - 9n)(5m + 9n)$.

✓ Solution

This fits the pattern.	$\left(\underset{5m-9n}{\overset{a\ -\ b}{}}\right)\left(\underset{5m+9n}{\overset{a\ +\ b}{}}\right)$
Use the pattern.	$\overset{a^2\ -\ b^2}{(5m)^2-(9n)^2}$
Simplify.	$25m^2-81n^2$

> **TRY IT ::** 6.109   Find the product: $(4p-7q)(4p+7q)$.

> **TRY IT ::** 6.110   Find the product: $(3x-y)(3x+y)$.

EXAMPLE 6.56

Find the product: $(cd-8)(cd+8)$.

✓ Solution

This fits the pattern.	$\left(\underset{cd-8}{\overset{a\ -\ b}{}}\right)\left(\underset{cd+8}{\overset{a\ +\ b}{}}\right)$
Use the pattern.	$\overset{a^2\ -\ b^2}{(cd)^2-(8)^2}$
Simplify.	c^2d^2-64

> **TRY IT ::** 6.111   Find the product: $(xy-6)(xy+6)$.

> **TRY IT ::** 6.112   Find the product: $(ab-9)(ab+9)$.

EXAMPLE 6.57

Find the product: $\left(6u^2-11v^5\right)\left(6u^2+11v^5\right)$.

✓ Solution

This fits the pattern.	$\left(\underset{6u^2-11v^5}{\overset{a\ -\ b}{}}\right)\left(\underset{6u^2+11v^5}{\overset{a\ +\ b}{}}\right)$
Use the pattern.	$\overset{a^2\ -\ b^2}{(6u^2)^2-(11v^5)^2}$
Simplify.	$36u^4-121v^{10}$

> **TRY IT ::** 6.113   Find the product: $\left(3x^2-4y^3\right)\left(3x^2+4y^3\right)$.

> **TRY IT ::** 6.114   Find the product: $\left(2m^2-5n^3\right)\left(2m^2+5n^3\right)$.

Recognize and Use the Appropriate Special Product Pattern

We just developed special product patterns for Binomial Squares and for the Product of Conjugates. The products look similar, so it is important to recognize when it is appropriate to use each of these patterns and to notice how they differ. Look at the two patterns together and note their similarities and differences.

Comparing the Special Product Patterns

Binomial Squares	Product of Conjugates
$(a + b)^2 = a^2 + 2ab + b^2$	$(a - b)(a + b) = a^2 - b^2$
$(a - b)^2 = a^2 - 2ab + b^2$	
- Squaring a binomial	- Multiplying conjugates
- Product is a **trinomial**	- Product is a **binomial**
- Inner and outer terms with FOIL are **the same.**	- Inner and outer terms with FOIL are **opposites.**
- Middle term is **double the product** of the terms.	- There is **no** middle term.

EXAMPLE 6.58

Choose the appropriate pattern and use it to find the product:

ⓐ $(2x - 3)(2x + 3)$ ⓑ $(5x - 8)^2$ ⓒ $(6m + 7)^2$ ⓓ $(5x - 6)(6x + 5)$

✓ Solution

ⓐ $(2x - 3)(2x + 3)$ These are conjugates. They have the same first numbers, and the same last numbers, and one binomial is a sum and the other is a difference. It fits the Product of Conjugates pattern.

This fits the pattern.	$\left(\overset{a\ -\ b}{2x - 3}\right)\left(\overset{a\ +\ b}{2x + 3}\right)$
Use the pattern.	$\overset{a^2\ -\ b^2}{(2x)^2 - 3^2}$
Simplify.	$4x^2 - 9$

ⓑ $(8x - 5)^2$ We are asked to square a binomial. It fits the **binomial squares** pattern.

$$\left(\overset{a\ -\ b}{8x - 5}\right)^2$$

Use the pattern.	$\overset{a^2\ -\ \ 2ab\ \ +\ b^2}{(8x)^2 - 2 \cdot 8x \cdot 5 + 5^2}$
Simplify.	$64x^2 - 80x + 25$

ⓒ $(6m + 7)^2$ Again, we will square a binomial so we use the **binomial squares** pattern.

$$\left(\overset{a\ +\ b}{6m + 7}\right)^2$$

Use the pattern.	$\overset{a^2\ +\ \ 2ab\ \ +\ b^2}{(6m)^2 + 2 \cdot 6m \cdot 7 + 7^2}$
Simplify.	$36m^2 + 84m + 49$

ⓓ $(5x - 6)(6x + 5)$ This product does not fit the patterns, so we will use FOIL.

$$(5x - 6)(6x + 5)$$

Use FOIL.	$30x^2 + 25x - 36x - 30$
Simplify.	$30x^2 - 11x - 30$

> **TRY IT : : 6.115**   Choose the appropriate pattern and use it to find the product:

ⓐ $(9b - 2)(2b + 9)$ ⓑ $(9p - 4)^2$ ⓒ $(7y + 1)^2$ ⓓ $(4r - 3)(4r + 3)$

> **TRY IT : : 6.116**   Choose the appropriate pattern and use it to find the product:

ⓐ $(6x + 7)^2$ ⓑ $(3x - 4)(3x + 4)$ ⓒ $(2x - 5)(5x - 2)$ ⓓ $(6n - 1)^2$

▶ **MEDIA : :**

Access these online resources for additional instruction and practice with special products:

- **Special Products (https://openstax.org/l/25Specialprod)**

6.4 EXERCISES

Practice Makes Perfect

Square a Binomial Using the Binomial Squares Pattern

In the following exercises, square each binomial using the Binomial Squares Pattern.

302. $(w + 4)^2$

303. $(q + 12)^2$

304. $\left(y + \frac{1}{4}\right)^2$

305. $\left(x + \frac{2}{3}\right)^2$

306. $(b - 7)^2$

307. $(y - 6)^2$

308. $(m - 15)^2$

309. $(p - 13)^2$

310. $(3d + 1)^2$

311. $(4a + 10)^2$

312. $\left(2q + \frac{1}{3}\right)^2$

313. $\left(3z + \frac{1}{5}\right)^2$

314. $(3x - y)^2$

315. $(2y - 3z)^2$

316. $\left(\frac{1}{5}x - \frac{1}{7}y\right)^2$

317. $\left(\frac{1}{8}x - \frac{1}{9}y\right)^2$

318. $(3x^2 + 2)^2$

319. $(5u^2 + 9)^2$

320. $(4y^3 - 2)^2$

321. $(8p^3 - 3)^2$

Multiply Conjugates Using the Product of Conjugates Pattern

In the following exercises, multiply each pair of conjugates using the Product of Conjugates Pattern.

322. $(m - 7)(m + 7)$

323. $(c - 5)(c + 5)$

324. $\left(x + \frac{3}{4}\right)\left(x - \frac{3}{4}\right)$

325. $\left(b + \frac{6}{7}\right)\left(b - \frac{6}{7}\right)$

326. $(5k + 6)(5k - 6)$

327. $(8j + 4)(8j - 4)$

328. $(11k + 4)(11k - 4)$

329. $(9c + 5)(9c - 5)$

330. $(11 - b)(11 + b)$

331. $(13 - q)(13 + q)$

332. $(5 - 3x)(5 + 3x)$

333. $(4 - 6y)(4 + 6y)$

334. $(9c - 2d)(9c + 2d)$

335. $(7w + 10x)(7w - 10x)$

336. $\left(m + \frac{2}{3}n\right)\left(m - \frac{2}{3}n\right)$

337. $\left(p + \frac{4}{5}q\right)\left(p - \frac{4}{5}q\right)$

338. $(ab - 4)(ab + 4)$

339. $(xy - 9)(xy + 9)$

340. $\left(uv - \frac{3}{5}\right)\left(uv + \frac{3}{5}\right)$

341. $\left(rs - \frac{2}{7}\right)\left(rs + \frac{2}{7}\right)$

342. $(2x^2 - 3y^4)(2x^2 + 3y^4)$

343. $(6m^3 - 4n^5)(6m^3 + 4n^5)$

344.
$(12p^3 - 11q^2)(12p^3 + 11q^2)$

345. $(15m^2 - 8n^4)(15m^2 + 8n^4)$

Recognize and Use the Appropriate Special Product Pattern

In the following exercises, find each product.

346.

ⓐ $(p - 3)(p + 3)$

ⓑ $(t - 9)^2$

ⓒ $(m + n)^2$

ⓓ $(2x + y)(x - 2y)$

347.

ⓐ $(2r + 12)^2$

ⓑ $(3p + 8)(3p - 8)$

ⓒ $(7a + b)(a - 7b)$

ⓓ $(k - 6)^2$

348.

ⓐ $(a^5 - 7b)^2$

ⓑ $(x^2 + 8y)(8x - y^2)$

ⓒ $(r^6 + s^6)(r^6 - s^6)$

ⓓ $(y^4 + 2z)^2$

349.

ⓐ $(x^5 + y^5)(x^5 - y^5)$

ⓑ $(m^3 - 8n)^2$

ⓒ $(9p + 8q)^2$

ⓓ $(r^2 - s^3)(r^3 + s^2)$

Everyday Math

350. Mental math You can use the product of conjugates pattern to multiply numbers without a calculator. Say you need to multiply 47 times 53. Think of 47 as $50 - 3$ and 53 as $50 + 3$.

ⓐ Multiply $(50 - 3)(50 + 3)$ by using the product of conjugates pattern, $(a - b)(a + b) = a^2 - b^2$.

ⓑ Multiply $47 \cdot 53$ without using a calculator.

ⓒ Which way is easier for you? Why?

351. Mental math You can use the binomial squares pattern to multiply numbers without a calculator. Say you need to square 65. Think of 65 as $60 + 5$.

ⓐ Multiply $(60 + 5)^2$ by using the binomial squares pattern, $(a + b)^2 = a^2 + 2ab + b^2$.

ⓑ Square 65 without using a calculator.

ⓒ Which way is easier for you? Why?

Writing Exercises

352. How do you decide which pattern to use?

353. Why does $(a + b)^2$ result in a trinomial, but $(a - b)(a + b)$ result in a binomial?

354. Marta did the following work on her homework paper:

$$(3 - y)^2$$
$$3^2 - y^2$$
$$9 - y^2$$

Explain what is wrong with Marta's work.

355. Use the order of operations to show that $(3 + 5)^2$ is 64, and then use that numerical example to explain why $(a + b)^2 \neq a^2 + b^2$.

Self Check

ⓐ *After completing the exercises, use this checklist to evaluate your mastery of the objectives of this section.*

I can...	Confidently	With some help	No-I don't get it!
square a binomial using the binomial squares pattern.			
multiply conjugates using the product of conjugates pattern.			
recognize and use the appropriate special product pattern.			

ⓑ *On a scale of 1-10, how would you rate your mastery of this section in light of your responses on the checklist? How can you improve this?*

6.5 **Divide Monomials**

Learning Objectives

By the end of this section, you will be able to:

> Simplify expressions using the Quotient Property for Exponents
> Simplify expressions with zero exponents
> Simplify expressions using the quotient to a Power Property
> Simplify expressions by applying several properties
> Divide monomials

Be Prepared!

Before you get started, take this readiness quiz.

1. Simplify: $\frac{8}{24}$.

   If you missed this problem, review **Example 1.65**.

2. Simplify: $\left(2m^3\right)^5$.

   If you missed this problem, review **Example 6.23**.

3. Simplify: $\frac{12x}{12y}$.

   If you missed this problem, review **Example 1.67**.

Simplify Expressions Using the Quotient Property for Exponents

Earlier in this chapter, we developed the properties of exponents for multiplication. We summarize these properties below.

Summary of Exponent Properties for Multiplication

If a and b are real numbers, and m and n are whole numbers, then

$$\begin{aligned}
\textbf{Product Property} && a^m \cdot a^n &= a^{m+n} \\
\textbf{Power Property} && (a^m)^n &= a^{m \cdot n} \\
\textbf{Product to a Power} && (ab)^m &= a^m b^m
\end{aligned}$$

Now we will look at the exponent properties for division. A quick memory refresher may help before we get started. You have learned to simplify fractions by dividing out common factors from the numerator and denominator using the Equivalent Fractions Property. This property will also help you work with algebraic fractions—which are also quotients.

Equivalent Fractions Property

If a, b, and c are whole numbers where $b \neq 0$, $c \neq 0$,

$$\text{then} \quad \frac{a}{b} = \frac{a \cdot c}{b \cdot c} \quad \text{and} \quad \frac{a \cdot c}{b \cdot c} = \frac{a}{b}$$

As before, we'll try to discover a property by looking at some examples.

Consider	$\dfrac{x^5}{x^2}$ and	$\dfrac{x^2}{x^3}$
What do they mean?	$\dfrac{x \cdot x \cdot x \cdot x \cdot x}{x \cdot x}$	$\dfrac{x \cdot x}{x \cdot x \cdot x}$
Use the Equivalent Fractions Property.	$\dfrac{\cancel{x} \cdot \cancel{x} \cdot x \cdot x \cdot x}{\cancel{x} \cdot \cancel{x}}$	$\dfrac{\cancel{x} \cdot \cancel{x} \cdot 1}{\cancel{x} \cdot \cancel{x} \cdot x}$
Simplify.	x^3	$\dfrac{1}{x}$

Notice, in each case the bases were the same and we subtracted exponents.

When the larger exponent was in the numerator, we were left with factors in the numerator.

When the larger exponent was in the denominator, we were left with factors in the denominator—notice the numerator of 1.

We write:

$$\frac{x^5}{x^2} \qquad \frac{x^2}{x^3}$$

$$x^{5-2} \qquad \frac{1}{x^{3-2}}$$

$$x^3 \qquad \frac{1}{x}$$

This leads to the *Quotient Property for Exponents*.

Quotient Property for Exponents

If a is a real number, $a \neq 0$, and m and n are whole numbers, then

$$\frac{a^m}{a^n} = a^{m-n}, \, m > n \quad \text{and} \quad \frac{a^m}{a^n} = \frac{1}{a^{n-m}}, \, n > m$$

A couple of examples with numbers may help to verify this property.

$$\frac{3^4}{3^2} = 3^{4-2} \qquad\qquad \frac{5^2}{5^3} = \frac{1}{5^{3-2}}$$

$$\frac{81}{9} = 3^2 \qquad\qquad \frac{25}{125} = \frac{1}{5^1}$$

$$9 = 9 \checkmark \qquad\qquad \frac{1}{5} = \frac{1}{5} \checkmark$$

EXAMPLE 6.59

Simplify: ⓐ $\dfrac{x^9}{x^7}$ ⓑ $\dfrac{3^{10}}{3^2}$.

⊘ Solution

To simplify an expression with a quotient, we need to first compare the exponents in the numerator and denominator.

ⓐ

Since 9 > 7, there are more factors of *x* in the numerator.	$\dfrac{x^9}{x^7}$
Use the Quotient Property, $\dfrac{a^m}{a^n} = a^{m-n}$.	x^{9-7}
Simplify.	x^2

ⓑ

Since 10 > 2, there are more factors of *x* in the numerator.	$\dfrac{3^{10}}{3^2}$
Use the Quotient Property, $\dfrac{a^m}{a^n} = a^{m-n}$.	3^{10-2}
Simplify.	3^8

Notice that when the larger exponent is in the numerator, we are left with factors in the numerator.

> **TRY IT :: 6.117**   Simplify: ⓐ $\dfrac{x^{15}}{x^{10}}$ ⓑ $\dfrac{6^{14}}{6^5}$.

> **TRY IT :: 6.118**   Simplify: ⓐ $\dfrac{y^{43}}{y^{37}}$ ⓑ $\dfrac{10^{15}}{10^{7}}$.

EXAMPLE 6.60

Simplify: ⓐ $\dfrac{b^8}{b^{12}}$ ⓑ $\dfrac{7^3}{7^5}$.

⊘ **Solution**

To simplify an expression with a quotient, we need to first compare the exponents in the numerator and denominator.

ⓐ

Since 12 > 8, there are more factors of b in the denominator.	$\dfrac{b^8}{b^{12}}$
Use the Quotient Property, $\dfrac{a^m}{a^n} = \dfrac{1}{a^{n-m}}$.	$\dfrac{1}{b^{12-8}}$
Simplify.	$\dfrac{1}{b^4}$

ⓑ

Since 5 > 3, there are more factors of 3 in the denominator.	$\dfrac{7^3}{7^5}$
Use the Quotient Property, $\dfrac{a^m}{a^n} = \dfrac{1}{a^{n-m}}$.	$\dfrac{1}{7^{5-3}}$
Simplify.	$\dfrac{1}{7^2}$
Simplify.	$\dfrac{1}{49}$

Notice that when the larger exponent is in the denominator, we are left with factors in the denominator.

> **TRY IT :: 6.119**   Simplify: ⓐ $\dfrac{x^{18}}{x^{22}}$ ⓑ $\dfrac{12^{15}}{12^{30}}$.

> **TRY IT :: 6.120**   Simplify: ⓐ $\dfrac{m^7}{m^{15}}$ ⓑ $\dfrac{9^8}{9^{19}}$.

Notice the difference in the two previous examples:

- If we start with more factors in the numerator, we will end up with factors in the numerator.
- If we start with more factors in the denominator, we will end up with factors in the denominator.

The first step in simplifying an expression using the Quotient Property for Exponents is to determine whether the exponent is larger in the numerator or the denominator.

EXAMPLE 6.61

Simplify: ⓐ $\dfrac{a^5}{a^9}$ ⓑ $\dfrac{x^{11}}{x^7}$.

⊘ Solution

ⓐ Is the exponent of a larger in the numerator or denominator? Since $9 > 5$, there are more a's in the denominator and so we will end up with factors in the denominator.

$$\frac{a^5}{a^9}$$

Use the Quotient Property, $\dfrac{a^m}{a^n} = \dfrac{1}{a^{n-m}}$.	$\dfrac{1}{a^{9-5}}$
Simplify.	$\dfrac{1}{a^4}$

ⓑ Notice there are more factors of x in the numerator, since $11 > 7$. So we will end up with factors in the numerator.

$$\frac{x^{11}}{x^7}$$

Use the Quotient Property, $\dfrac{a^m}{a^n} = \dfrac{1}{a^{n-m}}$.	x^{11-7}
Simplify.	x^4

> **TRY IT : : 6.121** Simplify: ⓐ $\dfrac{b^{19}}{b^{11}}$ ⓑ $\dfrac{z^5}{z^{11}}$.

> **TRY IT : : 6.122** Simplify: ⓐ $\dfrac{p^9}{p^{17}}$ ⓑ $\dfrac{w^{13}}{w^9}$.

Simplify Expressions with an Exponent of Zero

A special case of the Quotient Property is when the exponents of the numerator and denominator are equal, such as an expression like $\dfrac{a^m}{a^m}$. From your earlier work with fractions, you know that:

$$\frac{2}{2} = 1 \qquad \frac{17}{17} = 1 \qquad \frac{-43}{-43} = 1$$

In words, a number divided by itself is 1. So, $\dfrac{x}{x} = 1$, for any x $(x \neq 0)$, since any number divided by itself is 1.

The Quotient Property for Exponents shows us how to simplify $\dfrac{a^m}{a^n}$ when $m > n$ and when $n < m$ by subtracting exponents. What if $m = n$?

Consider $\dfrac{8}{8}$, which we know is 1.

$$\frac{8}{8} = 1$$

Write 8 as 2^3.
$$\frac{2^3}{2^3} = 1$$

Subtract exponents.
$$2^{3-3} = 1$$

Simplify.
$$2^0 = 1$$

Now we will simplify $\dfrac{a^m}{a^m}$ in two ways to lead us to the definition of the zero exponent. In general, for $a \neq 0$:

$$\frac{a^m}{a^m} \qquad \frac{a^m}{a^m}$$

$$\overbrace{}^{m \text{ factors}}$$

$$a^{m-m} \qquad \frac{\cancel{a} \cdot \cancel{a} \cdot \ldots \cdot \cancel{a}}{\cancel{a} \cdot \cancel{a} \cdot \ldots \cdot \cancel{a}}$$

$$\underbrace{}_{m \text{ factors}}$$

$$a^0 \qquad 1$$

We see $\frac{a^m}{a^m}$ simplifies to a^0 and to 1. So $a^0 = 1$.

Zero Exponent

If a is a non-zero number, then $a^0 = 1$.

Any nonzero number raised to the zero power is 1.

In this text, we assume any variable that we raise to the zero power is not zero.

EXAMPLE 6.62

Simplify: ⓐ 9^0 ⓑ n^0.

✓ Solution

The definition says any non-zero number raised to the zero power is 1.

ⓐ

	9^0
Use the definition of the zero exponent.	1

ⓑ

	n^0
Use the definition of the zero exponent.	1

> **TRY IT : : 6.123**   Simplify: ⓐ 15^0 ⓑ m^0.

> **TRY IT : : 6.124**   Simplify: ⓐ k^0 ⓑ 29^0.

Now that we have defined the zero exponent, we can expand all the Properties of Exponents to include whole number exponents.

What about raising an expression to the zero power? Let's look at $(2x)^0$. We can use the product to a power rule to rewrite this expression.

	$(2x)^0$
Use the product to a power rule.	$2^0 x^0$
Use the zero exponent property.	$1 \cdot 1$
Simplify.	1

This tells us that any nonzero expression raised to the zero power is one.

EXAMPLE 6.63

Simplify: ⓐ $(5b)^0$ ⓑ $\left(-4a^2 b\right)^0$.

✓ Solution

ⓐ

$$(5b)^0$$

Use the definition of the zero exponent. 1

ⓑ

$$\left(-4a^2 b\right)^0$$

Use the definition of the zero exponent. 1

> **TRY IT :: 6.125**
>
> Simplify: ⓐ $(11z)^0$ ⓑ $\left(-11pq^3\right)^0$.

> **TRY IT :: 6.126**
>
> Simplify: ⓐ $(-6d)^0$ ⓑ $\left(-8m^2 n^3\right)^0$.

Simplify Expressions Using the Quotient to a Power Property

Now we will look at an example that will lead us to the Quotient to a Power Property.

$$\left(\frac{x}{y}\right)^3$$

This means: $\frac{x}{y} \cdot \frac{x}{y} \cdot \frac{x}{y}$

Multiply the fractions. $\frac{x \cdot x \cdot x}{y \cdot y \cdot y}$

Write with exponents. $\frac{x^3}{y^3}$

Notice that the exponent applies to both the numerator and the denominator.

We see that $\left(\frac{x}{y}\right)^3$ is $\frac{x^3}{y^3}$.

We write: $\left(\frac{x}{y}\right)^3$

$\frac{x^3}{y^3}$

This leads to the *Quotient to a Power Property for Exponents*.

Quotient to a Power Property for Exponents

If a and b are real numbers, $b \neq 0$, and m is a counting number, then

$$\left(\frac{a}{b}\right)^m = \frac{a^m}{b^m}$$

To raise a fraction to a power, raise the numerator and denominator to that power.

An example with numbers may help you understand this property:

$$\left(\frac{2}{3}\right)^3 = \frac{2^3}{3^3}$$

$$\frac{2}{3} \cdot \frac{2}{3} \cdot \frac{2}{3} = \frac{8}{27}$$

$$\frac{8}{27} = \frac{8}{27} ✓$$

EXAMPLE 6.64

Simplify: ⓐ $\left(\frac{3}{7}\right)^2$ ⓑ $\left(\frac{b}{3}\right)^4$ ⓒ $\left(\frac{k}{j}\right)^3$.

✓ **Solution**

ⓐ

$$\left(\frac{3}{7}\right)^2$$

Use the Quotient Property, $\left(\frac{a}{b}\right)^m = \frac{a^m}{b^m}$.	$\frac{3^2}{7^2}$
Simplify.	$\frac{9}{49}$

ⓑ

$$\left(\frac{b}{3}\right)^4$$

Use the Quotient Property, $\left(\frac{a}{b}\right)^m = \frac{a^m}{b^m}$.	$\frac{b^4}{3^4}$
Simplify.	$\frac{b^4}{81}$

ⓒ

$$\left(\frac{k}{j}\right)^3$$

Raise the numerator and denominator to the third power.	$\frac{k^3}{j^3}$

> **TRY IT : : 6.127**   Simplify: ⓐ $\left(\frac{5}{8}\right)^2$ ⓑ $\left(\frac{p}{10}\right)^4$ ⓒ $\left(\frac{m}{n}\right)^7$.

> **TRY IT : : 6.128**   Simplify: ⓐ $\left(\frac{1}{3}\right)^3$ ⓑ $\left(\frac{-2}{q}\right)^3$ ⓒ $\left(\frac{w}{x}\right)^4$.

Simplify Expressions by Applying Several Properties

We'll now summarize all the properties of exponents so they are all together to refer to as we simplify expressions using several properties. Notice that they are now defined for whole number exponents.

Summary of Exponent Properties

If a and b are real numbers, and m and n are whole numbers, then

Product Property	$a^m \cdot a^n = a^{m+n}$
Power Property	$(a^m)^n = a^{m \cdot n}$
Product to a Power	$(ab)^m = a^m b^m$
Quotient Property	$\dfrac{a^m}{b^m} = a^{m-n}, a \neq 0, m > n$
	$\dfrac{a^m}{a^n} = \dfrac{1}{a^{n-m}}, a \neq 0, n > m$
Zero Exponent Definition	$a^o = 1, a \neq 0$
Quotient to a Power Property	$\left(\dfrac{a}{b}\right)^m = \dfrac{a^m}{b^m}, b \neq 0$

EXAMPLE 6.65

Simplify: $\dfrac{\left(y^4\right)^2}{y^6}$.

✓ **Solution**

$$\dfrac{\left(y^4\right)^2}{y^6}$$

Multiply the exponents in the numerator.

$$\dfrac{y^8}{y^6}$$

Subtract the exponents.

$$y^2$$

> **TRY IT : : 6.129**

Simplify: $\dfrac{\left(m^5\right)^4}{m^7}$.

> **TRY IT : : 6.130**

Simplify: $\dfrac{\left(k^2\right)^6}{k^7}$.

EXAMPLE 6.66

Simplify: $\dfrac{b^{12}}{\left(b^2\right)^6}$.

✓ **Solution**

$$\dfrac{b^{12}}{\left(b^2\right)^6}$$

Multiply the exponents in the denominator.

$$\dfrac{b^{12}}{b^{12}}$$

Subtract the exponents.

$$b^0$$

Simplify.

$$1$$

Notice that after we simplified the denominator in the first step, the numerator and the denominator were equal. So the final value is equal to 1.

> **TRY IT : : 6.131**

Simplify: $\dfrac{n^{12}}{\left(n^3\right)^4}$.

> **TRY IT : : 6.132**

Simplify: $\dfrac{x^{15}}{\left(x^3\right)^5}$.

EXAMPLE 6.67

Simplify: $\left(\dfrac{y^9}{y^4}\right)^2$.

⊘ **Solution**

$$\left(\dfrac{y^9}{y^4}\right)^2$$

Remember parentheses come before exponents.
Notice the bases are the same, so we can simplify
inside the parentheses. Subtract the exponents.

$$\left(y^5\right)^2$$

Multiply the exponents.

$$y^{10}$$

> **TRY IT : : 6.133**

Simplify: $\left(\dfrac{r^5}{r^3}\right)^4$.

> **TRY IT : : 6.134**

Simplify: $\left(\dfrac{v^6}{v^4}\right)^3$.

EXAMPLE 6.68

Simplify: $\left(\dfrac{j^2}{k^3}\right)^4$.

⊘ **Solution**

Here we cannot simplify inside the parentheses first, since the bases are not the same.

$$\left(\dfrac{j^2}{k^3}\right)^4$$

Raise the numerator and denominator to the third power
using the Quotient to a Power Property, $\left(\dfrac{a}{b}\right)^m = \dfrac{a^m}{b^m}$.

$$\dfrac{\left(j^2\right)^4}{\left(k^3\right)^4}$$

Use the Power Property and simplify.

$$\dfrac{j^8}{k^{12}}$$

> **TRY IT : : 6.135**

Simplify: $\left(\dfrac{a^3}{b^2}\right)^4$.

> **TRY IT ::** 6.136

Simplify: $\left(\dfrac{q^7}{r^5}\right)^3$.

EXAMPLE 6.69

Simplify: $\left(\dfrac{2m^2}{5n}\right)^4$.

 Solution

$$\left(\dfrac{2m^2}{5n}\right)^4$$

Raise the numerator and denominator to the fourth power, using the Quotient to a Power Property, $\left(\dfrac{a}{b}\right)^m = \dfrac{a^m}{b^m}$.

$$\dfrac{(2m^2)^4}{(5n)^4}$$

Raise each factor to the fourth power.

$$\dfrac{2^4(m^2)^4}{5^4 n^4}$$

Use the Power Property and simplify.

$$\dfrac{16m^8}{625n^4}$$

> **TRY IT ::** 6.137

Simplify: $\left(\dfrac{7x^3}{9y}\right)^2$.

> **TRY IT ::** 6.138

Simplify: $\left(\dfrac{3x^4}{7y}\right)^2$.

EXAMPLE 6.70

Simplify: $\dfrac{\left(x^3\right)^4 \left(x^2\right)^5}{\left(x^6\right)^5}$.

 Solution

$$\dfrac{\left(x^3\right)^4 \left(x^2\right)^5}{\left(x^6\right)^5}$$

Use the Power Property, $(a^m)^n = a^{m \cdot n}$.

$$\dfrac{\left(x^{12}\right)\left(x^{10}\right)}{\left(x^{30}\right)}$$

Add the exponents in the numerator.

$$\dfrac{x^{22}}{x^{30}}$$

Use the Quotient Property, $\dfrac{a^m}{a^n} = \dfrac{1}{a^{n-m}}$.

$$\dfrac{1}{x^8}$$

> **TRY IT ::** 6.139

Simplify: $\dfrac{\left(a^2\right)^3 \left(a^2\right)^4}{\left(a^4\right)^5}$.

> TRY IT :: 6.140

Simplify: $\dfrac{\left(p^3\right)^4\left(p^5\right)^3}{\left(p^7\right)^6}$.

EXAMPLE 6.71

Simplify: $\dfrac{\left(10p^3\right)^2}{(5p)^3\left(2p^5\right)^4}$.

✓ Solution

	$\dfrac{\left(10p^3\right)^2}{(5p)^3\left(2p^5\right)^4}$
Use the Product to a Power Property, $(ab)^m = a^m b^m$.	$\dfrac{(10)^2\left(p^3\right)^2}{(5)^3\,(p)^3\,(2)^4\left(p^5\right)^4}$
Use the Power Property, $(a^m)^n = a^{m \cdot n}$.	$\dfrac{100p^6}{125p^3 \cdot 16p^{20}}$
Add the exponents in the denominator.	$\dfrac{100p^6}{125 \cdot 16p^{23}}$
Use the Quotient Property, $\dfrac{a^m}{a^n} = \dfrac{1}{a^{n-m}}$.	$\dfrac{100}{125 \cdot 16p^{17}}$
Simplify.	$\dfrac{1}{20p^{17}}$

> TRY IT :: 6.141

Simplify: $\dfrac{\left(3r^3\right)^2\left(r^3\right)^7}{\left(r^3\right)^3}$.

> TRY IT :: 6.142

Simplify: $\dfrac{\left(2x^4\right)^5}{\left(4x^3\right)^2\left(x^3\right)^5}$.

Divide Monomials

You have now been introduced to all the properties of exponents and used them to simplify expressions. Next, you'll see how to use these properties to divide monomials. Later, you'll use them to divide polynomials.

EXAMPLE 6.72

Find the quotient: $56x^7 \div 8x^3$.

✅ Solution

$$56x^7 \div 8x^3$$

Rewrite as a fraction.

$$\frac{56x^7}{8x^3}$$

Use fraction multiplication.

$$\frac{56}{8} \cdot \frac{x^7}{x^3}$$

Simplify and use the Quotient Property.

$$7x^4$$

> **TRY IT : : 6.143**   Find the quotient: $42y^9 \div 6y^3$.

> **TRY IT : : 6.144**   Find the quotient: $48z^8 \div 8z^2$.

EXAMPLE 6.73

Find the quotient: $\dfrac{45a^2 b^3}{-5ab^5}$.

✅ Solution

When we divide monomials with more than one variable, we write one fraction for each variable.

$$\frac{45a^2 b^3}{-5ab^5}$$

Use fraction multiplication.

$$\frac{45}{-5} \cdot \frac{a^2}{a} \cdot \frac{b^3}{b^5}$$

Simplify and use the Quotient Property.

$$-9 \cdot a \cdot \frac{1}{b^2}$$

Multiply.

$$-\frac{9a}{b^2}$$

> **TRY IT : : 6.145**   Find the quotient: $\dfrac{-72a^7 b^3}{8a^{12} b^4}$.

> **TRY IT : : 6.146**   Find the quotient: $\dfrac{-63c^8 d^3}{7c^{12} d^2}$.

EXAMPLE 6.74

Find the quotient: $\dfrac{24a^5 b^3}{48ab^4}$.

✅ Solution

$$\frac{24a^5 b^3}{48ab^4}$$

Use fraction multiplication.

$$\frac{24}{48} \cdot \frac{a^5}{a} \cdot \frac{b^3}{b^4}$$

Simplify and use the Quotient Property.

$$\frac{1}{2} \cdot a^4 \cdot \frac{1}{b}$$

Multiply.

$$\frac{a^4}{2b}$$

> **TRY IT : : 6.147**   Find the quotient: $\dfrac{16a^7b^6}{24ab^8}$.

> **TRY IT : : 6.148**   Find the quotient: $\dfrac{27p^4q^7}{-45p^{12}q}$.

Once you become familiar with the process and have practiced it step by step several times, you may be able to simplify a fraction in one step.

EXAMPLE 6.75

Find the quotient: $\dfrac{14x^7y^{12}}{21x^{11}y^6}$.

⊘ **Solution**

Be very careful to simplify $\frac{14}{21}$ by dividing out a common factor, and to simplify the variables by subtracting their exponents.

$$\frac{14x^7y^{12}}{21x^{11}y^6}$$

Simplify and use the Quotient Property. $\dfrac{2y^6}{3x^4}$

> **TRY IT : : 6.149**   Find the quotient: $\dfrac{28x^5y^{14}}{49x^9y^{12}}$.

> **TRY IT : : 6.150**   Find the quotient: $\dfrac{30m^5n^{11}}{48m^{10}n^{14}}$.

In all examples so far, there was no work to do in the numerator or denominator before simplifying the fraction. In the next example, we'll first find the product of two monomials in the numerator before we simplify the fraction. This follows the order of operations. Remember, a fraction bar is a grouping symbol.

EXAMPLE 6.76

Find the quotient: $\dfrac{\left(6x^2y^3\right)\left(5x^3y^2\right)}{\left(3x^4y^5\right)}$.

⊘ **Solution**

$$\frac{\left(6x^2y^3\right)\left(5x^3y^2\right)}{\left(3x^4y^5\right)}$$

Simplify the numerator. $\dfrac{30x^5y^5}{3x^4y^5}$

Simplify. $10x$

> **TRY IT : : 6.151**   Find the quotient: $\dfrac{\left(6a^4b^5\right)\left(4a^2b^5\right)}{12a^5b^8}$.

 **TRY IT : :** 6.152

Find the quotient: $\dfrac{\left(-12x^6 y^9\right)\left(-4x^5 y^8\right)}{-12x^{10}y^{12}}$.

▶ **MEDIA : :**

Access these online resources for additional instruction and practice with dividing monomials:

- **Rational Expressions (https://openstax.org/l/25RationalExp)**
- **Dividing Monomials (https://openstax.org/l/25DivideMono)**
- **Dividing Monomials 2 (https://openstax.org/l/25DivideMono2)**

6.5 EXERCISES

Practice Makes Perfect

Simplify Expressions Using the Quotient Property for Exponents

In the following exercises, simplify.

356. ⓐ $\dfrac{x^{18}}{x^3}$ ⓑ $\dfrac{5^{12}}{5^3}$

357. ⓐ $\dfrac{y^{20}}{y^{10}}$ ⓑ $\dfrac{7^{16}}{7^2}$

358. ⓐ $\dfrac{p^{21}}{p^7}$ ⓑ $\dfrac{4^{16}}{4^4}$

359. ⓐ $\dfrac{u^{24}}{u^3}$ ⓑ $\dfrac{9^{15}}{9^5}$

360. ⓐ $\dfrac{q^{18}}{q^{36}}$ ⓑ $\dfrac{10^2}{10^3}$

361. ⓐ $\dfrac{t^{10}}{t^{40}}$ ⓑ $\dfrac{8^3}{8^5}$

362. ⓐ $\dfrac{b}{b^9}$ ⓑ $\dfrac{4}{4^6}$

363. ⓐ $\dfrac{x}{x^7}$ ⓑ $\dfrac{10}{10^3}$

Simplify Expressions with Zero Exponents

In the following exercises, simplify.

364.
ⓐ 20^0
ⓑ b^0

365.
ⓐ 13^0
ⓑ k^0

366.
ⓐ -27^0
ⓑ $-\left(27^0\right)$

367.
ⓐ -15^0
ⓑ $-\left(15^0\right)$

368.
ⓐ $(25x)^0$
ⓑ $25x^0$

369.
ⓐ $(6y)^0$
ⓑ $6y^0$

370.
ⓐ $(12x)^0$
ⓑ $\left(-56p^4q^3\right)^0$

371.
ⓐ $7y^0\,(17y)^0$
ⓑ $\left(-93c^7d^{15}\right)^0$

372.
ⓐ $12n^0 - 18m^0$
ⓑ $(12n)^0 - (18m)^0$

373.
ⓐ $15r^0 - 22s^0$
ⓑ $(15r)^0 - (22s)^0$

Simplify Expressions Using the Quotient to a Power Property

In the following exercises, simplify.

374. ⓐ $\left(\dfrac{3}{4}\right)^3$ ⓑ $\left(\dfrac{p}{2}\right)^5$ ⓒ $\left(\dfrac{x}{y}\right)^6$

375. ⓐ $\left(\dfrac{2}{5}\right)^2$ ⓑ $\left(\dfrac{x}{3}\right)^4$ ⓒ $\left(\dfrac{a}{b}\right)^5$

376. ⓐ $\left(\dfrac{a}{3b}\right)^4$ ⓑ $\left(\dfrac{5}{4m}\right)^2$

377. ⓐ $\left(\dfrac{x}{2y}\right)^3$ ⓑ $\left(\dfrac{10}{3q}\right)^4$

Simplify Expressions by Applying Several Properties

In the following exercises, simplify.

378. $\dfrac{\left(a^2\right)^3}{a^4}$

379. $\dfrac{\left(p^3\right)^4}{p^5}$

380. $\dfrac{\left(y^3\right)^4}{y^{10}}$

381. $\dfrac{\left(x^4\right)^5}{x^{15}}$

382. $\dfrac{u^6}{\left(u^3\right)^2}$

383. $\dfrac{v^{20}}{\left(v^4\right)^5}$

384. $\dfrac{m^{12}}{\left(m^8\right)^3}$

385. $\dfrac{n^8}{\left(n^6\right)^4}$

386. $\left(\dfrac{p^9}{p^3}\right)^5$

387. $\left(\dfrac{q^8}{q^2}\right)^3$

388. $\left(\dfrac{r^2}{r^6}\right)^3$

389. $\left(\dfrac{m^4}{m^7}\right)^4$

390. $\left(\dfrac{p}{r^{11}}\right)^2$

391. $\left(\dfrac{a}{b^6}\right)^3$

392. $\left(\dfrac{w^5}{x^3}\right)^8$

393. $\left(\dfrac{y^4}{z^{10}}\right)^5$

394. $\left(\dfrac{2j^3}{3k}\right)^4$

395. $\left(\dfrac{3m^5}{5n}\right)^3$

396. $\left(\dfrac{3c^2}{4d^6}\right)^3$

397. $\left(\dfrac{5u^7}{2v^3}\right)^4$

398. $\left(\dfrac{k^2 k^8}{k^3}\right)^2$

399. $\left(\dfrac{j^2 j^5}{j^4}\right)^3$

400. $\dfrac{\left(t^2\right)^5 \left(t^4\right)^2}{\left(t^3\right)^7}$

401. $\dfrac{\left(q^3\right)^6 \left(q^2\right)^3}{\left(q^4\right)^8}$

402. $\dfrac{\left(-2p^2\right)^4 \left(3p^4\right)^2}{\left(-6p^3\right)^2}$

403. $\dfrac{\left(-2k^3\right)^2 \left(6k^2\right)^4}{\left(9k^4\right)^2}$

404. $\dfrac{\left(-4m^3\right)^2 \left(5m^4\right)^3}{\left(-10m^6\right)^3}$

405. $\dfrac{\left(-10n^2\right)^3 \left(4n^5\right)^2}{\left(2n^8\right)^2}$

Divide Monomials

In the following exercises, divide the monomials.

406. $56b^8 \div 7b^2$

407. $63v^{10} \div 9v^2$

408. $-88y^{15} \div 8y^3$

409. $-72u^{12} \div 12u^4$

410. $\dfrac{45a^6 b^8}{-15a^{10} b^2}$

411. $\dfrac{54x^9 y^3}{-18x^6 y^{15}}$

412. $\dfrac{15r^4 s^9}{18r^9 s^2}$ **413.** $\dfrac{20m^8 n^4}{30m^5 n^9}$ **414.** $\dfrac{18a^4 b^8}{-27a^9 b^5}$

415. $\dfrac{45x^5 y^9}{-60x^8 y^6}$ **416.** $\dfrac{64q^{11} r^9 s^3}{48q^6 r^8 s^5}$ **417.** $\dfrac{65a^{10} b^8 c^5}{42a^7 b^6 c^8}$

418. $\dfrac{\left(10m^5 n^4\right)\left(5m^3 n^6\right)}{25m^7 n^5}$ **419.** $\dfrac{\left(-18p^4 q^7\right)\left(-6p^3 q^8\right)}{-36p^{12} q^{10}}$ **420.** $\dfrac{\left(6a^4 b^3\right)\left(4ab^5\right)}{\left(12a^2 b\right)\left(a^3 b\right)}$

421. $\dfrac{\left(4u^2 v^5\right)\left(15u^3 v\right)}{\left(12u^3 v\right)\left(u^4 v\right)}$

Mixed Practice

422.
ⓐ $24a^5 + 2a^5$
ⓑ $24a^5 - 2a^5$
ⓒ $24a^5 \cdot 2a^5$
ⓓ $24a^5 \div 2a^5$

423.
ⓐ $15n^{10} + 3n^{10}$
ⓑ $15n^{10} - 3n^{10}$
ⓒ $15n^{10} \cdot 3n^{10}$
ⓓ $15n^{10} \div 3n^{10}$

424.
ⓐ $p^4 \cdot p^6$
ⓑ $\left(p^4\right)^6$

425.
ⓐ $q^5 \cdot q^3$
ⓑ $\left(q^5\right)^3$

426.
ⓐ $\dfrac{y^3}{y}$
ⓑ $\dfrac{y}{y^3}$

427.
ⓐ $\dfrac{z^6}{z^5}$
ⓑ $\dfrac{z^5}{z^6}$

428. $\left(8x^5\right)(9x) \div 6x^3$ **429.** $(4y)\left(12y^7\right) \div 8y^2$ **430.** $\dfrac{27a^7}{3a^3} + \dfrac{54a^9}{9a^5}$

431. $\dfrac{32c^{11}}{4c^5} + \dfrac{42c^9}{6c^3}$ **432.** $\dfrac{32y^5}{8y^2} - \dfrac{60y^{10}}{5y^7}$ **433.** $\dfrac{48x^6}{6x^4} - \dfrac{35x^9}{7x^7}$

434. $\dfrac{63r^6 s^3}{9r^4 s^2} - \dfrac{72r^2 s^2}{6s}$ **435.** $\dfrac{56y^4 z^5}{7y^3 z^3} - \dfrac{45y^2 z^2}{5y}$

Everyday Math

436. Memory One megabyte is approximately 10^6 bytes. One gigabyte is approximately 10^9 bytes. How many megabytes are in one gigabyte?

437. Memory One gigabyte is approximately 10^9 bytes. One terabyte is approximately 10^{12} bytes. How many gigabytes are in one terabyte?

Writing Exercises

438. Jennifer thinks the quotient $\dfrac{a^{24}}{a^6}$ simplifies to a^4. What is wrong with her reasoning?

439. Maurice simplifies the quotient $\dfrac{d^7}{d}$ by writing $\dfrac{\cancel{d}^7}{\cancel{d}} = 7$. What is wrong with his reasoning?

440. When Drake simplified -3^0 and $(-3)^0$ he got the same answer. Explain how using the Order of Operations correctly gives different answers.

441. Robert thinks x^0 simplifies to 0. What would you say to convince Robert he is wrong?

Self Check

ⓐ *After completing the exercises, use this checklist to evaluate your mastery of the objectives of this section.*

I can...	Confidently	With some help	No-I don't get it!
simplify expressions using the Quotient Property for Exponents.			
simplify expressions with zero exponents.			
simplify expressions using the Quotient to a Power Property.			
simplify expressions by applying several properties.			
divide monomials.			

ⓑ *On a scale of 1-10, how would you rate your mastery of this section in light of your responses on the checklist? How can you improve this?*

6.6 Divide Polynomials

Learning Objectives

By the end of this section, you will be able to:

› Divide a polynomial by a monomial
› Divide a polynomial by a binomial

Be Prepared!

Before you get started, take this readiness quiz.

1. Add: $\frac{3}{d} + \frac{x}{d}$.
   If you missed this problem, review **Example 1.77**.

2. Simplify: $\frac{30xy^3}{5xy}$.
   If you missed this problem, review **Example 6.72**.

3. Combine like terms: $8a^2 + 12a + 1 + 3a^2 - 5a + 4$.
   If you missed this problem, review **Example 1.24**.

Divide a Polynomial by a Monomial

In the last section, you learned how to divide a monomial by a monomial. As you continue to build up your knowledge of polynomials the next procedure is to divide a polynomial of two or more terms by a monomial.

The method we'll use to divide a polynomial by a monomial is based on the properties of fraction addition. So we'll start with an example to review fraction addition.

The sum, $\qquad \frac{y}{5} + \frac{2}{5}$,

simplifies to $\qquad \frac{y+2}{5}$.

Now we will do this in reverse to split a single fraction into separate fractions.

We'll state the fraction addition property here just as you learned it and in reverse.

Fraction Addition

If a, b, and c are numbers where $c \neq 0$, then

$$\frac{a}{c} + \frac{b}{c} = \frac{a+b}{c} \quad \text{and} \quad \frac{a+b}{c} = \frac{a}{c} + \frac{b}{c}$$

We use the form on the left to add fractions and we use the form on the right to divide a polynomial by a monomial.

For example, $\qquad \frac{y+2}{5}$

can be written $\qquad \frac{y}{5} + \frac{2}{5}$.

We use this form of fraction addition to divide polynomials by monomials.

Division of a Polynomial by a Monomial

To divide a polynomial by a monomial, divide each term of the polynomial by the monomial.

EXAMPLE 6.77

Find the quotient: $\frac{7y^2 + 21}{7}$.

Solution

$$\frac{7y^2 + 21}{7}$$

Divide each term of the numerator by the denominator. $\frac{7y^2}{7} + \frac{21}{7}$

Simplify each fraction. $y^2 + 3$

> **TRY IT : : 6.153**   Find the quotient: $\frac{8z^2 + 24}{4}$.

> **TRY IT : : 6.154**   Find the quotient: $\frac{18z^2 - 27}{9}$.

Remember that division can be represented as a fraction. When you are asked to divide a polynomial by a monomial and it is not already in fraction form, write a fraction with the polynomial in the numerator and the monomial in the denominator.

EXAMPLE 6.78

Find the quotient: $\left(18x^3 - 36x^2\right) \div 6x$.

Solution

$$\left(18x^3 - 36x^2\right) \div 6x$$

Rewrite as a fraction. $\frac{18x^3 - 36x^2}{6x}$

Divide each term of the numerator by the denominator. $\frac{18x^3}{6x} - \frac{36x^2}{6x}$

Simplify. $3x^2 - 6x$

> **TRY IT : : 6.155**   Find the quotient: $\left(27b^3 - 33b^2\right) \div 3b$.

> **TRY IT : : 6.156**   Find the quotient: $\left(25y^3 - 55y^2\right) \div 5y$.

When we divide by a negative, we must be extra careful with the signs.

EXAMPLE 6.79

Find the quotient: $\frac{12d^2 - 16d}{-4}$.

Solution

$$\frac{12d^2 - 16d}{-4}$$

Divide each term of the numerator by the denominator. $\frac{12d^2}{-4} - \frac{16d}{-4}$

Simplify. Remember, subtracting a negative is like adding a positive! $-3d^2 + 4d$

> **TRY IT : : 6.157**   Find the quotient: $\frac{25y^2 - 15y}{-5}$.

> **TRY IT ::** 6.158

Find the quotient: $\dfrac{42b^2 - 18b}{-6}$.

EXAMPLE 6.80

Find the quotient: $\dfrac{105y^5 + 75y^3}{5y^2}$.

⊘ **Solution**

$$\dfrac{105y^5 + 75y^3}{5y^2}$$

Separate the terms. $\dfrac{105y^5}{5y^2} + \dfrac{75y^3}{5y^2}$

Simplify. $21y^3 + 15y$

> **TRY IT ::** 6.159

Find the quotient: $\dfrac{60d^7 + 24d^5}{4d^3}$.

> **TRY IT ::** 6.160

Find the quotient: $\dfrac{216p^7 - 48p^5}{6p^3}$.

EXAMPLE 6.81

Find the quotient: $\left(15x^3 y - 35xy^2\right) \div (-5xy)$.

⊘ **Solution**

$$\left(15x^3 y - 35xy^2\right) \div (-5xy)$$

Rewrite as a fraction. $\dfrac{15x^3 y - 35xy^2}{-5xy}$

Separate the terms. Be careful with the signs! $\dfrac{15x^3 y}{-5xy} - \dfrac{35xy^2}{-5xy}$

Simplify. $-3x^2 + 7y$

> **TRY IT ::** 6.161

Find the quotient: $\left(32a^2 b - 16ab^2\right) \div (-8ab)$.

> **TRY IT ::** 6.162

Find the quotient: $\left(-48a^8 b^4 - 36a^6 b^5\right) \div \left(-6a^3 b^3\right)$.

EXAMPLE 6.82

Find the quotient: $\dfrac{36x^3 y^2 + 27x^2 y^2 - 9x^2 y^3}{9x^2 y}$.

⊘ Solution

$$\frac{36x^3y^2 + 27x^2y^2 - 9x^2y^3}{9x^2y}$$

Separate the terms.

$$\frac{36x^3y^2}{9x^2y} + \frac{27x^2y^2}{9x^2y} - \frac{9x^2y^3}{9x^2y}$$

Simplify.

$$4xy + 3y - y^2$$

> **TRY IT :: 6.163**

Find the quotient: $\dfrac{40x^3y^2 + 24x^2y^2 - 16x^2y^3}{8x^2y}$.

> **TRY IT :: 6.164**

Find the quotient: $\dfrac{35a^4b^2 + 14a^4b^3 - 42a^2b^4}{7a^2b^2}$.

EXAMPLE 6.83

Find the quotient: $\dfrac{10x^2 + 5x - 20}{5x}$.

⊘ Solution

$$\frac{10x^2 + 5x - 20}{5x}$$

Separate the terms.

$$\frac{10x^2}{5x} + \frac{5x}{5x} - \frac{20}{5x}$$

Simplify.

$$2x + 1 + \frac{4}{x}$$

> **TRY IT :: 6.165**

Find the quotient: $\dfrac{18c^2 + 6c - 9}{6c}$.

> **TRY IT :: 6.166**

Find the quotient: $\dfrac{10d^2 - 5d - 2}{5d}$.

Divide a Polynomial by a Binomial

To divide a polynomial by a binomial, we follow a procedure very similar to long division of numbers. So let's look carefully the steps we take when we divide a 3-digit number, 875, by a 2-digit number, 25.

We write the long division	$25\overline{)875}$

We divide the first two digits, 87, by 25.	$\begin{array}{r} 3 \\ 25\overline{)875} \end{array}$

We multiply 3 times 25 and write the product under the 87.	$\begin{array}{r} 3 \\ 25\overline{)875} \\ 75 \end{array}$

Now we subtract 75 from 87.	$\begin{array}{r} 3 \\ 25\overline{)875} \\ -75 \\ \hline 12 \end{array}$

Then we bring down the third digit of the dividend, 5.	$\begin{array}{r} 3 \\ 25\overline{)875} \\ -75 \\ \hline 125 \end{array}$

Repeat the process, dividing 25 into 125.	$\begin{array}{r} 35 \\ 25\overline{)875} \\ -75 \\ \hline 125 \\ -125 \end{array}$

We check division by multiplying the quotient by the divisor.

If we did the division correctly, the product should equal the dividend.

$$35 \cdot 25$$
$$875 \checkmark$$

Now we will divide a trinomial by a binomial. As you read through the example, notice how similar the steps are to the numerical example above.

EXAMPLE 6.84

Find the quotient: $\left(x^2 + 9x + 20\right) \div (x + 5)$.

✓ **Solution**

	$(x^2 + 9x + 20) \div (x + 5)$

Write it as a long division problem.	
Be sure the dividend is in standard form.	$x + 5\overline{)x^2 + 9x + 20}$

Divide x^2 by x. It may help to ask yourself, "What do I need to multiply x by to get x^2?"	
Put the answer, x, in the quotient over the x term.	$\begin{array}{r} x \\ x + 5\overline{)x^2 + 9x + 20} \end{array}$

Multiply x times x + 5. Line up the like terms under the dividend.	$\begin{array}{r} x \\ x+5\overline{)\ x^2+9x+20} \\ x^2+5x \end{array}$
Subtract $x^2 + 5x$ from $x^2 + 9x$.	
You may find it easier to **change the signs** and then add. Then bring down the last term, 20.	$\begin{array}{r} x \\ x+5\overline{)\ x^2+9x+20} \\ -x^2+(-5x) \\ \hline 4x+20 \end{array}$
Divide $4x$ by x. It may help to ask yourself, "What do I need to multiply x by to get $4x$?"	
Put the answer, 4, in the quotient over the constant term.	$\begin{array}{r} x+4 \\ x+5\overline{)\ x^2+9x+20} \\ -x^2+(-5x) \\ \hline 4x+20 \end{array}$
Multiply 4 times $x + 5$.	$\begin{array}{r} x+4 \\ x+5\overline{)\ x^2+9x+20} \\ -x^2+(-5x) \\ \hline 4x+20 \\ 4x+20 \\ \hline \end{array}$
Subtract $4x + 20$ from $4x + 20$.	$\begin{array}{r} x+4 \\ x+5\overline{)\ x^2+9x+20} \\ -x^2+(-5x) \\ \hline 4x+20 \\ -4x+(-20) \\ \hline 0 \end{array}$
Check:	
Multiply the quotient by the divisor.	
$(x + 4)(x + 5)$	
You should get the dividend.	
$x^2 + 9x + 20$ ✓	

> **TRY IT :: 6.167**   Find the quotient: $\left(y^2 + 10y + 21\right) \div (y + 3)$.

> **TRY IT :: 6.168**   Find the quotient: $\left(m^2 + 9m + 20\right) \div (m + 4)$.

When the divisor has subtraction sign, we must be extra careful when we multiply the partial quotient and then subtract. It may be safer to show that we change the signs and then add.

EXAMPLE 6.85

Find the quotient: $\left(2x^2 - 5x - 3\right) \div (x - 3)$.

✓ Solution

	$(2x^2 - 5x - 3) \div (x - 3)$
Write it as a long division problem.	
Be sure the dividend is in standard form.	$x - 3{\overline{\smash{\big)}\,2x^2 - 5x - 3}}$
Divide $2x^2$ by x. Put the answer, $2x$, in the quotient over the x term.	$\begin{array}{r} 2x \\ x - 3{\overline{\smash{\big)}\,2x^2 - 5x - 3}} \end{array}$
Multiply $2x$ times $x - 3$. Line up the like terms under the dividend.	$\begin{array}{r} 2x \\ x - 3{\overline{\smash{\big)}\,2x^2 - 5x - 3}} \\ \underline{2x^2 - 6x } \end{array}$
Subtract $2x^2 - 6x$ from $2x^2 - 5x$. Change the signs and then add. Then bring down the last term.	$\begin{array}{r} 2x \\ x - 3{\overline{\smash{\big)}\,2x^2 - 5x - 3}} \\ \underline{-2x^2 + 6x } \\ x - 3 \end{array}$
Divide x by x. Put the answer, 1, in the quotient over the constant term.	$\begin{array}{r} 2x + 1 \\ x - 3{\overline{\smash{\big)}\,2x^2 - 5x - 3}} \\ \underline{-2x^2 - (-6x) } \\ x - 3 \end{array}$
Multiply 1 times $x - 3$.	$\begin{array}{r} 2x + 1 \\ x - 3{\overline{\smash{\big)}\,2x^2 - 5x - 3}} \\ \underline{-2x^2 + 6x } \\ x - 3 \\ \underline{x - 3} \end{array}$
Subtract $x - 3$ from $x - 3$ by changing the signs and adding.	$\begin{array}{r} 2x + 1 \\ x - 3{\overline{\smash{\big)}\,2x^2 - 5x - 3}} \\ \underline{-2x^2 + 6x } \\ x - 3 \\ \underline{-x + 3} \\ 0 \end{array}$
To check, multiply $(x - 3)(2x + 1)$.	
The result should be $2x^2 - 5x - 3$.	

> **TRY IT :: 6.169** Find the quotient: $\left(2x^2 - 3x - 20\right) \div (x - 4)$.

> **TRY IT :: 6.170** Find the quotient: $\left(3x^2 - 16x - 12\right) \div (x - 6)$.

When we divided 875 by 25, we had no remainder. But sometimes division of numbers does leave a remainder. The same is true when we divide polynomials. In **Example 6.86**, we'll have a division that leaves a remainder. We write the remainder as a fraction with the divisor as the denominator.

EXAMPLE 6.86

Find the quotient: $\left(x^3 - x^2 + x + 4\right) \div (x + 1)$.

✓ Solution

$(x^3 - x^2 + x + 4) \div (x + 1)$

Write it as a long division problem.	
Be sure the dividend is in standard form.	$x + 1 \overline{)\, x^3 - x^2 + x + 4}$

Divide x^3 by x.
Put the answer, x^2, in the quotient over the x^2 term.
Multiply x^2 times $x + 1$. Line up the like terms under the dividend.

$$\begin{array}{r} x^2 \\ x + 1 \overline{)\, x^3 - x^2 + x + 4} \\ \underline{x^3 + x^2} \end{array}$$

Subtract $x^3 + x^2$ from $x^3 - x^2$ by changing the signs and adding.
Then bring down the next term.

$$\begin{array}{r} x^2 \\ x + 1 \overline{)\, x^3 - x^2 + x + 4} \\ -x^3 + (-x^2) \\ \hline -2x^2 + x \end{array}$$

Divide $-2x^2$ by x.
Put the answer, $-2x$, in the quotient over the x term.
Multiply $-2x$ times $x + 1$. Line up the like terms under the dividend.

$$\begin{array}{r} x^2 - 2x \\ x + 1 \overline{)\, x^3 - x^2 + x + 4} \\ -x^3 + (-x^2) \\ \hline -2x^2 + x \\ \underline{-2x^2 - 2x} \end{array}$$

Subtract $-2x^2 - 2x$ from $-2x^2 + x$ by changing the signs and adding.
Then bring down the last term.

$$\begin{array}{r} x^2 - 2x \\ x + 1 \overline{)\, x^3 - x^2 + x + 4} \\ -x^3 + (-x^2) \\ \hline -2x^2 + x \\ +2x^2 + 2x \\ \hline 3x + 4 \end{array}$$

Divide $3x$ by x.
Put the answer, 3, in the quotient over the constant term.
Multiply 3 times $x + 1$. Line up the like terms under the dividend.

$$\begin{array}{r} x^2 - 2x + 3 \\ x + 1 \overline{)\, x^3 - x^2 + x + 4} \\ -x^3 + (-x^2) \\ \hline -2x^2 + x \\ +2x^2 + 2x \\ \hline 3x + 4 \\ 3x + 3 \\ \hline \end{array}$$

Subtract $3x + 3$ from $3x + 4$ by changing the signs and adding.
Write the remainder as a fraction with the divisor as the denominator.

$$\begin{array}{r} x^2 - 2x + 3 + \dfrac{1}{x + 1} \\ x + 1 \overline{)\, x^3 - x^2 + x + 4} \\ -x^3 + (-x^2) \\ \hline -2x^2 + x \\ +2x^2 + 2x \\ \hline 3x + 4 \\ -3x + (-3) \\ \hline 1 \end{array}$$

To check, multiply $(x + 1)\left(x^2 - 2x + 3 + \dfrac{1}{x + 1}\right)$.

The result should be $x^3 - x^2 + x + 4$.

> **TRY IT :: 6.171**   Find the quotient: $\left(x^3 + 5x^2 + 8x + 6\right) \div (x + 2)$.

TRY IT :: 6.172   Find the quotient: $\left(2x^3 + 8x^2 + x - 8\right) \div (x + 1)$.

Look back at the dividends in **Example 6.84**, **Example 6.85**, and **Example 6.86**. The terms were written in descending order of degrees, and there were no missing degrees. The dividend in **Example 6.87** will be $x^4 - x^2 + 5x - 2$. It is missing an x^3 term. We will add in $0x^3$ as a placeholder.

EXAMPLE 6.87

Find the quotient: $\left(x^4 - x^2 + 5x - 2\right) \div (x + 2)$.

✓ **Solution**

Notice that there is no x^3 term in the dividend. We will add $0x^3$ as a placeholder.

	$(x^4 - x^2 + 5x - 2) \div (x + 2)$
Write it as a long division problem. Be sure the dividend is in standard form with placeholders for missing terms.	$x + 2\overline{)x^4 - 0x^3 - x^2 + 5x - 2}$
Divide x^4 by x. Put the answer, x^3, in the quotient over the x^3 term. Multiply x^3 times $x + 2$. Line up the like terms. Subtract and then bring down the next term.	$\begin{array}{r} x^3 \\ x + 2\overline{)x^4 + 0x^3 - x^2 + 5x - 2} \\ \underline{-(x^4 + 2x^3)} \\ -2x^3 - x^2 \end{array}$ It may be helpful to change the signs and add.
Divide $-2x^3$ by x. Put the answer, $-2x^2$, in the quotient over the x^2 term. Multiply $-2x^2$ times $x + 1$. Line up the like terms. Subtract and bring down the next term.	$\begin{array}{r} x^3 - 2x^2 \\ x + 2\overline{)x^4 + 0x^3 - x^2 + 5x - 2} \\ \underline{-(x^4 + 2x^3)} \\ -2x^3 - x^2 \\ \underline{-(-2x^3 - 4x^2)} \\ 3x^2 + 5x \end{array}$ It may be helpful to change the signs and add.
Divide $3x^2$ by x. Put the answer, $3x$, in the quotient over the x term. Multiply $3x$ times $x + 1$. Line up the like terms. Subtract and bring down the next term.	$\begin{array}{r} x^3 - 2x^2 + 3x \\ x + 2\overline{)x^4 + 0x^3 - x^2 + 5x - 2} \\ \underline{-(x^4 + 2x^3)} \\ -2x^3 - x^2 \\ \underline{-(-2x^3 - 4x^2)} \\ 3x^2 + 5x \\ \underline{-(3x^2 + 6x)} \\ -x - 2 \end{array}$ It may be helpful to change the signs and add.
Divide $-x$ by x. Put the answer, -1, in the quotient over the constant term. Multiply -1 times $x + 1$. Line up the like terms. Change the signs, add.	$\begin{array}{r} x^3 - 2x^2 + 3x - 1 \\ x + 2\overline{)x^4 + 0x^3 - x^2 + 5x - 2} \\ \underline{-(x^4 + 2x^3)} \\ -2x^3 - x^2 \\ \underline{-(-2x^3 - 4x^2)} \\ 3x^2 + 5x \\ \underline{-(3x^2 + 6x)} \\ -x - 2 \\ \underline{-(-x - 2)} \\ 0 \end{array}$ It may be helpful to change the signs and add.

To check, multiply $(x + 2)\left(x^3 - 2x^2 + 3x - 1\right)$.

The result should be $x^4 - x^2 + 5x - 2$.

> **TRY IT :: 6.173**   Find the quotient: $\left(x^3 + 3x + 14\right) \div (x + 2)$.

> **TRY IT :: 6.174**   Find the quotient: $\left(x^4 - 3x^3 - 1000\right) \div (x + 5)$.

In **Example 6.88**, we will divide by $2a - 3$. As we divide we will have to consider the constants as well as the variables.

EXAMPLE 6.88

Find the quotient: $\left(8a^3 + 27\right) \div (2a + 3)$.

⊘ Solution

This time we will show the division all in one step. We need to add two placeholders in order to divide.

$$(8a^3 + 27) \div (2a + 3)$$

$$
\begin{array}{r}
4a^2 - 6a + 9 \\
2a + 3 \overline{)\ 8a^3 + 0a^2 + 0a + 27} \\
-(8a^3 + 12a^2) \qquad \longleftarrow 4a^2(2a + 3) \\
\overline{\qquad -12a^2 + 0a} \\
-(-12a^2 - 18a) \qquad \longleftarrow 6a(2a + 3) \\
\overline{\qquad 18a + 27} \\
-(18a + 27) \qquad \longleftarrow 9(2a + 3) \\
\overline{\qquad 0}
\end{array}
$$

To check, multiply $(2a + 3)\left(4a^2 - 6a + 9\right)$.

The result should be $8a^3 + 27$.

> **TRY IT :: 6.175**   Find the quotient: $\left(x^3 - 64\right) \div (x - 4)$.

> **TRY IT :: 6.176**   Find the quotient: $\left(125x^3 - 8\right) \div (5x - 2)$.

> **MEDIA ::**
> Access these online resources for additional instruction and practice with dividing polynomials:
> - **Divide a Polynomial by a Monomial (https://openstax.org/l/25DividePolyMo1)**
> - **Divide a Polynomial by a Monomial 2 (https://openstax.org/l/25DividePolyMo2)**
> - **Divide Polynomial by Binomial (https://openstax.org/l/25DividePolyBin)**

 **6.6 EXERCISES**

Practice Makes Perfect

In the following exercises, divide each polynomial by the monomial.

442. $\dfrac{45y + 36}{9}$

443. $\dfrac{30b + 75}{5}$

444. $\dfrac{8d^2 - 4d}{2}$

445. $\dfrac{42x^2 - 14x}{7}$

446. $\left(16y^2 - 20y\right) \div 4y$

447. $\left(55w^2 - 10w\right) \div 5w$

448. $\left(9n^4 + 6n^3\right) \div 3n$

449. $\left(8x^3 + 6x^2\right) \div 2x$

450. $\dfrac{18y^2 - 12y}{-6}$

451. $\dfrac{20b^2 - 12b}{-4}$

452. $\dfrac{35a^4 + 65a^2}{-5}$

453. $\dfrac{51m^4 + 72m^3}{-3}$

454. $\dfrac{310y^4 - 200y^3}{5y^2}$

455. $\dfrac{412z^8 - 48z^5}{4z^3}$

456. $\dfrac{46x^3 + 38x^2}{2x^2}$

457. $\dfrac{51y^4 + 42y^2}{3y^2}$

458. $\left(24p^2 - 33p\right) \div (-3p)$

459. $\left(35x^4 - 21x\right) \div (-7x)$

460. $\left(63m^4 - 42m^3\right) \div \left(-7m^2\right)$

461. $\left(48y^4 - 24y^3\right) \div \left(-8y^2\right)$

462. $\left(63a^2 b^3 + 72ab^4\right) \div (9ab)$

463. $\left(45x^3 y^4 + 60xy^2\right) \div (5xy)$

464. $\dfrac{52p^5 q^4 + 36p^4 q^3 - 64p^3 q^2}{4p^2 q}$

465. $\dfrac{49c^2 d^2 - 70c^3 d^3 - 35c^2 d^4}{7cd^2}$

466. $\dfrac{66x^3 y^2 - 110x^2 y^3 - 44x^4 y^3}{11x^2 y^2}$

467. $\dfrac{72r^5 s^2 + 132r^4 s^3 - 96r^3 s^5}{12r^2 s^2}$

468. $\dfrac{4w^2 + 2w - 5}{2w}$

469. $\dfrac{12q^2 + 3q - 1}{3q}$

470. $\dfrac{10x^2 + 5x - 4}{-5x}$

471. $\dfrac{20y^2 + 12y - 1}{-4y}$

472. $\dfrac{36p^3 + 18p^2 - 12p}{6p^2}$

473. $\dfrac{63a^3 - 108a^2 + 99a}{9a^2}$

Divide a Polynomial by a Binomial

In the following exercises, divide each polynomial by the binomial.

474. $\left(y^2 + 7y + 12\right) \div (y + 3)$

475. $\left(d^2 + 8d + 12\right) \div (d + 2)$

476. $\left(x^2 - 3x - 10\right) \div (x + 2)$

477. $\left(a^2 - 2a - 35\right) \div (a + 5)$

478. $\left(t^2 - 12t + 36\right) \div (t - 6)$

479. $\left(x^2 - 14x + 49\right) \div (x - 7)$

480. $\left(6m^2 - 19m - 20\right) \div (m - 4)$

481. $\left(4x^2 - 17x - 15\right) \div (x - 5)$

482. $\left(q^2 + 2q + 20\right) \div (q + 6)$

483. $\left(p^2 + 11p + 16\right) \div (p + 8)$ **484.** $\left(y^2 - 3y - 15\right) \div (y - 8)$ **485.** $\left(x^2 + 2x - 30\right) \div (x - 5)$

486. $\left(3b^3 + b^2 + 2\right) \div (b + 1)$ **487.** $\left(2n^3 - 10n + 24\right) \div (n + 3)$ **488.** $\left(2y^3 - 6y - 36\right) \div (y - 3)$

489. $\left(7q^3 - 5q - 2\right) \div (q - 1)$ **490.** $\left(z^3 + 1\right) \div (z + 1)$ **491.** $\left(m^3 + 1000\right) \div (m + 10)$

492. $\left(a^3 - 125\right) \div (a - 5)$ **493.** $\left(x^3 - 216\right) \div (x - 6)$ **494.** $\left(64x^3 - 27\right) \div (4x - 3)$

495. $\left(125y^3 - 64\right) \div (5y - 4)$

Everyday Math

496. Average cost Pictures Plus produces digital albums. The company's average cost (in dollars) to make x albums is given by the expression $\frac{7x + 500}{x}$.

  ⓐ Find the quotient by dividing the numerator by the denominator.

  ⓑ What will the average cost (in dollars) be to produce 20 albums?

497. Handshakes At a company meeting, every employee shakes hands with every other employee. The number of handshakes is given by the expression $\frac{n^2 - n}{2}$, where n represents the number of employees. How many handshakes will there be if there are 10 employees at the meeting?

Writing Exercises

498. James divides $48y + 6$ by 6 this way: $\frac{48y + \cancel{6}}{\cancel{6}} = 48y$. What is wrong with his reasoning?

499. Divide $\frac{10x^2 + x - 12}{2x}$ and explain with words how you get each term of the quotient.

Self Check

ⓐ After completing the exercises, use this checklist to evaluate your mastery of the objectives of this section.

I can...	Confidently	With some help	No-I don't get it!
divide a polynomial by a monomial.			
divide a polynomial by a binomial.			

ⓑ After reviewing this checklist, what will you do to become confident for all goals?

6.7 Integer Exponents and Scientific Notation

Learning Objectives

By the end of this section, you will be able to:

> Use the definition of a negative exponent
> Simplify expressions with integer exponents
> Convert from decimal notation to scientific notation
> Convert scientific notation to decimal form
> Multiply and divide using scientific notation

Be Prepared!

Before you get started, take this readiness quiz.

1. What is the place value of the 6 in the number 64,891 ?
   If you missed this problem, review Example 1.1.
2. Name the decimal: 0.0012.
   If you missed this problem, review Example 1.91.
3. Subtract: $5 - (-3)$.
   If you missed this problem, review Example 1.42.

Use the Definition of a Negative Exponent

We saw that the Quotient Property for Exponents introduced earlier in this chapter, has two forms depending on whether the exponent is larger in the numerator or the denominator.

Quotient Property for Exponents

If a is a real number, $a \neq 0$, and m and n are whole numbers, then

$$\frac{a^m}{a^n} = a^{m-n}, \, m > n \quad \text{and} \quad \frac{a^m}{a^n} = \frac{1}{a^{n-m}}, \, n > m$$

What if we just subtract exponents regardless of which is larger?

Let's consider $\frac{x^2}{x^5}$.

We subtract the exponent in the denominator from the exponent in the numerator.

$$\frac{x^2}{x^5}$$
$$x^{2-5}$$
$$x^{-3}$$

We can also simplify $\frac{x^2}{x^5}$ by dividing out common factors:

$$\frac{x \cdot x}{x \cdot x \cdot x \cdot x \cdot x}$$
$$\frac{1}{x^3}$$

This implies that $x^{-3} = \frac{1}{x^3}$ and it leads us to the definition of a *negative exponent*.

Negative Exponent

If n is an integer and $a \neq 0$, then $a^{-n} = \frac{1}{a^n}$.

The negative exponent tells us we can re-write the expression by taking the reciprocal of the base and then changing the sign of the exponent.

Any expression that has negative exponents is not considered to be in simplest form. We will use the definition of a negative exponent and other properties of exponents to write the expression with only positive exponents.

For example, if after simplifying an expression we end up with the expression x^{-3}, we will take one more step and write $\frac{1}{x^3}$. The answer is considered to be in simplest form when it has only positive exponents.

EXAMPLE 6.89

Simplify: ⓐ 4^{-2} ⓑ 10^{-3}.

⊘ **Solution**

ⓐ

	4^{-2}
Use the definition of a negative exponent, $a^{-n} = \frac{1}{a^n}$.	$\frac{1}{4^2}$
Simplify.	$\frac{1}{16}$

ⓑ

	10^{-3}
Use the definition of a negative exponent, $a^{-n} = \frac{1}{a^n}$.	$\frac{1}{10^3}$
Simplify.	$\frac{1}{1000}$

> **TRY IT :: 6.177**   Simplify: ⓐ 2^{-3} ⓑ 10^{-7}.

> **TRY IT :: 6.178**   Simplify: ⓐ 3^{-2} ⓑ 10^{-4}.

In **Example 6.89** we raised an integer to a negative exponent. What happens when we raise a fraction to a negative exponent? We'll start by looking at what happens to a fraction whose numerator is one and whose denominator is an integer raised to a negative exponent.

	$\frac{1}{a^{-n}}$
Use the definition of a negative exponent, $a^{-n} = \frac{1}{a^n}$.	$\frac{1}{\frac{1}{a^n}}$
Simplify the complex fraction.	$1 \cdot \frac{a^n}{1}$
Multiply.	a^n

This leads to the Property of Negative Exponents.

Property of Negative Exponents

If n is an integer and $a \neq 0$, then $\frac{1}{a^{-n}} = a^n$.

EXAMPLE 6.90

Simplify: ⓐ $\frac{1}{y^{-4}}$ ⓑ $\frac{1}{3^{-2}}$.

✓ **Solution**

ⓐ

$$\frac{1}{y^{-4}}$$

Use the property of a negative exponent, $\frac{1}{a^{-n}} = a^n$.

$$y^4$$

ⓑ

$$\frac{1}{3^{-2}}$$

Use the property of a negative exponent, $\frac{1}{a^{-n}} = a^n$.

$$3^2$$

Simplify.

$$9$$

> **TRY IT :: 6.179**   Simplify: ⓐ $\dfrac{1}{p^{-8}}$ ⓑ $\dfrac{1}{4^{-3}}$.

> **TRY IT :: 6.180**   Simplify: ⓐ $\dfrac{1}{q^{-7}}$ ⓑ $\dfrac{1}{2^{-4}}$.

Suppose now we have a fraction raised to a negative exponent. Let's use our definition of negative exponents to lead us to a new property.

$$\left(\frac{3}{4}\right)^{-2}$$

Use the definition of a negative exponent, $a^{-n} = \frac{1}{a^n}$.

$$\frac{1}{\left(\frac{3}{4}\right)^2}$$

Simplify the denominator.

$$\frac{1}{\frac{9}{16}}$$

Simplify the complex fraction.

$$\frac{16}{9}$$

But we know that $\frac{16}{9}$ is $\left(\frac{4}{3}\right)^2$.

This tells us that:

$$\left(\frac{3}{4}\right)^{-2} = \left(\frac{4}{3}\right)^2$$

To get from the original fraction raised to a negative exponent to the final result, we took the reciprocal of the base—the fraction—and changed the sign of the exponent.

This leads us to the *Quotient to a Negative Power Property*.

Quotient to a Negative Exponent Property

If a and b are real numbers, $a \neq 0$, $b \neq 0$, and n is an integer, then $\left(\frac{a}{b}\right)^{-n} = \left(\frac{b}{a}\right)^n$.

EXAMPLE 6.91

Simplify: ⓐ $\left(\frac{5}{7}\right)^{-2}$ ⓑ $\left(-\frac{2x}{y}\right)^{-3}$.

✓ Solution

ⓐ

$$\left(\frac{5}{7}\right)^{-2}$$

Use the Quotient to a Negative Exponent Property, $\left(\frac{a}{b}\right)^{-n} = \left(\frac{b}{a}\right)^{n}$.

Take the reciprocal of the fraction and change the sign of the exponent.

$$\left(\frac{7}{5}\right)^{2}$$

Simplify.

$$\frac{49}{25}$$

ⓑ

$$\left(-\frac{2x}{y}\right)^{-3}$$

Use the Quotient to a Negative Exponent Property, $\left(\frac{a}{b}\right)^{-n} = \left(\frac{b}{a}\right)^{n}$.

Take the reciprocal of the fraction and change the sign of the exponent.

$$\left(-\frac{y}{2x}\right)^{3}$$

Simplify.

$$-\frac{y^3}{8x^3}$$

> **TRY IT : : 6.181** Simplify: ⓐ $\left(\frac{2}{3}\right)^{-4}$ ⓑ $\left(-\frac{6m}{n}\right)^{-2}$.

> **TRY IT : : 6.182** Simplify: ⓐ $\left(\frac{3}{5}\right)^{-3}$ ⓑ $\left(-\frac{a}{2b}\right)^{-4}$.

When simplifying an expression with exponents, we must be careful to correctly identify the base.

EXAMPLE 6.92

Simplify: ⓐ $(-3)^{-2}$ ⓑ -3^{-2} ⓒ $\left(-\frac{1}{3}\right)^{-2}$ ⓓ $-\left(\frac{1}{3}\right)^{-2}$.

✓ Solution

ⓐ Here the exponent applies to the base -3.

$$(-3)^{-2}$$

Take the reciprocal of the base and change the sign of the exponent.

$$\frac{1}{(-3)^{-2}}$$

Simplify.

$$\frac{1}{9}$$

ⓑ The expression -3^{-2} means "find the opposite of 3^{-2}". Here the exponent applies to the base 3.

$$-3^{-2}$$

Rewrite as a product with -1.

$$-1 \cdot 3^{-2}$$

Take the reciprocal of the base and change the sign of the exponent.

$$-1 \cdot \frac{1}{3^{2}}$$

Simplify.

$$-\frac{1}{9}$$

ⓒ Here the exponent applies to the base $\left(-\frac{1}{3}\right)$.

$$\left(-\frac{1}{3}\right)^{-2}$$

Take the reciprocal of the base and change the sign of the exponent. $\left(-\frac{3}{1}\right)^{2}$

Simplify.   9

ⓓ The expression $-\left(\frac{1}{3}\right)^{-2}$ means "find the opposite of $\left(\frac{1}{3}\right)^{-2}$ ". Here the exponent applies to the base $\left(\frac{1}{3}\right)$.

$$-\left(\frac{1}{3}\right)^{-2}$$

Rewrite as a product with -1. $-1 \cdot \left(\frac{1}{3}\right)^{-2}$

Take the reciprocal of the base and change the sign of the exponent. $-1 \cdot \left(\frac{3}{1}\right)^{2}$

Simplify. -9

> **TRY IT : : 6.183**

Simplify: ⓐ $(-5)^{-2}$ ⓑ -5^{-2} ⓒ $\left(-\frac{1}{5}\right)^{-2}$ ⓓ $-\left(\frac{1}{5}\right)^{-2}$.

> **TRY IT : : 6.184**

Simplify: ⓐ $(-7)^{-2}$ ⓑ -7^{-2}, ⓒ $\left(-\frac{1}{7}\right)^{-2}$ ⓓ $-\left(\frac{1}{7}\right)^{-2}$.

We must be careful to follow the Order of Operations. In the next example, parts (a) and (b) look similar, but the results are different.

EXAMPLE 6.93

Simplify: ⓐ $4 \cdot 2^{-1}$ ⓑ $(4 \cdot 2)^{-1}$.

⊘ **Solution**

ⓐ

Do exponents before multiplication. $4 \cdot 2^{-1}$

Use $a^{-n} = \frac{1}{a^n}$. $4 \cdot \frac{1}{2^1}$

Simplify. 2

ⓑ

                                                                                       $(4 \cdot 2)^{-1}$

Simplify inside the parentheses first. $(8)^{-1}$

Use $a^{-n} = \frac{1}{a^n}$. $\frac{1}{8^1}$

Simplify. $\frac{1}{8}$

> **TRY IT : : 6.185**   Simplify: ⓐ $6 \cdot 3^{-1}$ ⓑ $(6 \cdot 3)^{-1}$.

> **TRY IT : : 6.186**   Simplify: ⓐ $8 \cdot 2^{-2}$ ⓑ $(8 \cdot 2)^{-2}$.

When a variable is raised to a negative exponent, we apply the definition the same way we did with numbers. We will assume all variables are non-zero.

EXAMPLE 6.94

Simplify: ⓐ x^{-6} ⓑ $\left(u^4\right)^{-3}$.

✓ Solution

ⓐ

		x^{-6}
Use the definition of a negative exponent, $a^{-n} = \frac{1}{a^n}$.		$\frac{1}{x^6}$

ⓑ

		$\left(u^4\right)^{-3}$
Use the definition of a negative exponent, $a^{-n} = \frac{1}{a^n}$.		$\frac{1}{\left(u^4\right)^3}$
Simplify.		$\frac{1}{u^{12}}$

> **TRY IT :: 6.187** Simplify: ⓐ y^{-7} ⓑ $\left(z^3\right)^{-5}$.

> **TRY IT :: 6.188** Simplify: ⓐ p^{-9} ⓑ $\left(q^4\right)^{-6}$.

When there is a product and an exponent we have to be careful to apply the exponent to the correct quantity. According to the Order of Operations, we simplify expressions in parentheses before applying exponents. We'll see how this works in the next example.

EXAMPLE 6.95

Simplify: ⓐ $5y^{-1}$ ⓑ $(5y)^{-1}$ ⓒ $(-5y)^{-1}$.

✓ Solution

ⓐ

	$5y^{-1}$
Notice the exponent applies to just the base y. Take the reciprocal of y and change the sign of the exponent.	$5 \cdot \frac{1}{y^1}$
Simplify.	$\frac{5}{y}$

ⓑ

	$(5y)^{-1}$
Here the parentheses make the exponent apply to the base $5y$. Take the reciprocal of $5y$ and change the sign of the exponent.	$\frac{1}{(5y)^1}$
Simplify.	$\frac{1}{5y}$

ⓒ

	$(-5y)^{-1}$
The base here is $-5y$.	
Take the reciprocal of $-5y$ and change the sign of the exponent.	$\dfrac{1}{(-5y)^1}$
Simplify.	$\dfrac{1}{-5y}$
Use $\dfrac{a}{-b} = -\dfrac{a}{b}$.	$-\dfrac{1}{5y}$

> **TRY IT : : 6.189** Simplify: ⓐ $8p^{-1}$ ⓑ $(8p)^{-1}$ ⓒ $(-8p)^{-1}$.

> **TRY IT : : 6.190** Simplify: ⓐ $11q^{-1}$ ⓑ $(11q)^{-1}$ $-(11q)^{-1}$ ⓒ $(-11q)^{-1}$.

With negative exponents, the Quotient Rule needs only one form $\dfrac{a^m}{a^n} = a^{m-n}$, for $a \neq 0$. When the exponent in the denominator is larger than the exponent in the numerator, the exponent of the quotient will be negative.

Simplify Expressions with Integer Exponents

All of the exponent properties we developed earlier in the chapter with whole number exponents apply to integer exponents, too. We restate them here for reference.

Summary of Exponent Properties

If a and b are real numbers, and m and n are integers, then

Product Property	$a^m \cdot a^n = a^{m+n}$
Power Property	$(a^m)^n = a^{m \cdot n}$
Product to a Power	$(ab)^m = a^m b^m$
Quotient Property	$\dfrac{a^m}{a^n} = a^{m-n},\ a \neq 0$
Zero Exponent Property	$a^0 = 1,\ a \neq 0$
Quotient to a Power Property	$\left(\dfrac{a}{b}\right)^m = \dfrac{a^m}{b^m},\ b \neq 0$
Properties of Negative Exponents	$a^{-n} = \dfrac{1}{a^n}$ and $\dfrac{1}{a^{-n}} = a^n$
Quotient to a Negative Exponent	$\left(\dfrac{a}{b}\right)^{-n} = \left(\dfrac{b}{a}\right)^n$

EXAMPLE 6.96

Simplify: ⓐ $x^{-4} \cdot x^6$ ⓑ $y^{-6} \cdot y^4$ ⓒ $z^{-5} \cdot z^{-3}$.

✓ **Solution**

ⓐ

	$x^{-4} \cdot x^6$
Use the Product Property, $a^m \cdot a^n = a^{m+n}$.	x^{-4+6}
Simplify.	x^2

ⓑ

	$y^{-6} \cdot y^4$
Notice the same bases, so add the exponents.	y^{-6+4}
Simplify.	y^{-2}
Use the definition of a negative exponent, $a^{-n} = \frac{1}{a^n}$.	$\frac{1}{y^2}$

ⓒ

	$z^{-5} \cdot z^{-3}$
Add the exponents, since the bases are the same.	z^{-5-3}
Simplify.	z^{-8}
Take the reciprocal and change the sign of the exponent, using the definition of a negative exponent.	$\frac{1}{z^8}$

> **TRY IT : : 6.191**   Simplify: ⓐ $x^{-3} \cdot x^7$ ⓑ $y^{-7} \cdot y^2$ ⓒ $z^{-4} \cdot z^{-5}$.

> **TRY IT : : 6.192**   Simplify: ⓐ $a^{-1} \cdot a^6$ ⓑ $b^{-8} \cdot b^4$ ⓒ $c^{-8} \cdot c^{-7}$.

In the next two examples, we'll start by using the Commutative Property to group the same variables together. This makes it easier to identify the like bases before using the Product Property.

EXAMPLE 6.97

Simplify: $\left(m^4 n^{-3}\right)\left(m^{-5} n^{-2}\right)$.

✓ **Solution**

	$\left(m^4 n^{-3}\right)\left(m^{-5} n^{-2}\right)$
Use the Commutative Property to get like bases together.	$m^4 m^{-5} \cdot n^{-2} n^{-3}$
Add the exponents for each base.	$m^{-1} \cdot n^{-5}$
Take reciprocals and change the signs of the exponents.	$\frac{1}{m^1} \cdot \frac{1}{n^5}$
Simplify.	$\frac{1}{mn^5}$

> **TRY IT : : 6.193**   Simplify: $\left(p^6 q^{-2}\right)\left(p^{-9} q^{-1}\right)$.

> **TRY IT : : 6.194**   Simplify: $\left(r^5 s^{-3}\right)\left(r^{-7} s^{-5}\right)$.

If the monomials have numerical coefficients, we multiply the coefficients, just like we did earlier.

EXAMPLE 6.98

Simplify: $\left(2x^{-6} y^8\right)\left(-5x^5 y^{-3}\right)$.

✓ Solution

$$(2x^{-6}y^8)(-5x^5y^{-3})$$

Rewrite with the like bases together.

$$2(-5)\cdot(x^{-6}x^5)\cdot(y^8y^{-3})$$

Multiply the coefficients and add the exponents of each variable.

$$-10\cdot x^{-1}\cdot y^5$$

Use the definition of a negative exponent, $a^{-n} = \frac{1}{a^n}$.

$$-10\cdot\frac{1}{x^1}\cdot y^5$$

Simplify.

$$\frac{-10y^5}{x}$$

> **TRY IT : : 6.195**   Simplify: $(3u^{-5}v^7)(-4u^4v^{-2})$.

> **TRY IT : : 6.196**   Simplify: $(-6c^{-6}d^4)(-5c^{-2}d^{-1})$.

In the next two examples, we'll use the Power Property and the Product to a Power Property.

EXAMPLE 6.99

Simplify: $(6k^3)^{-2}$.

✓ Solution

$$(6k^3)^{-2}$$

Use the Product to a Power Property, $(ab)^m = a^m b^m$.

$$(6)^{-2}(k^3)^{-2}$$

Use the Power Property, $(a^m)^n = a^{m\cdot n}$.

$$6^{-2}k^{-6}$$

Use the Definition of a Negative Exponent, $a^{-n} = \frac{1}{a^n}$.

$$\frac{1}{6^2}\cdot\frac{1}{k^6}$$

Simplify.

$$\frac{1}{36k^6}$$

> **TRY IT : : 6.197**   Simplify: $(-4x^4)^{-2}$.

> **TRY IT : : 6.198**   Simplify: $(2b^3)^{-4}$.

EXAMPLE 6.100

Simplify: $(5x^{-3})^2$.

✓ Solution

$$\left(5x^{-3}\right)^2$$

Use the Product to a Power Property, $(ab)^m = a^m b^m$.	$5^2\left(x^{-3}\right)^2$
Simplify 5^2 and multiply the exponents of x using the Power Property, $(a^m)^n = a^{m \cdot n}$.	$25 \cdot x^{-6}$
Rewrite x^{-6} by using the Definition of a Negative Exponent, $a^{-n} = \dfrac{1}{a^n}$.	$25 \cdot \dfrac{1}{x^6}$
Simplify.	$\dfrac{25}{x^6}$

> **TRY IT :: 6.199** Simplify: $\left(8a^{-4}\right)^2$.

> **TRY IT :: 6.200** Simplify: $\left(2c^{-4}\right)^3$.

To simplify a fraction, we use the Quotient Property and subtract the exponents.

EXAMPLE 6.101

Simplify: $\dfrac{r^5}{r^{-4}}$.

✓ Solution

$$\dfrac{r^5}{r^{-4}}$$

Use the Quotient Property, $\dfrac{a^m}{a^n} = a^{m-n}$.	$r^{5-(-4)}$
Simplify.	r^9

> **TRY IT :: 6.201** Simplify: $\dfrac{x^8}{x^{-3}}$.

> **TRY IT :: 6.202** Simplify: $\dfrac{y^8}{y^{-6}}$.

Convert from Decimal Notation to Scientific Notation

Remember working with place value for whole numbers and decimals? Our number system is based on powers of 10. We use tens, hundreds, thousands, and so on. Our decimal numbers are also based on powers of tens—tenths, hundredths, thousandths, and so on. Consider the numbers 4,000 and 0.004. We know that 4,000 means $4 \times 1,000$ and 0.004 means $4 \times \dfrac{1}{1,000}$.

If we write the 1000 as a power of ten in exponential form, we can rewrite these numbers in this way:

$$4{,}000 \qquad\qquad 0.004$$

$$4 \times 1{,}000 \qquad\qquad 4 \times \frac{1}{1{,}000}$$

$$4 \times 10^3 \qquad\qquad 4 \times \frac{1}{10^3}$$

$$4 \times 10^{-3}$$

When a number is written as a product of two numbers, where the first factor is a number greater than or equal to one but less than 10, and the second factor is a power of 10 written in exponential form, it is said to be in *scientific notation*.

Scientific Notation

A number is expressed in **scientific notation** when it is of the form

$$a \times 10^n \text{ where } 1 \le a < 10 \text{ and } n \text{ is an integer}$$

It is customary in scientific notation to use as the $\times$ multiplication sign, even though we avoid using this sign elsewhere in algebra.

If we look at what happened to the decimal point, we can see a method to easily convert from decimal notation to scientific notation.

$$4000. = 4 \times 10^3 \qquad\qquad 0.004 = 4 \times 10^{-3}$$

$$4000. = 4 \times 10^3 \qquad\qquad 0.004 = 4 \times 10^{-3}$$

Moved the decimal point 3   Moved the decimal point 3
places to the left.   places to the right.

In both cases, the decimal was moved 3 places to get the first factor between 1 and 10.

The power of 10 is positive when the number is larger than 1: $\qquad 4{,}000 = 4 \times 10^3$

The power of 10 is negative when the number is between 0 and 1: $\quad 0.004 = 4 \times 10^{-3}$

EXAMPLE 6.102   HOW TO CONVERT FROM DECIMAL NOTATION TO SCIENTIFIC NOTATION

Write in scientific notation: 37,000.

✓ **Solution**

Step 1. Move the decimal point so that the first factor is greater than or equal to 1 but less than 10.	Remember, there is a decimal at the end of 37,000. Move the decimal after the 3. 3.700 is between 1 and 10.	37,000.
Step 2. Count the number of decimal places, n, that the decimal point was moved.	The decimal point was moved 4 places to the left.	37000.
Step 3. Write the number as a product with a power of 10. If the original number is: Greater than 1, the power of 10 will be 10^n. Between 0 and 1, the power of 10 will be 10^{-n}.	37,000 is greater than 1 so the power of 10 will have exponent 4.	3.7×10^4
Step 4. Check.	Check to see if your answer makes sense. 	10^4 is 10,000 and 10,000 times 3.7 will be 37,000. $37{,}000 = 3.7 \times 10^4$

> **TRY IT : : 6.203**   Write in scientific notation: 96,000.

> **TRY IT :: 6.204**   Write in scientific notation: 48,300.

HOW TO :: CONVERT FROM DECIMAL NOTATION TO SCIENTIFIC NOTATION

Step 1.   Move the decimal point so that the first factor is greater than or equal to 1 but less than 10.

Step 2.   Count the number of decimal places, n, that the decimal point was moved.

Step 3.   Write the number as a product with a power of 10.
If the original number is:
- greater than 1, the power of 10 will be 10^n.
- between 0 and 1, the power of 10 will be 10^{-n}.

Step 4.   Check.

EXAMPLE 6.103

Write in scientific notation: 0.0052.

✓ **Solution**

The original number, 0.0052, is between 0 and 1 so we will have a negative power of 10.

	0.0052
Move the decimal point to get 5.2, a number between 1 and 10.	0.0052
Count the number of decimal places the point was moved.	3 places
Write as a product with a power of 10.	5.2×10^{-3}
Check.	

$$5.2 \times 10^{-3}$$
$$5.2 \times \frac{1}{10^3}$$
$$5.2 \times \frac{1}{1000}$$
$$5.2 \times 0.001$$

0.0052	$0.0052 = 5.2 \times 10^{-3}$

> **TRY IT :: 6.205**   Write in scientific notation: 0.0078.

> **TRY IT :: 6.206**   Write in scientific notation: 0.0129.

Convert Scientific Notation to Decimal Form

How can we convert from scientific notation to decimal form? Let's look at two numbers written in scientific notation and see.

$$9.12 \times 10^4 \qquad\qquad 9.12 \times 10^{-4}$$
$$9.12 \times 10,000 \qquad\qquad 9.12 \times 0.0001$$
$$91,200 \qquad\qquad 0.000912$$

If we look at the location of the decimal point, we can see an easy method to convert a number from scientific notation to decimal form.

$$9.12 \times 10^4 = 91{,}200 \qquad\qquad 9.12 \times 10^{-4} = 0.000912$$

9.12 × 10⁴ = 91,200   9.12 × 10⁴ = 0.000912

9.12⌣⌣⌣ × 10⁴ = 91,200   ⌣⌣⌣9.12 × 10⁴ = 0.000912

Move the decimal   Move the decimal
point 4 places to   point 4 places to
the right.   the left.

In both cases the decimal point moved 4 places. When the exponent was positive, the decimal moved to the right. When the exponent was negative, the decimal point moved to the left.

EXAMPLE 6.104   HOW TO CONVERT SCIENTIFIC NOTATION TO DECIMAL FORM

Convert to decimal form: 6.2×10^3.

✓ **Solution**

Step 1. Determine the exponent, n, on the factor 10.	The exponent is 3.	6.2×10^3
Step 2. Move the decimal n places, adding zeros if needed. If the exponent is positive, move the decimal point n places to the right. If the exponent is negative, move the decimal point $\lvert n \rvert$ places to the left.	The exponent is positive, so move the decimal point 3 places to the right. We need to add 2 zeros as placeholders.	6.200. 6,200
Step 3. Check to see if your answer makes sense.		10^3 is 1000 and 1000 times 6.2 will be 6,200. $6.2 \times 10^3 = 6{,}200$

> **TRY IT : : 6.207**   Convert to decimal form: 1.3×10^3.

> **TRY IT : : 6.208**   Convert to decimal form: 9.25×10^4.

The steps are summarized below.

HOW TO : : CONVERT SCIENTIFIC NOTATION TO DECIMAL FORM.

To convert scientific notation to decimal form:

Step 1.   Determine the exponent, n, on the factor 10.

Step 2.   Move the decimal n places, adding zeros if needed.

     ◦   If the exponent is positive, move the decimal point n places to the right.

     ◦   If the exponent is negative, move the decimal point $\lvert n \rvert$ places to the left.

Step 3.   Check.

EXAMPLE 6.105

Convert to decimal form: 8.9×10^{-2}.

Solution

	8.9×10^{-2}
Determine the exponent, *n*, on the factor 10.	The exponent is –2.
Since the exponent is negative, move the decimal point 2 places to the left.	8 . 9
Add zeros as needed for placeholders.	$8.9 \times 10^{-2} = 0.089$

> **TRY IT :: 6.209**   Convert to decimal form: 1.2×10^{-4}.

> **TRY IT :: 6.210**   Convert to decimal form: 7.5×10^{-2}.

Multiply and Divide Using Scientific Notation

Astronomers use very large numbers to describe distances in the universe and ages of stars and planets. Chemists use very small numbers to describe the size of an atom or the charge on an electron. When scientists perform calculations with very large or very small numbers, they use scientific notation. Scientific notation provides a way for the calculations to be done without writing a lot of zeros. We will see how the Properties of Exponents are used to multiply and divide numbers in scientific notation.

EXAMPLE 6.106

Multiply. Write answers in decimal form: $\left(4 \times 10^5\right)\left(2 \times 10^{-7}\right)$.

Solution

	$\left(4 \times 10^5\right)\left(2 \times 10^{-7}\right)$
Use the Commutative Property to rearrange the factors.	$4 \cdot 2 \cdot 10^5 \cdot 10^{-7}$
Multiply.	8×10^{-2}
Change to decimal form by moving the decimal two places left.	0.08

> **TRY IT :: 6.211**   Multiply $\left(3 \times 10^6\right)\left(2 \times 10^{-8}\right)$. Write answers in decimal form.

> **TRY IT :: 6.212**   Multiply $\left(3 \times 10^{-2}\right)\left(3 \times 10^{-1}\right)$. Write answers in decimal form.

EXAMPLE 6.107

Divide. Write answers in decimal form: $\dfrac{9 \times 10^3}{3 \times 10^{-2}}$.

⊘ **Solution**

$$\frac{9 \times 10^3}{3 \times 10^{-2}}$$

Separate the factors, rewriting as the product of two fractions. $\frac{9}{3} \times \frac{10^3}{10^{-2}}$

Divide. 3×10^5

Change to decimal form by moving the decimal five places right.   300,000

> | **TRY IT ::** 6.213   Divide $\frac{8 \times 10^4}{2 \times 10^{-1}}$. Write answers in decimal form.

> | **TRY IT ::** 6.214   Divide $\frac{8 \times 10^2}{4 \times 10^{-2}}$. Write answers in decimal form.

▶ | **MEDIA ::**

Access these online resources for additional instruction and practice with integer exponents and scientific notation:

- **Negative Exponents (https://openstax.org/l/25Negexponents)**
- **Scientific Notation (https://openstax.org/l/25Scientnot1)**
- **Scientific Notation 2 (https://openstax.org/l/25Scientnot2)**

6.7 EXERCISES

Practice Makes Perfect

Use the Definition of a Negative Exponent

In the following exercises, simplify.

500.
ⓐ 4^{-2}

ⓑ 10^{-3}

501.
ⓐ 3^{-4}

ⓑ 10^{-2}

502.
ⓐ 5^{-3}

ⓑ 10^{-5}

503.
ⓐ 2^{-8}

ⓑ 10^{-2}

504.
ⓐ $\dfrac{1}{c^{-5}}$

ⓑ $\dfrac{1}{3^{-2}}$

505.
ⓐ $\dfrac{1}{c^{-5}}$

ⓑ $\dfrac{1}{5^{-2}}$

506.
ⓐ $\dfrac{1}{q^{-10}}$

ⓑ $\dfrac{1}{10^{-3}}$

507.
ⓐ $\dfrac{1}{t^{-9}}$

ⓑ $\dfrac{1}{10^{-4}}$

508.
ⓐ $\left(\dfrac{5}{8}\right)^{-2}$

ⓑ $\left(-\dfrac{3m}{n}\right)^{-2}$

509.
ⓐ $\left(\dfrac{3}{10}\right)^{-2}$

ⓑ $\left(-\dfrac{2}{cd}\right)^{-3}$

510.
ⓐ $\left(\dfrac{4}{9}\right)^{-3}$

ⓑ $\left(-\dfrac{u^2}{2v}\right)^{-5}$

511.
ⓐ $\left(\dfrac{7}{2}\right)^{-3}$

ⓑ $\left(-\dfrac{3}{xy^2}\right)^{-3}$

512.
ⓐ $(-5)^{-2}$

ⓑ -5^{-2}

ⓒ $\left(-\dfrac{1}{5}\right)^{-2}$

ⓓ $-\left(\dfrac{1}{5}\right)^{-2}$

513.
ⓐ $(-7)^{-2}$

ⓑ -7^{-2}

ⓒ $\left(-\dfrac{1}{7}\right)^{-2}$

ⓓ $-\left(\dfrac{1}{7}\right)^{-2}$

514.
ⓐ -3^{-3}

ⓑ $\left(-\dfrac{1}{3}\right)^{-3}$

ⓒ $-\left(\dfrac{1}{3}\right)^{-3}$

ⓓ $(-3)^{-3}$

515.
ⓐ -5^{-3}

ⓑ $\left(-\dfrac{1}{5}\right)^{-3}$

ⓒ $-\left(\dfrac{1}{5}\right)^{-3}$

ⓓ $(-5)^{-3}$

516.
ⓐ $3 \cdot 5^{-1}$

ⓑ $(3 \cdot 5)^{-1}$

517.
ⓐ $2 \cdot 5^{-1}$

ⓑ $(2 \cdot 5)^{-1}$

518.
ⓐ $4 \cdot 5^{-2}$

ⓑ $(4 \cdot 5)^{-2}$

519.
ⓐ $3 \cdot 4^{-2}$

ⓑ $(3 \cdot 4)^{-2}$

520.
ⓐ m^{-4}

ⓑ $\left(x^3\right)^{-4}$

521.
ⓐ b^{-5}

ⓑ $\left(k^2\right)^{-5}$

522.
ⓐ p^{-10}

ⓑ $\left(q^6\right)^{-8}$

523.
ⓐ s^{-8}

ⓑ $\left(a^9\right)^{-10}$

524.
ⓐ $7n^{-1}$

ⓑ $(7n)^{-1}$

ⓒ $(-7n)^{-1}$

525.
ⓐ $6r^{-1}$

ⓑ $(6r)^{-1}$

ⓒ $(-6r)^{-1}$

526.
ⓐ $(3p)^{-2}$

ⓑ $3p^{-2}$

ⓒ $-3p^{-2}$

527.
ⓐ $(2q)^{-4}$

ⓑ $2q^{-4}$

ⓒ $-2q^{-4}$

Simplify Expressions with Integer Exponents

In the following exercises, simplify.

528.
ⓐ $b^4 b^{-8}$

ⓑ $r^{-2} r^5$

ⓒ $x^{-7} x^{-3}$

529.
ⓐ $s^3 \cdot s^{-7}$

ⓑ $q^{-8} \cdot q^3$

ⓒ $y^{-2} \cdot y^{-5}$

530.
ⓐ $a^3 \cdot a^{-3}$

ⓑ $a \cdot a^3$

ⓒ $a \cdot a^{-3}$

531.
ⓐ $y^5 \cdot y^{-5}$

ⓑ $y \cdot y^5$

ⓒ $y \cdot y^{-5}$

532. $p^5 \cdot p^{-2} \cdot p^{-4}$

533. $x^4 \cdot x^{-2} \cdot x^{-3}$

534. $\left(w^4 x^{-5}\right)\left(w^{-2} x^{-4}\right)$

535. $\left(m^3 n^{-3}\right)\left(m^{-5} n^{-1}\right)$

536. $\left(uv^{-2}\right)\left(u^{-5} v^{-3}\right)$

537. $\left(pq^{-4}\right)\left(p^{-6} q^{-3}\right)$

538. $\left(-6c^{-3} d^9\right)\left(2c^4 d^{-5}\right)$

539. $\left(-2j^{-5} k^8\right)\left(7j^2 k^{-3}\right)$

540. $\left(-4r^{-2} s^{-8}\right)\left(9r^4 s^3\right)$

541. $\left(-5m^4 n^6\right)\left(8m^{-5} n^{-3}\right)$

542. $\left(5x^2\right)^{-2}$

543. $\left(4y^3\right)^{-3}$

544. $\left(3z^{-3}\right)^2$

545. $\left(2p^{-5}\right)^2$

546. $\dfrac{t^9}{t^{-3}}$

547. $\dfrac{n^5}{n^{-2}}$

548. $\dfrac{x^{-7}}{x^{-3}}$

549. $\dfrac{y^{-5}}{y^{-10}}$

Convert from Decimal Notation to Scientific Notation

In the following exercises, write each number in scientific notation.

550. 57,000

551. 340,000

552. 8,750,000

553. 1,290,000

554. 0.026

555. 0.041

556. 0.00000871

557. 0.00000103

Convert Scientific Notation to Decimal Form

In the following exercises, convert each number to decimal form.

558. 5.2×10^2

559. 8.3×10^2

560. 7.5×10^6

561. 1.6×10^{10}

562. 2.5×10^{-2}

563. 3.8×10^{-2}

564. 4.13×10^{-5}

565. 1.93×10^{-5}

Multiply and Divide Using Scientific Notation

In the following exercises, multiply. Write your answer in decimal form.

566. $\left(3 \times 10^{-5}\right)\left(3 \times 10^9\right)$

567. $\left(2 \times 10^2\right)\left(1 \times 10^{-4}\right)$

568. $\left(7.1 \times 10^{-2}\right)\left(2.4 \times 10^{-4}\right)$

569. $\left(3.5 \times 10^{-4}\right)\left(1.6 \times 10^{-2}\right)$

In the following exercises, divide. Write your answer in decimal form.

570. $\dfrac{7 \times 10^{-3}}{1 \times 10^{-7}}$

571. $\dfrac{5 \times 10^{-2}}{1 \times 10^{-10}}$

572. $\dfrac{6 \times 10^4}{3 \times 10^{-2}}$

573. $\dfrac{8 \times 10^6}{4 \times 10^{-1}}$

Everyday Math

574. The population of the United States on July 4, 2010 was almost 310,000,000. Write the number in scientific notation.

575. The population of the world on July 4, 2010 was more than 6,850,000,000. Write the number in scientific notation

576. The average width of a human hair is 0.0018 centimeters. Write the number in scientific notation.

577. The probability of winning the 2010 Megamillions lottery was about 0.0000000057. Write the number in scientific notation.

578. In 2010, the number of Facebook users each day who changed their status to 'engaged' was 2×10^4. Convert this number to decimal form.

579. At the start of 2012, the US federal budget had a deficit of more than $\$1.5 \times 10^{13}$. Convert this number to decimal form.

580. The concentration of carbon dioxide in the atmosphere is 3.9×10^{-4}. Convert this number to decimal form.

581. The width of a proton is 1×10^{-5} of the width of an atom. Convert this number to decimal form.

582. Health care costs The Centers for Medicare and Medicaid projects that consumers will spend more than $4 trillion on health care by 2017.

ⓐ Write 4 trillion in decimal notation.

ⓑ Write 4 trillion in scientific notation.

583. Coin production In 1942, the U.S. Mint produced 154,500,000 nickels. Write 154,500,000 in scientific notation.

584. Distance The distance between Earth and one of the brightest stars in the night star is 33.7 light years. One light year is about 6,000,000,000,000 (6 trillion), miles.

ⓐ Write the number of miles in one light year in scientific notation.

ⓑUse scientific notation to find the distance between Earth and the star in miles. Write the answer in scientific notation.

585. Debt At the end of fiscal year 2015 the gross United States federal government debt was estimated to be approximately $18,600,000,000,000 ($18.6 trillion), according to the Federal Budget. The population of the United States was approximately 300,000,000 people at the end of fiscal year 2015.

ⓐ Write the debt in scientific notation.

ⓑ Write the population in scientific notation.

ⓒ Find the amount of debt per person by using scientific notation to divide the debt by the population. Write the answer in scientific notation.

Writing Exercises

586.

ⓐ Explain the meaning of the exponent in the expression 2^3.

ⓑ Explain the meaning of the exponent in the expression 2^{-3}.

587. When you convert a number from decimal notation to scientific notation, how do you know if the exponent will be positive or negative?

Self Check

ⓐ After completing the exercises, use this checklist to evaluate your mastery of the objectives of this section.

I can...	Confidently	With some help	No-I don't get it!
use the definition of a negative exponent.			
simplify expressions with integer exponents.			
convert from decimal notation to scientific notation.			
convert scientific notation to decimal form.			
multiply and divide using scientific notation.			

ⓑ Overall, after looking at the checklist, do you think you are well-prepared for the next section? Why or why not?

CHAPTER 6 REVIEW

KEY TERMS

binomial A binomial is a polynomial with exactly two terms.

conjugate pair A conjugate pair is two binomials of the form $(a - b)$, $(a + b)$; the pair of binomials each have the same first term and the same last term, but one binomial is a sum and the other is a difference.

degree of a constant The degree of any constant is 0.

degree of a polynomial The degree of a polynomial is the highest degree of all its terms.

degree of a term The degree of a term is the exponent of its variable.

monomial A monomial is a term of the form ax^m, where a is a constant and m is a whole number; a monomial has exactly one term.

negative exponent If n is a positive integer and $a \neq 0$, then $a^{-n} = \frac{1}{a^n}$.

polynomial A polynomial is a monomial, or two or more monomials combined by addition or subtraction.

scientific notation A number is expressed in scientific notation when it is of the form $a \times 10^n$ where $a \geq 1$ and $a < 10$ and n is an integer.

standard form A polynomial is in standard form when the terms of a polynomial are written in descending order of degrees.

trinomial A trinomial is a polynomial with exactly three terms.

KEY CONCEPTS

6.1 Add and Subtract Polynomials

- **Monomials**
  - A monomial is a term of the form ax^m, where a is a constant and m is a whole number

- **Polynomials**
  - **polynomial**—A monomial, or two or more monomials combined by addition or subtraction is a polynomial.
  - **monomial**—A polynomial with exactly one term is called a monomial.
  - **binomial**—A polynomial with exactly two terms is called a binomial.
  - **trinomial**—A polynomial with exactly three terms is called a trinomial.

- **Degree of a Polynomial**
  - The **degree of a term** is the sum of the exponents of its variables.
  - The **degree of a constant** is 0.
  - The **degree of a polynomial** is the highest degree of all its terms.

6.2 Use Multiplication Properties of Exponents

- **Exponential Notation**

  a^m ← exponent, base ; a^m means multiply m factors of a

  $$a^m = a \cdot a \cdot a \cdot \ldots \cdot a$$
  m factors

- **Properties of Exponents**
  - If a, b are real numbers and m, n are whole numbers, then

Product Property	$a^m \cdot a^n$	$= a^{m+n}$
Power Property	$(a^m)^n$	$= a^{m \cdot n}$
Product to a Power	$(ab)^m$	$= a^m b^m$

6.3 Multiply Polynomials

- **FOIL Method for Multiplying Two Binomials**—To multiply two binomials:

  Step 1. Multiply the **First** terms.

  Step 2. Multiply the **Outer** terms.

  Step 3. Multiply the **Inner** terms.

  Step 4. Multiply the **Last** terms.

- **Multiplying Two Binomials**—To multiply binomials, use the:
  - Distributive Property (**Example 6.34**)
  - FOIL Method (**Example 6.39**)
  - Vertical Method (**Example 6.44**)

- **Multiplying a Trinomial by a Binomial**—To multiply a trinomial by a binomial, use the:
  - Distributive Property (**Example 6.45**)
  - Vertical Method (**Example 6.46**)

6.4 Special Products

- **Binomial Squares Pattern**
  - If a, b are real numbers,

$$\underbrace{(a+b)^2}_{(binomial)^2} = \underbrace{a^2}_{(first\ term)^2} + \underbrace{2ab}_{2(product\ of\ terms)} + \underbrace{b^2}_{(last\ term)^2}$$

  - $(a+b)^2 = a^2 + 2ab + b^2$
  - $(a-b)^2 = a^2 - 2ab + b^2$
  - To square a binomial: square the first term, square the last term, double their product.

- **Product of Conjugates Pattern**
  - If a, b are real numbers,

$$\underbrace{(a-b)(a+b)}_{conjugates} = \underbrace{a^2}_{squares} - \underbrace{b^2}_{squares}$$

  - $(a-b)(a+b) = a^2 - b^2$
  - The product is called a difference of squares.

- **To multiply conjugates:**
  - **square the first term square the last term** write it as a difference of squares

6.5 Divide Monomials

- **Quotient Property for Exponents:**
  - If a is a real number, $a \neq 0$, and m, n are whole numbers, then:

  $\frac{a^m}{a^n} = a^{m-n}, m > n$ and $\frac{a^m}{a^n} = \frac{1}{a^{m-n}}, n > m$

- **Zero Exponent**
  - If a is a non-zero number, then $a^0 = 1$.

- **Quotient to a Power Property for Exponents**:
  - ◦ If a and b are real numbers, $b \neq 0,$ and m is a counting number, then:

  $$\left(\frac{a}{b}\right)^m = \frac{a^m}{b^m}$$

  - ◦ To raise a fraction to a power, raise the numerator and denominator to that power.

- **Summary of Exponent Properties**
  - ◦ If a, b are real numbers and m, n are whole numbers, then

Product Property	$a^m \cdot a^n$	$= a^{m+n}$
Power Property	$(a^m)^n$	$= a^{m \cdot n}$
Product to a Power	$(ab)^m$	$= a^m b^m$
Quotient Property	$\dfrac{a^m}{b^m}$	$= a^{m-n}, a \neq 0, m > n$
	$\dfrac{a^m}{a^n}$	$= \dfrac{1}{a^{n-m}}, a \neq 0, n > m$
Zero Exponent Definition	a^o	$= 1, a \neq 0$
Quotient to a Power Property	$\left(\dfrac{a}{b}\right)^m$	$= \dfrac{a^m}{b^m}, b \neq 0$

6.6 Divide Polynomials

- **Fraction Addition**
  - ◦ If a, b, and c are numbers where $c \neq 0$, then

  $$\frac{a}{c} + \frac{b}{c} = \frac{a+b}{c} \text{ and } \frac{a+b}{c} = \frac{a}{c} + \frac{b}{c}$$

- **Division of a Polynomial by a Monomial**
  - ◦ To divide a polynomial by a monomial, divide each term of the polynomial by the monomial.

6.7 Integer Exponents and Scientific Notation

- **Property of Negative Exponents**
  - ◦ If n is a positive integer and $a \neq 0$, then $\dfrac{1}{a^{-n}} = a^n$

- **Quotient to a Negative Exponent**
  - ◦ If a, b are real numbers, $b \neq 0$ and n is an integer, then $\left(\dfrac{a}{b}\right)^{-n} = \left(\dfrac{b}{a}\right)^n$

- **To convert a decimal to scientific notation:**

  Step 1. Move the decimal point so that the first factor is greater than or equal to 1 but less than 10.

  Step 2. Count the number of decimal places, n, that the decimal point was moved.

  Step 3. Write the number as a product with a power of 10. If the original number is:
  - greater than 1, the power of 10 will be 10^n
  - between 0 and 1, the power of 10 will be 10^{-n}

  Step 4. Check.

- **To convert scientific notation to decimal form:**

  Step 1. Determine the exponent, n, on the factor 10.

Step 2. Move the decimal n places, adding zeros if needed.

- If the exponent is positive, move the decimal point n places to the right.
- If the exponent is negative, move the decimal point $|n|$ places to the left.

Step 3. Check.

REVIEW EXERCISES

6.1 Section 6.1 Add and Subtract Polynomials

Identify Polynomials, Monomials, Binomials and Trinomials

In the following exercises, determine if each of the following polynomials is a monomial, binomial, trinomial, or other polynomial.

588.
ⓐ $11c^4 - 23c^2 + 1$
ⓑ $9p^3 + 6p^2 - p - 5$
ⓒ $\frac{3}{7}x + \frac{5}{14}$
ⓓ 10
ⓔ $2y - 12$

589.
ⓐ $a^2 - b^2$
ⓑ $24d^3$
ⓒ $x^2 + 8x - 10$
ⓓ $m^2 n^2 - 2mn + 6$
ⓔ $7y^3 + y^2 - 2y - 4$

Determine the Degree of Polynomials

In the following exercises, determine the degree of each polynomial.

590.
ⓐ $3x^2 + 9x + 10$
ⓑ $14a^2 bc$
ⓒ $6y + 1$
ⓓ $n^3 - 4n^2 + 2n - 8$
ⓔ -19

591.
ⓐ
$5p^3 - 8p^2 + 10p - 4$
ⓑ $-20q^4$
ⓒ $x^2 + 6x + 12$
ⓓ $23r^2 s^2 - 4rs + 5$
ⓔ 100

Add and Subtract Monomials

In the following exercises, add or subtract the monomials.

592. $5y^3 + 8y^3$

593. $-14k + 19k$

594. $12q - (-6q)$

595. $-9c - 18c$

596. $12x - 4y - 9x$

597. $3m^2 + 7n^2 - 3m^2$

598. $6x^2 y - 4x + 8xy^2$

599. $13a + b$

Add and Subtract Polynomials

In the following exercises, add or subtract the polynomials.

600.
$(5x^2 + 12x + 1) + (6x^2 - 8x + 3)$

601. $(9p^2 - 5p + 3) + (4p^2 - 4)$

602.
$(10m^2 - 8m - 1) - (5m^2 + m - 2)$

603. $(7y^2 - 8y) - (y - 4)$

604. Subtract
$(3s^2 + 10)$ from $(15s^2 - 2s + 8)$

605. Find the sum of $(a^2 + 6a + 9)$ and $(5a^3 - 7)$

Evaluate a Polynomial for a Given Value of the Variable

In the following exercises, evaluate each polynomial for the given value.

606. Evaluate $3y^2 - y + 1$ when:

ⓐ $y = 5$ ⓑ $y = -1$

ⓒ $y = 0$

607. Evaluate $10 - 12x$ when:

ⓐ $x = 3$ ⓑ $x = 0$

ⓒ $x = -1$

608. Randee drops a stone off the 200 foot high cliff into the ocean. The polynomial $-16t^2 + 200$ gives the height of a stone t seconds after it is dropped from the cliff. Find the height after $t = 3$ seconds.

609. A manufacturer of stereo sound speakers has found that the revenue received from selling the speakers at a cost of p dollars each is given by the polynomial $-4p^2 + 460p$. Find the revenue received when $p = 75$ dollars.

6.2 Section 6.2 Use Multiplication Properties of Exponents

Simplify Expressions with Exponents

In the following exercises, simplify.

610. 10^4

611. 17^1

612. $\left(\dfrac{2}{9}\right)^2$

613. $(0.5)^3$

614. $(-2)^6$

615. -2^6

Simplify Expressions Using the Product Property for Exponents

In the following exercises, simplify each expression.

616. $x^4 \cdot x^3$

617. $p^{15} \cdot p^{16}$

618. $4^{10} \cdot 4^6$

619. $8 \cdot 8^5$

620. $n \cdot n^2 \cdot n^4$

621. $y^c \cdot y^3$

Simplify Expressions Using the Power Property for Exponents

In the following exercises, simplify each expression.

622. $(m^3)^5$

623. $(5^3)^2$

624. $\left(y^4\right)^x$

625. $(3^r)^s$

Simplify Expressions Using the Product to a Power Property

In the following exercises, simplify each expression.

626. $(4a)^2$

627. $(-5y)^3$

628. $(2mn)^5$

629. $(10xyz)^3$

Simplify Expressions by Applying Several Properties

In the following exercises, simplify each expression.

630. $\left(p^2\right)^5 \cdot \left(p^3\right)^6$

631. $(4a^3b^2)^3$

632. $(5x)^2(7x)$

633. $(2q^3)^4(3q)^2$

634. $\left(\frac{1}{3}x^2\right)^2\left(\frac{1}{2}x\right)^3$

635. $\left(\frac{2}{5}m^2n\right)^3$

Multiply Monomials

In the following exercises 8, multiply the monomials.

636. $\left(-15x^2\right)\left(6x^4\right)$

637. $(-9n^7)(-16n)$

638. $(7p^5q^3)(8pq^9)$

639. $\left(\frac{5}{9}ab^2\right)\left(27ab^3\right)$

6.3 Section 6.3 Multiply Polynomials

Multiply a Polynomial by a Monomial

In the following exercises, multiply.

640. $7(a+9)$

641. $-4(y+13)$

642. $-5(r-2)$

643. $p(p+3)$

644. $-m(m+15)$

645. $-6u(2u+7)$

646. $9(b^2+6b+8)$

647. $3q^2(q^2-7q+6)$ 3

648. $(5z-1)z$

649. $(b-4)\cdot 11$

Multiply a Binomial by a Binomial

In the following exercises, multiply the binomials using: ⓐ the Distributive Property, ⓑ the FOIL method, ⓒ the Vertical Method.

650. $(x-4)(x+10)$

651. $(6y-7)(2y-5)$

In the following exercises, multiply the binomials. Use any method.

652. $(x+3)(x+9)$

653. $(y-4)(y-8)$

654. $(p-7)(p+4)$

655. $(q+16)(q-3)$

656. $(5m-8)(12m+1)$

657. $(u^2+6)(u^2-5)$

658. $(9x-y)(6x-5)$

659. $(8mn+3)(2mn-1)$

Multiply a Trinomial by a Binomial

In the following exercises, multiply using ⓐ the Distributive Property, ⓑ the Vertical Method.

660. $(n+1)(n^2+5n-2)$

661. $(3x-4)(6x^2+x-10)$

In the following exercises, multiply. Use either method.

662. $(y-2)(y^2-8y+9)$

663. $(7m+1)(m^2-10m-3)$

6.4 Section 6.4 Special Products

Square a Binomial Using the Binomial Squares Pattern

In the following exercises, square each binomial using the Binomial Squares Pattern.

664. $(c+11)^2$

665. $(q-15)^2$

666. $\left(x+\frac{1}{3}\right)^2$

667. $(8u + 1)^2$

668. $(3n^3 - 2)^2$

669. $(4a - 3b)^2$

Multiply Conjugates Using the Product of Conjugates Pattern

In the following exercises, multiply each pair of conjugates using the Product of Conjugates Pattern.

670. $(s - 7)(s + 7)$

671. $\left(y + \frac{2}{5}\right)\left(y - \frac{2}{5}\right)$

672. $(12c + 13)(12c - 13)$

673. $(6 - r)(6 + r)$

674. $\left(u + \frac{3}{4}v\right)\left(u - \frac{3}{4}v\right)$

675. $(5p^4 - 4q^3)(5p^4 + 4q^3)$

Recognize and Use the Appropriate Special Product Pattern

In the following exercises, find each product.

676. $(3m + 10)^2$

677. $(6a + 11)(6a - 11)$

678. $(5x + y)(x - 5y)$

679. $(c^4 + 9d)^2$

680. $(p^5 + q^5)(p^5 - q^5)$

681. $(a^2 + 4b)(4a - b^2)$

6.5 Section 6.5 Divide Monomials

Simplify Expressions Using the Quotient Property for Exponents

In the following exercises, simplify.

682. $\dfrac{u^{24}}{u^6}$

683. $\dfrac{10^{25}}{10^5}$

684. $\dfrac{3^4}{3^6}$

685. $\dfrac{v^{12}}{v^{48}}$

686. $\dfrac{x}{x^5}$

687. $\dfrac{5}{5^8}$

Simplify Expressions with Zero Exponents

In the following exercises, simplify.

688. 75^0

689. x^0

690. -12^0

691. $\left(-12^0\right)\left(-12\right)^0$

692. $25x^0$

693. $(25x)^0$

694. $19n^0 - 25m^0$

695. $(19n)^0 - (25m)^0$

Simplify Expressions Using the Quotient to a Power Property

In the following exercises, simplify.

696. $\left(\dfrac{2}{5}\right)^3$

697. $\left(\dfrac{m}{3}\right)^4$

698. $\left(\dfrac{r}{s}\right)^8$

699. $\left(\dfrac{x}{2y}\right)^6$

Simplify Expressions by Applying Several Properties

In the following exercises, simplify.

700. $\dfrac{\left(x^3\right)^5}{x^9}$

701. $\dfrac{n^{10}}{\left(n^5\right)^2}$

702. $\left(\dfrac{q^6}{q^8}\right)^3$

703. $\left(\dfrac{r^8}{r^3}\right)^4$

704. $\left(\dfrac{c^2}{d^5}\right)^9$

705. $\left(\dfrac{3x^4}{2y^2}\right)^5$

706. $\left(\dfrac{v^3 v^9}{v^6}\right)^4$

707. $\dfrac{\left(3n^2\right)^4\left(-5n^4\right)^3}{\left(-2n^5\right)^2}$

Divide Monomials

In the following exercises, divide the monomials.

708. $-65y^{14} \div 5y^2$

709. $\dfrac{64a^5 b^9}{-16a^{10} b^3}$

710. $\dfrac{144x^{15} y^8 z^3}{18x^{10} y^2 z^{12}}$

711. $\dfrac{\left(8p^6 q^2\right)\left(9p^3 q^5\right)}{16p^8 q^7}$

6.6 Section 6.6 Divide Polynomials

Divide a Polynomial by a Monomial

In the following exercises, divide each polynomial by the monomial.

712. $\dfrac{42z^2 - 18z}{6}$

713. $\left(35x^2 - 75x\right) \div 5x$

714. $\dfrac{81n^4 + 105n^2}{-3}$

715. $\dfrac{550p^6 - 300p^4}{10p^3}$

716. $\left(63xy^3 + 56x^2 y^4\right) \div (7xy)$

717.
$\dfrac{96a^5 b^2 - 48a^4 b^3 - 56a^2 b^4}{8ab^2}$

718. $\dfrac{57m^2 - 12m + 1}{-3m}$

719. $\dfrac{105y^5 + 50y^3 - 5y}{5y^3}$

Divide a Polynomial by a Binomial

In the following exercises, divide each polynomial by the binomial.

720. $\left(k^2 - 2k - 99\right) \div (k + 9)$

721. $\left(v^2 - 16v + 64\right) \div (v - 8)$

722. $\left(3x^2 - 8x - 35\right) \div (x - 5)$

723. $\left(n^2 - 3n - 14\right) \div (n + 3)$

724. $\left(4m^3 + m - 5\right) \div (m - 1)$

725. $\left(u^3 - 8\right) \div (u - 2)$

6.7 Section 6.7 Integer Exponents and Scientific Notation

Use the Definition of a Negative Exponent

In the following exercises, simplify.

726. 9^{-2}

727. $(-5)^{-3}$

728. $3 \cdot 4^{-3}$

729. $(6u)^{-3}$

730. $\left(\dfrac{2}{5}\right)^{-1}$

731. $\left(\dfrac{3}{4}\right)^{-2}$

Simplify Expressions with Integer Exponents

In the following exercises, simplify.

732. $p^{-2} \cdot p^8$

733. $q^{-6} \cdot q^{-5}$

734. $\left(c^{-2}d\right)\left(c^{-3}d^{-2}\right)$

735. $\left(y^8\right)^{-1}$

736. $\left(q^{-4}\right)^{-3}$

737. $\dfrac{a^8}{a^{12}}$

738. $\dfrac{n^5}{n^{-4}}$

739. $\dfrac{r^{-2}}{r^{-3}}$

Convert from Decimal Notation to Scientific Notation

In the following exercises, write each number in scientific notation.

740. 8,500,000

741. 0.00429

742. The thickness of a dime is about 0.053 inches.

743. In 2015, the population of the world was about 7,200,000,000 people.

Convert Scientific Notation to Decimal Form

In the following exercises, convert each number to decimal form.

744. 3.8×10^5

745. 1.5×10^{10}

746. 9.1×10^{-7}

747. 5.5×10^{-1}

Multiply and Divide Using Scientific Notation

In the following exercises, multiply and write your answer in decimal form.

748. $\left(2 \times 10^5\right)\left(4 \times 10^{-3}\right)$

749. $\left(3.5 \times 10^{-2}\right)\left(6.2 \times 10^{-1}\right)$

In the following exercises, divide and write your answer in decimal form.

750. $\dfrac{8 \times 10^5}{4 \times 10^{-1}}$

751. $\dfrac{9 \times 10^{-5}}{3 \times 10^2}$

PRACTICE TEST

752. For the polynomial
$10x^4 + 9y^2 - 1$

ⓐ Is it a monomial, binomial, or trinomial?

ⓑ What is its degree?

In the following exercises, simplify each expression.

753.
$\left(12a^2 - 7a + 4\right) + \left(3a^2 + 8a - 10\right)$

754. $\left(9p^2 - 5p + 1\right) - \left(2p^2 - 6\right)$

755. $\left(-\frac{2}{5}\right)^3$

756. $u \cdot u^4$

757. $\left(4a^3 b^5\right)^2$

758. $\left(-9r^4 s^5\right)\left(4rs^7\right)$

759. $3k\left(k^2 - 7k + 13\right)$

760. $(m + 6)(m + 12)$

761. $(v - 9)(9v - 5)$

762. $(4c - 11)(3c - 8)$

763. $(n - 6)\left(n^2 - 5n + 4\right)$

764. $(2x - 15y)(5x + 7y)$

765. $(7p - 5)(7p + 5)$

766. $(9v - 2)^2$

767. $\dfrac{3^8}{3^{10}}$

768. $\left(\dfrac{m^4 \cdot m}{m^3}\right)^6$

769. $\left(87x^{15} y^3 z^{22}\right)^0$

770. $\dfrac{80c^8 d^2}{16cd^{10}}$

771. $\dfrac{12x^2 + 42x - 6}{2x}$

772. $\left(70xy^4 + 95x^3 y\right) \div 5xy$

773. $\dfrac{64x^3 - 1}{4x - 1}$

774. $\left(y^2 - 5y - 18\right) \div (y + 3)$

775. 5^{-2}

776. $(4m)^{-3}$

777. $q^{-4} \cdot q^{-5}$

778. $\dfrac{n^{-2}}{n^{-10}}$

779. Convert 83,000,000 to scientific notation.

780. Convert 6.91×10^{-5} to decimal form.

In the following exercises, simplify, and write your answer in decimal form.

781. $\left(3.4 \times 10^9\right)\left(2.2 \times 10^{-5}\right)$

782. $\dfrac{8.4 \times 10^{-3}}{4 \times 10^3}$

783. A helicopter flying at an altitude of 1000 feet drops a rescue package. The polynomial $-16t^2 + 1000$ gives the height of the package t seconds a after it was dropped. Find the height when $t = 6$ seconds.

7   FACTORING

Figure 7.1 The Sydney Harbor Bridge is one of Australia's most photographed landmarks. It is the world's largest steel arch bridge with the top of the bridge standing 134 meters above the harbor. Can you see why it is known by the locals as the "Coathanger"?

Chapter Outline

7.1 Greatest Common Factor and Factor by Grouping

7.2 Factor Quadratic Trinomials with Leading Coefficient 1

7.3 Factor Quadratic Trinomials with Leading Coefficient Other than 1

7.4 Factor Special Products

7.5 General Strategy for Factoring Polynomials

7.6 Quadratic Equations

Introduction

Quadratic expressions may be used to model physical properties of a large bridge, the trajectory of a baseball or rocket, and revenue and profit of a business. By factoring these expressions, specific characteristics of the model can be identified. In this chapter, you will explore the process of factoring expressions and see how factoring is used to solve certain types of equations.

7.1   Greatest Common Factor and Factor by Grouping

Learning Objectives

By the end of this section, you will be able to:

> Find the greatest common factor of two or more expressions
> Factor the greatest common factor from a polynomial
> Factor by grouping

Be Prepared!

Before you get started, take this readiness quiz.

1. Factor 56 into primes.
   If you missed this problem, review **Example 1.7**.

2. Find the least common multiple of 18 and 24.
   If you missed this problem, review **Example 1.10**.

3. Simplify $-3(6a + 11)$.
   If you missed this problem, review **Example 1.135**.

Find the Greatest Common Factor of Two or More Expressions

Earlier we multiplied factors together to get a product. Now, we will be reversing this process; we will start with a product and then break it down into its factors. Splitting a product into factors is called **factoring**.

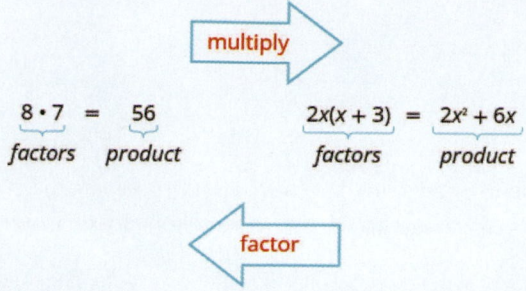

We have learned how to factor numbers to find the least common multiple (LCM) of two or more numbers. Now we will factor expressions and find the **greatest common factor** of two or more expressions. The method we use is similar to what we used to find the LCM.

Greatest Common Factor

The greatest common factor (GCF) of two or more expressions is the largest expression that is a factor of all the expressions.

First we'll find the GCF of two numbers.

EXAMPLE 7.1   HOW TO FIND THE GREATEST COMMON FACTOR OF TWO OR MORE EXPRESSIONS

Find the GCF of 54 and 36.

⊘ **Solution**

Step 1. Factor each coefficient into primes. Write all variables with exponents in expanded form.	Factor 54 and 36.	54 9 6 3 3 2 3 36 6 6 2 3 2 3
Step 2. In each column, circle the common factors.	Circle the 2, 3, and 3 that are shared by both numbers.	$36 = 2 \cdot 2 \cdot 3 \cdot 3$ $18 = 2 \cdot \quad 3 \cdot 3 \cdot 3$
Step 3. Bring down the common factors that all expressions share.	Bring down the 2, 3, and 3, and then multiply.	$GCF = 2 \cdot \quad 3 \cdot 3$
Step 4. Multiply the factors.		$GCF = 18$ The GCF of 54 and 36 is 18.

Notice that, because the GCF is a factor of both numbers, 54 and 36 can be written as multiples of 18.

$$54 = 18 \cdot 3$$
$$36 = 18 \cdot 2$$

> **TRY IT :: 7.1**   Find the GCF of 48 and 80.

> **TRY IT :: 7.2**   Find the GCF of 18 and 40.

We summarize the steps we use to find the GCF below.

HOW TO :: FIND THE GREATEST COMMON FACTOR (GCF) OF TWO EXPRESSIONS.

Step 1.   Factor each coefficient into primes. Write all variables with exponents in expanded form.

Step 2.   List all factors—matching common factors in a column. In each column, circle the common factors.

Step 3.   Bring down the common factors that all expressions share.

Step 4.   Multiply the factors.

In the first example, the GCF was a constant. In the next two examples, we will get variables in the greatest common factor.

EXAMPLE 7.2

Find the greatest common factor of $27x^3$ and $18x^4$.

✓ **Solution**

Factor each coefficient into primes and write the variables with exponents in expanded form. Circle the common factors in each column.

$$27x^3 = 3 \cdot 3 \cdot 3 \cdot x \cdot x \cdot x$$
$$18x^4 = 2 \cdot 3 \cdot 3 \cdot x \cdot x \cdot x \cdot x$$

Bring down the common factors.

$$GCF = 3 \cdot 3 \cdot x \cdot x \cdot x$$

Multiply the factors.

$$GCF = 9x^3$$

The GCF of $27x^3$ and $18x^4$ is $9x^3$.

> **TRY IT :: 7.3**   Find the GCF: $12x^2$, $18x^3$.

> **TRY IT :: 7.4**   Find the GCF: $16y^2$, $24y^3$.

EXAMPLE 7.3

Find the GCF of $4x^2 y$, $6xy^3$.

✓ **Solution**

Factor each coefficient into primes and write the variables with exponents in expanded form. Circle the common factors in each column.

$$4x^2y = 2 \cdot 2 \cdot x \cdot x \cdot y$$
$$6xy^3 = 2 \cdot 3 \cdot x \cdot y \cdot y \cdot y$$

Bring down the common factors.

$$GCF = 2 \cdot x \cdot y$$

Multiply the factors.

$$GCF = 2xy$$

The GCF of $4x^2y$ and $6xy^3$ is $2xy$.

> **TRY IT ::7.5**   Find the GCF: $6ab^4$, $8a^2b$.

> **TRY IT ::7.6**   Find the GCF: $9m^5n^2$, $12m^3n$.

EXAMPLE 7.4

Find the GCF of: $21x^3$, $9x^2$, $15x$.

⊘ **Solution**

Factor each coefficient into primes and write the variables with exponents in expanded form. Circle the common factors in each column.	$21x^3 = 3 \cdot 7 \cdot x \cdot x \cdot x$ $9x^2 = 3 \cdot 3 \cdot x \cdot x$ $15x = 3 \cdot 5 \cdot x$
Bring down the common factors.	$\text{GCF} = 3 \cdot \qquad x$
Multiply the factors.	$\text{GCF} = 3x$

The GCF of $21x^3$, $9x^2$ and $15x$ is $3x$.

> **TRY IT ::7.7**   Find the greatest common factor: $25m^4$, $35m^3$, $20m^2$.

> **TRY IT ::7.8**   Find the greatest common factor: $14x^3$, $70x^2$, $105x$.

Factor the Greatest Common Factor from a Polynomial

Just like in arithmetic, where it is sometimes useful to represent a number in factored form (for example, 12 as $2 \cdot 6$ or $3 \cdot 4$), in algebra, it can be useful to represent a polynomial in factored form. One way to do this is by finding the GCF of all the terms. Remember, we multiply a polynomial by a monomial as follows:

$$2(x + 7) \qquad \text{factors}$$
$$2 \cdot x + 2 \cdot 7$$
$$2x + 14 \qquad \text{product}$$

Now we will start with a product, like $2x + 14$, and end with its factors, $2(x + 7)$. To do this we apply the Distributive Property "in reverse."

We state the Distributive Property here just as you saw it in earlier chapters and "in reverse."

Distributive Property

If a, b, c are real numbers, then

$$a(b + c) = ab + ac \qquad \text{and} \qquad ab + ac = a(b + c)$$

The form on the left is used to multiply. The form on the right is used to factor.

So how do you use the Distributive Property to factor a polynomial? You just find the GCF of all the terms and write the polynomial as a product!

EXAMPLE 7.5 HOW TO FACTOR THE GREATEST COMMON FACTOR FROM A POLYNOMIAL

Factor: $4x + 12$.

✓ **Solution**

Step 1. Find the GCF of all the terms of the polynomial.	Find the GCF of $4x$ and 12.	$4x = 2 \cdot 2 \cdot x$ $12 = 2 \cdot 2 \cdot 3$ ――――――― $GCF = 2 \cdot 2$ $GCF = 4$
Step 2. Rewrite each term as a product using the GCF.	Rewrite $4x$ and 12 as products of their GCF, 4. $4x = 4 \cdot x$ $12 = 4 \cdot 3$	$4x + 12$ $4 \cdot x + 4 \cdot 3$
Step 3. Use the "reverse" Distributive Property to factor the expression.		$4(x + 3)$
Step 4. Check by multiplying the factors.		$4(x + 3)$ $4 \cdot x + 4 \cdot 3$ $4x + 12 ✓$

> **TRY IT ::** 7.9   Factor: $6a + 24$.

> **TRY IT ::** 7.10   Factor: $2b + 14$.

HOW TO :: FACTOR THE GREATEST COMMON FACTOR FROM A POLYNOMIAL.

Step 1.   Find the GCF of all the terms of the polynomial.

Step 2.   Rewrite each term as a product using the GCF.

Step 3.   Use the "reverse" Distributive Property to factor the expression.

Step 4.   Check by multiplying the factors.

Factor as a Noun and a Verb

We use "factor" as both a noun and a verb.

> Noun   7 is a factor of 14
>
> Verb   factor 3 from $3a + 3$

EXAMPLE 7.6

Factor: $5a + 5$.

✓ Solution

Find the GCF of $5a$ and 5.

$$5a = \boxed{5} \cdot a$$
$$5 = \boxed{5}$$
$$\overline{GCF = 5}$$

$$5a + 5$$

Rewrite each term as a product using the GCF.	$5 \cdot a + 5 \cdot 1$
Use the Distributive Property "in reverse" to factor the GCF.	$5(a + 1)$
Check by mulitplying the factors to get the orginal polynomial.	

$5(a + 1)$

$5 \cdot a + 5 \cdot 1$

$5a + 5$ ✓

> **TRY IT :: 7.11**   Factor: $14x + 14$.

> **TRY IT :: 7.12**   Factor: $12p + 12$.

The expressions in the next example have several factors in common. Remember to write the GCF as the product of all the common factors.

EXAMPLE 7.7

Factor: $12x - 60$.

✓ Solution

Find the GCF of $12x$ and 60.

$$12x = \boxed{2} \cdot \boxed{2} \cdot \boxed{3} \cdot \ x$$
$$60 = \boxed{2} \cdot \boxed{2} \cdot \boxed{3} \cdot 5$$
$$\overline{GCF = 2 \cdot 2 \cdot 3}$$
$$GCF = 12$$

$$12x - 60$$

Rewrite each term as a product using the GCF.	$12 \cdot x - 12 \cdot 5$
Factor the GCF.	$12(x - 5)$
Check by mulitplying the factors.	

$12(x - 5)$

$12 \cdot x - 12 \cdot 5$

$12x - 60$ ✓

> **TRY IT :: 7.13**   Factor: $18u - 36$.

> **TRY IT : : 7.14**   Factor: $30y - 60$.

Now we'll factor the greatest common factor from a trinomial. We start by finding the GCF of all three terms.

EXAMPLE 7.8

Factor: $4y^2 + 24y + 28$.

⊘ **Solution**

We start by finding the GCF of all three terms.

Find the GCF of $4y^2$, $24y$ and 28.	$4y^2 = 2 \cdot 2 \cdot \quad y \cdot y$ $24y = 2 \cdot 2 \cdot 2 \cdot 3 \cdot y$ $28 = 2 \cdot 2 \cdot \quad 7$ ——————————— $GCF = 2 \cdot 2$ $GCF = 4$
	$4y^2 + 24y + 28$
Rewrite each term as a product using the GCF.	$4 \cdot y^2 + 4 \cdot 6y + 4 \cdot 7$
Factor the GCF.	$4(y^2 + 6y + 7)$
Check by mulitplying.	

$$4\left(y^2 + 6y + 7\right)$$

$$4 \cdot y^2 + 4 \cdot 6y + 4 \cdot 7$$

$$4y^2 + 24y + 28 ✓$$

> **TRY IT : : 7.15**   Factor: $5x^2 - 25x + 15$.

> **TRY IT : : 7.16**   Factor: $3y^2 - 12y + 27$.

EXAMPLE 7.9

Factor: $5x^3 - 25x^2$.

⊘ **Solution**

Find the GCF of $5x^3$ and $25x^2$.	$5x^3 = 5 \cdot \quad x \cdot x \cdot x$ $25x^2 = 5 \cdot 5 \cdot x \cdot x$ ——————————— $GCF = 5 \cdot \quad x \cdot x$ $GCF = 5x^2$
	$5x^3 - 25x^2$
Rewrite each term.	$5x^2 \cdot x - 5x^2 \cdot 5$

Factor the GCF.	$5x^2(x-5)$
Check.	

$$5x^2(x-5)$$

$$5x^2 \cdot x - 5x^2 \cdot 5$$

$$5x^3 - 25x^2 \checkmark$$

> **TRY IT :: 7.17**   Factor: $2x^3 + 12x^2$.

> **TRY IT :: 7.18**   Factor: $6y^3 - 15y^2$.

EXAMPLE 7.10

Factor: $21x^3 - 9x^2 + 15x$.

⊘ Solution

In a previous example we found the GCF of $21x^3$, $9x^2$, $15x$ to be $3x$.

	$21x^3 - 9x^2 + 15x$
Rewrite each term using the GCF, $3x$.	$3x \cdot 7x^2 - 3x \cdot 3x + 3x \cdot 5$
Factor the GCF.	$3x(7x^2 - 3x + 5)$
Check.	

$$3x(7x^2 - 3x + 5)$$

$$3x \cdot 7x^2 - 3x \cdot 3x + 3x \cdot 5$$

$$21x^3 - 9x^2 + 15x \checkmark$$

> **TRY IT :: 7.19**   Factor: $20x^3 - 10x^2 + 14x$.

> **TRY IT :: 7.20**   Factor: $24y^3 - 12y^2 - 20y$.

EXAMPLE 7.11

Factor: $8m^3 - 12m^2 n + 20mn^2$.

⊘ Solution

Find the GCF of $8m^3$, $12m^2n$, $20mn^2$.

$$8m^3 = 2 \cdot 2 \cdot 2 \quad m \cdot m \cdot m$$
$$12m^2n = 2 \cdot 2 \cdot 3 \cdot \quad m \cdot m \cdot n$$
$$20mn^2 = 2 \cdot 2 \cdot \quad 5 \cdot m \cdot \quad n \cdot n$$

$$\overline{}$$

$$GCF = 2 \cdot 2 \cdot \quad m$$
$$GCF = 4m$$

$$8m^3 - 12m^2n + 20mn^2$$

Rewrite each term.	$4m \cdot 2m^2 - 4m \cdot 3m\,n + 4m \cdot 5n^2$
Factor the GCF.	$4m(2m^2 - 3m\,n + 5n^2)$
Check.	

$$4m\left(2m^2 - 3mn + 5n^2\right)$$

$$4m \cdot 2m^2 - 4m \cdot 3mn + 4m \cdot 5n^2$$

$$8m^3 - 12m^2n + 20mn^2 \checkmark$$

> **TRY IT :: 7.21**   Factor: $9xy^2 + 6x^2y^2 + 21y^3$.

> **TRY IT :: 7.22**   Factor: $3p^3 - 6p^2q + 9pq^3$.

When the leading coefficient is negative, we factor the negative out as part of the GCF.

EXAMPLE 7.12

Factor: $-8y - 24$.

⊘ Solution

When the leading coefficient is negative, the GCF will be negative.

Ignoring the signs of the terms, we first find the GCF of 8y and 24 is 8. Since the expression –8y – 24 has a negative leading coefficient, we use –8 as the GCF.

$$8y = 2 \cdot 2 \cdot 2 \cdot \quad y$$
$$24 = 2 \cdot 2 \cdot 2 \cdot 3$$

$$\overline{}$$

$$GCF = 2 \cdot 2 \cdot 2$$
$$GCF = 8$$

Rewrite each term using the GCF.	$-8y - 24$ $-8 \cdot y + (-8) \cdot 3$
Factor the GCF.	$-8(y + 3)$
Check.	

$$-8(y + 3)$$

$$-8 \cdot y + (-8) \cdot 3$$

$-8y - 24$ ✓

> **TRY IT : : 7.23**   Factor: $-16z - 64$.

> **TRY IT : : 7.24**   Factor: $-9y - 27$.

EXAMPLE 7.13

Factor: $-6a^2 + 36a$.

✓ **Solution**

The leading coefficient is negative, so the GCF will be negative.?

Since the leading coefficient is negative, the GCF is negative, $-6a$.

$$6a^2 = 2 \cdot 3 \cdot a \cdot a$$
$$36a = 2 \cdot 2 \cdot 3 \cdot 3 \cdot a$$
$$\overline{}$$
$$GCF = 2 \cdot \quad 3 \cdot \quad a$$
$$GCF = 6a$$
$$-6a^2 + 36a$$

Rewrite each term using the GCF.	$-6a \cdot a - (-6a) \cdot 6$
Factor the GCF.	$-6a(a - 6)$
Check.	

$-6a(a - 6)$

$-6a \cdot a + (-6a)(-6)$

$-6a^2 + 36a$ ✓

> **TRY IT : : 7.25**   Factor: $-4b^2 + 16b$.

> **TRY IT : : 7.26**   Factor: $-7a^2 + 21a$.

EXAMPLE 7.14

Factor: $5q(q + 7) - 6(q + 7)$.

✓ **Solution**

The GCF is the binomial $q + 7$.

$$5q(q + 7) - 6(q + 7)$$

Factor the GCF, $(q + 7)$.	$(q + 7)(5q - 6)$
Check on your own by multiplying.	

> **TRY IT :: 7.27**   Factor: $4m(m + 3) - 7(m + 3)$.

> **TRY IT :: 7.28**   Factor: $8n(n - 4) + 5(n - 4)$.

Factor by Grouping

When there is no common factor of all the terms of a polynomial, look for a common factor in just some of the terms. When there are four terms, a good way to start is by separating the polynomial into two parts with two terms in each part. Then look for the GCF in each part. If the polynomial can be factored, you will find a common factor emerges from both parts.

(Not all polynomials can be factored. Just like some numbers are prime, some polynomials are prime.)

EXAMPLE 7.15   HOW TO FACTOR BY GROUPING

Factor: $xy + 3y + 2x + 6$.

⊘ **Solution**

Step 1. Group terms with common factors.	Is there a greatest common factor of all four terms?	$xy + 3y + 2x + 6$
	No, so let's separate the first two terms from the second two.	$\underline{xy + 3y} + \underline{2x + 6}$
Step 2. Factor out the common factor in each group.	Factor the GCF from the first two terms.	$\underline{y(x + 3)} + 2x + 6$
	Factor the GCF from the second two terms.	$y(x + 3) + 2(x + 3)$
Step 3. Factor the common factor from the expression.	Notice that each term has a common factor of $(x + 3)$.	$y(x + 3) + 2(x + 3)$
	Factor out the common factor.	$(x + 3)(y + 2)$
Step 4. Check.	Multiply $(x + 3)(y + 2)$. Is the product the original expression?	$(x + 3)(y + 2)$
		$xy + 2x + 3y + 6$
		$xy + 3y + 2x + 6$ ✓

> **TRY IT :: 7.29**   Factor: $xy + 8y + 3x + 24$.

> **TRY IT :: 7.30**   Factor: $ab + 7b + 8a + 56$.

HOW TO :: FACTOR BY GROUPING.

Step 1.   Group terms with common factors.

Step 2.   Factor out the common factor in each group.

Step 3.   Factor the common factor from the expression.

Step 4.   Check by multiplying the factors.

EXAMPLE 7.16

Factor: $x^2 + 3x - 2x - 6$.

⊘ Solution

There is no GCF in all four terms.

$$x^2 + 3x \quad -2x - 6$$

Separate into two parts.

$$\underline{x^2 + 3x} \quad \underline{-2x - 6}$$

Factor the GCF from both parts. Be careful with the signs when factoring the GCF from the last two terms.

$$x(x + 3) - 2(x + 3)$$
$$(x + 3)(x - 2)$$

Check on your own by multiplying.

> **TRY IT ∷ 7.31**   Factor: $x^2 + 2x - 5x - 10$.

> **TRY IT ∷ 7.32**   Factor: $y^2 + 4y - 7y - 28$.

▶ **MEDIA ∷**

Access these online resources for additional instruction and practice with greatest common factors (GFCs) and factoring by grouping.

- **Greatest Common Factor (GCF) (https://openstax.org/l/25GCF1)**
- **Factoring Out the GCF of a Binomial (https://openstax.org/l/25GCF2)**
- **Greatest Common Factor (GCF) of Polynomials (https://openstax.org/l/25GCF3)**

7.1 EXERCISES
Practice Makes Perfect

Find the Greatest Common Factor of Two or More Expressions
In the following exercises, find the greatest common factor.

1. 8, 18
2. 24, 40
3. 72, 162

4. 150, 275
5. $10a$, 50
6. $5b$, 30

7. $3x$, $10x^2$
8. $21b^2$, $14b$
9. $8w^2$, $24w^3$

10. $30x^2$, $18x^3$
11. $10p^3 q$, $12pq^2$
12. $8a^2 b^3$, $10ab^2$

13. $12m^2 n^3$, $30m^5 n^3$
14. $28x^2 y^4$, $42x^4 y^4$
15. $10a^3$, $12a^2$, $14a$

16. $20y^3$, $28y^2$, $40y$
17. $35x^3$, $10x^4$, $5x^5$
18. $27p^2$, $45p^3$, $9p^4$

Factor the Greatest Common Factor from a Polynomial
In the following exercises, factor the greatest common factor from each polynomial.

19. $4x + 20$
20. $8y + 16$
21. $6m + 9$

22. $14p + 35$
23. $9q + 9$
24. $7r + 7$

25. $8m - 8$
26. $4n - 4$
27. $9n - 63$

28. $45b - 18$
29. $3x^2 + 6x - 9$
30. $4y^2 + 8y - 4$

31. $8p^2 + 4p + 2$
32. $10q^2 + 14q + 20$
33. $8y^3 + 16y^2$

34. $12x^3 - 10x$
35. $5x^3 - 15x^2 + 20x$
36. $8m^2 - 40m + 16$

37. $12xy^2 + 18x^2 y^2 - 30y^3$
38. $21pq^2 + 35p^2 q^2 - 28q^3$
39. $-2x - 4$

40. $-3b + 12$
41. $5x(x + 1) + 3(x + 1)$
42. $2x(x - 1) + 9(x - 1)$

43. $3b(b - 2) - 13(b - 2)$
44. $6m(m - 5) - 7(m - 5)$

Factor by Grouping
In the following exercises, factor by grouping.

45. $xy + 2y + 3x + 6$
46. $mn + 4n + 6m + 24$
47. $uv - 9u + 2v - 18$

48. $pq - 10p + 8q - 80$
49. $b^2 + 5b - 4b - 20$
50. $m^2 + 6m - 12m - 72$

51. $p^2 + 4p - 9p - 36$
52. $x^2 + 5x - 3x - 15$

Mixed Practice

In the following exercises, factor.

53. $-20x - 10$

54. $5x^3 - x^2 + x$

55. $3x^3 - 7x^2 + 6x - 14$

56. $x^3 + x^2 - x - 1$

57. $x^2 + xy + 5x + 5y$

58. $5x^3 - 3x^2 - 5x - 3$

Everyday Math

59. Area of a rectangle The area of a rectangle with length 6 less than the width is given by the expression $w^2 - 6w$, where $w =$ width. Factor the greatest common factor from the polynomial.

60. Height of a baseball The height of a baseball t seconds after it is hit is given by the expression $-16t^2 + 80t + 4$. Factor the greatest common factor from the polynomial.

Writing Exercises

61. The greatest common factor of 36 and 60 is 12. Explain what this means.

62. What is the GCF of y^4, y^5, and y^{10}? Write a general rule that tells you how to find the GCF of y^a, y^b, and y^c.

Self Check

ⓐ *After completing the exercises, use this checklist to evaluate your mastery of the objectives of this section.*

I can...	Confidently	With some help	No-I don't get it!
find the greatest common factor of two or more expressions.			
factor the greatest common factor from a polynomial.			
factor by grouping.			

ⓑ *If most of your checks were:*

...confidently. *Congratulations! You have achieved your goals in this section! Reflect on the study skills you used so that you can continue to use them. What did you do to become confident of your ability to do these things? Be specific!*

...with some help. *This must be addressed quickly as topics you do not master become potholes in your road to success. Math is sequential—every topic builds upon previous work. It is important to make sure you have a strong foundation before you move on. Who can you ask for help? Your fellow classmates and instructor are good resources. Is there a place on campus where math tutors are available? Can your study skills be improved?*

...no - I don't get it! *This is critical and you must not ignore it. You need to get help immediately or you will quickly be overwhelmed. See your instructor as soon as possible to discuss your situation. Together you can come up with a plan to get you the help you need.*

7.2 | Factor Quadratic Trinomials with Leading Coefficient 1

Learning Objectives

By the end of this section, you will be able to:

> Factor trinomials of the form $x^2 + bx + c$
> Factor trinomials of the form $x^2 + bxy + cy^2$

Be Prepared!

Before you get started, take this readiness quiz.

1. Multiply: $(x + 4)(x + 5)$.
   If you missed this problem, review **Example 6.38**.

2. Simplify: ⓐ $-9 + (-6)$ ⓑ $-9 + 6$.
   If you missed this problem, review **Example 1.37**.

3. Simplify: ⓐ $-9(6)$ ⓑ $-9(-6)$.
   If you missed this problem, review **Example 1.46**.

4. Simplify: ⓐ $|-5|$ ⓑ $|3|$.
   If you missed this problem, review **Example 1.33**.

Factor Trinomials of the Form $x^2 + bx + c$

You have already learned how to multiply binomials using FOIL. Now you'll need to "undo" this multiplication—to start with the product and end up with the factors. Let's look at an example of multiplying binomials to refresh your memory.

$$(x + 2)(x + 3) \text{ factors}$$
$$F \quad O \quad I \quad L$$
$$x^2 + 3x + 2x + 6$$
$$x^2 + 5x + 6 \quad \text{product}$$

To factor the trinomial means to start with the product, $x^2 + 5x + 6$, and end with the factors, $(x + 2)(x + 3)$. You need to think about where each of the terms in the trinomial came from.

The *first term* came from multiplying the first term in each binomial. So to get x^2 in the product, each binomial must start with an x.

$$x^2 + 5x + 6$$
$$(x \quad)(x \quad)$$

The *last term* in the trinomial came from multiplying the last term in each binomial. So the last terms must multiply to 6.

What two numbers multiply to 6?

The factors of 6 could be 1 and 6, or 2 and 3. How do you know which pair to use?

Consider the *middle term*. It came from adding the outer and inner terms.

So the numbers that must have a product of 6 will need a sum of 5. We'll test both possibilities and summarize the results in **Table 7.13**—the table will be very helpful when you work with numbers that can be factored in many different ways.

Factors of 6	Sum of factors
1, 6	$1 + 6 = 7$
2, 3	$2 + 3 = 5$

Table 7.13

We see that 2 and 3 are the numbers that multiply to 6 and add to 5. So we have the factors of $x^2 + 5x + 6$. They are $(x + 2)(x + 3)$.

$$x^2 + 5x + 6 \qquad \text{product}$$
$$(x + 2)(x + 3) \qquad \text{factors}$$

You should check this by multiplying.

Looking back, we started with $x^2 + 5x + 6$, which is of the form $x^2 + bx + c$, where $b = 5$ and $c = 6$. We factored it into two binomials of the form $(x + m)$ and $(x + n)$.

$$x^2 + 5x + 6 \qquad x^2 + bx + c$$
$$(x + 2)(x + 3) \qquad (x + m)(x + n)$$

To get the correct factors, we found two numbers m and n whose product is c and sum is b.

EXAMPLE 7.17 HOW TO FACTOR TRINOMIALS OF THE FORM $x^2 + bx + c$

Factor: $x^2 + 7x + 12$.

✓ **Solution**

Step 1. Write the factors as two binomials with first terms x.	Write two sets of parentheses and put x as the first term.	$x^2 + 7x + 12$ $(x \quad)(x \quad)$
Step 2. Find two numbers m and n that multiply to c, $\quad m \cdot n = c$ add to b, $\quad m + n = b$	Find two numbers that multiply to 12 and add to 7. <table><tr><td>Factors of 12</td><td>Sum of factors</td></tr><tr><td>1, 12</td><td>$1 + 12 = 13$</td></tr><tr><td>2, 6</td><td>$2 + 6 = 8$</td></tr><tr><td>3, 4</td><td>$3 + 4 = 7*$</td></tr></table>	
Step 3. Use m and n as the last terms of the factors.	Use 3 and 4 as the last terms of the binomials.	$(x + 3)(x + 4)$
Step 4. Check by multiplying the factors.		$(x + 3)(x + 4)$ $x^2 + 4x + 3x + 12$ $x^2 + 7x + 12$ ✓

> **TRY IT :: 7.33** Factor: $x^2 + 6x + 8$.

> **TRY IT :: 7.34** Factor: $y^2 + 8y + 15$.

Let's summarize the steps we used to find the factors.

HOW TO :: FACTOR TRINOMIALS OF THE FORM $x^2 + bx + c$.

Step 1. Write the factors as two binomials with first terms x: $(x \quad)(x \quad)$.

Step 2. Find two numbers m and n that
   Multiply to c, $m \cdot n = c$
   Add to b, $m + n = b$

Step 3. Use m and n as the last terms of the factors: $(x + m)(x + n)$.

Step 4. Check by multiplying the factors.

EXAMPLE 7.18

Factor: $u^2 + 11u + 24$.

⊘ Solution

Notice that the variable is u, so the factors will have first terms u.

$$u^2 + 11u + 24$$

Write the factors as two binomials with first terms u. $(u \quad)(u \quad)$

Find two numbers that: multiply to 24 and add to 11.

Factors of 24	Sum of factors
1, 24	$1 + 24 = 25$
2, 12	$2 + 12 = 14$
3, 8	$3 + 8 = 11*$
4, 6	$4 + 6 = 10$

Use 3 and 8 as the last terms of the binomials. $(u + 3)(u + 8)$

Check.

$(u + 3)(u + 8)$
$u^2 + 3u + 8u + 24$
$u^2 + 11u + 24$ ✓

| > | **TRY IT : : 7.35** | Factor: $q^2 + 10q + 24$. |

| > | **TRY IT : : 7.36** | Factor: $t^2 + 14t + 24$. |

EXAMPLE 7.19

Factor: $y^2 + 17y + 60$.

⊘ Solution

$$y^2 + 17y + 60$$

Write the factors as two binomials with first terms y. $(y \quad)(y \quad)$

Find two numbers that multiply to 60 and add to 17.

Factors of 60	Sum of factors
1, 60	$1 + 60 = 61$
2, 30	$2 + 30 = 32$
3, 20	$3 + 20 = 23$
4, 15	$4 + 15 = 19$
5, 12	$5 + 12 = 17*$
6, 10	$6 + 10 = 16$

Use 5 and 12 as the last terms. $(y + 5)(y + 12)$

Check.

$$(y + 5)(y + 12)$$
$$(y^2 + 12y + 5y + 60)$$
$$(y^2 + 17y + 60) \checkmark$$

> **TRY IT :: 7.37**   Factor: $x^2 + 19x + 60$.

> **TRY IT :: 7.38**   Factor: $v^2 + 23v + 60$.

Factor Trinomials of the Form $x^2 + bx + c$ with b Negative, c Positive

In the examples so far, all terms in the trinomial were positive. What happens when there are negative terms? Well, it depends which term is negative. Let's look first at trinomials with only the middle term negative.

Remember: To get a negative sum and a positive product, the numbers must both be negative.

Again, think about FOIL and where each term in the trinomial came from. Just as before,

- the first term, x^2, comes from the product of the two first terms in each binomial factor, x and y;
- the positive last term is the product of the two last terms
- the negative middle term is the sum of the outer and inner terms.

How do you get a *positive product* and a *negative sum*? With two negative numbers.

EXAMPLE 7.20

Factor: $t^2 - 11t + 28$.

⊘ **Solution**

Again, with the positive last term, 28, and the negative middle term, $-11t$, we need two negative factors. Find two numbers that multiply 28 and add to -11.

$$t^2 - 11t + 28$$

Write the factors as two binomials with first terms t. $(t\quad)(t\quad)$

Find two numbers that: multiply to 28 and add to -11.

Factors of 28	Sum of factors
$-1, -28$	$-1 + (-28) = -29$
$-2, -14$	$-2 + (-14) = -16$
$-4, -7$	$-4 + (-7) = -11*$

Use $-4, -7$ as the last terms of the binomials. $(t - 4)(t - 7)$

Check.

$$(t - 4)(t - 7)$$
$$t^2 - 7t - 4t + 28$$
$$t^2 - 11t + 28 \checkmark$$

> **TRY IT :: 7.39**   Factor: $u^2 - 9u + 18$.

> **TRY IT :: 7.40**   Factor: $y^2 - 16y + 63$.

Factor Trinomials of the Form $x^2 + bx + c$ with c Negative

Now, what if the last term in the trinomial is negative? Think about FOIL. The last term is the product of the last terms in the two binomials. A negative product results from multiplying two numbers with opposite signs. You have to be very careful to choose factors to make sure you get the correct sign for the middle term, too.

Remember: To get a negative product, the numbers must have different signs.

EXAMPLE 7.21

Factor: $z^2 + 4z - 5$.

⊘ Solution

To get a negative last term, multiply one positive and one negative. We need factors of -5 that add to positive 4.

Factors of -5	Sum of factors
$1, -5$	$1 + (-5) = -4$
$-1, 5$	$-1 + 5 = 4*$

Notice: We listed both $1, -5$ and $-1, 5$ to make sure we got the sign of the middle term correct.

$$z^2 + 4z - 5$$

Factors will be two binomials with first terms z. $(z\quad)(z\quad)$

Use $-1, 5$ as the last terms of the binomials. $(z - 1)(z + 5)$

Check.

$$(z - 1)(z + 5)$$
$$z^2 + 5z - 1z - 5$$
$$z^2 + 4z - 5 \checkmark$$

> **TRY IT : : 7.41**   Factor: $h^2 + 4h - 12$.

> **TRY IT : : 7.42**   Factor: $k^2 + k - 20$.

Let's make a minor change to the last trinomial and see what effect it has on the factors.

EXAMPLE 7.22

Factor: $z^2 - 4z - 5$.

⊘ **Solution**

This time, we need factors of -5 that add to -4.

Factors of -5	Sum of factors
$1, -5$	$1 + (-5) = -4*$
$-1, 5$	$-1 + 5 = 4$

$$z^2 - 4z - 5$$

Factors will be two binomials with first terms z. $(z\quad)(z\quad)$

Use $1, -5$ as the last terms of the binomials. $(z + 1)(z - 5)$

Check.

$(z + 1)(z - 5)$

$z^2 - 5z + 1z - 5$

$z^2 - 4z - 5$ ✓

Notice that the factors of $z^2 - 4z - 5$ are very similar to the factors of $z^2 + 4z - 5$. It is very important to make sure you choose the factor pair that results in the correct sign of the middle term.

> **TRY IT : : 7.43**   Factor: $x^2 - 4x - 12$.

> **TRY IT : : 7.44**   Factor: $y^2 - y - 20$.

EXAMPLE 7.23

Factor: $q^2 - 2q - 15$.

⊘ **Solution**

$$q^2 - 2q - 15$$

Factors will be two binomials with first terms q. $(q\quad)(q\quad)$

You can use $3, -5$ as the last terms of the $(q + 3)(q - 5)$
binomials.

Factors of -15	Sum of factors
$1, -15$	$1 + (-15) = -14$
$-1, 15$	$-1 + 15 = 14$
$3, -5$	$3 + (-5) = -2*$
$-3, 5$	$-3 + 5 = 2$

Check.

$$(q + 3)(q - 5)$$

$$q^2 - 5q + 3q - 15$$

$$q^2 - 2q - 15 \checkmark$$

> **TRY IT : : 7.45**   Factor: $r^2 - 3r - 40$.

> **TRY IT : : 7.46**   Factor: $s^2 - 3s - 10$.

Some trinomials are prime. The only way to be certain a trinomial is prime is to list all the possibilities and show that none of them work.

EXAMPLE 7.24

Factor: $y^2 - 6y + 15$.

⊘ **Solution**

$$y^2 - 6y + 15$$

Factors will be two binomials with first terms y.

$$(y \quad)(y \quad)$$

Factors of 15	Sum of factors
$-1, -15$	$-1 + (-15) = -16$
$-3, -5$	$-3 + (-5) = -8$

As shown in the table, none of the factors add to -6; therefore, the expression is prime.

> **TRY IT : : 7.47**   Factor: $m^2 + 4m + 18$.

> **TRY IT : : 7.48**   Factor: $n^2 - 10n + 12$.

EXAMPLE 7.25

Factor: $2x + x^2 - 48$.

⊘ **Solution**

$$2x + x^2 - 48$$

First we put the terms in decreasing degree order. $x^2 + 2x - 48$

Factors will be two binomials with first terms x. $(x\quad)(x\quad)$

As shown in the table, you can use $-6,\ 8$ as the last terms of the binomials.

$$(x - 6)(x + 8)$$

Factors of -48	Sum of factors
$-1,\ 48$	$-1 + 48 = 47$
$-2,\ 24$	$-2 + 24 = 22$
$-3,\ 16$	$-3 + 16 = 13$
$-4,\ 12$	$-4 + 12 = 8$
$-6,\ 8$	$-6 + 8 = 2$

Check.

$(x - 6)(x + 8)$

$x^2 - 6q + 8q - 48$

$x^2 + 2x - 48$ ✓

> **TRY IT :: 7.49**   Factor: $9m + m^2 + 18$.

> **TRY IT :: 7.50**   Factor: $-7n + 12 + n^2$.

Let's summarize the method we just developed to factor trinomials of the form $x^2 + bx + c$.

HOW TO :: FACTOR TRINOMIALS.

When we factor a trinomial, we look at the signs of its terms first to determine the signs of the binomial factors.

$$x^2 + bx + c$$
$$(x + m)(x + n)$$

When c is positive, m and n have the same sign.

b positive	b negative
m, n positive	m, n negative
$x^2 + 5x + 6$	$x^2 - 6x + 8$
$(x + 2)(x + 3)$	$(x - 4)(x - 2)$
same signs	same signs

When c is negative, m and n have opposite signs.

$x^2 + x - 12$	$x^2 - 2x - 15$
$(x + 4)(x - 3)$	$(x - 5)(x + 3)$
opposite signs	opposite signs

Notice that, in the case when m and n have opposite signs, the sign of the one with the larger absolute value matches the sign of b.

Factor Trinomials of the Form $x^2 + bxy + cy^2$

Sometimes you'll need to factor trinomials of the form $x^2 + bxy + cy^2$ with two variables, such as $x^2 + 12xy + 36y^2$.

The first term, x^2, is the product of the first terms of the binomial factors, $x \cdot x$. The y^2 in the last term means that the second terms of the binomial factors must each contain y. To get the coefficients b and c, you use the same process summarized in the previous objective.

EXAMPLE 7.26

Factor: $x^2 + 12xy + 36y^2$.

 Solution

$$x^2 + 12xy + 36y^2$$

Note that the first terms are x, last terms contain y.

$$(x\_y)(x\_y)$$

Find the numbers that multiply to 36 and add to 12.

Factors of 36	Sum of factors
1, 36	$1 + 36 = 37$
2, 18	$2 + 18 = 20$
3, 12	$3 + 12 = 15$
4, 9	$4 + 9 = 13$
6, 6	$6 + 6 = 12*$

Use 6 and 6 as the coefficients of the last terms. $(x + 6y)(x + 6y)$

Check your answer.

$$(x + 6y)(x + 6y)$$
$$x^2 + 6xy + 6xy + 36y^2$$
$$x^2 + 12xy + 36y^2 \ \checkmark$$

> | **TRY IT ::** 7.51   Factor: $u^2 + 11uv + 28v^2$.

> | **TRY IT ::** 7.52   Factor: $x^2 + 13xy + 42y^2$.

EXAMPLE 7.27

Factor: $r^2 - 8rx - 9s^2$.

⊘ **Solution**

We need r in the first term of each binomial and s in the second term. The last term of the trinomial is negative, so the factors must have opposite signs.

$$r^2 - 8rx - 9s^2$$

Note that the first terms are r, last terms contain s. $(r\_s)(r\_s)$

Find the numbers that multiply to -9 and add to -8.

Factors of -9	Sum of factors
1, -9	$1 + (-9) = -8*$
-1, 9	$-1 + 9 = 8$
3, -3	$3 + (-3) = 0$

Use 1, -9 as coefficients of the last terms. $(r + s)(r - 9s)$

Check your answer.

$$(r - 9s)(r + s)$$
$$r^2 + rs - 9rs - 9s^2$$
$$r^2 - 8rs - 9s^2 \ \checkmark$$

> | **TRY IT ::** 7.53   Factor: $a^2 - 11ab + 10b^2$.

> | **TRY IT ::** 7.54   Factor: $m^2 - 13mn + 12n^2$.

EXAMPLE 7.28

Factor: $u^2 - 9uv - 12v^2$.

⊘ Solution

We need u in the first term of each binomial and v in the second term. The last term of the trinomial is negative, so the factors must have opposite signs.

$$u^2 - 9uv - 12v^2$$

Note that the first terms are u, last terms contain v. $(u\_v)(u\_v)$

Find the numbers that multiply to -12 and add to -9.

Factors of -12	Sum of factors
1, −12	$1 + (-12) = -11$
−1, 12	$-1 + 12 = 11$
2, −6	$2 + (-6) = -4$
−2, 6	$-2 + 6 = 4$
3, −4	$3 + (-4) = -1$
−3, 4	$-3 + 4 = 1$

Note there are no factor pairs that give us -9 as a sum. The trinomial is prime.

> **TRY IT : : 7.55**   Factor: $x^2 - 7xy - 10y^2$.

> **TRY IT : : 7.56**   Factor: $p^2 + 15pq + 20q^2$.

 **7.2 EXERCISES**

Practice Makes Perfect

Factor Trinomials of the Form $x^2 + bx + c$

In the following exercises, factor each trinomial of the form $x^2 + bx + c$.

63. $x^2 + 4x + 3$

64. $y^2 + 8y + 7$

65. $m^2 + 12m + 11$

66. $b^2 + 14b + 13$

67. $a^2 + 9a + 20$

68. $m^2 + 7m + 12$

69. $p^2 + 11p + 30$

70. $w^2 + 10x + 21$

71. $n^2 + 19n + 48$

72. $b^2 + 14b + 48$

73. $a^2 + 25a + 100$

74. $u^2 + 101u + 100$

75. $x^2 - 8x + 12$

76. $q^2 - 13q + 36$

77. $y^2 - 18x + 45$

78. $m^2 - 13m + 30$

79. $x^2 - 8x + 7$

80. $y^2 - 5y + 6$

81. $p^2 + 5p - 6$

82. $n^2 + 6n - 7$

83. $y^2 - 6y - 7$

84. $v^2 - 2v - 3$

85. $x^2 - x - 12$

86. $r^2 - 2r - 8$

87. $a^2 - 3a - 28$

88. $b^2 - 13b - 30$

89. $w^2 - 5w - 36$

90. $t^2 - 3t - 54$

91. $x^2 + x + 5$

92. $x^2 - 3x - 9$

93. $8 - 6x + x^2$

94. $7x + x^2 + 6$

95. $x^2 - 12 - 11x$

96. $-11 - 10x + x^2$

Factor Trinomials of the Form $x^2 + bxy + cy^2$

In the following exercises, factor each trinomial of the form $x^2 + bxy + cy^2$.

97. $p^2 + 3pq + 2q^2$

98. $m^2 + 6mn + 5n^2$

99. $r^2 + 15rs + 36s^2$

100. $u^2 + 10uv + 24v^2$

101. $m^2 - 12mn + 20n^2$

102. $p^2 - 16pq + 63q^2$

103. $x^2 - 2xy - 80y^2$

104. $p^2 - 8pq - 65q^2$

105. $m^2 - 64mn - 65n^2$

106. $p^2 - 2pq - 35q^2$

107. $a^2 + 5ab - 24b^2$

108. $r^2 + 3rs - 28s^2$

109. $x^2 - 3xy - 14y^2$

110. $u^2 - 8uv - 24v^2$

111. $m^2 - 5mn + 30n^2$

112. $c^2 - 7cd + 18d^2$

Mixed Practice

In the following exercises, factor each expression.

113. $u^2 - 12u + 36$

114. $w^2 + 4w - 32$

115. $x^2 - 14x - 32$

116. $y^2 + 41y + 40$

117. $r^2 - 20rs + 64s^2$

118. $x^2 - 16xy + 64y^2$

119. $k^2 + 34k + 120$

120. $m^2 + 29m + 120$

121. $y^2 + 10y + 15$

122. $z^2 - 3z + 28$

123. $m^2 + mn - 56n^2$

124. $q^2 - 29qr - 96r^2$

125. $u^2 - 17uv + 30v^2$

126. $m^2 - 31mn + 30n^2$

127. $c^2 - 8cd + 26d^2$

128. $r^2 + 11rs + 36s^2$

Everyday Math

129. Consecutive integers Deirdre is thinking of two consecutive integers whose product is 56. The trinomial $x^2 + x - 56$ describes how these numbers are related. Factor the trinomial.

130. Consecutive integers Deshawn is thinking of two consecutive integers whose product is 182. The trinomial $x^2 + x - 182$ describes how these numbers are related. Factor the trinomial.

Writing Exercises

131. Many trinomials of the form $x^2 + bx + c$ factor into the product of two binomials $(x + m)(x + n)$. Explain how you find the values of *m* and *n*.

132. How do you determine whether to use plus or minus signs in the binomial factors of a trinomial of the form $x^2 + bx + c$ where *b* and *c* may be positive or negative numbers?

133. Will factored $x^2 - x - 20$ as $(x + 5)(x - 4)$. Bill factored it as $(x + 4)(x - 5)$. Phil factored it as $(x - 5)(x - 4)$. Who is correct? Explain why the other two are wrong.

134. Look at **Example 7.19**, where we factored $y^2 + 17y + 60$. We made a table listing all pairs of factors of 60 and their sums. Do you find this kind of table helpful? Why or why not?

Self Check

ⓐ *After completing the exercises, use this checklist to evaluate your mastery of the objectives of this section.*

I can...	Confidently	With some help	No-I don't get it!
factor trinomials of the form $x^2 + bx + c$.			
factor trinomials of the form $x^2 + bxy + cy^2$.			

ⓑ *After reviewing this checklist, what will you do to become confident for all goals?*

7.3 Factor Quadratic Trinomials with Leading Coefficient Other than 1

Learning Objectives

By the end of this section, you will be able to:

› Recognize a preliminary strategy to factor polynomials completely
› Factor trinomials of the form $ax^2 + bx + c$ with a GCF
› Factor trinomials using trial and error
› Factor trinomials using the 'ac' method

Be Prepared!

Before you get started, take this readiness quiz.

1. Find the GCF of $45p^2$ and $30p^6$.
   If you missed this problem, review **Example 7.2**.

2. Multiply $(3y + 4)(2y + 5)$.
   If you missed this problem, review **Example 6.40**.

3. Combine like terms $12x^2 + 3x + 5x + 9$.
   If you missed this problem, review **Example 1.24**.

Recognize a Preliminary Strategy for Factoring

Let's summarize where we are so far with factoring polynomials. In the first two sections of this chapter, we used three methods of factoring: factoring the GCF, factoring by grouping, and factoring a trinomial by "undoing" FOIL. More methods will follow as you continue in this chapter, as well as later in your studies of algebra.

How will you know when to use each factoring method? As you learn more methods of factoring, how will you know when to apply each method and not get them confused? It will help to organize the factoring methods into a strategy that can guide you to use the correct method.

As you start to factor a polynomial, always ask first, "Is there a greatest common factor?" If there is, factor it first.

The next thing to consider is the type of polynomial. How many terms does it have? Is it a binomial? A trinomial? Or does it have more than three terms?

If it is a trinomial where the leading coefficient is one, $x^2 + bx + c$, use the "undo FOIL" method.

If it has more than three terms, try the grouping method. This is the only method to use for polynomials of more than three terms.

Some polynomials cannot be factored. They are called "prime."

Below we summarize the methods we have so far. These are detailed in **Choose a strategy to factor polynomials completely**.

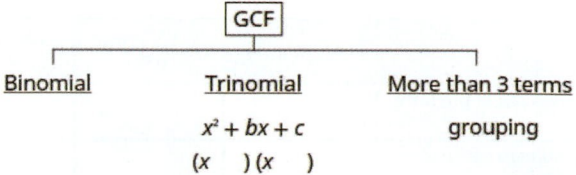

HOW TO :: CHOOSE A STRATEGY TO FACTOR POLYNOMIALS COMPLETELY.

Step 1. Is there a greatest common factor?

   ◦ Factor it out.

Step 2. Is the polynomial a binomial, trinomial, or are there more than three terms?

   ◦ If it is a binomial, right now we have no method to factor it.

   ◦ If it is a trinomial of the form $x^2 + bx + c$: Undo FOIL $(x \quad)(x \quad)$

   ◦ If it has more than three terms: Use the grouping method.

Step 3. Check by multiplying the factors.

Use the preliminary strategy to completely factor a polynomial. A polynomial is factored completely if, other than monomials, all of its factors are prime.

EXAMPLE 7.29

Identify the best method to use to factor each polynomial.

  ⓐ $6y^2 - 72$ ⓑ $r^2 - 10r - 24$ ⓒ $p^2 + 5p + pq + 5q$

✅ **Solution**

ⓐ

	$6y^2 - 72$
Is there a greatest common factor?	Yes, 6.
Factor out the 6.	$6(y^2 - 12)$
Is it a binomial, trinomial, or are there more than 3 terms?	Binomial, we have no method to factor binomials yet.

ⓑ

	$r^2 - 10r - 24$
Is there a greatest common factor?	No, there is no common factor.
Is it a binomial, trinomial, or are there more than three terms?	Trinomial, with leading coefficient 1, so "undo" FOIL.

ⓒ

	$p^2 + 5p + pq + 5q$
Is there a greatest common factor?	No, there is no common factor.
Is it a binomial, trinomial, or are there more than three terms?	More than three terms, so factor using grouping.

> **TRY IT :: 7.57** Identify the best method to use to factor each polynomial:
>
>   ⓐ $4y^2 + 32$ ⓑ $y^2 + 10y + 21$ ⓒ $yz + 2y + 3z + 6$

> **TRY IT :: 7.58** Identify the best method to use to factor each polynomial:
>
>   ⓐ $ab + a + 4b + 4$ ⓑ $3k^2 + 15$ ⓒ $p^2 + 9p + 8$

Factor Trinomials of the form $ax^2 + bx + c$ with a GCF

Now that we have organized what we've covered so far, we are ready to factor trinomials whose leading coefficient is not 1, trinomials of the form $ax^2 + bx + c$.

Remember to always check for a GCF first! Sometimes, after you factor the GCF, the leading coefficient of the trinomial becomes 1 and you can factor it by the methods in the last section. Let's do a few examples to see how this works.

Watch out for the signs in the next two examples.

EXAMPLE 7.30

Factor completely: $2n^2 - 8n - 42$.

⊘ **Solution**

Use the preliminary strategy.

Is there a greatest common factor?	$2n^2 - 8n - 42$
Yes, GCF = 2. Factor it out.	$2(n^2 - 4n - 21)$

Inside the parentheses, is it a binomial, trinomial, or are there more than three terms?

It is a trinomial whose coefficient is 1, so undo FOIL.	$2(n\quad)(n\quad)$
Use 3 and −7 as the last terms of the binomials.	$2(n + 3)(n - 7)$

Factors of −21	Sum of factors
1, −21	$1 + (-21) = -20$
3, −7	$3 + (-7) = -4$*

Check.

$$2(n + 3)(n - 7)$$

$$2(n^2 - 7n + 3n - 21)$$

$$2(n^2 - 4n - 21)$$

$$2n^2 - 8n - 42 \checkmark$$

> **TRY IT :: 7.59**   Factor completely: $4m^2 - 4m - 8$.

> **TRY IT :: 7.60**   Factor completely: $5k^2 - 15k - 50$.

EXAMPLE 7.31

Factor completely: $4y^2 - 36y + 56$.

⊘ **Solution**

Use the preliminary strategy.

Is there a greatest common factor? $4y^2 - 36y + 56$

    Yes, GCF = 4. Factor it. $4\left(y^2 - 9y + 14\right)$

Inside the parentheses, is it a binomial, trinomial, or are there more than three terms?

    It is a trinomial whose coefficient is 1. So undo FOIL. $4(y\quad)(y\quad)$

Use a table like the one below to find two numbers that multiply to 14 and add to -9.

Both factors of 14 must be negative. $4(y - 2)(y - 7)$

Factors of 14	Sum of factors
$-1, -14$	$-1 + (-14) = -15$
$-2, -7$	$-2 + (-7) = -9*$

Check.

    $4(y - 2)(y - 7)$

    $4(y^2 - 7y - 2y + 14)$

    $4(y^2 - 9y + 14)$

    $4y^2 - 36y + 42$ ✓

> **TRY IT : : 7.61**   Factor completely: $3r^2 - 9r + 6$.

> **TRY IT : : 7.62**   Factor completely: $2t^2 - 10t + 12$.

In the next example the GCF will include a variable.

EXAMPLE 7.32

Factor completely: $4u^3 + 16u^2 - 20u$.

✓ **Solution**

Use the preliminary strategy.

Is there a greatest common factor?	$4u^3 + 16u^2 - 20u$
Yes, GCF $= 4u$. Factor it.	$4u(u^2 + 4u - 5)$

Binomial, trinomial, or more than three terms?	
It is a trinomial. So "undo FOIL."	$4u(u\ \ )(u\ \ )$

Use a table like the table below to find two numbers that multiply to -5 and add to 4.

$4u(u - 1)(u + 5)$

Factors of -5	Sum of factors
$-1, 5$	$-1 + 5 = 4*$
$1, -5$	$1 + (-5) = -4$

Check.

$4u(u - 1)(u + 5)$

$4u(u^2 + 5u - u - 5)$

$4u(u^2 + 4u - 5)$

$4u^3 + 16u^2 - 20u$ ✓

> **TRY IT :: 7.63**   Factor completely: $5x^3 + 15x^2 - 20x$.

> **TRY IT :: 7.64**   Factor completely: $6y^3 + 18y^2 - 60y$.

Factor Trinomials using Trial and Error

What happens when the leading coefficient is not 1 and there is no GCF? There are several methods that can be used to factor these trinomials. First we will use the Trial and Error method.

Let's factor the trinomial $3x^2 + 5x + 2$.

From our earlier work we expect this will factor into two binomials.

$$3x^2 + 5x + 2$$
$$(\ \ )(\ \ )$$

We know the first terms of the binomial factors will multiply to give us $3x^2$. The only factors of $3x^2$ are $1x, 3x$. We can place them in the binomials.

$$3x^2 + 5x + 2$$
$$1x, 3x$$
$$(x\ \ )(3x\ \ )$$

Check. Does $1x \cdot 3x = 3x^2$?

We know the last terms of the binomials will multiply to 2. Since this trinomial has all positive terms, we only need to consider positive factors. The only factors of 2 are 1 and 2. But we now have two cases to consider as it will make a difference if we write 1, 2, or 2, 1.

$$3x^2 + 5x + 2 \qquad 3x^2 + 5x + 2$$
1x, 3x   1, 2   1x, 3x   1, 2

$$(x + 1)(3x + 2) \quad \text{or} \quad (x + 2)(3x + 1)$$

Which factors are correct? To decide that, we multiply the inner and outer terms.

$$3x^2 + 5x + 2 \qquad 3x^2 + 5x + 2$$
1x, 3x   1, 2   1x, 3x   1, 2

$$(x + 1)(3x + 2) \quad \text{or} \quad (x + 2)(3x + 1)$$

3x   6x

2x   1x

5x   7x

Since the middle term of the trinomial is 5x, the factors in the first case will work. Let's FOIL to check.

$$(x + 1)(3x + 2)$$
$$3x^2 + 2x + 3x + 2$$
$$3x^2 + 5x + 2 \checkmark$$

Our result of the factoring is:

$$3x^2 + 5x + 2$$
$$(x + 1)(3x + 2)$$

EXAMPLE 7.33   HOW TO FACTOR TRINOMIALS OF THE FORM $ax^2 + bx + c$ USING TRIAL AND ERROR

Factor completely: $3y^2 + 22y + 7$.

⊘ Solution

Step 1. Write the trinomial in descending order.	The trinomial is already in descending order.	$3y^2 + 22y + 7$
Step 2. Find all the factor pairs of the first term.	The only factors of $3y^2$ are $1y, 3y$ Since there is only one pair, we can put them in the parentheses.	$3y^2 + 22y + 7$ 1y, 3y $3y^2 + 22y + 7$ 1y, 3y $(y \quad)(3y \quad)$
Step 3. Find all the factor pairs of the third term.	The only factors of 7 are 1, 7.	$3y^2 + 22y + 7$ 1y, 3y 1, 7 $(y \quad)(3y \quad)$
Step 4. Test all the possible combinations of the factors until the correct product is found.	$3y^2 + 22y + 7$ 1y, 3y 1, 7 $(y + 1)(3y + 7)$ 3y 7y 10y No. We need 22y $3y^2 + 22y + 7$ 1y, 3y 1, 7 $(y + 7)(3y + 1)$ 21y y 22y	<table><tr><td colspan="2" align="center">$3y^2 + 22y + 7$</td></tr><tr><td>Possible factors</td><td>Product</td></tr><tr><td>$(y + 1)(3y + 7)$</td><td>$3y^2 + 10y + 7$</td></tr><tr><td>$(y + 7)(3y + 1)$</td><td>$3y^2 + 22y + 7$</td></tr></table> $(y + 7)(3y + 1)$

| **Step 5.** Check by multiplying. | $(y+7)(3y+1)$
 $3y^2 + 22y + 7 ✓$ |

> **TRY IT :: 7.65**   Factor completely: $2a^2 + 5a + 3$.

> **TRY IT :: 7.66**   Factor completely: $4b^2 + 5b + 1$.

HOW TO :: FACTOR TRINOMIALS OF THE FORM $ax^2 + bx + c$ **USING TRIAL AND ERROR.**

Step 1.   Write the trinomial in descending order of degrees.

Step 2.   Find all the factor pairs of the first term.

Step 3.   Find all the factor pairs of the third term.

Step 4.   Test all the possible combinations of the factors until the correct product is found.

Step 5.   Check by multiplying.

When the middle term is negative and the last term is positive, the signs in the binomials must both be negative.

EXAMPLE 7.34

Factor completely: $6b^2 - 13b + 5$.

⊘ **Solution**

The trinomial is already in descending order.	$6b^2 - 13b + 5$
Find the factors of the first term.	$6b^2 - 13b + 5$ $1b \cdot 6b$ $2b \cdot 3b$
Find the factors of the last term. Consider the signs. Since the last term, 5 is positive its factors must both be positive or both be negative. The coefficient of the middle term is negative, so we use the negative factors.	$6b^2 - 13b + 5$ $1b \cdot 6b$ $-1, -5$ $2b \cdot 3b$

Consider all the combinations of factors.

$6b^2 - 13b + 5$	
Possible factors	**Product**
$(b-1)(6b-5)$	$6b^2 - 11b + 5$
$(b-5)(6b-1)$	$6b^2 - 31b + 5$
$(2b-1)(3b-5)$	$6b^2 - 13b + 5$ *
$(2b-5)(3b-1)$	$6b^2 - 17b + 5$

The correct factors are those whose product
is the original trinomial.

$(2b - 1)(3b - 5)$

Check by multiplying.

$(2b - 1)(3b - 5)$
$6b^2 - 10b - 3b + 5$
$6b^2 - 13b + 5$ ✓

> **TRY IT : : 7.67**   Factor completely: $8x^2 - 13x + 3$.

> **TRY IT : : 7.68**   Factor completely: $10y^2 - 37y + 7$.

When we factor an expression, we always look for a greatest common factor first. If the expression does not have a greatest common factor, there cannot be one in its factors either. This may help us eliminate some of the possible factor combinations.

EXAMPLE 7.35

Factor completely: $14x^2 - 47x - 7$.

⊘ Solution

The trinomial is already in descending order.	$14x^2 - 47x - 7$
Find the factors of the first term.	$14x^2 - 47x - 7$ $1x \cdot 14x$ $2x \cdot 7x$
Find the factors of the last term. Consider the signs. Since it is negative, one factor must be positive and one negative.	$14x^2 - 47x - 7$ $1x \cdot 14x$ $1, -7$ $2x \cdot 7x$ $-1, 7$

Consider all the combinations of factors. We use each pair of the factors of $14x^2$ with each pair of factors of -7.

Factors of $14x^2$	Pair with	Factors of -7
$x, 14x$		$1, -7$ $-7, 1$ (reverse order)
$x, 14x$		$-1, 7$ $7, -1$ (reverse order)
$2x, 7x$		$1, -7$ $-7, 1$ (reverse order)
$2x, 7x$		$-1, 7$ $7, -1$ (reverse order)

These pairings lead to the following eight combinations.

$14x^2 - 47x - 7$	
Possible factors	**Product**
$(x + 1)(14x - 7)$	Not an option
$(x - 7)(14x + 1)$	$14x^2 - 97x - 7$
$(x - 1)(14x + 7)$	Not an option
$(x + 7)(14x - 1)$	$14x^2 + 97x - 7$
$(2x + 1)(7x - 7)$	Not an option
$(2x - 7)(7x + 1)$	$14x^2 - 47x - 7*$
$(2x - 1)(7x + 7)$	Not an option
$(2x + 7)(7x - 1)$	$14x^2 + 47x - 7$

If the trinomial has no common factors, then neither factor can contain a common factor. That means each of these combinations is not an option.

The correct factors are those whose product is the original trinomial.

$(2x - 7)(7x + 1)$

Check by multiplying.

$(2x - 7)(7x + 1)$

$14x^2 + 2x - 49x - 7$

$14x^2 - 47x - 7$ ✓

> **TRY IT ::** 7.69   Factor completely: $8a^2 - 3a - 5$.

> **TRY IT ::** 7.70   Factor completely: $6b^2 - b - 15$.

EXAMPLE 7.36

Factor completely: $18n^2 - 37n + 15$.

⊘ **Solution**

The trinomial is already in descending order.	$18n^2 - 37n + 15$

Find the factors of the first term.	$18n^2 - 37n + 15$ $1n \cdot 18n$ $2n \cdot 9n$ $3n \cdot 6n$

Find the factors of the last term. Consider the signs. Since 15 is positive and the coefficient of the middle term is negative, we use the negative facotrs.	$18n^2 - 37n + 15$ $1n \cdot 18n$ $-1(-15)$ $2n \cdot 9n$ $-3(-5)$ $3n \cdot 6n$

Consider all the combinations of factors.

$18n^2 - 37n + 15$	
Possible factors	**Product**
$(n - 1)(18n - 15)$	Not an option
$(n - 15)(18n - 1)$	$18n^2 - 271n + 15$
$(n - 3)(18n - 5)$	$18n^2 - 59n + 15$
$(n - 5)(18n - 3)$	Not an option
$(2n - 1)(9n - 15)$	Not an option
$(2n - 15)(9n - 1)$	$18n^2 - 137n + 15$
$(2n - 3)(9n - 5)$	$18n^2 - 37n + 15*$
$(2n - 5)(9n - 3)$	Not an option
$(3n - 1)(6n - 15)$	Not an option
$(3n - 15)(6n - 1)$	Not an option
$(3n - 3)(6n - 5)$	Not an option
$(3n - 5)(6n - 3)$	Not an option

If the trinomial has no common factors, then neither factor can contain a common factor. That means this combination is not an option.

The correct factors are those whose product is
the original trinomial.

$(2n - 3)(9n - 5)$

Check by multiplying.

$(2n - 3)(9n - 5)$
$18n^2 - 10n - 27n + 15$
$18n^2 - 37n + 15$ ✓

> | **TRY IT :: 7.71**   Factor completely: $18x^2 - 3x - 10$.

> | **TRY IT :: 7.72**   Factor completely: $30y^2 - 53y - 21$.

Don't forget to look for a GCF first.

EXAMPLE 7.37

Factor completely: $10y^4 + 55y^3 + 60y^2$.

⊘ **Solution**

$$10y^4 + 55y^3 + 60y^2$$

Notice the greatest common factor, and factor it first. $15y^2(2y^2 + 11y + 12)$

Factor the trinomial.

$5y^2(2y^2 + 11y + 12)$
$y \cdot 2y$
$1 \cdot 12$
$2 \cdot 6$
$3 \cdot 4$

Consider all the combinations.

2y² + 11y + 12	
Possible factors	**Product**
(y + 1) (2y + 12)	Not an option
(y + 12) (2y + 1)	2y² + 25y + 12
(y + 2) (2y + 6)	Not an option
(y + 6) (2y + 2)	Not an option
(y + 3) (2y + 4)	Not an option
(y + 4) (2y + 3)	2y² + 11y + 12*

If the trinomial has no common factors, then neither factor can contain a common factor. That means this combination is not an option.

The correct factors are those whose product
is the original trinomial. Remember to include
the factor $5y^2$.

$5y^2(y + 4)(2y + 3)$

Check by multiplying.

$5y^2(y + 4)(2y + 3)$
$5y^2(2y^2 + 8y + 3y + 12)$
$10y^4 + 55y^3 + 60y^2$ ✓

> | **TRY IT :: 7.73**   Factor completely: $15n^3 - 85n^2 + 100n$.

> TRY IT :: 7.74   Factor completely: $56q^3 + 320q^2 - 96q$.

Factor Trinomials using the "ac" Method

Another way to factor trinomials of the form $ax^2 + bx + c$ is the "ac" method. (The "ac" method is sometimes called the grouping method.) The "ac" method is actually an extension of the methods you used in the last section to factor trinomials with leading coefficient one. This method is very structured (that is step-by-step), and it always works!

EXAMPLE 7.38   HOW TO FACTOR TRINOMIALS USING THE "AC" METHOD

Factor: $6x^2 + 7x + 2$.

✓ **Solution**

Step 1. Factor any GCF.	Is there a greatest common factor? No.	$6x^2 + 7x + 2$
Step 2. Find the product *ac*.	$a \cdot c$ $6 \cdot 2$ 12	$ax^2 + bx + c$ $6x^2 + 7x + 2$
Step 3. Find two numbers *m* and *n* that: Multiply to *ac* $m \cdot n = a \cdot c$ Add to *b* $m + n = b$	Find two numbers that multiply to 12 and add to 7. Both factors must be positive. $3 \cdot 4 = 12$ $3 + 4 = 7$	
Step 4. Split the middle term using *m*, and *n* $ax^2 + bx + c$ bx $ax^2 + mx + nx + c$	Rewrite 7x as 3x + 4x. Notice that $6x^2 + 3x + 4x + 2$ is equal to $6x^2 + 7x + 2$. We just split the middle term to get a more useful form.	$6x^2 + 7x + 2$ $6x^2 + 3x + 4x + 2$
Step 5. Factor by grouping.		$3x(2x + 1) + 2(2x + 1)$ $(2x + 1)(3x + 2)$
Step 6. Check by multiplying.		$(2x + 1)(3x + 2)$ $6x^2 + 4x + 3x + 2$ $6x^2 + 7x + 2$ ✓

> TRY IT :: 7.75   Factor: $6x^2 + 13x + 2$.

> TRY IT :: 7.76   Factor: $4y^2 + 8y + 3$.

HOW TO :: FACTOR TRINOMIALS OF THE FORM USING THE "AC" METHOD.

Step 1.   Factor any GCF.

Step 2.   Find the product ac.

Step 3.   Find two numbers m and n that:
Multiply to ac $m \cdot n = a \cdot c$
Add to b $m + n = b$

Step 4.   Split the middle term using m and n:

$$ax^2 + bx + c$$
$$ax^2 + \overbrace{mx + nx}^{bx} + c$$

Step 5.   Factor by grouping.

Step 6.   Check by multiplying the factors.

When the third term of the trinomial is negative, the factors of the third term will have opposite signs.

EXAMPLE 7.39

Factor: $8u^2 - 17u - 21$.

✓ **Solution**

Is there a greatest common factor? No.

$$\overset{ax^2 \ + \ bx \ + \ c}{8u^2 - 17u - 21}$$

Find $a \cdot c$.

$$a \cdot c$$

$$8(-21)$$

$$-168$$

Find two numbers that multiply to -168 and add to -17. The larger factor must be negative.

Factors of -168	Sum of factors
1, −168	$1 + (-168) = -167$
2, −84	$2 + (-84) = -82$
3, −56	$3 + (-56) = -53$
4, −42	$4 + (-42) = -38$
6, −28	$6 + (-28) = -22$
7, −24	$7 + (-24) = -17*$
8, −21	$8 + (-21) = -13$

Split the middle term using $7u$ and $-24u$.

$$8u^2 - 17u - 21$$

$$\underline{8u^2 + 7u} \underline{-24u - 21}$$

Factor by grouping.

$$u(8u + 7) - 3(8u + 7)$$

$$(8u + 7)(u - 3)$$

Check by multiplying.

$$(8u + 7)(u - 3)$$
$$8u^2 - 24u + 7u - 21$$
$$8u^2 - 17u - 21 \checkmark$$

> **TRY IT :: 7.77**   Factor: $20h^2 + 13h - 15$.

> **TRY IT :: 7.78**   Factor: $6g^2 + 19g - 20$.

EXAMPLE 7.40

Factor: $2x^2 + 6x + 5$.

⊘ **Solution**

Is there a greatest common factor? No.

$$\overset{ax^2 \;+\; bx \;+\; c}{2x^2 + 6x + 5}$$

Find $a \cdot c$.	$a \cdot c$
	$2(5)$
	10

Find two numbers that multiply to 10 and add to 6.

Factors of 10	Sum of factors
1, 10	$1 + 10 = 11$
2, 5	$2 + 5 = 7$

There are no factors that multiply to 10 and add to 6. The polynomial is prime.

> **TRY IT :: 7.79**   Factor: $10t^2 + 19t - 15$.

> **TRY IT :: 7.80**   Factor: $3u^2 + 8u + 5$.

Don't forget to look for a common factor!

EXAMPLE 7.41

Factor: $10y^2 - 55y + 70$.

✓ Solution

Is there a greatest common factor? Yes. The GCF is 5.	$10y^2 - 55y + 70$
Factor it. Be careful to keep the factor of 5 all the way through the solution!	$5(2y^2 - 11y + 14)$
The trinomial inside the parentheses has a leading coefficient that is not 1.	$ax^2 + bx + c$ $5(2y^2 - 11y + 14)$
Factor the trinomial.	$5(y - 2)(2y - 7)$
Check by mulitplying all three factors.	

$$5\left(2y^2 - 2y - 4y + 14\right)$$

$$5\left(2y^2 - 11y + 14\right)$$

$$10y^2 - 55y + 70 \checkmark$$

> **TRY IT :: 7.81** Factor: $16x^2 - 32x + 12$.

> **TRY IT :: 7.82** Factor: $18w^2 - 39w + 18$.

We can now update the Preliminary Factoring Strategy, as shown in **Figure 7.2** and detailed in **Choose a strategy to factor polynomials completely (updated)**, to include trinomials of the form $ax^2 + bx + c$. Remember, some polynomials are prime and so they cannot be factored.

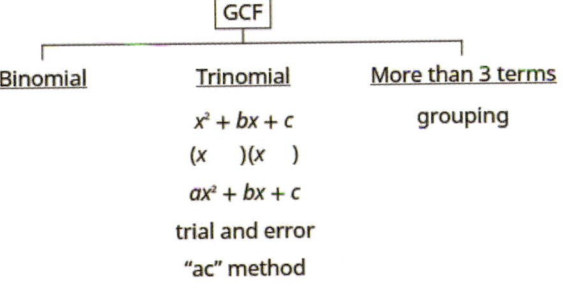

Figure 7.2

HOW TO :: CHOOSE A STRATEGY TO FACTOR POLYNOMIALS COMPLETELY (UPDATED).

Step 1. Is there a greatest common factor?
- Factor it.

Step 2. Is the polynomial a binomial, trinomial, or are there more than three terms?
- If it is a binomial, right now we have no method to factor it.

- If it is a trinomial of the form $x^2 + bx + c$
  Undo FOIL $(x \quad)(x \quad)$.

- If it is a trinomial of the form $ax^2 + bx + c$
  Use Trial and Error or the "ac" method.

- If it has more than three terms
  Use the grouping method.

Step 3. Check by multiplying the factors.

 MEDIA ::

Access these online resources for additional instruction and practice with factoring trinomials of the form $ax^2 + bx + c$.

- **Factoring Trinomials, a is not 1 (https://openstax.org/l/25FactorTrinom)**

7.3 EXERCISES

Practice Makes Perfect

Recognize a Preliminary Strategy to Factor Polynomials Completely

In the following exercises, identify the best method to use to factor each polynomial.

135.
ⓐ $10q^2 + 50$

ⓑ $a^2 - 5a - 14$

ⓒ $uv + 2u + 3v + 6$

136.
ⓐ $n^2 + 10n + 24$

ⓑ $8u^2 + 16$

ⓒ $pq + 5p + 2q + 10$

137.
ⓐ $x^2 + 4x - 21$

ⓑ $ab + 10b + 4a + 40$

ⓒ $6c^2 + 24$

138.
ⓐ $20x^2 + 100$

ⓑ $uv + 6u + 4v + 24$

ⓒ $y^2 - 8y + 15$

Factor Trinomials of the form $ax^2 + bx + c$ with a GCF

In the following exercises, factor completely.

139. $5x^2 + 35x + 30$

140. $12s^2 + 24s + 12$

141. $2z^2 - 2z - 24$

142. $3u^2 - 12u - 36$

143. $7v^2 - 63v + 56$

144. $5w^2 - 30w + 45$

145. $p^3 - 8p^2 - 20p$

146. $q^3 - 5q^2 - 24q$

147. $3m^3 - 21m^2 + 30m$

148. $11n^3 - 55n^2 + 44n$

149. $5x^4 + 10x^3 - 75x^2$

150. $6y^4 + 12y^3 - 48y^2$

Factor Trinomials Using Trial and Error

In the following exercises, factor.

151. $2t^2 + 7t + 5$

152. $5y^2 + 16y + 11$

153. $11x^2 + 34x + 3$

154. $7b^2 + 50b + 7$

155. $4w^2 - 5w + 1$

156. $5x^2 - 17x + 6$

157. $6p^2 - 19p + 10$

158. $21m^2 - 29m + 10$

159. $4q^2 - 7q - 2$

160. $10y^2 - 53y - 11$

161. $4p^2 + 17p - 15$

162. $6u^2 + 5u - 14$

163. $16x^2 - 32x + 16$

164. $81a^2 + 153a - 18$

165. $30q^3 + 140q^2 + 80q$

166. $5y^3 + 30y^2 - 35y$

Factor Trinomials using the 'ac' Method

In the following exercises, factor.

167. $5n^2 + 21n + 4$

168. $8w^2 + 25w + 3$

169. $9z^2 + 15z + 4$

170. $3m^2 + 26m + 48$ **171.** $4k^2 - 16k + 15$ **172.** $4q^2 - 9q + 5$

173. $5s^2 - 9s + 4$ **174.** $4r^2 - 20r + 25$ **175.** $6y^2 + y - 15$

176. $6p^2 + p - 22$ **177.** $2n^2 - 27n - 45$ **178.** $12z^2 - 41z - 11$

179. $3x^2 + 5x + 4$ **180.** $4y^2 + 15y + 6$ **181.** $60y^2 + 290y - 50$

182. $6u^2 - 46u - 16$ **183.** $48z^3 - 102z^2 - 45z$ **184.** $90n^3 + 42n^2 - 216n$

185. $16s^2 + 40s + 24$ **186.** $24p^2 + 160p + 96$ **187.** $48y^2 + 12y - 36$

188. $30x^2 + 105x - 60$

Mixed Practice

In the following exercises, factor.

189. $12y^2 - 29y + 14$ **190.** $12x^2 + 36y - 24z$ **191.** $a^2 - a - 20$

192. $m^2 - m - 12$ **193.** $6n^2 + 5n - 4$ **194.** $12y^2 - 37y + 21$

195. $2p^2 + 4p + 3$ **196.** $3q^2 + 6q + 2$ **197.** $13z^2 + 39z - 26$

198. $5r^2 + 25r + 30$ **199.** $x^2 + 3x - 28$ **200.** $6u^2 + 7u - 5$

201. $3p^2 + 21p$ **202.** $7x^2 - 21x$ **203.** $6r^2 + 30r + 36$

204. $18m^2 + 15m + 3$ **205.** $24n^2 + 20n + 4$ **206.** $4a^2 + 5a + 2$

207. $x^2 + 2x - 24$ **208.** $2b^2 - 7b + 4$

Everyday Math

209. Height of a toy rocket The height of a toy rocket launched with an initial speed of 80 feet per second from the balcony of an apartment building is related to the number of seconds, t, since it is launched by the trinomial $-16t^2 + 80t + 96$. Completely factor the trinomial.

210. Height of a beach ball The height of a beach ball tossed up with an initial speed of 12 feet per second from a height of 4 feet is related to the number of seconds, t, since it is tossed by the trinomial $-16t^2 + 12t + 4$. Completely factor the trinomial.

Writing Exercises

211. List, in order, all the steps you take when using the "ac" method to factor a trinomial of the form $ax^2 + bx + c$.

212. How is the "ac" method similar to the "undo FOIL" method? How is it different?

213. What are the questions, in order, that you ask yourself as you start to factor a polynomial? What do you need to do as a result of the answer to each question?

214. On your paper draw the chart that summarizes the factoring strategy. Try to do it without looking at the book. When you are done, look back at the book to finish it or verify it.

Self Check

ⓐ *After completing the exercises, use this checklist to evaluate your mastery of the objectives of this section.*

I can...	Confidently	With some help	No-I don't get it!
recognize a preliminary strategy to factor polynomials completely.			
factor trinomials of the form $ax^2 + bx + c$ with a GCF.			
factor trinomials using trial and error.			
factor trinomials using the "ac" method.			

ⓑ *What does this checklist tell you about your mastery of this section? What steps will you take to improve?*

7.4 Factor Special Products

Learning Objectives

By the end of this section, you will be able to:

> Factor perfect square trinomials
> Factor differences of squares
> Factor sums and differences of cubes
> Choose method to factor a polynomial completely

Be Prepared!

Before you get started, take this readiness quiz.

1. Simplify: $(12x)^2$.
   If you missed this problem, review **Example 6.23**.

2. Multiply: $(m + 4)^2$.
   If you missed this problem, review **Example 6.47**.

3. Multiply: $(p - 9)^2$.
   If you missed this problem, review **Example 6.48**.

4. Multiply: $(k + 3)(k - 3)$.
   If you missed this problem, review **Example 6.52**.

The strategy for factoring we developed in the last section will guide you as you factor most binomials, trinomials, and polynomials with more than three terms. We have seen that some binomials and trinomials result from special products—squaring binomials and multiplying conjugates. If you learn to recognize these kinds of polynomials, you can use the special products patterns to factor them much more quickly.

Factor Perfect Square Trinomials

Some trinomials are perfect squares. They result from multiplying a binomial times itself. You can square a binomial by using FOIL, but using the Binomial Squares pattern you saw in a previous chapter saves you a step. Let's review the Binomial Squares pattern by squaring a binomial using FOIL.

$$(3x + 4)^2$$

$$(3x + 4)(3x + 4)$$

$$\overset{F \qquad O \qquad I \qquad L}{9x^2 + 12x + 12x + 16}$$

$$9x^2 + 24x + 16$$

The first term is the square of the first term of the binomial and the last term is the square of the last. The middle term is twice the product of the two terms of the binomial.

$$(3x)^2 + 2(3x \cdot 4) + 4^2$$
$$9x^2 + 24x + 16$$

The trinomial $9x^2 + 24 + 16$ is called a perfect square trinomial. It is the square of the binomial $3x+4$.

We'll repeat the Binomial Squares Pattern here to use as a reference in factoring.

Binomial Squares Pattern

If a and b are real numbers,

$$(a + b)^2 = a^2 + 2ab + b^2 \qquad (a - b)^2 = a^2 - 2ab + b^2$$

When you square a binomial, the product is a perfect square trinomial. In this chapter, you are learning to factor—now, you will start with a perfect square trinomial and factor it into its prime factors.

You could factor this trinomial using the methods described in the last section, since it is of the form $ax^2 + bx + c$. But if you recognize that the first and last terms are squares and the trinomial fits the **perfect square trinomials pattern**, you will save yourself a lot of work.

Here is the pattern—the reverse of the binomial squares pattern.

Perfect Square Trinomials Pattern

If a and b are real numbers,

$$a^2 + 2ab + b^2 = (a+b)^2 \qquad a^2 - 2ab + b^2 = (a-b)^2$$

To make use of this pattern, you have to recognize that a given trinomial fits it. Check first to see if the leading coefficient is a perfect square, a^2. Next check that the last term is a perfect square, b^2. Then check the middle term—is it twice the product, $2ab$? If everything checks, you can easily write the factors.

EXAMPLE 7.42   HOW TO FACTOR PERFECT SQUARE TRINOMIALS

Factor: $9x^2 + 12x + 4$.

✓ **Solution**

Step 1. Does the trinomial fit the perfect square trinomials pattern, $a^2 + 2ab + b^2$?		
• Is the first term a perfect square? Write it as a square, a^2.	Is $9x^2$ a perfect square? Yes—write it as $(3x)^2$.	$9x^2 + 12x + 4$ $(3x)^2$
• Is the last term a perfect square? Write it as a square, b^2.	Is 4 a perfect square? Yes—write it as $(2)^2$.	$(3x)^2 \qquad (2)^2$
• Check the middle term. Is it $2ab$?	Is $12x$ twice the product of $3x$ and 2? Does it match? Yes, so we have a perfect square trinomial!	$(3x)^2 \qquad (2)^2$ $2(3x)(2)$ $12x$
Step 2. Write the square of the binomial.	Write it as the square of a binomial.	$9x^2 + 12x + 4$ $a^2 + 2 \cdot a \cdot b + b^2$ $(3x)^2 + 2 \cdot 3x \cdot 2 + 2^2$ $(a + b)^2$ $(3x + 2)^2$
Step 3. Check. $(3x + 2)^2$ $(3x)^2 + 2 \cdot 3x \cdot 2 + 2^2$ $9x^2 + 12x + 4$ ✓		

> **TRY IT :: 7.83**   Factor: $4x^2 + 12x + 9$.

> **TRY IT :: 7.84**   Factor: $9y^2 + 24y + 16$.

The sign of the middle term determines which pattern we will use. When the middle term is negative, we use the pattern $a^2 - 2ab + b^2$, which factors to $(a-b)^2$.

The steps are summarized here.

HOW TO :: FACTOR PERFECT SQUARE TRINOMIALS.

Step 1. Does the trinomial fit the pattern? $a^2 + 2ab + b^2$ $a^2 - 2ab + b^2$

- Is the first term a perfect square? $(a)^2$ $(a)^2$
  Write it as a square.

- Is the last term a perfect square? $(a)^2$ $(b)^2$ $(a)^2$ $(b)^2$
  Write it as a square.

- Check the middle term. Is it $2ab$? $(a)^2 \searrow_{2 \cdot a \cdot b} \nearrow (b)^2$ $(a)^2 \searrow_{2 \cdot a \cdot b} \nearrow (b)^2$

Step 2. Write the square of the binomial. $(a + b)^2$ $(a - b)^2$
Step 3. Check by multiplying.

We'll work one now where the middle term is negative.

EXAMPLE 7.43

Factor: $81y^2 - 72y + 16$.

⊘ **Solution**

The first and last terms are squares. See if the middle term fits the pattern of a perfect square trinomial. The middle term is negative, so the binomial square would be $(a - b)^2$.

	$81y^2 - 72y + 16$
Are the first and last terms perfect squares?	$(9y)^2$ $(4)^2$
Check the middle term.	$(9y)^2$ $(4)^2$ $2(9y)(4)$ $72y$
Does is match $(a - b)^2$? Yes.	$a^2 - 2 \cdot a \cdot b + b^2$ $(9y)^2 - 2 \cdot 9y \cdot 4 + 4^2$
Write the square of a binomial.	$(9y - 4)^2$
Check by mulitplying.	
$(9y - 4)^2$	
$(9y)^2 - 2 \cdot 9y \cdot 4 + 4^2$	
$81y^2 - 72y + 16$ ✓	

> **TRY IT :: 7.85**   Factor: $64y^2 - 80y + 25$.

> **TRY IT :: 7.86**   Factor: $16z^2 - 72z + 81$.

The next example will be a perfect square trinomial with two variables.

EXAMPLE 7.44

Factor: $36x^2 + 84xy + 49y^2$.

✓ Solution

	$36x^2 + 84xy + 49y^2$
Test each term to verify the pattern.	$\overset{a^2}{(6x)^2} + \overset{+2\quad a\quad b}{2 \cdot 6x \cdot 7y} + \overset{+\ b^2}{(7y)^2}$
Factor.	$(6x + 7y)^2$
Check by mulitplying.	
$(6x + 7y)^2$	
$(6x)^2 + 2 \cdot 6x \cdot 7y + (7y)^2$	
$36x^2 + 84xy + 49y^2$ ✓	

> | **TRY IT ::** 7.87   Factor: $49x^2 + 84xy + 36y^2$.

> | **TRY IT ::** 7.88   Factor: $64m^2 + 112mn + 49n^2$.

EXAMPLE 7.45

Factor: $9x^2 + 50x + 25$.

✓ Solution

	$9x^2 + 50x + 25$
Are the first and last terms perfect squares?	$(3x)^2 \qquad (5)^2$
Check the middle term—is it $2ab$?	$(3x)^2 \underset{30x}{\searrow 2(3x)(5) \nearrow} (5)^2$
No! $30x \neq 50x$	This does not fit the pattern!
Factor using the "ac" method.	$9x^2 + 50x + 25$
Notice: $\overset{ac}{9 \cdot 25}$ and $\begin{array}{l} 5 \cdot 45 = 225 \\ 5 + 45 = 50 \end{array}$	
	225

Split the middle term.	$9x^2 + 5x + 45x + 25$
Factor by grouping.	$x(9x + 5) + 5(9x + 5)$
	$(9x + 5)(x + 5)$

Check.
$(9x + 5)(x + 5)$
$9x^2 + 45x + 5x + 25$
$9x^2 + 50x + 25$ ✓

> | **TRY IT ::** 7.89   Factor: $16r^2 + 30rs + 9s^2$.

> **TRY IT ::** 7.90   Factor: $9u^2 + 87u + 100$.

Remember the very first step in our Strategy for Factoring Polynomials? It was to ask "is there a greatest common factor?" and, if there was, you factor the GCF before going any further. Perfect square trinomials may have a GCF in all three terms and it should be factored out first. And, sometimes, once the GCF has been factored, you will recognize a perfect square trinomial.

EXAMPLE 7.46

Factor: $36x^2y - 48xy + 16y$.

⊘ Solution

	$36x^2y - 48xy + 16y$
Is there a GCF? Yes, 4y, so factor it out.	$4y\left(9x^2 - 12x + 4\right)$
Is this a perfect square trinomial?	
Verify the pattern.	$\overset{a^2 \quad -2\ a\ b\ +\ b^2}{4y[(3x)^2 - 2\cdot 3x\cdot 2 + 2^2]}$
Factor.	$4y(3x - 2)^2$
Remember: Keep the factor 4y in the final product.	
Check.	

$4y(3x - 2)^2$

$4y\left[(3x)^2 - 2\cdot 3x \cdot 2 + 2^2\right]$

$4y(9x)^2 - 12x + 4$

$36x^2y - 48xy + 16y$ ✓

> **TRY IT ::** 7.91   Factor: $8x^2y - 24xy + 18y$.

> **TRY IT ::** 7.92   Factor: $27p^2q + 90pq + 75q$.

Factor Differences of Squares

The other special product you saw in the previous was the Product of Conjugates pattern. You used this to multiply two binomials that were conjugates. Here's an example:

$$(3x - 4)(3x + 4)$$
$$9x^2 - 16$$

Remember, when you multiply conjugate binomials, the middle terms of the product add to 0. All you have left is a binomial, the difference of squares.

Multiplying conjugates is the only way to get a binomial from the product of two binomials.

Product of Conjugates Pattern

If a and b are real numbers

$$(a - b)(a + b) = a^2 - b^2$$

The product is called a difference of squares.

To factor, we will use the product pattern "in reverse" to factor the difference of squares. A **difference of squares** factors to a product of conjugates.

Difference of Squares Pattern

If a and b are real numbers,

$$a^2 - b^2 = (a - b)(a + b) \qquad a^2 \underset{\text{squares}}{\overset{\text{difference}}{-}} b^2 = (a - b)(a + b)$$

squares   conjugates

Remember, "difference" refers to subtraction. So, to use this pattern you must make sure you have a binomial in which two squares are being subtracted.

EXAMPLE 7.47   HOW TO FACTOR DIFFERENCES OF SQUARES

Factor: $x^2 - 4$.

✓ **Solution**

Step 1. Does the binomial fit the pattern?		$x^2 - 4$
• Is this a difference?	Yes	$x^2 - 4$
• Are the first and last terms perfect squares?	Yes	
Step 2. Write them as squares.	Write them as x^2 and 2^2.	$a^2 - b^2$ $(x)^2 - 2^2$
Step 3. Write the product of conjugates.		$(a - b)(a + b)$ $(x - 2)(x + 2)$
Step 4. Check.		$(x - 2)(x + 2)$ $x^2 - 4$ ✓

> **TRY IT :: 7.93**   Factor: $h^2 - 81$.

> **TRY IT :: 7.94**   Factor: $k^2 - 121$.

HOW TO :: FACTOR DIFFERENCES OF SQUARES.

Step 1. Does the binomial fit the pattern? $a^2 - b^2$
    • Is this a difference? $\underline{\quad} - \underline{\quad}$
    • Are the first and last terms perfect squares?

Step 2. Write them as squares. $(a)^2 - (b)^2$
Step 3. Write the product of conjugates. $(a - b)(a + b)$
Step 4. Check by multiplying.

It is important to remember that *sums of squares do not factor into a product of binomials*. There are no binomial factors that multiply together to get a sum of squares. After removing any GCF, the expression $a^2 + b^2$ is prime!

Don't forget that 1 is a perfect square. We'll need to use that fact in the next example.

EXAMPLE 7.48

Factor: $64y^2 - 1$.

⊘ Solution

	$64y^2 - 1$
Is this a difference? Yes.	$64y^2 - 1$
Are the first and last terms perfect squares?	
Yes - write them as squares.	$\overset{a^2 - b^2}{(8y)^2 - 1^2}$
Factor as the product of conjugates.	$\overset{(a - b)\ (a + b)}{(8y - 1)(8y + 1)}$
Check by multiplying.	
$(8y - 1)(8y + 1)$	
$64y^2 - 1$ ✓	

> **TRY IT ::** 7.95   Factor: $m^2 - 1$.

> **TRY IT ::** 7.96   Factor: $81y^2 - 1$.

EXAMPLE 7.49

Factor: $121x^2 - 49y^2$.

⊘ Solution

$$121x^2 - 49y^2$$

Is this a difference of squares? Yes. $(11x)^2 - (7y)^2$

Factor as the product of conjugates. $(11x - 7y)(11x + 7y)$

Check by multiplying.

$(11x - 7y)(11x + 7y)$
$121x^2 - 49y^2$ ✓

> **TRY IT ::** 7.97   Factor: $196m^2 - 25n^2$.

> **TRY IT ::** 7.98   Factor: $144p^2 - 9q^2$.

The binomial in the next example may look "backwards," but it's still the difference of squares.

Factor: $100 - h^2$.

⊘ **Solution**

$$100 - h^2$$

Is this a difference of squares? Yes. $(10)^2 - (h)^2$

Factor as the product of conjugates. $(10 - h)(10 + h)$

Check by multiplying.

$$(10 - h)(10 + h)$$
$$100 - h^2 \checkmark$$

Be careful not to rewrite the original expression as $h^2 - 100$.

Factor $h^2 - 100$ on your own and then notice how the result differs from $(10 - h)(10 + h)$.

> **TRY IT : : 7.99**   Factor: $144 - x^2$.

> **TRY IT : : 7.100**   Factor: $169 - p^2$.

To completely factor the binomial in the next example, we'll factor a difference of squares twice!

Factor: $x^4 - y^4$.

⊘ **Solution**

$$x^4 - y^4$$

Is this a difference of squares? Yes.

$$\left(x^2\right)^2 - \left(y^2\right)^2$$

Factor it as the product of conjugates.

$$\left(x^2 - y^2\right)\left(x^2 + y^2\right)$$

Notice the first binomial is also a difference of squares!

$$\left((x)^2 - (y)^2\right)\left(x^2 + y^2\right)$$

Factor it as the product of conjugates. The last factor, the sum of squares, cannot be factored.

$$(x - y)(x + y)\left(x^2 + y^2\right)$$

Check by multiplying.

$$(x - y)(x + y)\left(x^2 + y^2\right)$$
$$[(x - y)(x + y)]\left(x^2 + y^2\right)$$
$$\left(x^2 - y^2\right)\left(x^2 + y^2\right)$$
$$x^4 - y^4 \checkmark$$

> TRY IT :: 7.101   Factor: $a^4 - b^4$.

> TRY IT :: 7.102   Factor: $x^4 - 16$.

As always, you should look for a common factor first whenever you have an expression to factor. Sometimes a common factor may "disguise" the difference of squares and you won't recognize the perfect squares until you factor the GCF.

EXAMPLE 7.52

Factor: $8x^2 y - 18y$.

✓ Solution

$$8x^2 y - 98y$$

Is there a GCF? Yes, $2y$—factor it out! $2y\left(4x^2 - 49\right)$

Is the binomial a difference of squares? Yes. $2y\left((2x)^2 - (7)^2\right)$

Factor as a product of conjugates. $2y(2x - 7)(2x + 7)$

Check by multiplying.

$2y(2x - 7)(2x + 7)$
$2y[(2x - 7)(2x + 7)]$
$2y\left(4x^2 - 49\right)$
$8x^2 y - 98y$ ✓

> TRY IT :: 7.103   Factor: $7xy^2 - 175x$.

> TRY IT :: 7.104   Factor: $45a^2 b - 80b$.

EXAMPLE 7.53

Factor: $6x^2 + 96$.

✓ Solution

$$6x^2 + 96$$

Is there a GCF? Yes, 6—factor it out! $6\left(x^2 + 16\right)$

Is the binomial a difference of squares? No, it
is a sum of squares. Sums of squares do not factor!

Check by multiplying.

$6\left(x^2 + 16\right)$
$6x^2 + 96$ ✓

> **TRY IT ::** 7.105   Factor: $8a^2 + 200$.

> **TRY IT ::** 7.106   Factor: $36y^2 + 81$.

Factor Sums and Differences of Cubes

There is another special pattern for factoring, one that we did not use when we multiplied polynomials. This is the pattern for the sum and difference of cubes. We will write these formulas first and then check them by multiplication.

$$a^3 + b^3 = (a + b)\left(a^2 - ab + b^2\right)$$
$$a^3 - b^3 = (a - b)\left(a^2 + ab + b^2\right)$$

We'll check the first pattern and leave the second to you.

$$(a + b)\,(a^2 - ab + b^2)$$

Distribute.	$a(a^2 - ab + b^2) + b(a^2 - ab + b^2)$
Multiply.	$a^3 - a^2 b + ab^2 + a^2 b - ab^2 + b^3$
Combine like terms.	$a^3 + b^3$

Sum and Difference of Cubes Pattern

$$a^3 + b^3 = (a + b)\left(a^2 - ab + b^2\right)$$
$$a^3 - b^3 = (a - b)\left(a^2 + ab + b^2\right)$$

The two patterns look very similar, don't they? But notice the signs in the factors. The sign of the binomial factor matches the sign in the original binomial. And the sign of the middle term of the trinomial factor is the opposite of the sign in the original binomial. If you recognize the pattern of the signs, it may help you memorize the patterns.

$$a^3 + b^3 = (a + b)\,(a^2 - ab + b^2)$$

same sign
opposite signs

$$a^3 - b^3 = (a - b)\,(a^2 + ab + b^2)$$

same sign
opposite signs

The trinomial factor in the sum and difference of cubes pattern cannot be factored.

It can be very helpful if you learn to recognize the cubes of the integers from 1 to 10, just like you have learned to recognize squares. We have listed the cubes of the integers from 1 to 10 in **Figure 7.3**.

n	1	2	3	4	5	6	7	8	9	10
n^3	1	8	27	64	125	216	343	512	729	1000

Figure 7.3

EXAMPLE 7.54   HOW TO FACTOR THE SUM OR DIFFERENCE OF CUBES

Factor: $x^3 + 64$.

✅ Solution

Step 1. Does the binomial fit the sum or difference of cubes pattern?		$x^3 + 64$
• Is it a sum or difference?	This is a sum.	$x^3 + 64$
• Are the first and last terms perfect cubes?	Yes	
Step 2. Write the terms as cubes.	Write them as x^3 and 4^3	$a^3 + b^3$ $x^3 + 4^3$
Step 3. Use either the sum or difference of cubes pattern.	This is a sum of cubes.	$\underset{(x+4)}{(a+b)}\underset{(x^2 - 4x + 4^2)}{(a^2 - ab + b^2)}$
Step 4. Simplify inside the parentheses.	Simplify 4^2.	$(x+4)(x^2 - 4x + 16)$
Step 5. Check by multiplying the factors.		$\begin{array}{r} x^2 - 4x + 16 \\ x + 4 \\ \hline 4x^2 - 16x + 64 \\ x^3 - 4x^2 + 16x \\ \hline x^3 + 64 \checkmark \end{array}$

> **TRY IT :: 7.107**   Factor: $x^3 + 27$.

> **TRY IT :: 7.108**   Factor: $y^3 + 8$.

HOW TO :: FACTOR THE SUM OR DIFFERENCE OF CUBES.

To factor the sum or difference of cubes:

Step 1.   Does the binomial fit the sum or difference of cubes pattern?
   ◦ Is it a sum or difference?
   ◦ Are the first and last terms perfect cubes?
Step 2.   Write them as cubes.
Step 3.   Use either the sum or difference of cubes pattern.
Step 4.   Simplify inside the parentheses
Step 5.   Check by multiplying the factors.

EXAMPLE 7.55

Factor: $x^3 - 1000$.

✅ Solution

	$x^3 - 1000$
This binomial is a difference. The first and last terms are perfect cubes.	
Write the terms as cubes.	$a^3 - b^3$ $x^3 - 10^3$

Use the difference of cubes pattern.

$$\left(\overset{a\ -\ b}{x-10}\right)\left(\overset{a^2\ +\ \ ab\ \ +\ b^2}{x^2+10\cdot x+10^2}\right)$$

Simplify.

$$\left(\overset{a\ -\ b}{x-10}\right)\left(\overset{a^2+\ \ ab\ \ +\ b^2}{x^2+10x+100}\right)$$

Check by multiplying.

$(x-10)\ (x^2+10x+100)$

$x^2+10x+100$
$\quad\quad\quad x-100$
_____
x^3+10x^2+100x
$\quad -10x^2-100x-1000\ \checkmark$
_____
$x^3\quad\quad\quad\quad -1000$

 **TRY IT :: 7.109** Factor: u^3-125.

 **TRY IT :: 7.110** Factor: v^3-343.

Be careful to use the correct signs in the factors of the sum and difference of cubes.

EXAMPLE 7.56

Factor: $512-125p^3$.

✓ **Solution**

$512-125p^3$

This binomial is a difference. The first and last terms are perfect cubes.

Write the terms as cubes.

$$\overset{a^3\ -\ \ b^3}{8^3-(5p)^3}$$

Use the difference of cubes pattern.

$$\left(\overset{a\ -\ b}{8-5p}\right)\left(\overset{a^2\ +\ \ ab\ \ +\ b^2}{8^2+8\cdot 5p+(5p)^2}\right)$$

Simplify.

$$\left(\overset{a\ -\ b}{8-5p}\right)\left(\overset{a^2\ +\ \ ab\ \ +\ b^2}{64+40p+25p^2}\right)$$

Check by multiplying.

We'll leave the check to you.

 **TRY IT :: 7.111** Factor: $64-27x^3$.

 **TRY IT :: 7.112** Factor: $27-8y^3$.

EXAMPLE 7.57

Factor: $27u^3-125v^3$.

✓ Solution

	$27u^3 - 125v^3$
This binomial is a difference. The first and last terms are perfect cubes.	
Write the terms as cubes.	$(3u)^3 - (5v)^3$
Use the difference of cubes pattern.	$(3u - 5v)\left((3u)^2 + 3u \cdot 5v + (5v)^2\right)$
Simplify.	$(3u - 5v)\left(9u^2 + 15uv + 25v^2\right)$
Check by multiplying.	We'll leave the check to you.

> **TRY IT ::** 7.113   Factor: $8x^3 - 27y^3$.

> **TRY IT ::** 7.114   Factor: $1000m^3 - 125n^3$.

In the next example, we first factor out the GCF. Then we can recognize the sum of cubes.

EXAMPLE 7.58

Factor: $5m^3 + 40n^3$.

✓ Solution

	$5m^3 + 40n^3$
Factor the common factor.	$5(m^3 + 8n^3)$
This binomial is a sum. The first and last terms are perfect cubes.	
Write the terms as cubes.	$5\left(m^3 + (2n)^3\right)$
Use the sum of cubes pattern.	$5(m + 2n)\left(m^2 - m \cdot 2n + (2n)^2\right)$
Simplify.	$5(m + 2n)\left(m^2 - 2mn + 4n^2\right)$

Check. To check, you may find it easier to multiply the sum of cubes factors first, then multiply that product by 5. We'll leave the multiplication for you.

$5(m + 2n)\left(m^2 - 2mn + 4n^2\right)$

> **TRY IT ::** 7.115   Factor: $500p^3 + 4q^3$.

> **TRY IT ::** 7.116   Factor: $432c^3 + 686d^3$.

 **MEDIA : :**

Access these online resources for additional instruction and practice with factoring special products.

- **Sum of Difference of Cubes (https://openstax.org/l/25SumCubes)**
- **Difference of Cubes Factoring (https://openstax.org/l/25DiffCubes)**

7.4 EXERCISES

Practice Makes Perfect

Factor Perfect Square Trinomials

In the following exercises, factor.

215. $16y^2 + 24y + 9$

216. $25v^2 + 20v + 4$

217. $36s^2 + 84s + 49$

218. $49s^2 + 154s + 121$

219. $100x^2 - 20x + 1$

220. $64z^2 - 16z + 1$

221. $25n^2 - 120n + 144$

222. $4p^2 - 52p + 169$

223. $49x^2 - 28xy + 4y^2$

224. $25r^2 - 60rs + 36s^2$

225. $25n^2 + 25n + 4$

226. $100y^2 - 52y + 1$

227. $64m^2 - 34m + 1$

228. $100x^2 - 25x + 1$

229. $10k^2 + 80k + 160$

230. $64x^2 - 96x + 36$

231. $75u^3 - 30u^2v + 3uv^2$

232. $90p^3 + 300p^2q + 250pq^2$

Factor Differences of Squares

In the following exercises, factor.

233. $x^2 - 16$

234. $n^2 - 9$

235. $25v^2 - 1$

236. $169q^2 - 1$

237. $121x^2 - 144y^2$

238. $49x^2 - 81y^2$

239. $169c^2 - 36d^2$

240. $36p^2 - 49q^2$

241. $4 - 49x^2$

242. $121 - 25s^2$

243. $16z^4 - 1$

244. $m^4 - n^4$

245. $5q^2 - 45$

246. $98r^3 - 72r$

247. $24p^2 + 54$

248. $20b^2 + 140$

Factor Sums and Differences of Cubes

In the following exercises, factor.

249. $x^3 + 125$

250. $n^3 + 512$

251. $z^3 - 27$

252. $v^3 - 216$

253. $8 - 343t^3$

254. $125 - 27w^3$

255. $8y^3 - 125z^3$

256. $27x^3 - 64y^3$

257. $7k^3 + 56$

258. $6x^3 - 48y^3$

259. $2 - 16y^3$

260. $-2x^3 - 16y^3$

Mixed Practice

In the following exercises, factor.

261. $64a^2 - 25$

262. $121x^2 - 144$

263. $27q^2 - 3$

264. $4p^2 - 100$

265. $16x^2 - 72x + 81$

266. $36y^2 + 12y + 1$

267. $8p^2 + 2$

268. $81x^2 + 169$

269. $125 - 8y^3$

270. $27u^3 + 1000$

271. $45n^2 + 60n + 20$

272. $48q^3 - 24q^2 + 3q$

Everyday Math

273. Landscaping Sue and Alan are planning to put a 15 foot square swimming pool in their backyard. They will surround the pool with a tiled deck, the same width on all sides. If the width of the deck is w, the total area of the pool and deck is given by the trinomial $4w^2 + 60w + 225$. Factor the trinomial.

274. Home repair The height a twelve foot ladder can reach up the side of a building if the ladder's base is b feet from the building is the square root of the binomial $144 - b^2$. Factor the binomial.

Writing Exercises

275. Why was it important to practice using the binomial squares pattern in the chapter on multiplying polynomials?

276. How do you recognize the binomial squares pattern?

277. Explain why $n^2 + 25 \neq (n + 5)^2$. Use algebra, words, or pictures.

278. Maribel factored $y^2 - 30y + 81$ as $(y - 9)^2$. Was she right or wrong? How do you know?

Self Check

ⓐ After completing the exercises, use this checklist to evaluate your mastery of the objectives of this section.

I can...	Confidently	With some help	No-I don't get it!
factor perfect square trinomials.			
factor differences of squares.			
factor sums and differences of cubes.			

ⓑ On a scale of 1–10, how would you rate your mastery of this section in light of your responses on the checklist? How can you improve this?

7.5 General Strategy for Factoring Polynomials

Learning Objectives

By the end of this section, you will be able to:

> Recognize and use the appropriate method to factor a polynomial completely

---

Be Prepared!

Before you get started, take this readiness quiz.

1. Factor $y^2 - 2y - 24$.
   If you missed this problem, review **Example 7.23**.

2. Factor $3t^2 + 17t + 10$.
   If you missed this problem, review **Example 7.38**.

3. Factor $36p^2 - 60p + 25$.
   If you missed this problem, review **Example 7.42**.

4. Factor $5x^2 - 80$.
   If you missed this problem, review **Example 7.52**.

Recognize and Use the Appropriate Method to Factor a Polynomial Completely

You have now become acquainted with all the methods of factoring that you will need in this course. (In your next algebra course, more methods will be added to your repertoire.) The figure below summarizes all the factoring methods we have covered. **Factor polynomials.** outlines a strategy you should use when factoring polynomials.

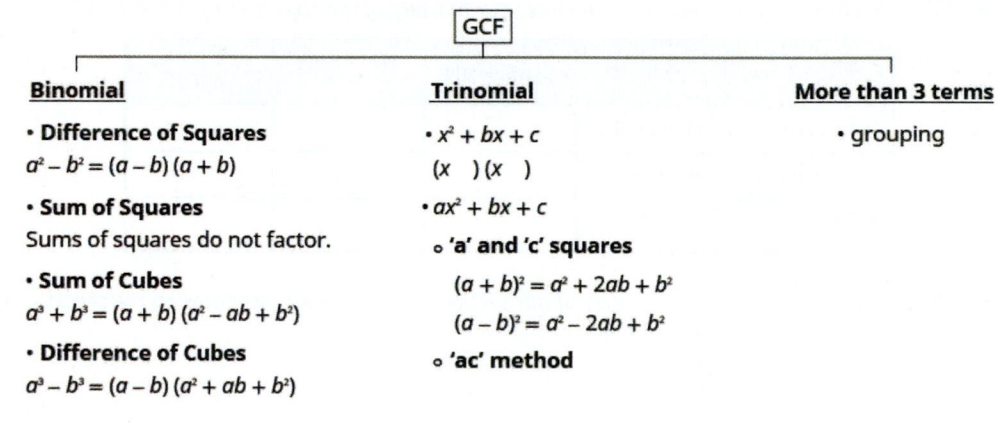

General Strategy for Factoring Polynomials

GCF

| **Binomial** | **Trinomial** | **More than 3 terms** |

- **Difference of Squares**
$a^2 - b^2 = (a - b)(a + b)$

- **Sum of Squares**
Sums of squares do not factor.

- **Sum of Cubes**
$a^3 + b^3 = (a + b)(a^2 - ab + b^2)$

- **Difference of Cubes**
$a^3 - b^3 = (a - b)(a^2 + ab + b^2)$

- $x^2 + bx + c$
$(x\ \ )(x\ \ )$

- $ax^2 + bx + c$
  o 'a' and 'c' squares
    $(a + b)^2 = a^2 + 2ab + b^2$
    $(a - b)^2 = a^2 - 2ab + b^2$
  o 'ac' method

- grouping

Figure 7.4

HOW TO :: FACTOR POLYNOMIALS.

Step 1.   Is there a greatest common factor?

   ◦ Factor it out.

Step 2.   Is the polynomial a binomial, trinomial, or are there more than three terms?

   ◦ If it is a binomial:
   Is it a sum?

      ▪ Of squares? Sums of squares do not factor.

      ▪ Of cubes? Use the sum of cubes pattern.

   Is it a difference?

      ▪ Of squares? Factor as the product of conjugates.

      ▪ Of cubes? Use the difference of cubes pattern.

   ◦ If it is a trinomial:
   Is it of the form $x^2 + bx + c$? Undo FOIL.

   Is it of the form $ax^2 + bx + c$?

      ▪ If a and c are squares, check if it fits the trinomial square pattern.

      ▪ Use the trial and error or "ac" method.

   ◦ If it has more than three terms:
   Use the grouping method.

Step 3.   Check.

   ◦ Is it factored completely?

   ◦ Do the factors multiply back to the original polynomial?

Remember, a polynomial is completely factored if, other than monomials, its factors are prime!

EXAMPLE 7.59

Factor completely: $4x^5 + 12x^4$.

✓ **Solution**

Is there a GCF?	Yes, $4x^4$.	$4x^5 + 12x^4$
	Factor out the GCF.	$4x^4(x+3)$
In the parentheses, is it a binomial, a trinomial, or are there more than three terms?	Binomial.	
Is it a sum?	Yes.	
Of squares? Of cubes?	No.	
Check.		
Is the expression factored completely?	Yes.	
Multiply.		

$$4x^4(x+3)$$
$$4x^4 \cdot x + 4x^4 \cdot 3$$
$$4x^5 + 12x^4 \checkmark$$

 **TRY IT :: 7.117**   Factor completely: $3a^4 + 18a^3$.

> **TRY IT :: 7.118**   Factor completely: $45b^6 + 27b^5$.

EXAMPLE 7.60

Factor completely: $12x^2 - 11x + 2$.

⊘ **Solution**

	$12x^2 - 11x + 2$
Is there a GCF?	No.
Is it a binomial, trinomial, or are there more than three terms?	Trinomial.
Are a and c perfect squares?	No, $a = 12$, not a perfect square.

Use trial and error or the "ac" method. We will use trial and error here.	$12x^2 - 11x + 2$ 1x, 12x −1, −2 2x, 6x 3x, 4x

$12x^2 - 11x + 2$	
Possible factors	**Product**
$(x - 1)(12x - 2)$	Not an option
$(x - 2)(12x - 1)$	$12x^2 - 25x + 2$
$(2x - 1)(6x - 2)$	Not an option
$(2x - 2)(6x - 1)$	Not an option
$(3x - 1)(4x - 2)$	Not an option
$(3x - 2)(4x - 1)$	$12x^2 - 11x + 2$

If the trinomial has no common factors, then neither factor can contain a common factor. That means each of these combinations is not an option.

Check.

$$(3x - 2)(4x - 1)$$

$$12x^2 - 3x - 8x + 2$$

$$12x^2 - 11x + 2 \checkmark$$

> **TRY IT :: 7.119**   Factor completely: $10a^2 - 17a + 6$.

> **TRY IT :: 7.120**   Factor completely: $8x^2 - 18x + 9$.

EXAMPLE 7.61

Factor completely: $g^3 + 25g$.

Solution

Is there a GCF?	Yes, g.	$g^3 + 25g$
Factor out the GCF.		$g(g^2 + 25)$

In the parentheses, is it a binomial, trinomial, or are there more than three terms?	Binomial.	
Is it a sum ? Of squares?	Yes.	Sums of squares are prime.

Check.

Is the expression factored completely?	Yes.
Multiply.	

$$g(g^2 + 25)$$

$$g^3 + 25g \checkmark$$

> **TRY IT : : 7.121**   Factor completely: $x^3 + 36x$.

> **TRY IT : : 7.122**   Factor completely: $27y^2 + 48$.

EXAMPLE 7.62

Factor completely: $12y^2 - 75$.

Solution

Is there a GCF?	Yes, 3.	$12y^2 - 75$
Factor out the GCF.		$3(4y^2 - 25)$

In the parentheses, is it a binomial, trinomial, or are there more than three terms?	Binomial.	
Is it a sum?	No.	
Is it a difference? Of squares or cubes?	Yes, squares.	$3\left((2y)^2 - (5)^2\right)$
Write as a product of conjugates.		$3(2y - 5)(2y + 5)$

Check.

Is the expression factored completely?	Yes.
Neither binomial is a difference of squares.	
Multiply.	

$$3(2y - 5)(2y + 5)$$

$$3(4y^2 - 25)$$

$$12y^2 - 75 \checkmark$$

> **TRY IT : : 7.123**   Factor completely: $16x^3 - 36x$.

> **TRY IT : : 7.124**   Factor completely: $27y^2 - 48$.

EXAMPLE 7.63

Factor completely: $4a^2 - 12ab + 9b^2$.

✓ **Solution**

Is there a GCF?	No.	$4a^2 - 12ab + 9b^2$
Is it a binomial, trinomial, or are there more terms?		
Trinomial with $a \neq 1$. But the first term is a perfect square.		
Is the last term a perfect square?	Yes.	$(2a)^2 - 12ab + (3b)^2$
Does it fit the pattern, $a^2 - 2ab + b^2$?	Yes.	$(2a)^2 - 12ab + (3b)^2$ $-2(2a)(3b)$ $12ab$
Write it as a square.		$(2a - 3b)^2$
Check your answer.		
Is the expression factored completely?		
Yes.		
The binomial is not a difference of squares.		
Multiply.		

$$(2a - 3b)^2$$

$$(2a)^2 - 2 \cdot 2a \cdot 3b + (3b)^2$$

$$4a^2 - 12ab + 9b^2 ✓$$

> **TRY IT :: 7.125** Factor completely: $4x^2 + 20xy + 25y^2$.

> **TRY IT :: 7.126** Factor completely: $9m^2 + 42mn + 49n^2$.

EXAMPLE 7.64

Factor completely: $6y^2 - 18y - 60$.

✓ Solution

Is there a GCF?	Yes, 6.	$6y^2 - 18y - 60$
Factor out the GCF.	Trinomial with leading coefficient 1.	$6(y^2 - 3y - 10)$
In the parentheses, is it a binomial, trinomial, or are there more terms?		
"Undo" FOIL.	$6(y\ \ )(y\ \ )$	$6(y + 2)(y - 5)$
Check your answer.		
Is the expression factored completely?		Yes.
Neither binomial is a difference of squares.		
Multiply.		

$$6(y + 2)(y - 5)$$
$$6(y^2 - 5y + 2y - 10)$$
$$6(y^2 - 3y - 10)$$
$$6y^2 - 18y - 60 \ \checkmark$$

> **TRY IT ::: 7.127**   Factor completely: $8y^2 + 16y - 24$.

> **TRY IT ::: 7.128**   Factor completely: $5u^2 - 15u - 270$.

EXAMPLE 7.65

Factor completely: $24x^3 + 81$.

✓ Solution

Is there a GCF?	Yes, 3.	$24x^3 + 81$
Factor it out.		$3(8x^3 + 27)$
In the parentheses, is it a binomial, trinomial, or are there more than three terms?	Binomial.	
Is it a sum or difference?	Sum.	
Of squares or cubes?	Sum of cubes.	$3\left(\overset{a^3}{(2x)^3} + \overset{b^3}{(3)^3}\right)$
Write it using the sum of cubes pattern.		$3\left(\overset{a}{2x} + \overset{b}{3}\right)\left(\overset{a^2}{(2x)^2} - \overset{ab}{2x \cdot 3} + \overset{b^2}{3^2}\right)$
Is the expression factored completely?	Yes.	$3(2x + 3)(4x^2 - 6x + 9)$
Check by multiplying.		We leave the check to you.

> **TRY IT ::: 7.129**   Factor completely: $250m^3 + 432$.

> **TRY IT :: 7.130**   Factor completely: $81q^3 + 192$.

EXAMPLE 7.66

Factor completely: $2x^4 - 32$.

✓ **Solution**

Is there a GCF?	Yes, 2.	$2x^4 - 32$
Factor it out.		$2(x^4 - 16)$
In the parentheses, is it a binomial, trinomial, or are there more than three terms?	Binomial.	
Is it a sum or difference?	Yes.	
Of squares or cubes?	Difference of squares.	$2\left((x^2)^2 - (4)^2\right)$
Write it as a product of conjugates.		$2(x^2 - 4)(x^2 + 4)$
The first binomial is again a difference of squares.		$2\left((x)^2 - (2)^2\right)(x^2 + 4)$
Write it as a product of conjugates.		$2(x - 2)(x + 2)(x^2 + 4)$
Is the expression factored completely?	Yes.	
None of these binomials is a difference of squares.		

Check your answer.

Multiply.

$2(x - 2)(x + 2)(x^2 + 4)$

$2(x^2 - 4)(x^2 + 4)$

$2(x^4 - 16)$

$2x^4 - 32$ ✓

> **TRY IT :: 7.131**   Factor completely: $4a^4 - 64$.

> **TRY IT :: 7.132**   Factor completely: $7y^4 - 7$.

EXAMPLE 7.67

Factor completely: $3x^2 + 6bx - 3ax - 6ab$.

✓ Solution

Is there a GCF?	Yes, 3.	$3x^2 + 6bx - 3ax - 6ab$
Factor out the GCF.		$3\left(x^2 + 2bx - ax - 2ab\right)$
In the parentheses, is it a binomial, trinomial, or are there more terms?	More than 3 terms.	
Use grouping.		$3[x(x + 2b) - a(x + 2b)]$ $3(x + 2b)(x - a)$
Check your answer.		

Is the expression factored completely? Yes.
Multiply.

$3(x + 2b)(x - a)$
$3\left(x^2 - ax + 2bx - 2ab\right)$
$3x^2 - 3ax + 6bx - 6ab$ ✓

> **TRY IT :: 7.133**   Factor completely: $6x^2 - 12xc + 6bx - 12bc$.

> **TRY IT :: 7.134**   Factor completely: $16x^2 + 24xy - 4x - 6y$.

EXAMPLE 7.68

Factor completely: $10x^2 - 34x - 24$.

✓ Solution

Is there a GCF?	Yes, 2.	$10x^2 - 34x - 24$
Factor out the GCF.		$2\left(5x^2 - 17x - 12\right)$
In the parentheses, is it a binomial, trinomial, or are there more than three terms?	Trinomial with $a \neq 1$.	
Use trial and error or the "ac" method.		$2(5x^2 - 17x - 12)$ $2(5x + 3)(x - 4)$

Check your answer. Is the expression factored completely? Yes.

Multiply.

$2(5x + 3)(x - 4)$
$2\left(5x^2 - 20x + 3x - 12\right)$
$2\left(5x^2 - 17x - 12\right)$
$10x^2 - 34x - 24$ ✓

> **TRY IT : :** 7.135   Factor completely: $4p^2 - 16p + 12$.

> **TRY IT : :** 7.136   Factor completely: $6q^2 - 9q - 6$.

7.5 EXERCISES

Practice Makes Perfect

Recognize and Use the Appropriate Method to Factor a Polynomial Completely

In the following exercises, factor completely.

279. $10x^4 + 35x^3$

280. $18p^6 + 24p^3$

281. $y^2 + 10y - 39$

282. $b^2 - 17b + 60$

283. $2n^2 + 13n - 7$

284. $8x^2 - 9x - 3$

285. $a^5 + 9a^3$

286. $75m^3 + 12m$

287. $121r^2 - s^2$

288. $49b^2 - 36a^2$

289. $8m^2 - 32$

290. $36q^2 - 100$

291. $25w^2 - 60w + 36$

292. $49b^2 - 112b + 64$

293. $m^2 + 14mn + 49n^2$

294. $64x^2 + 16xy + y^2$

295. $7b^2 + 7b - 42$

296. $3n^2 + 30n + 72$

297. $3x^3 - 81$

298. $5t^3 - 40$

299. $k^4 - 16$

300. $m^4 - 81$

301. $15pq - 15p + 12q - 12$

302. $12ab - 6a + 10b - 5$

303. $4x^2 + 40x + 84$

304. $5q^2 - 15q - 90$

305. $u^5 + u^2$

306. $5n^3 + 320$

307. $4c^2 + 20cd + 81d^2$

308. $25x^2 + 35xy + 49y^2$

309. $10m^4 - 6250$

310. $3v^4 - 768$

Everyday Math

311. Watermelon drop A springtime tradition at the University of California San Diego is the Watermelon Drop, where a watermelon is dropped from the seventh story of Urey Hall.

ⓐ The binomial $-16t^2 + 80$ gives the height of the watermelon t seconds after it is dropped. Factor the greatest common factor from this binomial.

ⓑ If the watermelon is thrown down with initial velocity 8 feet per second, its height after t seconds is given by the trinomial $-16t^2 - 8t + 80$. Completely factor this trinomial.

312. Pumpkin drop A fall tradition at the University of California San Diego is the Pumpkin Drop, where a pumpkin is dropped from the eleventh story of Tioga Hall.

ⓐ The binomial $-16t^2 + 128$ gives the height of the pumpkin t seconds after it is dropped. Factor the greatest common factor from this binomial.

ⓑ If the pumpkin is thrown down with initial velocity 32 feet per second, its height after t seconds is given by the trinomial $-16t^2 - 32t + 128$. Completely factor this trinomial.

Writing Exercises

313. The difference of squares $y^4 - 625$ can be factored as $\left(y^2 - 25\right)\left(y^2 + 25\right)$. But it is not *completely* factored. What more must be done to completely factor it?

314. Of all the factoring methods covered in this chapter (GCF, grouping, undo FOIL, 'ac' method, special products) which is the easiest for you? Which is the hardest? Explain your answers.

Self Check

ⓐ *After completing the exercises, use this checklist to evaluate your mastery of the objectives of this section.*

I can...	Confidently	With some help	No-I don't get it!
recognize and use the appropriate method to factor a polynomial completely.			

ⓑ *Overall, after looking at the checklist, do you think you are well-prepared for the next section? Why or why not?*

7.6 Quadratic Equations

Learning Objectives

By the end of this section, you will be able to:

- Solve quadratic equations by using the Zero Product Property
- Solve quadratic equations factoring
- Solve applications modeled by quadratic equations

Be Prepared!

Before you get started, take this readiness quiz.

1. Solve: $5y - 3 = 0$.
   If you missed this problem, review **Example 2.27**.

2. Solve: $10a = 0$.
   If you missed this problem, review **Example 2.13**.

3. Combine like terms: $12x^2 - 6x + 4x$.
   If you missed this problem, review **Example 1.24**.

4. Factor $n^3 - 9n^2 - 22n$ completely.
   If you missed this problem, review **Example 7.32**.

We have already solved linear equations, equations of the form $ax + by = c$. In linear equations, the variables have no exponents. Quadratic equations are equations in which the variable is squared. Listed below are some examples of quadratic equations:

$$x^2 + 5x + 6 = 0 \quad 3y^2 + 4y = 10 \quad 64u^2 - 81 = 0 \quad n(n + 1) = 42$$

The last equation doesn't appear to have the variable squared, but when we simplify the expression on the left we will get $n^2 + n$.

The general form of a quadratic equation is $ax^2 + bx + c = 0$, with $a \neq 0$.

Quadratic Equation

An equation of the form $ax^2 + bx + c = 0$ is called a quadratic equation.

$$a, b, \text{ and } c \text{ are real numbers and } a \neq 0$$

To solve quadratic equations we need methods different than the ones we used in solving linear equations. We will look at one method here and then several others in a later chapter.

Solve Quadratic Equations Using the Zero Product Property

We will first solve some quadratic equations by using the Zero Product Property. The **Zero Product Property** says that if the product of two quantities is zero, it must be that at least one of the quantities is zero. The only way to get a product equal to zero is to multiply by zero itself.

Zero Product Property

If $a \cdot b = 0$, then either $a = 0$ or $b = 0$ or both.

We will now use the Zero Product Property, to solve a quadratic equation.

EXAMPLE 7.69 HOW TO USE THE ZERO PRODUCT PROPERTY TO SOLVE A QUADRATIC EQUATION

Solve: $(x + 1)(x - 4) = 0$.

Solution

Step 1. Set each factor equal to zero.	The product equals zero, so at least one factor must equal zero.	$(x+1)(x-4) = 0$ $x+1 = 0$ or $x-4 = 0$
Step 2. Solve the linear equations.	Solve each equation.	$x = -1$ or $\qquad x = 4$
Step 3. Check.	Substitute each solution separately into the original equation.	$x = -1$ $(x+1)(x-4) = 0$ $(-1+1)(-1-4) \overset{?}{=} 0$ $(0)(-5) \overset{?}{=} 0$ $0 = 0 \checkmark$ $x = 4$ $(x+1)(x-4) = 0$ $(4+1)(4-4) \overset{?}{=} 0$ $(5)(0) \overset{?}{=} 0$ $0 = 0 \checkmark$

> **TRY IT : : 7.137**   Solve: $(x-3)(x+5) = 0$.

> **TRY IT : : 7.138**   Solve: $(y-6)(y+9) = 0$.

We usually will do a little more work than we did in this last example to solve the linear equations that result from using the Zero Product Property.

EXAMPLE 7.70

Solve: $(5n-2)(6n-1) = 0$.

Solution

$$(5n-2)(6n-1) = 0$$

Use the Zero Product Property to set each factor to 0.	$5n-2 = 0 \qquad 6n-1 = 0$
Solve the equations.	$n = \dfrac{2}{5} \qquad n = \dfrac{1}{6}$
Check your answers.	

$$n = \frac{2}{5} \qquad\qquad n = \frac{1}{6}$$

$$(5n - 2)(6n - 1) = 0 \qquad\qquad (5n - 2)(6n - 1) = 0$$

$$\left(5 \cdot \frac{2}{5} - 2\right)\left(6 \cdot \frac{2}{5} - 1\right) \overset{?}{=} 0 \qquad \left(5 \cdot \frac{1}{6} - 2\right)\left(6 \cdot \frac{1}{6} - 1\right) \overset{?}{=} 0$$

$$(2 - 2)\left(\frac{12}{5} - \frac{5}{5}\right) \overset{?}{=} 0 \qquad\qquad \left(\frac{5}{6} - \frac{12}{6}\right)(1 - 1) \overset{?}{=} 0$$

$$(0)\left(\frac{7}{5}\right) \overset{?}{=} 0 \qquad\qquad\qquad \left(-\frac{7}{6}\right)(0) \overset{?}{=} 0$$

$$0 = 0\ \checkmark \qquad\qquad\qquad\qquad 0 = 0\ \checkmark$$

> **TRY IT :: 7.139**   Solve: $(3m - 2)(2m + 1) = 0$.

> **TRY IT :: 7.140**   Solve: $(4p + 3)(4p - 3) = 0$.

Notice when we checked the solutions that each of them made just one factor equal to zero. But the product was zero for both solutions.

EXAMPLE 7.71

Solve: $3p(10p + 7) = 0$.

⊘ **Solution**

$$3p(10p + 7) = 0$$

Use the Zero Product Property to set each factor to 0.	$3p = 0 \qquad 10p + 7 = 0$
Solve the equations.	$p = 0 \qquad\quad 10p = -7$
	$p = -\frac{7}{10}$

Check your answers.

$$p = 0 \qquad\qquad\qquad p = -\frac{7}{10}$$

$$3p(10p + 7) = 0 \qquad\qquad 3p(10p + 7) = 0$$

$$3 \cdot 0(10 \cdot 0 + 7) \overset{?}{=} 0 \qquad 3\left(-\frac{7}{10}\right)10\left(-\frac{7}{10}\right) + 7 \overset{?}{=} 0$$

$$0(0 + 7) \overset{?}{=} 0 \qquad\qquad \left(-\frac{21}{10}\right)(-7 + 7) \overset{?}{=} 0$$

$$0(7) \overset{?}{=} 0 \qquad\qquad\qquad \left(-\frac{21}{10}\right)(0) \overset{?}{=} 0$$

$$0 = 0\ \checkmark \qquad\qquad\qquad 0 = 0\ \checkmark$$

> **TRY IT :: 7.141**   Solve: $2u(5u - 1) = 0$.

> **TRY IT :: 7.142**   Solve: $w(2w + 3) = 0$.

It may appear that there is only one factor in the next example. Remember, however, that $(y - 8)^2$ means $(y - 8)(y - 8)$.

EXAMPLE 7.72

Solve: $(y - 8)^2 = 0$.

⊘ **Solution**

$$(y - 8)^2 = 0$$

Rewrite the left side as a product.	$(y - 8)(y - 8) = 0$
Use the Zero Product Property and set each factor to 0.	$y - 8 = 0 \qquad y - 8 = 0$
Solve the equations.	$y = 8 \qquad\qquad y = 8$
When a solution repeats, we call it a double root.	
Check your answer.	

$$y = 8$$
$$(y - 8)^2 = 0$$
$$(8 - 8)^2 \stackrel{?}{=} 0$$
$$(0)^2 \stackrel{?}{=} 0$$
$$0 = 0 ✓$$

> **TRY IT :: 7.143**   Solve: $(x + 1)^2 = 0$.

> **TRY IT :: 7.144**   Solve: $(v - 2)^2 = 0$.

Solve Quadratic Equations by Factoring

Each of the equations we have solved in this section so far had one side in factored form. In order to use the Zero Product Property, the quadratic equation must be factored, with zero on one side. So we be sure to start with the quadratic equation in standard form, $ax^2 + bx + c = 0$. Then we factor the expression on the left.

EXAMPLE 7.73   HOW TO SOLVE A QUADRATIC EQUATION BY FACTORING

Solve: $x^2 + 2x - 8 = 0$.

⊘ **Solution**

Step 1. Write the quadratic equation in standard form, $ax^2 + bx + c = 0$.	The equation is already in standard form.	$x^2 + 2x - 8 = 0$
Step 2. Factor the quadratic expression.	Factor $x^2 + 2x - 8$ $(x + 4)(x - 2)$	$(x + 4)(x - 2) = 0$

Step 3. Use the Zero Product Property.	Set each factor equal to zero.	$x+4=0$ or $x-2=0$
Step 4. Solve the linear equations.	We have two linear equations.	$x=-4$ or $x=2$
Step 5. Check.	Substitute each solution separately into the original equation.	$x^2+2x-8=0$ $x=-4$ $(-4)^2-2(-4)-8\overset{?}{=}0$ $16+(-8)-8\overset{?}{=}0$ $0=0\checkmark$ $x^2+2x-8=0$ $x=2$ $2^2-2(2)-8\overset{?}{=}0$ $4+4-8\overset{?}{=}0$ $0=0\checkmark$

> **TRY IT :: 7.145**   Solve: $x^2-x-12=0$.

> **TRY IT :: 7.146**   Solve: $b^2+9b+14=0$.

HOW TO :: SOLVE A QUADRATIC EQUATION BY FACTORING.

Step 1. Write the quadratic equation in standard form, $ax^2+bx+c=0$.

Step 2. Factor the quadratic expression.

Step 3. Use the Zero Product Property.

Step 4. Solve the linear equations.

Step 5. Check.

Before we factor, we must make sure the quadratic equation is in standard form.

EXAMPLE 7.74

Solve: $2y^2=13y+45$.

⊘ **Solution**

$$2y^2=13y+45$$

Write the quadratic equation in standard form.	$2y^2-13y-45=0$	
Factor the quadratic expression.	$(2y+5)(y-9)=0$	
Use the Zero Product Property to set each factor to 0.	$2y+5=0$	$y-9=0$
Solve each equation.	$y=-\dfrac{5}{2}$	$y=9$

Check your answers.

$y = -\dfrac{5}{2}$	$y = 9$
$2y^2 = 13y + 45$	$2y^2 = 13y + 45$
$2\left(-\dfrac{5}{2}\right)^2 \overset{?}{=} 13\left(-\dfrac{5}{2}\right) + 45$	$2(9)^2 \overset{?}{=} 13(9) + 45$
$2\left(\dfrac{25}{4}\right) \overset{?}{=} \left(-\dfrac{65}{2}\right) + \dfrac{90}{2}$	$2(81) \overset{?}{=} 117 + 45$
$\dfrac{25}{2} = \dfrac{25}{2}\ \checkmark$	$162 = 162\ \checkmark$

> **TRY IT ::** 7.147   Solve: $3c^2 = 10c - 8$.

> **TRY IT ::** 7.148   Solve: $2d^2 - 5d = 3$.

EXAMPLE 7.75

Solve: $5x^2 - 13x = 7x$.

⊘ **Solution**

	$5x^2 - 13x = 7x$
Write the quadratic equation in standard form.	$5x^2 - 20x = 0$
Factor the left side of the equation.	$5x(x - 4) = 0$
Use the Zero Product Property to set each factor to 0.	$5x = 0 \qquad x - 4 = 0$
Solve each equation.	$x = 0 \qquad\quad x = 4$

Check your answers.

$x = 0$	$x = 4$
$5x^2 - 13x = 7x$	$5x^2 - 13x = 7x$
$5(0)^2 - 13(0) \overset{?}{=} 7(0)$	$5(4)^2 - 13(4) \overset{?}{=} 7(4)$
$0 - 0 \overset{?}{=} 0$	$5(16) - 52 \overset{?}{=} 28$
$0 = 0\ \checkmark$	$28 = 28\ \checkmark$

> **TRY IT ::** 7.149   Solve: $6a^2 + 9a = 3a$.

> **TRY IT ::** 7.150   Solve: $45b^2 - 2b = -17b$.

Solving quadratic equations by factoring will make use of all the factoring techniques you have learned in this chapter! Do you recognize the special product pattern in the next example?

EXAMPLE 7.76

Solve: $144q^2 = 25$.

 **Solution**

$$144q^2 = 25$$

Write the quadratic equation in standard form.

$$144q^2 - 25 = 0$$

Factor. It is a difference of squares.

$$(12q - 5)(12q + 5) = 0$$

Use the Zero Product Property to set each factor to 0.

$$12q - 5 = 0 \qquad 12q + 5 = 0$$

Solve each equation.

$$12q = 5 \qquad 12q = -5$$
$$q = \frac{5}{12} \qquad q = -\frac{5}{12}$$

Check your answers.

> **TRY IT :: 7.151** Solve: $25p^2 = 49$.

> **TRY IT :: 7.152** Solve: $36x^2 = 121$.

The left side in the next example is factored, but the right side is not zero. In order to use the Zero Product Property, one side of the equation must be zero. We'll multiply the factors and then write the equation in standard form.

EXAMPLE 7.77

Solve: $(3x - 8)(x - 1) = 3x$.

 **Solution**

$$(3x - 8)(x - 1) = 3x$$

Multiply the binomials.

$$3x^2 - 11x + 8 = 3x$$

Write the quadratic equation in standard form.

$$3x^2 - 14x + 8 = 0$$

Factor the trinomial.

$$(3x - 2)(x - 4) = 0$$

Use the Zero Product Property to set each factor to 0.

$$3x - 2 = 0 \qquad x - 4 = 0$$

Solve each equation.

$$3x = 2 \qquad x = 4$$

$$x = \frac{2}{3}$$

Check your answers.   The check is left to you!

> **TRY IT :: 7.153**   Solve: $(2m + 1)(m + 3) = 12m$.

> **TRY IT : : 7.154**   Solve: $(k+1)(k-1) = 8$.

The Zero Product Property also applies to the product of three or more factors. If the product is zero, at least one of the factors must be zero. We can solve some equations of degree more than two by using the Zero Product Property, just like we solved quadratic equations.

EXAMPLE 7.78

Solve: $9m^3 + 100m = 60m^2$.

⊘ Solution

$$9m^3 + 100m = 60m^2$$

Bring all the terms to one side so that the other side is zero.

$$9m^3 - 60m^2 + 100m = 0$$

Factor the greatest common factor first.

$$m(9m^2 - 60m + 100) = 0$$

Factor the trinomial.

$$m(3m - 10)(3m - 10) = 0$$

Use the Zero Product Property to set each factor to 0.

$$m = 0 \qquad 3m - 10 = 0 \qquad 3m - 10 = 0$$

Solve each equation.

$$m = 0 \qquad m = \frac{10}{3} \qquad m = \frac{10}{3}$$

Check your answers.

The check is left to you.

> **TRY IT : : 7.155**   Solve: $8x^3 = 24x^2 - 18x$.

> **TRY IT : : 7.156**   Solve: $16y^2 = 32y^3 + 2y$.

When we factor the quadratic equation in the next example we will get three factors. However the first factor is a constant. We know that factor cannot equal 0.

EXAMPLE 7.79

Solve: $4x^2 = 16x + 84$.

⊘ Solution

$$4x^2 = 16x + 84$$

Write the quadratic equation in standard form.

$$4x^2 - 16x - 84 = 0$$

Factor the greatest common factor first.

$$4(x^2 - 4x - 21) = 0$$

Factor the trinomial.

$$4(x - 7)(x + 3) = 0$$

Use the Zero Product Property to set each factor to 0.

$$4 \neq 0 \qquad x - 7 = 0 \qquad x + 3 = 0$$

Solve each equation.

$$4 \neq 0 \qquad x = 7 \qquad x = -3$$

Check your answers.

The check is left to you.

> **TRY IT : : 7.157**   Solve: $18a^2 - 30 = -33a$.

TRY IT :: 7.158   Solve: $123b = -6 - 60b^2$.

Solve Applications Modeled by Quadratic Equations

The problem solving strategy we used earlier for applications that translate to linear equations will work just as well for applications that translate to quadratic equations. We will copy the problem solving strategy here so we can use it for reference.

> **HOW TO ::** USE A PROBLEM-SOLVING STRATEGY TO SOLVE WORD PROBLEMS.
>
> Step 1.   **Read** the problem. Make sure all the words and ideas are understood.
>
> Step 2.   **Identify** what we are looking for.
>
> Step 3.   **Name** what we are looking for. Choose a variable to represent that quantity.
>
> Step 4.   **Translate** into an equation. It may be helpful to restate the problem in one sentence with all the important information. Then, translate the English sentence into an algebra equation.
>
> Step 5.   **Solve** the equation using good algebra techniques.
>
> Step 6.   **Check** the answer in the problem and make sure it makes sense.
>
> Step 7.   **Answer** the question with a complete sentence.

We will start with a number problem to get practice translating words into a quadratic equation.

EXAMPLE 7.80

The product of two consecutive integers is 132. Find the integers.

 **Solution**

Step 1. Read the problem.

Step 2. Identify what we are looking for.	We are looking for two consecutive integers.
Step 3. Name what we are looking for.	Let $n =$ the first integer $n + 1 =$ the next consecutive integer
Step 4. Translate into an equation. Restate the problem in a sentence.	The product of the two consecutive integers is 132.
	The first integer times the next integer is 132.
Translate to an equation.	$n(n + 1) = 132$
Step 5. Solve the equation.	$n^2 + n = 132$
Bring all the terms to one side.	$n^2 + n - 132 = 0$
Factor the trinomial.	$(n - 11)(n + 12) = 0$
Use the zero product property.	$n - 11 = 0 \qquad n + 12 = 0$
Solve the equations.	$n = 11 \qquad n = -12$

There are two values for n that are solutions to this problem. So there are two sets of consecutive integers that will work.

If the first integer is $n = 11$	If the first integer is $n = -12$
then the next integer is $n + 1$	then the next integer is $n + 1$
$11 + 1$	$-12 + 1$
12	-11

Step 6. Check the answer.

The consecutive integers are 11, 12 and -11, -12. The product $11 \cdot 12 = 132$ and the product $-11(-12) = 132$. Both pairs of consecutive integers are solutions.

Step 7. Answer the question. The consecutive integers are 11, 12 and -11, -12.

> **TRY IT : : 7.159**   The product of two consecutive integers is 240. Find the integers.

> **TRY IT : : 7.160**   The product of two consecutive integers is 420. Find the integers.

Were you surprised by the pair of negative integers that is one of the solutions to the previous example? The product of the two positive integers and the product of the two negative integers both give 132.

In some applications, negative solutions will result from the algebra, but will not be realistic for the situation.

EXAMPLE 7.81

A rectangular garden has an area 15 square feet. The length of the garden is two feet more than the width. Find the length and width of the garden.

⊘ **Solution**

Step 1. Read the problem. In problems involving geometric figures, a sketch can help you visualize the situation.	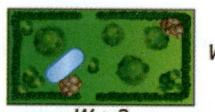
Step 2. Identify what you are looking for.	We are looking for the length and width.
Step 3. Name what you are looking for. The length is two feet more than width.	Let W = the width of the garden. $W + 2$ = the length of the garden
Step 4. Translate into an equation. Restate the important information in a sentence.	The area of the rectangular garden is 15 square feet.
Use the formula for the area of a rectangle.	$A = L \cdot W$
Substitute in the variables.	$15 = (W + 2)W$
Step 5. Solve the equation. Distribute first.	$15 = W^2 + 2W$
Get zero on one side.	$0 = W^2 + 2W - 15$
Factor the trinomial.	$0 = (W + 5)(W - 3)$
Use the Zero Product Property.	$0 = W + 5 \qquad 0 = W - 3$
Solve each equation.	$-5 = W \qquad\qquad 3 = W$

Since *W* is the width of the garden, it does not make sense for it to be negative. We eliminate that value for *W*.	$\cancel{-5 = W}$ $W = 3$	$3 = W$ Width is 3 feet.
Find the value of the length.	$W + 2 = \text{length}$	
	$3 + 2$	
	5	Length is 5 feet.

Step 6. Check the answer. Does the answer make sense?

W $\quad A = L \cdot W$
3 $\quad A = 3 \cdot 5$
$\quad\quad A = 15$

W + 2
3 + 2
5

| | Yes, this makes sense. |
| **Step 7. Answer** the question. | The width of the garden is 3 feet and the length is 5 feet. |

> **TRY IT : : 7.161**

> A rectangular sign has area 30 square feet. The length of the sign is one foot more than the width. Find the length and width of the sign.

> **TRY IT : : 7.162**

> A rectangular patio has area 180 square feet. The width of the patio is three feet less than the length. Find the length and width of the patio.

In an earlier chapter, we used the Pythagorean Theorem $\left(a^2 + b^2 = c^2\right)$. It gave the relation between the legs and the hypotenuse of a right triangle.

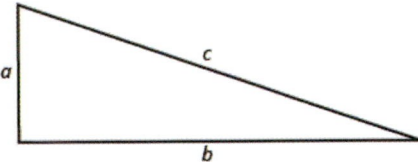

We will use this formula to in the next example.

Justine wants to put a deck in the corner of her backyard in the shape of a right triangle, as shown below. The hypotenuse will be 17 feet long. The length of one side will be 7 feet less than the length of the other side. Find the lengths of the sides of the deck.

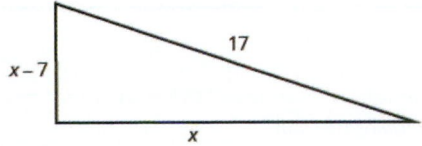

⊘ Solution

Step 1. Read the problem.

Step 2. Identify what you are looking for.	We are looking for the lengths of the sides of the deck.

Step 3. Name what you are looking for.
One side is 7 less than the other.

Let x = length of a side of the deck
$x - 7$ = length of other side

Step 4. Translate into an equation.
Since this is a right triangle we can use the
Pythagorean Theorem.

$$a^2 + b^2 = c^2$$

Substitute in the variables.

$$x^2 + (x - 7)^2 = 17^2$$

Step 5. Solve the equation.

$$x^2 + x^2 - 14x + 49 = 289$$

Simplify.

$$2x^2 - 14x + 49 = 289$$

It is a quadratic equation, so get zero on one side.

$$2x^2 - 14x - 240 = 0$$

Factor the greatest common factor.

$$2(x^2 - 7x - 120) = 0$$

Factor the trinomial.

$$2(x - 15)(x + 8) = 0$$

Use the Zero Product Property.

$2 \neq 0 \qquad x - 15 = 0 \qquad x + 8 = 0$

Solve.

$2 \neq 0 \qquad x = 15 \qquad x = -8$

Since x is a side of the triangle, $x = -8$ does not make sense.

$2 \neq 0 \qquad x = 15 \qquad \cancel{x = -8}$

Find the length of the other side.

If the length of one side is $\qquad x = 15$

then the length of the other side is $\qquad x - 7$

$15 - 7$

8 is the length of the other side.

Step 6. Check the answer.
Do these numbers make sense?

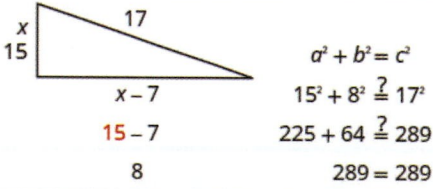

$a^2 + b^2 = c^2$
$15^2 + 8^2 \overset{?}{=} 17^2$
$225 + 64 \overset{?}{=} 289$
$289 = 289 \checkmark$

Step 7. Answer the question.

The sides of the deck are 8, 15, and 17 feet.

> **TRY IT : : 7.163**

A boat's sail is a right triangle. The length of one side of the sail is 7 feet more than the other side. The hypotenuse is 13. Find the lengths of the two sides of the sail.

TRY IT : : 7.164

A meditation garden is in the shape of a right triangle, with one leg 7 feet. The length of the hypotenuse is one more than the length of one of the other legs. Find the lengths of the hypotenuse and the other leg.

7.6 EXERCISES

Practice Makes Perfect

Use the Zero Product Property

In the following exercises, solve.

315. $(x-3)(x+7) = 0$

316. $(y-11)(y+1) = 0$

317. $(3a-10)(2a-7) = 0$

318. $(5b+1)(6b+1) = 0$

319. $6m(12m-5) = 0$

320. $2x(6x-3) = 0$

321. $(y-3)^2 = 0$

322. $(b+10)^2 = 0$

323. $(2x-1)^2 = 0$

324. $(3y+5)^2 = 0$

Solve Quadratic Equations by Factoring

In the following exercises, solve.

325. $x^2 + 7x + 12 = 0$

326. $y^2 - 8y + 15 = 0$

327. $5a^2 - 26a = 24$

328. $4b^2 + 7b = -3$

329. $4m^2 = 17m - 15$

330. $n^2 = 5 - 6n$ $n^2 = 5n - 6$

331. $7a^2 + 14a = 7a$

332. $12b^2 - 15b = -9b$

333. $49m^2 = 144$

334. $625 = x^2$

335. $(y-3)(y+2) = 4y$

336. $(p-5)(p+3) = -7$

337. $(2x+1)(x-3) = -4x$

338. $(x+6)(x-3) = -8$

339. $16p^3 = 24p^2 + 9p$

340. $m^3 - 2m^2 = -m$

341. $20x^2 - 60x = -45$

342. $3y^2 - 18y = -27$

Solve Applications Modeled by Quadratic Equations

In the following exercises, solve.

343. The product of two consecutive integers is 56. Find the integers.

344. The product of two consecutive integers is 42. Find the integers.

345. The area of a rectangular carpet is 28 square feet. The length is three feet more than the width. Find the length and the width of the carpet.

346. A rectangular retaining wall has area 15 square feet. The height of the wall is two feet less than its length. Find the height and the length of the wall.

347. A pennant is shaped like a right triangle, with hypotenuse 10 feet. The length of one side of the pennant is two feet longer than the length of the other side. Find the length of the two sides of the pennant.

348. A reflecting pool is shaped like a right triangle, with one leg along the wall of a building. The hypotenuse is 9 feet longer than the side along the building. The third side is 7 feet longer than the side along the building. Find the lengths of all three sides of the reflecting pool.

Mixed Practice

In the following exercises, solve.

349. $(x+8)(x-3) = 0$

350. $(3y-5)(y+7) = 0$

351. $p^2 + 12p + 11 = 0$

352. $q^2 - 12q - 13 = 0$

353. $m^2 = 6m + 16$

354. $4n^2 + 19n = 5$

355. $a^3 - a^2 - 42a = 0$

356. $4b^2 - 60b + 224 = 0$

357. The product of two consecutive integers is 110. Find the integers.

358. The length of one leg of a right triangle is three more than the other leg. If the hypotenuse is 15, find the lengths of the two legs.

Everyday Math

359. Area of a patio If each side of a square patio is increased by 4 feet, the area of the patio would be 196 square feet. Solve the equation $(s + 4)^2 = 196$ for s to find the length of a side of the patio.

360. Watermelon drop A watermelon is dropped from the tenth story of a building. Solve the equation $-16t^2 + 144 = 0$ for t to find the number of seconds it takes the watermelon to reach the ground.

Writing Exercises

361. Explain how you solve a quadratic equation. How many answers do you expect to get for a quadratic equation?

362. Give an example of a quadratic equation that has a GCF and none of the solutions to the equation is zero.

Self Check

ⓐ *After completing the exercises, use this checklist to evaluate your mastery of the objectives of this section.*

I can...	Confidently	With some help	No-I don't get it!
solve quadratic equations by using the Zero Product Property.			
solve quadratic equations by factoring.			
solve applications modeled by quadratic equations.			

ⓑ *Overall, after looking at the checklist, do you think you are well-prepared for the next section? Why or why not?*

CHAPTER 7 REVIEW

KEY TERMS

difference of squares pattern If a and b are real numbers,

$$a^2 - b^2 = (a-b)(a+b) \qquad a^2 \overset{\text{difference}}{-} b^2 = (a-b)(a+b)$$

$$\underset{\text{squares}}{} \qquad \underset{\text{conjugates}}{}$$

factoring Factoring is splitting a product into factors; in other words, it is the reverse process of multiplying.

greatest common factor The greatest common factor is the largest expression that is a factor of two or more expressions is the greatest common factor (GCF).

perfect square trinomials pattern If a and b are real numbers,

$$a^2 + 2ab + b^2 = (a+b)^2 \qquad a^2 - 2ab + b^2 = (a-b)^2$$

prime polynomials Polynomials that cannot be factored are prime polynomials.

quadratic equations are equations in which the variable is squared.

sum and difference of cubes pattern

$$a^3 + b^3 = (a+b)\left(a^2 - ab + b^2\right)$$
$$a^3 - b^3 = (a-b)\left(a^2 + ab + b^2\right)$$

Zero Product Property The Zero Product Property states that, if the product of two quantities is zero, at least one of the quantities is zero.

KEY CONCEPTS

7.1 Greatest Common Factor and Factor by Grouping

- **Finding the Greatest Common Factor (GCF):** To find the GCF of two expressions:

  Step 1. Factor each coefficient into primes. Write all variables with exponents in expanded form.

  Step 2. List all factors—matching common factors in a column. In each column, circle the common factors.

  Step 3. Bring down the common factors that all expressions share.

  Step 4. Multiply the factors as in **Example 7.2**.

- **Factor the Greatest Common Factor from a Polynomial:** To factor a greatest common factor from a polynomial:

  Step 1. Find the GCF of all the terms of the polynomial.

  Step 2. Rewrite each term as a product using the GCF.

  Step 3. Use the 'reverse' Distributive Property to factor the expression.

  Step 4. Check by multiplying the factors as in **Example 7.5**.

- **Factor by Grouping:** To factor a polynomial with 4 four or more terms

  Step 1. Group terms with common factors.

  Step 2. Factor out the common factor in each group.

  Step 3. Factor the common factor from the expression.

  Step 4. Check by multiplying the factors as in **Example 7.15**.

7.2 Factor Quadratic Trinomials with Leading Coefficient 1

- **Factor trinomials of the form** $x^2 + bx + c$

  Step 1. Write the factors as two binomials with first terms x: $(x \quad)(x \quad)$.

  Step 2. Find two numbers m and n that
  Multiply to c, $m \cdot n = c$
  Add to b, $m + n = b$

Step 3. Use m and n as the last terms of the factors: $(x+m)(x+n)$.

Step 4. Check by multiplying the factors.

7.3 Factor Quadratic Trinomials with Leading Coefficient Other than 1

- **Factor Trinomials of the Form $ax^2 + bx + c$ using Trial and Error:** See **Example 7.33**.

  Step 1. Write the trinomial in descending order of degrees.

  Step 2. Find all the factor pairs of the first term.

  Step 3. Find all the factor pairs of the third term.

  Step 4. Test all the possible combinations of the factors until the correct product is found.

  Step 5. Check by multiplying.

- **Factor Trinomials of the Form $ax^2 + bx + c$ Using the "ac" Method:** See **Example 7.38**.

  Step 1. Factor any GCF.

  Step 2. Find the product ac.

  Step 3. Find two numbers m and n that:

  | Multiply to ac | $m \cdot n = a \cdot c$ |
  | Add to b | $m + n = b$ |

  Step 4. Split the middle term using m and n:

  $$ax^2 + bx + c$$
  $$ax^2 + mx + nx + c$$

  Step 5. Factor by grouping.

  Step 6. Check by multiplying the factors.

- **Choose a strategy to factor polynomials completely (updated):**

  Step 1. Is there a greatest common factor? Factor it.

  Step 2. Is the polynomial a binomial, trinomial, or are there more than three terms?
  If it is a binomial, right now we have no method to factor it.

  If it is a trinomial of the form $x^2 + bx + c$
    Undo FOIL $(x\ \ )(x\ \ )$.

  If it is a trinomial of the form $ax^2 + bx + c$
    Use Trial and Error or the "ac" method.
  If it has more than three terms
    Use the grouping method.

  Step 3. Check by multiplying the factors.

7.4 Factor Special Products

- **Factor perfect square trinomials** See **Example 7.42**.

  Step 1. Does the trinomial fit the pattern? $a^2 + 2ab + b^2$ $a^2 - 2ab + b^2$

    Is the first term a perfect square? $(a)^2$ $(a)^2$
    Write it as a square.

    Is the last term a perfect square? $(a)^2$ $(b)^2$ $(a)^2$ $(b)^2$
    Write it as a square.

    Check the middle term. Is it $2ab$? $(a)^2 \searrow_{2 \cdot a \cdot b} \nearrow (b)^2$ $(a)^2 \searrow_{2 \cdot a \cdot b} \nearrow (b)^2$

  Step 2. Write the square of the binomial. $(a+b)^2$ $(a-b)^2$

  Step 3. Check by multiplying.

- **Factor differences of squares** See **Example 7.47**.

Step 1. Does the binomial fit the pattern? $\qquad a^2 - b^2$
Is this a difference?
Are the first and last terms perfect squares? $\underline{} - \underline{}$

Step 2. Write them as squares. $\qquad\qquad (a)^2 - (b)^2$

Step 3. Write the product of conjugates. $\qquad\quad (a-b)(a+b)$

Step 4. Check by multiplying.

- **Factor sum and difference of cubes** To factor the sum or difference of cubes: See **Example 7.54**.

  Step 1. Does the binomial fit the sum or difference of cubes pattern? Is it a sum or difference? Are the first and last terms perfect cubes?

  Step 2. Write them as cubes.

  Step 3. Use either the sum or difference of cubes pattern.

  Step 4. Simplify inside the parentheses

  Step 5. Check by multiplying the factors.

7.5 General Strategy for Factoring Polynomials

- **General Strategy for Factoring Polynomials** See **Figure 7.4**.
- **How to Factor Polynomials**

  Step 1. Is there a greatest common factor? Factor it out.

  Step 2. Is the polynomial a binomial, trinomial, or are there more than three terms?

  - If it is a binomial:
    Is it a sum?

    - Of squares? Sums of squares do not factor.
    - Of cubes? Use the sum of cubes pattern.

    Is it a difference?

    - Of squares? Factor as the product of conjugates.
    - Of cubes? Use the difference of cubes pattern.

  - If it is a trinomial:
    Is it of the form $x^2 + bx + c$? Undo FOIL.

    Is it of the form $ax^2 + bx + c$?

    - If 'a' and 'c' are squares, check if it fits the trinomial square pattern.
    - Use the trial and error or 'ac' method.

  - If it has more than three terms:
    Use the grouping method.

  Step 3. Check. Is it factored completely? Do the factors multiply back to the original polynomial?

7.6 Quadratic Equations

- **Zero Product Property** If $a \cdot b = 0$, then either $a = 0$ or $b = 0$ or both. See **Example 7.69**.
- **Solve a quadratic equation by factoring** To solve a quadratic equation by factoring: See **Example 7.73**.

  Step 1. Write the quadratic equation in standard form, $ax^2 + bx + c = 0$.

  Step 2. Factor the quadratic expression.

  Step 3. Use the Zero Product Property.

  Step 4. Solve the linear equations.

  Step 5. Check.

- **Use a problem solving strategy to solve word problems** See **Example 7.80**.

  Step 1. **Read** the problem. Make sure all the words and ideas are understood.

  Step 2. **Identify** what we are looking for.

Step 3. **Name** what we are looking for. Choose a variable to represent that quantity.

Step 4. **Translate** into an equation. It may be helpful to restate the problem in one sentence with all the important information. Then, translate the English sentence into an algebra equation.

Step 5. **Solve** the equation using good algebra techniques.

Step 6. **Check** the answer in the problem and make sure it makes sense.

Step 7. **Answer** the question with a complete sentence.

REVIEW EXERCISES

7.1 7.1 Greatest Common Factor and Factor by Grouping

Find the Greatest Common Factor of Two or More Expressions

In the following exercises, find the greatest common factor.

363. 42, 60

364. 450, 420

365. 90, 150, 105

366. 60, 294, 630

Factor the Greatest Common Factor from a Polynomial

In the following exercises, factor the greatest common factor from each polynomial.

367. $24x - 42$

368. $35y + 84$

369. $15m^4 + 6m^2 n$

370. $24pt^4 + 16t^7$

Factor by Grouping

In the following exercises, factor by grouping.

371. $ax - ay + bx - by$

372. $x^2 y - xy^2 + 2x - 2y$

373. $x^2 + 7x - 3x - 21$

374. $4x^2 - 16x + 3x - 12$

375. $m^3 + m^2 + m + 1$

376. $5x - 5y - y + x$

7.2 7.2 Factor Trinomials of the form $x^2 + bx + c$

Factor Trinomials of the Form $x^2 + bx + c$

In the following exercises, factor each trinomial of the form $x^2 + bx + c$.

377. $u^2 + 17u + 72$

378. $a^2 + 14a + 33$

379. $k^2 - 16k + 60$

380. $r^2 - 11r + 28$

381. $y^2 + 6y - 7$

382. $m^2 + 3m - 54$

383. $s^2 - 2s - 8$

384. $x^2 - 3x - 10$

Factor Trinomials of the Form $x^2 + bxy + cy^2$

In the following examples, factor each trinomial of the form $x^2 + bxy + cy^2$.

385. $x^2 + 12xy + 35y^2$

386. $u^2 + 14uv + 48v^2$

387. $a^2 + 4ab - 21b^2$

388. $p^2 - 5pq - 36q^2$

7.3 7.3 Factoring Trinomials of the form $ax^2 + bx + c$

Recognize a Preliminary Strategy to Factor Polynomials Completely

In the following exercises, identify the best method to use to factor each polynomial.

389. $y^2 - 17y + 42$

390. $12r^2 + 32r + 5$

391. $8a^3 + 72a$

392. $4m - mn - 3n + 12$

Factor Trinomials of the Form $ax^2 + bx + c$ with a GCF

In the following exercises, factor completely.

393. $6x^2 + 42x + 60$

394. $8a^2 + 32a + 24$

395. $3n^4 - 12n^3 - 96n^2$

396. $5y^4 + 25y^2 - 70y$

Factor Trinomials Using the "ac" Method

In the following exercises, factor.

397. $2x^2 + 9x + 4$

398. $3y^2 + 17y + 10$

399. $18a^2 - 9a + 1$

400. $8u^2 - 14u + 3$

401. $15p^2 + 2p - 8$

402. $15x^2 + 6x - 2$

403. $40s^2 - s - 6$

404. $20n^2 - 7n - 3$

Factor Trinomials with a GCF Using the "ac" Method

In the following exercises, factor.

405. $3x^2 + 3x - 36$

406. $4x^2 + 4x - 8$

407. $60y^2 - 85y - 25$

408. $18a^2 - 57a - 21$

7.4 7.4 Factoring Special Products

Factor Perfect Square Trinomials

In the following exercises, factor.

409. $25x^2 + 30x + 9$

410. $16y^2 + 72y + 81$

411. $36a^2 - 84ab + 49b^2$

412. $64r^2 - 176rs + 121s^2$

413. $40x^2 + 360x + 810$

414. $75u^2 + 180u + 108$

415. $2y^3 - 16y^2 + 32y$

416. $5k^3 - 70k^2 + 245k$

Factor Differences of Squares

In the following exercises, factor.

417. $81r^2 - 25$

418. $49a^2 - 144$

419. $169m^2 - n^2$

420. $64x^2 - y^2$

421. $25p^2 - 1$

422. $1 - 16s^2$

423. $9 - 121y^2$

424. $100k^2 - 81$

425. $20x^2 - 125$

426. $18y^2 - 98$ **427.** $49u^3 - 9u$ **428.** $169n^3 - n$

Factor Sums and Differences of Cubes

In the following exercises, factor.

429. $a^3 - 125$ **430.** $b^3 - 216$ **431.** $2m^3 + 54$

432. $81x^3 + 3$

7.5 7.5 General Strategy for Factoring Polynomials

Recognize and Use the Appropriate Method to Factor a Polynomial Completely

In the following exercises, factor completely.

433. $24x^3 + 44x^2$ **434.** $24a^4 - 9a^3$ **435.** $16n^2 - 56mn + 49m^2$

436. $6a^2 - 25a - 9$ **437.** $5r^2 + 22r - 48$ **438.** $5u^4 - 45u^2$

439. $n^4 - 81$ **440.** $64j^2 + 225$ **441.** $5x^2 + 5x - 60$

442. $b^3 - 64$ **443.** $m^3 + 125$ **444.** $2b^2 - 2bc + 5cb - 5c^2$

7.6 7.6 Quadratic Equations

Use the Zero Product Property

In the following exercises, solve.

445. $(a - 3)(a + 7) = 0$ **446.** $(b - 3)(b + 10) = 0$ **447.** $3m(2m - 5)(m + 6) = 0$

448. $7n(3n + 8)(n - 5) = 0$

Solve Quadratic Equations by Factoring

In the following exercises, solve.

449. $x^2 + 9x + 20 = 0$ **450.** $y^2 - y - 72 = 0$ **451.** $2p^2 - 11p = 40$

452. $q^3 + 3q^2 + 2q = 0$ **453.** $144m^2 - 25 = 0$ **454.** $4n^2 = 36$

Solve Applications Modeled by Quadratic Equations

In the following exercises, solve.

455. The product of two consecutive numbers is 462. Find the numbers.

456. The area of a rectangular shaped patio 400 square feet. The length of the patio is 9 feet more than its width. Find the length and width.

PRACTICE TEST

In the following exercises, find the Greatest Common Factor in each expression.

457. $14y - 42$

458. $-6x^2 - 30x$

459. $80a^2 + 120a^3$

460. $5m(m - 1) + 3(m - 1)$

In the following exercises, factor completely.

461. $x^2 + 13x + 36$

462. $p^2 + pq - 12q^2$

463. $3a^3 - 6a^2 - 72a$

464. $s^2 - 25s + 84$

465. $5n^2 + 30n + 45$

466. $64y^2 - 49$

467. $xy - 8y + 7x - 56$

468. $40r^2 + 810$

469. $9s^2 - 12s + 4$

470. $n^2 + 12n + 36$

471. $100 - a^2$

472. $6x^2 - 11x - 10$

473. $3x^2 - 75y^2$

474. $c^3 - 1000d^3$

475. $ab - 3b - 2a + 6$

476. $6u^2 + 3u - 18$

477. $8m^2 + 22m + 5$

In the following exercises, solve.

478. $x^2 + 9x + 20 = 0$

479. $y^2 = y + 132$

480. $5a^2 + 26a = 24$

481. $9b^2 - 9 = 0$

482. $16 - m^2 = 0$

483. $4n^2 + 19 + 21 = 0$

484. $(x - 3)(x + 2) = 6$

485. The product of two consecutive integers is 156. Find the integers.

486. The area of a rectangular place mat is 168 square inches. Its length is two inches longer than the width. Find the length and width of the placemat.

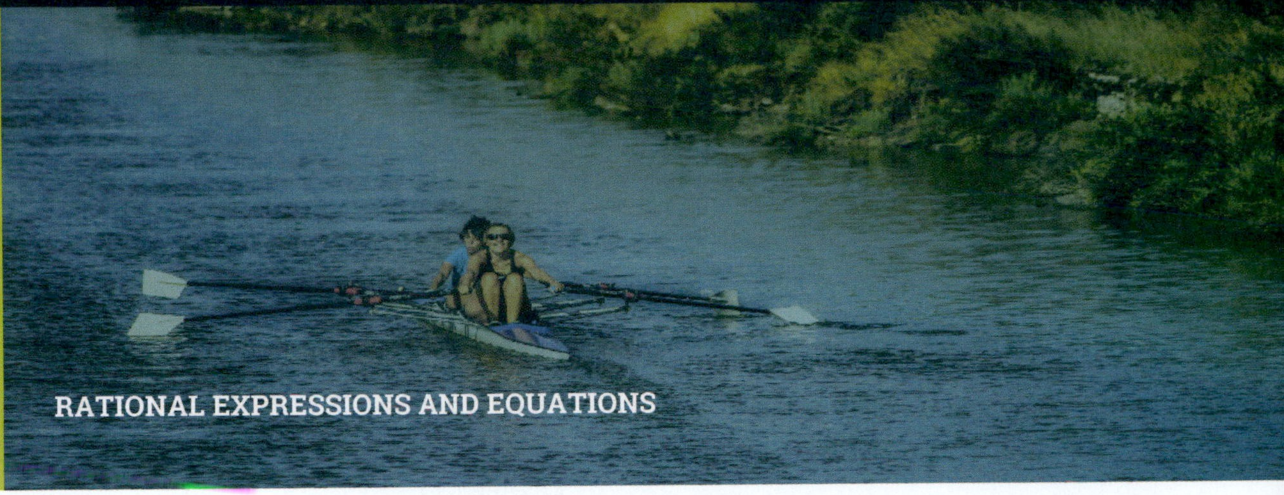

8 RATIONAL EXPRESSIONS AND EQUATIONS

Figure 8.1 Rowing a boat downstream can be very relaxing, but it takes much more effort to row the boat upstream.

Chapter Outline

8.1 Simplify Rational Expressions
8.2 Multiply and Divide Rational Expressions
8.3 Add and Subtract Rational Expressions with a Common Denominator
8.4 Add and Subtract Rational Expressions with Unlike Denominators
8.5 Simplify Complex Rational Expressions
8.6 Solve Rational Equations
8.7 Solve Proportion and Similar Figure Applications
8.8 Solve Uniform Motion and Work Applications
8.9 Use Direct and Inverse Variation

Introduction

Like rowing a boat, riding a bicycle is a situation in which going in one direction, downhill, is easy, but going in the opposite direction, uphill, can be more work. The trip to reach a destination may be quick, but the return trip whether upstream or uphill will take longer.

Rational equations are used to model situations like these. In this chapter, we will work with rational expressions, solve rational equations, and use them to solve problems in a variety of applications.

8.1 Simplify Rational Expressions

Learning Objectives

By the end of this section, you will be able to:

- Determine the values for which a rational expression is undefined
- Evaluate rational expressions
- Simplify rational expressions
- Simplify rational expressions with opposite factors

Be Prepared!

Before you get started, take this readiness quiz.

If you miss a problem, go back to the section listed and review the material.

1. Simplify: $\frac{90y}{15y^2}$.

If you missed this problem, review **Example 6.66**.

2. Factor: $6x^2 - 7x + 2$.

   If you missed this problem, review **Example 7.34**.

3. Factor: $n^3 + 8$.

   If you missed this problem, review **Example 7.54**.

In Chapter 1, we reviewed the properties of fractions and their operations. We introduced rational numbers, which are just fractions where the numerators and denominators are integers, and the denominator is not zero.

In this chapter, we will work with fractions whose numerators and denominators are polynomials. We call these rational expressions.

Rational Expression

A **rational expression** is an expression of the form $\dfrac{p(x)}{q(x)}$, where p and q are polynomials and $q \neq 0$.

Remember, division by 0 is undefined.

Here are some examples of rational expressions:

$$-\frac{13}{42} \qquad \frac{7y}{8z} \qquad \frac{5x+2}{x^2-7} \qquad \frac{4x^2+3x-1}{2x-8}$$

Notice that the first rational expression listed above, $-\dfrac{13}{42}$, is just a fraction. Since a constant is a polynomial with degree zero, the ratio of two constants is a rational expression, provided the denominator is not zero.

We will perform same operations with rational expressions that we do with fractions. We will simplify, add, subtract, multiply, divide, and use them in applications.

Determine the Values for Which a Rational Expression is Undefined

When we work with a numerical fraction, it is easy to avoid dividing by zero, because we can see the number in the denominator. In order to avoid dividing by zero in a rational expression, we must not allow values of the variable that will make the denominator be zero.

If the denominator is zero, the rational expression is undefined. The numerator of a rational expression may be 0—but not the denominator.

So before we begin any operation with a rational expression, we examine it first to find the values that would make the denominator zero. That way, when we solve a rational equation for example, we will know whether the algebraic solutions we find are allowed or not.

HOW TO :: DETERMINE THE VALUES FOR WHICH A RATIONAL EXPRESSION IS UNDEFINED.

Step 1. Set the denominator equal to zero.

Step 2. Solve the equation in the set of reals, if possible.

EXAMPLE 8.1

Determine the values for which the rational expression is undefined:

ⓐ $\dfrac{9y}{x}$ ⓑ $\dfrac{4b-3}{2b+5}$ ⓒ $\dfrac{x+4}{x^2+5x+6}$

⊘ **Solution**

The expression will be undefined when the denominator is zero.

(a)

	$\dfrac{9y}{x}$
Set the denominator equal to zero. Solve for the variable.	$x = 0$
	$\dfrac{9y}{x}$ is undefined for $x = 0$.

(b)

	$\dfrac{4b-3}{2b+5}$
	$2b + 5 \;=\; 0$
Set the denominator equal to zero. Solve for the variable.	$2b \;=\; -5$
	$b \;=\; -\dfrac{5}{2}$
	$\dfrac{4b-3}{2b+5}$ is undefined for $b = -\dfrac{5}{2}$.

(c)

	$\dfrac{x+4}{x^2+5x+6}$
	$x^2 + 5x + 6 \;=\; 0$
Set the denominator equal to zero. Solve for the variable.	$(x+2)(x+3) \;=\; 0$
	$x + 2 = 0 \text{ or } x + 3 \;=\; 0$
	$x = -2 \text{ or } x \;=\; -3$
	$\dfrac{x+4}{x^2+5x+6}$ is undefined for $x = -2$ or $x = -3$.

Saying that the rational expression $\dfrac{x+4}{x^2+5x+6}$ is undefined for $x = -2$ or $x = -3$ is similar to writing the phrase "void where prohibited" in contest rules.

> **TRY IT : : 8.1**   Determine the values for which the rational expression is undefined:

(a) $\dfrac{3y}{x}$ (b) $\dfrac{8n-5}{3n+1}$ (c) $\dfrac{a+10}{a^2+4a+3}$

> **TRY IT : : 8.2**   Determine the values for which the rational expression is undefined:

(a) $\dfrac{4p}{5q}$ (b) $\dfrac{y-1}{3y+2}$ (c) $\dfrac{m-5}{m^2+m-6}$

Evaluate Rational Expressions

To evaluate a rational expression, we substitute values of the variables into the expression and simplify, just as we have for many other expressions in this book.

EXAMPLE 8.2

Evaluate $\dfrac{2x+3}{3x-5}$ for each value:

(a) $x = 0$ (b) $x = 2$ (c) $x = -3$

✓ **Solution**

(a)

$$\frac{2x+3}{3x-5}$$

Substitute 0 for x.	$\frac{2(0)+3}{3(0)-5}$
Simplify.	$-\frac{3}{5}$

ⓑ

$$\frac{2x+3}{3x-5}$$

Substitute 2 for x.	$\frac{2(2)+3}{3(2)-5}$
Simplify.	$\frac{4+3}{6-5}$
	$\frac{7}{1}$
	7

ⓒ

$$\frac{2x+3}{3x-5}$$

Substitute –3 for x.	$\frac{2(-3)+3}{3(-3)-5}$
Simplify.	$\frac{-6+3}{-9-5}$
	$\frac{-3}{-14}$
	$\frac{3}{14}$

> **TRY IT :: 8.3**   Evaluate $\frac{y+1}{2y-3}$ for each value:

ⓐ $y=1$ ⓑ $y=-3$ ⓒ $y=0$

> **TRY IT :: 8.4**   Evaluate $\frac{5x-1}{2x+1}$ for each value:

ⓐ $x=1$ ⓑ $x=-1$ ⓒ $x=0$

EXAMPLE 8.3

Evaluate $\frac{x^2+8x+7}{x^2-4}$ for each value:

ⓐ $x=0$ ⓑ $x=2$ ⓒ $x=-1$

✓ Solution

ⓐ

	$\dfrac{x^2 + 8x + 7}{x^2 - 4}$
Substitute 0 for x.	$\dfrac{(0)^2 + 8(0) + 7}{(0)^2 - 4}$
Simplify.	$\dfrac{7}{-4}$
	$-\dfrac{7}{4}$

ⓑ

	$\dfrac{x^2 + 8x + 7}{x^2 - 4}$
Substitute 2 for x.	$\dfrac{(2)^2 + 8(2) + 7}{(2)^2 - 4}$
Simplify.	$\dfrac{4 + 16 + 7}{4 - 4}$
	$\dfrac{27}{0}$

This rational expression is undefined for $x = 2$.

ⓒ

	$\dfrac{x^2 + 8x + 7}{x^2 - 4}$
Substitute –1 for x.	$\dfrac{(-1)^2 + 8(-1) + 7}{(-1)^2 - 4}$
Simplify.	$\dfrac{1 - 8 + 7}{1 - 4}$
	$\dfrac{-7 + 7}{-3}$
	$\dfrac{0}{-3}$
	0

> **TRY IT : : 8.5** Evaluate $\dfrac{x^2 + 1}{x^2 - 3x + 2}$ for each value:
>
> ⓐ $x = 0$ ⓑ $x = -1$ ⓒ $x = 3$

> **TRY IT : : 8.6** Evaluate $\dfrac{x^2 + x - 6}{x^2 - 9}$ for each value:
>
> ⓐ $x = 0$ ⓑ $x = -2$ ⓒ $x = 1$

Remember that a fraction is simplified when it has no common factors, other than 1, in its numerator and denominator. When we evaluate a rational expression, we make sure to simplify the resulting fraction.

EXAMPLE 8.4

Evaluate $\dfrac{a^2 + 2ab + b^2}{3ab^3}$ for each value:

ⓐ $a = 1,\ b = 2$ ⓑ $a = -2,\ b = -1$ ⓒ $a = \frac{1}{3},\ b = 0$

✓ **Solution**

ⓐ

$$\dfrac{a^2 + 2ab + b^2}{3ab^2} \quad \text{when} \quad a = 1,\ b = 2.$$

Substitute 1 for *a* and 2 for *b*.	$\dfrac{(1)^2 + 2(1)(2) + (2)^2}{3(1)(2)^2}$
Simplify.	$\dfrac{1 + 4 + 4}{3(4)}$
	$\dfrac{9}{12}$
	$\dfrac{3}{4}$

ⓑ

$$\dfrac{a^2 + 2ab + b^2}{3ab^2} \quad \text{when} \quad a = -2,\ b = -1.$$

Substitute –2 for *a* and –1 for *b*.	$\dfrac{(-2)^2 + 2(-2)(-1) + (-1)^2}{3(-2)(-1)^2}$
Simplify.	$\dfrac{4 + 4 + 1}{-6}$
	$-\dfrac{9}{6}$
	$-\dfrac{3}{2}$

ⓒ

$$\dfrac{a^2 + 2ab + b^2}{3ab^2} \quad \text{when} \quad a = \frac{1}{3},\ b = 0.$$

Substitute $\frac{1}{3}$ for *a* and 0 for *b*.	$\dfrac{\left(\frac{1}{3}\right)^2 + 2\left(\frac{1}{3}\right)(0) + (0)^2}{3\left(\frac{1}{3}\right)(0)^2}$

Simplify.	$$\frac{\frac{1}{9}+0+0}{0}$$
	$$\frac{\frac{1}{9}}{0}$$
	The expression is undefined.

> **TRY IT ::** 8.7

Evaluate $\dfrac{2a^3 b}{a^2 + 2ab + b^2}$ for each value:

ⓐ $a = -1,\ b = 2$ ⓑ $a = 0,\ b = -1$ ⓒ $a = 1,\ b = \frac{1}{2}$

> **TRY IT ::** 8.8

Evaluate $\dfrac{a^2 - b^2}{8ab^3}$ for each value:

ⓐ $a = 1,\ b = -1$ ⓑ $a = \frac{1}{2},\ b = -1$ ⓒ $a = -2,\ b = 1$

Simplify Rational Expressions

Just like a fraction is considered simplified if there are no common factors, other than 1, in its numerator and denominator, a rational expression is *simplified* if it has no common factors, other than 1, in its numerator and denominator.

Simplified Rational Expression

A rational expression is considered simplified if there are no common factors in its numerator and denominator.

For example:

- $\frac{2}{3}$ is simplified because there are no common factors of 2 and 3.

- $\frac{2x}{3x}$ is not simplified because x is a common factor of 2x and 3x.

We use the Equivalent Fractions Property to simplify numerical fractions. We restate it here as we will also use it to simplify rational expressions.

Equivalent Fractions Property

If a, b, and c are numbers where $b \neq 0$, $c \neq 0$, then $\dfrac{a}{b} = \dfrac{a \cdot c}{b \cdot c}$ and $\dfrac{a \cdot c}{b \cdot c} = \dfrac{a}{b}$.

Notice that in the Equivalent Fractions Property, the values that would make the denominators zero are specifically disallowed. We see $b \neq 0$, $c \neq 0$ clearly stated. Every time we write a rational expression, we should make a similar statement disallowing values that would make a denominator zero. However, to let us focus on the work at hand, we will omit writing it in the examples.

Let's start by reviewing how we simplify numerical fractions.

EXAMPLE 8.5

Simplify: $-\dfrac{36}{63}$.

✓ **Solution**

	$-\dfrac{36}{63}$
Rewrite the numerator and denominator showing the common factors.	$-\dfrac{4\cdot9}{7\cdot9}$
Simplify using the Equivalent Fractions Property.	$-\dfrac{4}{7}$

Notice that the fraction $-\dfrac{4}{7}$ is simplified because there are no more common factors.

> **TRY IT :: 8.9**   Simplify: $-\dfrac{45}{81}$.

> **TRY IT :: 8.10**   Simplify: $-\dfrac{42}{54}$.

Throughout this chapter, we will assume that all numerical values that would make the denominator be zero are excluded. We will not write the restrictions for each rational expression, but keep in mind that the denominator can never be zero. So in this next example, $x \neq 0$ and $y \neq 0$.

EXAMPLE 8.6

Simplify: $\dfrac{3xy}{18x^2y^2}$.

✓ **Solution**

	$\dfrac{3xy}{18x^2y^2}$
Rewrite the numerator and denominator showing the common factors.	$\dfrac{1\cdot3xy}{6xy\cdot3xy}$
Simplify using the Equivalent Fractions Property.	$\dfrac{1}{6xy}$

Did you notice that these are the same steps we took when we divided monomials in **Polynomials**?

> **TRY IT :: 8.11**   Simplify: $\dfrac{4x^2y}{12xy^2}$.

> **TRY IT :: 8.12**   Simplify: $\dfrac{16x^2y}{2xy^2}$.

To simplify rational expressions we first write the numerator and denominator in factored form. Then we remove the common factors using the Equivalent Fractions Property.

Be very careful as you remove common factors. Factors are multiplied to make a product. You can remove a factor from a product. You cannot remove a term from a sum.

$$\frac{2 \cdot \cancel{3} \cdot \cancel{7}}{\cancel{3} \cdot 5 \cdot \cancel{7}} \qquad \frac{3x(x-9)}{5(x-9)} \qquad \frac{x+5}{x}$$

$$\frac{2}{5} \qquad \qquad \frac{3x}{5} \qquad \qquad \textbf{NO COMMON FACTORS}$$

We removed the common factors 3 and 7. They are factors of the product. | We removed the common factor $(x-9)$. It is a factor of the product. | While there is an x in both the numerator and denominator, the x in the numerator is a term of a sum!

Note that removing the x's from $\frac{x+5}{x}$ would be like cancelling the 2's in the fraction $\frac{2+5}{2}$!

EXAMPLE 8.7   HOW TO SIMPLIFY RATIONAL BINOMIALS

Simplify: $\frac{2x+8}{5x+20}$.

✓ **Solution**

Step 1. Factor the numerator and denominator completely.	Factor $2x+8$ and $5x-20$.	$\frac{2x+8}{5x+20}$ $\frac{2(x+4)}{5(x+4)}$
Step 2. Simplify by dividing out common factors.	Divide out the common factors.	$\frac{2\cancel{(x+4)}}{5\cancel{(x+4)}}$ $\frac{2}{5}$

> **TRY IT :: 8.13**   Simplify: $\frac{3x-6}{2x-4}$.

> **TRY IT :: 8.14**   Simplify: $\frac{7y+35}{5y+25}$.

We now summarize the steps you should follow to simplify rational expressions.

HOW TO :: SIMPLIFY A RATIONAL EXPRESSION.

Step 1.   Factor the numerator and denominator completely.

Step 2.   Simplify by dividing out common factors.

Usually, we leave the simplified rational expression in factored form. This way it is easy to check that we have removed all the common factors!

We'll use the methods we covered in Factoring to factor the polynomials in the numerators and denominators in the following examples.

EXAMPLE 8.8

Simplify: $\frac{x^2+5x+6}{x^2+8x+12}$.

Solution

$$\frac{x^2 + 5x + 6}{x^2 + 8x + 12}$$

Factor the numerator and denominator.

$$\frac{(x+2)(x+3)}{(x+2)(x+6)}$$

Remove the common factor $x + 2$ from the numerator and the denominator.

$$\frac{\cancel{(x+2)}(x+3)}{\cancel{(x+2)}(x+6)}$$

$$\frac{x+3}{x+6}$$

Can you tell which values of x must be excluded in this example?

> **TRY IT :: 8.15**   Simplify: $\dfrac{x^2 - x - 2}{x^2 - 3x + 2}$.

> **TRY IT :: 8.16**   Simplify: $\dfrac{x^2 - 3x - 10}{x^2 + x - 2}$.

EXAMPLE 8.9

Simplify: $\dfrac{y^2 + y - 42}{y^2 - 36}$.

Solution

$$\frac{y^2 + y - 42}{y^2 - 36}$$

Factor the numerator and denominator.

$$\frac{(y+7)(y-6)}{(y+6)(y-6)}$$

Remove the common factor $y - 6$ from the numerator and the denominator.

$$\frac{(y+7)\cancel{(y-6)}}{(y+6)\cancel{(y-6)}}$$

$$\frac{y+7}{y+6}$$

> **TRY IT :: 8.17**   Simplify: $\dfrac{x^2 + x - 6}{x^2 - 4}$.

> **TRY IT :: 8.18**   Simplify: $\dfrac{x^2 + 8x + 7}{x^2 - 49}$.

EXAMPLE 8.10

Simplify: $\dfrac{p^3 - 2p^2 + 2p - 4}{p^2 - 7p + 10}$.

Solution

$$\frac{p^3 - 2p^2 + 2p - 4}{p^2 - 7p + 10}$$

Factor the numerator and denominator, using grouping to factor the numerator.

$$\frac{p^2(p - 2) + 2(p - 2)}{(p - 5)(p - 2)}$$

$$\frac{(p^2 + 2)(p - 2)}{(p - 5)(p - 2)}$$

Remove the common factor of $p - 2$ from the numerator and the denominator.

$$\frac{(p^2 + 2)(\cancel{p - 2})}{(p - 5)(\cancel{p - 2})}$$

$$\frac{p^2 + 2}{p - 5}$$

> **TRY IT :: 8.19**   Simplify: $\dfrac{y^3 - 3y^2 + y - 3}{y^2 - y - 6}$.

> **TRY IT :: 8.20**   Simplify: $\dfrac{p^3 - p^2 + 2p - 2}{p^2 + 4p - 5}$.

EXAMPLE 8.11

Simplify: $\dfrac{2n^2 - 14n}{4n^2 - 16n - 48}$.

Solution

$$\frac{2n^2 - 14n}{4n^2 - 16n - 48}$$

Factor the numerator and denominator, first factoring out the GCF.

$$\frac{2n(n - 7)}{4(n^2 - 4n - 12)}$$

$$\frac{2n(n - 7)}{4(n - 6)(n + 2)}$$

Remove the common factor, 2.

$$\frac{\cancel{2}n(n - 7)}{\cancel{2} \cdot 2(n - 6)(n + 2)}$$

$$\frac{n(n - 7)}{2(n - 6)(n + 2)}$$

> **TRY IT :: 8.21**   Simplify: $\dfrac{2n^2 - 10n}{4n^2 - 16n - 20}$.

> **TRY IT :: 8.22**   Simplify: $\dfrac{4x^2 - 16x}{8x^2 - 16x - 64}$.

EXAMPLE 8.12

Simplify: $\dfrac{3b^2 - 12b + 12}{6b^2 - 24}$.

⊘ **Solution**

$$\dfrac{3b^2 - 12b + 12}{6b^2 - 24}$$

Factor the numerator and denominator, first factoring out the GCF.

$$\dfrac{3\left(b^2 - 4b + 4\right)}{6\left(b^2 - 4\right)}$$

$$\dfrac{3(b - 2)(b - 2)}{6(b + 2)(b - 2)}$$

Remove the common factors of $b - 2$ and 3.

$$\dfrac{\cancel{3}(b - 2)\cancel{(b - 2)}}{\cancel{3} \cdot 2(b + 2)\cancel{(b - 2)}}$$

$$\dfrac{b - 2}{2(b + 2)}$$

> **TRY IT ::** 8.23   Simplify: $\dfrac{2x^2 - 12x + 18}{3x^2 - 27}$.

> **TRY IT ::** 8.24   Simplify: $\dfrac{5y^2 - 30y + 25}{2y^2 - 50}$.

EXAMPLE 8.13

Simplify: $\dfrac{m^3 + 8}{m^2 - 4}$.

⊘ **Solution**

$$\dfrac{m^3 + 8}{m^2 - 4}$$

Factor the numerator and denominator, using the formulas for sum of cubes and difference of squares.

$$\dfrac{(m + 2)\left(m^2 - 2m + 4\right)}{(m + 2)(m - 2)}$$

Remove the common factor of $m + 2$.

$$\dfrac{\cancel{(m + 2)}\left(m^2 - 2m + 4\right)}{\cancel{(m + 2)}(m - 2)}$$

$$\dfrac{m^2 - 2m + 4}{m - 2}$$

> **TRY IT ::** 8.25   Simplify: $\dfrac{p^3 - 64}{p^2 - 16}$.

> **TRY IT ::** 8.26   Simplify: $\dfrac{x^3 + 8}{x^2 - 4}$.

Simplify Rational Expressions with Opposite Factors

Now we will see how to simplify a rational expression whose numerator and denominator have opposite factors. Let's

---

start with a numerical fraction, say $\frac{7}{-7}$. We know this fraction simplifies to -1. We also recognize that the numerator and denominator are opposites.

In **Foundations**, we introduced opposite notation: the opposite of a is $-a$. We remember, too, that $-a = -1 \cdot a$.

We simplify the fraction $\frac{a}{-a}$, whose numerator and denominator are opposites, in this way:

$$\frac{a}{-a}$$

We could rewrite this. $\qquad\qquad\qquad\qquad \dfrac{1 \cdot a}{-1 \cdot a}$

Remove the common factors. $\qquad\qquad \dfrac{1}{-1}$

Simplify. $\qquad\qquad\qquad\qquad\qquad -1$

So, in the same way, we can simplify the fraction $\frac{x-3}{-(x-3)}$:

We could rewrite this. $\qquad\qquad\qquad \dfrac{1 \cdot (x-3)}{-1 \cdot (x-3)}$

Remove the common factors. $\qquad\qquad \dfrac{1}{-1}$

Simplify. $\qquad\qquad\qquad\qquad\qquad -1$

But the opposite of $x - 3$ could be written differently:

$$-(x-3)$$

Distribute. $\qquad\qquad\qquad\qquad\quad -x + 3$

Rewrite. $\qquad\qquad\qquad\qquad\qquad 3 - x$

This means the fraction $\frac{x-3}{3-x}$ simplifies to -1.

In general, we could write the opposite of $a - b$ as $b - a$. So the rational expression $\frac{a-b}{b-a}$ simplifies to -1.

Opposites in a Rational Expression

The opposite of $a - b$ is $b - a$.

$$\frac{a-b}{b-a} = -1 \qquad a \neq b$$

An expression and its opposite divide to -1.

We will use this property to simplify rational expressions that contain opposites in their numerators and denominators.

EXAMPLE 8.14

Simplify: $\frac{x-8}{8-x}$.

⊘ Solution

$$\frac{x-8}{8-x}$$

Recognize that $x - 8$ and $8 - x$ are opposites. $\qquad -1$

> **TRY IT :: 8.27** Simplify: $\frac{y-2}{2-y}$.

> **TRY IT :: 8.28** Simplify: $\frac{n-9}{9-n}$.

Remember, the first step in simplifying a rational expression is to factor the numerator and denominator completely.

EXAMPLE 8.15

Simplify: $\dfrac{14 - 2x}{x^2 - 49}$.

✓ **Solution**

	$\dfrac{14 - 2x}{x^2 - 49}$
Factor the numerator and denominator.	$\dfrac{2(7 - x)}{(x + 7)(x - 7)}$
Recognize that $7 - x$ and $x - 7$ are opposites.	$\dfrac{2\cancel{(7 - x)}}{(x + 7)\cancel{(x - 7)}}(-1)$
Simplify.	$-\dfrac{2}{x + 7}$

> **TRY IT : : 8.29**   Simplify: $\dfrac{10 - 2y}{y^2 - 25}$.

> **TRY IT : : 8.30**   Simplify: $\dfrac{3y - 27}{81 - y^2}$.

EXAMPLE 8.16

Simplify: $\dfrac{x^2 - 4x - 32}{64 - x^2}$.

✓ **Solution**

	$\dfrac{x^2 - 4x - 32}{64 - x^2}$
Factor the numerator and denominator.	$\dfrac{(x - 8)(x + 4)}{(8 - x)(8 + x)}$
Recognize the factors that are opposites.	$(-1)\dfrac{\cancel{(x - 8)}(x + 4)}{\cancel{(8 - x)}(8 + x)}$
Simplify.	$-\dfrac{x + 4}{x + 8}$

> **TRY IT : : 8.31**   Simplify: $\dfrac{x^2 - 4x - 5}{25 - x^2}$.

> **TRY IT : : 8.32**   Simplify: $\dfrac{x^2 + x - 2}{1 - x^2}$.

8.1 EXERCISES

Practice Makes Perfect

In the following exercises, determine the values for which the rational expression is undefined.

1.

ⓐ $\dfrac{2x}{z}$

ⓑ $\dfrac{4p-1}{6p-5}$

ⓒ $\dfrac{n-3}{n^2+2n-8}$

2.

ⓐ $\dfrac{10m}{11n}$

ⓑ $\dfrac{6y+13}{4y-9}$

ⓒ $\dfrac{b-8}{b^2-36}$

3.

ⓐ $\dfrac{4x^2y}{3y}$

ⓑ $\dfrac{3x-2}{2x+1}$

ⓒ $\dfrac{u-1}{u^2-3u-28}$

4.

ⓐ $\dfrac{5pq^2}{9q}$

ⓑ $\dfrac{7a-4}{3a+5}$

ⓒ $\dfrac{1}{x^2-4}$

Evaluate Rational Expressions

In the following exercises, evaluate the rational expression for the given values.

5. $\dfrac{2x}{x-1}$

ⓐ $x=0$

ⓑ $x=2$

ⓒ $x=-1$

6. $\dfrac{4y-1}{5y-3}$

ⓐ $y=0$

ⓑ $y=2$

ⓒ $y=-1$

7. $\dfrac{2p+3}{p^2+1}$

ⓐ $p=0$

ⓑ $p=1$

ⓒ $p=-2$

8. $\dfrac{x+3}{2-3x}$

ⓐ $x=0$

ⓑ $x=1$

ⓒ $x=-2$

9. $\dfrac{y^2+5y+6}{y^2-1}$

ⓐ $y=0$

ⓑ $y=2$

ⓒ $y=-2$

10. $\dfrac{z^2+3z-10}{z^2-1}$

ⓐ $z=0$

ⓑ $z=2$

ⓒ $z=-2$

11. $\dfrac{a^2-4}{a^2+5a+4}$

ⓐ $a=0$

ⓑ $a=1$

ⓒ $a=-2$

12. $\dfrac{b^2+2}{b^2-3b-4}$

ⓐ $b=0$

ⓑ $b=2$

ⓒ $b=-2$

13. $\dfrac{x^2+3xy+2y^2}{2x^3y}$

ⓐ $x=1,\ y=-1$

ⓑ $x=2,\ y=1$

ⓒ $x=-1,\ y=-2$

14. $\dfrac{c^2 + cd - 2d^2}{cd^3}$

   ⓐ $c = 2, d = -1$

   ⓑ $c = 1, d = -1$

   ⓒ $c = -1, d = 2$

15. $\dfrac{m^2 - 4n^2}{5mn^3}$

   ⓐ $m = 2, n = 1$

   ⓑ $m = -1, n = -1$

   ⓒ $m = 3, n = 2$

16. $\dfrac{2s^2 t}{s^2 - 9t^2}$

   ⓐ $s = 4, t = 1$

   ⓑ $s = -1, t = -1$

   ⓒ $s = 0, t = 2$

Simplify Rational Expressions

In the following exercises, simplify.

17. $-\dfrac{4}{52}$

18. $-\dfrac{44}{55}$

19. $\dfrac{56}{63}$

20. $\dfrac{65}{104}$

21. $\dfrac{6ab^2}{12a^2 b}$

22. $\dfrac{15xy}{3x^3 y^3}$

23. $\dfrac{8m^3 n}{12mn^2}$

24. $\dfrac{36v^3 w^2}{27vw^3}$

25. $\dfrac{3a + 6}{4a + 8}$

26. $\dfrac{5b + 5}{6b + 6}$

27. $\dfrac{3c - 9}{5c - 15}$

28. $\dfrac{4d + 8}{9d + 18}$

29. $\dfrac{7m + 63}{5m + 45}$

30. $\dfrac{8n - 96}{3n - 36}$

31. $\dfrac{12p - 240}{5p - 100}$

32. $\dfrac{6q + 210}{5q + 175}$

33. $\dfrac{a^2 - a - 12}{a^2 - 8a + 16}$

34. $\dfrac{x^2 + 4x - 5}{x^2 - 2x + 1}$

35. $\dfrac{y^2 + 3y - 4}{y^2 - 6y + 5}$

36. $\dfrac{v^2 + 8v + 15}{v^2 - v - 12}$

37. $\dfrac{x^2 - 25}{x^2 + 2x - 15}$

38. $\dfrac{a^2 - 4}{a^2 + 6a - 16}$

39. $\dfrac{y^2 - 2y - 3}{y^2 - 9}$

40. $\dfrac{b^2 + 9b + 18}{b^2 - 36}$

41. $\dfrac{y^3 + y^2 + y + 1}{y^2 + 2y + 1}$

42. $\dfrac{p^3 + 3p^2 + 4p + 12}{p^2 + p - 6}$

43. $\dfrac{x^3 - 2x^2 - 25x + 50}{x^2 - 25}$

44. $\dfrac{q^3 + 3q^2 - 4q - 12}{q^2 - 4}$

45. $\dfrac{3a^2 + 15a}{6a^2 + 6a - 36}$

46. $\dfrac{8b^2 - 32b}{2b^2 - 6b - 80}$

47. $\dfrac{-5c^2 - 10c}{-10c^2 + 30c + 100}$

48. $\dfrac{4d^2 - 24d}{2d^2 - 4d - 48}$

49. $\dfrac{3m^2 + 30m + 75}{4m^2 - 100}$

50. $\dfrac{5n^2 + 30n + 45}{2n^2 - 18}$

51. $\dfrac{5r^2 + 30r - 35}{r^2 - 49}$

52. $\dfrac{3s^2 + 30s + 24}{3s^2 - 48}$

53. $\dfrac{t^3 - 27}{t^2 - 9}$

54. $\dfrac{v^3 - 1}{v^2 - 1}$

55. $\dfrac{w^3 + 216}{w^2 - 36}$

56. $\dfrac{v^3 + 125}{v^2 - 25}$

Simplify Rational Expressions with Opposite Factors

In the following exercises, simplify each rational expression.

57. $\dfrac{a - 5}{5 - a}$

58. $\dfrac{b - 12}{12 - b}$

59. $\dfrac{11 - c}{c - 11}$

60. $\dfrac{5 - d}{d - 5}$

61. $\dfrac{12 - 2x}{x^2 - 36}$

62. $\dfrac{20 - 5y}{y^2 - 16}$

63. $\dfrac{4v - 32}{64 - v^2}$

64. $\dfrac{7w - 21}{9 - w^2}$

65. $\dfrac{y^2 - 11y + 24}{9 - y^2}$

66. $\dfrac{z^2 - 9z + 20}{16 - z^2}$

67. $\dfrac{a^2 - 5z - 36}{81 - a^2}$

68. $\dfrac{b^2 + b - 42}{36 - b^2}$

Everyday Math

69. Tax Rates For the tax year 2015, the amount of tax owed by a single person earning between $37,450 and $90,750, can be found by evaluating the formula $0.25x - 4206.25$, where x is income. The average tax rate for this income can be found by evaluating the formula $\dfrac{0.25x - 4206.25}{x}$. What would be the average tax rate for a single person earning $50,000?

70. Work The length of time it takes for two people for perform the same task if they work together can be found by evaluating the formula $\dfrac{xy}{x + y}$. If Tom can paint the den in $x = 45$ minutes and his brother Bobby can paint it in $y = 60$ minutes, how many minutes will it take them if they work together?

Writing Exercises

71. Explain how you find the values of x for which the rational expression $\dfrac{x^2 - x - 20}{x^2 - 4}$ is undefined.

72. Explain all the steps you take to simplify the rational expression $\dfrac{p^2 + 4p - 21}{9 - p^2}$.

Self Check

ⓐ *After completing the exercises, use this checklist to evaluate your mastery of the objectives of this section.*

I can...	Confidently	With some help	No-I don't get it!
determine the values for which a rational expression is undefined.			
evaluate rational expressions.			
simplify rational expressions.			
simplify rational expressions with opposite factors.			

ⓑ *If most of your checks were:*

...confidently. *Congratulations! You have achieved your goals in this section! Reflect on the study skills you used so that you can continue to use them. What did you do to become confident of your ability to do these things? Be specific!*

...with some help. *This must be addressed quickly as topics you do not master become potholes in your road to success. Math is sequential - every topic builds upon previous work. It is important to make sure you have a strong foundation before you move on. Who can you ask for help? Your fellow classmates and instructor are good resources. Is there a place on campus where math* *tutors are available? Can your study skills be improved?*

...no - I don't get it! *This is critical and you must not ignore it. You need to get help immediately or you will quickly be overwhelmed. See your instructor as soon as possible to discuss your situation. Together you can come up with a plan to get you the help you need.*

8.2 Multiply and Divide Rational Expressions

Learning Objectives

By the end of this section, you will be able to:

> Multiply rational expressions
> Divide rational expressions

Be Prepared!

Before you get started, take this readiness quiz.

If you miss a problem, go back to the section listed and review the material.

1. Multiply: $\frac{14}{15} \cdot \frac{6}{35}$.

   If you missed this problem, review **Example 1.68**.

2. Divide: $\frac{14}{15} \div \frac{6}{35}$.

   If you missed this problem, review **Example 1.71**.

3. Factor completely: $2x^2 - 98$.
   If you missed this problem, review **Example 7.62**.

4. Factor completely: $10n^3 + 10$.
   If you missed this problem, review **Example 7.65**.

5. Factor completely: $10p^2 - 25pq - 15q^2$.

   If you missed this problem, review **Example 7.68**.

Multiply Rational Expressions

To multiply rational expressions, we do just what we did with numerical fractions. We multiply the numerators and multiply the denominators. Then, if there are any common factors, we remove them to simplify the result.

Multiplication of Rational Expressions

If p, q, r, s are polynomials where $q \neq 0$ and $s \neq 0$, then

$$\frac{p}{q} \cdot \frac{r}{s} = \frac{pr}{qs}$$

To multiply rational expressions, multiply the numerators and multiply the denominators.

We'll do the first example with numerical fractions to remind us of how we multiplied fractions without variables.

EXAMPLE 8.17

Multiply: $\frac{10}{28} \cdot \frac{8}{15}$.

✓ **Solution**

	$\frac{10}{28} \cdot \frac{8}{15}$
Multiply the numerators and denominators.	$\frac{10 \cdot 8}{28 \cdot 15}$
Look for common factors, and then remove them.	$\frac{2 \cdot \cancel{5} \cdot 2 \cdot \cancel{4}}{7 \cdot \cancel{4} \cdot 3 \cdot \cancel{5}}$
Simplify.	$\frac{4}{21}$

> **TRY IT :: 8.33**   Mulitply: $\dfrac{6}{10} \cdot \dfrac{15}{12}$.

> **TRY IT :: 8.34**   Mulitply: $\dfrac{20}{15} \cdot \dfrac{6}{8}$.

Remember, throughout this chapter, we will assume that all numerical values that would make the denominator be zero are excluded. We will not write the restrictions for each rational expression, but keep in mind that the denominator can never be zero. So in this next example, $x \neq 0$ and $y \neq 0$.

EXAMPLE 8.18

Mulitply: $\dfrac{2x}{3y^2} \cdot \dfrac{6xy^3}{x^2 y}$.

⊘ **Solution**

	$\dfrac{2x}{3y^2} \cdot \dfrac{6xy^3}{x^2 y}$
Multiply.	$\dfrac{2x \cdot 6xy^3}{3y^2 \cdot x^2 y}$
Factor the numerator and denominator completely, and then remove common factors.	$\dfrac{2 \cdot x \cdot 2 \cdot 3 \cdot x \cdot y \cdot y \cdot y}{3 \cdot y \cdot y \cdot x \cdot x \cdot y}$
Simplify.	4

> **TRY IT :: 8.35**   Mulitply: $\dfrac{3pq}{q^2} \cdot \dfrac{5p^2 q}{6pq}$.

> **TRY IT :: 8.36**   Mulitply: $\dfrac{6x^3 y}{7x^2} \cdot \dfrac{2xy^3}{x^2 y}$.

EXAMPLE 8.19   HOW TO MULTIPLY RATIONAL EXPRESSIONS

Mulitply: $\dfrac{2x}{x^2 + x + 12} \cdot \dfrac{x^2 - 9}{6x^2}$.

⊘ **Solution**

Step 1. Factor the numerator and denominator completely.	Factor $x^2 - 9$ and $x^2 + x + 12$.	$\dfrac{2x}{x^2 + x + 12} \cdot \dfrac{x^2 - 9}{6x^2}$ $\dfrac{2x}{(x-3)(x-4)} \cdot \dfrac{(x-3)(x+3)}{6x^2}$
Step 2. Multiply the numerators and denominators.	Multiply the numerators and denominators. It is helpful to write the monomials first.	$\dfrac{2x(x-3)(x+3)}{6x^2(x-3)(x-4)}$
Step 3. Simplify by dividing out common factors.	Divide out the common factors. Leave the denominator in factored form.	$\dfrac{2x\,(x-3)(x+3)}{2 \cdot 3 \cdot x \cdot x\,(x-3)(x-4)}$ $\dfrac{(x+3)}{3x(x-4)}$

> **TRY IT : : 8.37**

Mulitply: $\dfrac{5x}{x^2 + 5x + 6} \cdot \dfrac{x^2 - 4}{10x}$.

> **TRY IT : : 8.38**

Mulitply: $\dfrac{9x^2}{x^2 + 11x + 30} \cdot \dfrac{x^2 - 36}{3x^2}$.

HOW TO : : MULTIPLY A RATIONAL EXPRESSION.

Step 1.   Factor each numerator and denominator completely.

Step 2.   Multiply the numerators and denominators.

Step 3.   Simplify by dividing out common factors.

EXAMPLE 8.20

Multiply: $\dfrac{n^2 - 7n}{n^2 + 2n + 1} \cdot \dfrac{n + 1}{2n}$.

⊘ **Solution**

$$\dfrac{n^2 - 7n}{n^2 + 2n + 1} \cdot \dfrac{n + 1}{2n}$$

Factor each numerator and denominator.

$$\dfrac{n(n - 7)}{(n + 1)(n + 1)} \cdot \dfrac{n + 1}{2n}$$

Multiply the numerators and the denominators.

$$\dfrac{n(n - 7)(n + 1)}{(n + 1)(n + 1)2n}$$

Remove common factors.

$$\dfrac{\cancel{n}(n - 7)\cancel{(n + 1)}}{(n + 1)\cancel{(n + 1)}2\cancel{n}}$$

Simplify.

$$\dfrac{n - 7}{2(n + 1)}$$

> **TRY IT : : 8.39**

Multiply: $\dfrac{x^2 - 25}{x^2 - 3x - 10} \cdot \dfrac{x + 2}{x}$.

> **TRY IT : : 8.40**

Multiply: $\dfrac{x^2 - 4x}{x^2 + 5x + 6} \cdot \dfrac{x + 2}{x}$.

EXAMPLE 8.21

Multiply: $\dfrac{16 - 4x}{2x - 12} \cdot \dfrac{x^2 - 5x - 6}{x^2 - 16}$.

✓ **Solution**

$$\frac{16-4x}{2x-12} \cdot \frac{x^2-5x-6}{x^2-16}$$

Factor each numerator and denominator.

$$\frac{4(4-x)}{2(x-6)} \cdot \frac{(x-6)(x+1)}{(x-4)(x+4)}$$

Multiply the numerators and the denominators.

$$\frac{4(4-x)(x-6)(x+1)}{2(x-6)(x-4)(x+4)}$$

Remove common factors.

$$(-1)\frac{\cancel{2} \cdot 2(\cancel{4-x})(\cancel{x-6})(x+1)}{\cancel{2}(\cancel{x-6})(\cancel{x-4})(x+4)}$$

Simplify.

$$-\frac{2(x+1)}{(x+4)}$$

> **TRY IT :: 8.41**   Multiply: $\dfrac{12x-6x^2}{x^2+8x} \cdot \dfrac{x^2+11x+24}{x^2-4}$.

> **TRY IT :: 8.42**   Multiply: $\dfrac{9v-3v^2}{9v+36} \cdot \dfrac{v^2+7v+12}{v^2-9}$.

EXAMPLE 8.22

Multiply: $\dfrac{2x-6}{x^2-8x+15} \cdot \dfrac{x^2-25}{2x+10}$.

✓ **Solution**

	$\dfrac{2x-6}{x^2-8x+15} \cdot \dfrac{x^2-25}{2x+10}$
Factor each numerator and denominator.	$\dfrac{2(x-3)}{(x-3)(x-5)} \cdot \dfrac{(x-5)(x+5)}{2(x+5)}$
Multiply the numerators and denominators.	$\dfrac{2(x-3)(x-5)(x+5)}{2(x-3)(x-5)(x+5)}$
Remove common factors.	$\dfrac{\cancel{2}\cancel{(x-3)}\cancel{(x-5)}\cancel{(x+5)}}{\cancel{2}\cancel{(x-3)}\cancel{(x-5)}\cancel{(x+5)}}$
Simplify.	1

> **TRY IT :: 8.43**   Multiply: $\dfrac{3a-21}{a^2-9a+14} \cdot \dfrac{a^2-4}{3a+6}$.

> **TRY IT :: 8.44**   Multiply: $\dfrac{b^2-b}{b^2+9b-10} \cdot \dfrac{b^2-100}{b^2-10b}$.

Divide Rational Expressions

To divide rational expressions we multiply the first fraction by the reciprocal of the second, just like we did for numerical fractions.

Remember, the **reciprocal** of $\frac{a}{b}$ is $\frac{b}{a}$. To find the reciprocal we simply put the numerator in the denominator and the denominator in the numerator. We "flip" the fraction.

Division of Rational Expressions

If p, q, r, s are polynomials where $q \neq 0$, $r \neq 0$, $s \neq 0$, then

$$\frac{p}{q} \div \frac{r}{s} = \frac{p}{q} \cdot \frac{s}{r}$$

To divide rational expressions multiply the first fraction by the reciprocal of the second.

EXAMPLE 8.23   HOW TO DIVIDE RATIONAL EXPRESSIONS

Divide: $\frac{x+9}{6-x} \div \frac{x^2-81}{x-6}$.

✓ **Solution**

Step 1. Rewrite the division as the product of the first rational expression and the reciprocal of the second.		$\frac{x+9}{6-x} \div \frac{x^2-81}{x-6}$
	"Flip" the second fraction and change the division sign to multiplication.	$\frac{x+9}{6-x} \cdot \frac{x-6}{x^2-81}$
Step 2. Factor the numerators and denominators completely.	Factor $x^2 - 81$.	$\frac{x+9}{6-x} \cdot \frac{x-6}{(x-9)(x+9)}$
Step 3. Multiply the numerators and denominators.		$\frac{(x+9)(x-6)}{(6-x)(x-9)(x+9)}$
Step 4. Simplify by dividing out common factors.	Divide out the common factors. Remember opposites divide to –1.	$(-1)\dfrac{\cancel{(x+9)}\cancel{(x-6)}}{\cancel{(6-x)}(x-9)\cancel{(x+9)}}$ $-\dfrac{1}{(x-9)}$

> **TRY IT :: 8.45**   Divide: $\frac{c+3}{5-c} \div \frac{c^2-9}{c-5}$.

> **TRY IT :: 8.46**   Divide: $\frac{2-d}{d-4} \div \frac{4-d^2}{4-d}$.

HOW TO :: DIVIDE RATIONAL EXPRESSIONS.

Step 1.   Rewrite the division as the product of the first rational expression and the reciprocal of the second.

Step 2.   Factor the numerators and denominators completely.

Step 3.   Multiply the numerators and denominators together.

Step 4.   Simplify by dividing out common factors.

EXAMPLE 8.24

Divide: $\dfrac{3n^2}{n^2 - 4n} \div \dfrac{9n^2 - 45n}{n^2 - 7n + 10}$.

✓ **Solution**

	$\dfrac{3n^2}{n^2 - 4n} \div \dfrac{9n^2 - 45n}{n^2 - 7n + 10}$
Rewrite the division as the product of the first rational expression and the reciprocal of the second.	$\dfrac{3n^2}{n^2 - 4n} \cdot \dfrac{n^2 - 7n + 10}{9n^2 - 45n}$
Factor the numerators and denominators and then multiply.	$\dfrac{3 \cdot n \cdot n \cdot (n - 5)(n - 2)}{n(n - 4) \cdot 3 \cdot 3 \cdot n \cdot (n - 5)}$
Simplify by dividing out common factors.	$\dfrac{3 \cdot \cancel{n} \cdot \cancel{n}\cancel{(n-5)}(n - 2)}{\cancel{n}(n - 4)\cancel{3} \cdot 3 \cdot \cancel{n}\cancel{(n-5)}}$
	$\dfrac{n - 2}{3(n - 4)}$

> **TRY IT :: 8.47**   Divide: $\dfrac{2m^2}{m^2 - 8m} \div \dfrac{8m^2 + 24m}{m^2 + m - 6}$.

> **TRY IT :: 8.48**   Divide: $\dfrac{15n^2}{3n^2 + 33n} \div \dfrac{5n - 5}{n^2 + 9n - 22}$.

Remember, first rewrite the division as multiplication of the first expression by the reciprocal of the second. Then factor everything and look for common factors.

EXAMPLE 8.25

Divide: $\dfrac{2x^2 + 5x - 12}{x^2 - 16} \div \dfrac{2x^2 - 13x + 15}{x^2 - 8x + 16}$.

✓ **Solution**

	$\dfrac{2x^2 + 5x - 12}{x^2 - 16} \div \dfrac{2x^2 - 13x + 15}{x^2 - 8x + 16}$
Rewrite the division as multiplication of the first expression by the reciprocal of the second.	$\dfrac{2x^2 + 5x - 12}{x^2 - 16} \cdot \dfrac{x^2 - 8x + 16}{2x^2 - 13x + 15}$
Factor the numerators and denominators and then multiply.	$\dfrac{(2x - 3)(x + 4)(x - 4)(x - 4)}{(x - 4)(x + 4)(2x - 3)(x - 5)}$
Simplify by dividing out common factors.	$\dfrac{\cancel{(2x - 3)}\cancel{(x + 4)}\cancel{(x - 4)}(x - 4)}{\cancel{(x - 4)}\cancel{(x + 4)}\cancel{(2x - 3)}(x - 5)}$
Simplify.	$\dfrac{x - 4}{x - 5}$

> **TRY IT :: 8.49**   Divide: $\dfrac{3a^2 - 8a - 3}{a^2 - 25} \div \dfrac{3a^2 - 14a - 5}{a^2 + 10a + 25}$.

> TRY IT :: 8.50 Divide: $\dfrac{4b^2 + 7b - 2}{1 - b^2} \div \dfrac{4b^2 + 15b - 4}{b^2 - 2b + 1}$.

EXAMPLE 8.26

Divide: $\dfrac{p^3 + q^3}{2p^2 + 2pq + 2q^2} \div \dfrac{p^2 - q^2}{6}$.

✓ **Solution**

$$\dfrac{p^3 + q^3}{2p^2 + 2pq + 2q^2} \div \dfrac{p^2 - q^2}{6}$$

Rewrite the division as a multiplication of the first expression times the reciprocal of the second.

$$\dfrac{p^3 + q^3}{2p^2 + 2pq + 2q^2} \cdot \dfrac{6}{p^2 - q^2}$$

Factor the numerators and denominators and then multiply.

$$\dfrac{(p + q)\left(p^2 - pq + q^2\right)6}{2\left(p^2 + pq + q^2\right)(p - q)(p + q)}$$

Simplify by dividing out common factors.

$$\dfrac{\cancel{(p + q)}\left(p^2 - pq + q^2\right)\cancel{6}^{3}}{\cancel{2}\left(p^2 + pq + q^2\right)(p - q)\cancel{(p + q)}}$$

Simplify.

$$\dfrac{3\left(p^2 - pq + q^2\right)}{(p - q)\left(p^2 + pq + q^2\right)}$$

> TRY IT :: 8.51 Divide: $\dfrac{x^3 - 8}{3x^2 - 6x + 12} \div \dfrac{x^2 - 4}{6}$.

> TRY IT :: 8.52 Divide: $\dfrac{2z^2}{z^2 - 1} \div \dfrac{z^3 - z^2 + z}{z^3 - 1}$.

Before doing the next example, let's look at how we divide a fraction by a whole number. When we divide $\frac{3}{5} \div 4$, we first write 4 as a fraction so that we can find its reciprocal.

$$\frac{3}{5} \div 4$$
$$\frac{3}{5} \div \frac{4}{1}$$
$$\frac{3}{5} \cdot \frac{1}{4}$$

We do the same thing when we divide rational expressions.

EXAMPLE 8.27

Divide: $\dfrac{a^2 - b^2}{3ab} \div \left(a^2 + 2ab + b^2\right)$.

⊘ **Solution**

$$\frac{a^2 - b^2}{3ab} \div \left(a^2 + 2ab + b^2\right)$$

Write the second expression as a fraction.

$$\frac{a^2 - b^2}{3ab} \div \frac{a^2 + 2ab + b^2}{1}$$

Rewrite the division as the first
expression times the reciprocal of the
second expression.

$$\frac{a^2 - b^2}{3ab} \cdot \frac{1}{a^2 + 2ab + b^2}$$

Factor the numerators and the
denominators, and then multiply.

$$\frac{(a - b)(a + b) \cdot 1}{3ab \cdot (a + b)(a + b)}$$

Simplify by dividing out common factors.

$$\frac{(a - b)\cancel{(a + b)}}{3ab \cdot \cancel{(a + b)}(a + b)}$$

Simplify.

$$\frac{(a - b)}{3ab(a + b)}$$

> **TRY IT :: 8.53**   Divide: $\dfrac{2x^2 - 14x - 16}{4} \div (x^2 + 2x + 1)$.

> **TRY IT :: 8.54**   Divide: $\dfrac{y^2 - 6y + 8}{y^2 - 4y} \div (3y^2 - 12y)$.

Remember a fraction bar means division. A complex fraction is another way of writing division of two fractions.

EXAMPLE 8.28

Divide: $\dfrac{\dfrac{6x^2 - 7x + 2}{4x - 8}}{\dfrac{2x^2 - 7x + 3}{x^2 - 5x + 6}}$.

Solution

$$\frac{\dfrac{6x^2 - 7x + 2}{4x - 8}}{\dfrac{2x^2 - 7x + 3}{x^2 - 5x + 6}}$$

Rewrite with a division sign.

$$\frac{6x^2 - 7x + 2}{4x - 8} \div \frac{2x^2 - 7x + 3}{x^2 - 5x + 6}$$

Rewrite as product of first times reciprocal of second.

$$\frac{6x^2 - 7x + 2}{4x - 8} \cdot \frac{x^2 - 5x + 6}{2x^2 - 7x + 3}$$

Factor the numerators and the denominators, and then multiply.

$$\frac{(2x - 1)(3x - 2)(x - 2)(x - 3)}{4(x - 2)(2x - 1)(x - 3)}$$

Simplify by dividing out common factors.

$$\frac{\cancel{(2x - 1)}(3x - 2)\cancel{(x - 2)}\cancel{(x - 3)}}{4\cancel{(x - 2)}\cancel{(2x - 1)}\cancel{(x - 3)}}$$

Simplify.

$$\frac{3x - 2}{4}$$

> **TRY IT :: 8.55**   Divide: $\dfrac{\dfrac{3x^2 + 7x + 2}{4x + 24}}{\dfrac{3x^2 - 14x - 5}{x^2 + x - 30}}$.

> **TRY IT :: 8.56**   Divide: $\dfrac{\dfrac{y^2 - 36}{2y^2 + 11y - 6}}{\dfrac{2y^2 - 2y - 60}{8y - 4}}$.

If we have more than two rational expressions to work with, we still follow the same procedure. The first step will be to rewrite any division as multiplication by the reciprocal. Then we factor and multiply.

EXAMPLE 8.29

Divide: $\dfrac{3x - 6}{4x - 4} \cdot \dfrac{x^2 + 2x - 3}{x^2 - 3x - 10} \div \dfrac{2x + 12}{8x + 16}$.

Solution

$$\frac{3x - 6}{4x - 4} \cdot \frac{x^2 + 2x - 3}{x^2 - 3x - 10} \div \frac{2x + 12}{8x + 16}$$

Rewrite the division as multiplication by the reciprocal.

$$\frac{3x - 6}{4x - 4} \cdot \frac{x^2 + 2x - 3}{x^2 - 3x - 10} \cdot \frac{8x + 16}{2x + 12}$$

Factor the numerators and the denominators, and then multiply.

$$\frac{3 \cdot 8(x - 2)(x + 3)(x - 1)(x + 2)}{4 \cdot 2(x - 1)(x + 2)(x - 5)(x + 6)}$$

Simplify by dividing out common factors.

$$\frac{3 \cdot \cancel{8}(x - 2)(x + 3)\cancel{(x - 1)}\cancel{(x + 2)}}{\cancel{4} \cdot \cancel{2}\cancel{(x - 1)}\cancel{(x + 2)}(x - 5)(x + 6)}$$

Simplify.

$$\frac{3(x - 2)(x + 3)}{(x - 5)(x + 6)}$$

> **TRY IT ::** 8.57   Divide: $\dfrac{4m+4}{3m-15} \cdot \dfrac{m^2-3m-10}{m^2-4m-32} \div \dfrac{12m-36}{6m-48}$.

> **TRY IT ::** 8.58   Divide: $\dfrac{2n^2+10n}{n-1} \div \dfrac{n^2+10n+24}{n^2+8n-9} \cdot \dfrac{n+4}{8n^2+12n}$.

8.2 EXERCISES

Practice Makes Perfect

Multiply Rational Expressions

In the following exercises, multiply.

73. $\dfrac{12}{16} \cdot \dfrac{4}{10}$

74. $\dfrac{32}{5} \cdot \dfrac{16}{24}$

75. $\dfrac{18}{10} \cdot \dfrac{4}{30}$

76. $\dfrac{21}{36} \cdot \dfrac{45}{24}$

77. $\dfrac{5x^2 y^4}{12xy^3} \cdot \dfrac{6x^2}{20y^2}$

78. $\dfrac{8w^3 y}{9y^2} \cdot \dfrac{3y}{4w^4}$

79. $\dfrac{12a^3 b}{b^2} \cdot \dfrac{2ab^2}{9b^3}$

80. $\dfrac{4mn^2}{5n^3} \cdot \dfrac{mn^3}{8m^2 n^2}$

81. $\dfrac{5p^2}{p^2 - 5p - 36} \cdot \dfrac{p^2 - 16}{10p}$

82. $\dfrac{3q^2}{q^2 + q - 6} \cdot \dfrac{q^2 - 9}{9q}$

83. $\dfrac{4r}{r^2 - 3r - 10} \cdot \dfrac{r^2 - 25}{8r^2}$

84. $\dfrac{s}{s^2 - 9s + 14} \cdot \dfrac{s^2 - 49}{7s^2}$

85. $\dfrac{x^2 - 7x}{x^2 + 6x + 9} \cdot \dfrac{x + 3}{4x}$

86. $\dfrac{2y^2 - 10y}{y^2 + 10y + 25} \cdot \dfrac{y + 5}{6y}$

87. $\dfrac{z^2 + 3z}{z^2 - 3z - 4} \cdot \dfrac{z - 4}{z^2}$

88. $\dfrac{2a^2 + 8a}{a^2 - 9a + 20} \cdot \dfrac{a - 5}{a^2}$

89. $\dfrac{28 - 4b}{3b - 3} \cdot \dfrac{b^2 + 8b - 9}{b^2 - 49}$

90. $\dfrac{18c - 2c^2}{6c + 30} \cdot \dfrac{c^2 + 7c + 10}{c^2 - 81}$

91. $\dfrac{35d - 7d^2}{d^2 + 7d} \cdot \dfrac{d^2 + 12d + 35}{d^2 - 25}$

92.
$\dfrac{72m - 12m^2}{8m + 32} \cdot \dfrac{m^2 + 10m + 24}{m^2 - 36}$

93. $\dfrac{4n + 20}{n^2 + n - 20} \cdot \dfrac{n^2 - 16}{4n + 16}$

94. $\dfrac{6p^2 - 6p}{p^2 + 7p - 18} \cdot \dfrac{p^2 - 81}{3p^2 - 27p}$

95. $\dfrac{q^2 - 2q}{q^2 + 6q - 16} \cdot \dfrac{q^2 - 64}{q^2 - 8q}$

96. $\dfrac{2r^2 - 2r}{r^2 + 4r - 5} \cdot \dfrac{r^2 - 25}{2r^2 - 10r}$

Divide Rational Expressions

In the following exercises, divide.

97. $\dfrac{t - 6}{3 - t} \div \dfrac{t^2 - 9}{t - 5}$

98. $\dfrac{v - 5}{11 - v} \div \dfrac{v^2 - 25}{v - 11}$

99. $\dfrac{10 + w}{w - 8} \div \dfrac{100 - w^2}{8 - w}$

100. $\dfrac{7 + x}{x - 6} \div \dfrac{49 - x^2}{x + 6}$

101. $\dfrac{27y^2}{3y - 21} \div \dfrac{3y^2 + 18}{y^2 + 13y + 42}$

102. $\dfrac{24z^2}{2z - 8} \div \dfrac{4z - 28}{z^2 - 11z + 28}$

103. $\dfrac{16a^2}{4a + 36} \div \dfrac{4a^2 - 24a}{a^2 + 4a - 45}$

104. $\dfrac{24b^2}{2b - 4} \div \dfrac{12b^2 + 36b}{b^2 - 11b + 18}$

105. $\dfrac{5c^2 + 9c + 2}{c^2 - 25} \div \dfrac{3c^2 - 14c - 5}{c^2 + 10c + 25}$

106. $\dfrac{2d^2 + d - 3}{d^2 - 16} \div \dfrac{2d^2 - 9d - 18}{d^2 - 8d + 16}$

107. $\dfrac{6m^2 - 2m - 10}{9 - m^2} \div \dfrac{6m^2 + 29m - 20}{m^2 - 6m + 9}$

108. $\dfrac{2n^2 - 3n - 14}{25 - n^2} \div \dfrac{2n^2 - 13n + 21}{n^2 - 10n + 25}$

109. $\dfrac{3s^2}{s^2 - 16} \div \dfrac{s^3 - 4s^2 + 16s}{s^3 - 64}$

110. $\dfrac{r^2 - 9}{15} \div \dfrac{r^3 - 27}{5r^2 + 15r + 45}$

111. $\dfrac{p^3 + q^3}{3p^2 + 3pq + 3q^2} \div \dfrac{p^2 - q^2}{12}$

112. $\dfrac{v^3 - 8w^3}{2v^2 + 4vw + 8w^2} \div \dfrac{v^2 - 4w^2}{4}$

113. $\dfrac{t^2 - 9}{2t} \div (t^2 - 6t + 9)$

114. $\dfrac{x^2 + 3x - 10}{4x} \div (2x^2 + 20x + 50)$

115. $\dfrac{2y^2 - 10yz - 48z^2}{2y - 1} \div (4y^2 - 32yz)$

116. $\dfrac{2m^2 - 98n^2}{2m + 6} \div (m^2 - 7mn)$

117. $\dfrac{\dfrac{2a^2 - a - 21}{5a + 20}}{\dfrac{a^2 + 7a + 12}{a^2 + 8a + 16}}$

118. $\dfrac{\dfrac{3b^2 + 2b - 8}{12b + 18}}{\dfrac{3b^2 + 2b - 8}{2b^2 - 7b - 15}}$

119. $\dfrac{\dfrac{12c^2 - 12}{2c^2 - 3c + 1}}{\dfrac{4c + 4}{6c^2 - 13c + 5}}$

120. $\dfrac{\dfrac{4d^2 + 7d - 2}{35d + 10}}{\dfrac{d^2 - 4}{7d^2 - 12d - 4}}$

121. $\dfrac{10m^2 + 80m}{3m - 9} \cdot \dfrac{m^2 + 4m - 21}{m^2 - 9m + 20}$

$\div \dfrac{5m^2 + 10m}{2m - 10}$

122. $\dfrac{4n^2 + 32n}{3n + 2} \cdot \dfrac{3n^2 - n - 2}{n^2 + n - 30}$

$\div \dfrac{108n^2 - 24n}{n + 6}$

123. $\dfrac{12p^2 + 3p}{p + 3} \div \dfrac{p^2 + 2p - 63}{p^2 - p - 12}$

$\cdot \dfrac{p - 7}{9p^3 - 9p^2}$

124. $\dfrac{6q + 3}{9q^2 - 9q} \div \dfrac{q^2 + 14q + 33}{q^2 + 4q - 5}$

$\cdot \dfrac{4q^2 + 12q}{12q + 6}$

Everyday Math

125. Probability The director of large company is interviewing applicants for two identical jobs. If $w =$ the number of women applicants and $m =$ the number of men applicants, then the probability that two women are selected for the jobs is $\dfrac{w}{w + m} \cdot \dfrac{w - 1}{w + m - 1}$.

ⓐ Simplify the probability by multiplying the two rational expressions.

ⓑ Find the probability that two women are selected when $w = 5$ and $m = 10$.

126. Area of a triangle The area of a triangle with base b and height h is $\dfrac{bh}{2}$. If the triangle is stretched to make a new triangle with base and height three times as much as in the original triangle, the area is $\dfrac{9bh}{2}$.

Calculate how the area of the new triangle compares to the area of the original triangle by dividing $\dfrac{9bh}{2}$ by $\dfrac{bh}{2}$.

Writing Exercises

127.

ⓐ Multiply $\frac{7}{4} \cdot \frac{9}{10}$ and explain all your steps.

ⓑ Multiply $\frac{n}{n-3} \cdot \frac{9}{n+3}$ and explain all your steps.

ⓒ Evaluate your answer to part (b) when $n = 7$. Did you get the same answer you got in part (a)? Why or why not?

128.

ⓐ Divide $\frac{24}{5} \div 6$ and explain all your steps.

ⓑ Divide $\frac{x^2 - 1}{x} \div (x + 1)$ and explain all your steps.

ⓒ Evaluate your answer to part (b) when $x = 5$. Did you get the same answer you got in part (a)? Why or why not?

Self Check

ⓐ *After completing the exercises, use this checklist to evaluate your mastery of the objectives of this section.*

I can...	Confidently	With some help	No-I don't get it!
multiply rational expressions.			
divide rational expressions.			

ⓑ *After reviewing this checklist, what will you do to become confident for all objectives?*

8.3 Add and Subtract Rational Expressions with a Common Denominator

Learning Objectives

By the end of this section, you will be able to:

> Add rational expressions with a common denominator
> Subtract rational expressions with a common denominator
> Add and subtract rational expressions whose denominators are opposites

---

Be Prepared!

Before you get started, take this readiness quiz.

If you miss a problem, go back to the section listed and review the material.

1. Add: $\frac{y}{3} + \frac{9}{3}$.

    If you missed this problem, review **Example 1.77**.

2. Subtract: $\frac{10}{x} - \frac{2}{x}$.

    If you missed this problem, review **Example 1.79**.

3. Factor completely: $8n^5 - 20n^3$.

    If you missed this problem, review **Example 7.59**.

4. Factor completely: $45a^3 - 5ab^2$.

    If you missed this problem, review **Example 7.62**.

Add Rational Expressions with a Common Denominator

What is the first step you take when you add numerical fractions? You check if they have a common denominator. If they do, you add the numerators and place the sum over the common denominator. If they do not have a common denominator, you find one before you add.

It is the same with rational expressions. To add rational expressions, they must have a common denominator. When the denominators are the same, you add the numerators and place the sum over the common denominator.

Rational Expression Addition

If p, q, and r are polynomials where $r \neq 0$, then

$$\frac{p}{r} + \frac{q}{r} = \frac{p + q}{r}$$

To add rational expressions with a common denominator, add the numerators and place the sum over the common denominator.

We will add two numerical fractions first, to remind us of how this is done.

EXAMPLE 8.30

Add: $\frac{5}{18} + \frac{7}{18}$.

⊘ Solution

$$\frac{5}{18} + \frac{7}{18}$$

The fractions have a common denominator, so add the numerators and place the sum over the common denominator.	$\frac{5+7}{18}$
Add in the numerator.	$\frac{12}{18}$
Factor the numerator and denominator to show the common factors.	$\frac{6 \cdot 2}{6 \cdot 3}$
Remove common factors.	$\frac{\cancel{6} \cdot 2}{\cancel{6} \cdot 3}$
Simplify.	$\frac{2}{3}$

> **TRY IT :: 8.59**   Add: $\frac{7}{16} + \frac{5}{16}$.

> **TRY IT :: 8.60**   Add: $\frac{3}{10} + \frac{1}{10}$.

Remember, we do not allow values that would make the denominator zero. What value of y should be excluded in the next example?

EXAMPLE 8.31

Add: $\frac{3y}{4y-3} + \frac{7}{4y-3}$.

⊘ Solution

$$\frac{3y}{4y-3} + \frac{7}{4y-3}$$

The fractions have a common denominator, so add the numerators and place the sum over the common denominator.	$\frac{3y+7}{4y-3}$

The numerator and denominator cannot be factored. The fraction is simplified.

> **TRY IT :: 8.61**   Add: $\frac{5x}{2x+3} + \frac{2}{2x+3}$.

> **TRY IT :: 8.62**   Add: $\frac{x}{x-2} + \frac{1}{x-2}$.

EXAMPLE 8.32

Add: $\dfrac{7x+12}{x+3} + \dfrac{x^2}{x+3}$.

✓ Solution

$$\dfrac{7x+12}{x+3} + \dfrac{x^2}{x+3}$$

The fractions have a common denominator, so add the numerators and place the sum over the common denominator.

$$\dfrac{7x+12+x^2}{x+3}$$

Write the degrees in descending order.

$$\dfrac{x^2+7x+12}{x+3}$$

Factor the numerator.

$$\dfrac{(x+3)(x+4)}{x+3}$$

Simplify by removing common factors.

$$\dfrac{\cancel{(x+3)}(x+4)}{\cancel{x+3}}$$

Simplify.

$$x+4$$

> **TRY IT :: 8.63**   Add: $\dfrac{9x+14}{x+7} + \dfrac{x^2}{x+7}$.

> **TRY IT :: 8.64**   Add: $\dfrac{x^2+8x}{x+5} + \dfrac{15}{x+5}$.

Subtract Rational Expressions with a Common Denominator

To subtract rational expressions, they must also have a common denominator. When the denominators are the same, you subtract the numerators and place the difference over the common denominator.

Rational Expression Subtraction

If $p,\ q,$ and r are polynomials where $r \neq 0$, then

$$\dfrac{p}{r} - \dfrac{q}{r} = \dfrac{p-q}{r}$$

To subtract rational expressions, subtract the numerators and place the difference over the common denominator.

We always simplify rational expressions. Be sure to factor, if possible, after you subtract the numerators so you can identify any common factors.

EXAMPLE 8.33

Subtract: $\dfrac{n^2}{n-10} - \dfrac{100}{n-10}$.

✓ Solution

$$\frac{n^2}{n-10} - \frac{100}{n-10}$$

The fractions have a common denominator, so subtract the numerators and place the difference over the common denominator.

$$\frac{n^2 - 100}{n-10}$$

Factor the numerator.

$$\frac{(n-10)(n+10)}{n-10}$$

Simplify by removing common factors.

$$\frac{(\cancel{n-10})(n+10)}{\cancel{n-10}}$$

Simplify.

$$n + 10$$

> **TRY IT :: 8.65**   Subtract: $\dfrac{x^2}{x+3} - \dfrac{9}{x+3}$.

> **TRY IT :: 8.66**   Subtract: $\dfrac{4x^2}{2x-5} - \dfrac{25}{2x-5}$.

Be careful of the signs when you subtract a binomial!

EXAMPLE 8.34

Subtract: $\dfrac{y^2}{y-6} - \dfrac{2y+24}{y-6}$.

✓ Solution

$$\frac{y^2}{y-6} - \frac{2y+24}{y-6}$$

The fractions have a common denominator, so subtract the numerators and place the difference over the common denominator.

$$\frac{y^2 - (2y + 24)}{y-6}$$

Distribute the sign in the numerator.

$$\frac{y^2 - 2y - 24}{y-6}$$

Factor the numerator.

$$\frac{(y-6)(y+4)}{y-6}$$

Remove common factors.

$$\frac{(\cancel{y-6})(y+4)}{\cancel{y-6}}$$

Simplify.

$$y + 4$$

> **TRY IT :: 8.67**   Subtract: $\dfrac{n^2}{n-4} - \dfrac{n+12}{n-4}$.

> **TRY IT : : 8.68**
> Subtract: $\dfrac{y^2}{y-1} - \dfrac{9y-8}{y-1}$.

EXAMPLE 8.35

Subtract: $\dfrac{5x^2-7x+3}{x^2-3x+18} - \dfrac{4x^2+x-9}{x^2-3x+18}$.

✓ **Solution**

$$\dfrac{5x^2-7x+3}{x^2-3x+18} - \dfrac{4x^2+x-9}{x^2-3x+18}$$

Subtract the numerators and place the difference over the common denominator.

$$\dfrac{5x^2-7x+3-\left(4x^2+x-9\right)}{x^2-3x+18}$$

Distribute the sign in the numerator.

$$\dfrac{5x^2-7x+3-4x^2-x+9}{x^2-3x-18}$$

Combine like terms.

$$\dfrac{x^2-8x+12}{x^2-3x-18}$$

Factor the numerator and the denominator.

$$\dfrac{(x-2)(x-6)}{(x+3)(x-6)}$$

Simplify by removing common factors.

$$\dfrac{(x-2)\cancel{(x-6)}}{(x+3)\cancel{(x-6)}}$$

Simplify.

$$\dfrac{(x-2)}{(x+3)}$$

> **TRY IT : : 8.69**
> Subtract: $\dfrac{4x^2-11x+8}{x^2-3x+2} - \dfrac{3x^2+x-3}{x^2-3x+2}$.

> **TRY IT : : 8.70**
> Subtract: $\dfrac{6x^2-x+20}{x^2-81} - \dfrac{5x^2+11x-7}{x^2-81}$.

Add and Subtract Rational Expressions whose Denominators are Opposites

When the denominators of two rational expressions are opposites, it is easy to get a common denominator. We just have to multiply one of the fractions by $\dfrac{-1}{-1}$.

Let's see how this works.

$$\dfrac{7}{d} + \dfrac{5}{-d}$$

Multiply the second fraction by $\dfrac{-1}{-1}$. $\dfrac{7}{d} + \dfrac{(-1)5}{(-1)(-d)}$

The denominators are the same.	$\dfrac{7}{d} + \dfrac{-5}{d}$
Simplify.	$\dfrac{2}{d}$

EXAMPLE 8.36

Add: $\dfrac{4u-1}{3u-1} + \dfrac{u}{1-3u}$.

✓ **Solution**

	$\dfrac{4u-1}{3u-1} + \dfrac{u}{1-3u}$
The denominators are opposites, so multiply the second fraction by $\dfrac{-1}{-1}$.	$\dfrac{4u-1}{3u-1} + \dfrac{(-1)u}{(-1)(1-3u)}$
Simplify the second fraction.	$\dfrac{4u-1}{3u-1} + \dfrac{-u}{3u-1}$
The denominators are the same. Add the numerators.	$\dfrac{4u-1-u}{3u-1}$
Simplify.	$\dfrac{3u-1}{3u-1}$
Simplify.	1

> **TRY IT :: 8.71** Add: $\dfrac{8x-15}{2x-5} + \dfrac{2x}{5-2x}$.

> **TRY IT :: 8.72** Add: $\dfrac{6y^2+7y-10}{4y-7} + \dfrac{2y^2+2y+11}{7-4y}$.

EXAMPLE 8.37

Subtract: $\dfrac{m^2-6m}{m^2-1} - \dfrac{3m+2}{1-m^2}$.

✓ **Solution**

	$\dfrac{m^2-6m}{m^2-1} - \dfrac{3m+2}{1-m^2}$
The denominators are opposites, so multiply the second fraction by $\dfrac{-1}{-1}$.	$\dfrac{m^2-6m}{m^2-1} - \dfrac{-1(3m+2)}{-1(1-m^2)}$
Simplify the second fraction.	$\dfrac{m^2-6m}{m^2-1} - \dfrac{-3m-2}{m^2-1}$
The denominators are the same. Subtract the numerators.	$\dfrac{m^2-6m-(-3m-2)}{m^2-1}$

Distribute.	$\dfrac{m^2 - 6m + 3m + 2}{m^2 - 1}$
Combine like terms.	$\dfrac{m^2 - 3m + 2}{m^2 - 1}$
Factor the numerator and denominator.	$\dfrac{(m-1)(m-2)}{(m-1)(m+1)}$
Simplify by removing common factors.	$\dfrac{\cancel{(m-1)}(m-2)}{\cancel{(m-1)}(m+1)}$
Simplify.	$\dfrac{m-2}{m+1}$

> **TRY IT :: 8.73**   Subtract: $\dfrac{y^2 - 5y}{y^2 - 4} - \dfrac{6y - 6}{4 - y^2}$.

> **TRY IT :: 8.74**   Subtract: $\dfrac{2n^2 + 8n - 1}{n^2 - 1} - \dfrac{n^2 - 7n - 1}{1 - n^2}$.

8.3 EXERCISES

Practice Makes Perfect

Add Rational Expressions with a Common Denominator

In the following exercises, add.

129. $\dfrac{2}{15} + \dfrac{7}{15}$

130. $\dfrac{4}{21} + \dfrac{3}{21}$

131. $\dfrac{7}{24} + \dfrac{11}{24}$

132. $\dfrac{7}{36} + \dfrac{13}{36}$

133. $\dfrac{3a}{a-b} + \dfrac{1}{a-b}$

134. $\dfrac{3c}{4c-5} + \dfrac{5}{4c-5}$

135. $\dfrac{d}{d+8} + \dfrac{5}{d+8}$

136. $\dfrac{7m}{2m+n} + \dfrac{4}{2m+n}$

137. $\dfrac{p^2+10p}{p+2} + \dfrac{16}{p+2}$

138. $\dfrac{q^2+12q}{q+3} + \dfrac{27}{q+3}$

139. $\dfrac{2r^2}{2r-1} + \dfrac{15r-8}{2r-1}$

140. $\dfrac{3s^2}{3s-2} + \dfrac{13s-10}{3s-2}$

141. $\dfrac{8t^2}{t+4} + \dfrac{32t}{t+4}$

142. $\dfrac{6v^2}{v+5} + \dfrac{30v}{v+5}$

143. $\dfrac{2w^2}{w^2-16} + \dfrac{8w}{w^2-16}$

144. $\dfrac{7x^2}{x^2-9} + \dfrac{21x}{x^2-9}$

Subtract Rational Expressions with a Common Denominator

In the following exercises, subtract.

145. $\dfrac{y^2}{y+8} - \dfrac{64}{y+8}$

146. $\dfrac{z^2}{z+2} - \dfrac{4}{z+2}$

147. $\dfrac{9a^2}{3a-7} - \dfrac{49}{3a-7}$

148. $\dfrac{25b^2}{5b-6} - \dfrac{36}{5b-6}$

149. $\dfrac{c^2}{c-8} - \dfrac{6c+16}{c-8}$

150. $\dfrac{d^2}{d-9} - \dfrac{6d+27}{d-9}$

151. $\dfrac{3m^2}{6m-30} - \dfrac{21m-30}{6m-30}$

152. $\dfrac{2n^2}{4n-32} - \dfrac{30n-16}{4n-32}$

153.
$\dfrac{6p^2+3p+4}{p^2+4p-5} - \dfrac{5p^2+p+7}{p^2+4p-5}$

154.
$\dfrac{5q^2+3q-9}{q^2+6q+8} - \dfrac{4q^2+9q+7}{q^2+6q+8}$

155.
$\dfrac{5r^2+7r-33}{r^2-49} - \dfrac{4r^2-5r-30}{r^2-49}$

156. $\dfrac{7t^2-t-4}{t^2-25} - \dfrac{6t^2+2t-1}{t^2-25}$

Add and Subtract Rational Expressions whose Denominators are Opposites

In the following exercises, add.

157. $\dfrac{10v}{2v-1} + \dfrac{2v+4}{1-2v}$

158. $\dfrac{20w}{5w-2} + \dfrac{5w+6}{2-5w}$

159.
$\dfrac{10x^2+16x-7}{8x-3} + \dfrac{2x^2+3x-1}{3-8x}$

160.

$$\frac{6y^2 + 2y - 11}{3y - 7} + \frac{3y^2 - 3y + 17}{7 - 3y}$$

In the following exercises, subtract.

161. $\dfrac{z^2 + 6z}{z^2 - 25} - \dfrac{3z + 20}{25 - z^2}$ **162.** $\dfrac{a^2 + 3a}{a^2 - 9} - \dfrac{3a - 27}{9 - a^2}$ **163.**

$$\frac{2b^2 + 30b - 13}{b^2 - 49} - \frac{2b^2 - 5b - 8}{49 - b^2}$$

164.

$$\frac{c^2 + 5c - 10}{c^2 - 16} - \frac{c^2 - 8c - 10}{16 - c^2}$$

Everyday Math

165. Sarah ran 8 miles and then biked 24 miles. Her biking speed is 4 mph faster than her running speed. If r represents Sarah's speed when she ran, then her running time is modeled by the expression $\dfrac{8}{r}$ and her biking time is modeled by the expression $\dfrac{24}{r + 4}$.

Add the rational expressions $\dfrac{8}{r} + \dfrac{24}{r + 4}$ to get an expression for the total amount of time Sarah ran and biked.

166. If Pete can paint a wall in p hours, then in one hour he can paint $\dfrac{1}{p}$ of the wall. It would take Penelope 3 hours longer than Pete to paint the wall, so in one hour she can paint $\dfrac{1}{p + 3}$ of the wall. Add the rational expressions $\dfrac{1}{p} + \dfrac{1}{p + 3}$ to get an expression for the part of the wall Pete and Penelope would paint in one hour if they worked together.

Writing Exercises

167. Donald thinks that $\dfrac{3}{x} + \dfrac{4}{x}$ is $\dfrac{7}{2x}$. Is Donald correct? Explain.

168. Explain how you find the Least Common Denominator of $x^2 + 5x + 4$ and $x^2 - 16$.

Self Check

ⓐ *After completing the exercises, use this checklist to evaluate your mastery of the objectives of this section.*

I can...	Confidently	With some help	No-I don't get it!
add rational expressions with a common denominator.			
subtract rational expressions with a common denominator.			
add and subtract rational expressions whose denominators are opposites.			

ⓑ *What does this checklist tell you about your mastery of this section? What steps will you take to improve?*

8.4 ## Add and Subtract Rational Expressions with Unlike Denominators

Learning Objectives

By the end of this section, you will be able to:

> Find the least common denominator of rational expressions
> Find equivalent rational expressions
> Add rational expressions with different denominators
> Subtract rational expressions with different denominators

Be Prepared!

Before you get started, take this readiness quiz.

If you miss a problem, go back to the section listed and review the material.

1. Add: $\frac{7}{10} + \frac{8}{15}$.
   If you missed this problem, review **Example 1.81**.

2. Subtract: $6(2x + 1) - 4(x - 5)$.
   If you missed this problem, review **Example 1.139**.

3. Find the Greatest Common Factor of $9x^2 y^3$ and $12xy^5$.
   If you missed this problem, review **Example 7.3**.

4. Factor completely $-48n - 12$.
   If you missed this problem, review **Example 7.11**.

Find the Least Common Denominator of Rational Expressions

When we add or subtract rational expressions with unlike denominators we will need to get common denominators. If we review the procedure we used with numerical fractions, we will know what to do with rational expressions.

Let's look at the example $\frac{7}{12} + \frac{5}{18}$ from **Foundations**. Since the denominators are not the same, the first step was to find the least common denominator (LCD). Remember, the LCD is the least common multiple of the denominators. It is the smallest number we can use as a common denominator.

To find the LCD of 12 and 18, we factored each number into primes, lining up any common primes in columns. Then we "brought down" one prime from each column. Finally, we multiplied the factors to find the LCD.

$$12 = 2 \cdot 2 \cdot 3$$
$$\underline{18 = 2 \cdot 3 \cdot 3}$$
$$LCD = 2 \cdot 2 \cdot 3 \cdot 3$$
$$LCD = 36$$

We do the same thing for rational expressions. However, we leave the LCD in factored form.

 HOW TO :: FIND THE LEAST COMMON DENOMINATOR OF RATIONAL EXPRESSIONS.

Step 1.   Factor each expression completely.

Step 2.   List the factors of each expression. Match factors vertically when possible.

Step 3.   Bring down the columns.

Step 4.   Multiply the factors.

Remember, we always exclude values that would make the denominator zero. What values of x should we exclude in this next example?

EXAMPLE 8.38

Find the LCD for $\dfrac{8}{x^2-2x-3}$, $\dfrac{3x}{x^2+4x+3}$.

✓ Solution

Find the LCD for $\dfrac{8}{x^2-2x-3}$, $\dfrac{3x}{x^2+4x+3}$.

Factor each expression completely, lining up common factors. Bring down the columns.	$x^2-2x-3=(x+1)(x-2)$ $\dfrac{x^2+4x+3=(x+1)\quad(x+3)}{\text{LCD}=(x+1)(x-2)(x+3)}$
Multiply the factors.	The LCD is $(x+1)(x-3)(x+3)$.

> **TRY IT :: 8.75**   Find the LCD for $\dfrac{2}{x^2-x-12}$, $\dfrac{1}{x^2-16}$.

> **TRY IT :: 8.76**   Find the LCD for $\dfrac{x}{x^2+8x+15}$, $\dfrac{5}{x^2+9x+18}$.

Find Equivalent Rational Expressions

When we add numerical fractions, once we find the LCD, we rewrite each fraction as an equivalent fraction with the LCD.

$\dfrac{7}{12}+\dfrac{5}{18}$ $12=2\cdot2\cdot3$
$18=2\cdot\quad3\cdot3$
$\overline{\text{LCD}=2\cdot2\cdot3\cdot3}$
LCD$=36$

$\dfrac{7\cdot3}{12\cdot3}+\dfrac{5\cdot2}{18\cdot2}$

$\dfrac{21}{36}+\dfrac{10}{36}$

We will do the same thing for rational expressions.

EXAMPLE 8.39

Rewrite as equivalent rational expressions with denominator $(x+1)(x-3)(x+3)$: $\dfrac{8}{x^2-2x-3}$, $\dfrac{3x}{x^2+4x+3}$.

✓ Solution

	$\dfrac{8}{x^2-2x-3}$, $\dfrac{3x}{x^2+4x+3}$
Factor each denominator.	$\dfrac{8}{(x+1)(x-3)}$, $\dfrac{3x}{(x+1)(x+3)}$
Find the LCD.	$x^2-2x-3=(x+1)(x-3)\bigcirc$ $x^2+4x+3=(x+1)\bigcirc(x+3)$ $\text{LCD}=(x+1)(x-3)(x+3)$
Multiply each denominator by the 'missing' factor and multiply each numerator by the same factor.	$\dfrac{8(x+3)}{(x+1)(x-3)(x+3)}$, $\dfrac{3x(x-3)}{(x+1)(x+3)(x-3)}$
Simplify the numerators.	$\dfrac{8x+24}{(x+1)(x-3)(x+3)}$, $\dfrac{3x^2-9x}{(x+1)(x+3)(x-3)}$

> **TRY IT :: 8.77**   Rewrite as equivalent rational expressions with denominator $(x+3)(x-4)(x+4)$:

$$\frac{2}{x^2 - x - 12}, \frac{1}{x^2 - 16}.$$

> **TRY IT :: 8.78**   Rewrite as equivalent rational expressions with denominator $(x+3)(x+5)(x+6)$:

$$\frac{x}{x^2 + 8x + 15}, \frac{5}{x^2 + 9x + 18}.$$

Add Rational Expressions with Different Denominators

Now we have all the steps we need to add rational expressions with different denominators. As we have done previously, we will do one example of adding numerical fractions first.

EXAMPLE 8.40

Add: $\dfrac{7}{12} + \dfrac{5}{18}$.

⊘ Solution

$$\frac{7}{12} + \frac{5}{18}$$

Find the LCD of 12 and 18.	$\begin{aligned} 12 &= 2 \cdot 2 \cdot 3 \\ 18 &= 2 \cdot 3 \cdot 3 \\ \hline \text{LCD} &= 2 \cdot 2 \cdot 3 \cdot 3 \\ \text{LCD} &= 36 \end{aligned}$
Rewrite each fraction as an equivalent fraction with the LCD.	$\dfrac{7 \cdot 3}{12 \cdot 3} + \dfrac{5 \cdot 2}{18 \cdot 2}$
Add the fractions.	$\dfrac{21}{36} + \dfrac{10}{36}$
The fraction cannot be simplified.	$\dfrac{31}{36}$

> **TRY IT :: 8.79**   Add: $\dfrac{11}{30} + \dfrac{7}{12}$.

> **TRY IT :: 8.80**   Add: $\dfrac{3}{8} + \dfrac{9}{20}$.

Now we will add rational expressions whose denominators are monomials.

EXAMPLE 8.41

Add: $\dfrac{5}{12x^2 y} + \dfrac{4}{21xy^2}$.

✓ **Solution**

$$\frac{5}{12x^2y} + \frac{4}{21xy^2}$$

Find the LCD of $12x^2y$ and $21xy^2$.

$$\begin{aligned} 12x^2y &= 2 \cdot 2 \cdot 3 \cdot \quad x \cdot x \cdot y \\ 21xy^2 &= \qquad\quad 3 \cdot 7 \cdot x \cdot \quad y \cdot y \\ \hline LCD &= 2 \cdot 2 \cdot 3 \cdot 7 \cdot x \cdot x \cdot y \cdot y \\ LCD &= 84x^2y^2 \end{aligned}$$

$$\frac{5}{12x^2y} + \frac{4}{21xy^2}$$

Rewrite each rational expression as an equivalent fraction with the LCD.

$$\frac{5 \cdot 7y}{12x^2y \cdot 7y} + \frac{4 \cdot 4x}{21xy^2 \cdot 4x}$$

Simplify.

$$\frac{35y}{84x^2y^2} + \frac{16x}{84x^2y^2}$$

Add the rational expressions.

$$\frac{16x + 35y}{84x^2y^2}$$

There are no factors common to the numerator and denominator. The fraction cannot be simplified.

> **TRY IT :: 8.81**   Add: $\dfrac{2}{15a^2b} + \dfrac{5}{6ab^2}$.

> **TRY IT :: 8.82**   Add: $\dfrac{5}{16c} + \dfrac{3}{8cd^2}$.

Now we are ready to tackle polynomial denominators.

EXAMPLE 8.42   HOW TO ADD RATIONAL EXPRESSIONS WITH DIFFERENT DENOMINATORS

Add: $\dfrac{3}{x-3} + \dfrac{2}{x-2}$.

✓ **Solution**

Step 1. Determine if the expressions have a common denominator.
• Yes – Go to step 2.
• No – Rewrite each rational expression with the LCD.
  ◦ Find the LCD.
  ◦ Rewrite each rational expression as an equivalent rational expression with the LCD.
	No
	
	Find the LCD of $(x-3), (x-2)$
		$\begin{aligned} x-3 &: (x-3) \\ x-2 &: \qquad (x-2) \\ \hline LCD &: (x-3)(x-2) \end{aligned}$
		
		$\dfrac{3}{x-3} + \dfrac{2}{x-2}$

	Change into equivalent rational expressions with the LCD, $(x-3)(x-2)$.	$\dfrac{3(x-2)}{(x-3)(x-2)} + \dfrac{2(x-3)}{(x-2)(x-3)}$
	Keep the denominators factored!	$\dfrac{3x-6}{(x-3)(x-2)} + \dfrac{2x-6}{(x-2)(x-3)}$

Step 2. Add the rational expressions.	Add the numerators and place the sum over the common denominator.	$\dfrac{3x - 6 + 2x - 6}{(x-3)(x-2)}$ $\dfrac{5x - 12}{(x-3)(x-2)}$
Step 3. Simplify, if possible.	Because $5x - 12$ cannot be factored, the answer is simplified.	

> **TRY IT : :** 8.83   Add: $\dfrac{2}{x-2} + \dfrac{5}{x+3}$.

> **TRY IT : :** 8.84   Add: $\dfrac{4}{m+3} + \dfrac{3}{m+4}$.

The steps to use to add rational expressions are summarized in the following procedure box.

HOW TO : : ADD RATIONAL EXPRESSIONS.

Step 1.   Determine if the expressions have a common denominator.
Yes – go to step 2.
No – Rewrite each rational expression with the LCD.
Find the LCD.
Rewrite each rational expression as an equivalent rational expression with the LCD.

Step 2.   Add the rational expressions.

Step 3.   Simplify, if possible.

EXAMPLE 8.43

Add: $\dfrac{2a}{2ab + b^2} + \dfrac{3a}{4a^2 - b^2}$.

⊘ **Solution**

$$\frac{2a}{2ab + b^2} + \frac{3a}{4a^2 - b^2}$$

Do the expressions have a common denominator? No. Rewrite each expression with the LCD.	

Find the LCD.

$$\begin{aligned} 2ab + b^2 &= b(2a + b) \\ 4a^2 - b^2 &= (2a + b)(2a - b) \\ \hline \text{LCD} &= b(2a + b)(2a - b) \end{aligned}$$

Rewrite each rational expression as an equivalent rational expression with the LCD.

$$\frac{2a(2a - b)}{b(2a + b)(2a - b)} + \frac{3a \cdot b}{(2a + b)(2a - b) \cdot b}$$

Simplify the numerators.

$$\frac{4a^2 - 2ab}{b(2a + b)(2a - b)} + \frac{3ab}{b(2a + b)(2a - b)}$$

Add the rational expressions.

$$\frac{4a^2 - 2ab + 3ab}{b(2a + b)(2a - b)}$$

Simplify the numerator.

$$\frac{4a^2 + ab}{b(2a + b)(2a - b)}$$

Factor the numerator.

$$\frac{a(4a + b)}{b(2a + b)(2a - b)}$$

There are no factors common to the numerator and denominator. The fraction cannot be simplified.

> **TRY IT :: 8.85**   Add: $\dfrac{5x}{xy - y^2} + \dfrac{2x}{x^2 + y^2}$.

> **TRY IT :: 8.86**   Add: $\dfrac{7}{2m + 6} + \dfrac{4}{m^2 + 4m + 3}$.

Avoid the temptation to simplify too soon! In the example above, we must leave the first rational expression as $\dfrac{2a(2a - b)}{b(2a + b)(2a - b)}$ to be able to add it to $\dfrac{3a \cdot b}{(2a + b)(2a - b) \cdot b}$. Simplify only after you have combined the numerators.

EXAMPLE 8.44

Add: $\dfrac{8}{x^2 - 2x - 3} + \dfrac{3x}{x^2 + 4x + 3}$.

⊘ **Solution**

$$\frac{8}{x^2 - 2x - 3} + \frac{3x}{x^2 + 4x + 3}$$

Do the expressions have a common denominator? No. Rewrite each expression with the LCD.	

Find the LCD.

$$\begin{aligned} x^2 - 2x - 3 &= (x + 1)(x - 3) \\ x^2 + 4x + 3 &= (x + 1) \qquad (x + 3) \\ \hline \text{LCD} &= (x + 1)(x - 3)(x + 3) \end{aligned}$$

Rewrite each rational expression as an equivalent fraction with the LCD.

$$\frac{8(x + 3)}{(x + 1)(x - 3)(x + 3)} + \frac{3x(x - 3)}{(x + 1)(x + 3)(x - 3)}$$

Simplify the numerators.	$\dfrac{8x+24}{(x+1)(x-3)(x+3)} + \dfrac{3x^2-9x}{(x+1)(x+3)(x-3)}$
Add the rational expressions.	$\dfrac{8x+24+3x^2+9x}{(x+1)(x-3)(x+3)}$
Simplify the numerator.	$\dfrac{3x^2-x^2+24}{(x+1)(x-3)(x+3)}$

The numerator is prime, so there are no common factors.

> **TRY IT :: 8.87** Add: $\dfrac{1}{m^2-m-2} + \dfrac{5m}{m^2+3m+2}$.

> **TRY IT :: 8.88** Add: $\dfrac{2n}{n^2-3n-10} + \dfrac{6}{n^2+5n+6}$.

Subtract Rational Expressions with Different Denominators

The process we use to subtract rational expressions with different denominators is the same as for addition. We just have to be very careful of the signs when subtracting the numerators.

EXAMPLE 8.45   HOW TO SUBTRACT RATIONAL EXPRESSIONS WITH DIFFERENT DENOMINATORS

Subtract: $\dfrac{x}{x-3} - \dfrac{x-2}{x+3}$.

⊘ **Solution**

Step 1. Determine if the expressions have a common denominator. • Yes – Go to step 2. • No – Rewrite each rational expression with the LCD. ○ Find the LCD. ○ Rewrite each rational expression as an equivalent rational expression with the LCD.	No Find the LCD of $(x-3)$, $(x+3)$. Change into equivalent fractions with the LCD, $(x-3)(x+3)$. Keep the denominators factored!	$x-3:\ (x-3)$ $x+3:\ \qquad (x+3)$ $\overline{\text{LCD: }(x-3)(x+3)}$ $\dfrac{x}{x-3} - \dfrac{x-2}{x+3}$ $\dfrac{x(x+3)}{(x-3)(x+3)} - \dfrac{(x-2)(x-3)}{(x+3)(x-3)}$ $\dfrac{x^2+3x}{(x-3)(x+3)} - \dfrac{x^2-5x+6}{(x-3)(x+3)}$
Step 2. Subtract the rational expressions.	Subtract the numerators and place the difference over the common denominator. Be careful with the signs!	$\dfrac{x^2+3x-(x^2-5x+6)}{(x-3)(x+3)}$ $\dfrac{x^2+3x-x^2+5x-6}{(x-3)(x+3)}$ $\dfrac{8x-6}{(x-3)(x+3)}$
Step 3. Simplify, if possible.	The numerator and denominator have no factors in common. The answer is simplified.	$\dfrac{2(4x-3)}{(x-3)(x+3)}$

> **TRY IT :: 8.89** Subtract: $\dfrac{y}{y+4} - \dfrac{y-2}{y-5}$.

> **TRY IT :: 8.90**   Subtract: $\dfrac{z+3}{z+2} - \dfrac{z}{z+3}$.

The steps to take to subtract rational expressions are listed below.

HOW TO :: SUBTRACT RATIONAL EXPRESSIONS.

Step 1.   Determine if they have a common denominator.
           Yes – go to step 2.
           No – Rewrite each rational expression with the LCD.
           Find the LCD.
           Rewrite each rational expression as an equivalent rational expression with the LCD.

Step 2.   Subtract the rational expressions.

Step 3.   Simplify, if possible.

EXAMPLE 8.46

Subtract: $\dfrac{8y}{y^2 - 16} - \dfrac{4}{y - 4}$.

 **Solution**

	$\dfrac{8y}{y^2 - 16} - \dfrac{4}{y - 4}$
Do the expressions have a common denominator? No. Rewrite each expression with the LCD.	
Find the LCD. $\begin{aligned} y^2 - 16 &= (y-4)(y+4) \\ y - 4\ &= y - 4 \\ \hline \text{LCD}\ &= (y-4)(y+4) \end{aligned}$	
Rewrite each rational expression as an equivalent rational expression with the LCD.	$\dfrac{8y}{(y-4)(y+4)} - \dfrac{4(y+4)}{(y-4)(y+4)}$
Simplify the numerators.	$\dfrac{8y}{(y-4)(y+4)} - \dfrac{4y+16}{(y-4)(y+4)}$
Subtract the rational expressions.	$\dfrac{8y - 4y - 16}{(y-4)(y+4)}$
Simplify the numerators.	$\dfrac{4y - 16}{(y-4)(y+4)}$
Factor the numerator to look for common factors.	$\dfrac{4(y-4)}{(y-4)(y+4)}$
Remove common factors.	$\dfrac{4\cancel{(y-4)}}{\cancel{(y-4)}(y+4)}$
Simplify.	$\dfrac{4}{(y+4)}$

> **TRY IT :: 8.91**   Subtract: $\dfrac{2x}{x^2 - 4} - \dfrac{1}{x + 2}$.

> **TRY IT ::** 8.92 Subtract: $\dfrac{3}{z+3} - \dfrac{6z}{z^2-9}$.

There are lots of negative signs in the next example. Be extra careful!

EXAMPLE 8.47

Subtract: $\dfrac{-3n-9}{n^2+n-6} - \dfrac{n+3}{2-n}$.

✓ **Solution**

	$\dfrac{-3n-9}{n^2+n-6} - \dfrac{n+3}{2-n}$
Factor the denominator.	$\dfrac{-3n-9}{(n-2)(n+3)} - \dfrac{n+3}{2-n}$
Since $n-2$ and $2-n$ are opposites, we will mutliply the second rational expression by $\dfrac{-1}{-1}$.	$\dfrac{-3n-9}{(n-2)(n+3)} - \dfrac{(-1)(n+3)}{(-1)(2-n)}$
Simplify.	$\dfrac{-3n-9}{(n-2)(n+3)} + \dfrac{(n+3)}{(n-2)}$
Do the expressions have a common denominator? No.	
Find the LCD.	$n^2+n-6 = (n-2)(n+3)$ $\underline{\quad n-2 = (n-2) \qquad\qquad}$ $LCD = (n-2)(n+3)$
Rewrite each rational expression as an equivalent rational expression with the LCD.	$\dfrac{-3n-9}{(n-2)(n+3)} + \dfrac{(n+3)(n+3)}{(n-2)(n+3)}$
Simplify the numerators.	$\dfrac{-3n-9}{(n-2)(n+3)} + \dfrac{n^2+6n+9}{(n-2)(n+3)}$
Simplify the rational expressions.	$\dfrac{-3n-9+n^2+6n+9}{(n-2)(n+3)}$
Somplify the numerator.	$\dfrac{n^2+3n}{(n-2)(n+3)}$
Factor the numerator to look for common factors.	$\dfrac{n\cancel{(n+3)}}{(n-2)\cancel{(n+3)}}$
Simplify.	$\dfrac{n}{(n-2)}$

> **TRY IT ::** 8.93 Subtract: $\dfrac{3x-1}{x^2-5x-6} - \dfrac{2}{6-x}$.

> **TRY IT ::** 8.94 Subtract: $\dfrac{-2y-2}{y^2+2y-8} - \dfrac{y-1}{2-y}$.

When one expression is not in fraction form, we can write it as a fraction with denominator 1.

EXAMPLE 8.48

Subtract: $\dfrac{5c+4}{c-2} - 3$.

⊘ **Solution**

	$\dfrac{5c+4}{c-2} - 3$
Write 3 as $\frac{3}{1}$ to have 2 rational expressions.	$\dfrac{5c+4}{c-2} - \dfrac{3}{1}$
Do the rational expressions have a common denominator? No.	
Find the LCD of $c-2$ and 1. LCD = $c-2$.	
Rewrite $\frac{3}{1}$ as an equivalent rational expression with the LCD.	$\dfrac{5c+4}{c-2} - \dfrac{3(c-2)}{1(c-2)}$
Simplify.	$\dfrac{5c+4}{c-2} - \dfrac{3c-6}{c-2}$
Subtract the rational expressions.	$\dfrac{5c+4-(3c-6)}{c-2}$
Simplify.	$\dfrac{2c+10}{c-2}$
Factor to check for common factors.	$\dfrac{2(c+5)}{c-2}$
There are no common factors; the rational expression is simplified.	

> **TRY IT :: 8.95**   Subtract: $\dfrac{2x+1}{x-7} - 3$.

> **TRY IT :: 8.96**   Subtract: $\dfrac{4y+3}{2y-1} - 5$.

HOW TO :: ADD OR SUBTRACT RATIONAL EXPRESSIONS.

Step 1.   Determine if the expressions have a common denominator.
   Yes – go to step 2.
   No – Rewrite each rational expression with the LCD.
   Find the LCD.
   Rewrite each rational expression as an equivalent rational expression with the LCD.

Step 2.   Add or subtract the rational expressions.

Step 3.   Simplify, if possible.

We follow the same steps as before to find the LCD when we have more than two rational expressions. In the next example we will start by factoring all three denominators to find their LCD.

EXAMPLE 8.49

Simplify: $\dfrac{2u}{u-1} + \dfrac{1}{u} - \dfrac{2u-1}{u^2-u}$.

⊘ Solution

$$\frac{2u}{u-1} + \frac{1}{u} - \frac{2u-1}{u^2-u}$$

Do the rational expressions have a common denominator? No.

$$\begin{aligned} u - 1 &= u - 1 \\ u &= u \\ u^2 - u &= u(u-1) \end{aligned}$$

Find the LCD. $$\text{LCD} = u(u-1)$$

Rewrite each rational expression as an equivalent rational expression with the LCD.

$$\frac{2u \cdot u}{(u-1)u} + \frac{1 \cdot (u-1)}{u \cdot (u-1)} - \frac{2u-1}{u(u-1)}$$

$$\frac{2u^2}{(u-1)u} + \frac{u-1}{u \cdot (u-1)} - \frac{2u-1}{u(u-1)}$$

Write as one rational expression.

$$\frac{2u^2 + u - 1 - 2u + 1}{u(u-1)}$$

Simplify.

$$\frac{2u^2 - u}{u(u-1)}$$

Factor the numerator, and remove common factors.

$$\frac{\cancel{u}(2u-1)}{\cancel{u}(u-1)}$$

Simplify.

$$\frac{2u-1}{u-1}$$

> **TRY IT :: 8.97**   Simplify: $\dfrac{v}{v+1} + \dfrac{3}{v-1} - \dfrac{6}{v^2-1}$.

> **TRY IT :: 8.98**   Simplify: $\dfrac{3w}{w+2} + \dfrac{2}{w+7} - \dfrac{17w+4}{w^2+9w+14}$.

8.4 EXERCISES

Practice Makes Perfect

In the following exercises, find the LCD.

169. $\dfrac{5}{x^2 - 2x - 8}, \dfrac{2x}{x^2 - x - 12}$

170. $\dfrac{8}{y^2 + 12y + 35}, \dfrac{3y}{y^2 + y - 42}$

171. $\dfrac{9}{z^2 + 2z - 8}, \dfrac{4z}{z^2 - 4}$

172. $\dfrac{6}{a^2 + 14a + 45}, \dfrac{5a}{a^2 - 81}$

173. $\dfrac{4}{b^2 + 6b + 9}, \dfrac{2b}{b^2 - 2b - 15}$

174. $\dfrac{5}{c^2 - 4c + 4}, \dfrac{3c}{c^2 - 10c + 16}$

175.

$\dfrac{2}{3d^2 + 14d - 5}, \dfrac{5d}{3d^2 - 19d + 6}$

176.

$\dfrac{3}{5m^2 - 3m - 2}, \dfrac{6m}{5m^2 + 17m + 6}$

In the following exercises, write as equivalent rational expressions with the given LCD.

177. $\dfrac{5}{x^2 - 2x - 8}, \dfrac{2x}{x^2 - x - 12}$
LCD $(x - 4)(x + 2)(x + 3)$

178. $\dfrac{8}{y^2 + 12y + 35}, \dfrac{3y}{y^2 + y - 42}$
LCD $(y + 7)(y + 5)(y - 6)$

179. $\dfrac{9}{z^2 + 2z - 8}, \dfrac{4z}{z^2 - 4}$
LCD $(z - 2)(z + 4)(z + 2)$

180. $\dfrac{6}{a^2 + 14a + 45}, \dfrac{5a}{a^2 - 81}$
LCD $(a + 9)(a + 5)(a - 9)$

181. $\dfrac{4}{b^2 + 6b + 9}, \dfrac{2b}{b^2 - 2b - 15}$
LCD $(b + 3)(b + 3)(b - 5)$

182. $\dfrac{5}{c^2 - 4c + 4}, \dfrac{3c}{c^2 - 10c + 10}$
LCD $(c - 2)(c - 2)(c - 8)$

183.

$\dfrac{2}{3d^2 + 14d - 5}, \dfrac{5d}{3d^2 - 19d + 6}$
LCD $(3d - 1)(d + 5)(d - 6)$

184.

$\dfrac{3}{5m^2 - 3m - 2}, \dfrac{6m}{5m^2 + 17m + 6}$
LCD $(5m + 2)(m - 1)(m + 3)$

In the following exercises, add.

185. $\dfrac{5}{24} + \dfrac{11}{36}$

186. $\dfrac{7}{30} + \dfrac{13}{45}$

187. $\dfrac{9}{20} + \dfrac{11}{30}$

188. $\dfrac{8}{27} + \dfrac{7}{18}$

189. $\dfrac{7}{10x^2 y} + \dfrac{4}{15xy^2}$

190. $\dfrac{1}{12a^3 b^2} + \dfrac{5}{9a^2 b^3}$

191. $\dfrac{1}{2m} + \dfrac{7}{8m^2 n}$

192. $\dfrac{5}{6p^2 q} + \dfrac{1}{4p}$

193. $\dfrac{3}{r + 4} + \dfrac{2}{r - 5}$

194. $\dfrac{4}{s - 7} + \dfrac{5}{s + 3}$

195. $\dfrac{8}{t + 5} + \dfrac{6}{t - 5}$

196. $\dfrac{7}{v + 5} + \dfrac{9}{v - 5}$

197. $\dfrac{5}{3w - 2} + \dfrac{2}{w + 1}$

198. $\dfrac{4}{2x + 5} + \dfrac{2}{x - 1}$

199. $\dfrac{2y}{y + 3} + \dfrac{3}{y - 1}$

200. $\dfrac{3z}{z - 2} + \dfrac{1}{z + 5}$

201. $\dfrac{5b}{a^2 b - 2a^2} + \dfrac{2b}{b^2 - 4}$

202. $\dfrac{4}{cd + 3c} + \dfrac{1}{d^2 - 9}$

203. $\dfrac{2m}{3m-3} + \dfrac{5m}{m^2+3m-4}$

204. $\dfrac{3}{4n+4} + \dfrac{6}{n^2-n-2}$

205.

$\dfrac{3}{n^2+3n-18} + \dfrac{4n}{n^2+8n+12}$

206.

$\dfrac{6}{q^2-3q-10} + \dfrac{5q}{q^2-8q+15}$

207. $\dfrac{3r}{r^2+7r+6} + \dfrac{9}{r^2+4r+3}$

208. $\dfrac{2s}{s^2+2s-8} + \dfrac{4}{s^2+3s-10}$

In the following exercises, subtract.

209. $\dfrac{t}{t-6} - \dfrac{t-2}{t+6}$

210. $\dfrac{v}{v-3} - \dfrac{v-6}{v+1}$

211. $\dfrac{w+2}{w+4} - \dfrac{w}{w-2}$

212. $\dfrac{x-3}{x+6} - \dfrac{x}{x+3}$

213. $\dfrac{y-4}{y+1} - \dfrac{1}{y+7}$

214. $\dfrac{z+8}{z-3} - \dfrac{z}{z-2}$

215. $\dfrac{5a}{a+3} - \dfrac{a+2}{a+6}$

216. $\dfrac{3b}{b-2} - \dfrac{b-6}{b-8}$

217. $\dfrac{6c}{c^2-25} - \dfrac{3}{c+5}$

218. $\dfrac{4d}{d^2-81} - \dfrac{2}{d+9}$

219. $\dfrac{6}{m+6} - \dfrac{12m}{m^2-36}$

220. $\dfrac{4}{n+4} - \dfrac{8n}{n^2-16}$

221. $\dfrac{-9p-17}{p^2-4p-21} - \dfrac{p+1}{7-p}$

222. $\dfrac{7q+8}{q^2-2q-24} - \dfrac{q+2}{4-q}$

223. $\dfrac{-2r-16}{r^2+6r-16} - \dfrac{5}{2-r}$

224. $\dfrac{2t-30}{t^2+6t-27} - \dfrac{2}{3-t}$

225. $\dfrac{5v-2}{v+3} - 4$

226. $\dfrac{6w+5}{w-1} + 2$

227. $\dfrac{2x+7}{10x-1} + 3$

228. $\dfrac{8y-4}{5y+2} - 6$

In the following exercises, add and subtract.

229. $\dfrac{5a}{a-2} + \dfrac{9}{a} - \dfrac{2a+18}{a^2-2a}$

230. $\dfrac{2b}{b-5} + \dfrac{3}{2b} - \dfrac{2b-15}{2b^2-10b}$

231. $\dfrac{c}{c+2} + \dfrac{5}{c-2} - \dfrac{11c}{c^2-4}$

232.

$\dfrac{6d}{d-5} + \dfrac{1}{d+4} - \dfrac{7d-5}{d^2-d-20}$

In the following exercises, simplify.

233. $\dfrac{6a}{3ab+b^2} + \dfrac{3a}{9a^2-b^2}$

234. $\dfrac{2c}{2c+10} + \dfrac{7c}{c^2+9c+20}$

235. $\dfrac{6d}{d^2-64} - \dfrac{3}{d-8}$

236. $\dfrac{5}{n+7} - \dfrac{10n}{n^2-49}$

237.

$\dfrac{4m}{m^2+6m-7} + \dfrac{2}{m^2+10m+21}$

238.

$\dfrac{3p}{p^2+4p-12} + \dfrac{1}{p^2+p-30}$

239. $\dfrac{-5n-5}{n^2+n-6} + \dfrac{n+1}{2-n}$

240. $\dfrac{-4b-24}{b^2+b-30} + \dfrac{b+7}{5-b}$

241. $\dfrac{7}{15p} + \dfrac{5}{18pq}$

242. $\dfrac{3}{20a^2} + \dfrac{11}{12ab^2}$

243. $\dfrac{4}{x-2} + \dfrac{3}{x+5}$

244. $\dfrac{6}{m+4} + \dfrac{9}{m-8}$

245. $\dfrac{2q+7}{y+4} - 2$

246. $\dfrac{3y-1}{y+4} - 2$

247. $\dfrac{z+2}{z-5} - \dfrac{z}{z+1}$

248. $\dfrac{t}{t-5} - \dfrac{t-1}{t+5}$

249. $\dfrac{3d}{d+2} + \dfrac{4}{d} - \dfrac{d+8}{d^2+2d}$

250.

$$\dfrac{2q}{q+5} + \dfrac{3}{q-3} - \dfrac{13q+15}{q^2+2q-15}$$

Everyday Math

251. Decorating cupcakes Victoria can decorate an order of cupcakes for a wedding in t hours, so in 1 hour she can decorate $\dfrac{1}{t}$ of the cupcakes. It would take her sister 3 hours longer to decorate the same order of cupcakes, so in 1 hour she can decorate $\dfrac{1}{t+3}$ of the cupcakes.

ⓐ Find the fraction of the decorating job that Victoria and her sister, working together, would complete in one hour by adding the rational expressions $\dfrac{1}{t} + \dfrac{1}{t+3}$.

ⓑ Evaluate your answer to part (a) when $t=5$.

252. Kayaking When Trina kayaks upriver, it takes her $\dfrac{5}{3-c}$ hours to go 5 miles, where c is the speed of the river current. It takes her $\dfrac{5}{3+c}$ hours to kayak 5 miles down the river.

ⓐ Find an expression for the number of hours it would take Trina to kayak 5 miles up the river and then return by adding $\dfrac{5}{3-c} + \dfrac{5}{3+c}$.

ⓑ Evaluate your answer to part (a) when $c=1$ to find the number of hours it would take Trina if the speed of the river current is 1 mile per hour.

Writing Exercises

253. Felipe thinks $\dfrac{1}{x} + \dfrac{1}{y}$ is $\dfrac{2}{x+y}$.

ⓐ Choose numerical values for x and y and evaluate $\dfrac{1}{x} + \dfrac{1}{y}$.

ⓑ Evaluate $\dfrac{2}{x+y}$ for the same values of x and y you used in part (a).

ⓒ Explain why Felipe is wrong.

ⓓ Find the correct expression for $\dfrac{1}{x} + \dfrac{1}{y}$.

254. Simplify the expression $\dfrac{4}{n^2+6n+9} - \dfrac{1}{n^2-9}$ and explain all your steps.

Self Check

ⓐ *After completing the exercises, use this checklist to evaluate your mastery of the objectives of this section.*

I can...	Confidently	With some help	No-I don't get it!
find the least common denominator of rational expressions.			
find equivalent rational expressions.			
add rational expressions with different denominators.			
subtract rational expressions with different denominators.			

ⓑ *On a scale of 1-10, how would you rate your mastery of this section in light of your responses on the checklist? How can you improve this?*

8.5 **Simplify Complex Rational Expressions**

Learning Objectives

By the end of this section, you will be able to:

> Simplify a complex rational expression by writing it as division
> Simplify a complex rational expression by using the LCD

Be Prepared!

Before you get started, take this readiness quiz.

If you miss a problem, go back to the section listed and review the material.

1. Simplify: $\dfrac{\frac{3}{5}}{\frac{9}{10}}$.

   If you missed this problem, review **Example 1.72**.

2. Simplify: $\dfrac{1 - \frac{1}{3}}{4^2 + 4 \cdot 5}$.

   If you missed this problem, review **Example 1.74**.

Complex fractions are fractions in which the numerator or denominator contains a fraction. In Chapter 1 we simplified complex fractions like these:

$$\frac{\frac{3}{4}}{\frac{5}{8}} \qquad \frac{\frac{x}{2}}{\frac{xy}{6}}$$

In this section we will simplify *complex rational expressions*, which are rational expressions with rational expressions in the numerator or denominator.

Complex Rational Expression

A **complex rational expression** is a rational expression in which the numerator or denominator contains a rational expression.

Here are a few complex rational expressions:

$$\frac{\frac{4}{y-3}}{\frac{8}{y^2-9}} \qquad \frac{\frac{1}{x}+\frac{1}{y}}{\frac{x}{y}-\frac{y}{x}} \qquad \frac{\frac{2}{x+6}}{\frac{4}{x-6}-\frac{4}{x^2-36}}$$

Remember, we always exclude values that would make any denominator zero.

We will use two methods to simplify complex rational expressions.

Simplify a Complex Rational Expression by Writing it as Division

We have already seen this complex rational expression earlier in this chapter.

$$\frac{\frac{6x^2-7x+2}{4x-8}}{\frac{2x^2-8x+3}{x^2-5x+6}}$$

We noted that fraction bars tell us to divide, so rewrote it as the division problem

$$\left(\frac{6x^2-7x+2}{4x-8}\right) \div \left(\frac{2x^2-8x+3}{x^2-5x+6}\right)$$

Then we multiplied the first rational expression by the reciprocal of the second, just like we do when we divide two fractions.

This is one method to simplify rational expressions. We write it as if we were dividing two fractions.

EXAMPLE 8.50

Simplify: $\dfrac{\frac{4}{y-3}}{\frac{8}{y^2-9}}$.

✓ **Solution**

$$\dfrac{\frac{4}{y-3}}{\frac{8}{y^2-9}}$$

Rewrite the complex fraction as division.

$$\frac{4}{y-3} \div \frac{8}{y^2-9}$$

Rewrite as the product of first times the reciprocal of the second.

$$\frac{4}{y-3} \cdot \frac{y^2-9}{8}$$

Multiply.

$$\frac{4(y^2-9)}{8(y-3)}$$

Factor to look for common factors.

$$\frac{4(y-3)(y+3)}{4 \cdot 2(y-3)}$$

Remove common factors.

$$\frac{\cancel{4}(y-3)(y+3)}{\cancel{4} \cdot 2(y-3)}$$

Simplify.

$$\frac{y+3}{2}$$

Are there any value(s) of y that should not be allowed? The simplified rational expression has just a constant in the denominator. But the original complex rational expression had denominators of $y-3$ and y^2-9. This expression would be undefined if $y=3$ or $y=-3$.

> **TRY IT :: 8.99**

Simplify: $\dfrac{\frac{2}{x^2-1}}{\frac{3}{x+1}}$.

> **TRY IT :: 8.100**

Simplify: $\dfrac{\frac{1}{x^2-7x+12}}{\frac{2}{x-4}}$.

Fraction bars act as grouping symbols. So to follow the Order of Operations, we simplify the numerator and denominator as much as possible before we can do the division.

EXAMPLE 8.51

Simplify: $\dfrac{\frac{1}{3}+\frac{1}{6}}{\frac{1}{2}-\frac{1}{3}}$.

✓ Solution

<table>
<tr><td></td><td>$\dfrac{\dfrac{1}{3}+\dfrac{1}{6}}{\dfrac{1}{2}-\dfrac{1}{3}}$</td></tr>
<tr><td>Simplify the numerator and denominator.</td><td></td></tr>
<tr><td>Find the LCD and add the fractions in the numerator.
Find the LCD and add the fractions in the denominator.</td><td>$\dfrac{\dfrac{1\cdot 2}{3\cdot 2}+\dfrac{1}{6}}{\dfrac{1\cdot 3}{2\cdot 3}-\dfrac{1\cdot 2}{3\cdot 2}}$</td></tr>
<tr><td>Simplify the numerator and denominator.</td><td>$\dfrac{\dfrac{2}{6}+\dfrac{1}{6}}{\dfrac{3}{6}-\dfrac{2}{6}}$</td></tr>
<tr><td>Simplify the numerator and denominator, again.</td><td>$\dfrac{\dfrac{3}{6}}{\dfrac{1}{6}}$</td></tr>
<tr><td>Rewrite the complex rational expression as a division problem.</td><td>$\dfrac{3}{6}\div\dfrac{1}{6}$</td></tr>
<tr><td>Multiply the first times by the reciprocal of the second.</td><td>$\dfrac{3}{6}\cdot\dfrac{6}{1}$</td></tr>
<tr><td>Simplify.</td><td>3</td></tr>
</table>

> **TRY IT :: 8.101**
>
> Simplify: $\dfrac{\dfrac{1}{2}+\dfrac{2}{3}}{\dfrac{5}{6}+\dfrac{1}{12}}$.

> **TRY IT :: 8.102**
>
> Simplify: $\dfrac{\dfrac{3}{4}-\dfrac{1}{3}}{\dfrac{1}{8}+\dfrac{5}{6}}$.

EXAMPLE 8.52 HOW TO SIMPLIFY A COMPLEX RATIONAL EXPRESSION BY WRITING IT AS DIVISION

Simplify: $\dfrac{\dfrac{1}{x}+\dfrac{1}{y}}{\dfrac{x}{y}-\dfrac{y}{x}}$.

✓ **Solution**

Step 1. Simplify the numerator and denominator.	We will simplify the sum in the numerator and difference in the denominator.	$\dfrac{\frac{1}{x}+\frac{1}{y}}{\frac{x}{y}-\frac{y}{x}}$
	Find a common denominator and add the fractions in the numerator.	$\dfrac{\frac{1\cdot y}{x\cdot y}+\frac{1\cdot x}{y\cdot x}}{\frac{x\cdot x}{y\cdot x}-\frac{y\cdot y}{x\cdot y}}$
	Find a common denominator and subtract the fractions in the numerator.	$\dfrac{\frac{y}{xy}+\frac{x}{xy}}{\frac{x^2}{xy}-\frac{y^2}{xy}}$
	We now have just one rational expression in the numerator and one in the denominator.	$\dfrac{\frac{y+x}{xy}}{\frac{x^2-y^2}{xy}}$
Step 2. Rewrite the complex rational expression as a division problem.	We write the numerator divided by the denominator.	$\left(\dfrac{y+x}{xy}\right)\div\left(\dfrac{x^2-y^2}{xy}\right)$
Step 3. Divide the expressions.	Multiply the first by the reciprocal of the second.	$\left(\dfrac{y+x}{xy}\right)\cdot\left(\dfrac{xy}{x^2-y^2}\right)$
	Factor any expressions if possible.	$\dfrac{xy(y+x)}{xy(x-y)(x+y)}$
	Remove common factors.	$\dfrac{\cancel{xy}\cancel{(y+x)}}{\cancel{xy}(x-y)\cancel{(x+y)}}$
	Simplify.	$\dfrac{1}{x-y}$

> **TRY IT ::** 8.103

Simplify: $\dfrac{\frac{1}{x}+\frac{1}{y}}{\frac{1}{x}-\frac{1}{y}}$.

> **TRY IT ::** 8.104

Simplify: $\dfrac{\frac{1}{a}+\frac{1}{b}}{\frac{1}{a^2}-\frac{1}{b^2}}$.

HOW TO :: SIMPLIFY A COMPLEX RATIONAL EXPRESSION BY WRITING IT AS DIVISION.

Step 1.   Simplify the numerator and denominator.

Step 2.   Rewrite the complex rational expression as a division problem.

Step 3.   Divide the expressions.

EXAMPLE 8.53

Simplify: $\dfrac{n-\frac{4n}{n+5}}{\frac{1}{n+5}+\frac{1}{n-5}}$.

⊘ **Solution**

$$\frac{n - \dfrac{4n}{n+5}}{\dfrac{1}{n+5} + \dfrac{1}{n-5}}$$

Simplify the numerator and denominator.

Find the LCD and add the fractions in the numerator.
Find the LCD and add the fractions in the denominator.

$$\frac{\dfrac{n(n+5)}{1(n+5)} - \dfrac{4n}{n+5}}{\dfrac{1(n-5)}{(n+5)(n-5)} + \dfrac{1(n+5)}{(n-5)(n+5)}}$$

Simplify the numerators.

$$\frac{\dfrac{n^2+5n}{n+5} - \dfrac{4n}{n+5}}{\dfrac{n-5}{(n+5)(n-5)} + \dfrac{n+5}{(n-5)(n+5)}}$$

Subtract the rational expressions in the numerator and add in the denominator.

Simplify.

$$\frac{\dfrac{n^2+n}{n+5}}{\dfrac{2n}{(n+5)(n-5)}}$$

Rewrite as fraction division.

$$\frac{n^2+n}{n+5} \div \frac{2n}{(n+5)(n-5)}$$

Multiply the first times the reciprocal of the second.

$$\frac{n^2+n}{n+5} \cdot \frac{(n+5)(n-5)}{2n}$$

Factor any expressions if possible.

$$\frac{n(n+1)(n+5)(n-5)}{(n+5)2n}$$

Remove common factors.

$$\frac{\cancel{n}(n+1)\cancel{(n+5)}(n-5)}{\cancel{(n+5)}2\cancel{n}}$$

Simplify.

$$\frac{(n+1)(n-5)}{2}$$

> **TRY IT : :** 8.105

Simplify: $\dfrac{b - \dfrac{3b}{b+5}}{\dfrac{2}{b+5} + \dfrac{1}{b-5}}$.

> **TRY IT : :** 8.106

Simplify: $\dfrac{1 - \dfrac{3}{c+4}}{\dfrac{1}{c+4} + \dfrac{c}{3}}$.

Simplify a Complex Rational Expression by Using the LCD

We "cleared" the fractions by multiplying by the LCD when we solved equations with fractions. We can use that strategy here to simplify complex rational expressions. We will multiply the numerator and denominator by LCD of all the rational expressions.

Let's look at the complex rational expression we simplified one way in **Example 8.51**. We will simplify it here by multiplying the numerator and denominator by the LCD. When we multiply by $\dfrac{\text{LCD}}{\text{LCD}}$ we are multiplying by 1, so the value stays the same.

EXAMPLE 8.54

Simplify: $\dfrac{\frac{1}{3}+\frac{1}{6}}{\frac{1}{2}-\frac{1}{3}}$.

✓ **Solution**

	$\dfrac{\frac{1}{3}+\frac{1}{6}}{\frac{1}{2}-\frac{1}{3}}$
The LCD of all the fractions in the whole expression is 6.	
Clear the fractions by multiplying the numerator and denominator by that LCD.	$\dfrac{6\cdot\left(\frac{1}{3}+\frac{1}{6}\right)}{6\cdot\left(\frac{1}{2}-\frac{1}{3}\right)}$
Distribute.	$\dfrac{6\cdot\frac{1}{3}+6\cdot\frac{1}{6}}{6\cdot\frac{1}{2}-6\cdot\frac{1}{3}}$
Simplify.	$\dfrac{2+1}{3-2}$
	$\dfrac{3}{1}$
	3

> **TRY IT : : 8.107**
>
> Simplify: $\dfrac{\frac{1}{2}+\frac{1}{5}}{\frac{1}{10}+\frac{1}{5}}$.

> **TRY IT : : 8.108**
>
> Simplify: $\dfrac{\frac{1}{4}+\frac{3}{8}}{\frac{1}{2}-\frac{5}{16}}$.

EXAMPLE 8.55   HOW TO SIMPLIFY A COMPLEX RATIONAL EXPRESSION BY USING THE LCD

Simplify: $\dfrac{\frac{1}{x}+\frac{1}{y}}{\frac{x}{y}-\frac{y}{x}}$.

✓ **Solution**

Step 1. Find the LCD of all fractions in the complex rational expression.	The LCD of all the fractions is xy.	$\dfrac{\frac{1}{x}+\frac{1}{y}}{\frac{x}{y}-\frac{y}{x}}$
Step 2. Multiply the numerator and denominator by the LCD.	Multiply both the numerator and denominator by xy.	$\dfrac{xy\cdot\left(\frac{1}{x}+\frac{1}{y}\right)}{xy\cdot\left(\frac{x}{y}-\frac{y}{x}\right)}$

Step 3. Simplify the expression.	Distribute.	$$\dfrac{xy \cdot \dfrac{1}{x} + xy \cdot \dfrac{1}{y}}{xy \cdot \dfrac{x}{y} - xy \cdot \dfrac{y}{x}}$$
		$$\dfrac{y + x}{x^2 - y^2}$$
	Simplify.	$$\dfrac{\cancel{(y + x)}}{(x - y)\cancel{(x + y)}}$$
	Remove common factors.	$$\dfrac{1}{x - y}$$

> **TRY IT : : 8.109**

Simplify: $\dfrac{\dfrac{1}{a} + \dfrac{1}{b}}{\dfrac{a}{b} + \dfrac{b}{a}}$.

> **TRY IT : : 8.110**

Simplify: $\dfrac{\dfrac{1}{x^2} - \dfrac{1}{y^2}}{\dfrac{1}{x} + \dfrac{1}{y}}$.

HOW TO : : SIMPLIFY A COMPLEX RATIONAL EXPRESSION BY USING THE LCD.

Step 1.   Find the LCD of all fractions in the complex rational expression.

Step 2.   Multiply the numerator and denominator by the LCD.

Step 3.   Simplify the expression.

Be sure to start by factoring all the denominators so you can find the LCD.

Simplify: $\dfrac{\dfrac{2}{x + 6}}{\dfrac{4}{x - 6} - \dfrac{4}{x^2 - 36}}$.

✓ **Solution**

$$\dfrac{\dfrac{2}{x + 6}}{\dfrac{4}{x - 6} - \dfrac{4}{x^2 - 36}}$$

Find the LCD of all fractions in the complex rational expression. The LCD is $(x + 6)(x - 6)$.

Multiply the numerator and denominator by the LCD.

$$\dfrac{(x + 6)(x - 6)\dfrac{2}{x + 6}}{(x + 6)(x - 6)\left(\dfrac{4}{x - 6} - \dfrac{4}{(x + 6)(x - 6)}\right)}$$

Simplify the expression.

Distribute in the denominator.	$$\dfrac{(x+6)(x-6)\dfrac{2}{x+6}}{(x+6)(x-6)\left(\dfrac{4}{x-6}\right)-(x+6)(x-6)\left(\dfrac{4}{(x+6)(x-6)}\right)}$$
Simplify.	$$\dfrac{\cancel{(x+6)}(x-6)\dfrac{2}{\cancel{x+6}}}{(x+6)\cancel{(x-6)}\left(\dfrac{4}{\cancel{x-6}}\right)-\cancel{(x+6)(x-6)}\left(\dfrac{4}{\cancel{(x+6)(x-6)}}\right)}$$
Simplify.	$$\dfrac{2(x-6)}{4(x+6)-4}$$
To simplify the denominator, distribute and combine like terms.	$$\dfrac{2(x-6)}{4x+20}$$
Remove common factors.	$$\dfrac{\cancel{2}(x-6)}{\cancel{2}(2x+10)}$$
Simplify.	$$\dfrac{x-6}{2x+10}$$
Notice that there are no more factors common to the numerator and denominator.	

> **TRY IT : : 8.111**

Simplify: $\dfrac{\dfrac{3}{x+2}}{\dfrac{5}{x-2}-\dfrac{3}{x^2-4}}$.

> **TRY IT : : 8.112**

Simplify: $\dfrac{\dfrac{2}{x-7}-\dfrac{1}{x+7}}{\dfrac{6}{x+7}-\dfrac{1}{x^2-49}}$.

EXAMPLE 8.57

Simplify: $\dfrac{\dfrac{4}{m^2-7m+12}}{\dfrac{3}{m-3}-\dfrac{2}{m-4}}$.

✓ **Solution**

	$$\dfrac{\dfrac{4}{m^2-7m+12}}{\dfrac{3}{m-3}-\dfrac{2}{m-4}}$$
Find the LCD of all fractions in the complex rational expression. The LCD is $(m-3)(m-4)$.	
Multiply the numerator and denominator by the LCD.	$$\dfrac{(m-3)(m-4)\dfrac{4}{(m-3)(m-4)}}{(m-3)(m-4)\left(\dfrac{3}{m-3}-\dfrac{2}{m-4}\right)}$$
Simplify.	$$\dfrac{\cancel{(m-3)(m-4)}\dfrac{4}{\cancel{(m-3)(m-4)}}}{(m-3)(m-4)\left(\dfrac{3}{\cancel{m-3}}\right)-(m-3)(m-4)\left(\dfrac{2}{\cancel{m-4}}\right)}$$

Simplify.	$\dfrac{4}{3(m-4)-2\,(m-3)}$
Distribute.	$\dfrac{4}{3m-12-2m+6}$
Combine like terms.	$\dfrac{4}{m-6}$

> **TRY IT : : 8.113**

Simplify: $\dfrac{\dfrac{3}{x^2+7x+10}}{\dfrac{4}{x+2}+\dfrac{1}{x+5}}$.

> **TRY IT : : 8.114**

Simplify: $\dfrac{\dfrac{4y}{y+5}+\dfrac{2}{y+6}}{\dfrac{3y}{y^2+11y+30}}$.

EXAMPLE 8.58

Simplify: $\dfrac{\dfrac{y}{y+1}}{1+\dfrac{1}{y-1}}$.

✓ **Solution**

	$\dfrac{\dfrac{y}{y+1}}{1+\dfrac{1}{y-1}}$
Find the LCD of all fractions in the complex rational expression.	
The LCD is $(y+1)(y-1)$.	
Multiply the numerator and denominator by the LCD.	$\dfrac{(y+1)(y-1)\dfrac{y}{y+1}}{(y+1)(y-1)\left(1+\dfrac{1}{y-1}\right)}$
Distribute in the denominator and simplify.	$\dfrac{(y+1)(y-1)\left(\dfrac{y}{y+1}\right)}{(y+1)(y-1)(1)+(y+1)(y-1)\left(\dfrac{1}{y-1}\right)}$
Simplify.	$\dfrac{(y-1)y}{(y+1)(y-1)+(y+1)}$
Simplify the denominator, and leave the numerator factored.	$\dfrac{y(y-1)}{y^2-1+y+1}$
	$\dfrac{y(y-1)}{y^2+y}$
Factor the denominator, and remove factors common with the numerator.	$\dfrac{y(y-1)}{y(y+1)}$
Simplify.	$\dfrac{y-1}{y+1}$

> **TRY IT : : 8.115**   Simplify: $\dfrac{\dfrac{x}{x+3}}{1+\dfrac{1}{x+3}}$.

> **TRY IT : : 8.116**   Simplify: $\dfrac{1+\dfrac{1}{x-1}}{\dfrac{3}{x+1}}$.

8.5 EXERCISES

Practice Makes Perfect

Simplify a Complex Rational Expression by Writing It as Division

In the following exercises, simplify.

255. $\dfrac{\dfrac{2a}{a+4}}{\dfrac{4a^2}{a^2-16}}$

256. $\dfrac{\dfrac{3b}{b-5}}{\dfrac{b^2}{b^2-25}}$

257. $\dfrac{\dfrac{5}{c^2+5c-14}}{\dfrac{10}{c+7}}$

258. $\dfrac{\dfrac{8}{d^2+9d+18}}{\dfrac{12}{d+6}}$

259. $\dfrac{\dfrac{1}{2}+\dfrac{5}{6}}{\dfrac{2}{3}+\dfrac{7}{9}}$

260. $\dfrac{\dfrac{1}{2}+\dfrac{3}{4}}{\dfrac{3}{5}+\dfrac{7}{10}}$

261. $\dfrac{\dfrac{2}{3}-\dfrac{1}{9}}{\dfrac{3}{4}+\dfrac{5}{6}}$

262. $\dfrac{\dfrac{1}{2}-\dfrac{1}{6}}{\dfrac{2}{3}+\dfrac{3}{4}}$

263. $\dfrac{\dfrac{n}{m}+\dfrac{1}{n}}{\dfrac{1}{n}-\dfrac{n}{m}}$

264. $\dfrac{\dfrac{1}{p}+\dfrac{p}{q}}{\dfrac{q}{p}-\dfrac{1}{q}}$

265. $\dfrac{\dfrac{1}{r}+\dfrac{1}{t}}{\dfrac{1}{r^2}-\dfrac{1}{t^2}}$

266. $\dfrac{\dfrac{2}{v}+\dfrac{2}{w}}{\dfrac{1}{v^2}-\dfrac{1}{w^2}}$

267. $\dfrac{x-\dfrac{2x}{x+3}}{\dfrac{1}{x+3}+\dfrac{1}{x-3}}$

268. $\dfrac{y-\dfrac{2y}{y-4}}{\dfrac{2}{y-4}-\dfrac{2}{y+4}}$

269. $\dfrac{2-\dfrac{2}{a+3}}{\dfrac{1}{a+3}+\dfrac{a}{2}}$

270. $\dfrac{4-\dfrac{4}{b-5}}{\dfrac{1}{b-5}+\dfrac{b}{4}}$

Simplify a Complex Rational Expression by Using the LCD

In the following exercises, simplify.

271. $\dfrac{\dfrac{1}{3}+\dfrac{1}{8}}{\dfrac{1}{4}+\dfrac{1}{12}}$

272. $\dfrac{\dfrac{1}{4}+\dfrac{1}{9}}{\dfrac{1}{6}+\dfrac{1}{12}}$

273. $\dfrac{\dfrac{5}{6}+\dfrac{2}{9}}{\dfrac{7}{18}-\dfrac{1}{3}}$

274. $\dfrac{\dfrac{1}{6}+\dfrac{4}{15}}{\dfrac{3}{5}-\dfrac{1}{2}}$

275. $\dfrac{\dfrac{c}{d}+\dfrac{1}{d}}{\dfrac{1}{d}-\dfrac{d}{c}}$

276. $\dfrac{\dfrac{1}{m}+\dfrac{m}{n}}{\dfrac{n}{m}-\dfrac{1}{n}}$

277. $\dfrac{\dfrac{1}{p}+\dfrac{1}{q}}{\dfrac{1}{p^2}-\dfrac{1}{q^2}}$

278. $\dfrac{\dfrac{2}{r}+\dfrac{2}{t}}{\dfrac{1}{r^2}-\dfrac{1}{t^2}}$

279. $\dfrac{\dfrac{2}{x+5}}{\dfrac{3}{x-5}+\dfrac{1}{x^2-25}}$

280. $\dfrac{\dfrac{5}{y-4}}{\dfrac{3}{y+4}+\dfrac{2}{y^2-16}}$

281. $\dfrac{\dfrac{5}{z^2-64}+\dfrac{3}{z+8}}{\dfrac{1}{z+8}+\dfrac{2}{z-8}}$

282. $\dfrac{\dfrac{3}{s+6}+\dfrac{5}{s-6}}{\dfrac{1}{s^2-36}+\dfrac{4}{s+6}}$

283. $\dfrac{\dfrac{4}{a^2-2a-15}}{\dfrac{1}{a-5}+\dfrac{2}{a+3}}$

284. $\dfrac{\dfrac{5}{b^2-6b-27}}{\dfrac{3}{b-9}+\dfrac{1}{b+3}}$

285. $\dfrac{\dfrac{5}{c+2}-\dfrac{3}{c+7}}{\dfrac{5c}{c^2+9c+14}}$

286. $\dfrac{\dfrac{6}{d-4}-\dfrac{2}{d+7}}{\dfrac{2d}{d^2+3d-28}}$

287. $\dfrac{2+\dfrac{1}{p-3}}{\dfrac{5}{p-3}}$

288. $\dfrac{\dfrac{n}{n-2}}{3+\dfrac{5}{n-2}}$

289. $\dfrac{\dfrac{m}{m+5}}{4+\dfrac{1}{m-5}}$

290. $\dfrac{7+\dfrac{2}{q-2}}{\dfrac{1}{q+2}}$

Simplify

In the following exercises, use either method.

291. $\dfrac{\dfrac{3}{4}-\dfrac{2}{7}}{\dfrac{1}{2}+\dfrac{5}{14}}$

292. $\dfrac{\dfrac{v}{w}+\dfrac{1}{v}}{\dfrac{1}{v}-\dfrac{v}{w}}$

293. $\dfrac{\dfrac{2}{a+4}}{\dfrac{1}{a^2-16}}$

294. $\dfrac{\dfrac{3}{b^2-3b-40}}{\dfrac{5}{b+5}-\dfrac{2}{b-8}}$

295. $\dfrac{\dfrac{3}{m}+\dfrac{3}{n}}{\dfrac{1}{m^2}-\dfrac{1}{n^2}}$

296. $\dfrac{\dfrac{2}{r-9}}{\dfrac{1}{r+9}+\dfrac{3}{r^2-81}}$

297. $\dfrac{x-\dfrac{3x}{x+2}}{\dfrac{3}{x+2}+\dfrac{3}{x-2}}$

298. $\dfrac{\dfrac{y}{y+3}}{2+\dfrac{1}{y-3}}$

Everyday Math

299. Electronics The resistance of a circuit formed by connecting two resistors in parallel is $\dfrac{1}{\dfrac{1}{R_1}+\dfrac{1}{R_2}}$.

ⓐ Simplify the complex fraction $\dfrac{1}{\dfrac{1}{R_1}+\dfrac{1}{R_2}}$.

ⓑ Find the resistance of the circuit when $R_1 = 8$ and $R_2 = 12$.

300. Ironing Lenore can do the ironing for her family's business in h hours. Her daughter would take $h+2$ hours to get the ironing done. If Lenore and her daughter work together, using 2 irons, the number of hours it would take them to do all the ironing is

$$\dfrac{1}{\dfrac{1}{h}+\dfrac{1}{h+2}}.$$

ⓐ Simplify the complex fraction $\dfrac{1}{\dfrac{1}{h}+\dfrac{1}{h+2}}$.

ⓑ Find the number of hours it would take Lenore and her daughter, working together, to get the ironing done if $h = 4$.

Writing Exercises

301. In this section, you learned to simplify the complex fraction $\dfrac{\dfrac{3}{x+2}}{\dfrac{x}{x^2-4}}$ two ways:

rewriting it as a division problem

multiplying the numerator and denominator by the LCD

Which method do you prefer? Why?

302. Efraim wants to start simplifying the complex fraction $\dfrac{\dfrac{1}{a}+\dfrac{1}{b}}{\dfrac{1}{a}-\dfrac{1}{b}}$ by cancelling the variables from the numerator and denominator. Explain what is wrong with Efraim's plan.

Self Check

ⓐ *After completing the exercises, use this checklist to evaluate your mastery of the objectives of this section.*

I can...	Confidently	With some help	No-I don't get it!
simplify a complex rational expression by writing it as division.			
simplify a complex rational expression by using the LCD.			

ⓑ *After looking at the checklist, do you think you are well-prepared for the next section? Why or why not?*

8.6 Solve Rational Equations

Learning Objectives

By the end of this section, you will be able to:

> Solve rational equations
> Solve a rational equation for a specific variable

Be Prepared!

Before you get started, take this readiness quiz.

If you miss a problem, go back to the section listed and review the material.

1. Solve: $\frac{1}{6}x + \frac{1}{2} = \frac{1}{3}$.
   If you missed this problem, review **Example 2.48**.

2. Solve: $n^2 - 5n - 36 = 0$.
   If you missed this problem, review **Example 7.73**.

3. Solve for y in terms of x: $5x + 2y = 10$ for y.
   If you missed this problem, review **Example 2.65**.

After defining the terms *expression* and *equation* early in **Foundations**, we have used them throughout this book. We have *simplified* many kinds of *expressions* and *solved* many kinds of *equations*. We have simplified many rational expressions so far in this chapter. Now we will solve rational equations.

The definition of a rational equation is similar to the definition of equation we used in **Foundations**.

Rational Equation

A **rational equation** is two rational expressions connected by an equal sign.

You must make sure to know the difference between rational expressions and rational equations. The equation contains an equal sign.

Rational Expression	Rational Equation
$\frac{1}{8}x + \frac{1}{2}$	$\frac{1}{8}x + \frac{1}{2} = \frac{1}{4}$
$\frac{y+6}{y^2 - 36}$	$\frac{y+6}{y^2 - 36} = y + 1$
$\frac{1}{n-3} + \frac{1}{n+4}$	$\frac{1}{n-3} + \frac{1}{n+4} = \frac{15}{n^2 + n - 12}$

Solve Rational Equations

We have already solved linear equations that contained fractions. We found the LCD of all the fractions in the equation and then multiplied both sides of the equation by the LCD to "clear" the fractions.

Here is an example we did when we worked with linear equations:

	$\frac{1}{8}x + \frac{1}{2} = \frac{1}{4}$	LCD = 8
We multiplied both sides by the LCD.	$8\left(\frac{1}{8}x + \frac{1}{2}\right) = 8\left(\frac{1}{4}\right)$	
Then we distributed.	$8 \cdot \frac{1}{8}x + 8 \cdot \frac{1}{2} = 8 \cdot \frac{1}{4}$	
We simplified—and then we had an equation with no fractions.	$x + 4 = 2$	
Finally, we solved that equation.	$x + 4 - 4 = 2 - 4$	
	$x = -2$	

We will use the same strategy to solve rational equations. We will multiply both sides of the equation by the LCD. Then we will have an equation that does not contain rational expressions and thus is much easier for us to solve.

But because the original equation may have a variable in a denominator we must be careful that we don't end up with a solution that would make a denominator equal to zero.

So before we begin solving a rational equation, we examine it first to find the values that would make any denominators zero. That way, when we solve a rational equation we will know if there are any algebraic solutions we must discard.

An algebraic solution to a rational equation that would cause any of the rational expressions to be undefined is called an *extraneous solution*.

Extraneous Solution to a Rational Equation

An **extraneous solution to a rational equation** is an algebraic solution that would cause any of the expressions in the original equation to be undefined.

We note any possible extraneous solutions, c, by writing $x \neq c$ next to the equation.

EXAMPLE 8.59   HOW TO SOLVE EQUATIONS WITH RATIONAL EXPRESSIONS

Solve: $\frac{1}{x} + \frac{1}{3} = \frac{5}{6}$.

✓ **Solution**

Step 1. Note any value of the variable that would make any denominator zero.	If $x = 0$, then $\frac{1}{x}$ is undefined.	
	So we'll write $x \neq 0$ next to the equation.	$\frac{1}{x} + \frac{1}{3} = \frac{5}{6}$, $x \neq 0$
Step 2. Find the least common denominator of *all* denominators in the equation.	Find the LCD of $\frac{1}{x}$, $\frac{1}{3}$, and $\frac{5}{6}$.	The LCD is $6x$.

Step 3. Clear the fractions by multiplying both sides of the equation by the LCD.	Multiply both sides of the equation by the LCD, $6x$.	$6x \cdot \left(\dfrac{1}{x} + \dfrac{1}{3}\right) = 6x \cdot \left(\dfrac{5}{6}\right)$
	Use the Distributive Property.	$6x \cdot \dfrac{1}{x} + 6x \cdot \dfrac{1}{3} = 6x \cdot \left(\dfrac{5}{6}\right)$
	Simplify – and notice, no more fractions!	$6 + 2x = 5x$
Step 4. Solve the resulting equation.	Simplify.	$6 = 3x$ $2 = x$
Step 5. Check. • If any values found in Step 1 are algebraic solutions, discard them. • Check any remaining solutions in the original equation.	We did not get 0 as an algebraic solution. We substitute $x = 2$ into the original equation.	$\dfrac{1}{x} + \dfrac{1}{3} = \dfrac{5}{6}$ $\dfrac{1}{2} + \dfrac{1}{3} \overset{?}{=} \dfrac{5}{6}$ $\dfrac{3}{6} + \dfrac{2}{6} \overset{?}{=} \dfrac{5}{6}$ $\dfrac{5}{6} = \dfrac{5}{6} \checkmark$

> **TRY IT :: 8.117**   Solve: $\dfrac{1}{y} + \dfrac{2}{3} = \dfrac{1}{5}$.

> **TRY IT :: 8.118**   Solve: $\dfrac{2}{3} + \dfrac{1}{5} = \dfrac{1}{x}$.

The steps of this method are shown below.

HOW TO :: SOLVE EQUATIONS WITH RATIONAL EXPRESSIONS.

Step 1.   Note any value of the variable that would make any denominator zero.

Step 2.   Find the least common denominator of *all* denominators in the equation.

Step 3.   Clear the fractions by multiplying both sides of the equation by the LCD.

Step 4.   Solve the resulting equation.

Step 5.   Check.

- If any values found in Step 1 are algebraic solutions, discard them.
- Check any remaining solutions in the original equation.

We always start by noting the values that would cause any denominators to be zero.

EXAMPLE 8.60

Solve: $1 - \dfrac{5}{y} = -\dfrac{6}{y^2}$.

⊘ **Solution**

$$1 - \frac{5}{y} = -\frac{6}{y^2}$$

Note any value of the variable that would make any denominator zero.	$1 - \frac{5}{y} = -\frac{6}{y^2}, y \neq 0$
Find the least common denominator of all denominators in the equation. The LCD is y^2.	
Clear the fractions by multiplying both sides of the equation by the LCD.	$y^2\left(1 - \frac{5}{y}\right) = y^2\left(-\frac{6}{y^2}\right)$
Distribute.	$y^2 \cdot 1 - y^2\left(\frac{5}{y}\right) = y^2\left(-\frac{6}{y^2}\right)$
Multiply.	$y^2 - 5y = -6$
Solve the resulting equation. First write the quadratic equation in standard form.	$y^2 - 5y + 6 = 0$
Factor.	$(y - 2)(y - 3) = 0$
Use the Zero Product Property.	$y - 2 = 0 \text{ or } y - 3 = 0$
Solve.	$y = 2 \text{ or } y = 3$
Check.	

We did not get 0 as an algebraic solution.

Check $y = 2$ and $y = 3$ in the original equation.

$$1 - \frac{5}{y} = -\frac{6}{y^2} \qquad\qquad 1 - \frac{5}{y} = -\frac{6}{y^2}$$

$$1 - \frac{5}{2} \overset{?}{=} -\frac{6}{2^2} \qquad\qquad 1 - \frac{5}{3} \overset{?}{=} -\frac{6}{3^2}$$

$$1 - \frac{5}{2} \overset{?}{=} -\frac{6}{4} \qquad\qquad 1 - \frac{5}{3} \overset{?}{=} -\frac{6}{9}$$

$$\frac{2}{2} - \frac{5}{2} \overset{?}{=} -\frac{6}{4} \qquad\qquad \frac{3}{3} - \frac{5}{3} \overset{?}{=} -\frac{6}{9}$$

$$-\frac{3}{2} \overset{?}{=} -\frac{6}{4} \qquad\qquad -\frac{2}{3} \overset{?}{=} -\frac{6}{9}$$

$$-\frac{3}{2} = -\frac{3}{2} \checkmark \qquad\qquad -\frac{2}{3} = -\frac{2}{3} \checkmark$$

> **TRY IT :: 8.119**   Solve: $1 - \frac{2}{a} = \frac{15}{a^2}$.

> **TRY IT :: 8.120**   Solve: $1 - \frac{4}{b} = \frac{12}{b^2}$.

EXAMPLE 8.61

Solve: $\dfrac{5}{3u - 2} = \dfrac{3}{2u}$.

✓ **Solution**

$$\frac{5}{3u-2} = \frac{3}{2u}$$

Note any value of the variable that would make any denominator zero.	$\frac{5}{3u-2} = \frac{3}{2u}, \; u \neq \frac{2}{3}, \; u \neq 0$
Find the least common denominator of all denominators in the equation. The LCD is $2u(3u-2)$.	
Clear the fractions by multiplying both sides of the equation by the LCD.	$2u\,(3u-2)\left(\frac{5}{3u-2}\right) = 2u\,(3u-2)\left(\frac{3}{2u}\right)$
Remove common factors.	$2u\,\cancel{(3u-2)}\left(\frac{5}{\cancel{3u-2}}\right) = \cancel{2u}\,(3u-2)\left(\frac{3}{\cancel{2u}}\right)$
Simplify.	$2u\,(5) = (3u-2)(3)$
Multiply.	$10u = 9u - 6$
Solve the resulting equation.	$u = -6$

We did not get 0 or $\frac{2}{3}$ as algebraic solutions.

Check $u = -6$ in the original equation.

$$\frac{5}{3u-2} = \frac{3}{2u}$$

$$\frac{5}{3(-6)-2} \overset{?}{=} \frac{3}{2(-6)}$$

$$\frac{5}{-20} \overset{?}{=} \frac{3}{-12}$$

$$-\frac{1}{4} = -\frac{1}{4} \checkmark$$

> **TRY IT :: 8.121**   Solve: $\dfrac{1}{x-1} = \dfrac{2}{3x}$.

> **TRY IT :: 8.122**   Solve: $\dfrac{3}{5n+1} = \dfrac{2}{3n}$.

When one of the denominators is a quadratic, remember to factor it first to find the LCD.

EXAMPLE 8.62

Solve: $\dfrac{2}{p+2} + \dfrac{4}{p-2} = \dfrac{p-1}{p^2-4}$.

✓ **Solution**

$$\frac{2}{p+2} + \frac{4}{p-2} = \frac{p-1}{p^2-4}$$

Note any value of the variable that would make any denominator zero.	$\frac{2}{p+2} + \frac{4}{p-2} = \frac{p-1}{(p+2)(p+2)}, \; p \neq -2, p \neq 2$

Find the least common denominator of all denominators in the equation. The LCD is $(p+2)(p-2)$.	

Clear the fractions by multiplying both sides of the equation by the LCD.	$(p+2)(p-2)\left(\dfrac{2}{p+2}+\dfrac{4}{p-2}\right)=(p+2)(p-2)\left(\dfrac{p-1}{p^2-4}\right)$
Distribute.	$(p+2)(p-2)\dfrac{2}{p+2}+(p+2)(p-2)\dfrac{4}{p-2}=(p+2)(p-2)\left(\dfrac{p-1}{p^2-4}\right)$
Remove common factors.	$\cancel{(p+2)}(p-2)\dfrac{2}{\cancel{p+2}}+(p+2)\cancel{(p-2)}\dfrac{4}{\cancel{p-2}}=\cancel{(p+2)(p-2)}\left(\dfrac{p-1}{\cancel{p^2-4}}\right)$
Simplify.	$2(p-2)+4(p+2)=p-1$
Distribute.	$2p-4+4p+8=p-1$
Solve.	$6p+4=p-1$
	$5p=-5$
	$p=-1$

We did not get 2 or -2 as algebraic solutions.	

Check $p=-1$ in the original equation.

$$\frac{2}{p+2}+\frac{4}{p-2}=\frac{p-1}{p^2-4}$$

$$\frac{2}{(-1)+2}+\frac{4}{(-1)-2}\overset{?}{=}\frac{(-1)-1}{(-1)^2-4}$$

$$\frac{2}{1}+\frac{4}{-3}\overset{?}{=}\frac{-2}{-3}$$

$$\frac{6}{3}-\frac{4}{-3}\overset{?}{=}\frac{-2}{-3}$$

$$\frac{2}{3}=\frac{2}{3}\checkmark$$

> **TRY IT :: 8.123** Solve: $\dfrac{2}{x+1}+\dfrac{1}{x-1}=\dfrac{1}{x^2-1}$.

> **TRY IT :: 8.124** Solve: $\dfrac{5}{y+3}+\dfrac{2}{y-3}=\dfrac{5}{y^2-9}$.

EXAMPLE 8.63

Solve: $\dfrac{4}{q-4}-\dfrac{3}{q-3}=1$.

✓ Solution

	$\dfrac{4}{q-4} - \dfrac{3}{q-3} = 1$
Note any value of the variable that would make any denominator zero.	$\dfrac{4}{q-4} + \dfrac{3}{q-3} = 1, \qquad q \neq 4, q \neq 3$
Find the least common denominator of all denominators in the equation. The LCD is $(q-4)(q-3)$.	
Clear the fractions by multiplying both sides of the equation by the LCD.	$(q-4)(q-3)\left(\dfrac{4}{q-4} - \dfrac{3}{q-3}\right) = (q-4)(q-3)(1)$
Distribute.	$(q-4)(q-3)\left(\dfrac{4}{q-4}\right) - (q-4)(q-3)\left(\dfrac{3}{q-3}\right) = (q-4)(q-3)(1)$
Remove common factors.	$(q-4)(q-3)\left(\dfrac{4}{q-4}\right) - (q-4)(q-3)\left(\dfrac{3}{q-3}\right) = (q-4)(q-3)(1)$
Simplify.	$4(q-3) - 3(q-4) = (q-4)(q-3)$
Simplify.	$4q - 12 - 3q + 12 = q^2 - 7q + 12$
Combine like terms.	$q = q^2 - 7q + 12$
Solve. First write in standard form.	$0 = q^2 - 8q + 12$
Factor.	$0 = (q-2)(q-6)$
Use the Zero Product Property.	$q = 2 \text{ or } q = 6$

We did not get 4 or 3 as algebraic solutions.

Check $q = 2$ and $q = 6$ in the original equation.

$$\frac{4}{q-4} - \frac{3}{q-3} = 1 \qquad\qquad \frac{4}{q-4} - \frac{3}{q-3} = 1$$

$$\frac{4}{2-4} - \frac{3}{2-3} \overset{?}{=} 1 \qquad\qquad \frac{4}{6-4} - \frac{3}{6-3} \overset{?}{=} 1$$

$$\frac{4}{-2} - \frac{3}{-1} \overset{?}{=} 1 \qquad\qquad \frac{4}{2} - \frac{3}{1} \overset{?}{=} 1$$

$$-2 - (-3) \overset{?}{=} 1 \qquad\qquad 2 - 1 \overset{?}{=} 1$$

$$1 = 1 \checkmark \qquad\qquad\qquad 1 = 1 \checkmark$$

> **TRY IT :: 8.125**   Solve: $\dfrac{2}{x+5} - \dfrac{1}{x-1} = 1$.

> **TRY IT :: 8.126**   Solve: $\dfrac{3}{x+8} - \dfrac{2}{x-2} = 1$.

EXAMPLE 8.64

Solve: $\dfrac{m+11}{m^2 - 5m + 4} = \dfrac{5}{m-4} - \dfrac{3}{m-1}$.

✓ Solution

$$\frac{m+11}{m^2-5m+4} = \frac{5}{m-4} = \frac{3}{m-1}$$

Factor all the denominators, so we can note any value of the variable the would make any denominator zero.	$\dfrac{m+11}{(m-4)(m-1)} = \dfrac{5}{m-4} - \dfrac{3}{m-1}, \ m \neq 4, m \neq 1$
Find the least common denominator of all denominators in the equation. The LCD is $(m-4)(m-1)$.	
Clear the fractions.	$(m-4)(m-1)\left(\dfrac{m+11}{(m-4)(m-1)}\right) = (m-4)(m-1)\left(\dfrac{5}{m-4} - \dfrac{3}{m-1}\right)$
Distribute.	$(m-4)(m-1)\left(\dfrac{m+11}{(m-4)(m-1)}\right) = (m-4)(m-1)\dfrac{5}{m-4} - (m-4)(m-1)\dfrac{3}{m-1}$
Remove common factors.	$\cancel{(m-4)(m-1)}\left(\dfrac{m+11}{\cancel{(m-4)(m-1)}}\right) = \cancel{(m-4)}(m-1)\dfrac{5}{\cancel{m-4}} - (m-4)\cancel{(m-1)}\dfrac{3}{\cancel{m-1}}$
Simplify.	$m+11 = 5(m-1) - 3(m-4)$
Solve the resulting equation.	$m+11 = 5m - 5 - 3m + 12$
	$4 = m$
Check. The only algebraic solution was 4, but we said that 4 would make a denominator equal to zero. The algebraic solution is an extraneous solution. There is no solution to this equation.	

> **TRY IT :: 8.127** Solve: $\dfrac{x+13}{x^2-7x+10} = \dfrac{6}{x-5} - \dfrac{4}{x-2}$.

> **TRY IT :: 8.128** Solve: $\dfrac{y-14}{y^2+3y-4} = \dfrac{2}{y+4} + \dfrac{7}{y-1}$.

The equation we solved in **Example 8.64** had only one algebraic solution, but it was an extraneous solution. That left us with no solution to the equation. Some equations have no solution.

EXAMPLE 8.65

Solve: $\dfrac{n}{12} + \dfrac{n+3}{3n} = \dfrac{1}{n}$.

✓ Solution

	$\dfrac{n}{12} + \dfrac{n+3}{3n} = \dfrac{1}{n}$
Note any value of the variable that would make any denominator zero.	$\dfrac{n}{12} + \dfrac{n+3}{3n} = \dfrac{1}{n}, \ n \neq 0$

Find the least common denominator of all denominators in the equation. The LCD is $12n$.

Clear the fractions by multiplying both sides of the equation by the LCD.	$12n\left(\dfrac{n}{12}+\dfrac{n+3}{3n}\right)=12n\left(\dfrac{1}{n}\right)$
Distribute.	$12n\left(\dfrac{n}{12}\right)+12n\left(\dfrac{n+3}{3n}\right)=12n\left(\dfrac{1}{n}\right)$
Remove common factors.	$\cancel{12}n\left(\dfrac{n}{\cancel{12}}\right)+4\cdot\cancel{3n}\left(\dfrac{n+3}{\cancel{3n}}\right)=12\cancel{n}\left(\dfrac{1}{\cancel{n}}\right)$
Simplify.	$n\cdot n+4(n+3)=12\cdot 1$
Solve the resulting equation.	$n^2+4n+12=12$
	$n^2+4n=0$
	$n(n+4)=0$
	$n=0 \text{ or } n=-4$

Check.

$n=0$ is an extraneous solution.

Check $n=-4$ in the original equation.

$$\frac{n}{12}+\frac{n+3}{3n}=\frac{1}{n}$$

$$\frac{-4}{12}+\frac{-4+3}{3(-4)}\overset{?}{=}\frac{1}{-4}$$

$$-\frac{4}{12}+\frac{1}{12}\overset{?}{=}-\frac{1}{4}$$

$$-\frac{3}{12}\overset{?}{=}-\frac{1}{4}$$

$$-\frac{1}{4}=-\frac{1}{4}\ \checkmark$$

> **TRY IT :: 8.129** Solve: $\dfrac{x}{18}+\dfrac{x+6}{9x}=\dfrac{2}{3x}$.

> **TRY IT :: 8.130** Solve: $\dfrac{y+5}{5y}+\dfrac{y}{15}=\dfrac{1}{y}$.

EXAMPLE 8.66

Solve: $\dfrac{y}{y+6}=\dfrac{72}{y^2-36}+4$.

⊘ **Solution**

$$\frac{y}{y+6}=\frac{72}{y^2-36}+4$$

Factor all the denominators, so we can note any value of the variable that would make any denominator zero.	$\dfrac{y}{y+6}=\dfrac{72}{(y-6)(y+6)}+4,\ y\neq 6,\ y\neq -6$

Find the least common denominator. The LCD is $(y-6)(y+6)$.

Clear the fractions.	$(y-6)(y+6)\left(\dfrac{y}{y+6}\right) = (y-6)(y+6)\left(\dfrac{72}{(y-6)(y+6)} + 4\right)$
Simplify.	$(y-6) \cdot y = 72 + (y-6)(y+6) \cdot 4$
Simplify.	$y(y-6) = 72 + 4(y^2 - 36)$
Solve the resulting equation.	$y^2 - 6y = 72 + 4y^2 - 144$
	$0 = 3y^2 + 6y - 72$
	$0 = 3(y^2 + 2y - 24)$
	$0 = 3(y+6)(y-4)$
	$y = -6,\ y = 4$

Check.

$y = -6$ is an extraneous solution.

Check $y = 4$ in the original equation.

$$\frac{y}{y+6} = \frac{72}{y^2 - 36} + 4$$

$$\frac{4}{4+6} \overset{?}{=} \frac{72}{4^2 - 36} + 4$$

$$\frac{4}{10} \overset{?}{=} \frac{72}{-20} + 4$$

$$\frac{4}{10} \overset{?}{=} -\frac{36}{10} + \frac{40}{10}$$

$$\frac{4}{10} = \frac{4}{10} \checkmark$$

> **TRY IT :: 8.131**　　Solve: $\dfrac{x}{x+4} = \dfrac{32}{x^2 - 16} + 5$.

> **TRY IT :: 8.132**　　Solve: $\dfrac{y}{y+8} = \dfrac{128}{y^2 - 64} + 9$.

EXAMPLE 8.67

Solve: $\dfrac{x}{2x-2} - \dfrac{2}{3x+3} = \dfrac{5x^2 - 2x + 9}{12x^2 - 12}$.

⊘ **Solution**

$$\frac{x}{2x-2} - \frac{2}{3x+3} = \frac{5x^2-2x+9}{12x^2-12}$$

We will start by factoring all denominators, to make it easier to identify extraneous solutions and the LCD.	$\frac{x}{2(x-1)} - \frac{2}{3(x+1)} = \frac{5x^2-2x+9}{12(x-1)(x+1)}$
Note any value of the variable that would make any denominator zero.	$\frac{x}{2(x-1)} - \frac{2}{3(x+1)} = \frac{5x^2-2x+9}{12(x-1)(x+1)}, x \neq 1, x \neq -1$
Find the least common denominator.The LCD is $12(x-1)(x+1)$	
Clear the fractions.	$12(x-1)(x+1)\left(\frac{x}{2(x-1)} - \frac{2}{3(x+1)}\right) = 12(x-1)(x+1)\left(\frac{5x^2-2x+9}{12(x-1)(x+1)}\right)$
Simplify.	$6(x+1) \cdot x - 4(x-1) \cdot 2 = 5x^2 - 2x + 9$
Simplify.	$6x(x+1) - 4 \cdot 2(x-1) = 5x^2 - 2x + 9$
Solve the resulting equation.	$6x^2 + 6x - 8x + 8 = 5x^2 - 2x + 9$
	$x^2 - 1 = 0$
	$(x-1)(x+1) = 0$
	$x = 1 \text{ or } x = -1$
Check.	

$x = 1$ and $x = -1$ are extraneous solutions.
The equation has no solution.

> **TRY IT ∷ 8.133**

Solve: $\dfrac{y}{5y-10} - \dfrac{5}{3y+6} = \dfrac{2y^2-19y+54}{15y^2-60}$.

> **TRY IT ∷ 8.134**

Solve: $\dfrac{z}{2z+8} - \dfrac{3}{4z-8} = \dfrac{3z^2-16z-6}{8z^2+8z-64}$.

Solve a Rational Equation for a Specific Variable

When we solved linear equations, we learned how to solve a formula for a specific variable. Many formulas used in business, science, economics, and other fields use rational equations to model the relation between two or more variables. We will now see how to solve a rational equation for a specific variable.

We'll start with a formula relating distance, rate, and time. We have used it many times before, but not usually in this form.

EXAMPLE 8.68

Solve: $\dfrac{D}{T} = R$ for T.

⊘ Solution

	$\frac{D}{T} = R$ for T
Note any value of the variable that would make any denominator zero.	$\frac{D}{T} = R, T \neq 0$
Clear the fractions by multiplying both sides of the equations by the LCD, T.	$T\left(\frac{D}{T}\right) = T(R)$
Simplify.	$D = T \cdot R$
Divide both sides by R to isolate T.	$\frac{D}{R} = \frac{RT}{R}$
Simplify.	$\frac{D}{R} = T$

> **TRY IT ::** 8.135   Solve: $\frac{A}{L} = W$ for L.

> **TRY IT ::** 8.136   Solve: $\frac{E}{A} = M$ for A.

Example 8.69 uses the formula for slope that we used to get the point-slope form of an equation of a line.

EXAMPLE 8.69

Solve: $m = \frac{x-2}{y-3}$ for y.

⊘ Solution

	$m = \frac{x-2}{y-3}$ for y
Note any value of the variable that would make any denominator zero.	$m = \frac{x-2}{y-3}, \; y \neq 3$
Clear the fractions by multiplying both sides of the equations by the LCD, $y - 3$.	$(y-3)m = (y-3)\left(\frac{x-2}{y-3}\right)$
Simplify.	$ym - 3m = x - 2$
Isolate the term with y.	$ym = x - 2 + 3m$
Divide both sides by m to isolate y.	$\frac{ym}{m} = \frac{x-2+3m}{m}$
Simplify.	$y = \frac{x-2+3m}{m}$

> **TRY IT ::** 8.137   Solve: $\frac{y-2}{x+1} = \frac{2}{3}$ for x.

> **TRY IT ::** 8.138   Solve: $x = \frac{y}{1-y}$ for y.

Be sure to follow all the steps in Example 8.70. It may look like a very simple formula, but we cannot solve it instantly for either denominator.

EXAMPLE 8.70

Solve $\frac{1}{c} + \frac{1}{m} = 1$ for c.

✓ **Solution**

$$\frac{1}{c} + \frac{1}{m} = 1 \text{ for } c$$

Note any value of the variable that would make any denominator zero.	$\frac{1}{c} + \frac{1}{m} = 1, c \neq 0, m \neq 0$
Clear the fractions by multiplying both sides of the equations by the LCD, cm.	$cm\left(\frac{1}{c} + \frac{1}{m}\right) = cm(1)$
Distribute.	$cm\left(\frac{1}{c}\right) + cm\,\frac{1}{m} = cm(1)$
Simplify.	$m + c = cm$
Collect the terms with c to the right.	$m = cm - c$
Factor the expression on the right.	$m = c(m - 1)$
To isolate c, divide both sides by $m - 1$.	$\frac{m}{m-1} = \frac{c(m-1)}{m-1}$
Simplify by removing common factors.	$\frac{m}{m-1} = c$

Notice that even though we excluded $c = 0$ and $m = 0$ from the original equation, we must also now state that $m \neq 1$.

> **TRY IT : : 8.139**   Solve: $\frac{1}{a} + \frac{1}{b} = c$ for a.

> **TRY IT : : 8.140**   Solve: $\frac{2}{x} + \frac{1}{3} = \frac{1}{y}$ for y.

8.6 EXERCISES

Practice Makes Perfect

Solve Rational Equations

In the following exercises, solve.

303. $\frac{1}{a} + \frac{2}{5} = \frac{1}{2}$

304. $\frac{5}{6} + \frac{3}{b} = \frac{1}{3}$

305. $\frac{5}{2} - \frac{1}{c} = \frac{3}{4}$

306. $\frac{6}{3} - \frac{2}{d} = \frac{4}{9}$

307. $\frac{4}{5} + \frac{1}{4} = \frac{2}{v}$

308. $\frac{3}{7} + \frac{2}{3} = \frac{1}{w}$

309. $\frac{7}{9} + \frac{1}{x} = \frac{2}{3}$

310. $\frac{3}{8} + \frac{2}{y} = \frac{1}{4}$

311. $1 - \frac{2}{m} = \frac{8}{m^2}$

312. $1 + \frac{4}{n} = \frac{21}{n^2}$

313. $1 + \frac{9}{p} = \frac{-20}{p^2}$

314. $1 - \frac{7}{q} = \frac{-6}{q^2}$

315. $\frac{1}{r+3} = \frac{4}{2r}$

316. $\frac{3}{t-6} = \frac{1}{t}$

317. $\frac{5}{3v-2} = \frac{7}{4v}$

318. $\frac{8}{2w+1} = \frac{3}{w}$

319. $\frac{3}{x+4} + \frac{7}{x-4} = \frac{8}{x^2-16}$

320. $\frac{5}{y-9} + \frac{1}{y+9} = \frac{18}{y^2-81}$

321. $\frac{8}{z-10} + \frac{7}{z+10} = \frac{5}{z^2-100}$

322. $\frac{9}{a+11} + \frac{6}{a-11} = \frac{7}{a^2-121}$

323. $\frac{1}{q+4} - \frac{7}{q-2} = 1$

324. $\frac{3}{r+10} - \frac{4}{r-4} = 1$

325. $\frac{1}{t+7} - \frac{5}{t-5} = 1$

326. $\frac{2}{s+7} - \frac{3}{s-3} = 1$

327. $\frac{v-10}{v^2-5v+4} = \frac{3}{v-1} - \frac{6}{v-4}$

328. $\frac{w+8}{w^2-11w+28} = \frac{5}{w-7} + \frac{2}{w-4}$

329. $\frac{x-10}{x^2+8x+12} = \frac{3}{x+2} + \frac{4}{x+6}$

330. $\frac{y-3}{y^2-4y-5} = \frac{1}{y+1} + \frac{8}{y-5}$

331. $\frac{z}{16} + \frac{z+2}{4z} = \frac{1}{2z}$

332. $\frac{a}{9} + \frac{a+3}{3a} = \frac{1}{a}$

333. $\frac{b+3}{3b} + \frac{b}{24} = \frac{1}{b}$

334. $\frac{c+3}{12c} + \frac{c}{36} = \frac{1}{4c}$

335. $\frac{d}{d+3} = \frac{18}{d^2-9} + 4$

336. $\frac{m}{m+5} = \frac{50}{m^2-25} + 6$

337. $\frac{n}{n+2} = \frac{8}{m^2-4} + 3$

338. $\frac{p}{p+7} = \frac{98}{p^2-49} + 8$

339. $\frac{q}{3q-9} - \frac{3}{4q+12}$
$= \frac{7q^2+6q+63}{24q^2-216}$

340. $\frac{r}{3r-15} - \frac{1}{4r+20}$
$= \frac{3r^2+17r+40}{12r^2-300}$

341. $\frac{s}{2s+6} - \frac{2}{5s+5}$
$= \frac{5s^2-3s-7}{10s^2+40s+30}$

342. $\dfrac{t}{6t - 12} - \dfrac{5}{2t + 10}$

$= \dfrac{t^2 - 23t + 70}{12t^2 + 36t - 120}$

Solve a Rational Equation for a Specific Variable

In the following exercises, solve.

343. $\dfrac{C}{r} = 2\pi$ for r

344. $\dfrac{I}{r} = P$ for r

345. $\dfrac{V}{h} = lw$ for h

346. $\dfrac{2A}{b} = h$ for b

347. $\dfrac{v + 3}{w - 1} = \dfrac{1}{2}$ for w

348. $\dfrac{x + 5}{2 - y} = \dfrac{4}{3}$ for y

349. $a = \dfrac{b + 3}{c - 2}$ for c

350. $m = \dfrac{n}{2 - n}$ for n

351. $\dfrac{1}{p} + \dfrac{2}{q} = 4$ for p

352. $\dfrac{3}{s} + \dfrac{1}{t} = 2$ for s

353. $\dfrac{2}{v} + \dfrac{1}{5} = \dfrac{1}{2}$ for v

354. $\dfrac{6}{x} + \dfrac{2}{3} = \dfrac{1}{y}$ for y

355. $\dfrac{m + 3}{n - 2} = \dfrac{4}{5}$ for n

356. $\dfrac{E}{c} = m^2$ for c

357. $\dfrac{3}{x} - \dfrac{5}{y} = \dfrac{1}{4}$ for y

358. $\dfrac{R}{T} = W$ for T

359. $r = \dfrac{s}{3 - t}$ for t

360. $c = \dfrac{2}{a} + \dfrac{b}{5}$ for a

Everyday Math

361. House Painting Alain can paint a house in 4 days. Spiro would take 7 days to paint the same house. Solve the equation $\dfrac{1}{4} + \dfrac{1}{7} = \dfrac{1}{t}$ for t to find the number of days it would take them to paint the house if they worked together.

362. Boating Ari can drive his boat 18 miles with the current in the same amount of time it takes to drive 10 miles against the current. If the speed of the boat is 7 knots, solve the equation $\dfrac{18}{7 + c} = \dfrac{10}{7 - c}$ for c to find the speed of the current.

Writing Exercises

363. Why is there no solution to the equation $\dfrac{3}{x - 2} = \dfrac{5}{x - 2}$?

364. Pete thinks the equation $\dfrac{y}{y + 6} = \dfrac{72}{y^2 - 36} + 4$ has two solutions, $y = -6$ and $y = 4$. Explain why Pete is wrong.

Self Check

ⓐ *After completing the exercises, use this checklist to evaluate your mastery of the objectives of this section.*

I can...	Confidently	With some help	No-I don't get it!
solve rational equations.			
solve rational equations for a specific variable.			

ⓑ *After reviewing this checklist, what will you do to become confident for all objectives?*

8.7
8.7 Solve Proportion and Similar Figure Applications

Learning Objectives

By the end of this section, you will be able to:

> Solve proportions
> Solve similar figure applications

Be Prepared!

Before you get started, take this readiness quiz.

If you miss a problem, go back to the section listed and review the material.

1. Solve $\frac{n}{3} = 30$.

   If you missed this problem, review **Example 2.21**.

2. The perimeter of a triangular window is 23 feet. The lengths of two sides are ten feet and six feet. How long is the third side?
   If you missed this problem, review **Example 3.35**.

Solve Proportions

When two rational expressions are equal, the equation relating them is called a *proportion*.

Proportion

A **proportion** is an equation of the form $\frac{a}{b} = \frac{c}{d}$, where $b \neq 0$, $d \neq 0$.

The proportion is read "a is to b, as c is to d."

The equation $\frac{1}{2} = \frac{4}{8}$ is a proportion because the two fractions are equal. The proportion $\frac{1}{2} = \frac{4}{8}$ is read "1 is to 2 as 4 is to 8."

Proportions are used in many applications to 'scale up' quantities. We'll start with a very simple example so you can see how proportions work. Even if you can figure out the answer to the example right away, make sure you also learn to solve it using proportions.

Suppose a school principal wants to have 1 teacher for 20 students. She could use proportions to find the number of teachers for 60 students. We let x be the number of teachers for 60 students and then set up the proportion:

$$\frac{1 \text{ teacher}}{20 \text{ students}} = \frac{x \text{ teachers}}{60 \text{ students}}$$

We are careful to match the units of the numerators and the units of the denominators—teachers in the numerators, students in the denominators.

Since a proportion is an equation with rational expressions, we will solve proportions the same way we solved equations in **Solve Rational Equations**. We'll multiply both sides of the equation by the LCD to clear the fractions and then solve the resulting equation.

So let's finish solving the principal's problem now. We will omit writing the units until the last step.

$$\frac{1}{20} = \frac{x}{60}$$

Multiply both sides by the LCD, 60.	$\frac{1}{20} \cdot 60 = \frac{x}{60} \cdot 60$
Simplify.	$3 = x$
	The principal needs 3 teachers for 60 students.

Now we'll do a few examples of solving numerical proportions without any units. Then we will solve applications using proportions.

EXAMPLE 8.71

Solve the proportion: $\frac{x}{63} = \frac{4}{7}$.

⊘ Solution

$$\frac{x}{63} = \frac{4}{7}$$

To isolate x, multiply both sides by the LCD, 63.	$63\left(\frac{x}{63}\right) = 63\left(\frac{4}{7}\right)$
Simplify.	$x = \frac{9 \cdot 7 \cdot 4}{7}$
Divide the common factors.	$x = 36$
Check. To check our answer, we substitute into the original proportion.	
	$\frac{x}{63} = \frac{4}{7}$
Substitute $x = 36$.	$\frac{36}{63} \overset{?}{=} \frac{4}{7}$
Show common factors.	$\frac{4 \cdot 9}{7 \cdot 9} \overset{?}{=} \frac{4}{7}$
Simplify.	$\frac{4}{7} = \frac{4}{7} \checkmark$

> **TRY IT :: 8.141**   Solve the proportion: $\frac{n}{84} = \frac{11}{12}$.

> **TRY IT :: 8.142**   Solve the proportion: $\frac{y}{96} = \frac{13}{12}$.

When we work with proportions, we exclude values that would make either denominator zero, just like we do for all rational expressions. What value(s) should be excluded for the proportion in the next example?

EXAMPLE 8.72

Solve the proportion: $\frac{144}{a} = \frac{9}{4}$.

⊘ Solution

$$\frac{144}{a} = \frac{9}{4}$$

Multiply both sides by the LCD.	$\frac{144}{a} \cdot 4a = \frac{9}{4} \cdot 4a$
Remove common factors on each side.	$4 \cdot 144 = a \cdot 9$
Simplify.	$576 = 9a$
Divide both sides by 9.	$\frac{576}{9} = \frac{9a}{9}$

Simplify.	$64 = a$

Check.

$$\frac{144}{a} = \frac{9}{4}$$

Substitute $a = 64$.	$\frac{144}{64} \stackrel{?}{=} \frac{9}{4}$
Show common factors.	$\frac{9 \cdot 16}{4 \cdot 16} \stackrel{?}{=} \frac{9}{4}$
Simplify.	$\frac{9}{4} = \frac{9}{4} \checkmark$

> **TRY IT : : 8.143** Solve the proportion: $\frac{91}{b} = \frac{7}{5}$.

> **TRY IT : : 8.144** Solve the proportion: $\frac{39}{c} = \frac{13}{8}$.

EXAMPLE 8.73

Solve the proportion: $\frac{n}{n + 14} = \frac{5}{7}$.

⊘ **Solution**

$$\frac{n}{n + 14} = \frac{5}{7}$$

Multiply both sides by the LCD.	$7(n + 14)\left(\frac{n}{n + 14}\right) = 7(n + 14)\left(\frac{5}{7}\right)$
Remove common factors on each side.	$7n = 5(n + 14)$
Simplify.	$7n = 5n + 70$
Solve for n.	$2n = 70$
	$n = 35$

Check.

$$\frac{n}{n + 14} = \frac{5}{7}$$

Substitute $n = 35$.	$\frac{35}{35 + 14} \stackrel{?}{=} \frac{5}{7}$
Simplify.	$\frac{35}{49} \stackrel{?}{=} \frac{5}{7}$
Show common factors.	$\frac{5 \cdot 7}{7 \cdot 7} \stackrel{?}{=} \frac{5}{7}$
Simplify.	$\frac{5}{7} = \frac{5}{7} \checkmark$

> **TRY IT : : 8.145** Solve the proportion: $\frac{y}{y + 55} = \frac{3}{8}$.

> **TRY IT : : 8.146** Solve the proportion: $\dfrac{z}{z-84} = -\dfrac{1}{5}$.

EXAMPLE 8.74

Solve: $\dfrac{p+12}{9} = \dfrac{p-12}{6}$.

⊘ Solution

	$\dfrac{p+12}{9} = \dfrac{p-12}{6}$
Multiply both sides by the LCD, 18.	$18\left(\dfrac{p+12}{9}\right) = 18\left(\dfrac{p-12}{6}\right)$
Simplify.	$2(p+12) = 3(p-12)$
Distribute.	$2p + 24 = 3p - 36$
Solve for p.	$60 = p$
Check.	

	$\dfrac{p+12}{9} = \dfrac{p-12}{6}$
Substitute $p = 60$.	$\dfrac{60+12}{9} \overset{?}{=} \dfrac{60-12}{6}$
Simplify.	$\dfrac{72}{9} \overset{?}{=} \dfrac{48}{6}$
Divide.	$8 = 8 ✓$

> **TRY IT : : 8.147** Solve: $\dfrac{v+30}{8} = \dfrac{v+66}{12}$.

> **TRY IT : : 8.148** Solve: $\dfrac{2x+15}{9} = \dfrac{7x+3}{15}$.

To solve applications with proportions, we will follow our usual strategy for solving applications. But when we set up the proportion, we must make sure to have the units correct—the units in the numerators must match and the units in the denominators must match.

EXAMPLE 8.75

When pediatricians prescribe acetaminophen to children, they prescribe 5 milliliters (ml) of acetaminophen for every 25 pounds of the child's weight. If Zoe weighs 80 pounds, how many milliliters of acetaminophen will her doctor prescribe?

⊘ Solution

Identify what we are asked to find, and choose a variable to represent it.	How many ml of acetaminophen will the doctor prescribe?
	Let $a = $ ml of acetaminophen.
Write a sentence that gives the information to find it.	If 5 ml is prescribed for every 25 pounds, how much will be prescribed for 80 pounds?

Translate into a proportion–be careful of the units. $$\frac{\text{ml}}{\text{pounds}} = \frac{\text{ml}}{\text{pounds}}$$	$$\frac{5}{25} = \frac{a}{80}$$
Multiply both sides by the LCD, 400.	$$400\left(\frac{5}{25}\right) = 400\left(\frac{a}{80}\right)$$
Remove common factors on each side.	$$25 \cdot 16\left(\frac{5}{25}\right) = 80 \cdot 5\left(\frac{a}{80}\right)$$
Simplify, but don't multiply on the left. Notice what the next step will be.	$$16 \cdot 5 = 5a$$
Solve for a.	$$\frac{16 \cdot 5}{5} = \frac{5a}{5}$$
Check.	$$16 = a$$
Is the answer reasonable?	
Yes, since 80 is about 3 times 25, the medicine should be about 3 times 5. So 16 ml makes sense.	
Substitute $a = 16$ in the original proportion. $$\frac{5}{25} = \frac{a}{80}$$ $$\frac{5}{25} \overset{?}{=} \frac{16}{80}$$ $$\frac{1}{5} = \frac{1}{5} \checkmark$$	
Write a complete sentence.	The pediatrician would prescribe 16 ml of acetaminophen to Zoe.

> **TRY IT : : 8.149**

> Pediatricians prescribe 5 milliliters (ml) of acetaminophen for every 25 pounds of a child's weight. How many milliliters of acetaminophen will the doctor prescribe for Emilia, who weighs 60 pounds?

> **TRY IT : : 8.150**

> For every 1 kilogram (kg) of a child's weight, pediatricians prescribe 15 milligrams (mg) of a fever reducer. If Isabella weighs 12 kg, how many milligrams of the fever reducer will the pediatrician prescribe?

EXAMPLE 8.76

A 16-ounce iced caramel macchiato has 230 calories. How many calories are there in a 24-ounce iced caramel macchiato?

⊘ **Solution**

Identify what we are asked to find, and choose a variable to represent it.	How many calories are in a 24 ounce iced caramel macchiato?
	Let $c = $ calories in 24 ounces.
Write a sentence that gives the information to find it.	If there are 230 calories in 16 ounces, then how many calories are in 24 ounces?

Translate into a proportion–be careful of the units.

$$\frac{\text{calories}}{\text{ounce}} = \frac{\text{calories}}{\text{ounce}}$$

$$\frac{230}{16} = \frac{c}{24}$$

Multiply both sides by the LCD, 48.	$48\left(\frac{230}{16}\right) = 48\left(\frac{c}{24}\right)$
Remove common factors on each side.	$16 \cdot 3\left(\frac{230}{16}\right) = 24 \cdot 2\left(\frac{c}{24}\right)$
Simplify.	$690 = 2c$
Solve for c.	$\frac{690}{2} = \frac{2c}{2}$
	$345 = c$

Check.

Is the answer reasonable?

Yes, 345 calories for 24 ounces is more than 290 calories for 16 ounces, but not too much more.

Substitute $c = 345$ in the original proportion.

$$\frac{230}{16} = \frac{c}{24}$$

$$\frac{230}{16} \overset{?}{=} \frac{345}{24}$$

$$\frac{115}{8} = \frac{115}{8} \checkmark$$

Write a complete sentence.

There are 345 calories in a 24-ounce iced caramel macchiato.

> **TRY IT : : 8.151**

At a fast-food restaurant, a 22-ounce chocolate shake has 850 calories. How many calories are in their 12-ounce chocolate shake? Round your answer to nearest whole number.

> **TRY IT : : 8.152**

Yaneli loves Starburst candies, but wants to keep her snacks to 100 calories. If the candies have 160 calories for 8 pieces, how many pieces can she have in her snack?

EXAMPLE 8.77

Josiah went to Mexico for spring break and changed $325 dollars into Mexican pesos. At that time, the exchange rate had $1 US is equal to 12.54 Mexican pesos. How many Mexican pesos did he get for his trip?

✓ **Solution**

What are you asked to find?	How many Mexican pesos did Josiah get?
Assign a variable.	Let p = the number of Mexican pesos.

Write a sentence that gives the information to find it.	If \$1 US is equal to 12.54 Mexican pesos, then \$325 is how many pesos?
Translate into a proportion–be careful of the units.	
$\dfrac{\$}{\text{pesos}} = \dfrac{\$}{\text{pesos}}$	$\dfrac{1}{12.54} = \dfrac{325}{p}$
Multiply both sides by the LCD, $12.54p$.	$12.54p\left(\dfrac{1}{12.54}\right) = 12.54p\left(\dfrac{325}{p}\right)$
Remove common factors on each side.	$12.54p\left(\dfrac{1}{12.54}\right) = 12.54p\left(\dfrac{325}{p}\right)$
Simplify.	$p = 4075.5$
Check.	
Is the answer reasonable?	
Yes, \$100 would be 1,254 pesos. \$325 is a little more than 3 times this amount, so our answer of 4075.5 pesos makes sense.	
Substitute $p = 4075.5$ in the original proportion. Use a calculator. $\dfrac{1}{12.54} = \dfrac{325}{p}$ $\dfrac{1}{12.54} \overset{?}{=} \dfrac{325}{4075.5}$ $0.07874... = 0.07874...$ ✓	
Write a complete sentence.	Josiah got 4075.5 pesos for his spring break trip.

> **TRY IT : : 8.153**
>
> Yurianna is going to Europe and wants to change \$800 dollars into Euros. At the current exchange rate, \$1 US is equal to 0.738 Euro. How many Euros will she have for her trip?

> **TRY IT : : 8.154**
>
> Corey and Nicole are traveling to Japan and need to exchange \$600 into Japanese yen. If each dollar is 94.1 yen, how many yen will they get?

In the example above, we related the number of pesos to the number of dollars by using a proportion. We could say the number of pesos *is proportional to* the number of dollars. If two quantities are related by a proportion, we say that they are proportional.

Solve Similar Figure Applications

When you shrink or enlarge a photo on a phone or tablet, figure out a distance on a map, or use a pattern to build a bookcase or sew a dress, you are working with **similar figures**. If two figures have exactly the same shape, but different sizes, they are said to be *similar*. One is a scale model of the other. All their corresponding angles have the same measures and their corresponding sides are in the same ratio.

Similar Figures

Two figures are similar if the measures of their corresponding angles are equal and their corresponding sides are in the same ratio.

For example, the two triangles in **Figure 8.2** are similar. Each side of $\triangle ABC$ is 4 times the length of the corresponding side of $\triangle XYZ$.

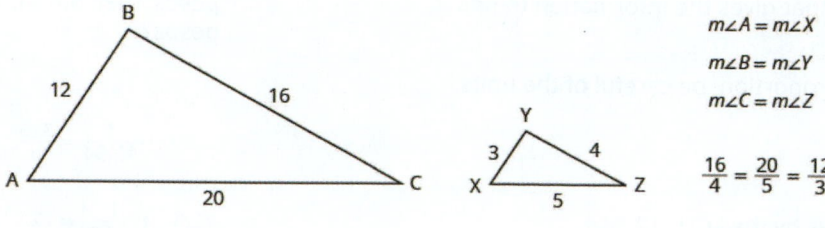

Figure 8.2

This is summed up in the Property of Similar Triangles.

Property of Similar Triangles

If $\triangle ABC$ is similar to $\triangle XYZ$, then their corresponding angle measure are equal and their corresponding sides are in the same ratio.

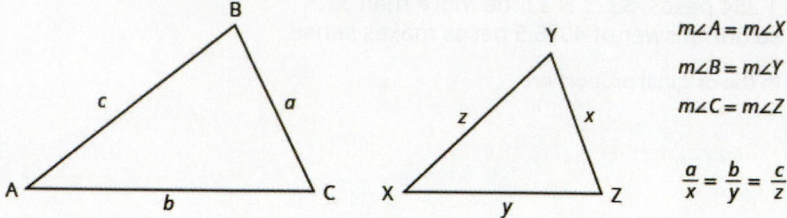

To solve applications with similar figures we will follow the Problem-Solving Strategy for Geometry Applications we used earlier.

HOW TO :: SOLVE GEOMETRY APPLICATIONS.

Step 1.   **Read** the problem and make all the words and ideas are understood. Draw the figure and label it with the given information.

Step 2.   **Identify** what we are looking for.

Step 3.   **Name** what we are looking for by choosing a variable to represent it.

Step 4.   **Translate** into an equation by writing the appropriate formula or model for the situation. Substitute in the given information.

Step 5.   **Solve the equation** using good algebra techniques.

Step 6.   **Check** the answer in the problem and make sure it makes sense.

Step 7.   **Answer** the question with a complete sentence.

EXAMPLE 8.78

$\triangle ABC$ is similar to $\triangle XYZ$. The lengths of two sides of each triangle are given. Find the lengths of the third sides.

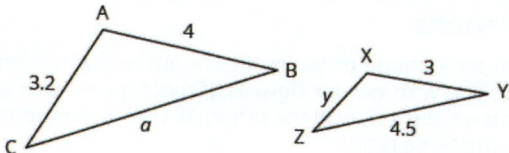

⊘ Solution

Step 1. Read the problem. Draw the figure and label it with the given information.	Figure is given.
Step 2. Identify what we are looking for.	the length of the sides of similar triangles
Step 3. Name the variables.	Let $a=$ length of the third side of $\triangle ABC$. $y=$ length of the third side of $\triangle XYZ$
Step 4. Translate.	Since the triangles are similar, the corresponding sides are proportional.

We need to write an equation that compares the side we are looking for to a known ratio. Since the side $AB = 4$ corresponds to the side $XY = 3$ we know $\frac{AB}{XY} = \frac{4}{3}$. So we write equations with $\frac{AB}{XY}$ to find the sides we are looking for. Be careful to match up corresponding sides correctly.

$$\frac{AB}{XY} = \frac{BC}{YZ} = \frac{AC}{XZ}.$$

	To find a:	To find y:
sides of large triangle $\longrightarrow$ sides of small triangle $\longrightarrow$	$\frac{AB}{XY} = \frac{BC}{YZ}$	$\frac{AB}{XY} = \frac{AC}{XZ}$

Substitute.	$\frac{4}{3} = \frac{a}{4.5}$	$\frac{4}{3} = \frac{3.2}{y}$
Step 5. Solve the equation.	$3a = 4(4.5)$	$4y = 3(3.2)$
	$a = 6$	$y = 2.4$

Step 6. Check.

$$\frac{4}{3} \overset{?}{=} \frac{6}{4.5} \qquad\qquad \frac{4}{3} \overset{?}{=} \frac{3.2}{2.4}$$

$$4(4.5) \overset{?}{=} 6(3) \qquad 4(2.4) \overset{?}{=} 3.2(3)$$

$$18 = 18 \checkmark \qquad\quad 9.6 = 9.6 \checkmark$$

Step 7. Answer the question.	The third side of $\triangle ABC$ is 6 and the third side of $\triangle XYZ$ is 2.4.

> **TRY IT : : 8.155**

$\triangle ABC$ is similar to $\triangle XYZ$. The lengths of two sides of each triangle are given in the figure.

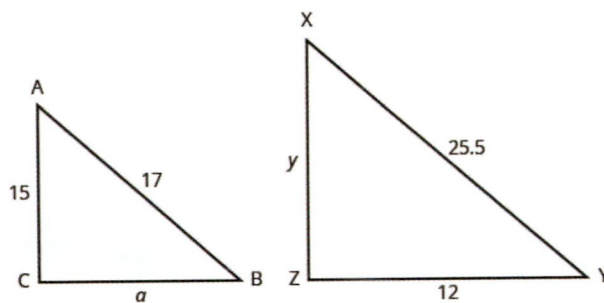

Find the length of side a.

> | **TRY IT :: 8.156**

$\triangle ABC$ is similar to $\triangle XYZ$. The lengths of two sides of each triangle are given in the figure.

Find the length of side y.

The next example shows how similar triangles are used with maps.

EXAMPLE 8.79

On a map, San Francisco, Las Vegas, and Los Angeles form a triangle whose sides are shown in the figure below. If the actual distance from Los Angeles to Las Vegas is 270 miles find the distance from Los Angeles to San Francisco.

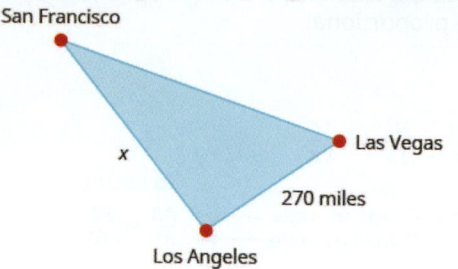 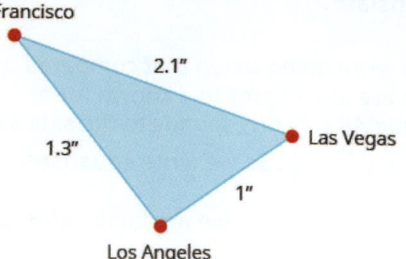

⊘ **Solution**

Read the problem. Draw the figures and label with the given information.	The figures are shown above.
Identify what we are looking for.	The actual distance from Los Angeles to San Francisco.
Name the variables.	Let $x =$ distance from Los Angeles to San Francisco.
Translate into an equation. Since the triangles are similar, the corresponding sides are proportional. We'll make the numerators "miles" and the denominators "inches."	$\dfrac{x \text{ miles}}{1.3 \text{ inches}} = \dfrac{270 \text{ miles}}{1 \text{ inch}}$
Solve the equation.	$1.3\left(\dfrac{x}{1.3}\right) = 1.3\left(\dfrac{270}{1}\right)$
	$x = 351$

Check.

On the map, the distance from Los Angeles to San Francisco is more than the distance from Los Angeles to Las Vegas. Since 351 is more than 270 the answer makes sense.

Check $x = 351$ in the original proportion. Use a calculator.

$$\frac{x \text{ miles}}{1.3 \text{ inches}} = \frac{270 \text{ miles}}{1 \text{ inch}}$$

$$\frac{351 \text{ miles}}{1.3 \text{ inches}} \overset{?}{=} \frac{270 \text{ miles}}{1 \text{ inch}}$$

$$\frac{270 \text{ miles}}{1 \text{ inch}} = \frac{270 \text{ miles}}{1 \text{ inch}} ✓$$

Answer the question.	The distance from Los Angeles to San Francisco is 351 miles.

> **TRY IT : : 8.157**

On the map, Seattle, Portland, and Boise form a triangle whose sides are shown in the figure below. If the actual distance from Seattle to Boise is 400 miles, find the distance from Seattle to Portland.

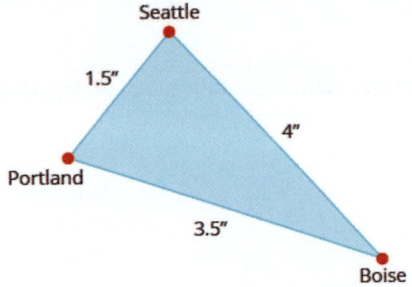

> **TRY IT : : 8.158**   Using the map above, find the distance from Portland to Boise.

We can use similar figures to find heights that we cannot directly measure.

EXAMPLE 8.80

Tyler is 6 feet tall. Late one afternoon, his shadow was 8 feet long. At the same time, the shadow of a tree was 24 feet long. Find the height of the tree.

⊘ **Solution**

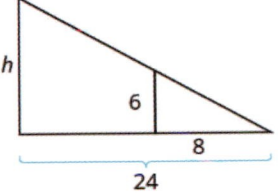

Read the problem and draw a figure.

We are looking for h, the height of the tree.

We will use similar triangles to write an equation.

The small triangle is similar to the large triangle. $\dfrac{h}{24} = \dfrac{6}{8}$

Solve the proportion. $24\left(\dfrac{6}{8}\right) = 24\left(\dfrac{h}{24}\right)$

Simplify. $18 = h$

Check.

Tyler's height is less than his shadow's length so it makes sense that the tree's height is less than the length of its shadow.

Check $h = 18$ in the original proportion.

$\dfrac{6}{8} = \dfrac{h}{24}$

$\dfrac{6}{8} \overset{?}{=} \dfrac{18}{24}$

$\dfrac{3}{4} = \dfrac{3}{4}$ ✓

> **TRY IT : : 8.159**

A telephone pole casts a shadow that is 50 feet long. Nearby, an 8 foot tall traffic sign casts a shadow that is 10 feet long. How tall is the telephone pole?

> **TRY IT : : 8.160**

A pine tree casts a shadow of 80 feet next to a 30-foot tall building which casts a 40 feet shadow. How tall is the pine tree?

8.7 EXERCISES

Practice Makes Perfect

Solve Proportions

In the following exercises, solve.

365. $\dfrac{x}{56} = \dfrac{7}{8}$

366. $\dfrac{n}{91} = \dfrac{8}{13}$

367. $\dfrac{49}{63} = \dfrac{z}{9}$

368. $\dfrac{56}{72} = \dfrac{y}{9}$

369. $\dfrac{5}{a} = \dfrac{65}{117}$

370. $\dfrac{4}{b} = \dfrac{64}{144}$

371. $\dfrac{98}{154} = \dfrac{-7}{p}$

372. $\dfrac{72}{156} = \dfrac{-6}{q}$

373. $\dfrac{a}{-8} = \dfrac{-42}{48}$

374. $\dfrac{b}{-7} = \dfrac{-30}{42}$

375. $\dfrac{2.7}{j} = \dfrac{0.9}{0.2}$

376. $\dfrac{2.8}{k} = \dfrac{2.1}{1.5}$

377. $\dfrac{a}{a+12} = \dfrac{4}{7}$

378. $\dfrac{b}{b-16} = \dfrac{11}{9}$

379. $\dfrac{c}{c-104} = -\dfrac{5}{8}$

380. $\dfrac{d}{d-48} = -\dfrac{13}{3}$

381. $\dfrac{m+90}{25} = \dfrac{m+30}{15}$

382. $\dfrac{n+10}{4} = \dfrac{40-n}{6}$

383. $\dfrac{2p+4}{8} = \dfrac{p+18}{6}$

384. $\dfrac{q-2}{2} = \dfrac{2q-7}{18}$

385. Pediatricians prescribe 5 milliliters (ml) of acetaminophen for every 25 pounds of a child's weight. How many milliliters of acetaminophen will the doctor prescribe for Jocelyn, who weighs 45 pounds?

386. Brianna, who weighs 6 kg, just received her shots and needs a pain killer. The pain killer is prescribed for children at 15 milligrams (mg) for every 1 kilogram (kg) of the child's weight. How many milligrams will the doctor prescribe?

387. A veterinarian prescribed Sunny, a 65 pound dog, an antibacterial medicine in case an infection emerges after her teeth were cleaned. If the dosage is 5 mg for every pound, how much medicine was Sunny given?

388. Belle, a 13 pound cat, is suffering from joint pain. How much medicine should the veterinarian prescribe if the dosage is 1.8 mg per pound?

389. A new energy drink advertises 106 calories for 8 ounces. How many calories are in 12 ounces of the drink?

390. One 12 ounce can of soda has 150 calories. If Josiah drinks the big 32 ounce size from the local mini-mart, how many calories does he get?

391. A new 7 ounce lemon ice drink is advertised for having only 140 calories. How many ounces could Sally drink if she wanted to drink just 100 calories?

392. Reese loves to drink healthy green smoothies. A 16 ounce serving of smoothie has 170 calories. Reese drinks 24 ounces of these smoothies in one day. How many calories of smoothie is he consuming in one day?

393. Janice is traveling to Canada and will change $250 US dollars into Canadian dollars. At the current exchange rate, $1 US is equal to $1.01 Canadian. How many Canadian dollars will she get for her trip?

394. Todd is traveling to Mexico and needs to exchange $450 into Mexican pesos. If each dollar is worth 12.29 pesos, how many pesos will he get for his trip?

395. Steve changed $600 into 480 Euros. How many Euros did he receive for each US dollar?

396. Martha changed $350 US into 385 Australian dollars. How many Australian dollars did she receive for each US dollar?

397. When traveling to Great Britain, Bethany exchanged her $900 into 570 British pounds. How many pounds did she receive for each American dollar?

398. A missionary commissioned to South Africa had to exchange his $500 for the South African Rand which is worth 12.63 for every dollar. How many Rand did he have after the exchange?

399. Ronald needs a morning breakfast drink that will give him at least 390 calories. Orange juice has 130 calories in one cup. How many cups does he need to drink to reach his calorie goal?

400. Sarah drinks a 32-ounce energy drink containing 80 calories per 12 ounce. How many calories did she drink?

401. Elizabeth is returning to the United States from Canada. She changes the remaining 300 Canadian dollars she has to $230.05 in American dollars. What was $1 worth in Canadian dollars?

402. Ben needs to convert $1000 to the Japanese Yen. One American dollar is worth 123.3 Yen. How much Yen will he have?

403. A golden retriever weighing 85 pounds has diarrhea. His medicine is prescribed as 1 teaspoon per 5 pounds. How much medicine should he be given?

404. Five-year-old Lacy was stung by a bee. The dosage for the anti-itch liquid is 150 mg for her weight of 40 pounds. What is the dosage per pound?

405. Karen eats $\frac{1}{2}$ cup of oatmeal that counts for 2 points on her weight loss program. Her husband, Joe, can have 3 points of oatmeal for breakfast. How much oatmeal can he have?

406. An oatmeal cookie recipe calls for $\frac{1}{2}$ cup of butter to make 4 dozen cookies. Hilda needs to make 10 dozen cookies for the bake sale. How many cups of butter will she need?

Solve Similar Figure Applications

In the following exercises, $\triangle ABC$ is similar to $\triangle XYZ$. Find the length of the indicated side.

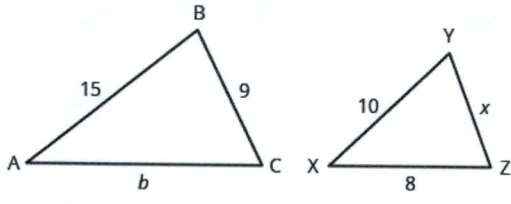

407. side b

408. side x

In the following exercises, $\triangle DEF$ is similar to $\triangle NPQ$.

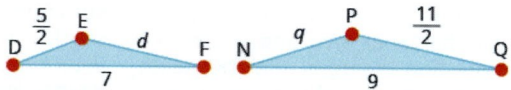

409. Find the length of side d.

410. Find the length of side q.

In the following two exercises, use the map shown. On the map, New York City, Chicago, and Memphis form a triangle whose sides are shown in the figure below. The actual distance from New York to Chicago is 800 miles.

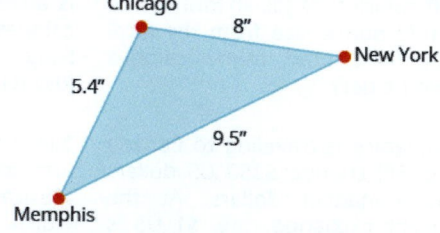

411. Find the actual distance from New York to Memphis.

412. Find the actual distance from Chicago to Memphis.

In the following two exercises, use the map shown. On the map, Atlanta, Miami, and New Orleans form a triangle whose sides are shown in the figure below. The actual distance from Atlanta to New Orleans is 420 miles.

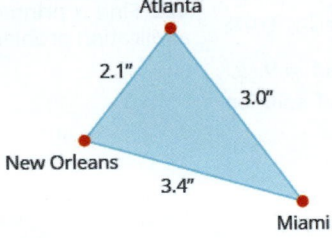

413. Find the actual distance from New Orleans to Miami.

414. Find the actual distance from Atlanta to Miami.

415. A 2 foot tall dog casts a 3 foot shadow at the same time a cat casts a one foot shadow. How tall is the cat?

416. Larry and Tom were standing next to each other in the backyard when Tom challenged Larry to guess how tall he was. Larry knew his own height is 6.5 feet and when they measured their shadows, Larry's shadow was 8 feet and Tom's was 7.75 feet long. What is Tom's height?

417. The tower portion of a windmill is 212 feet tall. A six foot tall person standing next to the tower casts a seven foot shadow. How long is the windmill's shadow?

418. The height of the Statue of Liberty is 305 feet. Nicole, who is standing next to the statue, casts a 6 foot shadow and she is 5 feet tall. How long should the shadow of the statue be?

Everyday Math

419. Heart Rate At the gym, Carol takes her pulse for 10 seconds and counts 19 beats.

ⓐ How many beats per minute is this?

ⓑ Has Carol met her target heart rate of 140 beats per minute?

420. Heart Rate Kevin wants to keep his heart rate at 160 beats per minute while training. During his workout he counts 27 beats in 10 seconds.

ⓐ How many beats per minute is this?

ⓑ Has Kevin met his target heart rate?

421. Cost of a Road Trip Jesse's car gets 30 miles per gallon of gas.

ⓐ If Las Vegas is 285 miles away, how many gallons of gas are needed to get there and then home?

ⓑ If gas is $3.09 per gallon, what is the total cost of the gas for the trip?

422. Cost of a Road Trip Danny wants to drive to Phoenix to see his grandfather. Phoenix is 370 miles from Danny's home and his car gets 18.5 miles per gallon.

ⓐ How many gallons of gas will Danny need to get to and from Phoenix?

ⓑ If gas is $3.19 per gallon, what is the total cost for the gas to drive to see his grandfather?

423. Lawn Fertilizer Phil wants to fertilize his lawn. Each bag of fertilizer covers about 4,000 square feet of lawn. Phil's lawn is approximately 13,500 square feet. How many bags of fertilizer will he have to buy?

424. House Paint April wants to paint the exterior of her house. One gallon of paint covers about 350 square feet, and the exterior of the house measures approximately 2000 square feet. How many gallons of paint will she have to buy?

425. Cooking Natalia's pasta recipe calls for 2 pounds of pasta for 1 quart of sauce. How many pounds of pasta should Natalia cook if she has 2.5 quarts of sauce?

426. Heating Oil A 275 gallon oil tank costs $400 to fill. How much would it cost to fill a 180 gallon oil tank?

Writing Exercises

427. Marisol solves the proportion $\frac{144}{a} = \frac{9}{4}$ by 'cross multiplying', so her first step looks like $4 \cdot 144 = 9 \cdot a$. Explain how this differs from the method of solution shown in **Example 8.72**.

428. Find a printed map and then write and solve an application problem similar to **Example 8.79**.

Self Check

ⓐ *After completing the exercises, use this checklist to evaluate your mastery of the objectives of this section.*

I can...	Confidently	With some help	No-I don't get it!
solve proportions.			
solve similar figure applications.			

ⓑ *What does this checklist tell you about your mastery of this section? What steps will you take to improve?*

8.8 Solve Uniform Motion and Work Applications

Learning Objectives

By the end of this section, you will be able to:

> Solve uniform motion applications
> Solve work applications

Be Prepared!

Before you get started, take this readiness quiz.

If you miss a problem, go back to the section listed and review the material.

1. An express train and a local bus leave Chicago to travel to Champaign. The express bus can make the trip in 2 hours and the local bus takes 5 hours for the trip. The speed of the express bus is 42 miles per hour faster than the speed of the local bus. Find the speed of the local bus.
    If you missed this problem, review **Example 3.48**.

2. Solve $\frac{1}{3}x + \frac{1}{4}x = \frac{5}{6}$.
    If you missed this problem, review **Example 3.49**.

3. Solve: $18t^2 - 30 = -33t$.
    If you missed this problem, review **Example 7.79**.

Solve Uniform Motion Applications

We have solved uniform motion problems using the formula $D = rt$ in previous chapters. We used a table like the one below to organize the information and lead us to the equation.

	Rate	•	Time	=	Distance

The formula $D = rt$ assumes we know r and t and use them to find D. If we know D and r and need to find t, we would solve the equation for t and get the formula $t = \frac{D}{r}$.

We have also explained how flying with or against a current affects the speed of a vehicle. We will revisit that idea in the next example.

EXAMPLE 8.81

An airplane can fly 200 miles into a 30 mph headwind in the same amount of time it takes to fly 300 miles with a 30 mph tailwind. What is the speed of the airplane?

⊘ Solution

This is a uniform motion situation. A diagram will help us visualize the situation.

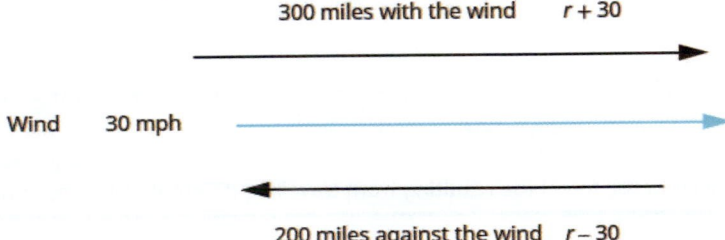

We fill in the chart to organize the information.

We are looking for the speed of the airplane.   Let $r =$ the speed of the airplane.

When the plane flies with the wind, the wind increases its speed and the rate is $r + 30$.

When the plane flies against the wind, the wind decreases its speed and the rate is $r - 30$.

Write in the rates.
Write in the distances.
Since $D = r \cdot t$, we solve for t and get $\frac{D}{r}$.

We divide the distance by the rate in each row, and place the expression in the time column.

	Rate •	Time	= Distance
Headwind	$r - 30$	$\frac{200}{r - 30}$	200
Tailwind	$r + 30$	$\frac{300}{r + 30}$	300

We know the times are equal and so we write our equation.

$$\frac{200}{r - 30} = \frac{300}{r + 30}$$

We multiply both sides by the LCD.
$200(r + 30) = 300(r - 30)$

$$(r + 30)(r - 30)\left(\frac{200}{r - 30}\right) = (r + 30)(r - 30)\left(\frac{300}{r + 30}\right)$$

Simplify.

$$(r + 30)(200) = (r - 30)(300)$$

$$200r + 6000 = 300r - 9000$$

Solve.

$$15000 = 100r$$
$$150 = r$$

Check.

Is 150 mph a reasonable speed for an airplane? Yes.
If the plane is traveling 150 mph and the wind is 30 mph:

Tailwind $150 + 30 = 180$mph $\frac{300}{180} = \frac{5}{3}$ hours

Headwind $150 - 30 = 120$mph $\frac{200}{120} = \frac{5}{3}$ hours

The times are equal, so it checks.   The plane was traveling 150 mph.

> **TRY IT : : 8.161**

   Link can ride his bike 20 miles into a 3 mph headwind in the same amount of time he can ride 30 miles with a 3 mph tailwind. What is Link's biking speed?

> **TRY IT : : 8.162**

   Judy can sail her boat 5 miles into a 7 mph headwind in the same amount of time she can sail 12 miles with a 7 mph tailwind. What is the speed of Judy's boat without a wind?

In the next example, we will know the total time resulting from travelling different distances at different speeds.

EXAMPLE 8.82

Jazmine trained for 3 hours on Saturday. She ran 8 miles and then biked 24 miles. Her biking speed is 4 mph faster than her running speed. What is her running speed?

⊘ Solution

This is a uniform motion situation. A diagram will help us visualize the situation.

We fill in the chart to organize the information.

		Rate	•	Time	=	Distance
Run		r		$\frac{8}{r}$		8
Bike		$r+4$		$\frac{24}{r+4}$		24
				3		

We are looking for Jazmine's running speed.

Let $r =$ Jazmine's running speed.

Her biking speed is 4 miles faster than her running speed.

$r+4 =$ her biking speed

The distances are given, enter them into the chart.

Since $D = r \cdot t$, we solve for t and get $t = \frac{D}{r}$.

We divide the distance by the rate in each row, and place the expression in the time column.

Write a word sentence.

Her time plus the time biking is 3 hours.

Translate the sentence to get the equation.

$$\frac{8}{r} + \frac{24}{r+4} = 3$$

Solve.

$$r(r+4)\left(\frac{8}{r} + \frac{24}{r+4}\right) = 3 \bullet r(r+4)$$
$$8(r+4) + 24r = 3r(r+4)$$
$$8r + 32 + 24r = 3r^2 + 12r$$
$$32 + 32r = 3r^2 + 12r$$
$$0 = 3r^2 - 20r - 32$$
$$0 = (3r+4)(r-8)$$

$$(3r+4) = 0 \quad (r-8) = 0$$

$$r = -\frac{4}{3} \quad r = 8$$

Check. $r = \cancel{-\frac{4}{3}}$ $r = 8$

A negative speed does not make sense in this problem, so $r = 8$ is the solution.

Is 8 mph a reasonable running speed? Yes.

Run 8 mph $\quad \frac{8 \text{ miles}}{8 \text{ mph}} = 1$ hour

Bike 12 mph $\quad \frac{24 \text{ miles}}{12 \text{ mph}} = 2$ hours

Total 3 hours $\qquad$ Jazmine's running speed is 8 mph.

> **TRY IT :: 8.163**

Dennis went cross-country skiing for 6 hours on Saturday. He skied 20 mile uphill and then 20 miles back downhill, returning to his starting point. His uphill speed was 5 mph slower than his downhill speed. What was Dennis' speed going uphill and his speed going downhill?

> **TRY IT :: 8.164**

Tony drove 4 hours to his home, driving 208 miles on the interstate and 40 miles on country roads. If he drove 15 mph faster on the interstate than on the country roads, what was his rate on the country roads?

Once again, we will use the uniform motion formula solved for the variable t.

EXAMPLE 8.83

Hamilton rode his bike downhill 12 miles on the river trail from his house to the ocean and then rode uphill to return home. His uphill speed was 8 miles per hour slower than his downhill speed. It took him 2 hours longer to get home than it took him to get to the ocean. Find Hamilton's downhill speed.

⊘ **Solution**

This is a uniform motion situation. A diagram will help us visualize the situation.

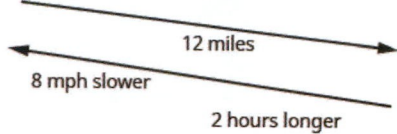

12 miles

8 mph slower

2 hours longer

We fill in the chart to organize the information.

We are looking for Hamilton's downhill speed.	Let $r =$ Hamilton's downhill speed.
His uphill speed is 8 miles per hour slower. Enter the rates into the chart.	$h - 8 =$ Hamilton's uphill speed

The distance is the same in both directions, 12 miles.
Since $D = r \cdot t$, we solve for t and get $t = \frac{D}{r}$.

We divide the distance by the rate in each row, and place the expression in the time column.

	Rate	•	Time	=	Distance
Downhill	h		$\frac{12}{h}$		12
Uphill	$h - 8$		$\frac{12}{h-8}$		12

Write a word sentence about the time.	He took 2 hours longer uphill than downhill. The uphill time is 2 more than the downhill time.

Translate the sentence to get the equation.

Solve.

$$\frac{12}{h-8} = \frac{12}{h} + 2$$

$$h(h-8)\left(\frac{12}{h-8}\right) = h(h-8)\left(\frac{12}{h} + 2\right)$$

$$12h = 12(h-8) + 2h(h-8)$$

$$12h = 12h - 96 + 2h^2 - 16h$$

$$0 = 2h^2 - 16h - 96$$

$$0 = 2(h^2 - 8h - 48)$$

$$0 = 2(h - 12)(h + 4)$$

$$h - 12 = 0 \quad h + 4 = 0$$

$$h = 12 \quad \cancel{h = -4}$$

Check. Is 12 mph a reasonable speed for biking downhill? Yes.

| Downhill | 12 mph | $\dfrac{12 \text{ miles}}{12 \text{ mph}} = 1$ hour |

| Uphill | $12 - 8 = 4$ mph | $\dfrac{12 \text{ miles}}{4 \text{ mph}} = 3$ hours |

The uphill time is 2 hours more than the downhill time. Hamilton's downhill speed is 12 mph.

> **TRY IT : : 8.165**
>
> Kayla rode her bike 75 miles home from college one weekend and then rode the bus back to college. It took her 2 hours less to ride back to college on the bus than it took her to ride home on her bike, and the average speed of the bus was 10 miles per hour faster than Kayla's biking speed. Find Kayla's biking speed.

> **TRY IT : : 8.166**
>
> Victoria jogs 12 miles to the park along a flat trail and then returns by jogging on an 18 mile hilly trail. She jogs 1 mile per hour slower on the hilly trail than on the flat trail, and her return trip takes her two hours longer. Find her rate of jogging on the flat trail.

Solve Work Applications

Suppose Pete can paint a room in 10 hours. If he works at a steady pace, in 1 hour he would paint $\frac{1}{10}$ of the room. If Alicia would take 8 hours to paint the same room, then in 1 hour she would paint $\frac{1}{8}$ of the room. How long would it take Pete and Alicia to paint the room if they worked together (and didn't interfere with each other's progress)?

This is a typical 'work' application. There are three quantities involved here – the time it would take each of the two people to do the job alone and the time it would take for them to do the job together.

Let's get back to Pete and Alicia painting the room. We will let t be the number of hours it would take them to paint the room together. So in 1 hour working together they have completed $\frac{1}{t}$ of the job.

	Rate	• Time	= Distance
Downhill	h	$\dfrac{12}{h}$	12
Uphill	$h - 8$	$\dfrac{12}{h-8}$	12

In one hour Pete did $\frac{1}{10}$ of the job. Alicia did $\frac{1}{8}$ of the job. And together they did $\frac{1}{t}$ of the job.

We can model this with the word equation and then translate to a rational equation. To find the time it would take them if they worked together, we solve for t.

$$\text{Pete's part} + \text{Alicia's part} = \text{part of total}$$

$$\frac{1}{10} \qquad \frac{1}{8} \qquad \frac{1}{t}$$

$$\frac{1}{10} + \frac{1}{8} = \frac{1}{t}$$

$$\frac{1}{10} + \frac{1}{8} = \frac{1}{t}$$

Multiply by the LCD, $40t$.	$40t\left(\frac{1}{10} + \frac{1}{8}\right) = 40t\left(\frac{1}{t}\right)$
Distribute.	$40t \cdot \frac{1}{10} + 40t \cdot \frac{1}{8} = 40t\left(\frac{1}{t}\right)$
Simplify and solve.	$4t + 5t = 40$
	$9t = 40$
	$t = \frac{40}{9}$
We'll write as a mixed number so that we can convert it to hours and minutes.	$t = 4\frac{4}{9}$ hours
Remember, 1 hour = 60 minutes.	$t = 4$ hours $+ \frac{4}{9}$(60 minutes)
Multiply, and then round to the nearest minute.	$t = 4$ hours $+ 27$ minutes
	It would take Pete and Alica about 4 hours and 27 minutes to paint the room.

Keep in mind, it should take less time for two people to complete a job working together than for either person to do it alone.

EXAMPLE 8.84

The weekly gossip magazine has a big story about the Princess' baby and the editor wants the magazine to be printed as soon as possible. She has asked the printer to run an extra printing press to get the printing done more quickly. Press #1 takes 6 hours to do the job and Press #2 takes 12 hours to do the job. How long will it take the printer to get the magazine printed with both presses running together?

⊘ **Solution**

This is a work problem. A chart will help us organize the information.

Let $t =$ the number of hours needed to complete the job together.

Enter the hours per job for Press #1, Press #2 and when they work together.

If a job on Press #1 takes 6 hours, then in 1 hour $\frac{1}{6}$ of the job is completed.
Similarly find the part of the job completed/hours for Press #2 and when they both work together.

	Number of hours to complete the job	Part of job completed/hour
Press #1	6	$\frac{1}{6}$
Press #2	12	$\frac{1}{12}$
Together	t	$\frac{1}{t}$

Write a word sentence.

The part completed by Press #1 plus the part completed by Press #2 equals the amount completed together.

Translate to an equation.

Work completed by
Press #1 + Press #2 = Together

$$\frac{1}{6} \quad + \quad \frac{1}{12} \quad = \quad \frac{1}{t}$$

Solve.

$$\frac{1}{6} + \frac{1}{12} = \frac{1}{t}$$

Multiply by the LCD, $12t$.

$$12t\left(\frac{1}{6} + \frac{1}{12}\right) = 12t\left(\frac{1}{t}\right)$$

Simplify.

$$2t + t = 12$$

$$3t = 12$$

$$t = 4$$

When both presses are running it takes 4 hours to do the job.

> **TRY IT : : 8.167**

One gardener can mow a golf course in 4 hours, while another gardener can mow the same golf course in 6 hours. How long would it take if the two gardeners worked together to mow the golf course?

> **TRY IT : : 8.168**

Carrie can weed the garden in 7 hours, while her mother can do it in 3. How long will it take the two of them working together?

EXAMPLE 8.85

Corey can shovel all the snow from the sidewalk and driveway in 4 hours. If he and his twin Casey work together, they can finish shoveling the snow in 2 hours. How many hours would it take Casey to do the job by himself?

⊘ Solution

This is a work application. A chart will help us organize the information.

We are looking for how many hours it would take Casey to complete the job by himself.

Let $t =$ the number of hours needed for Casey to complete.

Enter the hours per job for Corey, Casey, and when they work together.

If Corey takes 4 hours, then in 1 hour $\frac{1}{4}$ of the job is

completed. Similarly find the part of the job completed/ hours for Casey and when they both work together.

	Number of hours needed to complete the job	Part of job completed/ hour
Corey	4	$\frac{1}{4}$
Casey	t	$\frac{1}{t}$
Together	2	$\frac{1}{2}$

Write a word sentence.

The part completed by Corey plus the part completed by Casey equals the amount completed together.

Translate to an equation:

$$\text{Corey} + \text{Casey} = \text{Together}$$
$$\frac{1}{4} + \frac{1}{t} = \frac{1}{2}$$

Solve.

$$\frac{1}{4} + \frac{1}{t} = \frac{1}{2}$$

Multiply by the LCD, $4t$.

$$4t\left(\frac{1}{4} + \frac{1}{t}\right) = 4t\left(\frac{1}{2}\right)$$

Simplify.

$$t + 4 = 2t$$

$$4 = t$$

It would take Casey 4 hours to do the job alone.

> **TRY IT : : 8.169**

Two hoses can fill a swimming pool in 10 hours. It would take one hose 26 hours to fill the pool by itself. How long would it take for the other hose, working alone, to fill the pool?

> **TRY IT : : 8.170**

Cara and Cindy, working together, can rake the yard in 4 hours. Working alone, it takes Cindy 6 hours to rake the yard. How long would it take Cara to rake the yard alone?

8.8 EXERCISES

Practice Makes Perfect

Solve Uniform Motion Applications

In the following exercises, solve uniform motion applications

429. Mary takes a sightseeing tour on a helicopter that can fly 450 miles against a 35 mph headwind in the same amount of time it can travel 702 miles with a 35 mph tailwind. Find the speed of the helicopter.

430. A private jet can fly 1210 miles against a 25 mph headwind in the same amount of time it can fly 1694 miles with a 25 mph tailwind. Find the speed of the jet.

431. A boat travels 140 miles downstream in the same time as it travels 92 miles upstream. The speed of the current is 6mph. What is the speed of the boat?

432. Darrin can skateboard 2 miles against a 4 mph wind in the same amount of time he skateboards 6 miles with a 4 mph wind. Find the speed Darrin skateboards with no wind.

433. Jane spent 2 hours exploring a mountain with a dirt bike. When she rode the 40 miles uphill, she went 5 mph slower than when she reached the peak and rode for 12 miles along the summit. What was her rate along the summit?

434. Jill wanted to lose some weight so she planned a day of exercising. She spent a total of 2 hours riding her bike and jogging. She biked for 12 miles and jogged for 6 miles. Her rate for jogging was 10 mph less than biking rate. What was her rate when jogging?

435. Bill wanted to try out different water craft. He went 62 miles downstream in a motor boat and 27 miles downstream on a jet ski. His speed on the jet ski was 10 mph faster than in the motor boat. Bill spent a total of 4 hours on the water. What was his rate of speed in the motor boat?

436. Nancy took a 3 hour drive. She went 50 miles before she got caught in a storm. Then she drove 68 miles at 9 mph less than she had driven when the weather was good. What was her speed driving in the storm?

437. Chester rode his bike uphill 24 miles and then back downhill at 2 mph faster than his uphill. If it took him 2 hours longer to ride uphill than downhill, l, what was his uphill rate?

438. Matthew jogged to his friend's house 12 miles away and then got a ride back home. It took him 2 hours longer to jog there than ride back. His jogging rate was 25 mph slower than the rate when he was riding. What was his jogging rate?

439. Hudson travels 1080 miles in a jet and then 240 miles by car to get to a business meeting. The jet goes 300 mph faster than the rate of the car, and the car ride takes 1 hour longer than the jet. What is the speed of the car?

440. Nathan walked on an asphalt pathway for 12 miles. He walked the 12 miles back to his car on a gravel road through the forest. On the asphalt he walked 2 miles per hour faster than on the gravel. The walk on the gravel took one hour longer than the walk on the asphalt. How fast did he walk on the gravel?

441. John can fly his airplane 2800 miles with a wind speed of 50 mph in the same time he can travel 2400 miles against the wind. If the speed of the wind is 50 mph, find the speed of his airplane.

442. Jim's speedboat can travel 20 miles upstream against a 3 mph current in the same amount of time it travels 22 miles downstream with a 3 mph current speed. Find the speed of the Jim's boat.

443. Hazel needs to get to her granddaughter's house by taking an airplane and a rental car. She travels 900 miles by plane and 250 miles by car. The plane travels 250 mph faster than the car. If she drives the rental car for 2 hours more than she rode the plane, find the speed of the car.

444. Stu trained for 3 hours yesterday. He ran 14 miles and then biked 40 miles. His biking speed is 6 mph faster than his running speed. What is his running speed?

445. When driving the 9 hour trip home, Sharon drove 390 miles on the interstate and 150 miles on country roads. Her speed on the interstate was 15 more than on country roads. What was her speed on country roads?

446. Two sisters like to compete on their bike rides. Tamara can go 4 mph faster than her sister, Samantha. If it takes Samantha 1 hours longer than Tamara to go 80 miles, how fast can Samantha ride her bike?

Solve Work Applications

In the following exercises, solve work applications.

447. Mike, an experienced bricklayer, can build a wall in 3 hours, while his son, who is learning, can do the job in 6 hours. How long does it take for them to build a wall together?

448. It takes Sam 4 hours to rake the front lawn while his brother, Dave, can rake the lawn in 2 hours. How long will it take them to rake the lawn working together?

449. Mary can clean her apartment in 6 hours while her roommate can clean the apartment in 5 hours. If they work together, how long would it take them to clean the apartment?

450. Brian can lay a slab of concrete in 6 hours, while Greg can do it in 4 hours. If Brian and Greg work together, how long will it take?

451. Leeson can proofread a newspaper copy in 4 hours. If Ryan helps, they can do the job in 3 hours. How long would it take for Ryan to do his job alone?

452. Paul can clean a classroom floor in 3 hours. When his assistant helps him, the job takes 2 hours. How long would it take the assistant to do it alone?

453. Josephine can correct her students' test papers in 5 hours, but if her teacher's assistant helps, it would take them 3 hours. How long would it take the assistant to do it alone?

454. Washing his dad's car alone, eight year old Levi takes 2.5 hours. If his dad helps him, then it takes 1 hour. How long does it take the Levi's dad to wash the car by himself?

455. Jackson can remove the shingles off of a house in 7 hours, while Martin can remove the shingles in 5 hours. How long will it take them to remove the shingles if they work together?

456. At the end of the day Dodie can clean her hair salon in 15 minutes. Ann, who works with her, can clean the salon in 30 minutes. How long would it take them to clean the shop if they work together?

457. Ronald can shovel the driveway in 4 hours, but if his brother Donald helps it would take 2 hours. How long would it take Donald to shovel the driveway alone?

458. It takes Tina 3 hours to frost her holiday cookies, but if Candy helps her it takes 2 hours. How long would it take Candy to frost the holiday cookies by herself?

Everyday Math

459. Dana enjoys taking her dog for a walk, but sometimes her dog gets away and she has to run after him. Dana walked her dog for 7 miles but then had to run for 1 mile, spending a total time of 2.5 hours with her dog. Her running speed was 3 mph faster than her walking speed. Find her walking speed.

460. Ken and Joe leave their apartment to go to a football game 45 miles away. Ken drives his car 30 mph faster Joe can ride his bike. If it takes Joe 2 hours longer than Ken to get to the game, what is Joe's speed?

Writing Exercises

461. In **Example 8.83**, the solution $h = -4$ is crossed out. Explain why.

462. Paula and Yuki are roommates. It takes Paula 3 hours to clean their apartment. It takes Yuki 4 hours to clean the apartment. The equation $\frac{1}{3} + \frac{1}{4} = \frac{1}{t}$ can be used to find t, the number of hours it would take both of them, working together, to clean their apartment. Explain how this equation models the situation.

Self Check

ⓐ *After completing the exercises, use this checklist to evaluate your mastery of the objectives of this section.*

I can...	Confidently	With some help	No-I don't get it!
solve uniform motion applications.			
solve work applications.			

ⓑ *On a scale of 1–10, how would you rate your mastery of this section in light of your responses on the checklist? How can you improve this?*

8.9 Use Direct and Inverse Variation

Learning Objectives

By the end of this section, you will be able to:

› Solve direct variation problems
› Solve inverse variation problems

Be Prepared!

Before you get started, take this readiness quiz.

If you miss a problem, go back to the section listed and review the material.

1. Find the multiplicative inverse of -8.
   If you missed this problem, review **Example 1.126**.

2. Solve for n: $45 = 20n$.
   If you missed this problem, review **Example 2.13**.

3. Evaluate $5x^2$ when $x = 10$.
   If you missed this problem, review **Example 1.20**.

When two quantities are related by a proportion, we say they are *proportional* to each other. Another way to express this relation is to talk about the *variation* of the two quantities. We will discuss direct variation and inverse variation in this section.

Solve Direct Variation Problems

Lindsay gets paid $15 per hour at her job. If we let s be her salary and h be the number of hours she has worked, we could model this situation with the equation

$$s = 15h$$

Lindsay's salary is the product of a constant, 15, and the number of hours she works. We say that Lindsay's salary *varies directly* with the number of hours she works. Two variables vary directly if one is the product of a constant and the other.

Direct Variation

For any two variables x and y, y varies directly with x if

$$y = kx, \quad \text{where } k \neq 0$$

The constant k is called the constant of variation.

In applications using direct variation, generally we will know values of one pair of the variables and will be asked to find the equation that relates x and y. Then we can use that equation to find values of y for other values of x.

EXAMPLE 8.86   HOW TO SOLVE DIRECT VARIATION PROBLEMS

If y varies directly with x and $y = 20$ when $x = 8$, find the equation that relates x and y.

⊘ **Solution**

Step 1. Write the formula for direct variation.	The direct variation formula is $y = kx$.	$y = kx$
Step 2. Substitute the given values for the variables.	We are given $y = 20$, $x = 8$.	$20 = k \cdot 8$
Step 3. Solve for the constant of variation.	Divide both sides of the equation by 8, then simplify.	$\dfrac{20}{8} = k$ $k = 2.5$
Step 4. Write the equation that relates x and y.	Rewrite the general equation with the value we found for k.	$y = 2.5x$

> **TRY IT ::** 8.171   If y varies directly as x and $y = 3$, when $x = 10$. find the equation that relates x and y.

> **TRY IT ::** 8.172   If y varies directly as x and $y = 12$ when $x = 4$ find the equation that relates x and y.

We'll list the steps below.

HOW TO :: SOLVE DIRECT VARIATION PROBLEMS.

Step 1.   Write the formula for direct variation.
Step 2.   Substitute the given values for the variables.
Step 3.   Solve for the constant of variation.
Step 4.   Write the equation that relates x and y.

Now we'll solve a few applications of direct variation.

EXAMPLE 8.87

When Raoul runs on the treadmill at the gym, the number of calories, c, he burns varies directly with the number of minutes, m, he uses the treadmill. He burned 315 calories when he used the treadmill for 18 minutes.

      ⓐ Write the equation that relates c and m.

      ⓑ How many calories would he burn if he ran on the treadmill for 25 minutes?

✓ **Solution**

ⓐ

	The number of calories, c, varies directly with the number of minutes, m, on the treadmill, and $c = 315$ when $m = 18$.
Write the formula for direct variation.	$y = kx$
We will use c in place of y and m in place of x.	$c = km$
Substitute the given values for the variables.	$315 = k \cdot 18$
Solve for the constant of variation.	$\dfrac{315}{18} = \dfrac{k \cdot 18}{18}$
	$17.5 = k$
Write the equation that relates c and m.	$c = km$
Substitute in the constant of variation.	$c = 17.5m$

ⓑ

	Find c when $m = 25$.
Write the equation that relates c and m.	$c = 17.5m$
Substitute the given value for m.	$c = 17.5(25)$
Simplify.	$c = 437.5$

Raoul would burn 437.5 calories if he used the treadmill for 25 minutes.

> **TRY IT :: 8.173**

The number of calories, *c*, burned varies directly with the amount of time, *t*, spent exercising. Arnold burned 312 calories in 65 minutes exercising.

    ⓐ Write the equation that relates *c* and *t*.

    ⓑ How many calories would he burn if he exercises for 90 minutes?

> **TRY IT :: 8.174**

The distance a moving body travels, *d*, varies directly with time, *t*, it moves. A train travels 100 miles in 2 hours

    ⓐ Write the equation that relates *d* and *t*.

    ⓑ How many miles would it travel in 5 hours?

In the previous example, the variables *c* and *m* were named in the problem. Usually that is not the case. We will have to name the variables in the next example as part of the solution, just like we do in most applied problems.

EXAMPLE 8.88

The number of gallons of gas Eunice's car uses varies directly with the number of miles she drives. Last week she drove 469.8 miles and used 14.5 gallons of gas.

    ⓐ Write the equation that relates the number of gallons of gas used to the number of miles driven.

    ⓑ How many gallons of gas would Eunice's car use if she drove 1000 miles?

 Solution

	The number of gallons of gas varies directly with the number of miles driven.
First we will name the variables.	Let $g =$ number of gallons of gas. $m =$ number of miles driven
Write the formula for direct variation.	$y = kx$
We will use g in place of y and m in place of x.	$g = km$
Substitute the given values for the variables.	$g = 14.5$ when $m = 469.8$
	$14.5 = k(469.8)$
Solve for the constant of variation.	$\dfrac{14.5}{469.8} = \dfrac{k(469.8)}{469.8}$
We will round to the nearest thousandth.	$0.031 = k$
Write the equation that relates g and m.	$g = km$
Substitute in the constant of variation.	$g = 0.031m$

ⓑ

	Find g when $m = 1000$.
Write the equation that relates g and m.	$g = 0.031m$
Substitute the given value for m.	$g = 0.031(1000)$
Simplify.	$g = 31$
	Eunice's car would use 31 gallons of gas if she drove it 1,000 miles.

Notice that in this example, the units on the constant of variation are gallons/mile. In everyday life, we usually talk about miles/gallon.

> **TRY IT :: 8.175**

   The distance that Brad travels varies directly with the time spent traveling. Brad travelled 660 miles in 12 hours,

   ⓐ Write the equation that relates the number of miles travelled to the time.

   ⓑ How many miles could Brad travel in 4 hours?

> **TRY IT :: 8.176**

   The weight of a liquid varies directly as its volume. A liquid that weighs 24 pounds has a volume of 4 gallons.

   ⓐ Write the equation that relates the weight to the volume.

   ⓑ If a liquid has volume 13 gallons, what is its weight?

In some situations, one variable varies directly with the square of the other variable. When that happens, the equation of direct variation is $y = kx^2$. We solve these applications just as we did the previous ones, by substituting the given values into the equation to solve for k.

EXAMPLE 8.89

The maximum load a beam will support varies directly with the square of the diagonal of the beam's cross-section. A beam with diagonal 4" will support a maximum load of 75 pounds.

   ⓐ Write the equation that relates the maximum load to the cross-section.

   ⓑ What is the maximum load that can be supported by a beam with diagonal 8"?

⊘ **Solution**

	The maximum load varies directly with the square of the diagonal of the cross-section.
Name the variables.	Let $L =$ maximum load. $c =$ the diagonal of the cross-section
Write the formula for direct variation, where y varies directly with the square of x.	$y = kx^2$
We will use L in place of y and c in place of x.	$L = kc^2$
Substitute the given values for the variables.	$L = 75$ when $c = 4$
	$75 = k \cdot 4^2$

Solve for the constant of variation.	$\frac{75}{16} = \frac{k \cdot 16}{16}$
	$4.6875 = k$
Write the equation that relates L and c.	$L = kc^2$
Substitute in the constant of variation.	$L = 4.6875c^2$

ⓑ

	Find L when $c = 8$.
Write the equation that relates L and c.	$L = 4.6875c^2$
Substitute the given value for c.	$L = 4.6875(8)^2$
Simplify.	$L = 300$

A beam with diagonal 8" could support a maximum load of 300 pounds.

> **TRY IT : : 8.177**

The distance an object falls is directly proportional to the square of the time it falls. A ball falls 144 feet in 3 seconds.

ⓐ Write the equation that relates the distance to the time.

ⓑ How far will an object fall in 4 seconds?

> **TRY IT : : 8.178**

The area of a circle varies directly as the square of the radius. A circular pizza with a radius of 6 inches has an area of 113.04 square inches.

ⓐ Write the equation that relates the area to the radius.

ⓑ What is the area of a pizza with a radius of 9 inches?

Solve Inverse Variation Problems

Many applications involve two variable that *vary inversely*. As one variable increases, the other decreases. The equation that relates them is $y = \frac{k}{x}$.

Inverse Variation

For any two variables x and y, y varies inversely with x if

$$y = \frac{k}{x}, \text{ where } k \neq 0$$

The constant k is called the constant of variation.

The word 'inverse' in inverse variation refers to the multiplicative inverse. The multiplicative inverse of x is $\frac{1}{x}$.

We solve inverse variation problems in the same way we solved direct variation problems. Only the general form of the equation has changed. We will copy the procedure box here and just change 'direct' to 'inverse'.

HOW TO :: SOLVE INVERSE VARIATION PROBLEMS.

Step 1.   Write the formula for inverse variation.

Step 2.   Substitute the given values for the variables.

Step 3.   Solve for the constant of variation.

Step 4.   Write the equation that relates x and y.

EXAMPLE 8.90

If y varies inversely with x and $y = 20$ when $x = 8$, find the equation that relates x and y.

⊘ **Solution**

Write the formula for inverse variation.	$y = \frac{k}{x}$
Substitute the given values for the variables.	$y = 20$ when $x = 8$
	$20 = \frac{k}{8}$
Solve for the constant of variation.	$8(20) = 8\left(\frac{k}{8}\right)$
	$160 = k$
Write the equation that relates x and y.	$y = \frac{k}{x}$
Substitute in the constant of variation.	$y = \frac{160}{x}$

> **TRY IT :: 8.179**
>
> If p varies inversely with q and $p = 30$ when $q = 12$ find the equation that relates p and q.

> **TRY IT :: 8.180**
>
> If y varies inversely with x and $y = 8$ when $x = 2$ find the equation that relates x and y.

EXAMPLE 8.91

The fuel consumption (mpg) of a car varies inversely with its weight. A car that weighs 3100 pounds gets 26 mpg on the highway.

ⓐ Write the equation of variation.

ⓑ What would be the fuel consumption of a car that weighs 4030 pounds?

⊘ **Solution**

ⓐ

The fuel consumption varies inversely with the weight.

First we will name the variables.	Let $f =$ fuel consumption. $w =$ weight
Write the formula for inverse variation.	$y = \frac{k}{x}$
We will use f in place of y and w in place of x.	$f = \frac{k}{w}$
Substitute the given values for the variables.	$f = 26$ when $w = 3100$
	$26 = \frac{k}{3100}$
Solve for the constant of variation.	$3100(26) = 3100\left(\frac{k}{3100}\right)$
	$80{,}600 = k$
Write the equation that relates f and w.	$f = \frac{k}{w}$
Substitute in the constant of variation.	$f = \frac{80{,}600}{w}$

ⓑ

	Find f when $w = 4030$.
Write the equation that relates f and w.	$f = \frac{80{,}600}{w}$
Substitute the given value for w.	$f = \frac{80{,}600}{4030}$
Simplify.	$f = 20$

A car that weighs 4030 pounds would have fuel consumption of 20 mpg.

> **TRY IT : : 8.181**   A car's value varies inversely with its age. Elena bought a two-year-old car for $20,000.

ⓐ Write the equation of variation.

ⓑ What will be the value of Elena's car when it is 5 years old?

> **TRY IT : : 8.182**

The time required to empty a pool varies inversely as the rate of pumping. It took Lucy 2.5 hours to empty her pool using a pump that was rated at 400 gpm (gallons per minute).

ⓐ Write the equation of variation.

ⓑ How long will it take her to empty the pool using a pump rated at 500 gpm?

EXAMPLE 8.92

The frequency of a guitar string varies inversely with its length. A 26" long string has a frequency of 440 vibrations per second.

ⓐ Write the equation of variation.

ⓑ How many vibrations per second will there be if the string's length is reduced to 20" by putting a finger on a fret?

✓ **Solution**

ⓐ

	The frequency varies inversely with the length.
Name the variables.	Let $f=$ frequency. $L=$ length
Write the formula for inverse variation.	$y=\frac{k}{x}$
We will use f in place of y and L in place of x.	$f=\frac{k}{L}$
Substitute the given values for the variables.	$f=440$ when $L=26$
	$440=\frac{k}{26}$
Solve for the constant of variation.	$26(440)=26\left(\frac{k}{26}\right)$
	$11{,}440=k$
Write the equation that relates f and L.	$f=\frac{k}{L}$
Substitute in the constant of variation.	$f=\frac{11{,}440}{L}$

ⓑ

Find f when $L=20$.

Write the equation that relates f and L. $\qquad f=\frac{11{,}440}{L}$

Substitute the given value for L. $\qquad f=\frac{11{,}440}{20}$

Simplify. $\qquad f=572$

A 20" guitar string has frequency 572 vibrations per second.

> **TRY IT :: 8.183**

The number of hours it takes for ice to melt varies inversely with the air temperature. Suppose a block of ice melts in 2 hours when the temperature is 65 degrees.

    ⓐ Write the equation of variation.

    ⓑ How many hours would it take for the same block of ice to melt if the temperature was 78 degrees?

> **TRY IT :: 8.184**

The force needed to break a board varies inversely with its length. Richard uses 24 pounds of pressure to break a 2-foot long board.

    ⓐ Write the equation of variation.

    ⓑ How many pounds of pressure is needed to break a 5-foot long board?

8.9 EXERCISES

Practice Makes Perfect

Solve Direct Variation Problems

In the following exercises, solve.

463. If y varies directly as x and $y = 14$, when $x = 3$, find the equation that relates x and y.

464. If p varies directly as q and $p = 5$, when $q = 2$, find the equation that relates p and q.

465. If v varies directly as w and $v = 24$, when $w = 8$, find the equation that relates v and w.

466. If a varies directly as b and $a = 16$, when $b = 4$, find the equation that relates a and b.

467. If p varies directly as q and $p = 9.6$, when $q = 3$, find the equation that relates p and q.

468. If y varies directly as x and $y = 12.4$, when $x = 4$, find the equation that relates x and y

469. If a varies directly as b and $a = 6$, when $b = \frac{1}{3}$, find the equation that relates a and b.

470. If v varies directly as w and $v = 8$, when $w = \frac{1}{2}$, find the equation that relates v and w.

471. The amount of money Sally earns, P, varies directly with the number, n, of necklaces she sells. When Sally sells 15 necklaces she earns $150.

ⓐ Write the equation that relates P and n.

ⓑ How much money would she earn if she sold 4 necklaces?

472. The price, P, that Eric pays for gas varies directly with the number of gallons, g, he buys. It costs him $50 to buy 20 gallons of gas.

ⓐ Write the equation that relates P and g.

ⓑ How much would 33 gallons cost Eric?

473. Terri needs to make some pies for a fundraiser. The number of apples, a, varies directly with number of pies, p. It takes nine apples to make two pies.

ⓐ Write the equation that relates a and p.

ⓑ How many apples would Terri need for six pies?

474. Joseph is traveling on a road trip. The distance, d, he travels before stopping for lunch varies directly with the speed, v, he travels. He can travel 120 miles at a speed of 60 mph.

ⓐ Write the equation that relates d and v.

ⓑ How far would he travel before stopping for lunch at a rate of 65 mph?

475. The price of gas that Jesse purchased varies directly to how many gallons he purchased. He purchased 10 gallons of gas for $39.80.

ⓐ Write the equation that relates the price to the number of gallons.

ⓑ How much will it cost Jesse for 15 gallons of gas?

476. The distance that Sarah travels varies directly to how long she drives. She travels 440 miles in 8 hours.

ⓐ Write the equation that relates the distance to the number of hours.

ⓑ How far can Sally travel in 6 hours?

477. The mass of a liquid varies directly with its volume. A liquid with mass 16 kilograms has a volume of 2 liters.

ⓐ Write the equation that relates the mass to the volume.

ⓑ What is the volume of this liquid if its mass is 128 kilograms?

478. The length that a spring stretches varies directly with a weight placed at the end of the spring. When Sarah placed a 10 pound watermelon on a hanging scale, the spring stretched 5 inches.

ⓐ Write the equation that relates the length of the spring to the weight.

ⓑ What weight of watermelon would stretch the spring 6 inches?

479. The distance an object falls varies directly to the square of the time it falls. A ball falls 45 feet in 3 seconds.

ⓐ Write the equation that relates the distance to the time.

ⓑ How far will the ball fall in 7 seconds?

480. The maximum load a beam will support varies directly with the square of the diagonal of the beam's cross-section. A beam with diagonal 6 inch will support a maximum load of 108 pounds.

ⓐ Write the equation that relates the load to the diagonal of the cross-section.

ⓑ What load will a beam with a 10 inch diagonal support?

481. The area of a circle varies directly as the square of the radius. A circular pizza with a radius of 6 inches has an area of 113.04 square inches.

ⓐ Write the equation that relates the area to the radius.

ⓑ What is the area of a personal pizza with a radius 4 inches?

482. The distance an object falls varies directly to the square of the time it falls. A ball falls 72 feet in 3 seconds,

ⓐ Write the equation that relates the distance to the time.

ⓑ How far will the ball have fallen in 8 seconds?

Solve Inverse Variation Problems

In the following exercises, solve.

483. If y varies inversely with x and $y = 5$ when $x = 4$ find the equation that relates x and y.

484. If p varies inversely with q and $p = 2$ when $q = 1$ find the equation that relates p and q.

485. If v varies inversely with w and $v = 6$ when $w = \frac{1}{2}$ find the equation that relates v and w.

486. If a varies inversely with b and $a = 12$ when $b = \frac{1}{3}$ find the equation that relates a and b.

Write an inverse variation equation to solve the following problems.

487. The fuel consumption (mpg) of a car varies inversely with its weight. A Toyota Corolla weighs 2800 pounds and gets 33 mpg on the highway.

ⓐ Write the equation that relates the mpg to the car's weight.

ⓑ What would the fuel consumption be for a Toyota Sequoia that weighs 5500 pounds?

488. A car's value varies inversely with its age. Jackie bought a 10 year old car for $2,400.

ⓐ Write the equation that relates the car's value to its age.

ⓑ What will be the value of Jackie's car when it is 15 years old ?

489. The time required to empty a tank varies inversely as the rate of pumping. It took Janet 5 hours to pump her flooded basement using a pump that was rated at 200 gpm (gallons per minute),

ⓐ Write the equation that relates the number of hours to the pump rate.

ⓑ How long would it take Janet to pump her basement if she used a pump rated at 400 gpm?

490. The volume of a gas in a container varies inversely as pressure on the gas. A container of helium has a volume of 370 cubic inches under a pressure of 15 psi.

ⓐ Write the equation that relates the volume to the pressure.

ⓑ What would be the volume of this gas if the pressure was increased to 20 psi?

491. On a string instrument, the length of a string varies inversely as the frequency of its vibrations. An 11-inch string on a violin has a frequency of 400 cycles per second.

ⓐ Write the equation that relates the string length to its frequency.

ⓑ What is the frequency of a 10-inch string?

492. Paul, a dentist, determined that the number of cavities that develops in his patient's mouth each year varies inversely to the number of minutes spent brushing each night. His patient, Lori, had 4 cavities when brushing her teeth 30 seconds (0.5 minutes) each night.

ⓐ Write the equation that relates the number of cavities to the time spent brushing.

ⓑ How many cavities would Paul expect Lori to have if she had brushed her teeth for 2 minutes each night?

493. The number of tickets for a sports fundraiser varies inversely to the price of each ticket. Brianna can buy 25 tickets at $5each.

ⓐ Write the equation that relates the number of tickets to the price of each ticket.

ⓑ How many tickets could Brianna buy if the price of each ticket was $2.50?

494. Boyle's Law states that if the temperature of a gas stays constant, then the pressure varies inversely to the volume of the gas. Braydon, a scuba diver, has a tank that holds 6 liters of air under a pressure of 220 psi.

ⓐ Write the equation that relates pressure to volume.

ⓑ If the pressure increases to 330 psi, how much air can Braydon's tank hold?

Mixed Practice

495. If y varies directly as x and $y = 5,$ when $x = 3.$, find the equation that relates x and y.

496. If v varies directly as w and $v = 21,$ when $w = 8.$ find the equation that relates v and w.

497. If p varies inversely with q and $p = 5$ when $q = 6$, find the equation that relates p and q.

498. If y varies inversely with x and $y = 11$ when $x = 3$ find the equation that relates x and y.

499. If p varies directly as q and $p = 10,$ when $q = 2.$ find the equation that relates p and q.

500. If v varies inversely with w and $v = 18$ when $w = \frac{1}{3}$ find the equation that relates v and w.

501. The force needed to break a board varies inversely with its length. If Tom uses 20 pounds of pressure to break a 1.5-foot long board, how many pounds of pressure would he need to use to break a 6 foot long board?

502. The number of hours it takes for ice to melt varies inversely with the air temperature. A block of ice melts in 2.5 hours when the temperature is 54 degrees. How long would it take for the same block of ice to melt if the temperature was 45 degrees?

503. The length a spring stretches varies directly with a weight placed at the end of the spring. When Meredith placed a 6-pound cantaloupe on a hanging scale, the spring stretched 2 inches. How far would the spring stretch if the cantaloupe weighed 9 pounds?

504. The amount that June gets paid varies directly the number of hours she works. When she worked 15 hours, she got paid $111. How much will she be paid for working 18 hours?

505. The fuel consumption (mpg) of a car varies inversely with its weight. A Ford Focus weighs 3000 pounds and gets 28.7 mpg on the highway. What would the fuel consumption be for a Ford Expedition that weighs 5,500 pounds? Round to the nearest tenth.

506. The volume of a gas in a container varies inversely as the pressure on the gas. If a container of argon has a volume of 336 cubic inches under a pressure of 2,500 psi, what will be its volume if the pressure is decreased to 2,000 psi?

507. The distance an object falls varies directly to the square of the time it falls. If an object falls 52.8 feet in 4 seconds, how far will it fall in 9 seconds?

508. The area of the face of a Ferris wheel varies directly with the square of its radius. If the area of one face of a Ferris wheel with diameter 150 feet is 70,650 square feet, what is the area of one face of a Ferris wheel with diameter of 16 feet?

Everyday Math

509. Ride Service It costs $35 for a ride from the city center to the airport, 14 miles away.

ⓐ Write the equation that relates the cost, *c*, with the number of miles, *m*.

ⓑ What would it cost to travel 22 miles with this service?

510. Road Trip The number of hours it takes Jack to drive from Boston to Bangor is inversely proportional to his average driving speed. When he drives at an average speed of 40 miles per hour, it takes him 6 hours for the trip.

ⓐ Write the equation that relates the number of hours, *h*, with the speed, *s*.

ⓑ How long would the trip take if his average speed was 75 miles per hour?

Writing Exercises

511. In your own words, explain the difference between direct variation and inverse variation.

512. Make up an example from your life experience of inverse variation.

Self Check

ⓐ *After completing the exercises, use this checklist to evaluate your mastery of the objectives of this section.*

I can...	Confidently	With some help	No-I don't get it!
solve direct variation problems.			
solve inverse variation problems.			

ⓑ *After looking at the checklist, do you think you are well-prepared for the next chapter? Why or why not?*

CHAPTER 8 REVIEW

KEY TERMS

complex rational expression A complex rational expression is a rational expression in which the numerator or denominator contains a rational expression.

extraneous solution to a rational equation An extraneous solution to a rational equation is an algebraic solution that would cause any of the expressions in the original equation to be undefined.

proportion A proportion is an equation of the form $\frac{a}{b} = \frac{c}{d}$, where $b \neq 0$, $d \neq 0$. The proportion is read " a is to b, as c is to d."

rational equation A rational equation is two rational expressions connected by an equal sign.

rational expression A rational expression is an expression of the form $\frac{p}{q}$, where p and q are polynomials and $q \neq 0$.

similar figures Two figures are similar if the measures of their corresponding angles are equal and their corresponding sides are in the same ratio.

KEY CONCEPTS

8.1 Simplify Rational Expressions

- **Determine the Values for Which a Rational Expression is Undefined**

  Step 1. Set the denominator equal to zero.

  Step 2. Solve the equation, if possible.

- **Simplified Rational Expression**

  ○ A rational expression is considered simplified if there are no common factors in its numerator and denominator.

- **Simplify a Rational Expression**

  Step 1. Factor the numerator and denominator completely.

  Step 2. Simplify by dividing out common factors.

- **Opposites in a Rational Expression**

  ○ The opposite of $a - b$ is $b - a$.

  $$\frac{a-b}{b-a} = -1 \qquad a \neq 0,\ b \neq 0,\ a \neq b$$

8.2 Multiply and Divide Rational Expressions

- **Multiplication of Rational Expressions**

  ○ If $p,\ q,\ r,\ s$ are polynomials where $q \neq 0$, $s \neq 0$, then $\frac{p}{q} \cdot \frac{r}{s} = \frac{pr}{qs}$.

  ○ To multiply rational expressions, multiply the numerators and multiply the denominators

- **Multiply a Rational Expression**

  Step 1. Factor each numerator and denominator completely.

  Step 2. Multiply the numerators and denominators.

  Step 3. Simplify by dividing out common factors.

- **Division of Rational Expressions**

  ○ If $p,\ q,\ r,\ s$ are polynomials where $q \neq 0$, $r \neq 0$, $s \neq 0$, then $\frac{p}{q} \div \frac{r}{s} = \frac{p}{q} \cdot \frac{s}{r}$.

  ○ To divide rational expressions multiply the first fraction by the reciprocal of the second.

- **Divide Rational Expressions**

  Step 1. Rewrite the division as the product of the first rational expression and the reciprocal of the second.

  Step 2. Factor the numerators and denominators completely.

Step 3. Multiply the numerators and denominators together.

Step 4. Simplify by dividing out common factors.

8.3 Add and Subtract Rational Expressions with a Common Denominator

- **Rational Expression Addition**
  - If p, q, and r are polynomials where $r \neq 0$, then

$$\frac{p}{r} + \frac{q}{r} = \frac{p+q}{r}$$

  - To add rational expressions with a common denominator, add the numerators and place the sum over the common denominator.
- **Rational Expression Subtraction**
  - If p, q, and r are polynomials where $r \neq 0$, then

$$\frac{p}{r} - \frac{q}{r} = \frac{p-q}{r}$$

  - To subtract rational expressions, subtract the numerators and place the difference over the common denominator.

8.4 Add and Subtract Rational Expressions with Unlike Denominators

- **Find the Least Common Denominator of Rational Expressions**

  Step 1. Factor each expression completely.

  Step 2. List the factors of each expression. Match factors vertically when possible.

  Step 3. Bring down the columns.

  Step 4. Multiply the factors.

- **Add or Subtract Rational Expressions**

  Step 1. Determine if the expressions have a common denominator.
  Yes – go to step 2.
  No – Rewrite each rational expression with the LCD.

  - Find the LCD.
  - Rewrite each rational expression as an equivalent rational expression with the LCD.

  Step 2. Add or subtract the rational expressions.

  Step 3. Simplify, if possible.

8.5 Simplify Complex Rational Expressions

- **To Simplify a Rational Expression by Writing it as Division**

  Step 1. Simplify the numerator and denominator.

  Step 2. Rewrite the complex rational expression as a division problem.

  Step 3. Divide the expressions.

- **To Simplify a Complex Rational Expression by Using the LCD**

  Step 1. Find the LCD of all fractions in the complex rational expression.

  Step 2. Multiply the numerator and denominator by the LCD.

  Step 3. Simplify the expression.

8.6 Solve Rational Equations

- **Strategy to Solve Equations with Rational Expressions**

  Step 1. Note any value of the variable that would make any denominator zero.

  Step 2. Find the least common denominator of *all* denominators in the equation.

  Step 3. Clear the fractions by multiplying both sides of the equation by the LCD.

  Step 4. Solve the resulting equation.

Step 5. Check.
- If any values found in Step 1 are algebraic solutions, discard them.
- Check any remaining solutions in the original equation.

8.7 Solve Proportion and Similar Figure Applications

- **Property of Similar Triangles**
  - If $\triangle ABC$ is similar to $\triangle XYZ$, then their corresponding angle measures are equal and their corresponding sides are in the same ratio.
- **Problem Solving Strategy for Geometry Applications**

  Step 1. **Read** the problem and make sure all the words and ideas are understood. Draw the figure and label it with the given information.

  Step 2. **Identify** what we are looking for.

  Step 3. **Name** what we are looking for by choosing a variable to represent it.

  Step 4. **Translate** into an equation by writing the appropriate formula or model for the situation. Substitute in the given information.

  Step 5. **Solve the equation** using good algebra techniques.

  Step 6. **Check** the answer in the problem and make sure it makes sense.

  Step 7. **Answer** the question with a complete sentence.

REVIEW EXERCISES

8.1 Simplify Rational Expressions

Determine the Values for Which a Rational Expression is Undefined

In the following exercises, determine the values for which the rational expression is undefined.

513. $\dfrac{2a+1}{3a-2}$

514. $\dfrac{b-3}{b^2-16}$

515. $\dfrac{3xy^2}{5y}$

516. $\dfrac{u-3}{u^2-u-30}$

Evaluate Rational Expressions

In the following exercises, evaluate the rational expressions for the given values.

517. $\dfrac{4p-1}{p^2+5}$ when $p=-1$

518. $\dfrac{q^2-5}{q+3}$ when $q=7$

519. $\dfrac{y^2-8}{y^2-y-2}$ when $y=1$

520. $\dfrac{z^2+2}{4z-z^2}$ when $z=3$

Simplify Rational Expressions

In the following exercises, simplify.

521. $\dfrac{10}{24}$

522. $\dfrac{8m^4}{16mn^3}$

523. $\dfrac{14a-14}{a-1}$

524. $\dfrac{b^2+7b+12}{b^2+8b+16}$

Simplify Rational Expressions with Opposite Factors

In the following exercises, simplify.

525. $\dfrac{c^2 - c - 2}{4 - c^2}$

526. $\dfrac{d - 16}{16 - d}$

527. $\dfrac{7v - 35}{25 - v^2}$

528. $\dfrac{w^2 - 3w - 28}{49 - w^2}$

8.2 Multiply and Divide Rational Expressions

Multiply Rational Expressions

In the following exercises, multiply.

529. $\dfrac{3}{8} \cdot \dfrac{2}{15}$

530. $\dfrac{2xy^2}{8y^3} \cdot \dfrac{16y}{24x}$

531. $\dfrac{3a^2 + 21a}{a^2 + 6a - 7} \cdot \dfrac{a - 1}{ab}$

532. $\dfrac{5z^2}{5z^2 + 40z + 35} \cdot \dfrac{z^2 - 1}{3z}$

Divide Rational Expressions

In the following exercises, divide.

533. $\dfrac{t^2 - 4t + 12}{t^2 + 8t + 12} \div \dfrac{t^2 - 36}{6t}$

534. $\dfrac{r^2 - 16}{4} \div \dfrac{r^3 - 64}{2r^2 - 8r + 32}$

535. $\dfrac{11 + w}{w - 9} \div \dfrac{121 - w^2}{9 - w}$

536.
$\dfrac{3y^2 - 12y - 63}{4y + 3} \div (6y^2 - 42y)$

537. $\dfrac{\dfrac{c^2 - 64}{3c^2 + 26c + 16}}{\dfrac{c^2 - 4c - 32}{15c + 10}}$

538.
$\dfrac{8m^2 - 8m}{m - 4} \cdot \dfrac{m^2 + 2m - 24}{m^2 + 7m + 10} \div \dfrac{2m^2 - 6m}{m + 5}$

8.3 Add and Subtract Rational Expressions with a Common Denominator

Add Rational Expressions with a Common Denominator

In the following exercises, add.

539. $\dfrac{3}{5} + \dfrac{2}{5}$

540. $\dfrac{4a^2}{2a - 1} - \dfrac{1}{2a - 1}$

541. $\dfrac{p^2 + 10p}{p + 5} + \dfrac{25}{p + 5}$

542. $\dfrac{3x}{x - 1} + \dfrac{2}{x - 1}$

Subtract Rational Expressions with a Common Denominator

In the following exercises, subtract.

543. $\dfrac{d^2}{d + 4} - \dfrac{3d + 28}{d + 4}$

544. $\dfrac{z^2}{z + 10} - \dfrac{100}{z + 10}$

545. $\dfrac{4q^2 - q + 3}{q^2 + 6q + 5} - \dfrac{3q^2 - q - 6}{q^2 + 6q + 5}$

546.
$\dfrac{5t + 4t + 3}{t^2 - 25} - \dfrac{4t^2 - 8t - 32}{t^2 - 25}$

Add and Subtract Rational Expressions whose Denominators are Opposites

In the following exercises, add and subtract.

547. $\dfrac{18w}{6w-1}+\dfrac{3w-2}{1-6w}$

548. $\dfrac{a^2+3a}{a^2-4}-\dfrac{3a-8}{4-a^2}$

549.
$\dfrac{2b^2+3b-15}{b^2-49}-\dfrac{b^2+16b-1}{49-b^2}$

550.

$\dfrac{8y^2-10y+7}{2y-5}+\dfrac{2y^2+7y+2}{5-2y}$

8.4 Add and Subtract Rational Expressions With Unlike Denominators

Find the Least Common Denominator of Rational Expressions

In the following exercises, find the LCD.

551.

$\dfrac{4}{m^2-3m-10},\dfrac{2m}{m^2-m-20}$

552. $\dfrac{6}{n^2-4},\dfrac{2n}{n^2-4n+4}$

553.

$\dfrac{5}{3p^2+17p-6},\dfrac{2m}{3p^2-23p-8}$

Find Equivalent Rational Expressions

In the following exercises, rewrite as equivalent rational expressions with the given denominator.

554. Rewrite as equivalent rational expressions with denominator $(m+2)(m-5)(m+4)$:

$\dfrac{4}{m^2-3m-10},\dfrac{2m}{m^2-m-20}.$

555. Rewrite as equivalent rational expressions with denominator $(n-2)(n-2)(n+2)$:

$\dfrac{6}{n^2-4n+4},\dfrac{2n}{n^2-4}.$

556. Rewrite as equivalent rational expressions with denominator $(3p+1)(p+6)(p+8)$:

$\dfrac{5}{3p^2+19p+6},\dfrac{7p}{3p^2+25p+8}$

Add Rational Expressions with Different Denominators

In the following exercises, add.

557. $\dfrac{2}{3}+\dfrac{3}{5}$

558. $\dfrac{7}{5a}+\dfrac{3}{2b}$

559. $\dfrac{2}{c-2}+\dfrac{9}{c+3}$

560. $\dfrac{3d}{d^2-9}+\dfrac{5}{d^2+6d+9}$

561.

$\dfrac{2x}{x^2+10x+24}+\dfrac{3x}{x^2+8x+16}$

562. $\dfrac{5q}{p^2q-p^2}+\dfrac{4q}{q^2-1}$

Subtract Rational Expressions with Different Denominators

In the following exercises, subtract and add.

563. $\dfrac{3v}{v+2}-\dfrac{v+2}{v+8}$

564. $\dfrac{-3w-15}{w^2+w-20}-\dfrac{w+2}{4-w}$

565. $\dfrac{7m+3}{m+2}-5$

566. $\dfrac{n}{n+3}+\dfrac{2}{n-3}-\dfrac{n-9}{n^2-9}$

567. $\dfrac{8d}{d^2-64}-\dfrac{4}{d+8}$

568. $\dfrac{5}{12x^2y}+\dfrac{7}{20xy^3}$

8.5 Simplify Complex Rational Expressions

Simplify a Complex Rational Expression by Writing it as Division

In the following exercises, simplify.

569. $\dfrac{\frac{5a}{a+2}}{\frac{10a^2}{a^2-4}}$

570. $\dfrac{\frac{2}{5}+\frac{5}{6}}{\frac{1}{3}+\frac{1}{4}}$

571. $\dfrac{x-\frac{3x}{x+5}}{\frac{1}{x+5}+\frac{1}{x-5}}$

572. $\dfrac{\frac{2}{m}+\frac{m}{n}}{\frac{n}{m}-\frac{1}{n}}$

Simplify a Complex Rational Expression by Using the LCD

In the following exercises, simplify.

573. $\dfrac{6+\frac{2}{q-4}}{\frac{5}{q+4}}$

574. $\dfrac{\frac{3}{a^2}-\frac{1}{b}}{\frac{1}{a}+\frac{1}{b^2}}$

575. $\dfrac{\frac{2}{z^2-49}+\frac{1}{z+7}}{\frac{9}{z+7}+\frac{12}{z-7}}$

576. $\dfrac{\frac{3}{y^2-4y-32}}{\frac{2}{y-8}+\frac{1}{y+4}}$

8.6 Solve Rational Equations

Solve Rational Equations

In the following exercises, solve.

577. $\dfrac{1}{2}+\dfrac{2}{3}=\dfrac{1}{x}$

578. $1-\dfrac{2}{m}=\dfrac{8}{m^2}$

579. $\dfrac{1}{b-2}+\dfrac{1}{b+2}=\dfrac{3}{b^2-4}$

580. $\dfrac{3}{q+8}-\dfrac{2}{q-2}=1$

581.
$\dfrac{v-15}{v^2-9v+18}=\dfrac{4}{v-3}+\dfrac{2}{v-6}$

582. $\dfrac{z}{12}+\dfrac{z+3}{3z}=\dfrac{1}{z}$

Solve a Rational Equation for a Specific Variable

In the following exercises, solve for the indicated variable.

583. $\dfrac{V}{l}=hw$ for l

584. $\dfrac{1}{x}-\dfrac{2}{y}=5$ for y

585. $x=\dfrac{y+5}{z-7}$ for z

586. $P=\dfrac{k}{V}$ for V

8.7 Solve Proportion and Similar Figure Applications Similarity

Solve Proportions

In the following exercises, solve.

587. $\dfrac{x}{4}=\dfrac{3}{5}$

588. $\dfrac{3}{y}=\dfrac{9}{5}$

589. $\dfrac{s}{s+20}=\dfrac{3}{7}$

590. $\dfrac{t-3}{5}=\dfrac{t+2}{9}$

In the following exercises, solve using proportions.

591. Rachael had a 21 ounce strawberry shake that has 739 calories. How many calories are there in a 32 ounce shake?

592. Leo went to Mexico over Christmas break and changed $525 dollars into Mexican pesos. At that time, the exchange rate had $1 US is equal to 16.25 Mexican pesos. How many Mexican pesos did he get for his trip?

Solve Similar Figure Applications

In the following exercises, solve.

593. ΔABC is similar to ΔXYZ. The lengths of two sides of each triangle are given in the figure. Find the lengths of the third sides.

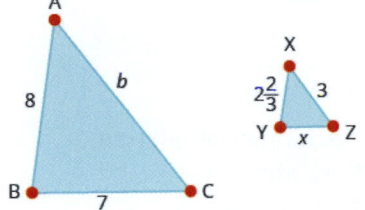

594. On a map of Europe, Paris, Rome, and Vienna form a triangle whose sides are shown in the figure below. If the actual distance from Rome to Vienna is 700 miles, find the distance from

ⓐ Paris to Rome

ⓑ Paris to Vienna

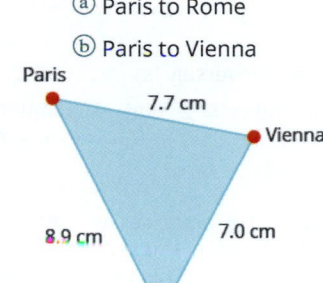

595. Tony is 5.75 feet tall. Late one afternoon, his shadow was 8 feet long. At the same time, the shadow of a nearby tree was 32 feet long. Find the height of the tree.

596. The height of a lighthouse in Pensacola, Florida is 150 feet. Standing next to the statue, 5.5 foot tall Natalie cast a 1.1 foot shadow How long would the shadow of the lighthouse be?

8.8 Solve Uniform Motion and Work Applications Problems

Solve Uniform Motion Applications

In the following exercises, solve.

597. When making the 5-hour drive home from visiting her parents, Lisa ran into bad weather. She was able to drive 176 miles while the weather was good, but then driving 10 mph slower, went 81 miles in the bad weather. How fast did she drive when the weather was bad?

598. Mark is riding on a plane that can fly 490 miles with a tailwind of 20 mph in the same time that it can fly 350 miles against a tailwind of 20 mph. What is the speed of the plane?

599. John can ride his bicycle 8 mph faster than Luke can ride his bike. It takes Luke 3 hours longer than John to ride 48 miles. How fast can John ride his bike?

600. Mark was training for a triathlon. He ran 8 kilometers and biked 32 kilometers in a total of 3 hours. His running speed was 8 kilometers per hour less than his biking speed. What was his running speed?

Solve Work Applications

In the following exercises, solve.

601. Jerry can frame a room in 1 hour, while Jake takes 4 hours. How long could they frame a room working together?

602. Lisa takes 3 hours to mow the lawn while her cousin, Barb, takes 2 hours. How long will it take them working together?

603. Jeffrey can paint a house in 6 days, but if he gets a helper he can do it in 4 days. How long would it take the helper to paint the house alone?

604. Sue and Deb work together writing a book that takes them 90 days. If Sue worked alone it would take her 120 days. How long would it take Deb to write the book alone?

8.9 Use Direct and Inverse Variation

Solve Direct Variation Problems

In the following exercises, solve.

605. If y varies directly as x, when $y = 9$ and $x = 3$, find x when $y = 21$.

606. If y varies inversely as x, when $y = 20$ and $x = 2$ find y when $x = 4$.

607. If m varies inversely with the square of n, when $m = 4$ and $n = 6$ find m when $n = 2$.

608. Vanessa is traveling to see her fiancé. The distance, d, varies directly with the speed, v, she drives. If she travels 258 miles driving 60 mph, how far would she travel going 70 mph?

609. If the cost of a pizza varies directly with its diameter, and if an 8" diameter pizza costs $12, how much would a 6" diameter pizza cost?

610. The distance to stop a car varies directly with the square of its speed. It takes 200 feet to stop a car going 50 mph. How many feet would it take to stop a car going 60 mph?

Solve Inverse Variation Problems

In the following exercises, solve.

611. The number of tickets for a music fundraiser varies inversely with the price of the tickets. If Madelyn has just enough money to purchase 12 tickets for $6, how many tickets can Madelyn afford to buy if the price increased to $8?

612. On a string instrument, the length of a string varies inversely with the frequency of its vibrations. If an 11-inch string on a violin has a frequency of 360 cycles per second, what frequency does a 12 inch string have?

PRACTICE TEST

In the following exercises, simplify.

613. $\dfrac{3a^2 b}{6ab^2}$

614. $\dfrac{5b - 25}{b^2 - 25}$

In the following exercises, perform the indicated operation and simplify.

615. $\dfrac{4x}{x + 2} \cdot \dfrac{x^2 + 5x + 6}{12x^2}$

616. $\dfrac{5y}{4y - 8} \cdot \dfrac{y^2 - 4}{10}$

617. $\dfrac{4}{pq} + \dfrac{5}{p}$

618. $\dfrac{1}{z - 9} - \dfrac{3}{z + 9}$

619. $\dfrac{\frac{2}{3} + \frac{3}{5}}{\frac{2}{5}}$

620. $\dfrac{\frac{1}{m} - \frac{1}{n}}{\frac{1}{n} + \frac{1}{m}}$

In the following exercises, solve each equation.

621. $\dfrac{1}{2} + \dfrac{2}{7} = \dfrac{1}{x}$

622. $\dfrac{5}{y - 6} = \dfrac{3}{y + 6}$

623. $\dfrac{1}{z - 5} + \dfrac{1}{z + 5} = \dfrac{1}{z^2 - 25}$

624. $\dfrac{t}{4} = \dfrac{3}{5}$

625. $\dfrac{2}{r - 2} = \dfrac{3}{r - 1}$

In the following exercises, solve.

626. If y varies directly with x, and $x = 5$ when $y = 30$, find x when $y = 42$.

627. If y varies inversely with x and $x = 6$ when $y = 20$, find y when $x = 2$.

628. If y varies inversely with the square of x and $x = 3$ when $y = 9$, find y when $x = 4$.

629. The recommended erythromycin dosage for dogs, is 5 mg for every pound the dog weighs. If Daisy weighs 25 pounds, how many milligrams of erythromycin should her veterinarian prescribe?

630. Julia spent 4 hours Sunday afternoon exercising at the gym. She ran on the treadmill for 10 miles and then biked for 20 miles. Her biking speed was 5 mph faster than her running speed on the treadmill. What was her running speed?

631. Kurt can ride his bike for 30 miles with the wind in the same amount of time that he can go 21 miles against the wind. If the wind's speed is 6 mph, what is Kurt's speed on his bike?

632. Amanda jogs to the park 8 miles using one route and then returns via a 14-mile route. The return trip takes her 1 hour longer than her jog to the park. Find her jogging rate.

633. An experienced window washer can wash all the windows in Mike's house in 2 hours, while a new trainee can wash all the windows in 7 hours. How long would it take them working together?

634. Josh can split a truckload of logs in 8 hours, but working with his dad they can get it done in 3 hours. How long would it take Josh's dad working alone to split the logs?

635. The price that Tyler pays for gas varies directly with the number of gallons he buys. If 24 gallons cost him $59.76, what would 30 gallons cost?

636. The volume of a gas in a container varies inversely with the pressure on the gas. If a container of nitrogen has a volume of 29.5 liters with 2000 psi, what is the volume if the tank has a 14.7 psi rating? Round to the nearest whole number.

637. The cities of Dayton, Columbus, and Cincinnati form a triangle in southern Ohio, as shown on the figure below, that gives the map distances between these cities in inches.

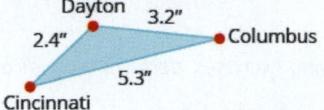

The actual distance from Dayton to Cincinnati is 48 miles. What is the actual distance between Dayton and Columbus?

Figure 9.1 Square roots are used to determine the time it would take for a stone falling from the edge of this cliff to hit the land below.

Chapter Outline

9.1 Simplify and Use Square Roots

9.2 Simplify Square Roots

9.3 Add and Subtract Square Roots

9.4 Multiply Square Roots

9.5 Divide Square Roots

9.6 Solve Equations with Square Roots

9.7 Higher Roots

9.8 Rational Exponents

Introduction

Suppose a stone falls from the edge of a cliff. The number of feet the stone has dropped after t seconds can be found by multiplying 16 times the square of t. But to calculate the number of seconds it would take the stone to hit the land below, we need to use a square root. In this chapter, we will introduce and apply the properties of square roots, and extend these concepts to higher order roots and rational exponents.

9.1 Simplify and Use Square Roots

Learning Objectives

By the end of this section, you will be able to:

- Simplify expressions with square roots
- Estimate square roots
- Approximate square roots
- Simplify variable expressions with square roots

Be Prepared!

Before you get started, take this readiness quiz.

1. Simplify: ⓐ 9^2 ⓑ $(-9)^2$ ⓒ -9^2.
   If you missed this problem, review **Example 1.50**.

2. Round 3.846 to the nearest hundredth.
   If you missed this problem, review **Example 1.94**.

3. For each number, identify whether it is a real number or not a real number:
ⓐ $-\sqrt{100}$ ⓑ $\sqrt{-100}$.
If you missed this problem, review **Example 1.113.**

Simplify Expressions with Square Roots

Remember that when a number n is multiplied by itself, we write n^2 and read it "n squared." For example, 15^2 reads as "15 squared," and 225 is called the square of 15, since $15^2 = 225$.

Square of a Number

If $n^2 = m$, then m is the square of n .

Sometimes we will need to look at the relationship between numbers and their squares in reverse. Because 225 is the square of 15, we can also say that 15 is a square root of 225. A number whose square is m is called a *square root* of m .

Square Root of a Number

If $n^2 = m$, then n is a square root of m .

Notice $(-15)^2 = 225$ also, so -15 is also a square root of 225. Therefore, both 15 and -15 are square roots of 225.

So, every positive number has two square roots—one positive and one negative. What if we only wanted the positive square root of a positive number? The *radical sign,* $\sqrt{m}$, denotes the positive square root. The positive square root is also called the principal square root.

We also use the radical sign for the square root of zero. Because $0^2 = 0$, $\sqrt{0} = 0$. Notice that zero has only one square root.

Square Root Notation

$$\text{radical sign} \longrightarrow \sqrt{m} \longleftarrow \text{radicand}$$

$\sqrt{m}$ is read as "the square root of m ."

If $m = n^2$, then $\sqrt{m} = n$, for $n \geq 0$.

The square root of m , $\sqrt{m}$, is the positive number whose square is m .

Since 15 is the positive square root of 225, we write $\sqrt{225} = 15$. Fill in **Figure 9.2** to make a table of square roots you can refer to as you work this chapter.

$\sqrt{1}$	$\sqrt{4}$	$\sqrt{9}$	$\sqrt{16}$	$\sqrt{25}$	$\sqrt{36}$	$\sqrt{49}$	$\sqrt{64}$	$\sqrt{81}$	$\sqrt{100}$	$\sqrt{121}$	$\sqrt{144}$	$\sqrt{169}$	$\sqrt{196}$	$\sqrt{225}$
														15

Figure 9.2

We know that every positive number has two square roots and the radical sign indicates the positive one. We write $\sqrt{225} = 15$. If we want to find the negative square root of a number, we place a negative in front of the radical sign. For example, $-\sqrt{225} = -15$.

EXAMPLE 9.1

Simplify: ⓐ $\sqrt{36}$ ⓑ $\sqrt{196}$ ⓒ $-\sqrt{81}$ ⓓ $-\sqrt{289}$.

⊘ **Solution**

ⓐ

	$\sqrt{36}$
Since $6^2 = 36$	6

ⓑ

	$\sqrt{196}$
Since $14^2 = 196$	14

ⓒ

	$-\sqrt{81}$
The negative is in front of the radical sign.	-9

ⓓ

	$-\sqrt{289}$
The negative is in front of the radical sign.	-17

> **TRY IT :: 9.1**   Simplify: ⓐ $-\sqrt{49}$ ⓑ $\sqrt{225}$.

> **TRY IT :: 9.2**   Simplify: ⓐ $\sqrt{64}$ ⓑ $-\sqrt{121}$.

EXAMPLE 9.2

Simplify: ⓐ $\sqrt{-169}$ ⓑ $-\sqrt{64}$.

⊘ **Solution**

ⓐ

$$\sqrt{-169}$$

There is no real number whose square is -169. $\sqrt{-169}$ is not a real number.

ⓑ

$$-\sqrt{64}$$

The negative is in front of the radical. -8

> **TRY IT :: 9.3**   Simplify: ⓐ $\sqrt{-196}$ ⓑ $-\sqrt{81}$.

> **TRY IT :: 9.4**   Simplify: ⓐ $-\sqrt{49}$ ⓑ $\sqrt{-121}$.

When using the order of operations to simplify an expression that has square roots, we treat the radical as a grouping symbol.

EXAMPLE 9.3

Simplify: ⓐ $\sqrt{25} + \sqrt{144}$ ⓑ $\sqrt{25 + 144}$.

⊘ **Solution**

ⓐ

$$\sqrt{25} + \sqrt{144}$$

Use the order of operations.	$5 + 12$
Simplify.	17

ⓑ

$$\sqrt{25 + 144}$$

Simplify under the radical sign.	$\sqrt{169}$
Simplify.	13

Notice the different answers in parts ⓐ and ⓑ!

> **TRY IT :: 9.5**   Simplify: ⓐ $\sqrt{9} + \sqrt{16}$ ⓑ $\sqrt{9 + 16}$.

> **TRY IT :: 9.6**   Simplify: ⓐ $\sqrt{64 + 225}$ ⓑ $\sqrt{64} + \sqrt{225}$.

Estimate Square Roots

So far we have only considered square roots of perfect square numbers. The square roots of other numbers are not whole numbers. Look at **Table 9.1** below.

Number	Square Root
4	$\sqrt{4} = 2$
5	$\sqrt{5}$
6	$\sqrt{6}$
7	$\sqrt{7}$
8	$\sqrt{8}$
9	$\sqrt{9} = 3$

Table 9.1

The square roots of numbers between 4 and 9 must be between the two consecutive whole numbers 2 and 3, and they are not whole numbers. Based on the pattern in the table above, we could say that $\sqrt{5}$ must be between 2 and 3. Using inequality symbols, we write:

$$2 < \sqrt{5} < 3$$

EXAMPLE 9.4

Estimate $\sqrt{60}$ between two consecutive whole numbers.

⊘ **Solution**

Think of the perfect square numbers closest to 60. Make a small table of these perfect squares and their squares roots.

Number	Square root
36	6
49	7
64	8
81	9

60 → → √60

Locate 60 between two consecutive perfect squares.	$49 < 60 < 64$
$\sqrt{60}$ is between their square roots.	$7 < \sqrt{60} < 8$

> **TRY IT : : 9.7**   Estimate the square root $\sqrt{38}$ between two consecutive whole numbers.

> **TRY IT : : 9.8**   Estimate the square root $\sqrt{84}$ between two consecutive whole numbers.

Approximate Square Roots

There are mathematical methods to approximate square roots, but nowadays most people use a calculator to find them. Find the $\sqrt{x}$ key on your calculator. You will use this key to approximate square roots.

When you use your calculator to find the square root of a number that is not a perfect square, the answer that you see is not the exact square root. It is an approximation, accurate to the number of digits shown on your calculator's display. The symbol for an approximation is $\approx$ and it is read 'approximately.'

Suppose your calculator has a 10-digit display. You would see that

$$\sqrt{5} \approx 2.236067978$$

If we wanted to round $\sqrt{5}$ to two decimal places, we would say

$$\sqrt{5} \approx 2.24$$

How do we know these values are approximations and not the exact values? Look at what happens when we square them:

$$(2.236067978)^2 = 5.000000002$$
$$(2.24)^2 = 5.0176$$

Their squares are close to 5, but are not exactly equal to 5.

Using the square root key on a calculator and then rounding to two decimal places, we can find:

$$\sqrt{4} = 2$$
$$\sqrt{5} \approx 2.24$$
$$\sqrt{6} \approx 2.45$$
$$\sqrt{7} \approx 2.65$$
$$\sqrt{8} \approx 2.83$$
$$\sqrt{9} = 3$$

EXAMPLE 9.5

Round $\sqrt{17}$ to two decimal places.

⊘ **Solution**

	$\sqrt{17}$
Use the calculator square root key.	4.123105626...
Round to two decimal places.	4.12
	$\sqrt{17} \approx 4.12$

> **TRY IT : : 9.9**   Round $\sqrt{11}$ to two decimal places.

> **TRY IT : : 9.10**   Round $\sqrt{13}$ to two decimal places.

Simplify Variable Expressions with Square Roots

What if we have to find a square root of an expression with a variable? Consider $\sqrt{9x^2}$. Can you think of an expression whose square is $9x^2$?

$$(?)^2 = 9x^2$$
$$(3x)^2 = 9x^2, \quad \text{so } \sqrt{9x^2} = 3x$$

When we use the radical sign to take the square root of a variable expression, we should specify that $x \geq 0$ to make sure we get the *principal square root*.

However, in this chapter we will assume that each variable in a square-root expression represents a non-negative number and so we will not write $x \geq 0$ next to every radical.

What about square roots of higher powers of variables? Think about the Power Property of Exponents we used in Chapter 6.

$$(a^m)^n = a^{m \cdot n}$$

If we square a^m, the exponent will become $2m$.

$$(a^m)^2 = a^{2m}$$

How does this help us take square roots? Let's look at a few:

$$\sqrt{25u^8} = 5u^4 \text{ because } \left(5u^4\right)^2 = 25u^8$$
$$\sqrt{16r^{20}} = 4r^{10} \text{ because } \left(4r^{10}\right)^2 = 16r^{20}$$
$$\sqrt{196q^{36}} = 14q^{18} \text{ because } \left(14q^{18}\right)^2 = 196q^{36}$$

EXAMPLE 9.6

Simplify: ⓐ $\sqrt{x^6}$ ⓑ $\sqrt{y^{16}}$.

⊘ **Solution**

ⓐ

$$\sqrt{x^6}$$

Since $\left(x^3\right)^2 = x^6$. x^3

ⓑ

$$\sqrt{y^{16}}$$

Since $\left(y^8\right)^2 = y^{16}$. y^8

> **TRY IT : : 9.11**   Simplify: ⓐ $\sqrt{y^8}$ ⓑ $\sqrt{z^{12}}$.

> **TRY IT : : 9.12**   Simplify: ⓐ $\sqrt{m^4}$ ⓑ $\sqrt{b^{10}}$.

EXAMPLE 9.7

Simplify: $\sqrt{16n^2}$.

⊘ **Solution**

$$\sqrt{16n^2}$$

Since $(4n)^2 = 16n^2$.　　　　$4n$

> **TRY IT :: 9.13** 　Simplify: $\sqrt{64x^2}$.

> **TRY IT :: 9.14** 　Simplify: $\sqrt{169y^2}$.

EXAMPLE 9.8

Simplify: $-\sqrt{81c^2}$.

⊘ **Solution**

$$-\sqrt{81c^2}$$

Since $(9c)^2 = 81c^2$.　　　　$-9c$

> **TRY IT :: 9.15** 　Simplify: $-\sqrt{121y^2}$.

> **TRY IT :: 9.16** 　Simplify: $-\sqrt{100p^2}$.

EXAMPLE 9.9

Simplify: $\sqrt{36x^2y^2}$.

⊘ **Solution**

$$\sqrt{36x^2y^2}$$

Since $(6xy)^2 = 36x^2y^2$.　　　　$6xy$

> **TRY IT :: 9.17** 　Simplify: $\sqrt{100a^2b^2}$.

> **TRY IT :: 9.18** 　Simplify: $\sqrt{225m^2n^2}$.

EXAMPLE 9.10

Simplify: $\sqrt{64p^{64}}$.

⊘ **Solution**

$$\sqrt{64p^{64}}$$

Since $\left(8p^{32}\right)^2 = 64p^{64}$.　　　　$8p^{32}$

> **TRY IT : : 9.19**   Simplify: $\sqrt{49x^{30}}$.

> **TRY IT : : 9.20**   Simplify: $\sqrt{81w^{36}}$.

EXAMPLE 9.11

Simplify: $\sqrt{121a^6 b^8}$

⊘ **Solution**

	$\sqrt{121a^6 b^8}$
Since $\left(11a^3 b^4\right)^2 = 121a^6 b^8$.	$11a^3 b^4$

> **TRY IT : : 9.21**   Simplify: $\sqrt{169x^{10} y^{14}}$.

> **TRY IT : : 9.22**   Simplify: $\sqrt{144p^{12} q^{20}}$.

▶ **MEDIA : :**

Access this online resource for additional instruction and practice with square roots.

• **Square Roots (https://openstax.org/l/25SquareRoots)**

9.1 EXERCISES

Practice Makes Perfect

Simplify Expressions with Square Roots

In the following exercises, simplify.

1. $\sqrt{36}$

2. $\sqrt{4}$

3. $\sqrt{64}$

4. $\sqrt{169}$

5. $\sqrt{9}$

6. $\sqrt{16}$

7. $\sqrt{100}$

8. $\sqrt{144}$

9. $-\sqrt{4}$

10. $-\sqrt{100}$

11. $-\sqrt{1}$

12. $-\sqrt{121}$

13. $\sqrt{-121}$

14. $\sqrt{-36}$

15. $\sqrt{-9}$

16. $\sqrt{-49}$

17. $\sqrt{9+16}$

18. $\sqrt{25+144}$

19. $\sqrt{9}+\sqrt{16}$

20. $\sqrt{25}+\sqrt{144}$

Estimate Square Roots

In the following exercises, estimate each square root between two consecutive whole numbers.

21. $\sqrt{70}$

22. $\sqrt{55}$

23. $\sqrt{200}$

24. $\sqrt{172}$

Approximate Square Roots

In the following exercises, approximate each square root and round to two decimal places.

25. $\sqrt{19}$

26. $\sqrt{21}$

27. $\sqrt{53}$

28. $\sqrt{47}$

Simplify Variable Expressions with Square Roots

In the following exercises, simplify.

29. $\sqrt{y^2}$

30. $\sqrt{b^2}$

31. $\sqrt{a^{14}}$

32. $\sqrt{w^{24}}$

33. $\sqrt{49x^2}$

34. $\sqrt{100y^2}$

35. $\sqrt{121m^{20}}$

36. $\sqrt{25h^{44}}$

37. $\sqrt{81x^{36}}$

38. $\sqrt{144z^{84}}$

39. $-\sqrt{81x^{18}}$

40. $-\sqrt{100m^{32}}$

41. $-\sqrt{64a^2}$

42. $-\sqrt{25x^2}$

43. $\sqrt{144x^2y^2}$

44. $\sqrt{196a^2b^2}$

45. $\sqrt{169w^8y^{10}}$

46. $\sqrt{81p^{24}q^6}$

47. $\sqrt{9c^8 d^{12}}$

48. $\sqrt{36r^6 s^{20}}$

Everyday Math

49. Decorating Denise wants to have a square accent of designer tiles in her new shower. She can afford to buy 625 square centimeters of the designer tiles. How long can a side of the accent be?

50. Decorating Morris wants to have a square mosaic inlaid in his new patio. His budget allows for 2025 square inch tiles. How long can a side of the mosaic be?

Writing Exercises

51. Why is there no real number equal to $\sqrt{-64}$?

52. What is the difference between 9^2 and $\sqrt{9}$?

Self Check

ⓐ *After completing the exercises, use this checklist to evaluate your mastery of the objectives of this section.*

I can...	Confidently	With some help	No-I don't get it!
simplify expressions with square roots.			
estimate square roots.			
approximate square roots.			
simplify variable expressions with square roots.			

ⓑ *On a scale of 1–10, how would you rate your mastery of this section in light of your responses on the checklist? How can you improve this?*

9.2   **Simplify Square Roots**

Learning Objectives

By the end of this section, you will be able to:
> Use the Product Property to simplify square roots
> Use the Quotient Property to simplify square roots

Be Prepared!

Before you get started take this readiness quiz.

1. Simplify: $\dfrac{80}{176}$.

   If you missed this problem, review **Example 1.65**.

2. Simplify: $\dfrac{n^9}{n^3}$.

   If you missed this problem, review **Example 6.59**.

3. Simplify: $\dfrac{q^4}{q^{12}}$.

   If you missed this problem, review **Example 6.60**.

In the last section, we estimated the square root of a number between two consecutive whole numbers. We can say that $\sqrt{50}$ is between 7 and 8. This is fairly easy to do when the numbers are small enough that we can use **Figure 9.2**.

But what if we want to estimate $\sqrt{500}$? If we simplify the square root first, we'll be able to estimate it easily. There are other reasons, too, to simplify square roots as you'll see later in this chapter.

A square root is considered *simplified* if its radicand contains no perfect square factors.

Simplified Square Root

$\sqrt{a}$ is considered simplified if a has no perfect square factors.

So $\sqrt{31}$ is simplified. But $\sqrt{32}$ is not simplified, because 16 is a perfect square factor of 32.

Use the Product Property to Simplify Square Roots

The properties we will use to simplify expressions with square roots are similar to the properties of exponents. We know that $(ab)^m = a^m b^m$. The corresponding property of square roots says that $\sqrt{ab} = \sqrt{a} \cdot \sqrt{b}$.

Product Property of Square Roots

If a, b are non-negative real numbers, then $\sqrt{ab} = \sqrt{a} \cdot \sqrt{b}$.

We use the Product Property of Square Roots to remove all perfect square factors from a radical. We will show how to do this in **Example 9.12**.

EXAMPLE 9.12   HOW TO USE THE PRODUCT PROPERTY TO SIMPLIFY A SQUARE ROOT

Simplify: $\sqrt{50}$.

 Solution

Step 1. Find the largest perfect square factor of the radicand.	25 is the largest perfect square factor of 50.	$\sqrt{50}$
Rewrite the radicand as a product using the perfect square factor.	$50 = 25 \cdot 2$	
	Always write the perfect square factor first.	$\sqrt{25 \cdot 2}$

Step 2. Use the product rule to rewrite the radical as the product of two radicals.		$\sqrt{25} \cdot \sqrt{2}$
Step 3. Simplify the square root of the perfect square.		$5\sqrt{2}$

> **TRY IT :: 9.23**   Simplify: $\sqrt{48}$.

> **TRY IT :: 9.24**   Simplify: $\sqrt{45}$.

Notice in the previous example that the simplified form of $\sqrt{50}$ is $5\sqrt{2}$, which is the product of an integer and a square root. We always write the integer in front of the square root.

 HOW TO :: SIMPLIFY A SQUARE ROOT USING THE PRODUCT PROPERTY.

Step 1.   Find the largest perfect square factor of the radicand. Rewrite the radicand as a product using the perfect-square factor.

Step 2.   Use the product rule to rewrite the radical as the product of two radicals.

Step 3.   Simplify the square root of the perfect square.

EXAMPLE 9.13

Simplify: $\sqrt{500}$.

 Solution

	$\sqrt{500}$
Rewrite the radicand as a product using the largest perfect square factor.	$\sqrt{100 \cdot 5}$
Rewrite the radical as the product of two radicals.	$\sqrt{100} \cdot \sqrt{5}$
Simplify.	$10\sqrt{5}$

> **TRY IT :: 9.25**   Simplify: $\sqrt{288}$.

> **TRY IT :: 9.26**   Simplify: $\sqrt{432}$.

We could use the simplified form $10\sqrt{5}$ to estimate $\sqrt{500}$. We know 5 is between 2 and 3, and $\sqrt{500}$ is $10\sqrt{5}$. So $\sqrt{500}$ is between 20 and 30.

The next example is much like the previous examples, but with variables.

EXAMPLE 9.14

Simplify: $\sqrt{x^3}$.

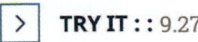 **Solution**

	$\sqrt{x^3}$
Rewrite the radicand as a product using the largest perfect square factor.	$\sqrt{x^2 \cdot x}$
Rewrite the radical as the product of two radicals.	$\sqrt{x^2} \cdot \sqrt{x}$
Simplify.	$x\sqrt{x}$

> **TRY IT :: 9.27**   Simplify: $\sqrt{b^5}$.

> **TRY IT :: 9.28**   Simplify: $\sqrt{p^9}$.

We follow the same procedure when there is a coefficient in the radical, too.

EXAMPLE 9.15

Simplify: $\sqrt{25y^5}$.

 Solution

	$\sqrt{25y^5}$
Rewrite the radicand as a product using the largest perfect square factor.	$\sqrt{25y^4 \cdot y}$
Rewrite the radical as the product of two radicals.	$\sqrt{25y^4} \cdot \sqrt{y}$
Simplify.	$5y^2\sqrt{y}$

> **TRY IT :: 9.29**   Simplify: $\sqrt{16x^7}$.

> **TRY IT :: 9.30**   Simplify: $\sqrt{49v^9}$.

In the next example both the constant and the variable have perfect square factors.

EXAMPLE 9.16

Simplify: $\sqrt{72n^7}$.

 **Solution**

	$\sqrt{72n^7}$
Rewrite the radicand as a product using the largest perfect square factor.	$\sqrt{36n^6 \cdot 2n}$
Rewrite the radical as the product of two radicals.	$\sqrt{36n^6} \cdot \sqrt{2n}$
Simplify.	$6n^3\sqrt{2n}$

> **TRY IT :: 9.31**   Simplify: $\sqrt{32y^5}$.

> **TRY IT : : 9.32**   Simplify: $\sqrt{75a^9}$.

EXAMPLE 9.17

Simplify: $\sqrt{63u^3 v^5}$.

⊘ **Solution**

$$\sqrt{63u^3 v^5}$$

Rewrite the radicand as a product using the largest perfect square factor.

$$\sqrt{9u^2 v^4 \cdot 7uv}$$

Rewrite the radical as the product of two radicals.

$$\sqrt{9u^2 v^4} \cdot \sqrt{7uv}$$

Simplify.

$$3uv^2 \sqrt{7uv}$$

> **TRY IT : : 9.33**   Simplify: $\sqrt{98a^7 b^5}$.

> **TRY IT : : 9.34**   Simplify: $\sqrt{180m^9 n^{11}}$.

We have seen how to use the Order of Operations to simplify some expressions with radicals. To simplify $\sqrt{25} + \sqrt{144}$ we must simplify each square root separately first, then add to get the sum of 17.

The expression $\sqrt{17} + \sqrt{7}$ cannot be simplified—to begin we'd need to simplify each square root, but neither 17 nor 7 contains a perfect square factor.

In the next example, we have the sum of an integer and a square root. We simplify the square root but cannot add the resulting expression to the integer.

EXAMPLE 9.18

Simplify: $3 + \sqrt{32}$.

⊘ **Solution**

$$3 + \sqrt{32}$$

Rewrite the radicand as a product using the largest perfect square factor.

$$3 + \sqrt{16 \cdot 2}$$

Rewrite the radical as the product of two radicals.

$$3 + \sqrt{16} \cdot \sqrt{2}$$

Simplify.

$$3 + 4\sqrt{2}$$

The terms are not like and so we cannot add them. Trying to add an integer and a radical is like trying to add an integer and a variable—they are not like terms!

> **TRY IT : : 9.35**   Simplify: $5 + \sqrt{75}$.

> **TRY IT : : 9.36**   Simplify: $2 + \sqrt{98}$.

The next example includes a fraction with a radical in the numerator. Remember that in order to simplify a fraction you need a common factor in the numerator and denominator.

EXAMPLE 9.19

Simplify: $\dfrac{4 - \sqrt{48}}{2}$.

⊘ **Solution**

$$\dfrac{4 - \sqrt{48}}{2}$$

Rewrite the radicand as a product using the largest perfect square factor.

$$\dfrac{4 - \sqrt{16 \cdot 3}}{2}$$

Rewrite the radical as the product of two radicals.

$$\dfrac{4 - \sqrt{16} \cdot \sqrt{3}}{2}$$

Simplify.

$$\dfrac{4 - 4\sqrt{3}}{2}$$

Factor the common factor from the numerator.

$$\dfrac{4(1 - \sqrt{3})}{2}$$

Remove the common factor, 2, from the numerator and denominator.

$$\dfrac{\cancel{2} \cdot 2(1 - \sqrt{3})}{\cancel{2}}$$

Simplify.

$$2(1 - \sqrt{3})$$

> **TRY IT :: 9.37**   Simplify: $\dfrac{10 - \sqrt{75}}{5}$.

> **TRY IT :: 9.38**   Simplify: $\dfrac{6 - \sqrt{45}}{3}$.

Use the Quotient Property to Simplify Square Roots

Whenever you have to simplify a square root, the first step you should take is to determine whether the radicand is a perfect square. A *perfect square fraction* is a fraction in which both the numerator and the denominator are perfect squares.

EXAMPLE 9.20

Simplify: $\sqrt{\dfrac{9}{64}}$.

⊘ **Solution**

$$\sqrt{\dfrac{9}{64}}$$

Since $\left(\dfrac{3}{8}\right)^2 = \dfrac{9}{64}$ $\dfrac{3}{8}$

> **TRY IT :: 9.39**   Simplify: $\sqrt{\dfrac{25}{16}}$.

> **TRY IT :: 9.40**   Simplify: $\sqrt{\dfrac{49}{81}}$.

If the numerator and denominator have any common factors, remove them. You may find a perfect square fraction!

EXAMPLE 9.21

Simplify: $\sqrt{\dfrac{45}{80}}$.

✓ **Solution**

$$\sqrt{\dfrac{45}{80}}$$

Simplify inside the radical first. Rewrite
showing the common factors of the
numerator and denominator.

$$\sqrt{\dfrac{5 \cdot 9}{5 \cdot 16}}$$

Simplify the fraction by removing common
factors.

$$\sqrt{\dfrac{9}{16}}$$

Simplify. $\left(\dfrac{3}{4}\right)^2 = \dfrac{9}{16}$

$$\dfrac{3}{4}$$

> **TRY IT : : 9.41**   Simplify: $\sqrt{\dfrac{75}{48}}$.

> **TRY IT : : 9.42**   Simplify: $\sqrt{\dfrac{98}{162}}$.

In the last example, our first step was to simplify the fraction under the radical by removing common factors. In the next example we will use the Quotient Property to simplify under the radical. We divide the like bases by subtracting their exponents, $\dfrac{a^m}{a^n} = a^{m-n}$, $a \neq 0$.

EXAMPLE 9.22

Simplify: $\sqrt{\dfrac{m^6}{m^4}}$.

✓ **Solution**

$$\sqrt{\dfrac{m^6}{m^4}}$$

Simplify the fraction inside the radical first.

Divide the like bases by subtracting the
exponents.

$$\sqrt{m^2}$$

Simplify.

$$m$$

> **TRY IT : : 9.43**   Simplify: $\sqrt{\dfrac{a^8}{a^6}}$.

> **TRY IT : : 9.44**   Simplify: $\sqrt{\dfrac{x^{14}}{x^{10}}}$.

EXAMPLE 9.23

Simplify: $\sqrt{\dfrac{48p^7}{3p^3}}$.

✓ **Solution**

$$\sqrt{\dfrac{48p^7}{3p^3}}$$

Simplify the fraction inside the radical first. $\sqrt{16p^4}$

Simplify. $4p^2$

> **TRY IT : : 9.45**   Simplify: $\sqrt{\dfrac{75x^5}{3x}}$.

> **TRY IT : : 9.46**   Simplify: $\sqrt{\dfrac{72z^{12}}{2z^{10}}}$.

Remember the Quotient to a Power Property? It said we could raise a fraction to a power by raising the numerator and denominator to the power separately.

$$\left(\dfrac{a}{b}\right)^m = \dfrac{a^m}{b^m}, \, b \neq 0$$

We can use a similar property to simplify a square root of a fraction. After removing all common factors from the numerator and denominator, if the fraction is not a perfect square we simplify the numerator and denominator separately.

Quotient Property of Square Roots

If a, b are non-negative real numbers and $b \neq 0$, then

$$\sqrt{\dfrac{a}{b}} = \dfrac{\sqrt{a}}{\sqrt{b}}$$

EXAMPLE 9.24

Simplify: $\sqrt{\dfrac{21}{64}}$.

✓ **Solution**

$$\sqrt{\dfrac{21}{64}}$$

We cannot simplify the fraction inside the radical. Rewrite using the quotient property. $\dfrac{\sqrt{21}}{\sqrt{64}}$

Simplify the square root of 64. The numerator cannot be simplified. $\dfrac{\sqrt{21}}{8}$

> **TRY IT : : 9.47**   Simplify: $\sqrt{\dfrac{19}{49}}$.

> **TRY IT : : 9.48**   Simplify: $\sqrt{\dfrac{28}{81}}$.

EXAMPLE 9.25   HOW TO USE THE QUOTIENT PROPERTY TO SIMPLIFY A SQUARE ROOT

Simplify: $\sqrt{\dfrac{27m^3}{196}}$.

✓ Solution

Step 1. Simplify the fraction in the radicand, if possible.	$\dfrac{27m^3}{196}$ cannot be simplified.	$\sqrt{\dfrac{27m^3}{196}}$
Step 2. Use the Quotient Property to rewrite the radical as the quotient of two radicals.	We rewrite $\sqrt{\dfrac{27m^3}{196}}$ as the quotient of $\sqrt{27m^3}$ and $\sqrt{196}$.	$\dfrac{\sqrt{27m^3}}{\sqrt{196}}$
Step 3. Simplify the radicals in the numerator and the denominator.	$9m^2$ and 196 are perfect squares.	$\dfrac{\sqrt{9m^2}\cdot\sqrt{3m}}{\sqrt{196}}$ $\dfrac{3m\sqrt{3m}}{14}$

> **TRY IT :: 9.49**
>
> Simplify: $\sqrt{\dfrac{24p^3}{49}}$.

> **TRY IT :: 9.50**
>
> Simplify: $\sqrt{\dfrac{48x^5}{100}}$.

HOW TO :: SIMPLIFY A SQUARE ROOT USING THE QUOTIENT PROPERTY.

Step 1.   Simplify the fraction in the radicand, if possible.

Step 2.   Use the Quotient Property to rewrite the radical as the quotient of two radicals.

Step 3.   Simplify the radicals in the numerator and the denominator.

EXAMPLE 9.26

Simplify: $\sqrt{\dfrac{45x^5}{y^4}}$.

✓ Solution

$$\sqrt{\dfrac{45x^5}{y^4}}$$

We cannot simplify the fraction in the radicand. Rewrite using the Quotient Property.

$$\dfrac{\sqrt{45x^5}}{\sqrt{y^4}}$$

Simplify the radicals in the numerator and the denominator.

$$\dfrac{\sqrt{9x^4}\cdot\sqrt{5x}}{y^2}$$

Simplify.

$$\dfrac{3x^2\sqrt{5x}}{y^2}$$

> **TRY IT :: 9.51**
>
> Simplify: $\sqrt{\dfrac{80m^3}{n^6}}$.

> **TRY IT :: 9.52**
>
> Simplify: $\sqrt{\dfrac{54u^7}{v^8}}$.

Be sure to simplify the fraction in the radicand first, if possible.

EXAMPLE 9.27

Simplify: $\sqrt{\dfrac{81d^9}{25d^4}}$.

✓ **Solution**

$$\sqrt{\dfrac{81d^9}{25d^4}}$$

Simplify the fraction in the radicand.

$$\sqrt{\dfrac{81d^5}{25}}$$

Rewrite using the Quotient Property.

$$\dfrac{\sqrt{81d^5}}{\sqrt{25}}$$

Simplify the radicals in the numerator and the denominator.

$$\dfrac{\sqrt{81d^4}\cdot\sqrt{d}}{5}$$

Simplify.

$$\dfrac{9d^2\sqrt{d}}{5}$$

> **TRY IT :: 9.53**
>
> Simplify: $\sqrt{\dfrac{64x^7}{9x^3}}$.

> **TRY IT :: 9.54**
>
> Simplify: $\sqrt{\dfrac{16a^9}{100a^5}}$.

EXAMPLE 9.28

Simplify: $\sqrt{\dfrac{18p^5 q^7}{32pq^2}}$.

✓ **Solution**

$$\sqrt{\dfrac{18p^5 q^7}{32pq^2}}$$

Simplify the fraction in the radicand, if possible.

$$\sqrt{\dfrac{9p^4 q^5}{16}}$$

Rewrite using the Quotient Property.

$$\dfrac{\sqrt{9p^4 q^5}}{\sqrt{16}}$$

Simplify the radicals in the numerator and the denominator.

$$\dfrac{\sqrt{9p^4 q^4}\cdot\sqrt{q}}{4}$$

Simplify.

$$\dfrac{3p^2 q^2 \sqrt{q}}{4}$$

> **TRY IT ::** 9.55

Simplify: $\sqrt{\dfrac{50x^5 y^3}{72x^4 y}}$.

> **TRY IT ::** 9.56

Simplify: $\sqrt{\dfrac{48m^7 n^2}{125m^5 n^9}}$.

9.2 EXERCISES

Practice Makes Perfect

Use the Product Property to Simplify Square Roots

In the following exercises, simplify.

53. $\sqrt{27}$

54. $\sqrt{80}$

55. $\sqrt{125}$

56. $\sqrt{96}$

57. $\sqrt{200}$

58. $\sqrt{147}$

59. $\sqrt{450}$

60. $\sqrt{252}$

61. $\sqrt{800}$

62. $\sqrt{288}$

63. $\sqrt{675}$

64. $\sqrt{1250}$

65. $\sqrt{x^7}$

66. $\sqrt{y^{11}}$

67. $\sqrt{p^3}$

68. $\sqrt{q^5}$

69. $\sqrt{m^{13}}$

70. $\sqrt{n^{21}}$

71. $\sqrt{r^{25}}$

72. $\sqrt{s^{33}}$

73. $\sqrt{49n^{17}}$

74. $\sqrt{25m^9}$

75. $\sqrt{81r^{15}}$

76. $\sqrt{100s^{19}}$

77. $\sqrt{98m^5}$

78. $\sqrt{32n^{11}}$

79. $\sqrt{125r^{13}}$

80. $\sqrt{80s^{15}}$

81. $\sqrt{200p^{13}}$

82. $\sqrt{128q^3}$

83. $\sqrt{242m^{23}}$

84. $\sqrt{175n^{13}}$

85. $\sqrt{147m^7 n^{11}}$

86. $\sqrt{48m^7 n^5}$

87. $\sqrt{75r^{13} s^9}$

88. $\sqrt{96r^3 s^3}$

89. $\sqrt{300p^9 q^{11}}$

90. $\sqrt{192q^3 r^7}$

91. $\sqrt{242m^{13} n^{21}}$

92. $\sqrt{150m^9 n^3}$

93. $5 + \sqrt{12}$

94. $8 + \sqrt{96}$

95. $1 + \sqrt{45}$

96. $3 + \sqrt{125}$

97. $\dfrac{10 - \sqrt{24}}{2}$

98. $\dfrac{8 - \sqrt{80}}{4}$

99. $\dfrac{3 + \sqrt{90}}{3}$

100. $\dfrac{15 + \sqrt{75}}{5}$

Use the Quotient Property to Simplify Square Roots

In the following exercises, simplify.

101. $\sqrt{\dfrac{49}{64}}$

102. $\sqrt{\dfrac{100}{36}}$

103. $\sqrt{\dfrac{121}{16}}$

104. $\sqrt{\dfrac{144}{169}}$

105. $\sqrt{\dfrac{72}{98}}$

106. $\sqrt{\dfrac{75}{12}}$

107. $\sqrt{\dfrac{45}{125}}$

108. $\sqrt{\dfrac{300}{243}}$

109. $\sqrt{\dfrac{x^{10}}{x^6}}$

110. $\sqrt{\dfrac{p^{20}}{p^{10}}}$

111. $\sqrt{\dfrac{y^4}{y^8}}$

112. $\sqrt{\dfrac{q^8}{q^{14}}}$

113. $\sqrt{\dfrac{200x^7}{2x^3}}$

114. $\sqrt{\dfrac{98y^{11}}{2y^5}}$

115. $\sqrt{\dfrac{96p^9}{6p}}$

116. $\sqrt{\dfrac{108q^{10}}{3q^2}}$

117. $\sqrt{\dfrac{36}{35}}$

118. $\sqrt{\dfrac{144}{65}}$

119. $\sqrt{\dfrac{20}{81}}$

120. $\sqrt{\dfrac{21}{196}}$

121. $\sqrt{\dfrac{96x^7}{121}}$

122. $\sqrt{\dfrac{108y^4}{49}}$

123. $\sqrt{\dfrac{300m^5}{64}}$

124. $\sqrt{\dfrac{125n^7}{169}}$

125. $\sqrt{\dfrac{98r^5}{100}}$

126. $\sqrt{\dfrac{180s^{10}}{144}}$

127. $\sqrt{\dfrac{28q^6}{225}}$

128. $\sqrt{\dfrac{150r^3}{256}}$

129. $\sqrt{\dfrac{75r^9}{s^8}}$

130. $\sqrt{\dfrac{72x^5}{y^6}}$

131. $\sqrt{\dfrac{28p^7}{q^2}}$

132. $\sqrt{\dfrac{45r^3}{s^{10}}}$

133. $\sqrt{\dfrac{100x^5}{36x^3}}$

134. $\sqrt{\dfrac{49r^{12}}{16r^6}}$

135. $\sqrt{\dfrac{121p^5}{81p^2}}$

136. $\sqrt{\dfrac{25r^8}{64r}}$

137. $\sqrt{\dfrac{32x^5y^3}{18x^3y}}$

138. $\sqrt{\dfrac{75r^6s^8}{48rs^4}}$

139. $\sqrt{\dfrac{27p^2q}{108p^5q^3}}$

140. $\sqrt{\dfrac{50r^5s^2}{128r^2s^5}}$

Everyday Math

141.

ⓐ Elliott decides to construct a square garden that will take up 288 square feet of his yard. Simplify $\sqrt{288}$ to determine the length and the width of his garden. Round to the nearest tenth of a foot.

ⓑ Suppose Elliott decides to reduce the size of his square garden so that he can create a 5-foot-wide walking path on the north and east sides of the garden. Simplify $\sqrt{288} - 5$ to determine the length and width of the new garden. Round to the nearest tenth of a foot.

142.

ⓐ Melissa accidentally drops a pair of sunglasses from the top of a roller coaster, 64 feet above the ground. Simplify $\sqrt{\dfrac{64}{16}}$ to determine the number of seconds it takes for the sunglasses to reach the ground.

ⓑ Suppose the sunglasses in the previous example were dropped from a height of 144 feet. Simplify $\sqrt{\dfrac{144}{16}}$ to determine the number of seconds it takes for the sunglasses to reach the ground.

Writing Exercises

143. Explain why $\sqrt{x^4} = x^2$. Then explain why $\sqrt{x^{16}} = x^8$.

144. Explain why $7 + \sqrt{9}$ is not equal to $\sqrt{7+9}$.

Self Check

ⓐ After completing the exercises, use this checklist to evaluate your mastery of the objectives of this section.

I can...	Confidently	With some help	No-I don't get it!
use the Product Property to simplify square roots.			
use the Quotient Property to simplify square roots.			

ⓑ After reviewing this checklist, what will you do to become confident for all objectives?

9.3 | Add and Subtract Square Roots

Learning Objectives

By the end of this section, you will be able to:

› Add and subtract like square roots
› Add and subtract square roots that need simplification

> **Be Prepared!**
>
> Before you get started, take this readiness quiz.
>
> 1. Add: ⓐ $3x + 9x$ ⓑ $5m + 5n$.
>   If you missed this problem, review **Example 1.24**.
>
> 2. Simplify: $\sqrt{50x^3}$.
>   If you missed this problem, review **Example 9.16**.

We know that we must follow the order of operations to simplify expressions with square roots. The radical is a grouping symbol, so we work inside the radical first. We simplify $\sqrt{2+7}$ in this way:

$$\sqrt{2+7}$$

Add inside the radical. $\sqrt{9}$
Simplify. 3

So if we have to add $\sqrt{2} + \sqrt{7}$, we must not combine them into one radical.

$$\sqrt{2} + \sqrt{7} \neq \sqrt{2+7}$$

Trying to add square roots with different radicands is like trying to add unlike terms.

But, just like we can add $x + x$,   we can add $\sqrt{3} + \sqrt{3}$.

$$x + x = 2x \qquad\qquad \sqrt{3} + \sqrt{3} = 2\sqrt{3}$$

Adding square roots with the same radicand is just like adding like terms. We call square roots with the same radicand like square roots to remind us they work the same as like terms.

> **Like Square Roots**
>
> Square roots with the same radicand are called **like square roots**.

We add and subtract like square roots in the same way we add and subtract like terms. We know that $3x + 8x$ is $11x$. Similarly we add $3\sqrt{x} + 8\sqrt{x}$ and the result is $11\sqrt{x}$.

Add and Subtract Like Square Roots

Think about adding like terms with variables as you do the next few examples. When you have like radicands, you just add or subtract the coefficients. When the radicands are not like, you cannot combine the terms.

EXAMPLE 9.29

Simplify: $2\sqrt{2} - 7\sqrt{2}$.

⊘ Solution

	$2\sqrt{2} - 7\sqrt{2}$
Since the radicals are like, we subtract the coefficients.	$-5\sqrt{2}$

 **TRY IT : : 9.57**   Simplify: $8\sqrt{2} - 9\sqrt{2}$.

 **TRY IT :: 9.58**   Simplify: $5\sqrt{3} - 9\sqrt{3}$.

EXAMPLE 9.30

Simplify: $3\sqrt{y} + 4\sqrt{y}$.

⊘ **Solution**

$$3\sqrt{y} + 4\sqrt{y}$$

Since the radicals are like, we add the coefficients.

$$7\sqrt{y}$$

 **TRY IT :: 9.59**   Simplify: $2\sqrt{x} + 7\sqrt{x}$.

> **TRY IT :: 9.60**   Simplify: $5\sqrt{u} + 3\sqrt{u}$.

EXAMPLE 9.31

Simplify: $4\sqrt{x} - 2\sqrt{y}$.

⊘ **Solution**

$$4\sqrt{x} - 2\sqrt{y}$$

Since the radicals are not like, we cannot subtract them. We leave the expression as is.

$$4\sqrt{x} - 2\sqrt{y}$$

> **TRY IT :: 9.61**   Simplify: $7\sqrt{p} - 6\sqrt{q}$.

> **TRY IT :: 9.62**   Simplify: $6\sqrt{a} - 3\sqrt{b}$.

EXAMPLE 9.32

Simplify: $5\sqrt{13} + 4\sqrt{13} + 2\sqrt{13}$.

⊘ **Solution**

$$5\sqrt{13} + 4\sqrt{13} + 2\sqrt{13}$$

Since the radicals are like, we add the coefficients.

$$11\sqrt{13}$$

> **TRY IT :: 9.63**   Simplify: $4\sqrt{11} + 2\sqrt{11} + 3\sqrt{11}$.

> **TRY IT :: 9.64**   Simplify: $6\sqrt{10} + 2\sqrt{10} + 3\sqrt{10}$.

EXAMPLE 9.33

Simplify: $2\sqrt{6} - 6\sqrt{6} + 3\sqrt{3}$.

Solution

Since the first two radicals are like, we subtract their coefficients.

$$2\sqrt{6} - 6\sqrt{6} + 3\sqrt{3}$$

$$-4\sqrt{6} + 3\sqrt{3}$$

> **TRY IT : : 9.65** Simplify: $5\sqrt{5} - 4\sqrt{5} + 2\sqrt{6}$.

> **TRY IT : : 9.66** Simplify: $3\sqrt{7} - 8\sqrt{7} + 2\sqrt{5}$.

EXAMPLE 9.34

Simplify: $2\sqrt{5n} - 6\sqrt{5n} + 4\sqrt{5n}$.

Solution

$$2\sqrt{5n} - 6\sqrt{5n} + 4\sqrt{5n}$$

Since the radicals are like, we combine them. $0\sqrt{5n}$

Simplify. 0

> **TRY IT : : 9.67** Simplify: $\sqrt{7x} - 7\sqrt{7x} + 4\sqrt{7x}$.

> **TRY IT : : 9.68** Simplify: $4\sqrt{3y} - 7\sqrt{3y} + 2\sqrt{3y}$.

When radicals contain more than one variable, as long as all the variables and their exponents are identical, the radicals are like.

EXAMPLE 9.35

Simplify: $\sqrt{3xy} + 5\sqrt{3xy} - 4\sqrt{3xy}$.

Solution

$$\sqrt{3xy} + 5\sqrt{3xy} - 4\sqrt{3xy}$$

Since the radicals are like, we combine them. $2\sqrt{3xy}$

> **TRY IT : : 9.69** Simplify: $\sqrt{5xy} + 4\sqrt{5xy} - 7\sqrt{5xy}$.

> **TRY IT : : 9.70** Simplify: $3\sqrt{7mn} + \sqrt{7mn} - 4\sqrt{7mn}$.

Add and Subtract Square Roots that Need Simplification

Remember that we always simplify square roots by removing the largest perfect-square factor. Sometimes when we have to add or subtract square roots that do not appear to have like radicals, we find like radicals after simplifying the square roots.

EXAMPLE 9.36

Simplify: $\sqrt{20} + 3\sqrt{5}$.

Solution

$$\sqrt{20} + 3\sqrt{5}$$

Simplify the radicals, when possible.

$$\sqrt{4} \cdot \sqrt{5} + 3\sqrt{5}$$
$$2\sqrt{5} + 3\sqrt{5}$$

Combine the like radicals.

$$5\sqrt{5}$$

> **TRY IT :: 9.71**   Simplify: $\sqrt{18} + 6\sqrt{2}$.

> **TRY IT :: 9.72**   Simplify: $\sqrt{27} + 4\sqrt{3}$.

EXAMPLE 9.37

Simplify: $\sqrt{48} - \sqrt{75}$.

Solution

$$\sqrt{48} - \sqrt{75}$$

Simplify the radicals.

$$\sqrt{16} \cdot \sqrt{3} - \sqrt{25} \cdot \sqrt{3}$$
$$4\sqrt{3} - 5\sqrt{3}$$

Combine the like radicals.

$$-\sqrt{3}$$

> **TRY IT :: 9.73**   Simplify: $\sqrt{32} - \sqrt{18}$.

> **TRY IT :: 9.74**   Simplify: $\sqrt{20} - \sqrt{45}$.

Just like we use the Associative Property of Multiplication to simplify $5(3x)$ and get $15x$, we can simplify $5(3\sqrt{x})$ and get $15\sqrt{x}$. We will use the Associative Property to do this in the next example.

EXAMPLE 9.38

Simplify: $5\sqrt{18} - 2\sqrt{8}$.

Solution

$$5\sqrt{18} - 2\sqrt{8}$$

Simplify the radicals.

$$5 \cdot \sqrt{9} \cdot \sqrt{2} - 2 \cdot \sqrt{4} \cdot \sqrt{2}$$

$$5 \cdot 3 \cdot \sqrt{2} - 2 \cdot 2 \cdot \sqrt{2}$$
$$15\sqrt{2} - 4\sqrt{2}$$

Combine the like radicals.

$$11\sqrt{2}$$

> **TRY IT :: 9.75**   Simplify: $4\sqrt{27} - 3\sqrt{12}$.

> **TRY IT :: 9.76**   Simplify: $3\sqrt{20} - 7\sqrt{45}$.

EXAMPLE 9.39

Simplify: $\frac{3}{4}\sqrt{192} - \frac{5}{6}\sqrt{108}$.

⊘ **Solution**

$$\frac{3}{4}\sqrt{192} - \frac{5}{6}\sqrt{108}$$

Simplify the radicals. $\frac{3}{4}\sqrt{64} \cdot \sqrt{3} - \frac{5}{6}\sqrt{36} \cdot \sqrt{3}$

$$\frac{3}{4} \cdot 8 \cdot \sqrt{3} - \frac{5}{6} \cdot 6 \cdot \sqrt{3}$$

$$6\sqrt{3} - 5\sqrt{3}$$

Combine the like radicals. $\sqrt{3}$

> **TRY IT : : 9.77**   Simplify: $\frac{2}{3}\sqrt{108} - \frac{5}{7}\sqrt{147}$.

> **TRY IT : : 9.78**   Simplify: $\frac{3}{5}\sqrt{200} - \frac{3}{4}\sqrt{128}$.

EXAMPLE 9.40

Simplify: $\frac{2}{3}\sqrt{48} - \frac{3}{4}\sqrt{12}$.

⊘ **Solution**

$$\frac{2}{3}\sqrt{48} - \frac{3}{4}\sqrt{12}$$

Simplify the radicals. $\frac{2}{3}\sqrt{16} \cdot \sqrt{3} - \frac{3}{4}\sqrt{4} \cdot \sqrt{3}$

$$\frac{2}{3} \cdot 4 \cdot \sqrt{3} - \frac{3}{4} \cdot 2 \cdot \sqrt{3}$$

$$\frac{8}{3}\sqrt{3} - \frac{3}{2}\sqrt{3}$$

Find a common denominator to subtract the $\frac{16}{6}\sqrt{3} - \frac{9}{6}\sqrt{3}$
coefficients of the like radicals.

Simplify. $\frac{7}{6}\sqrt{3}$

> **TRY IT : : 9.79**   Simplify: $\frac{2}{5}\sqrt{32} - \frac{1}{3}\sqrt{8}$.

> **TRY IT : : 9.80**   Simplify: $\frac{1}{3}\sqrt{80} - \frac{1}{4}\sqrt{125}$.

In the next example, we will remove constant and variable factors from the square roots.

EXAMPLE 9.41

Simplify: $\sqrt{18n^5} - \sqrt{32n^5}$.

✓ **Solution**

$$\sqrt{18n^5} - \sqrt{32n^5}$$

Simplify the radicals. $\sqrt{9n^4} \cdot \sqrt{2n} - \sqrt{16n^4} \cdot \sqrt{2n}$

$3n^2\sqrt{2n} - 4n^2\sqrt{2n}$

Combine the like radicals. $-n^2\sqrt{2n}$

> **TRY IT :: 9.81**   Simplify: $\sqrt{32m^7} - \sqrt{50m^7}$.

> **TRY IT :: 9.82**   Simplify: $\sqrt{27p^3} - \sqrt{48p^3}$.

EXAMPLE 9.42

Simplify: $9\sqrt{50m^2} - 6\sqrt{48m^2}$.

✓ **Solution**

$$9\sqrt{50m^2} - 6\sqrt{48m^2}$$

Simplify the radicals. $9\sqrt{25m^2} \cdot \sqrt{2} - 6\sqrt{16m^2} \cdot \sqrt{3}$

$9 \cdot 5m \cdot \sqrt{2} - 6 \cdot 4m \cdot \sqrt{3}$

$45m\sqrt{2} - 24m\sqrt{3}$

The radicals are not like and so cannot be combined.

> **TRY IT :: 9.83**   Simplify: $5\sqrt{32x^2} - 3\sqrt{48x^2}$.

> **TRY IT :: 9.84**   Simplify: $7\sqrt{48y^2} - 4\sqrt{72y^2}$.

EXAMPLE 9.43

Simplify: $2\sqrt{8x^2} - 5x\sqrt{32} + 5\sqrt{18x^2}$.

✓ **Solution**

$$2\sqrt{8x^2} - 5x\sqrt{32} + 5\sqrt{18x^2}$$

Simplify the radicals. $2\sqrt{4x^2} \cdot \sqrt{2} - 5x\sqrt{16} \cdot \sqrt{2} + 5\sqrt{9x^2} \cdot \sqrt{2}$

$2 \cdot 2x \cdot \sqrt{2} - 5x \cdot 4 \cdot \sqrt{2} + 5 \cdot 3x \cdot \sqrt{2}$

$4x\sqrt{2} - 20x\sqrt{2} + 15x\sqrt{2}$

Combine the like radicals. $-x\sqrt{2}$

> **TRY IT :: 9.85**   Simplify: $3\sqrt{12x^2} - 2x\sqrt{48} + 4\sqrt{27x^2}$.

> **TRY IT : : 9.86** Simplify: $3\sqrt{18x^2} - 6x\sqrt{32} + 2\sqrt{50x^2}$.

▶ **MEDIA : :**

Access this online resource for additional instruction and practice with the adding and subtracting square roots.

- **Adding/Subtracting Square Roots (https://openstax.org/l/25AddSubtrSR)**

9.3 EXERCISES

Practice Makes Perfect

Add and Subtract Like Square Roots

In the following exercises, simplify.

145. $8\sqrt{2} - 5\sqrt{2}$

146. $7\sqrt{2} - 3\sqrt{2}$

147. $3\sqrt{5} + 6\sqrt{5}$

148. $4\sqrt{5} + 8\sqrt{5}$

149. $9\sqrt{7} - 10\sqrt{7}$

150. $11\sqrt{7} - 12\sqrt{7}$

151. $7\sqrt{y} + 2\sqrt{y}$

152. $9\sqrt{n} + 3\sqrt{n}$

153. $\sqrt{a} - 4\sqrt{a}$

154. $\sqrt{b} - 6\sqrt{b}$

155. $5\sqrt{c} + 2\sqrt{c}$

156. $7\sqrt{d} + 2\sqrt{d}$

157. $8\sqrt{a} - 2\sqrt{b}$

158. $5\sqrt{c} - 3\sqrt{d}$

159. $5\sqrt{m} + \sqrt{n}$

160. $\sqrt{n} + 3\sqrt{p}$

161. $8\sqrt{7} + 2\sqrt{7} + 3\sqrt{7}$

162. $6\sqrt{5} + 3\sqrt{5} + \sqrt{5}$

163. $3\sqrt{11} + 2\sqrt{11} - 8\sqrt{11}$

164. $2\sqrt{15} + 5\sqrt{15} - 9\sqrt{15}$

165. $3\sqrt{3} - 8\sqrt{3} + 7\sqrt{5}$

166. $5\sqrt{7} - 8\sqrt{7} + 6\sqrt{3}$

167. $6\sqrt{2} + 2\sqrt{2} - 3\sqrt{5}$

168. $7\sqrt{5} + \sqrt{5} - 8\sqrt{10}$

169. $3\sqrt{2a} - 4\sqrt{2a} + 5\sqrt{2a}$

170. $\sqrt{11b} - 5\sqrt{11b} + 3\sqrt{11b}$

171. $8\sqrt{3c} + 2\sqrt{3c} - 9\sqrt{3c}$

172. $3\sqrt{5d} + 8\sqrt{5d} - 11\sqrt{5d}$

173. $5\sqrt{3ab} + \sqrt{3ab} - 2\sqrt{3ab}$

174. $8\sqrt{11cd} + 5\sqrt{11cd} - 9\sqrt{11cd}$

175. $2\sqrt{pq} - 5\sqrt{pq} + 4\sqrt{pq}$

176. $11\sqrt{2rs} - 9\sqrt{2rs} + 3\sqrt{2rs}$

Add and Subtract Square Roots that Need Simplification

In the following exercises, simplify.

177. $\sqrt{50} + 4\sqrt{2}$

178. $\sqrt{48} + 2\sqrt{3}$

179. $\sqrt{80} - 3\sqrt{5}$

180. $\sqrt{28} - 4\sqrt{7}$

181. $\sqrt{27} - \sqrt{75}$

182. $\sqrt{72} - \sqrt{98}$

183. $\sqrt{48} + \sqrt{27}$

184. $\sqrt{45} + \sqrt{80}$

185. $2\sqrt{50} - 3\sqrt{72}$

186. $3\sqrt{98} - \sqrt{128}$

187. $2\sqrt{12} + 3\sqrt{48}$

188. $4\sqrt{75} + 2\sqrt{108}$

189. $\frac{2}{3}\sqrt{72} + \frac{1}{5}\sqrt{50}$

190. $\frac{2}{5}\sqrt{75} + \frac{3}{4}\sqrt{48}$

191. $\frac{1}{2}\sqrt{20} - \frac{2}{3}\sqrt{45}$

192. $\frac{2}{3}\sqrt{54} - \frac{3}{4}\sqrt{96}$

193. $\frac{1}{6}\sqrt{27} - \frac{3}{8}\sqrt{48}$

194. $\frac{1}{8}\sqrt{32} - \frac{1}{10}\sqrt{50}$

195. $\frac{1}{4}\sqrt{98} - \frac{1}{3}\sqrt{128}$

196. $\frac{1}{3}\sqrt{24} + \frac{1}{4}\sqrt{54}$

197. $\sqrt{72a^5} - \sqrt{50a^5}$

198. $\sqrt{48b^5} - \sqrt{75b^5}$

199. $\sqrt{80c^7} - \sqrt{20c^7}$

200. $\sqrt{96d^9} - \sqrt{24d^9}$

201. $9\sqrt{80p^4} - 6\sqrt{98p^4}$

202. $8\sqrt{72q^6} - 3\sqrt{75q^6}$

203. $2\sqrt{50r^8} + 4\sqrt{54r^8}$

204. $5\sqrt{27s^6} + 2\sqrt{20s^6}$

205. $3\sqrt{20x^2} - 4\sqrt{45x^2} + 5x\sqrt{80}$

206. $2\sqrt{28x^2} - \sqrt{63x^2} + 6x\sqrt{7}$

207.
$3\sqrt{128y^2} + 4y\sqrt{162} - 8\sqrt{98y^2}$

208. $3\sqrt{75y^2} + 8y\sqrt{48} - \sqrt{300y^2}$

Mixed Practice

209. $2\sqrt{8} + 6\sqrt{8} - 5\sqrt{8}$

210. $\frac{2}{3}\sqrt{27} + \frac{3}{4}\sqrt{48}$

211. $\sqrt{175k^4} - \sqrt{63k^4}$

212. $\frac{5}{6}\sqrt{162} + \frac{3}{16}\sqrt{128}$

213. $2\sqrt{363} - 2\sqrt{300}$

214. $\sqrt{150} + 4\sqrt{6}$

215. $9\sqrt{2} - 8\sqrt{2}$

216. $5\sqrt{x} - 8\sqrt{y}$

217. $8\sqrt{13} - 4\sqrt{13} - 3\sqrt{13}$

218. $5\sqrt{12c^4} - 3\sqrt{27c^6}$

219. $\sqrt{80a^5} - \sqrt{45a^5}$

220. $\frac{3}{5}\sqrt{75} - \frac{1}{4}\sqrt{48}$

221. $21\sqrt{19} - 2\sqrt{19}$

222. $\sqrt{500} + \sqrt{405}$

223. $\frac{5}{6}\sqrt{27} + \frac{5}{8}\sqrt{48}$

224. $11\sqrt{11} - 10\sqrt{11}$

225. $\sqrt{75} - \sqrt{108}$

226. $2\sqrt{98} - 4\sqrt{72}$

227. $4\sqrt{24x^2} - \sqrt{54x^2} + 3x\sqrt{6}$

228. $8\sqrt{80y^6} - 6\sqrt{48y^6}$

Everyday Math

229. A decorator decides to use square tiles as an accent strip in the design of a new shower, but she wants to rotate the tiles to look like diamonds. She will use 9 large tiles that measure 8 inches on a side and 8 small tiles that measure 2 inches on a side. $9(8\sqrt{2}) + 8(2\sqrt{2})$. Determine the width of the accent strip by simplifying the expression $9(8\sqrt{2}) + 8(2\sqrt{2})$. (Round to the nearest tenth of an inch.)

230. Suzy wants to use square tiles on the border of a spa she is installing in her backyard. She will use large tiles that have area of 12 square inches, medium tiles that have area of 8 square inches, and small tiles that have area of 4 square inches. Once section of the border will require 4 large tiles, 8 medium tiles, and 10 small tiles to cover the width of the wall. Simplify the expression $4\sqrt{12} + 8\sqrt{8} + 10\sqrt{4}$ to determine the width of the wall.

Writing Exercises

231. Explain the difference between like radicals and unlike radicals. Make sure your answer makes sense for radicals containing both numbers and variables.

232. Explain the process for determining whether two radicals are like or unlike. Make sure your answer makes sense for radicals containing both numbers and variables.

Self Check

@ *After completing the exercises, use this checklist to evaluate your mastery of the objectives of this section.*

I can...	Confidently	With some help	No-I don't get it!
add and subtract like square roots.			
add and subtract square roots that need simplification.			

ⓑ *What does this checklist tell you about your mastery of this section? What steps will you take to improve?*

Multiply Square Roots

Learning Objectives

By the end of this section, you will be able to:
- Multiply square roots
- Use polynomial multiplication to multiply square roots

Be Prepared!

Before you get started, take this readiness quiz.

1. Simplify: $(3u)(8v)$.
   If you missed this problem, review **Example 6.26**.
2. Simplify: $6(12 - 7n)$.
   If you missed this problem, review **Example 6.28**.
3. Simplify: $(2 + a)(4 - a)$.
   If you missed this problem, review **Example 6.39**.

Multiply Square Roots

We have used the Product Property of Square Roots to simplify square roots by removing the perfect square factors. The Product Property of Square Roots says

$$\sqrt{ab} = \sqrt{a} \cdot \sqrt{b}$$

We can use the Product Property of Square Roots 'in reverse' to multiply square roots.

$$\sqrt{a} \cdot \sqrt{b} = \sqrt{ab}$$

Remember, we assume all variables are greater than or equal to zero.

We will rewrite the Product Property of Square Roots so we see both ways together.

Product Property of Square Roots

If a, b are nonnegative real numbers, then

$$\sqrt{ab} = \sqrt{a} \cdot \sqrt{b} \quad \text{and} \quad \sqrt{a} \cdot \sqrt{b} = \sqrt{ab}$$

So we can multiply $\sqrt{3} \cdot \sqrt{5}$ in this way:

$$\sqrt{3} \cdot \sqrt{5}$$
$$\sqrt{3 \cdot 5}$$
$$\sqrt{15}$$

Sometimes the product gives us a perfect square:

$$\sqrt{2} \cdot \sqrt{8}$$
$$\sqrt{2 \cdot 8}$$
$$\sqrt{16}$$
$$4$$

Even when the product is not a perfect square, we must look for perfect-square factors and simplify the radical whenever possible.

Multiplying radicals with coefficients is much like multiplying variables with coefficients. To multiply $4x \cdot 3y$ we multiply the coefficients together and then the variables. The result is $12xy$. Keep this in mind as you do these examples.

EXAMPLE 9.44

Simplify: ⓐ $\sqrt{2} \cdot \sqrt{6}$ ⓑ $(4\sqrt{3})(2\sqrt{12})$.

Solution

ⓐ

$$\sqrt{2} \cdot \sqrt{6}$$

Multiply using the Product Property.	$\sqrt{12}$
Simplify the radical.	$\sqrt{4} \cdot \sqrt{3}$
Simplify.	$2\sqrt{3}$

ⓑ

$$(4\sqrt{3})(2\sqrt{12})$$

Multiply using the Product Property.	$8\sqrt{36}$
Simplify the radical.	$8 \cdot 6$
Simplify.	48

Notice that in (b) we multiplied the coefficients and multiplied the radicals. Also, we did not simplify $\sqrt{12}$. We waited to get the product and then simplified.

> **TRY IT : : 9.87** Simplify: ⓐ $\sqrt{3} \cdot \sqrt{6}$ ⓑ $(2\sqrt{6})(3\sqrt{12})$.

> **TRY IT : : 9.88** Simplify: ⓐ $\sqrt{5} \cdot \sqrt{10}$ ⓑ $(6\sqrt{3})(5\sqrt{6})$.

EXAMPLE 9.45

Simplify: $(6\sqrt{2})(3\sqrt{10})$.

Solution

$$(6\sqrt{2})(3\sqrt{10})$$

Multiply using the Product Property.	$18\sqrt{20}$
Simplify the radical.	$18\sqrt{4} \cdot \sqrt{5}$
Simplify.	$18 \cdot 2 \cdot \sqrt{5}$
	$36\sqrt{5}$

> **TRY IT : : 9.89** Simplify: $(3\sqrt{2})(2\sqrt{30})$.

> **TRY IT : : 9.90** Simplify: $(3\sqrt{3})(3\sqrt{6})$.

When we have to multiply square roots, we first find the product and then remove any perfect square factors.

EXAMPLE 9.46

Simplify: ⓐ $\left(\sqrt{8x^3}\right)\left(\sqrt{3x}\right)$ ⓑ $\left(\sqrt{20y^2}\right)\left(\sqrt{5y^3}\right)$.

✓ **Solution**

ⓐ

$$\left(\sqrt{8x^3}\right)\left(\sqrt{3x}\right)$$

Multiply using the Product Property. $\sqrt{24x^4}$

Simplify the radical. $\sqrt{4x^4} \cdot \sqrt{6}$

Simplify. $2x^2\sqrt{6}$

ⓑ

$$\left(\sqrt{20y^2}\right)\left(\sqrt{5y^3}\right)$$

Multiply using the Product Property. $\sqrt{100y^5}$

Simplify the radical. $10y^2\sqrt{y}$

> **TRY IT :: 9.91**   Simplify: ⓐ $\left(\sqrt{6x^3}\right)\left(\sqrt{3x}\right)$ ⓑ $\left(\sqrt{2y^3}\right)\left(\sqrt{50y^2}\right)$.

> **TRY IT :: 9.92**   Simplify: ⓐ $\left(\sqrt{6x^5}\right)\left(\sqrt{2x}\right)$ ⓑ $\left(\sqrt{12y^2}\right)\left(\sqrt{3y^5}\right)$.

EXAMPLE 9.47

Simplify: $\left(10\sqrt{6p^3}\right)\left(3\sqrt{18p}\right)$.

✓ **Solution**

$$\left(10\sqrt{6p^3}\right)\left(3\sqrt{18p}\right)$$

Multiply. $30\sqrt{108p^4}$

Simplify the radical. $30\sqrt{36p^4} \cdot \sqrt{3}$

$30 \cdot 6p^2 \cdot \sqrt{3}$

$180p^2\sqrt{3}$

> **TRY IT :: 9.93**   Simplify: $\left(6\sqrt{2x^2}\right)\left(8\sqrt{45x^4}\right)$.

> **TRY IT :: 9.94**   Simplify: $\left(2\sqrt{6y^4}\right)\left(12\sqrt{30y}\right)$.

EXAMPLE 9.48

Simplify: ⓐ $(\sqrt{2})^2$ ⓑ $(-\sqrt{11})^2$.

✓ **Solution**

ⓐ

$$(\sqrt{2})^2$$

Rewrite as a product. $(\sqrt{2})(\sqrt{2})$

Multiply. $\sqrt{4}$

Simplify. 2

ⓑ

$$(-\sqrt{11})^2$$

Rewrite as a product. $(-\sqrt{11})(-\sqrt{11})$

Multiply. $\sqrt{121}$

Simplify. 11

> | **TRY IT ::** 9.95   Simplify: ⓐ $(\sqrt{12})^2$ ⓑ $(-\sqrt{15})^2$.

> | **TRY IT ::** 9.96   Simplify: ⓐ $(\sqrt{16})^2$ ⓑ $(-\sqrt{20})^2$.

The results of the previous example lead us to this property.

Squaring a Square Root

If a is a nonnegative real number, then

$$(\sqrt{a})^2 = a$$

By realizing that squaring and taking a square root are 'opposite' operations, we can simplify $(\sqrt{2})^2$ and get 2 right away. When we multiply the two like square roots in part (a) of the next example, it is the same as squaring.

EXAMPLE 9.49

Simplify: ⓐ $(2\sqrt{3})(8\sqrt{3})$ ⓑ $(3\sqrt{6})^2$.

✓ **Solution**

ⓐ

$$(2\sqrt{3})(8\sqrt{3})$$

Multiply. Remember, $(\sqrt{3})^2 = 3$. $16 \cdot 3$

Simplify. 48

ⓑ

$$(3\sqrt{6})^2$$

Multiply. $9 \cdot 6$

Simplify. 54

> **TRY IT : : 9.97**   Simplify: ⓐ $(6\sqrt{11})(5\sqrt{11})$ ⓑ $(5\sqrt{8})^2$.

> **TRY IT : : 9.98**   Simplify: ⓐ $(3\sqrt{7})(10\sqrt{7})$ ⓑ $(-4\sqrt{6})^2$.

Use Polynomial Multiplication to Multiply Square Roots

In the next few examples, we will use the Distributive Property to multiply expressions with square roots. We will first distribute and then simplify the square roots when possible.

EXAMPLE 9.50

Simplify: ⓐ $3(5 - \sqrt{2})$ ⓑ $\sqrt{2}(4 - \sqrt{10})$.

⊘ **Solution**

ⓐ
$$3(5 - \sqrt{2})$$

Distribute. $15 - 3\sqrt{2}$

ⓑ
$$\sqrt{2}(4 - \sqrt{10})$$

Distribute. $4\sqrt{2} - \sqrt{20}$

$$4\sqrt{2} - 2\sqrt{5}$$

> **TRY IT : : 9.99**   Simplify: ⓐ $2(3 - \sqrt{5})$ ⓑ $\sqrt{3}(2 - \sqrt{18})$.

> **TRY IT : : 9.100**   Simplify: ⓐ $6(2 + \sqrt{6})$ ⓑ $\sqrt{7}(1 + \sqrt{14})$.

EXAMPLE 9.51

Simplify: ⓐ $\sqrt{5}(7 + 2\sqrt{5})$ ⓑ $\sqrt{6}(\sqrt{2} + \sqrt{18})$.

⊘ **Solution**

ⓐ

$$\sqrt{5}(7 + 2\sqrt{5})$$

Multiply.

$$7\sqrt{5} + 2 \cdot 5$$

Simplify.

$$7\sqrt{5} + 10$$

$$10 + 7\sqrt{5}$$

ⓑ

$$\sqrt{6}(\sqrt{2} + \sqrt{18})$$

Multiply.

$$\sqrt{12} + \sqrt{108}$$

Simplify.

$$\sqrt{4} \cdot \sqrt{3} + \sqrt{36} \cdot \sqrt{3}$$
$$2\sqrt{3} + 6\sqrt{3}$$

Combine like radicals.

$$8\sqrt{3}$$

> **TRY IT ::** 9.101   Simplify: ⓐ $\sqrt{6}(1 + 3\sqrt{6})$ ⓑ $\sqrt{12}(\sqrt{3} + \sqrt{24})$.

> **TRY IT ::** 9.102   Simplify: ⓐ $\sqrt{8}(2 - 5\sqrt{8})$ ⓑ $\sqrt{14}(\sqrt{2} + \sqrt{42})$.

When we worked with polynomials, we multiplied binomials by binomials. Remember, this gave us four products before we combined any like terms. To be sure to get all four products, we organized our work—usually by the FOIL method.

EXAMPLE 9.52

Simplify: $(2 + \sqrt{3})(4 - \sqrt{3})$.

✓ Solution

$$(2 + \sqrt{3})(4 - \sqrt{3})$$

Multiply.

$$8 - 2\sqrt{3} + 4\sqrt{3} - 3$$

Combine like terms.

$$5 + 2\sqrt{3}$$

> **TRY IT ::** 9.103   Simplify: $(1 + \sqrt{6})(3 - \sqrt{6})$.

> **TRY IT ::** 9.104   Simplify: $(4 - \sqrt{10})(2 + \sqrt{10})$.

EXAMPLE 9.53

Simplify: $(3 - 2\sqrt{7})(4 - 2\sqrt{7})$.

✓ **Solution**

$$(3 - 2\sqrt{7})(4 - 2\sqrt{7})$$

Multiply. $12 - 6\sqrt{7} - 8\sqrt{7} + 4 \cdot 7$

Simplify. $12 - 6\sqrt{7} - 8\sqrt{7} + 28$

Combine like terms. $40 - 14\sqrt{7}$

> **TRY IT : : 9.105**   Simplify: $(6 - 3\sqrt{7})(3 + 4\sqrt{7})$.

> **TRY IT : : 9.106**   Simplify: $(2 - 3\sqrt{11})(4 - \sqrt{11})$.

EXAMPLE 9.54

Simplify: $(3\sqrt{2} - \sqrt{5})(\sqrt{2} + 4\sqrt{5})$.

✓ **Solution**

$$(3\sqrt{2} - \sqrt{5})(\sqrt{2} + 4\sqrt{5})$$

Multiply. $3 \cdot 2 + 12\sqrt{10} - \sqrt{10} - 4 \cdot 5$

Simplify. $6 + 12\sqrt{10} - \sqrt{10} - 20$

Combine like terms. $-14 + 11\sqrt{10}$

> **TRY IT : : 9.107**   Simplify: $(5\sqrt{3} - \sqrt{7})(\sqrt{3} + 2\sqrt{7})$.

> **TRY IT : : 9.108**   Simplify: $(\sqrt{6} - 3\sqrt{8})(2\sqrt{6} + \sqrt{8})$

EXAMPLE 9.55

Simplify: $(4 - 2\sqrt{x})(1 + 3\sqrt{x})$.

✓ **Solution**

$$(4 - 2\sqrt{x})(1 + 3\sqrt{x})$$

Multiply. $4 + 12\sqrt{x} - 2\sqrt{x} - 6x$

Combine like terms. $4 + 10\sqrt{x} - 6x$

> **TRY IT : : 9.109**   Simplify: $(6 - 5\sqrt{m})(2 + 3\sqrt{m})$.

> **TRY IT : : 9.110**   Simplify: $(10 + 3\sqrt{n})(1 - 5\sqrt{n})$.

Note that some special products made our work easier when we multiplied binomials earlier. This is true when we multiply square roots, too. The special product formulas we used are shown below.

Special Product Formulas

Binomial Squares	Product of Conjugates
$(a + b)^2 = a^2 + 2ab + b^2$	$(a - b)(a + b) = a^2 - b^2$
$(a - b)^2 = a^2 - 2ab + b^2$	

We will use the special product formulas in the next few examples. We will start with the Binomial Squares formula.

EXAMPLE 9.56

Simplify: ⓐ $(2 + \sqrt{3})^2$ ⓑ $(4 - 2\sqrt{5})^2$.

✓ Solution

Be sure to include the $2ab$ term when squaring a binomial.

ⓐ

$$\overset{(a\ +\ b)^2}{(2 + \sqrt{3})^2}$$

Multiply using the binomial square pattern.	$\overset{a^2\ +\qquad 2ab\qquad +\quad b^2}{2^2 + 2 \cdot 2 \cdot \sqrt{3} + (\sqrt{3})^2}$
Simplify.	$4 + 4\sqrt{3} + 3$
Combine like terms.	$7 + 4\sqrt{3}$

ⓑ

$$\overset{(a\ -\ b)^2}{(4 - 2\sqrt{5})^2}$$

Multiply using the binomial square pattern.	$\overset{a^2\ -\qquad 2ab\qquad +\quad b^2}{4^2 - 2 \cdot 4 \cdot 2\sqrt{5} + (2\sqrt{5})^2}$
Simplify.	$16 - 16\sqrt{5} + 4 \cdot 5$ $16 - 16\sqrt{5} + 20$
Combine like terms.	$36 - 16\sqrt{5}$

> **TRY IT :: 9.111** Simplify: ⓐ $(10 + \sqrt{2})^2$ ⓑ $(1 + 3\sqrt{6})^2$.

> **TRY IT :: 9.112** Simplify: ⓐ $(6 - \sqrt{5})^2$ ⓑ $(9 - 2\sqrt{10})^2$.

EXAMPLE 9.57

Simplify: $(1 + 3\sqrt{x})^2$.

⊘ **Solution**

$$\overset{(a\ +\ b)^2}{\left(1+3\sqrt{x}\right)^2}$$

Multiply using the binomial square pattern.	$\overset{a^2\ +\quad 2ab\quad\ +\quad b^2}{1^2+2\cdot1\cdot3\sqrt{x}+\left(3\sqrt{x}\right)^2}$
Simplify.	$1+6\sqrt{x}+9x$

> **TRY IT :: 9.113** Simplify: $\left(2+5\sqrt{m}\right)^2$.

> **TRY IT :: 9.114** Simplify: $\left(3-4\sqrt{n}\right)^2$.

In the next two examples, we will find the product of conjugates.

EXAMPLE 9.58

Simplify: $\left(4-\sqrt{2}\right)\left(4+\sqrt{2}\right)$.

⊘ **Solution**

$$\overset{(a\ -\ b)\ (a\ +\ b)}{\left(4-\sqrt{2}\right)\left(4+\sqrt{2}\right)}$$

Multiply using the binomial square pattern.	$\overset{a^2\ -\quad b^2}{4^2-\left(\sqrt{2}\right)^2}$
Simplify.	$16-2$ 14

> **TRY IT :: 9.115** Simplify: $\left(2-\sqrt{3}\right)\left(2+\sqrt{3}\right)$.

> **TRY IT :: 9.116** Simplify: $\left(1+\sqrt{5}\right)\left(1-\sqrt{5}\right)$.

EXAMPLE 9.59

Simplify: $\left(5-2\sqrt{3}\right)\left(5+2\sqrt{3}\right)$.

⊘ **Solution**

$$\overset{(a\ -\ b)\ (a\ +\ b)}{\left(5-2\sqrt{3}\right)\left(5+2\sqrt{3}\right)}$$

Multiply using the binomial square pattern.	$\overset{a^2\ -\quad b^2}{5^2-\left(2\sqrt{3}\right)^2}$
Simplify.	$25-4\cdot3$ 13

> **TRY IT :: 9.117** Simplify: $\left(3-2\sqrt{5}\right)\left(3+2\sqrt{5}\right)$.

> **TRY IT : :** 9.118   Simplify: $\left(4 + 5\sqrt{7}\right)\left(4 - 5\sqrt{7}\right)$.

▶ **MEDIA : :**

Access these online resources for additional instruction and practice with multiplying square roots.

- **Product Property (https://openstax.org/l/25ProductProp)**
- **Multiply Binomials with Square Roots (https://openstax.org/l/25MultPolySR)**

9.4 EXERCISES

Practice Makes Perfect

Multiply Square Roots

In the following exercises, simplify.

233.
ⓐ $\sqrt{2} \cdot \sqrt{8}$
ⓑ $(3\sqrt{3})(2\sqrt{18})$

234.
ⓐ $\sqrt{6} \cdot \sqrt{6}$
ⓑ $(3\sqrt{2})(2\sqrt{32})$

235.
ⓐ $\sqrt{7} \cdot \sqrt{14}$
ⓑ $(4\sqrt{8})(5\sqrt{8})$

236.
ⓐ $\sqrt{6} \cdot \sqrt{12}$
ⓑ $(2\sqrt{5})(2\sqrt{10})$

237. $(5\sqrt{2})(3\sqrt{6})$

238. $(2\sqrt{3})(4\sqrt{6})$

239. $(-2\sqrt{3})(3\sqrt{18})$

240. $(-4\sqrt{5})(5\sqrt{10})$

241. $(5\sqrt{6})(-\sqrt{12})$

242. $(6\sqrt{2})(-\sqrt{10})$

243. $(-2\sqrt{7})(-2\sqrt{14})$

244. $(-2\sqrt{11})(-4\sqrt{22})$

245.
ⓐ $(\sqrt{15y})(\sqrt{5y^3})$
ⓑ $(\sqrt{2n^2})(\sqrt{18n^3})$

246.
ⓐ $(\sqrt{14x^3})(\sqrt{7x^3})$
ⓑ $(\sqrt{3q^2})(\sqrt{48q^3})$

247.
ⓐ $(\sqrt{16y^2})(\sqrt{8y^4})$
ⓑ $(\sqrt{11s^6})(\sqrt{11s})$

248.
ⓐ $(\sqrt{8x^3})(\sqrt{3x})$
ⓑ $(\sqrt{7r})(\sqrt{7r^8})$

249. $(2\sqrt{5b^3})(4\sqrt{15b})$

250. $(3\sqrt{8c^5})(2\sqrt{6c^3})$

251. $(6\sqrt{3d^3})(4\sqrt{12d^5})$

252. $(2\sqrt{5b^3})(4\sqrt{15b})$

253. $(6\sqrt{3d^3})(4\sqrt{12d^5})$

254. $(-2\sqrt{7z^3})(3\sqrt{14z^8})$

255. $(4\sqrt{2k^5})(-3\sqrt{32k^6})$

256.
ⓐ $(\sqrt{7})^2$
ⓑ $(-\sqrt{15})^2$

257.
ⓐ $(\sqrt{11})^2$
ⓑ $(-\sqrt{21})^2$

258.
ⓐ $(\sqrt{19})^2$
ⓑ $(-\sqrt{5})^2$

259.
ⓐ $(\sqrt{23})^2$
ⓑ $(-\sqrt{3})^2$

260.
ⓐ $(4\sqrt{11})(-3\sqrt{11})$
ⓑ $(5\sqrt{3})^2$

261.
ⓐ $(2\sqrt{13})(-9\sqrt{13})$
ⓑ $(6\sqrt{5})^2$

262.
ⓐ $(-3\sqrt{12})(-2\sqrt{6})$
ⓑ $(-4\sqrt{10})^2$

263.
ⓐ $(-7\sqrt{5})(-3\sqrt{10})$
ⓑ $(-2\sqrt{14})^2$

Use Polynomial Multiplication to Multiply Square Roots

In the following exercises, simplify.

264.
ⓐ $3(4 - \sqrt{3})$

ⓑ $\sqrt{2}(4 - \sqrt{6})$

265.
ⓐ $4(6 - \sqrt{11})$

ⓑ $\sqrt{2}(5 - \sqrt{12})$

266.
ⓐ $5(3 - \sqrt{7})$

ⓑ $\sqrt{3}(4 - \sqrt{15})$

267.
ⓐ $7(-2 - \sqrt{11})$

ⓑ $\sqrt{7}(6 - \sqrt{14})$

268.
ⓐ $\sqrt{7}(5 + 2\sqrt{7})$

ⓑ $\sqrt{5}(\sqrt{10} + \sqrt{18})$

269.
ⓐ $\sqrt{11}(8 + 4\sqrt{11})$

ⓑ $\sqrt{3}(\sqrt{12} + \sqrt{27})$

270.
ⓐ $\sqrt{11}(-3 + 4\sqrt{11})$

ⓑ $\sqrt{3}(\sqrt{15} - \sqrt{18})$

271.
ⓐ $\sqrt{2}(-5 + 9\sqrt{2})$

ⓑ $\sqrt{7}(\sqrt{3} - \sqrt{21})$

272. $(8 + \sqrt{3})(2 - \sqrt{3})$

273. $(7 + \sqrt{3})(9 - \sqrt{3})$

274. $(8 - \sqrt{2})(3 + \sqrt{2})$

275. $(9 - \sqrt{2})(6 + \sqrt{2})$

276. $(3 - \sqrt{7})(5 - \sqrt{7})$

277. $(5 - \sqrt{7})(4 - \sqrt{7})$

278. $(1 + 3\sqrt{10})(5 - 2\sqrt{10})$

279. $(7 - 2\sqrt{5})(4 + 9\sqrt{5})$

280. $(\sqrt{3} + \sqrt{10})(\sqrt{3} + 2\sqrt{10})$

281. $(\sqrt{11} + \sqrt{5})(\sqrt{11} + 6\sqrt{5})$

282. $(2\sqrt{7} - 5\sqrt{11})(4\sqrt{7} + 9\sqrt{11})$

283. $(4\sqrt{6} + 7\sqrt{13})(8\sqrt{6} - 3\sqrt{13})$

284. $(5 - \sqrt{u})(3 + \sqrt{u})$

285. $(9 - \sqrt{w})(2 + \sqrt{w})$

286. $(7 + 2\sqrt{m})(4 + 9\sqrt{m})$

287. $(6 + 5\sqrt{n})(11 + 3\sqrt{n})$

288.
ⓐ $(3 + \sqrt{5})^2$

ⓑ $(2 - 5\sqrt{3})^2$

289.
ⓐ $(4 + \sqrt{11})^2$

ⓑ $(3 - 2\sqrt{5})^2$

290.
ⓐ $(9 - \sqrt{6})^2$

ⓑ $(10 + 3\sqrt{7})^2$

291.
ⓐ $(5 - \sqrt{10})^2$

ⓑ $(8 + 3\sqrt{2})^2$

292. $(3 - \sqrt{5})(3 + \sqrt{5})$

293. $(10 - \sqrt{3})(10 + \sqrt{3})$

294. $(4 + \sqrt{2})(4 - \sqrt{2})$

295. $(7 + \sqrt{10})(7 - \sqrt{10})$

296. $(4 + 9\sqrt{3})(4 - 9\sqrt{3})$

297. $(1 + 8\sqrt{2})(1 - 8\sqrt{2})$

298. $(12 - 5\sqrt{5})(12 + 5\sqrt{5})$

299. $(9 - 4\sqrt{3})(9 + 4\sqrt{3})$

Mixed Practice

In the following exercises, simplify.

300. $\sqrt{3} \cdot \sqrt{21}$

301. $(4\sqrt{6})(-\sqrt{18})$

302. $(-5 + \sqrt{7})(6 + \sqrt{21})$

303. $(-5\sqrt{7})(6\sqrt{21})$

304. $(-4\sqrt{2})(2\sqrt{18})$

305. $(\sqrt{35y^3})(\sqrt{7y^3})$

306. $(4\sqrt{12x^5})(2\sqrt{6x^3})$

307. $(\sqrt{29})^2$

308. $(-4\sqrt{17})(-3\sqrt{17})$

309. $(-4 + \sqrt{17})(-3 + \sqrt{17})$

Everyday Math

310. A landscaper wants to put a square reflecting pool next to a triangular deck, as shown below. The triangular deck is a right triangle, with legs of length 9 feet and 11 feet, and the pool will be adjacent to the hypotenuse.

ⓐ Use the Pythagorean Theorem to find the length of a side of the pool. Round your answer to the nearest tenth of a foot.

ⓑ Find the exact area of the pool.

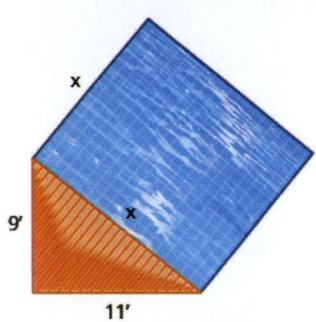

311. An artist wants to make a small monument in the shape of a square base topped by a right triangle, as shown below. The square base will be adjacent to one leg of the triangle. The other leg of the triangle will measure 2 feet and the hypotenuse will be 5 feet.

ⓐ Use the Pythagorean Theorem to find the length of a side of the square base. Round your answer to the nearest tenth of a foot.

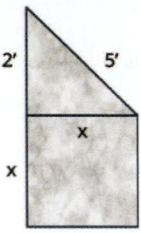

ⓑ Find the exact area of the face of the square base.

312. A square garden will be made with a stone border on one edge. If only $3 + \sqrt{10}$ feet of stone are available, simplify $(3 + \sqrt{10})^2$ to determine the area of the largest such garden.

313. A garden will be made so as to contain two square sections, one section with side length $\sqrt{5} + \sqrt{6}$ yards and one section with side length $\sqrt{2} + \sqrt{3}$ yards. Simplify $(\sqrt{5} + \sqrt{6})(\sqrt{2} + \sqrt{3})$ to determine the total area of the garden.

314. Suppose a third section will be added to the garden in the previous exercise. The third section is to have a width of $\sqrt{432}$ feet. Write an expression that gives the total area of the garden.

Writing Exercises

315.

ⓐ Explain why $(-\sqrt{n})^2$ is always positive, for $n \geq 0$.

ⓑ Explain why $(-\sqrt{n})^2$ is always negative, for $n \geq 0$.

316. Use the binomial square pattern to simplify $(3 + \sqrt{2})^2$. Explain all your steps.

Self Check

ⓐ *After completing the exercises, use this checklist to evaluate your mastery of the objectives of this section.*

I can...	Confidently	With some help	No-I don't get it!
multiply square roots.			
use polynomial multiplication to multiply square roots.			

ⓑ *On a scale of 1–10, how would you rate your mastery of this section in light of your responses on the checklist? How can you improve this?*

9.5 | Divide Square Roots

Learning Objectives

By the end of this section, you will be able to:

› Divide square roots
› Rationalize a one-term denominator
› Rationalize a two-term denominator

Be Prepared!

Before you get started, take this readiness quiz.

1. Find a fraction equivalent to $\frac{5}{8}$ with denominator 48.

   If you missed this problem, review **Example 1.64**.

2. Simplify: $\left(\sqrt{5}\right)^2$.

   If you missed this problem, review **Example 9.48**.

3. Multiply: $(7 + 3x)(7 - 3x)$.

   If you missed this problem, review **Example 6.54**.

Divide Square Roots

We know that we simplify fractions by removing factors common to the numerator and the denominator. When we have a fraction with a square root in the numerator, we first simplify the square root. Then we can look for common factors.

Common Factors	No common factors
$\dfrac{3\sqrt{2}}{3 \cdot 5}$	$\dfrac{2\sqrt{3}}{3 \cdot 5}$

EXAMPLE 9.60

Simplify: $\dfrac{\sqrt{54}}{6}$.

⊘ **Solution**

$$\frac{\sqrt{54}}{6}$$

Simplify the radical. $\dfrac{\sqrt{9} \cdot \sqrt{6}}{6}$

Simplify. $\dfrac{3\sqrt{6}}{6}$

Remove the common factors. $\dfrac{\cancel{3}\sqrt{6}}{\cancel{3} \cdot 2}$

Simplify. $\dfrac{\sqrt{6}}{2}$

 **TRY IT : : 9.119**   Simplify: $\dfrac{\sqrt{32}}{8}$.

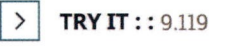 **TRY IT : : 9.120**   Simplify: $\dfrac{\sqrt{75}}{15}$.

EXAMPLE 9.61

Simplify: $\dfrac{6 - \sqrt{24}}{12}$.

✓ **Solution**

$$\dfrac{6 - \sqrt{24}}{12}$$

Simplify the radical.
$$\dfrac{6 - \sqrt{4} \cdot \sqrt{6}}{12}$$

Simplify.
$$\dfrac{6 - 2\sqrt{6}}{12}$$

Factor the common factor from the numerator.
$$\dfrac{2\left(3 - \sqrt{6}\right)}{2 \cdot 6}$$

Remove the common factors.
$$\dfrac{\cancel{2}\left(3 - \sqrt{6}\right)}{\cancel{2} \cdot 6}$$

Simplify.
$$\dfrac{3 - \sqrt{6}}{6}$$

> **TRY IT :: 9.121**   Simplify: $\dfrac{8 - \sqrt{40}}{10}$.

> **TRY IT :: 9.122**   Simplify: $\dfrac{10 - \sqrt{75}}{20}$.

We have used the Quotient Property of Square Roots to simplify square roots of fractions. The Quotient Property of Square Roots says

$$\sqrt{\dfrac{a}{b}} = \dfrac{\sqrt{a}}{\sqrt{b}},\ b \neq 0$$

Sometimes we will need to use the Quotient Property of Square Roots 'in reverse' to simplify a fraction with square roots.

$$\dfrac{\sqrt{a}}{\sqrt{b}} = \sqrt{\dfrac{a}{b}},\ b \neq 0$$

We will rewrite the Quotient Property of Square Roots so we see both ways together. Remember: we assume all variables are greater than or equal to zero so that their square roots are real numbers.

Quotient Property of Square Roots

If a, b are non-negative real numbers and $b \neq 0$, then

$$\sqrt{\dfrac{a}{b}} = \dfrac{\sqrt{a}}{\sqrt{b}} \quad \text{and} \quad \dfrac{\sqrt{a}}{\sqrt{b}} = \sqrt{\dfrac{a}{b}}$$

We will use the Quotient Property of Square Roots 'in reverse' when the fraction we start with is the quotient of two square roots, and neither radicand is a perfect square. When we write the fraction in a single square root, we may find common factors in the numerator and denominator.

EXAMPLE 9.62

Simplify: $\dfrac{\sqrt{27}}{\sqrt{75}}$.

⊘ **Solution**

$$\frac{\sqrt{27}}{\sqrt{75}}$$

Neither radicand is a perfect square, so rewrite using the quotient property of square roots.

$$\sqrt{\frac{27}{75}}$$

Remove common factors in the numerator and denominator.

$$\sqrt{\frac{\cancel{3}\cdot 9}{\cancel{3}\cdot 25}}$$

Simplify.

$$\sqrt{\frac{9}{25}}$$

$$\frac{3}{5}$$

> **TRY IT : : 9.123**   Simplify: $\dfrac{\sqrt{48}}{\sqrt{108}}$.

> **TRY IT : : 9.124**   Simplify: $\dfrac{\sqrt{96}}{\sqrt{54}}$.

We will use the Quotient Property for Exponents, $\dfrac{a^m}{a^n} = a^{m-n}$, when we have variables with exponents in the radicands.

EXAMPLE 9.63

Simplify: $\dfrac{\sqrt{6y^5}}{\sqrt{2y}}$.

⊘ **Solution**

$$\frac{\sqrt{6y^5}}{\sqrt{2y}}$$

Neither radicand is a perfect square, so rewrite using the quotient property of square roots.

$$\sqrt{\frac{6y^5}{2y}}$$

Remove common factors in the numerator and denominator.

$$\sqrt{\frac{\cancel{2}\cdot 3\cdot y^4\cdot \cancel{y}}{\cancel{2}\cdot \cancel{y}}}$$

Simplify.

$$\sqrt{3y^4}$$

Simplify the radical.

$$y^2\sqrt{3}$$

> **TRY IT : : 9.125**   Simplify: $\dfrac{\sqrt{12r^3}}{\sqrt{6r}}$.

> **TRY IT :: 9.126**

Simplify: $\dfrac{\sqrt{14p^9}}{\sqrt{2p^5}}$.

EXAMPLE 9.64

Simplify: $\dfrac{\sqrt{72x^3}}{\sqrt{162x}}$.

⊘ **Solution**

$$\dfrac{\sqrt{72x^3}}{\sqrt{162x}}$$

Rewrite using the quotient property of square roots.

$$\sqrt{\dfrac{72x^3}{162x}}$$

Remove common factors.

$$\sqrt{\dfrac{\cancel{18}\cdot 4\cdot x^2\cdot \cancel{x}}{\cancel{18}\cdot 9\cdot \cancel{x}}}$$

Simplify.

$$\sqrt{\dfrac{4x^2}{9}}$$

Simplify the radical.

$$\dfrac{2x}{3}$$

> **TRY IT :: 9.127**

Simplify: $\dfrac{\sqrt{50s^3}}{\sqrt{128s}}$.

> **TRY IT :: 9.128**

Simplify: $\dfrac{\sqrt{75q^5}}{\sqrt{108q}}$.

EXAMPLE 9.65

Simplify: $\dfrac{\sqrt{147ab^8}}{\sqrt{3a^3 b^4}}$.

⊘ **Solution**

$$\dfrac{\sqrt{147ab^8}}{\sqrt{3a^3 b^4}}$$

Rewrite using the quotient property of square roots.

$$\sqrt{\dfrac{147ab^8}{3a^3 b^4}}$$

Remove common factors.

$$\sqrt{\dfrac{49b^4}{a^2}}$$

Simplify the radical.

$$\dfrac{7b^2}{a}$$

> **TRY IT ::** 9.129

Simplify: $\dfrac{\sqrt{162x^{10}y^2}}{\sqrt{2x^6y^6}}$.

> **TRY IT ::** 9.130

Simplify: $\dfrac{\sqrt{300m^3n^7}}{\sqrt{3m^5n}}$.

Rationalize a One Term Denominator

Before the calculator became a tool of everyday life, tables of square roots were used to find approximate values of square roots. Figure 9.3 shows a portion of a table of squares and square roots. Square roots are approximated to five decimal places in this table.

n	n²	√n
200	40,000	14.14214
201	40,401	14.17745
202	40,804	14.21267
203	41,209	14.24781
204	41,616	14.28286
205	42,025	14.31782
206	42,436	14.35270
207	42,849	14.38749
208	43,264	14.42221
209	43,681	14.45683
210	44,100	14.49138

Figure 9.3 A table of square roots was used to find approximate values of square roots before there were calculators.

If someone needed to approximate a fraction with a square root in the denominator, it meant doing long division with a five decimal-place divisor. This was a very cumbersome process.

For this reason, a process called rationalizing the denominator was developed. A fraction with a radical in the denominator is converted to an equivalent fraction whose denominator is an integer. This process is still used today and is useful in other areas of mathematics, too.

Rationalizing the Denominator

The process of converting a fraction with a radical in the denominator to an equivalent fraction whose denominator is an integer is called **rationalizing the denominator**.

Square roots of numbers that are not perfect squares are irrational numbers. When we **rationalize the denominator**, we write an equivalent fraction with a rational number in the denominator.

Let's look at a numerical example.

Suppose we need an approximate value for the fraction. $\quad\dfrac{1}{\sqrt{2}}$

A five decimal place approximation to $\sqrt{2}$ is 1.41421. $\quad\dfrac{1}{1.41421}$

Without a calculator, would you want to do this division? $\quad 1.41421\overline{)1.0}$

But we can find a fraction equivalent to $\dfrac{1}{\sqrt{2}}$ by multiplying the numerator and denominator by $\sqrt{2}$.

$$\dfrac{1}{\sqrt{2}}$$

$$\dfrac{1\cdot\sqrt{2}}{\sqrt{2}\cdot\sqrt{2}}$$

$$\dfrac{\sqrt{2}}{2}$$

Now if we need an approximate value, we divide $2\overline{)1.41421}$. This is much easier.

Even though we have calculators available nearly everywhere, a fraction with a radical in the denominator still must be rationalized. It is not considered simplified if the denominator contains a square root.

Similarly, a square root is not considered simplified if the radicand contains a fraction.

Simplified Square Roots

A square root is considered simplified if there are

- no perfect-square factors in the radicand
- no fractions in the radicand
- no square roots in the denominator of a fraction

To rationalize a denominator, we use the property that $(\sqrt{a})^2 = a$. If we square an irrational square root, we get a rational number.

We will use this property to rationalize the denominator in the next example.

EXAMPLE 9.66

Simplify: $\dfrac{4}{\sqrt{3}}$.

⊘ **Solution**

To rationalize a denominator, we can multiply a square root by itself. To keep the fraction equivalent, we multiply both the numerator and denominator by the same factor.

$$\frac{4}{\sqrt{3}}$$

Multiply both the numerator and denominator by $\sqrt{3}$.

$$\frac{4 \cdot \sqrt{3}}{\sqrt{3} \cdot \sqrt{3}}$$

Simplify.

$$\frac{4\sqrt{3}}{3}$$

> **TRY IT : : 9.131**   Simplify: $\dfrac{5}{\sqrt{3}}$.

> **TRY IT : : 9.132**   Simplify: $\dfrac{6}{\sqrt{5}}$.

EXAMPLE 9.67

Simplify: $-\dfrac{8}{3\sqrt{6}}$.

⊘ **Solution**

To remove the square root from the denominator, we multiply it by itself. To keep the fractions equivalent, we multiply both the numerator and denominator by $\sqrt{6}$.

$$-\frac{8}{3\sqrt{6}}$$

Multiply both the numerator and the denominator by $\sqrt{6}$.

$$-\frac{8 \cdot \sqrt{6}}{3\sqrt{6} \cdot \sqrt{6}}$$

Simplify.	$\dfrac{8\sqrt{6}}{3 \cdot 6}$
Remove common factors.	$-\dfrac{4 \cdot \cancel{2}\sqrt{6}}{3 \cdot \cancel{2} \cdot 3}$
Simplify.	$-\dfrac{4\sqrt{6}}{9}$

> **TRY IT : : 9.133**   Simplify: $\dfrac{5}{2\sqrt{5}}$.

> **TRY IT : : 9.134**   Simplify: $-\dfrac{9}{4\sqrt{3}}$.

Always simplify the radical in the denominator first, before you rationalize it. This way the numbers stay smaller and easier to work with.

EXAMPLE 9.68

Simplify: $\sqrt{\dfrac{5}{12}}$.

✓ **Solution**

	$\sqrt{\dfrac{5}{12}}$
The fraction is not a perfect square, so rewrite using the Quotient Property.	$\dfrac{\sqrt{5}}{\sqrt{12}}$
Simplify the denominator	$\dfrac{\sqrt{5}}{2\sqrt{3}}$
Rationalize the denominator.	$\dfrac{\sqrt{5} \cdot \sqrt{3}}{2\sqrt{3} \cdot \sqrt{3}}$
Simplify.	$\dfrac{\sqrt{15}}{2 \cdot 3}$
Simplify.	$\dfrac{\sqrt{15}}{6}$

> **TRY IT : : 9.135**   Simplify: $\sqrt{\dfrac{7}{18}}$.

> **TRY IT : : 9.136**   Simplify: $\sqrt{\dfrac{3}{32}}$.

EXAMPLE 9.69

Simplify: $\sqrt{\dfrac{11}{28}}$.

✓ Solution

$$\sqrt{\frac{11}{28}}$$

Rewrite using the Quotient Property.	$\dfrac{\sqrt{11}}{\sqrt{28}}$
Simplify the denominator.	$\dfrac{\sqrt{11}}{2\sqrt{7}}$
Rationalize the denominator.	$\dfrac{\sqrt{11}\cdot\sqrt{7}}{2\sqrt{7}\cdot\sqrt{7}}$
Simplify.	$\dfrac{\sqrt{77}}{2\cdot7}$
Simplify.	$\dfrac{\sqrt{77}}{14}$

> **TRY IT :: 9.137**   Simplify: $\sqrt{\dfrac{3}{27}}$.

> **TRY IT :: 9.138**   Simplify: $\sqrt{\dfrac{10}{50}}$.

Rationalize a Two-Term Denominator

When the denominator of a fraction is a sum or difference with square roots, we use the Product of Conjugates pattern to rationalize the denominator.

$$\begin{array}{cc} (a-b)(a+b) & (2-\sqrt{5})(2+\sqrt{5}) \\ a^2-b^2 & 2^2-(\sqrt{5})^2 \\ & 4-5 \\ & -1 \end{array}$$

When we multiply a binomial that includes a square root by its conjugate, the product has no square roots.

EXAMPLE 9.70

Simplify: $\dfrac{4}{4+\sqrt{2}}$.

✓ Solution

$$\dfrac{4}{4+\sqrt{2}}$$

Multiply the numerator and denominator by the conjugate of the denominator.	$\dfrac{4(4-\sqrt{2})}{(4+\sqrt{2})(4-\sqrt{2})}$
Multiply the conjugates in the denominator.	$\dfrac{4(4-\sqrt{2})}{4^2-(\sqrt{2})^2}$
Simplify the denominator.	$\dfrac{4(4-\sqrt{2})}{16-2}$

| Simplify the denominator. | $\dfrac{4\left(4-\sqrt{2}\right)}{14}$ |

| Remove common factors from the numerator and denominator. | $\dfrac{2\left(4-\sqrt{2}\right)}{7}$ |

We leave the numerator in factored form to make it easier to look for common factors after we have simplified the denominator.

> **TRY IT : : 9.139**   Simplify: $\dfrac{2}{2+\sqrt{3}}$.

> **TRY IT : : 9.140**   Simplify: $\dfrac{5}{5+\sqrt{3}}$.

EXAMPLE 9.71

Simplify: $\dfrac{5}{2-\sqrt{3}}$.

⊘ **Solution**

| | $\dfrac{5}{2-\sqrt{3}}$ |

| Multiply the numerator and denominator by the conjugate of the denominator. | $\dfrac{5\left(2+\sqrt{3}\right)}{\left(2-\sqrt{3}\right)\left(2+\sqrt{3}\right)}$ |

| Multiply the conjugates in the denominator. | $\dfrac{5\left(2+\sqrt{3}\right)}{2^{2}-\left(\sqrt{3}\right)^{2}}$ |

| Simplify the denominator. | $\dfrac{5\left(2+\sqrt{3}\right)}{4-3}$ |

| Simplify the denominator. | $\dfrac{5\left(2+\sqrt{3}\right)}{1}$ |

| Simplify. | $5\left(2+\sqrt{3}\right)$ |

> **TRY IT : : 9.141**   Simplify: $\dfrac{3}{1-\sqrt{5}}$.

> **TRY IT : : 9.142**   Simplify: $\dfrac{2}{4-\sqrt{6}}$.

EXAMPLE 9.72

Simplify: $\dfrac{\sqrt{3}}{\sqrt{u}-\sqrt{6}}$.

⊘ Solution

$$\frac{\sqrt{3}}{\sqrt{u}-\sqrt{6}}$$

Multiply the numerator and denominator by the conjugate of the denominator.	$\dfrac{\sqrt{3}\left(\sqrt{u}+\sqrt{6}\right)}{\left(\sqrt{u}-\sqrt{6}\right)\left(\sqrt{u}+\sqrt{6}\right)}$
Multiply the conjugates in the denominator.	$\dfrac{\sqrt{3}\left(\sqrt{u}+\sqrt{6}\right)}{\left(\sqrt{u}\right)^2-\left(\sqrt{6}\right)^2}$
Simplify the denominator.	$\dfrac{\sqrt{3}\left(\sqrt{u}+\sqrt{6}\right)}{u-6}$

> **TRY IT : :** 9.143

Simplify: $\dfrac{\sqrt{5}}{\sqrt{x}+\sqrt{2}}$.

> **TRY IT : :** 9.144

Simplify: $\dfrac{\sqrt{10}}{\sqrt{y}-\sqrt{3}}$.

EXAMPLE 9.73

Simplify: $\dfrac{\sqrt{x}+\sqrt{7}}{\sqrt{x}-\sqrt{7}}$.

⊘ Solution

$$\frac{\sqrt{x}+\sqrt{7}}{\sqrt{x}-\sqrt{7}}$$

Multiply the numerator and denominator by the conjugate of the denominator.	$\dfrac{\left(\sqrt{x}+\sqrt{7}\right)\left(\sqrt{x}+\sqrt{7}\right)}{\left(\sqrt{x}-\sqrt{7}\right)\left(\sqrt{x}+\sqrt{7}\right)}$
Multiply the conjugates in the denominator.	$\dfrac{\left(\sqrt{x}+\sqrt{7}\right)\left(\sqrt{x}+\sqrt{7}\right)}{\left(\sqrt{x}\right)^2-\left(\sqrt{7}\right)^2}$
Simplify the denominator.	$\dfrac{\left(\sqrt{x}+\sqrt{7}\right)^2}{x-7}$

We do not square the numerator. In factored form, we can see there are no common factors to remove from the numerator and denominator.

> **TRY IT : :** 9.145

Simplify: $\dfrac{\sqrt{p}+\sqrt{2}}{\sqrt{p}-\sqrt{2}}$.

> **TRY IT : :** 9.146

Simplify: $\dfrac{\sqrt{q}-\sqrt{10}}{\sqrt{q}+\sqrt{10}}$.

MEDIA : :

Access this online resource for additional instruction and practice with dividing and rationalizing.

- **Dividing and Rationalizing (https://openstax.org/l/25DivideRation)**

9.5 EXERCISES

Practice Makes Perfect

Divide Square Roots

In the following exercises, simplify.

317. $\dfrac{\sqrt{27}}{6}$

318. $\dfrac{\sqrt{50}}{10}$

319. $\dfrac{\sqrt{72}}{9}$

320. $\dfrac{\sqrt{243}}{6}$

321. $\dfrac{2-\sqrt{32}}{8}$

322. $\dfrac{3+\sqrt{27}}{9}$

323. $\dfrac{6+\sqrt{45}}{6}$

324. $\dfrac{10-\sqrt{200}}{20}$

325. $\dfrac{\sqrt{80}}{\sqrt{125}}$

326. $\dfrac{\sqrt{72}}{\sqrt{200}}$

327. $\dfrac{\sqrt{128}}{\sqrt{72}}$

328. $\dfrac{\sqrt{48}}{\sqrt{75}}$

329. ⓐ $\dfrac{\sqrt{8x^6}}{\sqrt{2x^2}}$ ⓑ $\dfrac{\sqrt{200m^5}}{\sqrt{98m}}$

330. ⓐ $\dfrac{\sqrt{10y^3}}{\sqrt{5y}}$ ⓑ $\dfrac{\sqrt{108n^7}}{\sqrt{243n^3}}$

331. $\dfrac{\sqrt{75r^3}}{\sqrt{108r}}$

332. $\dfrac{\sqrt{196q^5}}{\sqrt{484q}}$

333. $\dfrac{\sqrt{108p^5q^2}}{\sqrt{34p^3q^6}}$

334. $\dfrac{\sqrt{98rs^{10}}}{\sqrt{2r^3s^4}}$

335. $\dfrac{\sqrt{320mn^5}}{\sqrt{45m^7n^3}}$

336. $\dfrac{\sqrt{810c^3d^7}}{\sqrt{1000c^5d}}$

337. $\dfrac{\sqrt{98}}{14}$

338. $\dfrac{\sqrt{72}}{18}$

339. $\dfrac{5+\sqrt{125}}{15}$

340. $\dfrac{6-\sqrt{45}}{12}$

341. $\dfrac{\sqrt{96}}{\sqrt{150}}$

342. $\dfrac{\sqrt{28}}{\sqrt{63}}$

343. $\dfrac{\sqrt{26y^7}}{\sqrt{2y}}$

344. $\dfrac{\sqrt{15x^3}}{\sqrt{3x}}$

Rationalize a One-Term Denominator

In the following exercises, simplify and rationalize the denominator.

345. $\dfrac{10}{\sqrt{6}}$

346. $\dfrac{8}{\sqrt{3}}$

347. $\dfrac{6}{\sqrt{7}}$

348. $\dfrac{4}{\sqrt{5}}$

349. $\dfrac{3}{\sqrt{13}}$

350. $\dfrac{10}{\sqrt{11}}$

351. $\dfrac{10}{3\sqrt{10}}$

352. $\dfrac{2}{5\sqrt{2}}$

353. $\dfrac{4}{9\sqrt{5}}$

354. $\dfrac{9}{2\sqrt{7}}$

355. $-\dfrac{9}{2\sqrt{3}}$

356. $-\dfrac{8}{3\sqrt{6}}$

357. $\sqrt{\dfrac{3}{20}}$

358. $\sqrt{\dfrac{4}{27}}$

359. $\sqrt{\dfrac{7}{40}}$

360. $\sqrt{\dfrac{8}{45}}$

361. $\sqrt{\dfrac{19}{175}}$

362. $\sqrt{\dfrac{17}{192}}$

Rationalize a Two-Term Denominator

In the following exercises, simplify by rationalizing the denominator.

363. ⓐ $\dfrac{3}{3+\sqrt{11}}$ ⓑ $\dfrac{8}{1-\sqrt{5}}$

364. ⓐ $\dfrac{4}{4+\sqrt{7}}$ ⓑ $\dfrac{7}{2-\sqrt{6}}$

365. ⓐ $\dfrac{5}{5+\sqrt{6}}$ ⓑ $\dfrac{6}{3-\sqrt{7}}$

366. ⓐ $\dfrac{6}{6+\sqrt{5}}$ ⓑ $\dfrac{5}{4-\sqrt{11}}$

367. $\dfrac{\sqrt{3}}{\sqrt{m}-\sqrt{5}}$

368. $\dfrac{\sqrt{5}}{\sqrt{n}-\sqrt{7}}$

369. $\dfrac{\sqrt{2}}{\sqrt{x}-\sqrt{6}}$

370. $\dfrac{\sqrt{7}}{\sqrt{y}+\sqrt{3}}$

371. $\dfrac{\sqrt{r}+\sqrt{5}}{\sqrt{r}-\sqrt{5}}$

372. $\dfrac{\sqrt{s}-\sqrt{6}}{\sqrt{s}+\sqrt{6}}$

373. $\dfrac{\sqrt{150x^2y^6}}{\sqrt{6x^4y^2}}$

374. $\dfrac{\sqrt{80p^3q}}{\sqrt{5pq^5}}$

375. $\dfrac{15}{\sqrt{5}}$

376. $\dfrac{3}{5\sqrt{8}}$

377. $\sqrt{\dfrac{8}{54}}$

378. $\sqrt{\dfrac{12}{20}}$

379. $\dfrac{3}{5+\sqrt{5}}$

380. $\dfrac{20}{4-\sqrt{3}}$

381. $\dfrac{\sqrt{2}}{\sqrt{x}-\sqrt{3}}$

382. $\dfrac{\sqrt{5}}{\sqrt{y}-\sqrt{7}}$

383. $\dfrac{\sqrt{x}+\sqrt{8}}{\sqrt{x}-\sqrt{8}}$

384. $\dfrac{\sqrt{m}-\sqrt{3}}{\sqrt{m}+\sqrt{3}}$

Everyday Math

385. A supply kit is dropped from an airplane flying at an altitude of 250 feet. Simplify $\sqrt{\dfrac{250}{16}}$ to determine how many seconds it takes for the supply kit to reach the ground.

386. A flare is dropped into the ocean from an airplane flying at an altitude of 1,200 feet. Simplify $\sqrt{\dfrac{1200}{16}}$ to determine how many seconds it takes for the flare to reach the ocean.

Writing Exercises

387.

ⓐ Simplify $\sqrt{\dfrac{27}{3}}$ and explain all your steps.

ⓑ Simplify $\sqrt{\dfrac{27}{5}}$ and explain all your steps.

ⓒ Why are the two methods of simplifying square roots different?

388.

ⓐ Approximate $\dfrac{1}{\sqrt{2}}$ by dividing $\dfrac{1}{1.414}$ using long division without a calculator.

ⓑ Rationalizing the denominator of $\dfrac{1}{\sqrt{2}}$ gives $\dfrac{\sqrt{2}}{2}$. Approximate $\dfrac{\sqrt{2}}{2}$ by dividing $\dfrac{1.414}{2}$ using long division without a calculator.

ⓒ Do you agree that rationalizing the denominator makes calculations easier? Why or why not?

Self Check

ⓐ *After completing the exercises, use this checklist to evaluate your mastery of the objectives of this section.*

I can...	Confidently	With some help	No-I don't get it!
divide square roots.			
rationalize a one-term denominator.			
rationalize a two-term denominator.			

ⓑ *After looking at the checklist, do you think you are well-prepared for the next section? Why or why not?*

9.6 Solve Equations with Square Roots

Learning Objectives

By the end of this section, you will be able to:

> Solve radical equations
> Use square roots in applications

Be Prepared!

Before you get started, take this readiness quiz.

1. Simplify: ⓐ $\sqrt{9}$ ⓑ 9^2.
   If you missed this problem, review **Example 9.1** and **Example 1.19**.

2. Solve: $5(x + 1) - 4 = 3(2x - 7)$.
   If you missed this problem, review **Example 2.42**.

3. Solve: $n^2 - 6n + 8 = 0$.
   If you missed this problem, review **Example 7.73**.

Solve Radical Equations

In this section we will solve equations that have the variable in the radicand of a square root. Equations of this type are called radical equations.

Radical Equation

An equation in which the variable is in the radicand of a square root is called a **radical equation**.

As usual, in solving these equations, what we do to one side of an equation we must do to the other side as well. Since squaring a quantity and taking a square root are 'opposite' operations, we will square both sides in order to remove the radical sign and solve for the variable inside.

But remember that when we write $\sqrt{a}$ we mean the principal square root. So $\sqrt{a} \geq 0$ always. When we solve radical equations by squaring both sides we may get an algebraic solution that would make $\sqrt{a}$ negative. This algebraic solution would not be a solution to the original radical equation; it is an *extraneous solution.* We saw extraneous solutions when we solved rational equations, too.

EXAMPLE 9.74

For the equation $\sqrt{x + 2} = x$:

ⓐ Is $x = 2$ a solution? ⓑ Is $x = -1$ a solution?

⊘ Solution

ⓐ Is $x = 2$ a solution?

	$\sqrt{x + 2} = x$
Let $x = 2$.	$\sqrt{2 + 2} \overset{?}{=} 2$
Simplify.	$\sqrt{4} \overset{?}{=} 2$
	$2 = 2 \checkmark$
	2 is a solution.

ⓑ Is $x = -1$ a solution?

$$\sqrt{x+2} = x$$

Let $x = -1$.	$\sqrt{-1+2} \stackrel{?}{=} -1$
Simplify.	$\sqrt{1} \stackrel{?}{=} -1$
	$1 \neq -1$
	-1 is not a solution.
	-1 is an extraneous solution to the equation.

> **TRY IT : : 9.147**   For the equation $\sqrt{x+6} = x$:
>
> ⓐ Is $x = -2$ a solution? ⓑ Is $x = 3$ a solution?

> **TRY IT : : 9.148**   For the equation $\sqrt{-x+2} = x$:
>
> ⓐ Is $x = -2$ a solution? ⓑ Is $x = 1$ a solution?

Now we will see how to solve a radical equation. Our strategy is based on the relation between taking a square root and squaring.

$$\text{For } a \geq 0, \quad (\sqrt{a})^2 = a$$

EXAMPLE 9.75   HOW TO SOLVE RADICAL EQUATIONS

Solve: $\sqrt{2x-1} = 7$.

⊘ **Solution**

Step 1. Isolate the radical on one side of the equation.	$\sqrt{2x-1}$ is already isolated on the left side.	$\sqrt{2x-1} = 7$
Step 2. Square both sides of the equation.	Remember, $(\sqrt{a})^2 = a$	$(\sqrt{2x-1})^2 = (7)^2$
Step 3. Solve the new equation.		$2x - 1 = 49$ $2x = 50$ $x = 25$
Step 4. Check the answer.		Check: $\sqrt{2x-1} = 7$ $\sqrt{2(25)-1} \stackrel{?}{=} 7$ $\sqrt{50-1} \stackrel{?}{=} 7$ $\sqrt{49} \stackrel{?}{=} 7$ $7 = 7 \checkmark$ The solution is $x = 25$.

> **TRY IT : : 9.149**   Solve: $\sqrt{3x-5} = 5$.

> **TRY IT : : 9.150**   Solve: $\sqrt{4x+8} = 6$.

HOW TO :: SOLVE A RADICAL EQUATION.

Step 1.   Isolate the radical on one side of the equation.

Step 2.   Square both sides of the equation.

Step 3.   Solve the new equation.

Step 4.   Check the answer.

EXAMPLE 9.76

Solve: $\sqrt{5n - 4} - 9 = 0$.

 Solution

$$\sqrt{5n - 4} - 9 = 0$$

To isolate the radical, add 9 to both sides.	$\sqrt{5n - 4} - 9 + 9 = 0 + 9$
Simplify.	$\sqrt{5n - 4} = 9$
Square both sides of the equation.	$\left(\sqrt{5n - 4}\right)^2 = (9)^2$
Solve the new equation.	$5n - 4 = 81$
	$5n = 85$
	$n = 17$

Check the answer.

$$\sqrt{5n - 4} - 9 = 0$$
$$\sqrt{5(17) - 4} - 9 \overset{?}{=} 0$$
$$\sqrt{85 - 4} - 9 \overset{?}{=} 0$$
$$\sqrt{81} - 9 \overset{?}{=} 0$$
$$9 - 9 \overset{?}{=} 0$$
$$0 = 0 ✓$$

The solution is $n = 17$.

> **TRY IT :: 9.151**   Solve: $\sqrt{3m + 2} - 5 = 0$.

> **TRY IT :: 9.152**   Solve: $\sqrt{10z + 1} - 2 = 0$.

EXAMPLE 9.77

Solve: $\sqrt{3y + 5} + 2 = 5$.

✓ Solution

$$\sqrt{3y+5}+2=5$$

To isolate the radical, subtract 2 from both sides.	$\sqrt{3y+5}+2-2=5-2$
Simplify.	$\sqrt{3y+5}=3$
Square both sides of the equation.	$\left(\sqrt{3y+5}\right)^2=(3)^2$
Solve the new equation.	$3y+5=9$
	$3y=4$
	$y=\dfrac{4}{3}$

Check the answer.

$$\sqrt{3y+5}+2=5$$

$$\sqrt{3\left(\frac{4}{3}\right)+5}+2\overset{?}{=}5$$

$$\sqrt{4+5}+2\overset{?}{=}5$$

$$\sqrt{9}+2\overset{?}{=}5$$

$$3+2\overset{?}{=}5$$

$$5=5\checkmark$$

The solution is $y=\dfrac{4}{3}$.

> **TRY IT :: 9.153**   Solve: $\sqrt{3p+3}+3=5$.

> **TRY IT :: 9.154**   Solve: $\sqrt{5q+1}+4=6$.

When we use a radical sign, we mean the principal or positive root. If an equation has a square root equal to a negative number, that equation will have no solution.

EXAMPLE 9.78

Solve: $\sqrt{9k-2}+1=0$.

✓ Solution

$$\sqrt{9k-2}+1=0$$

To isolate the radical, subtract 1 from both sides.	$\sqrt{9k-2}+1-1=0-1$
Simplify.	$\sqrt{9k-2}=-1$

Since the square root is equal to a negative number, the equation has no solution.

> **TRY IT :: 9.155**   Solve: $\sqrt{2r-3}+5=0$.

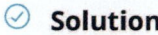 **TRY IT :: 9.156**   Solve: $\sqrt{7s-3}+2=0$.

If one side of the equation is a binomial, we use the binomial squares formula when we square it.

Binomial Squares
$$(a+b)^2 = a^2 + 2ab + b^2 \qquad (a-b)^2 = a^2 - 2ab + b^2$$

Don't forget the middle term!

EXAMPLE 9.79

Solve: $\sqrt{p-1}+1=p$.

⊘ **Solution**

	$\sqrt{p-1}+1=p$
To isolate the radical, subtract 1 from both sides.	$\sqrt{p-1}+1-1=p-1$
Simplify.	$\sqrt{p-1}=p-1$
Square both sides of the equation.	$\left(\sqrt{p-1}\right)^2 = (p-1)^2$
Simplify, then solve the new equation.	$p-1 = p^2 - 2p + 1$
It is a quadratic equation, so get zero on one side.	$0 = p^2 - 3p + 2$
Factor the right side.	$0 = (p-1)(p-2)$
Use the zero product property.	$0 = p-1 \quad 0 = p-2$
Solve each equation.	$p=1 \quad p=2$
Check the answers.	

$$p=1 \qquad \sqrt{p-1}+1=p \qquad\qquad p=2 \qquad \sqrt{p-1}+1=p$$

$$\sqrt{1-1}+1\overset{?}{=}1 \qquad\qquad\qquad \sqrt{2-1}+1\overset{?}{=}2$$

$$\sqrt{0}+1\overset{?}{=}1 \qquad\qquad\qquad\qquad \sqrt{1}+1\overset{?}{=}2$$

$$1=1 \checkmark \qquad\qquad\qquad\qquad\qquad 2=2 \checkmark$$

The solutions are $p = 1$, $p = 2$.

> **TRY IT :: 9.157**   Solve: $\sqrt{x-2}+2=x$.

> **TRY IT :: 9.158**   Solve: $\sqrt{y-5}+5=y$.

EXAMPLE 9.80

Solve: $\sqrt{r+4}-r+2=0$.

⊘ Solution

$$\sqrt{r+4} - r + 2 = 0$$

Isolate the radical.	$\sqrt{r+4} = r - 2$
Square both sides of the equation.	$\left(\sqrt{r+4}\right)^2 = (r-2)^2$
Solve the new equation.	$r + 4 = r^2 - 4r + 4$
It is a quadratic equation, so get zero on one side.	$0 = r^2 - 5r$
Factor the right side.	$0 = r(r-5)$
Use the zero product property.	$0 = r \quad 0 = r - 5$
Solve the equation.	$r = 0 \quad r = 5$
Check the answer.	

$r = 0 \quad \sqrt{r+4} - r + 2 = 0 \qquad r = 5 \quad \sqrt{r+4} - r + 2 = 0$

$\sqrt{0+4} - 0 + 2 \overset{?}{=} 0 \qquad\qquad \sqrt{5+4} - 5 + 2 \overset{?}{=} 0$

$\sqrt{4} + 2 \overset{?}{=} 0 \qquad\qquad\qquad \sqrt{9} - 3 \overset{?}{=} 0$

$4 \neq 0 \qquad\qquad\qquad\qquad\qquad 0 = 0 \checkmark \quad$ The solution is $r = 5$.

$r = 0$ is an extraneous solution.

> **TRY IT :: 9.159**   Solve: $\sqrt{m+9} - m + 3 = 0$.

> **TRY IT :: 9.160**   Solve: $\sqrt{n+1} - n + 1 = 0$.

When there is a coefficient in front of the radical, we must square it, too.

EXAMPLE 9.81

Solve: $3\sqrt{3x-5} - 8 = 4$.

⊘ **Solution**

$$3\sqrt{3x-5}-8=4$$

Isolate the radical.	$3\sqrt{3x-5}=12$
Square both sides of the equation.	$\left(3\sqrt{3x-5}\right)^2=(12)^2$
Simplify, then solve the new equation.	$9(3x-5)=144$
Distribute.	$27x-45=144$
Solve the equation.	$27x=189$
	$x=7$

Check the answer.

$x=7 \qquad 3\sqrt{3x-5}-8=4$

$\qquad\qquad 3\sqrt{3\,(7)-5}-8\overset{?}{=}4$

$\qquad\qquad 3\sqrt{21-5}-8\overset{?}{=}4$

$\qquad\qquad\quad 3\sqrt{16}-8\overset{?}{=}4$

$\qquad\qquad\quad\;\; 3(4)-8\overset{?}{=}4$

$\qquad\qquad\qquad\quad 4=4\checkmark \qquad\qquad$ The solution is $x=7$.

> **TRY IT :: 9.161**   Solve: $2\sqrt{4a+2}-16=16$.

> **TRY IT :: 9.162**   Solve: $3\sqrt{6b+3}-25=50$.

EXAMPLE 9.82

Solve: $\sqrt{4z-3}=\sqrt{3z+2}$.

⊘ **Solution**

$$\sqrt{4z-3}=\sqrt{3z+2}$$

The radical terms are isolated.	$\sqrt{4z-3}=\sqrt{3z+2}$
Square both sides of the equation.	$\left(\sqrt{4z-3}\right)^2=\left(\sqrt{3z+2}\right)^2$
Simplify, then solve the new equation.	$4z-3=3z+2$
	$z-3=2$
	$z=5$

Check the answer.
We leave it to you to show that 5 checks!   The solution is $z=5$.

> **TRY IT :: 9.163**   Solve: $\sqrt{2x-5}=\sqrt{5x+3}$.

> **TRY IT ::** 9.164   Solve: $\sqrt{7y+1} = \sqrt{2y-5}$.

Sometimes after squaring both sides of an equation, we still have a variable inside a radical. When that happens, we repeat Step 1 and Step 2 of our procedure. We isolate the radical and square both sides of the equation again.

EXAMPLE 9.83

Solve: $\sqrt{m} + 1 = \sqrt{m+9}$.

⊘ **Solution**

$$\sqrt{m} + 1 = \sqrt{m+9}$$

The radical on the right side is isolated. Square both sides. $$(\sqrt{m}+1)^2 = (\sqrt{m+9})^2$$

Simplify—be very careful as you multiply! $$m + 2\sqrt{m} + 1 = m + 9$$
There is still a radical in the equation.
So we must repeat the previous steps. Isolate the radical. $$2\sqrt{m} = 8$$

Square both sides. $$(2\sqrt{m})^2 = (8)^2$$

Simplify, then solve the new equation. $$4m = 64$$
$$m = 16$$

Check the answer.

We leave it to you to show that $m = 16$ checks!   The solution is $m = 16$.

> **TRY IT ::** 9.165   Solve: $\sqrt{x} + 3 = \sqrt{x+5}$.

> **TRY IT ::** 9.166   Solve: $\sqrt{m} + 5 = \sqrt{m+16}$.

EXAMPLE 9.84

Solve: $\sqrt{q-2} + 3 = \sqrt{4q+1}$.

⊘ **Solution**

$$\sqrt{q-2}+3 = \sqrt{4q+1}$$

The radical on the right side is isolated.
Square both sides.

$$\left(\sqrt{q-2}+3\right)^2 = \left(\sqrt{4q+1}\right)^2$$

Simplify.

$$q-2+6\sqrt{q-2}+9 = 4q+1$$

There is still a radical in the equation. So
we must repeat the previous steps. Isolate
the radical.

$$6\sqrt{q-2} = 3q-6$$

Square both sides.

$$\left(6\sqrt{q-2}\right)^2 = (3q-6)^2$$

Simplify, then solve the new equation.

$$36(q-2) = 9q^2 - 36q + 36$$

Distribute.

$$36q-72 = 9q^2 - 36q + 36$$

It is a quadratic equation, so get zero on
one side.

$$0 = 9q^2 - 72q + 108$$

Factor the right side.

$$0 = 9\left(q^2 - 8q + 12\right)$$
$$0 = 9(q-6)(q-2)$$

Use the zero product property.

$$q-6=0 \qquad q-2=0$$
$$q=6 \qquad\quad q=2$$

The checks are left to you. (Both solutions
should work.)

The solutions are $q=6$ and $q=2$.

> **TRY IT :: 9.167**   Solve: $\sqrt{y-3}+2 = \sqrt{4y+2}$.

> **TRY IT :: 9.168**   Solve: $\sqrt{n-4}+5 = \sqrt{3n+3}$.

Use Square Roots in Applications

As you progress through your college courses, you'll encounter formulas that include square roots in many disciplines.
We have already used formulas to solve geometry applications.

We will use our Problem Solving Strategy for Geometry Applications, with slight modifications, to give us a plan for solving
applications with formulas from any discipline.

HOW TO :: SOLVE APPLICATIONS WITH FORMULAS.

Step 1. **Read** the problem and make sure all the words and ideas are understood. When appropriate, draw a figure and label it with the given information.

Step 2. **Identify** what we are looking for.

Step 3. **Name** what we are looking for by choosing a variable to represent it.

Step 4. **Translate** into an equation by writing the appropriate formula or model for the situation. Substitute in the given information.

Step 5. **Solve the equation** using good algebra techniques.

Step 6. **Check** the answer in the problem and make sure it makes sense.

Step 7. **Answer** the question with a complete sentence.

We used the formula $A = L \cdot W$ to find the area of a rectangle with length L and width W. A square is a rectangle in which the length and width are equal. If we let s be the length of a side of a square, the area of the square is s^2.

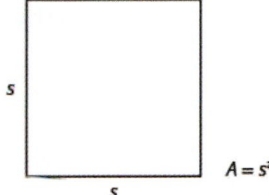

The formula $A = s^2$ gives us the area of a square if we know the length of a side. What if we want to find the length of a side for a given area? Then we need to solve the equation for s.

$$A = s^2$$

Take the square root of both sides. $\sqrt{A} = \sqrt{s^2}$

Simplify. $\sqrt{A} = s$

We can use the formula $s = \sqrt{A}$ to find the length of a side of a square for a given area.

Area of a Square

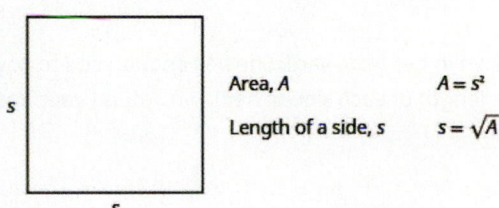

We will show an example of this in the next example.

EXAMPLE 9.85

Mike and Lychelle want to make a square patio. They have enough concrete to pave an area of 200 square feet. Use the formula $s = \sqrt{A}$ to find the length of each side of the patio. Round your answer to the nearest tenth of a foot.

✓ Solution

Step 1. **Read** the problem. Draw a figure and label it with the given information.	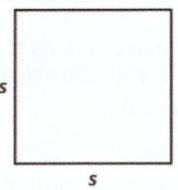
	$A = 200$ square feet
Step 2. **Identify** what you are looking for.	The length of a side of the square patio.
Step 3. **Name** what you are looking for by choosing a variable to represent it.	Let s = the length of a side.
Step 4. **Translate** into an equation by writing the appropriate formula or model for the situation. Substitute the given information.	$s = \sqrt{A}$, and $A = 200$ $s = \sqrt{200}$
Step 5. **Solve the equation** using good algebra techniques. Round to one decimal place.	$s = 14.14213...$ $s \approx 14.1$
Step 6. **Check** the answer in the problem and make sure it makes sense.	

$$14.1^2 \overset{?}{\approx} 200$$
$$14.1^2 \approx 198.81 \checkmark$$

This is close enough because we rounded the square root.
Is a patio with side 14.1 feet reasonable?
Yes.

Step 7. **Answer** the question with a complete sentence.	Each side of the patio should be 14.1 feet.

> **TRY IT : : 9.169**

Katie wants to plant a square lawn in her front yard. She has enough sod to cover an area of 370 square feet. Use the formula $s = \sqrt{A}$ to find the length of each side of her lawn. Round your answer to the nearest tenth of a foot.

> **TRY IT : : 9.170**

Sergio wants to make a square mosaic as an inlay for a table he is building. He has enough tile to cover an area of 2704 square centimeters. Use the formula $s = \sqrt{A}$ to find the length of each side of his mosaic. Round your answer to the nearest tenth of a foot.

Another application of square roots has to do with gravity.

Falling Objects

On Earth, if an object is dropped from a height of h feet, the time in seconds it will take to reach the ground is found by using the formula,

$$t = \frac{\sqrt{h}}{4}$$

For example, if an object is dropped from a height of 64 feet, we can find the time it takes to reach the ground by substituting $h = 64$ into the formula.

$$t = \frac{\sqrt{h}}{4}$$

$$t = \frac{\sqrt{64}}{4}$$

Take the square root of 64. $t = \frac{8}{4}$

Simplify the fraction. $t = 2$

It would take 2 seconds for an object dropped from a height of 64 feet to reach the ground.

EXAMPLE 9.86

Christy dropped her sunglasses from a bridge 400 feet above a river. Use the formula $t = \frac{\sqrt{h}}{4}$ to find how many seconds it took for the sunglasses to reach the river.

✓ **Solution**

Step 1. Read the problem.

Step 2. Identify what you are looking for.	The time it takes for the sunglasses to reach the river.
Step 3. Name what you are looking for by choosing a variable to represent it.	Let t = time.
Step 4. Translate into an equation by writing the appropriate formula or model for the situation. Substitute in the given information.	$t = \frac{\sqrt{h}}{4}$, and $h = 400$ $t = \frac{\sqrt{400}}{4}$
Step 5. Solve the equation using good algebra techniques.	$t = \frac{20}{4}$ $t = 5$

Step 6. Check the answer in the problem and make sure it makes sense.

$$5 \stackrel{?}{=} \frac{\sqrt{400}}{4}$$
$$5 \stackrel{?}{=} \frac{20}{4}$$
$$5 = 5 \checkmark$$

Does 5 seconds seem reasonable?
Yes.

Step 7. Answer the question with a complete sentence.	It will take 5 seconds for the sunglasses to hit the water.

> **TRY IT : : 9.171**

A helicopter dropped a rescue package from a height of 1,296 feet. Use the formula $t = \frac{\sqrt{h}}{4}$ to find how many seconds it took for the package to reach the ground.

> **TRY IT : : 9.172**

A window washer dropped a squeegee from a platform 196 feet above the sidewalk Use the formula $t = \frac{\sqrt{h}}{4}$ to find how many seconds it took for the squeegee to reach the sidewalk.

Police officers investigating car accidents measure the length of the skid marks on the pavement. Then they use square roots to determine the speed, in miles per hour, a car was going before applying the brakes.

Skid Marks and Speed of a Car

If the length of the skid marks is d feet, then the speed, s, of the car before the brakes were applied can be found by using the formula,

$$s = \sqrt{24d}$$

EXAMPLE 9.87

After a car accident, the skid marks for one car measured 190 feet. Use the formula $s = \sqrt{24d}$ to find the speed of the car before the brakes were applied. Round your answer to the nearest tenth.

⊘ **Solution**

Step 1. **Read** the problem.

Step 2. **Identify** what we are looking for.	The speed of a car.
Step 3. **Name** what we are looking for.	Let s = the speed.
Step 4. **Translate** into an equation by writing the appropriate formula.	$s = \sqrt{24d}$, and $d = 190$
Substitute the given information.	$s = \sqrt{24(190)}$
Step 5. **Solve the equation.**	$s = \sqrt{4560}$
	$s = 67.52777...$
Round to 1 decimal place.	$s \approx 67.5$

Step 6. **Check** the answer in the problem.

$67.5 \stackrel{?}{\approx} \sqrt{24(190)}$

$67.5 \stackrel{?}{\approx} \sqrt{4560}$

$67.5 \stackrel{?}{\approx} 67.5277...$

Is 67.5 mph a reasonable speed?	Yes.
Step 7. **Answer** the question with a complete sentence.	The speed of the car was approximately 67.5 miles per hour.

> **TRY IT : : 9.173**

An accident investigator measured the skid marks of the car. The length of the skid marks was 76 feet. Use the formula $s = \sqrt{24d}$ to find the speed of the car before the brakes were applied. Round your answer to the nearest tenth.

> **TRY IT : : 9.174**

The skid marks of a vehicle involved in an accident were 122 feet long. Use the formula $s = \sqrt{24d}$ to find the speed of the vehicle before the brakes were applied. Round your answer to the nearest tenth.

9.6 EXERCISES

Practice Makes Perfect

Solve Radical Equations

In the following exercises, check whether the given values are solutions.

389. For the equation $\sqrt{x+12} = x$:

ⓐ Is $x = 4$ a solution?

ⓑ Is $x = -3$ a solution?

390. For the equation $\sqrt{-y+20} = y$:

ⓐ Is $y = 4$ a solution?

ⓑ Is $y = -5$ a solution?

391. For the equation $\sqrt{t+6} = t$:

ⓐ Is $t = -2$ a solution?

ⓑ Is $t = 3$ a solution?

392. For the equation $\sqrt{u+42} = u$:

ⓐ Is $u = -6$ a solution?

ⓑ Is $u = 7$ a solution?

In the following exercises, solve.

393. $\sqrt{5y+1} = 4$

394. $\sqrt{7z+15} = 6$

395. $\sqrt{5x-6} = 8$

396. $\sqrt{4x-3} = 7$

397. $\sqrt{2m-3} - 5 = 0$

398. $\sqrt{2n-1} - 3 = 0$

399. $\sqrt{6v-2} - 10 = 0$

400. $\sqrt{4u+2} - 6 = 0$

401. $\sqrt{5q+3} - 4 = 0$

402. $\sqrt{4m+2} + 2 = 6$

403. $\sqrt{6n+1} + 4 = 8$

404. $\sqrt{2u-3} + 2 = 0$

405. $\sqrt{5v-2} + 5 = 0$

406. $\sqrt{3z-5} + 2 = 0$

407. $\sqrt{2m+1} + 4 = 0$

408.

ⓐ $\sqrt{u-3} + 3 = u$

ⓑ $\sqrt{x+1} - x + 1 = 0$

409.

ⓐ $\sqrt{v-10} + 10 = v$

ⓑ $\sqrt{y+4} - y + 2 = 0$

410.

ⓐ $\sqrt{r-1} - r = -1$

ⓑ $\sqrt{z+100} - z + 10 = 0$

411.

ⓐ $\sqrt{s-8} - s = -8$

ⓑ $\sqrt{w+25} - w + 5 = 0$

412. $3\sqrt{2x-3} - 20 = 7$

413. $2\sqrt{5x+1} - 8 = 0$

414. $2\sqrt{8r+1} - 8 = 2$

415. $3\sqrt{7y+1} - 10 = 8$

416. $\sqrt{3u-2} = \sqrt{5u+1}$

417. $\sqrt{4v+3} = \sqrt{v-6}$

418. $\sqrt{8+2r} = \sqrt{3r+10}$

419. $\sqrt{12c+6} = \sqrt{10-4c}$

420.

ⓐ $\sqrt{a+2} = \sqrt{a+4}$

ⓑ $\sqrt{b-2} + 1 = \sqrt{3b+2}$

421.

ⓐ $\sqrt{r+6} = \sqrt{r+8}$

ⓑ $\sqrt{s-3} + 2 = \sqrt{s+4}$

422.

ⓐ $\sqrt{u+1} = \sqrt{u+4}$

ⓑ $\sqrt{n-5} + 4 = \sqrt{3n+7}$

423.

ⓐ $\sqrt{x+10} = \sqrt{x+2}$

ⓑ $\sqrt{y-2} + 2 = \sqrt{2y+4}$

424. $\sqrt{2y+4} + 6 = 0$

425. $\sqrt{8u+1} + 9 = 0$

426. $\sqrt{a} + 1 = \sqrt{a+5}$

427. $\sqrt{d} - 2 = \sqrt{d-20}$

428. $\sqrt{6s+4} = \sqrt{8s-28}$

429. $\sqrt{9p+9} = \sqrt{10p-6}$

Use Square Roots in Applications

In the following exercises, solve. Round approximations to one decimal place.

430. Landscaping Reed wants to have a square garden plot in his backyard. He has enough compost to cover an area of 75 square feet. Use the formula $s = \sqrt{A}$ to find the length of each side of his garden. Round your answer to the nearest tenth of a foot.

431. Landscaping Vince wants to make a square patio in his yard. He has enough concrete to pave an area of 130 square feet. Use the formula $s = \sqrt{A}$ to find the length of each side of his patio. Round your answer to the nearest tenth of a foot.

432. Gravity While putting up holiday decorations, Renee dropped a light bulb from the top of a 64 foot tall tree. Use the formula $t = \frac{\sqrt{h}}{4}$ to find how many seconds it took for the light bulb to reach the ground.

433. Gravity An airplane dropped a flare from a height of 1024 feet above a lake. Use the formula $t = \frac{\sqrt{h}}{4}$ to find how many seconds it took for the flare to reach the water.

434. Gravity A hang glider dropped his cell phone from a height of 350 feet. Use the formula $t = \frac{\sqrt{h}}{4}$ to find how many seconds it took for the cell phone to reach the ground.

435. Gravity A construction worker dropped a hammer while building the Grand Canyon skywalk, 4000 feet above the Colorado River. Use the formula $t = \frac{\sqrt{h}}{4}$ to find how many seconds it took for the hammer to reach the river.

436. Accident investigation The skid marks for a car involved in an accident measured 54 feet. Use the formula $s = \sqrt{24d}$ to find the speed of the car before the brakes were applied. Round your answer to the nearest tenth.

437. Accident investigation The skid marks for a car involved in an accident measured 216 feet. Use the formula $s = \sqrt{24d}$ to find the speed of the car before the brakes were applied. Round your answer to the nearest tenth.

438. Accident investigation An accident investigator measured the skid marks of one of the vehicles involved in an accident. The length of the skid marks was 175 feet. Use the formula $s = \sqrt{24d}$ to find the speed of the vehicle before the brakes were applied. Round your answer to the nearest tenth.

439. Accident investigation An accident investigator measured the skid marks of one of the vehicles involved in an accident. The length of the skid marks was 117 feet. Use the formula $s = \sqrt{24d}$ to find the speed of the vehicle before the brakes were applied. Round your answer to the nearest tenth.

Writing Exercises

440. Explain why an equation of the form $\sqrt{x} + 1 = 0$ has no solution.

441.

ⓐ Solve the equation $\sqrt{r+4} - r + 2 = 0$.

ⓑ Explain why one of the "solutions" that was found was not actually a solution to the equation.

Self Check

ⓐ *After completing the exercises, use this checklist to evaluate your mastery of the objectives of this section.*

I can...	Confidently	With some help	No-I don't get it!
use square roots in applications.			

ⓑ *After reviewing this checklist, what will you do to become confident for all objectives?*

9.7 | # Higher Roots

Learning Objectives

By the end of this section, you will be able to:

> Simplify expressions with higher roots
> Use the Product Property to simplify expressions with higher roots
> Use the Quotient Property to simplify expressions with higher roots
> Add and subtract higher roots

Be Prepared!

Before you get started, take this readiness quiz.

1. Simplify: $y^5 y^4$.
   If you missed this problem, review **Example 6.18**.

2. Simplify: $\left(n^2\right)^6$.
   If you missed this problem, review **Example 6.22**.

3. Simplify: $\dfrac{x^8}{x^3}$.
   If you missed this problem, review **Example 6.59**.

Simplify Expressions with Higher Roots

Up to now, in this chapter we have worked with squares and square roots. We will now extend our work to include higher powers and higher roots.

Let's review some vocabulary first.

We write:	We say:
n^2	n squared
n^3	n cubed
n^4	n to the fourth
n^5	n to the fifth

The terms 'squared' and 'cubed' come from the formulas for area of a square and volume of a cube.

It will be helpful to have a table of the powers of the integers from -5 to 5. See **Figure 9.4**.

Number	Square	Cube	Fourth power	Fifth power
n	n^2	n^3	n^4	n^5
1	1	1	1	1
2	4	8	16	32
3	9	27	81	243
4	16	64	256	1024
5	25	125	625	3125
x	x^2	x^3	x^4	x^5
x^2	x^4	x^6	x^8	x^{10}

Number	Square	Cube	Fourth power	Fifth power
n	n^2	n^3	n^4	n^5
-1	1	-1	1	-1
-2	4	-8	16	-32
-3	9	-27	81	-243
-4	16	-64	256	-1024
-5	25	-125	625	-3125

Figure 9.4 First through fifth powers of integers from -5 to 5.

Notice the signs in **Figure 9.4**. All powers of positive numbers are positive, of course. But when we have a negative number, the even powers are positive and the odd powers are negative. We'll copy the row with the powers of -2 below to help you see this.

n	n^2	n^3	n^4	n^5
-2	4	-8	16	-32

Earlier in this chapter we defined the square root of a number.

$$\text{If } n^2 = m, \text{ then } n \text{ is a square root of } m.$$

And we have used the notation $\sqrt{m}$ to denote the **principal square root**. So $\sqrt{m} \geq 0$ always.

We will now extend the definition to higher roots.

---

nth Root of a Number

If $b^n = a$, then b is an **nth root of a number** a.

The principal nth root of a is written $\sqrt[n]{a}$.

n is called the **index** of the radical.

We do not write the index for a square root. Just like we use the word 'cubed' for b^3, we use the term 'cube root' for $\sqrt[3]{a}$.

We refer to **Figure 9.4** to help us find higher roots.

$$\begin{array}{rclcrcl}
4^3 & = & 64 & \qquad & \sqrt[3]{64} & = & 4 \\
3^4 & = & 81 & \qquad & \sqrt[4]{81} & = & 3 \\
(-2)^5 & = & -32 & \qquad & \sqrt[5]{-32} & = & -2
\end{array}$$

Could we have an even root of a negative number? No. We know that the square root of a negative number is not a real number. The same is true for any even root. Even roots of negative numbers are not real numbers. Odd roots of negative numbers are real numbers.

Properties of $\sqrt[n]{a}$

When n is an even number and

- $a \geq 0$, then $\sqrt[n]{a}$ is a real number

- $a < 0$, then $\sqrt[n]{a}$ is not a real number

When n is an odd number, $\sqrt[n]{a}$ is a real number for all values of a.

EXAMPLE 9.88

Simplify: ⓐ $\sqrt[3]{8}$ ⓑ $\sqrt[4]{81}$ ⓒ $\sqrt[5]{32}$.

✓ **Solution**

ⓐ

$$\sqrt[3]{8}$$

Since $(2)^3 = 8$. $\quad 2$

ⓑ

$$\sqrt[4]{81}$$

Since $(3)^4 = 81$. $\quad 3$

ⓒ

$$\sqrt[5]{32}$$

Since $(2)^5 = 32$. $\quad 2$

> **TRY IT : : 9.175**   Simplify: ⓐ $\sqrt[3]{27}$ ⓑ $\sqrt[4]{256}$ ⓒ $\sqrt[5]{243}$.

> **TRY IT : : 9.176**   Simplify: ⓐ $\sqrt[3]{1000}$ ⓑ $\sqrt[4]{16}$ ⓒ $\sqrt[5]{32}$.

EXAMPLE 9.89

Simplify: ⓐ $\sqrt[3]{-64}$ ⓑ $\sqrt[4]{-16}$ ⓒ $\sqrt[5]{-243}$.

✓ **Solution**

ⓐ

$$\sqrt[3]{-64}$$

Since $(-4)^3 = -64$.

$$-4$$

ⓑ

$$\sqrt[4]{-16}$$

Think, $(?)^4 = -16$. No real number raised to the fourth power is positive.

Not a real number.

ⓒ

$$\sqrt[5]{-243}$$

Since $(-3)^5 = -243$.

$$-3$$

> **TRY IT :: 9.177**   Simplify: ⓐ $\sqrt[3]{-125}$ ⓑ $\sqrt[4]{-16}$ ⓒ $\sqrt[5]{-32}$.

> **TRY IT :: 9.178**   Simplify: ⓐ $\sqrt[3]{-216}$ ⓑ $\sqrt[4]{-81}$ ⓒ $\sqrt[5]{-1024}$.

When we worked with square roots that had variables in the radicand, we restricted the variables to non-negative values. Now we will remove this restriction.

The odd root of a number can be either positive or negative. We have seen that $\sqrt[3]{-64} = -4$.

But the even root of a non-negative number is always non-negative, because we take the principal nth root.

Suppose we start with $a = -5$.

$$(-5)^4 = 625 \qquad\qquad \sqrt[4]{625} = 5$$

How can we make sure the fourth root of –5 raised to the fourth power, $(-5)^4$ is 5? We will see in the following property.

Simplifying Odd and Even Roots

For any integer $n \geq 2$,

$$\text{when } n \text{ is odd} \qquad \sqrt[n]{a^n} = a$$
$$\text{when } n \text{ is even} \qquad \sqrt[n]{a^n} = |a|$$

We must use the absolute value signs when we take an even root of an expression with a variable in the radical.

EXAMPLE 9.90

Simplify: ⓐ $\sqrt{x^2}$ ⓑ $\sqrt[3]{n^3}$ ⓒ $\sqrt[4]{p^4}$ ⓓ $\sqrt[5]{y^5}$.

✓ **Solution**

We use the absolute value to be sure to get the positive root.

ⓐ

Since $(x)^2 = x^2$ and we want the positive root.

$$\sqrt{x^2}$$
$$|x|$$

ⓑ

Since $(n)^3 = n^3$. It is an odd root so there is no need for an absolute value sign.

$$\sqrt[3]{n^3}$$
$$n$$

ⓒ

Since $(p)^4 = p^4$ and we want the positive root.

$$\sqrt[4]{p^4}$$
$$|p|$$

ⓓ

Since $(y)^5 = y^5$. It is an odd root so there is no need for an absolute value sign.

$$\sqrt[5]{y^5}$$
$$y$$

> **TRY IT :: 9.179**
>
> Simplify: ⓐ $\sqrt{b^2}$ ⓑ $\sqrt[3]{w^3}$ ⓒ $\sqrt[4]{m^4}$ ⓓ $\sqrt[5]{q^5}$.

> **TRY IT :: 9.180**
>
> Simplify: ⓐ $\sqrt{y^2}$ ⓑ $\sqrt[3]{p^3}$ ⓒ $\sqrt[4]{z^4}$ ⓓ $\sqrt[5]{q^5}$.

EXAMPLE 9.91

Simplify: ⓐ $\sqrt[3]{y^{18}}$ ⓑ $\sqrt[4]{z^8}$.

✓ **Solution**

ⓐ

Since $\left(y^6\right)^3 = y^{18}$.

$$\sqrt[3]{y^{18}}$$
$$\sqrt[3]{\left(y^6\right)^3}$$
$$y^6$$

ⓑ

Since $\left(z^2\right)^4 = z^8$.

$$\sqrt[4]{z^8}$$
$$\sqrt[4]{\left(z^2\right)^4}$$

Since z^2 is positive, we do not need an absolute value sign.

$$z^2$$

> **TRY IT :: 9.181**
>
> Simplify: ⓐ $\sqrt[4]{u^{12}}$ ⓑ $\sqrt[3]{v^{15}}$.

> **TRY IT : : 9.182**

Simplify: ⓐ $\sqrt[5]{c^{20}}$ ⓑ $\sqrt[6]{d^{24}}$.

EXAMPLE 9.92

Simplify: ⓐ $\sqrt[3]{64p^6}$ ⓑ $\sqrt[4]{16q^{12}}$.

✓ **Solution**

ⓐ

$$\sqrt[3]{64p^6}$$

Rewrite $64p^6$ as $\left(4p^2\right)^3$. $\sqrt[3]{\left(4p^2\right)^3}$

Take the cube root. $4p^2$

ⓑ

$$\sqrt[4]{16q^{12}}$$

Rewrite the radicand as a fourth power. $\sqrt[4]{\left(2q^3\right)^4}$

Take the fourth root. $2\left|q^3\right|$

> **TRY IT : : 9.183**

Simplify: ⓐ $\sqrt[3]{27x^{27}}$ ⓑ $\sqrt[4]{81q^{28}}$.

> **TRY IT : : 9.184**

Simplify: ⓐ $\sqrt[3]{125p^9}$ ⓑ $\sqrt[5]{243q^{25}}$.

Use the Product Property to Simplify Expressions with Higher Roots

We will simplify expressions with higher roots in much the same way as we simplified expressions with square roots. An nth root is considered simplified if it has no factors of m^n .

Simplified nth Root

$\sqrt[n]{a}$ is considered simplified if a has no factors of m^n .

We will generalize the Product Property of Square Roots to include any integer root $n \geq 2$.

Product Property of nth Roots

$$\sqrt[n]{ab} = \sqrt[n]{a} \cdot \sqrt[n]{b} \text{ and } \sqrt[n]{a} \cdot \sqrt[n]{b} = \sqrt[n]{ab}$$

when $\sqrt[n]{a}$ and $\sqrt[n]{b}$ are real numbers and for any integer $n \geq 2$

EXAMPLE 9.93

Simplify: ⓐ $\sqrt[3]{x^4}$ ⓑ $\sqrt[4]{x^7}$.

✓ **Solution**

ⓐ

$$\sqrt[3]{x^4}$$

Rewrite the radicand as a product using the largest perfect cube factor.

$$\sqrt[3]{x^3 \cdot x}$$

Rewrite the radical as the product of two radicals.

$$\sqrt[3]{x^3} \cdot \sqrt[3]{x}$$

Simplify.

$$x\sqrt[3]{x}$$

ⓑ

$$\sqrt[4]{x^7}$$

Rewrite the radicand as a product using the greatest perfect fourth power factor.

$$\sqrt[4]{x^4 \cdot x^3}$$

Rewrite the radical as the product of two radicals.

$$\sqrt[4]{x^4} \cdot \sqrt[4]{x^3}$$

Simplify.

$$|x|\sqrt[4]{x^3}$$

> **TRY IT :: 9.185**   Simplify: ⓐ $\sqrt[4]{y^6}$ ⓑ $\sqrt[3]{z^5}$.

> **TRY IT :: 9.186**   Simplify: ⓐ $\sqrt[5]{p^8}$ ⓑ $\sqrt[6]{q^{13}}$.

EXAMPLE 9.94

Simplify: ⓐ $\sqrt[3]{16}$ ⓑ $\sqrt[4]{243}$.

✓ **Solution**

ⓐ

$$\sqrt[3]{16}$$
$$\sqrt[3]{2^4}$$

Rewrite the radicand as a product using the greatest perfect cube factor.

$$\sqrt[3]{2^3 \cdot 2}$$

Rewrite the radical as the product of two radicals.

$$\sqrt[3]{2^3} \cdot \sqrt[3]{2}$$

Simplify.

$$2\sqrt[3]{2}$$

ⓑ

$$\sqrt[4]{243}$$
$$\sqrt[4]{3^5}$$

Rewrite the radicand as a product using the greatest perfect fourth power factor.

$$\sqrt[4]{3^4 \cdot 3}$$

Rewrite the radical as the product of two radicals.

$$\sqrt[4]{3^4} \cdot \sqrt[4]{3}$$

Simplify.

$$3\sqrt[4]{3}$$

> **TRY IT : : 9.187**

Simplify: ⓐ $\sqrt[3]{81}$ ⓑ $\sqrt[4]{64}$.

> **TRY IT : : 9.188**

Simplify: ⓐ $\sqrt[3]{625}$ ⓑ $\sqrt[4]{729}$.

Don't forget to use the absolute value signs when taking an even root of an expression with a variable in the radical.

EXAMPLE 9.95

Simplify: ⓐ $\sqrt[3]{24x^7}$ ⓑ $\sqrt[4]{80y^{14}}$.

✓ **Solution**

ⓐ

	$\sqrt[3]{24x^7}$
Rewrite the radicand as a product using perfect cube factors.	$\sqrt[3]{2^3 x^6 \cdot 3x}$
Rewrite the radical as the product of two radicals.	$\sqrt[3]{2^3 x^6} \cdot \sqrt[3]{3x}$
Rewrite the first radicand as $\left(2x^2\right)^3$.	$\sqrt[3]{\left(2x^2\right)^3} \cdot \sqrt[3]{3x}$
Simplify.	$2x^2\sqrt[3]{3x}$

ⓑ

	$\sqrt[4]{80y^{14}}$
Rewrite the radicand as a product using perfect fourth power factors.	$\sqrt[4]{2^4 y^{12} \cdot 5y^2}$
Rewrite the radical as the product of two radicals.	$\sqrt[4]{2^4 y^{12}} \cdot \sqrt[4]{5y^2}$
Rewrite the first radicand as $\left(2y^3\right)^4$.	$\sqrt[4]{\left(2y^3\right)^4} \cdot \sqrt[4]{5y^2}$
Simplify.	$2\left|y^3\right|\sqrt[4]{5y^2}$

> **TRY IT : : 9.189**

Simplify: ⓐ $\sqrt[3]{54p^{10}}$ ⓑ $\sqrt[4]{64q^{10}}$.

> **TRY IT : : 9.190**

Simplify: ⓐ $\sqrt[3]{128m^{11}}$ ⓑ $\sqrt[4]{162n^7}$.

EXAMPLE 9.96

Simplify: ⓐ $\sqrt[3]{-27}$ ⓑ $\sqrt[4]{-16}$.

✓ Solution

ⓐ

$$\sqrt[3]{-27}$$

Rewrite the radicand as a product using perfect cube factors.

$$\sqrt[3]{(-3)^3}$$

Take the cube root. -3

ⓑ

$$\sqrt[4]{-16}$$

There is no real number n where $n^4 = -16$.   Not a real number.

 TRY IT :: 9.191   Simplify: ⓐ $\sqrt[3]{-108}$ ⓑ $\sqrt[4]{-48}$.

 TRY IT :: 9.192   Simplify: ⓐ $\sqrt[3]{-625}$ ⓑ $\sqrt[4]{-324}$.

Use the Quotient Property to Simplify Expressions with Higher Roots

We can simplify higher roots with quotients in the same way we simplified square roots. First we simplify any fractions inside the radical.

EXAMPLE 9.97

Simplify: ⓐ $\sqrt[3]{\dfrac{a^8}{a^5}}$ ⓑ $\sqrt[4]{\dfrac{a^{10}}{a^2}}$.

✓ Solution

ⓐ

$$\sqrt[3]{\dfrac{a^8}{a^5}}$$

Simplify the fraction under the radical first. $\sqrt[3]{a^3}$
Simplify. a

ⓑ

$$\sqrt[4]{\dfrac{a^{10}}{a^2}}$$

Simplify the fraction under the radical first. $\sqrt[4]{a^8}$

Rewrite the radicand using perfect fourth power factors. $\sqrt[4]{(a^2)^4}$

Simplify. a^2

 **TRY IT :: 9.193**   Simplify: ⓐ $\sqrt[4]{\dfrac{x^7}{x^3}}$ ⓑ $\sqrt[4]{\dfrac{y^{17}}{y^5}}$.

TRY IT :: 9.194   Simplify: ⓐ $\sqrt[3]{\dfrac{m^{13}}{m^7}}$ ⓑ $\sqrt[5]{\dfrac{n^{12}}{n^2}}$.

Previously, we used the Quotient Property 'in reverse' to simplify square roots. Now we will generalize the formula to include higher roots.

Quotient Property of nth Roots

$$\sqrt[n]{\frac{a}{b}} = \frac{\sqrt[n]{a}}{\sqrt[n]{b}} \quad \text{and} \quad \frac{\sqrt[n]{a}}{\sqrt[n]{b}} = \sqrt[n]{\frac{a}{b}}$$

when $\sqrt[n]{a}$ and $\sqrt[n]{b}$ are real numbers, $b \neq 0$, and for any integer $n \geq 2$

EXAMPLE 9.98

Simplify: ⓐ $\dfrac{\sqrt[3]{-108}}{\sqrt[3]{2}}$ ⓑ $\dfrac{\sqrt[4]{96x^7}}{\sqrt[4]{3x^2}}$.

⊘ **Solution**

ⓐ

$$\frac{\sqrt[3]{-108}}{\sqrt[3]{2}}$$

Neither radicand is a perfect cube, so use the Quotient Property to write as one radical.	$\sqrt[3]{\dfrac{-108}{2}}$
Simplify the fraction under the radical.	$\sqrt[3]{-54}$
Rewrite the radicand as a product using perfect cube factors.	$\sqrt[3]{(-3)^3 \cdot 2}$
Rewrite the radical as the product of two radicals.	$\sqrt[3]{(-3)^3} \cdot \sqrt[3]{2}$
Simplify.	$-3\sqrt[3]{2}$

ⓑ

$$\frac{\sqrt[4]{96x^7}}{\sqrt[4]{3x^2}}$$

Neither radicand is a perfect fourth power, so use the Quotient Property to write as one radical.	$\sqrt[4]{\dfrac{96x^7}{3x^2}}$
Simplify the fraction under the radical.	$\sqrt[4]{32x^5}$
Rewrite the radicand as a product using perfect fourth power factors.	$\sqrt[4]{2^4 x^4 \cdot 2x}$
Rewrite the radical as the product of two radicals.	$\sqrt[4]{(2x)^4} \cdot \sqrt[4]{2x}$
Simplify.	$2|x|\sqrt[4]{2x}$

> **TRY IT : : 9.195**

Simplify: ⓐ $\dfrac{\sqrt[3]{-532}}{\sqrt[3]{2}}$ ⓑ $\dfrac{\sqrt[4]{486m^{11}}}{\sqrt[4]{3m^5}}$.

> **TRY IT : : 9.196**

Simplify: ⓐ $\dfrac{\sqrt[3]{-192}}{\sqrt[3]{3}}$ ⓑ $\dfrac{\sqrt[4]{324n^7}}{\sqrt[4]{2n^3}}$.

If the fraction inside the radical cannot be simplified, we use the first form of the Quotient Property to rewrite the expression as the quotient of two radicals.

EXAMPLE 9.99

Simplify: ⓐ $\sqrt[3]{\dfrac{24x^7}{y^3}}$ ⓑ $\sqrt[4]{\dfrac{48x^{10}}{y^8}}$.

⊘ **Solution**

ⓐ

$$\sqrt[3]{\dfrac{24x^7}{y^3}}$$

The fraction in the radicand cannot be simplified. Use the Quotient Property to write as two radicals.

$$\dfrac{\sqrt[3]{24x^7}}{\sqrt[3]{y^3}}$$

Rewrite each radicand as a product using perfect cube factors.

$$\dfrac{\sqrt[3]{8x^6 \cdot 3x}}{\sqrt[3]{y^3}}$$

Rewrite the numerator as the product of two radicals.

$$\dfrac{\sqrt[3]{\left(2x^2\right)^3}\,\sqrt[3]{3x}}{\sqrt[3]{y^3}}$$

Simplify.

$$\dfrac{2x^2\sqrt[3]{3x}}{y}$$

ⓑ
$$\sqrt[4]{\dfrac{48x^{10}}{y^8}}$$

The fraction in the radicand cannot be simplified. Use the Quotient Property to write as two radicals.

$$\dfrac{\sqrt[4]{48x^{10}}}{\sqrt[4]{y^8}}$$

Rewrite each radicand as a product using perfect fourth power factors.

$$\dfrac{\sqrt[4]{16x^8 \cdot 3x^2}}{\sqrt[4]{y^8}}$$

Rewrite the numerator as the product of two radicals.

$$\dfrac{\sqrt[4]{\left(2x^2\right)^4}\sqrt[4]{3x^2}}{\sqrt[4]{\left(y^2\right)^4}}$$

Simplify.

$$\dfrac{2x^2\sqrt[4]{3x^2}}{y^2}$$

> **TRY IT : :** 9.197

Simplify: ⓐ $\sqrt[3]{\dfrac{108c^{10}}{d^6}}$ ⓑ $\sqrt[4]{\dfrac{80x^{10}}{y^5}}$.

> **TRY IT : :** 9.198

Simplify: ⓐ $\sqrt[3]{\dfrac{40r^3}{s}}$ ⓑ $\sqrt[4]{\dfrac{162m^{14}}{n^{12}}}$.

Add and Subtract Higher Roots

We can add and subtract higher roots like we added and subtracted square roots. First we provide a formal definition of like radicals.

Like Radicals

Radicals with the same index and same radicand are called **like radicals**.

Like radicals have the same index and the same radicand.

- $9\sqrt[4]{42x}$ and $-2\sqrt[4]{42x}$ are like radicals.

- $5\sqrt[3]{125x}$ and $6\sqrt[3]{125y}$ are not like radicals. The radicands are different.

- $2\sqrt[5]{1000q}$ and $-4\sqrt[4]{1000q}$ are not like radicals. The indices are different.

We add and subtract like radicals in the same way we add and subtract like terms. We can add $9\sqrt[4]{42x} + \left(-2\sqrt[4]{42x}\right)$ and the result is $7\sqrt[4]{42x}$.

EXAMPLE 9.100

Simplify: ⓐ $\sqrt[3]{4x} + \sqrt[3]{4x}$ ⓑ $4\sqrt[4]{8} - 2\sqrt[4]{8}$.

⊘ **Solution**

ⓐ

$$\sqrt[3]{4x} + \sqrt[3]{4x}$$

The radicals are like, so we add the coefficients. $2\sqrt[3]{4x}$

ⓑ

$$4\sqrt[4]{8} - 2\sqrt[4]{8}$$

The radicals are like, so we subtract the coefficients. $2\sqrt[4]{8}$

> **TRY IT : : 9.199**   Simplify: ⓐ $\sqrt[5]{3x} + \sqrt[5]{3x}$ ⓑ $3\sqrt[3]{9} - \sqrt[3]{9}$.

> **TRY IT : : 9.200**   Simplify: ⓐ $\sqrt[4]{10y} + \sqrt[4]{10y}$ ⓑ $5\sqrt[6]{32} - 3\sqrt[6]{32}$.

When an expression does not appear to have like radicals, we will simplify each radical first. Sometimes this leads to an expression with like radicals.

EXAMPLE 9.101

Simplify: ⓐ $\sqrt[3]{54} - \sqrt[3]{16}$ ⓑ $\sqrt[4]{48} + \sqrt[4]{243}$.

⊘ **Solution**

ⓐ

$$\sqrt[3]{54} - \sqrt[3]{16}$$

Rewrite each radicand using perfect cube factors. $\sqrt[3]{27} \cdot \sqrt[3]{2} - \sqrt[3]{8} \cdot \sqrt[3]{2}$

Rewrite the perfect cubes. $\sqrt[3]{(3)^3}\sqrt[3]{2} - \sqrt[3]{(2)^3}\sqrt[3]{2}$

Simplify the radicals where possible. $3\sqrt[3]{2} - 2\sqrt[3]{2}$

Combine like radicals. $\sqrt[3]{2}$

ⓑ

$$\sqrt[4]{48} + \sqrt[4]{243}$$

Rewrite using perfect fourth power factors. $\sqrt[4]{16} \cdot \sqrt[4]{3} + \sqrt[4]{81} \cdot \sqrt[4]{3}$

Rewrite the perfect fourth powers. $\sqrt[4]{(2)^4}\sqrt[4]{3} + \sqrt[4]{(3)^4}\sqrt[4]{3}$

Simplify the radicals where possible. $2\sqrt[4]{3} + 3\sqrt[4]{3}$

Combine like radicals. $5\sqrt[4]{3}$

> **TRY IT : : 9.201**   Simplify: ⓐ $\sqrt[3]{192} - \sqrt[3]{81}$ ⓑ $\sqrt[4]{32} + \sqrt[4]{512}$.

> TRY IT : : 9.202   Simplify: ⓐ $\sqrt[3]{108} - \sqrt[3]{250}$ ⓑ $\sqrt[5]{64} + \sqrt[5]{486}$.

EXAMPLE 9.102

Simplify: ⓐ $\sqrt[3]{24x^4} - \sqrt[3]{-81x^7}$ ⓑ $\sqrt[4]{162y^9} + \sqrt[4]{516y^5}$.

✓ **Solution**

ⓐ

$$\sqrt[3]{24x^4} - \sqrt[3]{-81x^7}$$

Rewrite each radicand using perfect cube factors.

$$\sqrt[3]{8x^3} \cdot \sqrt[3]{3x} - \sqrt[3]{-27x^6} \cdot \sqrt[3]{3x}$$

Rewrite the perfect cubes.

$$\sqrt[3]{(2x)^3} \sqrt[3]{3x} - \sqrt[3]{\left(-3x^2\right)^3} \sqrt[3]{3x}$$

Simplify the radicals where possible.

$$2x\sqrt[3]{3x} - \left(-3x^2\sqrt[3]{3x}\right)$$

ⓑ

$$\sqrt[4]{162y^9} + \sqrt[4]{516y^5}$$

Rewrite each radicand using perfect fourth power factors.

$$\sqrt[4]{81y^8} \cdot \sqrt[4]{2y} + \sqrt[4]{256y^4} \cdot \sqrt[4]{2y}$$

Rewrite the perfect fourth powers.

$$\sqrt[4]{\left(3y^2\right)^4} \cdot \sqrt[4]{2y} + \sqrt[4]{(4y)^4} \cdot \sqrt[4]{2y}$$

Simplify the radicals where possible.

$$3y^2\sqrt[4]{2y} + 4|y|\sqrt[4]{2y}$$

> TRY IT : : 9.203   Simplify: ⓐ $\sqrt[3]{32y^5} - \sqrt[3]{-108y^8}$ ⓑ $\sqrt[4]{243r^{11}} + \sqrt[4]{768r^{10}}$.

> TRY IT : : 9.204   Simplify: ⓐ $\sqrt[3]{40z^7} - \sqrt[3]{-135z^4}$ ⓑ $\sqrt[4]{80s^{13}} + \sqrt[4]{1280s^6}$.

▶ **MEDIA : :**

Access these online resources for additional instruction and practice with simplifying higher roots.

- **Simplifying Higher Roots (https://openstax.org/l/25SimplifyHR)**
- **Add/Subtract Roots with Higher Indices (https://openstax.org/l/25AddSubtrHR)**

9.7 EXERCISES

Practice Makes Perfect

Simplify Expressions with Higher Roots

In the following exercises, simplify.

442.
ⓐ $\sqrt[3]{216}$
ⓑ $\sqrt[4]{256}$
ⓒ $\sqrt[5]{32}$

443.
ⓐ $\sqrt[3]{27}$
ⓑ $\sqrt[4]{16}$
ⓒ $\sqrt[5]{243}$

444.
ⓐ $\sqrt[3]{512}$
ⓑ $\sqrt[4]{81}$
ⓒ $\sqrt[5]{1}$

445.
ⓐ $\sqrt[3]{125}$
ⓑ $\sqrt[4]{1296}$
ⓒ $\sqrt[5]{1024}$

446.
ⓐ $\sqrt[3]{-8}$
ⓑ $\sqrt[4]{-81}$
ⓒ $\sqrt[5]{-32}$

447.
ⓐ $\sqrt[3]{-64}$
ⓑ $\sqrt[4]{-16}$
ⓒ $\sqrt[5]{-243}$

448.
ⓐ $\sqrt[3]{-125}$
ⓑ $\sqrt[4]{-1296}$
ⓒ $\sqrt[5]{-1024}$

449.
ⓐ $\sqrt[3]{-512}$
ⓑ $\sqrt[4]{-81}$
ⓒ $\sqrt[5]{-1}$

450.
ⓐ $\sqrt[5]{u^5}$
ⓑ $\sqrt[8]{v^8}$

451.
ⓐ $\sqrt[3]{a^3}$
ⓑ $\sqrt[12]{b^{12}}$

452.
ⓐ $\sqrt[4]{y^4}$
ⓑ $\sqrt[7]{m^7}$

453.
ⓐ $\sqrt[8]{k^8}$
ⓑ $\sqrt[6]{p^6}$

454.
ⓐ $\sqrt[3]{x^9}$
ⓑ $\sqrt[4]{y^{12}}$

455.
ⓐ $\sqrt[5]{a^{10}}$
ⓑ $\sqrt[3]{b^{27}}$

456.
ⓐ $\sqrt[4]{m^8}$
ⓑ $\sqrt[5]{n^{20}}$

457.
ⓐ $\sqrt[6]{r^{12}}$
ⓑ $\sqrt[3]{s^{30}}$

458.
ⓐ $\sqrt[4]{16x^8}$
ⓑ $\sqrt[6]{64y^{12}}$

459.
ⓐ $\sqrt[3]{-8c^9}$
ⓑ $\sqrt[3]{125d^{15}}$

460.
ⓐ $\sqrt[3]{216a^6}$
ⓑ $\sqrt[5]{32b^{20}}$

461.
ⓐ $\sqrt[7]{128r^{14}}$
ⓑ $\sqrt[4]{81s^{24}}$

Use the Product Property to Simplify Expressions with Higher Roots

In the following exercises, simplify.

462. ⓐ $\sqrt[3]{r^5}$ ⓑ $\sqrt[4]{s^{10}}$

463. ⓐ $\sqrt[5]{u^7}$ ⓑ $\sqrt[6]{v^{11}}$

464. ⓐ $\sqrt[4]{m^5}$ ⓑ $\sqrt[8]{n^{10}}$

465. ⓐ $\sqrt[5]{p^8}$ ⓑ $\sqrt[3]{q^8}$

466. ⓐ $\sqrt[4]{32}$ ⓑ $\sqrt[5]{64}$

467. ⓐ $\sqrt[3]{625}$ ⓑ $\sqrt[6]{128}$

468. ⓐ $\sqrt[5]{64}$ ⓑ $\sqrt[3]{256}$

469. ⓐ $\sqrt[4]{3125}$ ⓑ $\sqrt[3]{81}$

470. ⓐ $\sqrt[3]{108x^5}$ ⓑ $\sqrt[4]{48y^6}$

471. ⓐ $\sqrt[5]{96a^7}$ ⓑ $\sqrt[3]{375b^4}$

472. ⓐ $\sqrt[4]{405m^{10}}$ ⓑ $\sqrt[5]{160n^8}$

473. ⓐ $\sqrt[3]{512p^5}$ ⓑ $\sqrt[4]{324q^7}$

474. ⓐ $\sqrt[3]{-864}$ ⓑ $\sqrt[4]{-256}$

475. ⓐ $\sqrt[5]{-486}$ ⓑ $\sqrt[6]{-64}$

476. ⓐ $\sqrt[5]{-32}$ ⓑ $\sqrt[8]{-1}$

477. ⓐ $\sqrt[3]{-8}$ ⓑ $\sqrt[4]{-16}$

Use the Quotient Property to Simplify Expressions with Higher Roots

In the following exercises, simplify.

478. ⓐ $\sqrt[3]{\dfrac{p^{11}}{p^2}}$ ⓑ $\sqrt[4]{\dfrac{q^{17}}{q^{13}}}$

479. ⓐ $\sqrt[5]{\dfrac{d^{12}}{d^7}}$ ⓑ $\sqrt[8]{\dfrac{m^{12}}{m^4}}$

480. ⓐ $\sqrt[5]{\dfrac{u^{21}}{u^{11}}}$ ⓑ $\sqrt[6]{\dfrac{v^{30}}{v^{12}}}$

481. ⓐ $\sqrt[3]{\dfrac{r^{14}}{r^5}}$ ⓑ $\sqrt[4]{\dfrac{c^{21}}{c^9}}$

482. ⓐ $\dfrac{\sqrt[4]{64}}{\sqrt[4]{2}}$ ⓑ $\dfrac{\sqrt[5]{128x^8}}{\sqrt[5]{2x^2}}$

483. ⓐ $\dfrac{\sqrt[3]{-625}}{\sqrt[3]{5}}$ ⓑ $\dfrac{\sqrt[4]{80m^7}}{\sqrt[4]{5m}}$

484. ⓐ $\sqrt[3]{\dfrac{1050}{2}}$ ⓑ $\sqrt[4]{\dfrac{486y^9}{2y^3}}$

485. ⓐ $\sqrt[3]{\dfrac{162}{6}}$ ⓑ $\sqrt[4]{\dfrac{160r^{10}}{5r^3}}$

486. ⓐ $\sqrt[3]{\dfrac{54a^8}{b^3}}$ ⓑ $\sqrt[4]{\dfrac{64c^5}{d^2}}$

487. ⓐ $\sqrt[5]{\dfrac{96r^{11}}{s^3}}$ ⓑ $\sqrt[6]{\dfrac{128u^7}{v^3}}$

488. ⓐ $\sqrt[3]{\dfrac{81s^8}{t^3}}$ ⓑ $\sqrt[4]{\dfrac{64p^{15}}{q^{12}}}$

489. ⓐ $\sqrt[3]{\dfrac{625u^{10}}{v^3}}$ ⓑ $\sqrt[4]{\dfrac{729c^{21}}{d^8}}$

Add and Subtract Higher Roots

In the following exercises, simplify.

490.
ⓐ $\sqrt[7]{8p} + \sqrt[7]{8p}$
ⓑ $3\sqrt[3]{25} - \sqrt[3]{25}$

491.
ⓐ $\sqrt[3]{15q} + \sqrt[3]{15q}$
ⓑ $2\sqrt[4]{27} - 6\sqrt[4]{27}$

492.
ⓐ $3\sqrt[5]{9x} + 7\sqrt[5]{9x}$
ⓑ $8\sqrt[7]{3q} - 2\sqrt[7]{3q}$

493.
ⓐ $23\sqrt[13]{4y} + 19\sqrt[13]{4y}$
ⓑ $31\sqrt[19]{5z} - 17\sqrt[19]{5z}$

494.
ⓐ $\sqrt[3]{81} - \sqrt[3]{192}$
ⓑ $\sqrt[4]{512} - \sqrt[4]{32}$

495.
ⓐ $\sqrt[3]{250} - \sqrt[3]{54}$
ⓑ $\sqrt[4]{243} - \sqrt[4]{1875}$

496.
ⓐ $\sqrt[3]{128} + \sqrt[3]{250}$
ⓑ $\sqrt[5]{729} + \sqrt[5]{96}$

497.
ⓐ $\sqrt[4]{243} + \sqrt[4]{1250}$
ⓑ $\sqrt[3]{2000} + \sqrt[3]{54}$

498.
ⓐ $\sqrt[3]{64a^{10}} - \sqrt[3]{-216a^{12}}$
ⓑ $\sqrt[4]{486u^7} + \sqrt[4]{768u^3}$

499.
ⓐ $\sqrt[3]{80b^5} - \sqrt[3]{-270b^3}$
ⓑ $\sqrt[4]{160v^{10}} - \sqrt[4]{1280v^3}$

Mixed Practice

In the following exercises, simplify.

500. $\sqrt[4]{16}$

501. $\sqrt[6]{64}$

502. $\sqrt[3]{a^3}$

503. $\sqrt[12]{b^{12}}$

504. $\sqrt[3]{-8c^9}$

505. $\sqrt[3]{125d^{15}}$

506. $\sqrt[3]{r^5}$

507. $\sqrt[4]{s^{10}}$

508. $\sqrt[3]{108x^5}$

509. $\sqrt[4]{48y^6}$

510. $\sqrt[5]{-486}$

511. $\sqrt[6]{-64}$

512. $\dfrac{\sqrt[4]{64}}{\sqrt[4]{2}}$

513. $\dfrac{\sqrt[5]{128x^8}}{\sqrt[5]{2x^2}}$

514. $\sqrt[5]{\dfrac{96r^{11}}{s^3}}$

515. $\sqrt[6]{\dfrac{128u^7}{v^3}}$

516. $\sqrt[3]{81} - \sqrt[3]{192}$

517. $\sqrt[4]{512} - \sqrt[4]{32}$

518. $\sqrt[3]{64a^{10}} - \sqrt[3]{-216a^{12}}$

519. $\sqrt[4]{486u^7} + \sqrt[4]{768u^3}$

Everyday Math

520. Population growth The expression $10 \cdot x^n$ models the growth of a mold population after n generations. There were 10 spores at the start, and each had x offspring. So $10 \cdot x^n$ is the number of offspring at the fifth generation. At the fifth generation there were 10,240 offspring. Simplify the expression $\sqrt[5]{\dfrac{10,240}{10}}$ to determine the number of offspring of each spore.

521. Spread of a virus The expression $3 \cdot x^n$ models the spread of a virus after n cycles. There were three people originally infected with the virus, and each of them infected x people. So $3 \cdot x^4$ is the number of people infected on the fourth cycle. At the fourth cycle 1875 people were infected. Simplify the expression $\sqrt[4]{\dfrac{1875}{3}}$ to determine the number of people each person infected.

Writing Exercises

522. Explain how you know that $\sqrt[5]{x^{10}} = x^2$.

523. Explain why $\sqrt[4]{-64}$ is not a real number but $\sqrt[3]{-64}$ is.

Self Check

ⓐ *After completing the exercises, use this checklist to evaluate your mastery of the objectives of this section.*

I can...	Confidently	With some help	No-I don't get it!
simplify expressions with higher roots.			
use the Product Property to simplify expressions with higher roots.			
use the Quotient Property to simplify expressions with higher roots.			
add and subtract higher roots.			

ⓑ *What does this checklist tell you about your mastery of this section? What steps will you take to improve?*

9.8 | Rational Exponents

Learning Objectives

By the end of this section, you will be able to:

> Simplify expressions with $a^{\frac{1}{n}}$

> Simplify expressions with $a^{\frac{m}{n}}$

> Use the Laws of Exponents to simply expressions with rational exponents

---

Be Prepared!

Before you get started, take this readiness quiz.

1. Add: $\frac{7}{15} + \frac{5}{12}$.

   If you missed this problem, review **Example 1.81**.

2. Simplify: $\left(4x^2 y^5\right)^3$.

   If you missed this problem, review **Example 6.24**.

3. Simplify: 5^{-3}.

   If you missed this problem, review **Example 6.89**.

Simplify Expressions with $a^{\frac{1}{n}}$

Rational exponents are another way of writing expressions with radicals. When we use **rational exponents**, we can apply the properties of exponents to simplify expressions.

The Power Property for Exponents says that $(a^m)^n = a^{m \cdot n}$ when m and n are whole numbers. Let's assume we are now not limited to whole numbers.

Suppose we want to find a number p such that $(8^p)^3 = 8$. We will use the Power Property of Exponents to find the value of p.

$$(8^p)^3 = 8$$

Multiply the exponents on the left.	$8^{3p} = 8$
Write the exponent 1 on the right.	$8^{3p} = 8^1$
The exponents must be equal.	$3p = 1$
Solve for p.	$p = \frac{1}{3}$

$$\text{So} \left(8^{\frac{1}{3}}\right)^3 = 8.$$

But we know also $\left(\sqrt[3]{8}\right)^3 = 8$. Then it must be that $8^{\frac{1}{3}} = \sqrt[3]{8}$.

This same logic can be used for any positive integer exponent n to show that $a^{\frac{1}{n}} = \sqrt[n]{a}$.

Rational Exponent $a^{\frac{1}{n}}$

If $\sqrt[n]{a}$ is a real number and $n \geq 2$, $a^{\frac{1}{n}} = \sqrt[n]{a}$.

There will be times when working with expressions will be easier if you use rational exponents and times when it will be easier if you use radicals. In the first few examples, you'll practice converting expressions between these two notations.

EXAMPLE 9.103

Write as a radical expression: ⓐ $x^{\frac{1}{2}}$ ⓑ $y^{\frac{1}{3}}$ ⓒ $z^{\frac{1}{4}}$.

✓ **Solution**

We want to write each expression in the form $\sqrt[n]{a}$.

ⓐ

$$x^{\frac{1}{2}}$$

The denominator of the exponent is 2, so the index of the radical is 2. We do not show the index when it is 2.

$$\sqrt{x}$$

ⓑ

$$y^{\frac{1}{3}}$$

The denominator of the exponent is 3, so the index is 3.

$$\sqrt[3]{y}$$

ⓒ

$$z^{\frac{1}{4}}$$

The denominator of the exponent is 4, so the index is 4.

$$\sqrt[4]{z}$$

> **TRY IT : : 9.205**
>
> Write as a radical expression: ⓐ $t^{\frac{1}{2}}$ ⓑ $m^{\frac{1}{3}}$ ⓒ $r^{\frac{1}{4}}$.

> **TRY IT : : 9.206**
>
> Write as a radial expression: ⓐ $b^{\frac{1}{2}}$ ⓑ $z^{\frac{1}{3}}$ ⓒ $p^{\frac{1}{4}}$.

EXAMPLE 9.104

Write with a rational exponent: ⓐ $\sqrt{x}$ ⓑ $\sqrt[3]{y}$ ⓒ $\sqrt[4]{z}$.

✓ **Solution**

We want to write each radical in the form $a^{\frac{1}{n}}$.

ⓐ

$$\sqrt{x}$$

No index is shown, so it is 2.
The denominator of the exponent will be 2.

$$x^{\frac{1}{2}}$$

ⓑ

$$\sqrt[3]{y}$$

The index is 3, so the denominator of the exponent is 3.

$$y^{\frac{1}{3}}$$

ⓒ

The index is 4, so the denominator of the exponent is 4.

$$\sqrt[4]{z}$$
$$z^{\frac{1}{4}}$$

> **TRY IT : : 9.207**   Write with a rational exponent: ⓐ $\sqrt{s}$ ⓑ $\sqrt[3]{x}$ ⓒ $\sqrt[4]{b}$.

> **TRY IT : : 9.208**   Write with a rational exponent: ⓐ $\sqrt{v}$ ⓑ $\sqrt[3]{p}$ ⓒ $\sqrt[4]{p}$.

EXAMPLE 9.105

Write with a rational exponent: ⓐ $\sqrt{5y}$ ⓑ $\sqrt[3]{4x}$ ⓒ $3\sqrt[4]{5z}$.

⊘ Solution

We want to write each radical in the form $a^{\frac{1}{n}}$.

ⓐ

No index is shown, so it is 2.
The denominator of the exponent will be 2.

$$\sqrt{5y}$$
$$(5y)^{\frac{1}{2}}$$

ⓑ

The index is 3, so the denominator of the exponent is 3.

$$\sqrt[3]{4x}$$
$$(4x)^{\frac{1}{3}}$$

ⓒ

The index is 4, so the denominator of the exponent is 4.

$$3\sqrt[4]{5z}$$
$$3(5z)^{\frac{1}{4}}$$

> **TRY IT : : 9.209**   Write with a rational exponent: ⓐ $\sqrt{10m}$ ⓑ $\sqrt[5]{3n}$ ⓒ $3\sqrt[4]{6y}$.

> **TRY IT : : 9.210**   Write with a rational exponent: ⓐ $\sqrt[7]{3k}$ ⓑ $\sqrt[4]{5j}$ ⓒ $8\sqrt[3]{2a}$.

In the next example, you may find it easier to simplify the expressions if you rewrite them as radicals first.

EXAMPLE 9.106

Simplify: ⓐ $25^{\frac{1}{2}}$ ⓑ $64^{\frac{1}{3}}$ ⓒ $256^{\frac{1}{4}}$.

⊘ Solution

ⓐ

	$25^{\frac{1}{2}}$
Rewrite as a square root.	$\sqrt{25}$
Simplify.	5

ⓑ

	$64^{\frac{1}{3}}$
Rewrite as a cube root.	$\sqrt[3]{64}$
Recognize 64 is a perfect cube.	$\sqrt[3]{4^3}$
Simplify.	4

ⓒ

	$256^{\frac{1}{4}}$
Rewrite as a fourth root.	$\sqrt[4]{256}$
Recognize 256 is a perfect fourth power.	$\sqrt[4]{4^4}$
Simplify.	4

> **TRY IT :: 9.211**
> Simplify: ⓐ $36^{\frac{1}{2}}$ ⓑ $8^{\frac{1}{3}}$ ⓒ $16^{\frac{1}{4}}$.

> **TRY IT :: 9.212**
> Simplify: ⓐ $100^{\frac{1}{2}}$ ⓑ $27^{\frac{1}{3}}$ ⓒ $81^{\frac{1}{4}}$.

Be careful of the placement of the negative signs in the next example. We will need to use the property $a^{-n} = \frac{1}{a^n}$ in one case.

EXAMPLE 9.107

Simplify: ⓐ $(-64)^{\frac{1}{3}}$ ⓑ $-64^{\frac{1}{3}}$ ⓒ $(64)^{-\frac{1}{3}}$.

✓ **Solution**

ⓐ

	$(-64)^{\frac{1}{3}}$
Rewrite as a cube root.	$\sqrt[3]{-64}$
Rewrite -64 as a perfect cube.	$\sqrt[3]{(-4)^3}$
Simplify.	-4

ⓑ

	$-64^{\frac{1}{3}}$
The exponent applies only to the 64.	$-\left(64^{\frac{1}{3}}\right)$
Rewrite as a cube root.	$-\sqrt[3]{64}$
Rewrite 64 as 4^3.	$-\sqrt[3]{4^3}$
Simplify.	-4

ⓒ

	$(64)^{-\frac{1}{3}}$
Rewrite as a fraction with a positive exponent, using the property, $a^{-n} = \frac{1}{a^n}$.	$\frac{1}{\sqrt[3]{64}}$
Write as a cube root.	
Rewrite 64 as 4^3.	$\frac{1}{\sqrt[3]{4^3}}$
Simplify.	$\frac{1}{4}$

> **TRY IT ::** 9.213

Simplify: ⓐ $(-125)^{\frac{1}{3}}$ ⓑ $-125^{\frac{1}{3}}$ ⓒ $(125)^{-\frac{1}{3}}$.

> **TRY IT ::** 9.214

Simplify: ⓐ $(-32)^{\frac{1}{5}}$ ⓑ $-32^{\frac{1}{5}}$ ⓒ $(32)^{-\frac{1}{5}}$.

EXAMPLE 9.108

Simplify: ⓐ $(-16)^{\frac{1}{4}}$ ⓑ $-16^{\frac{1}{4}}$ ⓒ $(16)^{-\frac{1}{4}}$.

✓ **Solution**

ⓐ

	$(-16)^{\frac{1}{4}}$
Rewrite as a fourth root.	$\sqrt[4]{-16}$
There is no real number whose fourth power is -16.	

ⓑ

	$-16^{\frac{1}{4}}$
The exponent only applies to the 16. Rewrite as a fourth root.	$-\sqrt[4]{16}$
Rewrite 16 as 2^4.	$-\sqrt[4]{2^4}$
Simplify.	-2

ⓒ

	$(16)^{-\frac{1}{4}}$
Rewrite using the property $a^{-n} = \frac{1}{a^n}$.	$\dfrac{1}{(16)^{\frac{1}{4}}}$
Rewrite as a fourth root.	$\dfrac{1}{\sqrt[4]{16}}$
Rewrite 16 as 2^4.	$\dfrac{1}{\sqrt[4]{2^4}}$
Simplify.	$\dfrac{1}{2}$

> **TRY IT : : 9.215**
>
> Simplify: ⓐ $(-64)^{\frac{1}{2}}$ ⓑ $-64^{\frac{1}{2}}$ ⓒ $(64)^{-\frac{1}{2}}$.

> **TRY IT : : 9.216**
>
> Simplify: ⓐ $(-256)^{\frac{1}{4}}$ ⓑ $-256^{\frac{1}{4}}$ ⓒ $(256)^{-\frac{1}{4}}$.

Simplify Expressions with $a^{\frac{m}{n}}$

Let's work with the Power Property for Exponents some more.

Suppose we raise $a^{\frac{1}{n}}$ to the power m.

$$\left(a^{\frac{1}{n}}\right)^m$$

Multiply the exponents.	$a^{\frac{1}{n} \cdot m}$
Simplify.	$a^{\frac{m}{n}}$

So $a^{\frac{m}{n}} = (\sqrt[n]{a})^m$.

Now suppose we take a^m to the $\frac{1}{n}$ power.

$$(a^m)^{\frac{1}{n}}$$

Multiply the exponents.	$a^{m \cdot \frac{1}{n}}$
Simplify.	$a^{\frac{m}{n}}$

So $a^{\frac{m}{n}} = \sqrt[n]{a^m}$ also.

Which form do we use to simplify an expression? We usually take the root first—that way we keep the numbers in the radicand smaller.

> **Rational Exponent $a^{\frac{m}{n}}$**
>
> For any positive integers m and n,

$$a^{\frac{m}{n}} = (\sqrt[n]{a})^m$$

$$a^{\frac{m}{n}} = \sqrt[n]{a^m}$$

EXAMPLE 9.109

Write with a rational exponent: ⓐ $\sqrt{y^3}$ ⓑ $\sqrt[3]{x^2}$ ⓒ $\sqrt[4]{z^3}$.

✓ Solution

We want to use $a^{\frac{m}{n}} = \sqrt[n]{a^m}$ to write each radical in the form $a^{\frac{m}{n}}$.

ⓐ

The numerator of the exponent is the exponent of y, **3**.	$\sqrt{y^3}$
The denominator of the exponent is the index of the radical, **2**.	$y^{\frac{3}{2}}$

ⓑ

The numerator of the exponent is the exponent of x, **2**.	$\sqrt[3]{x^2}$
The denominator of the exponent is the index of the radical, **3**.	$x^{\frac{2}{3}}$

ⓒ

The numerator of the exponent is the exponent of z, **3**.	$\sqrt[4]{z^3}$
The denominator of the exponent Is the index of the radical, **4**.	$z^{\frac{3}{4}}$

> **TRY IT : : 9.217**
>
> Write with a rational exponent: ⓐ $\sqrt{x^5}$ ⓑ $\sqrt[4]{z^3}$ ⓒ $\sqrt[5]{y^2}$.

> **TRY IT : : 9.218**
>
> Write with a rational exponent: ⓐ $\sqrt[5]{a^2}$ ⓑ $\sqrt[3]{b^7}$ ⓒ $\sqrt[4]{m^5}$.

EXAMPLE 9.110

Simplify: ⓐ $9^{\frac{3}{2}}$ ⓑ $125^{\frac{2}{3}}$ ⓒ $81^{\frac{3}{4}}$.

✓ Solution

We will rewrite each expression as a radical first using the property, $a^{\frac{m}{n}} = (\sqrt[n]{a})^m$. This form lets us take the root first and so we keep the numbers in the radicand smaller than if we used the other form.

ⓐ

$$9^{\frac{3}{2}}$$

The power of the radical is the numerator of the exponent, 3. Since the denominator of the exponent is 2, this is a square root.	$(\sqrt{9})^3$
Simplify.	$(3)^3$
	27

ⓑ

$$125^{\frac{2}{3}}$$

The power of the radical is the numerator of the exponent, 2. The index of the radical is the denominator of the exponent, 3.

$$\left(\sqrt[3]{125}\right)^2$$

Simplify.

$$(5)^2$$

$$25$$

ⓒ

$$81^{\frac{3}{4}}$$

The power of the radical is the numerator of the exponent, 3. The index of the radical is the denominator of the exponent, 4.

$$\left(\sqrt[4]{81}\right)^3$$

Simplify.

$$(3)^3$$

$$27$$

> **TRY IT :: 9.219**

Simplify: ⓐ $4^{\frac{3}{2}}$ ⓑ $27^{\frac{2}{3}}$ ⓒ $625^{\frac{3}{4}}$.

> **TRY IT :: 9.220**

Simplify: ⓐ $8^{\frac{5}{3}}$ ⓑ $81^{\frac{3}{2}}$ ⓒ $16^{\frac{3}{4}}$.

Remember that $b^{-p} = \frac{1}{b^p}$. The negative sign in the exponent does not change the sign of the expression.

EXAMPLE 9.111

Simplify: ⓐ $16^{-\frac{3}{2}}$ ⓑ $32^{-\frac{2}{5}}$ ⓒ $4^{-\frac{5}{2}}$.

✓ **Solution**

We will rewrite each expression first using $b^{-p} = \frac{1}{b^p}$ and then change to radical form.

ⓐ

$$16^{-\frac{3}{2}}$$

Rewrite using $b^{-p} = \frac{1}{b^p}$.

$$\frac{1}{16^{\frac{3}{2}}}$$

Change to radical form. The power of the radical is the numerator of the exponent, 3. The index is the denominator of the exponent, 2.

$$\frac{1}{\left(\sqrt{16}\right)^3}$$

Simplify.

$$\frac{1}{4^3}$$

$$\frac{1}{64}$$

ⓑ

$$32^{-\frac{2}{5}}$$

Rewrite using $b^{-p} = \frac{1}{b^p}$.

$$\frac{1}{32^{\frac{2}{5}}}$$

Change to radical form.

$$\frac{1}{\left(\sqrt[5]{32}\right)^2}$$

Rewrite the radicand as a power.

$$\frac{1}{\left(\sqrt[5]{2^5}\right)^2}$$

Simplify.

$$\frac{1}{2^2}$$
$$\frac{1}{4}$$

ⓒ

$$4^{-\frac{5}{2}}$$

Rewrite using $b^{-p} = \frac{1}{b^p}$.

$$\frac{1}{4^{\frac{5}{2}}}$$

Change to radical form.

$$\frac{1}{(\sqrt{4})^5}$$

Simplify.

$$\frac{1}{2^5}$$
$$\frac{1}{32}$$

> **TRY IT :: 9.221**
>
> Simplify: ⓐ $8^{-\frac{5}{3}}$ ⓑ $81^{-\frac{3}{2}}$ ⓒ $16^{-\frac{3}{4}}$.

> **TRY IT :: 9.222**
>
> Simplify: ⓐ $4^{-\frac{3}{2}}$ ⓑ $27^{-\frac{2}{3}}$ ⓒ $625^{-\frac{3}{4}}$.

EXAMPLE 9.112

Simplify: ⓐ $-25^{\frac{3}{2}}$ ⓑ $-25^{-\frac{3}{2}}$ ⓒ $(-25)^{\frac{3}{2}}$.

✓ **Solution**

ⓐ

$$-25^{\frac{3}{2}}$$

Rewrite in radical form. $\quad -\left(\sqrt{25}\right)^3$

Simplify the radical. $\quad -(5)^3$

Simplify. $\quad -125$

ⓑ

$$-25^{-\frac{3}{2}}$$

Rewrite using $b^{-p} = \frac{1}{b^p}$. $\quad -\left(\dfrac{1}{25^{\frac{3}{2}}}\right)$

Rewrite in radical form. $\quad -\left(\dfrac{1}{\left(\sqrt{25}\right)^3}\right)$

Simplify the radical. $\quad -\left(\dfrac{1}{(5)^3}\right)$

Simplify. $\quad -\dfrac{1}{125}$

ⓒ

$$(-25)^{\frac{3}{2}}$$

Rewrite in radical form. $\quad \left(\sqrt{-25}\right)^3$

There is no real number whose square root is -25. $\quad$ Not a real number.

> **TRY IT :: 9.223**
> Simplify: ⓐ $-16^{\frac{3}{2}}$ ⓑ $-16^{-\frac{3}{2}}$ ⓒ $(-16)^{-\frac{3}{2}}$.

> **TRY IT :: 9.224**
> Simplify: ⓐ $-81^{\frac{3}{2}}$ ⓑ $-81^{-\frac{3}{2}}$ ⓒ $(-81)^{-\frac{3}{2}}$.

Use the Laws of Exponents to Simplify Expressions with Rational Exponents

The same laws of exponents that we already used apply to rational exponents, too. We will list the Exponent Properties here to have them for reference as we simplify expressions.

Summary of Exponent Properties

If a, b are real numbers and m, n are rational numbers, then

Product Property	$a^m \cdot a^n = a^{m+n}$
Power Property	$(a^m)^n = a^{m \cdot n}$
Product to a Power	$(ab)^m = a^m b^m$
Quotient Property	$\dfrac{a^m}{a^n} = a^{m-n}, \quad a \neq 0, \quad m > n$
	$\dfrac{a^m}{a^n} = \dfrac{1}{a^{n-m}}, \quad a \neq 0, \quad n > m$
Zero Exponent Definition	$a^0 = 1, \quad a \neq 0$
Quotient to a Power Property	$\left(\dfrac{a}{b}\right)^m = \dfrac{a^m}{b^m}, \quad b \neq 0$

When we multiply the same base, we add the exponents.

EXAMPLE 9.113

Simplify: ⓐ $2^{\frac{1}{2}} \cdot 2^{\frac{5}{2}}$ ⓑ $x^{\frac{2}{3}} \cdot x^{\frac{4}{3}}$ ⓒ $z^{\frac{3}{4}} \cdot z^{\frac{5}{4}}$.

✓ **Solution**

ⓐ

$$2^{\frac{1}{2}} \cdot 2^{\frac{5}{2}}$$

The bases are the same, so we add the exponents.

$$2^{\frac{1}{2} + \frac{5}{2}}$$

Add the fractions.

$$2^{\frac{6}{2}}$$

Simplify the exponent.

$$2^3$$

Simplify.

$$8$$

ⓑ

$$x^{\frac{2}{3}} \cdot x^{\frac{4}{3}}$$

The bases are the same, so we add the exponents.

$$x^{\frac{2}{3} + \frac{4}{3}}$$

Add the fractions.

$$x^{\frac{6}{3}}$$

Simplify.

$$x^2$$

ⓒ

$$z^{\frac{3}{4}} \cdot z^{\frac{5}{4}}$$

The bases are the same, so we add the exponents.

$$z^{\frac{3}{4} + \frac{5}{4}}$$

Add the fractions.

$$z^{\frac{8}{4}}$$

Simplify.

$$z^2$$

> **TRY IT :: 9.225**
>
> Simplify: ⓐ $3^{\frac{2}{3}} \cdot 3^{\frac{4}{3}}$ ⓑ $y^{\frac{1}{3}} \cdot y^{\frac{8}{3}}$ ⓒ $m^{\frac{1}{4}} \cdot m^{\frac{3}{4}}$.

> **TRY IT :: 9.226**

Simplify: ⓐ $5^{\frac{3}{5}} \cdot 5^{\frac{7}{5}}$ ⓑ $z^{\frac{1}{8}} \cdot z^{\frac{7}{8}}$ ⓒ $n^{\frac{2}{7}} \cdot n^{\frac{5}{7}}$.

We will use the Power Property in the next example.

EXAMPLE 9.114

Simplify: ⓐ $\left(x^4\right)^{\frac{1}{2}}$ ⓑ $\left(y^6\right)^{\frac{1}{3}}$ ⓒ $\left(z^9\right)^{\frac{2}{3}}$.

✓ **Solution**

ⓐ

$$\left(x^4\right)^{\frac{1}{2}}$$

To raise a power to a power, we multiply the exponents.

$$x^{4 \cdot \frac{1}{2}}$$

Simplify.

$$x^2$$

ⓑ

$$\left(y^6\right)^{\frac{1}{3}}$$

To raise a power to a power, we multiply the exponents.

$$y^{6 \cdot \frac{1}{3}}$$

Simplify.

$$y^2$$

ⓒ

$$\left(z^9\right)^{\frac{2}{3}}$$

To raise a power to a power, we multiply the exponents.

$$z^{9 \cdot \frac{2}{3}}$$

Simplify.

$$z^6$$

> **TRY IT :: 9.227**

Simplify: ⓐ $\left(p^{10}\right)^{\frac{1}{5}}$ ⓑ $\left(q^8\right)^{\frac{3}{4}}$ ⓒ $\left(x^6\right)^{\frac{4}{3}}$.

> **TRY IT :: 9.228**

Simplify: ⓐ $\left(r^6\right)^{\frac{5}{3}}$ ⓑ $\left(s^{12}\right)^{\frac{3}{4}}$ ⓒ $\left(m^9\right)^{\frac{2}{9}}$.

The Quotient Property tells us that when we divide with the same base, we subtract the exponents.

EXAMPLE 9.115

Simplify: ⓐ $\dfrac{x^{\frac{4}{3}}}{x^{\frac{1}{3}}}$ ⓑ $\dfrac{y^{\frac{3}{4}}}{y^{\frac{1}{4}}}$ ⓒ $\dfrac{z^{\frac{2}{3}}}{z^{\frac{5}{3}}}$.

✓ Solution

ⓐ

$$\dfrac{x^{\frac{4}{3}}}{x^{\frac{1}{3}}}$$

To divide with the same base, we subtract the exponents.

$$x^{\frac{4}{3}-\frac{1}{3}}$$

Simplify.

$$x$$

ⓑ

$$\dfrac{y^{\frac{3}{4}}}{y^{\frac{1}{4}}}$$

To divide with the same base, we subtract the exponents.

$$y^{\frac{3}{4}-\frac{1}{4}}$$

Simplify.

$$y^{\frac{1}{2}}$$

ⓒ

$$\dfrac{z^{\frac{2}{3}}}{z^{\frac{5}{3}}}$$

To divide with the same base, we subtract the exponents.

$$z^{\frac{2}{3}-\frac{5}{3}}$$

Rewrite without a negative exponent.

$$\dfrac{1}{z}$$

> **TRY IT : : 9.229**
>
> Simplify: ⓐ $\dfrac{u^{\frac{5}{4}}}{u^{\frac{1}{4}}}$ ⓑ $\dfrac{v^{\frac{3}{5}}}{v^{\frac{2}{5}}}$ ⓒ $\dfrac{x^{\frac{2}{3}}}{x^{\frac{5}{3}}}$.

> **TRY IT : : 9.230**
>
> Simplify: ⓐ $\dfrac{c^{\frac{12}{5}}}{c^{\frac{2}{5}}}$ ⓑ $\dfrac{m^{\frac{5}{4}}}{m^{\frac{9}{4}}}$ ⓒ $\dfrac{d^{\frac{1}{5}}}{d^{\frac{6}{5}}}$.

Sometimes we need to use more than one property. In the next two examples, we will use both the Product to a Power Property and then the Power Property.

EXAMPLE 9.116

Simplify: ⓐ $\left(27u^{\frac{1}{2}}\right)^{\frac{2}{3}}$ ⓑ $\left(8v^{\frac{1}{4}}\right)^{\frac{2}{3}}$.

✓ **Solution**

ⓐ

$$\left(27u^{\frac{1}{2}}\right)^{\frac{2}{3}}$$

First we use the Product to a Power Property.

$$(27)^{\frac{2}{3}}\left(u^{\frac{1}{2}}\right)^{\frac{2}{3}}$$

Rewrite 27 as a power of 3.

$$\left(3^3\right)^{\frac{2}{3}}\left(u^{\frac{1}{2}}\right)^{\frac{2}{3}}$$

To raise a power to a power, we multiply the exponents.

$$\left(3^2\right)\left(u^{\frac{1}{3}}\right)$$

Simplify.

$$9u^{\frac{1}{3}}$$

ⓑ

$$\left(8v^{\frac{1}{4}}\right)^{\frac{2}{3}}$$

First we use the Product to a Power Property.

$$(8)^{\frac{2}{3}}\left(v^{\frac{1}{4}}\right)^{\frac{2}{3}}$$

Rewrite 8 as a power of 2.

$$\left(2^3\right)^{\frac{2}{3}}\left(v^{\frac{1}{4}}\right)^{\frac{2}{3}}$$

To raise a power to a power, we multiply the exponents.

$$\left(2^2\right)\left(v^{\frac{1}{6}}\right)$$

Simplify.

$$4v^{\frac{1}{6}}$$

> **TRY IT :: 9.231**

Simplify: ⓐ $\left(32x^{\frac{1}{3}}\right)^{\frac{3}{5}}$ ⓑ $\left(64y^{\frac{2}{3}}\right)^{\frac{1}{3}}$.

> **TRY IT : : 9.232**

Simplify: ⓐ $\left(16m^{\frac{1}{3}}\right)^{\frac{3}{2}}$ ⓑ $\left(81n^{\frac{2}{5}}\right)^{\frac{3}{2}}$.

EXAMPLE 9.117

Simplify: ⓐ $\left(m^3 n^9\right)^{\frac{1}{3}}$ ⓑ $\left(p^4 q^8\right)^{\frac{1}{4}}$.

✓ **Solution**

ⓐ

$$\left(m^3 n^9\right)^{\frac{1}{3}}$$

First we use the Product to a Power Property.

$$\left(m^3\right)^{\frac{1}{3}} \left(n^9\right)^{\frac{1}{3}}$$

To raise a power to a power, we multiply the exponents.

$$mn^3$$

ⓑ

$$\left(p^4 q^8\right)^{\frac{1}{4}}$$

First we use the Product to a Power Property.

$$\left(p^4\right)^{\frac{1}{4}} \left(q^8\right)^{\frac{1}{4}}$$

To raise a power to a power, we multiply the exponents.

$$pq^2$$

We will use both the Product and Quotient Properties in the next example.

EXAMPLE 9.118

Simplify: ⓐ $\dfrac{x^{\frac{3}{4}} \cdot x^{-\frac{1}{4}}}{x^{-\frac{6}{4}}}$ ⓑ $\dfrac{y^{\frac{4}{3}} \cdot y}{y^{-\frac{2}{3}}}$.

✓ Solution

ⓐ

$$\frac{x^{\frac{3}{4}} \cdot x^{-\frac{1}{4}}}{x^{-\frac{6}{4}}}$$

Use the Product Property in the numerator, add the exponents.

$$\frac{x^{\frac{2}{4}}}{x^{-\frac{6}{4}}}$$

Use the Quotient Property, subtract the exponents.

$$x^{\frac{8}{4}}$$

Simplify.

$$x^2$$

ⓑ

$$\frac{y^{\frac{4}{3}} \cdot y}{y^{-\frac{2}{3}}}$$

Use the Product Property in the numerator, add the exponents.

$$\frac{y^{\frac{7}{3}}}{y^{-\frac{2}{3}}}$$

Use the Quotient Property, subtract the exponents.

$$y^{\frac{9}{3}}$$

Simplify.

$$y^3$$

> **TRY IT : : 9.233**

Simplify: ⓐ $\dfrac{m^{\frac{2}{3}} \cdot m^{-\frac{1}{3}}}{m^{-\frac{5}{3}}}$ ⓑ $\dfrac{n^{\frac{1}{6}} \cdot n}{n^{-\frac{11}{6}}}$.

> **TRY IT : : 9.234**

Simplify: ⓐ $\dfrac{u^{\frac{4}{5}} \cdot u^{-\frac{2}{5}}}{u^{-\frac{13}{5}}}$ ⓑ $\dfrac{v^{\frac{1}{2}} \cdot v}{v^{-\frac{7}{2}}}$.

9.8 EXERCISES

Practice Makes Perfect

Simplify Expressions with $a^{\frac{1}{n}}$

In the following exercises, write as a radical expression.

524.

ⓐ $x^{\frac{1}{2}}$

ⓑ $y^{\frac{1}{3}}$

ⓒ $z^{\frac{1}{4}}$

525.

ⓐ $r^{\frac{1}{2}}$

ⓑ $s^{\frac{1}{3}}$

ⓒ $t^{\frac{1}{4}}$

526.

ⓐ $u^{\frac{1}{5}}$

ⓑ $v^{\frac{1}{9}}$

ⓒ $w^{\frac{1}{20}}$

527.

ⓐ $g^{\frac{1}{7}}$

ⓑ $h^{\frac{1}{5}}$

ⓒ $j^{\frac{1}{25}}$

In the following exercises, write with a rational exponent.

528.

ⓐ $-\sqrt[7]{x}$

ⓑ $\sqrt[9]{y}$

ⓒ $\sqrt[5]{f}$

529.

ⓐ $\sqrt[8]{r}$

ⓑ $\sqrt[10]{s}$

ⓒ $\sqrt[4]{t}$

530.

ⓐ $\sqrt[3]{a}$

ⓑ $\sqrt[12]{b}$

ⓒ $\sqrt{c}$

531.

ⓐ $\sqrt[5]{u}$

ⓑ $\sqrt{v}$

ⓒ $\sqrt[10]{w}$

532.

ⓐ $\sqrt[3]{7c}$

ⓑ $\sqrt[7]{12d}$

ⓒ $3\sqrt[4]{5f}$

533.

ⓐ $\sqrt[4]{5x}$

ⓑ $\sqrt[8]{9y}$

ⓒ $7\sqrt[5]{3z}$

534.

ⓐ $\sqrt{21p}$

ⓑ $\sqrt[4]{8q}$

ⓒ $4\sqrt[6]{36r}$

535.

ⓐ $\sqrt[3]{25a}$

ⓑ $\sqrt{3b}$

ⓒ $\sqrt[10]{40c}$

In the following exercises, simplify.

536.

ⓐ $81^{\frac{1}{2}}$

ⓑ $125^{\frac{1}{3}}$

ⓒ $64^{\frac{1}{2}}$

537.

ⓐ $625^{\frac{1}{4}}$

ⓑ $243^{\frac{1}{5}}$

ⓒ $32^{\frac{1}{5}}$

538.

ⓐ $16^{\frac{1}{4}}$

ⓑ $16^{\frac{1}{2}}$

ⓒ $3125^{\frac{1}{5}}$

539.

ⓐ $216^{\frac{1}{3}}$

ⓑ $32^{\frac{1}{5}}$

ⓒ $81^{\frac{1}{4}}$

540.

ⓐ $(-216)^{\frac{1}{3}}$

ⓑ $-216^{\frac{1}{3}}$

ⓒ $(216)^{-\frac{1}{3}}$

541.

ⓐ $(-243)^{\frac{1}{5}}$

ⓑ $-243^{\frac{1}{5}}$

ⓒ $(243)^{-\frac{1}{5}}$

542.

ⓐ $(-1)^{\frac{1}{3}}$

ⓑ $-1^{\frac{1}{3}}$

ⓒ $(1)^{-\frac{1}{3}}$

543.

ⓐ $(-1000)^{\frac{1}{3}}$

ⓑ $-1000^{\frac{1}{3}}$

ⓒ $(1000)^{-\frac{1}{3}}$

544.

ⓐ $(-81)^{\frac{1}{4}}$

ⓑ $-81^{\frac{1}{4}}$

ⓒ $(81)^{-\frac{1}{4}}$

545.

ⓐ $(-49)^{\frac{1}{2}}$

ⓑ $-49^{\frac{1}{2}}$

ⓒ $(49)^{-\frac{1}{2}}$

546.

ⓐ $(-36)^{\frac{1}{2}}$

ⓑ $-36^{\frac{1}{2}}$

ⓒ $(36)^{-\frac{1}{2}}$

547.

ⓐ $(-1)^{\frac{1}{4}}$

ⓑ $(1)^{-\frac{1}{4}}$

ⓒ $-1^{\frac{1}{4}}$

548.

ⓐ $(-100)^{\frac{1}{2}}$

ⓑ $-100^{\frac{1}{2}}$

ⓒ $(100)^{-\frac{1}{2}}$

549.

ⓐ $(-32)^{\frac{1}{5}}$

ⓑ $(243)^{-\frac{1}{5}}$

ⓒ $-125^{\frac{1}{3}}$

Simplify Expressions with $a^{\frac{m}{n}}$

In the following exercises, write with a rational exponent.

550.

ⓐ $\sqrt{m^5}$

ⓑ $\sqrt[3]{n^2}$

ⓒ $\sqrt[4]{p^3}$

551.

ⓐ $\sqrt[4]{r^7}$

ⓑ $\sqrt[5]{s^3}$

ⓒ $\sqrt[3]{t^7}$

552.

ⓐ $\sqrt[5]{u^2}$

ⓑ $\sqrt[5]{v^8}$

ⓒ $\sqrt[9]{w^4}$

553.

ⓐ $\sqrt[3]{a}$

ⓑ $\sqrt{b^5}$

ⓒ $\sqrt[3]{c^5}$

In the following exercises, simplify.

554.

ⓐ $16^{\frac{3}{2}}$

ⓑ $8^{\frac{2}{3}}$

ⓒ $10,000^{\frac{3}{4}}$

555.

ⓐ $1000^{\frac{2}{3}}$

ⓑ $25^{\frac{3}{2}}$

ⓒ $32^{\frac{3}{5}}$

556.

ⓐ $27^{\frac{5}{3}}$

ⓑ $16^{\frac{5}{4}}$

ⓒ $32^{\frac{2}{5}}$

557.

ⓐ $16^{\frac{3}{2}}$

ⓑ $125^{\frac{5}{3}}$

ⓒ $64^{\frac{4}{3}}$

558.

ⓐ $32^{\frac{2}{5}}$

ⓑ $27^{-\frac{2}{3}}$

ⓒ $25^{-\frac{3}{2}}$

559.

ⓐ $64^{\frac{5}{2}}$

ⓑ $81^{-\frac{3}{2}}$

ⓒ $27^{-\frac{4}{3}}$

560.

ⓐ $25^{\frac{3}{2}}$

ⓑ $9^{-\frac{3}{2}}$

ⓒ $(-64)^{\frac{2}{3}}$

561.

ⓐ $100^{\frac{3}{2}}$

ⓑ $49^{-\frac{5}{2}}$

ⓒ $(-100)^{\frac{3}{2}}$

562.

ⓐ $-9^{\frac{3}{2}}$

ⓑ $-9^{-\frac{3}{2}}$

ⓒ $(-9)^{\frac{3}{2}}$

563.

ⓐ $-64^{\frac{3}{2}}$

ⓑ $-64^{-\frac{3}{2}}$

ⓒ $(-64)^{\frac{3}{2}}$

564.

ⓐ $-100^{\frac{3}{2}}$

ⓑ $-100^{-\frac{3}{2}}$

ⓒ $(-100)^{\frac{3}{2}}$

565.

ⓐ $-49^{\frac{3}{2}}$

ⓑ $-49^{-\frac{3}{2}}$

ⓒ $(-49)^{\frac{3}{2}}$

Use the Laws of Exponents to Simplify Expressions with Rational Exponents

In the following exercises, simplify.

566.

ⓐ $4^{\frac{5}{8}} \cdot 4^{\frac{11}{8}}$

ⓑ $m^{\frac{7}{12}} \cdot m^{\frac{17}{12}}$

ⓒ $p^{\frac{3}{7}} \cdot p^{\frac{18}{7}}$

567.

ⓐ $6^{\frac{5}{2}} \cdot 6^{\frac{1}{2}}$

ⓑ $n^{\frac{2}{10}} \cdot n^{\frac{8}{10}}$

ⓒ $q^{\frac{2}{5}} \cdot q^{\frac{13}{5}}$

568.

ⓐ $5^{\frac{1}{2}} \cdot 5^{\frac{7}{2}}$

ⓑ $c^{\frac{3}{4}} \cdot c^{\frac{9}{4}}$

ⓒ $d^{\frac{3}{5}} \cdot d^{\frac{2}{5}}$

569.

ⓐ $10^{\frac{1}{3}} \cdot 10^{\frac{5}{3}}$

ⓑ $x^{\frac{5}{6}} \cdot x^{\frac{7}{6}}$

ⓒ $y^{\frac{11}{8}} \cdot y^{\frac{21}{8}}$

570.

ⓐ $\left(m^6\right)^{\frac{5}{2}}$

ⓑ $\left(n^9\right)^{\frac{4}{3}}$

ⓒ $\left(p^{12}\right)^{\frac{3}{4}}$

571.

ⓐ $\left(a^{12}\right)^{\frac{1}{6}}$

ⓑ $\left(b^{15}\right)^{\frac{3}{5}}$

ⓒ $\left(c^{11}\right)^{\frac{1}{11}}$

572.

ⓐ $\left(x^{12}\right)^{\frac{2}{3}}$

ⓑ $\left(y^{20}\right)^{\frac{2}{5}}$

ⓒ $\left(z^{16}\right)^{\frac{1}{16}}$

573.

ⓐ $\left(h^{6}\right)^{\frac{4}{3}}$

ⓑ $\left(k^{12}\right)^{\frac{3}{4}}$

ⓒ $\left(j^{10}\right)^{\frac{7}{5}}$

574.

ⓐ $\dfrac{x^{\frac{7}{2}}}{x^{\frac{5}{2}}}$

ⓑ $\dfrac{y^{\frac{5}{2}}}{y^{\frac{1}{2}}}$

ⓒ $\dfrac{r^{\frac{4}{5}}}{r^{\frac{9}{5}}}$

575.

ⓐ $\dfrac{s^{\frac{11}{5}}}{s^{\frac{6}{5}}}$

ⓑ $\dfrac{z^{\frac{7}{3}}}{z^{\frac{1}{3}}}$

ⓒ $\dfrac{w^{\frac{2}{7}}}{w^{\frac{9}{7}}}$

576.

ⓐ $\dfrac{t^{\frac{12}{5}}}{t^{\frac{7}{5}}}$

ⓑ $\dfrac{x^{\frac{3}{2}}}{x^{\frac{1}{2}}}$

ⓒ $\dfrac{m^{\frac{13}{8}}}{m^{\frac{5}{8}}}$

577.

ⓐ $\dfrac{u^{\frac{13}{9}}}{u^{\frac{4}{9}}}$

ⓑ $\dfrac{r^{\frac{15}{7}}}{r^{\frac{8}{7}}}$

ⓒ $\dfrac{n^{\frac{3}{5}}}{n^{\frac{8}{5}}}$

578.

ⓐ $\left(9p^{\frac{2}{3}}\right)^{\frac{5}{2}}$

ⓑ $\left(27q^{\frac{3}{2}}\right)^{\frac{4}{3}}$

579.

ⓐ $\left(81r^{\frac{4}{5}}\right)^{\frac{1}{4}}$

ⓑ $\left(64s^{\frac{3}{7}}\right)^{\frac{1}{6}}$

580.

ⓐ $\left(16u^{\frac{1}{3}}\right)^{\frac{3}{4}}$

ⓑ $\left(100v^{\frac{2}{5}}\right)^{\frac{3}{2}}$

581.

ⓐ $\left(27m^{\frac{3}{4}}\right)^{\frac{2}{3}}$

ⓑ $\left(625n^{\frac{8}{3}}\right)^{\frac{3}{4}}$

582.

ⓐ $\left(x^{8}y^{10}\right)^{\frac{1}{2}}$

ⓑ $\left(a^{9}b^{12}\right)^{\frac{1}{3}}$

583.

ⓐ $\left(r^{8}s^{4}\right)^{\frac{1}{4}}$

ⓑ $\left(u^{15}v^{20}\right)^{\frac{1}{5}}$

584.

ⓐ $\left(a^{6}b^{16}\right)^{\frac{1}{2}}$

ⓑ $\left(j^{9}k^{6}\right)^{\frac{2}{3}}$

585.

ⓐ $\left(r^{16}s^{10}\right)^{\frac{1}{2}}$

ⓑ $\left(u^{10}v^{5}\right)^{\frac{4}{5}}$

586.

ⓐ $\dfrac{r^{\frac{5}{2}}\cdot r^{-\frac{1}{2}}}{r^{-\frac{3}{2}}}$

ⓑ $\dfrac{s^{\frac{1}{5}}\cdot s}{s^{-\frac{9}{5}}}$

587.

ⓐ $\dfrac{a^{\frac{3}{4}} \cdot a^{-\frac{1}{4}}}{a^{-\frac{10}{4}}}$

ⓑ $\dfrac{b^{\frac{2}{3}} \cdot b}{b^{-\frac{7}{3}}}$

588.

ⓐ $\dfrac{c^{\frac{5}{3}} \cdot c^{-\frac{1}{3}}}{c^{-\frac{2}{3}}}$

ⓑ $\dfrac{d^{\frac{3}{5}} \cdot d}{d^{-\frac{2}{5}}}$

589.

ⓐ $\dfrac{m^{\frac{7}{4}} \cdot m^{-\frac{5}{4}}}{m^{-\frac{2}{4}}}$

ⓑ $\dfrac{n^{\frac{3}{7}} \cdot n}{n^{-\frac{4}{7}}}$

590. $4^{\frac{5}{2}} \cdot 4^{\frac{1}{2}}$

591. $n^{\frac{2}{6}} \cdot n^{\frac{4}{6}}$

592. $\left(a^{24}\right)^{\frac{1}{6}}$

593. $\left(b^{10}\right)^{\frac{3}{5}}$

594. $\dfrac{w^{\frac{2}{5}}}{w^{\frac{7}{5}}}$

595. $\dfrac{z^{\frac{2}{3}}}{z^{\frac{8}{3}}}$

596. $\left(27r^{\frac{3}{5}}\right)^{\frac{1}{3}}$

597. $\left(64s^{\frac{3}{5}}\right)^{\frac{1}{6}}$

598. $\left(r^{9}s^{12}\right)^{\frac{1}{3}}$

599. $\left(u^{12}v^{18}\right)^{\frac{1}{6}}$

Everyday Math

600. Landscaping Joe wants to have a square garden plot in his backyard. He has enough compost to cover an area of 144 square feet. Simplify $144^{\frac{1}{2}}$ to find the length of each side of his garden.

601. Landscaping Elliott wants to make a square patio in his yard. He has enough concrete to pave an area of 242 square feet. Simplify $242^{\frac{1}{2}}$ to find the length of each side of his patio. Round to the nearest tenth of a foot.

602. Gravity While putting up holiday decorations, Bob dropped a decoration from the top of a tree that is 12 feet tall. Simplify $\dfrac{12^{\frac{1}{2}}}{16^{\frac{1}{2}}}$ to find how many seconds it took for the decoration to reach the ground. Round to the nearest tenth of a second.

603. Gravity An airplane dropped a flare from a height of 1024 feet above a lake. Simplify $\dfrac{1024^{\frac{1}{2}}}{16^{\frac{1}{2}}}$ to find how many seconds it took for the flare to reach the water.

Writing Exercises

604. Show two different algebraic methods to simplify $4^{\frac{3}{2}}$. Explain all your steps.

605. Explain why the expression $(-16)^{\frac{3}{2}}$ cannot be evaluated.

CHAPTER 9 REVIEW

KEY TERMS

index $\sqrt[n]{a}$ n is called the *index* of the radical.

like radicals Radicals with the same index and same radicand are called like radicals.

like square roots Square roots with the same radicand are called like square roots.

nth root of a number If $b^n = a$, then b is an *n*th root of a.

principal *n*th root The principal *n*th root of a is written $\sqrt[n]{a}$.

radical equation An equation in which the variable is in the radicand of a square root is called a radical equation

rational exponents

- If $\sqrt[n]{a}$ is a real number and $n \geq 2$, $a^{\frac{1}{n}} = \sqrt[n]{a}$.

- For any positive integers m and n, $a^{\frac{m}{n}} = (\sqrt[n]{a})^m$ and $a^{\frac{m}{n}} = \sqrt[n]{a^m}$.

rationalizing the denominator The process of converting a fraction with a radical in the denominator to an equivalent fraction whose denominator is an integer is called rationalizing the denominator.

square of a number

- If $n^2 = m$, then m is the square of n

square root notation

- If $m = n^2$, then $\sqrt{m} = n$. We read $\sqrt{m}$ as 'the square root of m.'

square root of a number

- If $n^2 = m$, then n is a square root of m

KEY CONCEPTS

9.1 Simplify and Use Square Roots

- Note that the square root of a negative number is not a real number.
- Every positive number has two square roots, one positive and one negative. The positive square root of a positive number is the principal square root.
- We can estimate square roots using nearby perfect squares.
- We can approximate square roots using a calculator.
- When we use the radical sign to take the square root of a variable expression, we should specify that $x \geq 0$ to make sure we get the principal square root.

9.2 Simplify Square Roots

- **Simplified Square Root** $\sqrt{a}$ is considered simplified if a has no perfect-square factors.
- **Product Property of Square Roots** If a, b are non-negative real numbers, then
$$\sqrt{ab} = \sqrt{a} \cdot \sqrt{b}$$
- **Simplify a Square Root Using the Product Property** To simplify a square root using the Product Property:
  Step 1. Find the largest perfect square factor of the radicand. Rewrite the radicand as a product using the perfect square factor.
  Step 2. Use the product rule to rewrite the radical as the product of two radicals.
  Step 3. Simplify the square root of the perfect square.
- **Quotient Property of Square Roots** If a, b are non-negative real numbers and $b \neq 0$, then
$$\sqrt{\frac{a}{b}} = \frac{\sqrt{a}}{\sqrt{b}}$$

- **Simplify a Square Root Using the Quotient Property** To simplify a square root using the Quotient Property:

  Step 1. Simplify the fraction in the radicand, if possible.

  Step 2. Use the Quotient Rule to rewrite the radical as the quotient of two radicals.

  Step 3. Simplify the radicals in the numerator and the denominator.

9.3 Add and Subtract Square Roots

- To add or subtract like square roots, add or subtract the coefficients and keep the like square root.
- Sometimes when we have to add or subtract square roots that do not appear to have like radicals, we find like radicals after simplifying the square roots.

9.4 Multiply Square Roots

- **Product Property of Square Roots** If a, b are nonnegative real numbers, then
$$\sqrt{ab} = \sqrt{a} \cdot \sqrt{b} \quad \text{and} \quad \sqrt{a} \cdot \sqrt{b} = \sqrt{ab}$$
- **Special formulas** for multiplying binomials and conjugates:
$$(a+b)^2 = a^2 + 2ab + b^2 \quad (a-b)(a+b) = a^2 - b^2$$
$$(a-b)^2 = a^2 - 2ab + b^2$$
- The FOIL method can be used to multiply binomials containing radicals.

9.5 Divide Square Roots

- **Quotient Property of Square Roots**

  ◦ If a, b are non-negative real numbers and $b \neq 0$, then
$$\sqrt{\frac{a}{b}} = \frac{\sqrt{a}}{\sqrt{b}} \quad \text{and} \quad \frac{\sqrt{a}}{\sqrt{b}} = \sqrt{\frac{a}{b}}$$

- **Simplified Square Roots**
  A square root is considered simplified if there are
  ◦ no perfect square factors in the radicand
  ◦ no fractions in the radicand
  ◦ no square roots in the denominator of a fraction

9.6 Solve Equations with Square Roots

- **To Solve a Radical Equation:**

  Step 1. Isolate the radical on one side of the equation.

  Step 2. Square both sides of the equation.

  Step 3. Solve the new equation.

  Step 4. Check the answer. Some solutions obtained may not work in the original equation.

- **Solving Applications with Formulas**

  Step 1. **Read** the problem and make sure all the words and ideas are understood. When appropriate, draw a figure and label it with the given information.

  Step 2. **Identify** what we are looking for.

  Step 3. **Name** what we are looking for by choosing a variable to represent it.

  Step 4. **Translate** into an equation by writing the appropriate formula or model for the situation. Substitute in the given information.

  Step 5. **Solve the equation** using good algebra techniques.

  Step 6. **Check** the answer in the problem and make sure it makes sense.

  Step 7. **Answer** the question with a complete sentence.

- **Area of a Square**

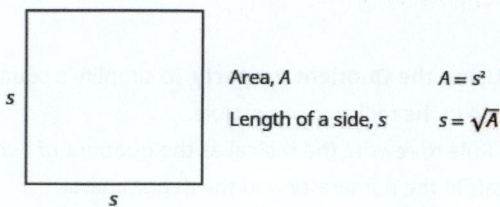

Area, A $A = s^2$

Length of a side, s $s = \sqrt{A}$

- **Falling Objects**
  - On Earth, if an object is dropped from a height of h feet, the time in seconds it will take to reach the ground is found by using the formula $t = \dfrac{\sqrt{h}}{4}$.

- **Skid Marks and Speed of a Car**
  - If the length of the skid marks is d feet, then the speed, s, of the car before the brakes were applied can be found by using the formula $s = \sqrt{24d}$.

9.7 Higher Roots

- **Properties of**
- $\sqrt[n]{a}$ when n is an even number and

  - $a \geq 0$, then $\sqrt[n]{a}$ is a real number

  - $a < 0$, then $\sqrt[n]{a}$ is not a real number

  - When n is an odd number, $\sqrt[n]{a}$ is a real number for all values of a.

  - For any integer $n \geq 2$, when n is odd $\sqrt[n]{a^n} = a$

  - For any integer $n \geq 2$, when n is even $\sqrt[n]{a^n} = |a|$

- $\sqrt[n]{a}$ is considered simplified if a has no factors of m^n.
- **Product Property of nth Roots**

$$\sqrt[n]{ab} = \sqrt[n]{a} \cdot \sqrt[n]{b} \ \text{ and } \ \sqrt[n]{a} \cdot \sqrt[n]{b} = \sqrt[n]{ab}$$

- **Quotient Property of nth Roots**

$$\sqrt[n]{\frac{a}{b}} = \frac{\sqrt[n]{a}}{\sqrt[n]{b}} \ \text{ and } \ \frac{\sqrt[n]{a}}{\sqrt[n]{b}} = \sqrt[n]{\frac{a}{b}}$$

- To combine like radicals, simply add or subtract the coefficients while keeping the radical the same.

9.8 Rational Exponents

- **Summary of Exponent Properties**
- If a, b are real numbers and m, n are rational numbers, then

  - **Product Property** $a^m \cdot a^n = a^{m+n}$

  - **Power Property** $(a^m)^n = a^{m \cdot n}$

  - **Product to a Power** $(ab)^m = a^m b^m$

  - **Quotient Property**:

$$\frac{a^m}{a^n} = a^{m-n}, \ \ a \neq 0, \ \ m > n$$

$$\frac{a^m}{a^n} = \frac{1}{a^{n-m}}, \ \ a \neq 0, \ \ n > m$$

  - **Zero Exponent Definition** $a^0 = 1$, $a \neq 0$

$$\circ \quad \textbf{Quotient to a Power Property} \left(\frac{a}{b}\right)^m = \frac{a^m}{b^m}, \quad b \neq 0$$

REVIEW EXERCISES

9.1 Section 9.1 Simplify and Use Square Roots

Simplify Expressions with Square Roots

In the following exercises, simplify.

606. $\sqrt{64}$

607. $\sqrt{144}$

608. $-\sqrt{25}$

609. $-\sqrt{81}$

610. $\sqrt{-9}$

611. $\sqrt{-36}$

612. $\sqrt{64} + \sqrt{225}$

613. $\sqrt{64 + 225}$

Estimate Square Roots

In the following exercises, estimate each square root between two consecutive whole numbers.

614. $\sqrt{28}$

615. $\sqrt{155}$

Approximate Square Roots

In the following exercises, approximate each square root and round to two decimal places.

616. $\sqrt{15}$

617. $\sqrt{57}$

Simplify Variable Expressions with Square Roots

In the following exercises, simplify.

618. $\sqrt{q^2}$

619. $\sqrt{64b^2}$

620. $-\sqrt{121a^2}$

621. $\sqrt{225m^2 n^2}$

622. $-\sqrt{100q^2}$

623. $\sqrt{49y^2}$

624. $\sqrt{4a^2 b^2}$

625. $\sqrt{121c^2 d^2}$

9.2 Section 9.2 Simplify Square Roots

Use the Product Property to Simplify Square Roots

In the following exercises, simplify.

626. $\sqrt{300}$

627. $\sqrt{98}$

628. $\sqrt{x^{13}}$

629. $\sqrt{y^{19}}$

630. $\sqrt{16m^4}$

631. $\sqrt{36n^{13}}$

632. $\sqrt{288m^{21}}$

633. $\sqrt{150n^7}$

634. $\sqrt{48r^5 s^4}$

635. $\sqrt{108r^5 s^3}$

636. $\dfrac{10 - \sqrt{50}}{5}$

637. $\dfrac{6 + \sqrt{72}}{6}$

Use the Quotient Property to Simplify Square Roots

In the following exercises, simplify.

638. $\sqrt{\dfrac{16}{25}}$

639. $\sqrt{\dfrac{81}{36}}$

640. $\sqrt{\dfrac{x^8}{x^4}}$

641. $\sqrt{\dfrac{y^6}{y^2}}$

642. $\sqrt{\dfrac{98p^6}{2p^2}}$

643. $\sqrt{\dfrac{72q^8}{2q^4}}$

644. $\sqrt{\dfrac{65}{121}}$

645. $\sqrt{\dfrac{26}{169}}$

646. $\sqrt{\dfrac{64x^4}{25x^2}}$

647. $\sqrt{\dfrac{36r^{10}}{16r^5}}$

648. $\sqrt{\dfrac{48p^3q^5}{27pq}}$

649. $\sqrt{\dfrac{12r^5s^7}{75r^2s}}$

9.3 Section 9.3 Add and Subtract Square Roots

Add and Subtract Like Square Roots

In the following exercises, simplify.

650. $3\sqrt{2} + \sqrt{2}$

651. $5\sqrt{5} + 7\sqrt{5}$

652. $4\sqrt{y} + 4\sqrt{y}$

653. $6\sqrt{m} - 2\sqrt{m}$

654. $-3\sqrt{7} + 2\sqrt{7} - \sqrt{7}$

655. $8\sqrt{13} + 2\sqrt{3} + 3\sqrt{13}$

656. $3\sqrt{5xy} - \sqrt{5xy} + 3\sqrt{5xy}$

657. $2\sqrt{3rs} + \sqrt{3rs} - 5\sqrt{rs}$

Add and Subtract Square Roots that Need Simplification

In the following exercises, simplify.

658. $\sqrt{32} + 3\sqrt{2}$

659. $\sqrt{8} + 3\sqrt{2}$

660. $\sqrt{72} + \sqrt{50}$

661. $\sqrt{48} + \sqrt{75}$

662. $3\sqrt{32} + \sqrt{98}$

663. $\dfrac{1}{3}\sqrt{27} - \dfrac{1}{8}\sqrt{192}$

664. $\sqrt{50y^5} - \sqrt{72y^5}$

665. $6\sqrt{18n^4} - 3\sqrt{8n^4} + n^2\sqrt{50}$

9.4 Section 9.4 Multiply Square Roots

Multiply Square Roots

In the following exercises, simplify.

666. $\sqrt{2} \cdot \sqrt{20}$

667. $2\sqrt{2} \cdot 6\sqrt{14}$

668. $\sqrt{2m^2} \cdot \sqrt{20m^4}$

669. $\left(6\sqrt{2y}\right)\left(3\sqrt{50y^3}\right)$

670. $\left(6\sqrt{3v^4}\right)\left(5\sqrt{30v}\right)$

671. $\left(\sqrt{8}\right)^2$

672. $\left(-\sqrt{10}\right)^2$

673. $\left(2\sqrt{5}\right)\left(5\sqrt{5}\right)$

674. $\left(-3\sqrt{3}\right)\left(5\sqrt{18}\right)$

Use Polynomial Multiplication to Multiply Square Roots

In the following exercises, simplify.

675. $10\left(2 - \sqrt{7}\right)$

676. $\sqrt{3}\left(4 + \sqrt{12}\right)$

677. $\left(5 + \sqrt{2}\right)\left(3 - \sqrt{2}\right)$

678. $\left(5 - 3\sqrt{7}\right)\left(1 - 2\sqrt{7}\right)$

679. $\left(1 - 3\sqrt{x}\right)\left(5 + 2\sqrt{x}\right)$

680. $\left(3 + 4\sqrt{y}\right)\left(10 - \sqrt{y}\right)$

681. $\left(1 + 6\sqrt{p}\right)^2$

682. $\left(2 - 6\sqrt{5}\right)^2$

683. $\left(3 + 2\sqrt{7}\right)\left(3 - 2\sqrt{7}\right)$

684. $\left(6 - \sqrt{11}\right)\left(6 + \sqrt{11}\right)$

9.5 Section 9.5 Divide Square Roots

Divide Square Roots

In the following exercises, simplify.

685. $\dfrac{\sqrt{75}}{10}$

686. $\dfrac{2-\sqrt{12}}{6}$

687. $\dfrac{\sqrt{48}}{\sqrt{27}}$

688. $\dfrac{\sqrt{75x^7}}{\sqrt{3x^3}}$

689. $\dfrac{\sqrt{20y^5}}{\sqrt{2y}}$

690. $\dfrac{\sqrt{98p^6q^4}}{\sqrt{2p^4q^8}}$

Rationalize a One Term Denominator

In the following exercises, rationalize the denominator.

691. $\dfrac{10}{\sqrt{15}}$

692. $\dfrac{6}{\sqrt{6}}$

693. $\dfrac{5}{3\sqrt{5}}$

694. $\dfrac{10}{2\sqrt{6}}$

695. $\sqrt{\dfrac{3}{28}}$

696. $\sqrt{\dfrac{9}{75}}$

Rationalize a Two Term Denominator

In the following exercises, rationalize the denominator.

697. $\dfrac{4}{4+\sqrt{27}}$

698. $\dfrac{5}{2-\sqrt{10}}$

699. $\dfrac{4}{2-\sqrt{5}}$

700. $\dfrac{5}{4-\sqrt{8}}$

701. $\dfrac{\sqrt{2}}{\sqrt{p}+\sqrt{3}}$

702. $\dfrac{\sqrt{x}-\sqrt{2}}{\sqrt{x}+\sqrt{2}}$

9.6 Section 9.6 Solve Equations with Square Roots

Solve Radical Equations

In the following exercises, solve the equation.

703. $\sqrt{7z+1}=6$

704. $\sqrt{4u-2}-4=0$

705. $\sqrt{6m+4}-5=0$

706. $\sqrt{2u-3}+2=0$

707. $\sqrt{u-4}+4=u$

708. $\sqrt{v-9}+9=0$

709. $\sqrt{r-4}-r=-10$

710. $\sqrt{s-9}-s=-9$

711. $2\sqrt{2x-7}-4=8$

712. $\sqrt{2-x}=\sqrt{2x-7}$

713. $\sqrt{a}+3=\sqrt{a+9}$

714. $\sqrt{r}+3=\sqrt{r+4}$

715. $\sqrt{u}+2=\sqrt{u+5}$

716. $\sqrt{n+11}-1=\sqrt{n+4}$

717. $\sqrt{y+5}+1=\sqrt{2y+3}$

Use Square Roots in Applications

In the following exercises, solve. Round approximations to one decimal place.

718. A pallet of sod will cover an area of about 600 square feet. Trinh wants to order a pallet of sod to make a square lawn in his backyard. Use the formula $s=\sqrt{A}$ to find the length of each side of his lawn.

719. A helicopter dropped a package from a height of 900 feet above a stranded hiker. Use the formula $t=\dfrac{\sqrt{h}}{4}$ to find how many seconds it took for the package to reach the hiker.

720. Officer Morales measured the skid marks of one of the cars involved in an accident. The length of the skid marks was 245 feet. Use the formula $s=\sqrt{24d}$ to find the speed of the car before the brakes were applied.

9.7 Section 9.7 Higher Roots

Simplify Expressions with Higher Roots

In the following exercises, simplify.

721.

ⓐ $\sqrt[6]{64}$

ⓑ $\sqrt[3]{64}$

722.

ⓐ $\sqrt[3]{-27}$

ⓑ $\sqrt[4]{-64}$

723.

ⓐ $\sqrt[9]{d^9}$

ⓑ $\sqrt[8]{v^8}$

724.

ⓐ $\sqrt[5]{a^{10}}$

ⓑ $\sqrt[3]{b^{27}}$

725.

ⓐ $\sqrt[4]{16x^8}$

ⓑ $\sqrt[6]{64y^{12}}$

726.

ⓐ $\sqrt[7]{128r^{14}}$

ⓑ $\sqrt[4]{81s^{24}}$

Use the Product Property to Simplify Expressions with Higher Roots

In the following exercises, simplify.

727.

ⓐ $\sqrt[9]{d^9}$

ⓑ $\sqrt[11]{m^{17}}$

728.

ⓐ $\sqrt[3]{54}$

ⓑ $\sqrt[4]{128}$

729.

ⓐ $\sqrt[5]{64c^8}$

ⓑ $\sqrt[4]{48d^7}$

730.

ⓐ $\sqrt[3]{343q^7}$

ⓑ $\sqrt[6]{192r^9}$

731.

ⓐ $\sqrt[3]{-500}$

ⓑ $\sqrt[4]{-16}$

Use the Quotient Property to Simplify Expressions with Higher Roots

In the following exercises, simplify.

732. $\sqrt[5]{\dfrac{r^{10}}{r^5}}$

733. $\sqrt[3]{\dfrac{w^{12}}{w^2}}$

734. $\sqrt[4]{\dfrac{64y^8}{4y^5}}$

735. $\sqrt[3]{\dfrac{54z^9}{2z^3}}$

736. $\sqrt[6]{\dfrac{64a^7}{b^2}}$

Add and Subtract Higher Roots

In the following exercises, simplify.

737. $4\sqrt[5]{20} - 2\sqrt[5]{20}$

738. $4\sqrt[3]{18} + 3\sqrt[3]{18}$

739. $\sqrt[4]{1250} - \sqrt[4]{162}$

740. $\sqrt[3]{640c^5} - \sqrt[3]{-80c^3}$

741. $\sqrt[5]{96t^8} + \sqrt[5]{486t^4}$

9.8 Section 9.8 Rational Exponents

Simplify Expressions with $a^{\frac{1}{n}}$

In the following exercises, write as a radical expression.

742. $r^{\frac{1}{8}}$

743. $s^{\frac{1}{10}}$

In the following exercises, write with a rational exponent.

744. $\sqrt[5]{u}$

745. $\sqrt[6]{v}$

746. $\sqrt[3]{9m}$

747. $\sqrt[6]{10z}$

In the following exercises, simplify.

748. $16^{\frac{1}{4}}$

749. $32^{\frac{1}{5}}$

750. $(-125)^{\frac{1}{3}}$

751. $(125)^{-\frac{1}{3}}$

752. $(-9)^{\frac{1}{2}}$

753. $(36)^{-\frac{1}{2}}$

Simplify Expressions with $a^{\frac{m}{n}}$

In the following exercises, write with a rational exponent.

754. $\sqrt[3]{q^5}$

755. $\sqrt[5]{n^8}$

In the following exercises, simplify.

756. $27^{-\frac{2}{3}}$

757. $64^{\frac{5}{2}}$

758. $36^{\frac{3}{2}}$

759. $81^{-\frac{5}{2}}$

Use the Laws of Exponents to Simplify Expressions with Rational Exponents

In the following exercises, simplify.

760. $3^{\frac{4}{5}} \cdot 3^{\frac{6}{5}}$

761. $\left(x^6\right)^{\frac{4}{3}}$

762. $\dfrac{z^{\frac{5}{2}}}{z^{\frac{7}{5}}}$

763. $\left(16s^{\frac{9}{4}}\right)^{\frac{1}{4}}$

764. $\left(m^8 n^{12}\right)^{\frac{1}{4}}$

765. $\dfrac{z^{\frac{2}{3}} \cdot z^{-\frac{1}{3}}}{z^{-\frac{5}{3}}}$

PRACTICE TEST

In the following exercises, simplify.

766. $\sqrt{81 + 144}$

767. $\sqrt{169m^4 n^2}$

768. $\sqrt{36n^{13}}$

769. $3\sqrt{13} + 5\sqrt{2} + \sqrt{13}$

770. $5\sqrt{20} + 2\sqrt{125}$

771. $\left(3\sqrt{6y}\right)\left(2\sqrt{50y^3}\right)$

772. $(2 - 5\sqrt{x})(3 + \sqrt{x})$

773. $(1 - 2\sqrt{q})^2$

774.

ⓐ $\sqrt[4]{a^{12}}$

ⓑ $\sqrt[3]{b^{21}}$

775.

ⓐ $\sqrt[4]{81x^{12}}$

ⓑ $\sqrt[6]{64y^{18}}$

776. $\sqrt{\dfrac{64r^{12}}{25r^6}}$

777. $\sqrt{\dfrac{14y^3}{7y}}$

778. $\dfrac{\sqrt[5]{256x^7}}{\sqrt[5]{4x^2}}$

779. $\sqrt[4]{512} - 2\sqrt[4]{32}$

780.

ⓐ $256^{\frac{1}{4}}$

ⓑ $243^{\frac{1}{5}}$

781. $49^{\frac{3}{2}}$

782. $25^{-\frac{5}{2}}$

783. $\dfrac{w^{\frac{3}{4}}}{w^{\frac{7}{4}}}$

784. $\left(27s^{\frac{3}{5}}\right)^{\frac{1}{3}}$

In the following exercises, rationalize the denominator.

785. $\dfrac{3}{2\sqrt{6}}$

786. $\dfrac{\sqrt{3}}{\sqrt{x} + \sqrt{5}}$

In the following exercises, solve.

787. $3\sqrt{2x - 3} - 20 = 7$

788. $\sqrt{3u - 2} = \sqrt{5u + 1}$

In the following exercise, solve.

789. A helicopter flying at an altitude of 600 feet dropped a package to a lifeboat. Use the formula $t = \dfrac{\sqrt{h}}{4}$ to find how many seconds it took for the package to reach the hiker. Round your answer to the nearest tenth of a second.

Figure 10.1 Fireworks accompany festive celebrations around the world. (Credit: modification of work by tlc, Flickr)

Chapter Outline

10.1 Solve Quadratic Equations Using the Square Root Property

10.2 Solve Quadratic Equations by Completing the Square

10.3 Solve Quadratic Equations Using the Quadratic Formula

10.4 Solve Applications Modeled by Quadratic Equations

10.5 Graphing Quadratic Equations

 ## Introduction

The trajectories of fireworks are modeled by quadratic equations. The equations can be used to predict the maximum height of a firework and the number of seconds it will take from launch to explosion. In this chapter, we will study the properties of quadratic equations, solve them, graph them, and see how they are applied as models of various situations.

 ## Solve Quadratic Equations Using the Square Root Property

Learning Objectives

By the end of this section, you will be able to:

› Solve quadratic equations of the form $ax^2 = k$ using the Square Root Property

› Solve quadratic equations of the form $a(x - h)^2 = k$ using the Square Root Property

---

Be Prepared!

Before you get started, take this readiness quiz.

1. Simplify: $\sqrt{75}$.
   If you missed this problem, review **Example 9.12**.

2. Simplify: $\sqrt{\dfrac{64}{3}}$.
   If you missed this problem, review **Example 9.67**.

3. Factor: $4x^2 - 12x + 9$.
   If you missed this problem, review **Example 7.43**.

Quadratic equations are equations of the form $ax^2 + bx + c = 0$, where $a \neq 0$. They differ from linear equations by including a term with the variable raised to the second power. We use different methods to solve **quadratic equations** than linear equations, because just adding, subtracting, multiplying, and dividing terms will not isolate the variable.

We have seen that some quadratic equations can be solved by factoring. In this chapter, we will use three other methods to solve quadratic equations.

Solve Quadratic Equations of the Form $ax^2 = k$ Using the Square Root Property

We have already solved some quadratic equations by factoring. Let's review how we used factoring to solve the quadratic equation $x^2 = 9$.

$$x^2 = 9$$

Put the equation in standard form.	$x^2 - 9 = 0$
Factor the left side.	$(x-3)(x+3) = 0$
Use the Zero Product Property.	$(x-3) = 0, \quad (x+3) = 0$
Solve each equation.	$x = 3, \quad x = -3$
Combine the two solutions into $\pm$ form.	$x = \pm 3$

(The solution is read 'x is equal to positive or negative 3.')

We can easily use factoring to find the solutions of similar equations, like $x^2 = 16$ and $x^2 = 25$, because 16 and 25 are perfect squares. But what happens when we have an equation like $x^2 = 7$? Since 7 is not a perfect square, we cannot solve the equation by factoring.

These equations are all of the form $x^2 = k$.
We defined the square root of a number in this way:

$$\text{If } n^2 = m, \text{ then } n \text{ is a square root of } m.$$

This leads to the **Square Root Property**.

Square Root Property

If $x^2 = k$, and $k \geq 0$, then $x = \sqrt{k}$ or $x = -\sqrt{k}$.

Notice that the Square Root Property gives two solutions to an equation of the form $x^2 = k$: the principal square root of k and its opposite. We could also write the solution as $x = \pm \sqrt{k}$.

Now, we will solve the equation $x^2 = 9$ again, this time using the Square Root Property.

$$x^2 = 9$$

Use the Square Root Property.	$x = \pm \sqrt{9}$
Simplify the radical.	$x = \pm 3$
Rewrite to show the two solutions.	$x = 3, \ x = -3$

What happens when the constant is not a perfect square? Let's use the Square Root Property to solve the equation $x^2 = 7$.

$$x^2 = 7$$

Use the Square Root Property.	$x = \pm \sqrt{7}$
Rewrite to show two solutions.	$x = \sqrt{7}, \quad x = -\sqrt{7}$

We cannot simplify $\sqrt{7}$, so we leave the answer as a radical.

EXAMPLE 10.1

Solve: $x^2 = 169$.

⊘ Solution

$$x^2 = 169$$

Use the Square Root Property.	$x = \pm \sqrt{169}$
Simplify the radical.	$x = \pm 13$
Rewrite to show two solutions.	$x = 13, \quad x = -13$

> **TRY IT :: 10.1**
Solve: $x^2 = 81$.

> **TRY IT :: 10.2**
Solve: $y^2 = 121$.

EXAMPLE 10.2 HOW TO SOLVE A QUADRATIC EQUATION OF THE FORM $ax^2 = k$ USING THE SQUARE ROOT PROPERTY

Solve: $x^2 - 48 = 0$.

✓ **Solution**

Step 1. Isolate the quadratic term and make its coefficient one.	Add 48 to both sides to get x^2 by itself.	$x^2 - 48 = 0$ $x^2 = 48$
Step 2. Use the Square Root Property.	Remember to add the ± symbol.	$x = \pm\sqrt{48}$
Step 3. Simplify the radical.		$x = \pm\sqrt{16} \cdot \sqrt{3}$ $x = \pm 4\sqrt{3}$ $x = 4\sqrt{3}, x = -4\sqrt{3}$
Step 4. Check the solutions.	Substitute in $x = 4\sqrt{3}$ and $x = -4\sqrt{3}$.	$x^2 - 48 = 0$ $\left(4\sqrt{3}\right)^2 - 48 \stackrel{?}{=} 0$ $16 \cdot 3 - 48 \stackrel{?}{=} 0$ $0 = 0 \checkmark$ $x^2 - 48 = 0$ $\left(-4\sqrt{3}\right)^2 - 48 \stackrel{?}{=} 0$ $16 \cdot 3 - 48 \stackrel{?}{=} 0$ $0 = 0 \checkmark$

> **TRY IT :: 10.3**
Solve: $x^2 - 50 = 0$.

> **TRY IT :: 10.4**
Solve: $y^2 - 27 = 0$.

HOW TO :: SOLVE A QUADRATIC EQUATION USING THE SQUARE ROOT PROPERTY.

Step 1. Isolate the quadratic term and make its coefficient one.

Step 2. Use Square Root Property.

Step 3. Simplify the radical.

Step 4. Check the solutions.

To use the Square Root Property, the coefficient of the variable term must equal 1. In the next example, we must divide both sides of the equation by 5 before using the Square Root Property.

EXAMPLE 10.3

Solve: $5m^2 = 80$.

⊘ Solution

The quadratic term is isolated.	$5m^2 = 80$
Divide by 5 to make its cofficient 1.	$\dfrac{5m^2}{5} = \dfrac{80}{5}$
Simplify.	$m^2 = 16$
Use the Square Root Property.	$m = \pm\sqrt{16}$
Simplify the radical.	$m = \pm 4$
Rewrite to show two solutions.	$m = 4,\ m = -4$

Check the solutions.

$$5m^2 = 80 \qquad\qquad 5m^2 = 80$$
$$5(4)^2 \overset{?}{=} 80 \qquad 5(-4)^2 \overset{?}{=} 80$$
$$5 \cdot 16 \overset{?}{=} 80 \qquad 5 \cdot 16 \overset{?}{=} 80$$
$$80 = 80 \checkmark \qquad\quad 80 = 80 \checkmark$$

> **TRY IT ::** 10.5   Solve: $2x^2 = 98$.

> **TRY IT ::** 10.6   Solve: $3z^2 = 108$.

The Square Root Property started by stating, 'If $x^2 = k$, and $k \geq 0$'. What will happen if $k < 0$? This will be the case in the next example.

EXAMPLE 10.4

Solve: $q^2 + 24 = 0$.

⊘ Solution

$$q^2 + 24 = 0$$

Isolate the quadratic term. $\qquad q^2 = -24$

Use the Square Root Property. $\qquad q = \pm\sqrt{-24}$

The $\sqrt{-24}$ is not a real number.   There is no real solution.

> **TRY IT ::** 10.7   Solve: $c^2 + 12 = 0$.

> **TRY IT ::** 10.8   Solve: $d^2 + 81 = 0$.

Remember, we first isolate the quadratic term and then make the coefficient equal to one.

EXAMPLE 10.5

Solve: $\dfrac{2}{3}u^2 + 5 = 17$.

⊘ Solution

$$\tfrac{2}{3}u^2 + 5 = 17$$

Isolate the quadratic term.	$\tfrac{2}{3}u^2 = 12$
Multiply by $\tfrac{3}{2}$ to make the coefficient 1.	$\tfrac{3}{2} \cdot \tfrac{2}{3}u^2 = \tfrac{3}{2} \cdot 12$
Simplify.	$u^2 = 18$
Use the Square Root Property.	$u = \pm\sqrt{18}$
Simplify the radical.	$u = \pm\sqrt{9}\sqrt{2}$
Simplify.	$u = \pm 3\sqrt{2}$
Rewrite to show two solutions.	$u = 3\sqrt{2},\ u = -3\sqrt{2}$

Check.

$$\tfrac{2}{3}u^2 + 5 = 17 \qquad\qquad \tfrac{2}{3}u^2 + 5 = 17$$

$$\tfrac{2}{3}(3\sqrt{2})^2 + 5 \overset{?}{=} 17 \qquad \tfrac{2}{3}(-3\sqrt{2})^2 + 5 \overset{?}{=} 17$$

$$\tfrac{2}{3}\cdot 18 + 5 \overset{?}{=} 17 \qquad\qquad \tfrac{2}{3}\cdot 18 + 5 \overset{?}{=} 17$$

$$12 + 5 \overset{?}{=} 17 \qquad\qquad\qquad 12 + 5 \overset{?}{=} 17$$

$$17 = 17\ \checkmark \qquad\qquad\qquad\quad 17 = 17\ \checkmark$$

> **TRY IT :: 10.9** Solve: $\tfrac{1}{2}x^2 + 4 = 24$.

> **TRY IT :: 10.10** Solve: $\tfrac{3}{4}y^2 - 3 = 18$.

The solutions to some equations may have fractions inside the radicals. When this happens, we must rationalize the denominator.

EXAMPLE 10.6

Solve: $2c^2 - 4 = 45$.

⊘ Solution

$$2c^2 - 4 = 45$$

Isolate the quadratic term.	$2c^2 = 49$
Divide by 2 to make the coefficient 1.	$\dfrac{2c^2}{2} = \dfrac{49}{2}$
Simplify.	$c^2 = \dfrac{49}{2}$
Use the Square Root Property.	$c = \pm\sqrt{\dfrac{49}{2}}$
Simplify the radical.	$c = \pm\dfrac{\sqrt{49}}{\sqrt{2}}$
Rationalize the denominator.	$c = \pm\dfrac{\sqrt{49}\cdot\sqrt{2}}{\sqrt{2}\cdot\sqrt{2}}$
Simplify.	$c = \pm\dfrac{7\sqrt{2}}{2}$
Rewrite to show two solutions.	$c = \dfrac{7\sqrt{2}}{2}, \quad c = -\dfrac{7\sqrt{2}}{2}$

Check. We leave the check for you.

> **TRY IT : : 10.11**   Solve: $5r^2 - 2 = 34$.

> **TRY IT : : 10.12**   Solve: $3t^2 + 6 = 70$.

Solve Quadratic Equations of the Form $a(x - h)^2 = k$ Using the Square Root Property

We can use the Square Root Property to solve an equation like $(x - 3)^2 = 16$, too. We will treat the whole binomial, $(x - 3)$, as the quadratic term.

EXAMPLE 10.7

Solve: $(x - 3)^2 = 16$.

⊘ Solution

$$(x - 3)^2 = 16$$

Use the Square Root Property.	$x - 3 = \pm\sqrt{16}$
Simplify.	$x - 3 = \pm 4$
Write as two equations.	$x - 3 = 4, \; x - 3 = -4$
Solve.	$x = 7, \; x = -1$

Check.

$$
\begin{array}{ll}
(7 - 3)^2 = 16 & (-1 - 3)^2 = 16 \\
(4)^2 = 16 & (-4)^2 = 16 \\
16 = 16 \checkmark & 16 = 16 \checkmark
\end{array}
$$

> **TRY IT : : 10.13**   Solve: $(q + 5)^2 = 1$.

TRY IT :: 10.14   Solve: $(r - 3)^2 = 25$.

EXAMPLE 10.8

Solve: $(y - 7)^2 = 12$.

✓ **Solution**

$$(y - 7)^2 = 12$$

Use the Square Root Property.	$y - 7 = \pm \sqrt{12}$
Simplify the radical.	$y - 7 = \pm 2\sqrt{3}$
Solve for y.	$y = 7 \pm 2\sqrt{3}$
Rewrite to show two solutions.	$y = 7 + 2\sqrt{3}, \; y = 7 - 2\sqrt{3}$

Check.

$(y - 7)^2 = 12$	$(y - 7)^2 = 12$
$(7 + 2\sqrt{3} - 7)^2 \overset{?}{=} 12$	$(7 - 2\sqrt{3} - 7)^2 \overset{?}{=} 12$
$(2\sqrt{3})^2 \overset{?}{=} 12$	$(-2\sqrt{3})^2 \overset{?}{=} 12$
$12 = 12 \checkmark$	$12 = 12 \checkmark$

 **TRY IT :: 10.15**   Solve: $(a - 3)^2 = 18$.

 **TRY IT :: 10.16**   Solve: $(b + 2)^2 = 40$.

Remember, when we take the square root of a fraction, we can take the square root of the numerator and denominator separately.

EXAMPLE 10.9

Solve: $\left(x - \frac{1}{2}\right)^2 = \frac{5}{4}$.

✓ **Solution**

$$\left(x - \frac{1}{2}\right)^2 = \frac{5}{4}$$

Use the Square Root Property.	$x - \frac{1}{2} = \pm \sqrt{\frac{5}{4}}$
Rewrite the radical as a fraction of square roots.	$x - \frac{1}{2} = \pm \frac{\sqrt{5}}{\sqrt{4}}$
Simplify the radical.	$x - \frac{1}{2} = \pm \frac{\sqrt{5}}{2}$
Solve for x.	$x = \frac{1}{2} \pm \frac{\sqrt{5}}{2}$
Rewrite to show two solutions.	$x = \frac{1}{2} + \frac{\sqrt{5}}{2}, \quad x = \frac{1}{2} - \frac{\sqrt{5}}{2}$

Check. We leave the check for you.

> **TRY IT : : 10.17**

Solve: $\left(x - \frac{1}{3}\right)^2 = \frac{5}{9}$.

> **TRY IT : : 10.18**

Solve: $\left(y - \frac{3}{4}\right)^2 = \frac{7}{16}$.

We will start the solution to the next example by isolating the binomial.

EXAMPLE 10.10

Solve: $(x - 2)^2 + 3 = 30$.

Solution

$$(x - 2)^2 + 3 = 30$$

Isolate the binomial term. $(x - 2)^2 = 27$

Use the Square Root Property. $x - 2 = \pm\sqrt{27}$

Simplify the radical. $x - 2 = \pm 3\sqrt{3}$

Solve for x. $x = 2 \pm 3\sqrt{3}$

Rewrite to show two solutions. $x = 2 + 3\sqrt{3}, \quad x = 2 - 3\sqrt{3}$

Check. We leave the check for you.

> **TRY IT : : 10.19**

Solve: $(a - 5)^2 + 4 = 24$.

> **TRY IT : : 10.20**

Solve: $(b - 3)^2 - 8 = 24$.

EXAMPLE 10.11

Solve: $(3v - 7)^2 = -12$.

Solution

Use the Square Root Property. $(3v - 7)^2 = -12$

     $3v - 7 = \pm\sqrt{-12}$

The $\sqrt{-12}$ is not a real number.   There is no real solution.

> **TRY IT : : 10.21**

Solve: $(3r + 4)^2 = -8$.

> **TRY IT : : 10.22**

Solve: $(2t - 8)^2 = -10$.

The left sides of the equations in the next two examples do not seem to be of the form $a(x - h)^2$. But they are perfect square trinomials, so we will factor to put them in the form we need.

EXAMPLE 10.12

Solve: $p^2 - 10p + 25 = 18$.

⊘ Solution

The left side of the equation is a perfect square trinomial. We will factor it first.

$$p^2 - 10p + 25 = 18$$

Factor the perfect square trinomial. $\qquad (p - 5)^2 = 18$

Use the Square Root Property. $\qquad p - 5 = \pm\sqrt{18}$

Simplify the radical. $\qquad p - 5 = \pm 3\sqrt{2}$

Solve for p. $\qquad p = 5 \pm 3\sqrt{2}$

Rewrite to show two solutions. $\qquad p = 5 + 3\sqrt{2}, \quad p = 5 - 3\sqrt{2}$

Check. We leave the check for you.

> **TRY IT :: 10.23**   Solve: $x^2 - 6x + 9 = 12$.

> **TRY IT :: 10.24**   Solve: $y^2 + 12y + 36 = 32$.

EXAMPLE 10.13

Solve: $4n^2 + 4n + 1 = 16$.

⊘ Solution

Again, we notice the left side of the equation is a perfect square trinomial. We will factor it first.

$$4n^2 + 4n + 1 = 16$$

Factor the perfect square trinomial.	$(2n + 1)^2 = 16$
Use the Square Root Property.	$2n + 1 = \pm\sqrt{16}$
Simplify the radical.	$2n + 1 = \pm 4$
Solve for n.	$2n = -1 \pm 4$
Divide each side by 2.	$\dfrac{2n}{2} = \dfrac{-1 \pm 4}{2}$ $n = \dfrac{-1 \pm 4}{2}$
Rewrite to show two solutions.	$n = \dfrac{-1 + 4}{2},\ n = \dfrac{-1 - 4}{2}$
Simplify each equation.	$n = \dfrac{3}{2},\ n = -\dfrac{5}{2}$

Check.

$$4n^2 + 4n + 1 = 16 \qquad\qquad 4n^2 + 4n + 1 = 16$$

$$4\left(\tfrac{3}{2}\right)^2 + 4\left(\tfrac{3}{2}\right) + 1 \overset{?}{=} 16 \qquad 4\left(-\tfrac{5}{2}\right)^2 + 4\left(-\tfrac{5}{2}\right) + 1 \overset{?}{=} 16$$

$$4\left(\tfrac{9}{4}\right) + 4\left(\tfrac{3}{2}\right) + 1 \overset{?}{=} 16 \qquad 4\left(\tfrac{25}{4}\right) + 4\left(-\tfrac{5}{2}\right) + 1 \overset{?}{=} 16$$

$$9 + 6 + 1 \overset{?}{=} 16 \qquad\qquad 25 - 10 + 1 \overset{?}{=} 16$$

$$16 = 16\ \checkmark \qquad\qquad\qquad 16 = 16\ \checkmark$$

> **TRY IT : :** 10.25   Solve: $9m^2 - 12m + 4 = 25$.

> **TRY IT : :** 10.26   Solve: $16n^2 + 40n + 25 = 4$.

▶ **MEDIA : :**

Access these online resources for additional instruction and practice with solving quadratic equations:

- **Solving Quadratic Equations: Solving by Taking Square Roots (https://openstax.org/l/25Solvebysqroot)**
- **Using Square Roots to Solve Quadratic Equations (https://openstax.org/l/25Usesqroots)**
- **Solving Quadratic Equations: The Square Root Method (https://openstax.org/l/25Sqrtproperty)**

10.1 EXERCISES

Practice Makes Perfect

Solve Quadratic Equations of the form $ax^2 = k$ Using the Square Root Property

In the following exercises, solve the following quadratic equations.

1. $a^2 = 49$

2. $b^2 = 144$

3. $r^2 - 24 = 0$

4. $t^2 - 75 = 0$

5. $u^2 - 300 = 0$

6. $v^2 - 80 = 0$

7. $4m^2 = 36$

8. $3n^2 = 48$

9. $x^2 + 20 = 0$

10. $y^2 + 64 = 0$

11. $\frac{2}{5}a^2 + 3 = 11$

12. $\frac{3}{2}b^2 - 7 = 41$

13. $7p^2 + 10 = 26$

14. $2q^2 + 5 = 30$

Solve Quadratic Equations of the Form $a(x - h)^2 = k$ Using the Square Root Property

In the following exercises, solve the following quadratic equations.

15. $(x + 2)^2 = 9$

16. $(y - 5)^2 = 36$

17. $(u - 6)^2 = 64$

18. $(v + 10)^2 = 121$

19. $(m - 6)^2 = 20$

20. $(n + 5)^2 = 32$

21. $\left(r - \frac{1}{2}\right)^2 = \frac{3}{4}$

22. $\left(t - \frac{5}{6}\right)^2 = \frac{11}{25}$

23. $(a - 7)^2 + 5 = 55$

24. $(b - 1)^2 - 9 = 39$

25. $(5c + 1)^2 = -27$

26. $(8d - 6)^2 = -24$

27. $m^2 - 4m + 4 = 8$

28. $n^2 + 8n + 16 = 27$

29. $25x^2 - 30x + 9 = 36$

30. $9y^2 + 12y + 4 = 9$

Mixed Practice

In the following exercises, solve using the Square Root Property.

31. $2r^2 = 32$

32. $4t^2 = 16$

33. $(a - 4)^2 = 28$

34. $(b + 7)^2 = 8$

35. $9w^2 - 24w + 16 = 1$

36. $4z^2 + 4z + 1 = 49$

37. $a^2 - 18 = 0$

38. $b^2 - 108 = 0$

39. $\left(p - \frac{1}{3}\right)^2 = \frac{7}{9}$

40. $\left(q - \frac{3}{5}\right)^2 = \frac{3}{4}$

41. $m^2 + 12 = 0$

42. $n^2 + 48 = 0$

43. $u^2 - 14u + 49 = 72$

44. $v^2 + 18v + 81 = 50$

45. $(m - 4)^2 + 3 = 15$

46. $(n-7)^2 - 8 = 64$

47. $(x+5)^2 = 4$

48. $(y-4)^2 = 64$

49. $6c^2 + 4 = 29$

50. $2d^2 - 4 = 77$

51. $(x-6)^2 + 7 = 3$

52. $(y-4)^2 + 10 = 9$

Everyday Math

53. Paola has enough mulch to cover 48 square feet. She wants to use it to make three square vegetable gardens of equal sizes. Solve the equation $3s^2 = 48$ to find s, the length of each garden side.

54. Kathy is drawing up the blueprints for a house she is designing. She wants to have four square windows of equal size in the living room, with a total area of 64 square feet. Solve the equation $4s^2 = 64$ to find s, the length of the sides of the windows.

Writing Exercises

55. Explain why the equation $x^2 + 12 = 8$ has no solution.

56. Explain why the equation $y^2 + 8 = 12$ has two solutions.

Self Check

ⓐ *After completing the exercises, use this checklist to evaluate your mastery of the objectives of this section.*

I can...	Confidently	With some help	No-I don't get it!
solve quadratic equations of the form $ax^2 = k$ using the square root property.			
solve quadratic equations of the form $a(x-h)^2 = k$ using the square root property.			

ⓑ *If most of your checks were:*

...confidently: *Congratulations! You have achieved the objectives in this section. Reflect on the study skills you used so that you can continue to use them. What did you do to become confident of your ability to do these things? Be specific.*

...with some help: *This must be addressed quickly because topics you do not master become potholes in your road to success. In math, every topic builds upon previous work. It is important to make sure you have a strong foundation before you move on. Who can you ask for help? Your fellow classmates and instructor are good resources. Is there a place on campus where math tutors are available? Can your study skills be improved?*

...no-I don't get it! *This is a warning sign and you must not ignore it. You should get help right away or you will quickly be overwhelmed. See your instructor as soon as you can to discuss your situation. Together you can come up with a plan to get you the help you need.*

10.2 Solve Quadratic Equations by Completing the Square

Learning Objectives

By the end of this section, you will be able to:

> Complete the square of a binomial expression

> Solve quadratic equations of the form $x^2 + bx + c = 0$ by completing the square

> Solve quadratic equations of the form $ax^2 + bx + c = 0$ by completing the square

Be Prepared!

Before you get started, take this readiness quiz. If you miss a problem, go back to the section listed and review the material.

1. Simplify $(x + 12)^2$.
    If you missed this problem, review **Example 6.47**.

2. Factor $y^2 - 18y + 81$.
    If you missed this problem, review **Example 7.42**.

3. Factor $5n^2 + 40n + 80$.
    If you missed this problem, review **Example 7.46**.

So far, we have solved quadratic equations by factoring and using the Square Root Property. In this section, we will solve quadratic equations by a process called 'completing the square.'

Complete The Square of a Binomial Expression

In the last section, we were able to use the Square Root Property to solve the equation $(y - 7)^2 = 12$ because the left side was a perfect square.

$$\begin{aligned} (y - 7)^2 &= 12 \\ y - 7 &= \pm\sqrt{12} \\ y - 7 &= \pm 2\sqrt{3} \\ y &= 7 \pm 2\sqrt{3} \end{aligned}$$

We also solved an equation in which the left side was a perfect square trinomial, but we had to rewrite it the form $(x - k)^2$ in order to use the square root property.

$$\begin{aligned} x^2 - 10x + 25 &= 18 \\ (x - 5)^2 &= 18 \end{aligned}$$

What happens if the variable is not part of a perfect square? Can we use algebra to make a perfect square?

Let's study the binomial square pattern we have used many times. We will look at two examples.

$$\begin{array}{ll} (x + 9)^2 & (y - 7)^2 \\ (x + 9)(x + 9) & (y - 7)(y - 7) \\ x^2 + 9x + 9x + 81 & y^2 - 7y - 7y + 49 \\ x^2 + 18x + 81 & y^2 - 14y + 49 \end{array}$$

Binomial Squares Pattern

If a, b are real numbers,

$$(a + b)^2 = a^2 + 2ab + b^2$$

$$(a+b)^2 = a^2 + 2ab + b^2$$

(binomial)²   (first term)²   2 × (product of terms)   (second term)²

$$(a-b)^2 = a^2 - 2ab + b^2$$

$$(a-b)^2 = a^2 - 2ab + b^2$$

(binomial)²   (first term)²   2 × (product of terms)   (second term)²

We can use this pattern to "make" a perfect square.

We will start with the expression $x^2 + 6x$. Since there is a plus sign between the two terms, we will use the $(a+b)^2$ pattern.

$$a^2 + 2ab + b^2 = (a+b)^2$$

Notice that the first term of $x^2 + 6x$ is a square, x^2.

We now know $a = x$.

What number can we add to $x^2 + 6x$ to make a perfect square trinomial?

$a^2 + 2ab + b^2$
$x^2 + 6x + \underline{}$

The middle term of the Binomial Squares Pattern, $2ab$, is twice the product of the two terms of the binomial. This means twice the product of x and some number is $6x$. So, two times some number must be six. The number we need is $\frac{1}{2} \cdot 6 = 3$. The second term in the binomial, b, must be 3.

$a^2 + \quad 2ab \quad + b^2$
$x^2 + 2 \cdot 3 \cdot x + \underline{}$

We now know $b = 3$.

Now, we just square the second term of the binomial to get the last term of the perfect square trinomial, so we square three to get the last term, nine.

$a^2 + 2ab + b^2$
$x^2 + 6x + 9$

We can now factor to

$(a+b)^2$
$(x+3)^2$

So, we found that adding nine to $x^2 + 6x$ 'completes the square,' and we write it as $(x+3)^2$.

HOW TO :: COMPLETE A SQUARE.

To complete the square of $x^2 + bx$:

Step 1.   Identify b, the coefficient of x.

Step 2.   Find $\left(\frac{1}{2}b\right)^2$, the number to complete the square.

Step 3.   Add the $\left(\frac{1}{2}b\right)^2$ to $x^2 + bx$.

EXAMPLE 10.14

Complete the square to make a perfect square trinomial. Then, write the result as a binomial square.

$x^2 + 14x$

✓ Solution

The coefficient of x is 14.	$x^2 +\ bx$ $x^2 + 14x$

Find $\left(\frac{1}{2}b\right)^2$.

$\left(\frac{1}{2} \cdot 14\right)^2$

$(7)^2$

49

Add 49 to the binomial to complete the square.	$x^2 + 14x + 49$
Rewrite as a binomial square.	$(x + 7)^2$

> **TRY IT : : 10.27**
>
> Complete the square to make a perfect square trinomial. Write the result as a binomial square.
>
> $y^2 + 12y$

> **TRY IT : : 10.28**
>
> Complete the square to make a perfect square trinomial. Write the result as a binomial square.
>
> $z^2 + 8z$

EXAMPLE 10.15

Complete the square to make a perfect square trinomial. Then, write the result as a binomial squared. $m^2 - 26m$

⊘ Solution

The coefficient of m is –26.	$x^2 - bx$ $m^2 - 26m$

Find $\left(\frac{1}{2}b\right)^2$.

$$\left(\frac{1}{2} \cdot \left(-26\right)\right)^2$$
$$(-13)^2$$
$$169$$

Add 169 to the binomial to complete the square.	$m^2 - 26m + 169$

Rewrite as a binomial square.	$(m - 13)^2$

> **TRY IT :: 10.29**
>
> Complete the square to make a perfect square trinomial. Write the result as a binomial square.
>
> $a^2 - 20a$

> **TRY IT :: 10.30**
>
> Complete the square to make a perfect square trinomial. Write the result as a binomial square.
>
> $b^2 - 4b$

EXAMPLE 10.16

Complete the square to make a perfect square trinomial. Then, write the result as a binomial squared.

$u^2 - 9u$

⊘ Solution

The coefficient of u is –9.	$x^2 + bx$ $u^2 - 9u$

Find $\left(\frac{1}{2}b\right)^2$.

$$\left(\frac{1}{2} \cdot \left(-9\right)\right)^2$$
$$\left(-\frac{9}{2}\right)^2$$
$$\frac{81}{4}$$

Add $\frac{81}{4}$ to the binomial to complete the square.	$u^2 - 9u + \frac{81}{4}$

Rewrite as a binomial square.	$\left(u - \frac{9}{2}\right)^2$

> **TRY IT : :** 10.31

Complete the square to make a perfect square trinomial. Write the result as a binomial square.

$m^2 - 5m$

> **TRY IT : :** 10.32

Complete the square to make a perfect square trinomial. Write the result as a binomial square.

$n^2 + 13n$

EXAMPLE 10.17

Complete the square to make a perfect square trinomial. Then, write the result as a binomial squared.

$p^2 + \frac{1}{2}p$

⊘ **Solution**

The coefficient of p is $\frac{1}{2}$.	$\overset{x^2 \ + \ bx}{p^2 + \frac{1}{2}p}$
Find $\left(\frac{1}{2}b\right)^2$.	
$\left(\frac{1}{2} \cdot \frac{1}{2}\right)^2$	
$\left(\frac{1}{4}\right)^2$	
$\frac{1}{16}$	
Add $\frac{1}{16}$ to the binomial to complete the square.	$p^2 + \frac{1}{2}p + \frac{1}{16}$
Rewrite as a binomial square.	$\left(p + \frac{1}{4}\right)^2$

> **TRY IT : :** 10.33

Complete the square to make a perfect square trinomial. Write the result as a binomial square.

$p^2 + \frac{1}{4}p$

> **TRY IT : :** 10.34

Complete the square to make a perfect square trinomial. Write the result as a binomial square.

$q^2 - \frac{2}{3}q$

Solve Quadratic Equations of the Form $x^2 + bx + c = 0$ by completing the square

In solving equations, we must always do the same thing to both sides of the equation. This is true, of course, when we solve a quadratic equation by completing the square, too. When we add a term to one side of the equation to make a perfect square trinomial, we must also add the same term to the other side of the equation.

For example, if we start with the equation $x^2 + 6x = 40$ and we want to complete the square on the left, we will add nine to both sides of the equation.

$$x^2 + 6x = 40$$
$$x^2 + 6x + \underline{} = 40 + \underline{}$$
$$x^2 + 6x + 9 = 40 + 9$$

Then, we factor on the left and simplify on the right.

$$(x + 3)^2 = 49$$

Now the equation is in the form to solve using the Square Root Property. Completing the square is a way to transform an equation into the form we need to be able to use the Square Root Property.

EXAMPLE 10.18   HOW TO SOLVE A QUADRATIC EQUATION OF THE FORM $x^2 + bx + c = 0$ BY COMPLETING THE SQUARE

Solve $x^2 + 8x = 48$ by completing the square.

✓ **Solution**

Step 1. Isolate the variable terms on one side and the constant terms on the other.	This equation has all the variables on the left.	$\overset{x^2 + bx \quad c}{x^2 + 8x = 48}$
Step 2. Find $\left(\frac{1}{2} \cdot b\right)^2$, the number to complete the square. Add it to both sides of the equation.	Take half of 8 and square it. $4^2 = 16$ Add 16 to BOTH sides of the equation.	$x^2 + 8x + \underset{\left(\frac{1}{2} \cdot 8\right)^2}{\underline{}} = 48$ $x^2 + 8x + 16 = 48 + 16$
Step 3. Factor the perfect square trinomial as a binomial square.	$x^2 + 8x + 16 = (x + 4)^2$ Add the terms on the right.	$(x + 4)^2 = 64$
Step 4. Use the Square Root Property.		$x + 4 = \pm\sqrt{64}$
Step 5. Simplify the radical and then solve the two resulting equations.		$x + 4 = \pm 8$ $x + 4 = 8 \quad x + 4 = -8$ $x = 4 \qquad x = -12$
Step 6. Check the solutions.	Put each answer in the original equation to check. Substitute $x = 4$.	$x^2 + 8x = 48$ $(4)^2 + 8(4) \overset{?}{=} 48$ $16 + 32 \overset{?}{=} 48$ $48 = 48 ✓$
	Substitute $x = -12$.	$x^2 + 8x = 48$ $(-12)^2 + 8(-12) \overset{?}{=} 48$ $144 - 96 \overset{?}{=} 48$ $48 = 48 ✓$

> **TRY IT : : 10.35**   Solve $c^2 + 4c = 5$ by completing the square.

> **TRY IT : : 10.36**   Solve $d^2 + 10d = -9$ by completing the square.

HOW TO :: SOLVE A QUADRATIC EQUATION OF THE FORM $x^2 + bx + c = 0$ **BY COMPLETING THE SQUARE.**

Step 1. Isolate the variable terms on one side and the constant terms on the other.

Step 2. Find $\left(\frac{1}{2} \cdot b\right)^2$, the number to complete the square. Add it to both sides of the equation.

Step 3. Factor the perfect square trinomial as a binomial square.

Step 4. Use the Square Root Property.

Step 5. Simplify the radical and then solve the two resulting equations.

Step 6. Check the solutions.

EXAMPLE 10.19

Solve $y^2 - 6y = 16$ by completing the square.

✓ **Solution**

The variable terms are on the left side.	$\overset{x^2-bx}{\overbrace{y^2 - 6y}} = \overset{c}{\overbrace{16}}$
Take half of -6 and square it. $\left(\frac{1}{2}(-6)\right)^2 = 9$	$y^2 - 6y + \underset{\left(\frac{1}{2}\cdot(-6)\right)^2}{\underline{\hspace{2cm}}} = 16$
Add 9 to both sides.	$y^2 - 6y + 9 = 16 + 9$
Factor the perfect square trinomial as a binomial square.	$(y - 3)^2 = 25$
Use the Square Root Property.	$y - 3 = \pm\sqrt{25}$
Simplify the radical.	$y - 3 = \pm 5$
Solve for y.	$y = 3 \pm 5$
Rewrite to show two solutions.	$y = 3 + 5,\ y = 3 - 5$
Solve the equations.	$y = 8,\ y = -2$

Check.

$$y^2 - 6y = 16 \qquad\qquad y^2 - 6y = 16$$
$$8^2 - 6 \cdot 8 \overset{?}{=} 16 \qquad (-2)^2 - 6(-2) \overset{?}{=} 16$$
$$64 - 48 \overset{?}{=} 16 \qquad\qquad 4 + 12 \overset{?}{=} 16$$
$$16 = 16 \checkmark \qquad\qquad\quad 16 = 16 \checkmark$$

> **TRY IT :: 10.37**   Solve $r^2 - 4r = 12$ by completing the square.

> **TRY IT :: 10.38**   Solve $t^2 - 10t = 11$ by completing the square.

EXAMPLE 10.20

Solve $x^2 + 4x = -21$ by completing the square.

✓ Solution

The variable terms are on the left side.	$\overset{x^2+bx\qquad c}{x^2+4x=-21}$
Take half of 4 and square it. $\left(\frac{1}{2}(4)\right)^2=4$	$x^2+4x+\underset{\left(\frac{1}{2}\cdot 4\right)^2}{\underline{\qquad}}=-21$
Add 4 to both sides.	$x^2+4x+4=-21+4$
Factor the perfect square trinomial as a binomial square.	$(x+2)^2=-17$
Use the Square Root Property.	$x+2=\pm\sqrt{-17}$
We cannot take the square root of a negative number.	There is no real solution.

> **TRY IT :: 10.39**   Solve $y^2-10y=-35$ by completing the square.

> **TRY IT :: 10.40**   Solve $z^2+8z=-19$ by completing the square.

In the previous example, there was no real solution because $(x+k)^2$ was equal to a negative number.

EXAMPLE 10.21

Solve $p^2-18p=-6$ by completing the square.

✓ Solution

The variable terms are on the left side.	$\overset{x^2+bx\qquad c}{p^2-18p=-6}$
Take half of -18 and square it. $\left(\frac{1}{2}(-18)\right)^2=81$	$p^2-18p+\underset{\left(\frac{1}{2}\cdot(-18)\right)^2}{\underline{\qquad}}=-6$
Add 81 to both sides.	$p^2-18p+81=-6+81$
Factor the perfect square trinomial as a binomial square.	$(p-9)^2=75$
Use the Square Root Property.	$p-9=\pm\sqrt{75}$
Simplify the radical.	$p-9=\pm 5\sqrt{3}$
Solve for p.	$p=9\pm 5\sqrt{3}$
Rewrite to show two solutions.	$p=9+5\sqrt{3},p=9-5\sqrt{3}$

Check.

$$p^2-18p=-6 \qquad\qquad p^2-18p=-6$$
$$\left(9+5\sqrt{3}\right)^2-18\left(9+5\sqrt{3}\right)\overset{?}{=}-6 \qquad \left(9-5\sqrt{3}\right)^2-18\left(9-5\sqrt{3}\right)\overset{?}{=}-6$$
$$81+90\sqrt{3}+75-162-90\sqrt{3}\overset{?}{=}-6 \qquad 81-90\sqrt{3}+75-162+90\sqrt{3}\overset{?}{=}-6$$
$$-6=-6\checkmark \qquad\qquad\qquad -6=-6\checkmark$$

Another way to check this would be to use a calculator. Evaluate p^2-18p for both of the solutions. The answer should be -6.

> **TRY IT :: 10.41**   Solve $x^2 - 16x = -16$ by completing the square.

> **TRY IT :: 10.42**   Solve $y^2 + 8y = 11$ by completing the square.

We will start the next example by isolating the variable terms on the left side of the equation.

EXAMPLE 10.22

Solve $x^2 + 10x + 4 = 15$ by completing the square.

✓ **Solution**

The variable terms are on the left side.	$x^2 + 10x + 4 = 15$
Subtract 4 to get the constant terms on the right side.	$x^2 + 10x \quad = 11$
Take half of 10 and square it. $\left(\frac{1}{2}(10)\right)^2 = 25$	$x^2 + 10x + \underline{\quad\left(\frac{1}{2}\cdot(10)\right)^2} = 11$
Add 25 to both sides.	$x^2 + 10x + 25 = 11 + 25$
Factor the perfect square trinomial as a binomial square.	$(x + 5)^2 = 36$
Use the Square Root Property.	$x + 5 = \pm\sqrt{36}$
Simplify the radical.	$x + 5 = \pm\sqrt{36}$
Solve for x.	$x = -5 \pm 6$
Rewrite to show two equations.	$x = -5 + 6,\ x = -5 - 6$
Solve the equations.	$x = 1,\ x = -11$

Check.

$x^2 + 10x + 4 = 15$ 　　　　 $x^2 + 10x + 4 = 15$

$(1)^2 + 10(1) + 4 \overset{?}{=} 15$ 　　 $(-11)^2 + 10(-11) + 4 \overset{?}{=} 15$

$1 + 10 + 4 \overset{?}{=} 15$ 　　　　 $121 - 110 + 4 \overset{?}{=} 15$

$15 = 15\ ✓$ 　　　　　　 $15 = 15\ ✓$

> **TRY IT :: 10.43**   Solve $a^2 + 4a + 9 = 30$ by completing the square.

> **TRY IT :: 10.44**   Solve $b^2 + 8b - 4 = 16$ by completing the square.

To solve the next equation, we must first collect all the variable terms to the left side of the equation. Then, we proceed as we did in the previous examples.

EXAMPLE 10.23

Solve $n^2 = 3n + 11$ by completing the square.

✓ Solution

$$n^2 = 3n + 11$$

Subtract $3n$ to get the variable terms on the left side.	$n^2 - 3n = 11$
Take half of -3 and square it. $\left(\frac{1}{2}(-3)\right)^2 = \frac{9}{4}$	$n^2 - 3n + \underline{} = 11$ $\left(\frac{1}{2} \cdot (-3)\right)^2$
Add $\frac{9}{4}$ to both sides.	$n^2 - 3n + \frac{9}{4} = 11 + \frac{9}{4}$
Factor the perfect square trinomial as a binomial square.	$\left(n - \frac{3}{2}\right)^2 = \frac{44}{4} + \frac{9}{4}$
Add the fractions on the right side.	$\left(n - \frac{3}{2}\right)^2 = \frac{53}{4}$
Use the Square Root Property.	$n - \frac{3}{2} = \pm\sqrt{\frac{53}{4}}$
Simplify the radical.	$n - \frac{3}{2} = \pm\frac{\sqrt{53}}{2}$
Solve for n.	$n = \frac{3}{2} + \frac{\sqrt{53}}{2}$
Rewrite to show two equations.	$n = \frac{3}{2} + \frac{\sqrt{53}}{2}, n = \frac{3}{2} - \frac{\sqrt{53}}{2}$
Check. We leave the check for you!	

> **TRY IT :: 10.45** Solve $p^2 = 5p + 9$ by completing the square.

> **TRY IT :: 10.46** Solve $q^2 = 7q - 3$ by completing the square.

Notice that the left side of the next equation is in factored form. But the right side is not zero, so we cannot use the Zero Product Property. Instead, we multiply the factors and then put the equation into the standard form to solve by completing the square.

EXAMPLE 10.24

Solve $(x - 3)(x + 5) = 9$ by completing the square.

✓ Solution

	$(x - 3)(x + 5) = 9$
We multiply binomials on the left.	$x^2 + 2x - 15 = 9$
Add 15 to get the variable terms on the left side.	$x^2 + 2x \qquad = 24$
Take half of 2 and square it. $\left(\frac{1}{2}(2)\right)^2 = 1$	$x^2 + 2x + \underset{\left(\frac{1}{2}\,\cdot\,(2)\right)^2}{\underline{\qquad}} = 24$
Add 1 to both sides.	$x^2 + 2x + 1 = 24 + 1$
Factor the perfect square trinomial as a binomial square.	$(x + 1)^2 = 25$
Use the Square Root Property.	$x + 1 = \pm\sqrt{25}$
Solve for x.	$x = -1 \pm 5$
Rewite to show two solutions.	$x = -1 + 5,\ x = -1 - 5$
Simplify.	$x = 4,\ x = -6$
Check. We leave the check for you!	

> **TRY IT : : 10.47**   Solve $(c - 2)(c + 8) = 7$ by completing the square.

> **TRY IT : : 10.48**   Solve $(d - 7)(d + 3) = 56$ by completing the square.

Solve Quadratic Equations of the form $ax^2 + bx + c = 0$ by completing the square

The process of completing the square works best when the leading coefficient is one, so the left side of the equation is of the form $x^2 + bx + c$. If the x^2 term has a coefficient, we take some preliminary steps to make the coefficient equal to one.

Sometimes the coefficient can be factored from all three terms of the trinomial. This will be our strategy in the next example.

EXAMPLE 10.25

Solve $3x^2 - 12x - 15 = 0$ by completing the square.

✓ Solution

To complete the square, we need the coefficient of x^2 to be one. If we factor out the coefficient of x^2 as a common factor, we can continue with solving the equation by completing the square.

$$3x^2 - 12x - 15 = 0$$

Factor out the greatest common factor.	$3(x^2 - 4x - 5) = 0$
Divide both sides by 3 to isolate the trinomial.	$\dfrac{3(x^2 - 4x - 5)}{3} = \dfrac{0}{3}$
Simplify.	$x^2 - 4x - 5 = 0$
Subtract 5 to get the constant terms on the right.	$x^2 - 4x = 5$
Take half of 4 and square it. $\left(\frac{1}{2}(4)\right)^2 = 4$	$x^2 - 4x + \underset{\left(\frac{1}{2}\,\cdot\,(4)\right)^2}{\underline{}} = 5$
Add 4 to both sides.	$x^2 - 4x + 4 = 5 + 4$
Factor the perfect square trinomial as a binomial square.	$(x - 2)^2 = 9$
Use the Square Root Property.	$x - 2 = \pm\sqrt{9}$
Solve for x.	$x - 2 = \pm 3$
Rewrite to show 2 solutions.	$x = 2 + 3,\, x = 2 - 3$
Simplify.	$x = 5,\, x = -1$

Check.

$x = 5$	$x = -1$
$3x^2 - 12x - 15 = 0$	$3x^2 - 12x - 15 = 0$
$3(5)^2 - 12(5) - 15 \overset{?}{=} 0$	$3(-1)^2 - 12(-1) - 15 \overset{?}{=} 0$
$75 - 60 - 15 \overset{?}{=} 0$	$3 + 12 - 15 \overset{?}{=} 0$
$0 = 0 \checkmark$	$0 = 0 \checkmark$

> **TRY IT :: 10.49**   Solve $2m^2 + 16m - 8 = 0$ by completing the square.

> **TRY IT :: 10.50**   Solve $4n^2 - 24n - 56 = 8$ by completing the square.

To complete the square, the leading coefficient must be one. When the leading coefficient is not a factor of all the terms, we will divide both sides of the equation by the leading coefficient. This will give us a fraction for the second coefficient. We have already seen how to complete the square with fractions in this section.

EXAMPLE 10.26

Solve $2x^2 - 3x = 20$ by completing the square.

⊘ **Solution**

Again, our first step will be to make the coefficient of x^2 be one. By dividing both sides of the equation by the coefficient of x^2, we can then continue with solving the equation by completing the square.

	$2x^2 - 3x = 20$
Divide both sides by 2 to get the coefficient of x^2 to be 1.	$\dfrac{2x^2 - 3x}{2} = \dfrac{20}{2}$
Simplify.	$x^2 - \dfrac{3}{2}x = 10$
Take half of $-\dfrac{3}{2}$ and square it. $\left(\dfrac{1}{2}\left(-\dfrac{3}{2}\right)\right)^2 = \dfrac{9}{16}$	$x^2 - \dfrac{3}{2}x + \underset{\left(\frac{1}{2}\cdot\left(-\frac{3}{2}\right)\right)^2}{\underline{}} = 10$
Add $\dfrac{9}{16}$ to both sides.	$x^2 - \dfrac{3}{2}x + \dfrac{9}{16} = 10 + \dfrac{9}{16}$
Factor the perfect square trinomial as a binomial square.	$\left(x - \dfrac{3}{4}\right)^2 = \dfrac{160}{16} + \dfrac{9}{16}$
Add the fractions on the right side.	$\left(x - \dfrac{3}{4}\right)^2 = \dfrac{169}{16}$
Use the Square Root Property.	$x - \dfrac{3}{4} = \pm\sqrt{\dfrac{169}{16}}$
Simplify the radical.	$x - \dfrac{3}{4} = \pm\dfrac{13}{4}$
Solve for x.	$x = \dfrac{3}{4} \pm \dfrac{13}{4}$
Rewrite to show 2 solutions.	$x = \dfrac{3}{4} + \dfrac{13}{4},\ x = \dfrac{3}{4} - \dfrac{13}{4}$
Simplify.	$x = 4,\ x = -\dfrac{5}{2}$
Check. We leave the check for you.	

> **TRY IT : : 10.51**   Solve $3r^2 - 2r = 21$ by completing the square.

> **TRY IT : : 10.52**   Solve $4t^2 + 2t = 20$ by completing the square.

EXAMPLE 10.27

Solve $3x^2 + 2x = 4$ by completing the square.

⊘ **Solution**

Again, our first step will be to make the coefficient of x^2 be one. By dividing both sides of the equation by the coefficient of x^2, we can then continue with solving the equation by completing the square.

	$3x^2 + 2x = 4$
Divide both sides by 3 to make the coefficient of x^2 equal 1.	$\dfrac{3x^2 + 2x}{3} = \dfrac{4}{3}$
Simplify.	$x^2 + \dfrac{2}{3}x = \dfrac{4}{3}$
Take half of $\dfrac{2}{3}$ and square it. $\left(\dfrac{1}{2} \cdot \dfrac{2}{3}\right)^2 = \dfrac{1}{9}$	$x^2 + \dfrac{2}{3}x + \underbrace{}_{\left(\frac{1}{2} \cdot \frac{2}{3}\right)^2} = \dfrac{4}{3}$
Add $\dfrac{1}{9}$ to both sides.	$x^2 + \dfrac{2}{3}x + \dfrac{1}{9} = \dfrac{4}{3} + \dfrac{1}{9}$
Factor the perfect square trinomial as a binomial square.	$\left(x + \dfrac{1}{3}\right)^2 = \dfrac{12}{9} + \dfrac{1}{9}$
Use the Square Root Property.	$x + \dfrac{1}{3} = \pm\sqrt{\dfrac{13}{9}}$
Simplify the radical.	$x + \dfrac{1}{3} = \pm\dfrac{\sqrt{13}}{3}$
Solve for x.	$x = -\dfrac{1}{3} \pm \dfrac{\sqrt{13}}{3}$
Rewrite to show 2 solutions.	$x = -\dfrac{1}{3} + \dfrac{\sqrt{13}}{3},\ x = -\dfrac{1}{3} - \dfrac{\sqrt{13}}{3}$
Check. We leave the check for you.	

> **TRY IT : : 10.53**   Solve $4x^2 + 3x = 12$ by completing the square.

> **TRY IT : : 10.54**   Solve $5y^2 + 3y = 10$ by completing the square.

▶ **MEDIA : :**

Access these online resources for additional instruction and practice with solving quadratic equations by completing the square:

- **Introduction to the method of completing the square (https://openstax.org/l/25Completethesq)**
- **How to Solve By Completing the Square (https://openstax.org/l/25Solvebycompsq)**

10.2 EXERCISES

Practice Makes Perfect

Complete the Square of a Binomial Expression

In the following exercises, complete the square to make a perfect square trinomial. Then, write the result as a binomial squared.

57. $a^2 + 10a$

58. $b^2 + 12b$

59. $m^2 + 18m$

60. $n^2 + 16n$

61. $m^2 - 24m$

62. $n^2 - 16n$

63. $p^2 - 22p$

64. $q^2 - 6q$

65. $x^2 - 9x$

66. $y^2 + 11y$

67. $p^2 - \frac{1}{3}p$

68. $q^2 + \frac{3}{4}q$

Solve Quadratic Equations of the Form $x^2 + bx + c = 0$ by Completing the Square

In the following exercises, solve by completing the square.

69. $v^2 + 6v = 40$

70. $w^2 + 8w = 65$

71. $u^2 + 2u = 3$

72. $z^2 + 12z = -11$

73. $c^2 - 12c = 13$

74. $d^2 - 8d = 9$

75. $x^2 - 20x = 21$

76. $y^2 - 2y = 8$

77. $m^2 + 4m = -44$

78. $n^2 - 2n = -3$

79. $r^2 + 6r = -11$

80. $t^2 - 14t = -50$

81. $a^2 - 10a = -5$

82. $b^2 + 6b = 41$

83. $u^2 - 14u + 12 = -1$

84. $z^2 + 2z - 5 = 2$

85. $v^2 = 9v + 2$

86. $w^2 = 5w - 1$

87. $(x + 6)(x - 2) = 9$

88. $(y + 9)(y + 7) = 79$

Solve Quadratic Equations of the Form $ax^2 + bx + c = 0$ by Completing the Square

In the following exercises, solve by completing the square.

89. $3m^2 + 30m - 27 = 6$

90. $2n^2 + 4n - 26 = 0$

91. $2c^2 + c = 6$

92. $3d^2 - 4d = 15$

93. $2p^2 + 7p = 14$

94. $3q^2 - 5q = 9$

Everyday Math

95. Rafi is designing a rectangular playground to have an area of 320 square feet. He wants one side of the playground to be four feet longer than the other side. Solve the equation $p^2 + 4p = 320$ for p, the length of one side of the playground. What is the length of the other side?

96. Yvette wants to put a square swimming pool in the corner of her backyard. She will have a 3 foot deck on the south side of the pool and a 9 foot deck on the west side of the pool. She has a total area of 1080 square feet for the pool and two decks. Solve the equation $(s + 3)(s + 9) = 1080$ for s, the length of a side of the pool.

Writing Exercises

97. Solve the equation $x^2 + 10x = -25$ ⓐ by using the Square Root Property and ⓑ by completing the square. ⓒ Which method do you prefer? Why?

98. Solve the equation $y^2 + 8y = 48$ by completing the square and explain all your steps.

Self Check

ⓐ *After completing the exercises, use this checklist to evaluate your mastery of the objectives of this section.*

I can...	Confidently	With some help	No-I don't get it!
complete the square of a binomial expression.			
solve quadratic equations of the form $x^2 + bx + c = 0$ by completing the square.			
solve quadratic equations of the form $ax^2 + bx + c = 0$ by completing the square.			

ⓑ *After reviewing this checklist, what will you do to become confident for all objectives?*

10.3 **Solve Quadratic Equations Using the Quadratic Formula**

Learning Objectives

By the end of this section, you will be able to:

> Solve quadratic equations using the quadratic formula
> Use the discriminant to predict the number of solutions of a quadratic equation
> Identify the most appropriate method to use to solve a quadratic equation

Be Prepared!

Before you get started, take this readiness quiz.

1. Simplify: $\frac{-20 - 5}{10}$.
   If you missed this problem, review **Example 1.74**.

2. Simplify: $4 + \sqrt{121}$.
   If you missed this problem, review **Example 9.29**.

3. Simplify: $\sqrt{128}$.
   If you missed this problem, review **Example 9.12**.

When we solved quadratic equations in the last section by completing the square, we took the same steps every time. By the end of the exercise set, you may have been wondering 'isn't there an easier way to do this?' The answer is 'yes.' In this section, we will derive and use a formula to find the solution of a quadratic equation.

We have already seen how to solve a formula for a specific variable 'in general' so that we would do the algebraic steps only once and then use the new formula to find the value of the specific variable. Now, we will go through the steps of completing the square in general to solve a quadratic equation for x. It may be helpful to look at one of the examples at the end of the last section where we solved an equation of the form $ax^2 + bx + c = 0$ as you read through the algebraic steps below, so you see them with numbers as well as 'in general.'

We start with the standard form of a quadratic equation and solve it for x by completing the square.

$$ax^2 + bx + c = 0 \qquad a \neq 0$$

Isolate the variable terms on one side.

$$ax^2 + bx = -c$$

Make leading coefficient 1, by dividing by a.

$$\frac{ax^2}{a} + \frac{b}{a}x = -\frac{c}{a}$$

Simplify.

$$x^2 + \frac{b}{a}x = -\frac{c}{a}$$

To complete the square, find $\left(\frac{1}{2} \cdot \frac{b}{a}\right)^2$ and add it to both

sides of the equation. $\left(\frac{1}{2}\frac{b}{a}\right)^2 = \frac{b^2}{4a^2}$

$$x^2 + \frac{b}{a}x + \frac{b^2}{4a^2} = -\frac{c}{a} + \frac{b^2}{4a^2}$$

The left side is a perfect square, factor it.

$$\left(x + \frac{b}{2a}\right)^2 = -\frac{c}{a} + \frac{b^2}{4a^2}$$

Find the common denominator of the right side and write equivalent fractions with the common denominator.

$$\left(x + \frac{b}{2a}\right)^2 = \frac{b^2}{4a^2} - \frac{c \cdot 4a}{a \cdot 4a}$$

Simplify.

$$\left(x + \frac{b}{2a}\right)^2 = \frac{b^2}{4a^2} - \frac{4ac}{4a^2}$$

Combine to one fraction.

$$\left(x + \frac{b}{2a}\right)^2 = \frac{b^2 - 4ac}{4a^2}$$

Use the square root property.

$$x + \frac{b}{2a} = \pm\sqrt{\frac{b^2 - 4ac}{4a^2}}$$

Simplify.

$$x + \frac{b}{2a} = \pm\frac{\sqrt{b^2 - 4ac}}{2a}$$

Add $-\frac{b}{2a}$ to both sides of the equation.

$$x = -\frac{b}{2a} \pm \frac{\sqrt{b^2 - 4ac}}{2a}$$

Combine the terms on the right side.

$$x = \frac{-b \pm \sqrt{b^2 - 4ac}}{2a}$$

This last equation is the Quadratic Formula.

Quadratic Formula

The solutions to a quadratic equation of the form $ax^2 + bx + c = 0$, $a \neq 0$ are given by the formula:

$$x = \frac{-b \pm \sqrt{b^2 - 4ac}}{2a}$$

To use the Quadratic Formula, we substitute the values of a, b, and c into the expression on the right side of the formula. Then, we do all the math to simplify the expression. The result gives the solution(s) to the quadratic equation.

EXAMPLE 10.28 HOW TO SOLVE A QUADRATIC EQUATION USING THE QUADRATIC FORMULA

Solve $2x^2 + 9x - 5 = 0$ by using the Quadratic Formula.

⊘ Solution

Step 1. Write the quadratic equation in standard form. Identify the a, b, c values.	This equation is in standard form.	$ax^2 + bx + c = 0$ $2x^2 + 9x - 5 = 0$ $a = 2, b = 9, c = -5$
Step 2. Write the quadratic formula. Then substitute in the values of a, b, c.	Substitute in $a = 2, b = 9, c = -5$	$x = \dfrac{-b \pm \sqrt{b^2 - 4ac}}{2a}$ $x = \dfrac{-9 \pm \sqrt{9^2 - 4 \cdot 2 \cdot (-5)}}{2 \cdot 2}$
Step 3. Simplify the fraction, and solve for x.		$x = \dfrac{-9 \pm \sqrt{81 - (-40)}}{4}$ $x = \dfrac{-9 \pm \sqrt{121}}{4}$ $x = \dfrac{-9 \pm 11}{4}$ $x = \dfrac{-9 + 11}{4} \qquad x = \dfrac{-9 - 11}{4}$ $x = \dfrac{2}{4} \qquad x = \dfrac{-20}{4}$ $x = \dfrac{1}{2} \qquad x = -5$
Step 4. Check the solutions.	Put each answer in the original equation to check. Substitute $x = \dfrac{1}{2}$. Substitute $x = -5$.	$2x^2 + 9x - 5 = 0$ $2\left(\dfrac{1}{2}\right)^2 + 9 \cdot \dfrac{1}{2} - 5 \overset{?}{=} 0$ $2 \cdot \dfrac{1}{4} + 9 \cdot \dfrac{1}{2} - 5 \overset{?}{=} 0$ $2 \cdot \dfrac{1}{4} + 9 \cdot \dfrac{1}{2} - 5 \overset{?}{=} 0$ $\dfrac{1}{2} + \dfrac{9}{2} - 5 \overset{?}{=} 0$ $\dfrac{10}{2} - 5 \overset{?}{=} 0$ $5 - 5 \overset{?}{=} 0$ $0 = 0 \checkmark$ $2x^2 + 9x - 5 = 0$ $2(-5)^2 + 9(-5) - 5 \overset{?}{=} 0$ $2 \cdot 25 - 45 - 5 \overset{?}{=} 0$ $50 - 45 - 5 \overset{?}{=} 0$ $0 = 0 \checkmark$

> **TRY IT : : 10.55** Solve $3y^2 - 5y + 2 = 0$ by using the Quadratic Formula.

> **TRY IT : : 10.56** Solve $4z^2 + 2z - 6 = 0$ by using the Quadratic Formula.

HOW TO :: SOLVE A QUADRATIC EQUATION USING THE QUADRATIC FORMULA.

Step 1. Write the Quadratic Formula in standard form. Identify the a, b, and c values.

Step 2. Write the Quadratic Formula. Then substitute in the values of a, b, and c.

Step 3. Simplify.

Step 4. Check the solutions.

If you say the formula as you write it in each problem, you'll have it memorized in no time. And remember, the Quadratic Formula is an equation. Be sure you start with '$x =$ '.

EXAMPLE 10.29

Solve $x^2 - 6x + 5 = 0$ by using the Quadratic Formula.

✓ **Solution**

$$x^2 - 6x + 5 = 0$$

This equation is in standard form.	$ax^2 + bx + c = 0$ $x^2 - 6x + 5 = 0$
Identify the a, b, c values.	$a = 1, b = -6, c = 5$
Write the Quadratic Formula.	$x = \dfrac{-b \pm \sqrt{b^2 - 4ac}}{2a}$
Then substitute in the values of a, b, c.	$x = \dfrac{-(-6) \pm \sqrt{(-6)^2 - 4 \cdot 1 \cdot (5)}}{2 \cdot 1}$
	$x = \dfrac{6 \pm \sqrt{36 - 20}}{2}$
Simplify.	$x = \dfrac{6 \pm \sqrt{16}}{2}$
	$x = \dfrac{6 \pm 4}{2}$
Rewrite to show two solutions.	$x = \dfrac{6 + 4}{2}, x = \dfrac{6 - 4}{2}$
Simplify.	$x = \dfrac{10}{2}, x = \dfrac{2}{2}$ $x = 5, x = 1$

Check.

$$x^2 - 6x + 5 = 0 \qquad x^2 - 6x + 5 = 0$$
$$5^2 - 6 \cdot 5 + 5 \overset{?}{=} 0 \qquad 1^2 - 6 \cdot 1 + 5 \overset{?}{=} 0$$
$$25 - 30 + 5 \overset{?}{=} 0 \qquad 1 - 6 + 5 \overset{?}{=} 0$$
$$0 = 0 ✓ \qquad 0 = 0 ✓$$

> **TRY IT :: 10.57**   Solve $a^2 - 2a - 15 = 0$ by using the Quadratic Formula.

> **TRY IT :: 10.58**   Solve $b^2 + 10b + 24 = 0$ by using the Quadratic Formula.

When we solved quadratic equations by using the Square Root Property, we sometimes got answers that had radicals. That can happen, too, when using the Quadratic Formula. If we get a radical as a solution, the final answer must have the radical in its simplified form.

EXAMPLE 10.30

Solve $4y^2 - 5y - 3 = 0$ by using the Quadratic Formula.

✓ Solution

We can use the Quadratic Formula to solve for the variable in a quadratic equation, whether or not it is named 'x'.

	$4y^2 - 5y - 3 = 0$
This equation is in standard form.	$ax^2 + bx + c = 0$ $4y^2 - 5y - 3 = 0$
Identify the a, b, c values.	$a = 4,\ b = -5,\ c = -3$
Write the Quadratic Formula.	$y = \dfrac{-b \pm \sqrt{b^2 - 4ac}}{2a}$
Then substitute in the values of a, b, c.	$y = \dfrac{-(-5) \pm \sqrt{(-5)^2 - 4 \cdot 4 \cdot (-3)}}{2 \cdot 4}$
Simplify.	$y = \dfrac{5 \pm \sqrt{25 + 48}}{8}$ $y = \dfrac{5 \pm \sqrt{73}}{8}$
Rewrite to show two solutions.	$y = \dfrac{5 + \sqrt{73}}{8},\ y = \dfrac{5 - \sqrt{73}}{8}$
Check. We leave the check to you.	

> **TRY IT :: 10.59**   Solve $2p^2 + 8p + 5 = 0$ by using the Quadratic Formula.

> **TRY IT :: 10.60**   Solve $5q^2 - 11q + 3 = 0$ by using the Quadratic Formula.

EXAMPLE 10.31

Solve $2x^2 + 10x + 11 = 0$ by using the Quadratic Formula.

✓ Solution

	$2x^2 + 10x + 11 = 0$
This equation is in standard form.	$ax^2 + bx + c = 0$ $2x^2 + 10x + 11 = 0$
Identify the a, b, c values.	$a = 2,\ b = 10,\ c = 11$
Write the Quadratic Formula.	$x = \dfrac{-b \pm \sqrt{b^2 - 4ac}}{2a}$
Then substitute in the values of a, b, c.	$x = \dfrac{-(10) \pm \sqrt{(10)^2 - 4 \cdot 2 \cdot (11)}}{2 \cdot 2}$
Simplify.	$x = \dfrac{-10 \pm \sqrt{100 - 88}}{4}$ $x = \dfrac{-10 \pm \sqrt{12}}{4}$
Simplify the radical.	$x = \dfrac{-10 \pm 4\sqrt{3}}{4}$

Factor out the common factor in the numerator.	$x = \dfrac{2(-5 \pm 2\sqrt{3})}{4}$
Remove the common factors.	$x = \dfrac{-5 \pm 2\sqrt{3}}{2}$
Rewrite to show two solutions.	$x = \dfrac{-5 + 2\sqrt{3}}{2}, \; x = \dfrac{-5 - 2\sqrt{3}}{2}$
Check. We leave the check to you.	

> **TRY IT :: 10.61**   Solve $3m^2 + 12m + 7 = 0$ by using the Quadratic Formula.

> **TRY IT :: 10.62**   Solve $5n^2 + 4n - 4 = 0$ by using the Quadratic Formula.

We cannot take the square root of a negative number. So, when we substitute a, b, and c into the Quadratic Formula, if the quantity inside the radical is negative, the quadratic equation has no real solution. We will see this in the next example.

EXAMPLE 10.32

Solve $3p^2 + 2p + 9 = 0$ by using the Quadratic Formula.

✓ **Solution**

This equation is in standard form.	$ax^2 + bx + c = 0$ $3p^2 + 2p + 9 = 0$
Identify the a, b, c values.	$a = 3, \; b = 2, \; c = 9$
Write the Quadratic Formula.	$p = \dfrac{-b \pm \sqrt{b^2 - 4ac}}{2a}$
Then substitute in the values of a, b, c.	$p = \dfrac{-(2) \pm \sqrt{(2)^2 - 4 \cdot 3 \cdot (9)}}{2 \cdot 3}$
Simplify.	$p = \dfrac{-2 \pm \sqrt{4 - 108}}{6}$
Simplify the radical.	$p = \dfrac{-2 \pm \sqrt{-104}}{6}$
We cannot take the square root of a negative number.	There is no real solution.

> **TRY IT :: 10.63**   Solve $4a^2 - 3a + 8 = 0$ by using the Quadratic Formula.

> **TRY IT :: 10.64**   Solve $5b^2 + 2b + 4 = 0$ by using the Quadratic Formula.

The quadratic equations we have solved so far in this section were all written in standard form, $ax^2 + bx + c = 0$. Sometimes, we will need to do some algebra to get the equation into standard form before we can use the Quadratic Formula.

EXAMPLE 10.33

Solve $x(x + 6) + 4 = 0$ by using the Quadratic Formula.

Solution

$$x(x + 6) + 4 = 0$$

Distribute to get the equation in standard form.	$x^2 + 6x + 4 = 0$
This equation is now in standard form.	$ax^2 + bx + c = 0$ $x^2 + 6x + 4 = 0$
Identify the a, b, c values.	$a = 1$, $b = 6$, $c = 4$
Write the Quadratic Formula.	$x = \dfrac{-b \pm \sqrt{b^2 - 4ac}}{2a}$
Then substitute in the values of a, b, c.	$x = \dfrac{-(6) \pm \sqrt{(6)^2 - 4 \cdot 1 \cdot (4)}}{2 \cdot 1}$
Simplify.	$x = \dfrac{-6 \pm \sqrt{36 - 16}}{2}$
Simplify inside the radical.	$x = \dfrac{-6 \pm \sqrt{20}}{2}$
Simplify the radical.	$x = \dfrac{-6 \pm 2\sqrt{5}}{2}$
Factor out the common factor in the numerator.	$x = \dfrac{2(-3 \pm 2\sqrt{5})}{2}$
Remove the common factors.	$x = -3 \pm 2\sqrt{5}$
Rewrite to show two solutions.	$x = -3 + 2\sqrt{5}$, $x = -3 - 2\sqrt{5}$
Check. We leave the check to you.	

> **TRY IT : : 10.65**   Solve $x(x + 2) - 5 = 0$ by using the Quadratic Formula.

> **TRY IT : : 10.66**   Solve $y(3y - 1) - 2 = 0$ by using the Quadratic Formula.

When we solved linear equations, if an equation had too many fractions we 'cleared the fractions' by multiplying both sides of the equation by the LCD. This gave us an equivalent equation—without fractions—to solve. We can use the same strategy with quadratic equations.

EXAMPLE 10.34

Solve $\frac{1}{2}u^2 + \frac{2}{3}u = \frac{1}{3}$ by using the Quadratic Formula.

Solution

$$\frac{1}{2}u^2 + \frac{2}{3}u = \frac{1}{3}$$

Multiply both sides by the LCD, 6, to clear the fractions.	$6\left(\frac{1}{2}u^2 + \frac{2}{3}u\right) = 6\left(\frac{1}{3}\right)$
Multiply.	$3u^2 + 4u = 2$
Subtract 2 to get the equation in standard form.	$ax^2 + bx + c = 0$ $3u^2 + 4u - 2 = 0$
Identify the a, b, c values.	$a = 3$, $b = 4$, $c = -2$

Write the Quadratic Formula.	$u = \dfrac{-b \pm \sqrt{b^2 - 4ac}}{2a}$
Then substitute in the values of a, b, c.	$u = \dfrac{-(4) \pm \sqrt{(4)^2 - 4 \cdot 3 \cdot (-2)}}{2 \cdot 3}$
Simplify.	$u = \dfrac{-4 \pm \sqrt{16 + 24}}{6}$
	$u = \dfrac{-4 \pm \sqrt{40}}{6}$
Simplify the radical.	$u = \dfrac{-4 \pm 2\sqrt{10}}{6}$
Factor out the common factor in the numerator.	$u = \dfrac{2(-2 \pm \sqrt{10})}{6}$
Remove the common factors.	$u = \dfrac{-2 \pm \sqrt{10}}{3}$
Rewrite to show two solutions.	$u = \dfrac{-2 \pm \sqrt{10}}{3}, \; u = \dfrac{-2 - \sqrt{10}}{3}$
Check. We leave the check to you.	

> **TRY IT :: 10.67**   Solve $\frac{1}{4}c^2 - \frac{1}{3}c = \frac{1}{12}$ by using the Quadratic Formula.

> **TRY IT :: 10.68**   Solve $\frac{1}{9}d^2 - \frac{1}{2}d = -\frac{1}{2}$ by using the Quadratic Formula.

Think about the equation $(x - 3)^2 = 0$. We know from the Zero Products Principle that this equation has only one solution: $x = 3$.

We will see in the next example how using the Quadratic Formula to solve an equation with a perfect square also gives just one solution.

EXAMPLE 10.35

Solve $4x^2 - 20x = -25$ by using the Quadratic Formula.

⊘ Solution

$$4x^2 - 20x = -25$$

Add 25 to get the equation in standard form.	$ax^2 + bx + c = 0$ $4x^2 - 20x + 25 = 0$
Identify the a, b, c values.	$a = 4$, $b = -20$, $c = 25$
Write the Quadratic Formula.	$x = \dfrac{-b \pm \sqrt{b^2 - 4ac}}{2a}$
Then substitute in the values of a, b, c.	$x = \dfrac{-(-20) \pm \sqrt{(-20)^2 - 4 \cdot 4 \cdot (25)}}{2 \cdot 4}$
Simplify.	$x = \dfrac{20 \pm \sqrt{400 - 400}}{8}$
	$x = \dfrac{20 \pm \sqrt{0}}{8}$
Simplify the radical.	$x = \dfrac{20}{8}$
Simplify the fraction.	$x = \dfrac{5}{2}$
Check. We leave the check to you.	

Did you recognize that $4x^2 - 20x + 25$ is a perfect square?

> **TRY IT :: 10.69**   Solve $r^2 + 10r + 25 = 0$ by using the Quadratic Formula.

> **TRY IT :: 10.70**   Solve $25t^2 - 40t = -16$ by using the Quadratic Formula.

Use the Discriminant to Predict the Number of Solutions of a Quadratic Equation

When we solved the quadratic equations in the previous examples, sometimes we got two solutions, sometimes one solution, sometimes no real solutions. Is there a way to predict the number of solutions to a quadratic equation without actually solving the equation?

Yes, the quantity inside the radical of the Quadratic Formula makes it easy for us to determine the number of solutions. This quantity is called the discriminant.

Discriminant

In the Quadratic Formula $x = \dfrac{-b \pm \sqrt{b^2 - 4ac}}{2a}$, the quantity $b^2 - 4ac$ is called the **discriminant**.

Let's look at the discriminant of the equations in Example 10.28, Example 10.32, and Example 10.35, and the number of solutions to those quadratic equations.

Quadratic Equation (in standard form)	Discriminant $b^2 - 4ac$	Sign of the Discriminant	Number of real solutions
Example 10.28 $2x^2 + 9x - 5 = 0$	$9^2 - 4 \cdot 2(-5) = 121$	+	2
Example 10.35 $4x^2 - 20x + 25 = 0$	$(-20)^2 - 4 \cdot 4 \cdot 25 = 0$	0	1
Example 10.32 $3p^2 + 2p + 9 = 0$	$2^2 - 4 \cdot 3 \cdot 9 = -104$	−	0

When the discriminant is **positive** $\left(x = \dfrac{-b \pm \sqrt{+}}{2a}\right)$ the quadratic equation has **two solutions**.

When the discriminant is **zero** $\left(x = \dfrac{-b \pm \sqrt{0}}{2a}\right)$ the quadratic equation has **one solution**.

When the discriminant is **negative** $\left(x = \dfrac{-b \pm \sqrt{-}}{2a}\right)$ the quadratic equation has **no real solutions**.

HOW TO :: USE THE DISCRIMINANT, $b^2 - 4ac$, TO DETERMINE THE NUMBER OF SOLUTIONS OF A QUADRATIC EQUATION.

For a quadratic equation of the form $ax^2 + bx + c = 0$, $a \neq 0$,

- if $b^2 - 4ac > 0$, the equation has two solutions.

- if $b^2 - 4ac = 0$, the equation has one solution.

- if $b^2 - 4ac < 0$, the equation has no real solutions.

EXAMPLE 10.36

Determine the number of solutions to each quadratic equation:

ⓐ $2v^2 - 3v + 6 = 0$ ⓑ $3x^2 + 7x - 9 = 0$ ⓒ $5n^2 + n + 4 = 0$ ⓓ $9y^2 - 6y + 1 = 0$

 Solution

To determine the number of solutions of each quadratic equation, we will look at its discriminant.

ⓐ

	$2v^2 - 3v + 6 = 0$
The equation is in standard form, identify a, b, c.	$a = 2,\ b = -3,\ c = 6$
Write the discriminant.	$b^2 - 4ac$
Substitute in the values of a, b, c.	$(3)^2 - 4 \cdot 2 \cdot 6$
Simplify.	$9 - 48$
	-39

Because the discriminant is negative, there are no real solutions to the equation.

ⓑ

	$3x^2 + 7x - 9 = 0$
The equation is in standard form, identify a, b, c.	$a = 3,\ b = 7,\ c = -9$
Write the discriminant.	$b^2 - 4ac$
Substitute in the values of a, b, c.	$(7)^2 - 4 \cdot 3 \cdot (-9)$
Simplify.	$49 + 108$
	157

Because the discriminant is positive, there are two solutions to the equation.

ⓒ

The equation is in standard form, identify a, b, and c.	$5n^2 + n + 4 = 0$ $a = 5,\ b = 1,\ c = 4$
Write the discriminant.	$b^2 - 4ac$
Substitute in the values of a, b, c.	$(1)^2 - 4 \cdot 5 \cdot 4$
Simplify.	$1 - 80$ -79

Because the discriminant is negative, there are no real solutions to the equation.

ⓓ

The equation is in standard form, identify a, b, c.	$9y^2 - 6y + 1 = 0$ $a = 9,\ b = -6,\ c = 1$
Write the discriminant.	$b^2 - 4ac$
Substitute in the values of a, b, c.	$(-6)^2 - 4 \cdot 9 \cdot 1$
Simplify.	$36 - 36$ 0

Because the discriminant is 0, there is one solution to the equation.

> **TRY IT :: 10.71**

Determine the number of solutions to each quadratic equation:

ⓐ $8m^2 - 3m + 6 = 0$ ⓑ $5z^2 + 6z - 2 = 0$ ⓒ $9w^2 + 24w + 16 = 0$ ⓓ $9u^2 - 2u + 4 = 0$

> **TRY IT :: 10.72**

Determine the number of solutions to each quadratic equation:

ⓐ $b^2 + 7b - 13 = 0$ ⓑ $5a^2 - 6a + 10 = 0$ ⓒ $4r^2 - 20r + 25 = 0$ ⓓ $7t^2 - 11t + 3 = 0$

Identify the Most Appropriate Method to Use to Solve a Quadratic Equation

We have used four methods to solve quadratic equations:

- Factoring
- Square Root Property
- Completing the Square
- Quadratic Formula

You can solve any quadratic equation by using the Quadratic Formula, but that is not always the easiest method to use.

HOW TO :: IDENTIFY THE MOST APPROPRIATE METHOD TO SOLVE A QUADRATIC EQUATION.

Step 1. Try **Factoring** first. If the quadratic factors easily, this method is very quick.

Step 2. Try the **Square Root Property** next. If the equation fits the form $ax^2 = k$ or $a(x - h)^2 = k$, it can easily be solved by using the Square Root Property.

Step 3. Use the **Quadratic Formula**. Any quadratic equation can be solved by using the Quadratic Formula.

What about the method of completing the square? Most people find that method cumbersome and prefer not to use it. We needed to include it in this chapter because we completed the square in general to derive the Quadratic Formula. You will also use the process of completing the square in other areas of algebra.

EXAMPLE 10.37

Identify the most appropriate method to use to solve each quadratic equation:

ⓐ $5z^2 = 17$ ⓑ $4x^2 - 12x + 9 = 0$ ⓒ $8u^2 + 6u = 11$

✓ **Solution**

ⓐ $5z^2 = 17$

Since the equation is in the $ax^2 = k$, the most appropriate method is to use the Square Root Property.

ⓑ $4x^2 - 12x + 9 = 0$

We recognize that the left side of the equation is a perfect square trinomial, and so Factoring will be the most appropriate method.

ⓒ $8u^2 + 6u = 11$

Put the equation in standard form. $8u^2 + 6u - 11 = 0$

While our first thought may be to try Factoring, thinking about all the possibilities for trial and error leads us to choose the Quadratic Formula as the most appropriate method

---

> **TRY IT : : 10.73** Identify the most appropriate method to use to solve each quadratic equation:

ⓐ $x^2 + 6x + 8 = 0$ ⓑ $(n-3)^2 = 16$ ⓒ $5p^2 - 6p = 9$

> **TRY IT : : 10.74** Identify the most appropriate method to use to solve each quadratic equation:

ⓐ $8a^2 + 3a - 9 = 0$ ⓑ $4b^2 + 4b + 1 = 0$ ⓒ $5c^2 = 125$

▶ **MEDIA : :**

Access these online resources for additional instruction and practice with using the Quadratic Formula:

- **Solving Quadratic Equations: Solving with the Quadratic Formula (https://openstax.org/l/25Quadformula)**

- **How to solve a quadratic equation in standard form using the Quadratic Formula (example) (https://openstax.org/l/25Usequadform)**

- **Solving Quadratic Equations using the Quadratic Formula—Example 3 (https://openstax.org/l/25Byquadformula)**

- **Solve Quadratic Equations using Quadratic Formula (https://openstax.org/l/25Solvequadform)**

10.3 EXERCISES

Practice Makes Perfect

Solve Quadratic Equations Using the Quadratic Formula

In the following exercises, solve by using the Quadratic Formula.

99. $4m^2 + m - 3 = 0$

100. $4n^2 - 9n + 5 = 0$

101. $2p^2 - 7p + 3 = 0$

102. $3q^2 + 8q - 3 = 0$

103. $p^2 + 7p + 12 = 0$

104. $q^2 + 3q - 18 = 0$

105. $r^2 - 8r - 33 = 0$

106. $t^2 + 13t + 40 = 0$

107. $3u^2 + 7u - 2 = 0$

108. $6z^2 - 9z + 1 = 0$

109. $2a^2 - 6a + 3 = 0$

110. $5b^2 + 2b - 4 = 0$

111. $2x^2 + 3x + 9 = 0$

112. $6y^2 - 5y + 2 = 0$

113. $v(v + 5) - 10 = 0$

114. $3w(w - 2) - 8 = 0$

115. $\frac{1}{3}m^2 + \frac{1}{12}m = \frac{1}{4}$

116. $\frac{1}{3}n^2 + n = -\frac{1}{2}$

117. $16c^2 + 24c + 9 = 0$

118. $25d^2 - 60d + 36 = 0$

119. $5m^2 + 2m - 7 = 0$

120. $8n^2 - 3n + 3 = 0$

121. $p^2 - 6p - 27 = 0$

122. $25q^2 + 30q + 9 = 0$

123. $4r^2 + 3r - 5 = 0$

124. $3t(t - 2) = 2$

125. $2a^2 + 12a + 5 = 0$

126. $4d^2 - 7d + 2 = 0$

127. $\frac{3}{4}b^2 + \frac{1}{2}b = \frac{3}{8}$

128. $\frac{1}{9}c^2 + \frac{2}{3}c = 3$

129. $2x^2 + 12x - 3 = 0$

130. $16y^2 + 8y + 1 = 0$

Use the Discriminant to Predict the Number of Solutions of a Quadratic Equation

In the following exercises, determine the number of solutions to each quadratic equation.

131.
ⓐ $4x^2 - 5x + 16 = 0$
ⓑ $36y^2 + 36y + 9 = 0$
ⓒ $6m^2 + 3m - 5 = 0$
ⓓ $18n^2 - 7n + 3 = 0$

132.
ⓐ $9v^2 - 15v + 25 = 0$
ⓑ $100w^2 + 60w + 9 = 0$
ⓒ $5c^2 + 7c - 10 = 0$
ⓓ $15d^2 - 4d + 8 = 0$

133.
ⓐ $r^2 + 12r + 36 = 0$
ⓑ $8t^2 - 11t + 5 = 0$
ⓒ $4u^2 - 12u + 9 = 0$
ⓓ $3v^2 - 5v - 1 = 0$

134.
ⓐ $25p^2 + 10p + 1 = 0$
ⓑ $7q^2 - 3q - 6 = 0$
ⓒ $7y^2 + 2y + 8 = 0$
ⓓ $25z^2 - 60z + 36 = 0$

Identify the Most Appropriate Method to Use to Solve a Quadratic Equation

In the following exercises, identify the most appropriate method (Factoring, Square Root, or Quadratic Formula) to use to solve each quadratic equation. Do not solve.

135.

ⓐ $x^2 - 5x - 24 = 0$

ⓑ $(y + 5)^2 = 12$

ⓒ $14m^2 + 3m = 11$

136.

ⓐ $(8v + 3)^2 = 81$

ⓑ $w^2 - 9w - 22 = 0$

ⓒ $4n^2 - 10 = 6$

137.

ⓐ $6a^2 + 14 = 20$

ⓑ $\left(x - \frac{1}{4}\right)^2 = \frac{5}{16}$

ⓒ $y^2 - 2y = 8$

138.

ⓐ $8b^2 + 15b = 4$

ⓑ $\frac{5}{9}v^2 - \frac{2}{3}v = 1$

ⓒ $\left(w + \frac{4}{3}\right)^2 = \frac{2}{9}$

Everyday Math

139. A flare is fired straight up from a ship at sea. Solve the equation $16\left(t^2 - 13t + 40\right) = 0$ for t, the number of seconds it will take for the flare to be at an altitude of 640 feet.

140. An architect is designing a hotel lobby. She wants to have a triangular window looking out to an atrium, with the width of the window 6 feet more than the height. Due to energy restrictions, the area of the window must be 140 square feet. Solve the equation $\frac{1}{2}h^2 + 3h = 140$ for h, the height of the window.

Writing Exercises

141. Solve the equation $x^2 + 10x = 200$

ⓐ by completing the square

ⓑ using the Quadratic Formula

ⓒ Which method do you prefer? Why?

142. Solve the equation $12y^2 + 23y = 24$

ⓐ by completing the square

ⓑ using the Quadratic Formula

ⓒ Which method do you prefer? Why?

Self Check

ⓐ *After completing the exercises, use this checklist to evaluate your mastery of the objectives of this section.*

I can...	Confidently	With some help	No-I don't get it!
solve quadratic equations using the quadratic formula.			
use the discriminant to predict the number of solutions of a quadratic equation.			
identify the most appropriate method to use to solve a quadratic equation.			

ⓑ *What does this checklist tell you about your mastery of this section? What steps will you take to improve?*

10.4 Solve Applications Modeled by Quadratic Equations

Learning Objectives

By the end of this section, you will be able to:

> Solve applications modeled by Quadratic Equations

Be Prepared!

Before you get started, take this readiness quiz.

1. The sum of two consecutive odd numbers is -100. Find the numbers.
   If you missed this problem, review **Example 3.10**.

2. The area of triangular mural is 64 square feet. The base is 16 feet. Find the height.
   If you missed this problem, review **Example 3.36**.

3. Find the length of the hypotenuse of a right triangle with legs 5 inches and 12 inches.
   If you missed this problem, review **Example 3.39**.

Solve Applications of the Quadratic Formula

We solved some applications that are modeled by quadratic equations earlier, when the only method we had to solve them was factoring. Now that we have more methods to solve quadratic equations, we will take another look at applications. To get us started, we will copy our usual Problem Solving Strategy here so we can follow the steps.

HOW TO :: USE THE PROBLEM SOLVING STRATEGY.

Step 1. **Read** the problem. Make sure all the words and ideas are understood.

Step 2. **Identify** what we are looking for.

Step 3. **Name** what we are looking for. Choose a variable to represent that quantity.

Step 4. **Translate** into an equation. It may be helpful to restate the problem in one sentence with all the important information. Then, translate the English sentence into an algebra equation.

Step 5. **Solve** the equation using good algebra techniques.

Step 6. **Check** the answer in the problem and make sure it makes sense.

Step 7. **Answer** the question with a complete sentence.

We have solved number applications that involved **consecutive even integers** and **consecutive odd integers** by modeling the situation with linear equations. Remember, we noticed each even integer is 2 more than the number preceding it. If we call the first one n, then the next one is $n + 2$. The next one would be $n + 2 + 2$ or $n + 4$. This is also true when we use odd integers. One set of even integers and one set of odd integers are shown below.

Consecutive even integers		Consecutive odd integers	
64, 66, 68		77, 79, 81	
n	1st even integer	n	1st odd integer
$n + 2$	2nd consecutive even integer	$n + 2$	2nd consecutive odd integer
$n + 4$	3rd consecutive even integer	$n + 4$	3rd consecutive odd integer

Some applications of consecutive odd integers or consecutive even integers are modeled by quadratic equations. The notation above will be helpful as you name the variables.

EXAMPLE 10.38

The product of two consecutive odd integers is 195. Find the integers.

⊘ Solution

Step 1. Read the problem.	
Step 2. Identify what we are looking for.	We are looking for two consecutive odd integers.
Step 3. Name what we are looking for.	Let $n =$ the first odd integer. $n + 2 =$ the next odd integer
Step 4. Translate into an equation. State the problem in one sentence.	"The product of two consecutive odd integers is 195." The product of the first odd integer and the second odd integer is 195.
Translate into an equation	$n(n + 2) = 195$
Step 5. Solve the equation. Distribute.	$n^2 + 2n = 195$
Subtract 195 to get the equation in standard form.	$ax^2 + bx + c = 0$ $n^2 + 2n - 195 = 0$
Identify the a, b, c values.	$a = 1, b = 2, c = -195$
Write the quadratic equation.	$n = \dfrac{-b \pm \sqrt{b^2 - 4ac}}{2a}$
Then substitute in the values of a, b, c..	$n = \dfrac{-2 \pm \sqrt{2^2 - 4 \cdot 1 \cdot (-195)}}{2 \cdot 1}$
Simplify.	$n = \dfrac{-2 \pm \sqrt{4 + 780}}{2}$ $n = \dfrac{-2 \pm \sqrt{784}}{2}$
Simplify the radical.	$n = \dfrac{-2 \pm 28}{2}$
Rewrite to show two solutions.	$n = \dfrac{-2 + 28}{2}, \; n = \dfrac{-2 - 28}{2}$
Solve each equation.	$n = \dfrac{26}{2}, \; n = \dfrac{-30}{2}$ $n = 13 \quad n = -15$
There are two values of n that are solutions. This will give us two pairs of consecutive odd integers for our solution.	First odd integer $n = 13$ next odd integer $n + 2$ $\qquad 13 + 2$ $\qquad\quad 15$
	First odd integer $n = -15$ next odd integer $n + 2$ $\qquad -15 + 2$ $\qquad\quad -13$
Step 6. Check the answer. Do these pairs work? Are they consecutive odd integers? Is their product 195?	$13, \; 15, \;$ yes $\quad -13, \; -15, \;$ yes $13 \cdot 15 = 195, \;$ yes $\quad -13(-15) = 195, \;$ yes
Step 7. Answer the question.	The two consecutive odd integers whose product is 195 are 13, 15, and −13, −15.

> **TRY IT ::** 10.75   The product of two consecutive odd integers is 99. Find the integers.

> **TRY IT ::** 10.76   The product of two consecutive odd integers is 168. Find the integers.

We will use the formula for the area of a triangle to solve the next example.

Area of a Triangle

For a triangle with base b and height h, the area, A, is given by the formula $A = \frac{1}{2}bh$.

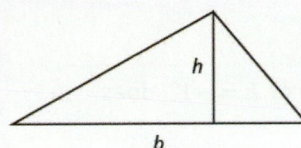

Recall that, when we solve geometry applications, it is helpful to draw the figure.

EXAMPLE 10.39

An architect is designing the entryway of a restaurant. She wants to put a triangular window above the doorway. Due to energy restrictions, the window can have an area of 120 square feet and the architect wants the width to be 4 feet more than twice the height. Find the height and width of the window.

⊘ **Solution**

Step 1. Read the problem. Draw a picture.	
Step 2. Identify what we are looking for.	We are looking for the height and width.
Step 3. Name what we are looking for.	Let $h =$ the height of the triangle. $2h + 4 =$ the width of the triangle
Step 4. Translate.	We know the area. Write the formula for the area of a triangle. $A = \frac{1}{2}bh$
Step 5. Solve the equation. Substitute in the values.	$120 = \frac{1}{2}(2h + 4)h$
Distribute.	$120 = h^2 + 2h$
This is a quadratic equation, rewrite it in standard form.	$ax^2 + bx + c = 0$ $h^2 + 2h - 120 = 0$
Solve the equation using the Quadratic Formula. Identify the a, b, c values.	$a = 1, b = 2, c = -120$
Write the quadratic equation.	$h = \dfrac{-b \pm \sqrt{b^2 - 4ac}}{2a}$
Then substitute in the values of a, b, c..	$h = \dfrac{-2 \pm \sqrt{2^2 - 4 \cdot 1 \cdot (-120)}}{2 \cdot 1}$

Simplify.	$h = \dfrac{-2 \pm \sqrt{4 + 480}}{2}$
	$h = \dfrac{-2 \pm \sqrt{484}}{2}$
Simplify the radical.	$h = \dfrac{-2 \pm 22}{2}$
Rewrite to show two solutions.	$h = \dfrac{-2 + 22}{2}$, $h = \dfrac{-2 - 22}{2}$
Simplify.	$h = \dfrac{20}{2}$, $h = \dfrac{-24}{2}$
Since h is the height of a window, a value of $h = -12$ does not make sense.	$h = 10$ $\cancel{h = -12}$

The height of the triangle: $h = 10$

The width of the triangle: $2h + 4$
$2 \cdot 10 + 4$
24

Step 6. Check the answer. Does a triangle with a height 10 and width 24 have area 120? Yes.

Step 7. Answer the question.

The height of the triangular window is 10 feet and the width is 24 feet.

Notice that the solutions were integers. That tells us that we could have solved the equation by factoring.

When we wrote the equation in standard form, $h^2 + 2h - 120 = 0$, we could have factored it. If we did, we would have solved the equation $(h + 12)(h - 10) = 0$.

> **TRY IT : :** 10.77

Find the dimensions of a triangle whose width is four more than six times its height and has an area of 208 square inches.

> **TRY IT : :** 10.78

If a triangle that has an area of 110 square feet has a height that is two feet less than twice the width, what are its dimensions?

In the two preceding examples, the number in the radical in the Quadratic Formula was a perfect square and so the solutions were rational numbers. If we get an irrational number as a solution to an application problem, we will use a calculator to get an approximate value.

The Pythagorean Theorem gives the relation between the legs and hypotenuse of a right triangle. We will use the Pythagorean Theorem to solve the next example.

Pythagorean Theorem

In any right triangle, where a and b are the lengths of the legs and c is the length of the hypotenuse, $a^2 + b^2 = c^2$

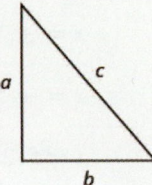

EXAMPLE 10.40

Rene is setting up a holiday light display. He wants to make a 'tree' in the shape of two right triangles, as shown below, and has two 10-foot strings of lights to use for the sides. He will attach the lights to the top of a pole and to two stakes on the ground. He wants the height of the pole to be the same as the distance from the base of the pole to each stake. How tall should the pole be?

⊘ **Solution**

Step 1. Read the problem. Draw a picture.	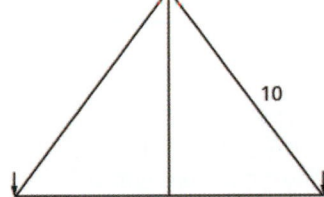
Step 2. Identify what we are looking for.	We are looking for the height of the pole.
Step 3. Name what we are looking for.	The distance from the base of the pole to either stake is the same as the height of the pole. Let $x =$ the height of the pole. $x =$ the distance from the pole to stake
Each side is a right triangle. We draw a picture of one of them.	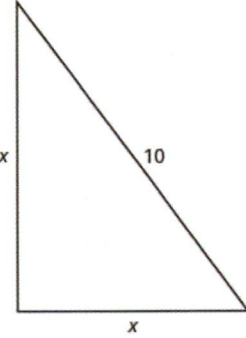
Step 4. Translate into an equation. We can use the Pythagorean Theorem to solve for x.	
Write the Pythagorean Theorem.	$a^2 + b^2 = c^2$
Step 5. Solve the equation. Substitute.	$x^2 + x^2 = 10^2$
Simplify.	$2x^2 = 100$
Divide by 2 to isolate the variable.	$\dfrac{2x^2}{2} = \dfrac{100}{2}$
Simplify.	$x^2 = 50$

Use the Square Root Property.	$x = \pm\sqrt{50}$
Simplify the radical.	$x = \pm 5\sqrt{2}$
Rewrite to show two solutions.	$x = 5\sqrt{2}$ $\cancel{x = -5\sqrt{2}}$
Approximate this number to the nearest tenth with a calculator.	$x \approx 7.1$
Step 6. Check the answer. Check on your own in the Pythagorean Theorem.	
Step 7. Answer the question.	The pole should be about 7.1 feet tall.

> **TRY IT : :** 10.79
>
> The sun casts a shadow from a flag pole. The height of the flag pole is three times the length of its shadow. The distance between the end of the shadow and the top of the flag pole is 20 feet. Find the length of the shadow and the length of the flag pole. Round to the nearest tenth of a foot.

> **TRY IT : :** 10.80
>
> The distance between opposite corners of a rectangular field is four more than the width of the field. The length of the field is twice its width. Find the distance between the opposite corners. Round to the nearest tenth.

EXAMPLE 10.41

Mike wants to put 150 square feet of artificial turf in his front yard. This is the maximum area of artificial turf allowed by his homeowners association. He wants to have a rectangular area of turf with length one foot less than three times the width. Find the length and width. Round to the nearest tenth of a foot.

⊘ **Solution**

Step 1. Read the problem. Draw a picture.	 w [rectangle] $3w - 1$
Step 2. Identify what we are looking for.	We are looking for the length and width.
Step 3. Name what we are looking for.	Let $w =$ the width of the rectangle. $3w - 1 =$ the length of the rectangle
Step 4. Translate into an equation. We know the area. Write the formula for the area of a rectangle.	$A = L \cdot W$
Step 5. Solve the equation. Substitute in the values.	$150 = (3w - 1)w$
Distribute.	$150 = 3w^2 - w$

This is a quadratic equation, rewrite it in standard form.	$ax^2 + bx + c = 0$ $3w^2 - w - 150 = 0$
Solve the equation using the Quadratic Formula.	
Identify the a, b, c values.	$a = 3$, $b = -1$, $c = -150$
Write the Quadratic Formula.	$w = \dfrac{-b \pm \sqrt{b^2 - 4ac}}{2a}$
Then substitute in the values of a, b, c.	$w = \dfrac{-(-1) \pm \sqrt{(-1)^2 - 4 \cdot 3 \cdot (-150)}}{2 \cdot 3}$
Simplify.	$w = \dfrac{1 \pm \sqrt{1 + 1800}}{6}$ $w = \dfrac{1 \pm \sqrt{1801}}{6}$
Rewrite to show two solutions.	$w = \dfrac{1 + \sqrt{1801}}{6}$, $w = \dfrac{1 - \sqrt{1801}}{6}$
Approximate the answers using a calculator. We eliminate the negative solution for the width.	$w \approx 7.2$, $\qquad$ $w \approx -6.9$ Width $w \approx 7.2$ Length $\approx 3w - 1$ $\approx 3(7.2) - 1$ ≈ 20.6
Step 6. Check the answer. Make sure that the answers make sense.	
Step 7. Answer the question.	The width of the rectangle is approximately 7.2 feet and the length 20.6 feet.

> **TRY IT : :** 10.81

The length of a 200 square foot rectangular vegetable garden is four feet less than twice the width. Find the length and width of the garden. Round to the nearest tenth of a foot.

> **TRY IT : :** 10.82

A rectangular tablecloth has an area of 80 square feet. The width is 5 feet shorter than the length. What are the length and width of the tablecloth? Round to the nearest tenth of a foot.

The height of a projectile shot upwards is modeled by a quadratic equation. The initial velocity, v_0, propels the object up until gravity causes the object to fall back down.

Projectile Motion

The height in feet, h, of an object shot upwards into the air with initial velocity, v_0, after t seconds is given by the formula:

$$h = -16t^2 + v_0 t$$

We can use the formula for projectile motion to find how many seconds it will take for a firework to reach a specific height.

EXAMPLE 10.42

A firework is shot upwards with initial velocity 130 feet per second. How many seconds will it take to reach a height of 260 feet? Round to the nearest tenth of a second.

✓ **Solution**

Step 1. Read the problem.

Step 2. Identify what we are looking for.	We are looking for the number of seconds, which is time.
Step 3. Name what we are looking for.	Let $t =$ the number of seconds.
Step 4. Translate into an equation.	Use the formula.

$$h = -16t^2 + v_0 t$$

Step 5. Solve the equation.
We know the velocity v_0 is 130 feet per second.

The height is 260 feet. Substitute the values.	$260 = -16t^2 + 130t$
This is a quadratic equation, rewrite it in standard form.	$ax^2 + bx + c = 0$ $16t^2 + 130t + 260 = 0$
Solve the equation using the Quadratic Formula.	
Identify the a, b, c values.	$a = 16, b = -130, c = 260$
Write the Quadratic Formula.	$t = \dfrac{-b \pm \sqrt{b^2 - 4ac}}{2a}$
Then substitute in the values of a, b, c.	$t = \dfrac{-(-130) \pm \sqrt{(-130)^2 - 4 \cdot 16 \cdot (260)}}{2 \cdot 16}$
Simplify.	$t = \dfrac{130 \pm \sqrt{16{,}900 - 16{,}640}}{32}$
	$t = \dfrac{130 \pm \sqrt{260}}{32}$
Rewrite to show two solutions.	$t = \dfrac{130 + \sqrt{260}}{32}, \ t = \dfrac{130 - \sqrt{260}}{32}$
Approximate the answers with a calculator.	$t \approx 4.6$ seconds, $t \approx 3.6$

Step 6. Check the answer.
The check is left to you.

Step 7. Answer the question.

The firework will go up and then fall back down. As the firework goes up, it will reach 260 feet after approximately 3.6 seconds. It will also pass that height on the way down at 4.6 seconds.

> **TRY IT :: 10.83**

An arrow is shot from the ground into the air at an initial speed of 108 ft/sec. Use the formula $h = -16t^2 + v_0 t$ to determine when the arrow will be 180 feet from the ground. Round the nearest tenth of a second.

> **TRY IT :: 10.84**

A man throws a ball into the air with a velocity of 96 ft/sec. Use the formula $h = -16t^2 + v_0 t$ to determine when the height of the ball will be 48 feet. Round to the nearest tenth of a second.

▶ **MEDIA ::**

Access these online resources for additional instruction and practice with solving word problems using the quadratic equation:

- **General Quadratic Word Problems (https://openstax.org/l/25Quadproblem)**
- **Word problem: Solve a projectile problem using a quadratic equation (https://openstax.org/l/25Projectile)**

10.4 EXERCISES

Practice Makes Perfect

Solve Applications of the Quadratic Formula

In the following exercises, solve by using methods of factoring, the square root principle, or the Quadratic Formula. Round your answers to the nearest tenth.

143. The product of two consecutive odd numbers is 255. Find the numbers.

144. The product of two consecutive even numbers is 360. Find the numbers.

145. The product of two consecutive even numbers is 624. Find the numbers.

146. The product of two consecutive odd numbers is 1023. Find the numbers.

147. The product of two consecutive odd numbers is 483. Find the numbers.

148. The product of two consecutive even numbers is 528. Find the numbers.

149. A triangle with area 45 square inches has a height that is two less than four times the width. Find the height and width of the triangle.

150. The width of a triangle is six more than twice the height. The area of the triangle is 88 square yards. Find the height and width of the triangle.

151. The hypotenuse of a right triangle is twice the length of one of its legs. The length of the other leg is three feet. Find the lengths of the three sides of the triangle.

152. The hypotenuse of a right triangle is 10 cm long. One of the triangle's legs is three times the length of the other leg. Round to the nearest tenth. Find the lengths of the three sides of the triangle.

153. A farmer plans to fence off sections of a rectangular corral. The diagonal distance from one corner of the corral to the opposite corner is five yards longer than the width of the corral. The length of the corral is three times the width. Find the length of the diagonal of the corral.

154. Nautical flags are used to represent letters of the alphabet. The flag for the letter O consists of a yellow right triangle and a red right triangle which are sewn together along their hypotenuse to form a square. The adjoining side of the two triangles is three inches longer than a side of the flag. Find the length of the side of the flag.

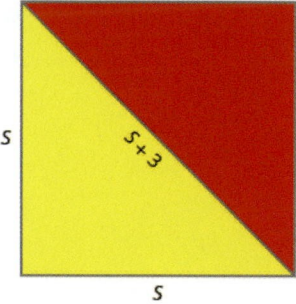

155. The length of a rectangular driveway is five feet more than three times the width. The area is 350 square feet. Find the length and width of the driveway.

156. A rectangular lawn has area 140 square yards. Its width that is six less than twice the length. What are the length and width of the lawn?

157. A firework rocket is shot upward at a rate of 640 ft/sec. Use the projectile formula $h = -16t^2 + v_0 t$ to determine when the height of the firework rocket will be 1200 feet.

158. An arrow is shot vertically upward at a rate of 220 feet per second. Use the projectile formula $h = -16t^2 + v_0 t$ to determine when height of the arrow will be 400 feet.

Everyday Math

159. A bullet is fired straight up from a BB gun with initial velocity 1120 feet per second at an initial height of 8 feet. Use the formula $h = -16t^2 + v_0 t + 8$ to determine how many seconds it will take for the bullet to hit the ground. (That is, when will $h = 0$?)

160. A city planner wants to build a bridge across a lake in a park. To find the length of the bridge, he makes a right triangle with one leg and the hypotenuse on land and the bridge as the other leg. The length of the hypotenuse is 340 feet and the leg is 160 feet. Find the length of the bridge.

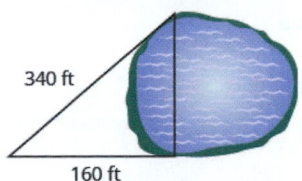

340 ft

160 ft

Writing Exercises

161. Make up a problem involving the product of two consecutive odd integers. Start by choosing two consecutive odd integers. ⓐ What are your integers? ⓑ What is the product of your integers? ⓒ Solve the equation $n(n + 2) = p$, where p is the product you found in part (b). ⓓ Did you get the numbers you started with?

162. Make up a problem involving the product of two consecutive even integers. Start by choosing two consecutive even integers. ⓐ What are your integers? ⓑ What is the product of your integers? ⓒ Solve the equation $n(n + 2) = p$, where p is the product you found in part (b). ⓓ Did you get the numbers you started with?

Self Check

ⓐ *After completing the exercises, use this checklist to evaluate your mastery of the objectives of this section.*

I can...	Confidently	With some help	No-I don't get it!
solve applications of the quadratic formula.			

ⓑ *On a scale of 1–10, how would you rate your mastery of this section in light of your responses on the checklist? How can you improve this?*

10.5 Graphing Quadratic Equations

Learning Objectives

By the end of this section, you will be able to:

> Recognize the graph of a quadratic equation in two variables
> Find the axis of symmetry and vertex of a parabola
> Find the intercepts of a parabola
> Graph quadratic equations in two variables
> Solve maximum and minimum applications

Be Prepared!

Before you get started, take this readiness quiz.

1. Graph the equation $y = 3x - 5$ by plotting points.
   If you missed this problem, review **Example 4.11**.

2. Evaluate $2x^2 + 4x - 1$ when $x = -3$.
   If you missed this problem, review **Example 1.57**.

3. Evaluate $-\frac{b}{2a}$ when $a = \frac{1}{3}$ and $b = \frac{5}{6}$.
   If you missed this problem, review **Example 1.89**.

Recognize the Graph of a Quadratic Equation in Two Variables

We have graphed equations of the form $Ax + By = C$. We called equations like this linear equations because their graphs are straight lines.

Now, we will graph equations of the form $y = ax^2 + bx + c$. We call this kind of equation a quadratic equation in two variables.

Quadratic Equation in Two Variables

A **quadratic equation in two variables**, where a, b, and c are real numbers and $a \neq 0$, is an equation of the form

$$y = ax^2 + bx + c$$

Just like we started graphing linear equations by plotting points, we will do the same for quadratic equations.

Let's look first at graphing the quadratic equation $y = x^2$. We will choose integer values of x between -2 and 2 and find their y values. See **Table 10.31**.

$y = x^2$	
x	y
0	0
1	1
-1	1
2	4
-2	4

Table 10.31

Notice when we let $x = 1$ and $x = -1$, we got the same value for y.

$$y = x^2 \qquad y = x^2$$

$$y = 1^2 \qquad y = (-1)^2$$

$$y = 1 \qquad y = 1$$

The same thing happened when we let $x = 2$ and $x = -2$.

Now, we will plot the points to show the graph of $y = x^2$. See **Figure 10.2**.

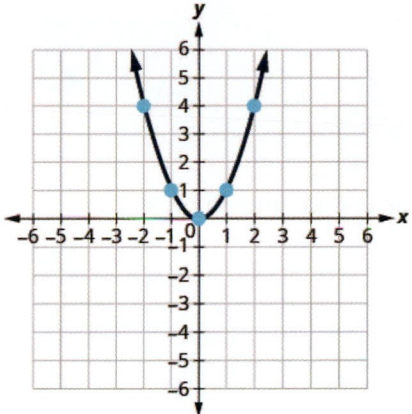

Figure 10.2

The graph is not a line. This figure is called a **parabola**. Every quadratic equation has a graph that looks like this. In **Example 10.43** you will practice graphing a parabola by plotting a few points.

EXAMPLE 10.43

Graph $y = x^2 - 1$.

⊘ **Solution**

We will graph the equation by plotting points.

Choose integers values for x, substitute them into the equation and solve for y.

Record the values of the ordered pairs in the chart.

$y = x^2 - 1$	
x	y
0	−1
1	0
−1	0
2	3
−2	3

Plot the points, and then connect them with a smooth curve. The result will be the graph of the equation $y = x^2 - 1$.

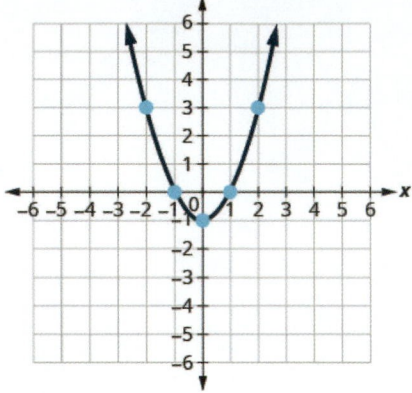

> **TRY IT : : 10.85**   Graph $y = -x^2$.

> **TRY IT : : 10.86**   Graph $y = x^2 + 1$.

How do the equations $y = x^2$ and $y = x^2 - 1$ differ? What is the difference between their graphs? How are their graphs the same?

All parabolas of the form $y = ax^2 + bx + c$ open upwards or downwards. See **Figure 10.3**.

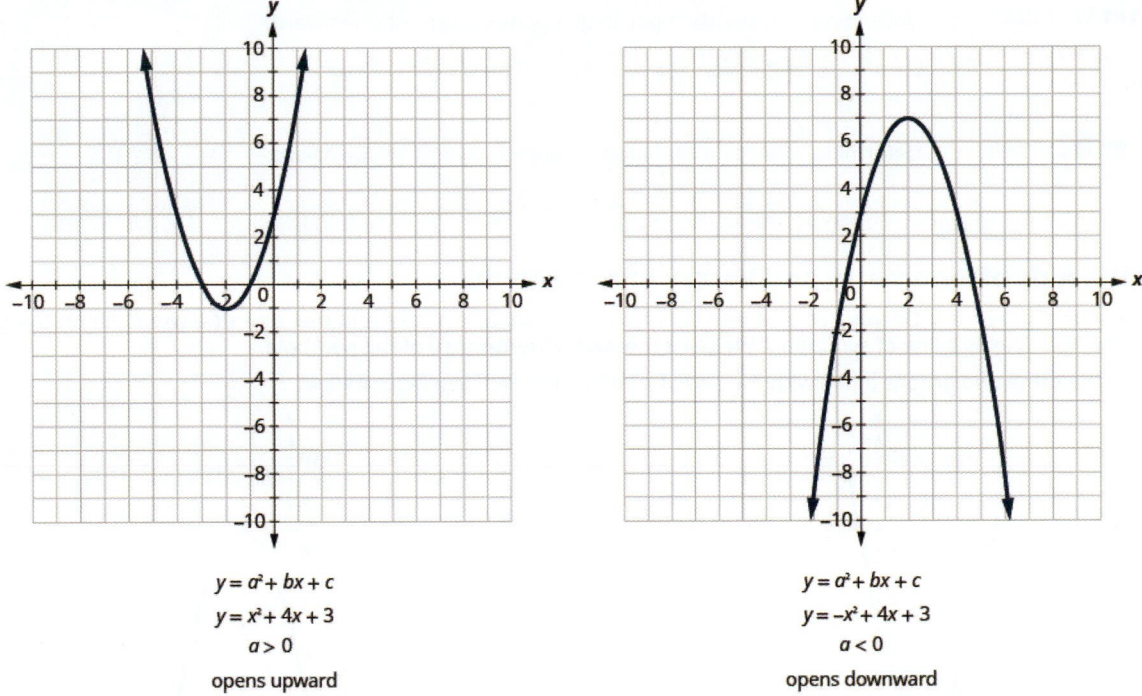

$$y = a^2 + bx + c$$
$$y = x^2 + 4x + 3$$
$$a > 0$$

opens upward

$$y = a^2 + bx + c$$
$$y = -x^2 + 4x + 3$$
$$a < 0$$

opens downward

Figure 10.3

Notice that the only difference in the two equations is the negative sign before the x^2 in the equation of the second graph in **Figure 10.3**. When the x^2 term is positive, the parabola opens upward, and when the x^2 term is negative, the parabola opens downward.

Parabola Orientation

For the quadratic equation $y = ax^2 + bx + c$, if:

- $a > 0$, the parabola opens upward
- $a < 0$, the parabola opens downward

EXAMPLE 10.44

Determine whether each parabola opens upward or downward:

ⓐ $y = -3x^2 + 2x - 4$ ⓑ $y = 6x^2 + 7x - 9$

⊘ **Solution**

ⓐ
Find the value of "a".

$$y = ax^2 + bx + c$$
$$y = -3x^2 + 2x - 4$$
$$a = -3$$

Since the "a" is negative, the parabola will open downward.

ⓑ
Find the value of "a".

$$y = ax^2 + bx + c$$
$$y = 6x^2 + 7x - 9$$
$$a = 6$$

Since the "a" is positive, the parabola will open upward.

> **TRY IT : : 10.87**   Determine whether each parabola opens upward or downward:

ⓐ $y = 2x^2 + 5x - 2$ ⓑ $y = -3x^2 - 4x + 7$

> **TRY IT : : 10.88**   Determine whether each parabola opens upward or downward:

ⓐ $y = -2x^2 - 2x - 3$ ⓑ $y = 5x^2 - 2x - 1$

Find the Axis of Symmetry and Vertex of a Parabola

Look again at **Figure 10.3**. Do you see that we could fold each parabola in half and that one side would lie on top of the other? The 'fold line' is a line of symmetry. We call it the **axis of symmetry** of the parabola.

We show the same two graphs again with the axis of symmetry in red. See **Figure 10.4**.

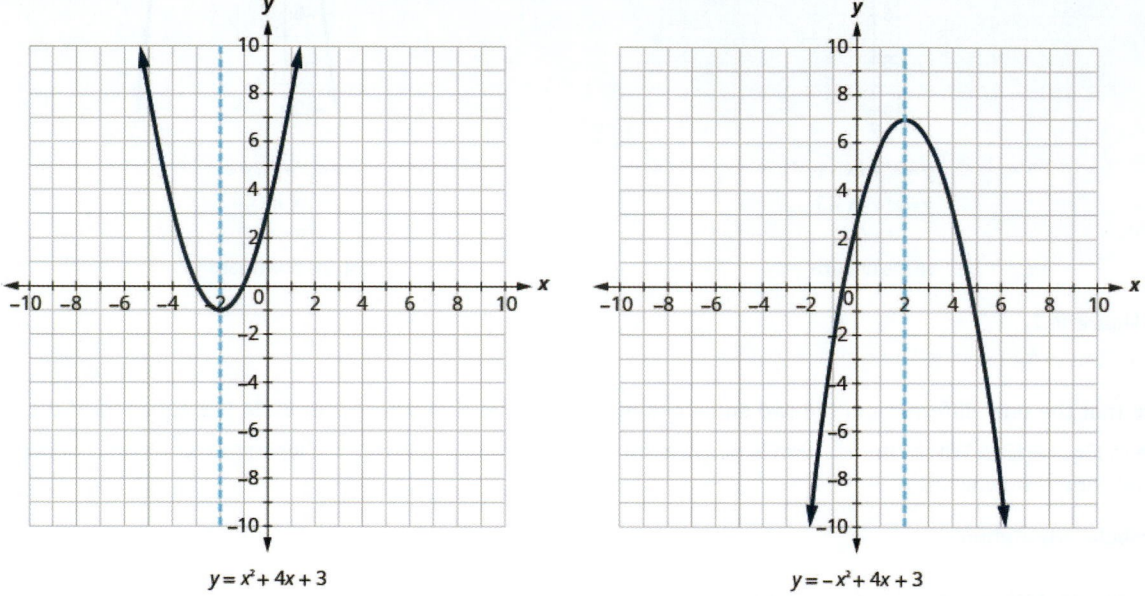

$y = x^2 + 4x + 3$ $y = -x^2 + 4x + 3$

Figure 10.4

The equation of the axis of symmetry can be derived by using the Quadratic Formula. We will omit the derivation here and proceed directly to using the result. The equation of the axis of symmetry of the graph of $y = ax^2 + bx + c$ is $x = -\dfrac{b}{2a}$.

So, to find the equation of symmetry of each of the parabolas we graphed above, we will substitute into the formula $x = -\dfrac{b}{2a}$.

$$\overset{\color{red}{ax^2 + bx + c}}{y = x^2 + 4x + 3} \qquad\qquad \overset{\color{red}{ax^2 + bx + c}}{y = -x^2 + 4x + 3}$$

axis of symmetry   axis of symmetry

$$x = -\frac{b}{2a} \qquad\qquad\qquad x = -\frac{b}{2a}$$

$$x = -\frac{4}{2 \cdot 1} \qquad\qquad\qquad x = -\frac{4}{2(-1)}$$

$$x = -2 \qquad\qquad\qquad\quad x = 2$$

Look back at **Figure 10.4**. Are these the equations of the dashed red lines?

The point on the parabola that is on the axis of symmetry is the lowest or highest point on the parabola, depending on whether the parabola opens upwards or downwards. This point is called the **vertex** of the parabola.

We can easily find the coordinates of the vertex, because we know it is on the axis of symmetry. This means its x-coordinate is $-\dfrac{b}{2a}$. To find the y-coordinate of the vertex, we substitute the value of the x-coordinate into the quadratic equation.

$$y = x^2 + 4x + 3$$

axis of symmetry is $x = -2$

vertex is $(-2, \_\_)$

$$y = x^2 + 4x + 3$$
$$y = (-2)^2 + 4(-2) + 3$$
$$y = -1$$

vertex is $(-2, -1)$

$$y = -x^2 + 4x + 3$$

axis of symmetry is $x = 2$

vertex is $(2, \_\_)$

$$y = -x^2 + 4x + 3$$
$$y = -(2)^2 + 4(2) + 3$$
$$y = 7$$

vertex is $(2, 7)$

Axis of Symmetry and Vertex of a Parabola

For a parabola with equation $y = ax^2 + bx + c$:

- The axis of symmetry of a parabola is the line $x = -\dfrac{b}{2a}$.

- The vertex is on the axis of symmetry, so its x-coordinate is $-\dfrac{b}{2a}$.

To find the y-coordinate of the vertex, we substitute $x = -\dfrac{b}{2a}$ into the quadratic equation.

EXAMPLE 10.45

For the parabola $y = 3x^2 - 6x + 2$ find: ⓐ the axis of symmetry and ⓑ the vertex.

✓ **Solution**

ⓐ
$$y = ax^2 + bx + c$$
$$y = 3x^2 - 6x + 2$$

The axis of symmetry is the line $x = -\dfrac{b}{2a}$.	$x = -\dfrac{b}{2a}$
Substitute the values of a, b into the equation.	$x = -\dfrac{-6}{2 \cdot 3}$
Simplify.	$x = 1$
	The axis of symmetry is the line $x = 1$.

ⓑ
$$y = 3x^2 - 6x + 2$$

The vertex is on the line of symmetry, so its x-coordinate will be $x = 1$.	
Substitute $x = 1$ into the equation and solve for y.	$y = 3(1)^2 - 6(1) + 2$
Simplify.	$y = 3 \cdot 1 - 6 + 2$
This is the y-coordinate.	$y = -1$
	The vertex is $(1, -1)$.

> **TRY IT :: 10.89**   For the parabola $y = 2x^2 - 8x + 1$ find: ⓐ the axis of symmetry and ⓑ the vertex.

> **TRY IT :: 10.90**   For the parabola $y = 2x^2 - 4x - 3$ find: ⓐ the axis of symmetry and ⓑ the vertex.

Find the Intercepts of a Parabola

When we graphed linear equations, we often used the x- and y-intercepts to help us graph the lines. Finding the coordinates of the intercepts will help us to graph parabolas, too.

Remember, at the **y-intercept** the value of x is zero. So, to find the y-intercept, we substitute $x = 0$ into the equation.

Let's find the y-intercepts of the two parabolas shown in the figure below.

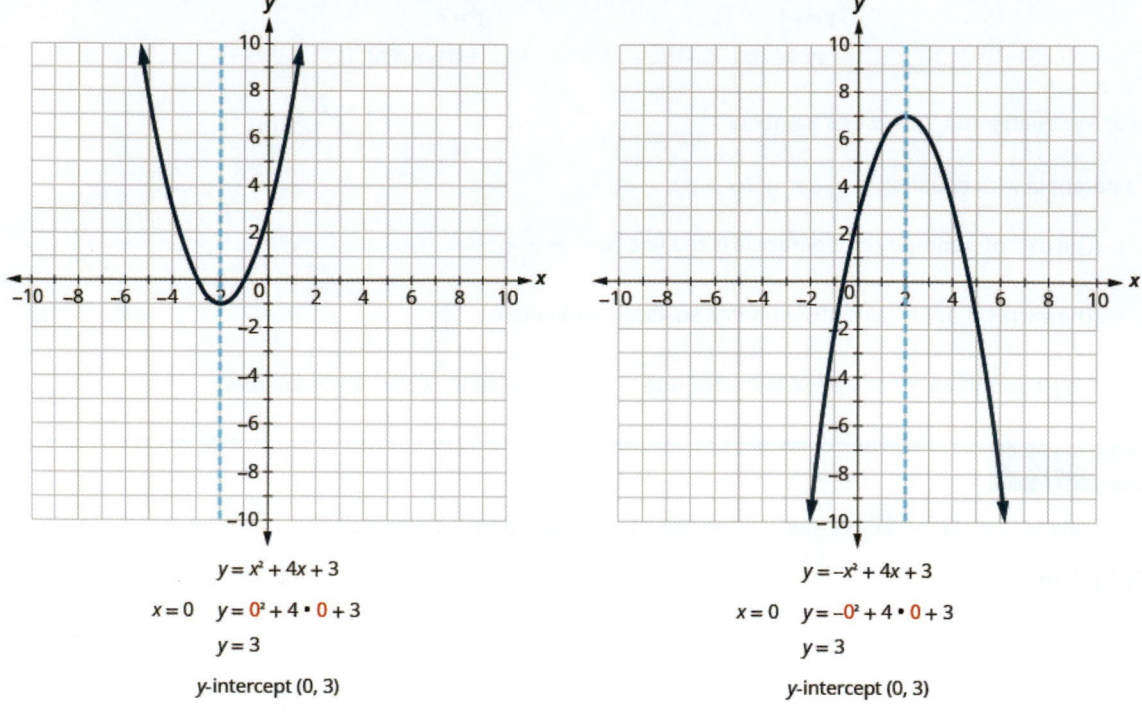

$$y = x^2 + 4x + 3$$
$$x = 0 \quad y = 0^2 + 4 \cdot 0 + 3$$
$$y = 3$$
$$y\text{-intercept } (0, 3)$$

$$y = -x^2 + 4x + 3$$
$$x = 0 \quad y = -0^2 + 4 \cdot 0 + 3$$
$$y = 3$$
$$y\text{-intercept } (0, 3)$$

Figure 10.5

At an **x-intercept**, the value of y is zero. To find an x-intercept, we substitute $y = 0$ into the equation. In other words, we will need to solve the equation $0 = ax^2 + bx + c$ for x.

$$y = ax^2 + bx + c$$
$$0 = ax^2 + bx + c$$

But solving quadratic equations like this is exactly what we have done earlier in this chapter.

We can now find the x-intercepts of the two parabolas shown in **Figure 10.5**.

First, we will find the x-intercepts of a parabola with equation $y = x^2 + 4x + 3$.

$$y = x^2 + 4x + 3$$

Let $y = 0$.	$0 = x^2 + 4x + 3$
Factor.	$0 = (x + 1)(x + 3)$
Use the zero product property.	$x + 1 = 0, x + 3 = 0$
Solve.	$x = -1, x = -3$
	The x intercepts are $(-1, 0)$ and $(-3, 0)$.

Now, we will find the *x*-intercepts of the parabola with equation $y = -x^2 + 4x + 3$.

$$y = -x^2 + 4x + 3$$

Let $y = 0$.	$0 = -x^2 + 4x + 3$
This quadratic does not factor, so we use the Quadratic Formula.	$x = \dfrac{-b \pm \sqrt{b^2 - 4ac}}{2a}$
$a = -1$, $b = 4$, $c = 3$	$x = \dfrac{-4 \pm \sqrt{4^2 - 4(-1)(3)}}{2(-1)}$
	$x = \dfrac{-4 \pm \sqrt{28}}{-2}$
Simplify.	$x = \dfrac{-4 \pm 2\sqrt{7}}{-2}$
	$x = \dfrac{-2(2 \pm \sqrt{7})}{-2}$
	$x = 2 \pm \sqrt{7}$
	The *x* intercepts are $\left(2 + \sqrt{7}, 0\right)$ and $\left(2 - \sqrt{7}, 0\right)$.

We will use the decimal approximations of the x-intercepts, so that we can locate these points on the graph.

$$\left(2 + \sqrt{7}, 0\right) \approx (4.6, 0) \qquad \left(2 - \sqrt{7}, 0\right) \approx (-0.6, 0)$$

Do these results agree with our graphs? See **Figure 10.6**.

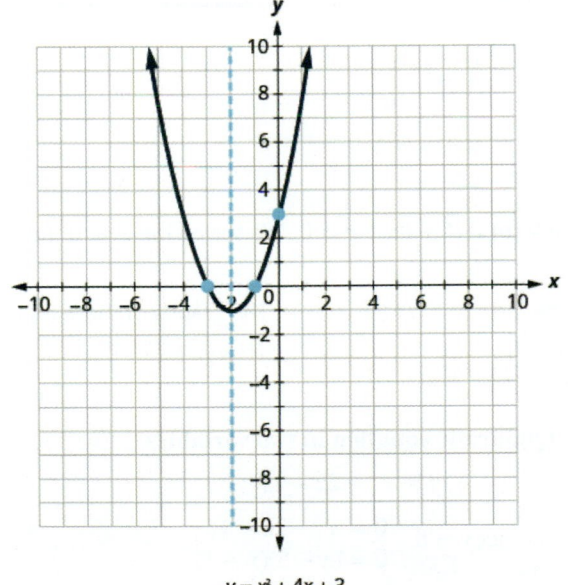

$y = x^2 + 4x + 3$

y-intercept (0, 3)

x-intercepts (−1, 0) and (−3, 0)

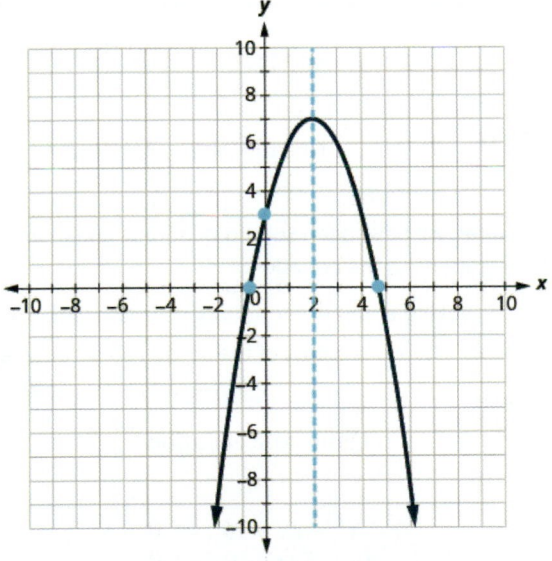

$y = -x^2 + 4x + 3$

y-intercept (0, 3)

x-intercepts $\left(2 + \sqrt{7}, 0\right) \approx (4.6, 0)$

$\left(2 - \sqrt{7}, 0\right) \approx (-0.6, 0)$

Figure 10.6

HOW TO :: FIND THE INTERCEPTS OF A PARABOLA.

To find the intercepts of a parabola with equation $y = ax^2 + bx + c$:

y-intercept	**x-intercepts**
Let $x = 0$ and solve for y.	Let $y = 0$ and solve for x.

EXAMPLE 10.46

Find the intercepts of the parabola $y = x^2 - 2x - 8$.

✓ **Solution**

$$y = x^2 - 2x - 8$$

To find the y-intercept, let $x = 0$ and solve for y.	$y = 0^2 - 2 \cdot 0 - 8$ $y = -8$

When $x = 0$, then $y = -8$.
The y-intercept is the point $(0, -8)$.

	$y = x^2 - 2x - 8$
To find the x-intercept, let $y = 0$ and solve for x.	$0 = x^2 - 2x - 8$
Solve by factoring.	$0 = (x - 4)(x + 2)$
	$0 = x - 4 \quad 0 = x + 2$ $4 = x \quad\quad -2 = x$

When $y = 0$, then $x = 4$ or $x = -2$. The x-intercepts are the points $(4, 0)$ and $(-2, 0)$.

> **TRY IT ::** 10.91   Find the intercepts of the parabola $y = x^2 + 2x - 8$.

> **TRY IT ::** 10.92   Find the intercepts of the parabola $y = x^2 - 4x - 12$.

In this chapter, we have been solving quadratic equations of the form $ax^2 + bx + c = 0$. We solved for x and the results were the solutions to the equation.

We are now looking at quadratic equations in two variables of the form $y = ax^2 + bx + c$. The graphs of these equations are parabolas. The x-intercepts of the parabolas occur where $y = 0$.

For example:

Quadratic equation	**Quadratic equation in two variables**
	$y = x^2 - 2x - 15$
$x^2 - 2x - 15 = 0$	let $y = 0 \quad\quad 0 = x^2 - 2x - 15$
$(x - 5)(x + 3) = 0$	$0 = (x - 5)(x + 3)$
$x - 5 = 0 \quad x + 3 = 0$	$x - 5 = 0 \quad x + 3 = 0$
$x = 5 \quad\quad x = -3$	$x = 5 \quad\quad x = -3$
	$(5, 0)$ and $(-3, 0)$
	x-intercepts

The solutions of the quadratic equation are the x values of the x-intercepts.

Earlier, we saw that quadratic equations have 2, 1, or 0 solutions. The graphs below show examples of parabolas for these three cases. Since the solutions of the equations give the x-intercepts of the graphs, the number of x-intercepts is the same as the number of solutions.

Previously, we used the discriminant to determine the number of solutions of a quadratic equation of the form $ax^2 + bx + c = 0$. Now, we can use the discriminant to tell us how many x-intercepts there are on the graph.

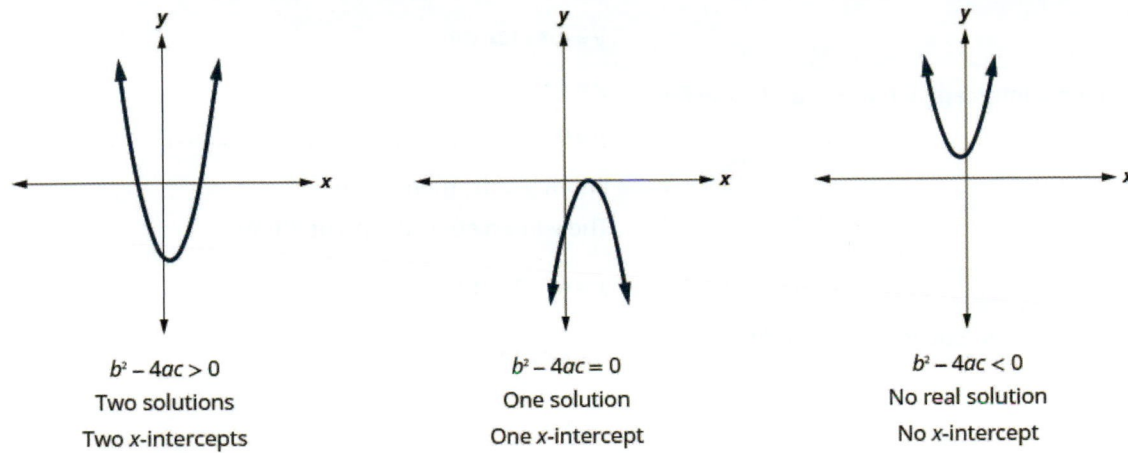

$b^2 - 4ac > 0$	$b^2 - 4ac = 0$	$b^2 - 4ac < 0$
Two solutions	One solution	No real solution
Two x-intercepts	One x-intercept	No x-intercept

Before you start solving the quadratic equation to find the values of the x-intercepts, you may want to evaluate the discriminant so you know how many solutions to expect.

EXAMPLE 10.47

Find the intercepts of the parabola $y = 5x^2 + x + 4$.

✓ Solution

$$y = 5x^2 + x + 4$$

To find the y-intercept, let $x = 0$ and solve for y.	$y = 5 \cdot 0^2 + 0 + 4$
	$y = 4$
	When $x = 0$, then $y = 4$. The y-intercept is the point $(0, 4)$.
	$y = 5x^2 + x + 4$
To find the x-intercept, let $y = 0$ and solve for x.	$0 = 5x^2 + x + 4$
Find the value of the discriminant to predict the number of solutions and so x-intercepts.	$b^2 \quad - \quad 4ac$ $1^2 \quad - \quad 4 \cdot 5 \cdot 4$ $1 \quad - \quad 80$ -79
Since the value of the discriminant is negative, there is no real solution to the equation.	There are no x-intercepts.

> **TRY IT :: 10.93**   Find the intercepts of the parabola $y = 3x^2 + 4x + 4$.

> **TRY IT :: 10.94**   Find the intercepts of the parabola $y = x^2 - 4x - 5$.

EXAMPLE 10.48

Find the intercepts of the parabola $y = 4x^2 - 12x + 9$.

✓ **Solution**

	$y = 4x^2 - 12x + 9$
To find the *y*-intercept, let $x = 0$ and solve for *y*.	$y = 4 \cdot 0^2 - 12 \cdot 0 + 9$ $y = 9$
	When $x = 0$, then $y = 9$. The *y*-intercept is the point $(0, 9)$.
	$y = 4x^2 - 12x + 9$
To find the *x*-intercept, let $y = 0$ and solve for *x*.	$0 = 4x^2 - 12x + 9$
Find the value of the discriminant to predict the number of solutions and so *x*-intercepts.	$b^2 - 4ac$ $2^2 - 4 \cdot 4 \cdot 9$ $144 - 144$ 0
	Since the value of the discriminant is 0, there is no real solution to the equation. So there is one *x*-intercept.
Solve the equation by factoring the perfect square trinomial.	$0 = (2x - 3)^2$
Use the Zero Product Property.	$0 = 2x - 3$
Solve for *x*.	$3 = 2x$ $\frac{3}{2} = x$
	When $y = 0$, then $\frac{3}{2} = x$.
	The *x*-intercept is the point $\left(\frac{3}{2}, 0\right)$.

> **TRY IT ::** 10.95   Find the intercepts of the parabola $y = -x^2 - 12x - 36$.

> **TRY IT ::** 10.96   Find the intercepts of the parabola $y = 9x^2 + 12x + 4$.

Graph Quadratic Equations in Two Variables

Now, we have all the pieces we need in order to graph a quadratic equation in two variables. We just need to put them together. In the next example, we will see how to do this.

EXAMPLE 10.49   HOW TO GRAPH A QUADRATIC EQUATION IN TWO VARIABLES

Graph $y = x^2 - 6x + 8$.

✓ Solution

Step 1. Write the quadratic equation with y on one side.	This equation has y on one side.	$y = x^2 - 6x + 8$ $a = 1, b = -6, c = 8$
Step 2. Determine whether the parabola opens upward or downward.	Look at a in the equation. $y = x^2 - 6x + 8$ Since a is positive, the parabola opens upward.	⋎ **The parabola opens upward.**
Step 3. Find the axis of symmetry.	$y = x^2 - 6x + 8$ The axis of symmetry is the line $x = -\dfrac{b}{2a}$.	Axis of Symmetry $x = -\dfrac{b}{2a}$ $x = -\dfrac{(-6)}{2 \cdot 1}$ $x = 3$ **The axis of symmetry is the line $x = 3$.**
Step 4. Find the vertex.	The vertex is on the axis of symmetry. Substitute $x = 3$ into the equation and solve for y.	Vertex $y = x^2 - 6x + 8$ $y = (3)^2 - 6(3) + 8$ $y = -1$ **The vertex is $(3, -1)$.**
Step 5. Find the y-intercept. Find the point symmetric to the y-intercept across the axis of symmetry.	We substitute $x = 0$ into the equation.	y-intercept $y = x^2 - 6x + 8$ $y = (0)^2 - 6(0) + 8$ $y = 8$ **The y-intercept is $(0, 8)$.**
	We use the axis of symmetry to find a point symmetric to the y-intercept. The y-intercept is 3 units left of the axis of symmetry, $x = 3$. A point 3 units to the right of the axis of symmetry has $x = 6$.	Point symmetric to y-intercept **The point is $(6, 8)$.**
Step 6. Find the x-intercepts.		x-intercept $y = x^2 - 6x + 8$
	We substitute $y = 0$ into the equation. We can solve this quadratic equation by factoring.	$0 = x^2 - 6x + 8$ $0 = (x - 2)(x - 4)$ $x = 2$ or $x = 4$ **The x-intercepts are $(2, 0)$ and $(4, 0)$.**

Step 7. Graph the parabola.	We graph the vertex, intercepts, and the point symmetric to the *y*-intercept. We connect these 5 points to sketch the parabola.	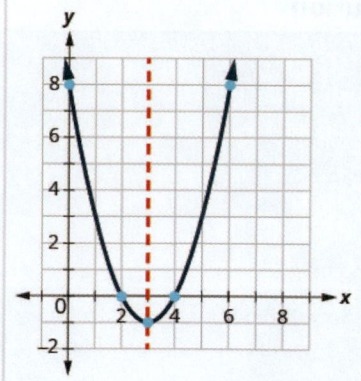

> **TRY IT :: 10.97** Graph the parabola $y = x^2 + 2x - 8$.

> **TRY IT :: 10.98** Graph the parabola $y = x^2 - 8x + 12$.

HOW TO :: GRAPH A QUADRATIC EQUATION IN TWO VARIABLES.

Step 1. Write the quadratic equation with y on one side.

Step 2. Determine whether the parabola opens upward or downward.

Step 3. Find the axis of symmetry.

Step 4. Find the vertex.

Step 5. Find the *y*-intercept. Find the point symmetric to the *y*-intercept across the axis of symmetry.

Step 6. Find the *x*-intercepts.

Step 7. Graph the parabola.

We were able to find the *x*-intercepts in the last example by factoring. We find the *x*-intercepts in the next example by factoring, too.

EXAMPLE 10.50

Graph $y = -x^2 + 6x - 9$.

 Solution

The equation *y* has on one side.	$y = ax^2 + bx + c$ $y = -x^2 + 6x - 9$

Since a is -1, the parabola opens downward.

To find the axis of symmetry, find $x = -\frac{b}{2a}$.

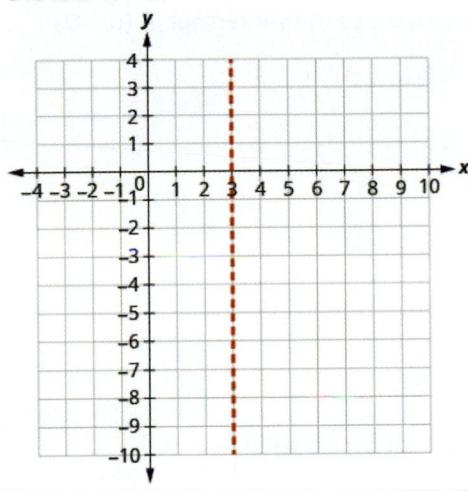

$x = -\frac{b}{2a}$

$x = -\frac{6}{2(-1)}$

$x = 3$

The axis of symmetry is $x = 3$. The vertex is on the line $x = 3$.

Find y when $x = 3$.

$y = -x^2 + 6x - 9$

$y = -3^2 + 6 \cdot 3 - 9$

$y = -9 + 18 - 9$

$y = 0$

The vertex is $(3, 0)$.

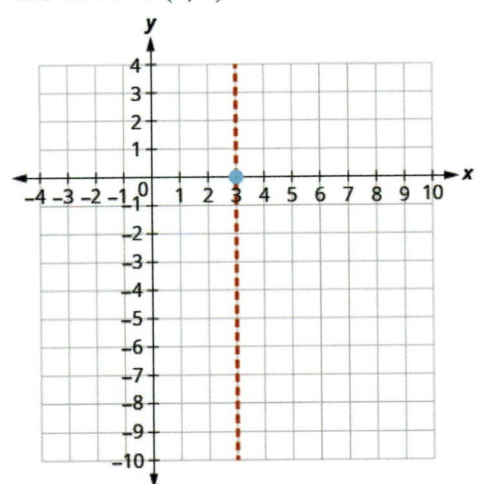

The y-intercept occurs when $x = 0$.

Substitute $x = 0$.

Simplify.

$y = -x^2 + 6x - 9$

$y = -0^2 + 6 \cdot 0 - 9$

$y = -9$

The point $(0, -9)$ is three units to the left of the line of symmetry.

The point three units to the right of the line of symmetry is $(6, -9)$.

Point symmetric to the y-intercept is $(6, -9)$

The y-intercept is $(0, -9)$.

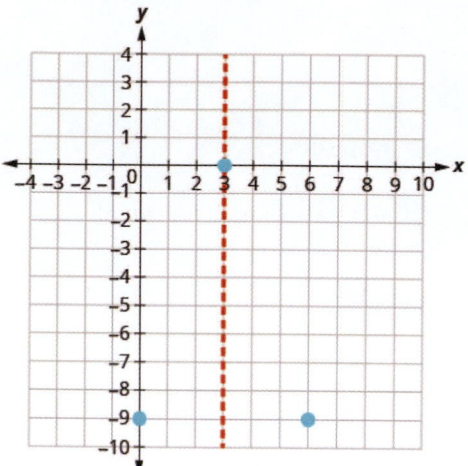

The x-intercept occurs when $y = 0$.		$y = -x^2 + 6x - 9$
Substitute $y = 0$.		$0 = -x^2 + 6x - 9$
Factor the GCF.		$0 = -(x^2 - 6x + 9)$
Factor the trinomial.		$0 = -(x - 3)^2$
Solve for x.		$x = 3$
Connect the points to graph the parabola.		

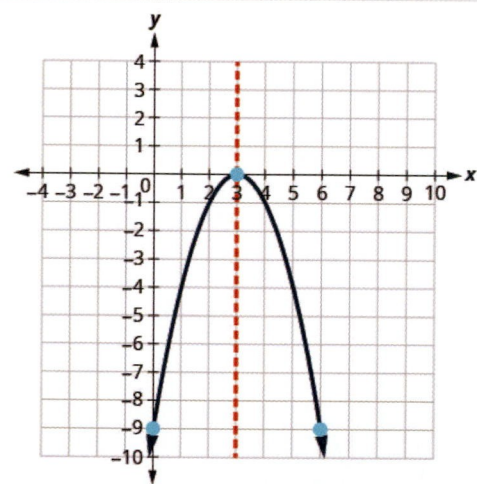

> **TRY IT : : 10.99**
Graph the parabola $y = -3x^2 + 12x - 12$.

> **TRY IT : : 10.100**
Graph the parabola $y = 25x^2 + 10x + 1$.

For the graph of $y = -x^2 + 6x - 9$, the vertex and the x-intercept were the same point. Remember how the discriminant determines the number of solutions of a quadratic equation? The discriminant of the equation $0 = -x^2 + 6x - 9$ is 0, so there is only one solution. That means there is only one x-intercept, and it is the vertex of the parabola.

How many x-intercepts would you expect to see on the graph of $y = x^2 + 4x + 5$?

EXAMPLE 10.51

Graph $y = x^2 + 4x + 5$.

⊘ Solution

The equation has y on one side.	$y = ax^2 + bx + c$ $y = x^2 + 4x + 5$
Since a is 1, the parabola opens upward.	
To find the axis of symmetry, find $x = -\dfrac{b}{2a}$.	$x = -\dfrac{b}{2a}$ $x = -\dfrac{4}{2(1)}$ $x = -2$ The axis of symmetry is $x = -2$.

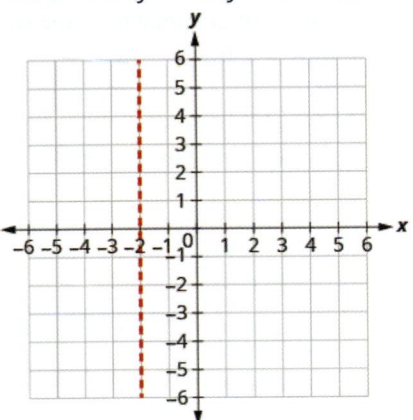

The vertex is on the line $x = -2$.

Find y when $x = -2$.

$$y = x^2 + 4x + 5$$

$$y = (-2)^2 + 4 \cdot (-2) + 5$$

$$y = 4 - 8 + 5$$

$$y = 1$$

The vertex is $(-2, 1)$.

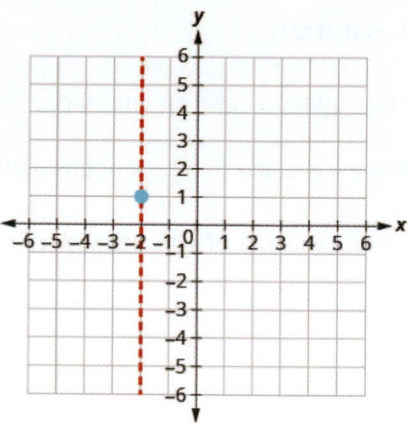

The y-intercept occurs when $x = 0$.

Substitute $x = 0$.

Simplify.

The point $(0, 5)$ is two units to the right of the line of symmetry.

The point two units to the left of the line of symmetry is $(-4, 5)$.

$$y = x^2 + 4x + 5$$

$$y = (0)^2 + 4(0) + 5$$

$$y = 5$$

The y-intercept is $(0, 5)$.

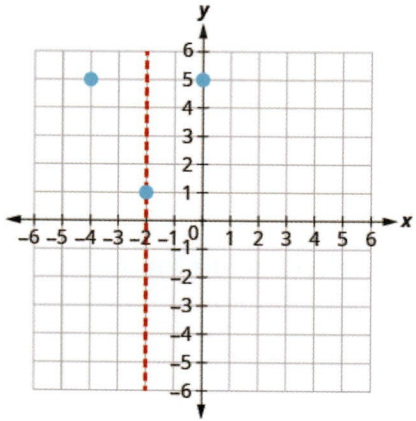

Point symmetric to the y-intercept is $(-4, 5)$.

The x-intercept occurs when $y = 0$.

Substitute $y = 0$.

Test the discriminant.

$$y = x^2 + 4x + 5$$

$$0 = x^2 + 4x + 5$$

$$b^2 - 4ac$$
$$4^2 - 4 \cdot 15$$
$$16 - 20$$
$$-4$$

Since the value of the discriminant is negative, there is no solution and so no *x*- intercept.
Connect the points to graph the parabola. You may want to choose two more points for greater accuracy.

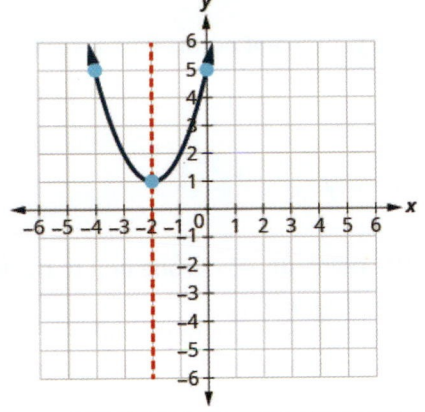

> **TRY IT : :** 10.101 Graph the parabola $y = 2x^2 - 6x + 5$.

> **TRY IT : :** 10.102 Graph the parabola $y = -2x^2 - 1$.

Finding the *y*-intercept by substituting $x = 0$ into the equation is easy, isn't it? But we needed to use the Quadratic Formula to find the *x*-intercepts in **Example 10.51**. We will use the Quadratic Formula again in the next example.

EXAMPLE 10.52

Graph $y = 2x^2 - 4x - 3$.

⊘ Solution

$$y = ax^2 + bx + c$$
$$y = 2x^2 - 4x - 3$$

The equation *y* has one side. Since *a* is 2, the parabola opens upward.	
To find the axis of symmetry, find $x = -\dfrac{b}{2a}$.	$x = -\dfrac{b}{2a}$ $x = -\dfrac{-4}{2 \cdot 2}$ $x = 1$ The axis of symmetry is $x = 1$.
The vertex on the line $x = 1$.	$y = 2x^2 - 4x - 3$

Find y when $x = 1$.	$y = 2(1)^2 - 4 \cdot (1) - 3$ $y = 2 - 4 - 3$ $y = -5$ The vertex is $(1, -5)$.
The y-intercept occurs when $x = 0$.	$y = 2x^2 - 4x - 3$
Substitute $x = 0$.	$y = 2 \cdot 0^2 - 4 \cdot 0 - 3$
Simplify.	$y = -3$ The y-intercept is $(0, -3)$.
The point $(0, -3)$ is one unit to the left of the line of symmetry. The point one unit to the right of the line of symmetry is $(2, -3)$	Point symmetric to the y-intercept is $(2, -3)$.
The x-intercept occurs when $y = 0$.	$y = 2x^2 - 4x - 3$
Substitute $y = 0$.	$0 = 2x^2 - 4x - 3$
Use the Quadratic Formula.	$x = \dfrac{-b \pm \sqrt{b^2 - 4ac}}{2a}$
Substitute in the values of a, b, c.	$x = \dfrac{-(-4) \pm \sqrt{(-4)^2 - 4 \cdot 2 \cdot (-3)}}{2 \cdot 2}$
Simplify.	$x = \dfrac{4 \pm \sqrt{16 + 24}}{4}$
Simplify inside the radical.	$x = \dfrac{4 \pm \sqrt{40}}{4}$
Simplify the radical.	$x = \dfrac{4 \pm 2\sqrt{10}}{4}$
Factor the GCF.	$x = \dfrac{2(2 \pm \sqrt{10})}{4}$
Remove common factors.	$x = \dfrac{2 \pm \sqrt{10}}{2}$
Write as two equations.	$x = \dfrac{2 + \sqrt{10}}{2}, x = \dfrac{2 - \sqrt{10}}{2}$
Approximate the values.	$x \approx 2.5, x \approx -0.6$
	The approximate values of the x-intercepts are $(2.5, 0)$ and $(-0.6, 0)$.

Graph the parabola using the points found.

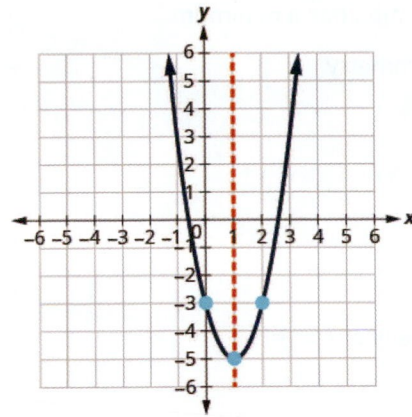

> **TRY IT : : 10.103**   Graph the parabola $y = 5x^2 + 10x + 3$.

> **TRY IT : : 10.104**   Graph the parabola $y = -3x^2 - 6x + 5$.

Solve Maximum and Minimum Applications

Knowing that the vertex of a parabola is the lowest or highest point of the parabola gives us an easy way to determine the minimum or maximum value of a quadratic equation. The y-coordinate of the vertex is the minimum y-value of a parabola that opens upward. It is the maximum y-value of a parabola that opens downward. See **Figure 10.7**.

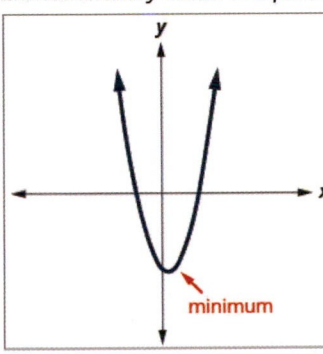
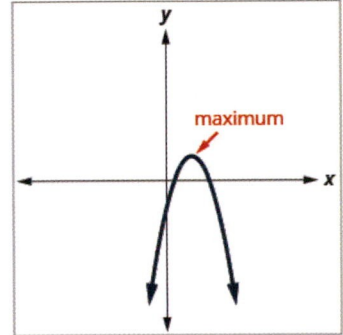

Figure 10.7

Minimum or Maximum Values of a Quadratic Equation

The **y-coordinate of the vertex** of the graph of a quadratic equation is the

- minimum value of the quadratic equation if the parabola opens upward.
- maximum value of the quadratic equation if the parabola opens downward.

EXAMPLE 10.53

Find the minimum value of the quadratic equation $y = x^2 + 2x - 8$.

⊘ Solution

$$y = x^2 + 2x - 8$$

Since a is positive, the parabola opens upward.

The quadratic equation has a minimum.

Find the axis of symmetry.	$x = -\dfrac{b}{2a}$
	$x = -\dfrac{2}{2 \cdot 1}$
	$x = -1$
	The axis of symmetry is $x = -1$.
The vertex is on the line $x = -1$.	$y = x^2 + 2x - 8$
Find y when $x = -1$.	$y = (-1)^2 + 2 \cdot (-1) - 8$
	$y = 1 - 2 - 8$
	$y = -9$
	The vertex is $(-1, -9)$.

Since the parabola has a minimum, the *y*-coordinate of the vertex is the minimum *y*-value of the quadratic equation.

The minimum value of the quadratic is -9 and it occurs when $x = -1$.

Show the graph to verify the result.

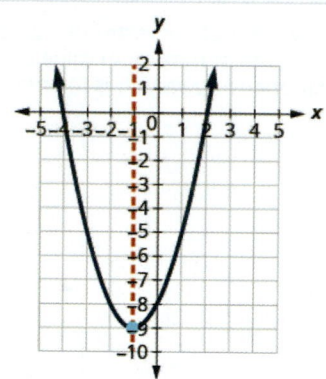

> **TRY IT :: 10.105**   Find the maximum or minimum value of the quadratic equation $y = x^2 - 8x + 12$.

> **TRY IT :: 10.106**   Find the maximum or minimum value of the quadratic equation $y = -4x^2 + 16x - 11$.

We have used the formula

$$h = -16t^2 + v_0 t + h_0$$

to calculate the height in feet, h, of an object shot upwards into the air with initial velocity, v_0, after t seconds.

This formula is a quadratic equation in the variable t, so its graph is a parabola. By solving for the coordinates of the vertex, we can find how long it will take the object to reach its maximum height. Then, we can calculate the maximum height.

EXAMPLE 10.54

The quadratic equation $h = -16t^2 + v_0 t + h_0$ models the height of a volleyball hit straight upwards with velocity 176 feet per second from a height of 4 feet.

ⓐ How many seconds will it take the volleyball to reach its maximum height?

ⓑ Find the maximum height of the volleyball.

✅ Solution

$h = -16t^2 + 176t + 4$

Since a is negative, the parabola opens downward.

The quadratic equation has a maximum.

ⓐ

Find the axis of symmetry.	$t = -\dfrac{b}{2a}$ $t = -\dfrac{176}{2(-16)}$ $t = 5.5$ The axis of symmetry is $t = 5.5$.
The vertex is on the line $t = 5.5$.	The maximum occurs when $t = 5.5$ seconds.

ⓑ

Find h when $t = 5.5$.	$h = -16t^2 + 176t + 4$ $h = -16(5.5)^2 + 176 \cdot (5.5) + 4$
Use a calculator to simplify.	$h = 488$
	The vertex is $(5.5,\ 488)$.
Since the parabola has a maximum, the h-coordinate of the vertex is the maximum y-value of the quadratic equation.	The maximum value of the quadratic is 488 feet and it occurs when $t = 5.5$ seconds.

> **TRY IT ∷ 10.107**
>
> The quadratic equation $h = -16t^2 + 128t + 32$ is used to find the height of a stone thrown upward from a height of 32 feet at a rate of 128 ft/sec. How long will it take for the stone to reach its maximum height? What is the maximum height? Round answers to the nearest tenth.

> **TRY IT ∷ 10.108**
>
> A toy rocket shot upward from the ground at a rate of 208 ft/sec has the quadratic equation of $h = -16t^2 + 208t$. When will the rocket reach its maximum height? What will be the maximum height? Round answers to the nearest tenth.

> ▶ **MEDIA ∷**
>
> Access these online resources for additional instruction and practice graphing quadratic equations:
> - **Graphing Quadratic Functions (https://openstax.org/l/25Graphquad1)**
> - **How do you graph a quadratic function? (https://openstax.org/l/25Graphquad2)**
> - **Graphing Quadratic Equations (https://openstax.org/l/25Graphquad3)**

10.5 EXERCISES

Practice Makes Perfect

Recognize the Graph of a Quadratic Equation in Two Variables

In the following exercises, graph:

163. $y = x^2 + 3$

164. $y = -x^2 + 1$

In the following exercises, determine if the parabola opens up or down.

165. $y = -2x^2 - 6x - 7$

166. $y = 6x^2 + 2x + 3$

167. $y = 4x^2 + x - 4$

168. $y = -9x^2 - 24x - 16$

Find the Axis of Symmetry and Vertex of a Parabola

In the following exercises, find ⓐ the axis of symmetry and ⓑ the vertex.

169. $y = x^2 + 8x - 1$

170. $y = x^2 + 10x + 25$

171. $y = -x^2 + 2x + 5$

172. $y = -2x^2 - 8x - 3$

Find the Intercepts of a Parabola

In the following exercises, find the x- and y-intercepts.

173. $y = x^2 + 7x + 6$

174. $y = x^2 + 10x - 11$

175. $y = -x^2 + 8x - 19$

176. $y = x^2 + 6x + 13$

177. $y = 4x^2 - 20x + 25$

178. $y = -x^2 - 14x - 49$

Graph Quadratic Equations in Two Variables

In the following exercises, graph by using intercepts, the vertex, and the axis of symmetry.

179. $y = x^2 + 6x + 5$

180. $y = x^2 + 4x - 12$

181. $y = x^2 + 4x + 3$

182. $y = x^2 - 6x + 8$

183. $y = 9x^2 + 12x + 4$

184. $y = -x^2 + 8x - 16$

185. $y = -x^2 + 2x - 7$

186. $y = 5x^2 + 2$

187. $y = 2x^2 - 4x + 1$

188. $y = 3x^2 - 6x - 1$

189. $y = 2x^2 - 4x + 2$

190. $y = -4x^2 - 6x - 2$

191. $y = -x^2 - 4x + 2$

192. $y = x^2 + 6x + 8$

193. $y = 5x^2 - 10x + 8$

194. $y = -16x^2 + 24x - 9$

195. $y = 3x^2 + 18x + 20$

196. $y = -2x^2 + 8x - 10$

Solve Maximum and Minimum Applications

In the following exercises, find the maximum or minimum value.

197. $y = 2x^2 + x - 1$

198. $y = -4x^2 + 12x - 5$

199. $y = x^2 - 6x + 15$

200. $y = -x^2 + 4x - 5$

201. $y = -9x^2 + 16$

202. $y = 4x^2 - 49$

In the following exercises, solve. Round answers to the nearest tenth.

203. An arrow is shot vertically upward from a platform 45 feet high at a rate of 168 ft/sec. Use the quadratic equation $h = -16t^2 + 168t + 45$ to find how long it will take the arrow to reach its maximum height, and then find the maximum height.

204. A stone is thrown vertically upward from a platform that is 20 feet high at a rate of 160 ft/sec. Use the quadratic equation $h = -16t^2 + 160t + 20$ to find how long it will take the stone to reach its maximum height, and then find the maximum height.

205. A computer store owner estimates that by charging x dollars each for a certain computer, he can sell $40 - x$ computers each week. The quadratic equation $R = -x^2 + 40x$ is used to find the revenue, R, received when the selling price of a computer is x. Find the selling price that will give him the maximum revenue, and then find the amount of the maximum revenue.

206. A retailer who sells backpacks estimates that, by selling them for x dollars each, he will be able to sell $100 - x$ backpacks a month. The quadratic equation $R = -x^2 + 100x$ is used to find the R received when the selling price of a backpack is x. Find the selling price that will give him the maximum revenue, and then find the amount of the maximum revenue.

207. A rancher is going to fence three sides of a corral next to a river. He needs to maximize the corral area using 240 feet of fencing. The quadratic equation $A = x(240 - 2x)$ gives the area of the corral, A, for the length, x, of the corral along the river. Find the length of the corral along the river that will give the maximum area, and then find the maximum area of the corral.

208. A veterinarian is enclosing a rectangular outdoor running area against his building for the dogs he cares for. He needs to maximize the area using 100 feet of fencing. The quadratic equation $A = x(100 - 2x)$ gives the area, A, of the dog run for the length, x, of the building that will border the dog run. Find the length of the building that should border the dog run to give the maximum area, and then find the maximum area of the dog run.

Everyday Math

209. In the previous set of exercises, you worked with the quadratic equation $R = -x^2 + 40x$ that modeled the revenue received from selling computers at a price of x dollars. You found the selling price that would give the maximum revenue and calculated the maximum revenue. Now you will look at more characteristics of this model.

ⓐ Graph the equation $R = -x^2 + 40x$. ⓑ Find the values of the *x*-intercepts.

210. In the previous set of exercises, you worked with the quadratic equation $R = -x^2 + 100x$ that modeled the revenue received from selling backpacks at a price of x dollars. You found the selling price that would give the maximum revenue and calculated the maximum revenue. Now you will look at more characteristics of this model.

ⓐ Graph the equation $R = -x^2 + 100x$. ⓑ Find the values of the *x*-intercepts.

Writing Exercises

211. For the revenue model in **Exercise 10.205** and **Exercise 10.209**, explain what the *x*-intercepts mean to the computer store owner.

212. For the revenue model in **Exercise 10.206** and **Exercise 10.210**, explain what the *x*-intercepts mean to the backpack retailer.

Self Check

ⓐ *After completing the exercises, use this checklist to evaluate your mastery of the objectives of this section.*

I can...	Confidently	With some help	No-I don't get it!
recognize the graph of a quadratic equation in two variables.			
find the axis of symmetry and vertex of a parabola.			
find the intercepts of a parabola.			
graph quadratic equations in two variables.			
solve maximum and minimum applications.			

ⓑ *What does this checklist tell you about your mastery of this section? What steps will you take to improve?*

CHAPTER 10 REVIEW

KEY TERMS

axis of symmetry The axis of symmetry is the vertical line passing through the middle of the parabola $y = ax^2 + bx + c$.

completing the square Completing the square is a method used to solve quadratic equations.

consecutive even integers Consecutive even integers are even integers that follow right after one another. If an even integer is represented by n, the next consecutive even integer is $n + 2$, and the next after that is $n + 4$.

consecutive odd integers Consecutive odd integers are odd integers that follow right after one another. If an odd integer is represented by n, the next consecutive odd integer is $n + 2$, and the next after that is $n + 4$.

discriminant In the Quadratic Formula, $x = \dfrac{-b \pm \sqrt{b^2 - 4ac}}{2a}$ the quantity $b^2 - 4ac$ is called the discriminant.

parabola The graph of a quadratic equation in two variables is a parabola.

quadratic equation A quadratic equation is an equation of the form $ax^2 + bx + c = 0$, where $a \neq 0$.

quadratic equation in two variables A quadratic equation in two variables, where a, b, and c are real numbers and $a \neq 0$ is an equation of the form $y = ax^2 + bx + c$.

Square Root Property The Square Root Property states that, if $x^2 = k$ and $k \geq 0$, then $x = \sqrt{k}$ or $x = -\sqrt{k}$.

vertex The point on the parabola that is on the axis of symmetry is called the *vertex* of the parabola; it is the lowest or highest point on the parabola, depending on whether the parabola opens upwards or downwards.

x-intercepts of a parabola The *x*-intercepts are the points on the parabola where $y = 0$.

y-intercept of a parabola The *y*-intercept is the point on the parabola where $x = 0$.

KEY CONCEPTS

10.1 Solve Quadratic Equations Using the Square Root Property

- Square Root Property
  If $x^2 = k$, and $k \geq 0$, then $x = \sqrt{k}$ or $x = -\sqrt{k}$.

10.2 Solve Quadratic Equations by Completing the Square

- Binomial Squares Pattern If a, b are real numbers,
  $$(a + b)^2 = a^2 + 2ab + b^2$$

$(a + b)^2$	$=$	a^2	$+$	$2ab$	$+$	b^2
(binomial)²		(first term)²		2 × (product of terms)		(second term)²

  $$(a - b)^2 = a^2 - 2ab + b^2$$

$(a - b)^2$	$=$	a^2	$-$	$2ab$	$+$	b^2
(binomial)²		(first term)²		2 × (product of terms)		(second term)²

- Complete a Square
  To complete the square of $x^2 + bx$:

  Step 1. Identify b, the coefficient of x.

  Step 2. Find $\left(\dfrac{1}{2}b\right)^2$, the number to complete the square.

Step 3.

Add the $\left(\frac{1}{2}b\right)^2$ to $x^2 + bx$.

10.3 Solve Quadratic Equations Using the Quadratic Formula

- **Quadratic Formula** The solutions to a quadratic equation of the form $ax^2 + bx + c = 0$, $a \neq 0$ are given by the formula:

$$x = \frac{-b \pm \sqrt{b^2 - 4ac}}{2a}$$

- **Solve a Quadratic Equation Using the Quadratic Formula**
  To solve a quadratic equation using the Quadratic Formula.

  Step 1. Write the quadratic formula in standard form. Identify the a, b, c values.

  Step 2. Write the quadratic formula. Then substitute in the values of a, b, c.

  Step 3. Simplify.

  Step 4. Check the solutions.

- **Using the Discriminant, $b^2 - 4ac$, to Determine the Number of Solutions of a Quadratic Equation**
  For a quadratic equation of the form $ax^2 + bx + c = 0$, $a \neq 0$,

  - if $b^2 - 4ac > 0$, the equation has 2 solutions.

  - if $b^2 - 4ac = 0$, the equation has 1 solution.

  - if $b^2 - 4ac < 0$, the equation has no real solutions.

- **To identify the most appropriate method to solve a quadratic equation:**

  Step 1. Try Factoring first. If the quadratic factors easily this method is very quick.

  Step 2. Try the Square Root Property next. If the equation fits the form $ax^2 = k$ or $a(x - h)^2 = k$, it can easily be solved by using the Square Root Property.

  Step 3. Use the Quadratic Formula. Any other quadratic equation is best solved by using the Quadratic Formula.

10.4 Solve Applications Modeled by Quadratic Equations

- **Area of a Triangle** For a triangle with base, b, and height, h, the area, A, is given by the formula: $A = \frac{1}{2}bh$

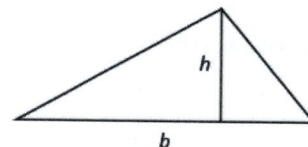

- **Pythagorean Theorem** In any right triangle, where a and b are the lengths of the legs, and c is the length of the hypthenuse, $a^2 + b^2 = c^2$

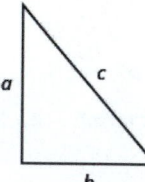

- **Projectile motion** The height in feet, h, of an object shot upwards into the air with initial velocity, v_0, after t seconds can be modeled by the formula:

$$h = -16t^2 + v_0 t$$

10.5 Graphing Quadratic Equations

- **The graph of every quadratic equation is a parabola.**
- **Parabola Orientation** For the quadratic equation $y = ax^2 + bx + c$, if
    - $a > 0$, the parabola opens upward.
    - $a < 0$, the parabola opens downward.
- **Axis of Symmetry and Vertex of a Parabola** For a parabola with equation $y = ax^2 + bx + c$:
    - The axis of symmetry of a parabola is the line $x = -\dfrac{b}{2a}$.
    - The vertex is on the axis of symmetry, so its x-coordinate is $-\dfrac{b}{2a}$.
    - To find the y-coordinate of the vertex we substitute $x = -\dfrac{b}{2a}$ into the quadratic equation.
- **Find the Intercepts of a Parabola** To find the intercepts of a parabola with equation $y = ax^2 + bx + c$:

y-intercept	x-intercepts
Let $x = 0$ and solve for y.	Let $y = 0$ and solve for x.

- **To Graph a Quadratic Equation in Two Variables**

    Step 1. Write the quadratic equation with y on one side.

    Step 2. Determine whether the parabola opens upward or downward.

    Step 3. Find the axis of symmetry.

    Step 4. Find the vertex.

    Step 5. Find the y-intercept. Find the point symmetric to the y-intercept across the axis of symmetry.

    Step 6. Find the x-intercepts.

    Step 7. Graph the parabola.
- **Minimum or Maximum Values of a Quadratic Equation**
    - The y-**coordinate of the vertex** of the graph of a quadratic equation is the
    - **minimum** value of the quadratic equation if the parabola opens upward.
    - **maximum** value of the quadratic equation if the parabola opens downward.

REVIEW EXERCISES

10.1 10.1 Solve Quadratic Equations Using the Square Root Property

In the following exercises, solve using the Square Root Property.

213. $x^2 = 100$

214. $y^2 = 144$

215. $m^2 - 40 = 0$

216. $n^2 - 80 = 0$

217. $4a^2 = 100$

218. $2b^2 = 72$

219. $r^2 + 32 = 0$

220. $t^2 + 18 = 0$

221. $\frac{4}{3}v^2 + 4 = 28$

222. $\frac{2}{3}w^2 - 20 = 30$

223. $5c^2 + 3 = 19$

224. $3d^2 - 6 = 43$

In the following exercises, solve using the Square Root Property.

225. $(p - 5)^2 + 3 = 19$

226. $(q + 4)^2 = 9$

227. $(u + 1)^2 = 45$

228. $(z-5)^2 = 50$

229. $\left(x-\frac{1}{4}\right)^2 = \frac{3}{16}$

230. $\left(y-\frac{2}{3}\right)^2 = \frac{2}{9}$

231. $(m-7)^2 + 6 = 30$

232. $(n-4)^2 - 50 = 150$

233. $(5c+3)^2 = -20$

234. $(4c-1)^2 = -18$

235. $m^2 - 6m + 9 = 48$

236. $n^2 + 10n + 25 = 12$

237. $64a^2 + 48a + 9 = 81$

238. $4b^2 - 28b + 49 = 25$

10.2 10.2 Solve Quadratic Equations Using Completing the Square

In the following exercises, complete the square to make a perfect square trinomial. Then write the result as a binomial squared.

239. $x^2 + 22x$

240. $y^2 + 6y$

241. $m^2 - 8m$

242. $n^2 - 10n$

243. $a^2 - 3a$

244. $b^2 + 13b$

245. $p^2 + \frac{4}{5}p$

246. $q^2 - \frac{1}{3}q$

In the following exercises, solve by completing the square.

247. $c^2 + 20c = 21$

248. $d^2 + 14d = -13$

249. $x^2 - 4x = 32$

250. $y^2 - 16y = 36$

251. $r^2 + 6r = -100$

252. $t^2 - 12t = -40$

253. $v^2 - 14v = -31$

254. $w^2 - 20w = 100$

255. $m^2 + 10m - 4 = -13$

256. $n^2 - 6n + 11 = 34$

257. $a^2 = 3a + 8$

258. $b^2 = 11b - 5$

259. $(u+8)(u+4) = 14$

260. $(z-10)(z+2) = 28$

261. $3p^2 - 18p + 15 = 15$

262. $5q^2 + 70q + 20 = 0$

263. $4y^2 - 6y = 4$

264. $2x^2 + 2x = 4$

265. $3c^2 + 2c = 9$

266. $4d^2 - 2d = 8$

10.3 10.3 Solve Quadratic Equations Using the Quadratic Formula

In the following exercises, solve by using the Quadratic Formula.

267. $4x^2 - 5x + 1 = 0$

268. $7y^2 + 4y - 3 = 0$

269. $r^2 - r - 42 = 0$

270. $t^2 + 13t + 22 = 0$

271. $4v^2 + v - 5 = 0$

272. $2w^2 + 9w + 2 = 0$

273. $3m^2 + 8m + 2 = 0$

274. $5n^2 + 2n - 1 = 0$

275. $6a^2 - 5a + 2 = 0$

276. $4b^2 - b + 8 = 0$

277. $u(u-10) + 3 = 0$

278. $5z(z-2) = 3$

279. $\frac{1}{8}p^2 - \frac{1}{5}p = -\frac{1}{20}$

280. $\frac{2}{5}q^2 + \frac{3}{10}q = \frac{1}{10}$

281. $4c^2 + 4c + 1 = 0$

282. $9d^2 - 12d = -4$

In the following exercises, determine the number of solutions to each quadratic equation.

283.

ⓐ $9x^2 - 6x + 1 = 0$

ⓑ $3y^2 - 8y + 1 = 0$

ⓒ $7m^2 + 12m + 4 = 0$

ⓓ $5n^2 - n + 1 = 0$

284.

ⓐ $5x^2 - 7x - 8 = 0$

ⓑ $7x^2 - 10x + 5 = 0$

ⓒ

$25x^2 - 90x + 81 = 0$

ⓓ $15x^2 - 8x + 4 = 0$

In the following exercises, identify the most appropriate method (Factoring, Square Root, or Quadratic Formula) to use to solve each quadratic equation.

285.

ⓐ $16r^2 - 8r + 1 = 0$

ⓑ $5t^2 - 8t + 3 = 9$

ⓒ $3(c + 2)^2 = 15$

286.

ⓐ

$4d^2 + 10d - 5 = 21$

ⓑ

$25x^2 - 60x + 36 = 0$

ⓒ $6(5v - 7)^2 = 150$

10.4 10.4 Solve Applications Modeled by Quadratic Equations

In the following exercises, solve by using methods of factoring, the square root principle, or the quadratic formula.

287. Find two consecutive odd numbers whose product is 323.

288. Find two consecutive even numbers whose product is 624.

289. A triangular banner has an area of 351 square centimeters. The length of the base is two centimeters longer than four times the height. Find the height and length of the base.

290. Julius built a triangular display case for his coin collection. The height of the display case is six inches less than twice the width of the base. The area of the of the back of the case is 70 square inches. Find the height and width of the case.

291. A tile mosaic in the shape of a right triangle is used as the corner of a rectangular pathway. The hypotenuse of the mosaic is 5 feet. One side of the mosaic is twice as long as the other side. What are the lengths of the sides? Round to the nearest tenth.

292. A rectangular piece of plywood has a diagonal which measures two feet more than the width. The length of the plywood is twice the width. What is the length of the plywood's diagonal? Round to the nearest tenth.

293. The front walk from the street to Pam's house has an area of 250 square feet. Its length is two less than four times its width. Find the length and width of the sidewalk. Round to the nearest tenth.

294. For Sophia's graduation party, several tables of the same width will be arranged end to end to give a serving table with a total area of 75 square feet. The total length of the tables will be two more than three times the width. Find the length and width of the serving table so Sophia can purchase the correct size tablecloth. Round answer to the nearest tenth.

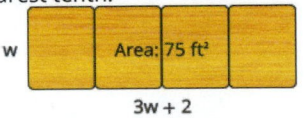

295. A ball is thrown vertically in the air with a velocity of 160 ft/sec. Use the formula $h = -16t^2 + v_0 t$ to determine when the ball will be 384 feet from the ground. Round to the nearest tenth.

296. A bullet is fired straight up from the ground at a velocity of 320 ft/sec. Use the formula $h = -16t^2 + v_0 t$ to determine when the bullet will reach 800 feet. Round to the nearest tenth.

10.5 10.5 Graphing Quadratic Equations in Two Variables

In the following exercises, graph by plotting point.

297. Graph $y = x^2 - 2$

298. Graph $y = -x^2 + 3$

In the following exercises, determine if the following parabolas open up or down.

299. $y = -3x^2 + 3x - 1$

300. $y = 5x^2 + 6x + 3$

301. $y = x^2 + 8x - 1$

302. $y = -4x^2 - 7x + 1$

In the following exercises, find ⓐ the axis of symmetry and ⓑ the vertex.

303. $y = -x^2 + 6x + 8$

304. $y = 2x^2 - 8x + 1$

In the following exercises, find the x- and y-intercepts.

305. $y = x^2 - 4x + 5$

306. $y = x^2 - 8x + 15$

307. $y = x^2 - 4x + 10$

308. $y = -5x^2 - 30x - 46$

309. $y = 16x^2 - 8x + 1$

310. $y = x^2 + 16x + 64$

In the following exercises, graph by using intercepts, the vertex, and the axis of symmetry.

311. $y = x^2 + 8x + 15$

312. $y = x^2 - 2x - 3$

313. $y = -x^2 + 8x - 16$

314. $y = 4x^2 - 4x + 1$

315. $y = x^2 + 6x + 13$

316. $y = -2x^2 - 8x - 12$

317. $y = -4x^2 + 16x - 11$

318. $y = x^2 + 8x + 10$

In the following exercises, find the minimum or maximum value.

319. $y = 7x^2 + 14x + 6$

320. $y = -3x^2 + 12x - 10$

In the following exercises, solve. Rounding answers to the nearest tenth.

321. A ball is thrown upward from the ground with an initial velocity of 112 ft/sec. Use the quadratic equation $h = -16t^2 + 112t$ to find how long it will take the ball to reach maximum height, and then find the maximum height.

322. A daycare facility is enclosing a rectangular area along the side of their building for the children to play outdoors. They need to maximize the area using 180 feet of fencing on three sides of the yard. The quadratic equation $A = -2x^2 + 180x$ gives the area, A, of the yard for the length, x, of the building that will border the yard. Find the length of the building that should border the yard to maximize the area, and then find the maximum area.

PRACTICE TEST

323. Use the Square Root Property to solve the quadratic equation: $3(w+5)^2 = 27$.

324. Use Completing the Square to solve the quadratic equation: $a^2 - 8a + 7 = 23$.

325. Use the Quadratic Formula to solve the quadratic equation: $2m^2 - 5m + 3 = 0$.

Solve the following quadratic equations. Use any method.

326. $8v^2 + 3 = 35$

327. $3n^2 + 8n + 3 = 0$

328. $2b^2 + 6b - 8 = 0$

329. $x(x+3) + 12 = 0$

330. $\frac{4}{3}y^2 - 4y + 3 = 0$

Use the discriminant to determine the number of solutions of each quadratic equation.

331. $6p^2 - 13p + 7 = 0$

332. $3q^2 - 10q + 12 = 0$

Solve by factoring, the Square Root Property, or the Quadratic Formula.

333. Find two consecutive even numbers whose product is 360.

334. The length of a diagonal of a rectangle is three more than the width. The length of the rectangle is three times the width. Find the length of the diagonal. (Round to the nearest tenth.)

For each parabola, find ⓐ which ways it opens, ⓑ the axis of symmetry, ⓒ the vertex, ⓓ the x- and y-intercepts, and ⓔ the maximum or minimum value.

335. $y = 3x^2 + 6x + 8$

336. $y = x^2 - 4$

337. $y = x^2 + 10x + 24$

338. $y = -3x^2 + 12x - 8$

339. $y = -x^2 - 8x + 16$

Graph the following parabolas by using intercepts, the vertex, and the axis of symmetry.

340. $y = 2x^2 + 6x + 2$

341. $y = 16x^2 + 24x + 9$

Solve.

342. A water balloon is launched upward at the rate of 86 ft/sec. Using the formula $h = -16t^2 + 86t$, find how long it will take the balloon to reach the maximum height and then find the maximum height. Round to the nearest tenth.

ANSWER KEY
Chapter 1
Try It

1.1. ⓐ ten millions ⓑ tens ⓒ hundred thousands ⓓ millions ⓔ ones

1.2. ⓐ billions ⓑ ten thousands ⓒ tens ⓓ hundred thousands ⓔ hundred millions

1.3. nine trillion, two hundred fifty-eight billion, one hundred thirty-seven million, nine hundred four thousand, sixty-one

1.4. seventeen trillion, eight hundred sixty-four billion, three hundred twenty-five million, six hundred nineteen thousand four

1.5. 2,466,714,051

1.6. 11,921,830,106

1.7. 17,900

1.8. 468,800

1.9. ⓐ 207,000 ⓑ 207,000 ⓒ 210,000

1.10. ⓐ 785,000 ⓑ 785,000 ⓒ 780,000

1.11. by 2, 3, and 6

1.12. by 3 and 5

1.13. $2 \cdot 2 \cdot 2 \cdot 2 \cdot 5$

1.14. $2 \cdot 2 \cdot 3 \cdot 5$

1.15. $2 \cdot 3 \cdot 3 \cdot 7$

1.16. $2 \cdot 3 \cdot 7 \cdot 7$

1.17. 36

1.18. 72

1.19. 36

1.20. 72

1.21. 84

1.22. 96

1.23. ⓐ 14 is less than or equal to 27 ⓑ 19 minus 2 is not equal to 8 ⓒ 12 is greater than 4 divided by 2 ⓓ x minus 7 is less than 1

1.24. ⓐ 19 is greater than or equal to 15 ⓑ 7 is equal to 12 minus 5 ⓒ 15 divided by 3 is less than 8 ⓓ y plus 3 is greater than 6

1.25. ⓐ equation ⓑ expression

1.26. ⓐ expression ⓑ equation

1.27. ⓐ 125 ⓑ 1

1.28. ⓐ 49 ⓑ 0

1.29. ⓐ 2 ⓑ 14

1.30. ⓐ 35 ⓑ 99

1.31. 16

1.32. 23

1.33. 86

1.34. 1

1.35. ⓐ 13 ⓑ 5

1.36. ⓐ 8 ⓑ 16

1.37. ⓐ 9 ⓑ 64

1.38. ⓐ 216 ⓑ 64

1.39. 40

1.40. 9

1.41. ⓐ 14 ⓑ 41 ⓒ 1

1.42. ⓐ 9 ⓑ 13 ⓒ 1

1.43. 9 and 15, y^2 and $11y^2$, $2x^3$ and $8x^3$

1.44. 19 and 24, $8x^2$ and $3x^2$, $4x^3$ and $6x^3$

1.45. $4x^2$, $5x$, 17

1.46. $5x$, $2y$

1.47. $10x^2 + 16x + 17$

1.48. $12y^2 + 9y + 7$

1.49. ⓐ $14x^2 - 13$ ⓑ $12x \div 2$

1.50. ⓐ $17y^2 + 19$ ⓑ $7y$

1.51. ⓐ $x + 11$ ⓑ $11a - 14$

1.52. ⓐ $z + 13$ ⓑ $8x - 18$

1.53. ⓐ $4(p + q)$ ⓑ $4p + q$

1.54. ⓐ $2x - 8$ ⓑ $2(x - 8)$

1.55. $w - 7$

1.56. $l - 6$

1.57. $4q - 8$

1.58. $7n + 3$

1.59. ⓐ > ⓑ < ⓒ > ⓓ >

1.60. ⓐ < ⓑ > ⓒ < ⓓ >

1.61. ⓐ -4 ⓑ 3 ⓒ 1

1.62. ⓐ -8 ⓑ 5 ⓒ 5

1.63. ⓐ -4 ⓑ 4

1.64. ⓐ -11 ⓑ 11

1.65. ⓐ 4 ⓑ 28 ⓒ 0

1.66. ⓐ 13 ⓑ 47 ⓒ 0

1.67. ⓐ > ⓑ > ⓒ < ⓓ >

1.68. ⓐ > ⓑ > ⓒ > ⓓ <

1.69. 16

1.70. 9

1.71. ⓐ 17 ⓑ 39 ⓒ -22 ⓓ -11

1.72. ⓐ 23 ⓑ 21 ⓒ -37 ⓓ -49

1.73. ⓐ 6 ⓑ -6

1.74. ⓐ 7 ⓑ -7

1.75. ⓐ 2 ⓑ -2

1.76. ⓐ 3 ⓑ −3 **1.77.** ⓐ −50 ⓑ −17 **1.78.** ⓐ −70 ⓑ −36

1.79. 13 **1.80.** 0 **1.81.** ⓐ 2 ⓑ −2

1.82. ⓐ 3 ⓑ −3 **1.83.** ⓐ −10 ⓑ 10 **1.84.** ⓐ −11 ⓑ 11

1.85. ⓐ 8 ⓑ −18 **1.86.** ⓐ 8 ⓑ −22 **1.87.** ⓐ 19 ⓑ −4

1.88. ⓐ 23 ⓑ 3 **1.89.** 3 **1.90.** 13

1.91. ⓐ −48 ⓑ 28 ⓒ −63 ⓓ 60 **1.92.** ⓐ −56 ⓑ 54 ⓒ −28 ⓓ 39 **1.93.** ⓐ −9 ⓑ 17

1.94. ⓐ −8 ⓑ 16 **1.95.** ⓐ −7 ⓑ 39 **1.96.** ⓐ −9 ⓑ 23

1.97. −63 **1.98.** −84 **1.99.** ⓐ 81 ⓑ −81

1.100. ⓐ 49 ⓑ −49 **1.101.** 29 **1.102.** 52

1.103. 4 **1.104.** 9 **1.105.** 21

1.106. 6 **1.107.** ⓐ −6 ⓑ 10 **1.108.** ⓐ −1 ⓑ 17

1.109. 196 **1.110.** 8 **1.111.** ⓐ −2 ⓑ 36

1.112. ⓐ −19 ⓑ 9 **1.113.** 39 **1.114.** 13

1.115. $(9 + (−16)) + 4 − 3$ **1.116.** $(−8 + (−12)) + 7 − 13$ **1.117.** ⓐ $14 − (−23); 37$ ⓑ $−17 − 21; −38$

1.118. ⓐ $11 − (−19); 30$ ⓑ $−11 − 18; −29$ **1.119.** $−5(12); −60$ **1.120.** $−8(13); −104$

1.121. $−63 ÷ (−9); 7$ **1.122.** $−72 ÷ (−9); 8$ **1.123.** The difference in temperatures was 45 degrees.

1.124. The difference in temperatures was 9 degrees. **1.125.** The Bears lost 105 yards. **1.126.** A \$16 fee was deducted from his checking account.

1.127. $\frac{6}{10}, \frac{9}{15}, \frac{12}{20}$; answers may vary **1.128.** $\frac{8}{10}, \frac{12}{15}, \frac{16}{20}$; answers may vary **1.129.** $−\frac{7}{9}$

1.130. $−\frac{5}{9}$ **1.131.** $−\frac{23}{40}$ **1.132.** $−\frac{5}{8}$

1.133. $\frac{x}{y}$ **1.134.** $\frac{a}{b}$ **1.135.** $−\frac{4}{21}$

1.136. $−\frac{3}{16}$ **1.137.** $−33a$ **1.138.** $−36b$

1.139. $−\frac{21}{5p}$ **1.140.** $−\frac{15}{8q}$ **1.141.** $\frac{4}{15}$

1.142. $\frac{2}{3}$ **1.143.** $\frac{4}{5}$ **1.144.** $\frac{11}{14}$

1.145. $\frac{3}{4b}$ **1.146.** $\frac{4}{2q}$ **1.147.** $−\frac{3}{4}$

1.148. $−\frac{2}{3}$ **1.149.** 4 **1.150.** 2

1.151. $\frac{a − b}{cd}$ **1.152.** $\frac{p + q}{r}$ **1.153.** $\frac{x + 3}{4}$

1.154. $\frac{y + 5}{8}$ **1.155.** $−\frac{26}{28}$ **1.156.** $−\frac{7}{8}$

1.157. $−\frac{16}{x}$ **1.158.** $−\frac{22}{a}$ **1.159.** $−1$

1.160. $−\frac{2}{3}$ **1.161.** $\frac{79}{60}$ **1.162.** $\frac{103}{60}$

1.163. $\frac{1}{96}$ **1.164.** $\frac{75}{224}$ **1.165.** $\frac{9y + 42}{54}$

1.166. $\frac{15x + 42}{135}$ **1.167.** ⓐ $\frac{27a − 32}{36}$ ⓑ $\frac{2a}{3}$ **1.168.** ⓐ $\frac{24k − 5}{30}$ ⓑ $\frac{2k}{15}$

1.169. $\frac{1}{90}$

1.170. 272

1.171. 2

1.172. $\frac{2}{7}$

1.173. ⓐ -1 ⓑ $-\frac{1}{2}$

1.174. ⓐ $\frac{7}{6}$ ⓑ $-\frac{1}{12}$

1.175. $-\frac{1}{4}$

1.176. $-\frac{17}{8}$

1.177. $-\frac{1}{2}$

1.178. $\frac{2}{3}$

1.179. $-\frac{5}{2}$

1.180. $\frac{3}{2}$

1.181. six and seven tenths

1.182. five and eight tenths

1.183. negative thirteen and four hundred sixty-one thousandths

1.184. negative two and fifty-three thousandths

1.185. 13.68

1.186. 5.94

1.187. 1.05

1.188. 9.17

1.189. ⓐ 6.58 ⓑ 6.6 ⓒ 7

1.190. ⓐ 15.218 ⓑ 15.22 ⓒ 15.2

1.191. 16.49

1.192. 23.593

1.193. 0.42

1.194. 12.58

1.195. -27.4815

1.196. -87.6148

1.197. ⓐ 25.8 ⓑ 258 ⓒ 2,580

1.198. ⓐ 142 ⓑ 1,420 ⓒ 14,200

1.199. 687.3

1.200. 34.25

1.201. $0.19

1.202. $0.42

1.203. $\frac{117}{500}$

1.204. $\frac{3}{125}$

1.205. -0.875

1.206. -0.375

1.207. $2.\overline{45}$

1.208. $2.3\overline{18}$

1.209. 5.275

1.210. 6.35

1.211. ⓐ 0.09 ⓑ 0.87 ⓒ 0.039

1.212. ⓐ 0.03 ⓑ 0.91 ⓒ 0.083

1.213. ⓐ 17% ⓑ 175% ⓒ 8.25%

1.214. ⓐ 41% ⓑ 225% ⓒ 9.25%

1.215. ⓐ 6 ⓑ 13

1.216. ⓐ 4 ⓑ 14

1.217. ⓐ -2 ⓑ -15

1.218. ⓐ -9 ⓑ -10

1.219. ⓐ $\frac{-24}{1}$ ⓑ $\frac{357}{100}$

1.220. ⓐ $\frac{-19}{1}$ ⓑ $\frac{841}{100}$

1.221. ⓐ $0.29,\ 0.81\overline{6}$ ⓑ $2.515115111\ldots$

1.222. ⓐ $2.6\overline{3},\ 0.125$ ⓑ $0.418302\ldots$

1.223. ⓐ rational ⓑ irrational

1.224. ⓐ irrational ⓑ rational

1.225. ⓐ not a real number ⓑ real number

1.226. ⓐ real number ⓑ not a real number

1.227. ⓐ $4,\ \sqrt{49}$ ⓑ $-3,\ 4,\ \sqrt{49}$ ⓒ $-3,\ 0.\overline{3},\ \frac{9}{5},\ 4,\ \sqrt{49}$ ⓓ $-\sqrt{2}$ ⓔ $-3,\ -\sqrt{2},\ 0.\overline{3},\ \frac{9}{5},\ 4,\ \sqrt{49}$

1.228. ⓐ $6,\ \sqrt{121}$ ⓑ $-\sqrt{25},\ -1,\ 6,\ \sqrt{121}$ ⓒ $-\sqrt{25},\ -\frac{3}{8},\ -1,\ 6,\ \sqrt{121}$ ⓓ $2.041975\ldots$ ⓔ $-\sqrt{25},\ -\frac{3}{8},\ -1,\ 6,\ \sqrt{121},\ 2.041975\ldots$

1.229.

1.230.

1.231. ⓐ > ⓑ > ⓒ < ⓓ <

1.232. ⓐ < ⓑ < ⓒ > ⓓ <

1.233.

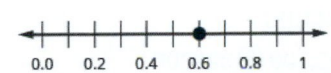

1.234.

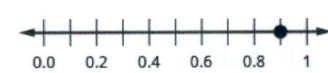

1.235.

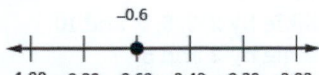

1.236.

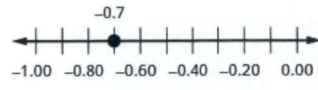

1.237. >

1.238. > **1.239.** > **1.240.** <

1.241. > **1.242.** > **1.243.** $32r + 29s$

1.244. $41m + 6n$

1.245. $1\frac{7}{15}$ **1.246.** $1\frac{2}{9}$

1.247. $32x$ **1.248.** $-63y$ **1.249.** ⓐ $-\frac{7}{9}$ ⓑ -1.2 ⓒ 14 ⓓ $\frac{9}{4}$

1.250. ⓐ $-\frac{7}{13}$ ⓑ -8.4 ⓒ 46 ⓓ $\frac{5}{2}$ **1.251.** ⓐ $\frac{1}{4}$ ⓑ -7 ⓒ $\frac{10}{3}$ **1.252.** ⓐ $\frac{1}{18}$ ⓑ $-\frac{5}{4}$ ⓒ $\frac{5}{3}$

1.253. ⓐ 0 ⓑ 0 ⓒ undefined **1.254.** ⓐ 0 ⓑ 0 ⓒ undefined **1.255.** $-48a$

1.256. $-92x$ **1.257.** $\frac{5}{49}$ **1.258.** $\frac{11}{25}$

1.259. ⓐ 0 ⓑ undefined **1.260.** ⓐ 0 ⓑ undefined **1.261.** $20y + 50$

1.262. $12z + 16$ **1.263.** $4x + 8$ **1.264.** $6x + 42$

1.265. $5y + 3$ **1.266.** $4n + 9$ **1.267.** $70 + 15p$

1.268. $4 + 35d$ **1.269.** $-18m - 15$ **1.270.** $-48n - 66$

1.271. $-10 + 15a$ **1.272.** $-56 + 105y$ **1.273.** $-z + 11$

1.274. $-x + 4$ **1.275.** $3 - 3x$ **1.276.** $2x - 20$

1.277. $5x - 66$ **1.278.** $7x - 13$ **1.279.** 2.5 feet

1.280. 54 feet **1.281.** 8,600 pounds **1.282.** 102,000,000 pounds

1.283. 440,000,000 yards **1.284.** 151,200 minutes **1.285.** 16 cups

1.286. 48 teaspoons **1.287.** 9 lbs. 8 oz **1.288.** 21 ft. 6 in.

1.289. 4 lbs. 8 oz. **1.290.** 11 gallons 2 qt. **1.291.** 5,000 meters

1.292. 250 centimeters **1.293.** 2.8 kilograms **1.294.** 4.5 kilograms

1.295. ⓐ 7,250 kiloliters ⓑ 6,300 milliliters **1.296.** ⓐ 35,000 liters ⓑ 410 centiliters **1.297.** 83 centimeters

1.298. 1.04 meters **1.299.** 2 liters **1.300.** 2.4 kilograms

1.301. 2.12 quarts **1.302.** 3.8 liters **1.303.** 19,335.6 feet

1.304. 8,993.46 km **1.305.** $15°C$ **1.306.** $5°C$

1.307. $59°F$ **1.308.** $50°F$

Section Exercises

1. ⓐ thousands ⓑ hundreds ⓒ tens ⓓ ten thousands ⓔ ones

3. ⓐ ones ⓑ ten thousands ⓒ hundred thousands ⓓ tens ⓔ hundreds

5. ⓐ ten millions ⓑ ten thousands ⓒ tens ⓓ thousands ⓔ millions

7. ⓐ trillions ⓑ billions ⓒ millions ⓓ tens ⓔ thousands

9. one thousand, seventy-eight

11. three hundred sixty-four thousand, five hundred ten

13. five million, eight hundred forty-six thousand, one hundred three

15. thirty-seven million, eight hundred eighty-nine thousand, five

17. 412

19. 35,975 **21.** 11,044,167 **23.** 3,226,512,017

25. ⓐ 390 ⓑ 2,930 **27.** ⓐ 13,700 ⓑ 391,800 **29.** ⓐ 1,490 ⓑ 1,500

31. ⓐ 64,000 ⓑ 63,900 **33.** ⓐ 392,500 ⓑ 393,000 ⓒ 390,000 **35.** ⓐ 2,587,000 ⓑ 2,587,000 ⓒ 2,590,000

37. divisible by 2, 3, and 6 **39.** divisible by 3 and 5 **41.** divisible by 2, 3, 5, 6, and 10

43. divisible by 2 **45.** divisible by 2, 3, and 10 **47.** divisible by 3 and 5

49. $2 \cdot 43$ **51.** $2 \cdot 2 \cdot 3 \cdot 11$ **53.** $3 \cdot 3 \cdot 7 \cdot 11$

55. $2 \cdot 2 \cdot 2 \cdot 2 \cdot 3 \cdot 3 \cdot 3$ **57.** $2 \cdot 2 \cdot 2 \cdot 2 \cdot 3 \cdot 3 \cdot 3 \cdot 5$ **59.** 24

61. 48

67. 420

63. 60

69. 440

65. 24

71. twenty-four thousand, four hundred ninety-three dollars

77. 80

73. ⓐ $24,490 ⓑ $24,500 ⓒ $24,000 ⓓ $20,000

81. Answers may vary.

87. 2 times 7, the product of two and seven

93. y minus 1 is greater than 6, the difference of y and one is greater than six

75. ⓐ 1,000,000,000 ⓑ 1,300,000,000 ⓒ 1,340,000,000

83. 16 minus 9, the difference of sixteen and nine

89. fourteen is less than twenty-one

95. 2 is less than or equal to 18 divided by 6; 2 is less than or equal to the quotient of eighteen and six

85. 28 divided by 4, the quotient of twenty-eight and four

91. thirty-six is greater than or equal to nineteen

97. equation

99. expression

105. 256

111. 34

117. 4

123. 149

129. 144

135. 9

141. 8

147. 16 and 4, $16b^2$ and $9b^2$

153. $13x$

159. $22a + 1$

165. $9 \cdot 7$

171. $\frac{y}{3}$

177. $2n - 7$

183. Answers may vary

189. 4

195. ⓐ 32 ⓑ 0 ⓒ 16

201. 56

207. ⓐ -19 ⓑ -33

213. -22

219. 6

225. ⓐ 16 ⓑ 16

231. -39

237. 9

243. 22

249. 6

255. ⓐ 20,329 ⓑ -282

261. Answers may vary

267. -63

273. -4

279. -47

285. 90

291. -9

297. ⓐ -47 ⓑ 16

101. expression

107. ⓐ 43 ⓑ 55

113. 58

119. 35

125. 50

131. 32

137. 73

143. 5

149. $15x^2$, $6x$, 2

155. $7c$

161. $17x^2 + 20x + 16$

167. $36 \div 9$

173. $8(y - 9)$

179. ⓐ $750 ⓑ $1,350

185. ⓐ $>$ ⓑ $<$ ⓒ $<$ ⓓ $>$

191. 15

197. ⓐ $<$ ⓑ $=$

203. 0

209. -80

215. 108

221. -9

227. ⓐ 45 ⓑ 45

233. -59

239. -2

245. 0

251. -8

257. ⓐ $540 million ⓑ $-$27 billion

263. Answers may vary

269. -6

275. 13

281. 64

287. 9

293. -29

299. ⓐ -4 ⓑ 10

103. 125

109. 5

115. 6

121. 58

127. 22

133. 21

139. 54

145. x^3 and $8x^3$, 14 and 5

151. $10y^3$, y, 2

157. $10u + 3$

163. $14 - 9$

169. $8x + 3x$

175. $b - 4$

181. Answers may vary

187. ⓐ -2 ⓑ 6

193. ⓐ -12 ⓑ 12

199. 5, -5

205. 8

211. 32

217. 29

223. 12

229. 27

235. -51

241. -2

247. 4

253. -11

259. -32

265. -32

271. 14

277. -12

283. -16

289. 41

295. 5

301. -8

303. -16

305. 121

307. (a) 1 (b) 33

309. (a) -5 (b) 25

311. 21

313. -56

315. $(3 + (-15)) + 7;\ -5$

317. $10 - (-18);\ 28$

319. $-5 - (-30);\ 25$

321. $-3 \cdot 15;\ -45$

323. $-60 \div (-20);\ 3$

325. $\dfrac{-6}{a+b}$

327. $-10(p - q)$

329. $96°$

331. 21

333. $-\$28$

335. $\$187$

337. $-\$3600$

339. Answers may vary

341. Answers may vary

343. $\dfrac{6}{16}, \dfrac{9}{24}, \dfrac{12}{32}$ answers may vary

345. $\dfrac{10}{18}, \dfrac{15}{27}, \dfrac{20}{36}$ answers may vary

347. $-\dfrac{5}{11}$

349. $-\dfrac{12}{7}$

351. $\dfrac{10}{21}$

353. $-\dfrac{x}{4y}$

355. $\dfrac{2x^2}{3y}$

357. $\dfrac{27}{40}$

359. $\dfrac{1}{4}$

361. $-\dfrac{1}{6}$

363. $-\dfrac{21}{50}$

365. $\dfrac{11}{30}$

367. $\dfrac{20}{11}$

369. $9n$

371. -34

373. $\dfrac{9}{8}$

375. 1

377. $\dfrac{33}{4x}$

379. $-\dfrac{4}{9}$

381. $\dfrac{10u}{9v}$

383. -10

385. $-\dfrac{1}{16}$

387. $-\dfrac{10}{9}$

389. $-\dfrac{2}{5}$

391. $\dfrac{2m}{3n}$

393. $\dfrac{5}{2}$

395. $\dfrac{16}{3}$

397. 0

399. $\dfrac{1}{3}$

401. $\dfrac{3}{5}$

403. $2\dfrac{8}{17}$

405. $\dfrac{3}{5}$

407. $\dfrac{5}{2}$

409. $\dfrac{11}{6}$

411. $\dfrac{5}{2}$

413. $\dfrac{r}{s+10}$

415. $\dfrac{x-y}{-8}$

417. (a) $1\dfrac{1}{2}$ cups (b) answers will vary

419. 20 bags

421. Answers may vary

423. Answers may vary

425. $\dfrac{11}{13}$

427. $\dfrac{x+3}{4}$

429. $-\dfrac{5}{8}$

431. $\dfrac{7}{17}$

433. $-\dfrac{16}{13}$

435. $\dfrac{4}{15}$

437. $\dfrac{1}{2}$

439. $\dfrac{5}{7}$

441. $\dfrac{5y-7}{8}$

443. $-\dfrac{38}{u}$

445. $\dfrac{1}{5}$

447. $-\dfrac{2}{9}$

449. $-\dfrac{1}{4}$

451. $\dfrac{n-4}{5}$

453. $-\dfrac{5}{24}$

455. $\dfrac{2}{9}$

457. $\dfrac{9}{14}$

459. $\dfrac{4}{9}$

461. $\dfrac{29}{24}$

463. $\dfrac{1}{48}$

465. $\dfrac{7}{24}$

467. $\dfrac{37}{120}$

469. $\dfrac{17}{105}$

471. $-\dfrac{53}{40}$

473. $\dfrac{1}{12}$

475. $\dfrac{15}{8}$

477. $\dfrac{4x+3}{12}$

479. $\dfrac{4y-12}{20}$

481. (a) $\dfrac{5}{6}$ (b) 4

483. (a) $\dfrac{25n}{16}$ (b) $\dfrac{25n-16}{30}$

485. $\dfrac{5}{4}$

487. $\dfrac{1}{24}$

489. $\dfrac{13}{18}$

491. $\dfrac{-28-15y}{60}$

493. $\dfrac{33}{64}$

495. 54

497. $\dfrac{49}{25}$

499. $\dfrac{15}{4}$

501. $\dfrac{5}{21}$

503. $\dfrac{7}{9}$

505. -5

507. $\dfrac{19}{12}$

509. $\dfrac{23}{24}$

511. $\dfrac{11}{5}$

513. 1

515. $\dfrac{13}{3}$

517. (a) $-\dfrac{1}{2}$ (b) -1

519. (a) $\dfrac{1}{5}$ (b) -1

521. (a) $\dfrac{1}{5}$ (b) $\dfrac{6}{5}$

523. $-\dfrac{1}{9}$

525. $-\dfrac{5}{11}$

527. $\dfrac{7}{8}$ yard

529. Answers may vary

531. 29.81

533. 0.7

535. 0.029

537. -11.0009

539. five and five tenths

541. eight and seventy-one hundredths

543. two thousandths

545. negative seventeen and nine tenths

547. 0.7

549. 2.8

551. 0.85

553. 0.30

555. 4.10

557. (a) 5.78 (b) 5.8 (c) 6

559. (a) 63.48 (b) 63.5 (c) 63

561. 24.48

563. -9.23

565. -40.91

567. -7.22

569. -27.5

571. 15.73

573. 102.212

575. 51.31

577. -4.89

579. 0.144

581. 42.008

583. -11.653

585. 337.8914

587. 1.305

589. 92.4

591. 55,200

593. 0.19

595. $2.44

597. 3

599. -4.8

601. 35

603. 2.08

605. 20

607. $\dfrac{1}{25}$

609. $\dfrac{13}{25}$

611. $\dfrac{5}{4}$

613. $\dfrac{3}{8}$

615. $\dfrac{19}{200}$

617. 0.85

619. 2.75

621. -12.4

623. $1.\overline{36}$

625. $0.\overline{135}$

627. 3.025

629. 0.011

631. 0.63

633. 1.5

635. 0.214

637. 0.078

639. 1%

641. 135%

643. 300%

645. 8.75%

647. 225.4%

649. (a) $58,966 (b) $59,000 (c) $60,000

651. (a) $142.19; (b) $142

653. (a) $243.57 (b) $79.35

655. Answers may vary

657. Answers may vary

659. 6

661. 8

663. 3

665. 10

667. -2

669. -1

671. (a) $\dfrac{5}{1}$ (b) $\dfrac{319}{100}$

673. (a) $\dfrac{-12}{1}$ (b) $\dfrac{9297}{1000}$

675. (a) $0.75,\ 0.22\overline{3}$ (b) $1.39174\ldots$

677. (a) $0.4\overline{5},\ 3.59$ (b) $1.919293\ldots$

679. (a) rational (b) irrational

681. (a) irrational (b) rational

683. (a) real number (b) not a real number

685. (a) not a real number (b) real number

687. ⓐ 0, √36, 9 ⓑ −8, √36, 9

ⓒ −8, 0, $\frac{12}{5}$, √36, 9 ⓓ

1.95286... ⓔ

−8, 0, 1.95286..., $\frac{12}{5}$, √36, 9

689. ⓐ none ⓑ −√100, −7, −1 ⓒ

−√100, −7, − $\frac{8}{3}$, −1, 0.77, $3\frac{1}{4}$ ⓓ

none ⓔ

−√100, −7, − $\frac{8}{3}$, −1, 0.77, $3\frac{1}{4}$

691.

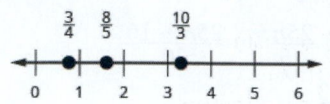

693.

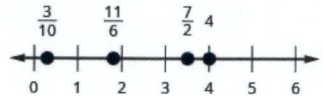

695.

697.

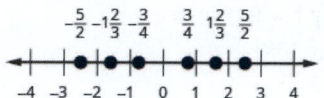

699. <
705. <

701. >
707.

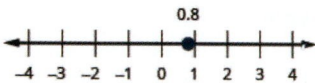

703. >
709.

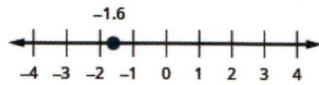

711. <
717. <

713. >
719. ⓐ 4 busses ⓑ answers may
vary ⓒ answers may vary

715. <
721. Answers may vary

723. $12x$

725. $y + 40$

727. $\frac{7}{8}$

729. $\frac{49}{11}$

731. -63

733. $1\frac{5}{6}$

735. 17
741. $-75m$

737. 14.88
743. $10p$

739. $28a$
745. $27m + (-21n)$

747. $\frac{5}{4}g + \frac{1}{2}h$

749. $2.43p + 8.26q$

751. ⓐ $-\frac{2}{5}$ ⓑ -4.3 ⓒ 8 ⓓ $\frac{10}{3}$

753. ⓐ $\frac{7}{6}$ ⓑ 0.075 ⓒ -23 ⓓ

$-\frac{1}{4}$

755. ⓐ $\frac{1}{6}$ ⓑ $-\frac{4}{3}$ ⓒ $\frac{10}{7}$

757. ⓐ $\frac{12}{11}$ ⓑ $-\frac{10}{11}$ ⓒ $-\frac{1}{4}$

759. 0
765. 0
771. 0
777. undefined
783. $6c - 78$

761. 0
767. 44
773. 0
779. $36d + 90$

785. $\frac{3}{4}q + 3$

763. 0
769. id
775. undefined
781. $32y + 72$

787. $5y - 3$

789. $3 + 8r$
795. $-28p - 7$
801. $-3y - 8$
807. $4m - 10$
813. $6c + 34$

791. $rs - 18r$
797. $-3x + 18$
803. $-33c + 26$
809. $22n + 9$
815. $12y + 63$

793. $yp + 4p$
799. $-3x + 7$
805. $-a + 19$
811. $72x - 25$
817. ⓐ $80 ⓑ $80 ⓒ answers will
vary

819. ⓐ $23.88 ⓑ no, the price is
the same
825. 72 inches

821. Answers may vary

827. 1.5 feet

823. Answers may vary

829. $53\frac{1}{3}$ yards

831. 7,920 feet

833. 9,200 pounds

835. $17\frac{1}{2}$ tons

837. 5,400 s

839. 96 teaspoons

841. 224 ounces

843. $1\frac{1}{4}$ gallons

845. 26 in.

847. 65 days

849. 115 ounces

851. 8 lbs. 13 oz.

853. 3.05 hours

855. 10 ft. 2 in.

857. 4 yards

859. 5,000 meters

861. 155 centimeters

863. 3.072 kilometers

865. 1.5 grams

867. 91,600 milligrams

869. 2,000 grams

871. 0.75 liters

873. 91 centimeters

875. 855 grams

877. 2.1 kilograms

879. 1.6 liters

881. 190.5 centimeters

883. 21.9 meters

885. 742.5 kilograms

887. 3.1 miles

889. 44 pounds

891. 53.2 liters

893. 30°C

895. 40°C

897. 22.2°C

899. −17.8°C

901. 41°F

903. 14°F

905. 71.6°F

907. 109.4°F

909. 14.6 kilograms

911. Answers may vary.

Review Exercises

913. ⓐ tens ⓑ ten thousands ⓒ hundreds ⓓ ones ⓔ thousands

915. ⓐ ten millions ⓑ tens ⓒ hundred thousands ⓓ millions ⓔ ten thousands

917. six thousand, one hundred four

919. three million, nine hundred seventy-five thousand, two hundred eighty-four

921. 315

923. 90,425,016

925. ⓐ 410 ⓑ 8,560

927. ⓐ 865,000 ⓑ 865,000 ⓒ 860,000

929. by 2, 3, 6

931. by 3, 5

933. by 2, 5, 10

935. $2 \cdot 2 \cdot 3 \cdot 5 \cdot 7$

937. $3 \cdot 3 \cdot 5 \cdot 5$

939. $2 \cdot 2 \cdot 2 \cdot 3 \cdot 5 \cdot 13$

941. $2 \cdot 2 \cdot 2 \cdot 3 \cdot 3$

943. $2 \cdot 2 \cdot 3 \cdot 3 \cdot 7$

945. 30

947. 120

949. 25 minus 7, the difference of twenty-five and seven

951. 45 divided by 5, the quotient of forty-five and five

953. forty-two is greater than or equal to twenty-seven

955. 3 is less than or equal to 20 divided by 4, three is less than or equal to the quotient of twenty and four

957. expression

959. 243

961. 13

963. 10

965. 41

967. 34

969. 81

971. 58

973. 12

975. 12 and 3, n^2 and $3n^2$

977. $11x^2$, $3x$, 6

979. $26a$

981. $12x + 8$

983. $12p + 2$

985. $8 + 12$

987. $x - 4$

989. $6y$

991. $b + 15$

993. ⓐ > ⓑ < ⓒ < ⓓ >

995. ⓐ 8 ⓑ −1

997. 19

999. ⓐ −3 ⓑ 3

1001. ⓐ 7 ⓑ 25 ⓒ 0

1003. ⓐ < ⓑ =

1005. 4

1007. 80

1009. ⓐ 28 ⓑ 15

1011. −135

1013. 0

1015. 132

1017. 6

1019. ⓐ 9 ⓑ 9

1021. ⓐ 17 ⓑ 17

1023. 29

1025. −48

1027. −42

1029. 18

1031. −20

1033. −35

1035. 36

1037. −4

1039. 6

1041. 43

1043. −125

1045. −88

1047. −5

1049. 55

1051. ⓐ −18 ⓑ −87

1053. ⓐ −5 ⓑ 17

1055. 8

1057. 38

1059. $(-4 + (-17)) + 32$; 11

1061. $\frac{-45}{-9}$; 5

1063. 84 degrees

1065. $\frac{2}{8}$, $\frac{3}{12}$, $\frac{4}{16}$ answers may vary

1067. $\frac{10}{12}$, $\frac{15}{18}$, $\frac{20}{24}$ answers may vary

1069. $\frac{1}{3}$

1071. $\frac{3}{4}$

1073. $-\frac{7}{8}$

1075. $\frac{x}{y}$

1077. $\frac{2}{15}$

1079. $-\frac{2}{9}$

1081. $7p$

1083. -42

1085. 2

1087. $-\frac{7}{5}$

1089. $\frac{25}{4a}$

1091. $\frac{2}{9}$

1093. $-\frac{1}{25}$

1095. $\frac{3}{4}$

1097. $-\frac{3}{10}$

1099. $\frac{3r}{5s}$

1101. $\frac{15}{8}$

1103. -6

1105. $\frac{4}{3}$

1107. -4

1109. $\frac{5}{2}$

1111. -2

1113. $\frac{c}{d+9}$

1115. $\frac{5}{9}$

1117. $\frac{y+2}{3}$

1119. $-\frac{1}{2}$

1121. $\frac{3}{5}$

1123. $\frac{y-9}{17}$

1125. $-\frac{11}{d}$

1127. $\frac{8}{15}$

1129. $\frac{3}{10}$

1131. $\frac{17}{12}$

1133. $\frac{13}{24}$

1135. $\frac{19}{80}$

1137. $\frac{11}{6}$

1139. $\frac{1}{275}$

1141. 14

1143. ⓐ $\frac{3}{8}$ ⓑ 0

1145. $\frac{5}{9}$

1147. -6

1149. 8.03

1151. 0.001

1153. seven and eight tenths

1155. five thousandths

1157. ⓐ 5.79 ⓑ 5.8 ⓒ 6

1159. ⓐ 12.48 ⓑ 12.5 ⓒ 12

1161. 27.73

1163. -5.53

1165. -13.5

1167. 35.8

1169. 42.51

1171. 0.12

1173. 26.7528

1175. 2.2302

1177. 0.03

1179. $0.71

1181. 150

1183. $\frac{2}{25}$

1185. $\frac{17}{40}$

1187. $\frac{7}{4}$

1189. 0.4

1191. -0.375

1193. $0.\overline{5}$

1195. 7

1197. 0.05

1199. 0.4

1201. 1.15

1203. 18%

1205. 0.9%

1207. 150%

1209. 8

1211. -5

1213. ⓐ $\frac{9}{1}$ ⓑ $\frac{847}{100}$

1215. ⓐ 0.84, $1.\overline{3}$ ⓑ 0.79132…,

1217. ⓐ rational ⓑ irrational

1219. ⓐ not a real number ⓑ real number

1221. ⓐ 0, $\sqrt{16}$ ⓑ -4, 0, $\sqrt{16}$ ⓒ -4, 0, $\frac{5}{6}$, $\sqrt{16}$ ⓓ $\sqrt{18}$, 5.2537… ⓔ -4, 0, $\frac{5}{6}$, $\sqrt{16}$, $\sqrt{18}$, 5.2537…

1223.

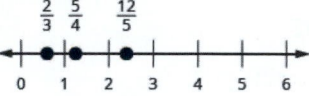

1225.

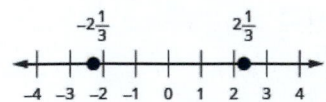

1227. <

1229. >

1231.

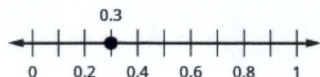

1233.

1235. >

1237. >

1239. $-48m$

1241. $a + 47$

1243. 37

1245. $\frac{35}{9}$

1247. $1\frac{7}{12}$

1249. $27x + 23y$

1251. ⓐ $-\frac{1}{3}$ ⓑ -5.1 ⓒ 14 ⓓ $\frac{8}{5}$

1253. ⓐ $\frac{1}{10}$ ⓑ $-\frac{9}{4}$ ⓒ $\frac{5}{3}$

1255. 0

1257. undefined

1259. 39

1261. 57

1263. 8

1265. $7x + 63$

1267. $-18m + 3$

1269. $5n - 2$

1271. $-5a - 13$

1273. 84 inches

1275. 64 inches

1277. 2.7 miles

1279. 105 minutes

1281. 94 ounces

1283. 10 lbs. 6 oz.

1285. 15 feet

1287. 170 centimeters

1289. 0.488 grams

1291. 2,900 grams

1293. 1.12 meter

1295. 0.9 kilograms

1297. 175.3 centimeters

1299. 1.6 miles

1301. 14.6 gallons

1303. 35° C

1305. −6.7° C

1307. 86° F

1309. 10.4° F

Practice Test

1311. 205,617

1313. 72

1315. -2

1317. $-7 - 20; \ -27$

1319. 677.13

1321. 185%

1323. 99

1325. -44

1327. 3

1329. $\frac{16}{9}$

1331. $\frac{m + 10}{7}$

1333. -10.5

1335. 2.2365

1337. $-10p$

1339. $15x + 4y$

1341. undefined

1343. 100 minutes

1345. 4.508 km

Chapter 2

Try It

2.1. no

2.2. yes

2.3. $x = -46$

2.4. $x = -50$

2.5. $n = -14$

2.6. $p = -32$

2.7. $p = \frac{9}{6} \ p = \frac{3}{2}$

2.8. $q = \frac{4}{3}$

2.9. $b = -1.63$

2.10. $c = -3.67$

2.11. $y = 15$

2.12. $z = 2$

2.13. $p = 5$

2.14. $q = -16$

2.15. $h = 6$

2.16. $x = 2$

2.17. $x + 10 = 41; x = 31$

2.18. $y - 12 = 51; y = 63$

2.19. $4x - 3x = 14; x = 14$

2.20. $7a - 6a = -8; a = -8$

2.21. 7 pounds

2.22. 42 books

2.23. $20,900

2.24. $11.00

2.25. $y = \frac{-41}{3}$

2.26. $z = \frac{-55}{4}$

2.27. $a = 294$

2.28. $b = 144$

2.29. $k = -8$

2.30. $g = -3$

2.31. $n = 35$

2.32. $y = 18$

2.33. $z = -\frac{9}{5}$

2.34. $r = -\frac{5}{16}$

2.35. $c = -3$

2.36. $x = -\frac{4}{7}$

2.37. $q = -6$

2.38. $r = -5$

2.39. $132 = -12y; y = -11$

2.40. $117 = -13z; z = -9$

2.41. $\frac{n}{7} = -21; n = -147$

2.42. $\frac{n}{8} = -56; n = -448$

2.43. $\frac{q}{-8} = 72$; $q = -576$

2.44. $\frac{p}{-9} = 81$; $p = -729$

2.45. $\frac{2}{5}f = 16$; $f = 40$

2.46. $\frac{3}{4}f = 21$; $f = 28$

2.47. $\frac{5}{8} + x = \frac{1}{4}$; $x = -\frac{3}{8}$

2.48. $\frac{3}{4} + x = \frac{5}{6}$; $x = \frac{1}{12}$

2.49. $0.39

2.50. $5.83

2.51. $120,000

2.52. 21 flats

2.53. $x = -4$

2.54. $a = -8$

2.55. $y = 5$

2.56. $m = 9$

2.57. $n = -10$

2.58. $c = -1$

2.59. $p = -7$

2.60. $m = -3$

2.61. $j = 2$

2.62. $h = 1$

2.63. $x = -1$

2.64. $y = 4$

2.65. $q = 1$

2.66. $n = 1$

2.67. $a = -5$

2.68. $k = -6$

2.69. $x = 10$

2.70. $y = -3$

2.71. $x = -5$

2.72. $y = -5$

2.73. $x = 4$

2.74. $y = 1$

2.75. $y = -6$

2.76. $z = 8$

2.77. $m = 2$

2.78. $n = 2$

2.79. $u = 2$

2.80. $x = 4$

2.81. $j = \frac{5}{3}$

2.82. $k = \frac{5}{2}$

2.83. $p = -2$

2.84. $q = -8$

2.85. $y = -\frac{17}{5}$

2.86. $z = 0$

2.87. $n = 1$

2.88. $m = -1$

2.89. identity; all real numbers

2.90. identity; all real numbers

2.91. conditional equation; $q = \frac{9}{11}$

2.92. conditional equation; $k = \frac{193}{14}$

2.93. contradiction; no solution

2.94. contradiction; no solution

2.95. $x = \frac{1}{2}$

2.96. $x = -2$

2.97. $x = 12$

2.98. $u = -12$

2.99. $x = -1$

2.100. $c = -2$

2.101. $p = -4$

2.102. $q = 2$

2.103. $n = 2$

2.104. $m = -1$

2.105. $y = 2$

2.106. $z = -2$

2.107. $b = -20$

2.108. $c = -6$

2.109. $r = 1$

2.110. $s = -8$

2.111. $h = 12$

2.112. $k = -1$

2.113. $n = 9$

2.114. $d = 16$

2.115. 330 miles

2.116. 7 miles

2.117. 11 hours

2.118. 56 mph

2.119. ⓐ $r = 45$ ⓑ $r = \frac{d}{t}$

2.120. ⓐ $r = 65$ ⓑ $r = \frac{d}{t}$

2.121. ⓐ $h = 20$ ⓑ $h = \frac{2A}{b}$

2.122. ⓐ $b = 4$ ⓑ $b = \frac{2A}{h}$

2.123. ⓐ $12,000 ⓑ $P = \frac{I}{rt}$

2.124. ⓐ $9,000 ⓑ $P = \frac{I}{rt}$

2.125. ⓐ $y = 1$ ⓑ $y = \frac{10 - 3x}{4}$

2.126. ⓐ $y = -1$ ⓑ $y = \frac{18 - 5x}{2}$

2.127. $b = P - a - c$

2.128. $c = P - a - b$

2.129. $y = \frac{9 - 4x}{7}$

2.130. $y = \frac{1 - 5x}{8}$

2.131.

ⓐ

ⓑ

ⓒ

2.132.

ⓐ

ⓑ

ⓒ

2.133.

(a)

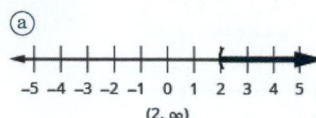

(2, ∞)

(b)

−1.5

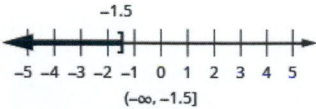

(−∞, −1.5]

(c)

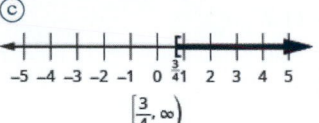

$\left[\frac{3}{4}, \infty\right)$

2.134.

(a)

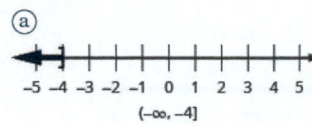

(−∞, −4]

(b)

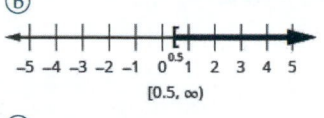

[0.5, ∞)

(c)

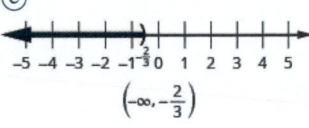

$\left(-\infty, -\frac{2}{3}\right)$

2.135.

$p \geq \frac{11}{12}$

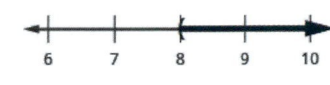

$\left[\frac{11}{12}, \infty\right)$

2.136.

$r \leq \frac{11}{12}$

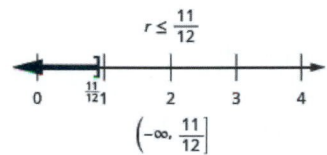

$\left(-\infty, \frac{11}{12}\right]$

2.137. $c > 8$

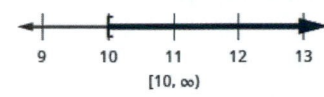

2.138. $d \leq 5$

2.139. $q > -4$

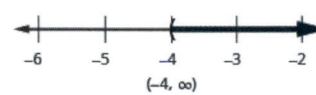

(−4, ∞)

2.140.

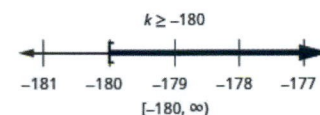

[10, ∞)

2.141.

$m \geq 64$

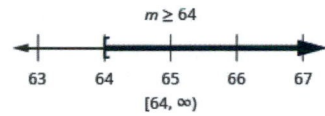

[64, ∞)

2.142.

$n > -18$

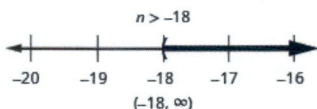

(−18, ∞)

2.143.

$k \geq -180$

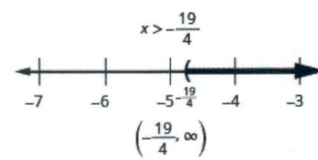

[−180, ∞)

2.144.

$u \leq 64$

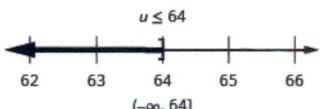

(−∞, 64]

2.145.

$q \leq \frac{23}{4}$

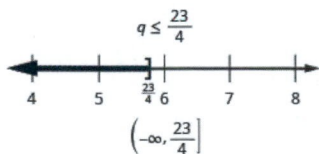

$\left(-\infty, \frac{23}{4}\right]$

2.146.

$x > -\frac{19}{4}$

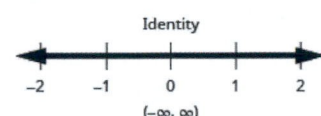

$\left(-\frac{19}{4}, \infty\right)$

2.147.

$y > -6$

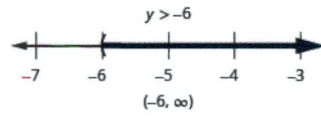

(−6, ∞)

2.148.

$u > 10$

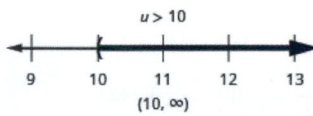

(10, ∞)

2.149.

Identity

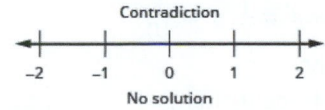

(−∞, ∞)

2.150.

Identity

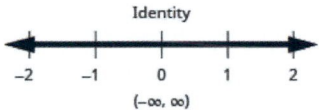

(−∞, ∞)

2.151.

Contradiction

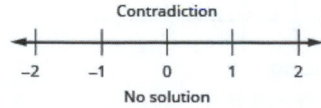

No solution

2.152.

Contradiction

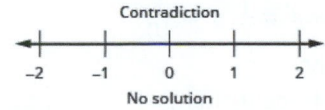

No solution

2.153.

$20y \leq 100$
$y \leq 5$

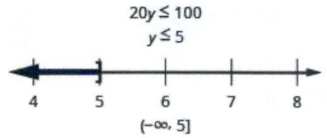

(−∞, 5]

2.154.

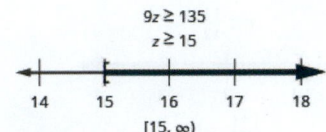

$9z \geq 135$
$z \geq 15$

14 15 16 17 18

$[15, \infty)$

2.155.

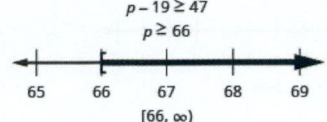

$p - 19 \geq 47$
$p \geq 66$

65 66 67 68 69

$[66, \infty)$

2.156.

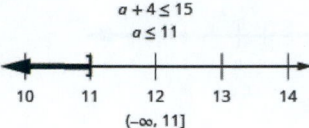

$a + 4 \leq 15$
$a \leq 11$

10 11 12 13 14

$(-\infty, 11]$

Section Exercises

1. yes

7. $y = -111$

13. $a = 121$

19. $y = 13.8$

25. $q = -\frac{1}{4}$

31. $x = 1$

37. $u = -7.1$

43. $n = -50$

49. $x = 13$

55. $y + (-30) = 40; y = 70$

61. $-4n + 5n = -82; -82$

67. 21 pounds

73. $d = \frac{17}{24}$ inch

79. $c = -11$

85. $z = 13$

91. $q = 100$

97. $y = -6$

103. $w = -64$

109. $p = -\frac{7}{9}$

115. $n = 88$

121. $x = -\frac{72}{5}$

127. $w = -198$

133. $m = 27$

139. $c = 51$

145. $-152 = 8q; q = -19$

151. $\frac{b}{-6} = 18; b = -108$

157. $\frac{5}{2}v = 50; v = 20$

163. 6 balloons

169. $195,000

175. $x = 6$

181. $b = -8$

3. no

9. $b = \frac{1}{2}$

15. $m = -182$

21. $x = -255$

27. $p = \frac{16}{15}$

33. $x = 0$

39. $a = -30$

45. $j = 8$

51. $x + 9 = 52; x = 43$

57. $9x - 8x = 107; 107$

63. 11 miles

69. 100.5 degrees

75. No. Justifications will vary.

81. $y = -\frac{809}{15}$

87. $x = 0$

93. $y = -144$

99. $v = 72$

105. $a = -4$

111. $u = \frac{1}{4}$

117. $d = 15$

123. $x = 100$

129. $a = \frac{9}{2}$

135. $p = 2130$

141. $x = 8$

147. $\frac{r}{12} = -48; r = -576$

153. $\frac{k}{22} = -66; k = -1,452$

159. $\frac{2}{5} + f = \frac{1}{2}; f = \frac{1}{10}$

165. $1.08

171. 15 49-cent stamps

177. $y = 6$

183. $x = -4$

5. $x = 11$

11. $p = -11.7$

17. $x = \frac{7}{3}$

23. $z = -9.02$

29. $c = 25$

35. $y = 8$ $y = 24$

41. $y = 28$

47. $a = -\frac{11}{4}$

53. $m - 10 = -14; m = -4$

59. $n - \frac{1}{6} = \frac{1}{2}; \frac{2}{3}$

65. 7 years old

71. $121.19

77. $x = 7$

83. $p = -\frac{541}{37}$

89. $x = 140$

95. $m = -540$

101. $y = 72$

107. $x = \frac{20}{3}$

113. $p = -12$

119. $q = -11$

125. $y = 9$

131. $x = -2$

137. $y = -7$

143. $133 = -19n; n = -7$

149. $\frac{j}{-20} = -80; j = 1,600$

155. $\frac{3}{10}x = 15; x = 50$

161. $q - \frac{1}{8} = \frac{3}{4}; q = \frac{7}{8}$

167. 36 mpg

173. Answers will vary.

179. $m = -8$

185. $q = -\frac{9}{7}$

187. $k = -11$

193. $z = 3$

189. $x = 9$

195. $x = -\dfrac{3}{4}$

191. $b = -3$

197. $r = -2$

199. $x = 19$

205. $c = -4$

201. $f = 7$

207. $x = 2$

203. $q = -5$

209. $p = -\dfrac{8}{5}$

211. $y = 4$

217. $a = -40$

223. $w = 60$

229. $x = 2$ Justifications will vary.

235. $n = -2$

213. $m = -6$

219. $p = 15$

225. $x = 23$

231. Yes. Justifications will vary.

237. $p = -\dfrac{4}{3}$

215. $a = 7$

221. $z = 3.46$

227. 30 feet

233. $y = 3$

239. $t = -9$

241. $b = 2$

247. $t = 1$

253. $d = 1$

259. $y = -4$

243. $m = 6$

249. $x = -2$

255. $r = -7$

261. $w = \dfrac{1}{2}$

245. $k = \dfrac{3}{2}$

251. $x = 5$

257. $m = -15$

263. $a = 4$

265. $r = 8$

271. $x = 34$

277. $a = -4$

283. $k = \dfrac{3}{5}$

267. $m = 3$

273. $x = -6$

279. $m = -4$

285. $n = -5$

269. $y = -3$

275. $n = -1$

281. $d = -3$

287. $v = 1$

289. $p = -34$

295. identity; all real numbers

291. $m = -4$

297. conditional equation; $d = \dfrac{2}{3}$

293. identity; all real numbers

299. conditional equation; $n = 7$

301. contradiction; no solution

303. contradiction; no solution

305. conditional equation; $k = 26$

307. contradiction; no solution

313. 8 nickels

319. $x = 1$

325. $x = 3$

331. $z = -2$

337. $x = 1$

309. identity; all real numbers

315. Answers will vary.

321. $y = -1$

327. $n = -24$

333. $a = 1$

339. $q = 7$

311. identity; all real numbers

317. Answers will vary.

323. $b = -2$

329. $x = -4$

335. $x = -6$

341. $x = -\dfrac{5}{2}$

343. $m = -1$

349. $d = -24$

355. $y = 15$

361. $x = 22$

367. $d = 10$

373. Answers will vary.

345. $p = -4$

351. $y = -1$

357. $k = 2$

363. $x = 8$

369. $q = 15$

375. 100. Justifications will vary.

347. $v = 20$

353. $v = 4$

359. $x = 20$

365. $n = 19$

371. $s = 35$

377. 290 miles

379. 30 miles

385. 75 mph

381. 5 hours

387. 75 mph

383. 11.25 hours

389. ⓐ $t = 4$ ⓑ $t = \dfrac{d}{r}$

391. ⓐ $t = 3.5$ ⓑ $t = \dfrac{d}{r}$

397. ⓐ $h = 16$ ⓑ $h = \dfrac{2A}{b}$

393. ⓐ $r = 70$ ⓑ $r = \dfrac{d}{t}$

399. ⓐ $b = 10$ ⓑ $b = \dfrac{2A}{h}$

395. ⓐ $r = 40$ ⓑ $r = \dfrac{d}{t}$

401. ⓐ $P = \$13,166.67$ ⓑ $P = \dfrac{I}{rt}$

403. ⓐ $t = 2$ years ⓑ $t = \dfrac{I}{Pr}$

405. ⓐ $y = -5$ ⓑ $y = \dfrac{10 - 5x}{2}$

407. ⓐ $y = 17$ ⓑ $y = 5 - 4x$

409. $a = 90 - b$

411. $c = 180 - a - b$

413. $y = 13 - 9x$

415. $y = -1 + 5x$

417. $y = \dfrac{11 - 3x}{4}$

419. $y = 3 + x$

421. $W = \dfrac{P - 2L}{2}$

423. $\pi = \dfrac{C}{d}$

425. $H = \dfrac{V}{LW}$

427. $10°C$

429. Answers will vary.

431.

(a)
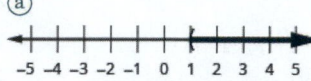

(b)
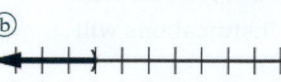

(c)

433.

(a)

(b)
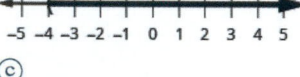

(c)

435.

(a)
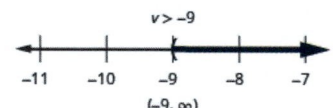
$(3, \infty)$

(b)
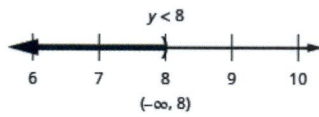
$(-\infty, -0.5]$

(c)

$\left[\dfrac{1}{3}, \infty\right)$

437.

(a)

$(-\infty, 5]$

(b)
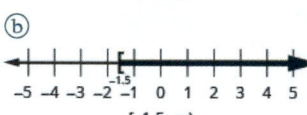
$[-1.5, \infty)$

(c)
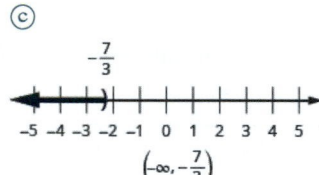
$\left(-\infty, -\dfrac{7}{3}\right)$

439.
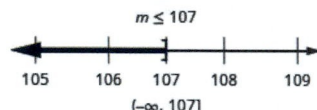
$(-\infty, 107]$

441.
$v > -9$
$(-9, \infty)$

443.
$b \geq -\dfrac{17}{24}$
$\left[-\dfrac{17}{24}, \infty\right)$

445.
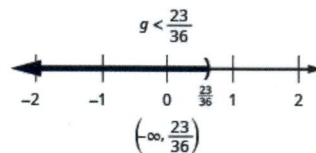
$g < \dfrac{23}{36}$
$\left(-\infty, \dfrac{23}{36}\right)$

447.
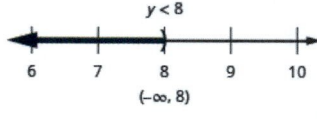
$y < 8$
$(-\infty, 8)$

449.
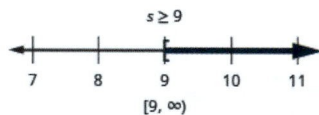
$s \geq 9$
$[9, \infty)$

451.
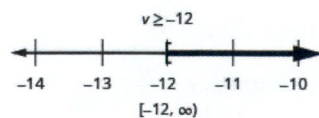
$v \geq -12$
$[-12, \infty)$

453.
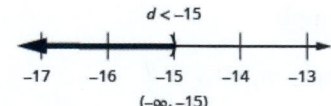
$d < -15$
$(-\infty, -15)$

455.
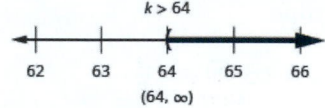
$k > 64$
$(64, \infty)$

457.

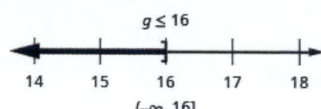

$g \le 16$

$(-\infty, 16]$

459.

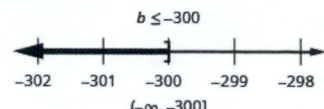

$b \le -300$

$(-\infty, -300]$

461.

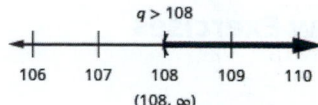

$q > 108$

$(108, \infty)$

463.

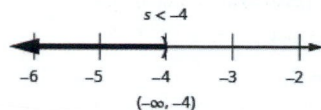

$s < -4$

$(-\infty, -4)$

465.

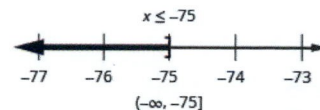

$x \le -75$

$(-\infty, -75]$

467.

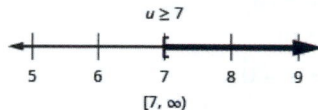

$u \ge 7$

$[7, \infty)$

469.

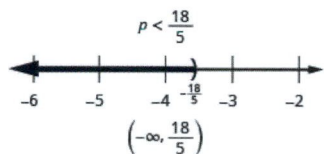

$p < \dfrac{18}{5}$

$\left(-\infty, \dfrac{18}{5}\right)$

471.

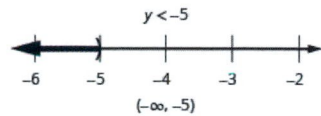

$y < -5$

$(-\infty, -5)$

473.

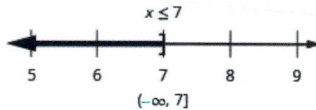

$x \le 7$

$(-\infty, 7]$

475.

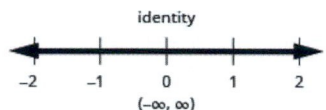

identity

$(-\infty, \infty)$

477.

Contradiction

No solution

479.

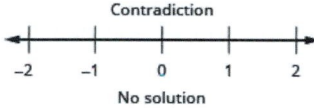

Contradiction

No solution

481.

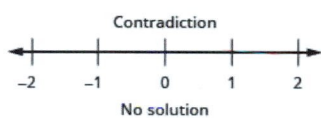

Contradiction

No solution

483.

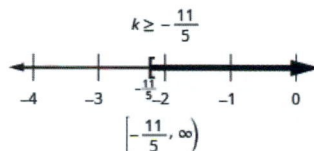

$k \ge -\dfrac{11}{5}$

$\left[-\dfrac{11}{5}, \infty\right)$

485.

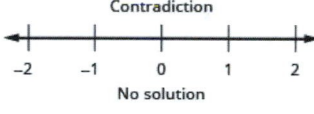

Contradiction

No solution

487.

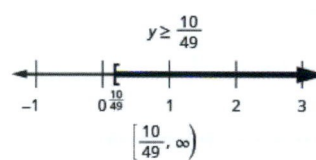

$y \ge \dfrac{10}{49}$

$\left[\dfrac{10}{49}, \infty\right)$

489.

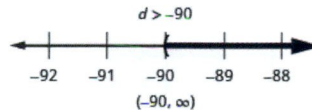

$d > -90$

$(-90, \infty)$

491.

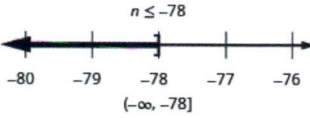

$n \le -78$

$(-\infty, -78]$

493.

$90c < 450$

$c < 5$

$(-\infty, 5)$

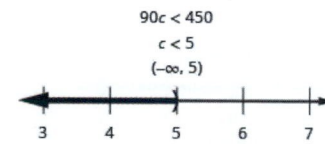

495.

$-10y \le -110$

$y \le -11$

$(-\infty, -11]$

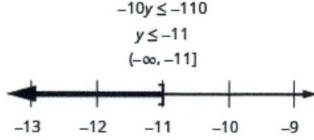

497.

$k + 6 > 25$

$k > 19$

$(19, \infty)$

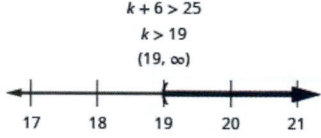

499.

$x - 12 \ge 21$

$x \ge 33$

$[33, \infty)$

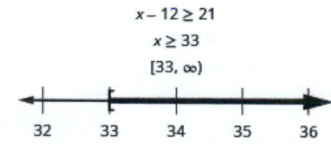

501.

$-2s < 56$

$s > -28$

$(-28, \infty)$

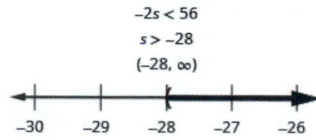

503.

$a - 15 \ge -7$

$a \ge 8$

$[8, \infty)$

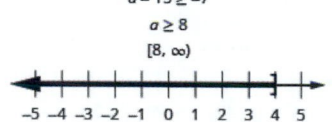

505. $h \le 77$

507. $n \le 8$

509. Answers will vary.

511. Answers will vary.

Review Exercises

513. no

515. yes

517. $y = -8$

519. $n = 1.5$

521. $x = 5$

523. $p = 18.8$

525. $y = -25$

527. $d = 12.1$

529. $x = -8$

531. $p = -28$

533. $n - 4 = 13;\ n = 17$

535. 161 pounds

537. $131.19

539. $a = -5$

541. $y = -4$

543. $y = -300$

545. $u = \dfrac{3}{2}$

547. $c = 324$

549. $y = \dfrac{11}{8}$

551. $x = -1$

553. $d = 3$

555. $\dfrac{b}{9} = -27;\ b = -243$

557. $s - \dfrac{1}{12} = \dfrac{1}{4};\ s = \dfrac{1}{3}$

559. $32

561. $w = 7$

563. $n = -4$

565. $a = -7$

567. $x = \dfrac{3}{8}$

569. $n = -5$

571. $c = 32$

573. $p = \dfrac{13}{2}$

575. $n = 12$

577. $m = -14$

579. $q = 18$

581. $x = -1$

583. $k = \dfrac{3}{4}$

585. contradiction; no solution

587. identity; all real numbers

589. $x = 15$

591. $k = 41$

593. $y = -1$

595. $u = -85$

597. $d = -20$

599. 5 hours

601. 37.5 miles

603. ⓐ $r = 82$ mph; ⓑ $r = \dfrac{D}{t}$

605. ⓐ $h = 17$ ⓑ $h = \dfrac{2A}{b}$

607. ⓐ $y = \dfrac{14}{3}$ ⓑ $y = \dfrac{6 - 4x}{3}$

609. $H = \dfrac{V}{LW}$

611.

ⓐ

ⓑ

ⓒ

613.

ⓐ

$(2, \infty)$

ⓑ

$(-\infty, -1.5]$

ⓒ

$\left[\dfrac{5}{3}, \infty\right)$

615.

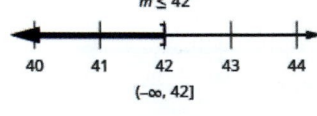

$m \le 42$

$(-\infty, 42]$

617.

$b \ge \dfrac{3}{8}$

$\left[\dfrac{3}{8}, \infty\right)$

619.

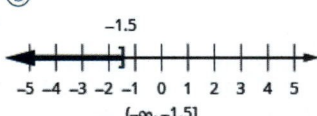

$d \ge -9$

$[-9, \infty)$

621.

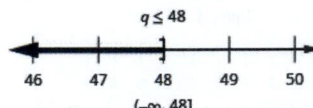

$q \le 48$

$(-\infty, 48]$

623.

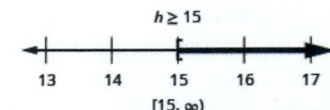

$h \ge 15$

$[15, \infty)$

625.

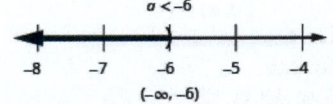

$a < -6$

$(-\infty, -6)$

627.

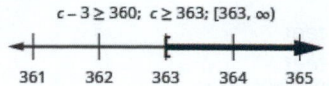

$c - 3 \geq 360;\ c \geq 363;\ [363, \infty)$

629.

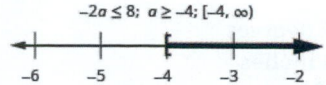

$-2a \leq 8;\ a \geq -4;\ [-4, \infty)$

Practice Test

631. ⓐ no ⓑ yes

633. $c = 32$

635. $x = -5$

637. $x = 9$

639. $y = -4$

641. $m = 9$

643. $d = -14$

645. $x = -\dfrac{1}{3}$

647. $p = \dfrac{10}{3}$

649. contradiction; no solution

651. ⓐ $y = 4$ ⓑ $y = \dfrac{5 - x}{2}$

653.

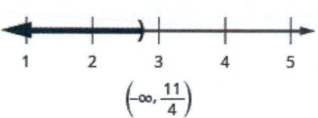

$\left(-\infty, \dfrac{11}{4}\right)$

655.

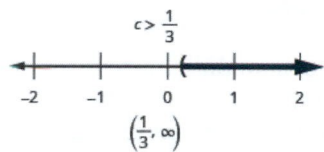

$\left(\dfrac{1}{3}, \infty\right)$

657. $n + 15 \geq 48;\ n \geq 33$

659. $120 = \dfrac{2}{3}p$; The original price was \$180.

Chapter 3

Try It

3.1. \$180

3.2. 40

3.3. 2

3.4. 7

3.5. 25

3.6. 4

3.7. 3

3.8. 6

3.9. 9, 15

3.10. 27, 31

3.11. $-15, -8$

3.12. $-29, 11$

3.13. $-4, 0$

3.14. $-3, -2$

3.15. 47, 48

3.16. $-16, -15$

3.17. $-33, -32, -31$

3.18. $-13, -12, -11$

3.19. 32, 34, 36

3.20. $-10, -8, -6$

3.21. \$5,000

3.22. \$19,300

3.23. 36

3.24. 33

3.25. \$26

3.26. \$36

3.27. 125%

3.28. 175%

3.29. \$14.67

3.30. \$2.16

3.31. 25 grams

3.32. 2,375 mg

3.33. 26%

3.34. 40%

3.35. 8.8%

3.36. 50%

3.37. 6.3%

3.38. 10%

3.39. \$142.50

3.40. \$7,020

3.41. 6%

3.42. 4%

3.43. \$11,450

3.44. \$9,600

3.45. ⓐ \$11.60 ⓑ \$17.40

3.46. ⓐ \$256.75 ⓑ \$138.25

3.47. ⓐ \$184.80 ⓑ 33%

3.48. ⓐ \$60 ⓑ 15%

3.49. ⓐ \$600 ⓑ \$1,800

3.50. ⓐ \$2,975 ⓑ \$11,475

3.51. 9 nickels, 16 dimes

3.52. 17 nickels, 5 quarters

3.53. 42 nickels, 21 dimes

3.54. 51 dimes, 17 quarters

3.55. 41 nickels, 18 quarters

3.56. 22 nickels, 59 dimes

3.57. 330 day passes, 367 tournament passes

3.58. 112 adult tickets, 199 senior/child tickets

3.59. 84 adult tickets, 31 student tickets

3.60. 26 full-fare, 14 reduced fare

3.61. 32 at \$0.41, 12 at \$0.02

3.62. 26 at \$0.41, 10 at \$0.20

3.63. 5 pounds cereal squares, 25 pounds nuts

3.64. 21 gallons of fruit punch, 7 gallons of soda

3.65. $4,200 at 4%, $9,800 at 7%

3.66. $2,500 at 3.2%, $5,500 at 8%

3.67. 21 degrees

3.68. 56 degrees

3.69. 8 feet

3.70. 6 feet

3.71. 14 inches

3.72. 6 feet

3.73. $34°$

3.74. $45°$

3.75. $20°, 70°, 90°$

3.76. $30°, 60°, 90°$

3.77. $c = 10$

3.78. $c = 13$

3.79. 8

3.80. 12

3.81. 12 feet

3.82. 8 feet

3.83. 340 yards

3.84. 220 feet

3.85. 26 feet

3.86. 29 meters

3.87. 15

3.88. 9

3.89. 18 meters, 11 meters

3.90. 19 feet, 11 feet

3.91. 24, 8

3.92. 5, 4

3.93. 70 feet, 30 feet

3.94. 90 yards, 60 yards

3.95. Wayne 21 mph, Dennis 28 mph

3.96. Jeromy 80 mph, mother 60 mph

3.97. Carina 50 mph, brother 65 mph

3.98. parents 55 mph, Ashley 62 mph

3.99. 3 hours

3.100. 2.2 hours

3.101. uphill 1.8 mph, downhill three mph

3.102. upstream 8 mph, downstream 12 mph

3.103. biking 16 mph, running 10 mph

3.104. uphill 12 mph, flat street 20 mph

3.105. There can be no more than 20 boxes.

3.106. A maximum of 14 people can safely ride in the elevator.

3.107. seven packs

3.108. 11 people

3.109. at least $4,000,000

3.110. at least $1,200,000

3.111. no more than 106 text messages

3.112. no more than 76 therms

3.113. at least 96 hours

3.114. at least 1,380 invitations

3.115. at least 23 hours

3.116. at least 20 driveways

Section Exercises

1. Answers will vary

3. 30

5. 125

7. 6

9. 58

11. $750

13. $180,000

15. 4

17. 15

19. 5

21. 12

23. -5

25. 18, 24

27. 8, 12

29. $-18, -27$

31. $-111, -205$

33. 32, 46

35. 23, 28

37. $-2, -3$

39. 4, 10

41. 38, 39

43. $-11, -12$

45. 25, 26, 27

47. $-11, -12, -13$

49. 84, 86, 88

51. 55, 57, 59

53. $-10, -12, -14$

55. $-69, -71, -73$

57. $45

59. $1.25

61. $1080

63. answers will vary

65. Consecutive odd integers are odd numbers that immediately follow each other. An example of three consecutive odd integers between 50 and 60 would be 51, 53, and 55.

67. 54

69. 26.88

71. 162.5

73. 18,000

75. 112

77. 108

79. $35

81. $940

83. 30%

85. 36%

87. 150%

89. 175%

91. $11.88

93. $259.80

95. 24.2 g

97. 2407 mg

99. 175 lb.

101. $120

103. 18%

105. 45%

107. 25%

109. 12%

111. 200%

113. 72.7%

115. 2.5%

117. 11%

119. 5.5%

121. $116

123. $4,836

125. 3%

127. 3.75%

129. $35,000

131. $3,345

133. $139

135. $125

137. ⓐ $26.97 ⓑ $17.98

139. ⓐ $128.37 ⓑ $260.63

141. ⓐ $332.48 ⓑ $617.47

143. ⓐ $576 ⓑ 30%

145. ⓐ $53.25 ⓑ 15%

147. ⓐ $370 ⓑ 43.5%

149. ⓐ $7.20 ⓑ $23.20

151. ⓐ $0.20 ⓑ $0.80

153. ⓐ $258.75 ⓑ $373.75

155. 21.2%

157. The number should be greater than 44. Since 80% equals 0.8 in decimal form, 0.8 is less than one, and we must multiply the number by 0.8 to get 44, the number must be greater than 44.

159. He meant that he should have packed half the shorts and twice the shirts.

161. 8 nickels, 22 dimes

163. 15 dimes, 8 quarters

165. 10 at $1, 7 at $5

167. 18 quarters, 36 nickels

169. 30 pennies, 90 dimes

171. 10 at $10, 5 at $5

173. 12 dimes and 27 nickels

175. 63 dimes, 20 quarters

177. 16 nickels, 12 dimes, 7 quarters

179. 30 child tickets, 50 adult tickets

181. 110 child tickets, 50 adult tickets

183. 34 general, 61 youth

185. 114 general, 246 student

187. 40 postcards, 100 stamps

189. 30 at $0.41, 10 at $0.02

191. 15 $10 shares, 5 $12 shares

193. 5 liters champagne, 10 liters orange juice

195. 7.5 lbs Maui beans, 17.5 Kona beans

197. 800 at $1.50, 200 at $9.00

199. $8000 at 4%, $17,000 at 9%

201. $10,000 in CD, $30,000 in mutual fund

203. 9.1%

205. 9 girls, 3 adults

207. Answers will vary.

209. Answers will vary.

211. 56 degrees

213. 44 degrees

215. 11 feet

217. 8 feet

219. 0.75 sq. ft.

221. 23 inches

223. 57

225. 67.5

227. 13 ft., 12 ft., 14 ft.

229. 3 ft., 6 ft., 8 ft.

231. 45°, 45°, 90°

233. 30°, 60°, 90°

235. 15

237. 25

239. 8

241. 12

243. 10.2

245. 9.8

247. 5 feet

249. 14.1 feet

251. 260 feet

253. 58 feet

255. 23 meters

257. 26 inches

259. 24

261. 4.9

263. 16 in., 7 in.

265. 17 m, 12 m

267. 13.5 m length, 12.8 m width

269. 50 ft., 25 ft.

271. 7 m width, 11 m length

273. 160 yd., 120 yd.

275. 85 ft., 40 ft.

277. 24 feet

279. area; answers will vary

281. ⓐ Answers will vary.

ⓑ Answers will vary.

ⓒ Answers will vary.

ⓓ The areas are the same.

ⓔ The 2x8 rectangle has a larger perimeter than the 4x4 square.

283. Mason 75 mph, train 60 mph

285. express bus 75mph, local 50 mph

287. Saul 82 mph, Erwin 74 mph

289. Helen 60 mph, Anne 56 mph

291. Aletheia 2.4 mph, Elvira 3 mph

293. 3 hours

295. 3.5 hours

297. walking 3 mph, jogging 4.5 mph

299. upstream 6 mph, downstream 9 mph

301. skateboarding 8 mph, walking 3.2 mph

303. heavy traffic 30 mph, light traffic 70 mph

305. ⓐ 36 mph ⓑ 25 mph ⓒ 3.05 hours ⓓ 29.5 mph

307. Answers will vary.

309. 15 children

311. 12 people

313. five drinks

315. 31 aprons

317. $15,000

319. $400,000

321. 16 ccf

323. nine movies

325. 48 cut & styles
331. 9 cars
337. Answers will vary.

327. 44 meals
333. 91 students

329. 49 cars
335. 169 guests

Review Exercises

339. answers will vary
345. $-18, -9$
351. 15, 17, 19
357. 120%
363. 2.5%
369. $63

341. 116
347. $-3, -10$
353. 167.5
359. $3.89
365. $428.22

343. 38
349. $-48, -47, -46$
355. 160
361. $875
367. $1,900

371. ⓐ $55.20 ⓑ $82.80

373. ⓐ $70 ⓑ 15.6%

375. ⓐ $175 ⓑ $225

377. 16 dimes, 11 quarters

379. six $5 bills, 11 $10 bills

381. 35 adults, 82 children

383. 57 students, 68 adults

385. three $0.26 stamps, eight $0.41 stamps

387. 2.2 lb. of raisins, 7.8 lb. of nuts
393. 600 square inches
399. 26
405. $6'$

389. $12,500 at 4.5%, $2,500 at 1.8%
395. $30°, 60°, 90°$

401. 8
407. ⓐ 110 ft. ⓑ 684 sq. ft.

391. $73°$

397. $22.5°, 67.5°, 90°$

403. 8.1
409. 62 m

411. 24.5 cm, 12.5 cm
417. flat road 11 mph, uphill 8 mph
423. at least 112 jobs

413. Gabe 70 mph, Elsa 77 mph
419. $33 per day

415. 2.5 hours
421. at least $300,000

Practice Test

425. 15

431. ⓐ $208.50 ⓑ 30%
437. 10
443. at most $44.44 per costume

427. $-49, -47$

433. 140 adult, 85 children
439. 127.3 ft.

429. 8%

435. $41.5°, 55.5°, 83°$

441. 2.5 hours

Chapter 4

Try It

4.1.

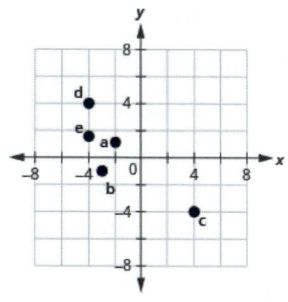

4.2.

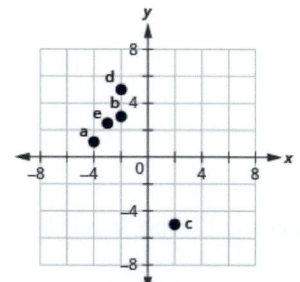

4.3.

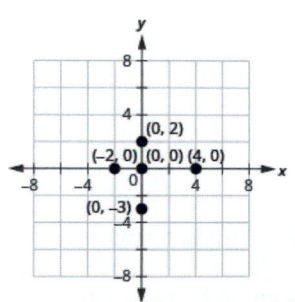

4.4.

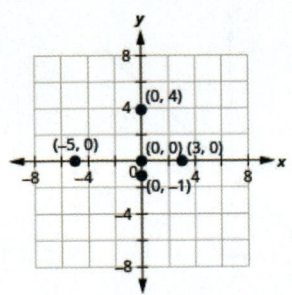

4.5. A: (5, 1)   B: (−2, 4)   C: (−5, −1)   D: (3, −2)   E: (0, −5)   F: (4, 0)

4.6. A: (4, 2)   B: (−2, 3)   C: (−4, −4)   D: (3, −5)   E: (−3, 0)   F: (0, 2)

4.7. a, c

4.8. b, c

4.9. b

4.10. a, b

4.11.

$y = 3x - 1$		
x	*y*	*(x, y)*
0	−1	(0, −1)
−1	−4	(−1, −4)
2	5	(2, 5)

4.12.

$y = 6x + 1$		
x	*y*	*(x, y)*
0	1	(0, 1)
1	7	(1, 7)
−2	−11	(−2, −11)

4.13.

$2x - 5y = 20$		
x	*y*	*(x, y)*
0	−4	(0, −4)
10	0	(10, 0)
−5	−6	(−5, −6)

4.14.

$3x - 4y = 12$		
x	*y*	*(x, y)*
0	−3	(0, −3)
4	0	(4, 0)
−4	−6	(−4, −6)

4.15. Answers will vary.

4.16. Answers will vary.

4.19. ⓐ yes, yes   ⓑ yes, yes

4.17. Answers will vary.

4.20. ⓐ no, no   ⓑ yes, yes

4.18. Answers will vary.

4.21.

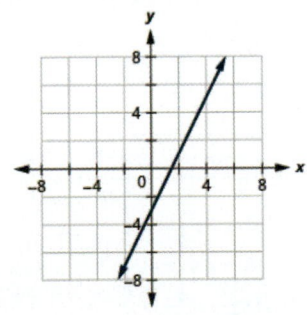

4.22.

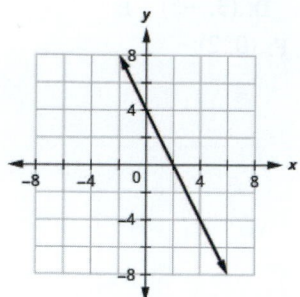

4.23.

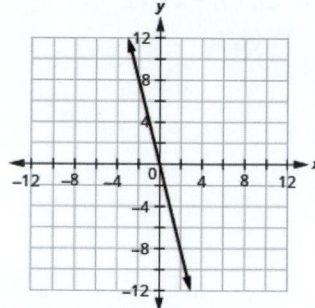

4.24.

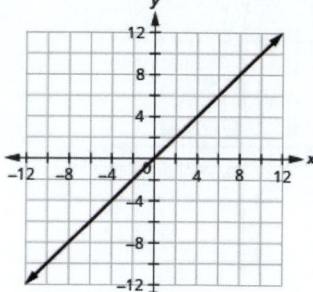

4.25.

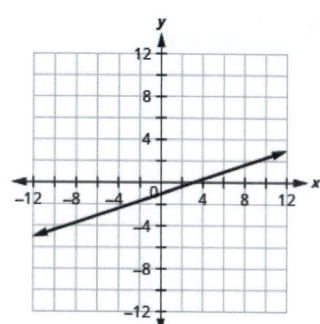

4.26.

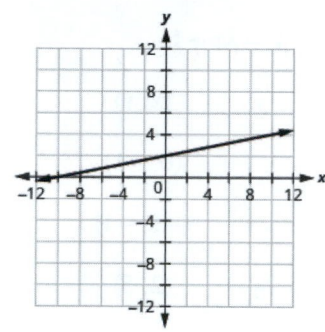

4.27.

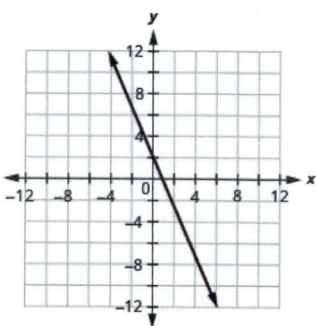

4.28.

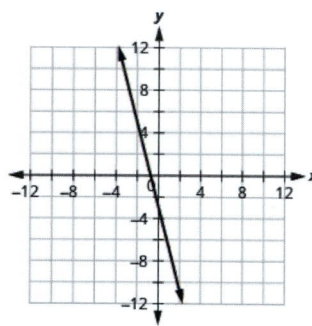

4.29.

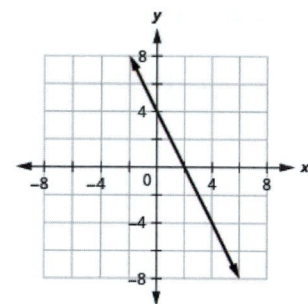

4.30.

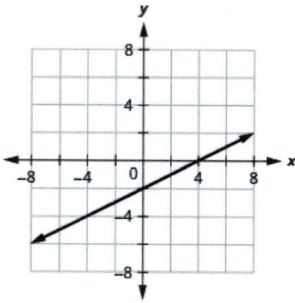

4.31.

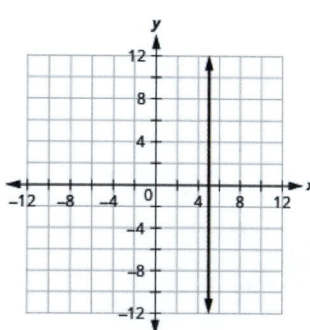

4.32.

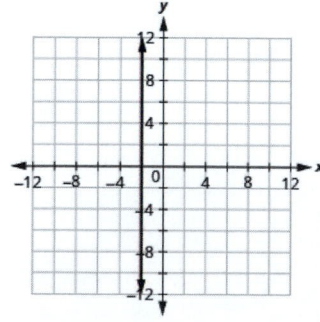

4.33.

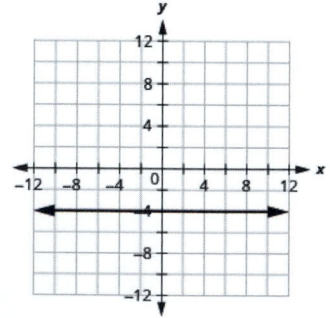

4.34.

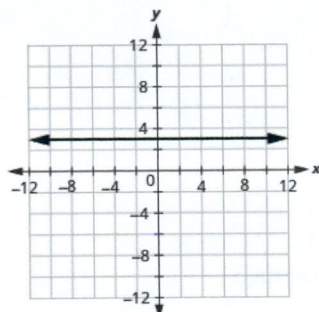

4.35.

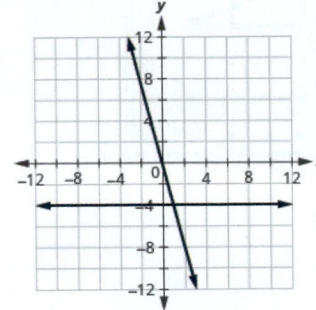

4.36.

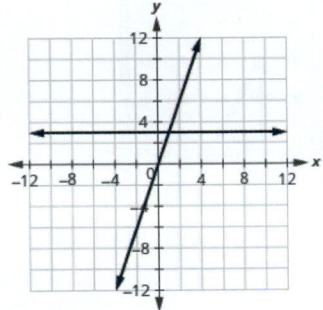

4.37. *x*- intercept: $(2, 0)$; *y*-intercept: $(0, -2)$

4.38. *x*- intercept: $(3, 0)$, *y*-intercept: $(0, 2)$

4.39. *x*- intercept: $(4, 0)$, *y*-intercept: $(0, 12)$

4.40. *x*- intercept: $(8, 0)$, *y*-intercept: $(0, 2)$

4.41. *x*- intercept: $(4, 0)$, *y*-intercept: $(0, -3)$

4.42. *x*- intercept: $(4, 0)$, *y*-intercept: $(0, -2)$

4.43.

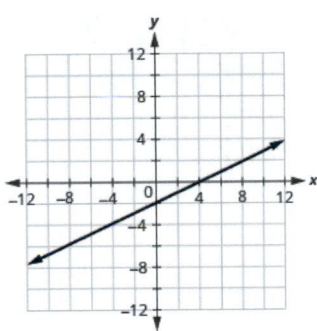

4.44.

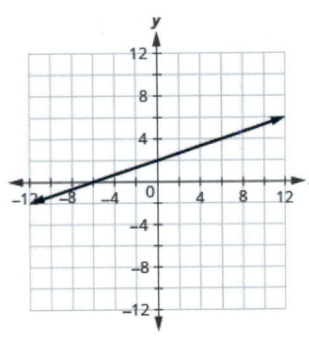

4.45.

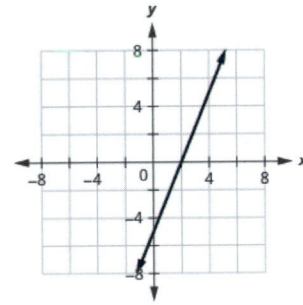

4.46.

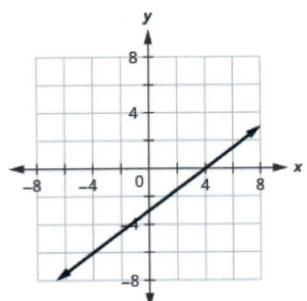

4.47.

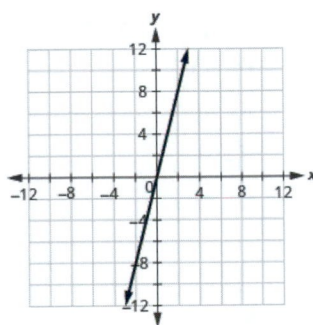

4.48.

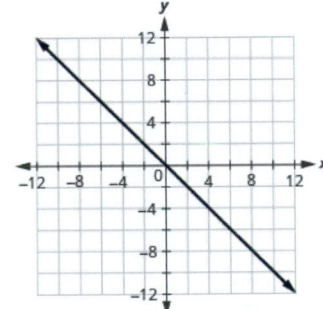

4.49. $\frac{4}{3}$

4.50. $\frac{1}{4}$

4.51. $-\frac{2}{3}$

4.52. $-\frac{4}{3}$

4.53.

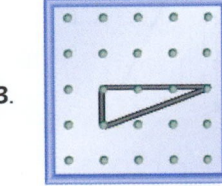

4.54.

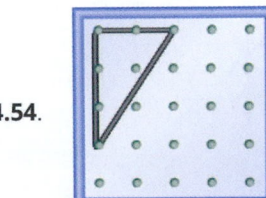

4.55.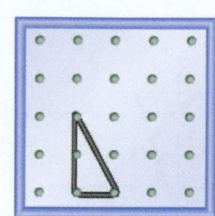

4.56.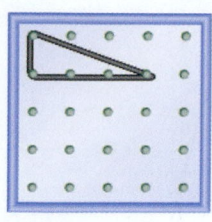

4.57. $\frac{2}{5}$

4.58. $\frac{3}{4}$

4.59. $-\frac{4}{3}$

4.60. $-\frac{3}{5}$

4.61. $\frac{5}{4}$

4.62. $\frac{3}{2}$

4.63. undefined

4.64. 0

4.65. 1

4.66. 1

4.67. -1

4.68. 10

4.69.

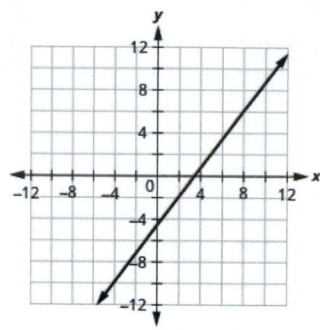

4.70.

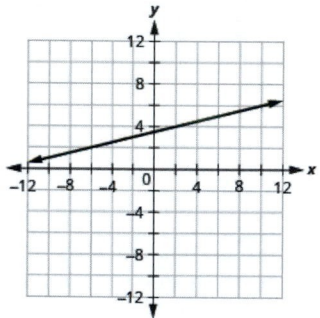

4.71.

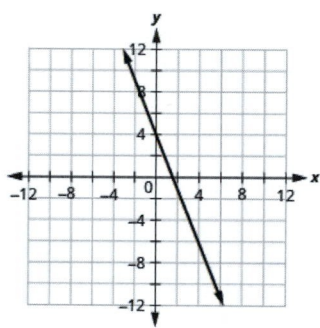

4.72.

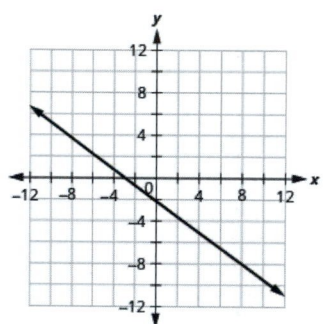

4.73.

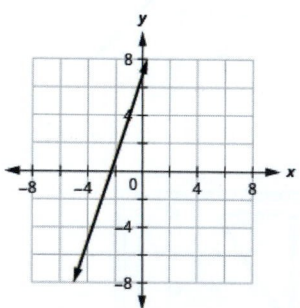

4.74.

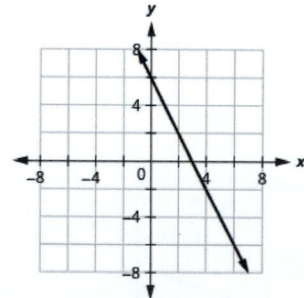

4.75. $\frac{7}{12}$

4.76. $\frac{5}{12}$

4.77. $-\frac{1}{36}$

4.78. $-\frac{1}{48}$

4.79. slope $m = \frac{2}{3}$ and y-intercept $(0, -1)$

4.80. slope $m = \frac{1}{2}$ and y-intercept $(0, 3)$

4.81. $\frac{2}{5}$; $(0, -1)$

4.82. $-\frac{4}{3}$; (0, 1)

4.83. $-\frac{1}{4}$; (0, 2)

4.84. $-\frac{3}{2}$; (0, 6)

4.85.

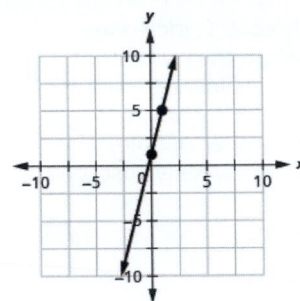

4.86.

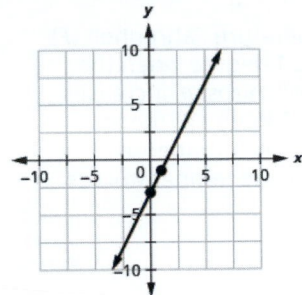

4.87.

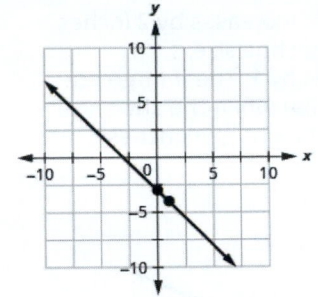

4.88.

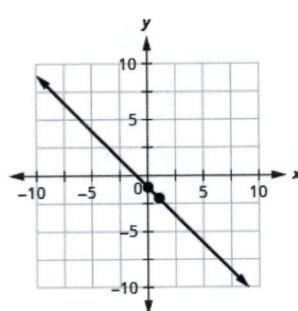

4.89.

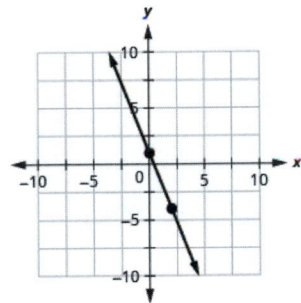

4.90.

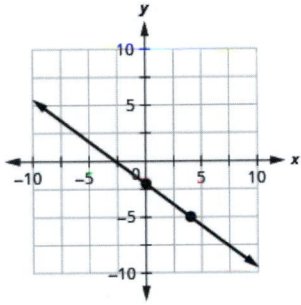

4.91.

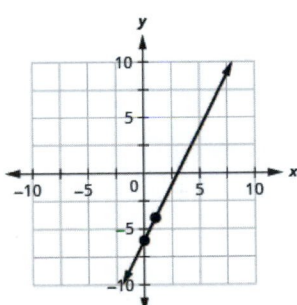

4.92.

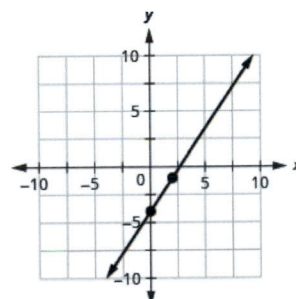

4.93.

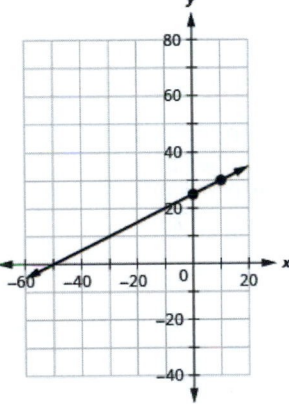

4.94.

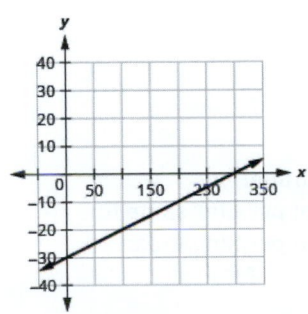

4.95. ⓐ intercepts ⓑ horizontal line ⓒ slope–intercept ⓓ vertical line

4.96. ⓐ vertical line ⓑ slope–intercept ⓒ horizontal line ⓓ intercepts

4.97.

ⓐ 50 inches ⓑ 66 inches

ⓒ The slope, 2, means that the height, h, increases by 2 inches when the shoe size, s, increases by 1. The h-intercept means that when the shoe size is 0, the height is 50 inches.

ⓓ

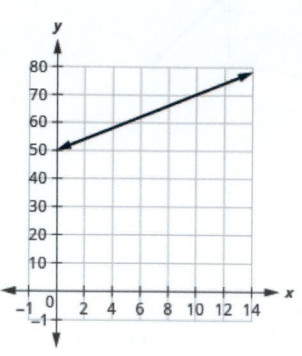

4.98.

ⓐ 40 degrees ⓑ 65 degrees

ⓒ The slope, $\frac{1}{4}$, means that the temperature Fahrenheit (F) increases 1 degree when the number of chirps, n, increases by 4. The T-intercept means that when the number of chirps is 0, the temperature is 40°.

ⓓ

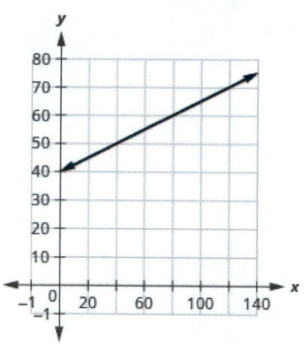

4.99.

ⓐ $60 ⓑ $185

ⓒ The slope, 0.5, means that the weekly cost, C, increases by $0.50 when the number of miles driven, n, increases by 1. The C-intercept means that when the number of miles driven is 0, the weekly cost is $60

ⓓ

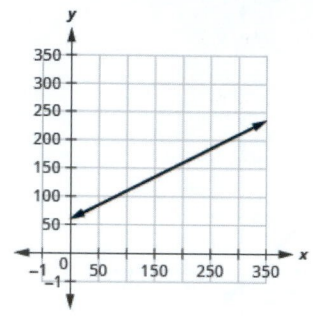

4.100.

ⓐ $35 ⓑ $170

ⓒ The slope, 1.8, means that the weekly cost, C, increases by $1.80 when the number of invitations, n, increases by 1.80.
The C-intercept means that when the number of invitations is 0, the weekly cost is $35.;

ⓓ

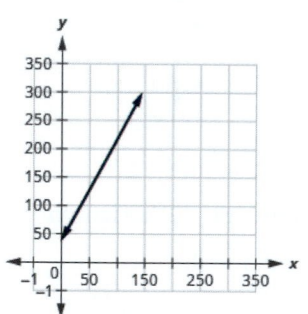

4.101. parallel

4.102. parallel

4.103. parallel
4.106. parallel
4.109. perpendicular
4.112. not perpendicular

4.115. $y = \frac{3}{5}x + 1$

4.104. parallel
4.107. not parallel; same line
4.110. perpendicular
4.113. $y = \frac{2}{5}x + 4$

4.116. $y = \frac{4}{3}x - 5$

4.105. parallel
4.108. not parallel; same line
4.111. not perpendicular
4.114. $y = -x - 3$

4.117. $y = \frac{5}{6}x - 2$

4.118. $y = \frac{2}{3}x - 4$

4.119. $y = -\frac{2}{5}x - 1$

4.120. $y = -\frac{3}{4}x - 4$

4.121. $y = 8$

4.122. $y = 4$

4.123. $y = \frac{5}{2}x - \frac{13}{2}$

4.124. $y = -\frac{2}{5}x + \frac{22}{5}$

4.125. $y = \frac{1}{3}x - \frac{10}{3}$

4.126. $y = -\frac{2}{5}x - \frac{23}{5}$

4.127. $x = 5$

4.128. $x = -4$

4.129. $y = 3x - 10$

4.130. $y = \frac{1}{2}x + 1$

4.131. $y = -\frac{1}{3}x + \frac{10}{3}$

4.132. $y = -2x + 16$

4.133. $y = -5$

4.134. $y = -1$

4.135. $x = -5$

4.136. $x = -4$

4.137. ⓐ yes ⓑ yes ⓒ yes ⓓ yes ⓔ no

4.138. ⓐ yes ⓑ yes ⓒ no ⓓ no ⓔ yes

4.139. $y \geq -2x + 3$

4.140. $y < \frac{1}{2}x - 4$

4.141. $x - 4y \leq 8$

4.142. $3x - y \leq 6$

4.143.

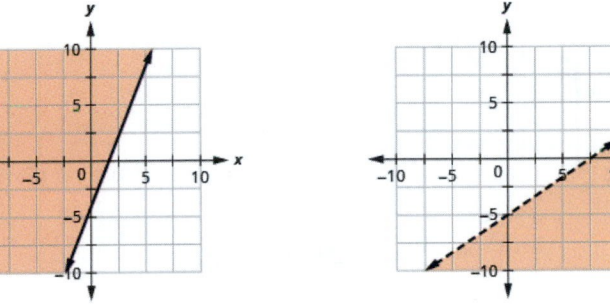

4.144.

4.145.

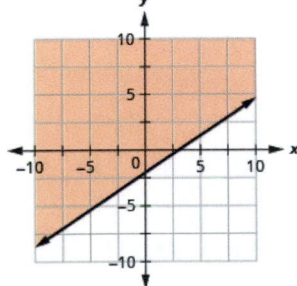

4.146.

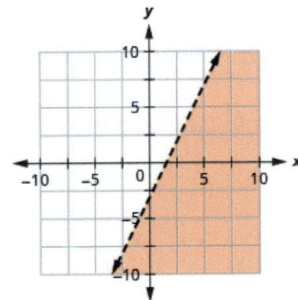

4.147.

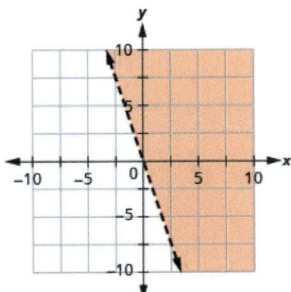

4.148.

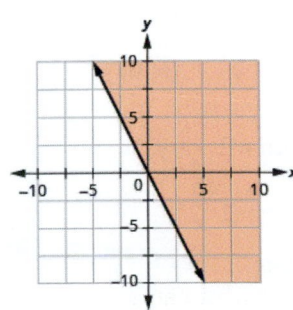

4.149.

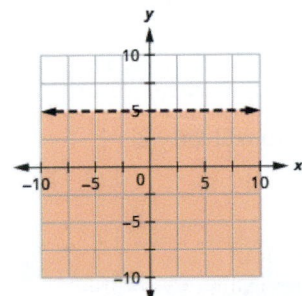

4.150.

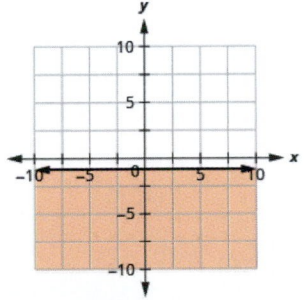

Section Exercises

1.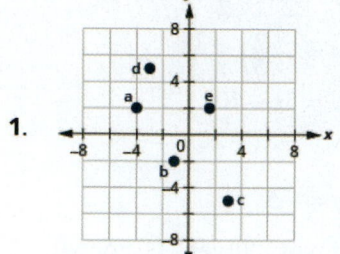

3.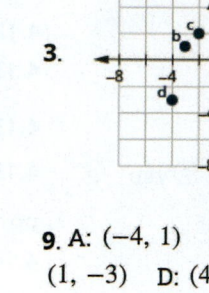

5.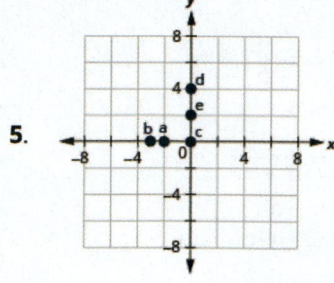

9. A: (−4, 1)   B: (−3, −4)   C: (1, −3)   D: (4, 3)

11. A: (0, −2)   B: (−2, 0)   C: (0, 5)   D: (5, 0)

7.

13. a, b   **15.** a, c   **17.** b, c
19. a, b   **21.**   **23.**

x	y	(x, y)
0	−4	(0, −4)
2	0	(2, 0)
−1	−6	(−1, −6)

x	y	(x, y)
0	5	(0, 5)
3	2	(3, 2)
−2	7	(−2, 7)

25.   **27.**   **29.**

x	y	(x, y)
0	1	(0, 1)
3	2	(3, 2)
6	3	(6, 3)

x	y	(x, y)
0	−2	(0, −2)
2	−5	(2, −5)
−2	1	(−2, 1)

x	y	(x, y)
0	2	(0, 2)
3	4	(3, 1)
6	0	(6, 0)

31.   **33.** Answers will vary.   **35.** Answers will vary.

x	y	(x, y)
0	−2	(0, −2)
10	2	(10, 2)
5	0	(5, 0)

37. Answers will vary.   **39.** Answers will vary.   **41.** Answers will vary.
43. Answers will vary.   **45.** Answers will vary.   **47.** Answers will vary.

49. ⓐ

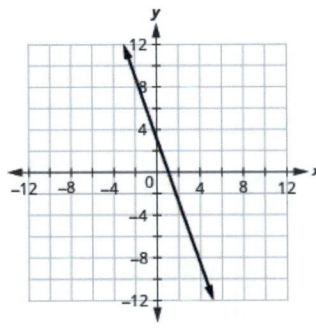

ⓑ Age and weight are only positive.

51. Answers will vary.

53. Answers will vary.

55. ⓐ yes; no ⓑ no; no ⓒ yes; yes ⓓ yes; yes

57. ⓐ yes; yes ⓑ yes; yes ⓒ yes; yes ⓓ no; no

59.

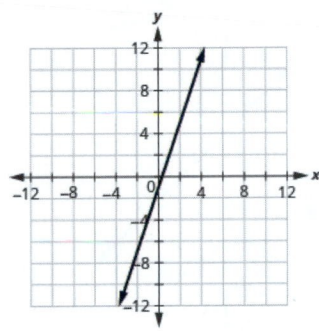

61.

63.

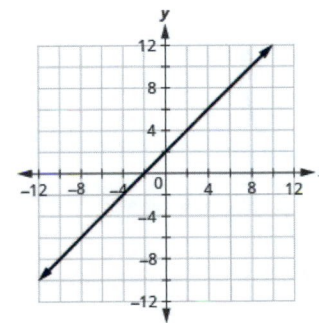

65.

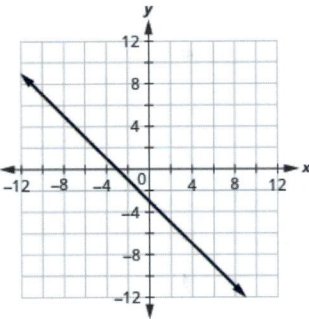

67.

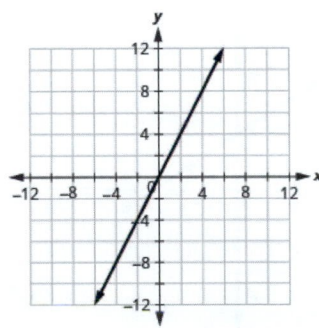

69.

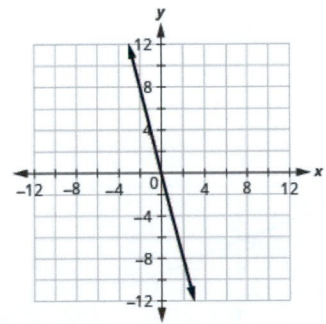

71.

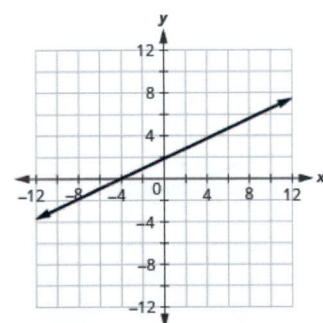

73.

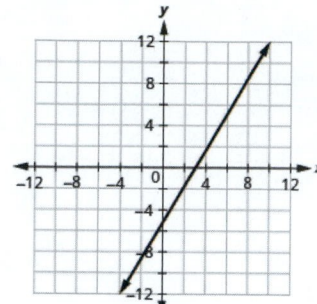

75.

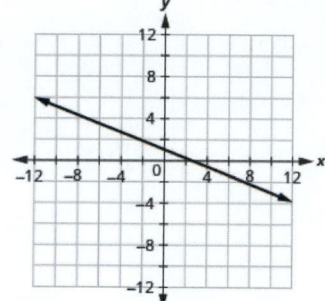

77.

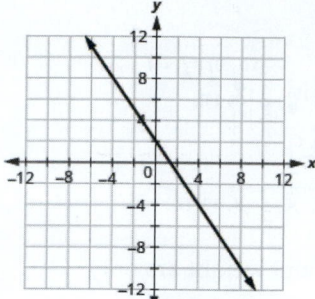

79.

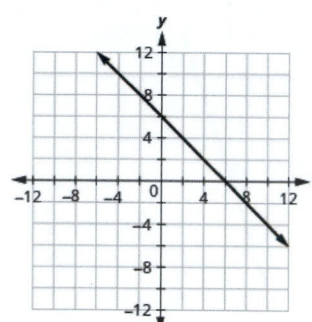

81.

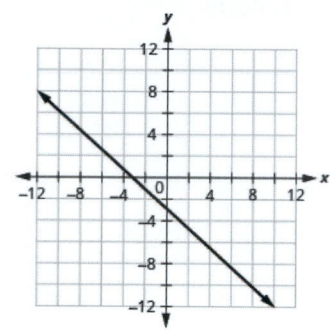

83.

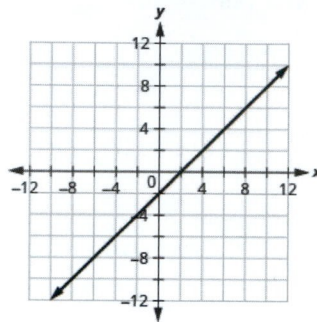

85.

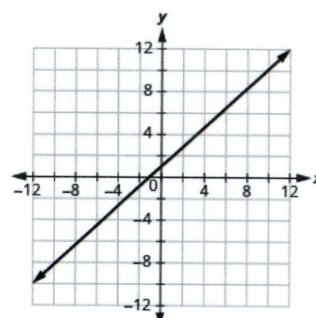

87.

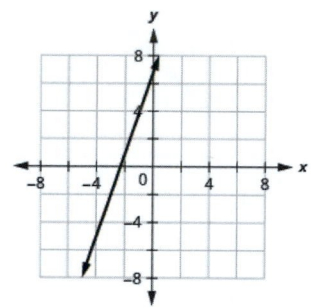

89.

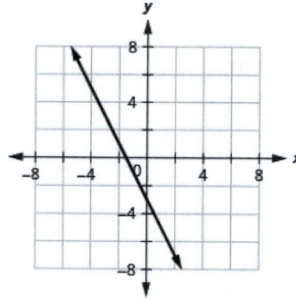

91.

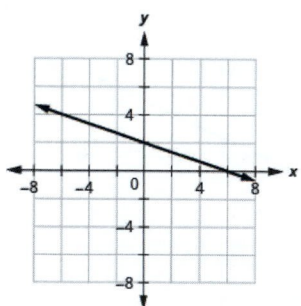

93.

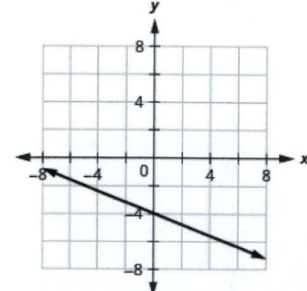

95.

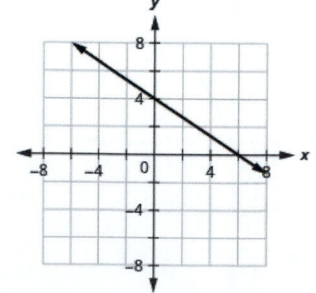

97.

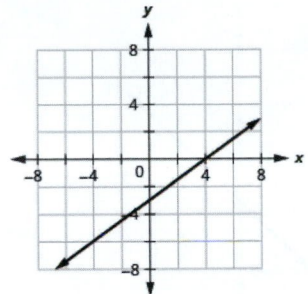

99.

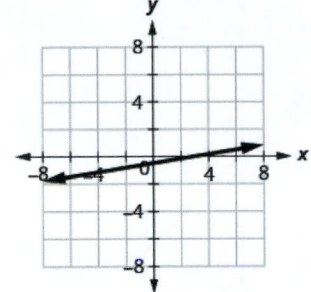

101.

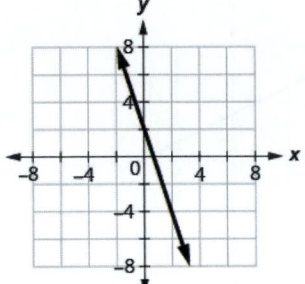

103.

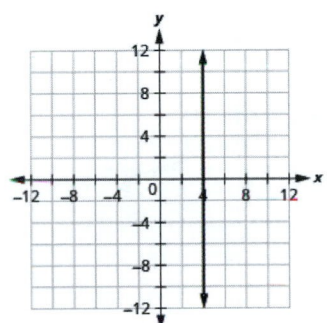

105.

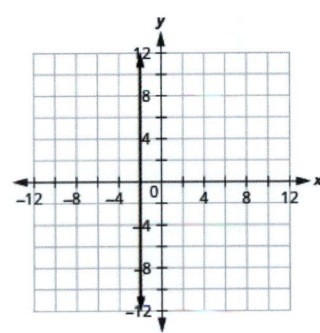

107.

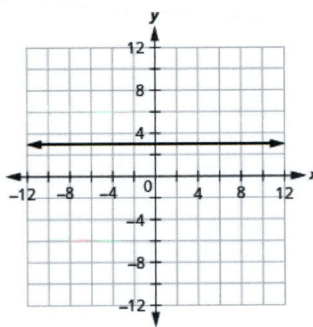

109.

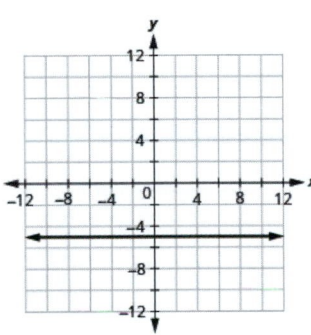

111.

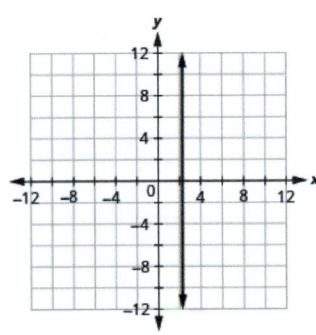

113.

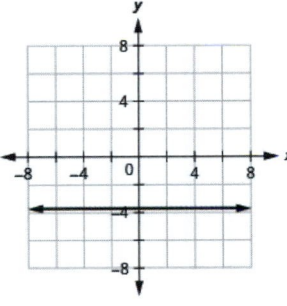

115.

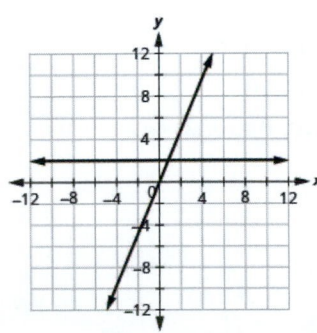

117.

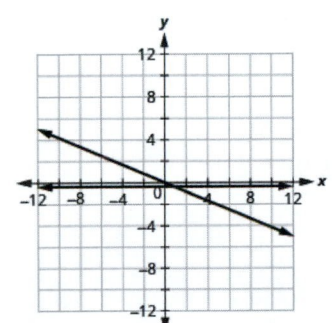

119.

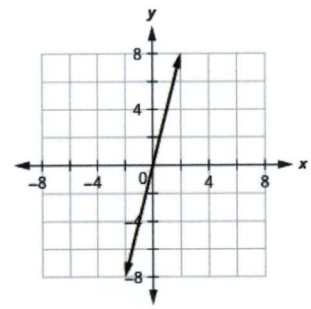

121.

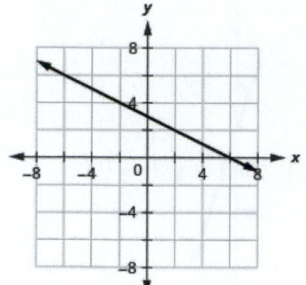

123.

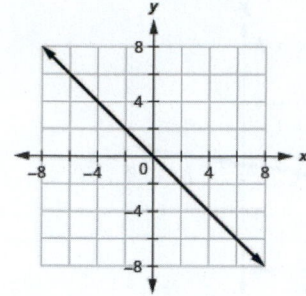

125.

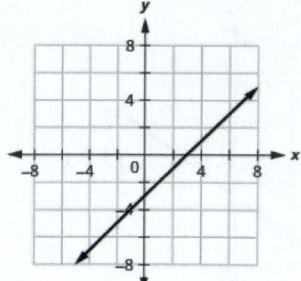

127.

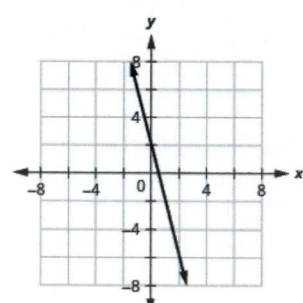

129.

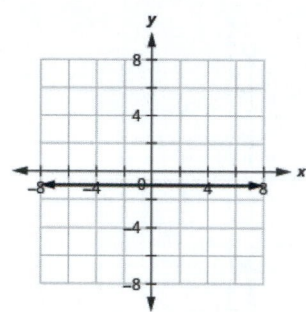

131.

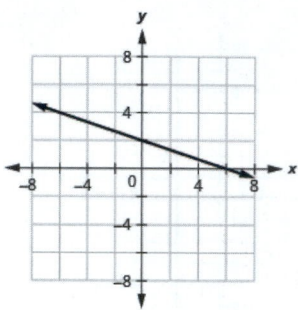

133.

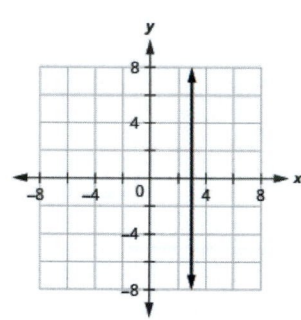

135. $722, $850, $978

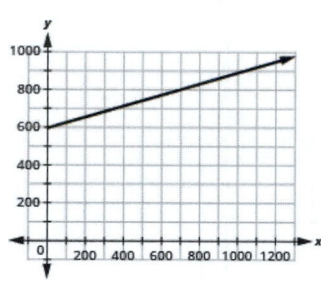
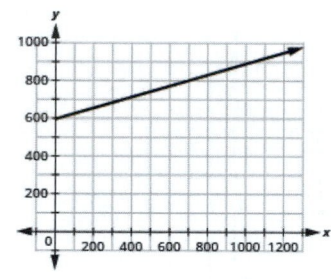
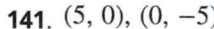

137. Answers will vary.

139. $(3, 0), (0, 3)$

141. $(5, 0), (0, -5)$

143. $(-2, 0), (0, -2)$

145. $(-1, 0), (0, 1)$

147. $(6, 0), (0, 3)$

149. $(0, 0)$

151. $(4, 0), (0, 4)$

153. $(-2, 0), (0, -2)$

155. $(5, 0), (0, -5)$

157. $(-3, 0), \ (0, 3)$

159. $(8, 0), (0, 4)$

161. $(2, 0), (0, 6)$

163. $(12, 0), (0, -4)$

165. $(2, 0), (0, -8)$

167. $(5, 0), (0, 2)$

169. $(4, 0), (0, -6)$

171. $(3, 0), (0, -1)$

173. $(-10, 0), (0, 2)$

175. $(0, 0)$

177. $(0, 0)$

179.

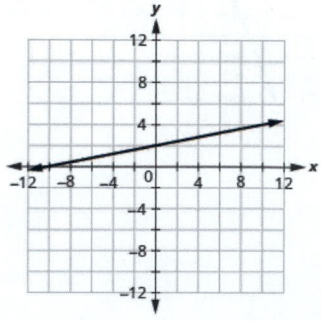

181.

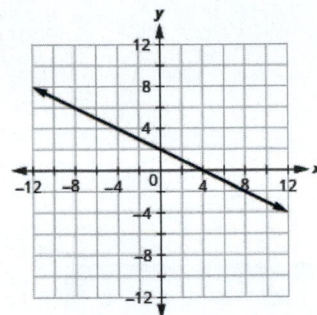

183.

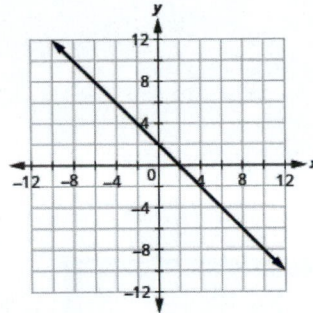

185.

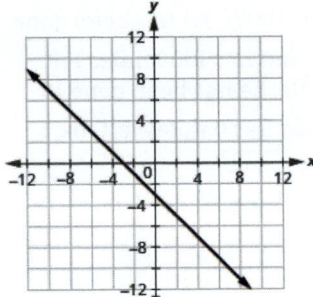

187.

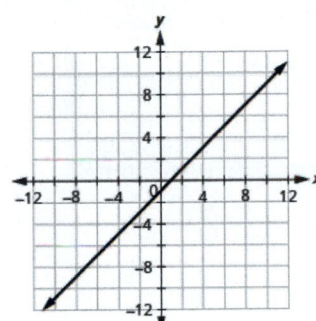

189.

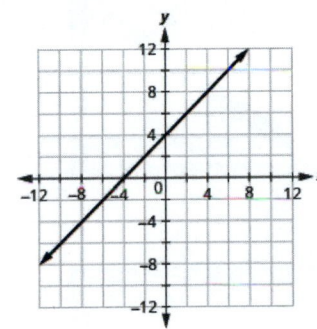

191.

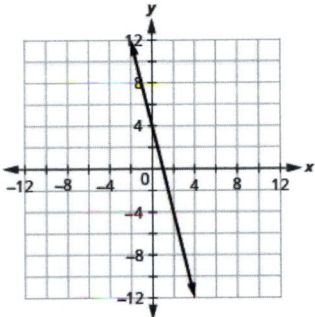

193.

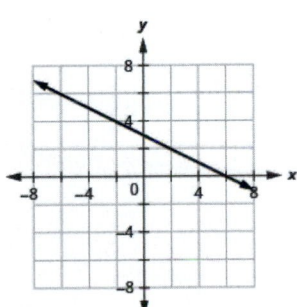

195.

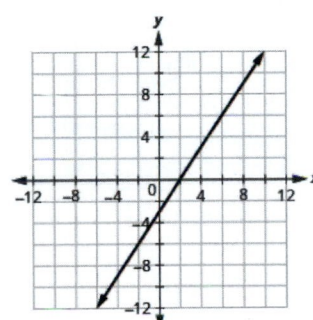

197.

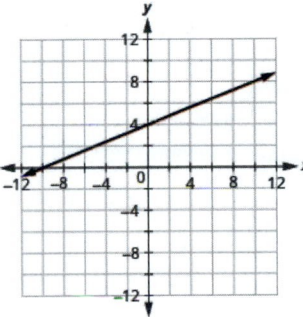

199.

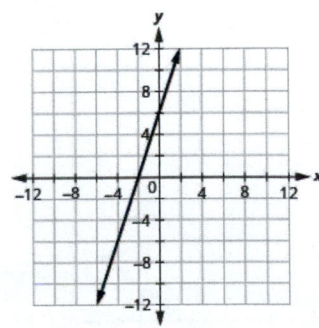

201.

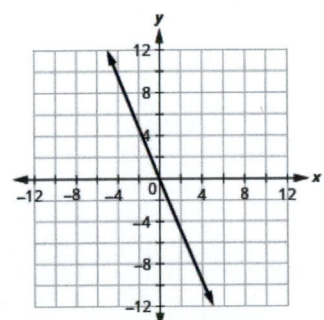

203.

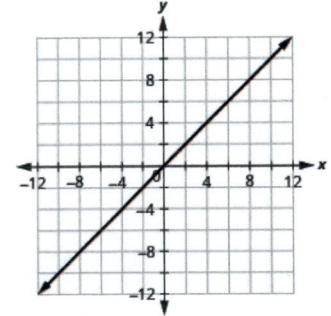

205. ⓐ $(0, 1000), (15, 0)$

ⓑ At $(0, 1000)$, he has been gone 0 hours and has 1000 miles left. At $(15, 0)$, he has been gone 15 hours and has 0 miles left to go.

207. Answers will vary.

209. Answers will vary.

211. $\frac{1}{4}$

213. $\frac{2}{3}$

215. $\frac{-3}{2} = -\frac{3}{2}$

217. $-\frac{2}{3}$

219.

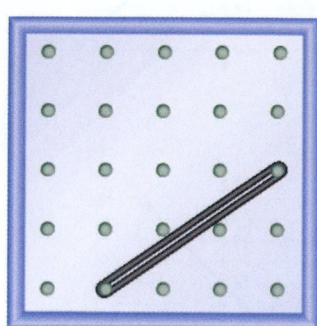

221.

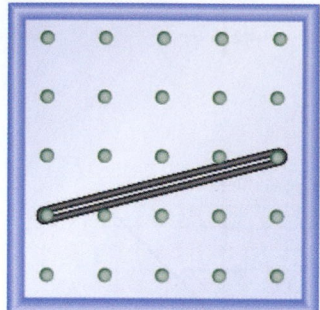

223.

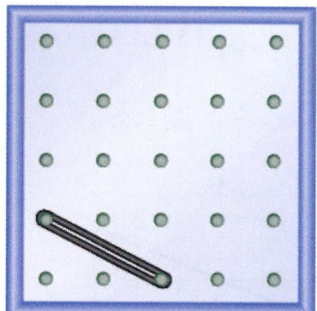

225.

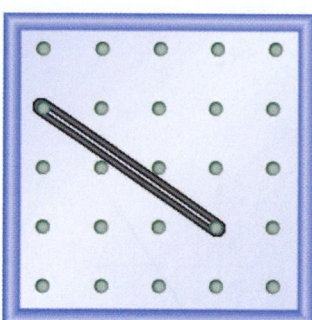

227. $\frac{2}{5}$

229. $\frac{5}{4}$

231. $-\frac{1}{3}$

233. $-\frac{3}{4}$

235. $\frac{3}{4}$

237. $-\frac{5}{2}$

239. $-\frac{2}{3}$

241. $\frac{1}{4}$

243. 0

245. undefined

247. 0

249. undefined

251. $\frac{5}{2}$

253. $\frac{3}{4}$

255. $-\frac{5}{2}$

257. $-\frac{8}{7}$

259. $\frac{7}{3}$

261. -1

263.

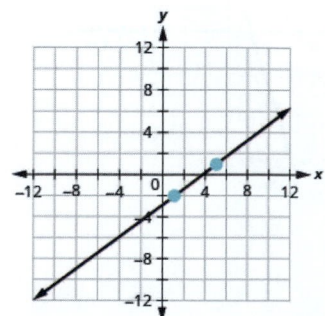

265.

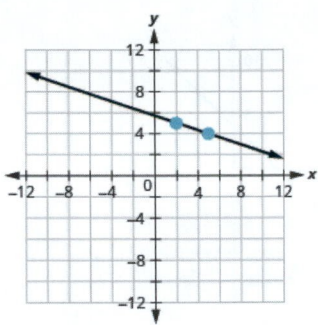

267.

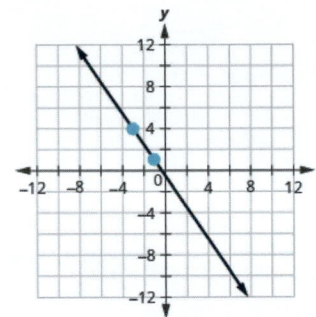

269.

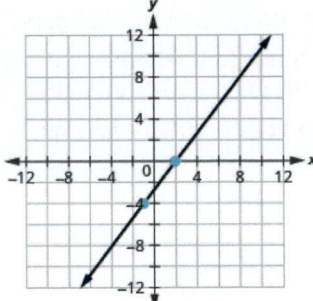

271.

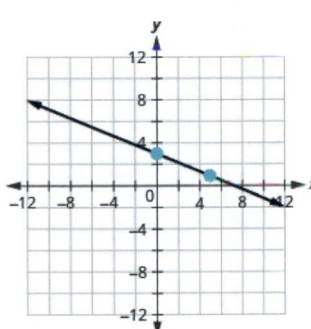

273.

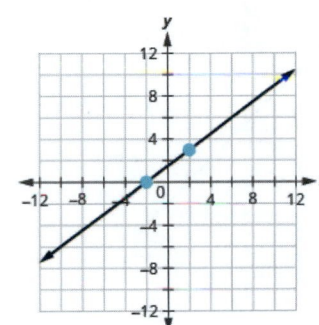

275.

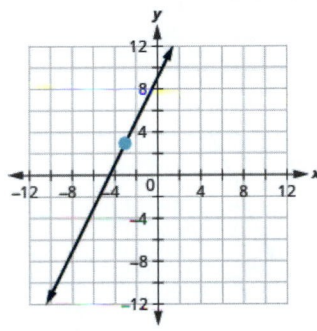

277.

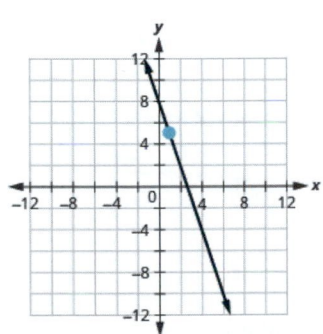

279. ⓐ $\frac{1}{3}$ ⓑ 4 12 pitch or 4-in-12 pitch

281. $\frac{3}{50}$; rise = 3, run = 50

283. ⓐ 288 inches (24 feet)   ⓑ Models will vary.

285. When the slope is a positive number the line goes up from left to right. When the slope is a negative number the line goes down from left to right.

287. A vertical line has 0 run and since division by 0 is undefined the slope is undefined.

289. slope $m = 4$ and y-intercept $(0, -2)$

291. slope $m = -3$ and y-intercept $(0, 1)$

293. slope $m = -\frac{2}{4}$ and y-intercept $(0, 3)$

295. -9; $(0, 7)$

297. 4; $(0, -10)$

299. -4; $(0, 8)$

301. $-\frac{8}{3}$; $(0, 4)$

303. $\frac{7}{3}$; $(0, -3)$

305.

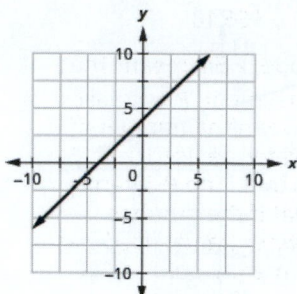

307.

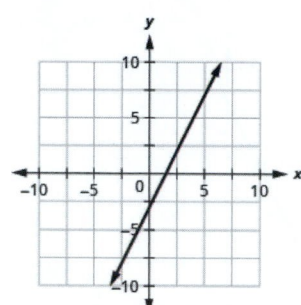

309.

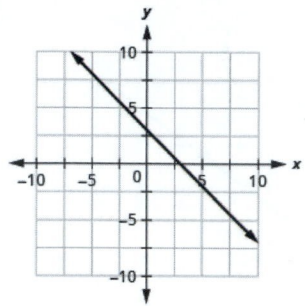

311.

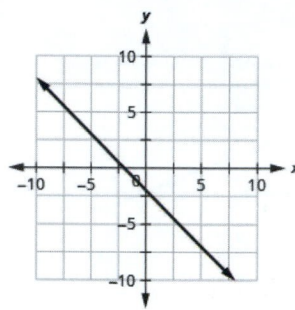

313.

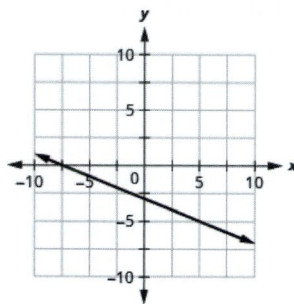

315.

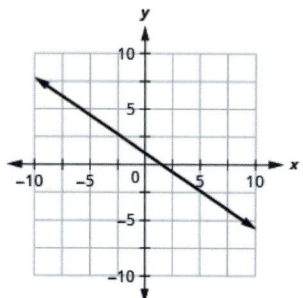

317.

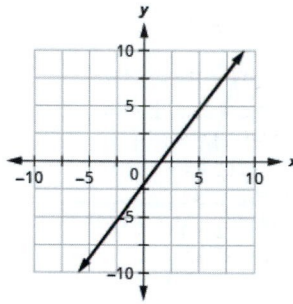

319.

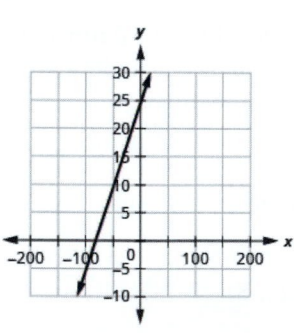

321. horizontal line

323. vertical line

325. slope−intercept

331. horizontal line

327. intercepts

333. intercepts

329. slope−intercept

335. slope−intercept

337.

ⓐ $28 ⓑ $66.10

ⓒ The slope, 2.54, means that Randy's payment, *P*, increases by $2.54 when the number of units of water he used, *w*, increases by 1. The *P*–intercept means that if the number units of water Randy used was 0, the payment would be $28.

ⓓ

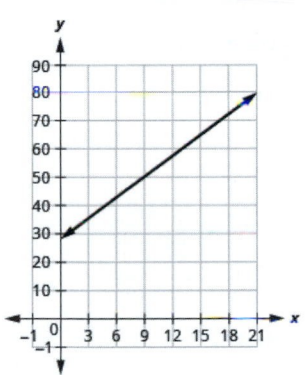

339.

ⓐ $15 ⓑ $143

ⓒ The slope, 0.32, means that the cost, *C*, increases by $0.32 when the number of miles driven, *m*, increases by 1. The *C*-intercept means that if Janelle drives 0 miles one day, the cost would be $15.

ⓓ

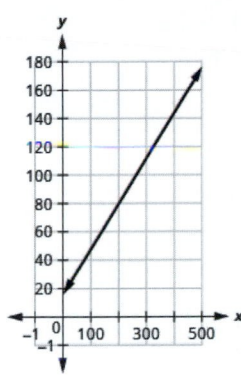

341.

ⓐ $750 ⓑ $2418.60

ⓒ The slope, 0.09, means that Patel's salary, *S*, increases by $0.09 for every $1 increase in his sales. The *S*-intercept means that when his sales are $0, his salary is $750.

ⓓ

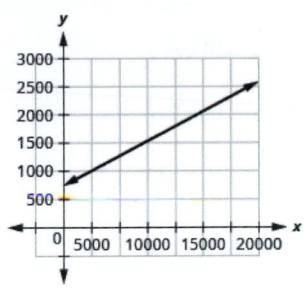

343.

ⓐ $2850 ⓑ $4950

ⓒ The slope, 42, means that the cost, *C*, increases by $42 for when the number of guests increases by 1. The *C*-intercept means that when the number of guests is 0, the cost would be $750.

ⓓ

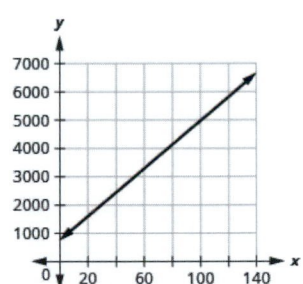

345. parallel

347. parallel

349. not parallel
355. parallel
361. not parallel
367. not parallel
373. perpendicular

351. parallel
357. parallel
363. not parallel
369. not parallel
375. not perpendicular

353. parallel
359. parallel
365. not parallel
371. perpendicular
377. not perpendicular

379. perpendicular

381. perpendicular

383.

ⓐ For every increase of one degree Fahrenheit, the number of chirps increases by four.

ⓑ There would be -160 chirps when the Fahrenheit temperature is $0°$. (Notice that this does not make sense; this model cannot be used for all possible temperatures.)

385. Answers will vary.

387. $y = 4x + 1$

389. $y = 8x - 6$

391. $y = -x + 7$

393. $y = -3x - 1$

395. $y = \frac{1}{5}x - 5$

397. $y = -\frac{2}{3}x - 3$

399. $y = 2$

401. $y = -4x$

403. $y = -2x + 4$

405. $y = \frac{3}{4}x + 2$

407. $y = -\frac{3}{2}x - 1$

409. $y = 6$

411. $y = \frac{3}{8}x - 1$

413. $y = \frac{5}{6}x + 2$

415. $y = -\frac{3}{5}x + 1$

417. $y = -\frac{1}{3}x - 11$

419. $y = 4$

421. $y = -7$

423. $y = -\frac{5}{2}x - 22$

425. $y = -4x - 11$

427. $y = -8$

429. $y = -4x + 13$

431. $y = x + 5$

433. $y = -\frac{1}{3}x - \frac{14}{3}$

435. $y = 7x + 22$

437. $y = -\frac{6}{7}x + \frac{4}{7}$

439. $y = \frac{1}{5}x - 2$

441. $x = 4$

443. $x = -2$

445. $y = 2$

447. $y = -3$

449. $y = 4x$

451. $y = \frac{1}{2}x + \frac{3}{2}$

453. $y = 5$

455. $y = 3x - 1$

457. $y = -3x + 3$

459. $y = 2x - 6$

461. $y = -\frac{2}{3}x + 5$

463. $x = -3$

465. $x = 4$

467. $y = -4$

469. $y = -1$

471. $y = x$

473. $y = -\frac{3}{2}x - 1$

475. $y = -\frac{3}{4}x - \frac{1}{4}$

477. $y = \frac{5}{4}x$

479. $x = -5$

481. $y = 1$

483. $y = x + 5$

485. $y = \frac{5}{6}x + 2$

487. $y = -\frac{2}{3}x + 5$

489. $y = -\frac{3}{5}x + 1$

491. $y = 2$

493. $y = x + 2$

495. $x = -3$

497. $y = -\frac{1}{3}x - \frac{14}{3}$

499. $y = \frac{3}{4}x$

501. $y = 1.2x + 5.2$

503. Answers will vary.

505. ⓐ yes ⓑ no ⓒ no ⓓ yes ⓔ no

507. ⓐ yes ⓑ no ⓒ no ⓓ yes ⓔ yes

509. ⓐ no ⓑ no ⓒ no ⓓ yes ⓔ yes

511. $y < 2x - 4$

513. $y \le -\frac{1}{3}x - 2$

515. $x + y \ge 3$

517. $x + 2y \ge -2$

519. $2x - y < 4$

521. $4x - 3y > 12$

523.

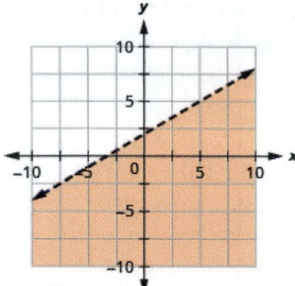

525.

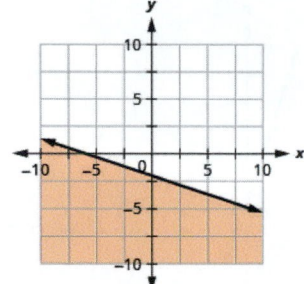

527.

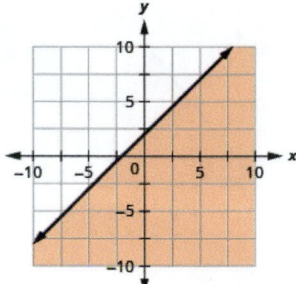

529.

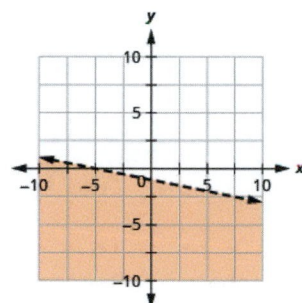

531.

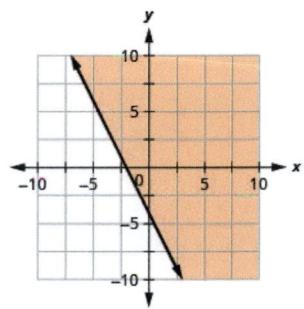

533.

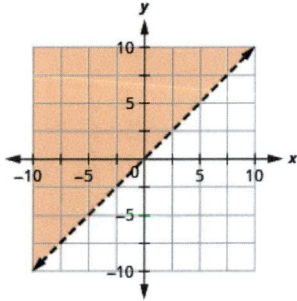

535.

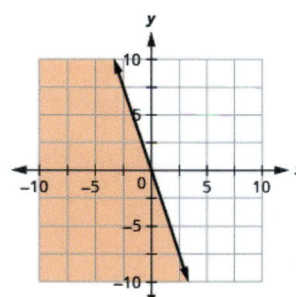

537.

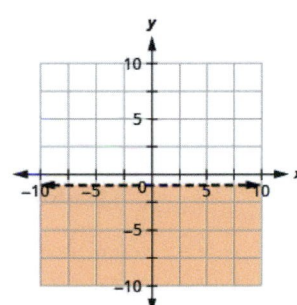

539.

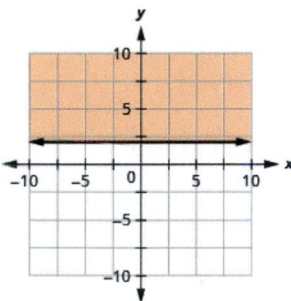

541.

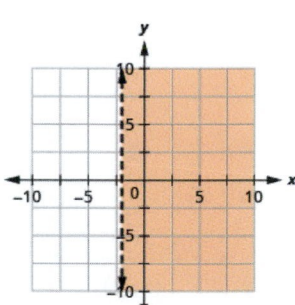

543.

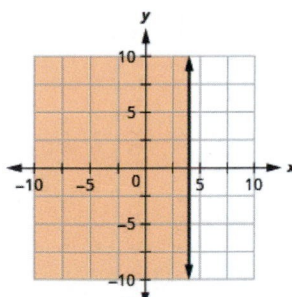

545.

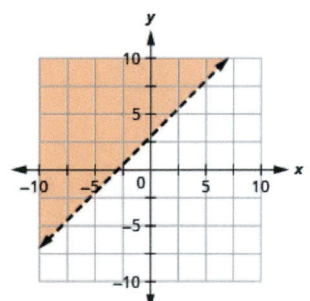

547.

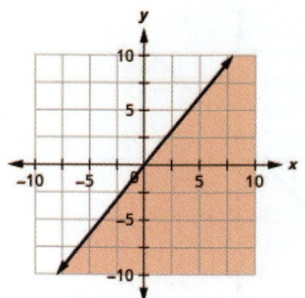

549.

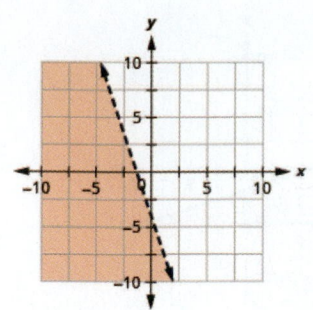

551.

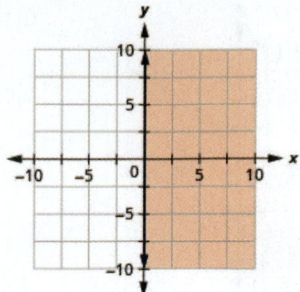

553.

ⓐ

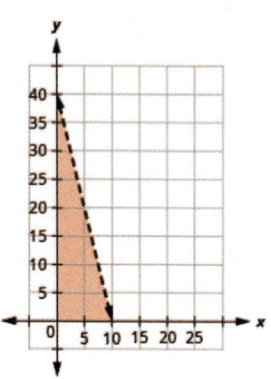

ⓑ Answers will vary.

Review Exercises

557.

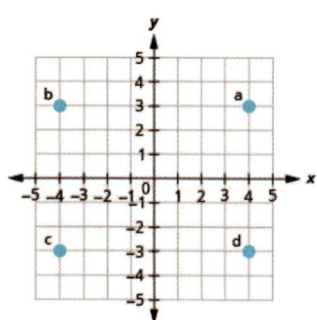

559.

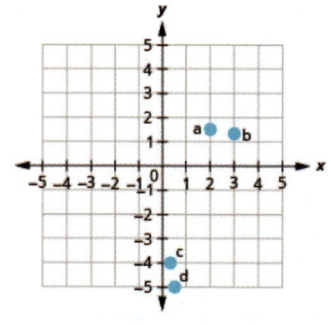

561. ⓐ (2, 0) ⓑ (0, −5) ⓒ
(−4.0) ⓓ (0, 3)

563. a, b

565.

x	y	(x, y)
0	3	(0, 3)
4	1	(4, 1)
−2	4	(−2, 4)

567.

x	y	(x, y)
0	−3	(0, −3)
2	0	(2, 0)
−2	−6	(−2, −6)

569. Answers will vary.

571. Answers will vary.

573. ⓐ yes; yes ⓑ yes; no

575.

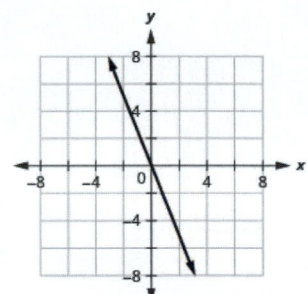

577.

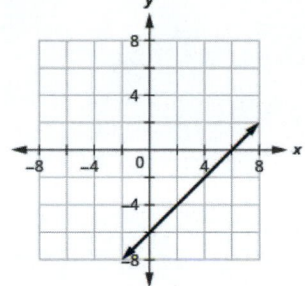

579.

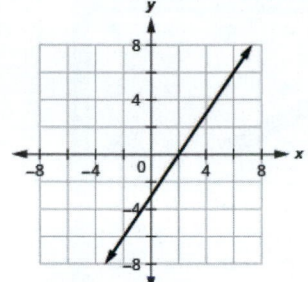

581.

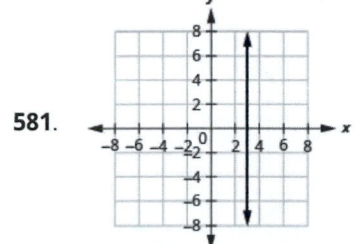

583.

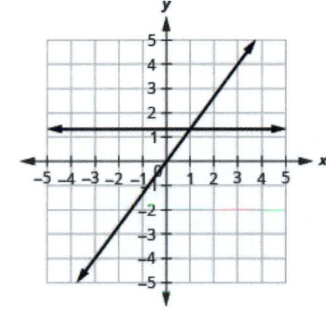

585. (3, 0), (0, 3)

587. (−1, 0), (0, 1)

589. (6, 0), (0, 4)

591. (0, 0)

593.

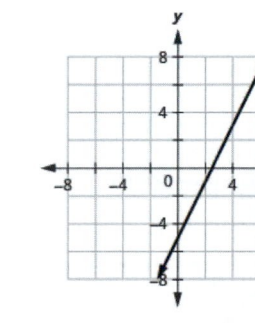

595.

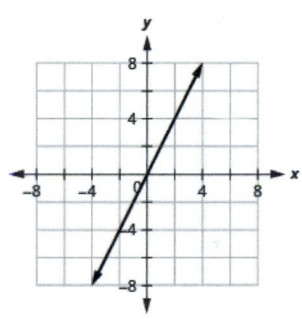

597.

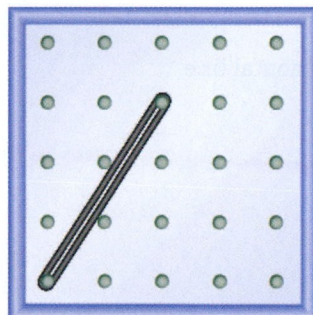

599. $\frac{4}{3}$

601. $-\frac{2}{3}$

603.

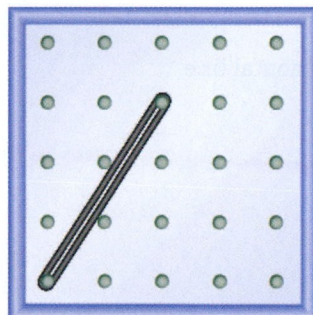

605.

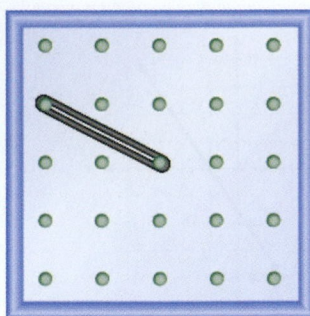

607. 1

609. $-\frac{1}{2}$

611. undefined

613. 0

615. -6

617. $\frac{5}{2}$

619.

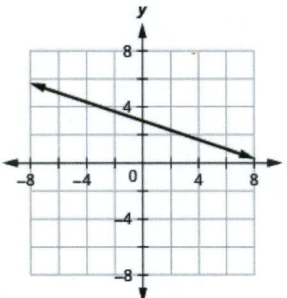

621.

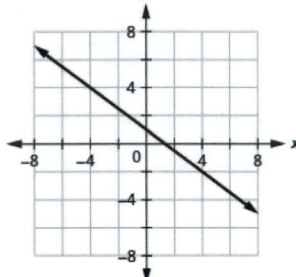

623. $\frac{1}{10}$

625. slope $m = -\frac{2}{3}$ and y-intercept $(0, 4)$

627. $\frac{5}{3}$; $(0, -6)$

629. $\frac{4}{5}$; $\left(0, -\frac{8}{5}\right)$

631.

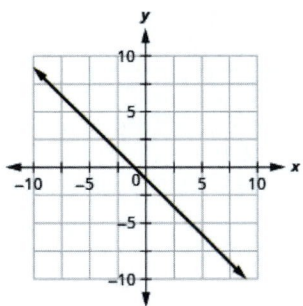

633.

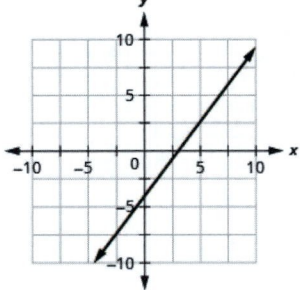

635. horizontal line

637. intercepts

639. plotting points

641. ⓐ –$250 ⓑ $450 ⓒ The slope, 35, means that Marjorie's weekly profit, P, increases by $35 for each additional student lesson she teaches. The P–intercept means that when the number of lessons is 0, Marjorie loses $250. ⓓ

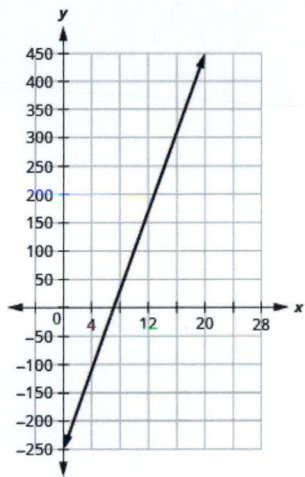

643. not parallel

645. perpendicular

647. $y = -5x - 3$

653. $y = -4$

659. $y = \frac{1}{2}x - \frac{5}{2}$

665. $y = 3$

649. $y = -2x$

655. $y = \frac{3}{5}x$

661. $y = 2$

667. $y = -\frac{3}{2}x - 6$

651. $y = -3x + 5$

657. $y = -2x - 5$

663. $y = -\frac{2}{5}x + 8$

669. $y = 1$

671. ⓐ yes ⓑ no ⓒ yes ⓓ yes ⓔ no

673. $y > \frac{2}{3}x - 3$

675. $x - 2y \geq 6$

677.

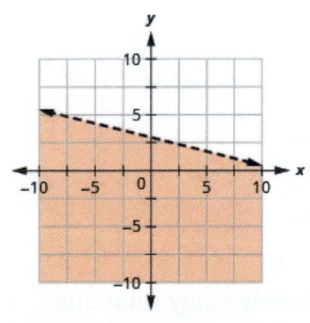

679.

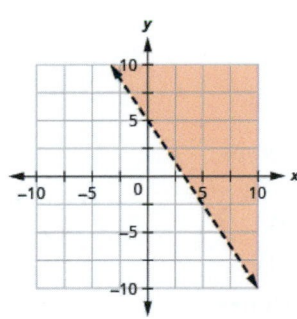

681.

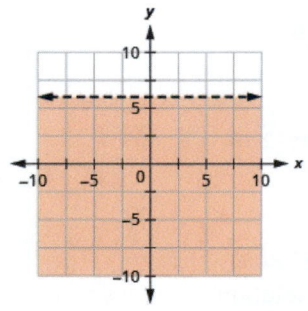

Practice Test

683. ⓐ yes ⓑ yes ⓒ no

685. $(3, 0), (0, -4)$

687. undefined

689. 1

691.

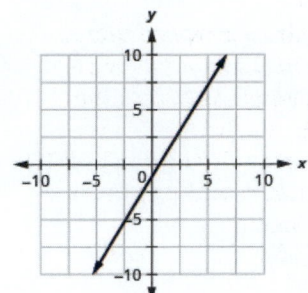

693.

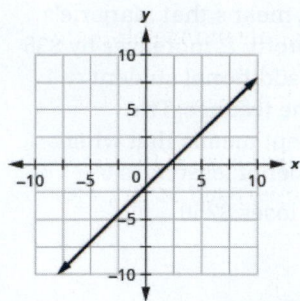

695.

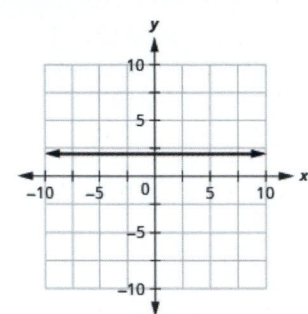

697. $y = -\frac{3}{4}x - 2$

699. $y = \frac{1}{2}x - 4$

701. $y = -\frac{4}{5}x - 5$

703.

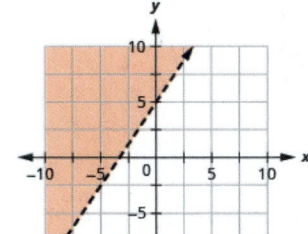

705.

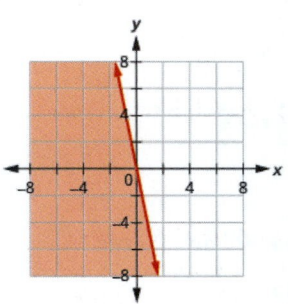

Chapter 5

Try It

5.1. ⓐ yes ⓑ no

5.2. ⓐ no ⓑ yes

5.3. $(3, 2)$

5.4. $(2, 3)$

5.5. $(-2, -2)$

5.6. $(1, 6)$

5.7. $(3, 4)$

5.8. $(5, -4)$

5.9. $(4, 2)$

5.10. $(5, -3)$

5.11. $(9, -1)$

5.12. $(4, -6)$

5.13. no solution

5.14. no solution

5.15. infinitely many solutions

5.16. infinitely many solutions

5.17. no solution, inconsistent, independent

5.18. no solution, inconsistent, independent

5.19. one solution, consistent, independent

5.20. one solution, consistent, independent

5.21. infinitely many solutions, consistent, dependent

5.22. infinitely many solutions, consistent, dependent

5.23. Manny needs 3 quarts juice concentrate and 9 quarts water.

5.24. Alisha needs 15 ounces of coffee and 3 ounces of milk.

5.25. $(6, 1)$

5.26. $(4, 2)$

5.27. $(2, 4)$

5.28. $(-1, -3)$

5.29. $(1, -2)$

5.30. $(2, 6)$

5.31. $(-2, -3)$

5.32. $(6, 2)$

5.33. $(6, 2)$

5.34. $(8, 2)$

5.35. $\left(2, \frac{3}{2}\right)$

5.36. $\left(-\frac{1}{2}, -2\right)$

5.37. infinitely many solutions

5.38. infinitely many solutions

5.39. no solution

5.40. no solution

5.41. The numbers are 3 and 7.

5.42. The numbers are 2 and −8.

5.43. The length is 12 and the width is 8.

5.44. The length is 23 and the width is 6.

5.45. The measure of the angles are 22 degrees and 68 degrees.

5.46. The measure of the angles are 36 degrees and 54 degrees.

5.47. There would need to be 160 policies sold to make the total pay the same.

5.48. Kenneth would need to sell 1,000 suits.

5.49. $(2, -1)$

5.50. $(-2, 3)$

5.51. $(3, -1)$

5.52. $(-2, 5)$

5.53. $(1, 1)$

5.54. $(-2, 4)$

5.55. $(1, 3)$

5.56. $(4, -3)$

5.57. $(6, 2)$

5.58. $(1, -2)$

5.59. infinitely many solutions

5.60. infinitely many solutions

5.61. no solution

5.62. no solution

5.63. The numbers are 25 and 17.

5.64. The numbers are −25 and 10.

5.65. The bag of diapers costs $11 and the can of formula costs $13.

5.66. There are 105 calories in a banana and 5 calories in a strawberry.

5.67. ⓐ Since both equations are in standard form, using elimination will be most convenient. ⓑ Since one equation is already solved for x, using substitution will be most convenient.

5.68. ⓐ Since one equation is already solved for y, using substitution will be most convenient; ⓑ Since both equations are in standard form, using elimination will be most convenient.

5.69. $\begin{cases} m + n = -23 \\ m = n - 7 \end{cases}$

5.70. $\begin{cases} m + n = -18 \\ m = n + 40 \end{cases}$

5.71. $\begin{cases} w + h = 84,000 \\ h = 2w - 18,000 \end{cases}$

5.72. $\begin{cases} s = 2n - 5 \\ s + n = 43 \end{cases}$

5.73. Ali is 28 and Jameela is 16.

5.74. Jake is 9 and his dad is 33.

5.75. Mark burned 11 calories for each minute of yoga and 7 calories for each minute of jumping jacks.

5.76. Erin burned 11 calories for each minute on the rowing machine and 5 calories for each minute of weight lifting.

5.77. The angle measures are 55 degrees and 35 degrees.

5.78. The angle measures are 5 degrees and 85 degrees.

5.79. The angle measures are 42 degrees and 138 degrees.

5.80. The angle measures are 66 degrees and 114 degrees.

5.81. The length is 60 feet and the width is 35 feet.

5.82. The length is 60 feet and the width is 38 feet.

5.83. It will take Clark 4 hours to catch Mitchell.

5.84. It will take Sally $1\frac{1}{2}$ hours to catch up to Charlie.

5.85. The rate of the boat is 11 mph and the rate of the current is 1 mph.

5.86. The speed of the canoe is 7 mph and the speed of the current is 1 mph.

5.87. The speed of the jet is 235 mph and the speed of the wind is 30 mph.

5.88. The speed of the jet is 408 mph and the speed of the wind is 24 mph.

5.89. There were 206 adult tickets sold and 347 children tickets sold.

5.90. There were 521 adult tickets sold and 842 children tickets sold.

5.91. Matilda has 13 dimes and 29 quarters.

5.92. Juan has 36 nickels and 63 dimes.

5.93. Greta should use 3 pounds of peanuts and 2 pounds of cashews.

5.94. Sammy should purchase 10 pounds of beans and 10 pounds of ground beef.

5.95. LeBron needs 120 ml of the 25% solution and 30 ml of the 50% solution.

5.96. Anatole should mix 125 ml of the 10% solution and 125 ml of the 40% solution.

5.97. Leon should put $42,000 in the stock fund and $8000 in the savings account.

5.98. Julius invested $1,750 at 11% and $5,250 at 13%.

5.99. The principal amount for the bank loan was $4,000. The principal amount for the federal loan was $14,000.

5.100. The principal amount for was $41,200 at 4.5%. The principal amount was, $24,000 at 7.2%.

5.101. ⓐ no ⓑ yes

5.102. ⓐ no ⓑ no

5.103.

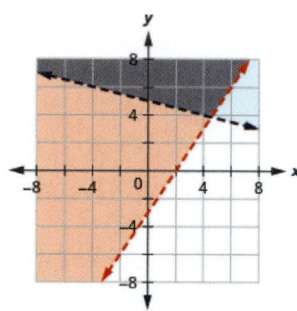

5.104.

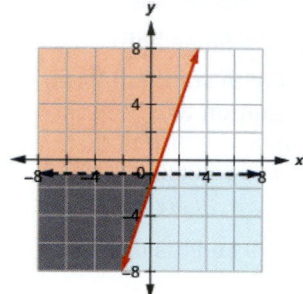

5.105.

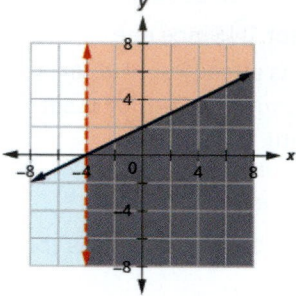

5.106.

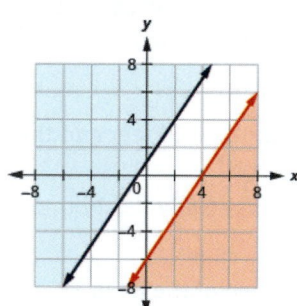

5.107.

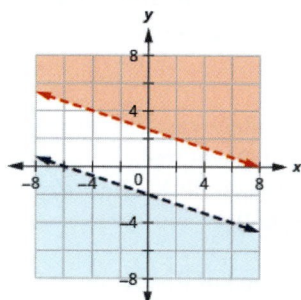

5.108.

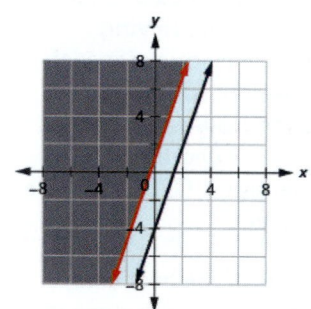

5.109. no solution

5.110. no solution

5.111. $y \geq 3x + 1$

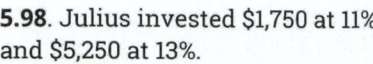

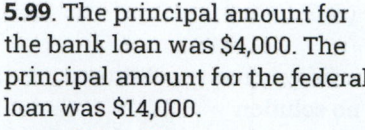

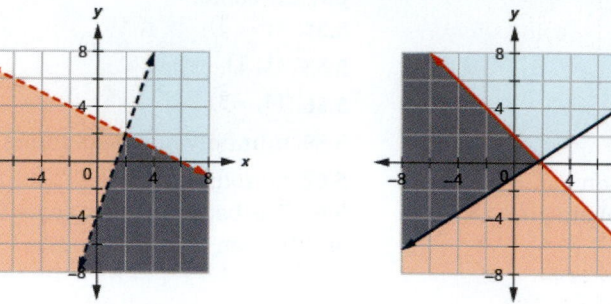

5.112. $x + 4y \leq 4$

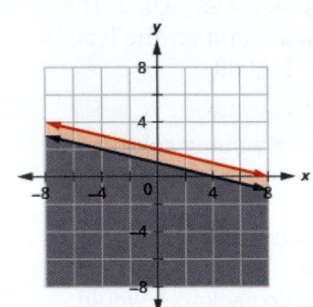

5.113.

ⓐ $\begin{cases} 30m + 20p \leq 160 \\ 2m + 3p \leq 15 \end{cases}$

ⓑ

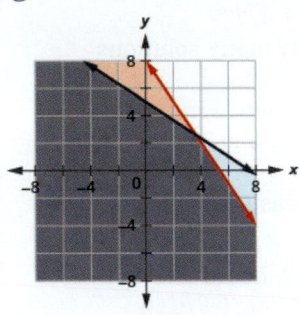

ⓒ yes　ⓓ no

5.114.

ⓐ $\begin{cases} a \geq p + 5 \\ a + 2p \leq 400 \end{cases}$

ⓑ

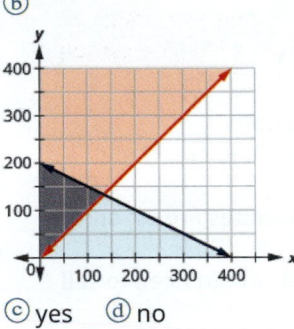

ⓒ yes　ⓓ no

5.115.

ⓐ $\begin{cases} 0.75d + 2e \leq 25 \\ 360d + 110e \geq 1000 \end{cases}$

ⓑ

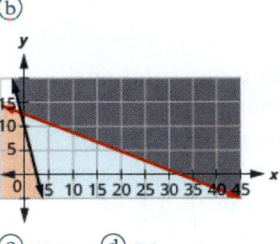

ⓒ yes　ⓓ no

5.116.

ⓐ $\begin{cases} 140p + 125j \geq 1000 \\ 1.80p + 1.25j \leq 12 \end{cases}$

ⓑ

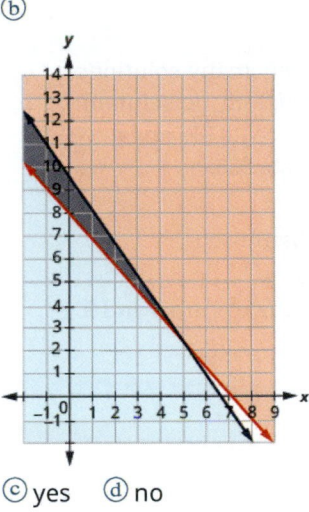

ⓒ yes　ⓓ no

Section Exercises

1. ⓐ yes ⓑ no

3. ⓐ yes ⓑ no

5. ⓐ yes ⓑ no

7. ⓐ no ⓑ yes

9. $(-2, 3)$

11. $(1, 2)$

13. $(0, 2)$

15. $(2, 4)$

17. $(2, -1)$

19. $(1, 2)$

21. $(3, 2)$

23. $(1, 1)$

25. $(-1, -4)$

27. $(3, 3)$

29. $(-5, 6)$

31. $(6, -2)$

33. $(3, 2)$

35. $(1, 3)$

37. $(-3, -1)$

39. no solution

41. no solution

43. no solution

45. infinitely many solutions

47. infinitely many solutions

49. $(2, 2)$

51. 0 solutions

53. 0 solutions

55. 0 solutions

57. consistent, 1 solution

59. infinitely many solutions

61. infinitely many solutions

63. Molly needs 16 ounces of strawberry juice and 48 ounces of water.

65. Enrique needs 8 ounces of nuts and 16 ounces of water.

67. Leo should plant 50 tulips and 300 daffodils.

69. There is an infinite number of possible solutions to the system of equations.

71. $(-2, 0)$

73. $(7, 6)$

75. $(0, 3)$

77. $(6, -3)$

79. $(3, -1)$

81. $(6, 6)$

83. $(5, 0)$

85. $(-2, 7)$

87. $(-5, 2)$

89. $(-1, 7)$

91. $(-3, 5)$

93. $(10, 12)$

95. $\left(\frac{1}{2}, 3\right)$

97. $\left(\frac{1}{2}, -\frac{3}{4}\right)$

99. Infinitely many solutions

101. Infinitely many solutions

103. No solution

105. No solution

107. The numbers are 13 and 17.

109. The numbers are -7 and -19.

111. The length is 20 and the width is 10.

113. The length is 34 and the width is 8.

115. The measures are 16° and 74°.

117. The measures are 45° and 45°.

119. 80 cable packages would need to be sold.

121. Mitchell would need to sell 120 stoves.

123. ⓐ $t = 2$ hours ⓑ $s = 2\frac{1}{2}$ hours

125. Answers will vary.

127. $(6, 9)$

129. $(-2, 1)$

131. $(-7, -1)$

133. $(-2, -4)$

135. $(-1, 3)$

137. $(-1, 2)$

139. $(-5, 9)$

141. $(6, 1)$

143. $(-2, 10)$

145. $(2, 3)$

147. $(-7, 6)$

149. $(-9, 3)$

151. $(9, 5)$

153. $(-3, 2)$

155. $(-2, 3)$

157. infinitely many solutions

159. infinitely many solutions

161. infinitely many solutions

163. inconsistent, no solution

165. inconsistent, no solution

167. The numbers are 20 and 45.

169. The numbers are 16 and -43.

171. A shirt costs $16 and a sweater costs $33.

173. There are 860 mg in a hot dog. There are 1,000 mg in a cup of cottage cheese.

175. ⓐ elimination ⓑ substitution

177. ⓐ substitution ⓑ elimination

179. ⓐ $r = 4$ ⓑ $c = 1$

181.

ⓐ $(8, 2)$

ⓑ

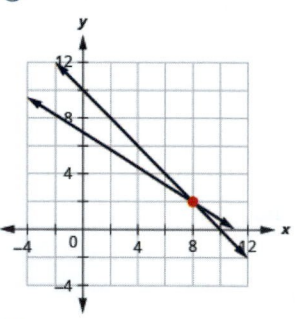

ⓒ Answers will vary.

183. The numbers are 6 and 9.

185. The numbers are -5 and -25.

187. The numbers are 5 and 4.

189. The numbers are 2 and 3.

191. $10,000

193. She put $15,000 into a CD and $35,000 in bonds.

195. The amount of the first year's loan was $30,000 and the amount of the second year's loan was $12,000.

197. Bethany is 16 years old and Alyssa is 28 years old.

199. Noelle is 20 years old and her dad is 54 years old.

201. The small container holds 20 gallons and the large container holds 30 gallons.

203. There were 10 calories burned jogging and 10 calories burned cycling.

205. Notebooks are $4 and thumb drives are $20.

207. The measures are 60 degrees and 30 degrees.

209. The measures are 125 degrees and 55 degrees.

211. 94 degrees and 86 degrees

213. 72.5 degrees and 17.5 degrees

215. The measures are 44 degrees and 136 degrees.

217. The measures are 34 degrees and 56 degrees.
223. It took Sarah's sister 12 hours.
229. The boat rate is 18 mph and the current rate is 2 mph.
235. $s = 183$, $a = 242$

241. Josie bought 40 adult tickets and 32 children tickets.
247. Peter had 11 dimes and 48 quarters.

253. Hannah needs 10 gallons of soda and 15 gallons of fruit drink.

259. The scientist should mix 15 liters of the 25% solution and 50 liters of the 12% solution.
265. Sam invested $28,000 at 10% and $20,000 at 6%.

271. 14 boys paid the full-year fee. 4 boys paid the partial-year fee,
277. ⓐ false ⓑ true
283.

219. The width is 10 feet and the length is 25 feet.
225. It took Lucy's friend 2 hours.
231. The jet rate is 240 mph and the wind speed is 28 mph.
237. Answers will vary.

243. There were 125 adult tickets and 128 children tickets sold.
249. The cashier has fourteen $10 bills and sixteen $20 bills.

255. Julia and her husband should buy 12 pounds of City Roast Columbian coffee and 8 pounds of French Roast Columbian coffee.
261. 160 liters of the 40% solution and 80 liters of the 70% solution will be used.
267. The federal loan is $62,500 and the bank loan is $3,300.

273. Answers will vary.

279. ⓐtrue ⓑfalse
285.

221. The width is 15 feet and the length is 15 feet.
227. The canoe rate is 5 mph and the current rate is 1 mph.
233. The jet rate is 415 mph and the wind speed is 19 mph.
239. There 1120 adult tickets and 530 child tickets sold.
245. Brandon has 12 quarters and 8 dimes.
251. Marissa should use 60 pounds of the $1.20/lb candy and 30 pounds of the $1.80/lb candy.
257. Jotham should mix 2 liters of the 30% solution and 28 liters of the 80% solution.

263. Hattie should invest $900 at 12% and $2,100 at 10%.

269. $12,000 should be invested at 5.25% and $13,000 should be invested at 4%.

275. ⓐ true ⓑ false
281. ⓐ true ⓑtrue
287.

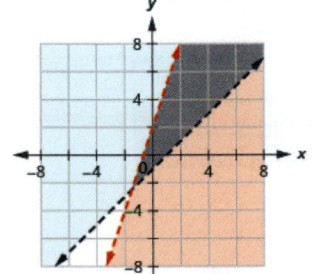

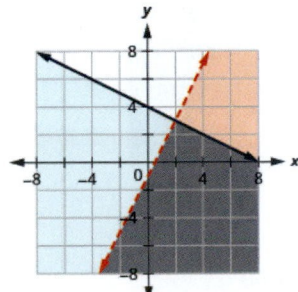

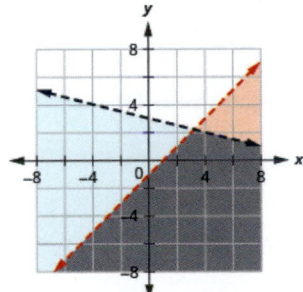

289.

291.

293.

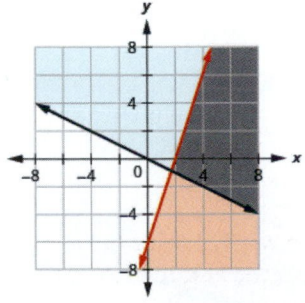

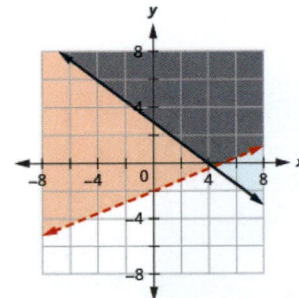

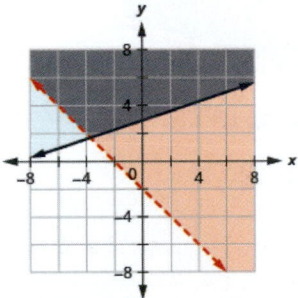

295.

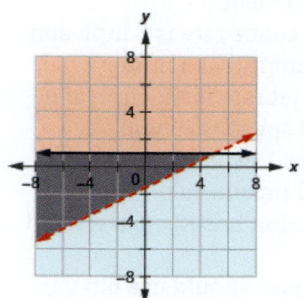

297.

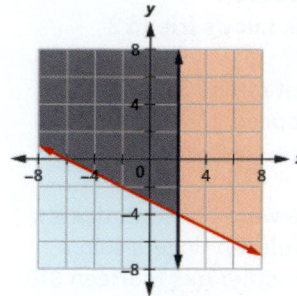

299.

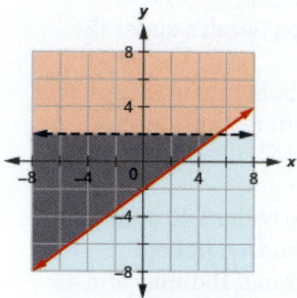

301.

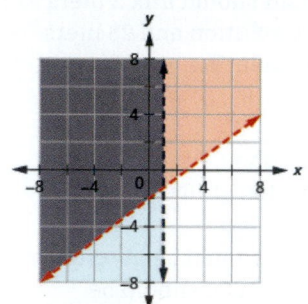

303.

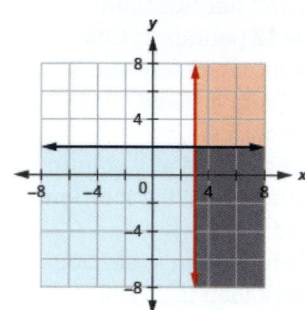

305. No solution

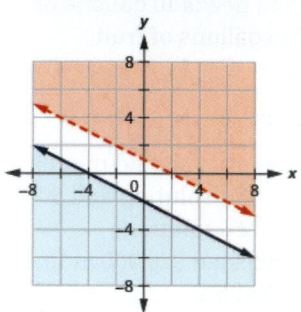

307. No solution

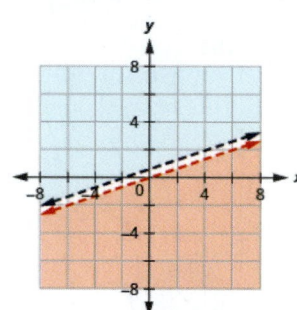

309.

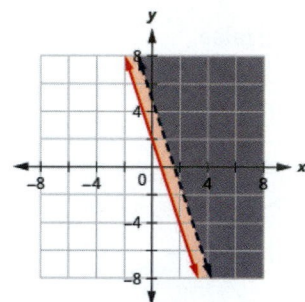

311. $x + 4y < 6$

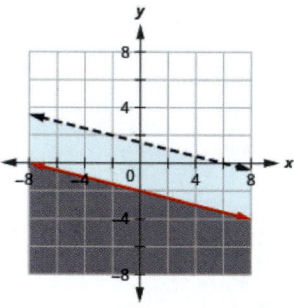

313. $-2x + 6y > 8$

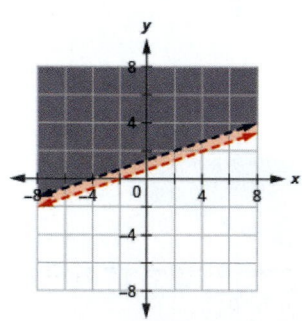

315.

ⓐ $\begin{cases} p + l \geq 60 \\ 15p + 10l \geq 800 \end{cases}$

ⓑ

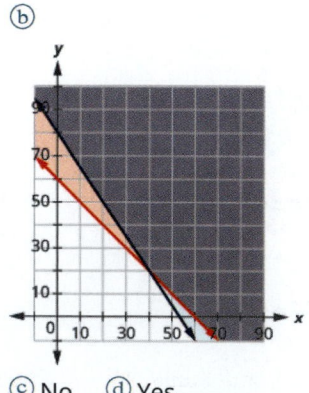

ⓒ No   ⓓ Yes

317.

ⓐ $\begin{cases} 7p + 3c \leq 500 \\ p \geq 2c + 4 \end{cases}$

ⓑ

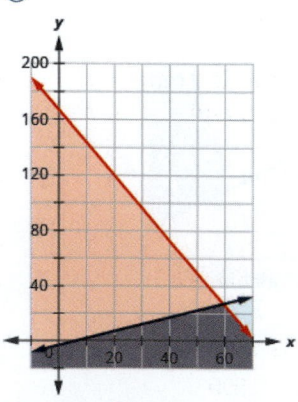

ⓒ Yes   ⓓ No

319.

ⓐ $\begin{cases} 90b + 150g \geq 500 \\ 0.35b + 2.50g \leq 15 \end{cases}$

ⓑ

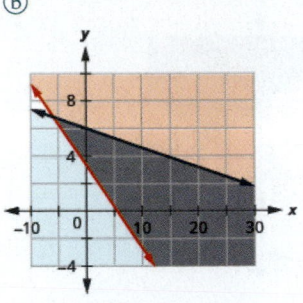

ⓒ No   ⓓ Yes

325. Answers will vary.

321.

ⓐ $\begin{cases} 7c + 11p \geq 35 \\ 110c + 22p \leq 200 \end{cases}$

ⓑ

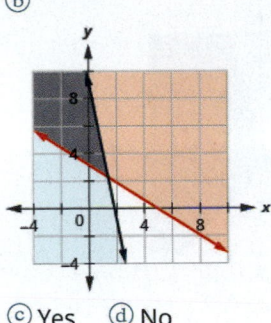

ⓒ Yes   ⓓ No

323.

ⓐ $\begin{cases} 3a + 3c < 75 \\ 2a + 4c < 62 \end{cases}$

ⓑ

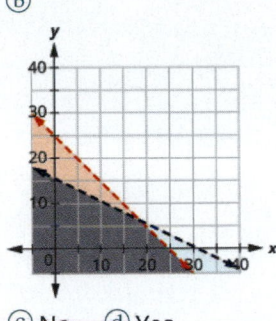

ⓒ No   ⓓ Yes

Review Exercises

327. ⓐ no ⓑ yes

329. $(3, -3)$

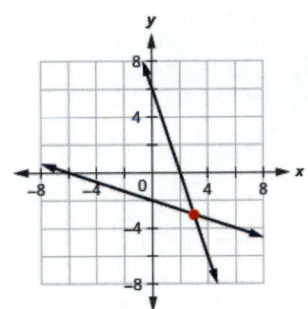

331. $(5, 4)$

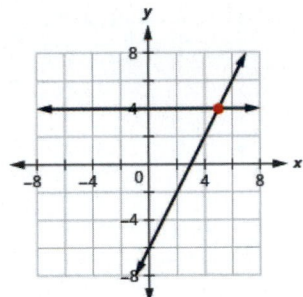

333. coincident lines

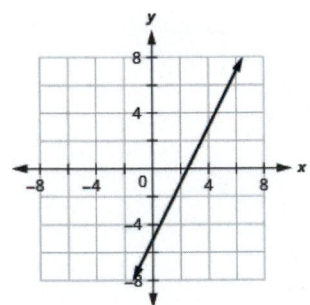

335. infinitely many solutions, consistent system, dependent equations

337. no solutions, inconsistent system, independent equations

339. LaVelle needs 8 ounces of chocolate syrup and 40 ounces of coffee.

345. no solution

351. $(1, 11)$

357. The numbers are -37 and -53.

363. $\begin{cases} j + d = 7200 \\ d = j + 1600 \end{cases}$

369. The pergola is 8 feet high and 12 feet wide.

341. $(-1, 2)$

347. The numbers are 22 and 33.

353. $(4, -1)$

359. elimination

365. Pam is 51 and Jan is 48.

371. It will take Lenore 3 hours.

343. $(-2, -2)$

349. The measures are 23 degrees and 67 degrees.

355. $(-2, 5)$

361. $\begin{cases} x + y = -32 \\ x = 2y - 2 \end{cases}$

367. The measures are 119 degrees and 61 degrees.

373. The rate of the boat is 10.5 mph. The rate of the current is 1.5 mph.

8.160. 60 feet

8.163. 10 mph

8.166. 6 mph

8.169. 16.25 hours

8.172. $y = 3x$

8.161. 15 mph

8.164. 50 mph

8.167. 2 hours and 24 minutes

8.170. 12 hours

8.162. 17 mph

8.165. 15 mph

8.168. 2 hours and 6 minutes

8.171. $y = \frac{3}{10}x$

8.173. ⓐ $c = 4.8t$ ⓑ 432 calories

8.175. ⓐ $m = 55h$ ⓑ 220 miles

8.178. ⓐ $A = 3.14r^2$ ⓑ 254.34 square inches

8.181. ⓐ $v = \frac{40,000}{a}$ ⓑ $8,000

8.184. ⓐ $F = \frac{48}{L}$ ⓑ 9.6 pounds

8.174. ⓐ $d = 50t$ ⓑ 250 miles

8.176. ⓐ $w = 6v$ ⓑ 78 pounds

8.179. $p = \frac{360}{q}$

8.182. ⓐ $t = \frac{1000}{r}$ ⓑ 2 hours

8.177. ⓐ $d = 16t^2$ ⓑ 256 feet

8.180. $y = \frac{16}{x}$

8.183. ⓐ $h = \frac{130}{t}$ ⓑ $1\frac{2}{3}$ hours

Section Exercises

1. ⓐ $z = 0$ ⓑ $p = \frac{5}{6}$
ⓒ $n = -4, n = 2$

3. ⓐ $y = 0$ ⓑ $x = -\frac{1}{2}$
ⓒ $u = -4, u = 7$

5. ⓐ 0 ⓑ 4 ⓒ 1

7. ⓐ 3 ⓑ $\frac{5}{2}$ ⓒ $-\frac{1}{5}$

9. ⓐ -6 ⓑ $\frac{20}{3}$ ⓒ 0

11. ⓐ -1 ⓑ $-\frac{3}{10}$ ⓒ 0

13. ⓐ 0 ⓑ $\frac{3}{4}$ ⓒ $\frac{15}{4}$

15. ⓐ 0 ⓑ $-\frac{3}{5}$ ⓒ $-\frac{7}{20}$

17. $-\frac{1}{13}$

19. $\frac{8}{9}$

21. $\frac{b}{2a}$

23. $\frac{2m^2}{3n}$

25. $\frac{3}{4}$

27. $\frac{3}{5}$

29. $\frac{7}{5}$

31. $\frac{12}{5}$

33. $\frac{a+3}{a-4}$

35. $\frac{y+4}{y-5}$

37. $\frac{x-5}{x-3}$

39. $\frac{y+1}{y+3}$

41. $\frac{y^2+1}{y+1}$

43. $x - 2$

45. $\frac{a(a+5)}{2(a+3)(a-2)}$

47. $\frac{c}{2(c-5)}$

49. $\frac{3(m+5)}{4(m-5)}$

51. $\frac{5(r-1)}{r+7}$

53. $\frac{t^2+3t+9}{t+3}$

55. $\frac{w^2-6w+36}{w-6}$

57. -1

59. -1

61. $-\frac{2}{x+6}$

63. $-\frac{4}{8+v}$

65. $-\frac{y-8}{3+y}$

67. $-\frac{a+4}{9+a}$

69. 16.5%

73. $\frac{3}{10}$

75. $\frac{6}{25}$

77. $\frac{x^3}{8y}$

79. $\frac{8ab}{3}$

81. $\frac{p(p-4)}{2(p-9)}$

83. $\frac{r+5}{2r(r+2)}$

85. $\frac{x-7}{4(x+3)}$

87. $\frac{z+3}{z(z+1)}$

89. $-\frac{4(b+9)}{3(b+7)}$

91. -7

93. 1

95. 1

97. $-\frac{2t}{t^3-5t-9}$

99. $-\frac{1}{10-w}$

101. $\frac{3y^2(y+6)(y+7)}{(y-7)(y^2+6)}$

103. $\frac{a(a-5)}{a-6}$

375. Lynn bought 227 student tickets and 34 adult tickets.

381. ⓐ yes ⓑ no

377. Yumi should use 4 cups of candies and 8 cups of nuts.

383.

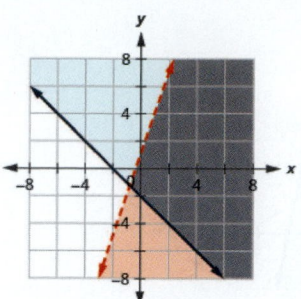

379. Jack should put $3600 into savings and $8400 into the CD.

385.

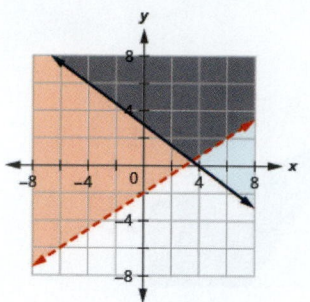

387. No solution

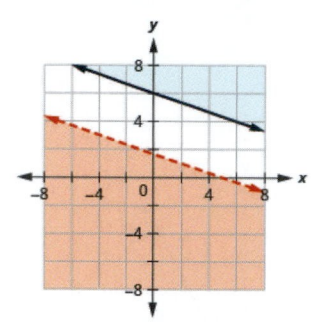

389.

ⓐ $\begin{cases} b + n \leq 40 \\ 12b + 18n \geq 500 \end{cases}$

ⓑ

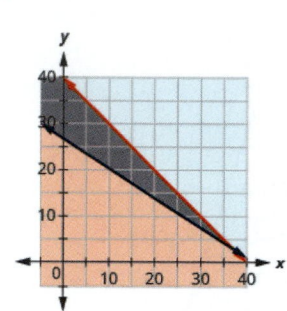

ⓒ yes

ⓓ no

Practice Test

391. ⓐ yes ⓑ no

393.

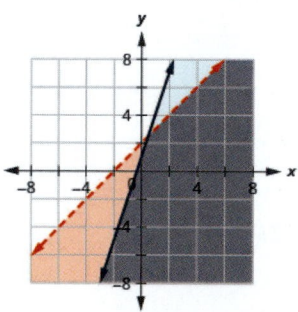

395. $(4, -7)$

397. infinitely many solutions

399. The numbers are 40 and 64

401. The measures of the angles are 28 degrees and 62 degrees.

403. It will take Kathy $\frac{1}{6}$ of an hour (or 10 minutes).

405. Liz bought 23 children's tickets and 5 adult tickets.

407.

ⓐ $\begin{cases} C + 0.5L \leq 50 \\ L \geq 3C \end{cases}$

ⓑ

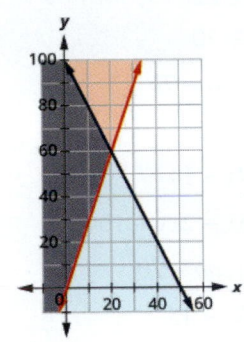

ⓒ No

ⓓ Yes

Chapter 6

Try It

6.1. ⓐ monomial ⓑ polynomial ⓒ trinomial ⓓ binomial ⓔ monomial

6.2. ⓐ binomial ⓑ trinomial ⓒ monomial ⓓ polynomial ⓔ monomial

6.3. ⓐ 1 ⓑ 4 ⓒ 12 ⓓ 3 ⓔ 0

6.4. ⓐ 0 ⓑ 5 ⓒ 1 ⓓ 2 ⓔ 3

6.5. $21q^2$

6.6. $-7c^2$

6.7. $13m$

6.8. $-10z^3$

6.9. $5y^2 + 3z^2$

6.10. $-4m^2 + n^2$

6.11. There are no like terms to combine.

6.12. There are no like terms to combine.

6.13. $8x^2 - 11x + 1$

6.14. $17y^2 + 14y + 1$

6.15. $15x^2 + 3x - 5$

6.16. $6b^2 + 3$

6.17. $2z^2 + 12z - 2$

6.18. $5x^2 + 14x + 7$

6.19. $5x^2 - 5xy + 5y^2$

6.20. $7x^2 - 6xy - 2y^2$

6.21. $-5ab - 5b^2$

6.22. $4n^2 + 7mn$

6.23. $x^3 - y^3$

6.24. $p^3 - 2p^2q + q^3$

6.25. ⓐ 18 ⓑ 50 ⓒ -15

6.26. ⓐ 18 ⓑ -4 ⓒ 14

6.27. 250

6.28. 106

6.29. $576

6.30. $750

6.31. ⓐ 216 ⓑ 15 ⓒ $\frac{9}{49}$ ⓓ 0.1849

6.32. ⓐ 32 ⓑ 21 ⓒ $\frac{8}{125}$ ⓓ 0.047524

6.33. ⓐ 81 ⓑ -81

6.34. ⓐ 169 ⓑ -169

6.35. b^{17}

6.36. x^{16}

6.37. ⓐ 5^6 ⓑ 4^{18}

6.38. ⓐ 7^{14} ⓑ 10^{11}

6.39. ⓐ p^6 ⓑ y^{43}

6.40. ⓐ z^8 ⓑ b^{49}

6.41. x^{18}

6.42. b^{19}

6.43. ⓐ b^{35} ⓑ 5^{12}

6.44. ⓐ z^{54} ⓑ 3^{49}

6.45. ⓐ $144y^2$ ⓑ $32w^5x^5$

6.46. ⓐ $125w^3x^3$ ⓑ $-27y^3$

6.47. ⓐ a^{48} ⓑ $-8c^{12}d^6$

6.48. ⓐ $81x^{24}y^{28}$ ⓑ q^{29}

6.49. ⓐ $75n^{12}$ ⓑ $81c^{24}d^{30}$

6.50. ⓐ $256a^{22}b^{24}$ ⓑ $40x^{10}$

6.51. $-35y^{11}$

6.52. $54b^9$

6.53. $6a^5b^6$

6.54. $8r^{11}s^8$

6.55. $5x + 35$

6.56. $3y + 39$

6.57. $x^2 - 7x$

6.58. $d^2 - 11d$

6.59. $5x^2 + 20xy$

6.60. $12p^2 + 2pr$

6.61. $-15y^3 - 24y^2 + 21y$

6.62. $8x^4 - 24x^3 + 20x^2$

6.63. $12x^3 - 20x^2 + 12x$

6.64. $-18a^5 + 12a^4 - 36a^3$

6.65. $xp + 8p$

6.66. $ap + 4p$

6.67. $x^2 + 17x + 72$

6.68. $20x^2 + 51x + 27$

6.69. $12b^2 + 38b + 30$

6.70. $a^2 + 17a + 70$

6.71. $30y^2 - 3y - 6$

6.72. $15c^2 + 14c - 8$

6.73. $a^2 - ab + 7a - 7b$

6.74. $x^2 - xy + 5x - 5y$

6.75. $x^2 + 14x + 48$

6.76. $y^2 + 20y + 51$

6.77. $x^2 - 2x - 35$

6.78. $b^2 + 3b - 18$

6.79. $15x^2 + 29x - 14$

6.80. $16y^2 - 20y - 50$

6.81. $10c^2 - 60c - cd + 6d$

6.82. $14x^2 - 35x - 2xy + 10y$

6.83. $x^3 - 8x^2 + 6x - 48$

6.84. $y^3 - 9y^2 + 7y - 63$

6.85. $8a^2b^2 + 12ab - 20$

6.86. $8x^2y^2 + 2xy - 15$

6.87. $15m^2 - 51m + 42$

6.88. $42b^2 - 53b + 15$

6.89. $y^3 - 8y^2 + 17y - 6$

6.90. $2x^3 + 5x^2 - 7x + 20$

6.91. $y^3 - 8y^2 + 17y - 6$

6.92. $2x^3 + 5x^2 - 7x + 20$

6.93. $x^2 + 18x + 81$

6.94. $y^2 + 22y + 121$

6.95. $x^2 - 18x + 81$

6.96. $p^2 - 26p + 169$

6.97. $36x^2 + 36x + 9$

6.98. $16x^2 + 72x + 81$

6.99. $4c^2 - 4cd + d^2$

6.100. $16x^2 - 40xy + 25y^2$

6.101. $4x^4 + 4x^2 + 1$

6.102. $9y^6 + 12y^3 + 4$

6.103. $x^2 - 25$

6.104. $w^2 - 9$

6.105. $36x^2 - 25$

6.106. $4x^2 - 49$

6.107. $49 - 16x^2$

6.108. $81 - 4y^2$

6.109. $16p^2 - 49q^2$

6.110. $9x^2 - y^2$

6.111. $x^2y^2 - 36$

6.112. $a^2b^2 - 81$

6.113. $9x^4 - 16y^6$

6.114. $4m^4 - 25n^6$

6.115. ⓐ FOIL; $18b^2 + 77b - 18$ ⓑ Binomial Squares; $81p^2 - 72p + 16$ ⓒ Binomial Squares; $49y^2 + 14y + 1$ ⓓ Product of Conjugates; $16r^2 - 9$

6.116. ⓐ Binomial Squares; $36x^2 + 84x + 49$ ⓑ Product of Conjugates; $9x^2 - 16$ ⓒ FOIL; $10x^2 - 29x + 10$ ⓓ Binomial Squares; $36n^2 - 12n + 1$

6.117. ⓐ x^5 ⓑ 6^9

6.118. ⓐ y^6 ⓑ 10^8

6.119. ⓐ $\dfrac{1}{x^4}$ ⓑ $\dfrac{1}{12^{15}}$

6.120. ⓐ $\dfrac{1}{m^8}$ ⓑ $\dfrac{1}{9^{11}}$

6.121. ⓐ b^8 ⓑ $\dfrac{1}{z^6}$

6.122. ⓐ $\dfrac{1}{p^8}$ ⓑ w^4

6.123. ⓐ 1 ⓑ 1

6.124. ⓐ 1 ⓑ 1

6.125. ⓐ 1 ⓑ 1

6.126. ⓐ 1 ⓑ 1

6.127. ⓐ $\dfrac{25}{64}$ ⓑ $\dfrac{p^4}{10,000}$ ⓒ $\dfrac{m^7}{n^7}$

6.128. ⓐ $\dfrac{1}{27}$ ⓑ $\dfrac{-8}{q^3}$ ⓒ $\dfrac{w^4}{x^4}$

6.129. m^{13}

6.130. k^5

6.131. 1

6.132. 1

6.133. r^8

6.134. v^6

6.135. $\dfrac{a^{12}}{b^8}$

6.136. $\dfrac{q^{21}}{r^{15}}$

6.137. $\dfrac{49x^6}{81y^2}$

6.138. $\dfrac{9x^8}{49y^2}$

6.139. $\dfrac{1}{a^6}$

6.140. $\dfrac{1}{p^{15}}$

6.141. $9r^{18}$

6.142. $\dfrac{2}{x}$

6.143. $7y^6$

6.144. $6z^6$

6.145. $-\dfrac{9}{a^5b}$

6.146. $\dfrac{-9d}{c^4}$

6.147. $\dfrac{2a^6}{3b^2}$

6.148. $-\dfrac{3q^6}{5p^8}$

6.149. $\dfrac{4y^2}{7x^4}$

6.150. $\dfrac{5}{8m^5n^3}$

6.151. $2ab^2$

6.152. $-4xy^5$

6.153. $2z^2 + 6$

6.154. $2z^2 - 3$

6.155. $9b^2 - 11b$

6.156. $5y^2 - 11y$

6.157. $-5y^2 + 3y$

6.158. $-7b^2 + 3b$

6.159. $15d^4 + 6d^2$

6.160. $36p^4 - 8p^2$

6.161. $-4a + 2b$

6.162. $8a^5b + 6a^3b^2$

6.163. $5xy + 3y - 2y^2$

6.164. $5a^2 + 2a^2b - 6b^2$

6.165. $3c + 1 - \dfrac{3}{2c}$

6.166. $2d - 1 - \dfrac{2}{5d}$

6.167. $y + 7$

6.168. $m + 5$

6.169. $2x + 5$

6.170. $3x + 2$

6.171. $x^2 + 3x + 2 + \dfrac{2}{x+2}$

6.172. $2x^2 + 6x - 5 - \dfrac{3}{x+1}$

6.173. $x^2 - 3x + 7 \quad x^2 - 2x + 7$

6.174. $x^3 - 8x^2 + 40x - 200$

6.175. $x^2 + 4x + 16$

6.176. $25x^2 + 10x + 4$

6.177. ⓐ $\dfrac{1}{8}$ ⓑ $\dfrac{1}{10^7}$

6.178. ⓐ $\dfrac{1}{9}$ ⓑ $\dfrac{1}{10,000}$

6.179. ⓐ p^8 ⓑ 64

6.180. ⓐ q^7 ⓑ 16

6.181. ⓐ $\dfrac{81}{16}$ ⓑ $\dfrac{n^2}{36m^2}$

6.182. ⓐ $\dfrac{125}{27}$ ⓑ $\dfrac{16b^4}{a^4}$

6.183. ⓐ $\dfrac{1}{25}$ ⓑ $-\dfrac{1}{25}$ ⓒ 25 ⓓ -25

6.184. ⓐ $\dfrac{1}{49}$ ⓑ $-\dfrac{1}{49}$ ⓒ 49 ⓓ -49

6.185. ⓐ 2 ⓑ $\dfrac{1}{18}$

6.186. ⓐ 2 ⓑ $\dfrac{1}{16}$

6.187. ⓐ $\dfrac{1}{y^7}$ ⓑ $\dfrac{1}{z^{15}}$

6.188. ⓐ $\dfrac{1}{p^9}$ ⓑ $\dfrac{1}{q^{24}}$

6.189. ⓐ $\dfrac{8}{p}$ ⓑ $\dfrac{1}{8p}$ ⓒ $-\dfrac{1}{8p}$

6.190. ⓐ $\dfrac{1}{11q}$ ⓑ $\dfrac{1}{11q}$ ⓒ $-\dfrac{1}{11q}$ $-\dfrac{1}{11q}$

6.191. ⓐ x^4 ⓑ $\dfrac{1}{y^5}$ ⓒ $\dfrac{1}{z^9}$

6.192. ⓐ a^5 ⓑ $\dfrac{1}{b^4}$ ⓒ $\dfrac{1}{c^{15}}$

6.193. $\dfrac{1}{p^3q^3}$

6.194. $\dfrac{1}{r^2s^8}$

6.195. $-\dfrac{12v^5}{u}$

6.196. $\dfrac{30d^3}{c^8}$

6.197. $\dfrac{1}{16x^8}$

6.198. $\dfrac{1}{16b^{12}}$

6.199. $\dfrac{64}{a^8}$

6.200. $\dfrac{8}{c^{12}}$

6.201. x^{11}

6.202. y^{13}

6.203. 9.6×10^4

6.204. 4.83×10^4

6.205. 7.8×10^{-3}

6.206. 1.29×10^{-2}

6.207. 1,300

6.208. 92,500

6.209. 0.00012

6.210. 0.075

6.211. 0.06

6.212. 0.009

6.213. 400,000

6.214. 20,000

Section Exercises

1. ⓐ trinomial ⓑ polynomial ⓒ binomial ⓓ monomial ⓔ binomial

3. ⓐ binomial ⓑ trinomial ⓒ polynomial ⓓ trinomial ⓔ monomial

5. ⓐ 2 ⓑ 4 ⓒ 1 ⓓ 3 ⓔ 0

7. ⓐ 1 ⓑ 2 ⓒ 3 ⓓ 3 ⓔ 0

9. $12x^2$

11. $6w$

13. $-5a$

15. $40x$

17. $-22b$

19. $-10a + 5b$

21. $-4a^2 + b^2$

23. $xy^2 - 5x - 5y^2$

25. $a^2b - 4a - 5ab^2$

27. $12a + 8b$

29. $-4a - 3b$

31. $-17x^6$

33. $11y^2 + 4y + 11$

35. $-3x^2 + 17x - 1$

37. $11x^2 - 5x + 5$

39. $6a^2 - 4a - 1$

41. $2m^2 - 7m + 4$

43. $5a + 3$

45. $12s^2 - 14s + 9$

47. $3x^2 - x + 4$

49. $w^2 + 3w + 4$

51. $2p^3 + p^2 + 9p + 10$

53. $8a^3 + a^2 - 2a + 12$

55. $11w - 64$

57. $12c - 45$

59. $10x^2 - 7xy + 6y^2$

61. $10m^2 + 3mn - 8n^2$

63. $-3ab + 3b^2$

65. $4uv + 2v^2$

67. $p^3 - 6p^2q + pq^2 + 4q^3$

69. $x^3 + 2x^2y - 5xy^2 + y^3$

71. ⓐ 187 ⓑ 46 ⓒ 2

73. ⓐ -104 ⓑ 4 ⓒ 40

75. 11

77. $10,800

79. 4

81. $58

83. 149

85. Answers will vary.

87. Answers will vary.

89. ⓐ 10,000 ⓑ 17 ⓒ $\frac{4}{81}$ ⓓ 0.125

91. ⓐ 512 ⓑ 8 ⓒ $\frac{27}{64}$

ⓓ 0.064

93. ⓐ 64 ⓑ -64

95. ⓐ $-\frac{4}{9}$

ⓑ $\frac{4}{9}$

97. ⓐ -0.001 ⓑ 0.001

99. x^6

101. q^{42}

103. ⓐ 3^{16} ⓑ 5^5

105. ⓐ w^6 ⓑ u^{94}

107. y^9

109. c^{18}

111. n^{y+2}

113. x^{p+q}

115. ⓐ b^{14} ⓑ 3^{16}

117. ⓐ x^{2y} ⓑ 7^{ab}

119. ⓐ $25x^2$ ⓑ $16a^2b^2$

121. ⓐ $-343n^3$ ⓑ $81x^4y^4z^4$

123. ⓐ w^{22} ⓑ $32x^5y^{20}$

125. ⓐ $-1000q^6p^{12}$ ⓑ n^{40}

127. ⓐ $48y^4$ ⓑ $25,000k^{24}$

129. ⓐ $432b^5$ ⓑ $\frac{1}{200}j^{16}$

131. ⓐ $128r^8$ ⓑ $\frac{1}{200}j^{16}$

133. $-18y^{11}$

135. $72u^7$

137. $4f^{11}$

139. $36a^5b^7$

141. $8r^2s^5$

143. $\frac{1}{2}x^3y^3$

145. x^{14}

147. a^{36}

149. $8m^{18}$

151. $1000x^6y^3$

153. $16a^{12}b^8$

155. $\frac{8}{27}x^6y^3$

157. $1024a^{10}$

159. $25000p^{24}$

161. $x^{18}y^{18}$

163. $144m^8n^{22}$

165. 1,000,000

167. $168.07

169. Answers will vary.

171. Answers will vary.

173. $4w + 40$

175. $-3a - 21$

177. $2x - 14$

179. $-3k + 12$

181. $q^2 + 5q$

183. $-b^2 - 9b$

185. $-x^2 + 10x$

187. $24r^2 + 6rs$

189. $12x^2 - 120x$

191. $-27a^2 - 45a$

193. $3p^2 + 30p + 75$

195. $-8x^3 - 16x^2 + 120x$

197. $5q^6 - 10q^4 + 30q^3$

199. $-8y^3 - 16y^2 + 120y$

201. $5q^5 - 10q^4 + 30q^3$

203. $-12z^4 - 48z^3 + 4z^2$

205. $2m^2 - 9m$

207. $8w - 48$

209. $4x + 40$

211. $15r - 360$

213. $-3m - 33$

215. $-8z + 40$

217. $u^2 + 5u$

219. $n^3 - 3n^2$

221. $24x^2 + 6xy$

223. $55p^2 - 25pq$

225. $3v^2 + 30v + 75$

227. $8n^3 - 8n^2 + 2n$

229. $-8y^3 - 16y^2 + 120y$

231. $5q^5 - 10q^4 + 30q^3$

233. $-12z^4 - 48z^3 + 4z^2$

235. $18y^2 - 9y$

237. $w^2 + 12w + 35$

239. $p^2 + 7p - 44$

241. $x^2 + 11x + 24$

243. $y^2 - 8y + 12$

245. $w^2 + 3w - 28$

247. $p^2 + 7p - 60$

249. $6p^2 + 11p + 5$

251. $20t^2 - 88t - 9$

253. $15x^2 - 3xy - 30x + 6y$

255. $2a^2 + 5ab + 3b^2$

257. $4z^2 - 24z - zy + 6y$

259. $x^3 + 2x^2 + 3x + 6$

261. $x^4 + 3x^2 - 40$

263. $10a^2 b^2 + 13ab - 3$

265. $24p^2 q^2 - 42pq + 15$

267. $x^3 + 9x^2 + 23x + 15$

269. $4y^3 + 33y^2 + y - 56$

271. $w^3 - 16w^2 + 73w - 70$

273. $3q^3 - 11q^2 - 19q - 5$

275. $14y - 13$

277. $-11x - 28$

279. $15q^3 - 30q^2 + 55q$

281. $s^2 + 2s - 63$

283. $y^3 - y^2 - 2y$

285. $3n^3 - n^2 - 25n + 28$

287. $49p^2 - 100$

289. $4m^4 - 3m^3 - 7m^2$

291. $25a^2 + 70ab + 49b^2$

293. $16y^2 - 144z^2$

295. ⓐ 306 ⓑ 306 ⓒ Answers will vary.

297. Answers will vary.

299. Answers may vary.

301. Answers may vary.

303. $q^2 + 24q + 144$

305. $x^2 + \frac{4}{3}x + \frac{4}{9}$

307. $y^2 - 12y + 36$

309. $p^2 - 26p + 169$

311. $16a^2 + 80a + 100$

313. $9z^2 + \frac{6}{5}z + \frac{1}{25}$

315. $4y^2 - 12yz + 9z^2$

317. $\frac{1}{64}x^2 - \frac{1}{36}xy + \frac{1}{81}y^2$

319. $25u^4 + 90u^2 + 81$

321. $64p^6 - 48p^3 + 9$

323. $c^2 - 25$

325. $b^2 - \frac{36}{49}$

327. $64j^2 - 16$

329. $81c^2 - 25$

331. $169 - q^2$

333. $16 - 36y^2$

335. $49w^2 - 100x^2$

337. $p^2 - \frac{16}{25}q^2$

339. $x^2 y^2 - 81$

341. $r^2 s^2 - \frac{4}{49}$

343. $36m^6 - 16n^{10}$

345. $225m^4 - 64n^8$

347. ⓐ $4r^2 + 48r + 144$ ⓑ $9p^2 - 64$ ⓒ $7a^2 - 48ab - 7b^2$ ⓓ $k^2 - 12k + 36$

349. ⓐ $x^{10} - y^{10}$ ⓑ $m^6 - 16m^3 n + 64n^2$ ⓒ $81p^2 + 144pq + 64q^2$ ⓓ $r^5 + r^2 s^2 - r^3 s^3 - s^5$

351. ⓐ 4,225 ⓑ 4,225 ⓒ Answers will vary.

353. Answers will vary.

355. Answers will vary.

357. ⓐ y^{10} ⓑ 7^{14}

359. ⓐ u^{21} ⓑ 9^{10}

361. ⓐ $\frac{1}{t^{30}}$ ⓑ $\frac{1}{64}$

363. ⓐ $\frac{1}{x^6}$ ⓑ $\frac{1}{100}$

365. ⓐ 1 ⓑ 1

367. ⓐ -1 ⓑ -1

369. ⓐ 1 ⓑ 6

371. ⓐ 7 ⓑ 1

373. ⓐ -7 ⓑ 0

375. ⓐ $\frac{4}{25}$ ⓑ $\frac{x^4}{81}$ ⓒ $\frac{a^5}{b^5}$

377. ⓐ $\frac{x^3}{8y^3}$ ⓑ $\frac{10,000}{81q^4}$

379. p^7

381. x^5

383. 1

385. $\frac{1}{n^{16}}$

387. q^{18}

389. $\frac{1}{m^{12}}$

391. $\frac{a^3}{b^{18}}$

393. $\frac{y^{20}}{z^{50}}$

395. $\frac{27m^{15}}{125n^3}$

397. $\frac{625u^{28}}{16v^{12}}$

399. j^9

401. $\frac{1}{q^8}$

403. $64k^6$

405. $-4,000$

407. $7v^8$

409. $-6u^8$

411. $-\frac{3x^3}{y^{12}}$

413. $\frac{-2m^3}{3n^5}$

415. $\dfrac{-3y^3}{4x^3}$

417. $\dfrac{65a^3 b^2}{42c^3}$

419. $\dfrac{-3q^5}{p^5}$

421. $\dfrac{5v^4}{u^2}$

423. ⓐ $18n^{10}$
ⓑ $12n^{10}$
ⓒ $45n^{20}$
ⓓ 5

425. ⓐ q^8
ⓑ q^{15}

427. ⓐ z ⓑ $\frac{1}{z}$

429. $6y^6$

431. $15c^6$

433. $3x^2$

435. $-yz^2$

437. 10^3

439. Answers will vary.

441. Answers will vary.

443. $6b + 15$

445. $6x^2 - 2x$

447. $11w - 2$

449. $4x^2 + 3x$

451. $-5b^2 + 3b$

453. $-17m^4 - 24m^3$

455. $103z^5 - 12z^2$

457. $17y^2 + 14$

459. $-5x^3 + 3$

461. $-6y^2 + 3y$

463. $9x^2 y^3 + 12y$

465. $7c - 10c^2 d - 5cd^2$

467. $6r^3 + 11r^2 s - 8rs^3$

469. $4q + 1 - \dfrac{1}{3q}$

471. $-5y - 3 + \dfrac{1}{4y}$

473. $7a - 12 + \dfrac{11}{a}$

475. $d + 6$

477. $a - 7$

479. $x - 7$

481. $4x + 3$

483. $p + 3 - \dfrac{8}{p + 8}$

485. $x + 7 + \dfrac{5}{x - 5}$

487. $2n^2 - 6n + 8$

489. $7q^2 + 7q + 2$

491. $m^2 - 10m + 100$

493. $x^2 + 6x + 36$

495. $25y^2 + 20x + 16$

497. 45

499. Answers will vary.

501. ⓐ $\frac{1}{81}$ ⓑ $\frac{1}{100}$

503. ⓐ $\frac{1}{256}$ ⓑ $\frac{1}{100}$

505. ⓐ c^5 ⓑ 25

507. ⓐ t^9 ⓑ 10000

509. ⓐ $\dfrac{100}{9}$ ⓑ $-\dfrac{c^3 d^3}{8}$

511. ⓐ $\dfrac{8}{343}$ ⓑ $-\dfrac{x^3 y^6}{27}$

513. ⓐ $\frac{1}{49}$ ⓑ $-\frac{1}{49}$ ⓒ 49 ⓓ -49

515. ⓐ $-\dfrac{1}{125}$ ⓑ -125 ⓒ -125 ⓓ $-\dfrac{1}{125}$

517. ⓐ $\frac{2}{5}$ ⓑ $\frac{1}{10}$

519. ⓐ $\frac{3}{16}$ ⓑ $\frac{1}{144}$

521. ⓐ $\frac{1}{b^5}$ ⓑ $\frac{1}{k^{10}}$

523. ⓐ $\frac{1}{s^8}$ ⓑ $\frac{1}{a^{90}}$

525. ⓐ $\frac{6}{r}$ ⓑ $\frac{1}{6r}$ ⓒ $-\frac{1}{6r}$

527. ⓐ $\dfrac{1}{16q^4}$ ⓑ $\dfrac{2}{q^4}$ ⓒ $-\dfrac{2}{q^4}$

529. ⓐ $\frac{1}{s^4}$ ⓑ $\frac{1}{q^5}$ ⓒ $\frac{1}{y^7}$

531. ⓐ 1 ⓑ y^6 ⓒ $\frac{1}{y^4}$

533. $\frac{1}{x}$

535. $\dfrac{1}{m^2 n^4}$

537. $\dfrac{1}{p^5 q^7}$

539. $-\dfrac{14k^5}{j^3}$

541. $-\dfrac{40n^3}{m}$

543. $\dfrac{1}{64y^9}$

545. $\dfrac{4}{p^{10}}$

547. n^7

549. y^5

551. 3.4×10^5

553. 1.29×10^6

555. 4.1×10^{-2}

557. 1.03×10^{-6}

559. 830

561. 16,000,000,000

563. 0.038

565. 0.0000193

567. 0.02

569. 5.6×10^{-6}

571. 500,000,000

573. 20,000,000

575. 6.85×10^9.

577. 5.7×10^{-10}

579. 15,000,000,000,000

581. 0.00001

583. 1.545×10^8

585. ⓐ 1.86×10^{13} ⓑ 3×10^8 ⓒ 6.2×10^4

587. answers will vary

Review Exercises

589. ⓐ binomial ⓑ monomial ⓒ trinomial ⓓ trinomial ⓔ other polynomial

591. ⓐ 3 ⓑ 4 ⓒ 2 ⓓ 4 ⓔ 0

593. $5k$

595. $-27c$

597. $7n^2$

599. $13a + b$

601. $13p^2 - 5p - 1$

603. $7y^2 - 9y + 4$

605. $5a^3 + a^2 + 6a + 2$

607. ⓐ -26 ⓑ 10 ⓒ 22

609. $12{,}000$

611. 17

613. 0.125

615. -64

617. p^{31}

619. 8^6

621. y^{c+3}

623. 5^6

625. 3^{rs}

627. $-125y^3$

629. $1000x^3 y^3 z^3$

631. $64a^9 b^6$

633. $48q^{14}$

635. $\frac{8}{125} m^6 n^3$

637. $144n^8$

639. $15a^2 b^5$

641. $-4y - 52$

643. $p^2 + 3p$

645. $-12u^2 - 42u$

647. $3q^4 - 21q^3 + 18q^2$

649. $11b - 44$

651. ⓐ $12y^2 - 44y + 35$ ⓑ $12y^2 - 44y + 35$ ⓒ $12y^2 - 44y + 35$

653. $y^2 - 12y + 32$

655. $q^2 + 13q - 48$

657. $u^4 + u^2 - 30$

659. $16m^2 n^2 - 2mn - 3$

661. ⓐ $18x^3 - 21x^2 - 34x + 40$ ⓑ $18x^3 - 21x^2 - 34x + 40$

663. $7m^3 - 69m^2 - 31m - 3$

665. $q^2 - 30q + 225$

667. $64u^2 + 16u + 1$

669. $16a^2 - 24ab + 9b^2$

671. $y^2 - \frac{4}{25}$

673. $36 - r^2$

675. $25p^8 - 16q^6$

677. $36a^2 - 121$

679. $c^8 + 18c^4 d + 81d^2$

681. $4a^3 + 3a^2 b - 4b^3$

683. 10^{20}

685. $\frac{1}{v^{36}}$

687. $\frac{1}{5^7}$

689. 1

691. 1

693. 1

695. 0

697. $\frac{m^4}{81}$

699. $\frac{x^6}{64y^6}$

701. 1

703. r^{20}

705. $\frac{343x^{20}}{32y^{10}}$

707. $-\frac{10{,}125n^{10}}{4}$

709. $-\frac{4b^6}{a^5}$

711. $\frac{9p}{2}$

713. $7x - 15$

715. $55p^3 - 30p$

717. $12a^4 - 6a^3 b - 7ab^2$

719. $21y^2 + 10 - \frac{1}{y^2}$

721. $v - 8$

723. $n - 6 + \frac{4}{n+3}$

725. $u^2 + 2u + 4$

727. $-\frac{1}{125}$

729. $\frac{1}{216u^3}$

731. $\frac{16}{9}$

733. $\frac{1}{q^{11}}$

735. $\frac{1}{y^8}$

737. $\frac{1}{a^4}$

739. r

741. 4.29×10^{-3}

743. 7.2×10^9

745. $15{,}000{,}000{,}000$

747. 0.55

749. 0.0217

751. 0.0000003

Practice Test

753. $15a^2 + a - 6$

755. $-\dfrac{8}{125}$

757. $16a^6 b^{10}$

759. $3k^3 - 21k^2 + 39k$

761. $9v^2 - 86v + 45$

763. $n^3 - 11n^2 + 34n - 24$

765. $49p^2 - 25$

767. $\dfrac{1}{9}$

769. 1

771. $6x + 21 - \dfrac{3}{x}$

773. $16x^2 + 4x + 1$

775. $\dfrac{1}{25}$

777. $\dfrac{1}{q^9}$

779. 8.3×10^7

781. 74,800

783. 424 feet

Chapter 7

Try It

7.1. 16

7.2. 2

7.3. $3x^2$

7.4. $8y^2$

7.5. $2ab$

7.6. $3m^3 n$

7.7. $5m^2$

7.8. $7x$

7.9. $6(a + 4)$

7.10. $2(b + 7)$

7.11. $14(x + 1)$

7.12. $12(p + 1)$

7.13. $8(u - 2)$

7.14. $30(y - 2)$

7.15. $5\left(x^2 - 5x + 3\right)$

7.16. $3\left(y^2 - 4y + 9\right)$

7.17. $2x^2 (x + 6)$

7.18. $3y^2 (2y - 5)$

7.19. $2x\left(10x^2 - 5x + 7\right)$

7.20. $4y\left(6y^2 - 3y - 5\right)$

7.21. $3y^2\left(3x + 2x^2 + 7y\right)$

7.22. $3p\left(p^2 - 2pq + 3q^2\right)$

7.23. $-8(8z + 8)$

7.24. $-9(y + 3)$

7.25. $-4b(b - 4)$

7.26. $-7a(a - 3)$

7.27. $(m + 3)(4m - 7)$

7.28. $(n - 4)(8n + 5)$

7.29. $(x + 8)(y + 3)$

7.30. $(a + 7)(b + 8)$

7.31. $(x - 5)(x + 2)$

7.32. $(y + 4)(y - 7)$

7.33. $(x + 2)(x + 4)$

7.34. $(y + 3)(y + 5)$

7.35. $(q + 4)(q + 6)$

7.36. $(t + 2)(t + 12)$

7.37. $(x + 4)(x + 15)$

7.38. $(v + 3)(v + 20)$

7.39. $(u - 3)(u - 6)$

7.40. $(y - 7)(y - 9)$

7.41. $(h - 2)(h + 6)$

7.42. $(k - 4)(k + 5)$

7.43. $(x + 2)(x - 6)$

7.44. $(y + 4)(y - 5)$

7.45. $(r + 5)(r - 8)$

7.46. $(s + 2)(s - 5)$

7.47. prime

7.48. prime

7.49. $(m + 3)(m + 6)$

7.50. $(n - 3)(n - 4)$

7.51. $(u + 4v)(u + 7v)$

7.52. $(x + 6y)(x + 7y)$

7.53. $(a - b)(a - 10b)$

7.54. $(m - n)(m - 12n)$

7.55. prime

7.56. prime

7.57. ⓐ no method ⓑ undo using FOIL ⓒ factor with grouping

7.58. ⓐ factor using grouping ⓑ no method ⓒ undo using FOIL

7.59. $4(m + 1)(m - 2)$

7.60. $5(k + 2)(k - 5)$

7.61. $3(r - 1)(r - 2)$

7.62. $2(t - 2)(t - 3)$

7.63. $5x(x - 1)(x + 4)$

7.64. $6y(y - 2)(y + 5)$

7.65. $(a + 1)(2a + 3)$

7.66. $(b + 1)(4b + 1)$

7.67. $(2x - 3)(4x - 1)$

7.68. $(2y - 7)(5y - 1)$

7.69. $(a - 1)(8a + 5)$

7.70. $(2b + 3)(3b - 5)$

7.71. $(3x + 2)(6x - 5)$

7.72. $(3y + 1)(10y - 21)$

7.73. $5n(n - 4)(3n - 5)$

7.74. $8q(q + 6)(7q - 2)$

7.75. $(x + 2)(6x + 1)$

7.76. $(2y + 1)(2y + 3)$

7.77. $(4n - 5)(5n + 3)$

7.78. $(q + 4)(6q - 5)$

7.79. $(2t + 5)(5t - 3)$

7.80. $(u + 1)(3u + 5)$

7.81. $4(2x - 3)(2x - 1)$

7.82. $3(3w - 2)(2w - 3)$

7.83. $(2x + 3)^2$

7.84. $(3y + 4)^2$

7.85. $(8y - 5)^2$

7.86. $(4z - 9)^2$

7.87. $(7x + 6y)^2$

7.88. $(8m + 7n)^2$

7.89. $(8r + 3s)(2r + 3s)$

7.90. $(3u + 4)(3u + 25)$

7.91. $2y(2x - 3)^2$

7.92. $3q(3p + 5)^2$

7.93. $(h - 9)(h + 9)$

7.94. $(k - 11)(k + 11)$

7.95. $(m - 1)(m + 1)$

7.96. $(9y - 1)(9y + 1)$

7.97. $(16m - 5n)(16m + 5n)$

7.98. $(12p - 3q)(12p + 3q)$

7.99. $(12 - x)(12 + x)$

7.100. $(13 - p)(13 + p)$

7.101. $(a^2 + b^2)(a + b)(a - b)$

7.102. $(x^2 + 4)(x + 2)(x - 2)$

7.103. $7x(y - 5)(y + 5)$

7.104. $5b(3a - 4)(3a + 4)$

7.105. $8(a^2 + 25)$

7.106. $9(4y^2 + 9)$

7.107. $(x + 3)(x^2 - 3x + 9)$

7.108. $(y + 2)(y^2 - 2y + 4)$

7.109. $(u - 5)(u^2 + 5u + 25)$

7.110. $(v - 7)(v^2 + 7v + 49)$

7.111. $(4 - 3x)(16 - 12x + 9x^2)$

7.112. $(3 - 2y)(9 - 6y + 4y^2)$

7.113. $(2x - 3y)(4x^2 - 6xy + 9y^2)$

7.114.
$(10m - 5n)(100m^2 - 50mn + 25n^2)$

7.115.
$4(5p + q)(25p^2 - 5pq + q^2)$

7.116.
$2(6c + 7d)(36c^2 - 42cd + 49d^2)$

7.117. $3a^3(a + 6)$

7.118. $9b^5(5b + 3)$

7.119. $(5a - 6)(2a - 1)$

7.120. $(2x - 3)(4x - 3)$

7.121. $x(x^2 + 36)$

7.122. $3(9y^2 + 16)$

7.123. $4x(2x - 3)(2x + 3)$

7.124. $3(3y - 4)(3y + 4)$

7.125. $(2x + 5y)^2$

7.126. $(3m + 7n)^2$

7.127. $8(y - 1)(y + 3)$

7.128. $5(u - 9)(u + 6)$

7.129.
$2(5m + 6)(25m^2 - 30m + 36)$

7.130. $81(q + 2)(q^2 - 2q + 4)$

7.131. $4(a^2 + 4)(a - 2)(a + 2)$

7.132. $7(y^2 + 1)(y - 1)(y + 1)$

7.133. $6(x + b)(x - 2c)$

7.134. $2(4x - 1)(x + 3y)$

7.135. $4(p - 1)(p - 3)$

7.136. $3(q - 2)(2q + 1)$

7.137. $x = 3, x = -5$

7.138. $y = 6, y = -9$

7.139. $m = \frac{2}{3}, m = -\frac{1}{2}$

7.140. $p = -\frac{3}{4}, p = \frac{3}{4}$

7.141. $u = 0, u = \frac{1}{5}$

7.142. $w = 0, w = -\frac{3}{2}$

7.143. $x = 1$

7.144. $v = 2$

7.145. $x = 4, x = -3$

7.146. $b = -2, b = -7$

7.147. $c = 0, c = \frac{4}{3}$

7.148. $d = 3, d = -\frac{1}{2}$

7.149. $a = 0, a = -1$

7.150. $b = 0, b = -\frac{1}{3}$

7.151. $p = \frac{7}{5}, p = -\frac{7}{5}$

7.152. $x = \frac{11}{6}, x = -\frac{11}{6}$

7.153. $m = 1, m = \frac{3}{2}$

7.154. $k = 3, k = -3$

7.155. $x = 0, x = \frac{3}{2}$

7.156. $y = 0, y = \frac{1}{4}$

7.157. $a = -\frac{5}{2}, a = \frac{2}{3}$

7.158. $b = 2, b = \frac{1}{20}$

7.159. $-15, -16$ and $15, 16$

7.160. $-21, -20$ and $20, 21$

7.161. 5 feet and 6 feet

7.162. 12 feet and 15 feet

7.163. 5 feet and 12 feet

7.164. 24 feet and 25 feet

Section Exercises

1. 2

3. 18

5. 10

7. x

9. $8w^2$

11. $2pq$

13. $6m^2 n^3$

15. $2a$

17. $5x^3$

19. $4(x + 5)$

21. $3(2m + 3)$

23. $9(q + 1)$

25. $8(m - 1)$

27. $9(n - 7)$

29. $3(x^2 + 2x - 3)$

31. $2(4p^2 + 2p + 1)$

33. $8y^2(y + 2)$

35. $5x(x^2 - 3x + 4)$

37. $6y^2(2x + 3x^2 - 5y)$

39. $-2(x + 4)$

41. $(x + 1)(5x + 3)$

43. $(b - 2)(3b - 13)$

45. $(y + 3)(x + 2)$

47. $(u + 2)(v - 9)$

49. $(b - 4)(b + 5)$

51. $(p - 9)(p + 4)$

53. $-10(2x + 1)$

55. $(x^2 + 2)(3x - 7)$

57. $(x + y)(x + 5)$

59. $w(w - 6)$

61. Answers will vary.

63. $(x + 1)(x + 3)$

65. $(m + 1)(m + 11)$

67. $(a + 4)(a + 5)$

69. $(p + 5)(p + 6)$

71. $(n + 3)(n + 16)$

73. $(a + 5)(a + 20)$

75. $(x - 2)(x - 6)$

77. $(y - 3)(y - 15)$

79. $(x - 1)(x - 7)$

81. $(p - 1)(p + 6)$

83. $(y + 1)(y - 7)$

85. $(x - 4)(x + 1)$ $(x - 4)(x + 3)$

87. $(a - 7)(a + 4)$

89. $(w - 9)(w + 4)$

91. prime

93. $(x - 4)(x - 2)$

95. $(x - 12)(x + 1)$

97. $(p + q)(p + 2q)$

99. $(r + 3s)(r + 12s)$

101. $(m - 2n)(m - 10n)$

103. $(x + 8y)(x - 10y)$

105. $(m + n)(m - 65n)$

107. $(a + 8b)(a - 3b)$

109. prime

111. prime

113. $(u - 6)(u - 6)$

115. $(x + 2)(x - 16)$

117. $(r - 4s)(r - 16s)$

119. $(k + 4)(k + 30)$

121. prime

123. $(m + 8n)(m - 7n)$

125. $(u - 15v)(u - 2v)$

127. prime

129. $(x + 8)(x - 7)$

131. Answers may vary

133. Answers may vary

135. ⓐ factor the GCF, binomial ⓑ Undo FOIL ⓒ factor by grouping

137. ⓐ undo FOIL ⓑ factor by grouping ⓒ factor the GCF, binomial

139. $5(x + 1)(x + 6)$

141. $2(z - 4)(z + 3)$

143. $7(v - 1)(v - 8)$

145. $p(p - 10)(p + 2)$

147. $3m(m - 5)(m - 2)$

149. $5x^2(x - 3)(x + 5)$

151. $(2t + 5)(t + 1)$

153. $(11x + 1)(x + 3)$

155. $(4w - 1)(w - 1)$

157. $(3p - 2)(2p - 5)$

159. $(4q + 1)(q - 2)$

161. $(4p - 3)(p + 5)$

163. $16(x - 1)(x - 1)$

165. $10q(3q + 2)(q + 4)$

167. $(5n + 1)(n + 4)$

169. $(3z + 1)(3z + 4)$

171. $(2k - 3)(2k - 5)$

173. $(5s - 4)(s - 1)$

175. $(3y + 5)(2y - 3)$

177. $(2n + 3)(n - 15)$

179. prime

181. $10(6y - 1)(y + 5)$

183. $3z(8z + 3)(2z - 5)$

185. $8(2s + 3)(s + 1)$

187. $12(4y - 3)(y + 1)$

189. $(4y - 7)(3y - 2)$

191. $(a - 5)(a + 4)$

193. $(2n - 1)(3n + 4)$

195. prime

197. $13(z^2 + 3z - 2)$

199. $(x + 7)(x - 4)$

201. $3p(p + 7)$

203. $6(r + 2)(r + 3)$

205. $4(2n + 1)(3n + 1)$

207. $(x + 6)(x - 4)$

209. $-16(t - 6)(t + 1)$

211. Answers may vary.

213. Answers may vary.

215. $(4y + 3)^2$

217. $(6s + 7)^2$

219. $(10x - 1)^2$

221. $(5n - 12)^2$

223. $(7x - 2y)^2$

225. $(5n + 4)(5n + 1)$

227. $(32m - 1)(2m - 1)$

229. $10(k + 4)^2$

231. $3u(5u - v)^2$

233. $(x - 4)(x + 4)$

235. $(5v - 1)(5v + 1)$

237. $(11x - 12y)(11x + 12y)$

239. $(13c - 6d)(13c + 6d)$

241. $(7x - 2)(7x + 2)$ $(2 - 7x)(2 + 7x)$

243. $(2z - 1)(2z + 1)(4z^2 + 1)$

245. $5(q - 3)(q + 3)$

247. $6(4p^2 + 9)$

249. $(x + 5)(x^2 - 5x + 25)$

251. $(z - 3)(z^2 + 3z + 9)$

253. $(2 - 7t)(4 + 14t + 49t^2)$

255. $(2y - 5z)(4y^2 + 10yz + 25z^2)$

257. $7(k + 2)(k^2 - 2k + 4)$

259. $2(1 - 2y)(1 + 2y + 4y^2)$

261. $(8a - 5)(8a + 5)$

263. $3(3q - 1)(3q + 1)$

265. $(4x - 9)^2$

267. $2(4p^2 + 1)$

269. $(5 - 2y)(25 + 10y + 4y^2)$

271. $5(3n+2)^2$

273. $(2w+15)^2$

275. Answers may vary.

277. Answers may vary.

279. $5x^3(2x+7)$

281. $(y-3)(y+13)$

283. $(2n-1)(n+7)$

285. $a^3(a^2+9)$

287. $(11r-s)(11r+s)$

289. $8(m-2)(m+2)$

291. $(5w-6)^2$

293. $(m+7n)^2$

295. $7(b+3)(b-2)$

297. $3(x-3)(x^2+3x+9)$

299. $(k-2)(k+2)(k^2+4)$

301. $3(5p+4)(q-1)$

303. $4(x+3)(x+7)$

305. $u^2(u+1)(u^2-u+1)$

307. prime

309. $10(m-5)(m+5)(m^2+25)$

311. ⓐ $-16(t^2-5)$ ⓑ $-8(2t+5)(t-2)$

315. $x=3, x=-7$

317. $a=10/3, a=7/2$

319. $m=0, m=5/12$

321. $y=3$

323. $x=1/2$

325. $x=3, x=4$
$x=-3, x=-4$

327. $a=-5/4, a=6$
$a=-4/5, a=6$

329. $m=5/4, m=3$

331. $a=-1, a=0$

333. $m=12/7, m=-12/7$

335. $y=-1, y=6$

337. $x=3/2, x=-1$

339. $p=0, p=\frac{3}{4}$

341. $x=-2/3 \quad x=3/2$

343. 7 and 8; -8 and -7

345. 4 feet and 7 feet

347. 6 feet and 8 feet

349. $x=-8, x=3$

351. $p=-1, p=-11$

353. $m=-2, m=8$

355. $a=0, a=-6, a=7$

357. 10 and 11; -11 and -10

359. 10 feet

361. Answers may vary.

Review Exercises

363. 6

365. 15

367. $6(4x-7)$

369. $3m^2(5m^2+2n)$

371. $(a+b)(x-y)$

373. $(x-3)(x+7)$

375. $(m^2+1)(m+1)$

377. $(u+8)(u+9)$

379. $(k-6)(k-10)$

381. $(y+7)(y-1)$

383. $(s-4)(s+2)$

385. $(x+5y)(x+7y)$

387. $(a+7b)(a-3b)$

389. Undo FOIL

391. Factor the GCF

393. $6(x+2)(x+5)$

395. $3n^2(n-8)(n+4)$

397. $(x+4)(2x+1)$

399. $(3a-1)(6a-1)$

401. $(5p+4)(3p-2)$

403. $(5s-2)(8s+3)$

405. $3(x+4)(x-3)$

407. $5(4y+1)(3y-5)$

409. $(5x+3)^2$

411. $(6a-7b)^2$

413. $10(2x+9)^2$

415. $2y(y-4)^2$

417. $(9r-5)(9r+5)$

419. $(13m+n)(13m-n)$

421. $(5p-1)(5p+1)$

423. $(3+11y)(3-11y)$

425. $5(2x-5)(2x+5)$

427. $u(7u+3)(7u-3)$

429. $(a-5)(a^2+5a+25)$

431. $2(m+3)(m^2-3m+9)$

433. $4x^2(6x+11)$

435. $(4n-7m)^2$

437. $(r+6)(5r-8)$

439. $(n^2+9)(n+3)(n-3)$

441. $5(x-3)(x+4)$

443. $(m+5)(m^2-5m+25)$

445. $a=3 \ a=-7$

447. $m=0 \ m=-3 \ m=\frac{5}{2}$

449. $x=-4, x=-5$

451. $p=-\frac{5}{2}, p=8$

453. $m=\frac{5}{12}, m=-\frac{5}{12}$

455. $-21, -22 \ 21, 22$

Practice Test

457. $7(y-6)$

459. $40a^2(2+3a)$

461. $(x+7)(x+6)$

463. $3a(a^2-2a-14)$

465. $5(n+1)(n+5)$

467. $(x-8)(y+7)$

469. $(3s-2)^2$

471. $(10-a)(10+a)$

473. $3(x+5y)(x-5y)$

475. $(a-3)(b-2)$

481. $b=1, b=-1$

477. $(4m+1)(2m+5)$

483. $n=-\frac{7}{4}, n=-3$

479. $y=-11, y=12$

485. 12 and 13; -13 and -12

Chapter 8

Try It

8.1. ⓐ $x=0$ ⓑ $n=-\frac{1}{3}$ ⓒ $a=-1, a=-3$

8.2. ⓐ $q=0$ ⓑ $y=-\frac{2}{3}$ ⓒ $m=2, m=-3$

8.3. ⓐ -2 ⓑ $\frac{2}{9}$ ⓒ $-\frac{1}{3}$

8.4. ⓐ $\frac{4}{3}$ ⓑ 6 ⓒ -1

8.5. ⓐ $\frac{1}{2}$ ⓑ $\frac{1}{3}$ ⓒ 2

8.6. ⓐ $\frac{2}{3}$ ⓑ $\frac{4}{5}$ ⓒ $\frac{1}{2}$

8.7. ⓐ -4 ⓑ 0 ⓒ $\frac{4}{9}$

8.8. ⓐ 0 ⓑ $\frac{3}{16}$ ⓒ $\frac{3}{16}$

8.9. $-\frac{5}{9}$.

8.10. $-\frac{7}{9}$

8.11. $\frac{x}{3y}$

8.12. $\frac{8x}{y}$

8.13. $\frac{3}{2}$

8.14. $\frac{7}{5}$

8.15. $\frac{x+1}{x-1}$

8.16. $\frac{x-5}{x-1}$

8.17. $\frac{x+3}{x+2}$

8.18. $\frac{x+1}{x-7}$

8.19. $\frac{y^2+1}{y+2}$

8.20. $\frac{p^2+2}{p+5}$

8.21. $\frac{n}{2(n+1)}$

8.22. $\frac{x}{2(x+2)}$

8.23. $\frac{2(x-3)}{3(x+3)}$

8.24. $\frac{5(x-1)}{2(x+5)}$

8.25. $\frac{p^2+4p+16}{p+4}$

8.26. $\frac{x^2-2x+4}{x-2}$

8.27. -1

8.28. -1

8.29. $-\frac{2}{y+5}$.

8.30. $-\frac{3}{9+y}$

8.31. $-\frac{x+1}{x+5}$

8.32. $-\frac{x+2}{x+1}$

8.33. $\frac{3}{4}$

8.34. 1

8.35. $\frac{5p^2}{q}$

8.36. $\frac{12y^3}{7}$

8.37. $\frac{x-2}{2(x+3)}$

8.38. $\frac{3(x-6)}{x+5}$

8.39. $\frac{x+5}{x}$

8.40. $\frac{x-4}{x+3}$

8.41. $-\frac{6(x+3)}{x+2}$

8.42. $-\frac{y}{3}$

8.43. 1

8.44. 1

8.45. $-\frac{1}{c-3}$

8.46. $-\frac{1}{2+d}$

8.47. $\frac{(m-2)}{4(m-8)}$

8.48. $\frac{n(n-2)}{n-1}$

8.49. $\frac{(a-3)(a+5)}{(a-5)(a-5)}$

8.50. $-\frac{(b+2)(b-1)}{(1+b)(b+4)}$

8.51. $\frac{2(x^2+2x+4)}{(x+2)(x^2-2x+4)}$

8.52. $\frac{2z(z^2+z+1)}{(z+1)(z^2-z+1)}$

8.53. $\frac{x-8}{2(x+1)}$

8.54. $\frac{y-2}{3y(y-4)}$

8.55. $\frac{x+2}{4}$

8.56. $\frac{2}{y+5}$

8.57. $\frac{2(m+1)(m+2)}{3(m+4)(m-3)}$

8.58. $\frac{(n+5)(n+9)}{2(n+6)(2n+3)}$

8.59. $\frac{3}{4}$

8.60. $\frac{2}{5}$

8.61. $\frac{5x+2}{2x+3}$.

8.62. $\frac{x+1}{x-2}$

8.63. $x+2$

8.64. $x+3$

8.65. $x-3$

8.66. $2x+5$

8.67. $n+3$

8.68. $y-8$

8.69. $\frac{x-11}{x-2}$

8.70. $\dfrac{x-3}{x+9}$

8.71. 3

8.72. $y+3$

8.73. $\dfrac{y+3}{y+2}$

8.74. $\dfrac{3n-2}{n-1}$

8.75. $(x-4)(x+4)(x+3)$

8.76. $(x+3)(x+6)(x+5)$

8.77. $\dfrac{2x+8}{(x-4)(x+3)(x+4)},$ $\dfrac{x+3}{(x-4)(x+3)(x+4)}$

8.78. $\dfrac{x^2+6x}{(x+3)(x+5)(x+6)},$ $\dfrac{x+3}{(x+3)(x+5)(x+6)}$

8.79. $\dfrac{57}{30}$

8.80. $\dfrac{33}{40}$

8.81. $\dfrac{4b+25a}{30a^2b^2}$

8.82. $\dfrac{5d^2+6}{16cd^2}$

8.83. $\dfrac{7x-4}{(x+3)(x-2)}$

8.84. $\dfrac{7m+25}{(m+3)(m+4)}$

8.85. $\dfrac{x(5x+7y)}{y(x-y)(x+y)}$

8.86. $\dfrac{7m+15}{2(m+3)(m+1)}$

8.87. $\dfrac{5m^2-9m+2}{(m-2)(m+1)(m+2)}$

8.88. $\dfrac{2(n^2+6n-15)}{(n+2)(n-5)(n+3)}$

8.89. $\dfrac{-7y+8}{(y+4)(y-5)}$

8.90. $\dfrac{4z+9}{(z+2)(z+3)}$

8.91. $\dfrac{1}{x-2}$

8.92. $\dfrac{-3}{z-3}$

8.93. $\dfrac{1}{x-6}$

8.94. $\dfrac{y+3}{y+4}$

8.95. $\dfrac{-x+22}{x-7}$

8.96. $\dfrac{-2(3y-4)}{2y-1}$

8.97. $\dfrac{v+3}{v+1}$

8.98. $\dfrac{3w}{w+7}$

8.99. $\dfrac{2}{3(x-1)}$

8.100. $\dfrac{1}{2(x-3)}$

8.101. $\dfrac{14}{11}$

8.102. $\dfrac{10}{23}$

8.103. $\dfrac{y+x}{y-x}$

8.104. $\dfrac{ab}{b-a}$

8.105. $b(b+2)$

8.106. $\dfrac{3}{c+3}$

8.107. $\dfrac{10}{3}$

8.108. $\dfrac{7}{3}$

8.109. $\dfrac{b+a}{a^2+b^2}$

8.110. $\dfrac{y-x}{xy}$

8.111. $\dfrac{3x-6}{5x+7}$

8.112. $\dfrac{x+21}{6x+43}$

8.113. $\dfrac{3}{5x+22}$

8.114. $\dfrac{6y+34}{3y}$

8.115. $\dfrac{x}{x+4}$

8.116. $\dfrac{x(x+1)}{3(x-1)}$

8.117. $-\dfrac{15}{7}$

8.118. $\dfrac{15}{13}$

8.119. $5, -3$

8.120. $6, -2$

8.121. -2

8.122. -2

8.123. $\dfrac{2}{3}$

8.124. 2

8.125. $-1, -2$

8.126. $-2, -3$

8.127. no solution

8.128. no solution

8.129. -2

8.130. -3

8.131. $-4, 3$

8.132. $-7, 8$

8.133. no solution

8.134. no solution

8.135. $w=\dfrac{A}{l}$

8.136. $A=\dfrac{F}{M}$

8.137. $x=\dfrac{3y-8}{2}$

8.138. $y=\dfrac{x}{1+x}$

8.139. $a=\dfrac{b}{cb-1}$

8.140. $y=\dfrac{3x}{6+x}$

8.141. 77

8.142. 104

8.143. 65

8.144. 24

8.145. 33

8.146. 14

8.147. 42

8.148. 6

8.149. 12 ml

8.150. 180 ml

8.151. 464 calories

8.152. 5 pieces

8.153. 590.4 Euros

8.154. $56,460$ yen

8.155. 8

8.156. 22.5

8.157. 150 miles

8.158. 350 miles

8.159. 40 feet

105. $\dfrac{(c+2)(c+2)}{(c-2)(c-3)}$

107. $-\dfrac{(m-2)(m-3)}{(3+m)(m+4)}$

109. $\dfrac{3s}{s+4}$

111. $\dfrac{4(p^2-pq+q^2)}{(p-q)(p^2+pq+q^2)}$

113. $\dfrac{t+3}{2t(t-3)}$

115. $\dfrac{y+3z}{2y(2y-1)}$

117. $\dfrac{2a-7}{5}$

119. $3(3c-5)$

121. $\dfrac{4(m+8)(m+7)}{3(m-4)(m+2)}$

123. $\dfrac{(4p+1)(p-7)}{3p(p+9)(p-1)}$

125. ⓐ $\dfrac{w(w-1)}{(w+m)(w+m-1)}$

ⓑ $\dfrac{2}{21}$

127. Answers will vary.

129. $\dfrac{3}{5}$

131. $\dfrac{3}{4}$

133. $\dfrac{3a+1}{a+b}$

135. $\dfrac{d+5}{d+8}$

137. $p+8$

139. $r+8$

141. $8t$

143. $\dfrac{2w}{w-4}$

145. $y-8$

147. $3a+7$

149. $c+2$

151. $\dfrac{m-2}{3}$

153. $\dfrac{p+3}{p+5}$

155. $\dfrac{r+9}{r+7}$

157. 4

159. $x+2$

161. $\dfrac{z+4}{z-5}$

163. $\dfrac{4b-3}{b-7}$

165. $\dfrac{32r+32}{r(r+4)}$

169. $(x-4)(x+2)(x+3)$

171. $(z-2)(z+4)(z+2)$

173. $(b+3)(b+3)(b-5)$

175. $(3d-1)(d+5)(d-6)$

177. $\dfrac{5x+15}{(x-4)(x+2)(x+3)}$,

$\dfrac{2x^2+4x}{(x-4)(x+2)(x+3)}$

179. $\dfrac{9z+18}{(z-2)(z+4)(z+2)}$,

$\dfrac{4z^2+16}{(z-2)(z+4)(z+2)}$

181. $\dfrac{4b-20}{(b+3)(b+3)(b-5)}$,

$\dfrac{2b^2+6b}{(b+3)(b+3)(b-5)}$

183. $\dfrac{2d-12}{(3d-1)(d+5)(d-6)}$,

$\dfrac{5d^2+25d}{(3d-1)(d+5)(d-6)}$

185. $\dfrac{37}{72}$

187. $\dfrac{49}{60}$

189. $\dfrac{21y+8x}{30x^2y^2}$

191. $\dfrac{mn+14}{16m^2n}$

193. $\dfrac{5r-7}{(r+4)(r-5)}$

195. $\dfrac{14t-10}{(t+5)(t-5)}$

197. $\dfrac{11w+1}{(3w-2)(w+1)}$

199. $\dfrac{2y^2+y+9}{(y+3)(y-1)}$

201. $\dfrac{b(5b+10+2a^2)}{a^2(b-2)(b+2)}$

203. $\dfrac{2m^2+23m}{3(m-1)(m+4)}$

205. $\dfrac{4p^2-9p+6}{(p+3)(p+6)(p+2)}$

207. $\dfrac{3(r^2+6r+18)}{(r+1)(r+6)(r+3)}$

209. $\dfrac{2(7t-6)}{(t-6)(t+6)}$

211. $\dfrac{-4(1+w)}{(w+5)(w-2)}$

213. $\dfrac{2(y+14)}{(y+1)(y+7)}$

215. $\dfrac{4a^2+25a-6}{(a+3)(a+6)}$

217. $\dfrac{3}{c-5}$

219. $\dfrac{-6}{m-6}$

221. $\dfrac{p+2}{p+3}$

223. $\dfrac{3}{r-2}$

225. $\dfrac{-v-14}{v+3}$

227. $\dfrac{4(8x+1)}{10x-1}$

229. $\dfrac{5a^2+7a-36}{a(a-2)}$

231. $\dfrac{c-5}{c+2}$

233. $\dfrac{3a(6a-b)}{b(3a+b)(3a-b)}$

235. $\dfrac{3}{d+8}$

237. $\dfrac{4m^2+14m-1}{(m+7)(m-1)(m+3)}$

239. $\dfrac{n+1}{n+3}$

241. $\dfrac{7q+5}{90pq}$

243. $\dfrac{7(x+2)}{(x-2)(x+5)}$

245. $\dfrac{17q+2}{3q-1}$

247. $\dfrac{8z+2}{(z-5)(z+1)}$

249. $\dfrac{3(d+1)}{d+2}$

251. $\dfrac{2t+3}{t(t+3)}$

$\dfrac{13}{40}$

253. Answers may vary.

255. $\dfrac{a-4}{2}$

257. $\dfrac{3}{2(c-2)}$

259. $\dfrac{24}{26}$

261. $\dfrac{20}{57}$

263. $\dfrac{n^2+m}{m-n^2}$

265. $\dfrac{rt}{t-r}$

267. $\dfrac{(x+1)(x-3)}{2}$

269. $\dfrac{4}{a+1}$

271. $\dfrac{11}{8}$

273. 19

275. $\dfrac{c^2+c}{c-d}$

277. $\dfrac{pq}{q-p}$

279. $\dfrac{2x-10}{3x+16}$

281. $\dfrac{3z-19}{3z+8}$

283. $\dfrac{4}{3a-2}$

285. $\dfrac{2c+29}{5c}$

287. $\dfrac{(2p-5)}{5}$

289. $\dfrac{m(m-5)}{4m^2+m-95}$

291. $\dfrac{13}{24}$

293. $2(a-4)$

295. $\dfrac{3mn}{n-m}$

297. $\dfrac{(x-1)(x-2)}{6}$

299. $\dfrac{R_1 R_2}{R_2+R_1}$

$\dfrac{24}{5}$

301. Answers will vary.

303. 10

305. $\dfrac{4}{7}$

307. $\dfrac{40}{21}$

309. -9

311. $-2, 4$

313. $-5, -4$

315. -6

317. 14

319. -2

321. $-\dfrac{1}{3}$

323. $-2, -1$

325. $-5, -1$

327. no solution

329. no solution

331. -4

333. -8

335. 2

337. 1

339. no solution

341. no solution

343. $r = \dfrac{C}{2\pi}$

345. $h = \dfrac{v}{lw}$

347. $w = 2v+7$

349. $c = \dfrac{b+3+2a}{a}$

351. $p = \dfrac{q}{4q-2}$

353. $w = \dfrac{15v}{10+v}$

355. $n = \dfrac{5m+23}{n}$

357. $y = \dfrac{20x}{12-x}$

359. $t = \dfrac{3r-s}{r}$

361. $2\dfrac{6}{11}$ days

363. Answers will vary.

365. 49

367. 7

369. 9

371. -11

373. 7

375. 0.6

377. $\dfrac{4}{7}$

379. $-\dfrac{5}{8}$

381. 60

383. 30

385. 9 ml

387. 325 mg

389. 159 calories

391. 5 oz

393. 252.5 Canadian dollars

395. 0.80 Euros

397. 0.63 British pounds

399. 3 cups

401. 1.30 Canadian dollars

403. 17 tsp

405. $\dfrac{3}{4}$ cup

407. 2

409. $\dfrac{77}{18}$

411. 950 miles

413. 680 miles

415. $\dfrac{2}{3}$ foot (8 in)

417. 247.3 feet

419. 114 beats per minute
no

421. 19 gallons
$58.71

423. 4 bags

425. 5

431. 29 mph

437. 4 mph

443. 50 mph

449. 2 hours and 44 minutes

455. 2 hours and 55 minutes

463. $y = \frac{14}{3}x$

469. $a = 18b$

475. (a) $p = 3.98g$ (b) \$59.70

481. (a) $A = 3.14r^2$ (b) 50.24 sq. in.

487. (a) $g = \frac{92,400}{w}$ (b) 16.8 mpg

493. (a) $t = \frac{125}{p}$ (b) 50 tickets

499. $p = 5q$

505. 15.6 mpg

511. Answers will vary.

427. Answers will vary.

433. 30 mph

439. 60 mph

445. 50 mph

451. 12 hours

457. 4 hours

465. $u = 3w$

471. (a) $P = 10n$ (b) \$40

477. (a) $m = 8v$ (b) 16 liters

483. $y = \frac{20}{x}$

489. (a) $t = \frac{1000}{r}$ (b) 2.5 hours

495. $y = \frac{5}{3}x$

501. 5 pounds

507. 267.3 feet

429. 160 mph

435. 20 mph

441. 650 mph

447. 2 hours

453. 7 hours and 30 minutes

459. 3 mph

467. $p = 3.2q$

473. (a) $a = 4.5p$ (b) 27 apples

479. (a) $d = 5t^2$ (b) 245 feet

485. $v = \frac{3}{w}$

491. (a) $L = \frac{4,400}{f}$ (b) 440 cycles per second

497. $p = \frac{30}{q}$

503. 3 inches

509. (a) $c = 2.5m$ (b) \$55

Review Exercises

513. $a \neq \frac{2}{3}$

519. $\frac{7}{2}$

525. $\frac{c+1}{c+2}$

531. $\frac{3}{b}$

537. $\frac{5}{c+4}$

543. $d + 7$

549. $\frac{3b-2}{b+7}$

555. $\frac{6n+12}{(n-2)(n-2)(n+2)}$, $\frac{2n^2-4n}{(n-2)(n-2)(n+2)}$

561. $\frac{5x^2+26x}{(x+4)(x+4)(x+6)}$

567. $\frac{4}{d-8}$

573. $\frac{(q-2)(q+4)}{5(q-4)}$

579. $\frac{3}{2}$

585. $z = \frac{y+5+7x}{x}$

591. 1161 calories

515. $y \neq 0$

521. $\frac{5}{12}$

527. $-\frac{7}{5+v}$

533. $\frac{6t}{(t+6)(t+6)}$

539. 1

545. $\frac{q-3}{q+5}$

551. $(m+2)(m-5)(m+4)$

557. $\frac{19}{15}$

563. $\frac{2(v^2+10v-2)}{(v+2)(v+8)}$

569. $\frac{a-2}{2a}$

575. $\frac{z-5}{23z+21}$

581. no solution

587. $\frac{12}{5}$

593. $b = 9; x = 2\frac{1}{3}$

517. $-\frac{5}{6}$

523. 14

529. $\frac{1}{20}$

535. $\frac{1}{11+w}$

541. $p + 5$

547. $\frac{15w+2}{6w-1}$

553. $(3p+1)(p+6)(p+8)$

559. $\frac{11c-12}{(c-2)(c+3)}$

565. $\frac{2m-7}{m+2}$

571. $\frac{(x-8)(x-5)}{2}$

577. $\frac{6}{7}$

583. $l = \frac{V}{hw}$

589. 15

595. 23 feet

597. 45 mph

599. 16 mph

601. $\frac{4}{5}$ hour

603. 12 days

605. 7

607. 36

609. $9

611. 97 tickets

Practice Test

613. $\frac{a}{2b}$

615. $\frac{x+3}{3x}$

617. $\frac{4+5q}{pq}$

619. $\frac{19}{16}$

621. $\frac{14}{11}$

623. $\frac{1}{2}$

625. 4

627. 60

629. 125 mg

631. 14 mph

633. $1\frac{5}{9}$ hours

635. $74.70

637. 64 miles

Chapter 9

Try It

9.1. ⓐ -7 ⓑ 15

9.2. ⓐ 8 ⓑ -11

9.3. ⓐ not a real number ⓑ -9

9.4. ⓐ -7 ⓑ not a real number

9.5. ⓐ 7 ⓑ 5

9.6. ⓐ 17 ⓑ 23

9.7. $6 < \sqrt{38} < 7$

9.8. $9 < \sqrt{84} < 10$

9.9. ≈ 3.32

9.10. ≈ 3.61

9.11. ⓐ y^4 ⓑ z^6

9.12. ⓐ m^2 ⓑ b^5

9.13. $8x$

9.14. $13y$

9.15. $-11y$

9.16. $-10p$

9.17. $10ab$

9.18. $15mn$

9.19. $7x^{15}$

9.20. $9w^{18}$

9.21. $13x^5 y^7$

9.22. $12p^6 q^{10}$

9.23. $4\sqrt{3}$

9.24. $3\sqrt{5}$

9.25. $12\sqrt{2}$

9.26. $12\sqrt{3}$

9.27. $b^2 \sqrt{b}$

9.28. $p^4 \sqrt{p}$

9.29. $4x^3 \sqrt{x}$

9.30. $7v^4 \sqrt{v}$

9.31. $4y^2 \sqrt{2y}$

9.32. $5a^4 \sqrt{3a}$

9.33. $7a^3 b^2 \sqrt{2ab}$

9.34. $6m^4 n^5 \sqrt{5mn}$

9.35. $5 + 5\sqrt{3}$

9.36. $2 + 7\sqrt{2}$

9.37. $2 - \sqrt{3}$

9.38. $2 - \sqrt{5}$

9.39. $\frac{5}{4}$

9.40. $\frac{7}{9}$

9.41. $\frac{5}{4}$

9.42. $\frac{7}{9}$

9.43. a

9.44. x^2

9.45. $5x^2$

9.46. $6z$

9.47. $\frac{\sqrt{19}}{7}$

9.48. $\frac{2\sqrt{7}}{9}$

9.49. $\frac{2p\sqrt{6p}}{7}$

9.50. $\frac{2x^2 \sqrt{3x}}{5}$

9.51. $\frac{4m\sqrt{5m}}{n^3}$

9.52. $\frac{3u^3 \sqrt{6u}}{v^4}$

9.53. $\frac{8x^2}{3}$

9.54. $\frac{2a^2}{5}$

9.55. $\frac{5y\sqrt{x}}{6}$

9.56. $\frac{4m\sqrt{3}}{5n^3 \sqrt{5n}}$

9.57. $-\sqrt{2}$

9.58. $-4\sqrt{3}$

9.59. $9\sqrt{x}$

9.60. $8\sqrt{u}$

9.61. $7\sqrt{p} - 6\sqrt{q}$

9.62. $6\sqrt{a} - 3\sqrt{b}$

9.63. $9\sqrt{11}$

9.64. $11\sqrt{10}$

9.65. $\sqrt{5} + 2\sqrt{6}$

9.66. $-5\sqrt{7} + 2\sqrt{5}$

9.67. $-2\sqrt{7x}$

9.68. $-\sqrt{3y}$

9.69. $-2\sqrt{5xy}$

9.70. 0

9.71. $9\sqrt{2}$

9.72. $7\sqrt{3}$

9.73. $\sqrt{2}$

9.74. $-\sqrt{5}$

9.75. $6\sqrt{3}$

9.76. $-15\sqrt{5}$

9.77. $-\sqrt{3}$

9.78. 0

9.79. $\frac{14}{15}\sqrt{2}$

9.80. $\frac{1}{12}\sqrt{5}$

9.81. $-m^3\sqrt{2m}$

9.82. $-p\sqrt{3p}$

9.83. $20x\sqrt{2} - 12x\sqrt{3}$

9.84. $28y\sqrt{3} - 24y\sqrt{2}$

9.85. $10x\sqrt{3}$

9.86. $-5x\sqrt{2}$

9.87. ⓐ $3\sqrt{2}$ ⓑ $36\sqrt{2}$

9.88. ⓐ $5\sqrt{2}$ ⓑ $90\sqrt{2}$

9.89. $12\sqrt{15}$

9.90. $27\sqrt{2}$

9.91. ⓐ $3x^2\sqrt{2}$ ⓑ $10y^2\sqrt{y}$

9.92. ⓐ $2x^3\sqrt{3}$ ⓑ $6y^2\sqrt{y}$

9.93. $144x^3\sqrt{10}$

9.94. $144y^2\sqrt{5y}$

9.95. ⓐ 12 ⓑ 15

9.96. ⓐ 16 ⓑ 20

9.97. ⓐ 330 ⓑ 200

9.98. ⓐ 210 ⓑ 96

9.99. ⓐ $6 - 2\sqrt{5}$ ⓑ $2\sqrt{3} - 3\sqrt{6}$

9.100. ⓐ $12 + 6\sqrt{6}$ ⓑ $\sqrt{7} + 7\sqrt{2}$

9.101. ⓐ $18 + \sqrt{6}$ ⓑ $6 + 12\sqrt{2}$

9.102. ⓐ $-40 + 4\sqrt{2}$ ⓑ $2\sqrt{7} + 14\sqrt{3}$

9.103. $-3 + 2\sqrt{6}$

9.104. $-2 + 2\sqrt{10}$

9.105. $-66 + 15\sqrt{7}$

9.106. $41 + 14\sqrt{11}$

9.107. $1 + 9\sqrt{21}$

9.108. $-12 - 20\sqrt{3}$

9.109. $12 - 8\sqrt{m} - 15m$

9.110. $10 - 47\sqrt{n} - 15n$

9.111. ⓐ $102 + 20\sqrt{2}$ ⓑ $55 + 6\sqrt{6}$

9.112. ⓐ $31 - 12\sqrt{5}$ ⓑ $121 - 36\sqrt{10}$

9.113. $4 + 20\sqrt{m} + 25m$

9.114. $9 - 24\sqrt{n} + 16n$

9.115. 1

9.116. -4

9.117. -11

9.118. -159

9.119. $\frac{\sqrt{2}}{2}$

9.120. $\frac{\sqrt{3}}{3}$

9.121. $\frac{4 - \sqrt{10}}{5}$

9.122. $\frac{5 - \sqrt{3}}{4}$

9.123. $\frac{2}{3}$

9.124. $\frac{4}{3}$

9.125. $n\sqrt{2}$

9.126. $p^2\sqrt{7}$

9.127. $\frac{5s}{8}$

9.128. $\frac{5q^2}{6}$

9.129. $\frac{9x^2}{y^2}$

9.130. $\frac{10n^3}{m}$

9.131. $\frac{5\sqrt{3}}{3}$

9.132. $\frac{6\sqrt{5}}{5}$

9.133. $\frac{\sqrt{5}}{2}$

9.134. $-\frac{3\sqrt{3}}{4}$

9.135. $\frac{\sqrt{14}}{6}$

9.136. $\frac{\sqrt{6}}{8}$

9.137. $\frac{1}{3}$

9.138. $\frac{\sqrt{5}}{5}$

9.139. $\frac{2(2 - \sqrt{3})}{1}$

9.140. $\frac{5(5 - \sqrt{3})}{22}$

9.141. $-\frac{3(1 + \sqrt{5})}{4}$

9.142. $\frac{4 + \sqrt{6}}{5}$

9.143. $\frac{\sqrt{5}(\sqrt{x} - \sqrt{2})}{x - 2}$

9.144. $\frac{\sqrt{10}(\sqrt{y} + \sqrt{3})}{y - 3}$

9.145. $\frac{(\sqrt{p} + \sqrt{2})^2}{p - 2}$

9.146. $\frac{(\sqrt{q} - \sqrt{10})^2}{q - 10}$

9.147. ⓐ no ⓑ yes

9.148. ⓐ no ⓑ yes

9.149. 10

9.150. 7

9.151. $\frac{23}{3}$

9.152. $\frac{3}{10}$

9.153. $\frac{1}{2}$

9.154. $\frac{3}{5}$

9.155. no solution

9.156. no solution

9.157. $2, 3$

9.158. $5, 6$

9.159. 7

9.160. 3

9.161. $\frac{127}{2}$

9.162. $\frac{311}{3}$

9.163. no solution

9.164. no solution

9.165. no solution

9.166. no solution   **9.167.** no solution   **9.168.** no solution

9.169. 19.2 yards   **9.170.** 52.0 cm   **9.171.** 9 seconds

9.172. 3.5 seconds   **9.173.** 42.7 feet   **9.174.** 54.1 feet

9.175. ⓐ 3 ⓑ 4 ⓒ 3   **9.176.** ⓐ 10 ⓑ 2 ⓒ 2   **9.177.** ⓐ -5 ⓑ not real ⓒ -2

9.178. ⓐ -6 ⓑ not real ⓒ -4 **9.179.** ⓐ $|b|$ ⓑ w ⓒ $|m|$ ⓓ q **9.180.** ⓐ $|y|$ ⓑ p ⓒ $|z|$ ⓓ q

9.181. ⓐ u^3 ⓑ v^5 **9.182.** ⓐ c^4 ⓑ d^4 **9.183.** ⓐ $3x^9$ ⓑ $3|q^7|$

9.184. ⓐ $5p^3$ ⓑ $3q^5$ **9.185.** ⓐ $|y|\sqrt[4]{y^2}$ ⓑ $z\sqrt[3]{z^2}$ **9.186.** ⓐ $p\sqrt[5]{p^3}$ ⓑ $q^2\sqrt[6]{q}$

9.187. ⓐ $3\sqrt[3]{4}$ ⓑ $2\sqrt[4]{4}$ **9.188.** ⓐ $5\sqrt[3]{5}$ ⓑ $3\sqrt[4]{9}$ **9.189.** ⓐ $3p^3\sqrt[3]{2p}$ ⓑ $2q^2\sqrt[4]{4q^2}$

9.190. ⓐ $4m^3\sqrt[3]{2m^2}$ ⓑ $3|n|\sqrt[4]{2n^3}$ **9.191.** ⓐ $-3\sqrt[3]{4}$ ⓑ not real   **9.192.** ⓐ $-5\sqrt[3]{5}$ ⓑ not real

9.193. ⓐ $|x|$ ⓑ y^3 **9.194.** ⓐ m^2 ⓑ n^2 **9.195.** ⓐ not real ⓑ $3|m|\sqrt[4]{2m^2}$

9.196. ⓐ -4 ⓑ $3|n|\sqrt[4]{2}$ **9.197.** ⓐ $\dfrac{3c^3\sqrt[3]{4c}}{d^2}$ ⓑ $\dfrac{x^2}{|y|}\sqrt[4]{\dfrac{80x^2}{y}}$ **9.198.** ⓐ $r\sqrt[3]{\dfrac{40}{s}}$ ⓑ $\dfrac{3m^3\sqrt[4]{2m^2}}{|n^3|}$

9.199. ⓐ $2\sqrt[5]{3x}$ ⓑ $2\sqrt[3]{9}$ **9.200.** ⓐ $2\sqrt[4]{10y}$ ⓑ $2\sqrt[6]{32}$ **9.201.** ⓐ $\sqrt[3]{3}$ ⓑ $6\sqrt[4]{2}$

9.202. ⓐ $-\sqrt[3]{2}$ ⓑ $5\sqrt[5]{2}$ **9.203.** ⓐ $2y\sqrt[3]{4y^2}+3y^2\sqrt[3]{4y^2}$ ⓑ $3r^2\sqrt[4]{3r^3}+4r^2\sqrt[4]{3r^2}$ **9.204.** ⓐ $2z^2\sqrt[3]{5z}+3z\sqrt[3]{5z}$ ⓑ $2|s^3|\sqrt[4]{5s}+4|s|\sqrt[4]{5s}$

9.205. ⓐ $\sqrt{t}$ ⓑ $\sqrt[3]{m}$ ⓒ $\sqrt[4]{r}$ **9.206.** ⓐ $\sqrt{b}$ ⓑ $\sqrt[3]{z}$ ⓒ $\sqrt[4]{p}$ **9.207.** ⓐ $s^{\frac{1}{2}}$ ⓑ $x^{\frac{1}{3}}$ ⓒ $b^{\frac{1}{4}}$

9.208. ⓐ $v^{\frac{1}{2}}$ ⓑ $p^{\frac{1}{3}}$ ⓒ $p^{\frac{1}{4}}$ **9.209.** ⓐ $(10m)^{\frac{1}{2}}$ ⓑ $(3n)^{\frac{1}{5}}$ ⓒ $(486y)^{\frac{1}{4}}$ **9.210.** ⓐ $(3k)^{\frac{1}{7}}$ ⓑ $(5j)^{\frac{1}{4}}$ ⓒ $(1024a)^{\frac{1}{3}}$

9.211. ⓐ 6 ⓑ 2 ⓒ 2   **9.212.** ⓐ 10 ⓑ 3 ⓒ 3   **9.213.** ⓐ -5 ⓑ -5 ⓒ $\dfrac{1}{5}$

9.214. ⓐ -2 ⓑ -2 ⓒ $\dfrac{1}{2}$ **9.215.** ⓐ -8 ⓑ -8 ⓒ $\dfrac{1}{8}$ **9.216.** ⓐ -4 ⓑ -4 ⓒ $\dfrac{1}{4}$

9.217. ⓐ $x^{\frac{5}{2}}$ ⓑ $z^{\frac{3}{4}}$ ⓒ $y^{\frac{2}{5}}$ **9.218.** ⓐ $a^{\frac{2}{5}}$ ⓑ $b^{\frac{7}{3}}$ ⓒ $m^{\frac{5}{4}}$ **9.219.** ⓐ 8 ⓑ 9 ⓒ 125

9.220. ⓐ 32 ⓑ 729 ⓒ 8   **9.221.** ⓐ $\dfrac{1}{32}$ ⓑ $\dfrac{1}{729}$ ⓒ $\dfrac{1}{8}$ **9.222.** ⓐ $\dfrac{1}{8}$ ⓑ $\dfrac{1}{9}$ ⓒ $\dfrac{1}{125}$

9.223. ⓐ -64 ⓑ $-\dfrac{1}{64}$ ⓒ not a real number   **9.224.** ⓐ -729 ⓑ $-\dfrac{1}{729}$ ⓒ not a real number   **9.225.** ⓐ 9 ⓑ y^3 ⓒ m

9.226. ⓐ 25 ⓑ z ⓒ n **9.227.** ⓐ p^2 ⓑ q^6 ⓒ x^8 **9.228.** ⓐ r^{10} ⓑ s^9 ⓒ m^2

9.229. ⓐ u ⓑ $\dfrac{1}{v^5}$ ⓒ $\dfrac{1}{x}$ **9.230.** ⓐ c^6 ⓑ $\dfrac{1}{m}$ ⓒ $\dfrac{1}{d}$ **9.231.** ⓐ $8x^{\frac{1}{5}}$ ⓑ $4y^{\frac{2}{9}}$

9.232. ⓐ $64m^{\frac{1}{2}}$ ⓑ $729n^{\frac{3}{5}}$ **9.233.** ⓐ m^2 ⓑ n^3 **9.234.** ⓐ u^3 ⓑ v^5

Section Exercises

1. 6   **3.** 8   **5.** 3

7. 10   **9.** -2 **11.** -1

13. not a real number   **15.** not a real number   **17.** 5

19. 7   **21.** $8 < \sqrt{70} < 9$ **23.** $14 < \sqrt{200} < 15$

25. 4.36   **27.** 7.28   **29.** y

31. a^7 **33.** $7x$ **35.** $11m^{10}$

37. $9x^{18}$ **39.** $-9x^9$ **41.** $-8a$

43. $12xy$

45. $13w^4 y^5$

47. $3c^4 d^6$

49. 25 centimeters

51. Answers will vary.

53. $3\sqrt{3}$

55. $5\sqrt{5}$

57. $10\sqrt{2}$

59. $15\sqrt{2}$

61. $20\sqrt{2}$

63. $15\sqrt{3}$

65. $x^3 \sqrt{x}$

67. $p\sqrt{p}$

69. $m^6 \sqrt{m}$

71. $r^{12} \sqrt{r}$

73. $7n^8 \sqrt{n}$

75. $9r^7 \sqrt{r}$

77. $7m^2 \sqrt{2m}$

79. $5r^6 \sqrt{5r}$

81. $10p^6 \sqrt{2p}$

83. $11m^{11} \sqrt{2m}$

85. $7m^3 n^5 \sqrt{3mn}$

87. $5r^6 s^4 \sqrt{3rs}$ 70)

89. $10p^4 q^5 \sqrt{3pq}$

91. $11m^6 n^{10} \sqrt{2mn}$

93. $5 + 2\sqrt{3}$

95. $1 + 3\sqrt{5}$

97. $5 - 2\sqrt{6}$

99. $1 + \sqrt{10}$

101. $\frac{7}{8}$

103. $\frac{11}{4}$

105. $\frac{6}{7}$

107. $\frac{3}{5}$

109. x^2

111. $\frac{1}{y^2}$

113. $10x^2$

115. $4p^4$

117. $\frac{6}{\sqrt{35}}$

119. $\frac{2\sqrt{5}}{9}$

121. $\frac{4x^3 \sqrt{6x}}{11}$

123. $\frac{10m^2 \sqrt{3m}}{8}$

125. $\frac{7r^2 \sqrt{2r}}{10}$

127. $\frac{2q^3 \sqrt{7}}{15}$

129. $\frac{5r^4 \sqrt{3r}}{s^4}$

131. $\frac{4p^3 \sqrt{7p}}{q}$

133. $\frac{5x}{3}$

135. $\frac{11p\sqrt{p}}{9}$

137. $\frac{4xy}{3}$

139. $\frac{1}{2pq\sqrt{p}}$

141. ⓐ 17.0 feet ⓑ 15.0 feet

143. Answers will vary.

145. $3\sqrt{2}$

147. $9\sqrt{5}$

149. $-\sqrt{7}$

151. $9\sqrt{y}$

153. $-3\sqrt{a}$

155. $7\sqrt{c}$

157. $6\sqrt{b}$

159. $5\sqrt{m} + \sqrt{n}$

161. $13\sqrt{7}$

163. $-3\sqrt{11}$

165. $-5\sqrt{3} + 7\sqrt{5}$

167. $8\sqrt{2} - 3\sqrt{5}$

169. $4\sqrt{2a}$

171. $\sqrt{3c}$

173. $4\sqrt{3ab}$

175. $\sqrt{pq}$

177. $9\sqrt{2}$

179. $\sqrt{5}$

181. $-2\sqrt{3}$

183. $7\sqrt{3}$

185. $-8\sqrt{2}$

187. $16\sqrt{3}$

189. $3\sqrt{2}$

191. $-\sqrt{5}$

193. $-\sqrt{3}$

195. $-\frac{3}{4}\sqrt{2}$

197. $-a^2 \sqrt{2a}$

199. $2c^3 \sqrt{5c}$

201. $36p^2 \sqrt{5} - 42p^2 \sqrt{2}$

203. $10r^4 \sqrt{2} + 12r^4 \sqrt{6}$

205. $14x\sqrt{5}$

207. $-12y\sqrt{2}$

209. $3\sqrt{8}$

211. $-2k^2 \sqrt{7}$

213. $2\sqrt{3}$

215. $\sqrt{2}$

217. $\sqrt{13}$

219. $a^2 \sqrt{5a}$

221. $19\sqrt{19}$

223. $5\sqrt{3}$

225. $-\sqrt{3}$

227. $8x\sqrt{6}$

229. 124.5 inches

231. Answers will vary.

233. ⓐ 4 ⓑ 36

235. ⓐ $7\sqrt{2}$ ⓑ 160

237. $30\sqrt{3}$

239. $-18\sqrt{6}$

241. $-30\sqrt{2}$

243. $28\sqrt{2}$

245. ⓐ $5y^2 \sqrt{3}$ ⓑ $6n^2 \sqrt{n}$

247. ⓐ $8y^3 \sqrt{2}$ ⓑ $11s^3 \sqrt{s}$

249. $40b^2 \sqrt{3}$

251. $144d^4$

253. $54y^4 \sqrt{y}$

255. $-96k^5 \sqrt{k}$

257. ⓐ 11 ⓑ 21

259. ⓐ 23 ⓑ 3

261. ⓐ -234 ⓑ 180

263. ⓐ $105\sqrt{2}$ ⓑ 56

265. ⓐ $24 - 4\sqrt{11}$ ⓑ $5\sqrt{2} - 4\sqrt{6}$ **267.** ⓐ $-14 - 7\sqrt{11}$ ⓑ $6\sqrt{7} - 7\sqrt{2}$ **269.** ⓐ $44 + 8\sqrt{11}$ ⓑ 15

271. ⓐ $18 - 5\sqrt{2}$ ⓑ $\sqrt{21} - 7\sqrt{3}$ **273.** $60 + 2\sqrt{3}$ **275.** $52 + 3\sqrt{2}$

277. $27 - 9\sqrt{7}$ **279.** $-62 + 55\sqrt{5}$ **281.** $161 + 7\sqrt{55}$

283. $-81 + 44\sqrt{78}$ **285.** $18 + 7\sqrt{w} - w$ **287.** $66 + 73\sqrt{n} + 15n$

289. ⓐ $27 + 8\sqrt{11}$ ⓑ $29 - 12\sqrt{5}$ **291.** ⓐ $35 - 10\sqrt{10}$ ⓑ $82 + 48\sqrt{2}$ **293.** 97

295. 39 **297.** -127 **299.** 33

301. $-24\sqrt{3}$ **303.** $-210\sqrt{3}$ **305.** $7y^3\sqrt{5}$

307. 29 **309.** $29 - 7\sqrt{17}$ **311.** ⓐ 4.6 feet ⓑ 21 sq. feet

315. ⓐ when squaring a negative, it becomes a positive ⓑ since the negative is not included in the parenthesis, it is not squared, and remains negative   **317.** $\dfrac{\sqrt{3}}{2}$ **319.** $\dfrac{2\sqrt{2}}{3}$

321. $\dfrac{1 - 2\sqrt{2}}{4}$ **323.** $\dfrac{2 + \sqrt{5}}{2}$ **325.** $\dfrac{4}{5}$

327. $\dfrac{4}{3}$ **329.** ⓐ $2x^2$ ⓑ $\dfrac{10m^2}{7}$ **331.** $\dfrac{5r}{6}$

333. $\dfrac{6p}{q^2}$ **335.** $\dfrac{8n}{3m^3}$ **337.** $\dfrac{\sqrt{2}}{2}$

339. $\dfrac{1 + \sqrt{3}}{3}$ **341.** $\dfrac{4}{5}$ **343.** $y^3\sqrt{13}$

345. $\dfrac{5\sqrt{6}}{3}$ **347.** $\dfrac{6\sqrt{7}}{7}$ **349.** $\dfrac{3\sqrt{13}}{13}$

351. $\dfrac{\sqrt{10}}{3}$ **353.** $\dfrac{4\sqrt{5}}{45}$ **355.** $-\dfrac{3\sqrt{3}}{2}$

357. $\dfrac{\sqrt{15}}{10}$ **359.** $\dfrac{\sqrt{70}}{20}$ **361.** $\dfrac{\sqrt{133}}{35}$

363. ⓐ $\dfrac{3(3 - \sqrt{11})}{-2}$ ⓑ $-2(1 + \sqrt{5})$ **365.** ⓐ $\dfrac{5(5 - \sqrt{6})}{19}$ ⓑ $3(3 + \sqrt{7})$ **367.** $\dfrac{\sqrt{3}(\sqrt{m} + \sqrt{5})}{m - 5}$

369. $\dfrac{\sqrt{2}(\sqrt{x} + \sqrt{3})}{x - 6}$ **371.** $\dfrac{(\sqrt{r} + \sqrt{5})^2}{r - 5}$ **373.** $\dfrac{5y^2}{x}$

375. $3\sqrt{5}$ **377.** $\dfrac{2\sqrt{3}}{9}$ **379.** $\dfrac{3(5 - \sqrt{5})}{20}$

381. $\dfrac{\sqrt{2}(\sqrt{x} + \sqrt{3})}{x - 3}$ **383.** $\dfrac{(\sqrt{x} + 2\sqrt{2})^2}{x - 8}$ **385.** $\dfrac{5\sqrt{10}}{4}$ seconds

387. Answers will vary.   **389.** ⓐ yes ⓑ no   **391.** ⓐ no ⓑ yes

393. 3 **395.** 14 **397.** 14

399. 17 **401.** $\dfrac{13}{5}$ **403.** $\dfrac{5}{2}$

405. no solution   **407.** no solution   **409.** ⓐ $10, 11$ ⓑ 5

411. ⓐ $8, 9$ ⓑ 11 **413.** 3 **415.** 5

417. not a real number   **419.** $\dfrac{1}{4}$ **421.** ⓐ no solution ⓑ $\dfrac{57}{16}$

423. ⓐ no solution ⓑ 6 **425.** no solution   **427.** 36

429. 15 **431.** 11.4 feet   **433.** 8 seconds

435. 15.8 seconds

437. 72 feet

439. 53.0 feet

441. Answers will vary.

443. ⓐ 3 ⓑ 2 ⓒ 3

445. ⓐ 5 ⓑ 6 ⓒ 4

447. ⓐ -4 ⓑ not real ⓒ -3

449. ⓐ -8 ⓑ not a real number ⓒ -1

451. ⓐ a ⓑ $|b|$

453. ⓐ $|k|$ ⓑ $|p|$

455. ⓐ a^2 ⓑ b^9

457. ⓐ r^2 ⓑ s^{10}

459. ⓐ $-2c^3$ ⓑ $5d^5$

461. ⓐ $2r^2$ ⓑ $3s^6$

463. ⓐ $u\sqrt[5]{u^2}$ ⓑ $v\sqrt[6]{v^5}$

465. ⓐ $p\sqrt[5]{p^3}$ ⓑ $q^2\sqrt[3]{q^2}$

467. ⓐ $5\sqrt[3]{5}$ ⓑ $2\sqrt[6]{2}$

469. ⓐ $5\sqrt[4]{5}$ ⓑ $3\sqrt[3]{3}$

471. ⓐ $2a\sqrt[5]{3a^2}$ ⓑ $5b\sqrt[3]{3b}$

473. ⓐ $8p\sqrt[3]{p^2}$ ⓑ $3q\sqrt[4]{4q^3}$

475. ⓐ $-3\sqrt[5]{2}$ ⓑ not real

477. ⓐ -2 ⓑ not real

479. ⓐ d ⓑ $|m|$

481. ⓐ r^2 ⓑ $|c^3|$

483. ⓐ -5 ⓑ $4m\sqrt[4]{m^2}$

485. ⓐ $3\sqrt[3]{6}$ ⓑ $2|r|\sqrt[4]{2r^3}$

487. ⓐ $\dfrac{2r^2\sqrt[3]{3r}}{s^3}$ ⓑ $\dfrac{2u^3\sqrt[6]{2uv3}}{v}$

489. ⓐ $\dfrac{5u^3\sqrt[3]{5u}}{v}$ ⓑ $\dfrac{3c^5\sqrt[4]{9c}}{d^2}$

491. ⓐ $2\sqrt[3]{15q}$ ⓑ $-4\sqrt[4]{27}$

493. ⓐ $42\sqrt[3]{4y}$ ⓑ $14\sqrt[4]{5z}$

495. ⓐ $5\sqrt[3]{5}-3\sqrt[3]{2}$ ⓑ $-2\sqrt[4]{3}$

497. ⓐ $3\sqrt[4]{3}+5\sqrt[4]{2}$ ⓑ $13\sqrt[3]{2}$

499. ⓐ $2b\sqrt[3]{10b^2}+3b\sqrt[3]{10}$ ⓑ $2v^2\sqrt[4]{10v^2}-4\sqrt[4]{5v^3}$

501. 2

503. $|b|$

505. $5d^5$

507. $s^2\sqrt[4]{s^2}$

509. $2y\sqrt[4]{3y^2}$

511. not real

513. $2x\sqrt[5]{2x}$

515. $\dfrac{2u^3\sqrt[6]{2uv3}}{v}$

517. $4\sqrt[4]{2}$

519. $3u\sqrt[4]{6u^3}+4\sqrt[4]{3u^3}$

521. 5

523. Answers may vary.

525. ⓐ $\sqrt{r}$ ⓑ $\sqrt[3]{s}$ ⓒ $\sqrt[4]{t}$

527. ⓐ $\sqrt[7]{g}$ ⓑ $\sqrt[5]{h}$ ⓒ $\sqrt[25]{j}$

529. ⓐ $r^{\frac{1}{8}}$ ⓑ $s^{\frac{1}{10}}$ ⓒ $t^{\frac{1}{4}}$

531. ⓐ $u^{\frac{1}{5}}$ ⓑ $v^{\frac{1}{2}}$ ⓒ $w^{\frac{1}{16}}$

533. ⓐ $(5x)^{\frac{1}{4}}$ ⓑ $(9y)^{\frac{1}{8}}$ ⓒ $7(3z)^{\frac{1}{5}}$

535. ⓐ $(25a)^{\frac{1}{3}}$ ⓑ $(3b)^{\frac{1}{2}}$ ⓒ $(40c)^{\frac{1}{10}}$

537. ⓐ 5 ⓑ 3 ⓒ 2

539. ⓐ 6 ⓑ 2 ⓒ 3

541. ⓐ -3 ⓑ -3 ⓒ $\dfrac{1}{3}$

543. ⓐ -10 ⓑ -10 ⓒ $\dfrac{1}{10}$

545. ⓐ not a real number ⓑ -7 ⓒ $\dfrac{1}{7}$

547. ⓐ not a real number ⓑ 1 ⓒ -1

549. ⓐ -2 ⓑ $\dfrac{1}{3}$ ⓒ -5

551. ⓐ $r^{\frac{7}{4}}$ ⓑ $s^{\frac{3}{5}}$ ⓒ $t^{\frac{7}{3}}$

553. ⓐ $a^{\frac{1}{3}}$ ⓑ $b^{\frac{1}{5}}$ ⓒ $c^{\frac{5}{3}}$

555. ⓐ 100 ⓑ 125 ⓒ 8

557. ⓐ 64 ⓑ 3125 ⓒ 256

559. ⓐ 32,768 ⓑ $\dfrac{1}{729}$ ⓒ $\dfrac{1}{81}$

561. ⓐ 1000 ⓑ $\dfrac{1}{16,807}$ ⓒ not a real number

563. ⓐ -512 ⓑ $-\dfrac{1}{512}$ ⓒ not a real number

565. ⓐ -343 ⓑ $-\dfrac{1}{343}$ ⓒ not a real number

567. ⓐ 216 ⓑ n ⓒ q^3

569. ⓐ 100 ⓑ x^2 ⓒ y^4

571. ⓐ a^2 ⓑ b^9 ⓒ c

573. ⓐ h^8 ⓑ k^9 ⓒ j^{14}

575. ⓐ s ⓑ z^2 ⓒ $\dfrac{1}{w}$

577. ⓐ u ⓑ r ⓒ $\dfrac{1}{n}$

579. ⓐ $3r^{\frac{1}{5}}$ ⓑ $2s^{\frac{1}{14}}$

581. ⓐ $9m^{\frac{1}{2}}$ ⓑ $125n^2$

583. ⓐ r^2s ⓑ u^3v^4

585. ⓐ $r^8 s^5$ ⓑ $u^8 v^4$

587. ⓐ a^3 ⓑ b^4

589. ⓐ m ⓑ n^2

591. n

593. b^6

595. $\dfrac{1}{z^2}$

597. $2s^{\frac{1}{10}}$

599. $u^2 v^3$

601. 15.6 feet

603. 8 seconds

Review Exercises

607. 12

609. -9

611. not a real number

613. 17

615. $12 < \sqrt{155} < 13$

617. 7.55

619. $8b$

621. $15mn$

623. $7y$

625. $11cd$

627. $7\sqrt{2}$

629. $y^9 \sqrt{y}$

631. $6n^6 \sqrt{n}$

633. $5n^3 \sqrt{6n}$

635. $6r^2 s\sqrt{3rs}$

637. $1 + \sqrt{2}$

639. $\dfrac{3}{2}$

641. y^2

643. $6q^2$

645. $\dfrac{\sqrt{26}}{13}$

647. $\dfrac{3r^2 \sqrt{r}}{2}$

649. $\dfrac{2rs^3 \sqrt{r}}{5}$

651. $12\sqrt{5}$

653. $4\sqrt{m}$

655. $11\sqrt{13} + 2\sqrt{3}$

657. $3\sqrt{3rs} - 5\sqrt{rs}$

659. $5\sqrt{2}$

661. $9\sqrt{3}$

663. 0

665. $17n^2 \sqrt{2}$

667. $24\sqrt{7}$

669. $180y^2$

671. 8

673. 50

675. $20 - 10\sqrt{7}$

677. $13 - 2\sqrt{2}$

679. $5 - 13\sqrt{x} - 6x$

681. $1 + 12\sqrt{p} + 36p$

683. -19

685. $\dfrac{\sqrt{3}}{2}$

687. $\dfrac{4}{3}$

689. $y^2 \sqrt{10}$

691. $\dfrac{2\sqrt{15}}{3}$

693. $\dfrac{\sqrt{5}}{3}$

695. $\dfrac{\sqrt{21}}{14}$

697. $\dfrac{16 - 12\sqrt{3}}{-11}$

699. $-8 - 4\sqrt{5}$

701. $\dfrac{\sqrt{2p} - \sqrt{6}}{p - 3}$

703. 5

705. $\dfrac{7}{2}$

707. no solution

709. 13

711. $\dfrac{43}{2}$

713. 0

715. $\dfrac{1}{16}$

717. 11

719. 7.5 seconds

721. ⓐ 2 ⓑ 4

723. ⓐ d ⓑ $|v|$

725. ⓐ $2x^2$ ⓑ $2y^2$

727. ⓐ d ⓑ $m\sqrt[11]{m^6}$

729. ⓐ $2c\sqrt[5]{2c^3}$ ⓑ $2d\sqrt[4]{3d^3}$

731. ⓐ $-5\sqrt[3]{4}$ ⓑ not a real number

733. $w^3 \sqrt[3]{w}$

735. $3z^2$

737. $2\sqrt[5]{20}$

739. $2\sqrt[4]{2}$

741. $2t\sqrt[5]{3t^3} + 3\sqrt[5]{2t^4}$

743. $\sqrt[10]{s}$

745. $v^{\frac{1}{6}}$

747. $(10z)^{\frac{1}{6}}$

749. 2

751. $\dfrac{1}{5}$

753. $\dfrac{1}{6}$

755. $n^{\frac{8}{5}}$

757. 32,768

759. $\dfrac{1}{59,049}$

761. x^8

763. $2s^{\frac{9}{16}}$

765. z^2

Practice Test

767. $13m^2 |n|$

769. $4\sqrt{13} + 5\sqrt{2}$

771. $180y^2 \sqrt{3}$

773. $1 - 4\sqrt{q} + 4q$

775. ⓐ $3x^3$ ⓑ $2y^3$

777. $y\sqrt{2}$

779. 0

781. 343

783. $\frac{1}{w}$

785. $\frac{\sqrt{6}}{4}$

787. 42

789. 6.1 seconds

Chapter 10

Try It

10.1. $x = 9, x = -9$

10.2. $y = 11, y = -11$

10.3. $x = 5\sqrt{2}, x = -5\sqrt{2}$

10.4. $y = 3\sqrt{3}, y = -3\sqrt{3}$

10.5. $x = 7, x = -7$

10.6. $z = 6, z = -6$

10.7. no real solution

10.8. no real solution

10.9. $x = 2\sqrt{10}, x = -2\sqrt{10}$

10.10. $y = 2\sqrt{7}, y = -2\sqrt{7}$

10.11. $r = \frac{6\sqrt{5}}{5}, r = -\frac{6\sqrt{5}}{5}$

10.12. $t = \frac{8\sqrt{3}}{3}, t = -\frac{8\sqrt{3}}{3}$

10.13. $q = -6, q = -4$

10.14. $r = 8, r = -2$

10.15. $a = 3 + 3\sqrt{2}, a = 3 - 3\sqrt{2}$

10.16. $b = -2 + 2\sqrt{10}, b = -2 - 2\sqrt{10}$

10.17. $x = \frac{1}{3} + \frac{\sqrt{5}}{3}, x = \frac{1}{3} - \frac{\sqrt{5}}{3}$

10.18. $y = \frac{3}{4} + \frac{\sqrt{7}}{4}, y = \frac{3}{4} - \frac{\sqrt{7}}{4}$

10.19. $a = 5 + 2\sqrt{5}, a = 5 - 2\sqrt{5}$

10.20. $b = 3 + 4\sqrt{2}, b = 3 - 4\sqrt{2}$

10.21. no real solution

10.22. no real solution

10.23. $x = 3 + 2\sqrt{3}, x = 3 - 2\sqrt{3}$

10.24. $y = -6 + 4\sqrt{2}, y = -6 - 4\sqrt{2}$

10.25. $m = 7, m = -3$

10.26. $n = -\frac{3}{4}, n = -\frac{7}{4}$

10.27. $(y + 6)^2$

10.28. $(z + 4)^2$

10.29. $(a - 10)^2$

10.30. $(b - 2)^2$

10.31. $\left(m - \frac{5}{2}\right)^2$

10.32. $\left(n + \frac{13}{2}\right)^2$

10.33. $\left(p + \frac{1}{8}\right)^2$

10.34. $\left(q - \frac{1}{3}\right)^2$

10.35. $c = -5, c = 1$

10.36. $d = -9, d = -1$

10.37. $r = -2, r = 6$

10.38. $t = -1, t = 11$

10.39. no real solution

10.40. no real solution

10.41. $x = 8 \pm 4\sqrt{3}$

10.42. $y = -4 \pm 3\sqrt{3}$

10.43. $a = -7, a = 3$

10.44. $b = -10, b = -2$

10.45. $p = \frac{5}{2} \pm \frac{\sqrt{61}}{2}$

10.46. $q = \frac{7}{2} \pm \frac{\sqrt{37}}{2}$

10.47. $c = -3 \pm 4\sqrt{2}$

10.48. $d = -7, d = 11$

10.49. $m = -4 \pm 2\sqrt{5}$

10.50. $n = -2, 8$

10.51. $r = -\frac{7}{3}, r = 3$

10.52. $t = -\frac{5}{2}, t = 2$

10.53. $x = -\frac{3}{8} \pm \frac{\sqrt{201}}{8}$

10.54. $y = -\frac{3}{10} \pm \frac{\sqrt{209}}{10}$

10.55. $y = \frac{2}{3}, y = 1$

10.56. $z = -\frac{3}{2}, z = 1$

10.57. $a = -3, a = 5$

10.58. $b = -6, b = -4$

10.59. $p = \frac{-4 \pm \sqrt{6}}{2}$

10.60. $q = \frac{11 \pm \sqrt{61}}{10}$

10.61. $m = \frac{-6 \pm \sqrt{15}}{3}$

10.62. $n = \frac{-2 \pm 2\sqrt{6}}{5}$

10.63. no real solution

10.64. no real solution

10.65. $x = -1 \pm \sqrt{6}$

10.66. $y = -\frac{2}{3}, y = 1$

10.67. $c = \frac{2 \pm \sqrt{7}}{3}$

10.68. $d = \frac{2}{3}, d = 0$

10.69. $r = -5$

10.70. $t = \frac{4}{5}$

10.71. ⓐ no real solutions ⓑ 2 ⓒ 1 ⓓ no real solutions

10.72. ⓐ 2 ⓑ no real solutions ⓒ 1 ⓓ 2

10.73. ⓐ factor ⓑ Square Root Property ⓒ Quadratic Formula

10.74. ⓐ Quadratic Formula ⓑ factoring ⓒ Square Root Property

10.75. Two consecutive odd numbers whose product is 99 are 9 and 11, and −9 and −11.

10.76. Two consecutive even numbers whose product is 168 are 12 and 14, and −12 and −14.

10.77. The height of the triangle is 8 inches and the width is 52 inches.

10.78. The height of the triangle is 20 feet and the width is 11 feet.

10.79. The length of the shadow is 6.3 feet and the length of the flag pole is 18.9 ft.

10.80. The distance to the opposite corner is 3.2.

10.81. The width of the garden is 11 feet and the length is 18 feet.

10.82. The width of the tablecloth is 6.8 feet and the length is 11.8 feet.

10.83. The arrow will reach 180 on its way up in 3 seconds, and on the way down in 3.8 seconds.

10.84. The ball will reach 48 feet on its way up in .6 seconds and on the way down in 5.5 seconds.

10.85.

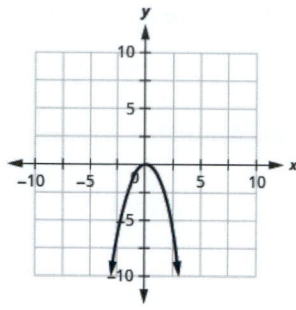

10.86.

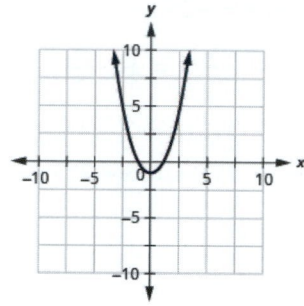

10.87. ⓐ up ⓑ down

10.88. ⓐ down ⓑ up

10.89. ⓐ $x = 2$ ⓑ $(2, -7)$

10.90. ⓐ $x = 1$ ⓑ $(1, -5)$

10.91.
y: $(0, -8)$; x: $(-4, 0)$, $(2, 0)$

10.92.
y: $(0, -12)$; x: $(6, 0)$, $(-2, 0)$

10.93. y: $(0, 4)$; x: none

10.94.
y: $(0, -5)$; x: $(5, 0)$ $(-1, 0)$

10.95. y: $(0, -36)$; x: $(-6, 0)$

10.96. y: $(0, 4)$; x: $\left(-\frac{2}{3}, 0\right)$

10.97. y: $(0, -8)$;
x: $(2, 0)$, $(-4, 0)$;
axis: $x = -1$; vertex: $(-1, -9)$;

10.98. y: $(0, 12)$; x: $(2, 0)$, $(6, 0)$;
axis: $x = 4$; vertex: $(4, -4)$;

10.99. y: $(0, -12)$; x: $(2, 0)$;
axis: $x = 2$; vertex: $(2, 0)$;

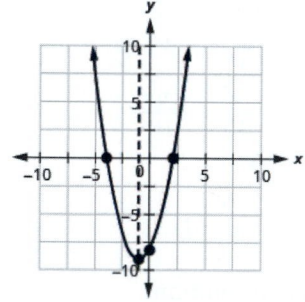

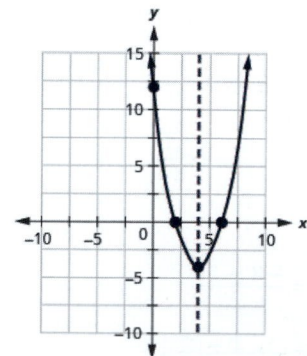

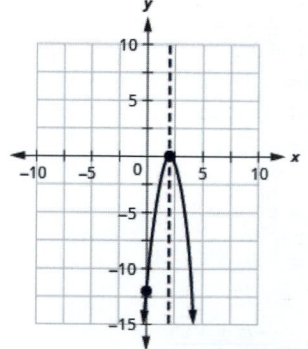

10.100. y: $(0, 1)$; x: $\left(-\frac{1}{5}, 0\right)$;

axis: $x = -\frac{1}{5}$; vertex: $\left(-\frac{1}{5}, 0\right)$;

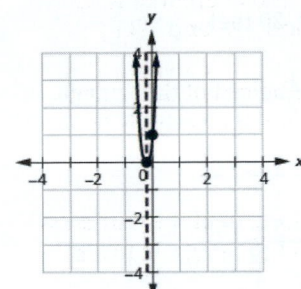

10.101. y: $(0, 5)$; x: none;

axis: $x = \frac{3}{2}$; vertex: $\left(\frac{3}{2}, \frac{1}{2}\right)$;

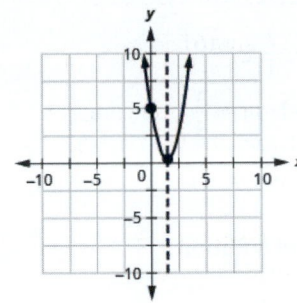

10.102. y: $(0, -1)$; x: none;

axis: $x = 0$; vertex: $(0, -1)$;

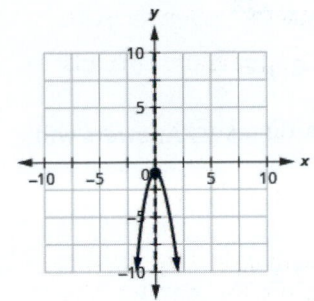

10.103.

y: $(0, 3)$; x: $(-1.6, 0)$, $(-0.4, 0)$;

axis: $x = -1$; vertex: $(-1, -2)$;

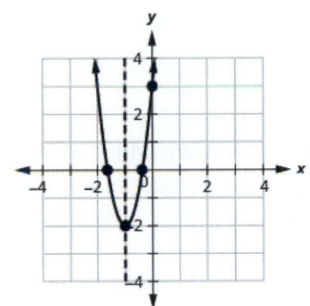

10.104.

y: $(0, 5)$; x: $(0.6, 0)$, $(-2.6, 0)$;

axis: $x = -1$; vertex: $(-1, 8)$;

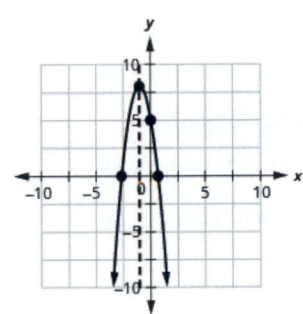

10.105. The minimum value is -4 when $x = 4$.

10.106. The maximum value is 5 when $x = 2$.

10.107. It will take 4 seconds to reach the maximum height of 288 feet.

10.108. It will take 6.5 seconds to reach the maximum height of 676 feet.

Section Exercises

1. $a = \pm\, 7$

7. $m = \pm\, 3$

13. $p = \pm\, \frac{4\sqrt{7}}{7}$

19. $m = 6 \pm 2\sqrt{5}$

25. no real solution

31. $r = \pm\, 4$

37. $a = \pm\, 3\sqrt{2}$

43. $u = 7 \pm 6\sqrt{2}$

49. $c = \pm\, \frac{5\sqrt{6}}{6}$

55. Answers will vary.

61. $(m - 12)^2$

67. $\left(p - \frac{1}{6}\right)^2$

3. $r = \pm\, 2\sqrt{6}$

9. no real solution

15. $x = 1$, $x = -5$

21. $r = \frac{1}{2} \pm \frac{\sqrt{3}}{2}$

27. $m = 2 \pm 2\sqrt{2}$

33. $a = 4 \pm 2\sqrt{7}$

39. $p = \frac{1}{3} \pm \frac{\sqrt{7}}{3}$

45. $m = 4 \pm 2\sqrt{3}$

51. no real solution

57. $(a + 5)^2$

63. $(p - 11)^2$

69. $v = -10$, $v = 4$

5. $u = \pm\, 10\sqrt{3}$

11. $a = \pm\, 2\sqrt{5}$

17. $u = 14$, $u = -2$

23. $a = 7 \pm 5\sqrt{2}$

29. $x = -\frac{3}{5}$, $x = \frac{9}{5}$

35. $w = 1$, $w = \frac{5}{3}$

41. no real solution

47. $x = -3$, $x = -7$

53. 4 feet

59. $(m + 9)^2$

65. $\left(x - \frac{9}{2}\right)^2$

71. $u = -3$, $u = 1$

73. $c = -1$, $c = 13$
79. no real solution

85. $v = \frac{9}{2} \pm \frac{\sqrt{89}}{2}$

91. $c = -2$, $c = \frac{3}{2}$

97. ⓐ -5 ⓑ -5 ⓒ Answers will
vary.
103. $p = -4$, $p = -3$

109. $a = \frac{3 \pm \sqrt{3}}{2}$

115. $m = -1$, $m = \frac{3}{4}$
121. $p = -3$, $p = 9$

127. $b = \frac{-2 \pm \sqrt{11}}{6}$

133. ⓐ 1 ⓑ no real solutions
ⓒ 1 ⓓ 2
139. 5 seconds, 8 seconds

145. Two consecutive even
numbers whose product is 624
are 24 and 26, and -26 and -24

151. The leg of the right triangle
is 1.7 feet and the hypotenuse is
3.4 feet.
157. The rocket will reach 1,200
feet on its way up in 2 seconds
and on the way down in 38
seconds.
163.

75. $x = -1$, $x = 21$
81. $a = 5 \pm 2\sqrt{5}$
87. $x = -7$, $x = 3$

93. $p = -\frac{7}{4} \pm \frac{\sqrt{161}}{4}$

99. $m = -1$, $m = \frac{3}{4}$

105. $r = -3$, $r = 11$

111. no real solution

117. $c = -\frac{3}{4}$

123. $r = \frac{-3 \pm \sqrt{89}}{8}$

129. $x = \frac{-6 \pm \sqrt{42}}{4}$

135. ⓐ factor ⓑ square root
ⓒ Quadratic Formula
141. ⓐ -20, 10 ⓑ -20, 10
ⓒ answers will vary

147. Two consecutive odd
numbers whose product is 483
are 21 and 23, and -21 and -23

153. The length of the fence is 7.1
units.

159. 70 seconds

165. down

77. no real solution
83. $u = 1$, $u = 13$

89. $m = -11$, $m = 1$

95. 16 feet, 20 feet

101. $p = \frac{1}{2}$, $p = 3$

107. $u = \frac{-7 \pm \sqrt{73}}{6}$

113. $v = \frac{-5 \pm \sqrt{65}}{2}$

119. $m = -\frac{7}{5}$, $m = 1$

125. $a = \frac{-6 \pm \sqrt{26}}{2}$

131. ⓐ no real solutions ⓑ 1
ⓒ 2 ⓓ no real solutions

137. ⓐ factor ⓑ square root
ⓒ factor
143. Two consecutive odd
numbers whose product is 255
are 15 and 17, and -15 and -17

149. The width of the triangle is 5
inches and the height is 18
inches.

155. The width of the driveway is
10 feet and its length is 35 feet.

161. ⓐ answers will vary
ⓑ answers will vary ⓒ answers
will vary ⓓ answers will vary
167. up

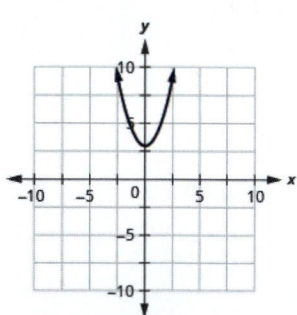

169. ⓐ $x = -4$ ⓑ $(-4, -17)$

171. ⓐ $x = 1$ ⓑ $(1, 6)$

173. y: $(0, 6)$; x: $(-1, 0)$, $(-6, 0)$

175. y: $(0, 19)$; x: none

177. y: $(0, 25)$; x: $(\frac{5}{2}, 0)$

179. y: $(0, 5)$; x: $(-1, 0)$, $(-5, 0)$;
axis: $x = -3$; vertex: $(-3, -4)$

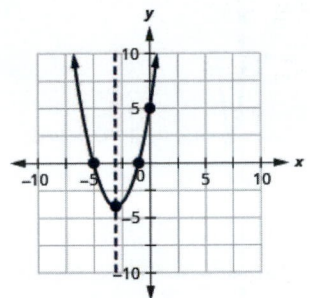

181. y: $(0, 3)$; x: $(-1, 0)$, $(-3, 0)$;
axis: $x = -2$; vertex: $(-2, -1)$

183. y: $(0, 4)$ x: $(-\frac{2}{3}, 0)$;
axis: $x = -\frac{2}{3}$; vertex: $(-\frac{2}{3}, 0)$

185. y: $(0, -7)$; x: none;
axis: $x = 1$; vertex: $(1, -6)$

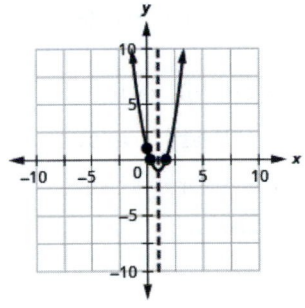

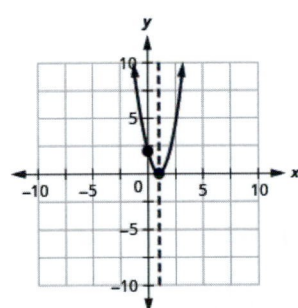

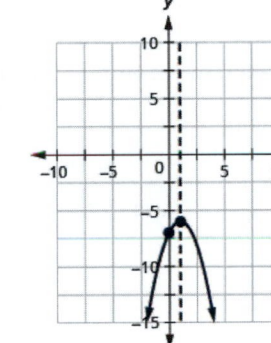

187. y: $(0, 1)$; x: $(1.7, 0)$, $(0.3, 0)$;
axis: $x = 1$; vertex: $(1, -1)$

189. y: $(0, 2)$ x: $(1, 0)$;
axis: $x = 1$; vertex: $(1, 0)$

191. y: $(0, 2)$ x: $(-4.4, 0)$, $(0.4, 0)$;
axis: $x = -2$; vertex: $(-2, 6)$

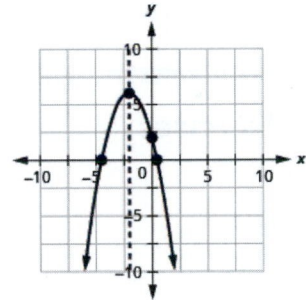

193. y: (0, 8); x: none; axis: $x = 1$; vertex: (1, 3)

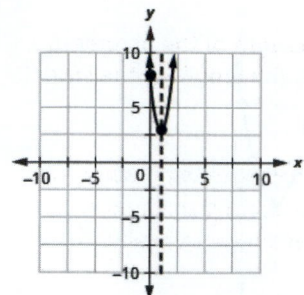

195.
y: (0, 20) x: (−4.5, 0), (−1.5, 0); axis: $x = -3$; vertex: (−3, −7)

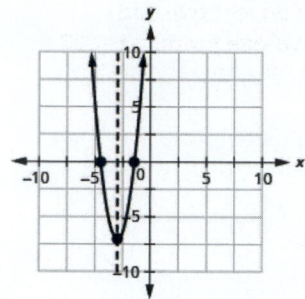

197. The minimum value is $-\frac{9}{8}$ when $x = -\frac{1}{4}$.

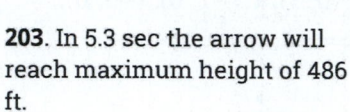

199. The minimum value is 6 when $x = 3$.

201. The maximum value is 16 when $x = 0$.

203. In 5.3 sec the arrow will reach maximum height of 486 ft.

205. 20 computers will give the maximum of $400 in receipts.

207. The length of the side along the river of the corral is 120 feet and the maximum area is 7,200 sq ft.

209.
ⓐ

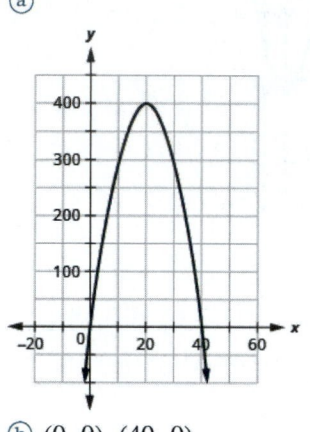

ⓑ (0, 0), (40, 0)

211. Answers will vary.

Review Exercises

213. $x = \pm\, 10$

219. no solution

225. $p = 1, 9$

231. $m = 7 \pm 2\sqrt{6}$

237. $a = -\frac{3}{2}, \frac{3}{4}$

243. $\left(a - \frac{3}{2}\right)^2$

249. $x = -4, 8$

255. $m = -9, -1$

261. $p = 0, 6$

267. $x = \frac{1}{4}, 1$

215. $m = \pm\, 2\sqrt{10}$

221. $v = \pm\, 3\sqrt{2}$

227. $u = -1 \pm 3\sqrt{5}$

233. no solution

239. $(x + 11)^2$

245. $\left(p + \frac{2}{5}\right)^2$

251. no solution

257. $a = \frac{3}{2} \pm \frac{\sqrt{41}}{2}$

263. $y = -\frac{1}{2}, 2$

269. $r = -6, 7$

217. $a = \pm\, 5$

223. $c = \pm\, \frac{4\sqrt{5}}{5}$

229. $x = \frac{1}{4} \pm \frac{\sqrt{3}}{4}$

235. $m = 3 \pm 4\sqrt{3}$

241. $(m - 4)^2$

247. $c = 1, -21$

253. $v = 7 \pm 3\sqrt{2}$

259. $u = -6 \pm 2\sqrt{2}$

265. $c = -\frac{1}{3} \pm \frac{2\sqrt{7}}{3}$

271. $v = -\frac{5}{4}, 1$

273. $m = \dfrac{-4 \pm \sqrt{10}}{3}$

279. $p = \dfrac{4 \pm \sqrt{6}}{5}$

285. ⓐ factor ⓑ Quadratic Formula ⓒ square root

291. The lengths of the sides of the mosaic are 2.2 and 4.4 feet.

297.

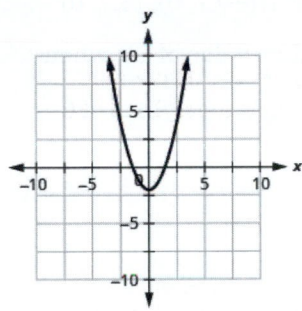

303. ⓐ $x = 3$ ⓑ $(3, 17)$

309. y: $(0, 1)$; x: $\left(\dfrac{1}{4}, 0\right)$

275. no real solution

281. $c = -\dfrac{1}{2}$

287. Two consecutive odd numbers whose product is 323 are 17 and 19, and -17 and -19.

293. The width of the front walk is 8.1 feet and its length is 30.8 feet.

299. down

305. y: $(0, 5)$; x: $(5, 0)$, $(-1, 0)$

311. y: $(0, 15)$; x: $(-3, 0)$, $(-5, 0)$; axis: $x = -4$; vertex: $(-4, -1)$

277. $u = 5 \pm \sqrt{22}$

283. ⓐ 1 ⓑ 2 ⓒ 2 ⓓ none

289. The height of the banner is 13 cm and the length of the side is 54 cm.

295. The ball will reach 384 feet on its way up in 4 seconds and on the way down in 6 seconds.

301. up

307. y: $(0, 10)$; x: none

313. y: $(0, -16)$; x: $(4, 0)$; axis: $x = 4$; vertex: $(4, 0)$

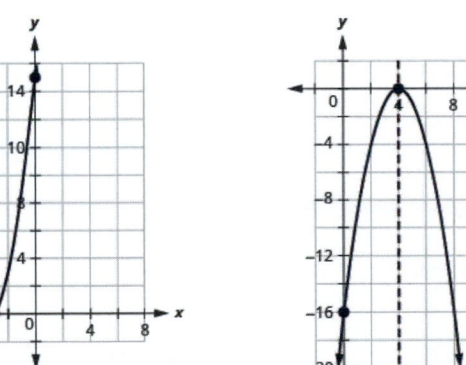

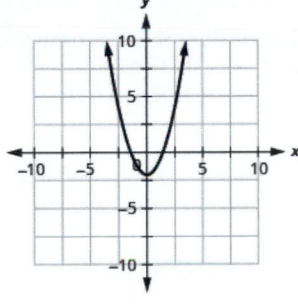

315. y: $(0, 13)$; x: none; axis: $x = -3$; vertex: $(-3, 4)$

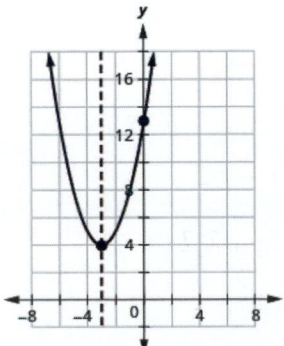

317.
y: $(0, -11)$ x: $(3.1, 0)$, $(0.9, 0)$; axis: $x = 2$; vertex: $(2, 5)$

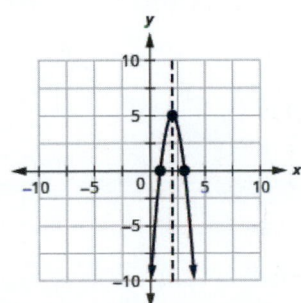

319. The minimum value is -1 when $x = -1$.

321. In 3.5 seconds the ball is at its maximum height of 196 feet.

Practice Test

323. $w = -2, -8$

325. $m = 1, \frac{3}{2}$

327. $n = \frac{-4 \pm \sqrt{7}}{3}$

329. no real solution

331. 2

333. Two consecutive even number are -20 and -18 and 18 and 20.

335. ⓐ up ⓑ $x = -1$ ⓒ $(-1, 5)$ ⓓ y: $(0, 8)$; x: none ⓔ minimum value of 5 when $x = -1$

337. ⓐ up ⓑ $x = -5$ ⓒ $(-5, -1)$ ⓓ y; $(0, 24)$; x: $(-6, 0)$, $(-4, 0)$ ⓔ minimum value of -5 when $x = -1$

339. ⓐ down ⓑ $x = -4$ ⓒ $(-4, 32)$ ⓓ y; $(0, 16)$; x: $(-9.7, 0)$, $(1.7, 0)$ ⓔ maximum value of 32 when $x = -4$

341. y: $(0, 9)$; x: $\left(-\frac{3}{4}, 0\right)$; axis: $x = -\frac{3}{4}$; vertex: $\left(-\frac{3}{4}, 0\right)$

INDEX

A

absolute value, **43**, **45**, **176**
addition, **21**
Addition Property of Equality, **200**, **200**
additive identity, **145**, **176**
additive inverse, **146**, **176**
amount of discount, **321**, **391**
an inconsistent system, **577**
area, **347**, **357**
associative property, **143**, **144**
axis of symmetry, **1194**, **1194**, **1215**

B

binomial, **674**, **701**, **703**, **711**, **717**, **721**, **751**, **779**
boundary line, **532**, **533**, **548**

C

coefficient, **31**, **176**
coincident lines, **576**, **663**
commutative property, **142**
complementary angles, **621**, **663**
completing the square, **1153**, **1215**
complex fraction, **84**, **176**
complex rational expression, **937**, **938**, **1003**
composite number, **13**, **176**
conditional equation, **242**, **285**
conjugate pair, **721**, **779**
consecutive even integers, **1179**, **1179**, **1215**
consecutive odd integers, **1179**, **1179**, **1215**
consistent system, **577**, **577**, **663**
constant, **21**, **21**, **176**
contradiction, **243**, **243**, **285**
counting number, **6**
counting numbers, **128**, **176**

D

decimal, **107**, **108**, **109**, **110**, **176**
degree of a constant, **675**, **779**
degree of a polynomial, **675**, **779**
degree of a term, **675**, **779**
denominator, **15**, **76**, **94**, **176**
dependent, **577**
dependent equations, **577**, **663**
difference of squares, **839**

difference of squares pattern, **876**
discount rate, **321**, **323**, **391**
discriminant, **1173**, **1173**, **1215**
distributive property, **151**, **153**
divisible, **11**
divisible by a number, **176**
division, **21**, **64**, **70**, **85**, **88**, **98**, **115**, **142**, **147**
Division Property of Equality, **213**, **213**

E

elapsed time, **376**
Elimination Method, **602**
equal sign, **22**, **24**, **26**
equality symbol, **176**
equals sign, **204**
equation, **24**, **176**
equivalent decimals, **136**, **176**
equivalent fractions, **77**, **176**
evaluate an expression, **29**, **176**
expression, **24**, **144**, **176**
extraneous solution to a rational equation, **951**, **1003**

F

factoring, **790**, **876**
factors, **12**, **12**, **161**, **176**
fraction, **107**, **132**, **176**
fraction bar, **86**
fraction operations, **96**

G

geoboard, **459**, **460**, **548**
graph of a linear equation, **426**, **548**
greatest common factor, **790**, **790**, **876**
grouping symbols, **23**

H

horizontal line, **437**, **548**

I

identity, **243**, **243**, **285**
Identity Property of Addition, **145**
Identity Property of Multiplication, **145**
improper fractions, **133**
inconsistent system, **577**, **663**
independent, **577**
independent equations, **577**, **663**

index, **1092**, **1128**
integers, **42**, **61**, **128**, **176**
intercepts of a line, **444**, **548**
interest, **318**, **391**
interval notation, **271**
Inverse Property of Addition, **146**
Inverse Property of Multiplication, **146**
irrational, **130**
irrational number, **130**, **176**

L

least common denominator, **94**, **176**
least common multiple, **15**, **15**, **176**
like radicals, **1038**, **1101**, **1101**, **1128**
like square roots, **1036**, **1049**, **1128**
like terms, **31**, **176**
linear equation, **409**, **409**, **548**
linear inequality, **530**, **538**, **548**
list price, **324**, **391**

M

mark-up, **324**, **391**
metric system, **164**
mixture problems, **330**, **391**
monomial, **674**, **675**, **695**, **701**, **703**, **748**, **748**, **779**
multiple of a number, **176**
multiples, **11**
multiplication, **21**, **61**, **70**
Multiplication Property of Equality, **214**
multiplicative identity, **145**, **176**
multiplicative inverse, **146**, **176**
multiplying fractions, **81**

N

negative exponent, **761**, **761**, **779**
negative numbers, **40**
negative slope, **463**, **548**
neutral pairs, **53**
*n*th root of a number, **1092**, **1128**
number line, **6**, **41**, **132**, **176**
number system, **6**
numerator, **76**, **176**

O

operations, 21
opposite, 41, 176
Order of Operations, 26
ordered pair, 404, 408, 410, 411, 548
origin, 176, 404, 404, 548
original cost, 324, 391

P

parabola, 1191, 1193, 1215
parallel lines, 500, 548
percent, 120, 176, 314
percent decrease, 317
percent increase, 316
perfect square trinomials pattern, 834, 876
perimeter, 347, 357
perpendicular, 504
perpendicular lines, 504, 548
place values, 6, 107, 111
point–slope form, 514, 548
point–slope method, 474
polynomial, 674, 679, 779
positive slope, 463, 548
prime factorization, 13, 13, 177
prime number, 13, 13, 79, 177
prime polynomials, 876
principal, 318, 391
principal nth root, 1093, 1128
principal square root, 127, 1092
product, 11, 13, 70, 81, 82, 113, 147
profit, 385
proper fractions, 117, 133
proportion, 965, 966, 1003
Pythagorean Theorem, 352

Q

quadrant, 405, 548
quadratic equation, 1137, 1215
quadratic equation in two variables, 1190, 1190, 1215
quadratic equations, 861, 876
quotient, 70, 88

R

radical equation, 1074, 1074, 1128
radical sign, 127, 177
rate of interest, 318, 391
ratio, 88
rational equation, 950, 951, 1003
rational exponents, 1107,

1108, 1128
rational expression, 884, 889, 1003
rational number, 128, 177
rationalize, 1141
rationalizing the denominator, 1064, 1128
real number, 131, 177
reciprocal, 82, 146, 177
rectangle, 357
rectangular coordinate system, 404, 404, 548
repeating decimal, 118, 177
right triangle, 350
rise, 460, 460, 548
run, 460, 460, 548

S

scientific notation, 770, 770, 779
similar figures, 971, 975, 1003
simple interest, 319, 341, 391
simplification, 202
simplified, 79
simplified fraction, 177
simplify an expression, 26, 177
slope formula, 473, 473, 548
slope of a line, 460, 548
slope-intercept form of an equation of a line, 548
slope–intercept form, 487
solution of a linear inequality, 530, 548
solution of an equation, 198, 285
solutions of a system of equations, 663
solutions to a system of equations, 566
square, 126
square and square root, 177
square of a number, 1128
square root, 126
square root notation, 1128
square root of a number, 1128
Square Root Property, 1138, 1138, 1215
standard form, 675, 779
subtraction, 21
subtraction property, 54
Subtraction Property of Equality, 199, 200
sum and difference of cubes pattern, 843, 876
supplementary angles, 621,

663
system of linear equations, 663
system of linear inequalities, 648, 663
systems of linear equations, 566

T

term, 31, 177
trinomial, 674, 706, 711, 712, 717, 752, 779

U

U.S. system, 160

V

variable, 21, 21, 43, 177
variables, 227
vertex, 1194, 1209, 1215
vertical line, 435, 548
vertices, 346

W

whole number, 6
whole numbers, 128, 177

X

x- intercept, 448, 548
x- intercept of a line, 445
x-coordinate, 404, 407, 548
x-intercept, 1196
x-intercepts of a parabola, 1196, 1215

Y

y- intercept, 448
y- intercept of a line, 446
y-coordinate, 404, 407, 548
y-intercept, 548, 1196
y-intercept of a parabola, 1215

Z

Zero Product Property, 861, 861, 876